An Introduction to
Medicinal Chemistry

THIRD EDITION

Graham L. Patrick

OXFORD

UNIVERSITY PRESS

OXFORD
UNIVERSITY PRESS

Great Clarendon Street, Oxford OX2 6DP

Oxford University Press is a department of the University of Oxford.
It furthers the University's objective of excellence in research, scholarship,
and education by publishing worldwide in

Oxford New York

Auckland Cape Town Dar es Salaam Hong Kong Karachi
Kuala Lumpur Madrid Melbourne Mexico City Nairobi
New Delhi Shanghai Taipei Toronto

With offices in

Argentina Austria Brazil Chile Czech Republic France Greece
Guatemala Hungary Italy Japan Poland Portugal Singapore
South Korea Switzerland Thailand Turkey Ukraine Vietnam

Oxford is a registered trade mark of Oxford University Press
in the UK and in certain other countries

Published in the United States
by Oxford University Press Inc., New York

© Graham L. Patrick, 2005

British Library Cataloguing in Publication Data
Data available

Library of Congress Cataloging in Publication Data
Data available

Typeset by Newgen Imaging Systems (P) Ltd., Chennai, India
Printed in Great Britain by
Ashford Colour Press Ltd, Gosport, Hampshire

ISBN 0-19-927500-9   978-0-19-927500-7

10 9 8 7 6 5 4 3 2 1

■ PREFACE

This text is aimed at undergraduates and postgraduates who have a basic grounding in chemistry and are studying a module or degree in medicinal chemistry. It attempts to convey, in a readable and interesting style, an understanding about drug design and the molecular mechanisms by which drugs act in the body. In so doing, it highlights the importance of medicinal chemistry in all our lives and the fascination of working in a field which overlaps the disciplines of chemistry, biochemistry, physiology, microbiology, cell biology, and pharmacology. Consequently, the book is of particular interest to students who might be considering a future career in the pharmaceutical industry.

Following the success of the first two editions, four new chapters have been included in the third edition, covering pharmacokinetics, antiviral agents, anticancer agents, and drug discovery. Other chapters have been expanded to include more recent advances in medicinal chemistry.

The book is divided into four parts:

- Part A covers pharmacodynamics in Chapters 1–7 and pharmacokinetics in Chapter 8. Pharmacodynamics includes the types of molecular targets used by drugs, the interactions which are involved when a drug meets that target, and the consequences of those interactions. Pharmacokinetics relates to the issues involved in a drug reaching its target in the first place.

- Part B covers the general principles and strategies involved in discovering and designing new drugs and developing them for the marketplace.

- Part C looks at particular 'tools of the trade' which are invaluable in those processes, i.e. QSAR, combinatorial synthesis, and computer aided design.

- Part D covers a selection of specific topics within medicinal chemistry—antibacterial, antiviral and anticancer agents, cholinergics and anticholinesterases, adrenergics, opiate analgesics and antiulcer agents. To some extent, those chapters reflect the changing emphasis in medicinal chemistry research. Antibacterial agents, cholinergics, adrenergics and opiates have long histories and much of the early development of these drugs relied heavily on random variations of lead compounds on a trial and error basis. This approach was wasteful but it led to the recognition of various design strategies which could be used in a more rational approach to drug design. The development of the antiulcer drug cimetidine (Chapter 22) represents one of the early examples of the rational approach to medicinal chemistry. However, the real revolution in drug design resulted from giant advances made in molecular biology and genetics which have provided a detailed understanding of drug targets and how they function at the molecular level. This, allied to the use of molecular modelling and X-ray crystallography, has revolutionized drug design. The development of protease inhibitors as antiviral agents (Chapter 17) is a prime example of the modern approach.

G. L. P.
January 2005

ABOUT THE BOOK

The third edition of *An Introduction to Medicinal Chemistry* and its accompanying companion web site contains many new learning features. This section illustrates each of these learning features and explains how they will help you to understand this fascinating subject.

Emboldened key words

Terminology is emboldened and defined in a glossary at the end of the book, helping you to become familiar with the language of medicinal chemistry.

Boxes

Boxes are used to present in-depth material and to explore how the concepts of medicinal chemistry are applied in practice.

Key points

Summaries at the end of major sections within chapters highlight and summarise key concepts and provide a basis for revision.

Questions

End of chapter questions allow you to test your understanding and apply concepts presented in the chapter.

Further reading

Selected references allow you to easily research those topics that are of particular interest to you.

Appendix

The appendix now includes an index of drug names and their corresponding trade names, and an expanded glossary.

---

404 CHAPTER 16: ANTIBACTERIAL AGENTS

and *Haemophilus* species. They show similar activity against Gram-negative aerobic rods such as *P. aeruginosa* but are generally more active against other Gram negative bacteria. Unfortunately, they have to be injected. Examples include azlocillin, which is 8–16 times more active than carbenicillin against *P. aeruginosa*, and is used primarily for the treatment of infections caused by that organism. It is susceptible to β-lactamases. **Mezlocillin** has a similar spectrum of activity to carbenicillin but is more active since it has a higher affinity for transpeptidases and can cross the outer membrane of Gram-negative bacteria more effectively. **Piperacillin** is similar to ampicillin in its activity against Gram-positive bacteria. It also has good activity against anaerobic species of both cocci and bacilli and can be used against a variety of infections. Piperacillin can be administered alongside **tazobactam** as **Tazocin** to widen its spectrum of activity (section 16.5.4.2 and Box 16.6).

BOX 16.6 SYNERGISM OF PENICILLINS WITH OTHER DRUGS

There are several examples in medicinal chemistry where the presence of one drug enhances the activity of another. In many cases this can be dangerous, leading to an effective overdose of the enhanced drug. In some cases, though, it can be useful. There are two interesting examples where the activity of penicillin has been enhanced by the presence of another drug.

One of these is the effect of clavulanic acid, described in section 16.5.4.1. The other is the administration of penicillins with a compound called **probenecid**. Probenecid is a moderately lipophilic carboxylic acid and as such is similar to penicillin. It is found that probenecid can block facilitated transport of penicillin through the kidney tubules. In other words, probenecid slows down the rate at which penicillin is excreted by competing with it in the excretion mechanism. Probenecid also competes with penicillin for binding sites on albumin. As a result, penicillin levels in the bloodstream are enhanced and the antibacterial activity increases—a useful tactic if faced with a particularly resistant bacterium.

Probenecid.

KEY POINTS

- Penicillins have a bicyclic structure consisting of a β-lactam ring fused to a thiazolidine ring. The strained β-lactam ring reacts irreversibly with the transpeptidase enzyme responsible for the final cross-linking of the bacterial cell wall.
- Penicillin analogues can be prepared by fermentation or by a semi-synthetic synthesis from 6-aminopenicillanic acid. Variation of the penicillin structure is limited to the acyl side chain.
- Penicillins can be made more resistant to acid conditions by incorporating an electron-withdrawing group into the acyl side chain.
- Steric shields can be added to penicillins to protect them from bacterial β-lactamase enzymes.
- Broad spectrum activity is associated with the presence of an α-hydrophilic group on the acyl side chain of penicillin.
- Prodrugs of penicillins are useful in masking polar groups and improving absorption from the gastrointestinal tract. Extended esters are used which undergo enzyme-catalysed hydrolysis to produce a product which spontaneously degrades to release the penicillin.
- Probenecid can be administered with penicillins to hinder the excretion of penicillins.

16.5.2 Cephalosporins

16.5.2.1 Cephalosporin C

Discovery and structure of cephalosporin C

The second major group of β-lactam antibiotics to be discovered were the cephalosporins. The first cephalosporin (**cephalosporin C**) was derived from a fungus obtained in the mid 1940s from sewer waters on the island of Sardinia. This was the work of an Italian professor, who noted that the waters surrounding the sewage outlet periodically cleared of microorganisms. He reasoned that an organism might be producing an antibacterial substance and so he collected samples and managed to isolate a fungus called *Cephalosporium acremonium* (now called *Acremonium chrysogenum*). The crude extract from this organism was shown to have antibacterial properties, and in 1948, workers at Oxford University isolated cephalosporin C, but it was not until 1961 that the structure was established by X-ray crystallography.

The structure of cephalosporin C (Fig. 16.33) has similarities to that of penicillin in that it has a bicyclic

FURTHER READING 641

QUESTIONS

1. Morphine is an example of a plant alkaloid. Alkaloids tend to be secondary metabolites that are not crucial to a plant's growth and are produced when the plant is mature. If that is the case, what role do you think these compounds have in plants, if any?

2. The synthesis in Box 21.1 shows that *N*-alkylated analogues can be synthesized by *N*-alkylation directly or by a two-stage process involving *N*-acylation. Why might a two-stage process be preferred to direct *N*-alkylation? What sort of products could not be synthesized by the two stage process? Is this likely to be a problem?

3. Show how you would synthesize nalorphine (Fig. 21.13).

4. Pethidine has been used in childbirth as it is short-acting and less hazardous than morphine in the newborn baby. Several drugs taken by the mother before giving birth can prove hazardous to the child after birth, but less so before birth. Why is this?

5. Show how you would synthesize diprenorphine and buprenorphine.

6. Why is buprenorphine considered the most lipophilic of the oripavine series of compounds?

7. Identify the potential hydrogen bond donors and acceptors in morphine. Structure-activity relationships reveal that one functional group in morphine is important as a hydrogen bond donor or as a hydrogen bond acceptor. Which group is that?

8. Propose the likely analgesic activity of 3-acetyl morphine relative to morphine, heroin, and 6-acetylmorphine.

9. Describe how you would synthesize the *N*-phenethyl analogue of morphine.

10. The *N*-phenethyl analogue of morphine is known as a semi-synthetic product. What does this mean?

11. Explain whether you think the following structures would act as agonists or antagonists.

12. Thebaine has no analgesic activity. Suggest why this might be so.

13. Morphine is the active principle of opium. What is meant by an active principle?

14. Identify the asymmetric centres in morphine.

15. How could heroin be synthesized from morphine, and what problems does this pose for drug regulation authorities?

FURTHER READING

Abraham, D. J. (ed.) (2003) Narcotic analgesics. Chapter 7 in: *Burger's medicinal chemistry and drug discovery*, 6th edn, Vol. 6. John Wiley and Sons, New York.

Feinberg, A. P., Creese, I., and Snyder, S. H. (1976) The opiate receptor: a model explaining structure-activity relationships of opiate agonists and antagonists. *Proceedings of the National Academy of Sciences of the USA*, 73, 4215–4219.

Hruby, V. J. (2002) Designing peptide receptor agonists and antagonists. *Nature Reviews Drug Discovery*, 1, 847–858.

Kreek, M. J., LaForge, K. S., and Butelman, E. (2002) Pharmacotherapy of addictions. *Nature Reviews Drug Discovery*, 1, 710–725.

Pouletty, P. (2002) Drug addictions: towards socially accepted and medically treatable diseases. *Nature Reviews Drug Discovery*, 1, 731–736.

Roberts, S. M. and Price, B. J. (eds.) (1985) Discovery of buprenorphine, a potent antagonist analgesic. Chapter 7 in: *Medicinal chemistry—the role of organic research in drug research*. Academic Press, New York.

Williams, D. A. and Lemke, T. L. (eds.) (2002) Opioid analgesics. Chapter 19 in: *Foye's principles of medicinal chemistry*, 5th. edn. Lippincott Williams and Wilkins, Philadelphia.

Titles for general further reading are listed on p. 711.

■ ABOUT THE COMPANION WEB SITE

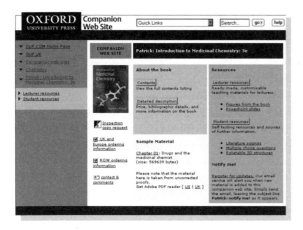

Companion web sites provide students and lecturers with ready-to-use teaching and learning resources. They are free-of-charge, designed to complement the textbook and offer additional materials which are suited to electronic delivery.

All these resources can be downloaded and are fully customisable allowing them to be incorporated into your institution's existing virtual learning environment.

You will find this material at:
www.oup.com/uk/booksites/chemistry

Student resources

Multiple choice questions

Self-test multiple choice questions are available for each chapter allowing you to test your knowledge and grasp of key concepts as you progress through the book. Each is submitted online giving you an immediate mark and instant feedback.

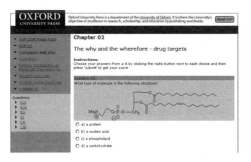

Rotatable 3D structures

JPEG and Chem3D versions of selected molecules in the book help you to visualise molecules. Full instructions and link to free software are provided.

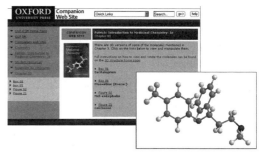

Literature sources

A selection of literature sources chosen by the author help you to explore the primary literature and the recent advances in the field.

Lecturer resources

Figures from the book

All of the figures from the textbook are available to download electronically for use in lectures and handouts.

PowerPoint slides

PowerPoint slides are provided to help teach selected topics from the book.

ACKNOWLEDGEMENTS

The author and Oxford University Press would like to thank the following people who gave advice on the second edition and reviewed draft chapters of the third edition:

Dr Lee Banting, School of Pharmacy and Biomedical Sciences, University of Portsmouth, UK

Professor Derek R. Boyd, Department of Chemistry, Queens University of Belfast, UK

Dr James Bruce, Department of Chemistry, Open University, UK

Professor Tim Bugg, Department of Chemistry, University of Warwick, UK

Dr Adrian Dobbs, Department of Chemistry, University of Exeter, UK

Dr Mark Edwards, School of Science, University of Greenwich at Medway, UK

Dr Don Green, Department of Health and Human Sciences, London Metropolitan University, UK

Dr Chris Hamilton, Department of Chemistry, Queens University of Belfast, UK

Dr Ian Hardcastle, Northern Institute for Cancer Research, University of Newcastle upon Tyne, UK

Professor Ray Jones, Department of Chemistry, Loughborough University, UK

Dr Barrie Kellam, School of Pharmacy, University of Nottingham, UK

Dr John Malone, Department of Chemistry, Queens University of Belfast, UK

Professor John Mann, Department of Chemistry, Queens University of Belfast, UK

Dr Susan Matthews, School of Chemical Sciences & Pharmacy, University of East Anglia, UK

Dr Marie Migaud, Department of Chemistry, Queens University of Belfast, UK

Dr Declan Naughton, School of Pharmacy & Biomolecular Science, University of Brighton, UK

Professor Michael Page, Department of Chemical and Biological sciences, University of Huddersfield, UK

Dr Fred Pearce, Department of Chemistry, UCL, UK

Dr Justin Perry, School of Applied Sciences, Northumbria University, UK

Dr Isabel Rozas, Department of Chemistry, Trinity College, University of Dublin, Ireland

Dr Derek Sharples, School of Pharmacy & Pharmaceutical Sciences, University of Manchester, UK

Dr Mike Southern, Department of Chemistry, Trinity College, University of Dublin, Ireland

Dr Dilys Thornton, School of Biosciences, University of the West of England, UK

Dr Stuart Warriner, School of Chemistry, University of Leeds, UK

Dr Alex White, Welsh School of Pharmacy, Cardiff University, UK

Professor Douglas Young, School of Chemistry, University of Sussex, UK

Dr Mikael Elofsson (Assistant Professor), Department of Chemistry, Umeå University, Sweden

Professor Rob Leurs, Department of Chemistry, Vrije Universiteit, Amsterdam, The Netherlands

Dr Ed Moret, Faculty of Pharmaceutical Sciences, Utrecht University, The Netherlands

Professor John Nielsen, Department of Natural Sciences, Royal Veterinary and Agricultural University, Denmark

Professor H. Timmerman, Department of Medicinal Chemistry, Vrije Universiteit, Amsterdam, The Netherlands

Professor Cliff Berkman, Dept. of Chemistry & Biochemistry, San Francisco State University, USA

Professor Nouri Neamati, School of Pharmacy, University of Southern California, USA

Professor John Warner, School of Health and the Environment, University of Massachusetts, Lowell, USA

The cartoons in the book were designed and drawn by Gary Learie, to whom the author is indebted. The author would also like to express his gratitude to Dr Stephen Bromidge of GlaxoSmithKline for permitting the description of his work on selective 5-HT2C antagonists, and for providing many of the diagrams for that case study. Thanks are also due to Cambridge Scientific, Oxford Molecular and Tripos for their advice and assistance in the writing of Chapter 15. Finally, the RSCB Protein Data Bank must be acknowledged for the use of images in Figures 3.9, 15.33 and 17.24.

■ BRIEF CONTENTS

■ DETAILED CONTENTS

■ LIST OF BOXES

■ ACRONYMS AND ABBREVIATIONS

ACE	angiotensin-converting enzyme	GCP	Good Clinical Practice
ADAPT	antibody-directed abzyme prodrug therapy	GDEPT	gene-directed enzyme prodrug therapy
		GEF	guanine nucleotide exchange factors
ADEPT	antibody-directed enzyme prodrug therapy	GGTase	geranylgeranyltransferase
ADH	alcohol dehydrogenase	GH	growth hormone
AIC	5-aminoimidazole-4-carboxamide	GIT	gastrointestinal tract
AIDS	acquired immune deficiency syndrome	GLP	Good Laboratory Practice
AML	acute myeloid leukaemia	GMP	Good Manufacturing Practice
AMP	adenosine 5′-monophosphate	GnRH	gonadotrophin-releasing hormone
ATP	adenosine triphosphate	HA	haemagglutinin
AUC	area under the curve	HAART	highly active antiretroviral therapy
CCK	cholecystokinin	HAMA	human anti-mouse antibodies
CDKs	cyclin dependent kinases	HIV	human immunodeficiency virus
cGMP	cyclic GMP	HOMO	highest occupied molecular orbital
CKIs	cyclin-dependent kinase inhibitors	HPLC	high-performance liquid chromatography
CML	chronic myeloid leukaemia	HPMA	N-(2-hydroxypropyl)methacrylamide
CMV	cytomegalovirus	HRV	human rhinoviruses
CNS	central nervous system	HTS	high-throughput screening
COMT	catechol O-methyltransferase	IGF-1R	insulin growth factor 1 receptor
COX	cyclooxygenase	IND	Investigational Exemption to a New Drug Application
CSD	Cambridge Structural Database		
CYP	enzymes that constitute the cytochrome P450 family	IP_3	inositol triphosphate
		IPER	International Preliminary Examination Report
DG	diacylglycerol		
DHFR	dihydrofolate reductase	IRB	Institutional Review Board
DNA	deoxyribonucleic acid	ISR	International Search Report
dTMP	deoxythymidine monophosphate	K_M	Michaelis constant
dUMP	deoxyuracil monophosphate	LH	luteinizing hormone
EC_{50}	concentration of drug required to produce 50% of the maximum possible effect	LHRH	luteinizing hormone-releasing hormones
		LUMO	lowest unoccupied molecular orbital
EGF	epidermal growth factor	MAA	Marketing Authorization Application
EMEA	European Agency for the Evaluation of Medicinal Products	MAb	monoclonal antibody
		MAO	monoamine oxidase
EPC	European Patent Convention	MAOI	monoamine oxidase inhibitor
EPO	European Patent Office	MAP	mitogen-activated protein
FDA	US Food and Drug Administration	MAPK	mitogen-activated protein kinases
FGF	fibroblast growth factor	MDR	multidrug resistance
FGF-R	fibroblast growth factor receptor	MEP	molecular electrostatic potential
FH_4	tetrahydrofolate	MMP	matrix metalloproteinase
FPGS	folylpolyglutamate synthetase	MMPI	matrix metalloproteinase inhibitor
FPP	farnesyl diphosphate	NA	neuraminidase noradrenaline
FT	farnesyl transferase	NAD^+/NADH	nicotinamide adenine dinucleotide
FTIs	farnesyl transferase inhibitors	NAG	N-acetylglucosamine
GABA	γ-aminobutyric acid	NAM	N-acetylmuramic acid
GAP	GTPase activating protein	NCE	new chemical entity

NDA	New Drug Application	QSARs	quantitative structure–activity relationships
NMDA	*N*-methyl-D-aspartate		
NME	new molecular entity	RES	reticuloendothelial system
NMR	nuclear magnetic resonance	RFC	reduced folate carrier
NNRTIs	non-nucleoside reverse transcriptase inhibitors	RMSD	root mean square distance
		RNA	ribonucleic acid
NO	nitric oxide	SAR	structure–activity relationship
NRTIs	nucleoside reverse transcriptase inhibitors	SCAL	safety catch linker
		SCF	stem cell factor
NSAID	non-steroidal anti-inflammatory drug	SCID	severe combined immunodeficiency disease
NVOC	nitroveratryloxycarbonyl		
PABA	*p*-aminobenzoic acid	SOPs	standard operating procedures
PBP	penicillin binding protein	SPA	scintillation proximity assay
PCP	phencyclidine, otherwise known as 'angel dust'	SPR	surface plasmon resonance
		TB	tuberculosis
PCT	Patent Cooperation Treaty	TFA	trifluoroacetic acid
PDB	Protein Data Bank	TGF-α	transforming growth factor α
PDGF	platelet-derived growth factor	TGF-β	transforming growth factor β
PDGF-R	platelet-derived growth factor receptor	THF	tetrahydrofuran
		TM	transmembrane
PDT	photodynamic therapy	TNF	tumour necrosis factor
PEG	polyethylene glycol	TNF-R	tumour necrosis factor receptors
PIP$_2$	phosphatidylinositol diphosphate	TNT	trinitrotoluene
PIs	protease inhibitors	TRAIL	TNF-related apoptosis-inducing ligand
PKA	protein kinase A	VEGF	vascular endothelial growth factor
PKB	protein kinase B	VEGF-R	vascular endothelial growth factor receptor
PKC	protein kinase C		
PLC	phospholipase C	VIP	vasoactive intestinal peptide
PLS	partial least squares	VOC–Cl	vinyloxycarbonyl chloride
PPIs	proton pump inhibitors	VRE	vancomycin-resistant enterococci
PPts	pyridinium 4-toluenesulfonate	VZV	varicella-zoster viruses

■ CLASSIFICATION OF DRUGS

The way drugs are classified or grouped can be confusing. Different textbooks group drugs in different ways.

1 By pharmacological effect

Drugs are grouped depending on the biological effect they have, e.g. analgesics, antipsychotics, antihypertensives, antiasthmatics, and antibiotics.

This is useful if one wishes to know the full scope of drugs available for a certain ailment. However, it should be emphasized that such groupings contain a large and extremely varied assortment of drugs. This is because there is very rarely one single way of dealing with a problem such as pain or heart disease. There are many biological mechanisms which the medicinal chemist can target to get the desired results, and so to expect all painkillers to look alike or to have some common thread running through them is not realistic.

A further point is that many drugs do not fit purely into one category or another and some drugs may have more than just one use. For example, a sedative might also have a use as an anticonvulsant. To pigeonhole a drug into one particular field and ignore other possible fields of action is folly.

The antibacterial agents (Chapter 16) are a group of drugs that are classified according to their pharmacological effect.

2 By chemical structure

Many drugs that have a common skeleton are grouped together, e.g. penicillins, barbiturates, opiates, steroids, catecholamines.

In some cases (e.g. penicillins) this is a useful classification, since the biological activity (e.g. antibiotic activity for penicillins) is the same. However, there is a danger that one could be misled into thinking that all compounds of a certain chemical group

have the same biological action. For example, barbiturates may look much alike and yet have completely different uses in medicine. The same holds true for steroids.

It is also important to consider that most drugs can act at several sites of the body and have several pharmacological effects.

The opiates (Chapter 21) are a group of drugs with similar chemical structures.

3 By target system

These are compounds that are classed according to whether they affect a certain target system in the body—usually involving a chemical messenger, e.g. antihistamines, cholinergics.

This classification is a bit more specific than the first, as it is identifying a system with which the drugs interact. It is, however, still a system with several stages and so the same point can be made as before—for example, one would not expect all antihistamines to be similar compounds because the system by which histamine is synthesized, released, interacts with its receptor, and is finally removed, can be attacked at all these stages.

4 By site of action

These are compounds that are grouped according to the enzyme or receptor with which they interact. For example, anticholinesterases (Chapter 19) are a group of drugs that act through inhibition of the enzyme acetylcholinesterase.

This is a more specific classification of drugs, since we have now identified the precise target at which the drugs act. In this situation, we might expect some common ground between the agents included, as a common mechanism of action is a reasonable though not inviolable assumption.

It is easy, however, to lose sight of the wood for the trees and forget why it is useful to have drugs that switch off a particular enzyme or receptor site.

For example, it is not intuitively obvious why an anticholinergic agent should paralyse muscle and why that should be useful.

FURTHER READING

Hansch, C., Sammes, P. G., and Taylor, J. B. (eds.) (1990) Classification of drugs. In: *Comprehensive medicinal chemistry*, 1, Chapter 3.1. Pergamon Press, Oxford.

Page, C., Curtis, M., Sutter, M., Walker, M., and Hoffman, B. (2002) Drug names and drug classification systems. In: *Integrated pharmacology*, 2nd edn, Chapter 2. Mosby, St. Louis.

■ NAMING OF DRUGS

The vast majority of chemicals that are synthesized in medicinal chemistry research never make it to the market place and it would be impractical to name them all. Instead, research groups refer to them by a code that usually consists of letters and numbers. The letters are specific to the research group undertaking the work and the number is specific for the compound. Thus, Ro31-8959, ABT-538, and MK-639 were compounds prepared by Roche, Abbott, and Merck pharmaceuticals respectively. If the compounds concerned show promise as therapeutic drugs they are taken into development and named. For example, the above compounds showed promise as anti-HIV drugs and were named saquinavir, ritonavir, and indinavir respectively. Finally, if the drugs prove successful and are marketed, they are given a proprietary, brand, or trade name that only the company can use. For example, the above compounds were marketed as Fortovase®, Norvir®, and Crixivan® respectively (note that brand names always start with a capital letter and have the symbol® to indicate that they are registered brand names). The proprietary names are also specific for the preparation or formulation of the drug. For example, Fortovase® (or Fortovase™) is a preparation containing 200 mg of saquinavir in a gel-filled beige capsule. If the formulation is changed, then a different name is used. For example, Roche sell a different preparation of saquinavir called Invirase® that consists of a brown/green capsule containing 200 mg of saquinavir as the mesilate salt. When a drug's patent has expired, it is possible for any pharmaceutical company to produce and sell that drug as a generic medicine. However, they are not allowed to use the trade name used by the company that originally invented it. European law requires that generic medicines are given a Recommended International Non-proprietary Name (rINN) which is usually identical to the name of the drug. In Britain, such drugs were given a British Approved Name (BAN), but these have now been modified in line with rINNs.

Because the naming of drugs is progressive, early research papers in the literature may use the original letter/number code as the name of the drug had not been allocated at the time of publication.

Throughout this text, the names of the active constituents are used rather than the trade names, although the trade name may be indicated if it is particularly well known. For example, it is indicated that sildenafil is Viagra® and that paclitaxel is Taxol®. The trade names for the drugs mentioned in the text are listed in Appendix 6, which also contains a list of trade names directing you to the relevant compound name. If you cannot find the drug you are looking for, you should refer to other textbooks or formularies such as the British National Formulary (see General Further Reading, p. 711).

PART A

Pharmacodynamics and pharmacokinetics

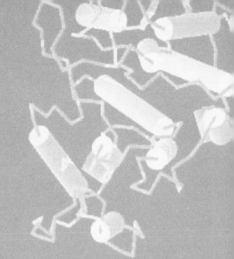

The role of the medicinal chemist is to design and synthesize new drugs. In order to carry out that role, it is important to identify the particular target for a specific drug, and to establish how the drug interacts with that target to produce a biological effect. In Chapters 2–7 we shall look at the various drug targets that are present in living systems, and the general mechanisms by which drugs can produce a pharmacological or biological effect. This is an area of study known as **pharmacodynamics**.

The major drug targets in the body are normally large molecules (macromolecules) such as proteins, nucleic acids, and carbohydrates. Knowing the structure, properties, and functions of a specific macromolecular target goes a long way to understanding how a drug works in the body. This is also crucial in helping to design better or novel drugs.

Drugs are normally small molecules with molecular weight less than 500 atomic mass units, much smaller than their macromolecular targets. As a result, they interact directly with only a small portion of the macromolecule. This is called a **binding site**. The binding site usually has a defined shape into which a drug must fit if it is to have an effect, and so it is important that the drug has the correct size and shape. However, there is more to drug action than just a good fit. Once an active drug enters a binding site, a variety of intermolecular bonding interactions are set up which hold it there and lead to further changes, culminating eventually in a biological effect. For this to occur, the drug must have the correct functional groups and molecular skeleton capable of participating in these interactions.

Pharmacodynamics is the study of how a drug interacts directly with its target, and optimizing the pharmacodynamics is clearly important if we are to design an effective drug. Having said that, there are examples of compounds which interact extremely well with their target but turn out to be useless in a clinical sense. That is because the compounds involved fail to reach their target in the body once they have been administered. There are various ways in which drugs can be administered, but generally the aim is to get the drug into the bloodstream so that it can then be carried to its particular target. Following the administration of a drug, there are a wide variety of hurdles and problems that have to be overcome. These include the efficiency with which a drug is absorbed into the bloodstream, how rapidly it is metabolized and excreted, and to what extent it is distributed round the body. This is an area of study known as **pharmacokinetics**, which is considered in Chapter 8.

1 Drugs and the medicinal chemist

In medicinal chemistry, the chemist attempts to design and synthesize a medicine or a pharmaceutical agent that will benefit humanity. Such a compound could also be called a 'drug', but this is a word that many scientists dislike because society views the term with suspicion. With media headlines such as 'Drugs Menace', or 'Drug Addiction Sweeps City Streets', this is hardly surprising. However, it suggests that a distinction can be drawn between drugs that are used in medicine and drugs that are abused. Is this really true? Can we draw a neat line between 'good drugs' like penicillin and 'bad drugs' like heroin? If so, how do we define what is meant by a good or a bad drug in the first place? Where would we place a so-called social drug like cannabis in this divide? What about nicotine, or alcohol?

The answers we get would almost certainly depend on who we were to ask. As far as the law is concerned, the dividing line is defined in black and white. As far as the partygoing teenager is concerned, the law is an ass. As far as we are concerned, the questions are irrelevant. Trying to divide drugs into two categories—safe or unsafe, good or bad—is futile and could even be dangerous.

First, let us consider the so-called 'good' drugs—the medicines. How 'good' are they? If a medicine is to be truly 'good' it would have to satisfy the following criteria. It would have to do what it is meant to do, have no toxic or unwanted side effects, and be easy to take.

How many medicines fit these criteria?

The short answer is 'none'. There is no pharmaceutical compound on the market today that can completely satisfy all these conditions. Admittedly, some come quite close to the ideal. Penicillin, for example, has been one of the most effective antibacterial agents ever discovered and has also been one

of the safest. Yet it too has drawbacks. It has never been able to kill all known bacteria and as the years have gone by, more and more bacterial strains have become resistant. Nor is penicillin totally safe: there are many examples of patients who show an allergic reaction to it and are required to take alternative antibacterial agents.

Penicillin is a relatively safe drug, but there are some medicines that are distinctly dangerous. Morphine is one such example. It is an excellent analgesic, yet it suffers from serious side effects such as tolerance, respiratory depression, and addiction. It can even kill if taken in excess.

Barbiturates are also known to be dangerous. At Pearl Harbor, American casualties were given barbiturates as general anaesthetics before surgery. However, because of a poor understanding about how barbiturates are stored in the body, many patients received sudden and fatal overdoses. In fact, it is thought that more casualties died at the hands of the anaesthetists at Pearl Harbor than died of their wounds.

To conclude, the 'good' drugs are not as perfect as one might think.

What about the 'bad' drugs, then? Is there anything good that can be said about them? Surely there is nothing we can say in defence of the highly addictive drug heroin?

Well, let us look at the facts about heroin. It is one of the best painkillers we know. In fact, it was named heroin at the end of the nineteenth century because it was thought to be the 'heroic' drug that would banish pain for good. Heroin went on the market in 1898, but 5 years later the true nature of its addictive properties became evident and the drug was speedily withdrawn from general distribution. However, heroin is still used in medicine today—under strict control, of course. The drug is called diamorphine and it is the drug of choice for treating patients dying of cancer. Not only does diamorphine reduce pain to acceptable levels, it also produces a euphoric effect that helps to counter the depression faced by patients close to death. Can we really condemn a drug which does that as being all 'bad'?

By now it should be evident that the division between good drugs and bad drugs is a woolly one and is not really relevant to our discussion of medicinal chemistry. All drugs have their good points and their bad points. Some have more good points than bad and vice versa, but, like people, they all have their own individual characteristics. So how are we to define a drug in general?

One definition could be to classify drugs as 'compounds which interact with a biological system to produce a biological response'. This definition covers all the drugs we have discussed so far, but it goes further. There are chemicals which we take every day and which have a biological effect on us. What are these everyday drugs?

One is contained in all the cups of tea, coffee, and cocoa that we consume. All of these beverages contain the stimulant caffeine. Whenever you take a cup of coffee, you are a drug user. We could go further. Whenever you crave a cup of coffee, you are a drug addict. Even kids are not immune. They get their caffeine 'shot' from Coke or Pepsi. Whether you like it or not, caffeine is a drug. When you take it, you experience a change of mood or feeling.

Caffeine

So too, if you are a worshipper of the 'nicotine stick'. The biological effect is different. In this case you crave sedation or a calming influence, and it is the nicotine in the cigarette smoke which induces that effect.

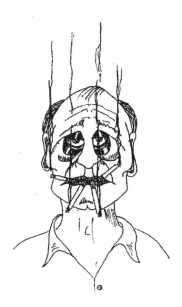

There can be little doubt that alcohol is a drug and, as such, causes society more problems than all other drugs put together. One only has to study road accident statistics to appreciate that fact. If alcohol was discovered today, it would probably be restricted in

exactly the same way as cocaine or cannabis. Considered in a purely scientific way, alcohol is a most unsatisfactory drug. As many will testify, it is notoriously difficult to judge the correct dose required to gain the beneficial effect of 'happiness' without drifting into the higher dose levels that produce unwanted side effects such as staggering down the street. Alcohol is also unpredictable in its biological effects. Either happiness or depression may result, depending on the user's state of mind. On a more serious note, addiction and tolerance in certain individuals have ruined the lives of addicts and relatives alike.

Our definition of a drug can also be used to include compounds which at first sight we might not consider to be drugs. Consider how the following examples fit our definition:

- morphine—reacts with the body to bring pain relief
- snake venom—reacts with the body to cause death!
- strychnine—reacts with the body to cause death!
- LSD—reacts with the body to produce hallucinations
- coffee—reacts with the body to wake you up
- penicillin—reacts with bacterial cells to kill them
- sugar—reacts with the tongue to produce a sense of taste.

All of these compounds fit our definition of drugs. It may seem strange to include poisons and snake

venoms as drugs, but they too react with a biological system and produce a biological response—a bit extreme perhaps, but a response all the same. The idea of poisons and venoms acting as drugs may not appear so strange if we consider penicillin. We have no problem in thinking of penicillin as a drug, but if we were to look closely at how penicillin works, then it is really a poison. It interacts with bacteria (the biological system) and kills them (the biological response). Fortunately for us, penicillin has no effect on human cells.

Even those medicinal drugs which do not act as poisons have the potential to become poisons—usually if they are taken in excess. We have already seen this with morphine. At low doses it is a painkiller. At high doses, it is a poison which kills by suffocation. Therefore, it is important that we treat all medicines as potential poisons and treat them with respect.

There is a term used in medicinal chemistry known as the **therapeutic index**, which indicates how safe a particular drug is. The therapeutic index is a measure of the drug's beneficial effects at a low dose versus its harmful effects at a high dose. To be more precise, the therapeutic index compares the drug dose levels leading to toxic effects in 50% of cases studied

to the dose levels leading to maximum therapeutic effects in 50% of cases studied. A high therapeutic index means that there is a large safety margin between beneficial and toxic doses. The values for cannabis and alcohol are 1000 and 10 respectively, which might imply that cannabis is safer and more predictable than alcohol. Indeed, it has been proposed that cannabis would be a useful medicine for the relief of chronic pain that is untreatable by normal analgesics (e.g. morphine). This may well be the case, but legalizing cannabis for medicinal use does not suddenly make it a safe drug. For example, the favourable therapeutic index of cannabis does not indicate its potential toxicity if it is taken over a long period (chronic use). For example, the various side effects of cannabis include dizziness, sedation, dry mouth, blurred vision, mental clouding, disconnected thought, muscle twitching, and impaired memory. More seriously, it can lead to panic attacks, paranoid delusions, and hallucinations. Clearly, the safety of drugs is a complex matter and it is not helped by media sensationalism.

If useful drugs can be poisons at high doses or over long periods of use, does the opposite hold true? Can a poison be a medicine at low doses? In certain cases, this is found to be the case.

Arsenic is well known as a poison, but arsenic-based compounds were used at the beginning of the twentieth century as antiprotozoal agents. Curare is a deadly poison which was used by the native people of South America to tip their arrows such that a minor arrow wound would be fatal, yet compounds based on the tubocurarine structure (the active principle of curare) have been used in surgical operations to relax muscles. Under proper control and in the correct dosage, a lethal poison may well have an important medical role. Alternatively, lethal poisons can be the starting point for the development of useful drugs. For example, ACE inhibitors are important cardiovascular drugs that were developed from a snake venom.

Since our definition covers any chemical that interacts with any biological system, we could include all pesticides and herbicides as drugs. They interact with bacteria, fungi, and insects, kill them and thus protect plants.

Sugar (or any sweet item for that matter) can be classed as a drug. It reacts with a biological system (the taste buds on the tongue) to produce a biological response (sense of sweetness). Even food can be a drug. Junk foods and fizzy drinks have been blamed for causing hyperactivity in children. It is believed that junk foods have high concentrations of certain amino acids which can be converted in the body to neurotransmitters. These are chemicals that pass messages between nerves. If an excess of these chemical messengers should accumulate, then too many messages are transmitted in the brain, leading to the disruptive behaviour observed in susceptible individuals. Allergies due to food additives and preservatives are also well recorded.

Some foods even contain toxic chemicals. Broccoli, cabbage, and cauliflower all contain high levels of

a chemical that can cause reproductive abnormalities in rats. Peanuts and maize sometimes contain fungal toxins and it is thought that fungal toxins in food were responsible for the biblical plagues. Basil contains over 50 compounds that are potentially carcinogenic, and other herbs contain some of the most potent carcinogens known. Carcinogenic compounds have also been identified in radishes, brown mustard, apricots, cherries, and plums. Such unpalatable facts might put you off your dinner, but take comfort—these chemicals are present in such small quantities that the risk is insignificant. Therein lies a great truth, which was recognized as long ago as the fifteenth century when it was stated that 'Everything is a poison, nothing is a poison. It is the dose that makes the poison'.

Almost anything taken in excess will be toxic. You can make yourself seriously ill by taking 100 aspirin tablets or a bottle of whisky or 9 kg of spinach. The choice is yours!

Having discussed what drugs are, we shall now consider why, where, and how drugs act.

KEY POINTS

- Drugs are compounds that interact with a biological system to produce a biological response.
- No drug is totally safe. Drugs vary in the side effects they might have.
- The dose level of a compound determines whether it will act as a medicine or as a poison.
- The therapeutic index is a measure of a drug's beneficial effect at a low dose versus its harmful effects at higher dose. A high therapeutic index indicates a large safety margin between beneficial and toxic doses.

FURTHER READING

Mann, J. (1992) *Murder, magic, and medicine*, Oxford University Press, Oxford.

Titles for general further reading are listed on p. 711.

2 The why and the wherefore: drug targets

2.1 Why should drugs work?

Why indeed? We take it for granted that they do, but why should chemicals, some of which have remarkably simple structures, have such an important effect on such a complicated and large structure as a human being? The answer lies in the way that the human body operates. If we could see inside our bodies to the molecular level, we would see a magnificent array of chemical reactions taking place, keeping the body healthy and functioning.

Drugs may be mere chemicals but they are entering a world of chemical reactions with which they interact. Therefore, there should be nothing odd in the fact that they can have an effect. The surprising thing might be that they can have such specific effects. This is more a result of *where* they act in the body.

2.2 Where do drugs work?

2.2.1 Cell structure

Since life is made up of cells, then quite clearly drugs must act on cells. The structure of a typical mammalian cell is shown in Fig. 2.1. All cells in the human body contain a boundary wall called the **cell membrane** which encloses the contents of the cell—the **cytoplasm**. The cell membrane seen under the electron microscope consists of two identifiable layers, each of which is made up of an ordered row of phosphoglyceride molecules such as **phosphatidylcholine (lecithin)** (Fig. 2.2). The outer layer of the membrane

is made up of phosphatidylcholine whereas the inner layer is made up of phosphatidylethanolamine, phosphatidylserine, and phosphatidylinositol. Each phosphoglyceride molecule consists of a small polar head group, and two long hydrophobic (water-hating) chains.

In the cell membrane, the two layers of phospholipids are arranged such that the hydrophobic tails point to each other and form a fatty, hydrophobic centre, while the ionic head groups are placed at the inner and outer surfaces of the cell membrane (Fig. 2.3). This is a stable structure because the ionic, hydrophilic head groups interact with the aqueous media inside and outside the cell, whereas the hydrophobic tails maximize hydrophobic interactions

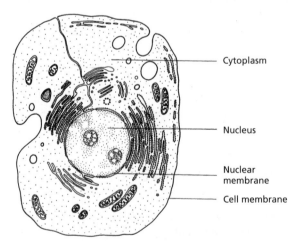

Figure 2.1 A typical mammalian cell. Taken from J. Mann, *Murder, magic, and medicine*, Oxford University Press (1992), with permission.

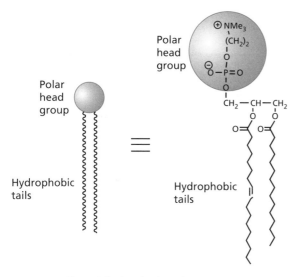

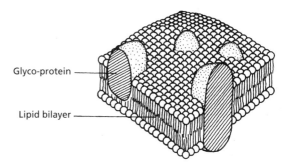

Figure 2.3 Cell membrane. Taken from J. Mann, *Murder, magic, and medicine*, Oxford University Press (1992), with permission.

Figure 2.2 Phosphoglyceride structure.

with each other and are kept away from the aqueous environments. The overall result of this structure is to construct a fatty barrier between the cell's interior and its surroundings.

The membrane is not just made up of phospholipids, however. There is a large variety of proteins situated in the cell membrane (Fig. 2.3). Some proteins lie attached to the inner or the outer surface of the membrane. Others are embedded in the membrane with part of their structure exposed to one surface or both. The extent to which these proteins are embedded within the cell membrane structure depends on the type of amino acid present. Portions of protein that are embedded in the cell membrane have a large number of hydrophobic amino acids, whereas those portions that stick out from the surface have a large number of hydrophilic amino acids. Many surface proteins also have short chains of carbohydrates attached to them and are thus classed as **glycoproteins**. These carbohydrate segments are important to cell–cell recognition.

Within the cytoplasm there are several structures, one of which is the **nucleus**. This acts as the 'control centre' for the cell. The nucleus contains the genetic code—the DNA—which acts as the blueprint for the construction of all the cell's proteins.

There are many other structures within a cell, such as the mitochondria, the Golgi apparatus, and the endoplasmic reticulum, but it is not the purpose of this book to look at the structure and function of these organelles. Suffice it to say that different drugs act on molecular targets at different locations in the cell.

2.2.2 Drug targets at the molecular level

We shall now move to the molecular level, because it is here that we can truly appreciate how drugs work. The four main molecular targets for drugs are lipids, carbohydrates, proteins, and nucleic acids. All of these are large molecules (**macromolecules**) having molecular weights measured in the order of several thousand atomic mass units. They are much bigger than the typical drug, which has a molecular weight in the order of a few hundred atomic mass units. Drugs interact with macromolecules in a process known as **binding**. There is usually a specific area of the macromolecule where this takes place, and this is known as the **binding site** (Fig. 2.4). Typically, this takes the form of a hollow or canyon on the surface of the macromolecule allowing the drug to sink into the body of the larger molecule. Some drugs react with the binding site and become permanently attached via a covalent bond that has a strength of 200–400 kJ mol^{-1}. However, most drugs interact through weaker forms of interaction known as **intermolecular bonds**. (It is also possible for these interactions to take place *within* a molecule, in which case they are called **intramolecular bonds**; see for example protein structure, sections 3.2 and 3.3.) None of these bonds is as strong as the covalent bonds that make up the skeleton of a molecule, and so they can be formed, then broken again. This means that an equilibrium takes place between the drug being bound and unbound to its target. The

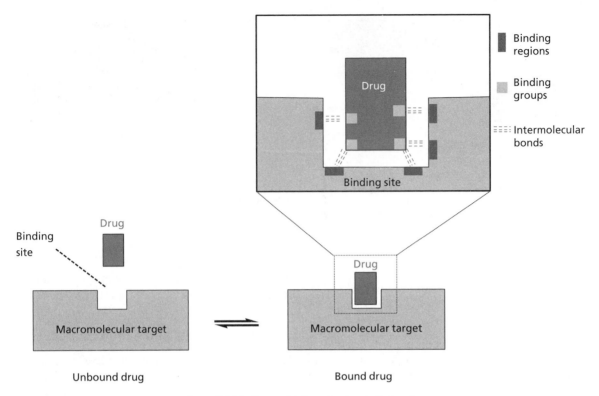

Figure 2.4 Binding equilibrium of a drug to its target.

binding forces are strong enough to hold the drug for a certain period of time to let it have an effect on the target, but weak enough to allow the drug to depart once it has done its job. The length of time the drug remains at its target will then depend on the number of intermolecular bonds involved in holding it there. Drugs having a large number of interactions are likely to remain bound longer than those that have only a few. The relative strength of the different intermolecular binding forces is also an important factor. Functional groups present in the drug can be important in forming intermolecular bonds with the target binding site. If they do so, they are called **binding groups**. However, the carbon skeleton of the drug also plays an important role in binding the drug to its target. As far as the target binding site is concerned, it too contains functional groups and carbon skeletons which can form intermolecular bonds with 'visiting' drugs. The specific regions where this takes place are known as **binding regions**. Let us now consider the types of intermolecular bond that are possible.

2.3 Intermolecular bonding forces

There are several types of intermolecular bonding interactions, which differ in their bond strengths. The number and types of these interactions depend on the structure of the drug and the functional groups that are present (section 10.1). Thus, each drug may use one or more of the following interactions but not necessarily all of them.

2.3.1 Electrostatic or ionic bonds

An ionic or electrostatic bond is the strongest of the intermolecular bonds $(20-40\,\text{kJ}\,\text{mol}^{-1})$ and takes place between groups having opposite charges such as a carboxylate ion and an ammonium ion (Fig. 2.5). The strength of the interaction is inversely proportional to the distance between the two charged atoms, and it is also dependent on the nature of the environment, being stronger in hydrophobic environments than in polar environments. Usually, the binding sites

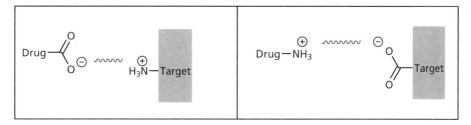

Figure 2.5 Electrostatic (ionic) interactions.

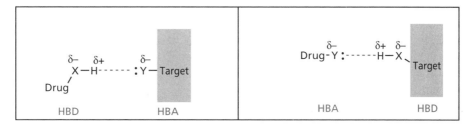

Figure 2.6 Hydrogen bonding shown by dotted line (X, Y = oxygen or nitrogen).

Figure 2.7 Orbital overlap in a hydrogen bond.

of macromolecules are hydrophobic in nature, and so this enhances the effect of an ionic interaction. The drop-off in ionic bonding strength with separation is less than in other intermolecular interactions, so if an ionic interaction is possible, it is likely to be the most important initial interaction as the drug enters the binding site.

2.3.2 Hydrogen bonds

A hydrogen bond can vary substantially in strength and normally takes place between an electron-rich heteroatom and an electron-deficient hydrogen (Fig. 2.6). The electron-rich heteroatom has to have a lone pair of electrons and is usually oxygen or nitrogen.

The electron-deficient hydrogen is usually linked by a covalent bond to an electronegative atom such as oxygen or nitrogen. Since the electronegative atom (X) has a greater attraction for electrons, the electron

distribution in the covalent bond (X–H) is weighted towards the more electronegative atom and so the hydrogen gains its slight positive charge. The functional group containing this feature is known as a **hydrogen bond donor** because it provides the hydrogen for the hydrogen bond. The functional group that provides the electron-rich atom to receive the hydrogen bond is known as the **hydrogen bond acceptor**.

Hydrogen bonds have been viewed as a weak form of electrostatic interaction, because the heteroatom is slightly negative and the hydrogen is slightly positive. However, there is more to hydrogen bonding than an attraction between partial charges. Unlike other intermolecular interactions, an interaction of orbitals takes place between the two molecules (Fig. 2.7). The orbital containing the lone pair of electrons on heteroatom (Y) interacts with the atomic orbitals normally involved in the covalent bond between X and H. This results in a weak form of sigma (σ) bonding and has an important directional consequence that is not evident in electrostatic bonds. The optimum orientation is where the X–H bond points directly to the lone pair on Y such that the angle formed between X, H, and Y is 180°. This is observed in very strong hydrogen bonds. However, the angle can vary between 130 and 180° for moderately strong hydrogen bonds and can be as low as 90° for weak hydrogen bonds. The lone pair orbital of Y also has a directional property, depending on its hybridization. For example, the nitrogen of a pyridine ring is sp^2 hybridized and so the lone pair points directly away from the ring and in the same plane (Fig. 2.8). The best location for a hydrogen bond donor would be the region of space indicated in the figure.

The strength of a hydrogen bond can vary widely, but most hydrogen bonds in drug–target interactions are moderate in strength, varying from 16 to 60 kJ mol^{-1}—approximately 10 times less than a covalent bond. The bond distance reflects this, and hydrogen bonds are typically 1.5–2.2 Å compared to 1.0–1.5 Å for a covalent bond. The strength of a hydrogen bond depends on how strong the hydrogen bond acceptor and the hydrogen bond donor are. A good hydrogen bond acceptor has to be electronegative and have a lone pair of electrons. Nitrogen and oxygen are the most common atoms involved as hydrogen bond acceptors in biological systems. Nitrogen has one lone pair of electrons and can act as an acceptor for one hydrogen bond; oxygen has two lone pairs of electrons and can act as an acceptor for two hydrogen bonds (Fig. 2.9).

Several drugs and macromolecular targets contain a sulfur atom, which is also electronegative. However, sulfur is a weak hydrogen bond acceptor because its lone pairs are in third-shell orbitals that are larger and more diffuse. This means that the orbitals concerned interact less efficiently with the small 1s orbitals of hydrogen atoms.

Fluorine, which is present in several drugs, is more electronegative than either oxygen or nitrogen. It also has three lone pairs of electrons, and this might suggest that it would make a good hydrogen bond acceptor. In fact, it is a weak hydrogen bond acceptor. It has been suggested that fluorine is so electronegative that it clings on tightly to its lone pairs of electrons, making them incapable of hydrogen bond interactions. This is in contrast to fluoride ions, which are very strong hydrogen bond acceptors.

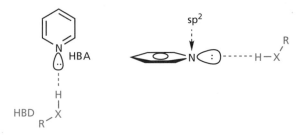

Figure 2.8 Directional influence of hybridization on hydrogen bonding.

Figure 2.9 Oxygen and nitrogen as hydrogen bond acceptors.

Any feature that affects the electron density of the hydrogen bond acceptor is likely to affect its ability to act as a hydrogen bond acceptor, since the greater the electron density of the heteroatom the greater its strength as a hydrogen bond acceptor. For example, the oxygen of a negatively charged carboxylate ion is a stronger hydrogen bond acceptor than the oxygen of the uncharged carboxylic acid (Fig. 2.10). Phosphate ions can also act as good hydrogen bond acceptors. Most hydrogen bond acceptors present in drugs and binding sites are neutral functional groups such as ethers, alcohols, phenols, amides, amines, and ketones. These groups will form moderately strong hydrogen bonds.

It has been proposed that the pi (π) systems present in alkynes and aromatic rings are regions of high electron density and can act as hydrogen bond acceptors. However, the electron density in these systems is diffuse, and so the hydrogen bonding interaction is much weaker than those involving oxygen or nitrogen. As a result, aromatic rings and alkynes are only likely to be significant hydrogen bond acceptors if they interact with a strong hydrogen bond donor such as a quaternary ammonium ion (NR_4^+).

More subtle effects can influence whether an atom is a good hydrogen bond acceptor or not. For example, the nitrogen atom of an aliphatic tertiary amine is a better hydrogen bond acceptor than the nitrogen of an

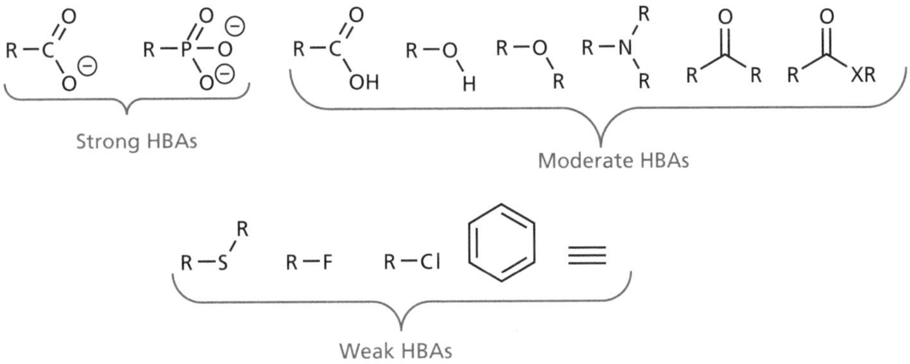

Figure 2.10 Relative strengths of hydrogen bond acceptors.

Tertiary amine—good HBA   Amide—N acts as poor HBA   Aniline—N acts as poor HBA

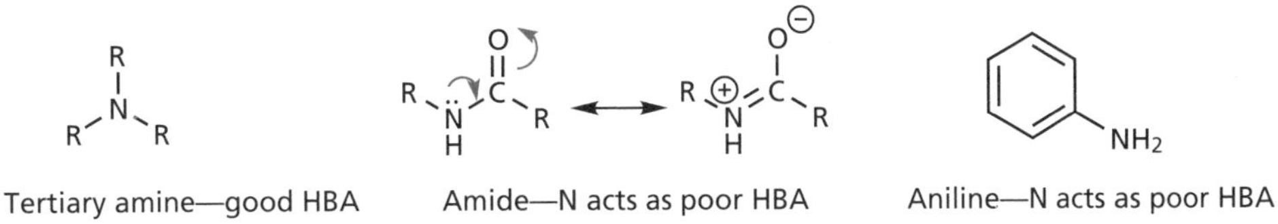

Figure 2.11 Comparison of different nitrogen containing functional groups as hydrogen bond acceptors.

amide or an aniline (Fig. 2.11). In the latter cases, the lone pair of the nitrogen can interact with neighbouring pi systems to form various resonance structures. As a result, it is less likely to take part in a hydrogen bond.

Similarly, the ability of a carbonyl group to act as a hydrogen bond acceptor varies depending on the functional group involved (Fig. 2.12).

Good hydrogen bond donors contain an electron-deficient proton linked to oxygen or nitrogen. The

Increasing strength of carbonyl oxygen as a hydrogen bond acceptor

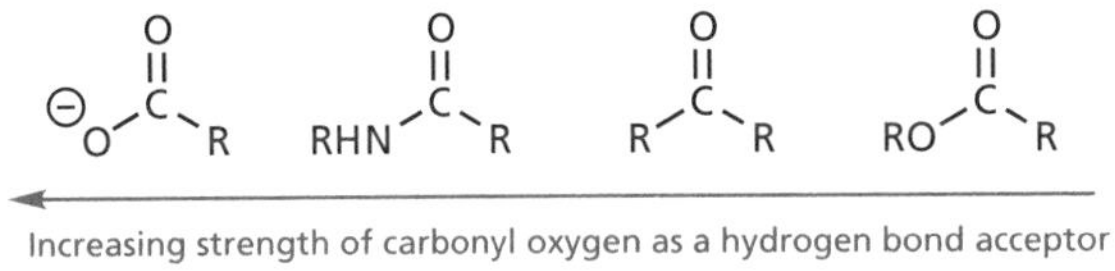

Figure 2.12 Carbonyl oxygens as hydrogen bond acceptors.

Quaternary ammonium ion (stronger HBD)   Secondary and primary amines

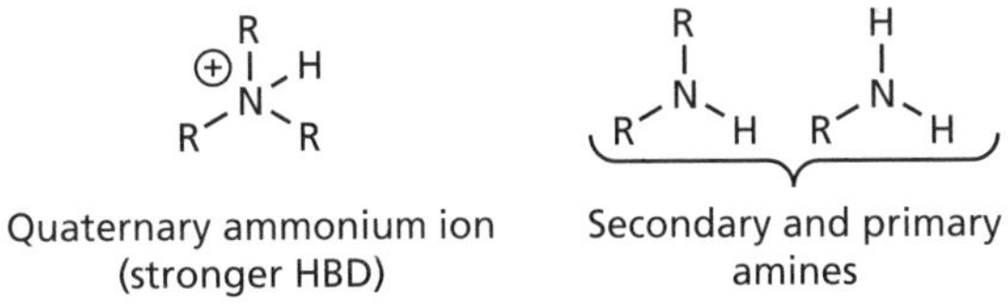

Figure 2.13 Comparison of hydrogen bond donors.

more electron-deficient the proton, the better it will act as a hydrogen bond donor. For example, a proton attached to a positively charged quaternary nitrogen atom acts as a stronger hydrogen bond donor than the proton of a primary or secondary amine (Fig. 2.13). Because the quaternary group is charged, the nitrogen has a greater pull on the electrons surrounding it, making attached protons even more electron deficient.

2.3.3 Van der Waals interactions

Van der Waals interactions are very weak interactions, typically 2–4 kJ mol^{-1} in strength, that involve interactions between hydrophobic regions of different molecules. These regions could be aliphatic substituents or the carbon skeleton of the molecules concerned. The electronic distribution in neutral, non-polar regions is never totally even or symmetrical and there are always transient areas of high and low electron densities leading to temporary dipoles. The dipoles in one molecule can induce dipoles in a neighbouring molecule, leading to weak interactions between the two molecules (Fig. 2.14). Thus, an area of high electron density on one molecule can have an attraction for an area of low electron density on another molecule. The strength of these interactions falls off rapidly the further the two molecules are apart, decreasing to the seventh power of the separation. Therefore, the drug has to be close to the target binding site before the interactions become important. Van der Waals interactions are also referred to as **London forces**. Although the interactions are individually weak, there may be many such interactions between a drug and its target and so the overall contribution of van der Waals interactions can often be crucial to binding. Hydrophobic forces are also important when the non-polar regions of molecules interact (section 2.3.6).

2.3.4 Dipole–dipole and ion–dipole interactions

Many molecules have a permanent dipole moment resulting from the different electronegativities of the atoms and functional groups present. For example, a

ketone has a dipole moment due to the different electronegativities of the carbon and oxygen making up the carbonyl bond. The binding site also contains functional groups, so it is inevitable that it too will have various local dipole moments. It is possible for the dipole moments of the drug and the binding site to interact as a drug approaches, aligning the drug such that the dipole moments are parallel and in opposite directions (Fig. 2.15). If this positions the drug such that other intermolecular interactions can take place between it and the target, the alignment is beneficial to both binding and activity. If not, then binding and activity may be weakened. The strength of dipole–dipole interactions reduces with the cube of the distance between the two dipoles. This means that dipole–dipole interactions fall away more quickly with distance than electrostatic interactions, but less quickly than van der Waals interactions.

An ion–dipole interaction is where a charged or ionic group in one molecule interacts with a dipole in a second molecule (Fig. 2.16). This is stronger than a

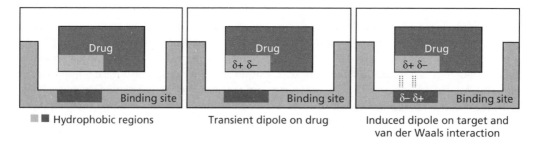

Figure 2.14 Van der Waals interactions.

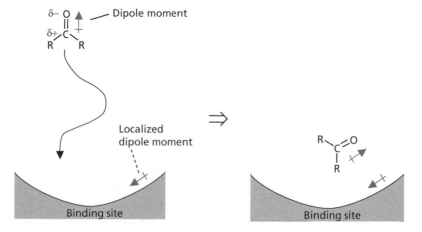

Figure 2.15 Dipole–dipole interactions.

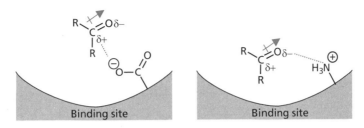

Figure 2.16 Ion–dipole interactions.

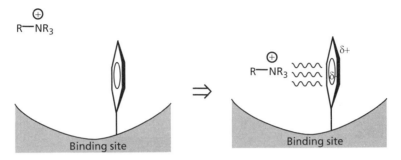

Figure 2.17 Induced dipole interaction.

dipole–dipole interaction and falls off less rapidly with separation (decreasing relative to the square of the separation).

Interactions involving an induced dipole moment have been proposed. There is evidence that an aromatic ring can interact with an ionic group such as a quaternary ammonium ion. Such an interaction is feasible if the positive charge of the quaternary ammonium group distorts the π electron cloud of the aromatic ring to produce a dipole moment where the face of the aromatic ring is electron rich and the edges are electron deficient (Fig. 2.17).

2.3.5 Repulsive interactions

So far we have talked only about attractive forces, which increase in strength the closer the molecules approach each other. Repulsive interactions are also important. Otherwise, there would be nothing to stop molecules trying to merge with each other! If molecules come too close, their molecular orbitals start to overlap and this results in repulsion. Other forms of repulsion are related to the types of groups present in both molecules. For example, two charged groups of identical charge are repelled.

2.3.6 The role of water and hydrophobic interactions

A crucial feature that is often overlooked when considering the interaction of a drug with its target is the role of water. The macromolecular targets in the body exist in an aqueous environment and the drug has to travel through that environment in order to reach its target, so both the drug and the macromolecule are solvated with water molecules before they meet each other. The water molecules surrounding the drug and the target binding site have to be stripped away before the interactions described above can take place (Fig. 2.18). This requires energy, and if the energy required to desolvate both the drug and the binding site is greater than the energy gained by the binding interactions, then the drug may be ineffective. In certain cases, it has been beneficial to remove a polar binding group from a drug if that group binds less strongly to the target binding site than to water. This was carried out during the development of the antiviral drug ritonavir (section 17.7.4.4).

It is not possible for water to solvate the non-polar or hydrophobic regions of a drug or its target binding site. Instead, the surrounding water molecules form

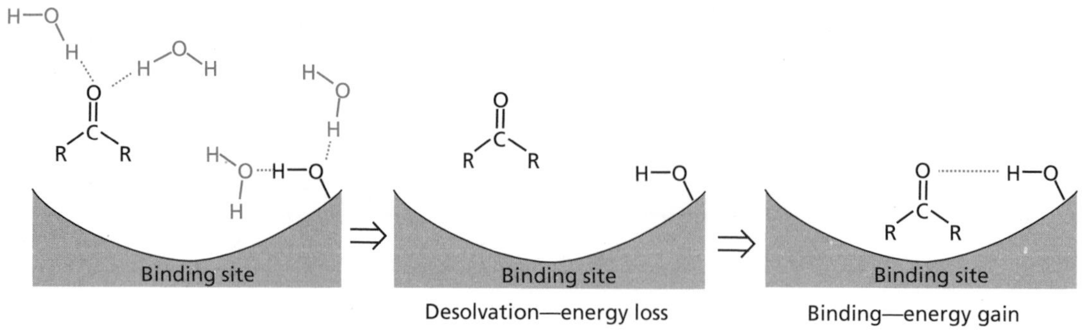

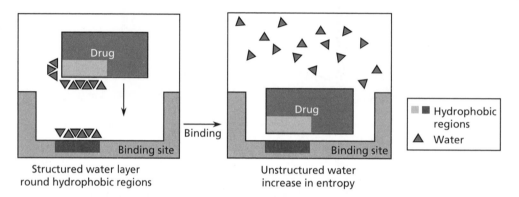

Figure 2.18 Desolvation.

Figure 2.19 Hydrophobic interactions.

stronger than usual interactions with each other, resulting in a more ordered layer of water next to the non-polar surface. This represents a negative entropy due to the increase in order. When the hydrophobic region of a drug interacts with a hydrophobic region of a binding site, these water molecules are freed and become less ordered (Fig. 2.19). This leads to an increase in entropy and a gain in binding energy.[1] The interactions involved are small—0.1 to 0.2 kJ mol^{-1} for each square angstrom of hydrophobic surface—but overall they can be substantial. Sometimes, a hydrophobic region in the drug may not be sufficiently close to a hydrophobic region in the binding site and water may be trapped between the two surfaces. The entropy increase is not so substantial in that case,

and there is a benefit in designing a better drug that fits more snugly.

KEY POINTS

- Drugs act on molecular targets located in the cell membrane of cells or within the cells themselves.

- Drug targets are macromolecules that have a binding site into which the drug fits and binds.

- Most drugs bind to their targets by means of intermolecular bonds.

- Electrostatic or ionic interactions occur between groups of opposite charge.

- Hydrogen bonds occur between an electron-rich heteroatom and an electron-deficient hydrogen.

- The functional group providing the hydrogen for a hydrogen bond is called a hydrogen bond donor. The functional group that interacts with the hydrogen in a hydrogen bond is called a hydrogen bond acceptor.

---

[1] The free energy gained by binding (ΔG) is related to the change in entropy (ΔS) by the equation $\Delta G = \Delta H - \Delta S$. If entropy increases, ΔS is positive which makes ΔG more negative. The more negative ΔG is, the more likely binding will take place.

- Van der Waals interactions take place between the non-polar regions of molecules and are caused by transient dipole–dipole interactions.

- Ion–dipole and dipole–dipole interactions are a weak form of electrostatic interaction.

- Hydrophobic interactions involve the displacement of ordered layers of water molecules which surround hydrophobic regions of molecules. The resulting increase in entropy contributes to the overall binding energy.

- Polar groups have to be desolvated before intermolecular interactions take place. This results in an energy penalty.

2.4 Drug targets

2.4.1 Lipids as drug targets

The number of drugs that interact with lipids is relatively small and, in general, they all act in the same way—by disrupting the lipid structure of cell membranes. For example, it has been proposed that general anaesthetics work by interacting with the lipids of cell membranes to alter the structure and conducting properties of the cell membrane.

The antifungal agent **amphotericin B** (Fig. 2.20) (used topically against athletes foot and systemically against life-threatening fungal diseases) interacts with the lipids of fungal cell membranes to build 'tunnels' through the membrane. Once in place, the contents of the cell are drained away and the cell is killed.

Amphotericin B is a fascinating molecule in that one half of the structure is made up of double bonds and is hydrophobic, whereas the other half contains a series of hydroxyl groups and is hydrophilic. It is

a molecule of extremes, and as such is ideally suited to act on the cell membrane in the way that it does. Several amphotericin molecules cluster together such that the alkene chains are to the exterior and interact favourably with the hydrophobic centre of the cell membrane. The tunnel resulting from this cluster is lined with the hydroxyl groups and so is hydrophilic, allowing the polar contents of the cell to escape (Fig. 2.21). The compound is a natural product derived from a microorganism (*Streptomyces nodosus*).

The antibiotics **valinomycin** and **gramicidin A** operate by acting within the cell membrane as ion carriers and ion channels respectively (section 16.6.1). The anticancer drug **cephalostatin 1** is thought to act by spanning the cell membrane and disrupting its structure (section 18.8.2).

2.4.2 Carbohydrates as drug targets

The term **glycomics** is used to describe the study of carbohydrates as drug targets or as drugs themselves. Carbohydrates are polyhydroxy structures, many of which have the general formula $C_nH_{2n}O_n$. Examples of some simple carbohydrate structures include **glucose**, **fructose**, and **ribose** (Fig. 2.22). These are called **monosaccharides**, because they can be viewed as the monomers required to make more complex polymeric carbohydrates. For example, glucose monomers are linked together to form the natural polymers **glycogen**, **cellulose** (Fig. 2.23), or **starch**.

Until relatively recently, carbohydrates were not considered useful drug targets. The main roles for carbohydrates in the cell were seen as energy storage (e.g. glycogen) or structural (e.g. starch and cellulose).

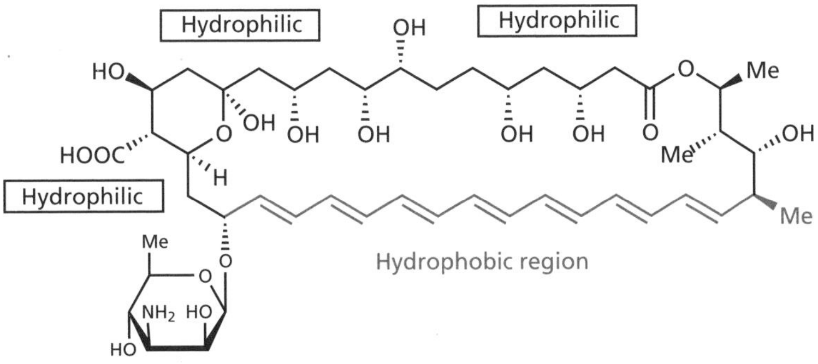

Figure 2.20 Amphotericin B.

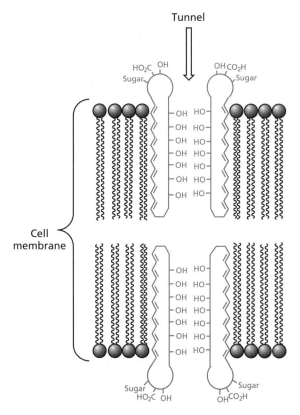

Figure 2.21 Amphotericin-formed channel through the cell membrane.

As a result, there are very few clinically useful drugs that act on carbohydrate targets. However, this is changing. It is now known that carbohydrates have important roles to play in various cellular processes such as cell recognition, regulation, and growth. Various disease states are associated with these cellular processes. For example, bacteria and viruses have to recognize host cells before they can infect them, and so the carbohydrate molecules involved in cell recognition are crucial to that process (sections 17.3, 17.7.1, and 17.8.1). Designing drugs to bind to these carbohydrates may well block the ability of bacteria and viruses to invade host cells. Alternatively, vaccines or drugs may be developed based on the structure of these important carbohydrates (section 17.8.3).

It has also been observed that autoimmune diseases and cancers are associated with changes in the structure of cell surface carbohydrates (section 18.1.10). Understanding how carbohydrates are involved in cell recognition and cell regulation may well allow the design of novel drugs to treat these diseases (section 18.9).

Many of the important cell recognition roles played by carbohydrates are not acted out by pure carbohydrates, but by carbohydrates linked to proteins (**glycoproteins** or **proteoglycans**) or lipids (**glycolipids**). Such

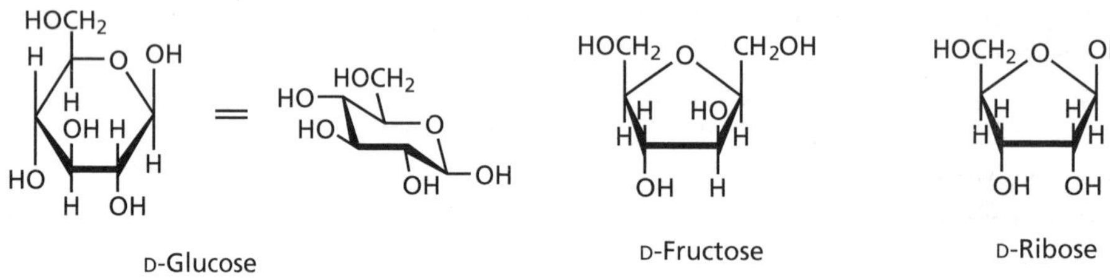

D-Glucose

D-Fructose

D-Ribose

Figure 2.22 Examples of monosaccharides.

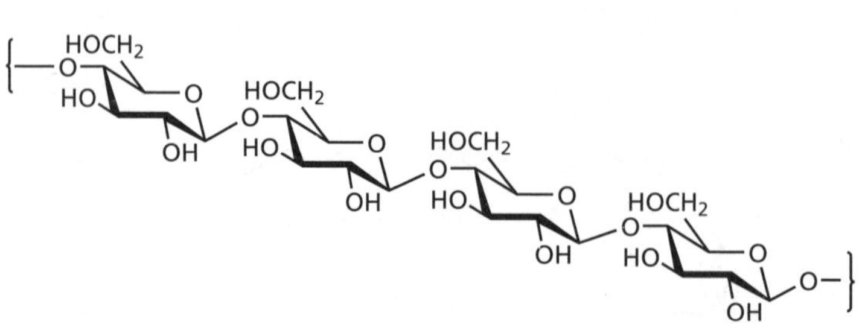

Figure 2.23 Cellulose where glucosyl units are linked β-1,4.

molecules are called **glycoconjugates** (see Box 2.1). Usually, the lipid or protein portion of the molecule is embedded within the cell membrane with the carbohydrate portion hanging free on the outside like the streamer of a kite. This allows the carbohydrate portion to serve the role of a molecular tag that labels and identifies the cell. The tag may also play the role of a receptor, binding other molecules or cells.

Such tags are frequently different between individuals and they can commonly act as **antigens** if the macromolecule or cell containing them is introduced into a different individual. The immune system recognizes the molecular tag as foreign and produces **antibodies** (section 3.9) which bind to it and trigger an immune response aimed at destroying the invader. This has important consequences when it comes to organ transplants, blood transfusions, and the use of proteins as drugs.

There is actually good sense in having a carbohydrate as a molecular tag rather than a peptide or a nucleic acid, because more structural variations are possible for carbohydrates than for other types of structure. For example, two molecules of alanine can only form one possible dipeptide, as there is only one way in which they can be linked (Fig. 2.24). However, because of the different hydroxyl groups on a carbohydrate, there are 11 possible disaccharides that can be formed from two glucose molecules (Fig. 2.25). This allows nature to create an almost infinite number of molecular tags based on different numbers and types of sugar units. Indeed, it has been calculated that 15 million possible structures can be derived from combining just four carbohydrate monomers.

Drugs which mimic natural carbohydrate tags have been used to block viral infection (Tamiflu and Relenza; section 17.8.3); others are being investigated which may block bacterial infection (e.g. *Helicobacter pylori*; section 22.4) or act as anticancer agents.

Several drugs are carbohydrates, or contain carbohydrates as part of their structure. Heparin, a large polysaccharide, is used as an anticoagulant to prevent blood clots after surgery. The important antibiotic **streptomycin** (section 16.7.1) contains carbohydrate units within its structure. Such carbohydrate units may have unusual structures and be important to the compound's activity, as well as lowering the chances of antibacterial resistance developing. An understanding of why these sugar units are important at the molecular level helps in the development of further antibacterial drugs.

The anti-HIV drug **zidovudine** contains an unusual sugar unit and the antiherpes drug **aciclovir** contains an incomplete sugar unit, both of which are important to the mechanism of action of these drugs (section 17.6.1).

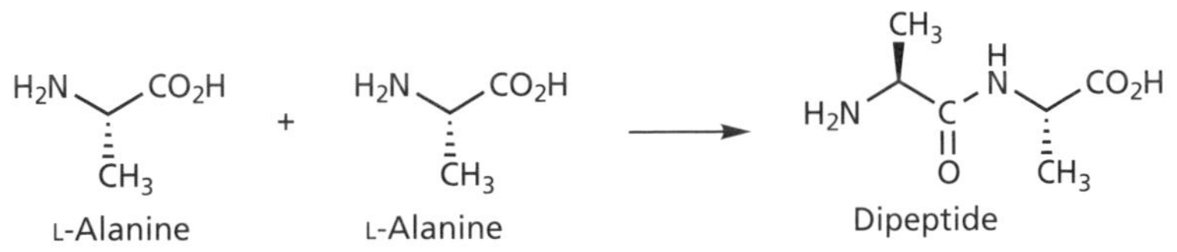

Figure 2.24 Dipeptide formed from linking two L-alanines.

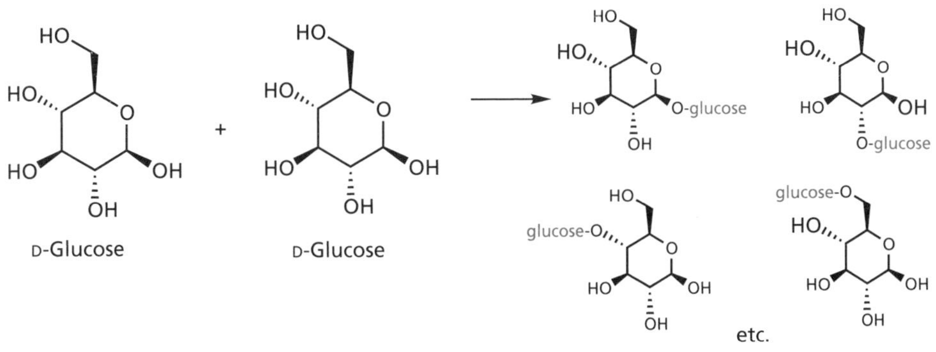

Figure 2.25 Variability in carbohydrate structures.

BOX 2.1 GLYCOSPHINGOLIPIDS

Glycosphingolipids are glycoconjugates which are thought to be important in the regulation of cell growth and consequently have a direct bearing on diseases such as cancer. They are also responsible for labelling red blood cells and hence identifying which blood group one belongs to (A, B, AB, or O). The glycosphingolipids are made up of three components—a carbohydrate structure, which can be highly variable and complex, a structure called sphingosine that consists of a 2-amino-1 3-diol unit

linked to a long-chain hydrocarbon, and a fatty acid (e.g. stearic acid). The portion of the molecule consisting of the sphingosine and the fatty acid is called a **ceramide** (Fig. 1).

The ceramide portion of the molecule is hydrophobic and is embedded within the cell membrane, thus acting as an anchor for the highly polar carbohydrate section. This portion lies outside the cell membrane and acts as the molecular tag for the cell (Fig. 2).

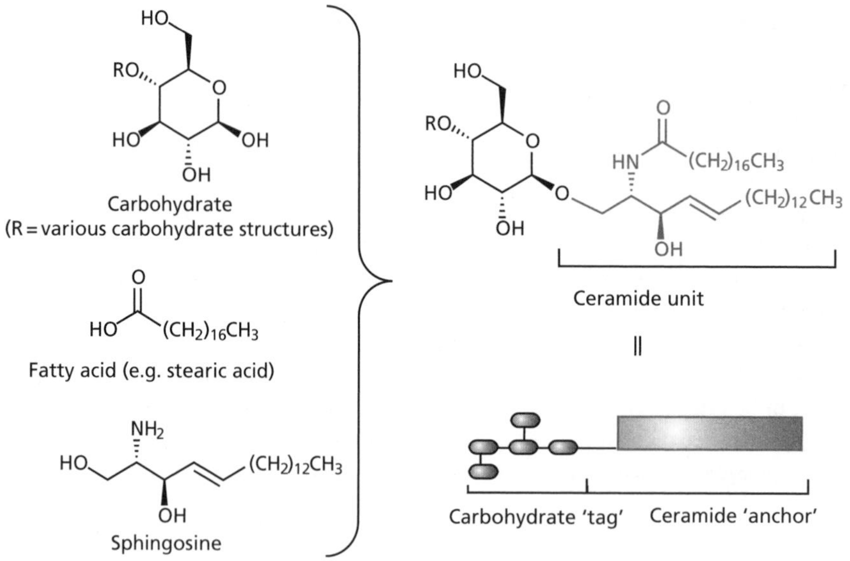

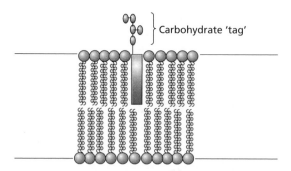

Figure 1 Structure of glycosphingolipids.

Figure 2 Glycosphingolipids as molecular tags.

The glycosides **digoxin** and **digitoxin** (Fig. 2.26) are used in cardiovascular medicine and contain three carbohydrate rings attached to a steroid nucleus. **Vancomycin** is a glycopeptide antibiotic used against antibiotic resistant infections and inhibits the production of a sugar component of the bacterial cell wall (section 16.5.5.2).

The therapeutic protein **erythropoietin** (used to stimulate red blood cell production in the treatment of anaemia) bears sugar molecules which are important in prolonging the drug's lifetime in the body.

Carbohydrate structures offer tremendous potential for novel drugs in the future. However, a relatively small amount of research has been carried out on carbohydrate-based drugs compared to peptide-based drugs. This is partly due to the greater complexity involved in synthesizing or modifying carbohydrates. Carbohydrates have large numbers of hydroxyl groups and asymmetric centres. For example, D-**glucose** has a cyclic structure containing five hydroxyl groups and five asymmetric centres. As a result, carbohydrate synthesis is a demanding activity, where the chemist is often required to carry out a reaction at a specific hydroxyl group while leaving the remaining hydroxyls unaffected. Moreover, reactions have to be done under mild conditions in order to avoid **epimerization**. The situation is made more complicated by the fact that many carbohydrates can easily form two different isomers called **anomers** in solution. For example, when D-glucose is dissolved in water, the cyclic structure opens up to form a linear structure containing an aldehyde functional group. The ring can then reform, but can do so in two ways to form two anomers (Fig. 2.27). One of these (the β-anomer) has the

hydroxyl group derived from the aldehyde in the equatorial position, whereas the other (the α-anomer) has the hydroxyl group in the axial position. An equilibrium is set up favouring the more stable anomer where the hydroxyl group is equatorial, resulting in partial epimerization of an asymmetric centre. Controlling a synthesis to produce a single anomer therefore becomes extremely difficult.

Despite these difficulties, the solid phase synthesis of carbohydrate monomers containing different substituents has proved possible, allowing the use of combinatorial synthesis (Chapter 14) to generate a range of such structures. With such techniques, there is the potential to synthesize linked carbohydrates, allowing access to a wider variety of structures than would be possible by linking amino acids together.

2.4.3 Proteins and nucleic acids as drug targets

We have seen examples of drugs which interact with the lipids of cell membranes, and of the growing importance of carbohydrates in medicinal chemistry. However, the vast majority of drugs used in medicine are targeted on proteins and nucleic acids, and these targets will be the subject of the next few chapters.

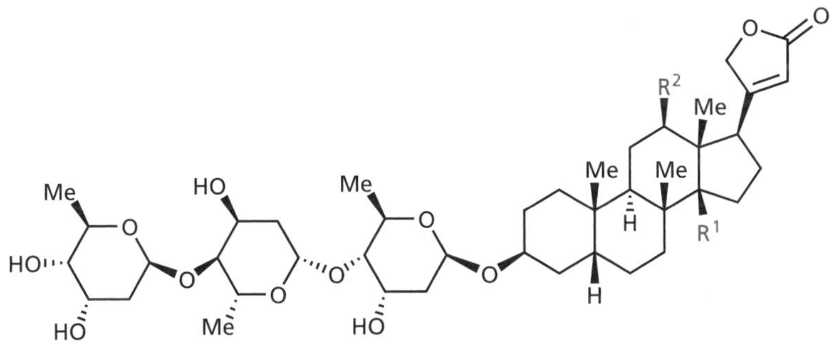

Figure 2.26 Digoxin: $R^1 = OH$, $R^2 = OH$; digitoxin: $R^1 = OH$, $R^2 = H$.

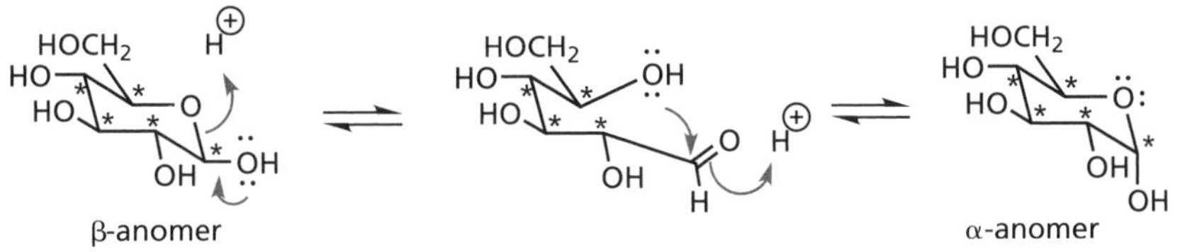

Figure 2.27 Reversible ring opening of D-glucose.

KEY POINTS

- The main molecular targets for drugs are lipids, carbohydrates, proteins, and nucleic acids. Most current drugs act on protein and nucleic acid targets.

- General anaesthetics and some antimicrobial agents target the phospholipid bilayer of cell membranes.

- Carbohydrates are of increasing importance as drugs or as drug targets in developing new therapies for infection, cancer, and autoimmune disease.

- Carbohydrates are more challenging to synthesize than peptides but offer a greater variety of potential novel structures.

QUESTIONS

1. If β-D-glucose is dissolved in water it has an optical rotation of +19°. However, with time, this increases to +52.5°. If α-D-glucose is dissolved in water, the initial optical rotation is +112°, but this eventually falls to +52.5°. When methyl β-D-glucose is dissolved in water, the optical rotation remains constant at −32.6°. Explain these observations.

2. The hormone adrenaline interacts with proteins located on the surface of cells and does not cross the cell membrane. However, larger steroid molecules such as oestrone cross cell membranes and interact with proteins located in the cell nucleus. Why is a large steroid molecule able to cross the cell membrane when a smaller molecule such as adrenaline cannot?

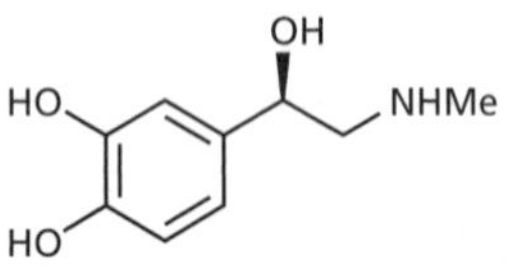

Adrenaline 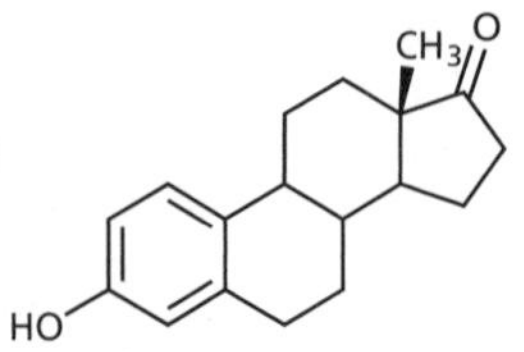   Oestrone

3. Valinomycin (Fig. 16.62) is an antibiotic which is able to transport ions across cell membranes and disrupt the ionic balance of the cell. Consider the structure of valinomycin and explain why it is able to carry out this task.

4. Archaea are microorganisms which can survive in extreme environments such as high temperature, low pH, or high salt concentration. It is observed that the cell membrane phospholipids in these organisms (see structure 1 below) are

markedly different from those in eukaryotic cell membranes. What differences are present and what function might they serve?

5. Teicoplanin (Fig. 16.59) is an antibiotic which inhibits the construction of bacterial cell walls. The cell wall is a barrier surrounding the cell membrane of bacterial cells and is not present in eukaryotic cells. Teicoplanin contains an alkyl chain highlighted in Fig. 16.59. If this chain is absent, activity drops. What role do you think this alkyl chain might serve?

6. The Ras protein is an important protein in signalling processes within the cell. It exists freely in the cell cytoplasm, but must become anchored to the inner surface of the cell membrane in order to carry out its function. What kind of modification to the protein might take place to allow this to happen?

7. Cholesterol is an important constituent of eukaryotic cell membranes and affects the fluidity of the membrane. Consider the structure of cholesterol (shown below) and suggest how it might be orientated in the membrane.

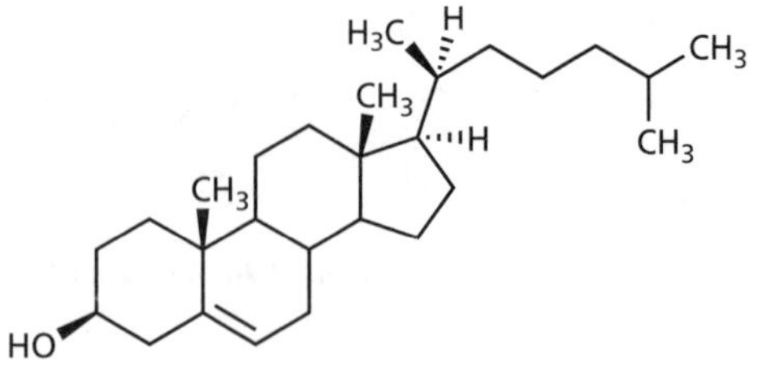

1

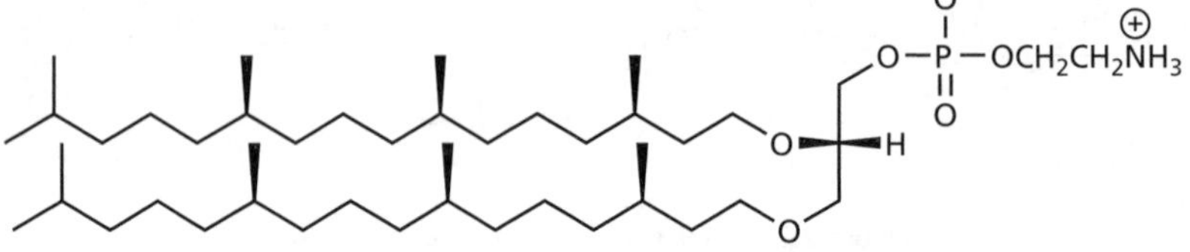

8. Most unsaturated alkyl chains in phospholipids are *cis* rather than *trans*. Consider the *cis*-unsaturated alkyl chain in the phospholipid shown in Fig. 2.2. Redraw this chain to give a better representation of its shape and compare it with the shape of its *trans* isomer. What conclusions can you make regarding the packing of such chains in the cell membrane and the effect on membrane fluidity?

9. The relative strength of carbonyl oxygens as hydrogen bond acceptors is shown in Fig. 2.12. Suggest why the order is as shown.

10. Consider the structures of adrenaline, oestrone, and cholesterol and suggest what kind of intermolecular interactions are possible for these molecules and where they occur.

FURTHER READING

Dwek, R. A. *et al.* (2002) Targeting glycosylation as a therapeutic approach. *Nature Reviews Drug Discovery,* **1**, 65–75.

Howard, J. A. K. *et al.* (1996) How good is fluorine as a hydrogen bond acceptor? *Tetrahedron,* **52**, 12613–12622.

Jeffrey, G. A. (1991) *Hydrogen bonding in biological structures.* Springer-Verlag, London.

Le, G. T. *et al.* (2003) Molecular diversity through sugar scaffolds. *Drug Discovery Today,* **8**, 701–709.

Maeder, T. (2002) Sweet medicines. *Scientific American,* July, 24–31.

Mann, J. (1992) *Murder, magic, and medicine,* Chapter 1. Oxford University Press, Oxford.

Meyer, E. G. *et al.* (1995) Backward binding and other structural surprises. *Perspectives in Drug Discovery and Design,* **3**, 168–195.

Wong, C. (2003) *Carbohydrate-based drug discovery.* John Wiley and Sons, Chichester.

Titles for general further reading are listed on p. 711.

3 Proteins as drug targets

In order to understand how drugs interact with proteins, it is necessary to understand protein structure. Proteins have four levels of structure—primary, secondary, tertiary, and quaternary.

3.1 Primary structure of proteins

The primary structure is the order in which the individual amino acids making up the protein are linked together through peptide bonds (Fig. 3.1). The 20 common amino acids found in humans are listed in Table 3.1, with the three-letter and one-letter codes frequently used to represent them. The structures of the amino acids are shown in Appendix 1. The primary structure of **Met-enkephalin** (one of the body's own painkillers) is shown in Fig. 3.2.

The peptide bond in proteins is planar in nature as a result of the resonance structure shown in Fig. 3.3. This gives the peptide bond a significant double bond character which prevents rotation. As a result bond

rotation in the protein backbone is only possible for the bonds on either side of each peptide bond. This has an important consequence for protein tertiary structure.

There are two possible configurations for the peptide bond (Fig. 3.4). The *trans* configuration is the one that is normally present in proteins, because the *cis* configuration leads to a steric clash between the residues. However, the *cis* configuration is possible for peptide bonds next to a proline residue.

Figure 3.1 Primary structure (R^1, R^2, and R^3 = amino acid residues).

Table 3.1 The 20 common amino acids found in humans

Synthesized in the human body			Essential to the diet		
Amino acid	Code		Amino acid	Code	
	3-letter	1-letter		3-letter	1-letter
Alanine	Ala	A	Histidine	His	H
Arginine	Arg	R	Isoleucine	Ile	I
Asparagine	Asn	N	Leucine	Leu	L
Aspartic acid	Asp	D	Lysine	Lys	K
Cysteine	Cys	C	Methionine	Met	M
Glutamic acid	Glu	E	Phenylalanine	Phe	F
Glutamine	Gln	Q	Threonine	Thr	T
Glycine	Gly	G	Tryptophan	Trp	W
Proline	Pro	P	Valine	Val	V
Serine	Ser	S			
Tyrosine	Tyr	Y			

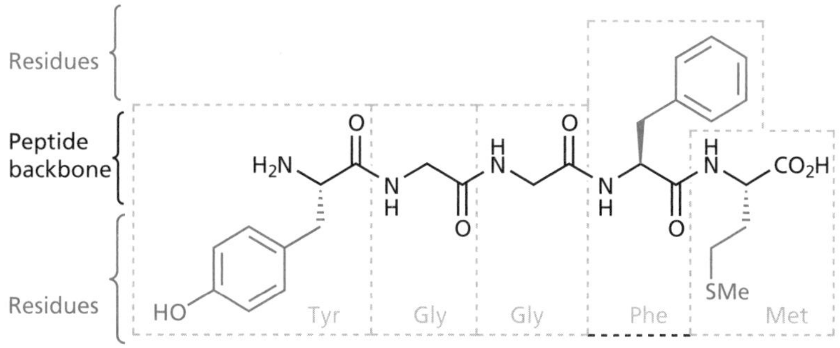

Figure 3.2 Met-enkephalin. The shorthand notation for this peptide is H-Tyr-Gly-Gly-Phe-Met-OH or YGGFM.

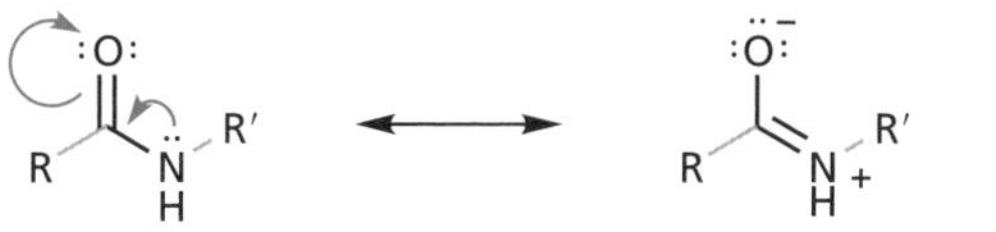

Figure 3.3 The planar peptide bond (bond rotation allowed for coloured bonds).

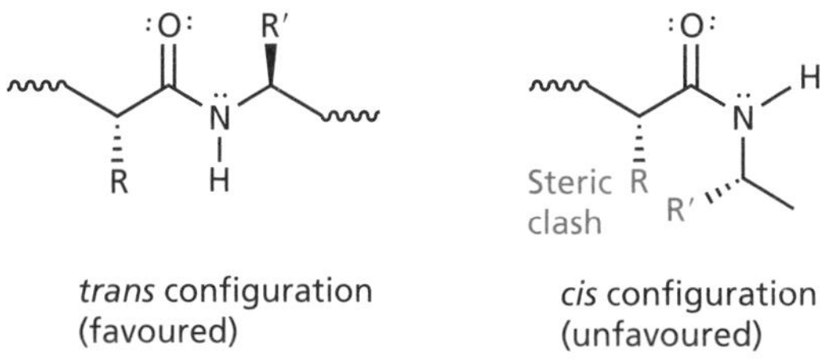

trans configuration (favoured)   *cis* configuration (unfavoured)

Figure 3.4 *Trans* and *cis* configurations of the peptide bond.

3.2 Secondary structure of proteins

The secondary structure of proteins consists of regions of ordered structures adopted by the protein chain. In structural proteins such as wool and silk, secondary structures are extensive and determine the overall shape and properties of such proteins. However, there are regions of secondary structure in most other proteins as well. There are three main structures—the α-helix, the β-pleated sheet, and the β-turn.

3.2.1 α-helix

The α-helix results from coiling of the protein chain such that the peptide bonds making up the backbone are able to form hydrogen bonds between each other. These hydrogen bonds are directed along the axis of the helix, as shown in Fig. 3.5. The residues of the component amino acids stick out at right angles from the helix, thus minimizing steric interactions and further stabilizing the structure.

3.2.2 β-pleated sheet

The β-pleated sheet is a layering of protein chains one on top of another, as shown in Fig. 3.6. Here too, the structure is held together by hydrogen bonds between the peptide chains. The residues are situated at right angles to the sheets, once again to reduce steric interactions. The chains in β-sheets can run in opposite directions (antiparallel) or in the same direction (parallel) (Fig. 3.7).

3.2.3 β-turn

A β-turn allows the polypeptide chain to turn abruptly and go in the opposite direction, and is important in allowing the protein to adopt a more globular compact shape. A hydrogen bonding interaction between the first and third peptide bond of the turn is important in stabilizing the turn (Fig. 3.8). Less abrupt changes in the direction of the polypeptide chain can also take place through longer loops which are less regular in their structure, but are often rigid and well defined.

3.3 Tertiary structure of proteins

The tertiary structure is the overall three-dimensional shape of a protein. Structural proteins are quite ordered in shape, whereas proteins such as enzymes and receptors fold up to form more complex

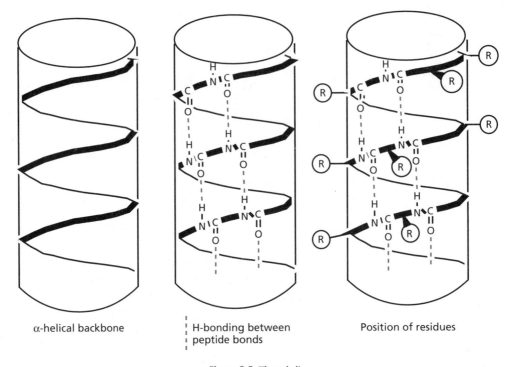

α-helical backbone   H-bonding between peptide bonds   Position of residues

Figure 3.5 The α-helix.

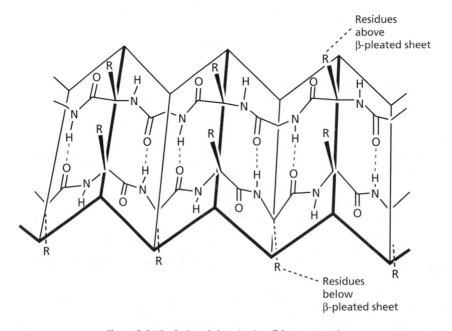

Residues above β-pleated sheet

Residues below β-pleated sheet

Figure 3.6 The β-pleated sheet (antiparallel arrangement).

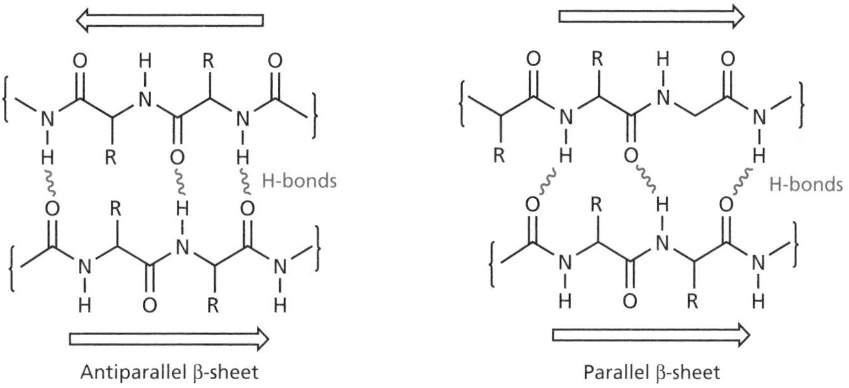

Figure 3.7 Hydrogen bonding in antiparallel and parallel β-sheets (arrows pointing to the C-terminal end of the chain).

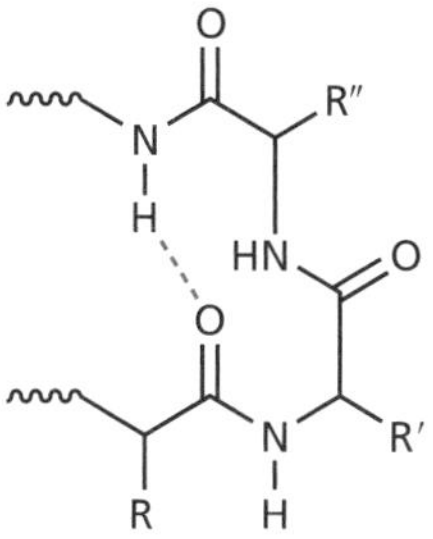

Figure 3.8 The β-turn.

structures. The tertiary structure of enzymes and receptors is crucial to their function and also to their interaction with drugs, so it is important to appreciate the forces that control tertiary structure.

Enzymes and receptors have a great variety of amino acids arranged in what appears to be a random fashion—the primary structure of the protein. Regions of secondary structure may also be present, and these vary from protein to protein. For example, **cyclin-dependent kinase 2** (a protein that catalyses phosphorylation reactions) has several regions of α-helices and β-pleated sheets (Fig. 3.9), whereas the digestive enzyme **chymotrypsin** has very little secondary structure. Nevertheless, the protein chains in both cyclin-dependent kinase 2 and chymotrypsin fold up to form a complex globular shape. How does this come about?

At first sight, the three-dimensional structure of cyclin-dependent kinase 2 looks like a ball of string after the cat has been at it. In fact, the structure shown

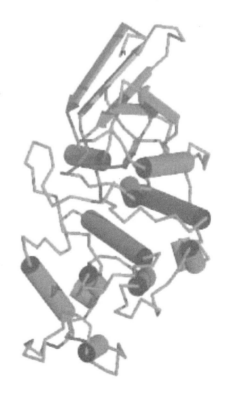

Figure 3.9 Human cyclin-dependent kinase 2 (CDK2) (cylinders represent α-helices, arrows represent β-sheets).

is a very precise shape which is taken up by every molecule of this protein synthesized in the body. There is no outside control or directing force that makes the protein take up this shape. It occurs spontaneously, as a consequence of the protein's primary structure—that is, the amino acids making up the protein and the order in which they are linked. This automatic folding of proteins takes place even as the protein is being synthesized in the cell (Fig. 3.10). Some proteins contain species known as cofactors (e.g. metal ions or small organic molecules) which also have an effect on tertiary structure.

Proteins are synthesized within cells on nucleic acid bodies called **ribosomes**. The ribosomes move along 'ticker-tape'-shaped molecules of another nucleic acid called **messenger RNA** (section 7.2). This messenger RNA contains the code for the protein and is called a messenger because it carries the message from the cell's DNA. The mechanism by which this takes place need not concern us here. We need only note that the ribosome holds on to the growing protein chain as the amino acids are added on one by one. As the chain grows, it automatically folds into its three-dimensional shape such that by the time the protein is fully synthesized and released from the ribosome, the three-dimensional shape is already adopted and is identical for every molecule of that protein synthesized. Indeed, it is possible to synthesize naturally occurring proteins in the laboratory which will adopt the same three-dimensional structure and function as the naturally occurring protein. The HIV-1 protease enzyme is a case in point (section 17.7.4.1).

This poses a problem. Why should a chain of amino acids take up such a precise three-dimensional shape? At first sight, it does not make sense. If we place a length of string on the table, it does not fold itself up into a precise complex shape. So why should a chain of amino acids do such a thing?

The answer lies in the fact that a protein is not just a bland length of string. It has a whole range of chemical functional groups attached along the length of its chain—the residues of each amino acid making up the chain. These can interact with each other. Some will attract each other. Others will repel. Thus the protein will twist and turn to minimize the unfavourable interactions and maximize the favourable ones until the most favourable shape (conformation) is found— the tertiary structure (Fig. 3.11).

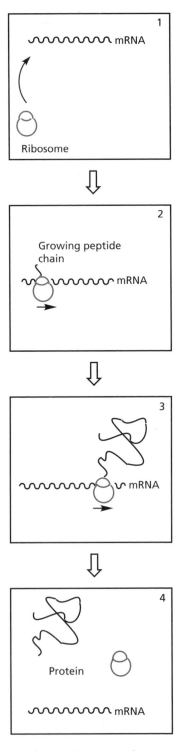

Figure 3.10 Protein synthesis.

What, then, are these important forces? Many of the bonding interactions involved were described in section 2.3 as **intermolecular bonding forces**. In protein tertiary structure, these bonds are taking place within the same molecule and so they are classed as **intramolecular bonds**. Nevertheless, the principles described in section 2.3 are the same regardless of whether the bond is intermolecular or intramolecular.

3.3.1 Covalent bonds

Cysteine has a residue containing a thiol group capable of forming a covalent bond in protein tertiary structure. When two such residues are close together, a covalent disulfide bond can be formed as a result of oxidation. A covalent bridge is thus formed between two different parts of the protein chain (Fig. 3.12).

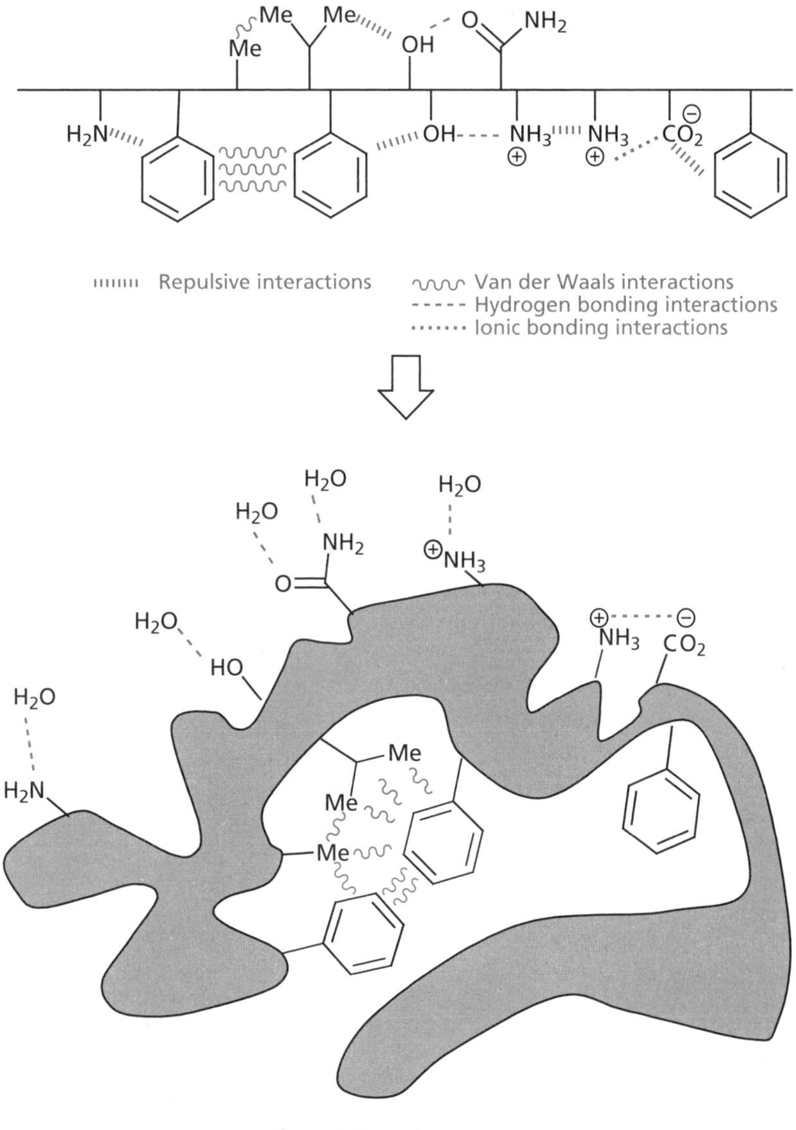

Figure 3.11 Tertiary structure.

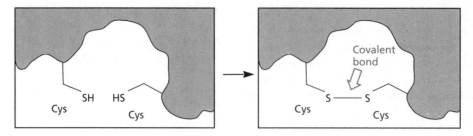

Figure 3.12 Covalent bonding.

3.3.2 Ionic bonds

An ionic bond can be formed between the carboxylate ion of an acidic residue such as aspartic acid or glutamic acid, and the ammonium ion of a basic residue such as lysine, arginine, or histidine (Fig. 3.13).

3.3.3 Hydrogen bonds

Hydrogen bonds can be formed between a large number of amino acid residues such as serine, threonine, aspartic acid, glutamic acid, glutamine, lysine, arginine, histidine, tryptophan, tyrosine, and asparagine. Two examples are shown in Fig. 3.14.

3.3.4 Van der Waals and hydrophobic interactions

Van der Waals interactions can take place between two hydrophobic regions of the protein. For example, they can take place between two aromatic rings or between two alkyl groups (Fig. 3.15). The amino acids alanine, valine, leucine, isoleucine, phenylalanine, and proline all have hydrophobic residues capable of interacting with each other by van der Waals interactions. The residues of other amino acids such as methionine, tryptophan, threonine, and tyrosine contain polar functional groups, but the residues also have a substantial hydrophobic character and so van der Waals interactions are possible for these amino acids as well. Hydrophobic interactions (section 2.3.6) are also important in the coming together of hydrophobic residues.

3.3.5 Relative importance of bonding interactions

We might expect the relative importance of the bonding interactions described above to follow the same order as their strengths: covalent, ionic,

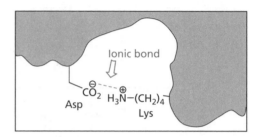

Figure 3.13 Ionic bonding.

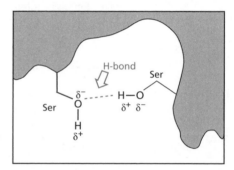

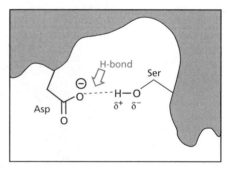

Figure 3.14 Hydrogen bonding.

hydrogen bonding, and finally van der Waals. In fact, the opposite is usually true. Usually the most important bonding interactions in tertiary structure are those due to van der Waals interactions and

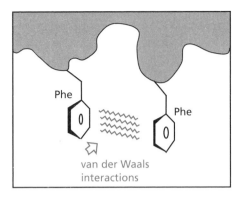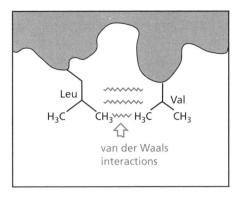

Figure 3.15 Van der Waals interactions.

hydrogen bonding, and the least important interactions are those due to covalent and ionic bonding. There are two reasons for this. First, in most proteins there are more possible opportunities for van der Waals and hydrogen bonding interactions than for covalent or ionic bonding. We only need to consider the types of amino acids in any typical globular protein to see why. The only covalent bond that can contribute to tertiary structure is a disulfide bond. Only cysteine can form such a bond, whereas there are several amino acids that can interact with each other through hydrogen bonding and van der Waals interactions.

Having said that, there *are* examples of proteins with a large number of disulfide bridges, where the relative importance of the covalent link to tertiary structure is more significant. Disulfide links are also more significant in small polypeptides such as the peptide hormones **vasopressin** and **oxytocin** (Fig. 3.16). Nevertheless, in most proteins, disulfide links play a minor role in controlling tertiary structure.

Only a limited number of amino acids have residues capable of forming ionic bonds, and so these too are outnumbered by the number of residues capable of forming hydrogen bonds or van der Waals interactions.

There is a second reason why van der Waals interactions are normally the most important form of bonding in tertiary structure. Proteins do not exist in a vacuum; they are surrounded by water. Water is a highly polar compound that interacts readily with polar, hydrophilic amino acid residues capable of forming hydrogen bonds (Fig. 3.17). The remaining non-polar, hydrophobic amino acid residues cannot

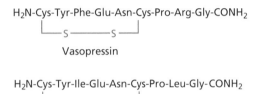

Figure 3.16 Vasopressin and oxytocin.

interact favourably with water, so the most stable tertiary structure will ensure that most of the hydrophilic groups are on the surface so that they interact with water, and most of the hydrophobic groups are in the centre so that they avoid water and interact with each other. Since the hydrophilic amino acids form hydrogen bonds with water, the number of ionic and hydrogen bonds contributing to the tertiary structure is reduced and this leaves hydrophobic and van der Waals interactions to largely determine the three-dimensional shape of the protein.

For the reasons stated above, the centre of the protein must be hydrophobic and non-polar. This has important consequences as far as the action of enzymes is concerned and helps to explain why reactions that should be impossible in an aqueous environment can take place in the presence of enzymes. The enzyme can provide a non-aqueous environment for the reaction to take place (Chapter 4).

There are also important consequences for drug design. Drugs interact with proteins by binding to binding sites which are normally hollows or canyons on the protein surface. These sites too tend to be hydrophobic in

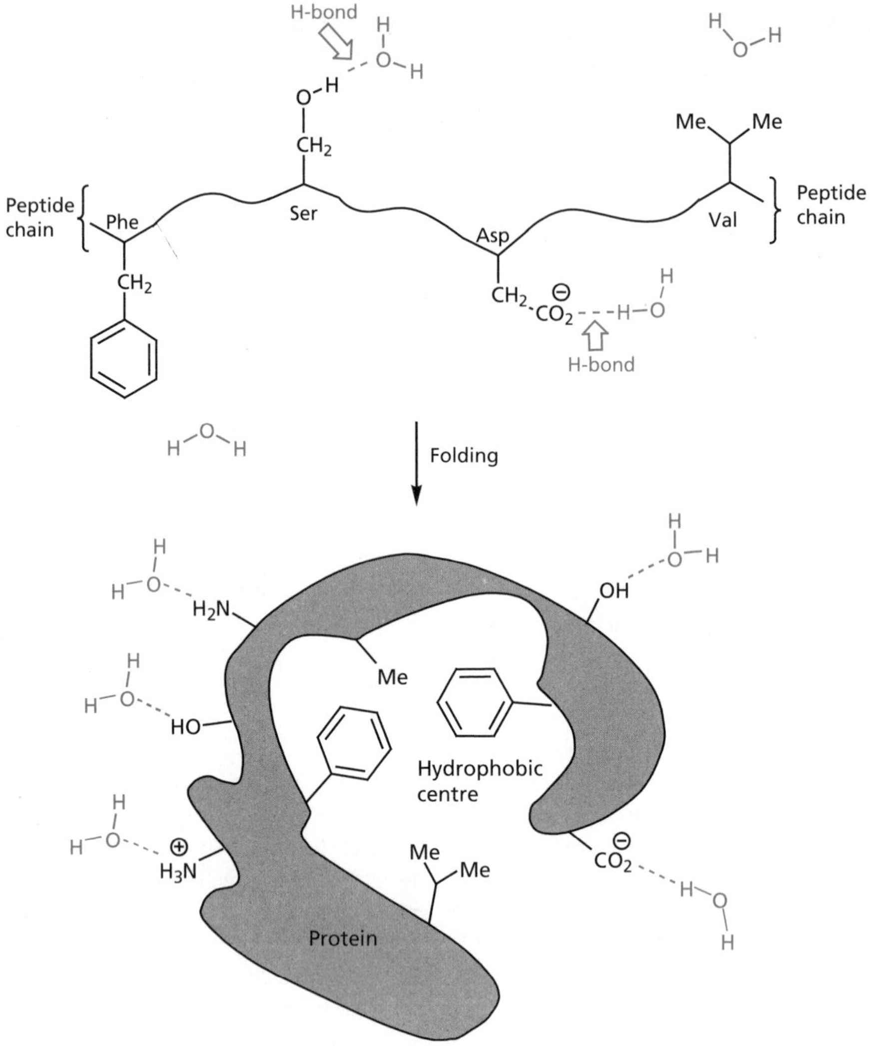

Figure 3.17 Bonding interactions with water.

character, and so van der Waals and hydrophobic interactions play an important role in the binding of a drug to its target, and consequently to its activity.

3.3.6 Role of the planar peptide bond

Planar peptide bonds indirectly play an important role in tertiary structure. Since bond rotation is not possible in peptide bonds, the number of possible conformations that a protein can adopt is significantly restricted, making it more likely that a specific conformation is adopted. Polymers without this restriction do not fold into a specific conformation, because the entropy change required to go from a highly disordered structure to an ordered structure is highly unfavourable.

Peptide bonds can also form hydrogen bonds with amino acid residues, and play a role in determining tertiary structure.

3.4 Quaternary structure of proteins

Only proteins that are made up of a number of protein subunits have quaternary structure. For example, **haemoglobin** is made up of four protein

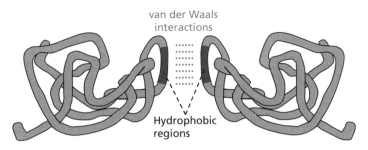

van der Waals
interactions

Hydrophobic
regions

Figure 3.18 Quaternary structure involving two protein subunits.

molecules—two identical alpha subunits and two identical beta subunits (not to be confused with the alpha and beta terminology used in secondary structure). The quaternary structure of haemoglobin is the way in which these four protein units associate with each other. Since this must inevitably involve interactions between the exterior surfaces of proteins, ionic bonding can be more important to quaternary structure than it is to tertiary structure. Nevertheless, hydrophobic (van der Waals) interactions have a role to play. It is not possible for a protein to fold up such that all hydrophobic groups are placed towards the centre. Some such groups may be stranded on the surface. If they form a small hydrophobic area on the protein surface, there is a distinct advantage for two protein molecules to form a dimer such that the two hydrophobic areas face each other rather than be exposed to an aqueous environment (Fig. 3.18).

3.5 Post-translational modifications

The process by which a protein is synthesized in the cell is called **translation** (section 7.2). Many proteins are modified following translation, and these modifications can have wide-ranging effects. For example, the *N*-terminals of many proteins are acetylated, making these proteins more resistant to degradation. Acetylation of proteins also has a role to play in the control of transcription, cell proliferation, and differentiation (section 18.7.4).

The fibres of **collagen** are stabilized by the hydroxylation of proline residues and insufficient hydroxylation results in scurvy (caused by a deficiency of vitamin C). The glutamate residues of **prothrombin**, a clotting protein, are carboxylated to form γ-carboxyglutamate structures. In cases of vitamin K deficiency, carboxylation does not occur and excessive bleeding results. The serine, threonine, and tyrosine residues of many proteins are phosphorylated and this plays an important role in signalling pathways within the cell (sections 6.5–6.7).

Many of the proteins present on the surface of cells are linked to carbohydrates through asparagine residues. Such carbohydrates are added as post-translational modifications and are important to cell–cell recognition, disease processes, and drug treatments (section 2.4.2).

Several proteins are cleaved into smaller proteins or peptides following translation. For example, the **enkephalins** are small peptides which are derived from proteins in this manner (section 21.6). Active enzymes are sometimes formed by cleaving a larger protein precursor. Often this serves to protect the cell from the indiscriminate action of an enzyme. For example, digestive enzymes are stored in the pancreas as inactive protein precursors and are only produced once the protein precursor is released into the intestine. In blood clotting, the soluble protein **fibrinogen** is cleaved to insoluble **fibrin** when the latter is required. Some polypeptide hormones are also produced from the cleavage of protein precursors. Finally, the cleavage of a viral polyprotein into constituent proteins is an important step in the life cycle of the HIV virus, and has proved a useful target for several drugs currently used to combat AIDS (section 17.7.4).

KEY POINTS

- The order in which amino acids are linked together in a protein is called the primary structure.

- The secondary structure of a protein refers to regions of ordered structure within the protein, such as α-helices, β-pleated sheets, or β-turns.

- The overall three-dimensional shape of a protein is called its tertiary structure.
- Proteins containing two or more subunits have a quaternary structure which defines how the subunits are arranged with respect to each other.
- Secondary, tertiary, and quaternary structures are formed to maximize favourable intra- and intermolecular bonds and to minimize unfavourable interactions.
- Amino acids with polar residues are favoured on the outer surface of a protein because this allows hydrogen bonding interactions with water. Amino acids with non-polar residues are favoured within the protein because this maximizes van der Waals and hydrophobic interactions.
- Many proteins undergo post-translational modifications.

3.6 Proteomics

A lot of publicity has been rightly accorded to the Human Genome Project, which has now been completed. The science behind this work is called **genomics** and involves the identification of the genetic code not only in humans but in other species as well. The success of this work has been hailed as a breakthrough that will lead to a new era in medicinal research. However, it is important to appreciate that this is only the start of this process. As we shall see in Chapter 7, DNA is the blueprint for the synthesis of proteins, and so the task is now to identify all the proteins present in each cell of the body and, more importantly, how they interact with each other—an area of science known as **proteomics**. Proteomics is far more challenging than genomics, because of the complexity of interactions that can take place between proteins (see Chapter 6). Moreover, the pattern and function of proteins present in a cell depend on the type of cell it is, and this pattern can alter in the diseased state. Nevertheless, the race is now on to analyse the structure and function of proteins, many of which are completely new to science, and to see whether they can act as novel drug targets for the future. This is no easy task, and it is made all the more difficult by the fact that it is not possible to simply derive the structure of proteins based on the known gene sequences. This is because different proteins can be derived from a single gene, and proteins are often modified following their synthesis (section 3.5). There are roughly 40 000 genes, whereas a typical cell contains hundreds of thousands of different proteins. Moreover, knowing the structure of a protein does not necessarily suggest its function or interactions.

Identifying the proteins present in a cell usually involves analysing the contents of the cell and separating out the proteins using a technique known as two-dimensional gel electrophoresis. Mass spectrometry can then be used to study the molecular weight of each protein. Assuming a pure sample of protein is obtained, its primary structure can be identified by traditional sequencing techniques. The analysis of secondary and tertiary structure is trickier. If the protein can be crystallized, then it is possible to determine its structure by X-ray crystallography. Not all proteins can be crystallized, though, and even if they are it is possible that the conformation in the crystal form is different from that in solution. In recent years nuclear magnetic resonance (NMR) spectroscopy has been successful in identifying the tertiary structure of some proteins.

There then comes the problem of identifying what role the protein has in the cell and whether it would serve as a useful drug target. If it does show promise as a target, the final problem is to discover or design a drug that will interact with it.

3.7 Drug action at proteins

We are now ready to discuss the various types of protein with which drugs interact—enzymes, receptors, carrier proteins, and structural proteins. Enzymes are the body's catalysts, and are discussed in Chapter 4. Receptors are the body's 'letter boxes' and are crucial to the communication between cells. They are discussed in Chapters 5 and 6.

3.7.1 Carrier proteins

Carrier proteins are present in the cell membrane and act as the cell's 'smugglers'—smuggling the important chemical building blocks of amino acids, sugars, and nucleic acids across the cell membrane such that the cell can synthesize its proteins, carbohydrates, and nucleic acids. They are also important in transporting important neurotransmitters (section 5.2) back into the nerve that released them so that the neurotransmitters only have a limited period of activity.

But why is this smuggling operation necessary? Why can't these molecules pass through the membrane by themselves?

Quite simply, the molecules concerned are polar structures and cannot pass through the hydrophobic cell membrane. The carrier proteins can float freely through the cell membrane because they have hydrophobic residues on the outside of the protein which interact favourably with the hydrophobic centre of the cell membrane. When they reach the outer surface, they bind the polar molecule (e.g. an amino acid), stow it away in a hydrophilic pocket and ferry it across the membrane to release it on the other side (Fig. 3.19).

Carrier proteins are not all identical; there are specific carrier ones for the different molecules that need to be smuggled across the membrane. These carrier proteins have a recognition site that allows them to bind and encapsulate their specific cargo. Some carrier proteins can be fooled, and it is possible to design drugs that are mistaken as the usual cargo molecules. As a result, the carrier protein accepts the drug and carries it across the cell membrane, delivering it into the cell. This is one way of delivering highly polar drugs into cells or across cell membranes (sections 8.2 and 11.6.1.3).

Carrier proteins can be viewed as drug targets when drugs prevent them from functioning properly. In such cases, drugs are tightly bound to the carrier protein once they are stowed away, and remain as permanent lodgers. The carrier protein can then no longer carry its usual guest across the cell membrane. Important drugs such as **cocaine** and the **tricyclic antidepressants** act in this way; they prevent neurotransmitters such as **noradrenaline** from re-entering nerve cells (section 20.12.4). This results in an increased level of the neurotransmitter at nerve synapses and has the same effect as adding drugs that mimic the neurotransmitter (Box 3.1).

3.7.2 Structural proteins

Structural proteins do not normally act as drug targets. However, the structural protein **tubulin** is an exception. Tubulin molecules polymerize to form small tubes called **microtubules** in the cell's cytoplasm (Fig. 3.20). These microtubules have various roles within the cell including the maintenance of shape, exocytosis, and release of neurotransmitters. They are also involved in the mobility of cells. For example, inflammatory cells called **neutrophils** are mobile cells which normally protect the body against infection.

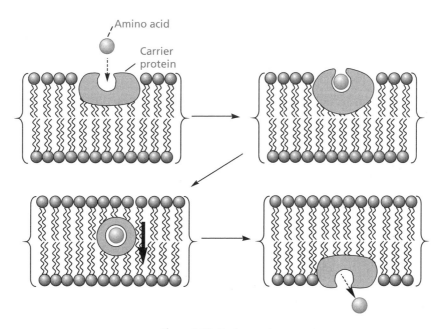

Figure 3.19 Carrier proteins.

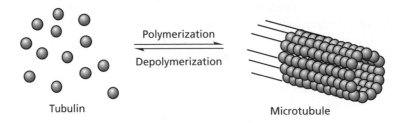

Figure 3.20 Polymerization of tubulin.

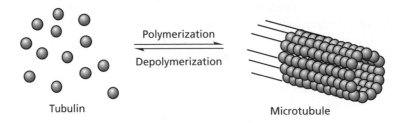

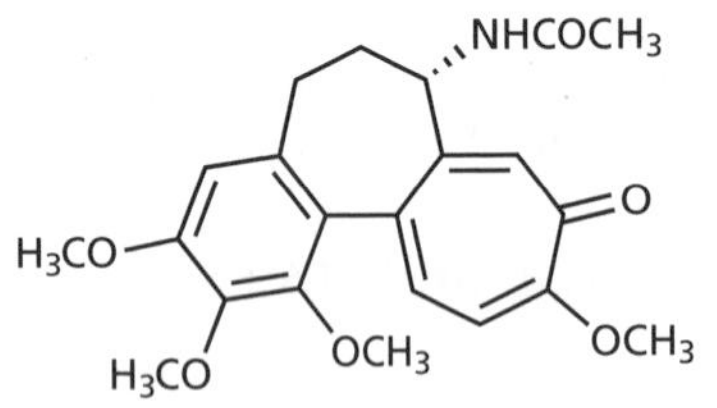

Figure 3.21 Colchicine.

However, they can also enter joints, leading to inflammation and arthritis. One way of treating this disease has been to administer **colchicine** (Fig. 3.21) which binds to tubulin and causes the microtubules to depolymerize. Once the microtubules have been broken down, the neutrophils lose their mobility and can no longer migrate into the joints. Unfortunately, colchicine has many side effects and is not an ideal drug for this treatment.

Tubulin is also crucial to cell division. When a cell is about to divide, its microtubules depolymerize to give tubulin. The tubulin is then repolymerized to form a structure called a **spindle** which then serves to push apart the two new cells and to act as a framework on which the chromosomes of the original cell are transferred to the nuclei of the daughter cells (Fig. 3.22). Several important anticancer drugs such as paclitaxel (Taxol) and vincristine work by binding to tubulin to prevent this polymerization/depolymeriza-

tion cycle taking place, and thus inhibit the growth of tumours (section 18.5).

3.8 Peptides or proteins as drugs

Many of the body's important hormones are peptides or proteins, and in certain diseases the body fails to produce sufficient quantities. Therefore, it is useful to provide such hormones as drugs to make up for any deficiency. Several such hormones have been introduced to the market (e.g. **insulin, human growth factor, interferon**, and **erythropoietin**) and their availability owes a great deal to genetic engineering (section 7.6). It is extremely tedious and expensive to obtain substantial quantities of these proteins by other means. For example, isolating and purifying a hormone from blood samples is doomed to failure because of the tiny quantities of hormone present. It is far more practical to use **recombinant DNA techniques**, whereby the human genes for the protein are cloned and then incorporated into the DNA of fast-growing bacterial, yeast, or mammalian cells. These cells then produce sufficient quantity of the protein.

Using these techniques, it is also possible to produce 'cut down' versions of important body proteins and polypeptides which can also be used therapeutically.

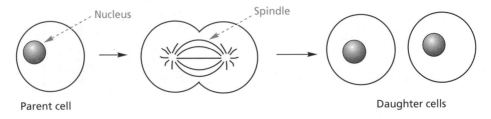

Figure 3.22 Cell division.

BOX 3.1 ANTIDEPRESSANT DRUGS ACTING ON
          CARRIER PROTEINS

The antidepressant drugs **fluoxetine** (Prozac), **citalopram**, and **escitalopram** selectively block the carrier protein responsible for the uptake of a neurotransmitter called **serotonin** from nerve synapses. A lack of serotonin in the brain has been linked with depression and by blocking its uptake, the serotonin that *is* released has a longer duration of action. Fluoxetine and citalopram are chiral molecules which are marketed as racemates. The *S*-enantiomer of citalopram is more active than the *R*-enantiomer and is now marketed as escitalopram. Replacing a racemic drug with a more effective enantiomer is known as **chiral switching**.

Fluoxetine (Prozac)

Citalopram

S-Citalopram (Escitalopram)

Antidepressant drugs acting to block the uptake of serotonin.

Escherichia coli. It consists of 34 amino acids that represent the *N*-terminal end of human parathyroid hormone (consisting of 84 amino acids). Another recombinant protein that has been approved is **etanercept**, which is used for the treatment of rheumatoid arthritis. More than 80 polypeptide drugs have reached the market as a result of the biotechnology revolution, with more to come.

However, as drugs, peptides and proteins have severe disadvantages. They are generally poorly absorbed from the gut and have to be injected, are rapidly metabolized and cleared from the body, and can potentially cause an immune response. This problem is considered in more detail in section 11.8.

3.9 Monoclonal antibodies in medicinal chemistry

Before leaving this chapter, it is worth mentioning another important group of proteins that are present in the body—antibodies (Fig. 3.23). Antibodies are crucial to the recognition and destruction of foreign cells and macromolecules by the immune response. They are Y-shaped molecules, made up of two heavy and two light peptide chains. At the *N*-terminals of these chains there is a highly variable region of amino acids, which differs from antibody to antibody. It is this region that recognizes and binds to particular chemical groupings on the invader. These

For example, **teriparatide** is a polypeptide which has been approved for the treatment of osteoporosis and which was produced by recombinant DNA technology using a genetically modified strain of the bacterium

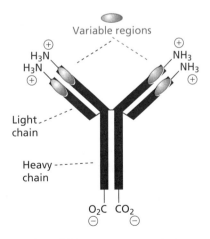

Figure 3.23 Structure of an antibody.

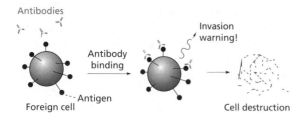

Figure 3.24 Role of antibodies in cell destruction.

groupings are called **antigens** and are normally the glycoconjugates on the outside of the cell (section 2.4.2). Once an antigen is recognized, the antibody binds to it and recruits the body's immune response to destroy the foreign cell (Fig. 3.24). All cells (including our own) have antigens on their outer surface. They act as a molecular signature for different cells, allowing the body to distinguish between its own cells and 'foreigners'. Fortunately, the body does not normally produce antibodies against its own cells and so we are safe from attack.

Because antibodies can recognize the chemical signature of a particular cell, they have great potential in targeting the immune response against cancer cells. Alternatively, they could be used to carry drugs or poisons to cancer cells (see section 11.4.1). Antibodies that recognize a particular antigen are generated by exposing a mouse to the antigen so that the mouse produces the desired antibodies (known as murine antibodies). However, the antibodies themselves are not isolated. Antibodies are produced by cells called B lymphocytes, and it is a mixture of B lymphocytes that is isolated from the mouse. The next task is to find the B lymphocyte responsible for producing the desired antibody. This is done by fusing the mixture with immortal (cancerous) human B lymphocytes to produce cells called **hybridomas**. These are then separated and cultured. The culture that produces the desired antibody can then be identified by its ability to bind to the antigen, and is then used to produce antibody on a large scale. Since all the cells in this culture are identical, the antibodies produced are also identical and are called **monoclonal antibodies**.

There was great excitement when this technology appeared in the 1980s, and an expectation that antibodies would be the magic bullet to tackle many

diseases. Unfortunately, the early antibodies failed to reach the clinic because they triggered an immune response in patients which resulted in antibodies being generated against the antibodies! In hindsight, this is not surprising; the antibodies were mouse-like in character and were identified as 'foreign' by the human immune system, resulting in the production of human anti-mouse antibodies (the **HAMA response**).

In order to tackle this problem, **chimeric antibodies** have been produced which are part human (66%) and part mouse in origin, to make them less 'foreign'. Genetic engineering has also been used to generate 'humanized' antibodies which are 90% human in nature. Genetic engineering has also been used to insert the human genes responsible for antibodies into mice such that the mice (transgenic mice) produce human antibodies rather than murine antibodies when they are exposed to the antigen. As a result of these efforts, 10 antibodies had reached the clinic in 2002 and are used as immune suppressants, antiviral agents (section 17.10.5), and anticancer agents (section 18.9.1). Many others are in the pipeline. **Omalizumab** is an example of a recombinant humanized monoclonal antibody which targets **immunoglobulin E** (IgE) and was approved in 2003 for the treatment of allergic asthmatic disease. It is known that exposure to allergens results in increased levels of IgE, which triggers the release of many of the chemicals responsible for the symptoms of asthma. Omalizumab works by binding to IgE and thus preventing it from acting in this way. Another recent example is **adalimumab**, which was launched in 2003 and is the first fully humanized antibody to be approved. It is used for the treatment of rheumatoid arthritis and works by binding to an inflammatory molecule called a cytokine, specifically one called tumour necrosis factor (TNF)-α. Molecules such as these are endogenous chemicals but are overproduced in arthritis, leading to chronic inflammation. By binding to the cytokine, the antibody prevents it interacting with its receptor. The antibody can also tag cells that are producing the chemical messenger leading to the cell's destruction by the body's immune system. **Infliximab** is another monoclonal antibody that targets TNF-α, but this is a chimeric monoclonal antibody and there is greater chance of the body developing an immune response against it during long-term use.

Work on the large-scale production of antibodies has also been continuing. They have traditionally been produced using hybridoma cells in bioreactors, but more recently companies have been looking at the possibility of using transgenic animals so that the antibodies can be collected in milk. Another possibility is to harvest transgenic plants which produce the antibody in their leaves or seeds.

KEY POINTS

- Carrier proteins, enzymes, and receptors are common drug targets.

- Carrier proteins transport essential polar molecules across the hydrophobic cell membrane. Drugs can be designed to take advantage of this transport system in order to gain access to cells, or to block the carrier protein.

- Tubulin is a structural protein which is crucial to cell division and cell mobility, and which is the target for several anticancer drugs.

- Many of the body's hormones are peptides and proteins and can be produced by recombinant DNA techniques. However, there are several disadvantages in using such compounds as drugs.

- Antibodies are proteins which are important to the body's immune response and which can identify foreign cells or macromolecules, marking them for destruction. They have been used therapeutically and can also be used to carry drugs to specific targets.

QUESTIONS

1. Draw the full structure of L-alanyl-L-phenylalanyl-glycine.

2. What is unique about glycine compared to other naturally occurring amino acids?

3. Identify the type of intermolecular interactions that are possible for the residues of the following amino acids; serine, phenylalanine, glycine, lysine, aspartic acid.

4. The chains of several cell membrane-bound proteins are known to wind back and forth through the cell membrane, such that some parts of the protein structure are extracellular, some parts are intracellular, and some parts lie within the cell membrane. How might the primary structure of such a protein help in distinguishing the portions of the protein embedded within the cell membrane from those positioned intracellularly or extracellularly?

5. What problems might you foresee if you tried to synthesize L-alanyl-L-valine directly from its two component amino acids?

6. The tertiary structure of many enzymes is significantly altered by the phosphorylation of serine, threonine, or tyrosine residues. Identify the functional groups that are involved in these phosphorylations and suggest why phosphorylation should have such an effect on tertiary structure.

7. What are the one-letter and three-letter codes code for the polypeptide Glu-Leu-Pro-Asp-Val-Val-Ala-Phe-Lys-Ser-Gly-Gly-Thr?

FURTHER READING

Berg, C., Neumeyer, K., and Kirkpatrick, P. (2003) Teriparatide. *Nature Reviews Drug Discovery*, **2**, 257–258.

Burke, M. (2002) Pharmas market. *Chemistry in Britain*, June, 30–32 (antibodies).

Darby, N. J. and Creighton, T. E. (1993) *Protein structure*, IRL Press, Oxford.

Dobson, C. M. (2003) Protein folding and disease: a view from the first Horizon symposium. *Nature Reviews Drug Discovery*, 2, 154–160.

Ezzell, C. (2001) Magic bullets fly again. *Scientific American*, October, 28–35 (antibodies).

Ezzell, C. (2002) Proteins rule. *Scientific American*, April, 7–33 (proteomics).

Harris, J. M. and Chess, R. B. (2003) Effect of pegylation on pharmaceuticals. *Nature Reviews Drug Discovery*, **2**, 214–221.

Jones, J. (1992) *Amino acid and peptide synthesis.* Oxford University Press, Oxford.

Stevenson, R. (2002) Proteomic analysis honoured. *Chemistry in Britain*, November, 21–23.

Teague, S. J. (2003) Implications of protein flexibility for drug discovery. *Nature Reviews Drug Discovery*, **2**, 527–541.

Titles for general further reading are listed on p. 711.

4 Proteins as drug targets: enzymes

4.1 Enzymes as catalysts

Enzymes are the body's catalysts—agents that speed up a chemical reaction without being consumed themselves. Without them, the cell's chemical reactions would either be too slow or not take place at all. An example of an enzyme catalysed reaction is the reduction of **pyruvic acid** to **lactic acid**, which takes place when muscles are overexercised and is catalysed by an enzyme called **lactate dehydrogenase** (Fig. 4.1).

Note that the reaction in Fig. 4.1 is shown as an equilibrium. It is therefore more correct to describe an enzyme as an agent that speeds up the approach to equilibrium, because the enzyme speeds up the reverse reaction just as efficiently as the forward reaction. The final equilibrium concentrations of the starting materials and products are unaffected by the presence of an enzyme.

How do enzymes affect the rate of a reaction without affecting the equilibrium? The answer lies in the existence of a high-energy intermediate or **transition state** that must be formed before the starting material can be converted to the product. The difference in energy between the transition state and the starting material is the **activation energy**, and it is the size of this activation energy that determines the rate of a reaction rather than the difference in energy between the starting material and the product (Fig. 4.2).

An enzyme acts to lower the activation energy by helping to stabilize the transition state. The energy of the starting material and products are unaffected, and therefore the equilibrium ratio of starting material to product is unaffected.

We can relate energy to the rate and equilibrium constants with the following equations:

$$\text{Energy difference} = \Delta G = -RT \ln K$$

where K is the equilibrium constant ($=$ [products]/[reactants]), R is the gas constant ($= 8.314$ J mol^{-1} K^{-1}), and T is the absolute temperature.

$$\text{Rate constant} = k = Ae^{-E/RT}$$

where E is the activation energy and A is the frequency factor.

Note that the rate constant k does not depend on the equilibrium constant K.

We have stated that enzymes speed up reaction rates by lowering the activation energy, but we have still to explain how.

Figure 4.1 Reaction catalysed by lactate dehydrogenase.

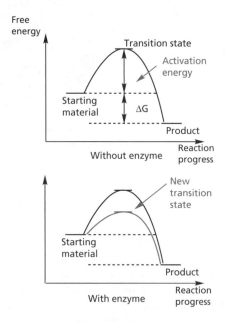

Figure 4.2 Activation energy.

An enzyme catalyses a reaction by providing a surface to which a substrate can bind, resulting in the weakening of high-energy bonds. The binding also holds the substrate in the correct orientation to increase the chances of reaction. The reaction takes place, aided by the enzyme, to give a product which is then released (Fig. 4.3). Note again that it is a reversible process. Enzymes can catalyse both forward and backward reactions. The final equilibrium mixture will, however, be the same, regardless of whether we supply the enzyme with substrate or product. Substrates bind to and react at a specific area of the enzyme called the **active site**—usually quite a small part of the overall protein structure.

4.2 How do enzymes lower activation energies?

There are several factors at work, which are summarized below and will be discussed in more detail in sections 4.2–4.5.

- Enzymes provide a reaction surface and suitable environment.
- Enzymes bring reactants together and position them correctly so that they easily attain their transition-state configurations.
- Enzymes weaken bonds in the reactants.
- Enzymes may participate in the mechanism.

4.3 The active site of an enzyme

The active site of an enzyme (Fig. 4.4) has to be on or near the surface of the enzyme if a substrate is to reach it. However, the site could be a groove, hollow, or gully allowing the substrate to sink into the enzyme. Normally the active site is hydrophobic in character, providing a suitable environment for many reactions that would be difficult or impossible to carry out in an aqueous environment.

Because of the overall folding of the enzyme, the amino acid residues that are close together in the active site may be far apart in the primary structure. For example, the important amino acids in the active site of lactate dehydrogenase are shown in Fig. 4.5. The numbers refer to their positions in the primary structure of the enzyme.

The amino acids present in the active site play an important role in enzyme function, and this can be demonstrated by comparing the primary structures of

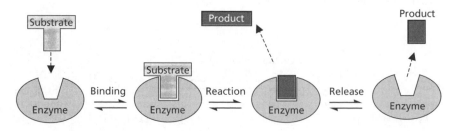

Figure 4.3 Enzyme catalysis.

the same enzyme from different organisms. Here, the primary structure differs from species to species as a result of mutations happening over millions of years, and the variability is proportional to how far apart the organisms are on the evolutionary ladder. However, there are certain amino acids that remain constant, no matter the source of the enzyme. These are amino acids that are crucial to the enzyme's function and, as such, are often the amino acids that make up the active site. If one of these amino acids should be altered through mutation, the enzyme could become useless and the cell bearing this mutation would have a poor chance of survival. Thus, the mutation would not be preserved. (The only exception would be if the mutation either introduced an amino acid which could perform the same task as the original amino acid, or improved substrate binding.) This consistency

of amino acids in the active site can often help scientists determine which amino acids are present in an active site, if this is not known already.

Amino acids present in the active site can have one of two roles:

● binding—the amino acid residue is involved in binding the substrate to the active site

● catalytic—the amino acid is involved in the mechanism of the reaction.

We shall study these in turn.

4.4 Substrate binding at an active site

The interactions which bind substrates to the active sites of enzymes include ionic bonds, hydrogen bonds, dipole–dipole, and ion–dipole interactions, as well as van der Waals and hydrophobic interactions (section 2.3). These binding interactions are the same bonding interactions responsible for the tertiary structure of proteins, but their relative importance differs. Ionic bonding plays a relatively minor role in protein tertiary structure compared to hydrogen

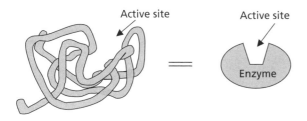

Figure 4.4 The active site of an enzyme.

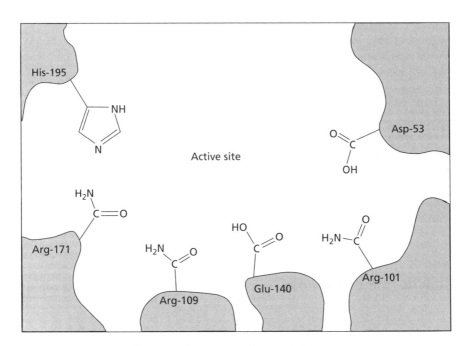

Figure 4.5 The active site of lactate dehydrogenase.

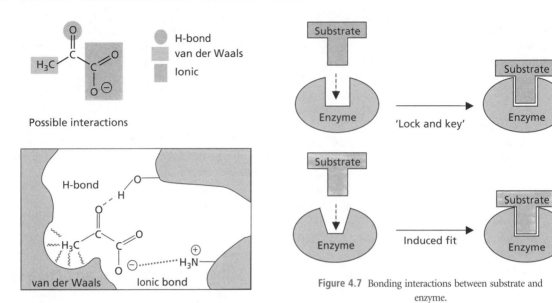

Figure 4.6 Interactions between pyruvic acid and lactate dehydrogenase.

Figure 4.7 Bonding interactions between substrate and enzyme.

bonding or van der Waals interactions, but it can play a crucial role in the binding of a substrate to an active site—not too surprising, because active sites are located on or near the surface of the enzyme.

Since we know the bonding forces involved in substrate binding, it is possible to look at the structure of a substrate and postulate the probable interactions that it will have with its active site. As an example, consider **pyruvic acid**—the substrate for **lactate dehydrogenase** (Fig. 4.6). If we look at the structure of pyruvic acid, we can propose three possible interactions by which it might bind to its active site—an ionic interaction involving the ionized carboxylate group, a hydrogen bond involving the ketonic oxygen, and a van der Waals interaction involving the methyl group. If these postulates are correct, it means that within the active site there must be **binding regions** containing suitable amino acids that can take part in these bonds. Lysine, serine, and phenylalanine residues respectively would fit the bill.

4.5 The catalytic role of enzymes

We now move on to consider the mechanism of enzymes, and how they catalyse reactions. In general, enzymes catalyse reactions by providing binding interactions, acid–base catalysis, nucleophilic groups, and/or cofactors.

4.5.1 Binding interactions

As mentioned previously, the rate of a reaction is increased if the energy of the transition state is lowered. This results from the bonding interactions between substrate and enzyme.

In the past, it was thought that a substrate fitted its active site in a similar way to a key fitting a lock (**Fischer's lock and key hypothesis**). Both the enzyme and the substrate were seen as rigid structures, with the substrate (the key) fitting perfectly into the active site (the lock) (Fig. 4.7). However, such a scenario does not explain how some enzymes can catalyse a reaction on a range of different substrates. It implies instead that an enzyme has an optimum substrate that fits it perfectly and can be catalysed very efficiently, whereas all other substrates are catalysed less efficiently. This is not the case, so the lock and key analogy is invalid.

It is now proposed that the substrate is not quite the ideal shape for the active site and when it enters the active site, it forces the latter to change shape—a kind of moulding process. This theory is known as **Koshland's theory of induced fit**, since the substrate induces the active site to take up the ideal shape to accommodate it (Fig. 4.7).

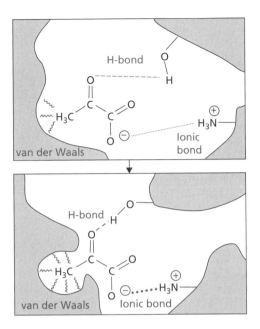

Figure 4.8 Example of induced fit.

For example, a substrate such as **pyruvic acid** might interact with specific binding regions in the active site of lactate dehydrogenase via one hydrogen bond, one ionic bond, and one van der Waals interaction (Fig. 4.8). However, if the fit is not perfect, the three bonding interactions are not ideal either. For example, the binding groups may be slightly too far away from the corresponding binding regions in the active site. In order to maximize the strength of these bonds, the enzyme changes shape such that the amino acid residues involved in the binding move closer to the substrate.

This theory of induced fit helps to explain why enzymes can catalyse reactions involving a wide range of substrates. Each substrate induces the active site into a shape that is ideal for it, and as long as the moulding process does not distort the active site so

much that the reaction mechanism proves impossible, the reaction can proceed. The range of substrates that can be accepted depends on the substrates being the correct size to fit the active site, and having the correct binding groups in the correct relative positions (compare section 5.6—the design of receptor agonists).

But note this. The substrate is not a passive spectator to the moulding process going on around it. As the enzyme changes shape to maximize bonding interactions, the same thing is happening to the substrate. It too will alter shape. Bond rotation may occur to fix the substrate in a particular conformation—not necessarily the most stable conformation. Bonds may even be stretched and weakened. Consequently, this moulding process designed to maximize binding interactions may force the substrate into the ideal conformation for the reaction to follow and may also weaken the very bonds that have to be broken.

Once bound to an active site, the substrate is now held ready for the reaction to follow. Binding has fixed the 'victim' so that it cannot evade attack, and this same binding has weakened its defences (bonds) so that reaction is easier (lower activation energy).

4.5.2 Acid–base catalysis

Usually acid–base catalysis is provided by the amino acid **histidine.** Histidine has an imidazole ring as part of its side chain that acts as a weak base. This means that it exists in equilibrium between its protonated and free base forms (Fig. 4.9), and can act as a proton bank; that is, it has the capability to accept and donate protons in a reaction mechanism. This is important, as active sites are frequently hydrophobic and will therefore have a low concentration of water and an even lower concentration of protons.

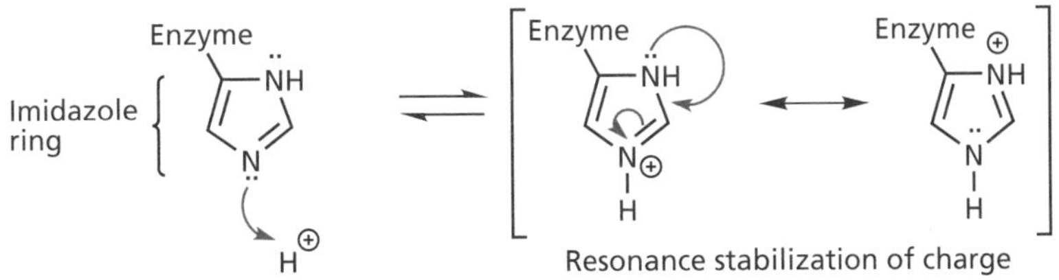

Resonance stabilization of charge

Figure 4.9 Histidine.

4.5.3 Nucleophilic groups

The amino acids **serine** and **cysteine** are commonly present in active sites. These amino acids have nucleophilic residues (OH and SH respectively) which are able to participate in the reaction mechanism. They do this by reacting with the substrate to form intermediates, which would not be formed in the uncatalysed reaction. These intermediates offer an alternative reaction pathway that may avoid a high-energy transition state and hence increase the rate of the reaction.

Normally, an alcoholic OH group such as that on serine is not a good nucleophile. However, there is usually a histidine residue close by to catalyse the reaction. For example, the mechanism by which chymotrypsin hydrolyses peptide bonds (Fig. 4.10) involves a catalytic triad of amino acids—serine, histidine, and aspartic acid. Serine and histidine participate in the mechanism as a nucleophile and

acid–base catalyst respectively. The aspartate group interacts with the histidine ring and serves to orient the ring correctly for the mechanism and also to activate it.

The presence of a nucleophilic serine residue means that water is not required in the initial stages of the mechanism. This is important, because water is a poor nucleophile and may also find it difficult to penetrate a hydrophobic active site. Secondly, a water molecule would have to drift into the active site, and search out the carboxyl group before it could attack it. This would be something similar to a game of blind man's buff. The enzyme, on the other hand, can provide a serine OH group, positioned in exactly the right spot to react with the substrate. Therefore, the nucleophile has no need to search for its substrate. The substrate has been delivered to it.

Water is eventually required to hydrolyse the acyl group attached to the serine residue. However, this is a

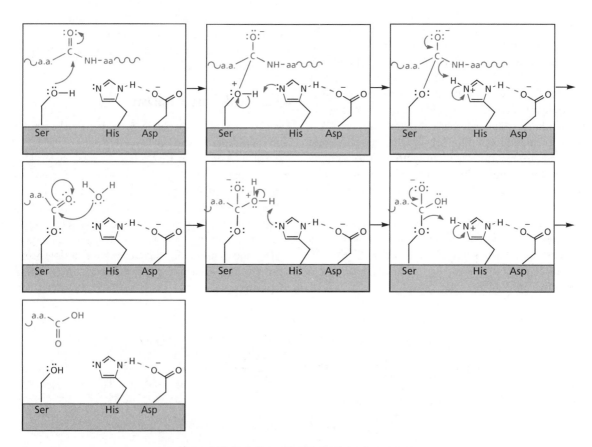

Figure 4.10 Hydrolysis of peptide bonds by chymotrypsin.

much easier step than the hydrolysis of a peptide link, as esters are more reactive than amides. Furthermore, the hydrolysis of the peptide link means that one half of the peptide can drift away from the active site and leave room for a water molecule to enter. A similar enzymatic mechanism is involved in the action of the enzyme **acetylcholinesterase** described in section 19.16.3.

4.5.4 Cofactors

Many enzymes require additional non-protein substances called cofactors for the reaction to take place. Deficiency of cofactors can arise from a poor diet and leads to loss of enzyme activity and subsequent disease

(e.g. scurvy). Cofactors are either metal ions (e.g. zinc) or small organic molecules called **coenzymes** (e.g. NAD^+, pyridoxal phosphate). Most coenzymes are bound by ionic bonds and other non-covalent bonding interactions, but some are bound covalently and are called **prosthetic groups**. Coenzymes are derived from water-soluble vitamins and act as the body's chemical reagents. For example, **lactate dehydrogenase** requires the coenzyme **nicotinamide adenine dinucleotide** (**NAD^+**) (Fig. 4.11) in order to catalyse the dehydrogenation of **lactic acid** to **pyruvic acid**. NAD^+ is bound to the active site along with lactic acid and acts as the oxidizing agent. During the reaction it is converted to its reduced form (NADH) (Fig. 4.12). Conversely, NADH can bind to the enzyme and act as a reducing agent when the enzyme catalyses the reverse reaction. $NADP^+$ and NADPH are phosphorylated analogues of NAD^+ and NADH respectively and carry out redox reactions by the same mechanism. NADPH is used almost exclusively for reductive biosynthesis, whereas NADH is used primarily for the generation of ATP.

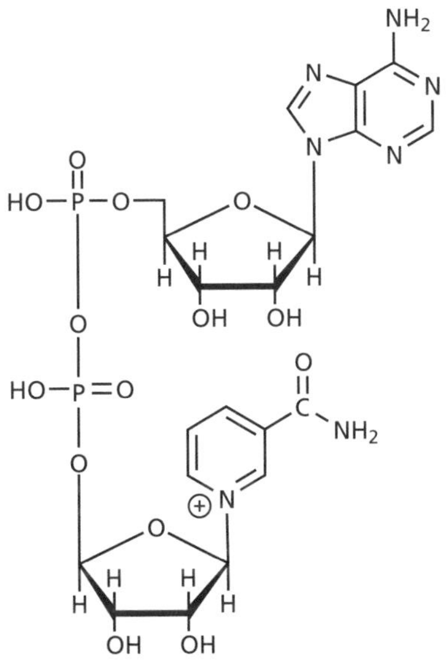

A knowledge of how the coenzyme binds to the enzyme allows the possibility of designing drugs that will fit the region of the active site normally occupied by the coenzyme and inhibit the enzyme (section 15.15).

4.5.5 Naming and classification of enzymes

The name of an enzyme reflects the types of reaction catalysed and has the suffix -ase to indicate it is an enzyme (e.g. oxidase or hydrolase). It is important to appreciate that enzymes can catalyse the forward and back reactions of an equilibrium reaction. This means

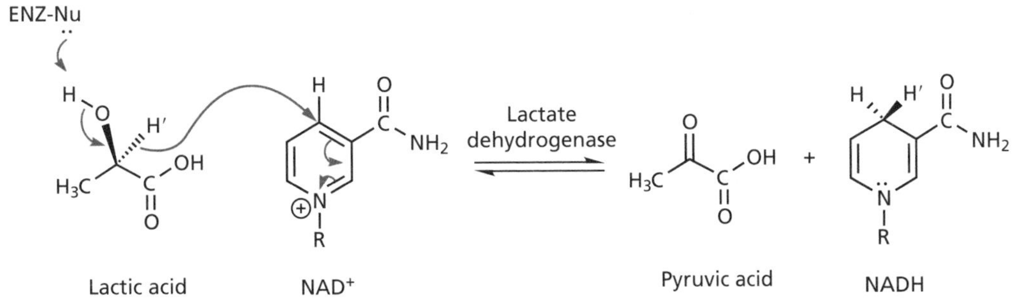

Figure 4.11 Nicotinamide adenine dinucleotide (NAD^+).

Figure 4.12 NAD^+ as a coenzyme.

that an oxidase enzyme can catalyse reductions as well as oxidations. The reaction catalysed depends on the nature of the substrate; that is whether it is in the reduced or oxidized form.

Enzymes are classified according to the general class of reaction they catalyse (Table 4.1).

4.6 Regulation of enzymes

Virtually all enzymes are controlled by agents which can either enhance or inhibit catalytic activity. Such control reflects the local conditions within the cell. For example, the enzyme **phosphorylase *a*** catalyses the breakdown of **glycogen** (a polymer of glucose monomers) to **glucose-1-phosphate** subunits (Fig. 4.13). It is stimulated by **adenosine 5′-monophosphate (AMP)** and inhibited by glucose-1-phosphate. Thus, rising levels of the product (glucose-1-phosphate) act as a self-regulating 'brake' on the enzyme.

But how does this control take place?

The answer is that many enzymes have a binding site which is separate from the active site and is called the **allosteric** binding site (Fig. 4.14). This is where the agents controlling the activity of the enzyme bind. By binding at this site, an induced fit takes place which alters not only the allosteric binding site but the active

Table 4.1 Classification of enzymes

EC number	Enzyme class	Type of reaction
EC.1.x.x.x	Oxidoreductases	Oxidations and reductions
EC.2.x.x.x	Transferases	Group transfer reactions
EC.3.x.x.x	Hydrolases	Hydrolysis reactions
EC.4.x.x.x	Lyases	Addition or removal of groups to form double bonds
EC.5.x.x.x	Isomerases	Isomerizations and intramolecular group transfers
EC.6.x.x.x	Ligases	Joining two substrates at the expense of ATP hydrolysis

Notes: EC stands for Enzyme Commission, a body set up by the International Union of Biochemistry (as it then was) in 1955.

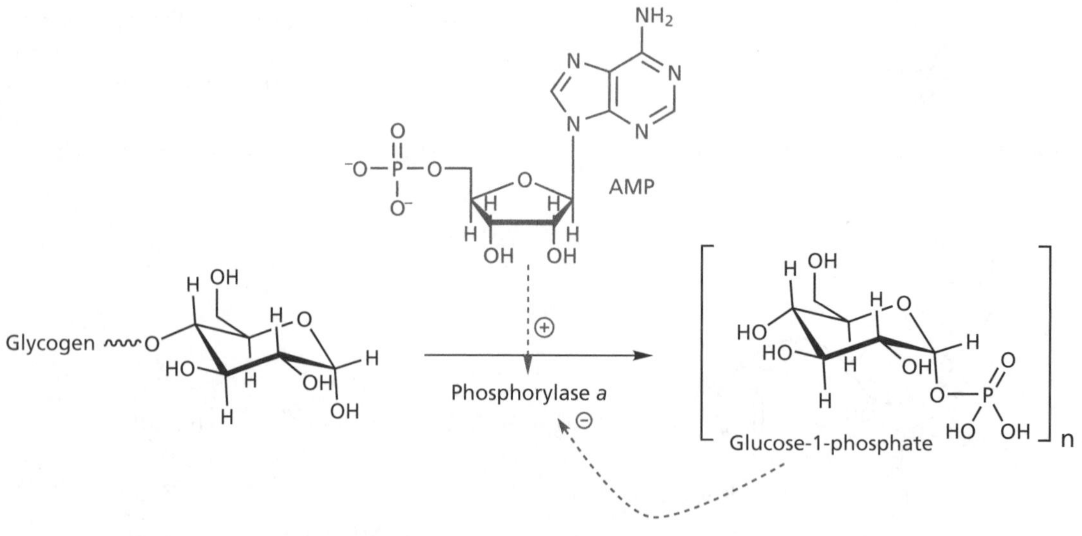

Figure 4.13 Internal control of phosphorylase *a*.

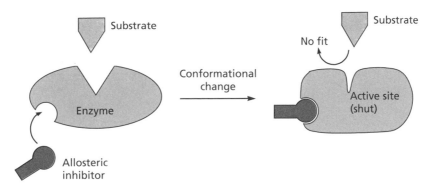

Figure 4.14 Non-competitive, reversible (allosteric) inhibition.

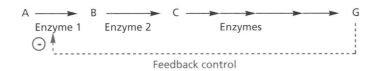

Figure 4.15 Feedback control.

site as well. Agents that inhibit the enzyme produce an induced fit that makes the active site unrecognizable to the substrate.

We might wonder why an agent inhibiting the enzyme has to bind to a separate, allosteric binding site and not at the active site itself. After all, if the agent could bind to the active site, it would block the natural substrate from entering. There are two explanations for this.

First, many of the enzymes that are under allosteric control are at the start of a biosynthetic pathway (Fig. 4.15). A biosynthetic pathway involves a series of enzymes, all working efficiently to produce a final product. Eventually, the cell will have enough of the required material and will need to stop production. The most common control mechanism is known as **feedback control**, where the final product controls its own synthesis by inhibiting the first enzyme in the biochemical pathway. When there are low levels of final product in the cell, the first enzyme in the pathway is not inhibited and works normally. As the levels of final product increase, more and more of the enzyme is blocked and the rate of synthesis drops

off in a graded fashion. Crucially, the final product has undergone many transformations from the original starting material and so it is no longer recognized by the active site of the first enzyme. A separate allosteric binding site is therefore needed which recognizes the final product. The biosynthesis of noradrenaline in section 20.4 is an example of a biosynthetic pathway under feedback control.

Second, binding of the final product to the active site would not be a very efficient method of feedback control, as the product would have to compete with the enzyme's substrate. If levels of the latter increased, then the inhibitor would be displaced and feedback control would fail.

Many enzymes can also be regulated externally (Box 4.1). We shall look at this in more detail in Chapters 5 and 6, but in essence, cells receive chemical messages from their environment which trigger a cascade of signals within the cell. These in turn ultimately activate a set of enzymes known as **protein kinases**. The protein kinases play an important part in controlling enzyme activity within the cell by phosphorylating amino acids such as **serine, threonine**, or

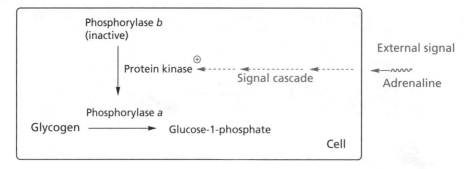

Figure 4.16 External control of phosphorylase *a*.

tyrosine in target enzymes—a covalent modification. For example, the hormone **adrenaline** is an external messenger which triggers a signalling sequence resulting in the activation of a protein kinase enzyme. Once activated, the protein kinase phosphorylates an inactive enzyme called **phosphorylase b** (Fig. 4.16). This enzyme now becomes active and is called **phosphorylase a**. It catalyses the breakdown of glycogen and remains active until it is dephosphorylated back to phosphorylase *b*.

In this case, phosphorylation of the target enzyme leads to activation. Other enzymes may be deactivated by phosphorylation. For example, **glycogen synthase**—the enzyme that catalyses the synthesis of glycogen from glucose-1-phosphate—is inactivated by phosphorylation and activated by dephosphorylation. The latter is effected by the hormone **insulin**, which triggers a different signalling cascade from that of adrenaline.

4.7 Isozymes

Enzymes having a quaternary structure are made up of different subunits, and the combination of these subunits can differ in different tissues. Such variations are called isozymes. Isozymes catalyse the same reaction but differ in their properties. For example, there are different isozymes of **lactate dehydrogenase**. This enzyme is made up of four identical subunits, but the subunits that make up lactate dehydrogenase in muscle are different in amino acid composition from those

making up lactate dehydrogenase in the heart. Both enzymes catalyse the conversion of lactic acid to pyruvic acid, but the isozyme in muscle is twice as active. Moreover, the isozyme in the heart is inhibited by excess pyruvic acid whereas the isozyme in muscle is not.

KEY POINTS

- Enzymes are proteins that act as the body's catalysts by binding substrates and participating in the reaction mechanism.

- The active site of an enzyme is usually a hydrophobic hollow or cleft in the protein. There are important amino acids present in the active site that either bind substrates or participate in the reaction mechanism.

- Binding of substrate or product to an active site involves intermolecular bonds.

- Substrate binding involves an induced fit where the shape of the active site alters to maximize binding interactions. The binding process also orientates the substrate correctly and may weaken crucial bonds in the substrate to facilitate the reaction mechanism.

- The amino acid histidine is often present in active sites and acts as an acid–base catalyst.

- The amino acids serine and cysteine are commonly present in active sites and act as nucleophiles in the reaction mechanism.

- Cofactors are metal ions or small organic molecules (coenzymes) which are required by many enzymes. Coenzymes can be viewed as the body's chemical reagents.

BOX 4.1 THE EXTERNAL CONTROL OF ENZYMES BY NITRIC OXIDE

The external control of enzymes is usually initiated by external chemical messengers which do not enter the cell. However, there is an exception to this. In recent years it has been discovered that cells can generate the gas **nitric oxide** by the reaction sequence shown in Fig. 1.

Because nitric oxide is a gas, it can easily diffuse through cell membranes into target cells. There, it activates enzymes called **cyclases** to generate **cyclic GMP** from **GTP** (Fig. 2).

Cyclic GMP then acts as a secondary messenger to influence other reactions within the cell (see Chapter 6). By this process, nitric oxide has an influence on a diverse range of physiological processes including blood pressure, neurotransmission and immunological defence mechanisms. The well-publicized drug **sildenafil (Viagra)** prevents the decomposition of cyclic GMP and so prolongs the signal initiated by nitric oxide (Fig. 3).

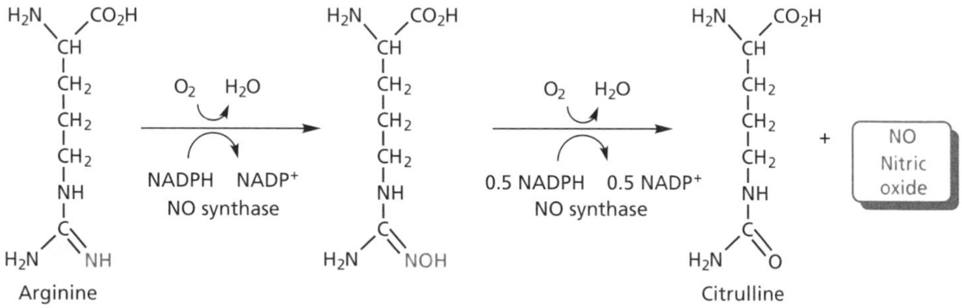

Figure 1 Synthesis of nitric oxide.

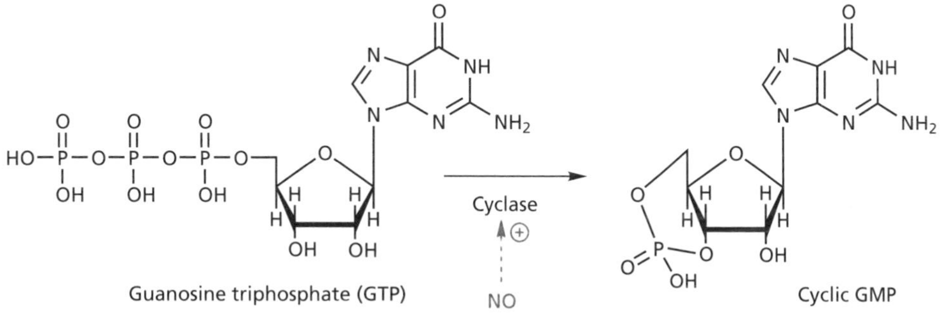

Figure 2 Activation of cyclase enzymes by NO.

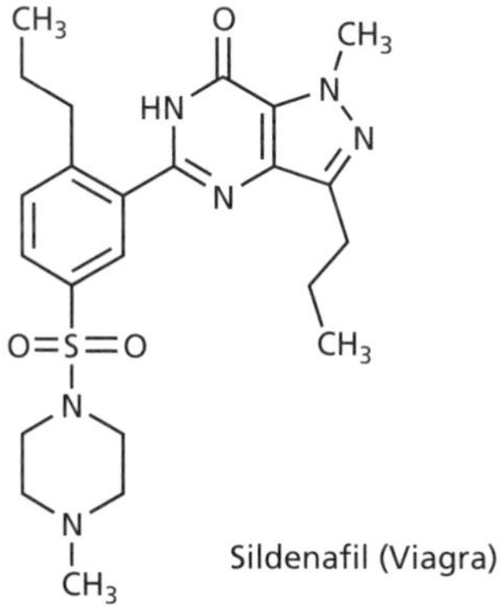

Figure 3 Sildenafil (Viagra).

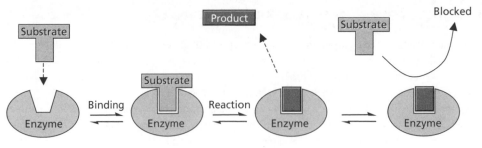

Figure 4.17 Enzyme 'clogging'.

- Prosthetic groups are coenzymes which are bound covalently to an enzyme.
- Enzymes are regulated by internal and/or external control.
- External control involves regulation initiated by a chemical messenger from outside the cell, and which ultimately involves the phosphorylation of enzymes.
- Allosteric inhibitors bind to a different binding site from the active site and alter the shape of the enzyme such that the active site is no longer recognizable. Allosteric inhibitors are usually involved in the feedback control of biosynthetic pathways.
- Isozymes are variations of the same enzyme. They catalyse the same reaction but differ in their primary structure, substrate specificity, and tissue distribution.

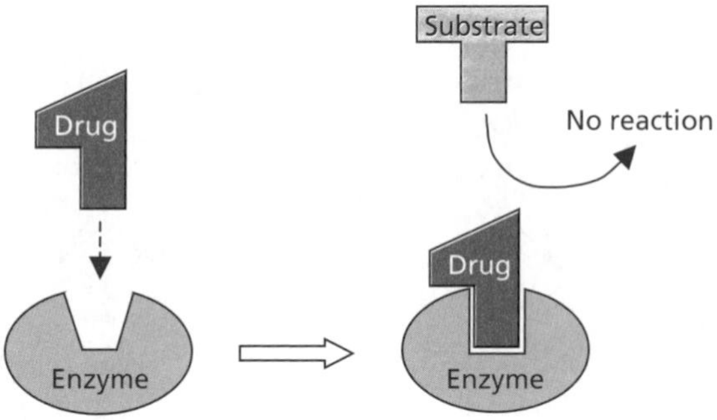

Figure 4.18 Competitive inhibition.

4.8 Enzyme inhibitors

4.8.1 Competitive (reversible) inhibitors

Binding interactions between substrate and enzyme are clearly important. If there were no interactions holding the substrate to the active site, then the substrate would drift in and drift out again before there was a chance for it to react. Therefore, the more binding interactions there are, the better the substrate will bind, and the better the chance of reaction. But there is a catch! What would happen if a substrate or product bound so strongly to the active site that it was not released again (Fig. 4.17)?

The answer is that the enzyme would become clogged up and would be unable to accept any more substrate. Therefore, the bonding interactions between the substrate or the product with the enzyme have to be properly balanced such that they are strong enough to keep the substrate at the active site to allow reaction, but weak enough to allow the product to depart. This bonding balancing act can be turned to great advantage if the medicinal chemist wishes to inhibit a particular enzyme, or to switch it off altogether. A molecule can be designed which is similar to the natural substrate or product, and can fit the active site, but which binds more strongly. It may not undergo any reaction when it is in the active site, but as long as it is there, it blocks access to the natural substrate and the enzymatic reaction stops (Fig. 4.18). This is known as competitive inhibition, as the drug is competing with the natural substrate for the active site. The longer the inhibitor is present in the active site, the greater the inhibition. Therefore, if the medicinal chemist knows which binding groups are present in an active site and where they are, a range of molecules can be designed with different inhibitory strengths.

There are many examples of useful drugs which act as competitive inhibitors (Box 4.2). For example, the **sulfonamides** act as antibacterial agents by inhibiting a bacterial enzyme in this fashion (section 16.4.1). Many diuretics used to control blood pressure are competitive inhibitors, as are many antidepressive agents.

BOX 4.2 A CURE FOR ANTIFREEZE POISONING

Competitive reversible inhibitors can generally be displaced by increasing the level of natural substrate. This feature has been useful in the treatment of accidental poisoning by antifreeze. The main constituent of antifreeze is **ethylene glycol**, which is oxidized in a series of enzymatic reactions to **oxalic acid**, which is toxic (Fig. 1). Blocking the synthesis of oxalic acid will lead to recovery.

The first step in this enzymatic process is the oxidation of ethylene glycol by **alcohol dehydrogenase (ADH)**. Ethylene glycol is acting here as a substrate, but we can view it as a competitive inhibitor because it is competing with the natural substrate for the enzyme (Fig. 2a). If the levels of natural substrate are increased, it will compete far better with ethylene glycol and prevent it from reacting (Fig. 2b). Toxic oxalic acid would no longer be formed and the unreacted ethylene glycol would eventually be excreted from the body. The cure, then, is to administer high doses of the natural substrate—alcohol!

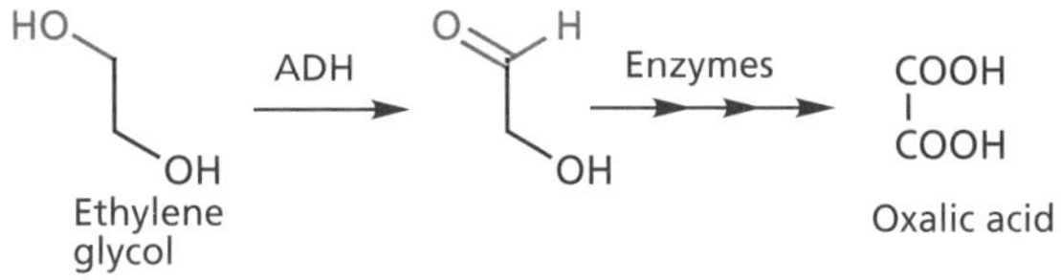

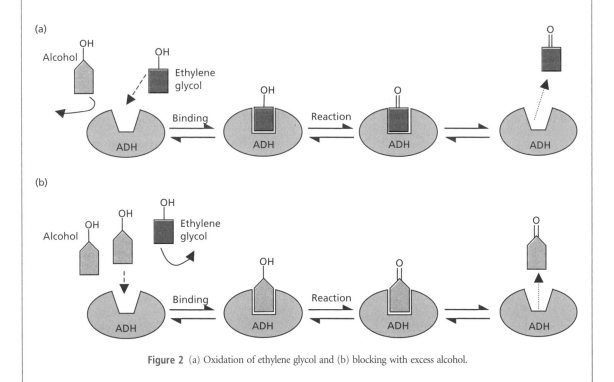

Figure 1 Formation of oxalic acid from ethylene glycol.

Figure 2 (a) Oxidation of ethylene glycol and (b) blocking with excess alcohol.

4.8.2 Non-competitive (irreversible) inhibitors

To stop an enzyme altogether, the chemist can design a drug that binds irreversibly to the active site and blocks it permanently. This would be a non-competitive form of inhibition, because increased levels of natural substrate would not be able to budge the unwanted squatter. The most effective irreversible inhibitors are those that can react with an amino acid at the active site to form a covalent bond. Amino acids such as **serine** and **cysteine** which bear nucleophilic residues (OH and SH respectively) are commonly present in enzyme active sites, as they are frequently involved in the mechanism of the enzyme reaction (section 4.5.3). By designing an electrophilic drug that fits the active site, it is possible to alkylate these nucleophilic groups and hence permanently clog up the active site (Fig. 4.19). The nerve gases (section 19.17.2) are irreversible inhibitors of mammalian enzymes and are therefore highly toxic. In the same way, penicillins (section 16.5.1) are highly toxic to bacteria by irreversibly inhibiting an enzyme crucial to cell wall synthesis.

Not all irreversible inhibitors are highly toxic. For example, **aspirin's** anti-inflammatory activity is due to irreversible inhibition of the **cyclooxygenase** (COX) enzyme, which is required for **prostaglandin** synthesis. Aspirin inhibits the enzyme by acetylating a serine residue in the active site.

4.8.3 Non-competitive, reversible (allosteric) inhibitors

So far we have discussed inhibitors that bind to the active site and prevent the natural substrate from binding. We would therefore expect these inhibitors to have some sort of structural similarity to the natural substrate. We would also expect reversible inhibitors to be displaced by increased levels of natural substrate.

However, there are many enzyme inhibitors that appear to have no structural similarity to the natural substrate. Furthermore, increasing the amount of natural substrate has no effect on the inhibition. Such inhibitors are therefore non-competitive inhibitors, but unlike the non-competitive inhibitors mentioned above, the inhibition can be reversible or irreversible.

Non-competitive or allosteric inhibitors bind to the allosteric binding site described in section 4.6 and are therefore not competing with the substrate for the active site. On binding to the allosteric binding site, an induced fit takes place (Fig. 4.14) which makes the active site unrecognizable to the substrate, so the substrate can no longer react. Adding more substrate will not reverse the situation, but that does not necessarily mean that the inhibition is irreversible. If the inhibitor uses non-covalent bonds to bind to the allosteric binding site, it will eventually depart in its own good time. If the inhibitor uses covalent bonds, it will bind irreversibly.

An enzyme with an allosteric binding site offers the medicinal chemist an extra option in designing an inhibitor. The chemist can not only design drugs which are based on the structure of the substrate and which bind directly to the active site, but can also design drugs based on the structure of the compound that binds to the allosteric binding site and controls the activity of the enzyme.

The drug **6-mercaptopurine**, used in the treatment of leukaemia (Fig. 4.20), is an example of an allosteric inhibitor. It inhibits the first enzyme involved in the synthesis of purines (section 7.1) and therefore blocks purine synthesis. This in turn blocks DNA synthesis.

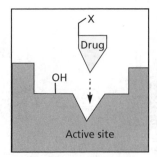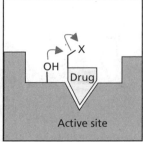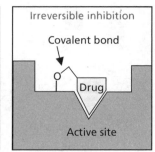

Figure 4.19 Irreversible inhibition. (X = halogen leaving group.)

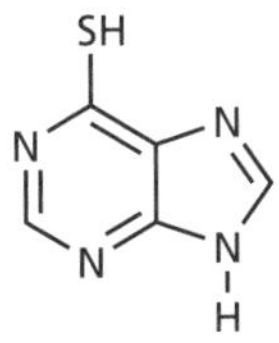

Figure 4.20 6-Mercaptopurine.

4.8.4 Transition-state analogues—renin inhibitors

An understanding of an enzyme mechanism can help medicinal chemists design more powerful inhibitors. For example, it is possible to design inhibitors which bind so strongly to the active site (using non-covalent forces) that they are effectively irreversible inhibitors—a bit like inviting the mother-in-law for dinner and finding that she has moved in on a permanent basis. One way of doing this is to design a drug that resembles the transition state for the catalysed reaction. Such a drug should bind more strongly than

either the substrate or the product, and be a strong inhibitor as a result. Such compounds are known as **transition-state analogues or inhibitors**.

The use of transition-state analogues has been particularly effective in the development of renin inhibitors (Fig. 4.21). Renin is a protease enzyme which is responsible for hydrolysing a specific peptide bond in the protein angiotensinogen to form angiotensin I. Angiotensin I is further converted to angiotensin II, which acts to constrict blood vessels and retain fluid in the kidneys, both of which lead to a rise in blood pressure. Therefore, an inhibitor of renin should act as an antihypertensive agent (i.e. lower blood pressure) by preventing the first stage in this process.

Renin contains two aspartyl residues and a bridging water molecule in the active site which are crucial to the mechanism of action (Fig. 4.22). As far as the protein substrate is concerned, a tetrahedral intermediate is formed during the reaction mechanism. The transition state for the reaction will be closer in character to the intermediate than either the substrate or the product and therefore a transition-state

Figure 4.21 Synthesis of angiotensin I and angiotensin II.

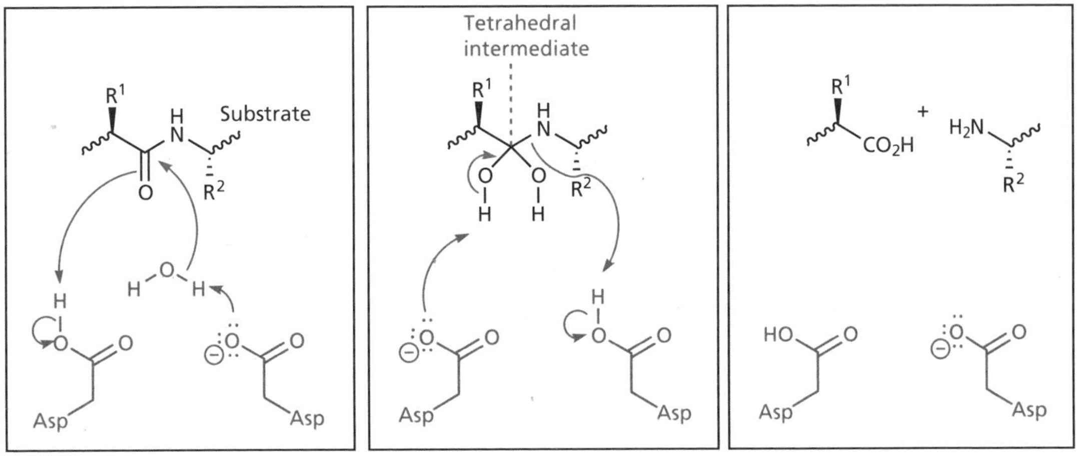

Figure 4.22 Mechanism of renin-catalysed hydrolysis.

analogue should mimic the intermediate. The intermediate itself cannot be used, as it is reactive and easily cleaved. Therefore, an analogue has to be used which mimics the tetrahedral intermediate, will bind more strongly than the substrate or the product, but will also be stable to hydrolysis. A variety of mimics has been tried, but a hydroxyethylene moiety has proved effective (e.g. CGP 38560; Fig. 4.23). The hydroxyethylene group mimics the transition state by having a tetrahedral geometry and one of the two hydroxyl groups. It is stable to hydrolysis because there is no leaving group present.

Similar strategies have been successfully used to design antiviral agents which act as transition-state analogue inhibitors for the HIV protease enzyme (section 17.7.4).

4.8.5 Suicide substrates

Transition-state analogues can be viewed as *bona fide* visitors to an enzyme's active site that become stubborn squatters once they have arrived. Other apparently harmless visitors can turn into lethal assassins

once they have bound to their target enzyme. Such agents are designed to undergo an enzyme-catalysed transformation which converts them into a highly reactive species that forms a covalent bond to the active site. One example of this is the unnatural amino acid **trifluoroalanine** which irreversibly inhibits the enzyme **alanine transaminase** (Fig. 4.24).

The normal mechanism for the transamination reaction is shown in Fig. 4.25 (R = H) and involves the condensation of alanine and **pyridoxal phosphate** to give an imine. A proton is lost from the imine to give a dihydropyridine intermediate. This reaction is catalysed by a basic amino acid provided by the enzyme, and by the electron-withdrawing effects of the protonated pyridine ring. The dihydropyridine structure now formed is hydrolysed to give the products.

Trifluoroalanine contains three fluorine atoms, which are very similar in size to the hydrogen atoms in alanine. The molecule is therefore able to fit into the active site of the enzyme and take alanine's place. The reaction mechanism proceeds as before to give the dihydropyridine intermediate. However, at this stage, an alternative mechanism now becomes possible

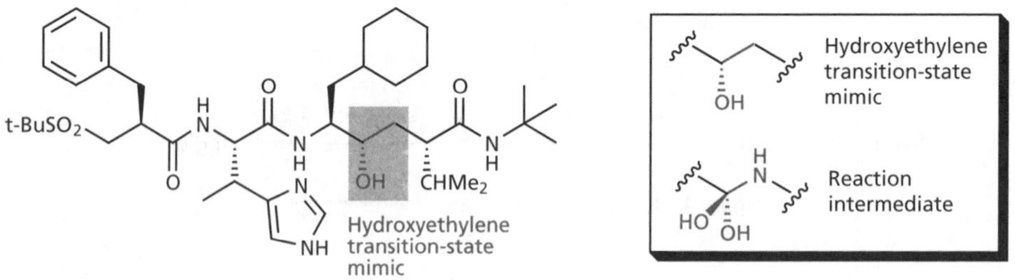

Figure 4.23 CGP 38560.

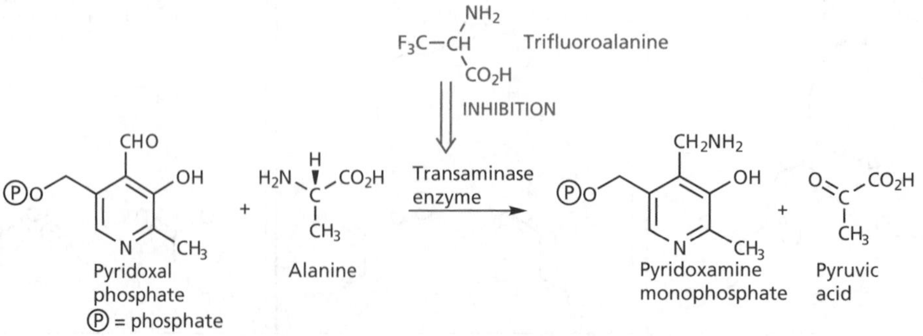

Figure 4.24 Irreversible inhibition of the enzyme alanine transaminase.

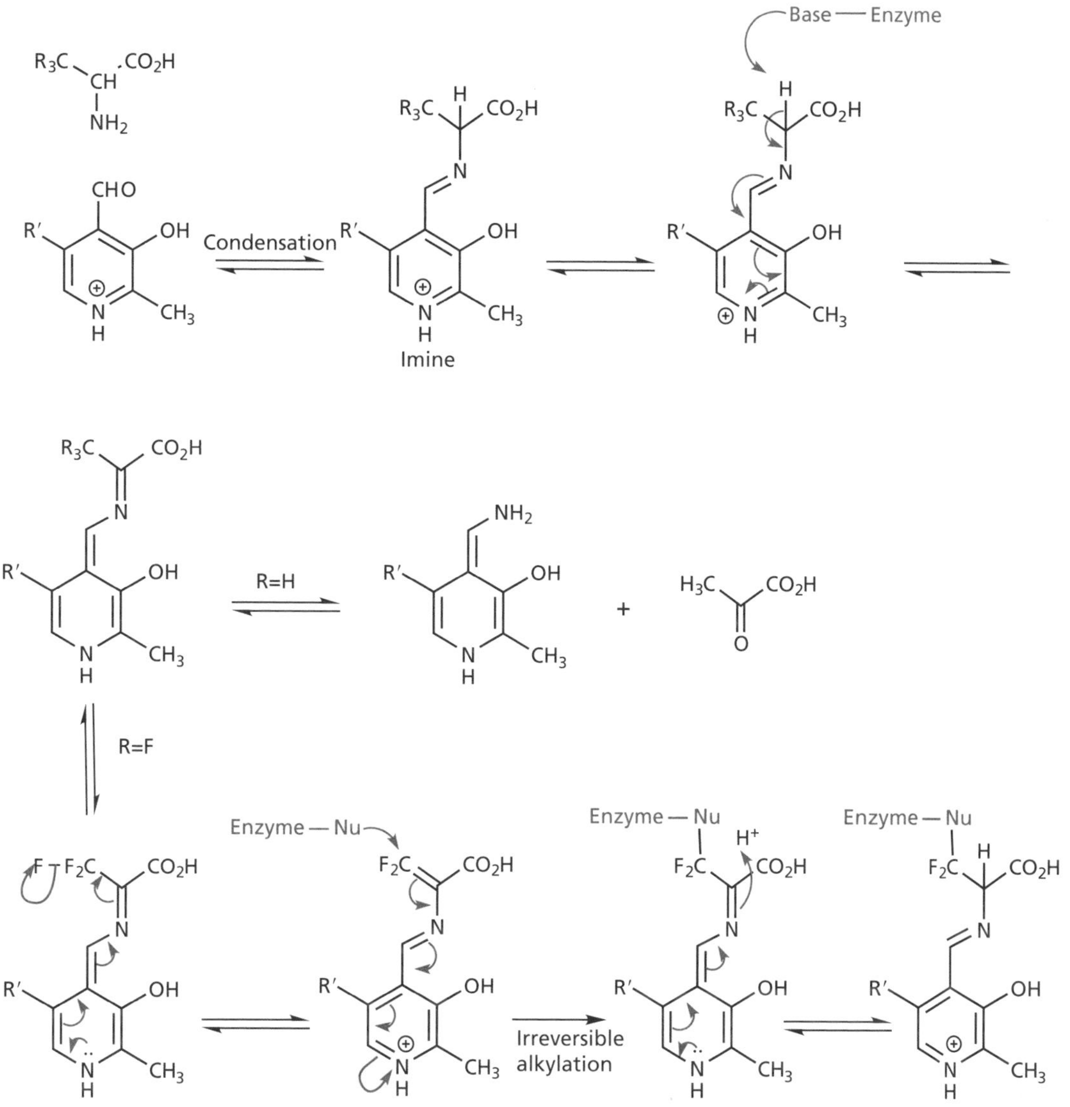

Figure 4.25 Mechanism for the transamination reaction and its inhibition.

(Fig. 4.25; R = F). A fluoride atom is electronegative and can therefore act as a leaving group. When this happens, a powerful alkylating agent is formed which can irreversibly alkylate any nucleophilic group present in the enzyme's active site. A covalent bond is now formed and the active site is unable to accept further substrate. As a result, the enzyme is irreversibly inhibited.

Drugs that operate in this way are often called **suicide substrates** because the enzyme is committing suicide by reacting with them (Box 4.3). The great advantage of this approach is that the alkylating agent is generated at the site where it is meant to act and is therefore highly selective for the target enzyme. If the alkylating group had not been disguised in this way, the drug would have alkylated the first nucleophilic group it met in the body and would have shown little or no selectivity. The uses of alkylating agents and the problems associated with them are also discussed in section 18.2.3.

Inhibiting the transaminase enzyme has no medicinal application; the enzyme is crucial to mammalian biochemistry, and inhibiting it would be toxic to the host. The main use for suicide substrates has been in labelling specific enzymes. The substrates can be labelled with radioactive isotopes and reacted with their target enzyme in order to locate the enzyme in tissue preparations. However, the suicide substrate approach has potential therapeutic applications against enzymes

that are unique to foreign invaders such as bacteria, protozoa, and fungi. **Clavulanic acid** (section 16.5.4.1) can also be classed as a suicide substrate.

One interesting example of a suicide substrate is **5-fluorodeoxyuracil monophosphate** (5-FdUMP). The anticancer agent **5-fluorouracil** is used to treat breast, liver, and skin cancers, and is converted to 5-FdUMP in the body. This then acts as the suicide substrate (section 18.3.2).

BOX 4.3 SUICIDE SUBSTRATES

Suicide substrates are agents which are converted to highly reactive species when they undergo an enzyme-catalysed reaction. They form covalent bonds to the enzyme and inhibit it irreversibly. In some cases, this can cause toxicity. For example, the diuretic agent **tienilic acid** had to be withdrawn from the market because it was found to act as a suicide substrate for the cytochrome P450 enzymes involved in drug metabolism (section 8.4). Unfortunately, the

metabolic reaction carried out by these enzymes converted tienilic acid to a thiophene sulfoxide which proved highly electrophilic. This encouraged a Michael reaction leading to alkylation of a thiol group in the enzyme's active site. Loss of water from the thiophene sulfoxide restored the thiophene ring and resulted in the formation of a covalent link to the enzyme, thus inhibiting the enzyme irreversibly.

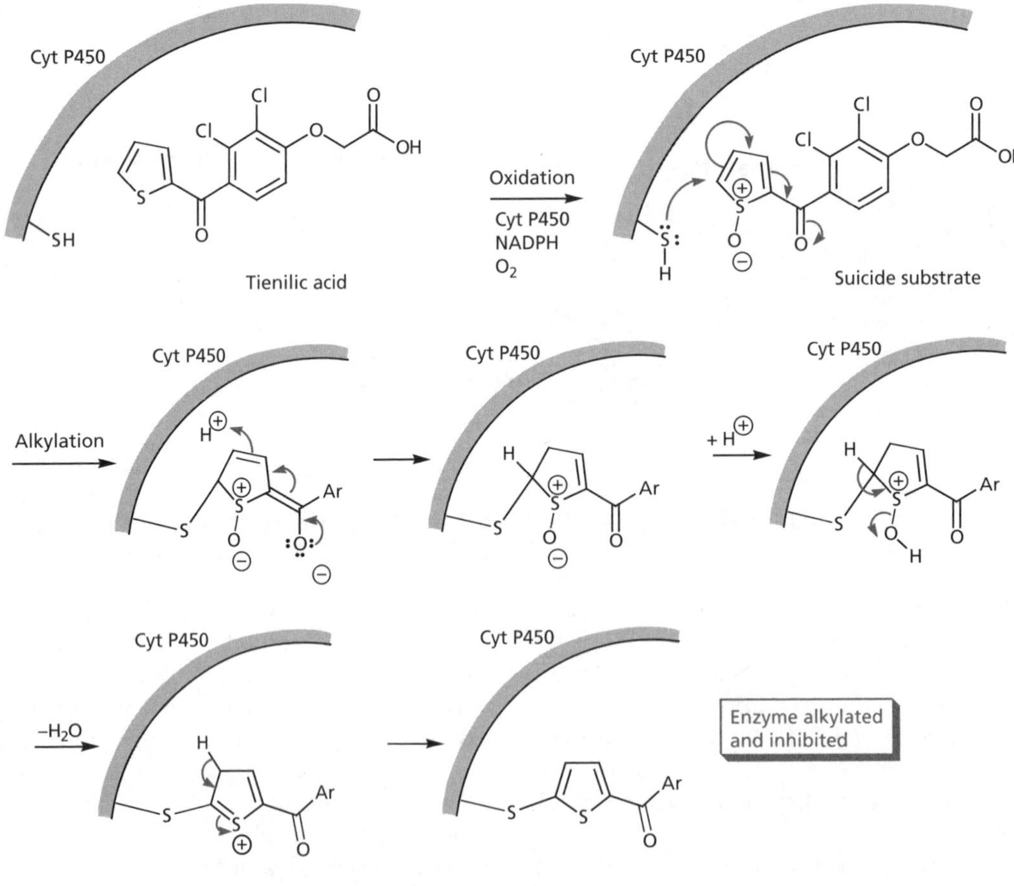

Inhibition of cytochrome P450 by tienilic acid.

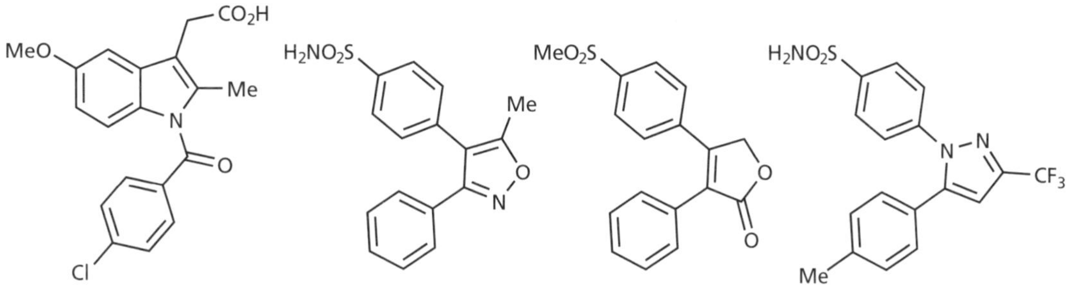

Figure 4.26 Cyclooxygenase inhibitors.

4.8.6 Isozyme selectivity of inhibitors

Identification of isozymes that are selective for certain tissues allows the possibility of designing tissue-selective enzyme inhibitors (see Box 4.4).

For example, the non-steroidal anti-inflammatory drug (NSAID) **indometacin** (Fig. 4.26) is used to treat inflammatory diseases such as rheumatoid arthritis, and works by inhibiting the enzyme **cyclooxygenase**. This enzyme is involved in the biosynthesis of **prostaglandins**—agents which are responsible for the pain and inflammation of rheumatoid arthritis. Inhibiting the enzyme lowers prostaglandin levels and alleviates the symptoms of the disease. However, the drug also inhibits the synthesis of beneficial prostaglandins in the gastrointestinal tract and the kidney. It has been discovered that cyclooxygenase has two isozymes, COX-1 and COX-2. Both isozymes carry out the same reactions, but COX-1 is the isozyme that is active under normal healthy conditions. In rheumatoid arthritis, the normally dormant COX-2 becomes activated and produces excess inflammatory prostaglandins. Therefore, drugs such as **valdecoxib, rofecoxib**, and **celecoxib** have been developed to be selective for the COX-2 isozyme, so that only the production of inflammatory prostaglandins is reduced. (Rofecoxib was authorized in 1999, but had to be withdrawn in September 2004 as it was linked to an increased risk of heart attack and stroke when taken over a period of 18 months or so.)

4.8.7 Medicinal uses of enzyme inhibitors

4.8.7.1 Enzyme inhibitors used against microorganisms

Inhibitors of enzymes have been extremely successful in the war against infection. If an enzyme is crucial to a microorganism, then switching it off will clearly kill the cell or prevent it from growing. Ideally, the enzyme chosen should be one that is not present in our own bodies, and fortunately such enzymes exist because of the significant biochemical differences between bacterial cells and our own. Nature, of course, is well ahead in this game. For example, many fungal strains produce metabolites such as **penicillin** which are toxic to bacteria but not to themselves. This gives fungi an advantage over their microbiological competitors when competing for nutrients.

Although it is preferable to target enzymes that are unique to the foreign invader, it is still possible to target enzymes that are present in both bacterial and mammalian cells, as long as there are significant differences between them. Such differences are perfectly feasible. Although the enzymes in both species may have derived from a common ancestral protein, they have evolved and mutated separately over several million years. Identifying these differences allows the medicinal chemist to design drugs that will bind and act selectively against the bacterial enzyme. Chapter 16 covers antibacterial agents such as the sulfonamides, penicillins, and cephalosporins, all of which act by inhibiting enzymes.

4.8.7.2 Enzyme inhibitors used against viruses

Enzyme inhibitors are also extremely important in the battle against viral infections (e.g. herpesvirus and HIV). Successful antiviral drugs include **aciclovir** for herpes, and drugs such as **zidovudine** and **saquinavir** for HIV (see Chapter 17).

BOX 4.4 DESIGNING DRUGS TO BE ISOZYME SELECTIVE

Designing drugs to be isozyme selective means that they can be designed to act on different diseases despite acting on the same enzyme. This is because isozymes differ in substrate specificity and are distributed differently in the body. **Monoamine oxidase (MAO)** is one of the enzymes responsible for the metabolism of important neurotransmitters such as **dopamine, noradrenaline**, and **serotonin** (Chapter 5) and exists in two isozymic forms (MAO-A and MAO-B). These isozymes differ in substrate specificity, tissue distribution, and primary structure but carry out the same reaction by the same mechanism (Fig. 1). MAO-A is

selective for noradrenaline and serotonin whereas MAO-B is selective for dopamine. MAO-A inhibitors such as **clorgiline** are used clinically as antidepressants; MAO-B inhibitors such as **selegiline** are administered with **levodopa** for the treatment of Parkinson's disease. MAO-B inhibition protects levodopa from metabolism. Clorgiline and selegiline are thought to act as suicide substrates where they are converted by the enzyme to reactive species which react with the enzyme and form covalent bonds. The amine and alkyne functional groups present in both drugs are crucial to this process (Fig. 2).

Figure 1 Reaction catalysed by MAO.

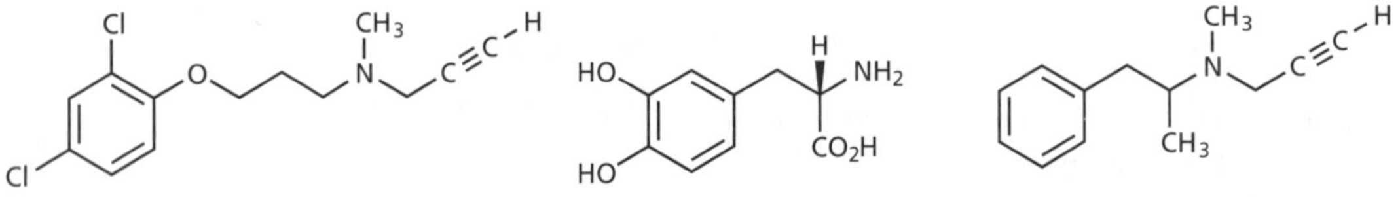

Figure 2 Clorgiline, levodopa, and selegiline.

4.8.7.3 Enzyme inhibitors used against the body's own enzymes

Drugs that act against the body's own enzymes are important in medicine. Some examples are given in Table 4.2. Agents known as **anticholinesterases**, which are used in a variety of diseases including Alzheimer's disease, are considered in sections 19.16–19.17. Proton pump inhibitors, used as antiulcer agents, are discussed in section 22.3.

The search continues for new enzyme inhibitors, especially those that are selective for a specific isozyme or act against recently discovered enzymes. Some current research projects include investigations into inhibitors of the **COX-2 isozyme** (sections 4.8.6 and 18.7.2), **matrix metalloproteinases** (antiarthritic and anticancer drugs; section 18.7.1), **aromatases** (anticancer agents; section 18.4.8), and **caspases.** The

caspases are implicated in the processes leading to cell death. Inhibitors of caspases may have potential in the treatment of stroke victims. A vast amount of research is also taking place on inhibitors of a family of enzymes known as **kinases**. These enzymes catalyse the phosphorylation of proteins and play an important role in signalling pathways within cells (see also Chapter 6 and section 18.6.2).

KEY POINTS

- Enzyme inhibitors binding to the active site are classed as competitive (reversible) or non-competitive (irreversible). Increasing the substrate concentration displaces the former but not the latter.

- Allosteric inhibitors bind to a different binding site from the active site and alter the shape of the enzyme such that the active site is no longer recognizable.

- Transition-state analogues are enzyme inhibitors which are designed to mimic the transition state of an enzyme-catalysed reaction mechanism. They bind more strongly than either the substrate or the product.

- Suicide substrates are molecules which act as substrates for a target enzyme but which are converted into highly reactive species as a result of the enzyme-catalysed reaction mechanism. These species react with amino acid residues present in the active site to form covalent bonds, and act as irreversible inhibitors.

- Drugs that selectively inhibit isozymes are less likely to have side effects and be more selective in their effect.

- Enzyme inhibitors are used in a wide variety of medicinal applications.

4.9 Enzyme kinetics

Studies of enzyme kinetics are extremely useful in determining the properties of an enzyme inhibitor (e.g. whether an inhibitor is competitive or non-competitive).

4.9.1 The Michaelis–Menten equation

The Michaelis–Menten equation holds for an enzyme (E) which combines with its substrate (S) to form an enzyme–substrate complex (ES). The enzyme–substrate complex can then either dissociate back to E and S, or go on to form a product (P). It is assumed that formation of the product is irreversible.

$$\text{E} + \text{S} \underset{k_2}{\overset{k_1}{\rightleftharpoons}} \text{ES} \overset{k_3}{\longrightarrow} \text{E} + \text{P}$$

where k_1, k_2, and k_3 are rate constants.

Table 4.2 Medicinally useful drugs that act against enzymes in the body

Drug	Target enzyme	Field of therapy
Aspirin	Cyclooxygenase	Anti-inflammatory
Captopril, enalapril	Angiotensin-converting enzyme (ACE)	Antihypertension
Simvastatin	HMG-CoA reductase	Lowering of cholesterol levels
Desipramine	Monoamine oxidase	Antidepressant
Clorgiline	Monoamine oxidase-A	Antidepressant
Selegiline	Monoamine oxidase-B	Treatment of Parkinson's disease
Methotrexate	Dihydrofolate reductase	Anticancer
5-Fluorouracil	Thymidylate synthase	Anticancer
Gefitinib, imatinib	Tyrosine kinases	Anticancer
Sildenafil (Viagra)	Phosphodiesterase enzyme (PDE5)	Treatment of male erectile dysfunction
Allopurinol	Xanthine oxidase	Treatment of gout
Zidovudine	HIV reverse transcriptase	AIDS therapy
Saquinavir	HIV protease	AIDS therapy
Aciclovir	Viral DNA polymerase	Treatment of herpes
Penicillins and cephalosporins	Bacterial transpeptidase	Antibacterial
Clavulanic acid	Bacterial β-lactamases	Antibacterial
Sulfonamides	Dihydropteroate synthetase	Antibacterial
Fluoroquinolones	Bacterial topoisomerases	Antibacterial
Ro41-0960	Catechol-O-methyltransferase	Treatment of Parkinson's disease
Omeprazole	H^+/K^+ ATPase proton pump	Ulcer therapy
Organophosphates	Acetylcholinesterase	Treatment of myasthenia gravis, glaucoma, and Alzheimer's disease
Acetazolamide	Carbonic anhydrase	Diuretic
Zileutin	5-Lipoxygenase	Antiasthmatic

For enzymes such as these, plotting the rate of enzyme reaction versus substrate concentration [S] gives a curve like that shown in Fig. 4.27. This curve shows that at low substrate concentrations the rate of reaction increases almost proportionally to the substrate concentration, whereas at high substrate concentration the rate becomes almost constant and approaches a maximum rate (rate$_{max}$), which is independent of substrate concentration. This reflects a situation where there is more substrate present than active sites available, and so increasing the amount of substrate will have little effect.

The Michaelis–Menten equation relates the rate of reaction to the substrate concentration for the curve in Fig. 4.27:

$$\text{rate} = \text{rate}_{max} \frac{[S]}{[S] + K_M}$$

The derivation of this equation is not covered here but can be found in most biochemistry textbooks. The constant K_M is known as the **Michaelis constant** and is equal to the substrate concentration at which the reaction rate is half of its maximum value. This can be demonstrated as follows. If $K_M = [S]$, then the Michaelis–Menten equation becomes:

$$\text{rate} = \text{rate}_{max} \frac{[S]}{[S] + [S]} = \text{rate}_{max} \frac{[S]}{2[S]} = \text{rate}_{max} \times \frac{1}{2}$$

The K_M of an enzyme is significant because it measures the concentration of substrate at which half the active sites in the enzyme are filled. This in turn provides a measure of the substrate concentration required for significant catalysis to occur.

K_M is also related to the rate constants of the enzyme-catalysed reaction:

$$K_M = \frac{k_2 + k_3}{k_1}$$

Consider the situation where there is rapid equilibration between S and ES, and a slower formation of P. This means that the substrate binds to the active site and departs several times before it is finally converted to product.

$$E + S \underset{k_2 \ \text{fast}}{\overset{k_1 \ \text{fast}}{\rightleftharpoons}} ES \xrightarrow[\text{slow}]{k_3} E + P$$

Under these conditions, the dissociation rate (k_2) of ES is much greater than the rate of formation of product (k_3). k_3 becomes insignificant relative to k_2, and the equation simplifies to the following:

$$K_M = \frac{k_2 + k_3}{k_1} = \frac{k_2}{k_1}$$

This corresponds to the following dissociation constant for ES:

$$[ES] \rightleftharpoons [E] + [S] \qquad \text{dissociation constant} = \frac{[E][S]}{[ES]}$$

In this situation, K_M effectively equals the dissociation constant of ES, and can be taken as a measure of how strongly the substrate binds to the enzyme. A high value of K_M indicates weak binding because the equilibrium is pushed to the right; a low K_M indicates strong binding because the equilibrium is to the left. K_M is also dependent on the particular substrate and on environmental conditions such as pH, temperature, and ionic strength.

The maximum rate is related to the total concentration of enzyme ($[E]_{total} = [E] + [ES]$) as follows:

$$\text{rate}_{max} = k_3 [E]_{total}$$

A knowledge of the maximum rate and the enzyme concentration allows the determination of k_3. For example, the enzyme carbonic anhydrase catalyses the formation of hydrogen carbonate and does so at a maximum rate of 0.6 M hydrogen carbonate molecules formed per second for a 10^{-6} M solution of the enzyme. Altering the above equation, k_3 can be determined as follows:

$$k_3 = \frac{\text{rate}_{max}}{[E]_{total}} = \frac{0.6 \ \text{Ms}^{-1}}{10^{-6} \ \text{M}} = 600\,000 \ \text{s}^{-1}$$

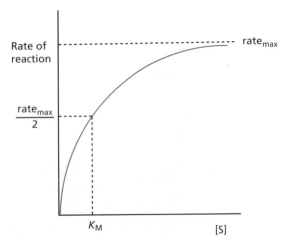

Figure 4.27 Reaction rate versus substrate concentration.

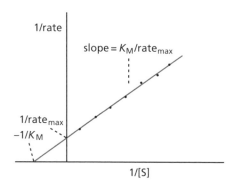

Figure 4.28 Lineweaver–Burk plot.

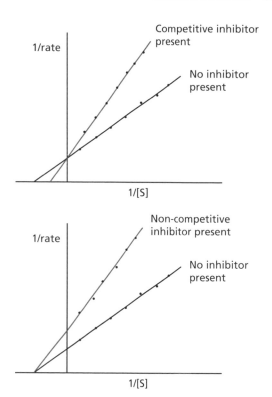

Figure 4.29 Distinguishing a competitive inhibitor from a non-competitive inhibitor.

Therefore, each enzyme is catalysing the formation of 600 000 hydrogen carbonate molecules per second. The turnover number is the time taken for each catalysed reaction to take place, i.e. $1/600\,000 = 1.7\,\mu s$.

4.9.2 Lineweaver–Burk plots

The Michaelis–Menten equation can be modified by taking its reciprocal such that a straight line plot is obtained (the Lineweaver–Burk plot) (Fig. 4.28):

$$\frac{1}{\text{rate}} = \frac{1}{\text{rate}_{max}} + \frac{K_M}{\text{rate}_{max}[S]}$$

Plots such as these can be used to determine whether an inhibitor is competitive or non-competitive (Fig. 4.29). The reaction rate of the enzyme reaction is measured with respect to varying substrate concentration, with and without the presence of the inhibitor.

In the case of competitive inhibition, the lines cross the y-axis at the same point (i.e. the maximum rate of the enzyme is unaffected), but the slopes are different. The fact that the maximum rate is unaffected reflects the fact that the inhibitor and substrate are competing for the same active site and that increasing the substrate concentration sufficiently will overcome the inhibition. The increase in the slope that results from adding an inhibitor is a measure of how strongly the inhibitor binds to the enzyme.

In the case of a non-competitive inhibitor, the lines have the same intercept point on the x-axis (i.e. K_M is unaffected), but have different slopes and different intercepts on the y-axis. Therefore, the maximum rate for the enzyme has been reduced.

Lineweaver–Burk plots are extremely useful in determining the nature of inhibition, but they have their limitations and are not applicable to enzymes that are under allosteric control.

4.9.3 Comparison of inhibitors

When comparing the activity of enzyme inhibitors, the IC_{50} value is often quoted. This is the concentration of inhibitor required to reduce the activity of the enzyme by 50%. Compounds with a high IC_{50} are less powerful inhibitors than those with a low IC_{50}, as a higher concentration of the former is required to attain the same level of inhibition.

K_i values are also reported in enzyme inhibition studies. K_i is the apparent inhibition constant for the following equilibrium between the enzyme and the inhibitor:

$$EI \rightleftharpoons E + I \qquad K_i = \frac{[E][I]}{[EI]}$$

It can be shown that $IC_{50} = K_i + [E]_{total}/2$. If the concentration of inhibitor required to inhibit the enzyme by 50% is much greater than the concentration of enzyme, then K_i is much larger than $[E]_{total}$ and this equation approximates to $IC_{50} = K_i$.

IC_{50} and K_i values are measured in assays involving isolated enzymes. However, it is often useful to carry out enzyme inhibition studies where the enzyme is present in whole cells or tissues. In these studies, a cellular effect resulting from enzyme activity is monitored. EC_{50} values represent the concentration of inhibitor required to reduce that particular cellular effect by 50%. It should be noted that the effect being measured may be several stages downstream from the enzyme reaction concerned.

KEY POINTS

- The Michaelis–Menten equation relates the rate of an enzyme-catalysed reaction to substrate concentration.

- Lineweaver–Burk plots are derived from the Michaelis–Menten equation and are used to determine whether inhibition is competitive or non-competitive.

- The activity of different enzymes can be compared by measuring values of EC_{50}, K_i, or IC_{50}.

QUESTIONS

1. Enzymes can be used in organic synthesis. For example, the reduction of an aldehyde is carried out using aldehyde dehydrogenase. Unfortunately, this reaction requires the use of the cofactor NADH, which is expensive and is used up in the reaction. If ethanol is added to the reaction, only catalytic amounts of cofactor are required. Why?

2. It is known that the amino acid at position 523 of the cyclooxygenase enzyme is part of the active site. In the iso-enzyme COX-1, this amino acid is isoleucine, whereas in COX-2 it is valine. Suggest how such information could be used in the design of drugs that selectively inhibit COX-2.

3. Acetylcholine is the substrate for the enzyme acetylcholinesterase. Suggest what sort of binding interactions could be involved in holding acetylcholine to the active site.

Acetylcholine

4. The ester bond of acetylcholine is hydrolysed by acetylcholinesterase. Suggest a mechanism by which the enzyme catalyses this reaction.

5. Suggest how binding interactions might make acetylcholine more susceptible to hydrolysis.

6. Neostigmine is an inhibitor of acetylcholinesterase. The enzyme attempts to catalyse the same reaction on neostigmine as it does with acetylcholine. However, a stable intermediate is formed which prevents completion of the process and which results in a molecule being covalently linked to the active site. Identify the stable intermediate and explain why it is stable.

Neostigmine

7. The human immunodeficiency virus contains a protease enzyme that is capable of hydrolysing the peptide bond of L-Phe-L-Pro. Structure I was designed as a transition-state inhibitor of the protease enzyme. What is a transition-state inhibitor, and how does structure I fit the description

L-Phe-L-Pro

I
IC_{50} 6500 nM

of a transition-state inhibitor? What is meant by IC_{50} 6500 nM?

8. Why should a transition state be bound more strongly to an enzyme than a substrate or a product?

FURTHER READING

Flower, R. J. (2003) The development of COX-2 inhibitors. *Nature Reviews Drug Discovery*, **2**, 179–191.

Knowles, J. R. (1991) Enzyme catalysis: not different, just better. *Science*, **350**, 121–124.

Maryanoff, B. E. and Maryanoff, C. A. (eds.) (1992) Some thoughts on enzyme inhibition and the quiescent affinity label concept. *Advances in Medicinal Chemistry*, **1**, 235–261.

Navia, M. A. and Murcko, M. A. (1992) Use of structural information in drug design. *Current Opinion in Structural Biology*, **2**, 202–216.

Teague, S. J. (2003) Implications of protein flexibility for drug discovery. *Nature Reviews Drug Discovery*, **2**, 527–541.

Titles for general further reading are listed on p. 711.

5

Proteins as drug targets: receptors

5.1 The receptor role

Drugs that interact with receptors are amongst the most important in medicine and provide treatment for ailments such as pain, depression, Parkinson's disease, psychosis, heart failure, asthma, and many other problems.

What are these receptors, and what do they do?

In a complex organism, cells have to get along with their neighbours. There has to be a communication system. After all, it would be pointless if individual heart cells were to contract at different times. The heart would then be a wobbly jelly and totally useless in its function as a pump. Communication is essential to ensure that all heart muscle cells contract at the same time. The same holds true for all the organs and functions of the body if it is to operate in a coordinated and controlled fashion.

Control and communication come primarily from the brain and spinal column (the **central nervous system**), which receives and sends messages via a vast network of nerves (Fig. 5.1). The detailed mechanism by which nerves transmit messages along their length need not concern us here (see Appendix 4). It is sufficient for our purposes to think of the message as being an electrical pulse which travels down the nerve cell (neuron) towards the target, whether that be a muscle cell or another nerve. If that was all there was to it, it would be difficult to imagine how drugs could affect this communication system. However, there is one important feature that is crucial to our understanding of drug action. Nerves do not connect directly to their target cells. They stop just short of the cell surface. The distance is minute, about 100 Å,

but it is a space that the electrical pulse is unable to jump.

Therefore, there has to be a method of carrying the message across the gap between the nerve ending and the target cell. The problem is solved by the release of a chemical messenger called a **neurotransmitter** from the nerve cell (Fig. 5.2). Once released, this chemical messenger diffuses across the gap to the target cell, where it binds and interacts with a specific protein (receptor) embedded in the cell membrane. This process of binding leads to a series or cascade of secondary effects, which results either in a flow of ions across the cell membrane or in the switching on (or off) of enzymes inside the target cell. A biological response then results, such as the contraction of a muscle cell or the activation of fatty acid metabolism in a fat cell.

We shall consider these secondary effects and how they result in a biological action in Chapter 6, but for the moment the important thing to note is that the communication system depends crucially on a chemical messenger. As a chemical process is involved, it should be possible for other chemicals (drugs) to interfere or interact with the process.

5.2 Neurotransmitters and hormones

There is a large variety of messengers that interact with receptors and that vary quite significantly in structure and complexity. Some neurotransmitters are simple in structure, such as monoamines (e.g. **acetylcholine**, **noradrenaline**, **dopamine**, and **serotonin**) or amino

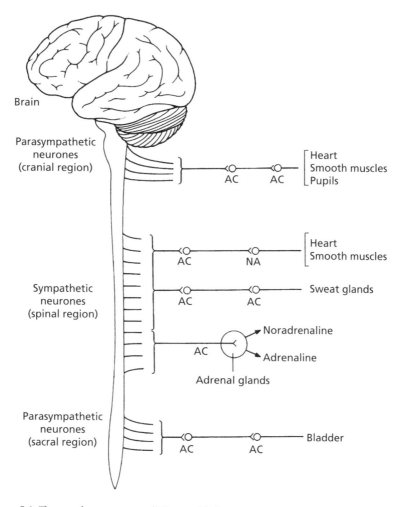

Figure 5.1 The central nervous system (AC = acetylcholine; NA = noradrenaline). Taken from J. Mann, *Murder, magic, and medicine*, Oxford University Press (1992), with permission.

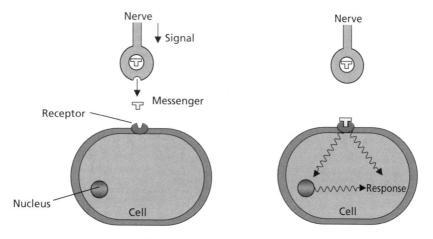

Figure 5.2 Neurotransmitter action.

acids (e.g. **γ-aminobutyric acid** [GABA], **glutamic acid**, and **glycine**) (Fig. 5.3). Even the calcium ion can act as a chemical messenger. Other chemical messengers are more complex in structure and include lipids such as **prostaglandins**; purines such as **adenosine** or **ATP** (Chapter 7); neuropeptides such as **endorphins** and **enkephalins** (section 21.6); peptide hormones such as **angiotensin** or **bradykinin**; and even enzymes such as **thrombin** (section 6.3.3).

In general, a nerve releases mainly one type of neurotransmitter, and the receptor which awaits it on the target cell will be specific for that messenger. However, that does not mean that the target cell has only one type of receptor protein. Each target cell has a large number of nerves communicating with it and they do not all use the same neurotransmitter (Fig. 5.4). Therefore, the target cell will have other types of receptors specific for those other neurotransmitters. It may also have receptors waiting to receive messages from chemical messengers that have longer distances to travel. These are the **hormones** released into the circulatory system by various glands in the body. The best known example of a hormone is **adrenaline**. When danger or exercise is anticipated, the adrenal medulla gland releases adrenaline into the bloodstream where it is carried round the body, preparing it for vigorous exercise.

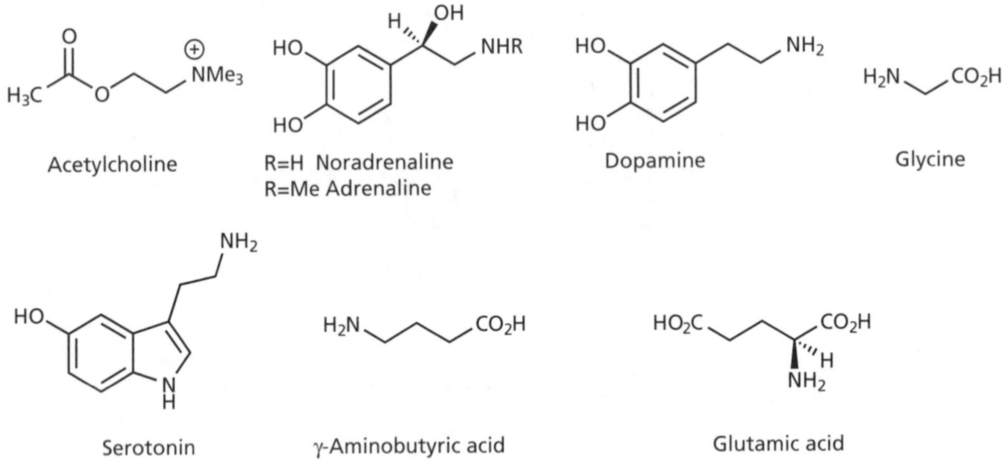

Acetylcholine

R=H Noradrenaline
R=Me Adrenaline

Dopamine

Glycine

Serotonin

γ-Aminobutyric acid

Glutamic acid

Figure 5.3 Examples of neurotransmitters.

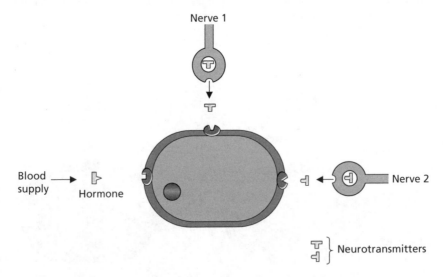

Figure 5.4 Target cell containing receptors specific to each type of messenger.

Hormones and neurotransmitters can be distinguished by the route they travel and by the way they are released, but their action when they reach the target cell is the same. They both interact with a receptor and a message is received. The cell responds to that message and adjusts its internal chemistry accordingly, and a biological response results.

Communication is clearly essential for the normal working of the human body, and if the communication should become faulty then it could lead to such ailments as depression, heart problems, schizophrenia, muscle fatigue, and many other problems. What sort of things *could* go wrong?

One problem would be if too many messengers were released. The target cell could (metaphorically) start to overheat. Alternatively, if too few messengers were sent out, the cell could become sluggish. It is at this point that drugs can play a role by either acting as replacement messengers (if there is a lack of the body's own messengers), or by blocking the receptors for the natural messengers (if there are too many host messengers). Drugs of the former type are known as **agonists**. Those of the latter type are known as **antagonists**.

What determines whether a drug is an agonist or an antagonist, and is it possible to predict whether a new drug will act as one or the other?

To answer this question, we have to move down to the molecular level and consider what happens when a small molecule such as a drug or a neurotransmitter interacts with a receptor protein. Let us first look at what happens when one of the body's own messengers (neurotransmitters or hormones) interacts with its receptor.

5.3 Receptors

A receptor is a protein molecule, usually embedded within the cell membrane with part of its structure facing the outside of the cell. The protein surface is a complicated shape containing hollows, ravines, and ridges, and somewhere within this complicated geography there is an area that has the correct shape to accept the incoming messenger. This area is known as the **binding site** and is analogous to the active site of an enzyme. When the chemical messenger fits into this site, it 'switches on' the receptor molecule and a message is received (Fig. 5.5). However, there is an important difference between enzymes and receptors in that the chemical messenger does not undergo a chemical reaction. It fits into the binding site of the receptor protein, passes on its message, and then leaves unchanged. If no reaction takes place, what *has* happened? How does the chemical messenger tell the receptor its message and how is this message conveyed to the cell?

5.4 How is the message received?

It all has to do with shape. Put simply, the messenger binds to the receptor and induces it to change shape. This change subsequently affects other components of the cell membrane and leads to a biological effect. Two main components are involved: ion channels and membrane-bound enzymes.

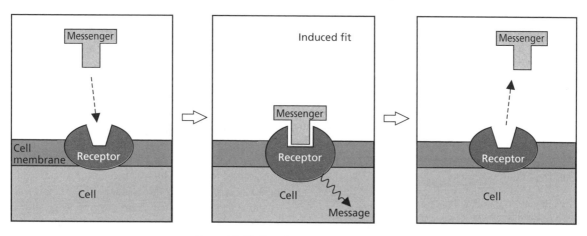

Figure 5.5 Binding of a messenger to a receptor.

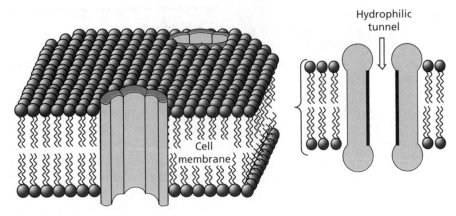

Figure 5.6 Ion channel protein structure. The bold lines show the hydrophilic sides of the channel.

5.4.1 Ion channels and their control

Some neurotransmitters operate by controlling ion channels. What are these ion channels and why are they necessary? Let us look again at the structure of the cell membrane.

As described in Chapter 2, the membrane is made up of a bilayer of phospholipid molecules, and so the middle of the cell membrane is 'fatty' and hydrophobic. Such a barrier makes it difficult for polar molecules or ions to move in or out of the cell. Yet it is important that these species should cross. For example, the movement of sodium and potassium ions across the membrane is crucial to the function of nerves (Appendix 4). It seems an intractable problem, but once again the ubiquitous proteins provide the answer by forming ion channels. These are protein complexes (Fig. 5.6) which traverse the cell membrane and consist of several protein subunits. The centre of the complex is hollow and is lined with polar amino acids to give a hydrophilic tunnel or pore.

Ions can now cross the fatty barrier of the cell membrane by moving through these hydrophilic channels or tunnels. But there has to be some control. In other words, there has to be a 'lock gate' that can be opened or closed as required. It makes sense that this lock gate should be controlled by a receptor protein sensitive to an external chemical messenger, and this is exactly what happens with many ion channels.[1] In fact,

the receptor protein is an integral part of the ion channel complex and is one or more of the constituent protein subunits. In the resting state, the ion channel is closed (i.e. the lock gate is shut). However, when a chemical messenger binds to the external binding site of the receptor protein, the receptor protein changes shape. This in turn causes the protein complex to change shape, opening up the lock gate and allowing ions to pass through the ion channel (Fig. 5.7).

It is worth emphasizing one important point at this stage. If the messenger is to make the receptor protein change shape, then the binding site cannot be just a negative image of the messenger molecule. This must be true, otherwise how could a change of shape result? Therefore, when the binding site receives its messenger it is moulded into the correct shape for the ideal fit. This theory of an **induced fit** has already been described in section 4.5.1 to explain the interaction between enzymes and their substrates.

The operation of an ion channel explains why the relatively small number of neurotransmitter molecules released by a nerve is able to have such a significant biological effect on the target cell. By opening a few ion channels, several thousand ions are mobilized for each neurotransmitter molecule involved.

5.4.2 Membrane-bound enzyme activation

The second way in which a receptor can pass on a message from a chemical messenger to the cell is if the receptor also doubles up as an enzyme (Fig. 5.8). In this situation, the receptor protein is embedded within the cell membrane, with part of its structure exposed

---

[1] Ion channels controlled by chemical messengers are called **ligand-gated ions channels**. Ion channels in excitable cell such as nerves are controlled by the membrane potential of the cell, and are called **voltage-gated ion channels** (see Appendix 4).

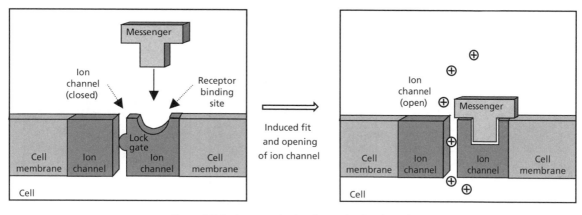

Figure 5.7 Lock-gate mechanism for opening ion channels.

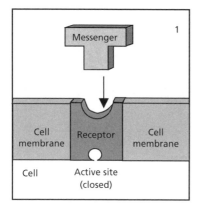
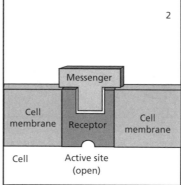
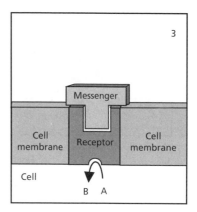

Figure 5.8 Enzyme activation.

on the outer surface of the cell and part exposed on the inner surface. The outer surface contains the binding site for the chemical messenger, and the inner surface has an active site that is closed in the resting state. When a chemical messenger binds to the receptor, it causes a change in shape which opens up the active site and leads to a chemical reaction within the cell. As long as the messenger is bound, the active site remains open and so a single neurotransmitter molecule can amplify its signal by binding long enough for several reactions to take place.

Some receptors control the activity of enzymes in an indirect manner, through the use of a protein messenger called a **G-protein** (Fig 5.9). Once again, the receptor is embedded within the membrane, with the binding site for the chemical messenger on the outer surface. On the inner surface, there is another binding site which is normally closed (frame 1). When the chemical messenger binds to its binding site, the

receptor protein changes shape, opening up the binding site on the inner surface (frame 2). This new binding site is recognized by the G-protein, which then binds (frame 3). The G-protein is made up of three protein subunits, but once it binds to the receptor the complex is destabilized and fragments (frame 4). One of the subunits then travels through the membrane to a membrane-bound enzyme and binds to an allosteric binding site (section 4.6) (frame 5). This causes the enzyme to alter shape, resulting in the opening of an active site and the start of a new reaction within the cell (frame 6).

There are several different G-proteins, which are recognized by different types of receptor. Some of these G-proteins have an inhibitory effect on a membrane-bound enzyme rather than a stimulatory effect. Nevertheless, the mechanism is much the same—the only difference is that the binding of the G-protein subunit to the enzyme leads to closure of the active site.

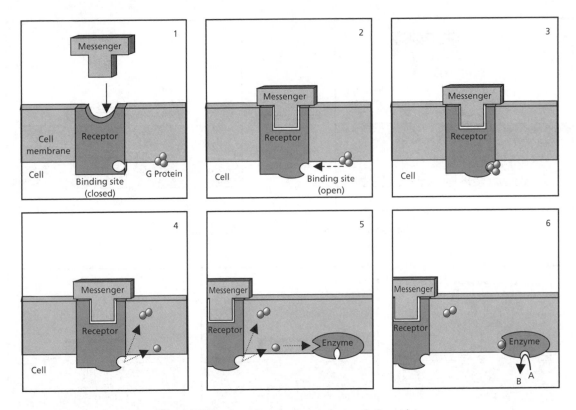

Figure 5.9 Membrane-bound enzyme activation via G-proteins.

At first sight, you might consider G-proteins to be unnecessary middlemen in the activation or deactivation of enzymes. However, they do have distinct advantages. For a start, each receptor will fragment several G-proteins for as long as the messenger molecule is bound to it. This results in an amplification of the signal initiated by the messenger, over and above any amplification resulting from activation of the enzyme.

Secondly, the existence of G-protein messengers allows different receptors to control the same membrane-bound enzyme. In other words, a receptor that interacts with one type of chemical messenger can activate a specific enzyme by involving one type of G-protein, while another receptor that interacts with a different chemical messenger could deactivate the same enzyme by involving a different type of G-protein. In this way, the same enzyme is under dual control. This is more efficient than having the enzyme specifically linked to each different type of receptor.

Regardless of the specific mechanism involved in the activation/deactivation of an enzyme, the overall result is the same. A change in receptor shape (or tertiary

structure) leads eventually to the activation or deactivation of enzymes. Since enzymes can catalyse the reaction of a large number of molecules, activation results in an amplification of the signal.

To conclude, the mechanisms by which neurotransmitters pass on their message involve changes in molecular shape rather than chemical reactions. These changes in shape ultimately lead to some sort of chemical reaction involving enzymes. This in turn results in an amplification of the original message, such that a relatively small number of neurotransmitter molecules can lead to a biological result. This topic is covered more fully in Chapter 6.

5.5 How does a receptor change shape?

We have seen already that it is the messenger molecule which induces the receptor to change shape, but how does it do that? It is not simply, as one might think, a moulding process whereby the receptor wraps itself around the shape of the messenger molecule. The answer lies rather in specific intermolecular binding interactions between messenger and receptor. These interactions— ionic bonding, hydrogen-bonding, van der Waals interactions, dipole–dipole interactions, ion–dipole interactions, induced dipole interactions and hydrophobic interactions—have already been described in section 2.3. The messenger and the receptor protein both take up conformations or shapes to maximize these bonding forces. As with enzyme–substrate binding, there is a fine balance involved in receptor–messenger binding. The bonding forces must be large enough to change the shape of the receptor in the first place, but not so strong that the messenger is unable to leave again. Most neurotransmitters bind quickly to their receptors, then depart again once the message has been received.

As an example of the various bonding forces involved, let us consider a hypothetical neurotransmitter and receptor as shown in Fig. 5.10a. The neurotransmitter has an aromatic ring that can take part in van der Waals interactions, an alcohol OH group that can take part in hydrogen bonding interactions, and a charged nitrogen centre that can take part in ionic or electrostatic interactions.

The hypothetical receptor protein has a binding site containing three complementary binding groups (binding regions) and is part of an ion channel. When the neurotransmitter is absent, the receptor is shaped such that it seals the ion channel (Fig. 5.10b).

As the binding site has complementary binding regions for the groups described above, the drug can fit into the binding site and bind strongly. This is all very well, but now that it has docked, how does it make the receptor change shape? As with enzyme–substrate binding we have to propose that the fit is not quite exact, or else there would be no reason for the receptor to change shape. In this example, we can envisage our messenger molecule fitting into the binding site and binding well to two of the three possible binding regions. The third binding region, however (the ionic one), is not quite in the right position (Fig. 5.10c). It is close enough to have a weak interaction, but not close enough for the optimum interaction. The receptor protein therefore alters shape to obtain the best binding interaction. The carboxylate group is pulled closer to the positively charged nitrogen on the messenger molecule and, as a result, the lock gate is opened and will remain open

until the messenger molecule detaches from the binding site, and allows the receptor to return to its original shape.

In reality, the conformational changes involved in the opening of an ion channel are quite complex and the lock gate is some distance from the receptor binding site. Nevertheless, the same principles hold. Binding of the chemical messenger induces a conformational change that causes a series of knock-on effects resulting in the final opening of the ion channel. We shall look at this again in Chapter 6.

KEY POINTS

- Most receptors are membrane-bound proteins which contain an external binding site for hormones or neurotransmitters. Binding results in an induced fit that changes the receptor conformation. This triggers a series of events that ultimately results in a change in cellular chemistry.

- Some receptors (e.g. steroid receptors) are intracellular.

- Neurotransmitters and hormones do not undergo a reaction when they bind to receptors. They depart the binding site unchanged once they have passed on their message.

- The interactions that bind a chemical messenger to the binding site must be strong enough to allow the chemical message to be received, but weak enough to allow the messenger to depart.

- Receptors controlling ion channels are an integral part of the ion channel. Binding of a messenger induces a change in shape, which results in the opening of the ion channel.

- Some membrane-bound receptors control the activation of signal proteins. Binding of a messenger results in the opening of a binding site for the signal protein. The latter binds and fragments, with one of the subunits departing to activate a membrane-bound enzyme.

- Some membrane-bound receptors contain an enzyme active site on the intracellular portion of their structure. Messenger binding results in the opening of the active site, allowing a catalytic reaction to take place in the cell.

5.6 The design of agonists

We are now at the stage of understanding how drugs might be designed in such a way that they mimic the natural neurotransmitters. Assuming that we know

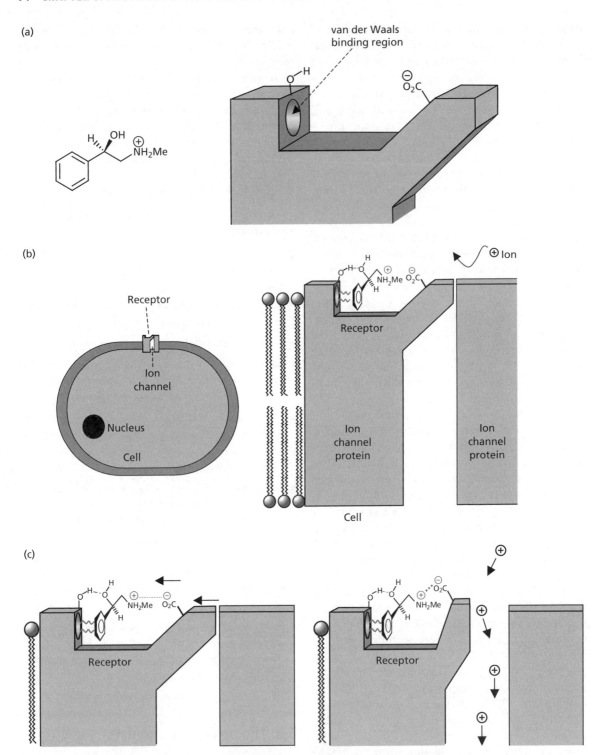

Figure 5.10 Hypothetical receptor controlling an ion channel: (a) A hypothetical neurotransmitter and receptor. (b) Receptor protein positioned in the cell membrane. (c) Opening of the 'lock gate'.

what binding regions are present in the receptor site and where they are, we can design drug molecules to interact with the receptor in the same way. Let us look at this more closely and consider the following requirements in turn.

- The drug must have the correct binding groups.
- The drug must have these binding groups correctly positioned.
- The drug must be the right size for the binding site.

5.6.1 Binding groups

If we consider our hypothetical receptor and its natural neurotransmitter, then we might reasonably predict which of a series of molecules would interact with the receptor and which would not. For example, consider the structures in Fig. 5.11. They all look different, but they all contain the necessary binding groups to interact with the receptor. Therefore, they may well be potential agonists or alternatives for the natural

neurotransmitter. However, the structures in Fig. 5.12 lack one or more of the required binding groups and might therefore be expected to have poor activity. We would expect them to drift into the receptor site and then drift back out again, binding only weakly if at all.

Of course, we are making an assumption here; that all three binding groups are essential. It might be argued that a compound such as structure II in Fig. 5.12 might be effective even though it lacks a suitable hydrogen bonding group. Why, for example, could it not bind initially by van der Waals interactions alone and then alter the shape of the receptor protein via ionic bonding?

In fact, this seems unlikely when we consider that neurotransmitters appear to bind, pass on their message and then leave the binding site relatively quickly. In order to do that, there must be a fine balance in the binding interactions between receptor and neurotransmitter. They must be strong enough to bind the neurotransmitter effectively such that the receptor

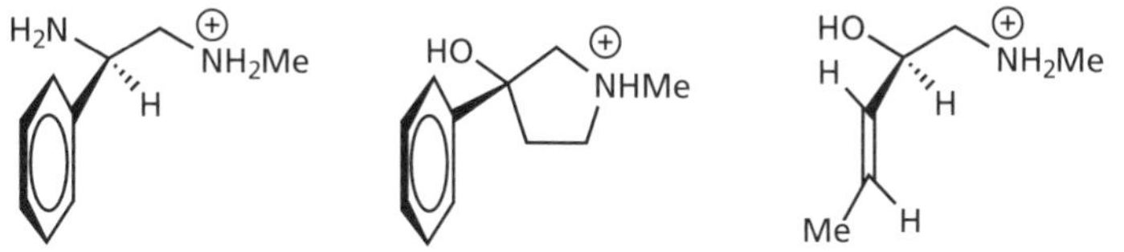

Figure 5.11 Possible agonists.

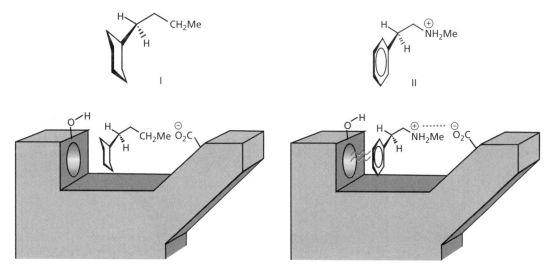

Figure 5.12 The weaker binding to the hypothetical receptor by structures I and II which possess fewer than the required binding groups.

changes shape. However, the binding interactions cannot be too strong, or else the neurotransmitter would not be able to leave and the receptor would not be able to return to its original shape. Therefore, it is reasonable to assume that a neurotransmitter needs all of its binding interactions to be effective. The lack of even one of these interactions would lead to a significant loss in activity.

5.6.2 Position of binding groups

The molecule may have the correct binding groups, but if they are in the wrong relative positions, they will not be able to form bonds at the same time. As a result, bonding would be too weak to be effective.

A molecule such as the one shown in Fig. 5.13 obviously has one of its binding groups (the hydroxyl group) in the wrong position, but there are more subtle examples of molecules that do not have the correct arrangement of binding groups. For example, the mirror image of our hypothetical neurotransmitter would not fit (Fig. 5.14). The structure has the same formula and the same constitutional structure as our original structure. It will have the same physical properties and undergo the same chemical reactions, but it is not the same shape. It is a non-superimposable mirror image and it cannot fit the receptor binding site (Fig. 5.15).

Compounds which have non-superimposable mirror images are termed **chiral** or **asymmetric**. There are only two detectable differences between the two mirror

images (or **enantiomers**) of a chiral compound. They rotate plane-polarized light in opposite directions, and they interact differently with other chiral systems such as enzymes. This has very important consequences for the pharmaceutical industry. Pharmaceutical agents are usually synthesized from simple starting materials using simple achiral (symmetrical) chemical reagents. These reagents are incapable of distinguishing between the two mirror images of a chiral compound. As a result, most chiral drugs used to be synthesized as a mixture of both mirror images (a **racemate**). However, we have seen from our own simple example that only one of these enantiomers is going to interact with a receptor. What happens to the other enantiomer? At best, it floats about in the body doing nothing. At worst, it interacts with a totally different receptor and results in an undesired side effect. Even if the 'wrong' enantiomer does not do any harm, it seems

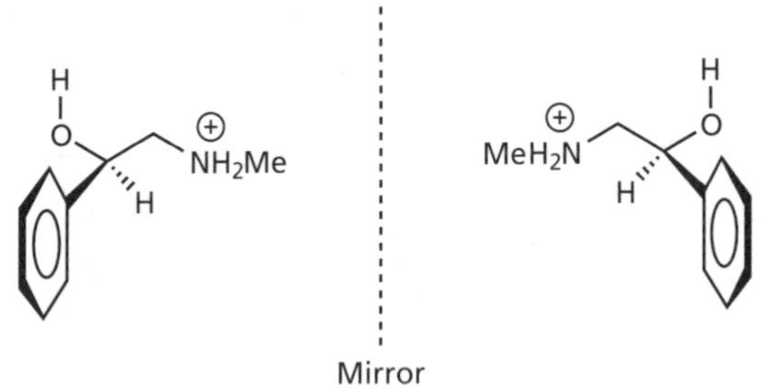

Figure 5.14 Mirror image of a hypothetical neurotransmitter.

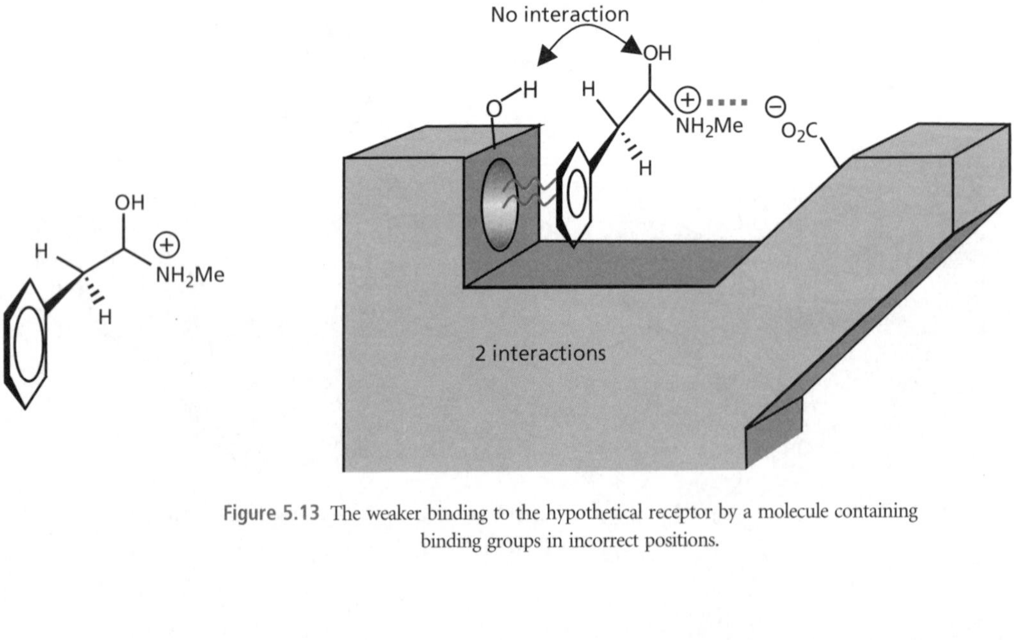

Figure 5.13 The weaker binding to the hypothetical receptor by a molecule containing binding groups in incorrect positions.

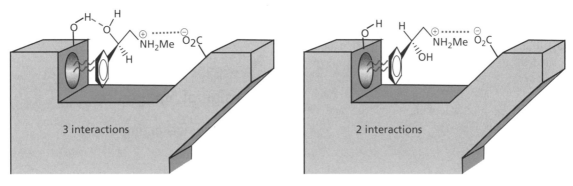

Figure 5.15 Interactions between the hypothetical neurotransmitter and its mirror image with the receptor binding site.

to be a great waste of time, money, and effort to synthesize drugs that are only 50% efficient. That is why one of the biggest areas of chemical research in recent years has been in **asymmetric synthesis**—the selective synthesis of a single enantiomer of a chiral compound.

Of course, nature has been at it for millions of years. Nature has chosen to work predominantly with the 'left-handed' enantiomer of amino acids,[2] so enzymes (made up of left-handed amino acids) are also present as a single mirror image and therefore catalyse **enantiospecific** reactions—reactions which give only one enantiomer. (This also means that the enantiomers of asymmetric enzyme inhibitors tend to be distinguished by the target enzyme, so that one enantiomer is more potent than the other.)

The importance of having binding groups in the correct position has led medicinal chemists to design drugs based on what is considered to be the important **pharmacophore** of the messenger molecule. In this approach, it is assumed that the correct positioning of the binding groups is what decides whether the drug will act as a messenger or not, and that the rest of the molecule serves only to hold the groups in these positions. Therefore, the activity of apparently disparate structures at a receptor can be explained if they all contain the correct binding groups at the correct positions. Totally novel structures or molecular frameworks could then be designed to hold these binding groups in the correct positions, leading to a new series of drugs. There is, however, a limiting factor to this, which will now be discussed.

5.6.3 Size and shape

It is possible for a compound to have the correct binding groups in the correct positions and yet fail to interact effectively if it has the wrong size or shape. As an example, consider the structure shown in Fig. 5.16 as a possible candidate for our hypothetical receptor. The structure has a *meta*-methyl group on the aromatic ring and a long alkyl chain attached to the nitrogen atom. Both of these features would prevent this molecule from binding effectively to the receptor. The *meta*-methyl group would act as a buffer and prevent the structure from sinking deep enough into the binding site for effective binding. Furthermore, the long alkyl chain on the nitrogen atom would make that part of the molecule too long for the space available to it. A thorough understanding of the space available in the binding site is therefore necessary when designing drugs to fit it.

5.6.4 Pharmacodynamics and pharmacokinetics

The study of how molecules interact with targets such as receptors or enzymes is called pharmacodynamics. Such studies can typically be carried out

---

[2] Naturally occurring asymmetric amino acids exist in mammals as the one enantiomer, termed the L-enantiomer. This terminology is historical, and defines the absolute configuration of the asymmetric carbon present at the head group of the amino acid. The current terminology for asymmetric centres is to define them as *R* or *S* according to a set of rules known as the Cahn–Ingold–Prelog rules. The L-amino acids exist as the (*S*)-configuration (except for cysteine, which is *R*), but the older terminology still dominates here. Experimentally, the L-amino acids are found to rotate plane-polarized light anticlockwise or to the left. It should be noted that D-amino acids do occur naturally, for example in bacteria (see for example section 16.5.5).

on the pure target protein, or on isolated cells or tissues which bear the target protein (*in vitro* studies), but it is important to appreciate that designing a drug to interact effectively with a protein *in vitro* does not guarantee a clinically useful drug. Studies should also be carried out concurrently to ensure that promising-looking drugs are active in whole organisms (*in vivo* studies). This is a field

known as pharmacokinetics and is covered in Chapters 8 and 11.

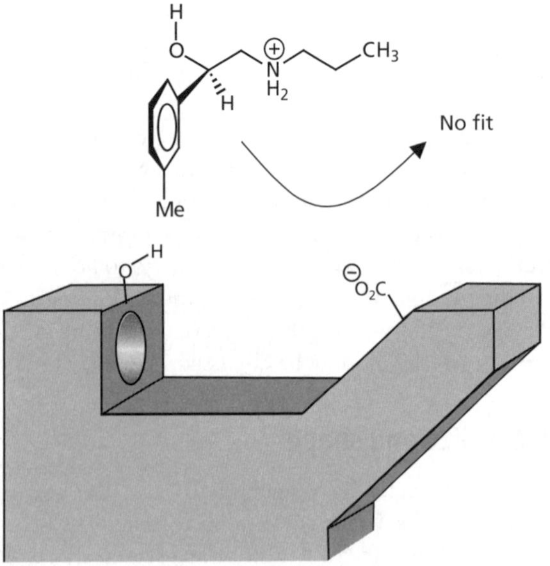

5.7 Design of antagonists

5.7.1 Antagonists acting at the binding site

We have seen how it might be possible to design drugs (agonists) to mimic natural neurotransmitters and how these would be useful in treating a shortage of the natural neurotransmitter. However, suppose that we have too much neurotransmitter operating in the body. How could a drug counteract that?

There are several strategies, but in theory we could design a drug (an antagonist) which would be the right shape to bind to the receptor site, but which would either fail to change the shape of the receptor protein or would distort it in the wrong way. Consider the following scenario.

The compound shown in Fig. 5.17 fits the binding site perfectly and, as a result, does not cause any change of shape. Therefore, there is no biological effect and the binding site is blocked to the natural neurotransmitter. The antagonist has to compete with the agonist for the receptor, but will usually get the better of this contest as it is usually designed to bind more strongly.

Figure 5.16 Failed interaction of a structure with a hypothetical receptor due to steric factors.

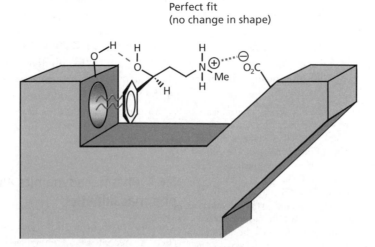

Figure 5.17 Compound acting as an antagonist at the binding site.

To sum up, if we know the shape and make-up of a receptor binding site, then we should be able to design drugs to act as agonists or antagonists. Unfortunately, it is not as straightforward as it sounds. Finding the receptor and determining the layout of its binding site is no easy task. In the past the theoretical shape of many receptor sites was hypothesized by synthesizing a large number of compounds and identifying those molecules that fitted and those that did not—a bit like a 3D jigsaw. Nowadays, the use of genetic engineering, computer-based molecular modelling, and X-ray crystallography allow a more accurate representation of proteins and their binding sites (Chapter 15) and has heralded a new phase of drug development known as structure-based drug design.

5.7.2 Antagonists acting outwith the binding site

Even if the layout of a binding site is known, it may not help in the design of antagonists. There are many examples of antagonists which bear no apparent structural similarity to the native neurotransmitter and could not possibly fit the geometrical requirements of the binding site. Such antagonists frequently contain one or more aromatic rings, suggesting that van der Waals interactions are important in their binding, yet there may not be any corresponding area in the binding site. How then do such antagonists work? There are two possible explanations.

- **Allosteric antagonists**: The antagonist may bind to a totally different part of the receptor. The process

of binding could alter the shape of the receptor protein such that the neurotransmitter binding site is distorted and is unable to recognize the natural neurotransmitter (Fig. 5.18). Therefore, binding between neurotransmitter and receptor would be prevented and the message lost. This form of antagonism is non-competitive, as the antagonist is not competing with the neurotransmitter for the same binding site (compare allosteric inhibitors of enzymes, section 4.6).

- **Antagonism by the 'umbrella effect'**: It has to be remembered that the receptor protein is bristling with amino acid residues, all of which are capable of interacting with a visiting molecule. It is unrealistic to think of the neurotransmitter binding site as an isolated landing pad surrounded by a bland, featureless no-go zone. There will almost certainly be regions close to the binding site which are capable of binding through van der Waals, ionic, or hydrogen bonding interactions. These regions may not be used by the natural neurotransmitter, but they can be used by other molecules. If these molecules bind to such regions and happen to overlap the neurotransmitter binding site, then they will act as antagonists and prevent the neurotransmitter reaching its binding site (Fig. 5.19). This form of antagonism has also been dubbed the **umbrella effect** and is a form of competitive antagonism because the normal binding site is directly affected. Many antagonists are capable of binding to both the normal binding site and neighbouring regions. Such antagonists will therefore bind far more strongly than agonists (Box 5.1).

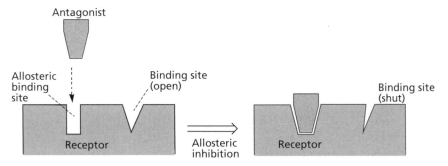
Figure 5.18 Allosteric antagonists.

BOX 5.1 ANTAGONISTS USED AS A MOLECULAR LABEL

Antagonists generally bind more strongly to receptors than agonists. As a result, antagonists have been useful in the isolation and identification of specific receptors present in tissues. A useful tactic in this respect is to incorporate a highly reactive chemical group—usually an electrophilic group—into the structure of a powerful antagonist. The electrophilic group will then react with any convenient nucleophilic group on the receptor surface and alkylate it to form a strong covalent bond. The antagonist will then be irreversibly tied to the receptor and can act as a molecular label. One example is tritium-labelled **propylbenzilylcholine mustard**— used to label the **muscarinic cholinergic receptor** (Chapter 19).

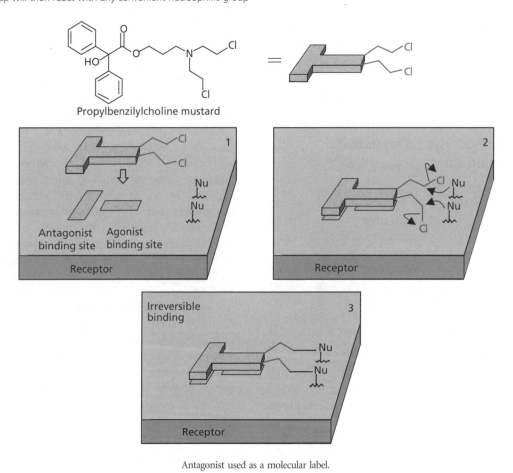

Propylbenzilylcholine mustard

Antagonist binding site Agonist binding site

Receptor

Irreversible binding

Receptor

Receptor

Antagonist used as a molecular label.

5.8 Partial agonists

Frequently a drug is discovered which cannot be defined as a pure antagonist or a pure agonist. The compound acts as an agonist and produces a biological effect, but that effect is not as great as one would get with a full agonist. Therefore the compound is called a **partial agonist**. There are several possible explanations for this.

- A partial agonist must bind to the receptor to have an agonist effect. However, it may be binding in such a way that the conformational change induced is not ideal and the subsequent effects of

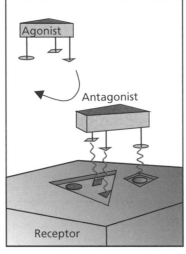

Binding regions   Binding groups

Figure 5.19 Antagonism by the 'umbrella effect'.

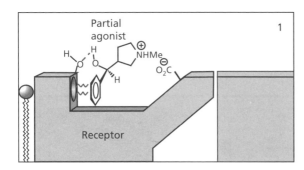
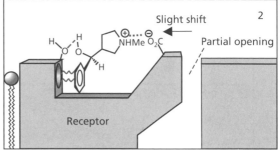

Figure 5.20 Partial agonism.

receptor activation are decreased. In our hypothetical situation (Fig. 5.20), we could imagine a partial agonist being a molecule which is almost a perfect fit for the binding site, such that binding results in only a very slight distortion of the receptor. This would then only partly open the ion channel.

- The partial agonist may be capable of binding to a receptor in two different ways by using different binding regions in the binding site. One method of binding would activate the receptor (agonist effect), but the other would not (antagonist effect). The balance of agonism versus antagonism would then depend on the relative proportions of molecules binding by either method.
- The partial agonist may be capable of distinguishing between different types of receptor (section 5.13).

Receptors which bind the same chemical messenger are not all the same. There are various types and subtypes. A partial agonist might distinguish between these, acting as an agonist at one subtype, but as an antagonist at another subtype.

Examples of partial agonists in the opiate and antihistamine fields are discussed in Chapters 21 and 22 respectively.

5.9 Inverse agonists

Many antagonists which bind to a receptor binding site are in fact more properly defined as inverse agonists. An inverse agonist has the same effect as an

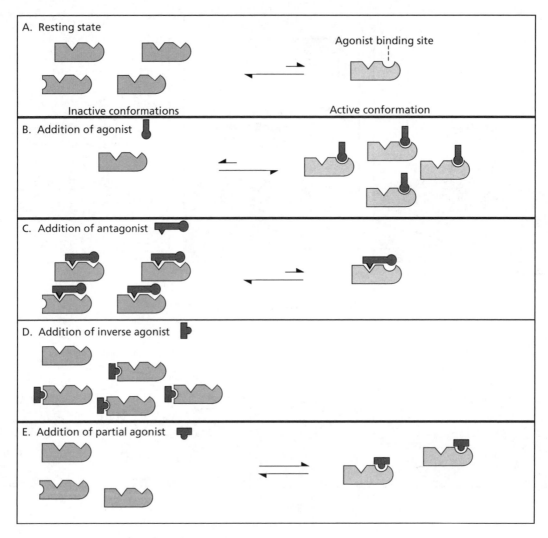

Figure 5.21 Equilibria between active and inactive receptor conformations.

antagonist in that it binds to a receptor, fails to activate it, and prevents the normal chemical messenger from binding. However, there is more to an inverse agonist than that. Some receptors (e.g. the **GABA** and **dihydropyridine receptors**) are found to have an inherent activity even in the absence of the chemical messenger. An inverse agonist is capable of preventing this activity as well.

The discovery that some receptors have an inherent activity has important implications for receptor theory. It suggests that these receptors do not have a 'fixed' inactive conformation, but are continually changing shape such that there is an equilibrium between the active conformation and different inactive conformations. In that equilibrium, most of the receptor population is in an inactive conformation but a small proportion of the receptors is in the active conformation. The action of agonists and antagonists is then explained by how that equilibrium is affected by binding preferences (Fig. 5.21).

If an agonist is introduced (frame B), it binds preferentially to the active conformation and stabilizes it, shifting the equilibrium to the active conformation and leading to an increase in the biological activity associated with the receptor.

In contrast, it is proposed that an antagonist binds equally well to all receptor conformations (both active and inactive) (frame C). There is no change in biological activity, as the equilibrium is unaffected. The introduction of an agonist has no effect either, because all the receptor binding sites are already occupied by the antagonist. Antagonists such as these will have some structural similarity to the natural agonist.

An inverse agonist is proposed to have a binding preference for an inactive conformation. This stabilizes the inactive conformation and shifts the equilibrium away from the active conformation leading to a drop in inherent biological activity (frame D). An inverse agonist need have no structural similarity to an agonist, as it could be binding to a different part of the receptor.

A partial agonist has a slight preference for the active conformation over any of the inactive conformations. The equilibrium is shifted to the active conformation but not to the same extent as with a full agonist and so the increase in biological activity is less (frame E).

5.10 Desensitization and sensitization

Some drugs bind relatively strongly to a receptor, switch it on, but then subsequently block the receptor after a certain period of time. Thus, they are acting as agonists, then antagonists. The mechanism of how this takes place is not clear, but it is believed that prolonged binding of the agonist to the receptor results in phosphorylation of hydroxyl or phenolic groups in the receptor. This causes the receptor to alter shape to an inactive conformation, despite the binding site being occupied (Fig. 5.22). This altered tertiary structure is then maintained as long as the binding site is occupied by the agonist. When the drug eventually leaves, the receptor is dephosphorylated and returns to its original resting shape.

On even longer exposure to a drug, the receptor–drug complex may be removed completely from the cell membrane by a process called **endocytosis**. Here, the relevant portion of the membrane is 'nipped out', absorbed into the cell, and metabolized.

Prolonged activation of a receptor can also result in the cell reducing its synthesis of the receptor protein. Consequently, it is generally true that the best agonists bind swiftly to the receptor, pass on their message, and then leave quickly. Antagonists, in contrast, tend to be slow to add and slow to leave. Prolonged exposure of a target receptor to an antagonist may lead to the opposite of desensitization (i.e. sensitization). This is where the cell synthesizes more receptors to compensate for the receptors that are blocked. This is known to happen when some β-blockers are given over long periods (section 20.11.3).

5.11 Tolerance and dependence

As mentioned above, depriving a target receptor of its neurotransmitter may induce that cell to synthesize more receptors. By doing so, the cell gains a greater sensitivity for what little neurotransmitter is left. This process can explain the phenomena of tolerance and dependence (Fig. 5.23).

Tolerance is a situation where higher levels of a drug are required to get the same biological response. If a drug is acting to suppress the binding of a neurotransmitter, then the cell may respond by increasing the number of receptors. This would require increasing the dose to regain the same level of antagonism.

If the drug is suddenly stopped, then all the receptors suddenly become available. There is now an excess of receptors, which would make the cell supersensitive to normal levels of neurotransmitter. This would be

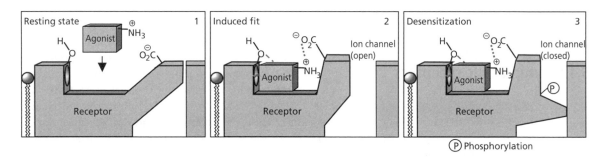

Figure 5.22 Desensitization.

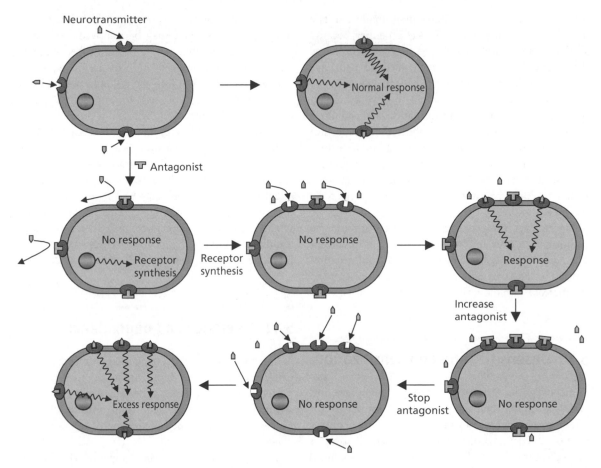

Figure 5.23 Increasing cell sensitivity.

equivalent to receiving a drug overdose. The resulting biological effects would explain the distressing withdrawal symptoms resulting from stopping certain drugs. These withdrawal symptoms would continue until the number of receptors returned to their original level. During this period, the patient may be tempted to take the drug again in order to 'return to normal' and will have then acquired a dependence on the drug.

5.12 Cytoplasmic receptors

Not all receptors are located in the cell membrane. Some receptors are within the cell, and their chemical messengers are required to pass through the cell membrane in order to reach them. Such messengers must be fatty in character in order to do this. The steroid hormones are

examples of chemical messengers which have to cross the cell membrane in order to reach their target receptor (e.g. the **oestrogen receptor**) (section 6.8).

5.13 Receptor types and subtypes

Receptors are identified by the specific neurotransmitter or hormone which activates them. Thus, the receptor activated by **dopamine** is called the **dopaminergic receptor**, the receptor activated by **acetylcholine** is called the **cholinergic receptor**, and the receptor activated by **adrenaline** or **noradrenaline** is called the **adrenergic receptor** or **adrenoceptor**.

However, not all receptors activated by the same chemical messenger are exactly the same throughout the body. For example, the adrenergic receptors in the lungs are slightly different from the adrenergic receptors in the heart. These differences arise from slight variations in amino acid composition, and if the variations are in the binding site, it allows medicinal chemists to design drugs which can distinguish between them. For example, adrenergic drugs can be designed to be 'lung' or 'heart' selective. In general, there are various **types** of a particular receptor and various **subtypes** of these, which are normally identified by numbers or letters. However, for historical reasons the main types of cholinergic receptor are named after natural products (section 19.6).

Some examples of receptor types and subtypes are given in Table 5.1. The identification of many of these subtypes is relatively recent, and the current emphasis in medicinal chemistry is to design drugs that are as selective as possible for receptor types and subtypes so that the drugs are tissue selective and have less side effects. For example, there are five types of dopaminergic receptor. All clinically effective antipsychotic agents (e.g. **clozapine**, **olanzapine**, and **risperidone**; Fig. 5.24) antagonize the dopaminergic receptors D_2 and D_3. However, blockade of D_2 receptors may lead to some of the side effects observed and a selective

Table 5.1 Some examples of receptor types and subtypes

Receptor	Type	Subtype	Examples of agonist therapies	Examples of antagonist therapies
Cholinergic	Nicotinic (N)	Nicotinic (4 subtypes)		Neuromuscular blockers and muscle relaxant (N)
	Muscarinic (M)	M_1–M_5	Stimulation of GIT motility (M_1) Glaucoma (M)	Peptic ulcers (M1) Motion sickness (M)
Adrenergic (adrenoceptors)	Alpha (α_1, α_2)	α_{1A}, α_{1B}, α_{1D}, α_{1A}–α_{2C}		
	Beta (β)	β_1, β_2, β_3	Anti-asthmatics (β_2)	β-Blockers (β_1)
Dopamine		D_1–D_5	Parkinson's disease	Antidepressant (D2/D3)
Histamine		H_1–H_3	Vasodilation (limited use)	Treatment of allergies, antiemetics, sedation (H_1) Antiulcer agents (H_2)
Opioid and opioid-like		μ, κ, δ, ORL1	Analgesics (κ)	Antidote to morphine overdose
5-Hydroxytryptamine (serotonin)	5-HT$_1$–5-HT$_7$	5-HT$_{1A-B}$, 5HT$_{1D-F}$ 5HT$_{2A-C}$, 5-HT$_{5A-B}$	Antimigraine (5-HT$_{1D}$) Stimulation of GIT motility (5-HT$_4$)	Anti-emetics (5-HT$_3$) Ketanserin (5-HT$_2$)
Oestrogen			Contraception	Breast cancer

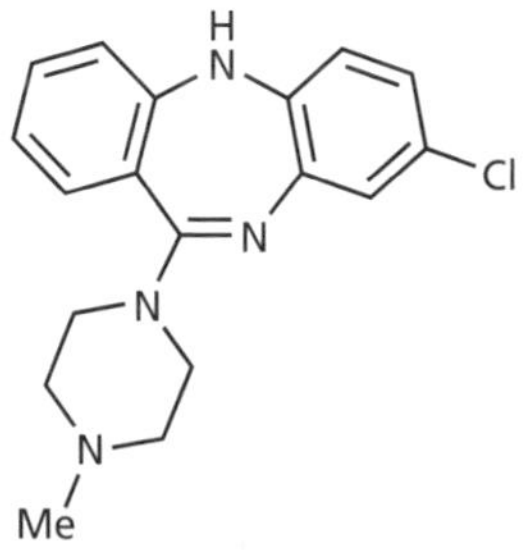

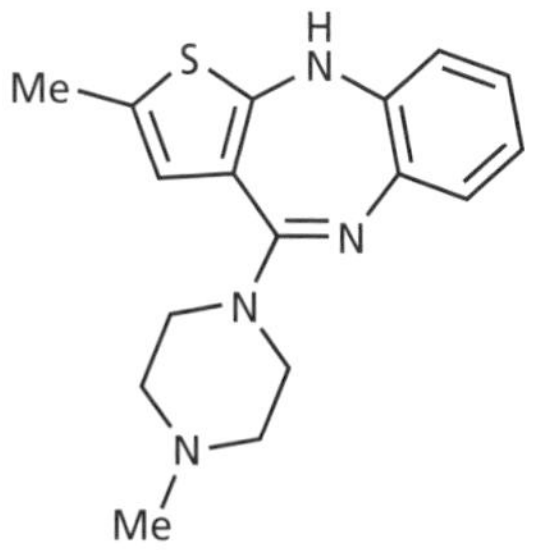

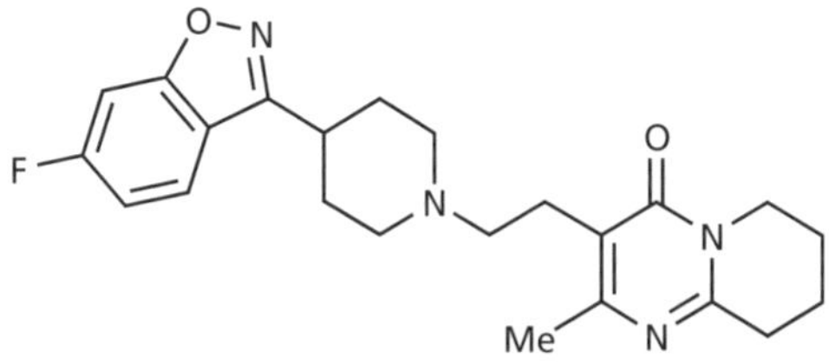

Figure 5.24 Antipsychotic agents.

D_3 antagonist may have better properties as an antipsychotic.

Other examples of current research projects include:

- muscarinic (M_2) agonists for the treatment of heart irregularities
- adrenergic (β_3) agonists for the treatment of obesity
- N-Methyl-D-aspartate (NMDA) antagonists for the treatment of stroke.
- cannabinoid (CB_1) antagonists for treatment of memory loss.

KEY POINTS

- Agonists are compounds which mimic the natural ligand for the receptor.

- Antagonists are agents which bind to the receptor, but which do not activate it. They block binding of the natural ligand.

- Agonists generally have a similar structure to the natural ligand.

- Antagonists bind differently to the natural ligand such that the receptor is not activated.

- Antagonists can bind to regions of the receptor that are not involved in binding the natural ligand. In general, antagonists tend to have more binding interactions than agonists and bind more strongly.

- Partial agonists induce a weaker effect than a full agonist.

- Inverse agonists act as antagonists, but also eliminate any resting activity associated with a receptor.

- Desensitization may occur when an agonist is bound to its receptor for a long period of time. Phosphorylation of the receptor results in a change of conformation.

- Sensitization can occur when an antagonist is bound to a receptor for a long period of time. The cell synthesizes more receptors to counter the antagonist effect.

- Tolerance is a situation where increased doses of a drug are required over time to achieve the same effect.

- Dependence is related to the body's ability to adapt to the presence of a drug. On stopping the drug, withdrawal symptoms occur as a result of abnormal levels of target receptor.

- There are several receptor types and subtypes, which vary in their distribution round the body. They also vary in their selectivity for agonists and antagonists.

- Pharmacodynamics is the study of how drugs interact with their targets. Pharmacokinetics is the study of how drugs reach their targets *in vivo*.

5.14 Affinity, efficacy, and potency

The affinity of a drug for a receptor is a measure of how strongly that drug binds to the receptor. Efficacy is a measure of the maximum biological effect that a drug can produce as a result of receptor binding. It is important to appreciate the distinction between affinity and efficacy. A compound with high affinity does not necessarily have high efficacy. For example, an antagonist can bind with high affinity but has no efficacy. The **potency** of a drug refers to the amount of drug required to achieve a defined biological effect. The smaller the dose required, the more potent the drug. It is possible for a drug to be potent (i.e. active in small doses) but have a low efficacy.

Affinity can be measured using a process known as **radioligand labelling**. A known antagonist (or ligand) for the target receptor is labelled with radioactivity and is added to cells or tissue such that it can bind to the receptors present. Once an equilibrium has been reached, the unbound ligands are removed by washing, filtration, or centrifugation. The extent of binding can then be measured by detecting the amount of radioactivity present in the cells or tissue, and the amount of radioactivity that was removed. The equilibrium constant for bound versus unbound radioligand is defined as the **dissociation binding constant (K_d)**.

$$[L] + [R] \rightleftharpoons \underset{\substack{\text{Receptor–ligand} \\ \text{complex}}}{[LR]} \qquad K_d = \frac{[L] \times [R]}{[LR]}$$

[L] and [LR] can be found by measuring the radioactivity of unbound ligand and bound ligand respectively. However, it is not possible to measure [R], and so we have to carry out some mathematical manipulations to remove [R] from the equation.

The total number of receptors present must equal the number of receptors occupied by the ligand ([LR]) and those that are unoccupied ([R]), i.e.

$$R_{tot} = [R] + [LR].$$

This means that the number of receptors unoccupied by a ligand is

$$[R] = R_{tot} - [LR].$$

Substituting this into the first equation and rearranging leads to the **Scatchard equation**, where both [LR] and [L] are measurable:

$$\frac{[\text{Bound ligand}]}{[\text{Free ligand}]} = \frac{[LR]}{[L]} = \frac{R_{tot} - [LR]}{K_d}$$

We are still faced with the problem that K_d and R_{tot} cannot be measured directly. However, these terms can be determined by drawing a graph based on a number of experiments where different concentrations of a known radioligand are used. [LR] and [L] are measured in each case and a **Scatchard plot** (Fig. 5.25) is drawn which compares the ratio [LR]/[L] versus [LR]. This gives a straight line, and the point where it meets the x-axis represents the total number of receptors available (R_{tot}) (line A; Fig. 5.25). The slope is a measure of the radioligand's affinity for the receptor and allows K_d to be determined.

We are now in the position to determine the affinity of a novel drug that is not radioactively labelled. This is done by repeating the radioligand experiments in the presence of the unlabelled test compound. The test compound competes with the radioligand for the receptor's binding sites and is called a **displacer**. The stronger the affinity of the test compound, the more effectively it will compete for binding sites and the less radioactivity will be measured for [LR]. This will result in a different line in the Scatchard plot.

If the test compound competes directly with the radiolabelled ligand for the same binding site on the receptor, then the slope is decreased but the intercept on the x-axis remains the same (line X in the graph). In other words, if the radioligand concentration is much greater than the test compound it will bind to all the receptors available.

Agents that bind to the receptor at an allosteric binding site do not compete with the radioligand for the same binding site and so cannot be displaced by high levels of radioligand. However, by binding to an allosteric site they make the normal binding site unrecognizable to the radioligand and so there are less receptors available. This results in a line with an identical slope to line A, but crossing the x-axis at a different point, thus indicating a lower total number of available receptors (line Y).

The data from these displacement experiments can be used to plot a different graph which compares the percentage of the radioligand that is bound to a receptor versus the concentration of the test compound. This results in a sigmoidal curve termed the **displacement** or **inhibition curve**, which can be used to identify the **IC$_{50}$ value** for the test compound (i.e. the concentration of compound that prevents 50% of the radioactive ligand being bound).

The **inhibitory** or **affinity constant** (K_i) for the test compound is the same as the [IC]$_{50}$ value if non-competitive interactions are involved. For compounds that *are* in competition with the radioligand for the binding site, the inhibitory constant depends on the level of radioligand present and is defined as

$$K_i = \frac{IC_{50}}{1 + [L]_{tot}/K_d}$$

where K_d is the dissociation constant for the radioactive ligand and $[L]_{tot}$ is the concentration of radioactive ligand used in the experiment.

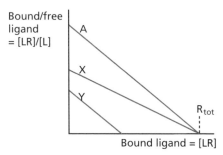
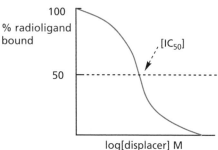

Figure 5.25 Scatchard plot. A, radioligand only; X, radioligand + competitive ligand; Y, radioligand + non-competitive ligand.

Efficacy is determined by measuring the maximum possible effect resulting from receptor–ligand binding. Potency can be determined by measuring the concentration of drug required to produce 50% of the maximum possible effect (EC_{50}) (Fig. 5.26). The smaller the value of EC_{50}, the more potent the drug. In practice, pD_2 is taken as the measure of potency where $pD_2 = -\log[EC]_{50}$.

A **Schild analysis** is used to determine the dissociation constant (K_d) of competitive antagonists (Fig. 5.27). An agonist is first used at different concentrations to activate the receptor, and an observable effect is measured at each concentration. The experiment is then repeated several times in the presence of different concentrations of antagonist. Comparing the effect ratio ($E_{observed}/E_{maximum}$) versus the log of the agonist concentration

(log[agonist]) produces a series of sigmoidal curves where the EC_{50} of the agonist increases with increasing antagonist concentration. In other words, greater concentrations of agonist are required to compete with the antagonist. A **Schild plot** is then constructed, which compares the reciprocal of the dose ratio versus the log of the antagonist concentration. (The **dose ratio** is the agonist concentration required to produce a specified level of effect when no antagonist is present, compared to the agonist concentration required to produce the same level in the presence of antagonist.) The line produced from these studies can be extended to the x-axis to find pA_2 ($= -\log K_d$), which represents the affinity of the competitive antagonist.

KEY POINTS

- Affinity is a measure of how strongly a drug binds to a receptor. Efficacy is a measure of the effect of that binding on the cell. Potency relates to how effective a drug is in producing a cellular effect.

- Affinity can be measured from Scatchard plots derived from radioligand displacement experiments.

- Efficacy is determined by the EC_{50} value—the concentration of agent required to produce 50% of the maximum possible effect resulting from receptor activation.

- A Schild analysis is used to determine the dissociation constant of competitive antagonists.

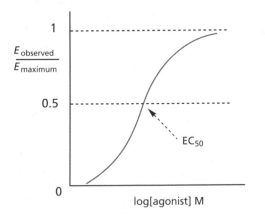

Figure 5.26 Measurement of EC_{50}.

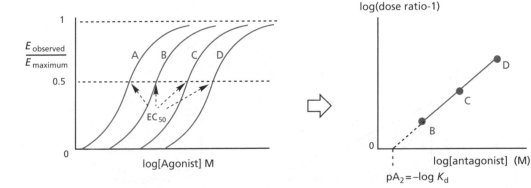

Figure 5.27 Schild analysis. A, no antagonist present; B–D, increasing concentrations of antagonist present.

QUESTIONS

1. Explain the distinction between a binding site and a binding region.

2. Consider the structures of the neurotransmitters shown in Fig. 5.2 and suggest what type of binding interactions could be involved in binding them to a receptor binding site. Identify possible amino acids in the binding site which could take part in each of these binding interactions.

3. Structure I is an agonist which binds to the cholinergic receptor and mimics the action of the natural ligand acetyl-choline. Structure II, on the other hand, shows no activity and does not bind to the receptor. Suggest why this might be the case.

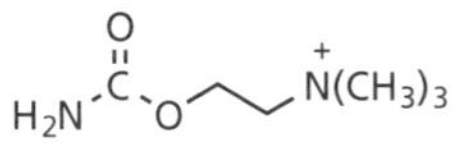

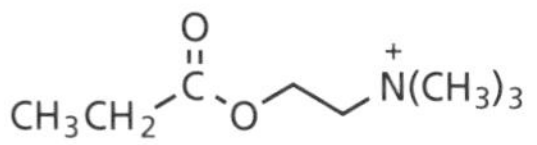

4. There are two main types of adrenergic receptor—the α- and β-adrenoceptors. Noradrenaline shows slight selectivity for the α-receptor, whereas isoprenaline shows selectivity for the β-adrenoceptor. Adrenaline shows no selectivity and binds equally well to both the α- and β-adrenoceptors. Suggest an explanation for these differences in selectivity.

5. Isoprenaline undergoes metabolism to give the inactive metabolite shown below. Suggest why this metabolite is inactive.

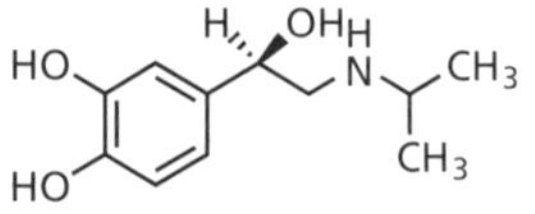

Isoprenaline 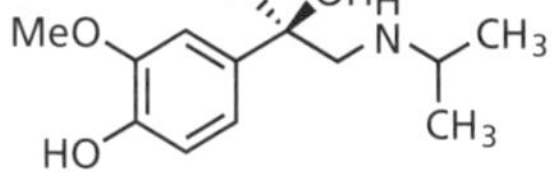 / Inactive metabolite

6. Salbutamol is an antiasthmatic agent which acts as an adrenergic agonist. Do you think it is likely to show any selectivity between the α- or β-adrenoceptors? Explain your answer.

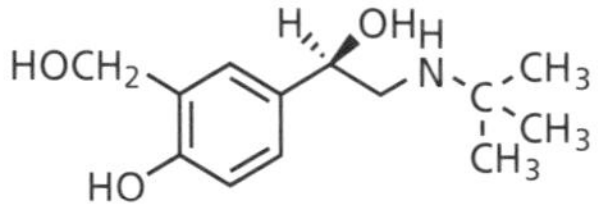

R-Salbutamol

7. Propranolol (Fig. 20.26) is an adrenergic antagonist. Compare the structure of propranolol with noradrenaline and identify which features are similar in both molecules. Suggest why this molecule might act as an antagonist rather than an agonist and whether it might show any selectivity between the different types of adrenergic receptor.

8. If you were asked to design drugs which acted as selective antagonists of the dopamine receptor, what structures might you consider synthesizing?

FURTHER READING

Alexander, S., Mead, A., and Peters, J. (eds.) (1998) *TiPS receptor and ion channel nomenclature supplement*, 9th edn. *Trends in Pharmacological Sciences*, **19**(3), Suppl. 1.

Chalmers, D. T. and Behan, D. P. (2002) The use of constitutively active GPCRs in drug discovery and functional genomics. *Nature Reviews Drug Discovery*, **1**, 599–608.

Christopoulis, A. (2002) Allosteric binding sites on cell surface receptors: novel targets for drug discovery. *Nature Reviews Drug Discovery*, **1**, 198–210.

Kreek, M. J., LaForge, K. S., and Butelman, E. (2002) Pharmacotherapy of addictions. *Nature Reviews Drug Discovery*, **1**, 710–725.

Maehle, A-H., Prull, C-R., and Halliwell, R. F. (2002) The emergence of the drug receptor theory. *Nature Reviews Drug Discovery*, **1**, 637–641.

Pouletty, P. (2002) Drug addictions: towards socially accepted and medically treatable diseases. *Nature Reviews Drug Discovery*, **1**, 731–736.

Titles for general further reading are listed on p. 711.

6 Proteins as drug targets: receptor structure and signal transduction

In Chapter 5, we discussed how neurotransmitters and hormones (ligands) interact with receptor proteins to change the tertiary structure of these proteins. This change of structure can result in the opening or closing of ion channels, or the activation or deactivation of enzymes, both of which lead to a change in chemistry within the cell. In Chapter 5, this was represented as a fairly simple process. In reality, the interaction of a receptor with its chemical messenger is the first step in a complex chain of events involving several secondary messengers, proteins, and enzymes. In this chapter, we shall look more closely at receptor structure and the events that occur after receptor–ligand binding. These events are referred to as **signal transduction**.

6.1 Receptor families

There are many different membrane-bound receptors, but they can be grouped into three **superfamilies**, on the basis of their structure and function:

- ion channel receptors (2-TM, 3-TM, and 4-TM receptors)
- G-protein-coupled receptors (7-TM receptors)
- kinase-linked receptors (1-TM receptors).

Here TM stands for transmembrane. Thus, 2-TM indicates that the proteins involved in these receptors contains two segments that transverse the cell membrane.

The structure and signalling pathways of these superfamilies differ as described in the following sections. They also differ in their speed of response. The signalling process involving ion channels is a rapid response, measured in a matter of milliseconds. These are the receptors involved in the rapid synaptic transmission between nerves. The signals arising from G-protein-coupled receptors result in a response measured in seconds, and the kinase-linked receptors lead to responses in a matter of minutes. There is a fourth superfamily of intracellular receptors which are not membrane-bound and which lead to responses in a matter of hours or days. These are the receptors in the cytoplasm which interact with steroid hormones.

6.2 Receptors that control ion channels (ligand-gated ion channel receptors)

Receptors that control ion channels are part of a five-protein ion channel structure (Fig. 6.1) and are called ligand-gated ion channel receptors. They are **glycoproteins** which traverse the cell membrane, and the fact that the receptor is part of the ion channel structure means that binding of a chemical messenger leads to the rapid response that is crucial to the speed and efficiency of nerve transmission. When the receptor binds a ligand, it changes shape and this has a knock-on effect on the protein complex which causes the ion channel to open—a process called **gating**.

Ion channels can select between different ions. There are cationic ion channels for sodium (Na^+), potassium (K^+), and calcium (Ca^{2+}) ions. When these channels are open, they are generally excitatory and lead to depolarization of the cell (see Appendix 4).

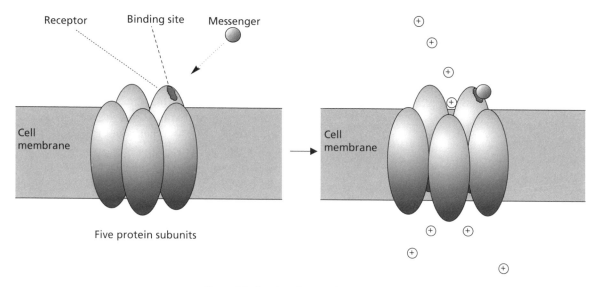

Figure 6.1 Opening of an ion channel—gating.

Some of the receptors involved in the control of cationic ion channels include **nicotinic** receptors (section 19.14), **glutamate** receptors, **5HT$_3$** receptors, and **ATP** receptors. There are also anionic ion channels for the chloride ion, controlled by the **GABA$_A$** and **glycine** receptors. These ion channels generally have an inhibitory effect when they are open.

The ion selectivity of different ion channels is dependent on the amino acids lining the ion channel and, interestingly enough, the mutation of one amino acid in this area is sufficient to change a cationic selective ion channel to an anionic selective channel.

There are at least three families of receptors involved in the control of ion channels and these are classified by the number of **transmembrane** domains they contain (4-TM, 3-TM, 2-TM).

6.2.1 Structure and function of 4-TM ion channel receptors

The 4-TM ion channel receptors include the **nicotinic acetylcholine** (nACh) receptor, the **serotonin** (5HT$_3$) receptor, the **glycine** receptor, and the **GABA$_A$** receptor. The ion channel controlled by the nicotinic acetylcholine receptor is made up five subunits of four different types (two α, β, γ, δ); the ion channel controlled by the glycine receptor is made up of five subunits of two different types (three α, two β) (Fig. 6.2).

The receptor protein in the ion channel controlled by glycine is the α-subunit. Three such subunits are

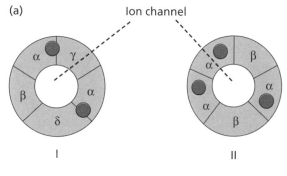

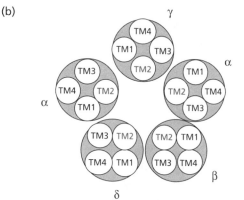

Figure 6.2 (a) Pentameric structure of ion channels (transverse view) I, ion channel controlled by a nicotinic receptor; II, ion channel controlled by a glycine receptor. The coloured circles indicate ligand binding sites. (b) Transverse view including TM regions.

present, all capable of interacting with glycine. However, the situation is slightly more complex in the ion channel controlled by acetylcholine. Most of the acetylcholine binding site is on the α-subunit but there is some involvement from neighbouring subunits. In this case, the ion channel complex as a whole might be viewed as the receptor.

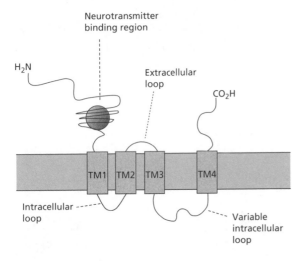

Figure 6.3 Structure of 4-TM receptor subunit.

Each protein subunit in the ion channel complex has a lengthy *N*-terminal extracellular chain which (in the case of the α-subunit) contains the ligand-binding site. There are also four hydrophobic sections which cross the membrane (the transmembrane domains) and which distinguish these ion channels (4-TM). Because the protein chain crosses the membrane four times, there is one extracellular and two intracellular loops connecting these regions, as well as an extra-cellular *C*-terminal chain. The transmembrane regions are numbered 1–4 from the *N*-terminal end, and at least one of these regions is an α-helix (Fig. 6.3).

The binding of a neurotransmitter to its binding site causes a conformational change in the receptor, which eventually opens up the central pore and allows ions to flow. This conformational change is quite complex, involving several knock-on effects from the initial binding process. This must be so, as the binding site is quite far from the lock gate. Studies have shown that the lock gate is made up of five kinked α-helices where one helix is contributed by each of the five protein subunits. In the closed state the kinks are pointing towards each other. The conformational change induced by ligand binding causes each of these helices to rotate such that the kink points the other way, thus opening up the pore (Fig. 6.4).

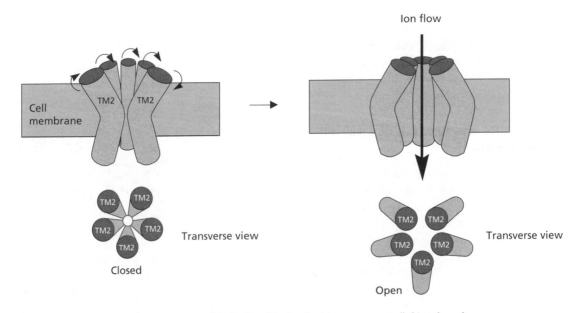

Figure 6.4 Opening of the 'lock gate' in the nicotinic receptor-controlled ion channel.

6.2.2 3-TM ion channel receptors

The ion channels controlled by these receptors also consist of five protein subunits. However, this time each protein subunit has three transmembrane segments, known as TM1, TM3, and TM4. This peculiar numbering is due to the presence of a hydrophobic segment that is thought to be embedded in the intracellular side of the membrane and is has been given the number TM2 (Fig. 6.5). The ligand binding site seems to involve both the N-terminal chain and the extracellular loop (regions A and B). The calcium ion channels controlled in the central nervous system by L-glutamate are examples of this type of ion channel, and are important in several processes including learning and memory.

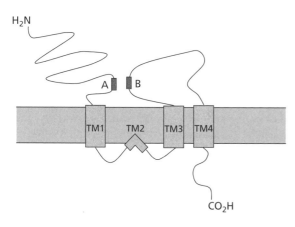

Figure 6.5 Structure of 3-TM ion channel receptor subunits. A and B are ligand-binding regions.

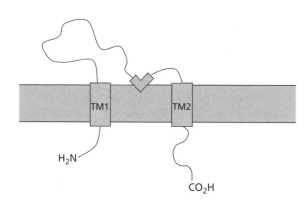

Figure 6.6 Structure of 2-TM ion channel receptor subunits.

6.2.3 2-TM ion channel receptors

2-TM ion channels consist of five protein subunits where each of the subunits contains two transmembrane segments. The *N*-terminal and the *C*-terminal chains are both inside the cell. However, most of the protein is extracellular and includes a hydrophobic region embedded in the outer surface of the cell membrane (Fig. 6.6). **Adenosine triphosphate** (ATP) is thought to control an ion channel of this type.

KEY POINTS

- The process by which receptor ligand binding leads to chemical effects in the cell is known as signal transduction.

- Membrane-bound receptors belong to three families—ion channels, G-protein-coupled receptors and kinase linked receptors.

- The signalling process is fastest for ion channels and slowest for intracellular receptors.

- Receptors controlling ion channels are called ligand-gated ion channel receptors. They consist of five protein subunits with the receptor binding site being present on one or more of the subunits. These receptors can be classified as 2-TM, 3-TM, or 4-TM depending on the number of transmembrane domains that are present in each subunit.

- Binding of a neurotransmitter to a 4-TM ion channel receptor causes a conformational change in the protein subunits of the ion channel where the second transmembrane domain of each subunit rotates to open the channel.

6.3 Structure of G-protein-coupled receptors

G-protein-coupled receptors do not directly affect ion channels or enzymes. Instead, they activate a signalling protein called a G-protein which then initiates a signalling cascade involving a variety of enzymes.

6.3.1 Structure of G-protein-coupled receptors

The G-protein-coupled receptors (or 7-TM receptors) are proteins which are embedded in the cell membrane, and have regions exposed to both the outside and the

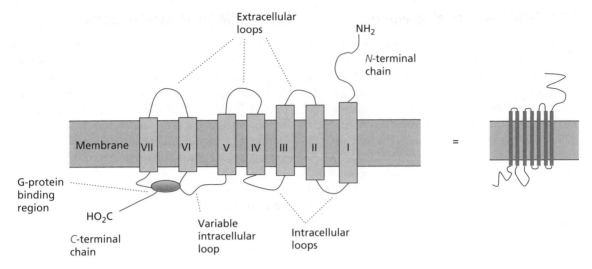

Figure 6.7 Structure of G-protein-coupled receptors.

inside of the cell (Fig. 6.7). The protein chain winds back and forth through the cell membrane seven times—hence the term 7-TM. Each of the seven transmembrane sections is hydrophobic and helical in shape, and it is usual to assign these helices roman numerals (I, II, etc.) starting from the N-terminal end of the protein. The winding of the protein back and forth through the membrane results in three extracellular and three intracellular loops. These loops are fairly constant in length except for the intracellular loop connecting helices V and VI, which varies depending on the specific receptor. The N-terminal chain is extracellular and is variable in length depending on the receptor, while the C-terminal chain is intracellular.

6.3.2 Neurotransmitters and hormones for G-protein-coupled receptors

There is a large number of different G-protein-coupled receptors, each of which interacts with a different neurotransmitter or hormone. The chemical messengers can be grouped by their structures as follows:

- monoamines (e.g. dopamine, histamine, serotonin, acetylcholine, noradrenaline)
- nucleotides
- lipids
- neuropeptides, peptide hormones, and protein hormones

- glycoprotein hormones
- glutamate
- calcium ions.

In general, these receptors mediate the action of hormones and slow-acting neurotransmitters. They include the **muscarinic** receptor (section 19.15) and **adrenergic** receptors (section 20.2).

6.3.3 Ligand binding

Despite the large variety of G-protein receptors, their overall structure is remarkably similar, and it was originally thought that there might be a common binding site into which the different chemical messengers (or ligands) would fit in different ways. However, it is now known that the different structural ligand groups fit their specific receptors in different areas as shown in Fig. 6.8.

- The binding site for monoamines, nucleotides, and lipids is a deep binding pocket between the transmembrane helices (A). The specific interactions involved depend on the nature of the ligand and the specific receptor.

- The binding site for neuropeptides, peptide hormones, and chemokines is closer to the surface of the cell membrane and involves the top of the transmembrane helices, some of the extracellular loops and the N-terminal chain (B).

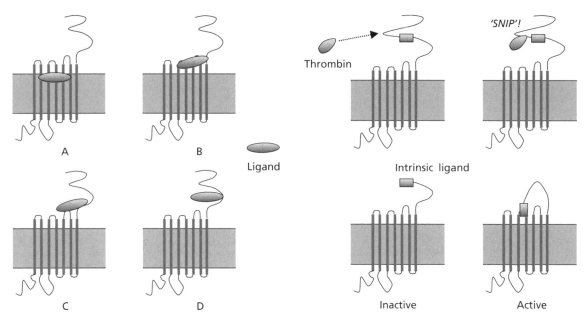

Figure 6.8 Ligand binding to G-protein-coupled receptors.

Figure 6.9 Action of thrombin.

- The binding site for glycoprotein hormones and some peptide hormones is mostly made up by the region of the *N*-terminal chain closest to the surface of the cell membrane, along with some regions of the extracellular loops (C).

- The binding site for glutamate and for calcium ions is made up entirely by the *N*-terminal chain above the surface of the cell membrane (D).

A curious case is the **thrombin** receptor, which has a 'hidden' intrinsic ligand already present in the *N*-terminal chain. This is revealed when thrombin (an enzyme) binds and cleaves off part of the *N*-terminal chain. This reveals a pentapeptide section of the remaining *N*-terminal chain which is then able to bind to the binding site in the extracellular loops of the receptor, resulting in receptor activation (Fig. 6.9).

6.3.4 The rhodopsin-like family of G-protein-coupled receptors

The G-protein-coupled receptors include the receptors for some of the best-known chemical messengers in medicinal chemistry (e.g. glutamic acid, GABA, noradrenaline, dopamine, acetylcholine, serotonin, prostaglandins, adenosine, endogenous opiates, angiotensin, bradykinin, and thrombin). Considering

the structural variety of the chemical messengers involved, it is remarkable that the overall structures of the G-protein-coupled receptors are so similar. Nevertheless, despite their similar overall structure, the amino acid sequences of the receptors vary quite significantly. This implies that these receptors have evolved over many millions of years from an ancient common ancestral protein. Comparing the amino acid sequences of the receptors allows us to construct an evolutionary tree and to group the receptors of this superfamily into various families. The most important of these as far as medicinal chemistry is concerned is the rhodopsin-like family—so called because the first receptor of this family to be studied in detail was the rhodopsin receptor itself, a receptor involved in the visual process. A study of the evolutionary tree of rhodopsin-like receptors throws up some interesting observations (Fig. 6.10). First of all, the evolutionary tree illustrates the similarity between different kinds of receptors based on their relative positions on the tree. Thus, the muscarinic, α-adrenergic, β-adrenergic, histamine, and dopamine receptors have evolved from a common branch of the evolutionary tree and have greater similarity to each other than to any receptors arising from an earlier evolutionary branch (e.g. the angiotensin receptor). Such receptor similarity may prove a problem in medicinal chemistry. Although

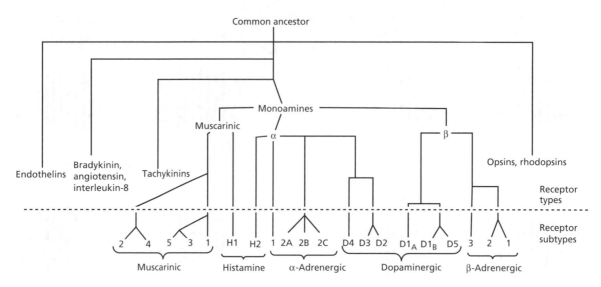

Figure 6.10 Evolutionary tree of G-protein-coupled receptors.

the receptors are distinguished by different neuro-transmitters or hormones in the body, a drug may not manage to make that distinction. Therefore, it is important to ensure that any new drug aimed at one kind of receptor (e.g. the dopamine receptor) does not interact with a similar kind of receptor (e.g. the muscarinic receptor).

Receptors have further evolved to give receptor **types** and **subtypes** which recognize the same chemical messenger, but are structurally different. For example, there are two types of adrenergic receptor (α and β), each of which has various subtypes (α_1, α_{2A}, α_{2B}, α_{2C}, β_1, β_2, β_3). There are two types of cholinergic receptor—nicotinic (an ion channel receptor) and muscarinic (a 7-TM receptor). Five subtypes of the muscarinic cholinergic receptor have been identified.

The existence of receptor subtypes allows the possibility of designing drugs that are selective for one receptor subtype over another. This is important, because one receptor subtype may be prevalent in one part of the body (e.g. the gut) while a different receptor subtype is prevalent in another part (e.g. the heart). Therefore, a drug that is designed to interact specifically with the receptor subtype in the gut is less likely to have side effects on the heart. Even if the different receptor subtypes are present in the same part of the body, it is still important to make drugs as specific as possible, because different receptor subtypes

frequently activate different signalling systems leading to different biological results.

A closer study of the evolutionary tree reveals some curious facts about the origins of receptor subtypes. As one might expect, various receptor subtypes have diverged from a common evolutionary branch (e.g. the dopamine subtypes D2, D3, D4). This is known as **divergent evolution** and there should be close structural similarity between these subtypes. However, receptor subtypes are also found in separate branches of the tree. For example, the dopamine receptor subtypes (D1$_A$, D1$_B$, and D5) have developed from a different evolutionary branch. In other words, the ability of a receptor to bind dopamine has developed in different evolutionary branches—an example of **convergent evolution**.

Consequently, there may sometimes be greater similarities between receptors which bind different ligands but which have evolved from the same branch of the tree, than there are between the various subtypes of receptors which bind the same ligand. For example, the histamine H$_1$ receptor resembles a muscarinic receptor more closely than it does the histamine H$_2$ receptor. This again has important consequences in drug design, since there is an increased possibility that a drug aimed at a muscarinic receptor may also interact with a histamine H$_1$ receptor and lead to unwanted side effects.

KEY POINTS

- G-protein-coupled receptors activate signal proteins called G-proteins.

- The G-protein-coupled receptors are membrane-bound proteins with seven transmembrane sections. The C-terminal chain lies within the cell and the N-terminal chain is extracellular.

- The location of the binding site differs between different G-protein-coupled receptors.

- The rhodopsin-like family of G-protein-coupled receptors includes many important receptors that are targets for currently available drugs.

- Receptor types and subtypes recognize the same chemical messenger, but have structural differences; this makes it possible to design drugs that are selective for one type (or subtype) of receptor over another.

- Receptor subtypes can arise from divergent evolution or convergent evolution.

6.4 Signal transduction pathways for G-protein-coupled receptors

The sequence of events leading from the combination of receptor and ligand (the chemical messenger) to the final activation of a target enzyme is quite lengthy and so we shall look at each stage of the process in turn.

6.4.1 Interaction of the 7-TM receptor–ligand complex with G-proteins

The first stage in the process is the binding of the ligand to the receptor, followed by the binding of a G-protein to the receptor–ligand complex (Fig. 6.11). G-proteins are membrane-bound proteins situated at the inner surface of the cell membrane and are made up of three protein subunits (α, β, and γ). The α-subunit has a binding pocket which can bind guanyl nucleotides (hence the name G-protein) and which

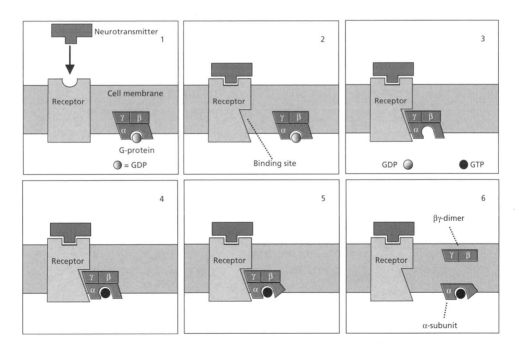

Figure 6.11 Interaction of 7-TM receptors with G-proteins.

binds **GDP (guanosine diphosphate)** when the G-protein is in the resting state. There are several types of G-protein (e.g. G_s, G_i/G_o, G_q/G_{11}) and several subtypes of these. Specific G-proteins are recognized by specific receptors. For example, G_s is recognized by the β-adrenoceptor, but not by the α-adrenoceptor. However, in all cases, the G-protein acts as a molecular 'relay runner', carrying the message received by the receptor to the next target in the signalling pathway.

We shall now look at what happens in detail.

First, the receptor binds its neurotransmitter or hormone (frame 1 in Fig. 6.11). As a result, the receptor changes shape and exposes a new binding site on its inner surface (frame 2). The newly exposed binding site now recognizes and binds a specific G-protein. Note that the cell membrane structure is a fluid structure and so it is possible for different proteins to 'float' through it. The binding process between the receptor and the G-protein causes the latter to change shape, which in turn changes the shape of the guanyl nucleotide binding site. This weakens the intermolecular bonding forces holding GDP and so GDP is released (frame 3). However, the binding pocket does not stay empty for long because it is now the right shape to bind **GTP (guanosine triphosphate)**. Therefore, GTP replaces GDP (frame 4).

Binding of GTP results in another conformational change in the G-protein (frame 5), which weakens the links between the protein subunits such that the α-subunit (with its GTP attached) splits off from the β- and γ-subunits (frame 6). Both the α-subunit and the βγ-dimer then depart the receptor.

The receptor–ligand complex is able to activate several G-proteins in this way before the ligand departs, switching off the receptor. This leads to an amplification of the signal.

Both the α-subunit and the βγ-dimer are now ready to enter the second stage of the signalling mechanism. We shall first consider what happens to the α-subunit.

6.4.2 Signal transduction pathways involving the α-subunit

The first stage of signal transduction (i.e. the splitting of a G-protein) is common to all of the 7-TM receptors. However, subsequent stages depend on what type of G-protein is involved and which specific α-subunit is formed (Fig. 6.12). Different α-subunits (there are at least 20 of them) have different targets and have different effects:

- α_s stimulates adenylate cyclase.
- α_i inhibits adenylate cyclase and may also activate potassium ion channels.

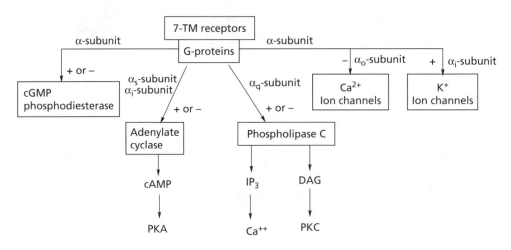

Figure 6.12 Signalling pathways arising from the splitting of G-proteins.

- α_o activates receptors that inhibit neuronal calcium ion channels.

- α_q activates phospholipase C.

We do not have the space to look at all of these pathways in detail. Instead, we shall concentrate on two—the activation of **adenylate cyclase** and the activation of **phospholipase C**.

6.5 Signal transduction involving G-protein-coupled receptors and cyclic AMP

6.5.1 Activation of adenylate cyclase by the α_s-subunit

The α_s-subunit binds to a membrane-bound enzyme called adenylate cyclase (or adenylyl cyclase) and 'switches' it on (Fig. 6.13). This enzyme now catalyses the synthesis of a molecule called cyclic AMP (cAMP)

(Fig. 6.14). cAMP is an example of a **secondary messenger** which moves into the cell's cytoplasm and carries the signal from the cell membrane into the cell itself. The enzyme will continue to be active as long as the α_s-subunit is bound, and this results in the synthesis of several hundred cyclic AMP molecules, representing another substantial amplification of the signal. However, the α_s-subunit has intrinsic GTP-ase activity (i.e. it can catalyse the hydrolysis of its bound GTP to GDP) and so it deactivates itself after a certain period and returns to the resting state. The α_s-subunit then departs the enzyme and recombines with the $\beta\gamma$-dimer to reform the G_s-protein, while the enzyme returns to its inactive conformation.

6.5.2 Activation of protein kinase A

cAMP now proceeds to activate an enzyme called protein kinase A (PKA) (Fig. 6.15). PKA belongs to a group of enzymes called the **serine-threonine kinases** which catalyse the phosphorylation of proteins and enzymes at their serine and threonine residues (Fig. 6.16). PKA is inactive in the resting state and

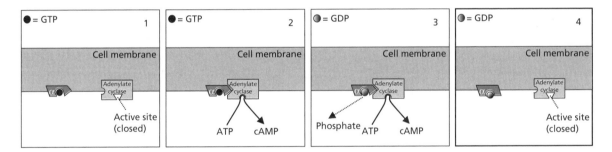

Figure 6.13 Interaction of α_s with adenylate cyclase.

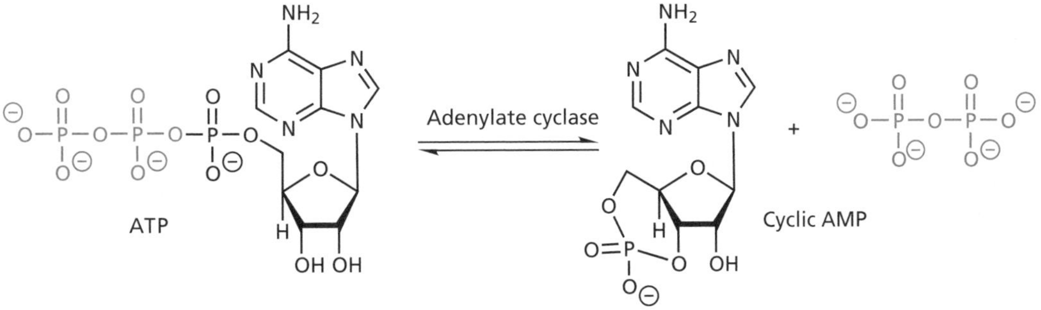

Figure 6.14 Synthesis of cAMP.

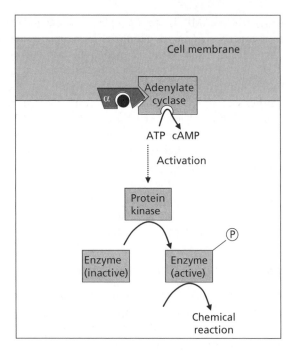

Figure 6.15 Activation of protein kinase A ((P) = phosphate).

Figure 6.16 Phosphorylation of serine and threonine residues.

consists of four protein subunits, two of which are regulatory and two of which are catalytic. Its activation occurs as shown in Fig. 6.17.

The regulatory subunits have two binding sites for cAMP, and when this binds, the catalytic subunits are split from the protein complex and become active. Once the protein kinase catalytic subunits become active, they catalyse the phosphorylation and activation of further enzymes with functions specific to the particular cell or organ in question. For example, a protein kinase activates lipase enzymes in fat cells that catalyse the breakdown of fat. The active site of a protein kinase has to be capable of binding the region of the protein substrate which is to be phosphorylated, as well as ATP which provides the necessary phosphate group.

There may be several more enzymes involved in the signalling pathway between the activation of PKA and the activation (or deactivation) of the target enzyme. For example, the enzymes involved in **glycogen** breakdown and glycogen synthesis in a liver cell are regulated as shown in Fig. 6.18.

Adrenaline is the initial hormone involved in the regulation process and is released when the body requires immediate energy in the form of **glucose**. The hormone initiates a signal at the **β-adrenoceptor** leading to the synthesis of cAMP and the activation of

PKA by the mechanism already discussed. The catalytic subunit of PKA now phosphorylates three enzymes within the cell, as follows:

- An enzyme called **phosphorylase kinase** is phosphorylated and is activated as a result. This enzyme then catalyses the phosphorylation of an inactive enzyme called **phosphorylase *b*** which is converted to its active form, **phosphorylase *a***. Phosphorylase *a* now catalyses the breakdown of glycogen by splitting off glucose-1-phosphate units.

- **Glycogen synthase** is phosphorylated to an inactive form, thus preventing the synthesis of glycogen.

- A molecule called **phosphorylase inhibitor** is phosphorylated. Once phosphorylated, it acts as an inhibitor for the **phosphatase** enzyme responsible for the conversion of phosphorylase *a* back to phosphorylase *b*. The lifetime of phosphorylase *a* is thereby prolonged.

The overall result of these different phosphorylations is a coordinated inhibition of glycogen synthesis and enhancement of glycogen metabolism to generate glucose in muscle cells. Note that the effect of adrenaline on other types of cell may be quite different. For example, adrenaline activates β-adrenoceptors in fat cells, leading to the activation of protein kinases as before. This time, however, phosphorylation activates

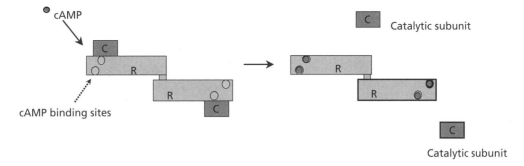

Figure 6.17 Activation of protein kinase A.

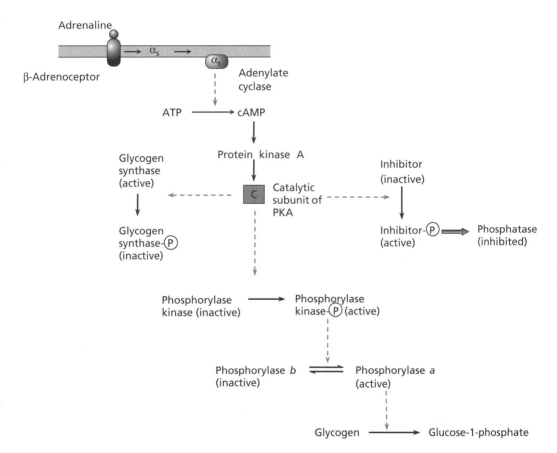

Figure 6.18 Regulation of glycogen synthesis and metabolism in a liver cell.

lipase enzymes which then catalyse the breakdown of fat to act as another source of glucose.

6.5.3 The G_i-protein

We have seen how the enzyme adenylate cyclase is activated by the α_s-subunit of the G_s-protein.

Adenylate cyclase can also be inhibited by a different G-protein—the G_i-protein. The G_i-protein interacts with different receptors from those that interact with the G_s-protein, but the mechanism leading to inhibition is the same as that leading to activation. The only difference is that the α_i-subunit released binds to adenylate cyclase to inhibit the enzyme rather than to activate it. Receptors that bind G_i-proteins include the **muscarinic M_2** receptor of cardiac muscle, **α_2-adrenoceptors** in smooth muscle, and **opioid receptors** in the central nervous system.

The existence of G_i- and G_s-proteins means that the generation of the secondary messenger cAMP is under the dual control of a brake and an accelerator, and this explains the process by which two different neurotransmitters can have opposing effects at a target cell. A neurotransmitter which stimulates the production of cAMP forms a receptor–ligand complex which activates a G_s-protein, whereas a neurotransmitter which inhibits the production of cAMP forms a receptor–ligand complex which activates a G_i-protein. For example, **noradrenaline** interacts with the **β-adrenoceptor** to activate a G_s-protein, whereas **acetylcholine** interacts with the muscarinic receptor to activate a G_i-protein.

Since there are various different types of receptor for a particular neurotransmitter, it is actually possible for that neurotransmitter to activate cAMP in one type of cell but inhibit it in another. For example, noradrenaline interacts with its β-adrenoceptor to activate adenylate cyclase because the β-adrenoceptor binds the G_s-protein. On the other hand, noradrenaline interacts with its α_2-adrenoceptor to inhibit adenylate cyclase because this receptor binds the G_i-protein. This example illustrates the point that it is the receptor, rather than the neurotransmitter or hormone, that determines which G-protein is activated.

It is also worth pointing out that enzymes such as adenylate cyclase and the kinases are never fully active or inactive. At any one time a certain proportion of the enzymes are active and the role of the G_s- and G_i-proteins is to either increase or decrease that proportion. In other words, the control is graded rather than all or nothing.

6.5.4 General points about the signalling cascade involving cAMP

The signalling cascade involving the G_s-protein, cAMP, and PKA appears very complex and you might wonder whether a simpler signalling process would be more efficient. There are several points worth noting about the process as it stands.

- First, the action of the G-protein and the generation of a secondary messenger explains how a message delivered to the outside of the cell surface can be transmitted to enzymes within the cell—enzymes that have no direct association with the cell membrane or the receptor. Such a signalling process avoids the difficulties involved in a messenger molecule (which is commonly hydrophilic) having to cross a hydrophobic cell membrane.

- Second, the process involves a molecular 'relay runner' (the G-protein) and several different enzymes in the signalling cascade. At each of these stages, the action of one protein or enzyme results in the activation of a much larger number of enzymes. Therefore, the effect of one neurotransmitter interacting with one receptor molecule results in a final effect several factors larger than one might expect. For example, each molecule of **adrenaline** is thought to generate 100 molecules of cAMP and each cAMP molecule starts off an amplification effect of its own within the cell.

- Third, there is an advantage in having the receptor, the G-protein, and adenylate cyclase as separate entities. The G-protein can bind to several different types of receptor–ligand complexes. This means that different neurotransmitters and hormones interacting with different receptors can switch on the same G-protein leading to activation of adenylate cyclase. Therefore, there is an economy of organization involved in the cellular signalling chemistry, as the adenylate cyclase signalling pathway can be used in many different cells, and yet respond to different signals. Moreover, different cellular effects will result depending on the type of cell involved (i.e. cells in different tissues will have different receptor types and subtypes and the signalling system will switch on different target enzymes). For example, **glucagon** activates G_s-linked receptors in the liver leading to gluconeogenesis in the liver, **adrenaline** activates G_s-linked β_2-adrenoceptors leading to lipolysis in fat cells, and **vasopressin** interacts with G_s-linked vasopressin (V_2) receptors in the kidney to affect sodium/water resorption. Adrenaline acts on $G_{i/o}$-linked α_2-adrenoceptors leading to contraction of smooth muscle, and **acetylcholine** acts on $G_{i/o}$-linked M_2 receptors leading to relaxation of heart muscle. All these effects are mediated by the cAMP signalling pathway.

- Finally, the dual of 'brake/accelerator' control provided by the G_s- and G_i-proteins allows fine control of adenylate cyclase activity.

6.5.5 The role of the βγ-dimer

If you've managed to follow the complexity of the G-protein signalling pathway so far, well done. Unfortunately, there's more! You may remember that when the G-protein binds to a receptor–ligand complex, it breaks up to form an α-subunit and a βγ-dimer. Until recently, the βγ-dimer was viewed merely as an anchor for the α-subunit to ensure that it remained in the cell membrane. However, it has now been found that the βγ-dimers from both the G_i- and the G_s-proteins can themselves activate or inhibit adenylate cyclase. There are actually six different types (or isozymes) of adenylate cyclase, and activation or inhibition depends on the isozyme involved. Moreover, adenylate cyclase is not the only enzyme that can be controlled by the βγ-dimer. The βγ-dimer is more promiscuous than the α-subunits and can affect several different targets, leading to a variety of different effects. This sounds like a recipe for anarchy. However, there is some advantage in the dimer having a signalling role, since it adds an extra subtlety to the signalling process. For example, it is found that higher concentrations of the dimer are required to result in any effect compared to the α-subunit. Therefore, regulation by the dimers becomes more important when a greater number of receptors are activated.

By now it should be becoming clear that the activation of a cellular process is more complicated than the interaction of one type of neurotransmitter interacting with one type of receptor. In reality, the cell is receiving a whole myriad of signals from different chemical messengers via various receptors and receptor–ligand interactions. The final signal depends on the number and type of G-proteins activated at any one time, as well as the various signal transduction pathways that these proteins initiate.

6.5.6 Phosphorylation

As we have seen above, phosphorylation is a key reaction in the activation or deactivation of enzymes. Phosphorylation requires ATP as a source for the phosphate group and occurs on the phenolic group of **tyrosine** residues when catalysed by **tyrosine kinases**, and on the alcohol groups of **serine** and **threonine** residues when catalysed by **serine–threonine kinases**. These functional groups are all capable of participating in hydrogen bonding, but if a bulky phosphate group is added to the OH group, hydrogen bonding is disrupted. Furthermore, the phosphate group is usually ionized at physiological pH and so phosphorylation

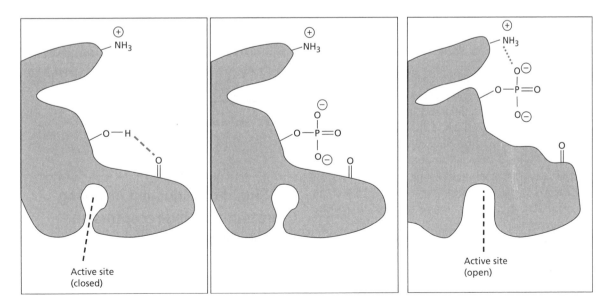

Figure 6.19 Conformational changes induced by phosphorylation.

BOX 6.1 PHOSPHODIESTERASE INHIBITORS

Some drugs and toxins are known to interact with the cAMP signal transduction process. For example, the bacterial toxin responsible for **cholera** causes persistent activation of adenylate cyclase. This leads to excessive secretion of fluid from epithelial cells in the gut, leading to diarrhoea. **Theophylline** and **caffeine** are thought to inhibit the **phosphodiesterases** responsible for metabolizing cAMP, thus prolonging the action of cAMP. There are several types of phosphodiesterase enzyme, and the pharmaceutical industry is currently looking at selective phosphodiesterase inhibitors as possible treatments for a variety of conditions such as inflammation and asthma.

Theophylline   Caffeine

Phosphodiesterase inhibitors.

introduces two negatively charged oxygens. These charged groups can now form strong ionic bonds with any positively charged residues in the protein, causing the enzyme to change its tertiary structure. This change in shape results in the exposure or closure of the active site (Fig. 6.19).

Phosphorylation by kinase enzymes also accounts for the desensitization of G-protein-linked receptors. Phosphorylation of serine and threonine residues occurs on the intracellular *C*-terminal chain after prolonged ligand binding. Since the *C*-terminal chain is involved in G-protein binding, phosphorylation changes the conformation of the protein in that region and prevents the G-protein from binding. Thus the receptor–ligand complex is no longer able to activate the G-protein.

KEY POINTS

- G-proteins consist of three protein subunits, with the α-subunit bound to GDP. There are several types of G-protein.

- Receptor–ligand binding opens a binding site for the G-protein. On binding, GDP is exchanged for GTP, and the G-protein fragments into an α-subunit (bearing GTP) and a βγ-dimer.

- G-proteins are bound and split for as long as the chemical messenger is bound to the receptor, resulting in a signal amplification.

- An α_s-subunit binds to adenylate cyclase and activates it such that it catalyses the formation of cAMP from ATP. The reaction proceeds for as long as the α_s-subunit is bound representing another signal amplification. An α_i-subunit inhibits adenylate cyclase.

- The α-subunits eventually exchange GTP for GDP and depart adenylate cyclase. They combine with their respective βγ-dimers to reform the original G-proteins.

- cAMP acts as a secondary messenger within the cell and activates PKA. PKA catalyses the phosphorylation of serine and threonine residues in other enzymes, leading to a biological effect determined by the type of cell involved.

- The signalling cascade initiated by receptor–ligand binding results in substantial signal amplification and does not require the original chemical messenger to enter the cell.

- The overall activity of adenylate cyclase is determined by the relevant proportions of G_s- and G_i-proteins that are split, which in turn depends on the types of receptors that are being activated.

- The βγ-dimer of G-proteins has a moderating role on the activity of adenylate cyclase and other enzymes when it is present in relatively high concentration.

- Tyrosine kinases are enzymes which phosphorylate the phenol group of tyrosine residues in enzyme substrates. Serine-threonine kinases phosphorylate the alcohol groups of serine and threonine in enzyme substrates. In both cases, phosphorylation results in conformational changes that affect the activity of the substrate enzyme.

- Kinases are involved in the desensitization of receptors.

6.6 Signal transduction involving G-protein-coupled receptors and phospholipase C

6.6.1 Activation by phospholipase C

Certain receptors bind G_s- or G_i-proteins and initiate a signalling pathway involving adenylate cyclase (section 6.5). Other 7-TM receptors bind a different G-protein called a G_q-**protein,** which initiates a different signalling pathway. This pathway involves the activation or deactivation of a membrane-bound enzyme called phospholipase C. The first part of the signalling mechanism is the interaction of the G-protein with a receptor–ligand complex as described previously in Fig. 6.11. This time however, the G-protein is a G_q-protein rather than a G_s- or G_i-protein and so an α_q-subunit is released. Depending on the nature of the released α_q-subunit, phospholipase C is activated or deactivated (Fig. 6.20). If activated, phospholipase C catalyses the hydrolysis of **phosphatidylinositol diphosphate** (PIP$_2$) (an integral part of the cell membrane structure) to generate the two secondary messengers **diacylglycerol** (DG) and **inositol triphosphate** (IP$_3$) (Fig. 6.21).

6.6.2 Action of the secondary messenger diacylglycerol

DG is a hydrophobic molecule and remains in the cell membrane once it is formed (Fig. 6.22). There, it activates an enzyme called **protein kinase C** (PKC) which moves from the cytoplasm to the cell membrane and then catalyses the phosphorylation of serine and threonine residues of enzymes within the cell. Once phosphorylated, these enzymes are activated and catalyse specific reactions within the cell. These induce effects such as tumour propagation, inflammatory responses, contraction or relaxation of smooth muscle, the increase or decrease of neurotransmitter release, the increase or decrease of neuronal excitability, and receptor desensitizations. Designing drugs to inhibit PKC may offer a means of tackling inflammatory or autoimmune diseases and could be used to prevent

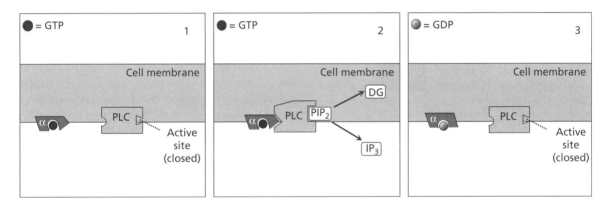

Figure 6.20 Activation of phospholipase C.

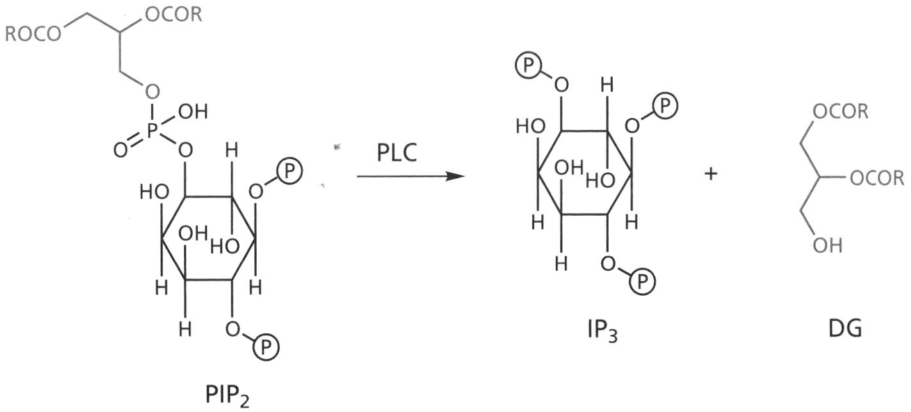

Figure 6.21 Hydrolysis of PIP$_2$ phosphate. Ⓟ $=$ phosphate.

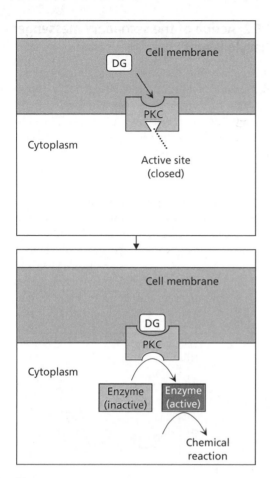

Figure 6.22 Activation of protein kinase C by diacylglycerol.

graft rejection. Inhibitors of PKC may also be useful in cancer therapy. For example, **bryostatin** (Fig. 6.23) is a complex natural product from a sea moss which inhibits PKC activity and has entered clinical trials as an anticancer agent.

So many extracellular and intracellular signals converge on this enzyme that inhibiting this one target may be better than blocking a range of separate receptors.

6.6.3 Action of the secondary messenger inositol triphosphate

Inositol triphosphate is a hydrophilic molecule and moves into the cytoplasm (Fig. 6.24). This messenger works by mobilizing calcium from calcium stores in the endoplasmic reticulum. It does so by binding to a receptor and opening up a calcium ion channel. Once the ion channel is open, calcium ions flood the cell and activate calcium-dependent protein kinases which in turn phosphorylate and activate cell-specific enzymes. The released calcium ions also bind to a calcium-binding protein called **calmodulin,** which then activates calmodulin-dependent protein kinases that phosphorylate and activate other cellular enzymes. Calcium has effects on contractile proteins and ion channels, but it is not possible to cover these effects in detail in this text. Suffice it to say that the release of calcium is crucial to a large

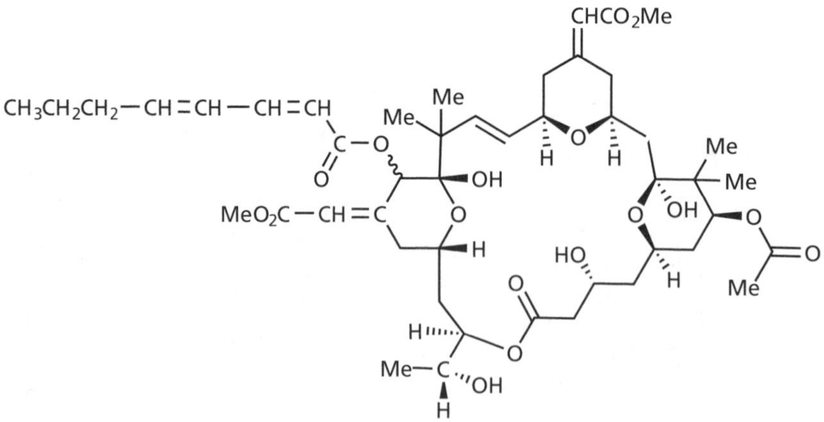

Figure 6.23 Bryostatin.

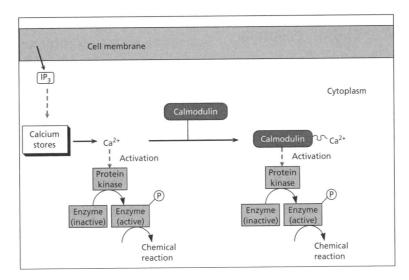

Figure 6.24 Action of inositol triphosphate.

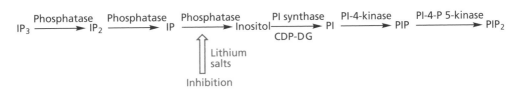

Figure 6.25 Recycling of IP_3 to PIP_2 (CDP-DG = cytidine diphosphate-diacylglycerol).

variety of cellular functions including smooth muscle and cardiac muscle contraction, secretion from exocrine glands, transmitter release from nerves, and hormone release.

6.6.4 Resynthesis of phosphatidylinositol diphosphate

Once IP_3 and DG have completed their tasks, they are recombined to form phosphatidylinositol diphosphate (PIP_2). Oddly enough, they cannot be linked directly and both molecules have to undergo several metabolic steps before resynthesis can occur. For example, IP_3 is dephosphorylated in three steps to inositol which is then used as one of the building blocks for the resynthesis of PIP_2 (Fig. 6.25). It is thought that **lithium** salts control the symptoms of manic depressive illness by interfering with this complex

synthesis, inhibiting the monophosphatase enzyme responsible for the final dephosphorylation leading to inositol.

KEY POINTS

- G_q-proteins are split in a similar manner to G_s- and G_i-proteins. The α_q-subunit affects the activity of phospholipase C which catalyses the hydrolysis of PIP_2 to form the secondary messengers IP_3 and DG.

- DG remains in the cell membrane and activates PKC which is a serine-threonine kinase.

- IP_3 is a polar molecule which moves into the cytoplasm and mobilizes calcium ions. The latter activate protein kinases both directly and via the calcium-binding protein calmodulin.

- IP_3 and DG are combined in a series of steps to reform PIP_2. Lithium salts are believed to interfere with this process.

6.7 Kinase-linked (1-TM) receptors

Kinase-linked receptors are a superfamily of receptors which activate enzymes directly and do not require a G-protein. They are activated by a large number of polypeptide hormones, growth factors, and cytokines. Loss of function of these receptors can lead to developmental defects or hormone resistance. Over-expression can result in malignant growth disorders. They have become important targets in the design of new anticancer drugs (section 18.6.2).

6.7.1 Structure of kinase-linked receptors

An important family of receptors belonging to this superfamily is the **tyrosine kinase receptors**. The basic structure of all the tyrosine kinase receptors consists of a single extracellular region (the *N*-terminal chain) that includes the binding site for the chemical messenger, a single hydrophobic region that traverses the membrane as an α-helix of seven turns (just sufficient to traverse the membrane), and the *C*-terminal chain on the inside of the cell membrane (Fig. 6.26). The *C*-terminal region contains a catalytic binding site, and so these receptors act as a receptor and an enzyme in the one molecule. These receptors include the receptor for **insulin,** and receptors for various **cytokines** and **growth factors**.

6.7.2 Signalling mechanism for the tyrosine kinase receptor family

As stated above, the tyrosine kinase receptors consist of membrane-bound proteins which have a dual role as receptors and enzymes. In the resting state, the receptor has no catalytic activity within the cell because the active site is hidden. However, when a ligand binds to the receptor, the receptor changes shape revealing the active site on the *C*-terminal chain. An enzymatic reaction can now take place within the cell. This reaction is yet another phosphorylation which specifically phosphorylates tyrosine residues, and so the receptor is classed as a tyrosine kinase receptor. As with other kinase enzymes, ATP is required to provide the necessary phosphate group. A specific example of a tyrosine kinase receptor is the receptor for a hormone called **epidermal growth factor** (EGF). EGF is a **bivalent** ligand which can bind to two receptors at the same time. This results in receptor dimerization as well as activation of enzymatic activity. The dimerization process is important because the active site on each half of the receptor dimer catalyses the phosphorylation of the accessible tyrosine residues on the other half (Fig. 6.27). If dimerization did not occur, no phosphorylation would take place. Note that these phosphorylations occur on the intracellular portion of the receptor protein chain.

Dimerization and autophosphorylation is a common theme for receptors in this family. However, some of

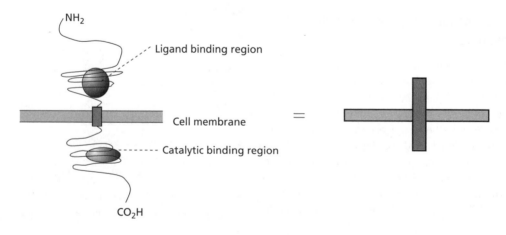

Figure 6.26 Structure of 1-TM receptors.

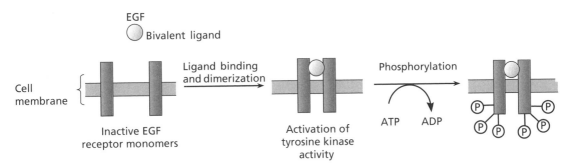

Figure 6.27 Signal mechanism for the EGF receptor.

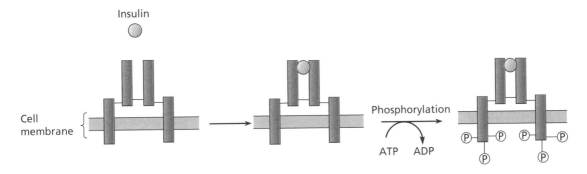

Figure 6.28 Ligand binding for the insulin receptor.

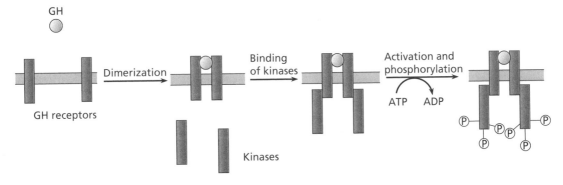

Figure 6.29 The GH receptor.

the receptors in this family already exist as dimers or tetramers and only require binding of the ligand. For example, the **insulin** receptor is a heterotetrameric complex (Fig. 6.28).

Some 1-TM kinase receptors bind ligands and dimerize in a similar fashion but do not have inherent catalytic activity in their *C*-terminal chain. However, once they have dimerized, they can bind and activate a tyrosine kinase enzyme from the cytoplasm. The **growth hormone** (GH) receptor is an example of this type of receptor (a tyrosine kinase-linked receptor) (Fig. 6.29).

6.7.3 Interaction of protein kinase receptors with signalling proteins

Once the kinase receptor (or its associated enzyme) has been phosphorylated, the phosphotyrosine groups and the regions around them act as binding sites for various signalling proteins or enzymes. Each phosphorylated tyrosine region can bind a specific signalling protein or enzyme. Some of these signalling proteins/enzymes become phosphorylated themselves once they are bound, and act as further binding sites for yet more signalling proteins (Fig. 6.30).

Not all of the phosphotyrosine binding regions can be occupied by signalling proteins at the one time and so the type of signalling that results depends on which signalling proteins *do* manage to bind to all the kinase receptors available. There is no room in an introductory text to consider what each and every signalling protein does, but most are the starting point for phosphorylation (kinase) cascades along the same principles as the cascades initiated by G-proteins (Fig. 6.31). Some growth factors activate a specific subtype of **phospholipase C** (PLCγ), which catalyses phospholipid breakdown leading to the generation of IP₃ and subsequent calcium release by the same mechanism described in section 6.6. Other signalling proteins are chemical 'adaptors', which serve to transfer a signal from the receptor to a wide variety of other proteins, including many involved in cell division and differentiation. For example, the principal action of **growth factors** is to stimulate transcription of particular genes through a kinase signalling cascade (Fig. 6.32). A signalling protein called **Grb2** binds to a specific phosphorylated site of the receptor–ligand complex and becomes phosphorylated itself. A membrane protein called **Ras** (with a bound molecule of **GDP**) interacts with the receptor–ligand–signal protein complex and functions in a similar way to a G-protein (i.e. GDP is lost and GTP is gained). Ras is now activated and activates a serine-threonine kinase called **Raf**, initiating a serine-threonine kinase cascade which finishes with the activation of mitogen activated protein **(MAP)-kinase**. This phosphorylates and activates proteins called **transcription factors** which enter the nucleus and initiate gene expression resulting in various responses including cell division. Many cancers can arise from malfunctions of this signalling cascade if the kinases become permanently activated despite the absence of the initial receptor signal, or are

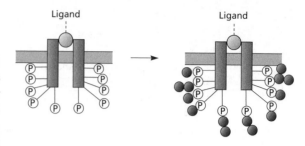

Figure 6.30 Binding of signalling proteins (indicated by dark circles).

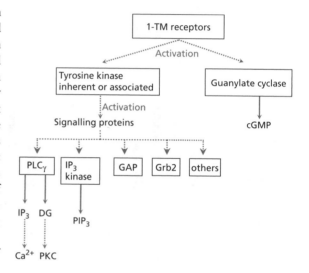

Figure 6.31 Signalling pathways from 1-TM receptors.

over-expressed. Consequently, inhibiting the kinase receptors or targeting the signalling pathway may lead to new drugs for the treatment of cancer (section 18.6).

6.7.4 Small G-proteins

The Ras signal protein described in section 6.7.3 is an example of a class of signalling proteins called the small G-proteins, so called because they are about two-thirds the size of the G-proteins described in sections 6.4–6.6. There are several subfamilies of small G-proteins (Ras, Rho, Arf, Rab, and Ran) and they can be viewed as being similar to the α-subunit of the larger G-proteins. Like the α-subunits they are able to bind either GDP in the resting state or GTP in the activated state. Unlike their larger cousins, the small G-proteins are not activated by direct interaction with a receptor, but are activated downstream of receptor activation through

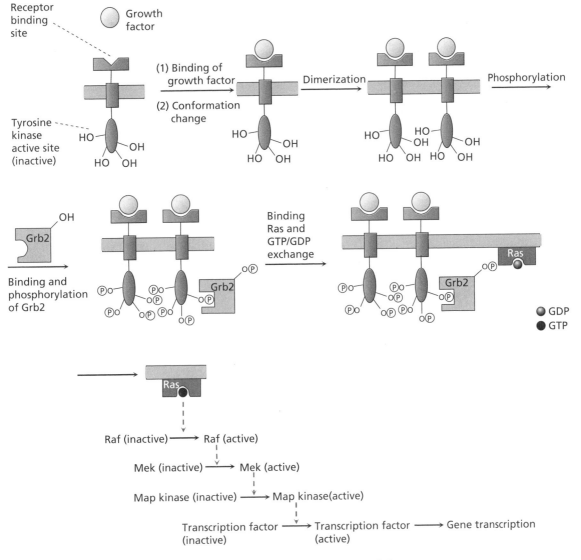

Figure 6.32 From growth factor to gene transcription.

intermediary proteins which are classed as guanine nucleotide exchange factors (GEF). For example, activation of Ras in Fig. 6.32 requires the prior involvement of the protein Grb2 following receptor activation. Like the α-subunits, small G-proteins can autocatalyse the hydrolysis of bound GTP to give bound GDP, resulting in a return to the resting state. However, this process can be accelerated by helper proteins known as **GTPase activating proteins** (GAPs). This means that the level of activity of small G-proteins is under simultaneous brake and accelerator control involving GAP and GEF respectively.

The small G-proteins are responsible for stimulating cell growth and differentiation through different signal transduction pathways. Many cancers are associated with defects in the small G-proteins such as the Ras protein. *Ras* is the gene coding for the Ras protein and is one of the genes most commonly mutated in human tumours. There are three Ras proteins in mammalian cells: H-, K-, and N-Ras. Mutations can occur which result in the inability of these proteins to autocatalyse the hydrolysis of bound GTP. As a result they remain permanently activated, leading in turn to permanent cell growth and division.

6.7.5 Activation of guanylate cyclase by 1-TM receptors

Some 1-TM receptors have the ability to catalyse the formation of **cyclic GMP** from **GTP**. Therefore, they are both receptor and enzyme (guanylate cyclase). The membrane-bound receptor/enzyme spans the cell membrane and has a single transmembrane segment. It has an extracellular receptor binding site and an intracellular guanylate cyclase active site. Its ligands are **α-atrial natriuretic peptide** and **brain natriuretic peptide**. Cyclic GMP appears to open sodium ion channels in the kidney, promoting the excretion of sodium.

KEY POINTS

- Kinase-linked receptors are receptors which are directly linked to kinase enzymes.

- Tyrosine kinase receptors have an extracellular binding site for a chemical messenger, and an intracellular enzymatic active site which catalyses the phosphorylation of tyrosine residues.

- Ligand binding to the EGF receptor results in dimerization and opening of the active sites. The active site on one half of the dimer catalyses the phosphorylation of tyrosine residues present on the C-terminal chain of the other half.

- The insulin receptor is a preformed heterotetrameric structure which acts as a tyrosine kinase receptor.

- The growth hormone receptor dimerizes on binding its ligand, then binds and activates tyrosine kinase enzymes from the cytoplasm.

- The phosphorylated tyrosine residues on activated receptors act as binding sites for various signalling proteins and enzymes which are activated in turn.

- Some 1-TM receptors have an intracellular active site capable of catalysing the formation of cyclic GMP from GTP.

6.8 Intracellular receptors

As discussed above, there are three superfamilies of membrane-bound receptors. There is a fourth superfamily of receptors, the members of which are not membrane-bound and are situated within the cell—the intracellular receptors. There are about 50 members of this group and they are particularly important in directly regulating gene transcription. As a result, they are often called **nuclear hormone receptors** or **nuclear transcription factors.** The chemical messengers for these receptors include steroid hormones, thyroid hormones, and retinoids. In all these cases, the messenger has to pass through the cell membrane in order to reach its receptor and so it has to be hydrophobic in nature.

The intracellular receptors all have a similar structure. They consist of a single protein containing a ligand binding site at the C-terminal end and a binding region for DNA near the centre (Fig. 6.33). The DNA binding region contains nine cysteine residues, eight of which are involved in binding two zinc ions. The zinc ions play a crucial role in stabilizing and determining the conformation of the DNA binding region. As a result, the stretches of protein concerned are called the **zinc finger domains.** The DNA binding region for each receptor can identify particular nucleotide sequences in DNA. For example, the zinc finger domains of the oestrogen receptor recognize the sequence 5'-AGGTCA-3' where A, G, and T are adenine, guanine, and thymine (Chapter 7).

The mechanism by which intracellular receptors work is also very similar (Fig. 6.34). Once the ligand has crossed the cell membrane, it seeks out its receptor and binds to it at the ligand binding site. An induced fit takes place which sees the receptor change shape. This in turn leads to a dimerization of the ligand–receptor complex. The dimer then binds to a protein called a **coactivator** and finally the whole complex binds to a particular region of the cell's DNA. Since

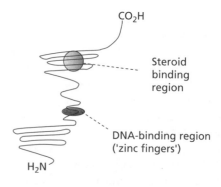

Figure 6.33 Structure of intracellular receptors.

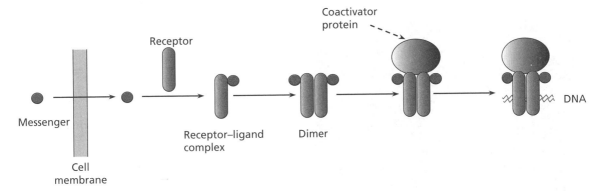

Figure 6.34 From messenger to control of gene transcription.

there are two receptors in the complex and two DNA binding regions, the complex recognizes two identical sequences of nucleotides in the DNA separated by a short distance. For example, the oestrogen ligand–receptor dimer (see Box 6.2) binds to a nucleotide sequence of 5'-AGGTCANNNTGACCT-3' where N can be any nucleic acid. Depending on the complex involved, binding of the complex to DNA either triggers or inhibits the start of transcription and the eventual synthesis of a protein.

BOX 6.2 OESTRADIOL AND THE OESTROGEN RECEPTOR

17β-Oestradiol (spelt estradiol in the USA) is a steroid hormone which affects the growth and development of a number of tissues. It does so by crossing cell membranes and interacting with the binding site of an oestrogen intracellular receptor. Oestradiol uses its alcohol and phenol groups to form hydrogen bonds with three amino acids in the binding site, while the hydrophobic skeleton of the molecule forms van der Waals and hydrophobic interactions with other regions (Fig. 1). The binding pocket is hydrophobic in nature and quite spacious, except for the region where the phenol ring binds. This is a narrow slot and will only accept a planar aromatic ring. Because of these constraints, the binding of oestradiol's phenolic ring determines the orientation for the rest of the molecule.

The binding of oestradiol induces a conformational change in the receptor which sees a helical section (known as H12) folding across the binding site like a lid (Fig. 2). This not only seals oestradiol into its binding site, it also exposes a hydrophobic region called the activating function (AF-2) region which acts as a binding site for a coactivator protein. Since dimerization has also taken place, there are two of these regions available and the coactivator binds to both to complete the nuclear transcription factor. This now binds to a specific region of DNA and switches on the transcription of a gene, resulting in the synthesis of a protein.

Raloxifene (Fig. 3) is an antagonist of the oestrogen receptor and is used for the treatment of hormone-dependent breast cancer. It is a synthetic agent which binds to the binding site without activating the receptor, and prevents oestradiol from binding. The molecule has two phenol groups which mimic the phenol and alcohol group of oestradiol. The skeleton is also hydrophobic and matches the hydrophobic character of oestradiol. So why does raloxifene not act as an agonist? The answer lies in a side chain. This side chain contains an amino group which is protonated and forms a hydrogen bond to Asp-351—an interaction that does not take place with oestradiol. In doing so, the side chain protrudes from the binding pocket and prevents the receptor helix H12 folding over as a lid. As a result, the AF-2 binding region is not exposed, the coactivator cannot bind and the transcription factor cannot be formed. Hence, the side chain is crucial to antagonism. It must contain an amine group of the correct basicity such that it ionizes and forms the interaction with Asp-351, and it must be of the correct length and flexibility to place the amine in the correct position for binding.

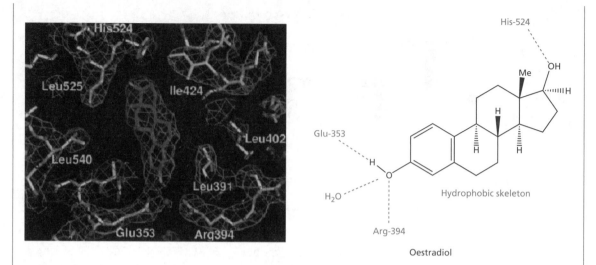

Figure 1 Binding mode of oestradiol with the oestrogen receptor.

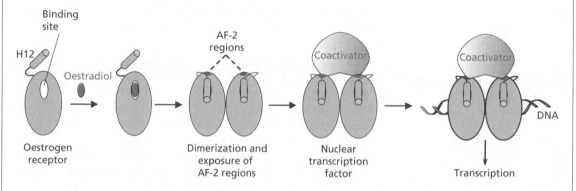

Figure 2 Control of transcription by the oestrogen receptor.

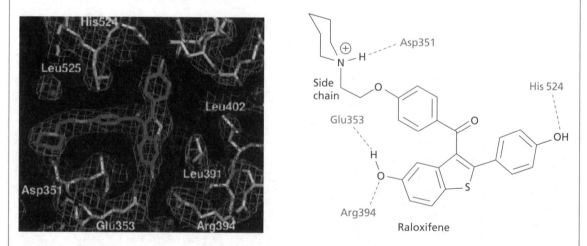

Figure 3 Binding mode of raloxifene with the oestrogen receptor.

KEY POINTS

● Intracellular receptors are located within the cell and are important in controlling transcription.

● The chemical messengers for intracellular receptors must be sufficiently hydrophobic to pass through the cell membrane.

● The binding of a ligand with an intracellular receptor results in dimerization and the formation of a transcription factor which binds to a specific nucleotide sequence on DNA.

QUESTIONS

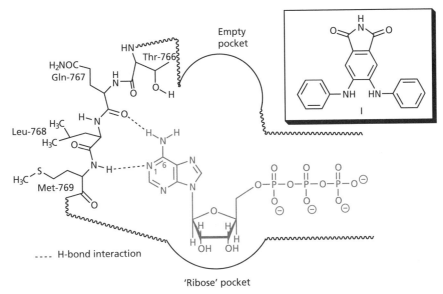

1. A model binding site for ATP was created for EGF receptor kinase, which demonstrates how ATP is bound (see above). Structure I is known to inhibit the binding of ATP. Suggest how structure I might bind.

2. Suggest why the transmembrane regions of many membrane-bound proteins are α-helices.

3. Tamoxifen acts as an antagonist for the oestrogen receptor. Suggest how it might bind to the receptor in order to do this.

4. The tamoxifen metabolite shown acts as an oestrogen agonist rather than an antagonist. Why?

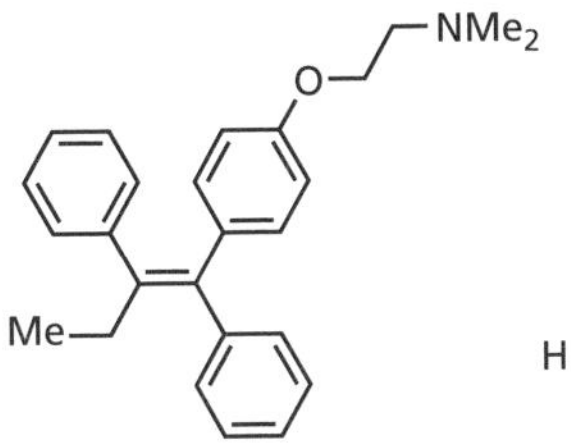

5. Small G-proteins like Ras have an autocatalytic property. What does this mean and what consequences would there be (if any) should that property be lost?

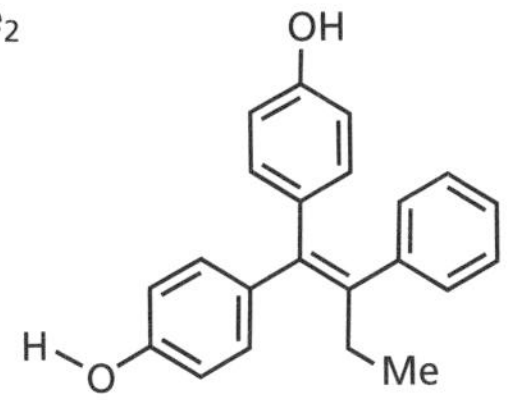

6. Farnesyl transferase is an enzyme which catalyses the attachment of a long hydrophobic chain to the Ras protein. What do you think is the purpose of this chain and what would be the effect if the enzyme was inhibited?

7. Consider the signal transduction pathways shown in Fig. 6.18 and identify where signal amplification takes place.

8. The enzyme cAMP phosphodiesterase hydrolyses cAMP to AMP. What effect would an inhibitor of this enzyme have on glucose-1-phosphate production (Fig. 6.18)?

9. An enzyme was produced by genetic engineering where several of the serine residues were replaced by glutamate residues. The mutated enzyme was permanently active, whereas the natural enzyme was only active in the presence of a serine-threonine protein kinase. Suggest an explanation.

Tamoxifen

Tamoxifen metabolite

10. Suggest why tyrosine kinases phosphorylate tyrosine residues in protein substrates, but not serine or threonine residues.

11. Antibodies have been generated to recognize the extracellular regions of growth-factor receptors. Binding of the antibody to the receptor should block the growth factor from reaching its binding site and block its signal. However, it has been observed that antibodies can sometimes trigger the same signal as the growth factor. Why should this occur? Consult section 3.9 to see the structure of an antibody.

FURTHER READING

Alexander, S., Mead, A., and Peters, J. (eds.) (1998) *TiPS receptor and ion channel nomenclature supplement*, 9th edn. *Trends in Pharmacological Sciences*, **19**(3), Suppl. 1.

Bikker, J. A., Trumpp-Kallmeyer, S., and Humblet, C. (1998) G-Protein coupled receptors: models, mutagenesis, and drug design. *Journal of Medicinal Chemistry*, **41**, 2911–2927.

Cohen, P. (2002) Protein kinases—the major drug targets of the twenty-first century? *Nature Reviews Drug Discovery*, **1**, 309–315.

Flower, D. (2000) Throwing light on GPCRs. *Chemistry in Britain*, November, 25.

Foreman, J. C. and Johansen, T. (eds.) (1996) *Textbook of receptor pharmacology*. CRC Press, Boca Raton.

George, S. R., O'Dowd, B. F., and Lee, S. P. (2002) G-Protein-coupled receptor oligomerization and its potential for drug discovery. *Nature Reviews Drug Discovery*, **1**, 808–820.

Kenakin, T. (2002) Efficacy at G-protein-coupled receptors. *Nature Reviews Drug Discovery*, **1**, 103–110.

Neubig, R. R. and Siderovski, D. P. (2002) Regulators of G-protein signalling as new central nervous system drug targets. *Nature Reviews Drug Discovery*, **1**, 187–197.

Schwarz, M. K. and Wells, T. N. C. (2002) New therapeutics that modulate chemokine networks. *Nature Reviews Drug Discovery*, **1**, 347–358.

Tai, H. J., Grossmann, M., and Ji, I. (1998) G-Protein-coupled receptors. *Journal of Biological Chemistry*, **273**, 17299–17302 and 17979–17982.

Takai, Y., Sasaki, T., and Matozaki, T. (2001) Small GTP-binding proteins. *Physiological Reviews*, **81**, 153–208.

Vlahos, C. J., McDowell, S. A., and Clerk, A. (2003) Kinases as therapeutic targets for heart failure. *Nature Reviews Drug Discovery*, **2**, 99–113.

Titles for general further reading are listed on p. 711.

7 Nucleic acids as drug targets

Although most drugs act on protein structures, there are several examples of important drugs which act directly on nucleic acids to disrupt its replication (section 7.1.2), transcription (section 7.2.2), and translation (section 7.2.2). There are two types of nucleic acid—DNA (deoxyribonucleic acid) and RNA (ribonucleic acid). We first consider the structure of DNA.

structure of DNA is far less varied than the primary structure of proteins. As a result, it was long thought that DNA only had a minor role to play in cell biochemistry, since it was hard to see how such an apparently simple molecule could have anything to do with the mysteries of the genetic code. The solution to this mystery lies in the secondary structure of DNA.

7.1 Structure of DNA

Like proteins, DNA has a primary, secondary, and tertiary structure.

7.1.1 The primary structure of DNA

The primary structure of DNA is the way in which the DNA building blocks are linked together. Whereas proteins have over 20 building blocks to choose from, DNA has only 4—the nucleosides **deoxyadenosine**, **deoxyguanosine**, **deoxycytidine**, and **deoxythymidine** (Fig. 7.1). Each nucleoside is constructed from two components—a **deoxyribose** sugar and a base. The sugar is the same in all four nucleosides and only the base is different. The four possible bases are two bicyclic purines (**adenine** and **guanine**) and two smaller pyrimidine structures (**cytosine** and **thymine**) (Fig. 7.2).

The nucleoside building blocks are joined together through phosphate groups which link the 5′-hydroxyl group of one nucleoside unit to the 3′-hydroxyl group of the next (Fig. 7.3). With only four types of building block available, the primary

Deoxyadenosine   Deoxyguanosine

Deoxythymidine   Deoxycytidine

Figure 7.1 Nucleosides—the building blocks of DNA.

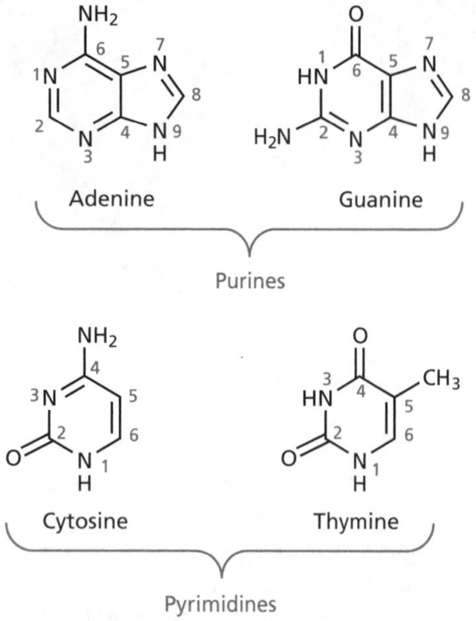

Figure 7.2 The nucleic acid bases for DNA.

7.1.2 The secondary structure of DNA

Watson and Crick solved the secondary structure of DNA by building a model that fitted all the known experimental results. The structure consists of two DNA chains arranged together in a double helix of constant diameter (Fig. 7.4). The double helix can be seen to have a major groove and a minor groove, which are of some importance to the action of several antibacterial agents (section 16.8).

The structure relies crucially on the pairing up of nucleic acid bases between the two chains. Adenine pairs only with thymine via two hydrogen bonds, whereas guanine pairs only with cytosine via three hydrogen bonds. Thus, a bicyclic purine base is always linked with a smaller monocyclic pyrimidine base, to allow the constant diameter of the double helix. The double helix is further stabilized by the fact that the base pairs are stacked one on top of each other, allowing hydrophobic interactions between the

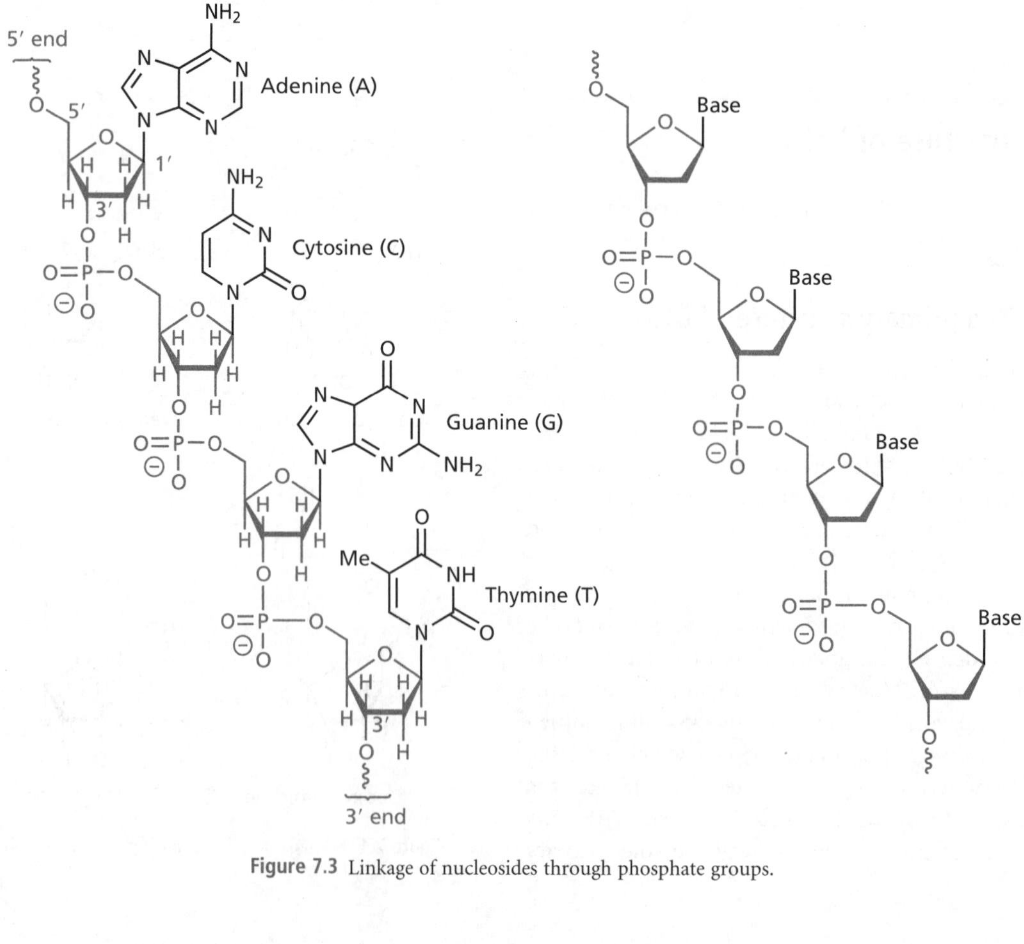

Figure 7.3 Linkage of nucleosides through phosphate groups.

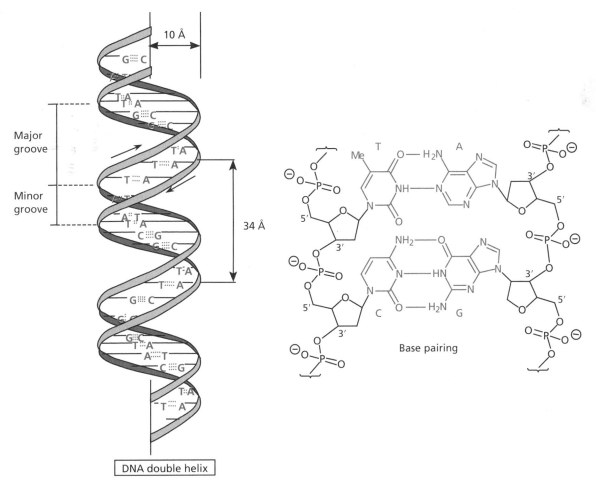

Figure 7.4 The secondary structure of DNA.

faces of the heterocyclic rings. The polar sugar phosphate backbone is placed to the outside of the structure and can form favourable polar interactions with water.

The fact that adenine always binds to thymine, and cytosine always binds to guanine, means that the chains are complementary to each other. In other words, one chain can be visualized as a complementary image of its partner. It is now possible to see how **replication** (the copying of genetic information) is feasible. If the double helix unravels, a new chain can be constructed on each of the original chains (Fig. 7.5). In other words, each of the original chains acts as a template for the construction of a new and identical double helix. The mechanism by which this takes place is shown in Figs. 7.6 and 7.7. The template chain has exposed bases which can base pair by hydrogen bonding with individual nucleotides in the form of triphosphates. Once a nucleotide has base paired, an enzyme-catalysed reaction takes place where the new nucleotide is spliced on to the growing complementary chain with loss of a diphosphate group, the latter acting as a good leaving group. Note that the process involves each new nucleotide reacting with the 3′ end of the growing chain.

We can now see how genetic information is passed on from generation to generation, but it is less obvious

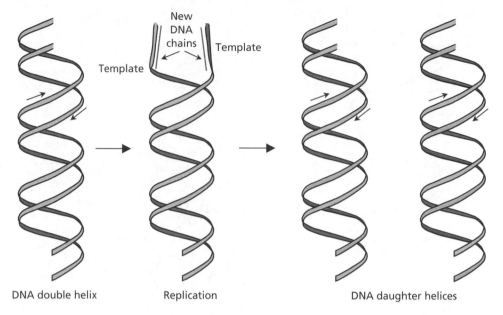

Figure 7.5 Replication of DNA chains.

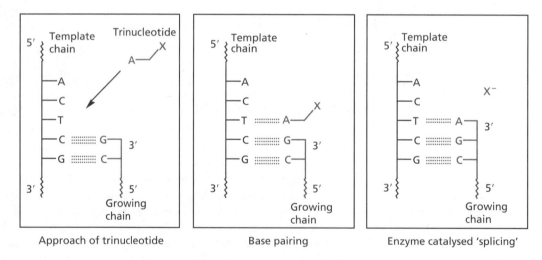

Figure 7.6 Base pairing and splicing.

how DNA codes for proteins. How can only 4 nucleotides code for over 20 amino acids? The answer lies in the **triplet code**. In other words, an amino acid is coded not by one nucleotide but by a set of 3. There are 64 (4^3) ways in which 4 nucleotides can be arranged in sets of 3—more than enough for the task required. Appendix 2 shows the standard genetic code for the various triplets. We shall look at how this code is interpreted to produce a protein in section 7.2.

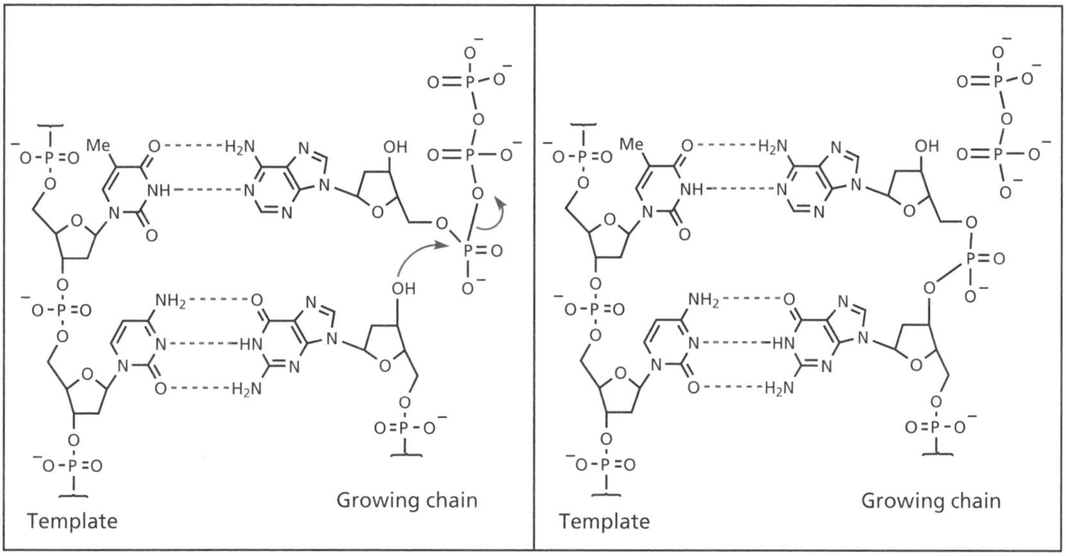

Figure 7.7 Splicing mechanism.

7.1.3 The tertiary structure of DNA

The tertiary structure of DNA is often neglected or ignored, but it is important to the action of the quinolone group of antibacterial agents (section 16.8.1) and to several anticancer agents (section 18.2.1). DNA is an extremely long molecule, so long in fact that it would not fit into the nucleus of the cell if it existed as a linear molecule. It has to be coiled into a more compact three-dimensional shape which *can* fit into the nucleus—a process known as **supercoiling**. This process requires the action of a family of enzymes called **topoisomerases**, which can catalyse the seemingly impossible act of passing one stretch of DNA helix across another stretch. They do this by temporarily cleaving one or both strands to create a temporary gap, then resealing the strand(s) once the crossover has taken place. This process is described in more detail in sections 16.8.1 and 18.2.1. Supercoiling allows the efficient storage of DNA, but the DNA has to be uncoiled again if replication and transcription (section 7.2.2) are to take place. The same topoisomerase enzymes are responsible for catalysing this process, so inhibition of these enzymes can effectively block transcription and replication.

KEY POINTS

- The primary structure of DNA consists of a sugar phosphate backbone with nucleic acid bases attached to each sugar moiety. The sugar is deoxyribose and the bases are adenine, thymine, cytosine, and guanine.

- The secondary structure of DNA is a double helix where the nucleic acid bases are stacked in the centre and paired up such that adenine pairs with thymine, and cytosine pairs with guanine. Hydrogen bonding is responsible for the base pairing and there are van der Waals interactions between the stacks of bases. Polar interactions occur between the sugar phosphate backbone and surrounding water.

- The DNA double helix is coiled up into a tertiary structure. The coiling and uncoiling of the double helix requires topoisomerase enzymes.

- The copying of DNA from one generation to the next is known as replication. Each strand of a parent DNA molecule acts as the template for a new daughter DNA molecule.

- The genetic code consists of nucleic acid bases which are read in sets of three during the synthesis of a protein. Each triplet of bases codes for a specific amino acid.

7.2 Ribonucleic acid and protein synthesis

7.2.1 Structure of RNA

The primary structure of RNA is the same as that of DNA, with two exceptions: **ribose** (Fig. 7.8) is the sugar component rather than **deoxyribose**, and **uracil** (Fig. 7.8) replaces thymine as one of the bases.

Base pairing between nucleic acid bases can occur in RNA with adenine pairing to uracil, and cytosine pairing to guanine. However, the pairing is between bases within the same chain, and it does not occur for the whole length of the molecule (e.g. Fig. 7.9). Therefore, RNA is not a double helix, but it does have regions of helical secondary structure.

Because the secondary structure is not uniform along the length of the RNA chain, more variety is allowed in RNA tertiary structure. There are three main types of RNA molecules with different cellular functions. The three are **messenger RNA** (mRNA), **transfer RNA** (tRNA), and **ribosomal RNA** (rRNA). These three molecules are crucial to the process by which protein synthesis takes place. Although DNA contains the genetic code for proteins, it cannot produce these proteins directly. Instead, RNA takes on that role, acting as the crucial 'middle man' between DNA and proteins. This has been termed the 'central dogma' of molecular biology.

The bases adenine, cytosine, guanine, and uracil are found in mRNA and are predominant in rRNA and tRNA. However, tRNA also contains a number of less common nucleic acids—see for example Fig. 7.9.

7.2.2 Transcription and translation

A molecule of mRNA represents a copy of the genetic information required to synthesize a single protein. Its role is to carry the required code out of the nucleus to a cellular organelle called the **endoplasmic reticulum**. This is where protein production takes place, on bodies called **ribosomes**. The segment of DNA which is copied is called a **gene** and the process involved is called **transcription**. The DNA double helix unravels and the stretch that is exposed acts as a template on which the mRNA can be built (Fig. 7.10). Once complete, the mRNA departs the nucleus to seek out a ribosome, while the DNA re-forms its double helix.

Ribosomal RNA is the most abundant of the three types of RNA and is the major component of ribosomes. These can be looked upon as the production site for protein synthesis—a process known as **translation**. The ribosome binds to one end of the mRNA molecule, then travels along it to the other end, allowing the triplet code to be read, and catalysing the construction of the protein molecule one amino acid at a time (Fig. 7.11). There are two segments to the mammalian ribosome, known as the 60S and 40S subunits. These combine to form an 80S ribosome. (In bacterial cells, the ribosomes are smaller and consist of 50S and 30S subunits combining to form a 70S ribosome. The terms 50S, etc. refer to the sedimentation properties of the various structures. These are related qualitatively to size and mass, but not quantitatively—that is why a 60S and a 40S subunit can combine to form an 80S ribosome.)

rRNA is the major component of each subunit, making up two thirds of the ribosome's mass. The 40S subunit contains one large rRNA molecule along with several proteins, whereas the 60S subunit contains three different-sized rRNAs, again with accompanying proteins. The secondary structure of rRNA includes extensive stretches of base pairing (duplex regions), resulting in a well-defined tertiary structure. At one time, it was thought that rRNA only played a structural role, and that the proteins were acting as enzymes to catalyse translation. The rRNA molecules certainly do have a crucial structural role, but it is now known that they, rather than the ribosomal proteins, have the major catalytic role. Indeed, the key sites in the ribosome where translation takes place are made up almost entirely of rRNA. The proteins are elongated structures which meander through the ribosome structure and are thought to have a fine-tuning effect on the translation process.

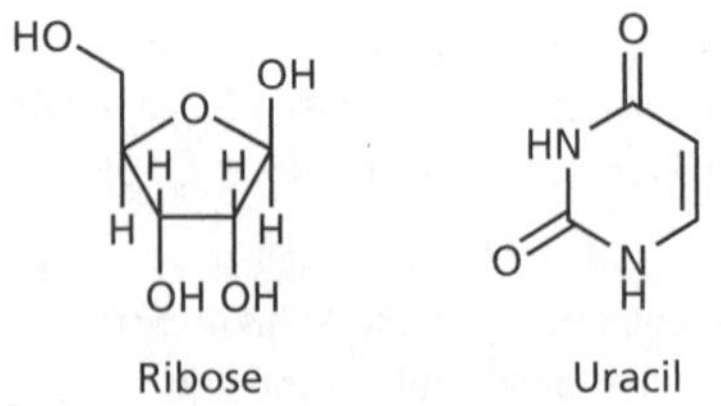

Figure 7.8 Ribose and uracil.

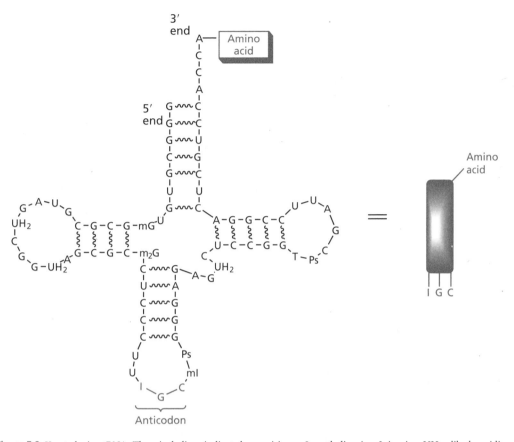

Figure 7.9 Yeast alanine tRNA. The wiggly lines indicate base pairing. mI, methylinosine; I, inosine; UH_2, dihydrouridine; T, ribothymidine; Ps, pseudouridine; mG, methylguanosine; m_2G, dimethylguanosine.

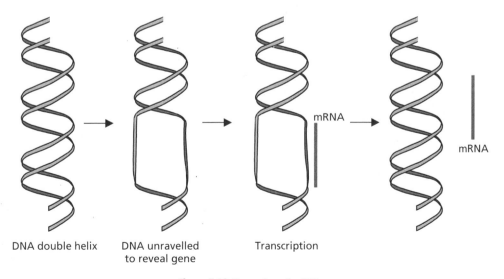

DNA double helix   DNA unravelled   Transcription
                    to reveal gene

mRNA

mRNA

Figure 7.10 Formation of mRNA.

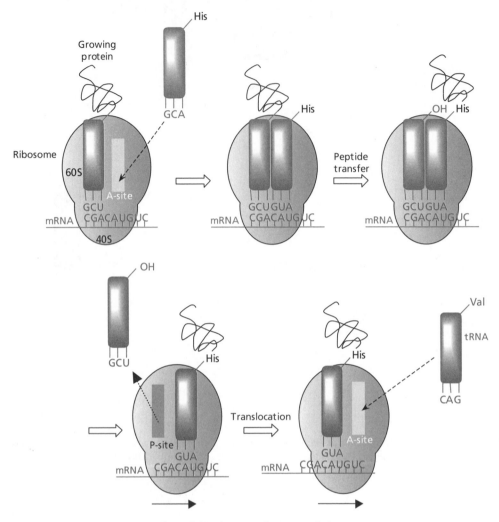

Figure 7.11 Protein synthesis—translation.

Transfer RNA is the crucial adaptor unit which links the triplet code on mRNA to a specific amino acid. This means there has to be a different tRNA for each amino acid. All the tRNAs are cloverleaf in shape, with two different binding regions at opposite ends of the molecule (see Fig. 7.9). One binding region is for the amino acid, where a specific amino acid is covalently linked to a terminal adenosyl residue. The other is a set of three nucleic acid bases (**anticodon**) which will base pair with a complementary triplet on the mRNA molecule. A tRNA having a particular anticodon will always have the same amino acid attached to it.

Let us now look at how translation takes place in more detail. As rRNA travels along mRNA, it reveals the triplet codes on mRNA one by one. For example, in Fig. 7.11 the triplet code CAU is revealed along with an associated binding site called the A site. The A stands for aminoacyl and refers to the attached amino acid on the incoming tRNA. tRNA molecules can enter this site, but they are accepted only if they have the necessary anticodon capable of base pairing with the exposed triplet on mRNA. In this case, tRNA having the anticodon GUA is accepted and brings with it the amino acid histidine. The peptide chain that has been created so far is attached to a tRNA molecule which is bound to the P binding site (standing for peptidyl). A grafting process then takes place, catalysed by rRNA, where the peptide chain is transferred to histidine (Fig. 7.12). The ribosome shifts along mRNA to reveal the next triplet (a process called translocation) and so the process continues until the whole strand is read. The new protein is then released from the ribosome, which is now available to start the process again. The overall process of transcription and translation is summarized in Fig. 7.13.

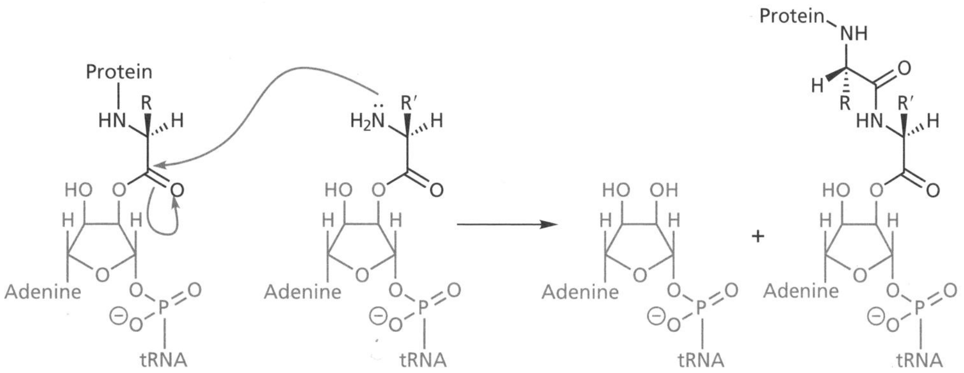

Figure 7.12 Mechanism of transfer of growing protein to next amino acid.

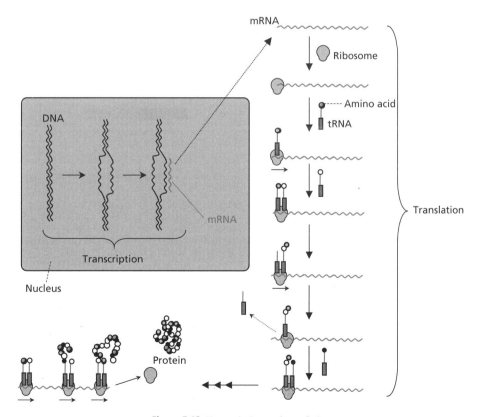

Figure 7.13 Transcription and translation.

7.2.3 Small nuclear RNA

After transcription, mRNA molecules are frequently modified before translation takes place. This involves a splicing operation where the middle section of mRNA (the intron) is excised and the ends of the mRNA molecule (the exons) are spliced together (Fig. 7.14).

Splicing requires the aid of an RNA–protein complex called a **spliceosome**. The RNA molecules involved in this complex are called small nuclear RNAs (snRNAs). As the name indicates, these are small RNA molecules with fewer than 300 nucleotides that occur in the nucleus of the cell. The role of the snRNAs in the spliceosome is to base pair with particular segments of mRNA such that the mRNA can be manipulated and aligned properly for the splicing process. Splice sites are recognized by their nucleotide sequences, but on occasions a mutation in DNA may introduce a new splice site somewhere else on mRNA. This results in faulty splicing, an altered mRNA, and a defective protein. About 15% of genetic diseases are thought to be due to mutations that result in defective splicing.

7.3 Drugs and nucleic acids

In general, we can classify the drugs which act on DNA as **intercalating agents, alkylating agents** or **chain cutters**.

● Intercalating drugs are compounds which are capable of slipping between the layers of nucleic acid base pairs and disrupting the shape of the double helix. This disruption prevents replication and transcription. Examples include the antibacterial agent **proflavine** (section 16.8.2) and the antitumour agents **dactinomycin** and **doxorubicin** (section 18.2.1). Drugs that work in this way must be flat in order to fit between the base pairs, and must therefore be aromatic or heteroaromatic in nature. Some drugs prefer to approach the helix via the major groove, whereas others prefer access via the minor groove. The highly effective antimalarial agent **chloroquine**—a drug developed from **quinine**—can attack the malarial parasite by blocking DNA transcription as part of its action. A flat heteroaromatic structure is present, which can intercalate DNA (Fig. 7.15).

Exon   Intron   Exon

mRNA

Transcription

Splicing

Excised segment

+

Processed mRNA

Translation

Figure 7.14 Splicing mRNA.

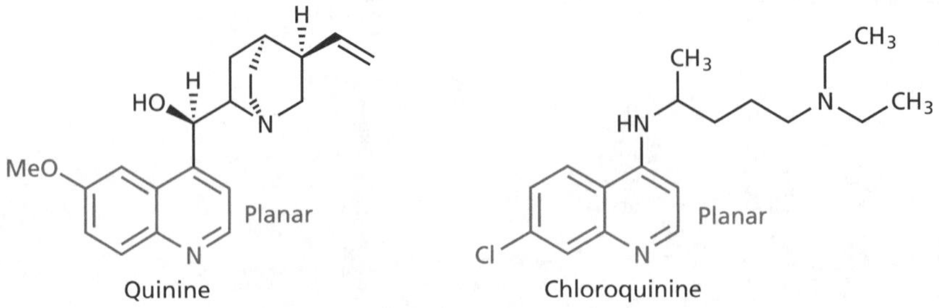

Quinine

Chloroquinine

Figure 7.15 Intercalating antimalarial drugs.

- Alkylating agents are highly electrophilic compounds which react with nucleophiles to form strong covalent bonds. There are several nucleophilic groups in DNA, and agents which alkylate these groups have been used in anticancer therapy (section 18.2.3).

- Chain cutters cut the strands of DNA and prevent the enzyme **DNA ligase** from repairing the damage. They appear to act by creating radicals on the DNA structure. These radicals react with oxygen to form peroxy species and the DNA chain fragments.

Agents such as these are also used in anticancer therapy (section 18.2.4).

The fluoroquinolone antibacterial agents (section 16.8.1) form complexes with DNA and an enzyme called **topoisomerase**, resulting in the inhibition of transcription and replication. Other antibiotic agents target RNA molecules and interfere with translation (Box 7.1). These are discussed in section 16.7.

Finally, there are several important drugs which are related structurally to nucleic acids or to the building

---

BOX 7.1 PUROMYCIN: MECHANISM OF ACTION

Puromycin is an antibiotic which disrupts translation by mimicking the aminoacyl terminus of an aminoacyl-tRNA (Fig. 1). Because puromycin resembles the aminoacyl and adenosine moieties of aminoacyl-tRNA, it is able to enter the A site of the ribosome and in doing so prevents aminoacyl-

tRNA molecules from binding. It has the amino group required for the transfer reaction and so the peptide chain is transferred from tRNA in the P binding site to puromycin in the A binding site (Fig. 2). Puromycin departs the ribosome carrying a stunted protein along with it.

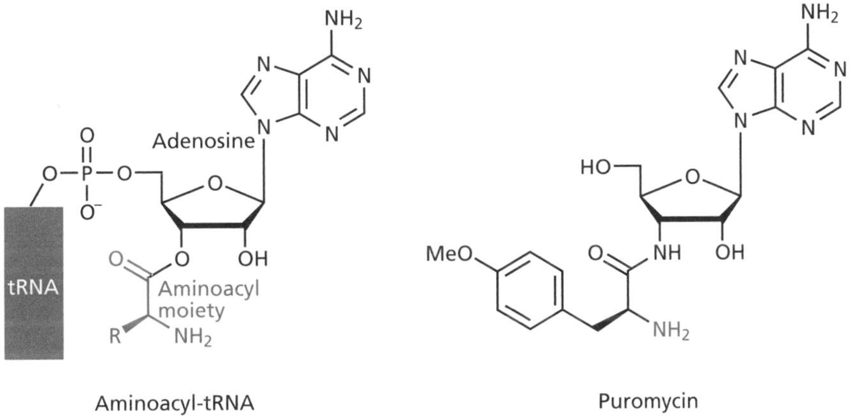

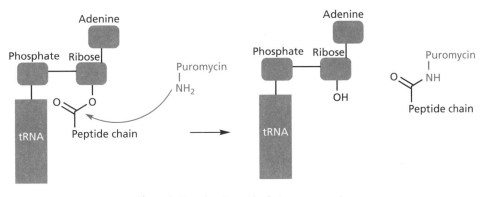

Figure 1 Comparison of puromycin and aminoacyl-tRNA.

Figure 2 Transfer of peptide chain to puromycin.

blocks for nucleic acids. Antisense oligonucleotides (section 7.4) and uracil mustard (section 18.2.3) are two such examples. Another important group of drugs that fall into this category are the various nucleoside like structures which are used as antiviral agents (sections 17.6.1 and 17.7.3).

7.4 Antisense therapy

A great deal of research has been carried out into the possibility of using **oligonucleotides** to block the coded messages carried by mRNA. This is an approach known as **antisense therapy** and has interesting potential for the future. The rationale is as follows (Fig. 7.16). Assuming that the primary sequence of a mRNA molecule is known, an oligonucleotide can be synthesized such that its nucleic acid bases are complementary to a specific stretch of the mRNA molecule. Since the oligonucleotide has a complementary base sequence, it is called an **antisense oligonucleotide**. When mixed with mRNA, the antisense oligonucleotide recognizes its complementary section in mRNA, interacts with it, and forms a duplex structure such that the bases pair up by hydrogen bonding. This section now acts as a barrier to the translation process and blocks protein synthesis.

There are several advantages to this approach. First of all, it can be highly specific. Statistically, an oligonucleotide of 17 nucleotides should be specific for a single mRNA molecule and block the synthesis of a single protein. (The number of possible oligonucleotides containing 17 nucleotides is 4^{17}, if we assume there are 4 different nucleic acids. Therefore the chances of the same segment being present in two different mRNA molecules is remote.) Secondly,

because one mRNA leads to several copies of the same protein, inhibiting mRNA should be more efficient than inhibiting the resulting protein. Both these factors should allow the antisense drug to be used in low doses and result in fewer side effects than conventional protein inhibition.

However, there are several difficulties involved in designing suitable antisense drugs. mRNA is a large molecule with a secondary and tertiary structure. Care has to be taken to choose a section that is exposed. There are also problems relating to the poor absorption of nucleotides and their susceptibility to metabolism (section 11.8.5).

Nevertheless, antisense oligonucleotides are potential antiviral and anticancer agents, as they should be capable of preventing the biosynthesis of 'rogue' proteins and have fewer side effects than currently used drugs. The first antisense oligonucleotide to be approved for the market, in 1998, was the antiviral agent **fomivirsen** (**Vitravene**) (section 17.6.3).

Antisense oligonucleotides are also being considered for the treatment of genetic diseases such as **muscular dystrophy** and **β-thalassaemia**. Abnormal mRNA is sometimes produced as a result of a faulty splicing mechanism (section 7.2.3). Designing an antisense molecule which binds to the faulty splice site might disguise that site and prevent the wrong splicing mechanism taking place.

7.5 Genetic illnesses

A number of genetic illnesses are due to genetic abnormalities that result in the non-expression of particular proteins, or the expression of defective

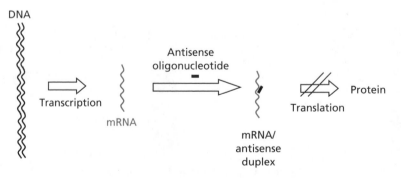

Figure 7.16 Antisense therapy.

proteins. For example, **albinism** is a condition where the skin, hair, and eyes lack pigment; it is associated with a deficiency of an enzyme called **tyrosinase**. This is a copper-containing enzyme that catalyses the first two stages in the synthesis of the pigment **melanin**. Over 90 mutations of the tyrosinase gene have been identified which lead to the expression of inactive enzyme. Mutations in the triplet code result in one or more amino acids being altered in the resulting protein, and if these amino acids are important to the activity of the enzyme, activity is lost. Mutations which alter amino acids in the active site are the ones most likely to result in loss of activity.

Phenylketonuria is a genetic disease caused by the absence or deficiency of an enzyme called **phenylalanine hydroxylase**. This enzyme normally converts phenylalanine to tyrosine, and in its absence the blood levels of phenylalanine rise substantially, along with alternative metabolic products such as phenylpyruvate. If left untreated, this disease results in severe mental retardation.

Haemophilias are inherited genetic diseases in which one of the blood coagulation factors is deficient. This results in uncontrolled bleeding after an injury. In the past, people with this disease were likely to die in their youth. Nowadays, with the proper treatment, affected individuals should have a normal life expectancy. Treatment in severe cases involves regular intravenous infusion with the relevant coagulation factor. In less severe cases, transfusions can be used when an injury has taken place. The coagulation factors used to be typically derived from blood plasma, but this meant that people with haemophilia were susceptible to infection from infected blood samples. For example, during the period 1979–1985 more than 1200 people in the UK were infected with HIV as a result of taking infected blood products. For the same reason, they were also prone to viral infections caused by hepatitis B and C. During the 1990s, recombinant DNA technology (section 7.6) successfully produced blood coagulation factors and these are now the agents of choice because they eliminate the risk of infection. Unfortunately, some victims produce an immune response to the infused factor, which can preclude their use. At present, clinical trials are under way to test whether gene therapy can be used as a treatment. This involves the introduction of a gene which will code for the normal coagulation factor so that it can be produced naturally in the body (section 7.6).

Muscular dystrophy is another genetic disease that affects 1 in every 3500 males and is characterized by the absence of a protein called **dystrophin**. This has an important structural role in cells and its absence results in muscle deterioration. Gene therapy is also being considered for this disease.

Many cancers are associated with genetic defects which result in molecular signalling defects in the cell. This is covered more fully in Chapter 18.

7.6 Molecular biology and genetic engineering

Over the last few years, rapid advances in molecular biology and genetic engineering have had important repercussions for medicinal chemistry. It is now possible to clone specific genes and to include these genes into the DNA of fast-growing cells such that the proteins encoded by these genes are expressed in the modified cell. As the cells are fast growing, this leads to a significant quantity of the desired protein which permits its isolation, purification, and structural determination. Before these techniques became available, it was extremely difficult to isolate and purify many proteins from their parent cells due to the small quantities present. Even if one was successful, the low yields inherent in the process made an analysis of the protein's structure and mechanism of action very difficult. Advances in molecular biology and recombinant DNA techniques have changed all that.

Recombinant DNA technology allows scientists to manipulate DNA sequences to produce modified DNA or completely novel DNA. The technology makes use of natural enzymes called **restriction enzymes** and **ligases** (Fig. 7.17). The restriction enzymes recognize a particular sequence of bases in each DNA molecule and split a specific sugar phosphate bond in each strand of the double helix. The break is not a clean one; there is an overlap between the two chains, resulting in a tail of unpaired bases on each side of the break. The bases on each tail are complementary and can still recognize each other, so they are described as 'sticky' ends. The same process is carried out on a different molecule of DNA and the molecules from both processes are mixed together. As these different molecules have the same sticky ends, they recognize each other

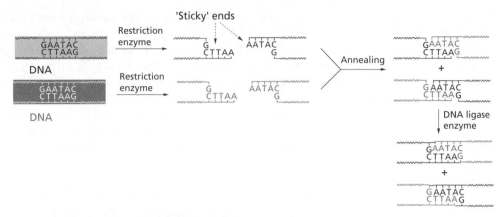

Figure 7.17 Recombinant DNA technology.

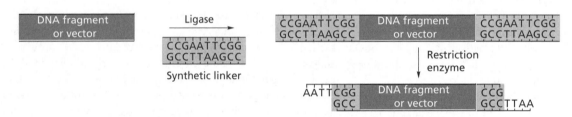

Figure 7.18 Attaching sequences recognized by restriction enzymes.

such that base pairing takes place in a process called **annealing**. Treatment with the ligase enzyme then repairs the sugar phosphate backbone and a new DNA molecule is formed. If the DNA molecule of interest does not have the required sequence recognized by the restriction enzyme, a synthetic DNA linker that *does* contain the sequence can be added to either end of the molecule using a ligase enzyme. This is then treated with the restriction enzyme as before (Fig. 7.18).

There are many applications for this technology, one of which is the ability to amplify and express the gene for a particular human protein in bacterial cells. In order to do this it is necessary to introduce the gene to the bacterial cell. This is done by using a suitable **vector** which will carry the gene into the cell. There are two suitable vectors—**plasmids** and **bacteriophages**. Plasmids are segments of circular DNA which are transferred naturally between bacterial cells and allow the sharing of genetic information. Because the DNA is circular, the DNA representing a human gene can be inserted into the vector's DNA by the same methods described above (Fig. 7.19). Bacteriophages (phages for short) are viruses which infect bacterial cells. There are a variety of

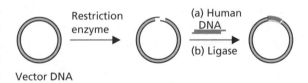

Figure 7.19 Inserting a human gene into a plasmid by recombinant DNA technology.

these, but the same recombinant DNA techniques can be used to insert human DNA into viral DNA.

Whichever vector is used, the modified DNA is introduced into the bacterial cell where it is cloned and amplified (Fig. 7.20). For example, once a phage containing modified nucleic acid infects a bacterial cell, the phage takes over the cell's biochemical machinery to produce multiple copies of itself and its nucleic acid.

Human genes can be introduced to bacterial cells such that the gene is incorporated into bacterial DNA and expressed as if it was the bacteria's own DNA. This allows the production of human proteins in much greater quantity than would be possible by any other means. Such proteins could then be used

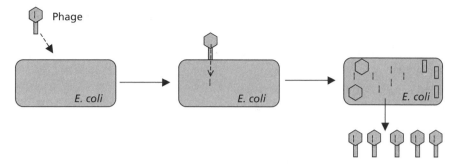

Figure 7.20 Infecting *E. coli* with a phage.

for medicinal purposes as described below. Modified genes can also be introduced and expressed to produce modified proteins to see what effect a mutation would have on the structure and function of a protein.

The following are some of the applications of genetic engineering to the medical field.

- **Harvesting important proteins:** The genes for important hormones or growth factors such as **insulin** and **human growth factor** have been included in fast-growing unicellular organisms. This allows the harvesting of these proteins in sufficient quantity that they can be marketed and administered to patients who are deficient in these important hormones. Genetic engineering has also been crucial in the production of monoclonal antibodies (section 3.9).

- **Genomics and the identification of new protein drug targets:** Nowadays, it is relatively easy to isolate and identify a range of signalling proteins, enzymes, and receptors by cloning techniques. This has led to the identification of a growing number of isozymes and receptor subtypes which offer potential drug targets for the future. The **Human Genome Project** involved the mapping of human DNA (completed in 2000) and has led to the discovery of new proteins previously unsuspected. These too may offer potential drug targets. The study of the structure and function of new proteins discovered from genomics is called **proteomics.**

- **Study of the molecular mechanism of target proteins:** Genetic engineering allows the controlled mutation of proteins such that specific amino acids are altered. This allows researchers to identify which amino acids are important to enzyme activity or to

receptor binding. This in turn leads to a greater understanding of how enzymes and receptors operate at the molecular level.

- **Somatic gene therapy:** Somatic gene therapy involves the use of a carrier virus to smuggle a healthy gene into cells in the body where the corresponding gene is defective. Once the virus has infected the cell, the healthy gene is inserted into the host DNA where it undergoes transcription and translation. This approach has great therapeutic potential for cancers, AIDS, and genetic abnormalities such as cystic fibrosis. However, the approach is still confined to research labs and there is still a long way to go before it is used clinically. There are several problems still to be tackled, such as how to target the viruses specifically to the defective cells, how to insert the gene into DNA in a controlled manner, how to regulate gene expression once it is in DNA, and how to avoid immune responses to the carrier virus. Progress in this field was set back significantly in 1999 as a result of a fatality to a teenage volunteer during a clinical trial in America—thought to be due to an over-reactive immune response to the carrier virus. Alternative non-viral delivery systems are also being studied, involving caged molecules called cyclodextrins. Lipids, polyaminoesters, and glycine polymers are also being investigated as carriers.

KEY POINTS

- The primary structure of RNA is similar to that of DNA, but it contains ribose instead of deoxyribose. Uracil is used as a base in place of thymine and other bases may be present in smaller quantities.

- Base pairing and sections of helical secondary structure are possible within the structure of RNA.

- There are three main types of RNA—messenger RNA, transfer RNA, and ribosomal RNA.

- Transcription is the process by which a segment of DNA is copied as mRNA. mRNA carries the genetic information required for the synthesis of a protein from the nucleus to the endoplasmic reticulum.

- rRNA is the main constituent of ribosomes where protein synthesis takes place. A ribosome moves along mRNA revealing each triplet of the genetic code in turn.

- tRNA interprets the coded message in mRNA. It contains an anticodon of three nucleic acid bases which binds to a complementary triplet on mRNA. Each tRNA carries a specific amino acid, the nature of which is determined by the anticodon.

- The process of protein synthesis is called translation. The growing protein chain is transferred from one tRNA to the amino acid on the next tRNA and is only released once the complete protein molecule has been synthesized.

- Antisense therapy involves the use of oligonucleotides which are complementary to small sections of mRNA. They form a duplex with mRNA and prevent translation taking place.

- Genetic engineering has been used in the production of important hormones for medicinal purposes, the identification of novel drug targets, the study of protein structure and function, and gene therapy.

QUESTIONS

1. Puromycin is an antibiotic which snatches a partially constructed protein from the ribosome (see Box 7.1). In this mechanism, the peptide chain is transferred to puromycin when it is bound to the A site. If the ribosome moves along mRNA, puromycin would be in the P site and the A site would be free to accept an aminoacyl-tRNA molecule. What would be the product if a transfer reaction now took place? Why is this reaction unlikely?

2. Proflavine is a topical antibacterial agent which intercalates bacterial DNA and was used to treat wounded soldiers in the Far East during the Second World War. What role (if any) is played by the tricyclic ring and the primary amino groups? The drug cannot be used systemically. Suggest why this is the case.

Proflavine

3. The following compounds are antiviral drugs which mimic natural nucleosides. What nucleosides do they mimic?

4. Adenine is an important component of several important biochemicals. It has been proposed that adenine was synthesized early on in the evolution of life when the Earth's atmosphere consisted of gases such as hydrogen cyanide and methane. It has also been possible to synthesize adenine from hydrogen cyanide. Consider the structure of adenine and identify how cyanide molecules might act as the building blocks for this molecule.

5. The genetic code involves three nucleic acid bases coding for a single amino acid (the triplet code). Therefore, a mutation to a particular triplet should result in a different amino acid. However, this is not always the case. For any triplet represented by XYZ, which mutation is least likely to result in a change in amino acid—X, Y, or Z?

6. The amino acids serine, glutamate, and phenylalanine were found to be important binding groups in a receptor binding site (see Appendix 1 for structures). The triplet codes for these amino acids in the mRNA for this receptor were AGU, GAA, and UUU respectively. Explain what effect the following mutations might have, if any:

AGU to ACU; AGU to GGU; AGU to AGC

GAA to GAU; GAA to AAA; GAA to GUA

UUU to UUC; UUU to UAU; UUU to AUU

FURTHER READING

Aldridge, S. (2003) The DNA story. *Chemistry in Britain*, April, 28–30.

Burke, M. (2003) On delivery. *Chemistry in Britain*, February, 36–38.

Johnson, I. S. (2003) The trials and tribulations of producing the first genetically engineered drug. *Nature Reviews Drug Discovery*, 2, 747–751.

Judson, H. F. (1979) *The Eighth Day of Creation*. Simon and Schuster, New York.

Langer, R. (2003) Where a pill won't reach. *Scientific American*, April, 32–39.

Lindpaintner, K. (2002) The impact of pharmacogenetics and pharmacogenomics on drug discovery. *Nature Reviews Drug Discovery*, 1, 463–469.

Opalinska, J. B. and Gewirtz, A. M. (2002) Nucleic-acid therapeutics: basic principles and recent applications. *Nature Reviews Drug Discovery*, 1, 503–514.

Petricoin, E. F. *et al.* (2002) Clinical proteomics. *Nature Reviews Drug Discovery*, 1, 683–695.

Winter, P. C., Hickey, G. I., and Fletcher, H. L. (1998) *Instant Notes Genetics*. Bios Scientific Publishers, Oxford.

Titles for general further reading are listed on p. 711.

8 Pharmacokinetics and related topics

8.1 Pharmacodynamics and pharmacokinetics

In Chapters 2–7, we looked at the molecular targets for drugs and the mechanisms by which drugs interact with these targets. In Chapter 10, we shall look at how drugs can be designed to optimize binding interactions with their targets. This is an area of medicinal chemistry known as **pharmacodynamics**. However, the compound with the best binding interactions for a target is not necessarily the best drug to use in medicine. This is because a clinically useful drug has to travel through the body in order to reach its target. There are many barriers and hurdles in its way and, as far as the drug is concerned, it is a long and arduous journey. The study of how a drug reaches it target, and what happens to it during that journey, is known as **pharmacokinetics**. When carrying out a drug design programme, it is important to study pharmacokinetics alongside pharmacodynamics. There is no point perfecting a compound with superb drug–target interactions if it has no chance of reaching its target. The four main topics to consider in pharmacokinetics are absorption, distribution, metabolism, and excretion (often abbreviated to ADME).

8.2 Drug absorption

Drug absorption refers to the route or method by which a drug reaches the blood supply. This in turn depends on how the drug is administered. The most common and preferred method of administering drugs is the oral route and so we shall first concentrate on the various barriers and problems associated with oral delivery.

An orally taken drug enters the **gastrointestinal tract** (GIT), which comprises the mouth, throat, stomach, and the upper and lower intestines. A certain amount of the drug may be absorbed through the mucosal membranes of the mouth, but the majority passes down into the stomach where it encounters gastric juices and hydrochloric acid. These chemicals aid in the digestion of food and will treat drugs in a similar fashion if the drug is susceptible to breakdown. For example, the first penicillin used clinically was broken down in the stomach and had to be administered by injection. Other acid-labile drugs, such as **local anaesthetics** or **insulin**, cannot be given orally. If the drug survives the stomach, it enters the upper intestine where it encounters digestive enzymes that serve to break down food. Assuming the drug survives this attack, it then has to pass through the cells lining the intestinal or gut wall. This means that the drug has to pass through a cell membrane on two occasions, first to enter the cell and then to exit it on the other side. Once the drug has passed through the cells of the gut wall, it can enter the blood supply relatively easily as the cells lining the blood vessels have pores between them through which most drugs can pass. In other words, drugs enter the blood vessels by passing between cells rather than through them.

The drug is now transported in the blood to the body's 'customs office'—the liver. The liver contains enzymes which are ready and waiting to intercept foreign chemicals, and modify them such that they are more easily excreted—a process called **drug metabolism** (section 8.4).

It can be seen that stringent demands are made on any orally taken drug. The drug must be chemically stable to survive the stomach acids, and metabolically stable to survive the digestive enzymes in the GIT as well as the metabolic enzymes in the liver. It must also have the correct balance of water versus fat solubility. If the drug is too polar (hydrophilic), it will fail to pass through the fatty cell membranes of the gut wall (section 2.2.1). On the other hand, if the drug is too fatty (hydrophobic), it will be poorly soluble in the gut and dissolve in fat globules. This means that there will be poor surface contact with the gut wall, resulting in poor absorption.

It is noticeable how many drugs contain an amine functional group. There are good reasons for this. Amines are often involved in a drug's binding interactions with its target. However, they are also an answer to the problem of balancing the dual requirements of water and fat solubility. Amines are weak bases, and in general it is found that many of the most effective drugs are amines having a pK_a value in the range 6–8. In other words, they are partially ionized at blood pH and can easily equilibrate between their ionized and non-ionized forms. This allows them to cross cell membranes in the non-ionized form, while the presence of the ionized form gives the drug good water solubility and permits good binding interactions with its target binding site (Fig. 8.1).

The extent of ionization at a particular pH can be determined by the **Henderson–Hasselbalch equation**:

$$pH = pK_a + \log\frac{[RNH_2]}{[RNH_3^+]}$$

where $[RNH_2]$ is the concentration of the free base and $[RNH_3^+]$ is the concentration of the ionized amine. K_a is the equilibrium constant for the equilibrium shown

in Fig. 8.1, and the Henderson–Hasselbalch equation can be derived from the equilibrium constant:

$$K_a = \frac{[H^+][RNH_2]}{[RNH_3^+]}$$

$$\text{Therefore}\quad pK_a = -\log\frac{[H^+][RNH_2]}{[RNH_3^+]}$$

$$= -\log[H^+] - \log\frac{[RNH_2]}{[RNH_3^+]}$$

$$= pH - \log\frac{[RNH_2]}{[RNH_3^+]}$$

$$\text{Therefore}\quad pH = pK_a + \log\frac{[RNH_2]}{[RNH_3^+]}$$

Note that when the concentration of the ionized and unionized amines are identical (i.e. when $[RNH_2] = [RNH_3^+]$), the ratio $([RNH_2]/[RNH_3^+])$ is 1. Since log 1 = 0, the Henderson–Hasselbalch equation will simplify to $pH = pK_a$. In other words, when the amine is 50% ionized, $pH = pK_a$. Therefore, drugs with a pK_a of 6–8 are approximately 50% ionized at blood pH (7.4).

The hydrophilic/hydrophobic character of the drug is the crucial factor affecting absorption through the gut wall, and the molecular weight of the drug should in theory be irrelevant. For example, **ciclosporin** is successfully absorbed through cell membranes, although it has a molecular weight of about 1200. In practice, however, larger molecules tend to be poorly absorbed, because they are likely to contain a large number of polar functional groups. As a rule of thumb, orally absorbed drugs tend to obey what is known as Lipinski's **rule of five**—so called because the numbers involved are multiples of 5:

- a molecular weight less than 500
- no more than 5 hydrogen bond donor groups
- no more than 10 hydrogen bond acceptor groups
- a calculated log P value less than +5 (log P is a measure of a drug's hydrophobicity—section 13.2.1).

Polar drugs that break these rules are usually poorly absorbed and have to be administered by injection. Nevertheless, some highly polar drugs can be absorbed from the digestive system. For example, there are polar drugs that can 'hijack' specific **carrier proteins** (section 3.7.1) in the cell membrane. Carrier proteins are essential to a cell's survival, as they transport the

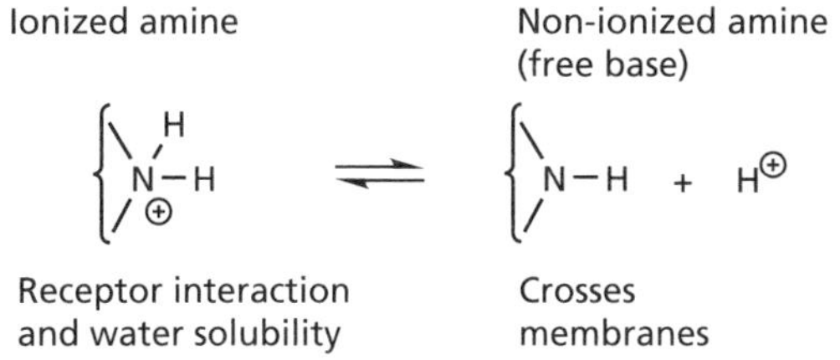

Ionized amine

Non-ionized amine (free base)

Receptor interaction and water solubility

Crosses membranes

Figure 8.1 Equilibrium between the ionized and non-ionized form of an amine.

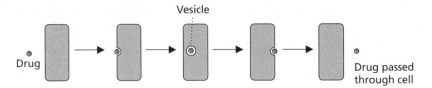

Figure 8.2 Pinocytosis.

highly polar building blocks required for various biosynthetic pathways (e.g. amino acids and nucleic acid bases). If the drug bears a structural resemblance to one of these building blocks, then it too may be smuggled into the cell. For example, **levodopa** is transported by the carrier protein for the amino acid phenylalanine, and **fluorouracil** is transported by carrier proteins for the nucleic acid bases thymine and uracil. The antihypertensive agent **lisinopril** is transported by carrier proteins for dipeptides, and the anticancer agent **methotrexate** and the antibiotic **erythromycin** are also absorbed by carrier proteins.

Other highly polar drugs can be absorbed into the blood supply if they have a low molecular weight (less than 200), as they can then pass through small pores between the cells lining the gut wall.

Occasionally, polar drugs with high molecular weight can cross the cells of the gut wall without actually passing through the membrane. This involves a process known as **pinocytosis** where the drug is engulfed by the cell membrane and a membrane-bound vesicle is pinched off to carry the drug across the cell (Fig. 8.2). The vesicle then fuses with the membrane to release the drug on the other side of the cell.

Sometimes drugs are deliberately designed to be highly polar so that they are not absorbed from the GIT. These are usually antibacterial agents targeted against gut infections. Making them highly polar ensures that the drug will reach the site of infection in higher concentration (section 16.4.1).

8.3 Drug distribution

Once a drug has been absorbed, it is rapidly distributed around the blood supply, then more slowly distributed to the various tissues and organs. The rate and extent of distribution depends on various factors, including the physical properties of the drug itself.

8.3.1 Distribution round the blood supply

The vessels carrying blood round the body are called **arteries**, **veins**, and **capillaries**. The heart is the pump that drives the blood through these vessels. The major artery carrying blood from the heart is called the **aorta**, and as it moves further from the heart, it divides into smaller and smaller arteries—similar to the limbs and branches radiating from the trunk of a tree. Eventually, the blood vessels divide to such an extent that they become extremely narrow—equivalent to the twigs of a tree. These blood vessels are called capillaries and it is from them that oxygen, nutrients, and drugs can escape in order to reach the tissues and organs of the body. At the same time, waste products such as cell breakdown products and carbon dioxide are transferred from the tissues into the capillaries to be carried away and disposed of. The capillaries now start uniting into bigger and bigger vessels, resulting in the formation of veins which return the blood to the heart.

Once a drug has been absorbed into the blood supply, it is rapidly and evenly distributed throughout the blood supply within a minute—the time taken for the blood volume to complete one circulation. However, this does not mean that the drug is evenly distributed around the body, because the blood supply is richer to some areas of the body than to others.

8.3.2 Distribution to tissues

Drugs do not stay confined to the blood supply. If they did, they would be of little use, since their targets are the cells of various organs and tissues. The drug has to leave the blood supply in order to reach those targets. The body has an estimated 10 billion capillaries with a total surface area of 200 m^2. They probe every part of the body, such that no cell is more than 20–30 μm away from a capillary. Each capillary is very narrow, not much wider than the red blood cells that pass through it. Its walls are made up of a thin single layer of cells packed tightly together. However, there are pores between the cells which are 90–150 Å in diameter—large enough to allow most drug-sized molecules to escape, but not large enough to allow the **plasma proteins** present in blood to escape. Therefore, drugs do not have to cross cell membranes in order to leave the blood system and can be freely and rapidly distributed into the aqueous fluid surrounding the various tissues and organs of the body. Having said that, some drugs are found to bind to plasma proteins in the blood. Since the plasma proteins cannot leave the capillaries, the proportion of drug bound to these proteins is also confined to the capillaries and cannot reach its target.

8.3.3 Distribution to cells

Once a drug has reached the tissues, it can immediately be effective if its target site is a receptor situated in a cell membrane. However, there are many drugs that have to enter the individual cells of tissues in order to reach their target. These include local anaesthetics, enzyme inhibitors, and drugs which act on nucleic acids or intracellular receptors. Such drugs must be hydrophobic enough to pass through the cell membrane, unless they are smuggled through by carrier proteins or taken in by pinocytosis.

8.3.4 Other distribution factors

The concentration levels of free drug circulating in the blood supply rapidly fall away after administration as a result of the distribution patterns described above, but there are other factors at work. Drugs that are excessively hydrophobic are often absorbed into fatty tissues and removed from the blood supply. This fat solubility can lead to problems. For example, obese patients undergoing surgery require a larger than normal volume of general anaesthetic because the gases used are particularly fat soluble. Unfortunately, once surgery is over and the patient has regained consciousness, the anaesthetics stored in the fat tissues will be released and may render the patient unconscious again. **Barbiturates** were once seen as potential intravenous anaesthetics which could replace the anaesthetic gases. Unfortunately, they too are fat soluble and as a result it is extremely difficult to estimate a sustained safe dosage. The initial dose can be estimated to allow for the amount of barbiturate taken up by fat cells, but further doses eventually lead to saturation of the fat depot, and result in a sudden and perhaps fatal increase of barbiturate levels in the blood supply.

Ionized drugs may be bound to various macromolecules and also removed from the blood supply. Drugs may also be bound reversibly to blood plasma proteins such as **albumin**, thus lowering the level of free drug. Therefore, only a small proportion of the drug that has been administered may actually reach the desired target.

8.3.5 Blood–brain barrier

The blood–brain barrier is an important barrier that drugs have to negotiate if they are to act in the brain. The blood capillaries feeding the brain are lined with tight-fitting cells which do not contain pores (unlike capillaries elsewhere in the body). Moreover, the capillaries are coated with a fatty layer formed from nearby cells, providing an extra fatty barrier through which drugs have to cross. Therefore, drugs entering the brain have to dissolve through the cell membranes of the capillaries and also through the fatty cells coating the capillaries. As a result, polar drugs such as **penicillin** do not easily enter the brain.

The existence of the blood–brain barrier makes it possible to design drugs which will act at various parts of the body (e.g. the heart) and have no activity in the brain, thus reducing any central nervous system (CNS) side effects. This is done by increasing the polarity of the drug such that it does not cross the blood–brain barrier. On the other hand, drugs that are intended to act in the brain must be designed such that they *are* able to cross the blood–brain barrier. This means that they must have a minimum number of polar groups, or have these groups temporarily masked (see prodrugs; section 11.6). Having said that, some polar drugs can cross the blood–brain barrier with the aid of carrier proteins and others (e.g. **insulin**) can cross by the process of pinocytosis previously described. The ability to cross the blood–brain barrier has an important bearing on the analgesic activity of opiates (section 21.2.3).

8.3.6 Placental barrier

The placental membranes separate a mother's blood from the blood of her fetus. The mother's blood provides the fetus with essential nutrients and carries away waste products, but these chemicals must pass through the placental barrier. Since food and waste products can pass through the placental barrier, it is perfectly feasible for drugs to pass through as well.

Drugs such as **alcohol, nicotine,** and **cocaine** can all pass into the fetal blood supply. Fat-soluble drugs will cross the barrier most easily, and drugs such as **barbiturates** will reach the same levels in fetal blood as in maternal blood. Such levels may have unpredictable effects on fetal development. They may also prove hazardous once the baby is born. Drugs and other toxins can be removed from fetal blood by the maternal blood and detoxified. Once the baby is born, it may have the same levels of drugs in its blood as the mother, but it does not have the same ability to detoxify or eliminate them. As a result, drugs will have a longer lifetime and may have fatal effects.

8.3.7 Drug–drug interactions

Drugs such as **warfarin** and **methotrexate** are bound to albumin and plasma proteins in the blood and are unavailable to interact with their targets. When another drug is taken which can compete for plasma protein binding (e.g. a sulfonamide), then a certain percentage of previously bound drug is released, increasing the concentration of the drug and its effect.

KEY POINTS

- Pharmacodynamics is the study of how drugs interact with a molecular target, whereas pharmacokinetics is the study of how a drug reaches its target in the body and how it is affected on that journey.

- The four main issues in pharmacokinetics are absorption, distribution, metabolism, and excretion.

- Orally taken drugs have to be chemically stable to survive the acidic conditions of the stomach, and metabolically stable to survive digestive and metabolic enzymes.

- Orally taken drugs must be sufficiently polar to dissolve in the GIT and blood supply, but sufficiently fatty to pass through cell membranes.

- Most orally taken drugs obey Lipinski's rule of five.

- Highly polar drugs can be orally active if they are small enough to pass between the cells of the gut wall, are recognized by carrier proteins, or are taken across the gut wall by pinocytosis.

- Distribution round the blood supply is rapid. Distribution to the interstitial fluid surrounding tissues and organs is rapid if the drug is not bound to plasma proteins.

- Some drugs have to enter cells in order to reach their target.

- A certain percentage of a drug may be absorbed into fatty tissue and/or bound to macromolecules.

- Drugs entering the CNS have to cross the blood–brain barrier. Polar drugs are unable to cross this barrier unless they make use of carrier proteins or are taken across by pinocytosis.

- Some drugs cross the placental barrier into the fetus and may harm development or prove toxic in newborn babies.

8.4 Drug metabolism

When drugs enter the body, they are subject to attack from a range of metabolic enzymes. The role of these enzymes is to degrade or modify the foreign structure, such that it can be more easily excreted. As a result, most drugs undergo some form of metabolic reaction resulting in structures known as metabolites. Very often these metabolites lose the activity of the original drug, but in some cases they may retain a certain level of activity. Some metabolites may possess a different activity from the parent drugs, resulting in side effects or toxicity. A knowledge of drug metabolism and its possible consequences can aid the medicinal chemist in designing new drugs which do not form unacceptable metabolites. Equally, it is possible to take advantage of drug metabolism to activate drugs in the body. This is known as a prodrug strategy (see section 11.6). It is now a requirement to identify all the metabolites of a new drug before it can be approved. The structure and stereochemistry of each metabolite has to be determined and the metabolite must be tested for biological activity (section 12.1.2).

8.4.1 Phase I and phase II metabolism

Drugs are foreign substances as far as the body is concerned, and the body has its own method of getting rid of such chemical invaders. If the drug is polar, it will be quickly excreted by the kidneys (section 8.5). However, non-polar drugs are not easily excreted and the purpose of drug metabolism is to convert such compounds into more polar molecules that *can* be easily excreted.

Non-specific enzymes (particularly **cytochrome P450 enzymes** in the liver) are able to add polar functional groups to a wide variety of drugs. Once the polar functional group has been added, the overall drug is more polar and water soluble, and is more likely to be excreted when it passes through the kidneys. An alternative set of enzymatic reactions can reveal masked polar functional groups which might already be present in a drug. For example, there are enzymes which can demethylate a methyl ether to reveal a more polar hydroxyl group. Once again, the more polar product (metabolite) is excreted more efficiently.

These reactions are classed as phase I reactions and generally involve oxidation, reduction, and hydrolysis (see Figs. 8.3–8.9). Most of these reactions occur in the liver but some (such as the hydrolysis of esters and amides) can also occur in the gut wall, blood plasma, and other tissues. Some of the structures most prone to oxidation are *N*-methyl groups, aromatic rings, the terminal positions of alkyl chains, and the least hindered positions of alicyclic rings. Nitro, azo, and carbonyl groups are prone to reduction by reductases, while amides and esters are prone to hydrolysis by esterases. For some drugs, two or more metabolic reactions might occur, resulting in different metabolites; other drugs may not be metabolized at all. A knowledge of the metabolic reactions that are possible for different functional groups allows the medicinal chemist to predict the likely metabolic products for any given drug, but only drug metabolism studies will establish whether these metabolites are really formed.

Drug metabolism has important implications when it comes to using chiral drugs, especially if the drug is to be used as a racemate. The enzymes involved in catalysing metabolic reactions will often distinguish between the two enantiomers of a chiral drug, such that one enantiomer undergoes different metabolic reactions from the other. As a result, both enantiomers of a chiral drug have to be tested separately to see what metabolites are formed. In practice, it is usually preferable to use a single enantiomer in medicine, or design the drug such that it is not asymmetric.

A series of metabolic reactions classed as phase II reactions also occur, mainly in the liver (see Figs. 8.10–8.16). Most of these reactions are **conjugation reactions**, whereby a polar molecule is attached to a suitable polar 'handle' that is already present on the drug or has been introduced by a phase I reaction. The resulting conjugate has greatly increased polarity, thus increasing its excretion rate in urine or bile even further.

Both phase I and phase II reactions can be species specific, which has implications for *in vivo* metabolic studies. In other words, the metabolites formed in an experimental animal may not necessarily be those formed in humans. A good knowledge of how metabolic reactions differ from species to species is important in determining which test animals are relevant for drug metabolism tests. Both sets of reactions can also be regioselective and stereoselective. This means that metabolic enzymes can distinguish between identical functional groups or alkyl groups located at different parts of the molecule (regioselectivity) as well as between different stereoisomers of chiral molecules (stereoselectivity).

8.4.2 Phase I transformations catalysed by cytochrome P450 enzymes

The enzymes that constitute the cytochrome P450 family are the most important metabolic enzymes and are located in liver cells. They are **haemoproteins** (containing haem and iron) and they catalyse a reaction that splits molecular oxygen, such that one of the oxygen atoms is introduced into the drug and the other ends up in water (Fig. 8.3). As a result they belong to a general class of enzymes called the **monooxygenases**.

There are at least 33 different P450 enzymes, grouped into four main families CYP1–CYP4. Within each family there are various subfamilies designated by a letter, and each enzyme within that subfamily is designated by a number. For example, CYP3A4 is enzyme 4 in the subfamily A of the main family 3. Most drugs in current use are metabolized by five primary CYP enzymes (CYP3A, CYP2D6, CYP2C, CYP1A2, and CYP2E1). The isozyme CYP3A4 is particularly important in drug metabolism and is

$$\text{Drug} - \text{H} + \text{O}_2 + \text{NADPH} + \text{H}^+ \xrightarrow{\text{Cytochrome P450 enzymes}} \text{Drug} - \text{OH} + \text{NADP} + \text{H}_2\text{O}$$

Figure 8.3 Oxidation by cytochrome P450 enzymes.

responsible for the metabolism of most drugs. The reactions catalysed by cytochrome P450 enzymes are shown in Figs. 8.4 and 8.5 and can involve the oxidation of carbon, nitrogen, phosphorus, sulphur, and other atoms.

Oxidation of carbon atoms can occur if the carbon atom is either exposed (i.e. easily accessible to the enzyme) or activated (see Fig. 8.4). For example, methyl substituents on the carbon skeleton of a drug are often easily accessible and are oxidized to form alcohols, which may be oxidized further to carboxylic acids. In the case of longer-chain substituents, the terminal carbon and the penultimate carbon are the most exposed carbons in the chain and are both susceptible to oxidation. If an aliphatic ring is present, the most exposed region is the part most likely to be oxidized.

Activated carbon atoms next to an sp^2 carbon centre (i.e. allylic or benzylic positions) or an sp carbon centre (i.e. a propynylic position) are more likely to be

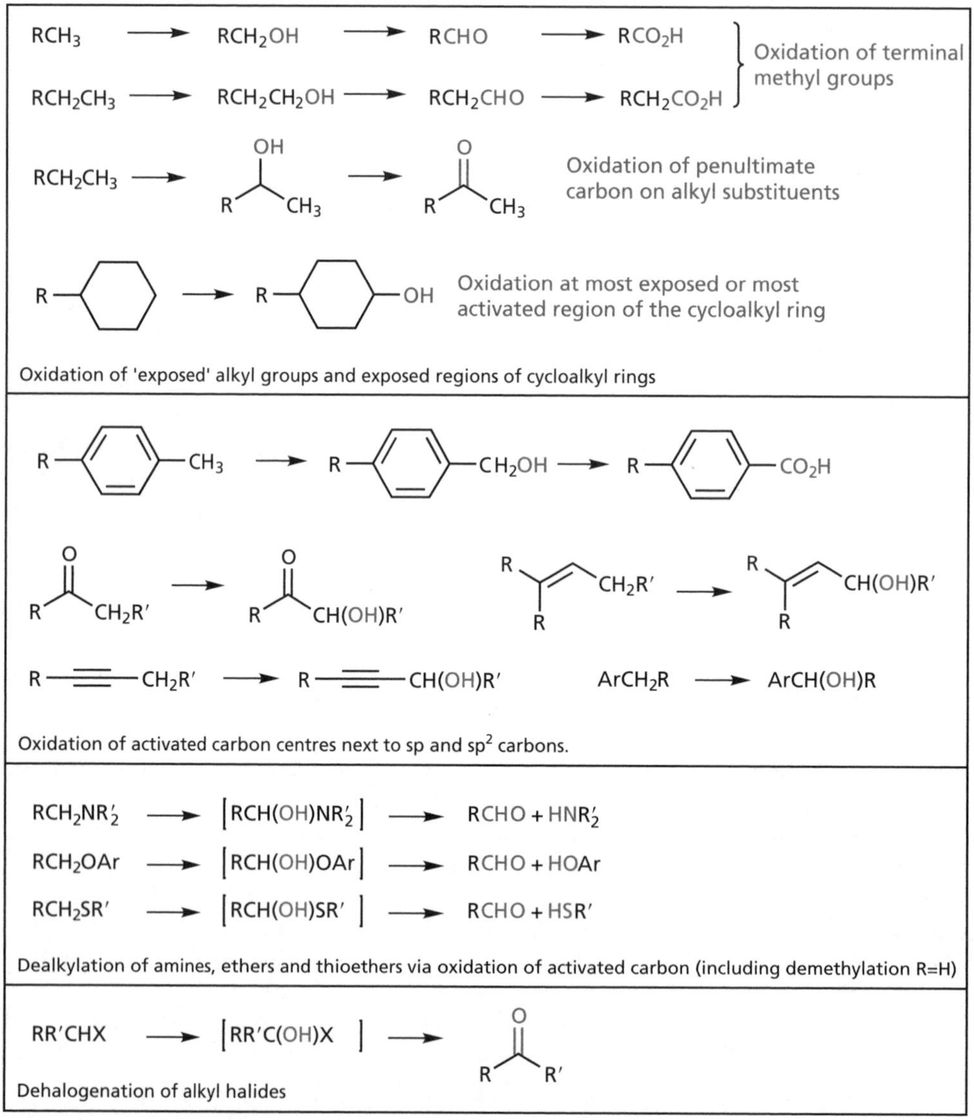

Figure 8.4 Oxidative reactions catalysed by cytochrome P450 enzymes on saturated carbon centres.

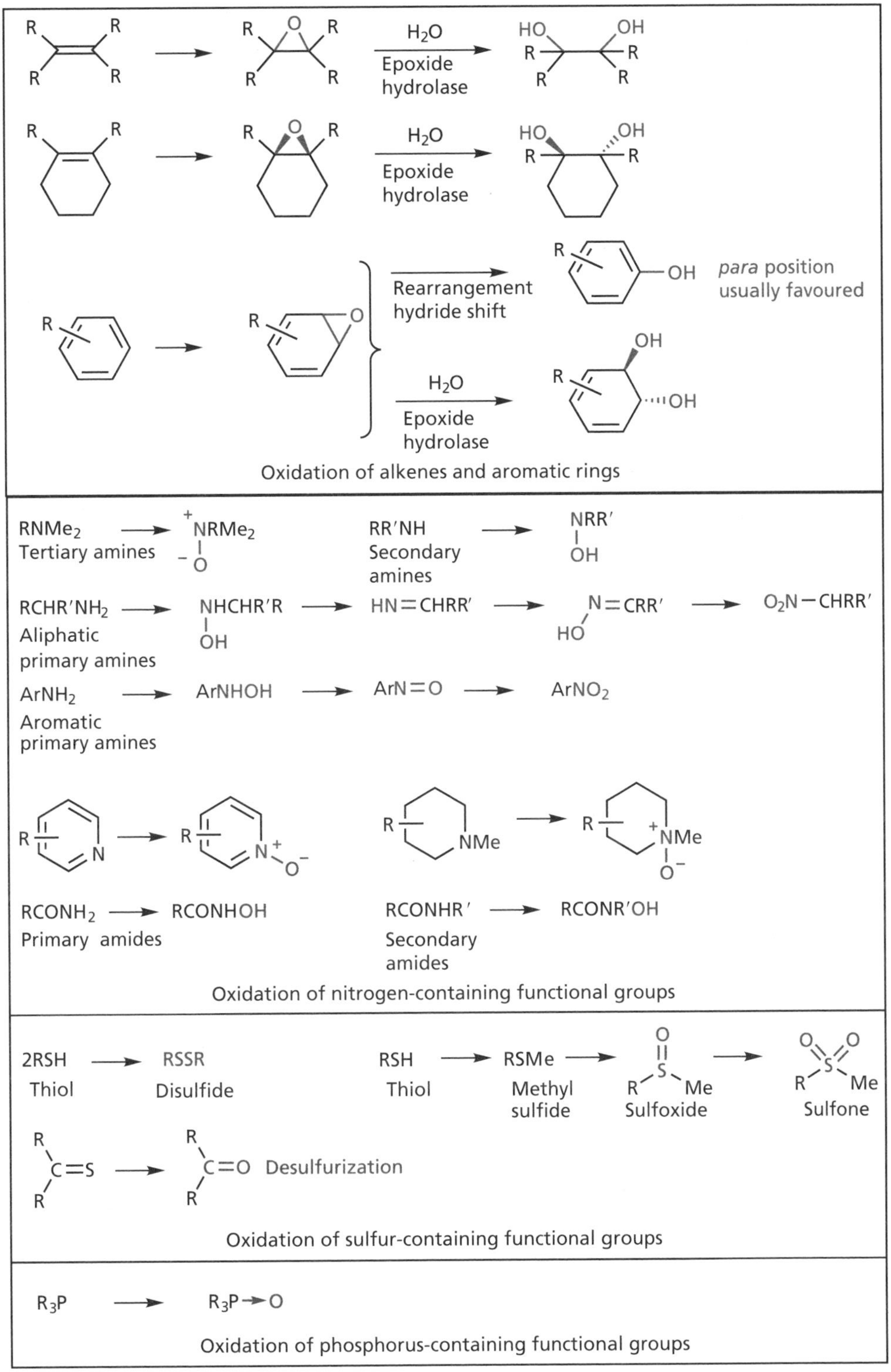

Oxidation of alkenes and aromatic rings

Oxidation of nitrogen-containing functional groups

Oxidation of sulfur-containing functional groups

Oxidation of phosphorus-containing functional groups

Figure 8.5 Oxidative reactions catalysed by cytochrome P450 enzymes on heteroatoms and unsaturated carbon centres.

oxidized than exposed carbon atoms (Fig. 8.4). Carbon atoms which are alpha to a heteroatom are also activated and prone to oxidation. In this case, hydroxylation results in an unstable metabolite that is immediately hydrolysed resulting in the dealkylation of amines, ethers, and thioethers or the dehalogenation of alkyl halides. The aldehydes which are formed from these reactions generally undergo further oxidation to carboxylic acids by aldehyde dehydrogenases (section 8.4.4). Tertiary amines are found to be more reactive to oxidative dealkylation than secondary amines because of their greater basicity, while *O*-demethylation of aromatic ethers is faster than *O*-dealkylation of larger alkyl groups. *O*-Demethylation is important to the analgesic activity of codeine (section 21.2.3).

Cytochrome P450 enzymes can catalyse the oxidation of unsaturated sp^2 and sp carbon centres present in alkenes, alkynes, and aromatic rings (Fig. 8.5). In the case of alkenes, a reactive epoxide is formed which is deactivated by the enzyme **epoxide hydrolase** to form a diol. In some cases, the epoxide may evade the enzyme. If this happens, it can act as an alkylating agent and react with nucleophilic groups present in proteins or nucleic acids, leading to toxicity. The oxidation of an aromatic ring results in a similarly reactive epoxide intermediate which can have several possible fates. It may undergo a rearrangement reaction involving a hydride transfer to form a phenol, normally at the *para* position. Alternatively, it may be deactivated by epoxide hydrolase to form a diol or react with **glutathione S-transferase** to form a conjugate (section 8.4.5). If the epoxide intermediate evades these enzymes, it may act as an alkylating agent and prove toxic. Electron-rich aromatic rings are likely to be epoxidized more quickly than those with electron-withdrawing substituents, and this has consequences for drug design (e.g. see section 15.18.10).

Tertiary amines are oxidized to *N*-oxides as long as the alkyl groups are not sterically demanding. Primary and secondary amines are also oxidized to *N*-oxides, but these are rapidly converted to hydroxylamines and beyond. Aromatic primary amines are also oxidized in stages to aromatic nitro groups—a process which is related to the toxicity of aromatic amines, since highly electrophilic intermediates are formed which can alkylate proteins or nucleic acids. Aromatic primary amines can also be methylated in a phase II reaction (section 8.4.5) to a secondary amine which can then

undergo phase I oxidation to produce formaldehyde and primary hydroxylamines. Primary and secondary amides can be oxidized to hydroxylamides. These functional groups have also been linked with toxicity and carcinogenicity. Thiols can be oxidized to disulfides. There is evidence that thiols can be methylated to methyl sulfides, which are then oxidized to sulfides and sulfones.

8.4.3 Phase I transformations catalysed by flavin-containing monooxygenases

Another group of metabolic enzymes present in the endoplasmic reticulum of liver cells is the flavin-containing monooxygenases. These enzymes are chiefly responsible for metabolic reactions involving oxidation at nucleophilic nitrogen, sulphur, and phosphorus atoms rather than at carbon atoms. Several examples are given in Fig. 8.6. Many of these reactions are also catalysed by cytochrome P450 enzymes.

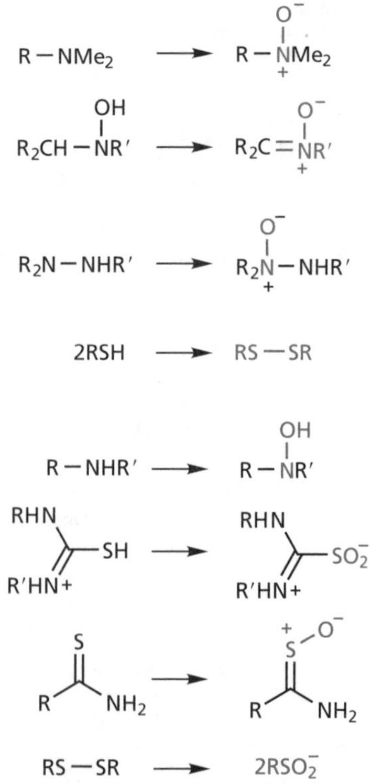

Figure 8.6 Phase I reactions catalysed by flavin monooxygenases.

8.4.4 Phase I transformations catalysed by other enzymes

There are several oxidative enzymes in various tissues around the body that are involved in the metabolism of endogenous compounds, but can also play a role in drug metabolism (Fig. 8.7). For example, **monoamine oxidases** are involved in the deamination of catecholamines (section 20.5), but have been observed to oxidize some drugs. Other important oxidative enzymes include alcohol dehydrogenases and aldehyde dehydrogenases. The aldehydes formed by the action of alcohol dehydrogenases on alcohols are usually not observed, as they are converted to carboxylic acids by aldehyde dehydrogenases.

Reductive phase I reactions are less common than oxidative reactions, but reductions of aldehyde, ketone, azo, and nitro functional groups have been observed in specific drugs (Fig. 8.8). Many of the oxidation reactions described for heteroatoms in Figs. 8.5–8.7 are reversible, and are catalysed by reductase enzymes. Cytochrome P450 enzymes are involved in catalysing some of these reactions. Remember that enzymes can catalyse a reaction in both directions, depending on the nature of the substrate, and so although cytochrome P450 enzymes are

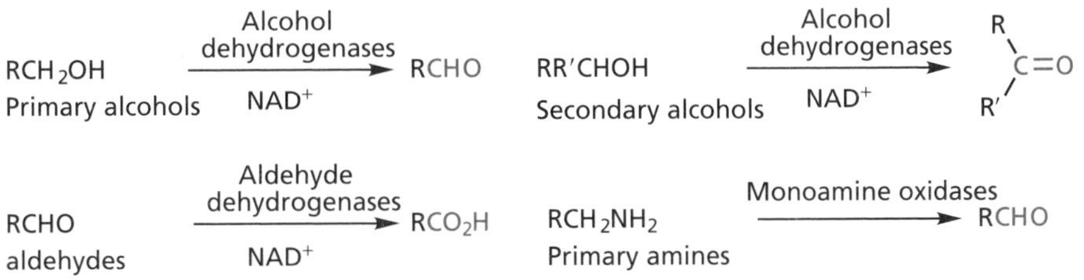

Figure 8.7 Phase I oxidative reactions catalysed by miscellaneous enzymes.

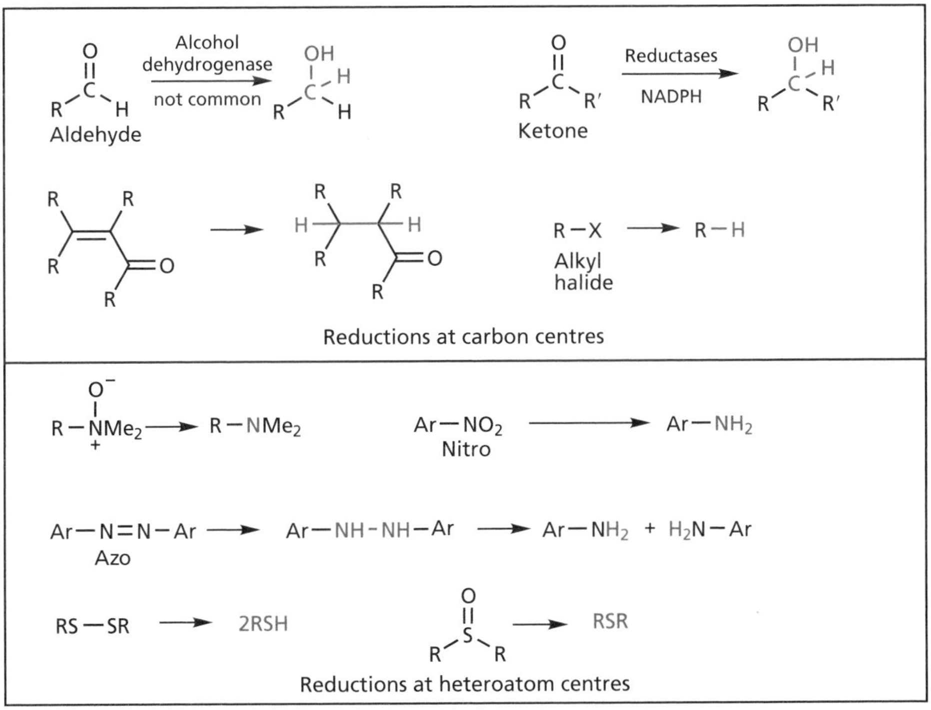

Figure 8.8 Phase I reductive reactions.

predominantly oxidative enzymes, it possible for them to catalyse some reductions.

The hydrolysis of esters and amides is a common metabolic reaction, catalysed by esterases and peptidases respectively (Fig. 8.9). These enzymes are present in various organs of the body including the liver. Amides tend to be hydrolysed more slowly than esters. The presence of electron-withdrawing groups can increase the susceptibility of both amides and esters to hydrolysis.

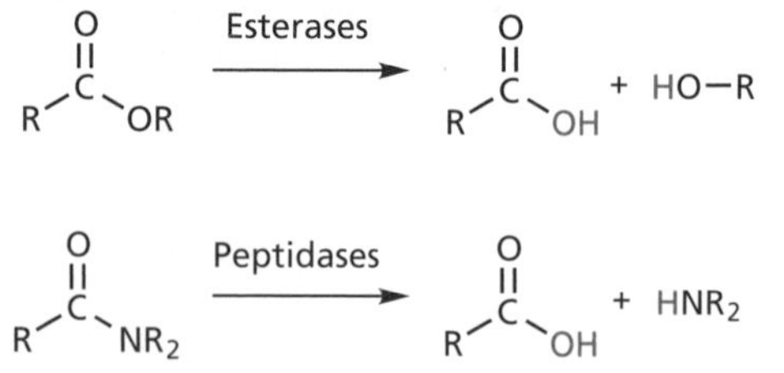

Figure 8.9 Hydrolysis of esters and amides.

8.4.5 Phase II transformations

Most phase II reactions are **conjugation reactions** catalysed by transferase enzymes. The resulting conjugates are usually inactive, but there are exceptions to this rule. Glucuronic acid conjugation is the most common of these reactions. Phenols, alcohols, hydroxylamines, and carboxylic acids form **O-glucuronides** by reaction with **UDFP-glucuronate** such that a highly polar glucuronic acid molecule is attached to the drug (Fig. 8.10). The resulting conjugate is excreted in the urine, but may also be excreted in the bile if the molecular weight is over 300.

A variety of other functional groups such as sulfonamides, amides, amines, and thiols (Fig. 8.11) can react to form *N*- or *S*-glucuronides. *C*-Glucuronides are also possible in situations where there is an activated carbon centre next to carbonyl groups.

Another form of conjugation is sulfate conjugation (Fig. 8.12). This is less common than glucuronation

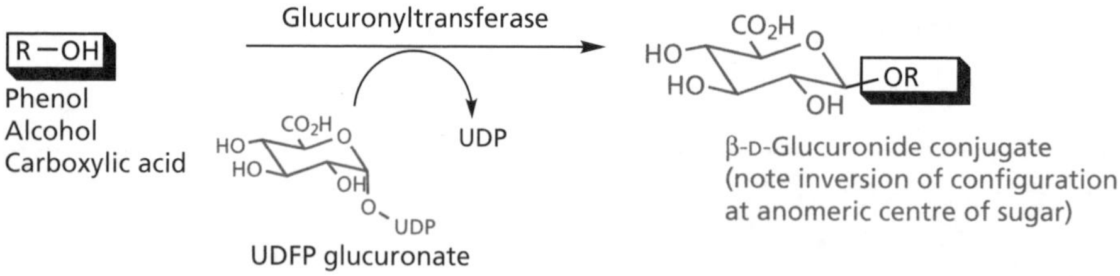

Figure 8.10 Glucuronidation of alcohols, phenols, and carboxylic acids.

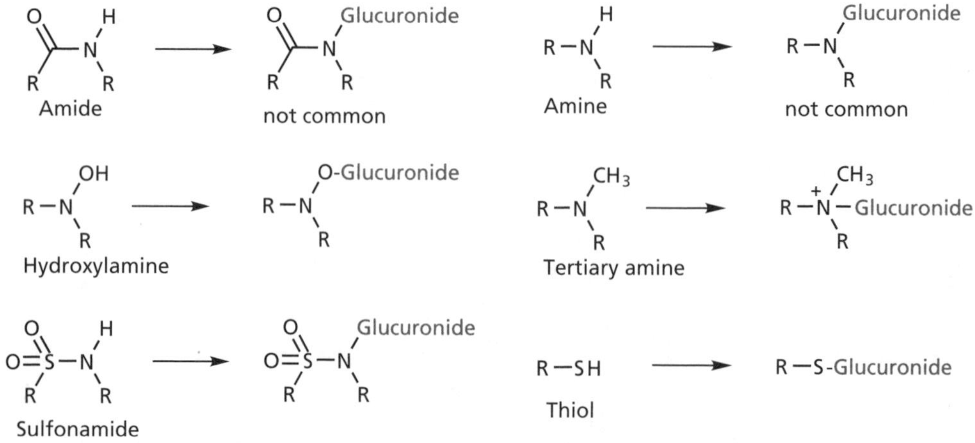

Figure 8.11 Glucuronidation of miscellaneous functional groups.

and is restricted mainly to phenols, alcohols, aryl-amines, and *N*-hydroxy compounds. The reaction is catalysed by **sulfotransferases** using the cofactor 3′-phosphoadenosine 5′-phosphosulfate as the sulfate source. Primary amines, secondary amines, secondary alcohols, and phenols form stable conjugates, whereas primary alcohols form reactive sulfates which can act as toxic alkylating agents. Aromatic hydroxylamines and hydroxylamides also form unstable sulfate conjugates that can be toxic.

Drugs bearing a carboxylic acid group can become conjugated to amino acids by the formation of a peptide link. In most animals glycine conjugates are generally formed, but L-glutamine is the most common amino acid used for conjugation in primates. The carboxylic acid present in the drug is first activated by formation of a coenzyme A thioester which is then linked to the amino acid (Fig. 8.13).

Electrophilic functional groups such as epoxides, alkyl halides, sulfonates, disulfides, and radical species can react with the nucleophilic thiol group of the tri-peptide **glutathione** to give glutathione conjugates which can be subsequently transformed to **mercapturic acids** (Fig. 8.14). The glutathione conjugation

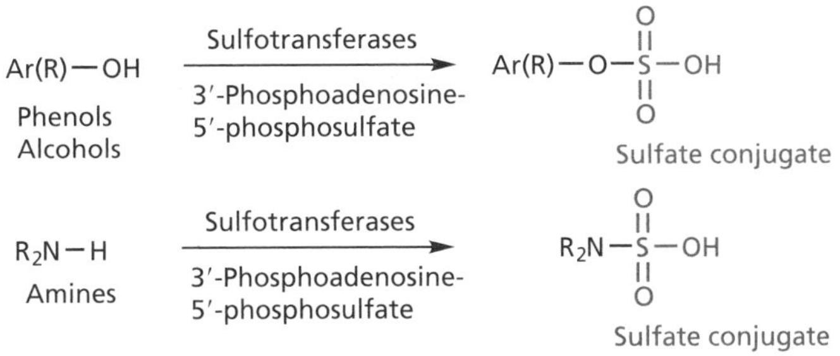

Figure 8.12 Examples of sulfoconjugation phase II reactions.

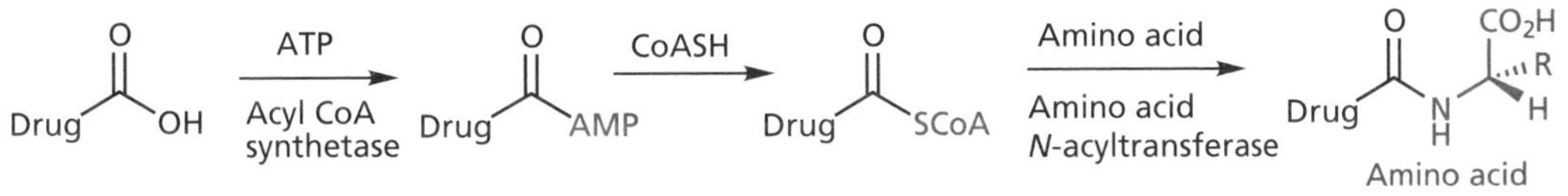

Figure 8.13 Formation of amino acid conjugates.

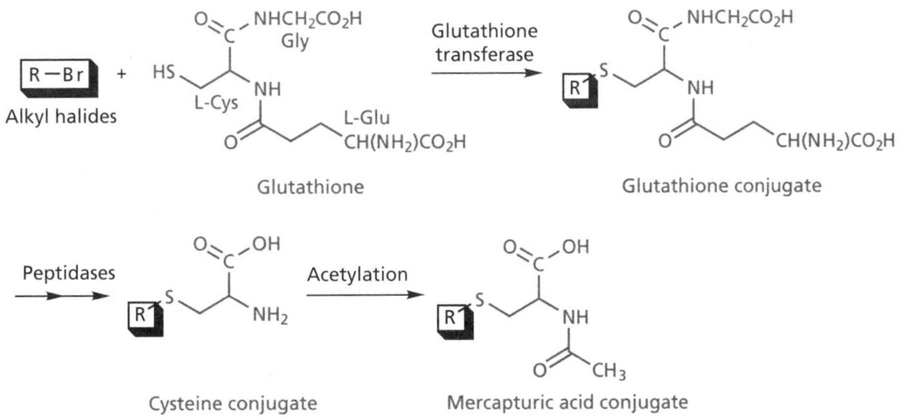

Figure 8.14 Formation of glutathione and mercapturic acid conjugates from an alkyl halide.

reaction can take place in most cells, especially those in the liver and kidney, and is catalysed by **glutathione transferase**. This conjugation reaction is important in detoxifying potentially dangerous environmental toxins or electrophilic alkylating agents formed by phase I reactions (Fig. 8.15). Glutathione conjugates are often excreted in the bile but are more usually converted to mercapturic acid conjugates before excretion.

Not all phase II reactions result in increased polarity. Methylation and acetylation are important phase II reactions which usually *decrease* the polarity of the drug (Fig. 8.16). An important exception is the methylation of pyridine rings, which leads to polar quaternary salts. The functional groups that are susceptible to methylation are phenols, amines, and thiols. Primary amines are also susceptible to acetylation. The enzyme cofactors involved in contributing the methyl group or acetyl

group are **S-adenosyl methionine** and **acetyl SCoA** respectively. Several methyltransferase enzymes are involved in the methylation reactions. The most important enzyme for *O*-methylations is catechol *O*-methyltransferase, which preferentially methylates the *meta* position of catechols (section 20.5). It should be pointed out, however, that methylation occurs less frequently than other conjugation reactions and is more important in biosynthetic pathways or the metabolism of endogenous compounds.

It is possible for drugs bearing carboxylic acids to become conjugated with cholesterol. Cholesterol conjugates can also be formed with drugs bearing an ester group by means of a transesterification reaction. Some drugs with an alcohol functional group form conjugates with fatty acids by means of an ester link.

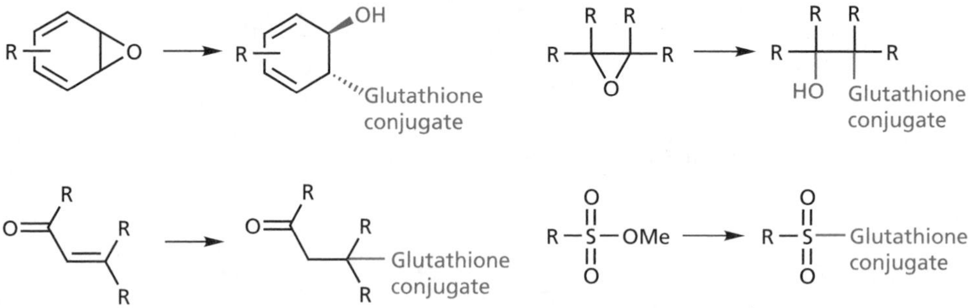

Figure 8.15 Formation of glutathione conjugates (Glu-Cys-Gly) with electrophilic groups.

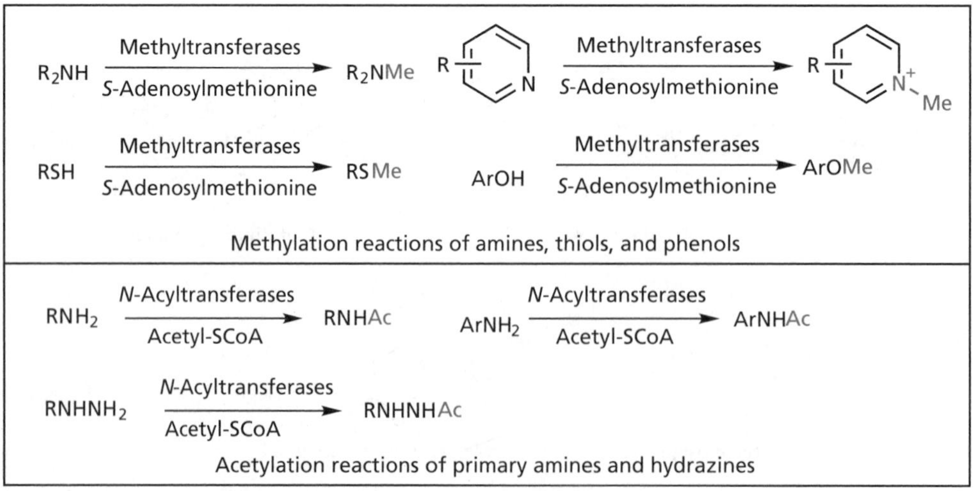

Figure 8.16 Methylation and acetylation.

BOX 8.1 METABOLISM OF AN ANTIVIRAL AGENT

Indinavir is an antiviral agent used in the treatment of HIV which is prone to metabolism, resulting in seven different metabolites (Fig. 1). Studies have shown that the CYP3A subfamily of cytochrome P450 enzymes is responsible for six of these metabolites. The metabolites concerned arise from *N*-dealkylation of the piperazine ring, *N*-oxidation of the pyridine ring, *para*-hydroxylation of the phenyl ring, and hydroxylation of the indane ring. The seventh metabolite is a glucuronide conjugate of the pyridine ring. All these reactions occur individually to produce five separate metabolites. The remaining two metabolites arise from two or more metabolic reactions taking place on the same molecule.

The major metabolites are those resulting from dealkylation. As a result, research has been carried out to try to design indinavir analogues that are resistant to this reaction. For

example, structures having two methyl substituents on the activated carbon next to pyridine have been effective in blocking dealkylation.

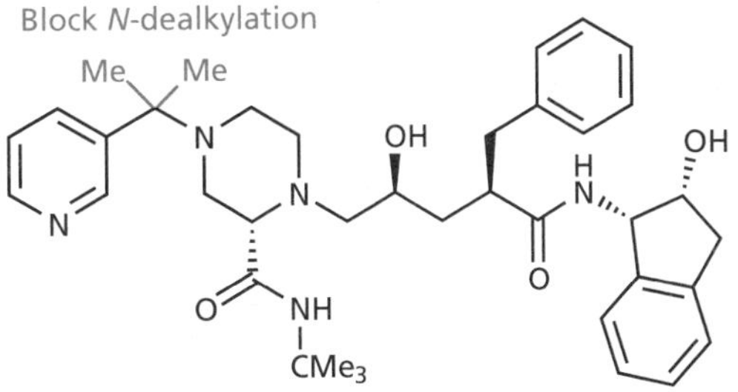

Figure 2 Analogue of indinavir resistant to *N*-dealkylation.

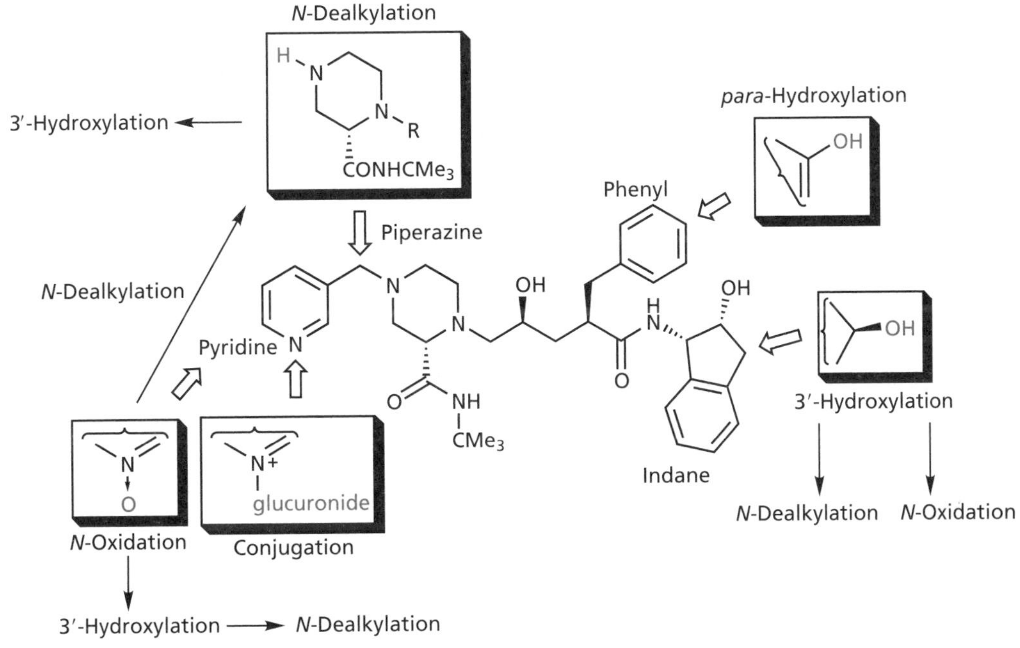

Figure 1 Metabolism of indinavir.

8.4.6 Metabolic stability

Ideally, a drug should be resistant to drug metabolism because the production of metabolites complicates drug therapy (see Box 8.1). For example, the metabolites

formed will usually have different properties from the original drug. In some cases, activity may be lost. In others, the metabolite may prove to be toxic. For example, the metabolites of **paracetamol** cause liver toxicity, and the carcinogenic properties of some polycyclic hydrocarbons are due to the formation of epoxides.

Another problem arises from the fact that the activity of metabolic enzymes varies from individual to individual. This is especially true of the cytochrome P450 enzymes. The profile of these enzymes in different patients can vary, resulting in a difference in the way a drug is metabolized. As a result, the amount of drug that can be safely administered also varies depending on the level of metabolism. Differences across populations can be quite significant, resulting in different countries having different recommended dose levels for particular drugs. For example, the rate at which the antibacterial agent **isoniazid** is acetylated and deactivated varies amongst populations. Asian populations acylate the drug at a fast rate, whereas 45–65% of Europeans and North Americans have a slow rate of acylation. **Pharmacogenomics** is the study of genetic variations between individuals and how these affect individual responses to drugs. In the future, it is possible that 'fingerprints' of an individual's genome may allow better prediction of which drugs would be suitable for that individual, and which drugs might produce unacceptable side effects. This in turn may avoid drugs having to be withdrawn from the market as a result of rare toxic side effects.

Another complication involving drug metabolism and drug therapy relates to the fact that cytochrome P450 activity can be affected by other chemicals. For example, certain foods have an influence. Brussels sprouts and cigarette smoke enhance activity, whereas grapefruit juice inhibits it. This can have a significant effect on the activity of drugs metabolized by cytochrome P450 enzymes. For example, the immuno-suppressant drug **ciclosporin** and the dihydropyridine hypotensive agents are more efficient when taken with grapefruit juice, as their metabolism is reduced. However, serious toxic effects can arise if the antihistamine agent **terfenadine** is taken with grapefruit juice. Terfenadine is actually inactive, but it is metabolized to an active form called **fexofenadine** (Fig. 8.17). If metabolism is inhibited by grapefruit juice, terfenadine persists in the body and can cause serious cardiac toxicity. As a result, fexofenadine itself is now favoured over terfenadine and is marketed as **Allegra**.

Certain drugs are also capable of inhibiting or promoting cytochrome P450 enzymes, leading to a phenomenon known as **drug–drug interaction** where

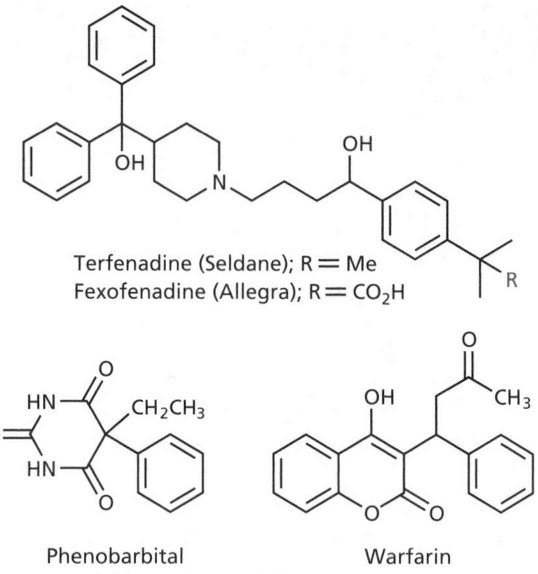

Figure 8.17 Drugs which are metabolized by cytochrome P450 enzymes or affect the activity of cytochrome P450 enzymes.

the presence of one drug affects the activity of another. For example, several antibiotics can act as cytochrome P450 inhibitors and will slow the metabolism of drugs metabolized by these enzymes. Other examples are the drug–drug interactions that occur between the anticoagulant **warfarin** and the barbiturate **phenobarbital** (Fig. 8.17), or between warfarin and the antiulcer drug **cimetidine** (section 22.2.7). Phenobarbitone stimulates cytochrome P450 enzymes and accelerates the metabolism of warfarin, making it less effective. On the other hand, cimetidine inhibits cytochrome P450 enzymes, thus slowing the metabolism of warfarin. Such drug–drug interactions affect the plasma levels of warfarin and could cause serious problems if the levels move outwith the normal therapeutic range.

Herbal medicine is not immune from this problem either. **St John's wort** is a popular remedy used for mild to moderate depression. However, it promotes the activity of cytochrome P450 enzymes and decreases the effectiveness of contraceptives and warfarin.

Because of the problems caused by cytochrome P450 activation or inhibition, new drugs are usually tested to check whether they have any effect on enzyme activity or are metabolized themselves by cytochrome P450 enzymes. Indeed, in many projects, an important goal is to ensure that such properties are lacking.

Drugs can be defined as hard or soft with respect to their metabolic susceptibility. In this context **hard drugs** are those that are resistant to metabolism and remain unchanged in the body. **Soft drugs** are designed to have a predictable, controlled metabolism where they are inactivated to non-toxic metabolites and excreted. A group is normally incorporated which is susceptible to metabolism but will ensure that the drug survives sufficiently long to achieve what it is meant to do before it is metabolized and excreted. Drugs such as these are also called **antedrugs**.

8.4.7 The first pass effect

Drugs that are taken orally pass directly to the liver once they enter the blood supply. Here, they are exposed to drug metabolism before they are distributed around the rest of the body, and so a certain percentage of the drug is transformed before it has the chance to reach its target. This is known as the first pass effect. Drugs that are administered in a different fashion (e.g. injection or inhalation) avoid the first pass effect and are distributed around the body before reaching the liver. Indeed, a certain proportion of the drug may not pass through the liver at all, but may be taken up in other tissues and organs *en route*.

8.5 Drug excretion

Drugs and their metabolites can be excreted from the body by a number of routes. Volatile or gaseous drugs are excreted through the lungs. Such drugs pass out of the capillaries that line the air sacs (alveoli) of the lungs, then diffuse through the cell membranes of the alveoli into the air sacs, from where they are exhaled. Gaseous **general anaesthetics** are excreted in this way and move down a concentration gradient from the blood supply into the lungs. They are also administered through the lungs, in which case the concentration gradient is in the opposite direction and the gas moves from the lungs to the blood supply.

The **bile duct** travels from the liver to the intestines and carries a greenish fluid called **bile** which contains bile acids and salts that are important to the digestion process. A small number of drugs are diverted from the blood supply back into the intestines by this route.

Since this happens from the liver, any drug eliminated in this way has not been distributed round the body. Therefore, the amount of drug distributed is less than that absorbed. However, once the drug has entered the intestine, it can be reabsorbed so it has another chance.

It is possible for as much as 10–15% of a drug to be lost through the skin in **sweat**. Drugs can also be excreted through saliva and breast milk, but these are minor excretion routes compared to the kidneys. There are concerns, however, that mothers may be passing on drugs such as **nicotine** to their baby through breast milk.

The **kidneys** are the principal route by which drugs and their metabolites are excreted (Fig. 8.18). The kidneys filter the blood of waste chemicals and these chemicals are subsequently removed in the urine. Drugs and their metabolites are excreted by the same mechanism.

Blood enters the kidneys by means of the **renal artery**. This divides into a large number of capillaries, each one of which forms a knotted structure called a **glomerulus** that fits into the opening of a duct called a **nephron**. The blood entering these glomeruli is under pressure and so plasma is forced through the pores in the capillary walls into the nephron, carrying with it any drugs and metabolites that might be present. Any compounds that are too big to pass

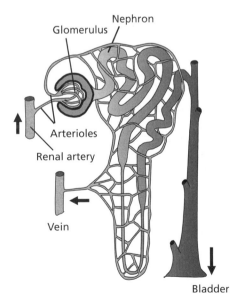

Figure 8.18 Excretion by the kidneys.

through the pores, such as plasma proteins and red blood cells, remain in the capillaries with the remaining plasma. Note that this is a filtration process, so it does not matter whether the drug is polar or hydrophobic: all drugs and drug metabolites will be passed equally efficiently into the nephron. However, this does not mean that every compound will be *excreted* equally efficiently, because there is more to the process than simple filtration.

The filtered plasma and chemicals now pass through the nephron on their route to the bladder. However, only a small proportion of what starts that journey actually finishes it. This is because the nephron is surrounded by a rich network of blood vessels carrying the filtered blood away from the glomerulus, permitting much of the contents of the nephron to be reabsorbed into the blood supply. Most of the water that was filtered into the nephron is quickly reabsorbed through pores in the nephron cell membrane which are specific for water molecules but bar the passage of ions or other molecules. These pores are made up of protein molecules called **aquaporins**. As water is reabsorbed, drugs and other agents are concentrated in the nephron and a concentration gradient is set up. There is now a driving force for compounds to move back into the blood supply down the concentration gradient. However, this can only happen if the drug is sufficiently hydrophobic to pass through the cell membranes of the nephron. This means that hydrophobic compounds are efficiently reabsorbed back into the blood, whereas polar compounds are not and are excreted. This process of excretion explains the importance of drug metabolism to drug excretion. Drug metabolism makes a drug more polar so that it is less likely to be reabsorbed from the nephrons.

Some drugs are actively transported from blood vessels into the nephrons. This process is called **facilitated transport**, and is important in the excretion of penicillins.

KEY POINTS

- Drugs are exposed to enzyme-catalysed reactions which modify their structure. This is called drug metabolism and can take place in various tissues. However, most reactions occur in the liver.

- Orally taken drugs are subject to the first pass effect.

- Drugs administered by methods other than the oral route avoid the first pass effect.

- Phase I metabolic reactions typically involve the addition or exposure of a polar functional group. Cytochrome P450 enzymes present in the liver carry out important phase I oxidation reactions. The types of cytochrome P450 enzymes present vary between individuals, leading to varying rates of drug metabolism.

- The activity of cytochrome P450 enzymes can be affected by food, chemicals, and drugs, resulting in drug–drug interactions and possible side effects.

- Phase II metabolic reactions involve the addition of a highly polar molecule to a functional group. The resulting conjugates are more easily excreted.

- Drug excretion can take place through sweat, exhaled air, or bile, but most excretion takes place through the kidneys.

- The kidneys filter blood such that drugs and their metabolites enter nephrons. Non-polar substances are reabsorbed into the blood supply, but polar substances are retained in the nephrons and excreted in the urine.

8.6 Drug administration

There is a large variety of ways in which drugs can be administered, and many of these avoid some of the problems associated with oral administration. The main routes are oral, sublingual, rectal, epithelial, inhalation, and injection. The method chosen will depend on the target organ and the pharmacokinetics of the drug.

8.6.1 Oral administration

Orally administered drugs are taken by mouth. This is the preferred option for most patients, so there is more chance that the patient will comply with the drug regime and complete the course. However, the oral route places the most demands on the chemical and physical properties of the drug as described above.

Drugs given orally can be taken as pills, capsules, or solutions. Drugs taken in solution are absorbed more quickly and a certain percentage may even be absorbed through the stomach wall. For example, approximately

25–33% of **alcohol** is absorbed into the blood supply from the stomach; the rest is absorbed from the upper intestine. Drugs taken as pills or capsules are mostly absorbed in the upper intestine. The rate of absorption is partly determined by the rate at which the pills and capsules dissolve. This in turn depends on such factors as particle size and crystal form. In general, about 75% of an orally administered drug is absorbed into the body within 1–3 hours. Specially designed pills and capsules can remain intact in the stomach to help protect acid-labile drugs from stomach acids. The containers then degrade once they reach the intestine.

Care has to be taken if drugs interact with food. For example, **tetracycline** binds strongly to calcium ions, which inhibits absorption, so foods such as milk should be avoided. Some drugs bind other drugs and prevent absorption. For example, **colestyramine** (used to lower cholesterol levels) binds to **warfarin** and also to the thyroid drug **levothyroxine sodium**, so these drugs should be taken separately.

8.6.2 Absorption through mucous membranes

Some drugs can be absorbed through the mucous membranes of the mouth or nose, thus avoiding the digestive and metabolic enzymes encountered during oral administration. For example, heart patients take **glyceryl trinitrate** (Fig. 8.19) by placing it under the tongue (sublingual administration). The opiate analgesic **fentanyl** (Fig. 8.19) has been given to children in the form of a lollipop and is absorbed through the mucous membranes of the mouth. The Incas absorbed **cocaine** sublingually by chewing coca leaves.

Nasal decongestants are absorbed through the mucous membranes of the nose. Cocaine powder is absorbed in this way when it is sniffed, as is **nicotine** in the form of snuff. Nasal sprays have been used to administer analogues of peptide hormones such as **antidiuretic hormone**. These drugs would be quickly degraded if taken orally.

Eye drops are used to administer drugs directly to the eye and thus reduce the possibility of side effects elsewhere in the body. For example, glaucoma is treated in this way. Nevertheless, some absorption into the blood supply can still occur and some asthmatic patients suffer bronchospasms when taking **timolol** eye drops.

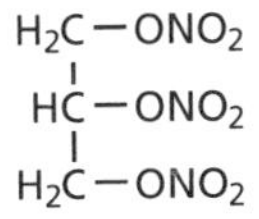

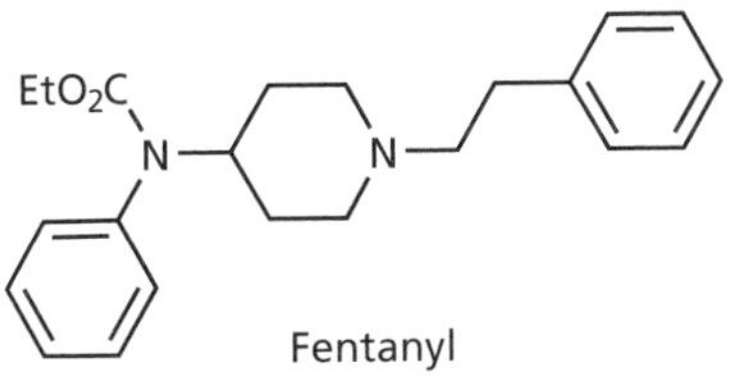

8.6.3 Rectal administration

Some drugs are administered rectally as **suppositories**, especially if the patient is unconscious, vomiting, or unable to swallow. However, there are several problems associated with rectal administration: the patient may suffer membrane irritation, and although the extent of drug absorption is efficient, it can be unpredictable. It is not the most popular of methods with patients either!

8.6.4 Topical administration

Topical drugs are those which are applied to the skin. For example, steroids are applied topically to treat local skin irritations. It is also possible for some of the drug to be absorbed through the skin (**transdermal absorption**) and to enter the blood supply, especially if the drug is lipophilic. **Nicotine patches** work in this fashion, as do hormone replacement therapies for **oestrogen**. Drugs are absorbed by this method at a steady rate, and avoid the acidity of the stomach or the enzymes in the gut or gut wall. Other drugs that have been applied in this way include the analgesic **fentanyl**, and the antihypertensive agent **clonidine**. Once applied, the drug is slowly released from the patch and absorbed through the skin into the blood supply over several days. As a result, the level of drug remains relatively constant over that period.

A technique known as **iontophoresis** is being investigated as a means of topical administration. Two miniature electrode patches are applied to the skin and linked to a reservoir of the drug. A painless

Figure 8.19 Glyceryl trinitrate, fentanyl, and methamphetamine.

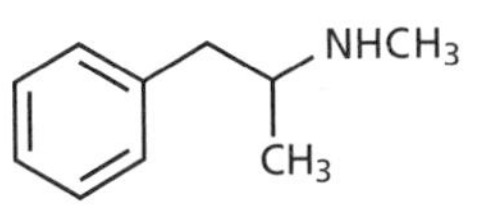

pulse of electricity is applied, which has the effect of making the skin more permeable to drug absorption. By timing the electrical pulses correctly, the drug can be administered such that fluctuations in blood levels are kept to a minimum. Similar devices are being investigated which use ultrasound to increase skin permeability.

8.6.5 Inhalation

Drugs administered by inhalation avoid the digestive and metabolic enzymes of the GIT or liver. Once inhaled, the drugs are absorbed through the cell linings of the respiratory tract into the blood supply. Assuming the drug is able to pass through the hydrophobic cell membranes, absorption is rapid and efficient because the blood supply is in close contact with the cell membranes of the lungs. For example, **general anaesthetic gases** are small, highly lipid-soluble molecules which are absorbed almost as fast as they are inhaled.

Non gaseous drugs can be administered as **aerosols**. This is how antiasthmatic drugs are administered, and it allows them to be delivered to the lungs in far greater quantities than if they were given orally or by injection. In the case of antiasthmatics, the drug is made sufficiently polar that it is poorly absorbed into the bloodstream. This localizes it in the airways, and lowers the possibility of side effects elsewhere in the body (e.g. action on the heart). However, a certain percentage of an inhaled drug is inevitably swallowed and can reach the blood supply by the oral route. This may lead to side effects. For example, tremor is a side effect of the antiasthmatic **salbutamol** as a result of the drug reaching the blood supply.

Several drugs of abuse are absorbed through inhalation or smoking (e.g. **nicotine, cocaine, marijuana, methamphetamine** (Fig. 8.19) and **heroin**). This is a particularly hazardous method of taking drugs. A normal cigarette is like a mini-furnace producing a complex mixture of potentially carcinogenic compounds, especially from the tars present in tobacco. These are not absorbed into the blood supply but coat the lung tissue, leading to long-term problems such as lung cancer. The tars in cannabis are considerably more dangerous than those in tobacco. If cannabis is to be used in medicine, safer methods of administration are desirable (i.e. inhalers).

8.6.6 Injection

Drugs can be introduced into the body by intravenous, intramuscular, subcutaneous, or intrathecal injection. Injection of a drug produces a much faster response than oral administration because the drug reaches the blood supply more quickly. The levels of drug administered are also more accurate, because absorption by the oral route has a level of unpredictability due to the first pass effect. Injecting a drug is potentially more hazardous, though. For example, some patients may have an unexpected reaction to a drug, and there is little one can do to reduce the levels once the drug has been injected. Such side effects would be more gradual and treatable if the drug was given orally. Secondly, sterile techniques are essential when giving injections, to avoid the risks of bacterial infection or of transmitting hepatitis or AIDS from a previous patient. Finally, there is a greater risk of receiving an overdose when injecting a drug.

The **intravenous** route involves injecting a solution of the drug directly into a vein. This method of administration is not particularly popular with patients, but it is a highly effective method of administering drugs in accurate doses and it is the fastest of the injection methods. However, it is also the most hazardous method of injection. Since its effects are rapid, the onset of any serious side effects or allergies is also rapid. It is therefore important to administer the drug as slowly as possible and to monitor the patient closely. An intravenous drip allows the drug to be administered in a controlled manner such that there is a steady level of drug in the system. The local anaesthetic **lidocaine** is given by intravenous injection. Drugs that are dissolved in oily liquids cannot be given by intravenous injection as this may result in the formation of blood clots.

The **intramuscular** route involves injecting drugs directly into muscle, usually in the arm, thigh, or buttocks. Drugs administered in this way do not pass round the body as rapidly as they would if given by intravenous injection, but they are still absorbed faster than by oral administration. The rate of absorption depends on various factors such as the diffusion of the drug, blood supply to the muscle, the solubility of the drug, and the volume of the injection. Local blood flow can be reduced by adding adrenaline to constrict blood vessels. Diffusion can be slowed by using a poorly absorbed salt, ester, or complex of the drug (see also section 11.6.2). The advantage of slowing down absorption is in prolonging activity. For example, oily suspensions of steroid hormone esters are used to slow absorption. Drugs are often administered by intramuscular injection when they are unsuitable for intravenous injection and so it is important to avoid injecting into a vein.

Subcutaneous injection involves injecting the drug under the surface of the skin. Absorption depends on factors such as how fast the drug diffuses, the level of blood supply to the skin, and the ability of the drug to enter the blood vessels. Absorption can be slowed by the same methods described for intramuscular injection. Drugs which can act as irritants should not be administered in this way as they can cause severe pain and may damage local tissues.

Intrathecal injection means that the drug is injected into the spinal cord. Antibacterial agents that do not normally cross the blood–brain barrier are often administered in this way. Intrathecal injections are also used to administer **methotrexate** in the treatment of childhood leukaemia in order to prevent relapse in the CNS.

Intraperitoneal injection involves injecting drugs directly into the abdominal cavity. This is very rarely used in medicine, but it is a method of injecting drugs into animals during preclinical tests.

8.6.7 Implants

Continuous osmotically driven minipumps for **insulin** have been developed which are implanted under the skin. The pumps monitor the level of insulin in the blood and release the hormone as required to keep levels constant. This avoids the problem of large fluctuations in insulin levels associated with regular injections.

Gliadel is a wafer that has been implanted into the brain to administer anticancer drugs direct to brain tumours, thus avoiding the blood–brain barrier.

Polymer-coated, drug-releasing stents have been used to keep blood vessels open after a clot clearing procedure called angioplasty.

Investigations are underway into the use of implantable microchips which could detect chemical signals in the body and release drugs in response to these signals.

KEY POINTS

- Oral administration is the preferred method of administering drugs, but it is also the most demanding on the drug.

- Drugs administered by methods other than the oral route avoid the first pass effect.

- Drugs can be administered such that they are absorbed through the mucous membranes of the mouth, nose, or eyes.

- Some drugs are administered rectally as suppositories.

- Topically administered drugs are applied to the skin. Some drugs are absorbed through the skin into the blood supply.

- Inhaled drugs are administered as gases or aerosols to act directly on the respiratory system. Some inhaled drugs are absorbed into the blood supply to act systemically.

- Polar drugs which are unable to cross cell membranes are given by injection.

- Injection is the most efficient method of administering a drug but it is also the most hazardous. Injection can be intravenous, intramuscular, subcutaneous, or intrathecal.

- Implants have been useful in providing controlled drug release such that blood concentrations of the drug remain as level as possible.

8.7 Drug dosing

Because of the number of pharmacokinetic variables involved, it can be difficult to estimate the correct dosing regime for a drug (i.e. the amount of drug used for each dose and the frequency of administration). There are other issues to consider as well. Ideally, the blood levels of any drug should be constant and controlled, but this would require a continuous, intravenous drip which is clearly impractical for most drugs. Therefore, drugs are usually taken at regular time intervals and the doses taken are designed to keep the blood levels of drug within a maximum and minimum level such that they are not too high to be toxic, yet not too low to be ineffective. In general, the concentration of free drug in the blood (i.e. not bound to plasma protein) is a good indication of the availability of that drug at its target site. This does not mean that blood concentration levels are the same as the concentration levels at the target site. However, any variations in blood concentration will result in similar fluctuations at the target site. Thus, blood concentration levels can be used to determine therapeutic and safe dosing levels for a drug.

Figure 8.20 shows two dosing regimes. Dosing regime A quickly reaches the therapeutic level but continues to rise to a steady state which is toxic. Dosing regime B involves half the amount of drug, provided with the same frequency. The time taken to reach the therapeutic level is certainly longer, but the steady state levels of the drug remain between the therapeutic and toxic levels—the **therapeutic window**.

Dosing regimes involving regular administration of a drug work well in most cases, especially if the size of each dose is less than 200 mg and doses are taken once or twice a day. However, there are certain situations where timed doses are not suitable. The treatment of diabetes with **insulin** is a case in point. Insulin is normally secreted continuously by the pancreas, so the injection of insulin at timed intervals is unnatural and can lead to a whole range of physiological complications.

Other dosing complications include differences of age, sex, and race. Diet, environment, and altitude also have an influence. Obese people present a particular problem, as it can be very difficult to estimate how much of a drug will be stored in fat tissue and how much will remain free in the blood supply. The precise time when drugs are taken may be important, because metabolic reaction rates can vary throughout the day.

Drugs can interact with other drugs. For example, some drugs used for diabetes are bound by plasma protein in the blood supply and are therefore not free to react with their targets. However, drugs such as **aspirin** may displace them from plasma protein, leading to a drug overdose. Aspirin has this same effect on anticoagulants.

Problems can also occur if a drug inhibits a metabolic reaction and is taken with a drug normally metabolized by that reaction. The latter is more slowly metabolized than normal, increasing the risk of an overdose. For example, the antidepressant drug **phenelzine** inhibits the metabolism of amines and should not be taken with drugs such as **amphetamines** or **pethidine**. Even amine-rich foods can lead to adverse effects, implying that cheese and wine parties are hardly the way to cheer the victim of depression. Other examples were described in section 8.4.6.

When one considers all these complications, it is hardly surprising that individual variability to drugs can vary by as much as a factor of 10.

8.7.1 Drug half-life

The half-life ($t_{1/2}$) of a drug is the time taken for the concentration of the drug in blood to fall by half. The removal or elimination of a drug takes place through both excretion and drug metabolism, and is not linear with time. Therefore, drugs can linger in the body for a significant period of time. For example, if a drug has a half-life of 1 hour, then there is 50% of it left after

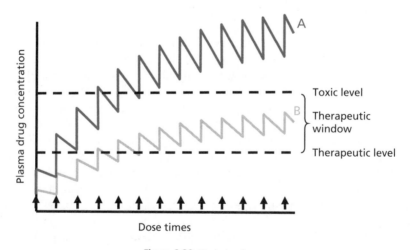

Figure 8.20 Dosing regimes.

1 hour. After 2 hours there is 25% of the original dose left, and after 3 hours 12.5% remains. It takes 7 hours for the level to fall below 1% of the original dose. Some drugs, such as the opiate analgesic **fentanyl**, have short half lives (45 minutes), whereas others such as **diazepam** (Valium) have a half-life measured in days. In the latter case, recovery from the drug may take a week or more.

8.7.2 Steady state concentration

Drugs are metabolized and eliminated as soon as they are administered, so it is necessary to provide regular doses in order to maintain therapeutic levels in the body. Therefore, it is important to know the half-life of the drug in order to calculate the frequency of dosing required to reach and maintain these levels. In general, the time taken to reach a **steady state concentration** is six times the drug's half-life. For example, the concentration levels of a drug with a half-life of 4 hours, supplied at 4-hourly intervals, is shown in Table 8.1 and Fig. 8.21. Note that there is a fluctuation in level in the period between each dose. The level is at a maximum after each dose, and falls to a minimum before the next dose is provided. It is important to ensure that the level does not drop below the therapeutic level, but does not rise to such a level that side effects are induced. The time taken to reach steady state concentration is not dependent on the size of the dose, but

the blood level achieved at steady state is. Therefore, the levels of drug present at steady state concentration depend on the size of each dose given, as well as the frequency of dosing. During clinical trials, blood samples are taken from patients at regular time intervals to determine the concentration of the drug in the blood. This helps determine the proper dosing regime in order to get the ideal blood levels.

The area under the plasma drug concentration curve (AUC) represents the total amount of drug that is available in the blood supply during the dosing regime.

8.7.3 Drug tolerance

With certain drugs, it is found that the effect of the drug diminishes after repeated doses, and it is necessary to increase the size of the dose in order to achieve the same results. This is known as drug tolerance. There are several mechanisms by which drug tolerance can occur. For example, the drug can induce the synthesis of metabolic enzymes which result in increased metabolism of the drug. **Pentobarbital** (Fig. 8.22) is a barbiturate sedative which induces enzymes in this fashion.

Alternatively, the target may adapt to the presence of a drug. Occupancy of a target receptor by an antagonist may induce cellular effects which result in the synthesis of more receptor (section 5.11). As a result, more drug will be needed in the next dose to antagonize all the receptors.

Physical dependence is usually associated with drug tolerance. Physical dependence is a state in which a patient becomes dependent on the drug in order to feel normal. If the drug is withdrawn, uncomfortable **withdrawal symptoms** may arise which can only be alleviated by re-taking the drug. These effects can be explained in part by the effects which lead to drug tolerance. For example, if cells have synthesized more receptors to counteract the presence of an antagonist, the removal of the antagonist means that the body will have too many receptors. This

Table 8.1 Fluctuation of drug concentration levels on regular dosing

Time of dosing (h)	0	4	8	12	16	20	24
Maximum level (µg/ml)	1.0	1.5	1.75	1.87	1.94	1.97	1.98
Minimum level (µg/ml)	0.5	0.75	0.87	0.94	0.97	0.98	0.99

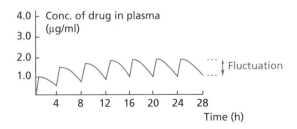

Figure 8.21 Graphical Representation of fluctuation of drug concentration levels on regular dosing.

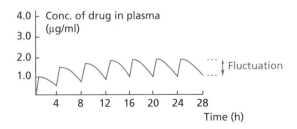

Figure 8.22 Pentobarbital.

results in a 'kickback' effect where the cell becomes oversensitive to the normal neurotransmitter or hormone, and this is what produces withdrawal symptoms. These will continue until the excess receptors have been broken down by normal cellular mechanisms—a process that may take several days or weeks (see also sections 5.10–5.11).

8.7.4 Bioavailability

Bioavailability refers to how quickly and how much of a particular drug reaches the blood supply once all the problems associated with absorption, distribution, metabolism, and excretion have been taken into account. Oral bioavailability (F) is the fraction of the ingested dose that survives to reach the blood supply. This is an important property when it comes to designing new drugs and should be considered alongside the pharmacodynamics of the drug—i.e. how effectively the drug interacts with its target.

8.8 Formulation

The way a drug is formulated can avoid some of the problems associated with oral administration. Drugs are normally taken orally as tablets or capsules. A tablet is usually a compressed preparation that contains 5–10% of the drug, 80% of fillers, disintegrants, lubricants, glidants, and binders, and 10% of compounds which ensure easy disintegration, disaggregation, and dissolution of the tablet in the stomach or the intestine. The disintegration time can be modified for a rapid effect or for sustained release. Special coatings can make the tablet resistant to the stomach acids such that it only disintegrates in the duodenum as a result of enzyme action or alkaline pH. Pills can also be coated with sugar, varnish, or wax to disguise taste. Some tablets are designed with an osmotically active core, surrounded by an impermeable membrane with a pore in it. This allows the drug to percolate out from the tablet at a constant rate as the tablet moves through the digestive tract.

A capsule is a gelatinous envelope enclosing the active substance. Capsules can be designed to remain intact for some hours after ingestion in order to delay absorption. They may also contain a mixture of slow- and fast-release particles to produce rapid and sustained absorption in the same dose.

The drug itself needs to be soluble in aqueous solution at a controlled rate. Such factors as particle size and crystal form can significantly affect dissolution. Fast dissolution is not always ideal. For example, slow dissolution rates can prolong the duration of action or avoid initial high plasma levels.

8.9 Drug delivery

The various aspects of drug delivery could fill a textbook in itself, so any attempt to cover the topic in a single section is merely tickling at the surface, let alone scratching it! However, it is worth appreciating that there are various methods by which drugs can be physically protected from degradation and/or targeted to treat particular diseases such as cancer and inflammation. One approach is to use a prodrug strategy (section 11.6), which involves chemical modifications to the drug. Another approach covered in this section is the use of water-soluble macromolecules to help the drug reach its target. The macromolecules concerned are many and varied and include synthetic polymers, proteins, liposomes, and antibodies. The drug itself may be covalently linked to the macromolecule or encapsulated within it. The following are some illustrations of drug delivery systems.

Antibodies were described in section 3.9 and have long been seen as a method of targeting drugs to cancer cells. Methods have been devised of linking anticancer drugs to antibodies to form **antibody–drug conjugates** that remain stable on their journey through the body but release the drug at the target cell. A lot of research has been carried out on these conjugates, and this is discussed in detail in section 18.9. However, there are problems associated with antibodies. The amount of drug that can be linked to the protein is quite limited and there is the risk of an immune reaction where the body identifies the antibody as foreign and tries to reject it.

A similar approach is to link the drugs to synthetic polymers such as polyethylene glycol (PEG), polyglutamate, or *N*-(2-hydroxypropyl)methacrylamide (HPMA) to form polymer–drug conjugates (Fig. 8.23). Again the amount of drug that can be linked is limited, but a variety of anticancer–polymer conjugates are currently undergoing clinical trials.

Protein-based polymers are being developed as drug delivery systems for the controlled release of ionized drugs. For example, the cationic drugs **Leu-enkephalin** or **naltrexone** could be delivered using polymers with anionic carboxylate groups. Ionic interactions between the drug and the protein result in folding and assembly of the protein polymer to form a protein–drug complex and the drug is then released at a slow and constant rate. The amount of drug carried could be predetermined by the density of carboxylate binding sites present and the accessible surface area of the vehicle. The rate of release could be controlled by varying the number of hydrophobic amino acids present. The greater the number of hydrophobic amino acids present, the weaker the affinity between the carboxylate binding groups and the drug. Once the drug is released, the protein carrier would be metabolized like any normal protein.

A physical method of protecting drugs from metabolic enzymes in the bloodstream and allowing a steady slow release of the drug is to encapsulate the drug within small vesicles called **liposomes** and then inject them into the blood supply (Fig. 8.24). These vesicles or globules consist of a bilayer of fatty phospholipid molecules (similar to a cell membrane) and will travel round the circulation, slowly leaking their contents. Liposomes are known to be concentrated in malignant tumours and this provides a possible method of delivering antitumour drugs to these cells. It is also found that liposomes can fuse with the plasma membranes of a variety of cells, allowing the delivery of drugs or DNA into these cells. As a result, they may be useful for gene therapy. The liposomes can be formed by sonicating a suspension of a phospholipid (e.g. phosphatidylcholine) in an aqueous solution of the drug.

Another future possibility for targeting liposomes is to incorporate antibodies into the liposome surface such that specific tissue antigens are recognized. Liposomes have a high drug-carrying capacity, but it can prove difficult to control the release of drug at the required rate and the slow leakage is a problem if the liposome is carrying a toxic anticancer drug such as **doxorubicin**. The liposomes can also be trapped by the reticuloendothelial system (RES) and removed from the blood supply. The RES is a network of cells which can be viewed as a kind of filter. One solution to this problem has been to PEGylate the liposome by attaching PEG polymers to the liposome (see also section 11.8.3). The tails of the PEG polymers project out from the liposome surface and act as a polar outer shell which both protects and shields the liposome from destructive enzymes and the reticuloendothelial system, increasing its lifetime significantly and reducing leakage of its passenger drug. **DOXIL** is a PEGylated liposome containing doxorubicin which is used successfully in anticancer therapy as a once-monthly infusion.

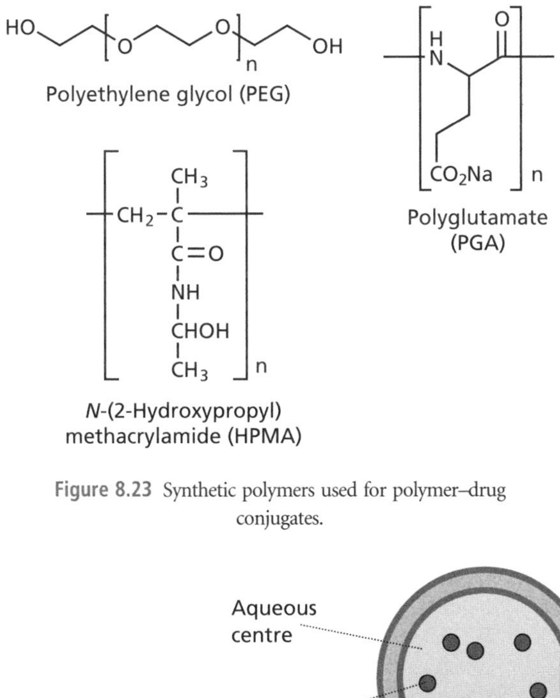

Figure 8.23 Synthetic polymers used for polymer–drug conjugates.

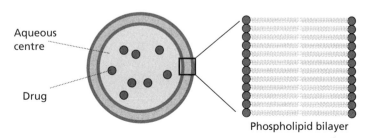

Figure 8.24 Liposome containing drug.

The use of injectable **microspheres** has been approved for the delivery of human growth hormone. The microspheres containing the drug are made up of a biologically degradable polymer and slowly release the hormone over a 4-week period.

A large number of important drugs have to be administered by injection because they are either susceptible to digestive enzymes or cannot cross the gut wall. This includes the ever-growing number of therapeutically useful peptides and proteins being generated by biotechnology companies using recombinant DNA technology. Drug delivery systems which could deliver these drugs orally would prove a huge step forward in medicine. For example, liposomes are currently being studied as possible oral delivery systems. Another approach currently being investigated is to link a therapeutic protein to a hydrophobic polymer such that it is more likely to be absorbed. However, it is important that the conjugate breaks up before the drug enters the blood supply or else it would have to be treated as a new drug and undergo expensive preclinical and clinical trials. **Hexyl-insulin monoconjugate 2** consists of a polymer linked to a lysine residue of insulin. It is currently being investigated as an oral delivery system for **insulin.**

Biologically erodable microspheres have also been designed to stick to the gut wall such that absorption of the drug within the sphere through the gut wall is increased. This has still to be used clinically, but has proved effective in enhancing the absorption of insulin and **plasmid DNA** in test animals. In a similar vein, drugs have been coated with bioadhesive polymers designed to adhere to the gut wall so that the drug has more chance of being absorbed. The use of anhydride polymers has the added advantage that these polymers are capable of crossing the gut wall and enter the bloodstream, taking their passenger drug with them. Emisphere Technologies Inc. have developed derivatives of amino acids and shown that they can enhance the absorption of specific proteins. It is thought that the amino acid derivatives interact with the protein and make it more lipophilic so that it can cross cell membranes directly.

KEY POINTS

- Drugs should be administered at the correct dose levels and frequency to ensure that blood concentrations remain within the therapeutic window.

- The half-life of a drug is the time taken for the blood concentration of the drug to fall by half. A knowledge of the half-life is required to calculate how frequently doses should be given to ensure a steady state concentration.

- Drug tolerance is where the effect of a drug diminishes after repeated doses. In physical dependence a patient becomes dependent on a drug and suffers withdrawal symptoms on stopping the treatment.

- Formulation refers to the method by which drugs are prepared for administration, whether by solution, pill, capsule, liposome, or microsphere. Suitable formulations can protect drugs from particular pharmacokinetic problems.

QUESTIONS

1. Benzene used to be a common solvent in organic chemistry but is no longer used because it is a suspect carcinogen. Benzene undergoes metabolic oxidation by cytochrome P450 enzymes to form an electrophilic epoxide which can alkylate proteins and DNA. Toluene is now used as a solvent in place of benzene. Toluene is also oxidized by cytochrome P450 enzymes, but the metabolite is less toxic and is rapidly excreted. Suggest what the metabolite might be, and why the metabolism of toluene is different from that of benzene.

2. The antipsychotic drug **fluphenazine** has a prolonged period of action when it is given by intramuscular injection, but

Fluphenazine

not when it is given by intravenous injection. Suggest why this is the case.

3. Morphine binds strongly to opiate receptors in the brain to produce analgesia. *In vitro* studies on opiate receptors show that the quaternary salt of morphine also binds strongly. However, the compound is inactive *in vivo* when injected intravenously. Explain this apparent contradiction.

Morphine; R=H
Quaternary salt; R=Me

4. The phenol group of morphine is important in binding morphine to opiate receptors and causing analgesia. Codeine has the same structure as morphine, but the phenol group is masked as a methyl ether. As a result, codeine binds poorly to opiate receptors and should show no analgesic activity. However, when it is taken *in vivo* it shows useful analgesic properties. Explain how this might occur.

5. The pK_a of histamine is 5.74. What is the ratio of ionized to unionized histamine (a) at pH 5.74 (b) at pH 7.4?

6. A drug contains an ionized carboxylate group and shows good activity against its target in *in vitro* tests. When *in vivo* tests were carried out, the drug showed poor activity when it

was administered orally but good activity when it was administered by intravenous injection. The same drug was converted to an ester, but proved inactive *in vitro*. Despite that, it proved to be active *in vivo* when it was administered orally. Explain these observations.

7. Atomoxetine and methylphenidate are used in the treatment of attention deficit hyperactivity disorder. Suggest possible metabolites for these structures.

Atomoxetine   Methylphenidate

8. Suggest metabolites for the proton pump inhibitor omeprazole.

Omeprazole

9. A drug has a half-life of 4 hours. How much of the drug remains after 24 hours?

FURTHER READING

Duncan, R. (2003) The dawning era of polymer therapeutics. *Nature Reviews Drug Discovery*, 2, 347–360.

Goldberg, M. and Gomez-Orellana, I. (2003) Challenges for the oral delivery of macromolecules. *Nature Reviews Drug Discovery*, 2, 257–258.

Guengerich, F. P. (2002) Cytochrome P450 enzymes in the generation of commercial products. *Nature Reviews Drug Discovery*, 1, 359–366.

Langer, R. (2003) Where a pill won't reach. *Scientific American*, April, 32–39.

LaVan, D. A., Lynn, D. M., and Langer, R. (2002) Moving smaller in drug discovery and delivery. *Nature Reviews Drug Discovery*, 1, 77–84.

Lindpaintner, K. (2002) The impact of pharmacogenetics and pharmacogenomics on drug discovery. *Nature Reviews Drug Discovery*, 1, 463–469.

Lipinski, C. A., Lombardo, F., Dominy, B. W., and Feeney, P. J. (1997) Experimental and computational approaches to estimate solubility and permeability in drug discovery and development settings. *Advanced Drug Delivery Reviews*, 23, 3–25 (rule of five).

Nicholson, J. K. and Wilson, I. D. (2003) Understanding global systems biology: metabonomics and the continuum of metabolism. *Nature Reviews Drug Discovery*, 2, 668–676.

Pardridge, W. M. (2002) Drug and gene targeting to the brain with molecular Trojan horses. *Nature Reviews Drug Discovery*, 1, 131–139.

Roden, D. M. and George, A. L. (2002) The genetic basis of variability in drug responses. *Nature Reviews Drug Discovery*, 1, 37–44.

Roses, A. D. (2002) Genome-based pharmacogenetics and the pharmaceutical industry. *Nature Reviews Drug Discovery*, 1, 541–549.

Rowland, M. and Tozer, T. N. (1980) *Clinical pharmacokinetics*. Lea and Febiger, Philadelphia.

Saltzman, W. M. and Olbricht, W. L. (2002) Building drug delivery into tissue engineering. *Nature Reviews Drug Discovery*, 1, 177–186.

Sam, A. P. and Tokkens, J. G. (eds.) (1996) *Innovations in drug delivery: impact on pharmacotherapy*. Anselmus Foundation, Hauten, The Netherlands.

Stevenson, R. (2003) Going with the flow. *Chemistry in Britain*, November, 18–20 (aquaporins).

van de Waterbeemd, H., Testa, B., and Folkers, G. (eds.) (1997) *Computer-assisted lead finding and optimisation*. Wiley-VCH, New York.

Willson, T. M. and Kliewer, S. A. (2002) PXR, CAR and drug metabolism. *Nature Reviews Drug Discovery*, 1, 259–266.

Titles for general further reading are listed on p. 711.

PART B

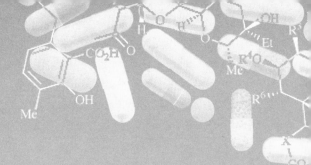

Drug discovery, design, and development

The past

Before the twentieth century, medicines consisted mainly of herbs and potions. It was not until the mid-nineteenth century that the first serious efforts were made to isolate and purify the **active principles** of these remedies (i.e. the pure chemicals responsible for the medicinal properties). The success of these efforts led to the birth of many of the pharmaceutical companies we know today. Since then, many naturally occurring drugs have been obtained and their structures determined (e.g. morphine from opium, cocaine from coca leaves, quinine from the bark of the cinchona tree).

These natural products sparked off a major synthetic effort where chemists made literally thousands of analogues in an attempt to improve on what nature had provided. Much of this work was carried out on a trial and error basis, but the results obtained revealed several general principles behind drug design. Many of these principles are described in Chapters 10 and 11.

An overall pattern for drug discovery and drug development also evolved, but there was still a high element of trial and error involved in the process. The mechanism by which a drug worked at the molecular level was rarely understood and drug research very much focused on what is known as the **lead compound** (pronounced 'leed')—an active principle isolated from a natural source or a synthetic compound prepared in the laboratory.

The present

In recent years, medicinal chemistry has undergone a revolutionary change. Rapid advances in the biological sciences have resulted in a much better understanding of how the body functions at the cellular and the molecular level. As a result, most research projects in the pharmaceutical industry or university sector now begin by identifying a suitable target in the body and designing a drug to interact with that target. An understanding of the structure and function of the target, as well as the mechanism by which it interacts with potential drugs is crucial to this approach. Generally, we can identify the following stages in drug discovery, design and development:

- Drug discovery—finding a lead
    - Choose a disease!
    - Choose a drug target.
    - Identify a bioassay.
    - Find a lead compound.
    - Isolate and purify the lead compound if necessary.
    - Determine the structure of the lead compound if necessary.

- Drug design
    - Identify structure–activity relationships (SARs).
    - Identify the pharmacophore.
    - Improve target interactions (pharmacodynamics).
    - Improve pharmacokinetic properties.

- Drug development
    - Patent the drug.
    - Carry out preclinical trials (drug metabolism, toxicology, formulation and stability tests, pharmacology studies, etc.).
    - Design a manufacturing process (chemical and process development).
    - Carry out clinical trials.
    - Register and market the drug.
    - Make money!

Many of these stages run concurrently and are dependent on each other. For example, preclinical trials are usually carried out in parallel with the development of a manufacturing process. Even so, the discovery, design and development of a new drug can take 15 years or more, involve the synthesis of over 10 000 compounds and cost in the region of $800 million or £450 million.

9 Drug discovery: finding a lead

In this chapter, we look at what happens when a pharmaceutical company or university research group initiates a new medicinal chemistry project through to the identification of a lead compound.

9.1 Choosing a disease

How does a pharmaceutical company decide which disease to target when designing a new drug? Clearly, it would make sense to concentrate on diseases where there is a need for new drugs. Pharmaceutical companies, however, have to consider economic factors as well as medical ones. A huge investment has to be made towards the research and development of a new drug. Therefore, companies must ensure that they get a good financial return for their investment. As a result, research projects tend to focus on diseases that are important in the developed world, because this is the market best able to afford new drugs. A great deal of research is carried out on ailments such as migraine, depression, ulcers, obesity, flu, cancer, and cardiovascular disease. Less is carried out on the tropical diseases of the developing world. Only when such diseases start to make an impact in richer countries do the pharmaceutical companies sit up and take notice. For example, there has been a noticeable increase in antimalarial research as a result of the increase in tourism to more exotic countries, and the spread of malaria into the southern states of the USA.

Choosing which disease to tackle is usually a matter for a company's market strategists. The science becomes important at the next stage.

9.2 Choosing a drug target

9.2.1 Drug targets

Once a therapeutic area has been identified, the next stage is to identify a suitable drug target (e.g. receptor, enzyme, or nucleic acid). An understanding of which biomacromolecules are involved in a particular disease state is clearly important (see Box 9.1). This allows the medicinal research team to identify whether agonists or antagonists should be designed for a particular receptor, or whether inhibitors should be designed for a particular enzyme. For example, agonists of serotonin receptors are useful for the treatment of migraine, and antagonists of dopamine receptors are useful as antidepressants. Sometimes it is not known for certain whether a particular target will be suitable or not. For example, **tricyclic antidepressants** such as **desipramine** (Fig. 9.1) are known to inhibit the uptake of the neurotransmitter **noradrenaline** from nerve synapses by inhibiting the carrier protein for noradrenaline (section 20.12.4). However, these drugs also inhibit uptake of a separate neurotransmitter called **serotonin,** and the possibility arose that inhibiting serotonin uptake might also be beneficial. A search for selective serotonin uptake inhibitors was initiated, which led to the discovery of the best-selling antidepressant drug **fluoxetine** (**Prozac**) (Fig. 9.1), but when this project was initiated it was not known for certain whether serotonin uptake inhibitors would be effective or not.

9.2.2 Discovering drug targets

If a drug or a poison produces a biological effect, there must be a molecular target for that agent in the body. In

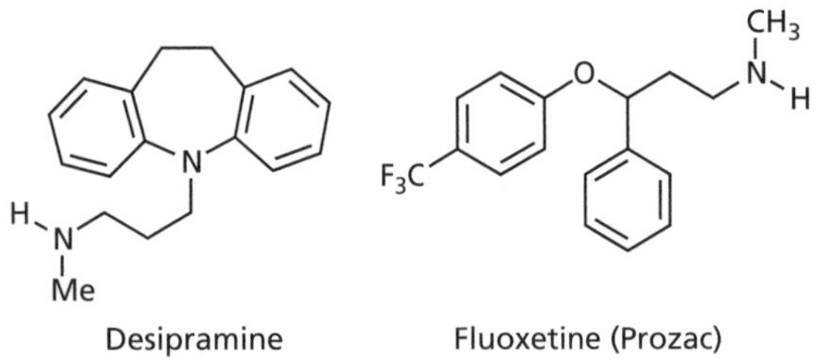

Figure 9.1 Antidepressant drugs.

the past, the discovery of drug targets depended on finding the drug first. Many early drugs such as the analgesic **morphine** are natural products derived from plants, and just happen to interact with a molecular target in the human body. As this involves coincidence more than design, the detection of drug targets was very much a hit and miss affair. Later, the body's own chemical messengers started to be discovered and pointed the finger at further targets. For example, since the 1970s a variety of peptides and proteins have been discovered which act as the body's own analgesics. Despite this, relatively few of the body's messengers were identified, either because they were present in such small quantities or because they were too short lived to be isolated. Indeed, many chemical messengers still

---

BOX 9.1 CASPASES: RECENTLY DISCOVERED TARGETS

The **caspases** are examples of recently discovered enzymes that may prove useful as drug targets. They are a family of protease enzymes which catalyse the hydrolysis of important cellular proteins and which have been found to play a role in inflammation and cell death. Cell death is a natural occurrence in the body, and cells are regularly recycled. Therefore, caspases should not necessarily be seen as 'bad' or 'undesirable' enzymes. Without them, cells could be more prone to unregulated growth resulting in diseases such as cancer.

The caspases catalyse the hydrolysis of particular target proteins such as those involved in DNA repair, and the regulation of cell cycles. By understanding how these enzymes operate, there is the possibility of producing new therapies for a variety of diseases. For example, agents that promote the activity of caspases and lead to more

rapid cell death might be useful in the treatment of diseases such as cancer, autoimmune disease and viral infections. Alternatively, agents that inhibit caspases and reduce the prevalence of cell death could provide novel treatments for trauma, neurodegenerative disease, and strokes. It is already known that the active site of caspases contains two amino acids that are crucial to the mechanism of hydrolysis: cysteine, which acts as a nucleophile, and histidine, which acts as an acid–base catalyst. The mechanism is similar to that used by acetylcholinesterase (section 19.16).

Caspases recognize aspartate groups within protein substrates and cleave the peptide link next to the aspartate group. Selective inhibitors have been developed which include aspartate or a mimic of it, but it remains to be seen whether such inhibitors have a clinical role.

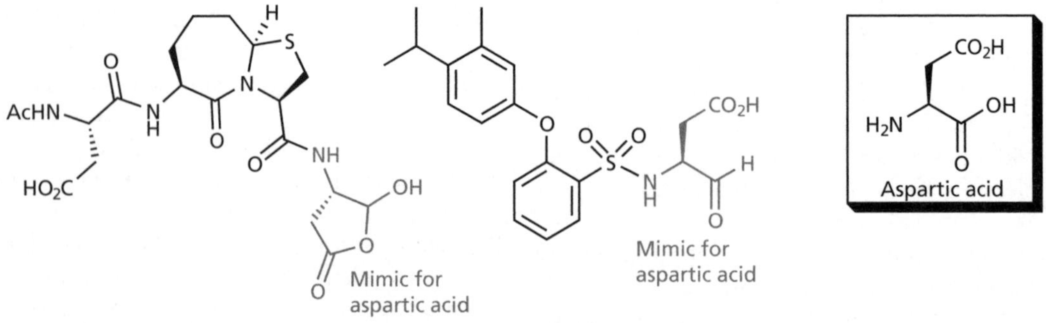

Selective caspase inhibitors.

remain undiscovered. This in turn means that many of the body's potential drug targets remain hidden. Or at least it did! The advances in genomics and proteomics have changed all that. The various genome projects which have mapped the DNA of humans and other life forms, along with the newer field of proteomics (section 3.6), are revealing an ever-increasing number of new proteins which are potential drug targets for the future. These targets have managed to stay hidden for so long that their natural chemical messengers are also unknown and, for the first time, medicinal chemistry is faced with new targets, but with no lead compounds to interact with them. Such targets have been defined as **orphan receptors**. The challenge is now to find a chemical that will interact with these targets in order to find out what their function is and whether they will be suitable as drug targets. This has been one of the main driving forces behind the rapidly expanding area of **combinatorial synthesis** (Chapter 14).

9.2.3 Target specificity and selectivity between species

Target specificity and selectivity is a crucial factor in modern medicinal chemistry research. The more selective a drug is for its target, the less chance that it will interact with different targets and have undesirable side effects.

In the field of antimicrobial agents, the best targets to choose are those that are unique to the microbe and are not present in humans. For example, **penicillin** targets an enzyme involved in bacterial cell wall biosynthesis. Mammalian cells do not have a cell wall, so this enzyme is absent in human cells and penicillin has few side effects (section 16.5). In a similar vein, sulfonamides inhibit a bacterial enzyme not present in human cells (section 16.4.1), and several agents used to treat AIDS inhibit an enzyme called retroviral reverse transcriptase which is unique to the infectious agent HIV (section 17.7.3).

Other cellular features that are unique to microorganisms could also be targeted. For example, the microorganisms which cause sleeping sickness in Africa are propelled by means of a tail-like structure called a **flagellum.** This feature is not present in mammalian cells, so designing drugs that bind to the proteins making up the flagellum and prevent it

Figure 9.2 Fluconazole.

from working could be potentially useful in treating that disease.

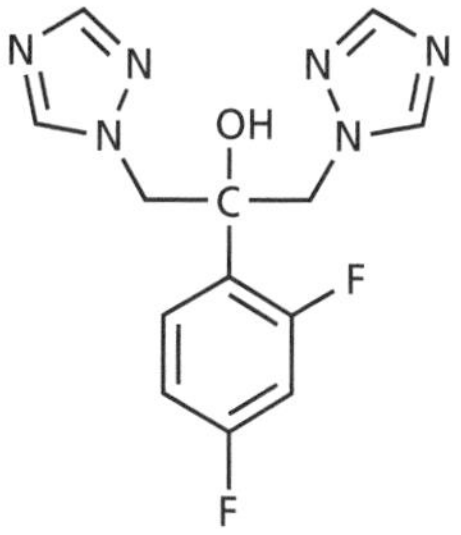

Having said all that, it is still possible to design drugs against targets which are present both in humans and in microbes, as long as the drugs show selectivity against the microbial target. Fortunately, this is perfectly feasible. An enzyme which catalyses a reaction in a bacterial cell differs significantly from the equivalent enzyme in a human cell. The enzymes may have been derived from an ancient common ancestor, but several million years of evolution have resulted in significant structural differences. For example, the antifungal agent **fluconazole** (Fig. 9.2) inhibits a fungal demethylase enzyme involved in steroid biosynthesis. This enzyme is also present in humans, but the structural differences between the two enzymes are significant enough that the antifungal agent is highly selective for the fungal enzyme. Other examples of bacterial or viral enzymes which are sufficiently different from their human equivalents are dihydrofolate reductase (section 16.4.2) and viral DNA polymerase (section 17.6.1).

9.2.4 Target specificity and selectivity within the body

Selectivity is also important for drugs acting on targets within the body. Enzyme inhibitors should only inhibit the target enzyme and not some other enzyme. Receptor agonists/antagonists should ideally interact with a specific kind of receptor (e.g. the adrenergic receptor) rather than a variety of different receptors. However nowadays, medicinal chemists aim for even higher standards of target selectivity. Ideally, enzyme inhibitors should show selectivity between the various isozymes of an enzyme. Receptor agonists and antagonists should not only show selectivity for a particular receptor (e.g. an adrenergic receptor) or even a particular receptor type (e.g. the β-adrenergic receptor), but also for a particular receptor subtype (e.g. the β_2-adrenergic receptor).

One of the current areas of research is to find antipsychotic agents with fewer side effects. Traditional antipsychotic agents act as antagonists of dopamine receptors. However, it has been found that there are five dopamine receptor subtypes and that traditional antipsychotic agents antagonize two of these (D_3 and D_2). There is good evidence that the D_2 receptor is responsible for the undesirable parkinsonian-type side effects of current drugs, and so research is now underway to find a selective D_3 antagonist.

9.2.5 Targeting drugs to specific organs and tissues

Targeting drugs against specific receptor subtypes often allows drugs to be targeted against specific organs or against specific areas of the brain. This is because the various receptor subtypes are not uniformly distributed around the body, but are often concentrated in particular tissues. For example, the β-adrenergic receptors in the heart are predominantly β_1 whereas those in the lungs are β_2. This makes it feasible to design drugs that will work on the lungs with a minimal side effect on the heart, and vice versa.

Attaining subtype selectivity is particularly important for drugs that are intended to mimic neurotransmitters. Neurotransmitters are released close to their target receptors and once they have passed on their message, they are quickly deactivated and do not have the opportunity to 'switch on' more distant receptors. Therefore, only those receptors which are fed by 'live' nerves are switched on.

In many diseases, there is a 'transmission fault' to a particular tissue or in a particular region of the brain. For example, in Parkinson's disease, **dopamine** transmission is deficient in certain regions of the brain although it is functioning normally elsewhere. A drug could be given to mimic dopamine in the brain. However, such a drug acts like a hormone rather than as a neurotransmitter because it has to travel round the body in order to reach its target. This means that the drug could potentially 'switch on' all the dopamine receptors around the body and not just the ones that are suffering the dopamine deficit. Such drugs would have a large

number of side effects, so it is important to make the drug as selective as possible for the particular type or subtype of dopamine receptor affected in the brain. This would target the drug more effectively to the affected area and reduce side effects elsewhere in the body.

Many research projects set out to discover new drugs with a defined profile of activity against a range of specific targets. For example, a research team may set out to find a drug that has agonist activity for one receptor subtype and antagonist activity at another. A further requirement may be that the drug does not inhibit metabolic enzymes (section 8.4).

9.2.6 Pitfalls

A word of caution! It is possible to identify whether a particular enzyme or receptor plays a role in a particular ailment. However, the body is a highly complex system. For any given function, there are usually several messengers, receptors, and enzymes involved in the process. For example, there is no one simple cause for hypertension (high blood pressure). This is illustrated by the variety of receptors and enzymes which can be targeted in its treatment. These include β_1-adrenoceptors, calcium ion channels, angiotensin-converting enzyme (ACE), and potassium ion channels.

As a result, more than one target may need to be addressed for a particular ailment. For example, most of the current therapies for asthma involve a combination of a bronchodilator (β_2-agonist) and an anti-inflammatory agent such as a corticosteroid.

Sometimes, drugs designed against a specific target become less effective over time. Because cells have a highly complex system of signalling mechanisms, it is possible that the blockade of one part of that system could be bypassed. This could be compared to blocking the main road into town to try to prevent congestion in the town centre. To begin with, the policy works, but in a day or two commuters discover alternative routes and congestion in the centre becomes as bad as ever (Fig. 9.3).

9.3 Identifying a bioassay

9.3.1 Choice of bioassay

Choosing the right bioassay or test system is crucial to the success of a drug research programme.

BOX 9.2 PITFALLS IN CHOOSING PARTICULAR TARGETS

Drugs are designed to interact with a particular target because it is believed that target is important to a particular disease process. Occasionally though, a particular target may not be so important to a disease as was first thought. For example, the dopamine D_2 receptor was thought to be involved in causing nausea. Therefore, the D_2 receptor antagonist **metoclopramide** was developed to as an antiemetic agent to prevent nausea. However, it was found that more potent D_2 antagonists were less effective, implying that a different receptor might be more important in producing nausea. Metoclopramide also antagonizes the 5-hydroxytryptamine ($5HT_3$) receptor, so antagonists for this receptor were studied, which led to the development of the antiemetic drugs **granisetron** and **ondansetron**.

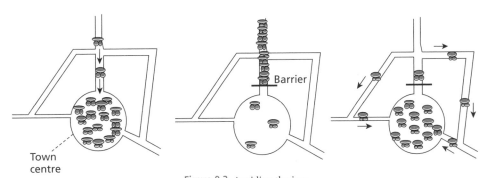

Metoclopramide   Granisetron   Ondansetron

Antiemetic agents.

Figure 9.3 Avoiding the jam.

The test should be simple, quick, and relevant, as there is usually a large number of compounds to be analysed. Human testing is not possible at such an early stage, so the test has to be done *in vitro* (i.e. on isolated cells, tissues, enzymes, or receptors) or *in vivo* (on animals). In general, *in vitro* tests are preferred over *in vivo* tests because they are cheaper, easier to carry out, less controversial, and they can be automated. However, *in vivo* tests are often needed to check whether drugs interacting with a specific target have the desired pharmacological activity, and also to monitor their pharmacokinetic properties. In modern medicinal chemistry, a variety of tests is usually carried out both *in vitro* and *in vivo* to determine not only whether the candidate drugs are acting at the desired target, but also whether they have activity at other undesired targets. The direction taken by projects is then determined by finding the drugs that have the best balance of good activity at the desired target and minimal activity at other targets.

9.3.2 *In vitro* tests

In vitro tests do not involve live animals. Instead, specific tissues, cells, or enzymes are used. Enzyme inhibitors can be tested on the pure enzyme in solution. In the past, it could be a major problem to isolate and purify sufficient enzyme to test, but nowadays genetic engineering can be used to incorporate the gene for a particular enzyme into fast-growing cells such as yeast or bacteria. These then produce the enzyme in larger quantities, making isolation easier. For example, **HIV protease** (section 17.7.4) has been cloned and expressed in the bacterium *E. coli*. A variety of experiments can be carried out on this enzyme to determine whether an enzyme inhibitor is competitive or non-competitive, and to determine IC_{50} values (section 4.9).

Receptor agonists and antagonists can be tested on isolated tissues or cells which express the target receptor on their surface. Sometimes these tissues can be used to test drugs for physiological effects. For example, bronchodilator activity can be tested by observing how well compounds inhibit contraction of isolated tracheal smooth muscle. Alternatively, the affinity of drugs for receptors (how strongly they bind) can be measured by **radioligand studies** (section 5.14). Many *in vitro* tests have been designed by genetic engineering where the gene coding for a specific receptor is identified, cloned, and expressed in fast-dividing cells such as bacterial, yeast, or tumour cells. For example, **Chinese Hamster Ovarian cells** (CHO cells) are commonly used for this purpose, as they express a large amount of the cloned receptor on their cell surface. *In vitro* studies on whole cells are useful because there are none of the complications of *in vivo* studies where the drug has to cross barriers such as the gut wall, or survive metabolic enzymes. The environment surrounding the cells can be easily controlled and both intracellular and intercellular events can be monitored, allowing a measurement of efficacy and potency (section 5.14). Primary cell cultures (i.e. cells that have not been modified) can be produced from embryonic tissues; transformed cell lines are derived from tumour tissue. Cells grown in this fashion are all identical.

Antibacterial drugs are tested *in vitro* by measuring how effectively they inhibit or kill bacterial cells in culture. It may seem strange to describe this as an *in vitro* test, as bacterial cells are living microorganisms.

However, *in vivo* tests are defined as those that are carried out on animals or humans to test whether antibacterial agents combat infection.

9.3.3 *In vivo* tests

In vivo tests on animals often involve inducing a clinical condition in the animal to produce observable symptoms. The animal is then treated to see whether the drug alleviates the problem by eliminating the observable symptoms. For example, the development of non-steroidal inflammatory drugs was carried out by inducing inflammation on test animals then testing drugs to see whether they relieved the inflammation.

Transgenic animals are often used in *in vivo* testing. These are animals whose genetic code has been altered. For example, it is possible to replace some mouse genes with human genes. The mouse produces the human receptor or enzyme and this allows *in vivo* testing against that target. Alternatively, the mouse's genes could be altered such that the animal becomes susceptible to a particular disease (e.g. breast cancer). Drugs can then be tested to see how well they prevent that disease.

There are several problems associated with *in vivo* testing. It is slow and it also causes animal suffering. There are the many problems of pharmacokinetics (Chapter 8), and so the results obtained may be misleading and difficult to rationalize if *in vivo* tests are carried out in isolation. For example, how can one tell whether a negative result is due to the drug failing to bind to its target or not reaching the target in the first place? Thus, *in vitro* tests are usually carried out first to determine whether a drug interacts with its receptor, and *in vivo* tests are then carried out to test pharmacokinetic properties.

Certain *in vivo* tests might turn out to be invalid. It is possible that the observed symptoms might be caused by a different physiological mechanism than the one intended. For example, many promising antiulcer drugs which proved effective in animal testing were ineffective in clinical trials. Finally, different results may be obtained in different animal species. For example, **penicillin methyl ester prodrugs** (Box 16.5) are hydrolysed in mice or rats to produce active penicillins, but are not hydrolysed in rabbits, dogs, or humans. Another example involves **thalidomide**, which is teratogenic in rabbits and humans but has no such effect in mice.

Despite these issues, *in vivo* testing is still crucial in identifying the particular problems which might be associated with using a drug *in vivo* and which cannot be picked up by *in vitro* tests.

9.3.4 Test validity

Sometimes the validity of testing procedures is easy and clear-cut. For example, an antibacterial agent can be tested *in vitro* by measuring how effectively it kills bacterial cells. A local anaesthetic can be tested *in vitro* on how well it blocks action potentials in isolated nerve tissue. In other cases, the testing procedure is more difficult. For example, how do you test a new anti-psychotic drug? There is no animal model for this condition and so a simple *in vivo* test is not possible. One way round this problem is to propose which receptor or receptors might be involved in a medical condition and to carry out *in vitro* tests against these in the expectation that the drug will have the desired activity when it comes to clinical trials. One problem with this approach is that it is not always clear-cut whether a specific receptor or enzyme is as important as one might think to the targeted disease (see Box 9.2).

9.3.5 High-throughput screening

Robotics and the miniaturization of *in vitro* tests on genetically modified cells has led to a process called high-throughput screening (HTS). This involves the automated testing of large numbers of compounds versus a large number of targets; typically, several thousand compounds can be tested at once in 30–50 biochemical tests. It is important that the test should produce an easily measurable effect which can be detected and measured automatically. This effect could be cell growth, an enzyme-catalysed reaction which produces a colour change, or displacement of radio-actively labelled ligands from receptors.

Receptor antagonists can be studied using modified cells which contain the target receptor in their cell membrane. Detection is possible by observing how effectively the test compounds inhibit the binding of a radiolabelled ligand. Another approach is to use yeast cells which have been modified such that activation of a target receptor results in the activation of an enzyme which, when supplied with a suitable substrate, catalyses the release of a dye. This produces an easily identifiable colour change.

9.3.6 Screening by NMR

NMR spectroscopy is an analytical tool which has been used for many years to determine the molecular structure of compounds. More recently it has been used to detect whether a compound binds to a protein target. In NMR spectroscopy, a compound is radiated with a short pulse of energy which excites the nuclei of specific atoms such as hydrogen, carbon, or nitrogen. Once the pulse of radiation has stopped, the excited nuclei slowly relax back to the ground state giving off energy as they do so. The time taken by different nuclei to give off this energy is called the **relaxation time** and this varies depending on the environment or position of each atom in the molecule. Therefore, a different signal will be obtained for each atom in the molecule and a spectrum is obtained which can be used to determine the structure.

The size of the molecule also plays an important role in the length of the relaxation time. Small molecules such as drugs have long relaxation times, whereas large molecules such as proteins have short relaxation times. Therefore it is possible to delay the measurement of energy emission, such that only small molecules are detected. This is the key to the detection of binding interactions between a protein and a test compound.

First of all, the NMR spectrum of the drug is taken, then the protein is added and the spectrum is re-run, introducing a delay in the measurement such that the protein signals are not detected. If the drug fails to bind to the protein, then its NMR spectrum will still be detected. If the drug bind to the protein, it essentially becomes part of the protein. As a result, its nuclei will have a shorter relaxation time and no NMR spectrum will be detected.

This screening method can also be applied to a mixture of compounds arising from a natural extract or from a combinatorial synthesis. If any of the compounds present bind to the protein, its relaxation time is shortened and so signals due to that compound will disappear from the spectrum. This will show that a component of the mixture is active and identify whether it is worthwhile separating the mixture or not.

There are several advantages in using NMR as a detection system:

- It is possible to screen 1000 small-molecular-weight compounds a day with one machine.

- The method can detect weak binding, which would be missed by conventional screening methods.

- It can identify the binding of small molecules to different regions of the binding site (section 9.4.10).

- It is complementary to HTS. The latter may give false-positive results, but these can be checked by NMR to ensure that the compounds concerned are binding in the correct binding site (section 9.4.10).

- The identification of weakly binding molecules allows the possibility of using them as building blocks for the construction of larger molecules that bind more strongly (section 9.4.10).

- Screening can be done on a new protein without needing to know its function.

NMR screening also has limitations, the main one being that at least 200 mg of the protein is required.

9.3.7 Affinity screening

A nice method of screening mixtures of compounds for active constituents is to take advantage of the binding affinity of compounds for the target. This not only detects the presence of such agents, but picks them out from the mixture. For example, the vancomycin family of antibacterial agents have a strong binding affinity for the dipeptide D-Ala-D-Ala (section 16.5.5.2). D-Ala-D-Ala was linked to sepharose resin and the resin was mixed with extracts from various microbes, which were known to have antibacterial activity. If an extract lost antibacterial activity as a result of this operation, it indicated that active compounds had bound to the resin. The resin could then be filtered off, and by changing the pH, the compounds could be released from the resin for identification.

9.3.8 Surface plasmon resonance

Surface plasmon resonance (SPR) is an optical method of detecting when a ligand binds to its target. The procedure is patented by Pharmacia Biosensor as **BIAcore** and makes use of a dextran-coated, gold-surfaced glass chip (Fig. 9.4). A ligand that is known to bind to the target is immobilized by linking it covalently to the dextran matrix, which is in a flow of buffer solution. Monochromatic, plane-polarized light is shone at an angle of incidence (α) from below the glass plate and is reflected back at the interface between the dense gold-coated glass and the less dense buffer solution. However, a component of the light called the evanescent wave penetrates a distance of about one wavelength into the buffer/dextran matrix. Normally, all of the light including the evanescent wave is reflected back, but if the gold film is very thin (a fraction of the evanescent wavelength), and the angle of incidence is exactly right, the evanescent wave interacts with free oscillating electrons called **plasmons** in the metal film. This is the surface plasmon resonance. Energy from the incident light is then lost to the gold film. As a result, there is a decrease in the reflected light intensity, which can be measured.

The angle of incidence when SPR occurs depends crucially on the refractive index of the buffer solution close to the metal film surface. This means that if the refractive index of the buffer changes, the angle of incidence at which SPR takes place changes as well. If the macromolecular target for the immobilized ligand is now introduced into the buffer flow, some of it will be bound by the immobilized ligand. This leads to a change of refractive index in the buffer solution close to the metal-coated surface, which can be detected by measuring the change in the angle of incidence required to get SPR. The technique allows the detection

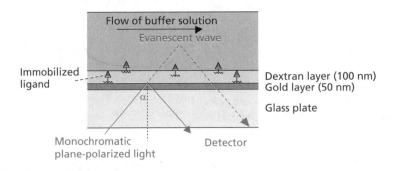

Figure 9.4 Surface plasmon resonance.

of ligand–target binding and can also be used to measure rate and equilibrium binding constants.

Suppose now we want to test whether a novel compound is binding to the target. This can be tested by introducing the novel compound into the buffer flow along with the target. If the test compound *does* bind to the target, less target will be available to bind to the immobilized ligands, so there will be a different change in refractive index and the change in the angle of incidence will also be different.

9.3.9 Scintillation proximity assay

Scintillation proximity assay (SPA) is a visual method of detecting whether a ligand binds to a target. It involves the immobilization of the target by linking it covalently to beads which are coated with a scintillant. A solution of a known ligand labelled with ^{125}I is then added to the beads. When the labelled ligand binds to the immobilized target, the ^{125}I acts as an energy donor and the scintillant-coated bead acts as an energy acceptor, resulting in an emission of light which can be detected. In order to find out whether a novel compound interacts with the target, the compound is added to the solution of the labelled ligand and the mixture is added to the beads. Successful binding by the novel compound will mean that less of the labelled ligand will bind, resulting in a reduction in the emission of light.

KEY POINTS

- Pharmaceutical companies tend to concentrate on developing drugs for diseases which are prevalent in developed countries, and aim to produce compounds with better properties than existing drugs.

- A molecular target is chosen which is believed to influence a particular disease when affected by a drug. The greater the selectivity that can be achieved, the less chance of side effects.

- A suitable bioassay must be devised which will demonstrate whether a drug has activity against a particular target. Bioassays can be carried out *in vitro* or *in vivo* and usually a combination of tests is used.

- HTS involves the miniaturization and automation of *in vitro* tests such that a large number of tests can be carried out in a short period of time.

- Compounds can be tested for their affinity to a macromolecular target by NMR spectroscopy. The relaxation times of ligands bound to a macromolecule are shorter than when they are unbound.

- SPR and SPA are two visual methods of detecting whether ligands bind to macromolecular targets.

9.4 Finding a lead compound

Once a target and a testing system have been chosen, the next stage is to find a **lead compound**—a compound which shows the desired pharmaceutical activity. The level of activity may not be very great and there may be undesirable side effects, but the lead compound provides a start for the drug design and development process. There are various ways in which a lead compound might be discovered as described in the following sections.

9.4.1 Screening of natural products

Natural products are a rich source of biologically active compounds. Many of today's medicines are either obtained directly from a natural source or were developed from a lead compound originally obtained from a natural source. Usually, the natural source has some form of biological activity and the compound responsible for that activity is known as the **active principle.** Such a structure can act as a lead compound. Most biologically active natural products are **secondary metabolites** with quite complex structures. This has an advantage in that they are extremely novel compounds. Unfortunately, this complexity also makes their synthesis difficult and the compound usually has to be extracted from its natural source—a slow, expensive, and inefficient process. As a result, there is usually an advantage in designing simpler analogues (section 10.3.8).

Many natural products have radically new chemical structures which no chemist would dream of synthesizing. For example, the antimalarial drug **artemisinin** (Fig. 9.5) is a natural product with an extremely unstable looking trioxane ring—one of the most unlikely structures to have appeared in recent years.

9.4.1.1 The plant kingdom

Plants have always been a rich source of lead compounds (e.g. **morphine, cocaine, digitalis, quinine,**

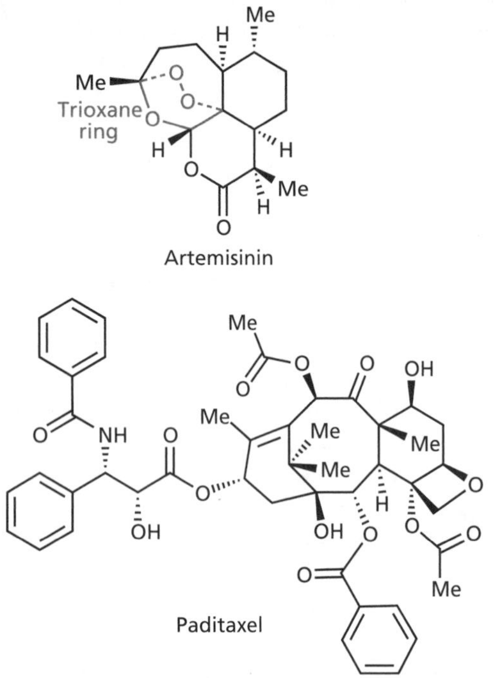

Artemisinin

Paditaxel

Figure 9.5 Plant natural products as drugs.

tubocurarine, nicotine, and **muscarine**). Many of these lead compounds are useful drugs in themselves (e.g. morphine and quinine), and others have been the basis for synthetic drugs (e.g. local anaesthetics developed from cocaine). Plants still remain a promising source of new drugs and will continue to be so. Clinically useful drugs which have recently been isolated from plants include the anticancer agent **paclitaxel (Taxol)** from the yew tree, and the antimalarial agent artemisinin from a Chinese plant (Fig. 9.5).

Plants provide a bank of rich, complex, and highly varied structures which are unlikely to be synthesized in laboratories. Furthermore, evolution has already carried out a screening process whereby plants are more likely to survive if they contain potent compounds which deter animals or insects from eating them. Considering the debt medicinal chemistry owes to the natural world, it is sobering to think that very few plants have been fully studied and the vast majority have not been studied at all. The rainforests of the world are particularly rich in plant species which have still to be discovered, let alone studied. Who knows how many exciting new lead compounds await discovery for the fight against cancer, AIDS, or any of the other myriad of human afflictions? This is one

reason why the destruction of rainforests and other ecosystems is so tragic: once these ecosystems are destroyed, unique plant species are lost to medicine for ever. For example, **silphion**—a plant cultivated near Cyrene in North Africa and famed as a contraceptive agent in ancient Greece—is now extinct. It is certain that many more useful plants have become extinct without medicine ever being aware of them.

9.4.1.2 The microbial world

Microorganisms such as bacteria and fungi have also provided rich pickings for drugs and lead compounds. These organisms produce a large variety of antimicrobial agents which have evolved to give their hosts an advantage over their competitors in the microbiological world. The screening of microorganisms became highly popular after the discovery of **penicillin.** Soil and water samples were collected from all round the world in order to study new fungal or bacterial strains, leading to an impressive arsenal of antibacterial agents such as the **cephalosporins, tetracyclines, aminoglycosides, rifamycins,** and **chloramphenicol** (Chapter 16). Although most of the drugs derived from microorganisms are used in antibacterial therapy, some microbial metabolites have provided lead compounds in other fields of medicine. For example, **asperlicin**—isolated from *Aspergillus alliaceus*—is a novel antagonist of a peptide hormone called **cholecystokinin** (CCK) which is involved in the control of appetite. CCK also acts as a neurotransmitter in the brain and is thought to be involved in panic attacks. Analogues of asperlicin may therefore have potential in treating anxiety.

Other examples include the fungal metabolite **lovastatin,** which was the lead compound for a series of drugs that lower cholesterol levels, and another fungal metabolite called **ciclosporin** (Fig. 9.6) which is used to suppress the immune response after transplantation operations.

9.4.1.3 The marine world

In recent years, there has been great interest in finding lead compounds from marine sources. Coral, sponges, fish, and marine microorganisms have a wealth of biologically potent chemicals with interesting inflammatory, antiviral, and anticancer activity. For example, **curacin A** (Fig. 9.7) is obtained from a marine cyanobacterium and shows potent antitumour activity. Other antitumour agents derived from marine sources include **eleutherobin, discodermolide, bryostatins,**

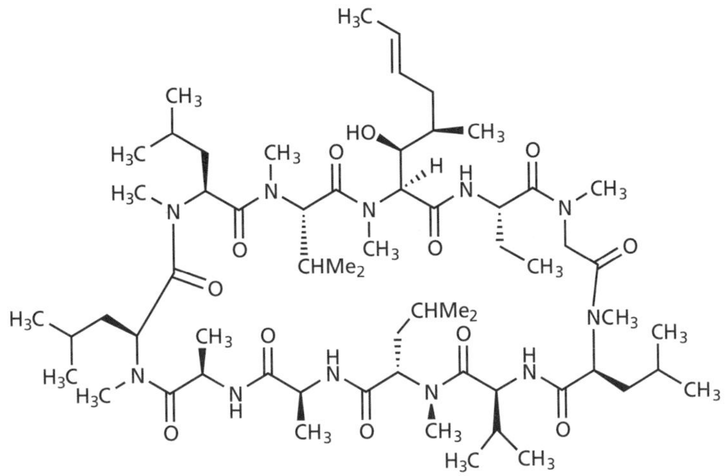

Figure 9.6 Ciclosporin.

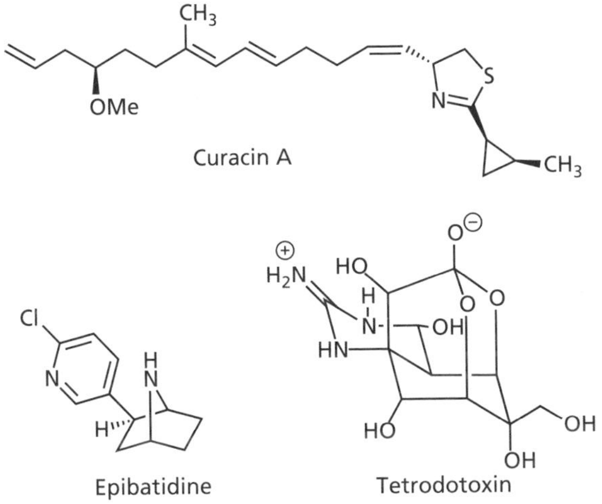

Figure 9.7 Natural products as drugs.

dolostatins, and **cephalostatins** (sections 18.5.2 and 18.8.2).

9.4.1.4 Animal sources

Animals can sometimes be a source of new lead compounds. For example, a series of antibiotic peptides were extracted from the skin of the African clawed frog and a potent analgesic compound called **epibatidine** (Fig. 9.7) was obtained from the skin extracts of the Ecuadorian poison frog.

9.4.1.5 Venoms and toxins

Venoms and toxins from animals, plants, snakes, spiders, scorpions, insects, and microorganisms are extremely potent because they often have very specific interactions with a macromolecular target in the body. As a result, they have proved important tools in studying receptors, ion channels, and enzymes. Many of these toxins are polypeptides (e.g. α-**bungarotoxin** from cobras). However, non-peptide toxins such as **tetrodotoxin** from the puffer fish (Fig. 9.7) are also extremely potent.

Venoms and toxins have been used as lead compounds in the development of novel drugs. For example, **teprotide,** a peptide isolated from the venom of the Brazilian viper, was the lead compound for the development of the antihypertensive agents **cilazapril** and **captopril** (section 10.3.11.2).

The neurotoxins from *Clostridium botulinum* are responsible for serious food poisoning (**botulism**), but they have a clinical use as well. They can be injected into specific muscles (such as those controlling the eyelid) to prevent muscle spasm. These toxins prevent cholinergic transmission (Chapter 19) and could well prove a lead for the development of novel anticholinergic drugs.

9.4.2 Medical folklore

In the past, ancient civilizations depended greatly on local flora and fauna for their survival. They would experiment with various berries, leaves, and roots to find out what effects they had. As a result, many brews were claimed by the local healer or shaman to have some medicinal use. More often than not, these concoctions were useless or downright dangerous, and if they worked at all, it was because the patient willed them to work—a **placebo effect**. However, some of these extracts may indeed have a real and beneficial effect, and a study of medical folklore can give clues as to which plants might be worth studying in more detail. **Rhubarb** root has been used as a purgative for many centuries,. In China, it was called 'The General' because of its 'galloping charge'! The most significant chemicals in rhubarb root are anthraquinones, which were used as the lead compounds in the design of the laxative **dantron** (Fig. 9.8).

The ancient records of Chinese medicine also provided the clue to the novel antimalarial drug **artemisinin** mentioned in section 9.4.1 above. The therapeutic properties of the opium poppy (active principle **morphine**) were known in Ancient Egypt, as were those of the *Solanaceae* plants in ancient Greece (active principles **atropine** and **hyoscine**; section 19.11.2). The snakeroot plant was well regarded in India (active principle **reserpine;** Fig. 9.8), and herbalists in medieval England used extracts from the willow tree (**salicin;** Fig. 9.8) and foxglove (active principle **digitalis**—a mixture of compounds such as digitoxin, digitonin, and digitalin). The Aztec and Mayan cultures of South America used extracts from a variety of bushes and trees including the ipecacuanha root (active principle **emetine;** Fig. 9.8), coca bush (active principle **cocaine**), and cinchona bark (active principle **quinine**).

9.4.3 Screening synthetic compound 'libraries'

The thousands of compounds which have been synthesized by the pharmaceutical companies over the years are another source of lead compounds. The vast majority of these compounds have never made the market place, but they have been stored in compound 'libraries' and are still available for testing. Pharmaceutical companies often screen their library of compounds whenever they study a new target. However, it has to be said that the vast majority of these compounds are merely variations on a theme; for example, 1000 or so different penicillin structures. This reduces the chances of finding a novel lead compound.

Pharmaceutical companies often try to diversify their range of structures by purchasing novel compounds prepared by research groups elsewhere—a useful source of revenue for hard-pressed university departments! These compounds may never have been synthesized with medicinal chemistry in mind and may be intermediates in a purely synthetic research project, but there is always the chance that they may have useful biological activity.

It can also be worth testing synthetic intermediates. For example, a series of thiosemicarbazones were synthesized and tested as antitubercular agents in the 1950s. They included isonicotinaldehyde thiosemicarbazone, the synthesis of which involved

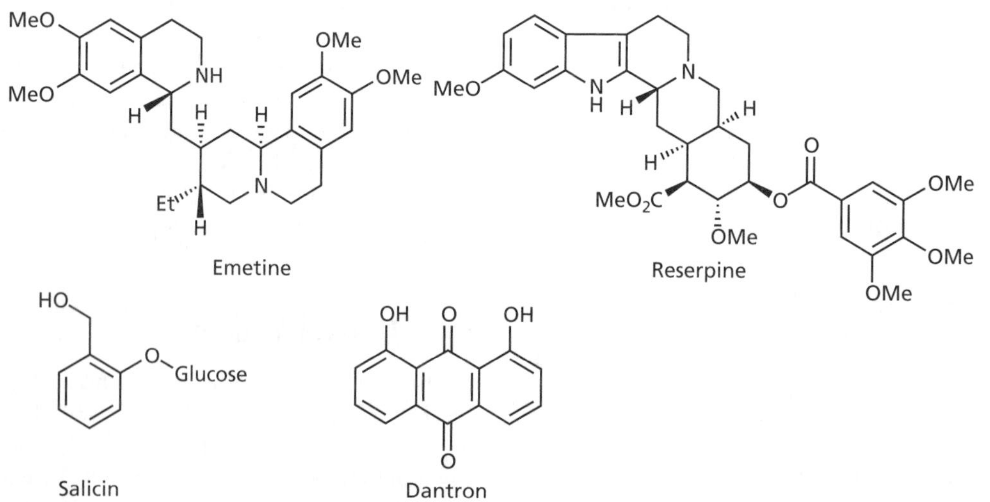

Figure 9.8 Active compounds resulting from studies of herbs and potions.

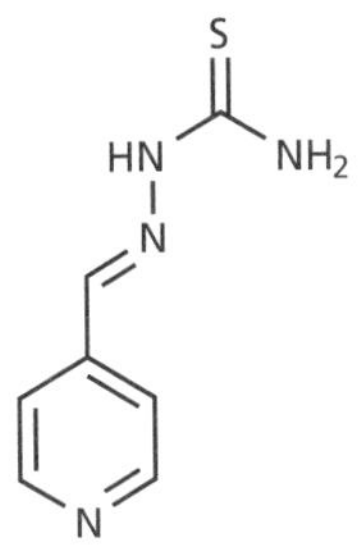

Figure 9.9 Pharmaceutically active compounds discovered from synthetic intermediates.

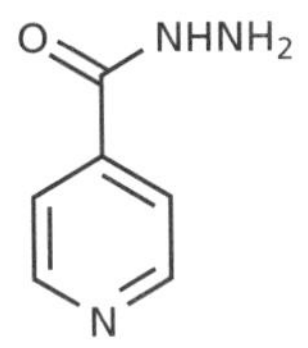

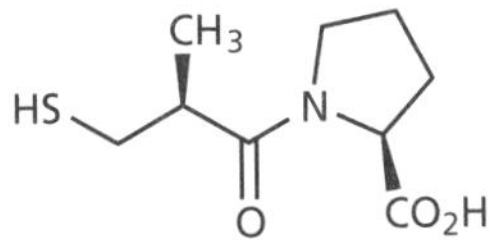

Figure 9.10 Captopril and 'me too' drugs.

the hydrazide structure **isoniazid** (Fig. 9.9) as a synthetic intermediate. It was subsequently found that isoniazid had greater activity than the target structure. Similarly, a series of **quinoline-3-carboxamide** intermediates (Fig. 9.9) were found to have antiviral activity.

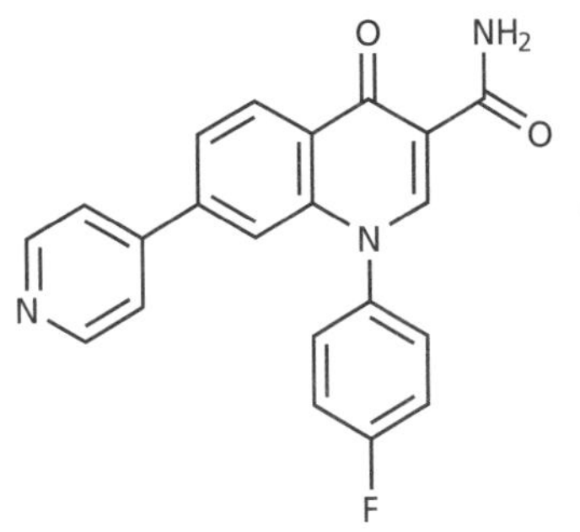

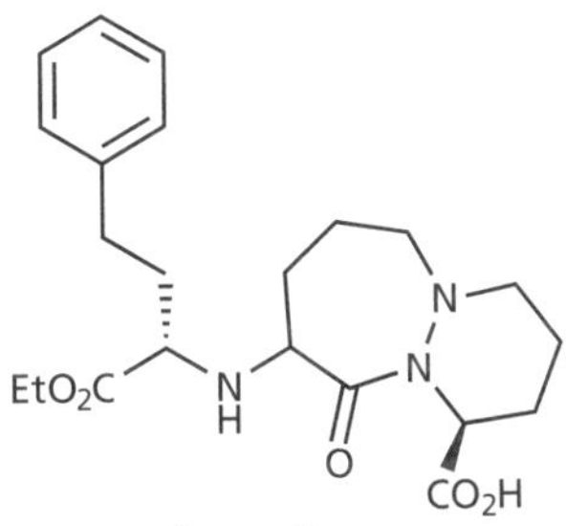

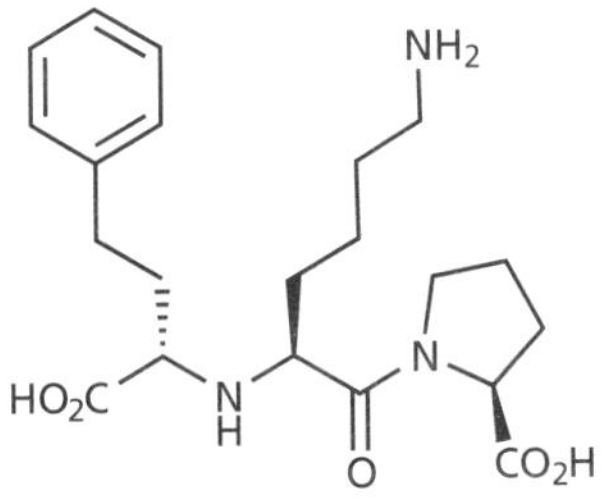

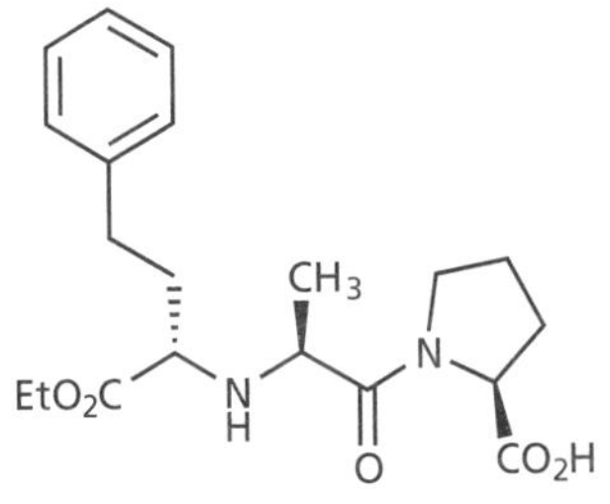

9.4.4 Existing drugs

9.4.4.1 'Me too' drugs

Many companies use established drugs from their competitors as lead compounds in order to design a drug that gives them a foothold in the same market area. The aim is to modify the structure sufficiently such that it avoids patent restrictions, retains activity, and ideally has improved therapeutic properties. For example, the antihypertensive drug captopril was used as a lead compound by various companies to produce their own antihypertensive agents (Fig. 9.10).

Although often disparaged as 'me too' drugs, they can often offer improvements over the original drug. For example, modern penicillins are more selective, more potent, and more stable than the original penicillins.

9.4.4.2 Enhancing a side effect

An existing drug may have a minor property or an undesirable side effect which might be of use in another area of medicine. As such, the drug could act as a lead compound on the basis of its side effects. The aim would then be to enhance the desired side effect and to eliminate the major biological activity.

For example, most sulfonamides were used as antibacterial agents. However, some sulfonamides with antibacterial activity could not be used clinically because they had convulsive side effects brought on by **hypoglycaemia** (lowered glucose levels in the blood). Clearly, this is an undesirable side effect for an antibacterial agent, but the ability to lower blood glucose levels would be useful in the treatment of diabetes. Therefore, structural alterations were made to the sulfonamides concerned in order to eliminate the antibacterial activity and to enhance the hypoglycaemic activity. This led to the antidiabetic agent **tolbutamide** (Fig. 9.11).

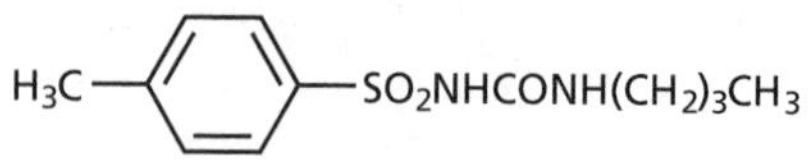

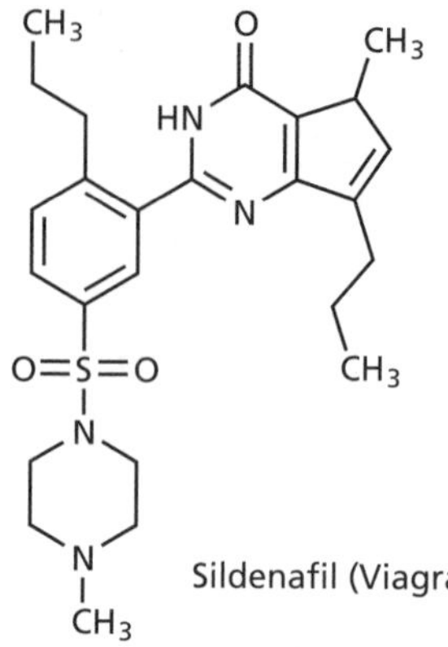

Figure 9.11 Tolbutamide and sildenafil (Viagra).

BOX 9.3 ENHANCING A SIDE EFFECT

Several drugs have been developed by enhancing the side effect of another drug. **Chlorpromazine** is used as a neuroleptic agent in psychiatry, but was developed from the antihistamine agent **promethazine**. This might appear an odd thing to do, but it is known that promethazine has sedative side effects and so medicinal chemists modified the structure to enhance the sedative effects at the expense of antihistamine activity. Similarly, the development of sulfonamide diuretics such as **chlorothiazide** arose from the observation that **sulfanilamide** has a diuretic effect in large doses (due to its action on an enzyme called **carbonic anhydrase**).

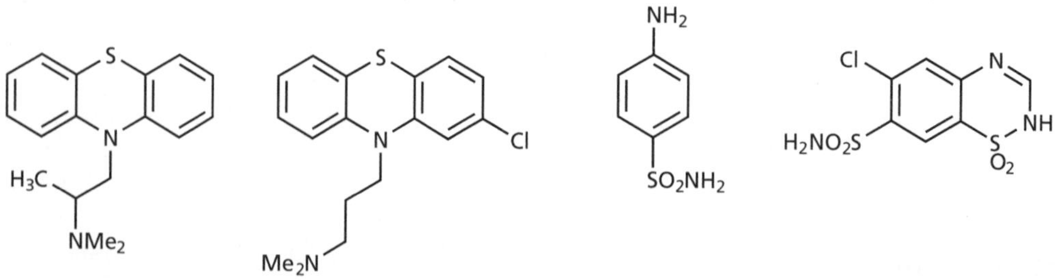

Drugs developed by enhancing a side effect.

In some cases, the side effect may be strong enough that the drug can be used without modification. For example, the anti-impotence drug **sildenafil** (Viagra) (Fig. 9.11) was originally designed as a vasodilator to treat angina and hypertension. During clinical trials, it was discovered that it acted as a vasodilator more effectively in the penis than in the heart, resulting in increased erectile function. The drug is now used to treat erectile dysfunction and sexual impotence. Another example is the antidepressant drug **bupropion**. Patients taking this drug reported that it helped them give up smoking and so the drug is now marketed as an antismoking aid (Zyban).

The moral of the story is that an unsatisfactory drug in one field may provide the lead compound in another field. Furthermore, one can fall into the trap of thinking that a structural group of compounds all have the same type of biological activity. The sulfonamides are generally thought of as antibacterial agents, but we have seen that they can have other properties as well (see Box 9.3).

9.4.5 Starting from the natural ligand or modulator

9.4.5.1 Natural ligands for receptors

The natural ligand of a target receptor has sometimes been used as the lead compound. The natural neurotransmitters **adrenaline** and **noradrenaline** were the starting points for the development of adrenergic β-agonists such as **salbutamol, dobutamine,** and **xamoterol** (section 20.10), and 5-hydroxytryptamine (5-HT) was the starting point for the development of the 5-HT$_1$ agonist **sumatriptan** (Fig. 9.12).

The natural ligand of a receptor can also be used as the lead compound in the design of an antagonist. For example, **histamine** was used as the original lead compound in the development of the H$_2$ histamine antagonist **cimetidine** (section 22.2). Turning an agonist into an antagonist is frequently achieved by adding extra binding groups to the lead structure. Other examples include the development of the adrenergic antagonist **pronethalol** (section 20.11), the H$_2$ antagonist **burimamide** (section 22.2), and the 5-HT$_3$ antagonists **ondansetron** and **granisetron** (Box. 9.2).

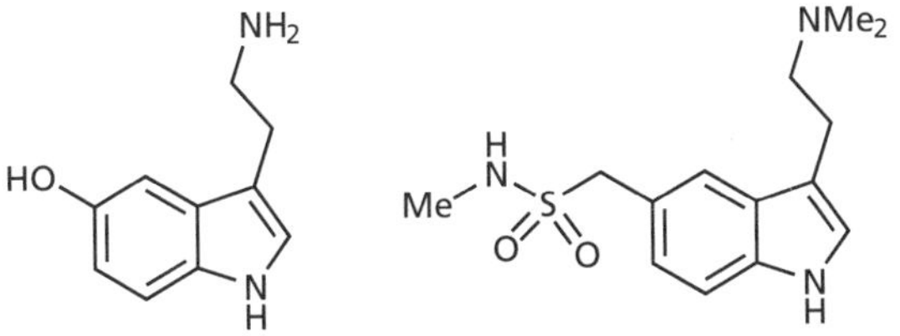

Sometimes the natural ligand for a receptor is not known (an **orphan receptor**) and the search for it can be a major project in itself. If the search is successful, however, it opens up a brand-new area of drug design (see Box 9.4). For example, the identification of the opiate receptors for **morphine** led to a search for endogenous opiates (natural body painkillers) which eventually led to the discovery of **endorphins, enkephalins**, and **endomorphins** as the natural ligands and their use as lead compounds (section 21.6).

9.4.5.2 Natural substrates for enzymes

The natural substrate for an enzyme can be used as the lead compound in the design of an enzyme inhibitor. For example, **enkephalins** have been used as lead compounds for the design of enkephalinase inhibitors. **Enkephalinases** are enzymes which metabolize enkephalins and their inhibition should prolong the activity of enkephalins (section 21.6.3).

BOX 9.4 NATURAL LIGANDS AS LEAD COMPOUNDS

The discovery of **cannabinoid** receptors in the early 1990s led to the discovery of two endogenous cannabinoid messengers—**arachidonylethanolamine** (**anandamide**) and **2-arachidonyl glycerol**. These have now been used as lead compounds for developing agents that will interact with cannabinoid receptors. Such agents may prove useful in suppressing nausea during chemotherapy or in stimulating appetite in AIDS patients.

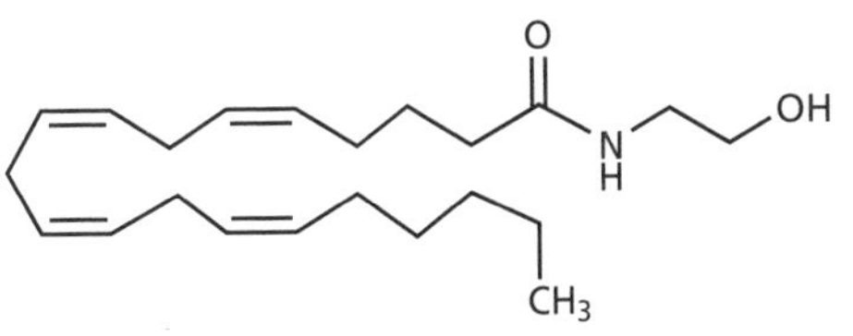

Anandamide.

5-Hydroxytryptamine   Sumatriptan (5-HT$_1$ agonist)

Figure 9.12 5-HT and sumatriptan.

9.4.5.3 Enzyme products as lead compounds

It should be remembered that enzymes catalyse a reaction in both directions and so the product of an enzyme-catalysed reaction can also be used as a lead compound for an inhibitor. For example, the design of the carboxypeptidase inhibitor L-**benzylsuccinic acid** was based on the products arising from the carboxypeptidase-catalysed hydrolysis of peptides (section 10.3.11.2).

9.4.5.4 Natural modulators as lead compounds

Many receptors and enzymes are under allosteric control (sections 4.6 and 5.7.2). The natural or endogenous chemicals that exert this control (modulators) could also serve as lead compounds.

In some cases, a modulator for an enzyme or receptor is suspected but has not yet been found. For example, the **benzodiazepines** are synthetic compounds that modulate the receptor for γ-**aminobutyric acid** (GABA) by binding to an allosteric binding site. The natural modulators for this allosteric site were not known at the time benzodiazepines were synthesized, but endogenous peptides called **endozepines** have since been discovered which bind to the same allosteric binding site and which may serve as lead compounds for novel drugs having the same activity as the benzodiazepines.

9.4.6 Combinatorial synthesis

The growing number of potentially new drug targets arising from genomic and proteomic projects has meant that there is an urgent need to find new lead compounds to interact with them. Unfortunately, the traditional sources of lead compounds have not managed to keep pace and in the last decade or so, research groups have invested greatly in a process called combinatorial synthesis in order to tackle this problem. Combinatorial synthesis is an automated solid-phase procedure aimed at producing as many different structures as possible in as short a time as possible. The reactions are carried out on very small scale, often in a way that will produce mixtures of compounds. In a sense, combinatorial synthesis aims to mimic what plants do, i.e. produce a pool of chemicals, one of which may prove to be a useful lead compound. Combinatorial science has developed so swiftly that it is almost a branch of chemistry in itself and a separate chapter is devoted to it (Chapter 14).

9.4.7 Computer-aided design

A detailed knowledge of a target binding site significantly aids in the design of novel lead compounds intended to bind with that target. In cases where enzymes or receptors can be crystallized, it is possible to determine the structure of the protein and its binding site by **X-ray crystallography**. Molecular modelling software programs can then be used to study the binding site, and to design molecules which will fit and bind to the site—*de novo* design (section 15.15).

In some cases, the enzyme or receptor cannot be crystallized and so X-ray crystallography cannot be carried out. However, if the structure of an analogous protein has been determined, this can be used as the basis for generating a computer model of the protein. This is covered in more detail in section 15.14. NMR spectroscopy has also been effective in determining the structure of proteins and can be used on proteins that cannot be studied by X-ray crystallography.

9.4.8 Serendipity and the prepared mind

Frequently, lead compounds are found as a result of serendipity (i.e. chance). However, it still needs someone with an inquisitive nature or a prepared mind to recognize the significance of chance discoveries and to take advantage of these events. The discovery of **cisplatin** (section 18.2.3) and **penicillin** (section 16.5.1) are two such examples, but there are many more (see Box 9.5).

Sometimes, the research carried out to improve a drug can have unexpected and beneficial spin offs. For example, **propranolol** and its analogues are effective β-blocking drugs (antagonists of β-adrenergic receptors) (section 20.11.3). However, they are also lipophilic, which means that they can enter the central nervous system and cause side effects. In an attempt to cut down entry into the central nervous system, it was decided to add a hydrophilic amide group to the molecule and so inhibit passage through the blood–brain barrier. One of the compounds made was **practolol**. As expected, this compound had fewer side effects acting on the central nervous system, but more importantly, it was found to be a selective antagonist for the β-receptors of the heart over β-receptors in other organs—a result that was highly

BOX 9.5 EXAMPLES OF SERENDIPITY

During the Second World War, an American ship carrying **mustard gas** exploded in an Italian harbour. It was observed that many of the survivors who had inhaled the gas lost their natural defences against microbes. Further study showed that their white blood cells had been destroyed. It is perhaps hard to see how a drug that weakens the immune system could be useful. However, there is one disease where this is the case—leukaemia. Leukaemia is a form of cancer which results in the excess proliferation of white blood cells, so a drug that kills these cells is potentially useful. As a result, a series of mustard-like drugs were developed based on the structure of the original mustard gas (section 18.2.3).

Another example involved the explosives industry, where it was quite common for workers to suffer severe headaches. These headaches resulted from dilatation of blood vessels in the brain, caused by handling trinitrotoluene (**TNT**). Once again, it is hard to see how such drugs could be useful. Certainly, the dilatation of blood vessels in the brain may not be particularly beneficial, but dilating the blood vessels in the heart could be useful in cardiovascular medicine. As a result, drugs were developed which dilated coronary blood vessels and alleviated the pain of angina.

Workers in the rubber industry found that they often acquired a distaste for **alcohol**! This was caused by an anti-oxidant used in the rubber manufacturing process which found its way into workers' bodies and prevented the normal oxidation of alcohol in the liver. As a result, there was a build up of **acetaldehyde** which was so unpleasant that workers preferred not to drink. The antioxidant became the lead compound for the development of **disulfiram (Antabuse)** used for the treatment of chronic alcoholism.

The following are further examples of lead compounds arising as a result of serendipity:

- **Clonidine** was originally designed to be a nasal vasoconstrictor to be used in nasal drops and shaving soaps. Clinical trials revealed that is caused a marked fall in blood pressure and so it became an important antihypertensive instead.

- **Imipramine** was synthesized as an analogue of chlorpromazine (Box 9.3), and was initially to be used as an antipsychotic. However, it was found to alleviate depression and this led to the development of a series of compounds classified as the **tricyclic antidepressants** (section 20.12.4).

- **Aminoglutethimide** was prepared as a potential anti-epileptic drug, but is now used as an anticancer agent (section 18.4.8).

- The anti-impotence drug **sildenafil (Viagra)** (Fig. 9.11) was discovered by chance from a project aimed at developing a new heart drug.

- **Isoniazid** (Fig. 9.9) was originally developed as an antituberculosis agent. Patients taking it proved remarkably cheerful and this led to the drug becoming the lead compound for a series of antidepressant drugs known as the **monoamine oxidase inhibitors** (MAOIs) (section 20.12.5).

- **Chlorpromazine** (Box 9.3) was synthesized as an antihistamine for possible use in preventing surgical shock, and was found to make patients relaxed and unconcerned. This led to the drug being tested in people with manic depression where it was found to have tranquillizing effects, resulting in it being the first of the neuroleptic drugs (major tranquillizers) used for schizophrenia.

- **Ciclosporin A** (Fig. 9.6) suppresses the immune system and is used during organ and bone marrow transplants to prevent the immune response rejecting the donor organs. The compound was isolated from a soil sample as part of a study aimed at finding new antibiotics. Fortunately, the compounds were more generally screened and ciclosporin A's immunosuppressant properties were identified.

- In a similar vein, the alkaloids **vincristine** and **vinblastine** (section 18.5) were discovered by chance when searching for compounds that could lower blood sugar levels. Vincristine is used in the treatment of Hodgkin's disease.

Drugs discovered by serendipity.

desirable, but not the one that was being looked for at the time.

Frequently, new lead compounds have arisen from research projects carried out in a totally different field of medicinal chemistry. This emphasizes the importance of keeping an open mind, especially when testing for biological activity. For example, we have already described the development of the anti-diabetic drug **tolbutamide** (section 9.4.4.2), based on the observation that some antibacterial sulfonamides could lower blood glucose levels.

9.4.9 Computerized searching of structural databases

New lead compounds can be found by carrying out computerized searches of structural databases. In order to carry out such a search, it is necessary to know the desired **pharmacophore** (section 10.2). This type of database searching is also known as **database mining** and is described in sections 15.11–15.13.

9.4.10 Designing lead compounds by NMR

So far we have described methods by which a lead compound can be discovered from a natural or synthetic source, but all these methods rely on an active compound being present. Unfortunately, there is no guarantee that this will be the case. Recently, NMR spectroscopy has been used to *design* a lead compound rather than to discover one (see Box 9.6). In essence, the method sets out to find small molecules (**epitopes**) which will bind to specific but different regions of a protein's binding site. These molecules will have no activity in themselves since they only bind to one part of the binding site, but if a larger molecule is designed which links these epitopes together, then a lead compound may be created which binds to the whole of the binding site and may have activity (Fig. 9.13).

Lead discovery by NMR can be applied to proteins of known structure which are labelled with ^{15}N such that each amide bond in the protein has an identifiable peak. To make identification of each peak easier, a 2D NMR spectrum is run matching ^{15}N with ^1H. Next, a range of low-molecular-weight compounds is screened to see whether any of them bind to a specific region of the binding site. Binding can be detected by observing a shift in any of the amide signals, which will not only show that binding is taking place, but will also reveal which part of the binding site is occupied. Once a compound (or ligand) has been found that binds to one region of the binding site, the process can be repeated to find another ligand that will bind to a different region of the binding site. This is usually done in the presence of the first ligand, to ensure that the second ligand does in fact bind to a distinct region. Once two ligands (or epitopes) have been identified, the structure of each can be optimized to find the best ligand for each of the binding regions, then a molecule can be designed where the two ligands are linked together.

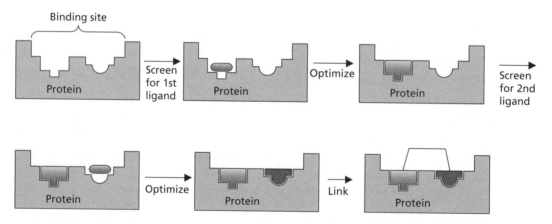

Figure 9.13 Epitope mapping.

There are several advantages to this approach. Since the individual ligands are optimized for each region of the binding site before synthesizing the overall structure, a lot of synthetic effort is spared. It is much easier to synthesize a series of small-molecular-weight compounds to optimize the interaction with specific parts of the binding site, than it is to synthesize a range of larger molecules to fit the overall binding site. A high level of diversity is also possible, as various combinations of fragments could be used.

Lead generation by NMR has advantages over combinatorial synthesis. In the latter, a large number of linked molecules are made from various building blocks requiring a large synthetic effort. In the former, the building blocks are optimized first and a small number of linked molecules are then prepared.

The method is limited to small protein targets (molecular weight >40 000) that can be obtained in quantities greater than 200 mg, and can be labelled with ^{15}N.

9.5 Isolation and purification

If the lead compound (or active principle) is present in a mixture of other compounds, whether the mixture be from a natural source or from a combinatorial synthesis (Chapter 14), it has to be isolated and purified. The ease with which the active principle can be isolated and purified depends very much on the structure, stability, and quantity of the compound. For example, Fleming recognized the antibiotic qualities of **penicillin** and its remarkable non-toxic nature to humans, but he disregarded it as a clinically useful drug because he was

---

BOX 9.6 THE USE OF NMR SPECTROSCOPY IN FINDING LEAD COMPOUNDS

NMR spectroscopy was used in the design of high-affinity ligands for the FK506 binding protein—a protein involved in the suppression of the immune response. Two optimized epitopes (A and B) were discovered, which bound to different regions of the binding site. Structure C was then synthesized,

where the two epitopes were linked by a propyl link. This compound had higher affinity than either of the individual epitopes and represents a lead compound for further development.

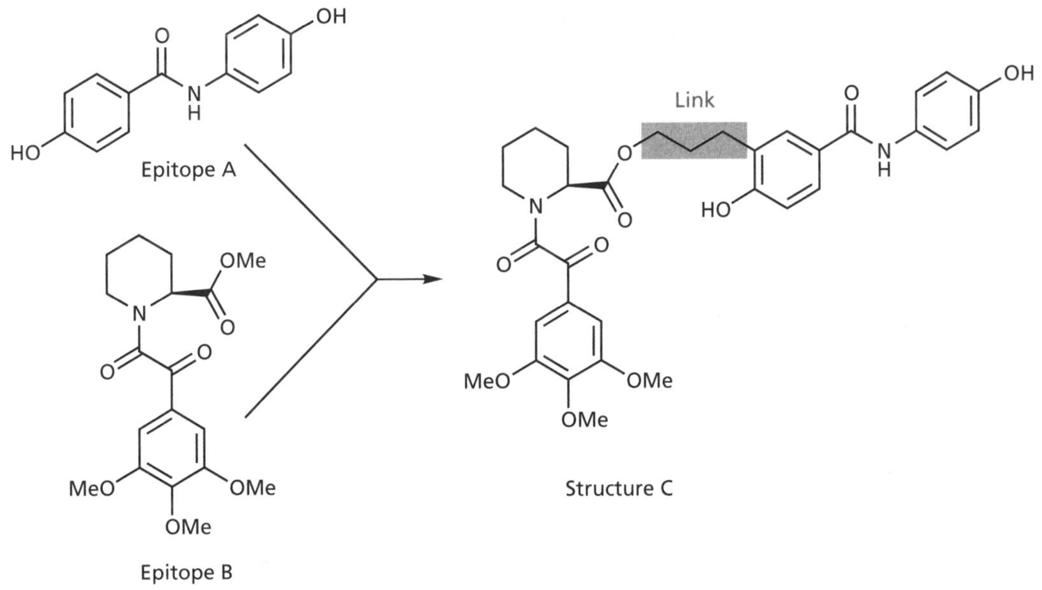

Design of a ligand for the FK 506 binding protein.

unable to purify it. He could isolate it in aqueous solution, but whenever he tried to remove the water, the drug was destroyed. It was not until the development of new experimental procedures such as freeze-drying and chromatography that the successful isolation and purification of penicillin and other natural products became feasible. A detailed description of the experimental techniques involved in the isolation and purification of compounds is outwith the scope of this textbook, and can be obtained from textbooks covering the practical aspects of chemistry.

9.6 Structure determination

It is sometimes hard for present-day chemists to appreciate how difficult structure determinations were before the days of NMR and IR spectroscopy. A novel structure which may now take a week's work to determine would have provided two or three decades of work in the past. For example, the microanalysis of **cholesterol** was carried out in 1888 to get its molecular formula, but its chemical structure was not fully established until an X-ray crystallographic study was carried out in 1932.

In the past, structures had to be degraded to simpler compounds, which were further degraded to recognizable fragments. From these scraps of evidence, a possible structure was proposed, but the only sure way of proving the proposal was to synthesize the structure and to compare its chemical and physical properties with those of the natural compound.

Today, structure determination is a relatively straightforward process and it is only when the natural product is obtained in minute quantities that a full synthesis is required to establish its structure. The most useful analytical techniques are **X-ray crystallography** and **NMR spectroscopy**. The former technique comes closest to giving a 'snapshot' of the molecule, but requires a suitable crystal of the sample. The latter technique is used more commonly as it can be carried out on any sample, whether it be a solid, oil, or liquid. There is a large variety of different NMR experiments that can be used to establish the structure of quite complex molecules. These include various 2D NMR experiments which involve a comparison of signals from different types of nuclei in the molecule

(e.g. carbon and hydrogen). Such experiments allow the chemist to build up a picture of the molecule atom by atom, and bond by bond.

In cases where there is not enough sample for an NMR analysis, mass spectrometry can be helpful. The fragmentation pattern can give useful clues about the structure, but it does not prove the structure. A full synthesis is still required as final proof.

KEY POINTS

- A lead compound is a structure which shows a useful pharmacological activity and can act as the starting point for drug design.

- Natural products are a rich source of lead compounds. The agent responsible for the biological activity of a natural extract is known as the active principle.

- Lead compounds have been isolated from plants, trees, microorganisms, animals, venoms, and toxins. A study of medical folklore indicates plants and herbs which may contain novel lead compounds.

- Lead compounds can be found by screening synthetic compounds obtained from combinatorial syntheses and other sources.

- Existing drugs can be used as lead compounds for the design of novel structures in the same therapeutic area. Alternatively, the side effects of an existing drug can be enhanced to design novel drugs in a different therapeutic area.

- The natural ligand, substrate, product, or modulator for a particular target can act as a lead compound.

- The ability to crystallize a molecular target allows the use of X-ray crystallography and molecular modelling to design lead compounds which will fit the relevant binding site.

- Serendipity has played a role in the discovery of new lead compounds.

- A knowledge of an existing drug's pharmacophore allows the computerized searching of structural databases to identify possible new lead compounds which share that pharmacophore.

- NMR spectroscopy can be used to identify whether small molecules (epitopes) bind to specific regions of a binding site. Epitopes can be optimized then linked together to give a lead compound.

- If a lead compound is present in a natural extract or a combinatorial synthetic mixture, it has to be isolated and purified such that its structure can be determined. X-ray crystallography and NMR spectroscopy are particularly important in structure determination.

9.7 Herbal medicine

We have described how useful drugs and lead compounds can be isolated from natural sources, so where does this place herbal medicine? Are there any advantages or disadvantages in using herbal medicines instead of the drugs developed from their active principles? There are no simple answers to this. Herbal medicines contain a large variety of different compounds, several of which may have biological activity, so there is a significant risk of side effects and even toxicity. The active principle is also present in small quantity, so the herbal medicine may be expected to be less active than the pure compound. Herbal medicines such as St John's wort can also interact with prescribed medicines (section 8.4.6), and in general there is a lack of regulation or control over their use.

Having said that, these same issues may actually be advantageous. If the herbal extract contains the active principle in small quantities, there is an inbuilt safety limit to the dose levels received. Different compounds within the extract may also have roles to play in the medicinal properties of the plant and enhance the effect of the active principle—a phenomenon known as **synergy.** Alternatively, some plant extracts have a wide variety of different active principles which act together to produce a beneficial effect. The **aloe plant** (the 'wand of heaven') is an example of this. It is a cactus-like plant found in the deserts of Africa and Arizona and has long been revered for its curative properties. Supporters of herbal medicine have proposed the use of aloe preparations to treat burns, irritable bowel syndrome, rheumatoid arthritis, asthma, chronic leg ulcers, itching, eczema, psoriasis, and acne, thus avoiding the undesirable side effects of long-term steroid use. The preparations are claimed to contain analgesic, anti-inflammatory, antimicrobial and many other agents which all contribute to the overall effect, and trying to isolate each active principle would detract from this. On the other hand, critics have stated that many of the beneficial effects claimed for aloe preparations have not been proven and that although the effects may be useful in some ailments, they are not very effective.

QUESTIONS

1. What is meant by target specificity and selectivity? Why is it important?

2. What are the advantages and disadvantages of natural products as lead compounds?

3. Fungi have been a richer source of antibacterial agents than bacteria. Suggest why this might be so.

4. Scuba divers and snorkellers are advised not to touch coral. Why do you think this might be? Why might it be of interest to medicinal chemists?

5. You are employed as a medicinal chemist and have been asked to initiate a research programme aimed at finding a drug which will prevent a novel tyrosine kinase receptor from functioning. There are no known lead compounds that have this property. What approaches can you make to establish a lead compound? (Consult section 6.7 to find out more about protein kinase receptors.)

6. A study was set up to look for agents that would inhibit the kinase active site of the epidermal growth factor receptor (section 6.7). Three assay methods were used: an assay carried out on a genetically engineered form of the protein that was water soluble and contained the kinase active site; a cell assay that measured total tyrosine phosphorylation in the presence of epidermal growth factor; and an *in vivo* study on mice that had tumours grafted onto their backs. How do you think these assays were carried out to measure the effect of an inhibitor? Why do you think three assays were necessary? What sort of information did they provide?

FURTHER READING

Bleicher, K. H. *et al.* (2003) Hit and lead generation: beyond high-throughput screening. *Nature Reviews Drug Discovery*, 2, 369–378.

Blundell, T. L., Jhoti, H., and Abell, C. (2002) High-throughput crystallography for lead discovery in drug design. *Nature Reviews Drug Discovery*, 1, 45–54.

Engel, L. W. and Straus, S. E. (2002) Development of therapeutics: opportunities within complementary and alternative medicine. *Nature Reviews Drug Discovery*, 1, 229–237.

Gershell, L. J. and Atkins, J. H. (2003) A brief history of novel drug discovery technologies. *Nature Reviews Drug Discovery*, 2, 321–327.

Hopkins, A. L. and Groom, C. R. (2002) The druggable genome. *Nature Reviews Drug Discovery*, 1, 727–730.

Lewis, R. J. and Garcia, M. L. (2003) Therapeutic potential of venom peptides. *Nature Reviews Drug Discovery*, 2, 790–802.

Lindsay, M. A. (2003) Target discovery. *Nature Reviews Drug Discovery*, 2, 831–838.

O'Shannessy, D. J. *et al.* (1993) Determination of rate and equilibrium binding constants for macromolecular interactions using surface plasmon resonance: use of non-linear least squares analysis methods. *Anal. Biochem.*, 212, 457–468.

Pellecchia, M., Sem, D. S., and Wuthrich, K. (2002) NMR in drug discovery. *Nature Reviews Drug Discovery*, 1, 211–219.

Sauter, G., Simon, R., and Hillan, K. (2003) Tissue micro-arrays in drug discovery. *Nature Reviews Drug Discovery*, 2, 962–972.

Shuker, S. B., Hajduk, P. J., Meadows, R. P., and Fesik, S. W. (1996) Discovering high-affinity ligands for proteins: SAR by NMR. *Science*, 274, 1531–1534.

Walters, W. P. and Namchuk, M. (2003) Designing screen: how to make your hits a hit. *Nature Reviews Drug Discovery*, 2, 259–266.

Titles for general further reading are listed on p. 711.

10 Drug design: optimizing target interactions

In Chapter 9, we looked at the various methods of discovering a lead compound. Once it *has* been discovered the lead compound can be used as the starting point for drug design. There are various aims in drug design. The eventual drug should have a good selectivity and level of activity for its target, and have minimal side effects. It should be easily synthesized and be chemically stable. Finally, it should have acceptable pharmacokinetic properties and be non-toxic. In this chapter, we concentrate on the design strategies that can be used to optimize the interaction of the drug with its target, in other words its **pharmacodynamic properties**. In Chapter 11, we look at the design strategies that can improve the drug's ability to reach its target and to have an acceptable lifetime, in other words its **pharmacokinetic properties**. Although these topics are in separate chapters, it would be wrong to think that they should be treated separately or in that order during a research programme. For example, it would be foolish to spend months or years perfecting a drug that interacts perfectly with its target, but has no chance of reaching that target because of adverse pharmacokinetic properties or metabolic instability. Pharmacodynamics and pharmacokinetics should have equal priority in influencing which strategies are used and which analogues are synthesized.

10.1 Structure–activity relationships

Once the structure of a lead compound is known, the medicinal chemist moves on to study its structure–activity relationships (SAR). The aim here is to discover which parts of the molecule are important to biological activity and which are not. If it is possible to crystallize the target with the drug bound to the binding site, the crystal structure of the complex could be solved by X-ray crystallography, then studied with molecular modelling software to identify important binding interactions. However, this may not be possible, either because the target structure cannot be crystallized or because the target structure has not been identified. If that is the case, it will be necessary to revert to the traditional method of synthesizing a selected number of compounds that vary slightly from the original structure, then studying what effect that has on the biological activity.

One can imagine the drug as a chemical knight entering the depths of a forest (the body) in order to make battle with an unseen dragon (the body's affliction) (Fig. 10.1). The knight (Sir Drugalot) is armed with a large variety of weapons and armour, but as his battle with the dragon goes unseen, it is impossible to tell which weapon he uses or whether his armour is essential to his survival. We only know of his success if he returns unscathed with the dragon slain. If the knight declines to reveal how he slew the dragon, then the only way to find out how he did it would be to remove some of his weapons and armour and to send him in against other dragons to see if he still succeeds.

As far as a drug is concerned, the weapons and armour are the various chemical functional groups and alkyl groups present in the structure that can bind to the drug's target (section 2.3), or assist and protect the drug on its journey through the body. Recognizing functional groups and the sort of intermolecular bonds they can form is important in understanding how a drug might bind to its target.

Figure 10.1

Let us imagine that we have isolated a natural product with the structure shown in Fig. 10.2. We shall name it glipine. There are a variety of functional groups present in the structure and the diagram shows the potential binding interactions that are possible with a target binding site. It is unlikely that all of these interactions take place, so we have to identify those that do. By synthesizing compounds (such as the examples shown in Fig. 10.3) where one particular functional group of the molecule is removed or altered, it is possible to find out which groups are essential and which are not. This involves testing all the analogues for biological activity and comparing them with the original compound. If an analogue shows a significantly lowered activity, then the group that has been modified must have been important. If the activity remains similar, then the group is not essential.

The ease with which this task is carried out depends on how easily we can synthesize the necessary analogues. It may be possible to modify some lead compounds directly to the required analogues, whereas the analogues of other lead compounds may best be prepared by total synthesis. Let us consider the binding interactions that are possible for different functional groups and the analogues that could be synthesized to establish whether they are involved in binding or not.

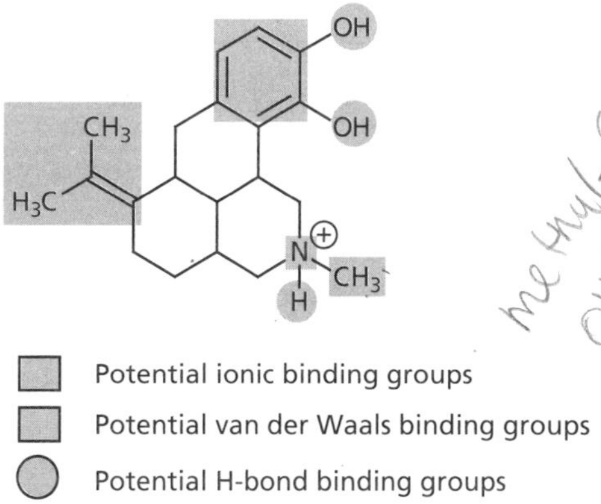

10.1.1 Binding role of alcohols and phenols

Alcohols and phenols are functional groups which are common in many drugs and are often involved in hydrogen bonding. The oxygen can act as a hydrogen bond acceptor, and the hydrogen can act as a hydrogen bond donor (Fig. 10.4). The directional preference for hydrogen bonding is indicated by the arrows in the figure, but it is important to realize that slight deviations are possible (section 2.3.2). One or all of these interactions may be important in binding the drug to the binding site. Synthesizing a methyl ether or an ester analogue would be relevant in testing this, as it is highly likely that the bonding would be disrupted in either analogue. Let us consider the methyl ether first.

There are two reasons why the ether might hinder or prevent the hydrogen bonding of the original alcohol or phenol. The obvious explanation is that the proton of the original hydroxyl group is involved as a hydrogen bond donor and if it is removed, the hydrogen bond is lost (Frames 1 and 2 in Fig. 10.5). However, suppose the oxygen atom is acting as a hydrogen bond acceptor (Frame 3 in Fig. 10.5). The oxygen is still present in the ether analogue, so could it still take part in hydrogen bonding? Well, it may, but possibly not to the same extent. The extra bulk of the methyl group should hinder the close approach that was previously attainable and should disrupt hydrogen bonding (Frame 4 in Fig. 10.5). The hydrogen bonding may not be completely destroyed, but we could reasonably expect it to be weakened.

The ester analogue cannot act as a hydrogen bond donor either. There is still the possibility of it acting as a hydrogen bond acceptor, but the extra bulk of the

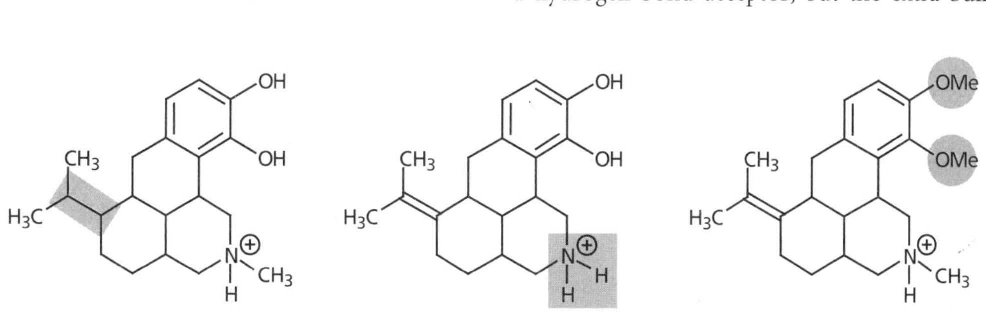

- ▢ Potential ionic binding groups
- ▢ Potential van der Waals binding groups
- ◯ Potential H-bond binding groups

Figure 10.2 Glipine.

Figure 10.3 Modifications of glipine.

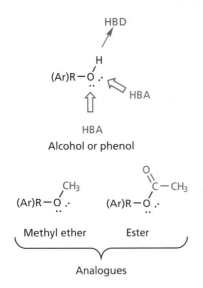

Figure 10.4 Possible hydrogen bonding interactions for an alcohol or phenol.

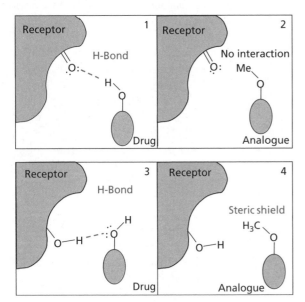

Figure 10.5 Possible hydrogen bond interactions.

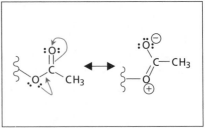

Electronic factor

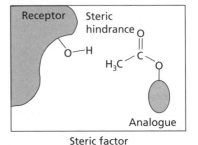

Steric factor

Figure 10.6 Factors by which an ester group can disrupt hydrogen bonding of the original hydroxyl group.

as a hydrogen bond acceptor. Of course, one could argue that the carbonyl oxygen may act as an alternative hydrogen bond acceptor, but it is in a different position relative to the rest of the molecule and it is unlikely that it would form a hydrogen bond to the same region of the binding site as the original oxygen.

It is relatively easy to acetylate alcohols and phenols to the corresponding esters and this was one of the early reactions that was carried out on natural products such as morphine (section 21.2.3).

Both phenols and alcohols can be converted to ethers and esters to see what effect that has on binding, but it should be remembered that phenols consist of an hydroxyl group linked to an aromatic ring. We have seen how an hydroxyl group can interact with a binding site, but it is also possible for the aromatic ring to take part in intermolecular interactions, as we shall now see.

10.1.2 Binding role of aromatic rings

Aromatic rings are planar, hydrophobic structures, commonly involved in van der Waals and hydrophobic interactions with flat hydrophobic regions of the binding site. An analogue containing a cyclohexane ring in place of the aromatic ring is less likely to bind so well, as the ring is no longer flat. The axial protons

acyl group is even greater than the methyl group of the ether, and this too should hinder the original hydrogen bonding interaction. There is also a difference between the electronic properties of an ester and an alcohol. The carboxyl group has a weak pull on the electrons from the neighbouring oxygen, giving the resonance structure shown in Fig. 10.6. Because the lone pair is involved in such an interaction, it will be less effective

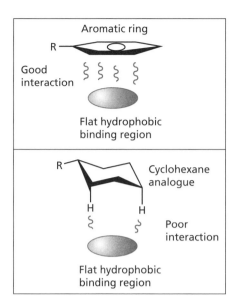

Figure 10.7 Binding comparison of an aromatic ring with a cyclohexyl ring.

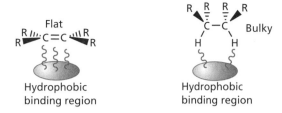

Fig. 10.8. Binding comparison of an alkene with an alkane.

can interact weakly, but they also serve as buffers to keep the rest of the cyclohexane ring at a distance (Fig. 10.7). The binding region for the aromatic ring may also be a narrow slot rather than a planar surface. In that scenario, the cyclohexane ring would be incapable of fitting into it, as it is a bulkier structure.

Although there are methods of converting aromatic rings to cyclohexane rings, they are unlikely to be successful with most lead compounds, and so such analogues would normally be prepared using a full synthesis.

Aromatic rings could also interact with an ammonium ion through induced dipole interactions or hydrogen bonding (sections 2.3.2 and 2.3.4). Such interactions would not be possible for the cyclohexyl analogue.

10.1.3 Binding role of alkenes

Like aromatic rings, alkenes are planar and hydrophobic so they too can interact with hydrophobic regions of the binding site through van der Waals and hydrophobic interactions. The activity of the equivalent saturated analogue would be worth testing, since the saturated alkyl region is bulkier and cannot approach the relevant region of the binding site so closely (Fig. 10.8). Alkenes are generally easier to reduce than aromatic rings, so it may be possible to

prepare the saturated analogue directly from the lead compound.

10.1.4 Binding role of ketones and aldehydes

A ketone group is not uncommon in many of the structures studied in medicinal chemistry. It is a planar group that can interact with a binding site through hydrogen bonding where the carbonyl oxygen acts as a hydrogen bond acceptor (Fig. 10.9). Two such interactions are possible, as two lone pairs of electrons are available on the carbonyl oxygen. The lone pairs are in sp^2 hybridized orbitals, which are in the same plane as the functional group. The carbonyl group also has a significant dipole moment and so a dipole–dipole interaction with the binding site is also possible.

It is relatively easy to reduce a ketone to an alcohol and it may be possible to carry out this reaction directly on the lead compound. This significantly changes the geometry of the functional group, from planar to tetrahedral. Such an alteration in geometry may well weaken any existing hydrogen bonding interactions and will certainly weaken any dipole–dipole interactions, as both the magnitude and orientation of the dipole moment will be altered (Fig. 10.10). If it was suspected that the oxygen present in the alcohol analogue might still be acting as a hydrogen bond acceptor then the ether or ester analogues could be studied as described above. Reactions are available that can reduce

Figure 10.9 Binding interactions that are possible for a carbonyl group.

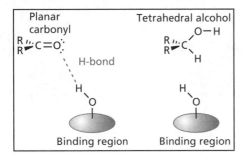

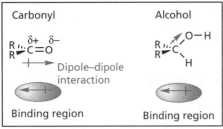

Figure 10.10 Effect on binding interactions following the reduction of a ketone or aldehyde.

a ketone completely to an alkane and remove the oxygen, but they are likely to be impractical for many of the lead compounds studied in medicinal chemistry.

Aldehydes are less common in drugs since they are more reactive and are susceptible to oxidation to carboxylic acids. However, they could interact in the same way as ketones, and similar analogues could be studied.

10.1.5 Binding role of amines

Amines are extremely important functional groups in medicinal chemistry and are present in many drugs. They may be involved in hydrogen bonding either as a hydrogen bond acceptor or a hydrogen bond donor (Fig. 10.11). The nitrogen atom has one lone pair of electrons and can act as a hydrogen bond acceptor for one hydrogen bond. Primary and secondary amines have N–H groups and can act as hydrogen bond donors. Aromatic and heteroaromatic amines act only as hydrogen bond donors, because the lone pair interacts with the aromatic or heteroaromatic ring.

In many cases, the amine may be protonated when it interacts with its target binding site, which means that it is ionized and cannot act as a hydrogen bond acceptor. However, it can still act as a hydrogen bond donor and will form a stronger hydrogen bond than if it was not ionized (Fig. 10.12). Alternatively, a strong

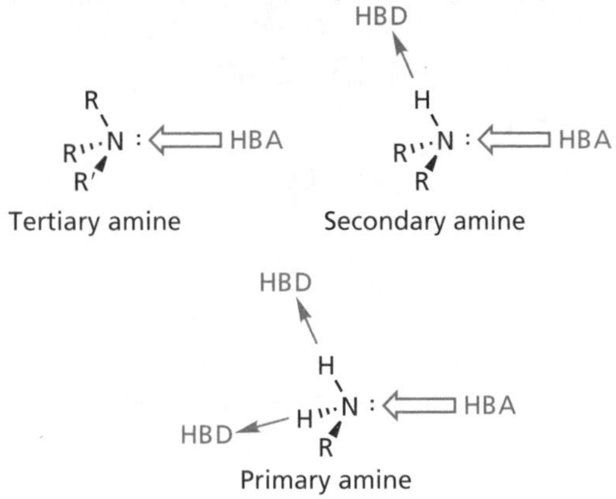

Figure 10.11 Possible binding interactions for amines.

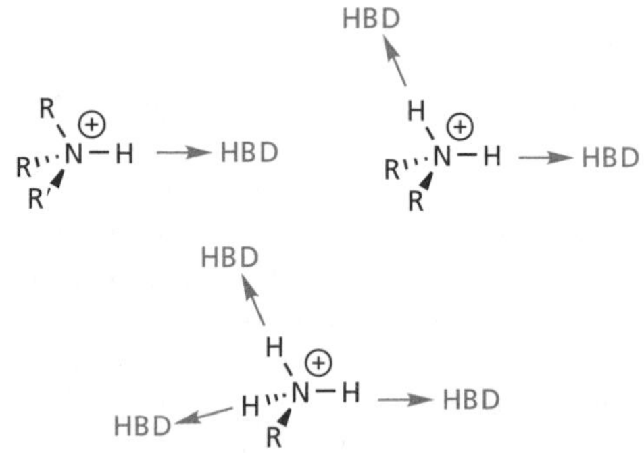

Figure 10.12 Possible hydrogen bonding interactions for ionized amines.

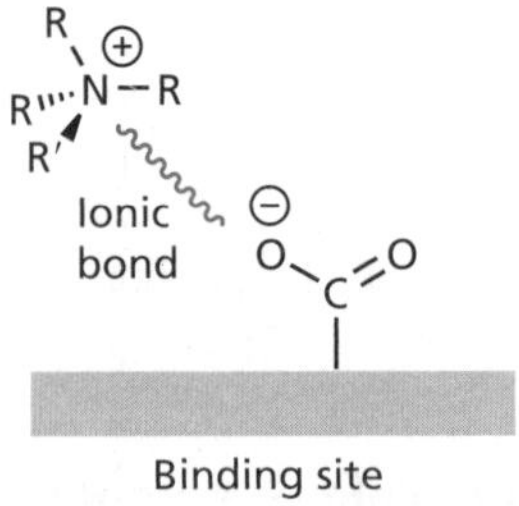

Figure 10.13 Ionic interaction between an ionized amine and a carboxylate ion.

ionic interaction may take place with a carboxylate ion in the binding site (Fig. 10.13).

To test whether ionic or hydrogen bonding interactions are taking place, an amide analogue could be studied. This will prevent the nitrogen acting as a hydrogen bond acceptor, as the nitrogen's lone pair

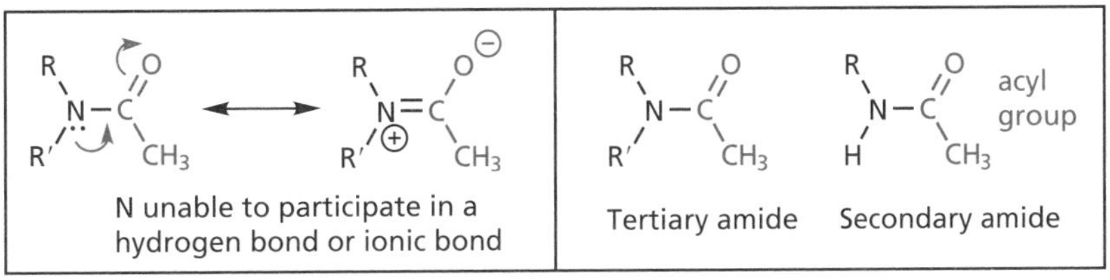

Figure 10.14 Interaction of the nitrogen lone pair with the neighbouring carbonyl group in amides.

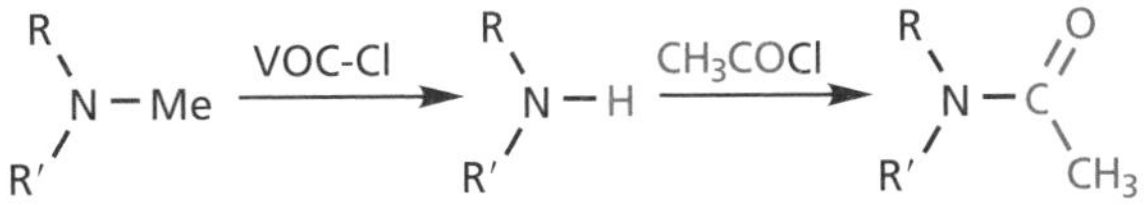

Figure 10.15 Dealkylation of a tertiary amine and formation of a secondary amide.

will interact with the neighbouring carbonyl group instead (Fig. 10.14). This interaction also prevents protonation of the nitrogen and rules out the possibility of ionic interactions. You might argue that the right-hand structure in Fig. 10.14 has a positive charge on the nitrogen and could still take part in an ionic interaction. However, this structure is not present as a distinct entity and the figure is only used to represent the fact that the lone pair can interact with the carbonyl group to produce double-bond character in the amide group. Any positive charge is extremely weak in comparison to that present in an ionized amine.

It is relatively easy to form secondary and tertiary amides from primary and secondary amines respectively, and it may be possible to carry out this reaction directly on the lead compound. A tertiary amide lacks the N–H group of the original secondary amine and would test whether this is involved as a hydrogen bond donor. The secondary amide formed from a primary amine still has an N–H group present, but the steric bulk of the acyl group should hinder it acting as a hydrogen bond donor.

Tertiary amines cannot be converted directly to amides, but if one of the alkyl groups is a methyl group it is often possible to remove it with vinyloxycarbonyl chloride (VOC–Cl) to form a secondary amine which could then be converted to the amide (Fig. 10.15). This demethylation reaction is extremely useful and has been used to good effect in the synthesis of morphine analogues (section 21.3.2).

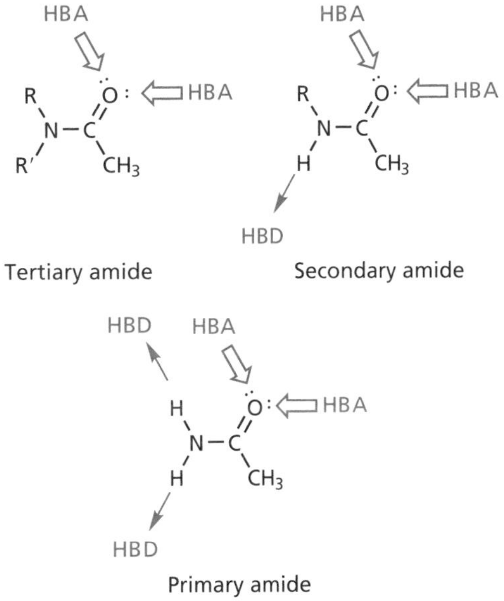

Figure 10.16 Possible hydrogen bonding interactions for amides.

10.1.6 Binding role of amides

Many of the lead compounds currently studied in medicinal chemistry are peptides or polypeptides, consisting of amino acids linked together by peptide or amide bonds (section 3.1). Amides are likely to interact with binding sites through hydrogen bonding (Fig. 10.16). The carbonyl oxygen atom can act as a hydrogen bond acceptor and has the potential to form

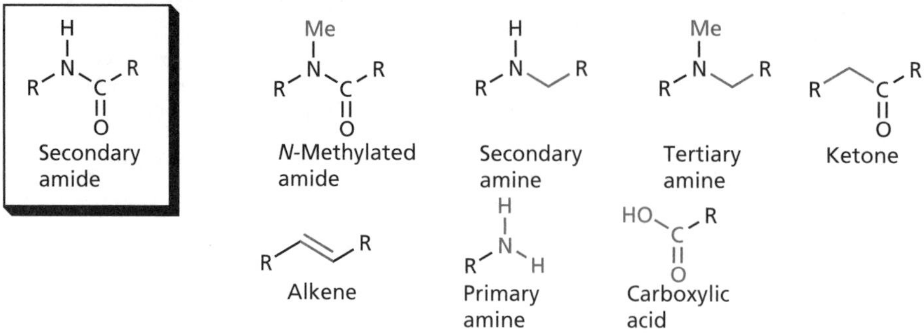

Figure 10.17 Possible analogues to test the binding interactions of a secondary amide.

two hydrogen bonds. Both the lone pairs involved are in sp² hybridized orbitals which are located in the same plane as the amide group. The nitrogen cannot act as a hydrogen bond acceptor because the lone pair interacts with the neighbouring carbonyl group, as described above. Primary and secondary amides have an N–H group, which allows the possibility of this group acting as a hydrogen bond donor.

The most common type of amide in peptide lead compounds is the secondary amide. Suitable analogues that could be prepared to test out possible binding interactions are shown in Fig. 10.17. All the analogues apart from the primary and secondary amines could be used to check whether the amide is acting as a hydrogen bond donor, and the alkenes and amines could be tested to see whether the amide is acting as a hydrogen bond acceptor. However, there are traps for the unwary. The amide group is planar and does not rotate, because of its partial double bond character. The ketone, the secondary amine, and the tertiary amine shown have a single bond at the equivalent position, which *can* rotate. This would alter the relative positions of any binding groups on either side of the amide group and lead to a loss of binding even if the amide itself was not involved in binding. Therefore, a loss of activity would not necessarily mean that the amide is important as a binding group. With these groups, it would only be safe to say that the amide group is not essential if activity is retained. Similarly, the primary amine and carboxylic acid may be found to have no activity, but this might be due to the loss of important binding groups in one half of the molecule. These particular analogues would only be worth considering if the amide group is peripheral to the molecule (e.g. R–NHCOMe or R–CONHMe) and not part of the main skeleton.

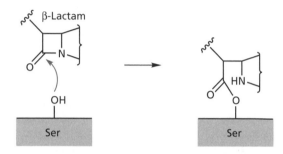

Figure 10.18 β-lactam ring acting as an acylating agent.

The alkene would be a particularly useful analogue to test, since it is planar, cannot rotate, and cannot act as a hydrogen bond donor or hydrogen bond acceptor. However, the synthesis of this analogue may not be simple. In fact, it is likely that all the analogues described would have to be prepared using a full synthesis. Amides are relatively stable functional groups and although several of the analogues described might be attainable from the lead compound directly, it is more likely that the lead compound would not survive the forcing conditions required.

Amides which are within a ring system are called **lactams**. They too can form intermolecular hydrogen bonds, as described above. However, if the ring is small and suffers ring strain, the lactam can undergo a chemical reaction with the target leading to the formation of a covalent bond. The best examples of this are the penicillins, which contain a four-membered β-lactam ring. This acts as an acylating agent and irreversibly inhibits a bacterial enzyme by acylating a serine residue in the active site (Fig. 10.18) (section 16.5.1.4).

10.1.7 Binding role of quaternary ammonium salts

Quaternary ammonium salts are ionized and can interact with carboxylate groups by ionic interactions (Fig. 10.19). Another possibility is an **induced dipole interaction** between the quaternary ammonium ion and any aromatic rings in the binding site. The positively charged nitrogen can distort the π electrons of the aromatic ring such that a dipole is induced, whereby the face of the ring is slightly negative and the edges are slightly positive. This allows an interaction between the slightly negative faces of the aromatic rings and the positive charge of the quaternary ammonium ion.

The importance of these interactions could be tested by synthesizing an analogue that has a tertiary amine group rather than the quaternary ammonium group. Of course, it is possible that such a group could become ionized by being protonated, and interact in the same way. Converting the amine to an amide would prevent this possibility. The neurotransmitter **acetylcholine** has a quaternary ammonium group which is thought to bind to the binding site of its target receptor by ionic bonding and/or induced dipole interactions (section 19.7).

10.1.8 Binding role of carboxylic acids

The carboxylic acid group is reasonably common in drugs. It can act as a hydrogen bond acceptor in various

ways, or as a hydrogen bond donor (Fig. 10.20). Alternatively, it may exist as the carboxylate ion. This allows the possibility of an ionic interaction or a strong hydrogen bond where the carboxylate ion acts as the hydrogen bond acceptor. The carboxylate ion has also been found to be a good ligand for zinc ions, which are present as cofactors in enzymes known as zinc metalloproteinases.

In order to test the possibility of such interactions, analogues such as esters, primary alcohols, and ketones could be synthesized and tested (Fig. 10.21). None of these functional groups can ionize, so a loss of activity could imply that an ionic bond is important. The primary alcohol could shed light on whether the carbonyl oxygen is involved in hydrogen bonding, whereas the ester and ketone could indicate whether the hydroxyl group of the carboxylic acid is involved in

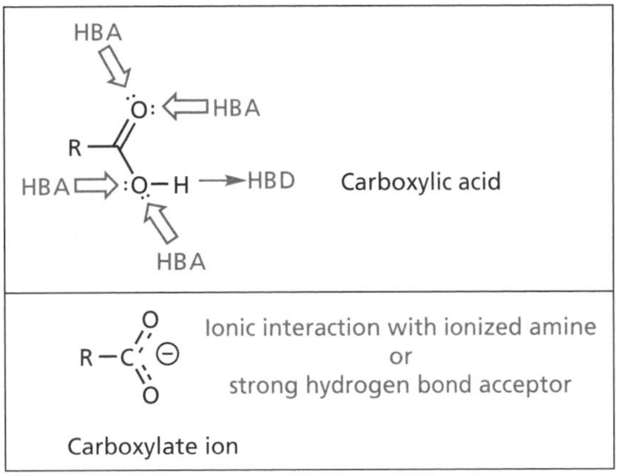

Figure 10.20 Possible binding interactions for a carboxylic acid and carboxylate ion.

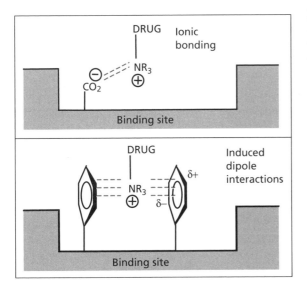

Figure 10.19 Possible binding interactions of a quaternary ammonium ion.

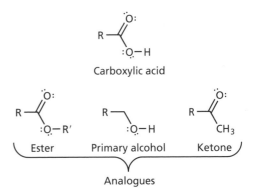

Figure 10.21 Analogues for a carboxylic acid.

hydrogen bonding. It may be possible to synthesize the ester analogue directly from the lead compound, but the reduction of a carboxylic acid to a primary alcohol requires harsher conditions and this sort of analogue would normally be prepared by a full synthesis. The ketone would also have to be prepared by a full synthesis.

10.1.9 Binding role of esters

An ester functional group has the potential to interact with a binding site as a hydrogen bond acceptor only (Fig. 10.22). The carbonyl oxygen is more likely to act as the hydrogen bond acceptor than the alkoxy oxygen, as it is sterically less hindered and has a greater electron density. The importance or otherwise of the carbonyl group could be judged by testing an equivalent ether, which would require a full synthesis.

Esters are susceptible to hydrolysis *in vivo* by metabolic enzymes called **esterases**. This may pose a problem if the lead compound contains an ester that is

important to binding, as it means the drug might have a short lifetime *in vivo*. Having said that, there are several drugs that *do* contain esters and are relatively stable to metabolism, thanks to either electronic factors that stabilize the ester or steric factors that protect it.

Esters that are susceptible to metabolic hydrolysis are sometimes used deliberately to mask a polar functional group such as a carboxylic acid, alcohol or phenol in order to achieve better absorption from the gastrointestinal tract. Once in the blood supply, the ester is hydrolysed to release the active drug. This is known as a **prodrug** strategy (section 11.6).

Special mention should be made of the ester group in aspirin. Aspirin has an anti-inflammatory action resulting from its ability to inhibit an enzyme called cyclooxygenase (COX) (section 4.8.2). In this case the ester is not taking part in intermolecular bonding but acts as an acylating agent where an acetyl group is covalently attached to a serine residue in the active site (Fig. 10.23).

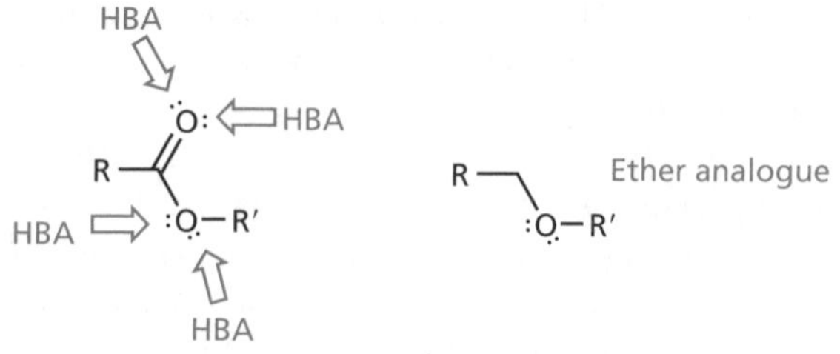

10.1.10 Binding role of alkyl and aryl halides

Alkyl halides involving chlorine, bromine, or iodine tend to be chemically reactive, since the halide ion is a good leaving group. As a result, a drug containing an alkyl halide is likely to react with any nucleophilic group that it encounters and become permanently linked to that group by a covalent bond—an alkylation reaction (Fig. 10.24). This poses a problem, as the drug is likely to alkylate a large variety of macromolecules which have nucleophilic groups, especially proteins

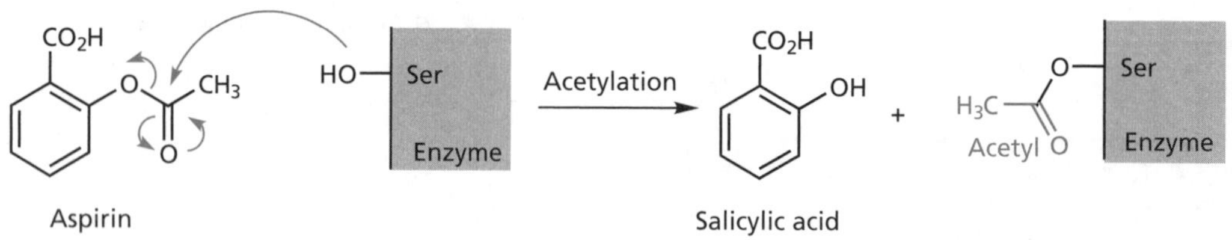

Figure 10.22 Possible binding interactions for an ester.

Figure 10.23 Aspirin acting as an acylating agent.

Figure 10.24 Alkylation of macromolecular targets by alkyl halides.

and nucleic acids. It is possible to moderate the reactivity to some extent, but selectivity is still a problem and leads to severe side effects. These drugs are therefore reserved for life-threatening diseases such as cancer (section 18.2.3). Alkyl fluorides, on the other hand, are not alkylating agents, because the C–F bond is a strong one and is not easily broken. Fluorine is commonly used to replace a proton as it is approximately the same size, but has different electronic properties. It may also protect the molecule from metabolism (sections 10.3.7 and 11.2.4).

Aryl halides do not act as alkylating agents and pose less of a problem in that respect. As the halogen substituents are electron-withdrawing groups, they affect the electron density of the aromatic ring and this may have an influence on the binding of the aromatic ring. The halogen substituents chlorine and bromine are hydrophobic in nature and may interact favourably with hydrophobic pockets in a binding site. Hydrogen bonding is not important. Although halide ions are strong hydrogen bond acceptors, halogen substituents are poor hydrogen bond acceptors.

Aliphatic and aromatic analogues lacking the halogen substituent could be prepared by a full synthesis to test whether the halogen has any importance in the activity of the lead compound.

10.1.11 Binding role of thiols

The thiol group (S–H) is known to be a good ligand for a zinc ion and has been incorporated into several drugs designed to inhibit enzymes containing a zinc cofactor. Such enzymes are known as zinc metalloproteinases (sections 10.3.11.2 and 18.7.1). If the lead compound has a thiol group, the corresponding alcohol could be tested. This would have a far weaker interaction with a transition metal such as zinc.

10.1.12 Binding role of other functional groups

A wide variety of other functional groups may be present in lead compounds that have no direct binding role, but could be important in other respects. Some may influence the electronic properties of the molecule (e.g. nitro groups or nitriles). Others may restrict the shape or conformation of a molecule (e.g. alkynes) (Box 10.3). Functional groups may also act as metabolic blockers (e.g. alkynes, aryl halides) (section 11.2.4).

10.1.13 Binding role of alkyl groups and the carbon skeleton

The alkyl substituents and carbon skeleton of a lead compound are hydrophobic and may bind with hydrophobic regions of the binding site through van der Waals and hydrophobic interactions. The relevance of an alkyl substituent to binding can be determined by synthesizing an analogue which lacks the substituent. Such analogues generally have to be synthesized using a full synthesis if they are attached to the carbon skeleton of the molecule. However, if the alkyl group is attached to nitrogen or oxygen, it may be possible to remove the group from the lead compound as shown in Fig. 10.25. The analogues obtained may then be expected to have less activity if the alkyl group was involved in important hydrophobic interactions.

10.1.14 Binding role of heterocycles

There is a large diversity of heterocycles to be found in lead compounds. Heterocycles are cyclic structures that contain one or more heteroatoms such as oxygen, nitrogen, or sulfur. Nitrogen-containing heterocycles are particularly prevalent. The heterocycles can be aliphatic or aromatic in character, and have the potential to interact with binding sites through a variety of bonding forces. For example, the overall heterocycle can interact through van der Waals and hydrophobic interactions, while the individual heteroatoms present in the structure could interact by hydrogen bonding or ionic bonding.

As far as hydrogen bonding is concerned, there is an important directional aspect. The position of the heteroatom in the ring and the orientation of the ring in the binding site can be crucial in determining whether or not a good interaction takes place. For example, a purine ring can take part in six hydrogen bonding interactions, three as a hydrogen bond donor and three as a hydrogen bond acceptor. The ideal

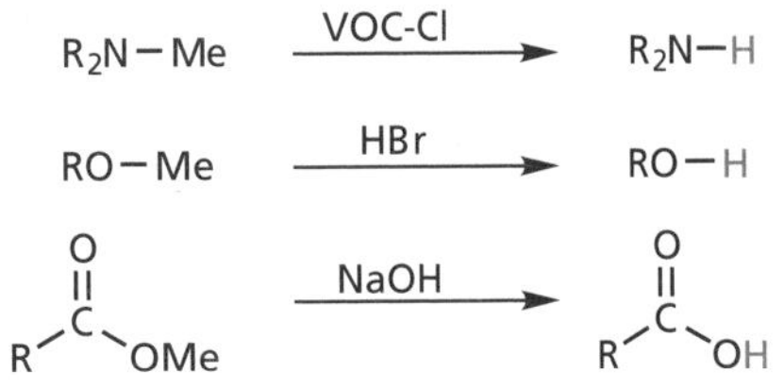

Figure 10.25 Removal of alkyl groups from heteroatoms.

directions for these interactions are shown in Fig. 10.26. Van der Waals interactions are also possible, to regions of the binding site above and below the plane of the ring system.

Heterocycles can be involved in quite intricate hydrogen bonding networks within a binding site. For example, the anticancer drug **methotrexate** contains a pteridine ring system that interacts with its binding site as shown in Fig. 10.27.

If the lead compound contains a heterocyclic ring, it is worth synthesizing analogues containing an aromatic ring or different heterocyclic rings to explore whether all the heteroatoms present are really necessary.

A complication with heterocycles is the possibility of **tautomers**. This played an important role in determining

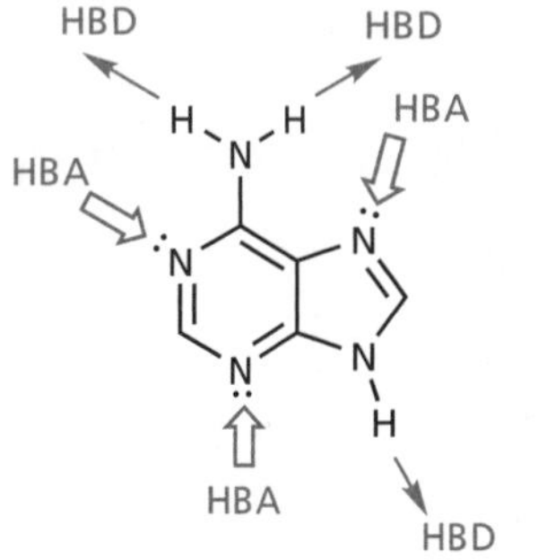

the structure of DNA (section 7.1). The structure of DNA consists of a double helix with base pairing between two sets of heterocyclic nucleic acid bases. Base pairing involves three hydrogen bonds between the base pair guanine and cytosine, and two hydrogen bonds between the base pair adenine and thymine (Fig. 10.28). The rings involved in the base pairing are coplanar, allowing the optimum orientation for the hydrogen bond donors and hydrogen bond acceptors. This in turn means that the base pairs are stacked above each other, allowing van der Waals interactions between the faces of each base pair. However, when Watson and Crick originally tried to devise a model for DNA, they incorrectly assumed that the preferred tautomers for the nucleic acid bases were as shown in the bottom part of Fig. 10.28. With these tautomers, the required hydrogen bonding is not possible and would not explain the base pairing observed in the structure of DNA.

In a similar vein, knowing the preferred tautomers of heterocycles can be important in understanding how drugs interact with their binding sites. This is amply illustrated in the design of the antiulcer agent cimetidine (section 22.2).

With heterocyclic compounds, it is possible for a hydrogen bond donor and a hydrogen bond acceptor to be part of a conjugated system. Polarization of the electrons in the conjugated system permits π-**bond cooperativity**, where the strength of the hydrogen bond

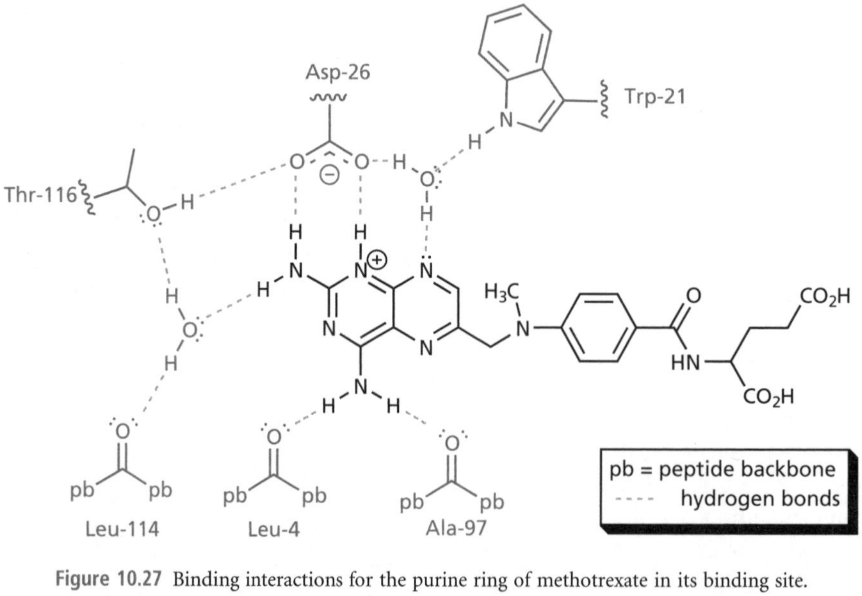

Figure 10.26 Possible hydrogen bonding interactions for adenine.

Figure 10.27 Binding interactions for the purine ring of methotrexate in its binding site.

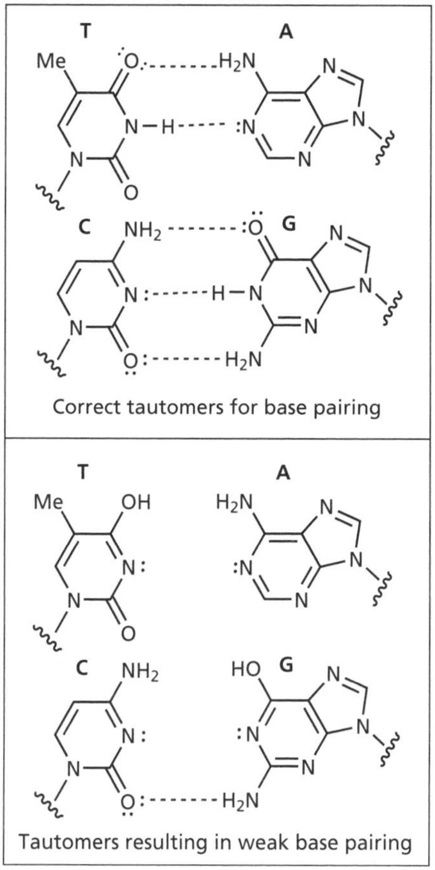

Figure 10.28 Base pairing in DNA and the importance of tautomers.

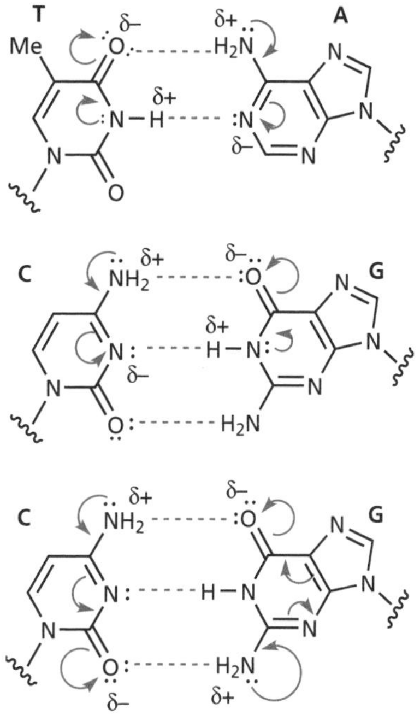

Figure 10.29 π-Bond cooperativity in hydrogen bonding.

donor is enhanced by the hydrogen bond acceptor and vice versa. This has also been called **resonance-assisted hydrogen bonding**. This type of hydrogen bonding is possible for the hydrogen bond donors and acceptors for the nucleic acid base pairs (Fig. 10.29).

10.1.15 Isosteres

Isosteres are atoms or groups of atoms which have the same valency (or number of outer shell electrons), and which have chemical or physical similarities (Fig. 10.30).

For example, SH, NH_2, and CH_3 are isosteres of OH, whereas S, NH, and CH_2 are isosteres of O. Isosteres can be used to determine whether a particular group is an important binding group or not, by altering the character of the molecule in as controlled a way as possible. Replacing O with CH_2, for example, will make little difference to the size of the analogue, but will have

a marked effect on its polarity, electronic distribution, and bonding. Replacing OH with the larger SH may not have such an influence on the electronic character, but steric factors become more significant.

Isosteric groups could be used to determine whether a particular group is involved in hydrogen bonding. For example, replacing OH with CH_3 would completely eliminate hydrogen bonding, whereas replacing OH with NH_2 would not.

The β-blocker propranolol has an ether linkage (Fig. 10.31). Replacement of the OCH_2 segment with the isosteres CH=CH, SCH_2, or CH_2CH_2 eliminates activity, whereas replacement with $NHCH_2$ retains activity (though reduced). These results show that the ether oxygen is important to the activity of the drug and suggests that it is involved in hydrogen bonding with the receptor.

The use of isosteres in drug design is described in section 10.3.7.

10.1.16 Testing procedures

When investigating structure–activity relationships for drug–target binding interactions, biological testing should involve *in vitro* tests; for example inhibition studies on isolated enzymes or binding studies on

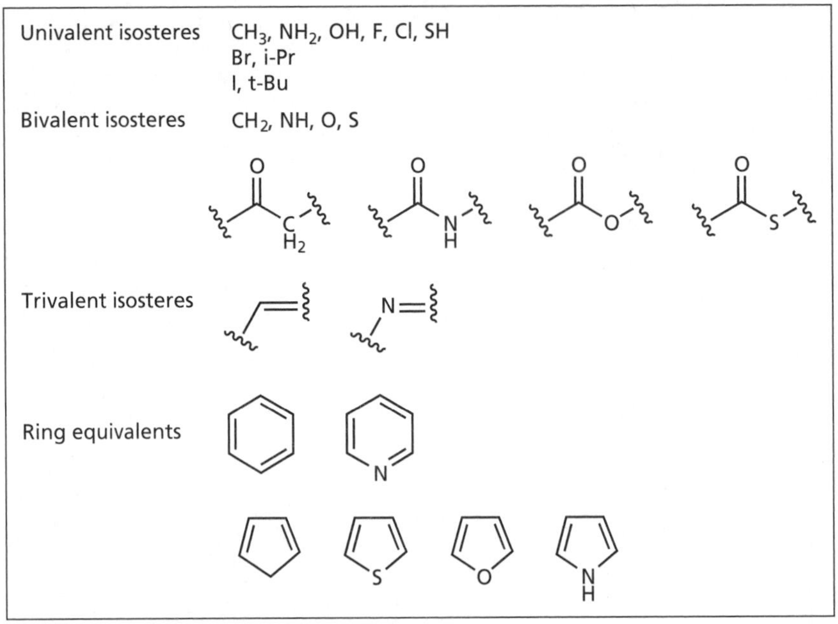

Univalent isosteres	CH$_3$, NH$_2$, OH, F, Cl, SH
	Br, i-Pr
	I, t-Bu
Bivalent isosteres	CH$_2$, NH, O, S

Figure 10.30 Examples of classical isosteres.

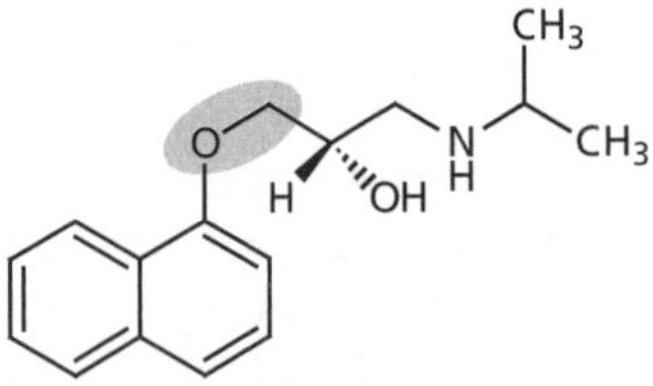

Figure 10.31 Propranolol.

membrane-bound receptors. The results then show conclusively which binding groups are important in drug–target interactions. If *in vivo* testing is carried out, the results are less clear-cut because loss of activity may be due to the inability of the drug to reach its target rather than reduced drug–target interactions. On the other hand, *in vivo* testing may reveal groups that are important in protecting or assisting the drug in its passage through the body, which would not be revealed by *in vitro* testing.

10.2 Identification of a pharmacophore

Once it is established which groups are important for a drug's activity, it is possible to move on to the next stage—the identification of the pharmacophore. The

pharmacophore summarizes the important binding groups which are required for activity, and their relative positions in space with respect to each other. For example, if we discover that the important binding groups for our hypothetical drug glipine are the two phenol groups, the aromatic ring, and the nitrogen atom, then the pharmacophore is as shown in Fig. 10.32. Structure I shows the 2D pharmacophore and structure II shows the 3D pharmacophore. The latter specifies the relative positions of the important groups in space. In this case, the nitrogen atom is 5.063 Å from the centre of the phenolic ring and lies at an angle of 18° from the plane of the ring. Note that it is not necessary to show the specific skeleton connecting the important groups. Indeed, there are benefits in not doing so, since it is easier to compare the 3D pharmacophores from different structural classes of compound to see if they share a common pharmacophore. 3D Pharmacophores can be defined using molecular modelling (section 15.11) which allows the definition of 'dummy bonds' such as the one in Fig. 10.32 between nitrogen and the centre of the aromatic ring. The centre of the ring can be defined by a dummy atom called a **centroid** (not shown).

An even more general type of 3D pharmacophore is the one shown in structure III—a bonding-type pharmacophore. Here the bonding characteristics of each

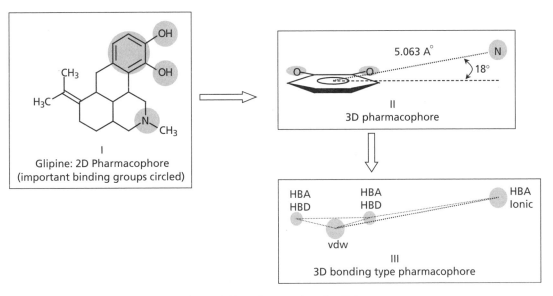

Figure 10.32 A pharmacophore for glipine.

functional group are defined rather than the group itself. Note also that the groups are defined as points in space. This includes the aromatic ring, which is defined by a point at the centre of the ring—the centroid. All the points are connected by pharmacophoric triangles to define their positions. This allows the comparison of molecules which may have the same pharmacophore and binding interactions, but which use different functional groups to achieve these interactions. In this case, the phenol groups can act as hydrogen bond donors or acceptors, the aromatic ring can participate in van der Waals interactions, and the amine can act as a hydrogen bond acceptor or an ionic centre if it is protonated. We shall return to the concept and use of 3D pharmacophores in sections 13.10 and 15.11.

Identifying 3D pharmacophores is relatively easy for rigid cyclic structures such as the hypothetical glipine. With more flexible structures, it is not so straightforward because the molecule can adopt a large number of shapes or conformations which place the important binding groups in different positions relative to each other. Normally, only one of these conformations is recognized and bound by the binding site. This conformation is known as the **active conformation**. In order to identify the 3D pharmacophore, it is necessary to know the active conformation. There are various ways in which this might be done. Rigid analogues of the flexible compound could be synthesized and tested to see whether activity is retained (section 10.3.9).

It may be possible to crystallize the target with the compound bound to the binding site and derive the structure of the complex and the active conformation by X-ray crystallography (section 15.10). More recently, progress has been made in using NMR spectroscopy to solve the active conformation of isotopically labelled molecules bound to their binding sites.

Finally, a warning! A drawback with pharmacophores is their unavoidable emphasis on functional groups as the crucial binding groups. In many situations this is certainly true, but in other situations it is not. It is not uncommon to find compounds that have the correct pharmacophore but nevertheless show disappointing activity and poor binding. It is important to realize that the overall skeleton of the molecule is involved in interactions with the binding site through van der Waals interactions and hydrophobic interactions. The strength of these interactions can sometimes be crucial in whether a drug binds effectively or not, and the 3D pharmacophore does not take this into account.

KEY POINTS

- SARs define the functional groups or regions of a lead compound which are important to its biological activity.

- Functional groups such as alcohols, amines, esters, amides, carboxylic acids, phenols, and ketones can interact with binding sites by means of hydrogen bonding.

- Functional groups such as amines, quaternary ammonium salts, and carboxylic acids can interact with binding sites by ionic bonding.

- Functional groups such as alkenes and aromatic rings can interact with binding sites by means of van der Waals interactions.

- Alkyl substituents and the carbon skeleton of the lead compound can interact with hydrophobic regions of binding sites by means of van der Waals interactions.

- Interactions involving dipole moments or induced dipole moments may play a role in binding a lead compound to a binding site.

- Reactive functional groups such as alkyl halides may lead to irreversible covalent bonds being formed between a lead compound and its target.

- The relevance of a functional group to binding can be determined by preparing analogues where the functional group is modified or removed in order to see whether activity is affected by such a change.

- Some functional groups can be important to the activity of a lead compound for reasons other than target binding. They may play a role in the electronic or stereochemical properties of the compound, or they may have an important pharmacokinetic role.

- Replacing a group in the lead compound with an isostere (a group having the same valency) makes it easier to determine whether a particular property such as hydrogen bonding is important.

- *In vitro* testing procedures should be used to determine the SAR for target binding.

- The pharmacophore summarizes the groups which are important in the binding of a lead compound to its target, as well as their relative positions in three dimensions.

10.3 Drug optimization: strategies in drug design

Once the important binding groups and pharmacophore of the lead compound have been identified, it is possible to synthesize analogues that contain the same pharmacophore. But why is this necessary? If the lead compound has useful biological activity,

why bother making analogues? The answer is that very few lead compounds are ideal. Most are likely to have low activity, poor selectivity, and significant side effects. They may also be difficult to synthesize, so there is an advantage in finding analogues with improved properties. We look now at strategies that can be used to optimize the interactions of a drug with its target to allow higher activity and selectivity.

10.3.1 Variation of substituents

Varying easily accessible substituents is a common method of fine tuning the binding interactions of a drug.

10.3.1.1 Alkyl substituents

Certain alkyl substituents can be varied more easily than others. For example, the alkyl substituents of ethers, amines, esters, and amides are easily varied as shown in Fig. 10.33. In these cases, the alkyl substituent already present can be removed and replaced by another substituent. Alkyl substituents which are part of the carbon skeleton of the molecule are not easily removed and it is usually necessary to carry out a full synthesis in order to vary them.

If alkyl groups are interacting with a hydrophobic pocket in the binding site, then varying the length and bulk of the alkyl group (e.g. methyl, ethyl, propyl, butyl, isopropyl, isobutyl, or t-butyl) allows one to probe the depth and width of the pocket. Choosing a substituent that will fill the pocket will then increase the binding interaction (Fig. 10.34).

Larger alkyl groups may also confer selectivity on the drug. For example, in the case of a compound that interacts with two different receptors, a bulkier alkyl substituent may prevent the drug from binding to one of those receptors and so cut down side effects (Fig. 10.35). **Isoprenaline** (Fig. 10.36) is an analogue of **adrenaline** where a methyl group was replaced by an isopropyl group, resulting in selectivity for adrenergic β-receptors over adrenergic α-receptors (section 20.11.3).

10.3.1.2 Aromatic substitutions

If a drug contains an aromatic ring, the position of substituents can be varied to find better binding interactions, resulting in increased activity (Fig. 10.37).

Figure 10.33 Alkyl modifications.

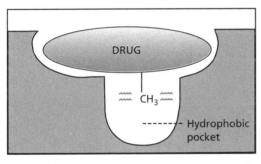

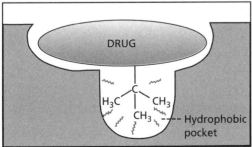

〜〜 van der Waals interactions

Figure 10.34 Variation of alkyl chain to fill a hydrophobic pocket.

For example, the antiarrythmic activity of a series of benzopyrans was found to be best when the sulfonamide substituent was at position 7 of the aromatic ring (Fig. 10.38).

Changing the position of one substituent may have an important effect on another. For example, an electron-withdrawing nitro group will affect the basicity of an aromatic amine more significantly if it is in the *para* position rather than the *meta* position (Fig. 10.39). At the *para* position, the nitro group will make the amine a weaker base and less liable to protonate. This would decrease the amine's ability to interact with ionic binding groups in the binding site, and decrease activity.

If the substitution pattern is ideal, then we can try varying the substituents themselves. Substituents having different steric, hydrophobic, and electronic properties are usually tried to see if these factors have any effect on activity. It may be that activity is improved by having a more electron-withdrawing substituent, in which case a chloro substituent might be tried in place of a methyl substituent.

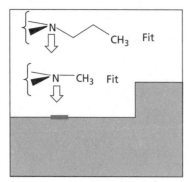

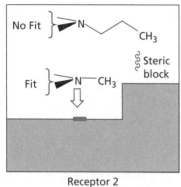

Figure 10.35 Use of a larger alkyl group to confer selectivity on a drug.

Isoprenaline

Adrenaline

Figure 10.36 Introducing selectivity for β-adrenoceptors over α-adrenoceptors.

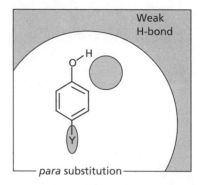

Figure 10.37 Aromatic substitutions.

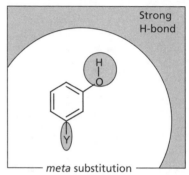

Figure 10.38 Benzopyrans.

The chemistry involved in these procedures is usually straightforward, so these analogues are made as a matter of course whenever a novel drug structure is developed. Furthermore, the variation of aromatic or aliphatic substituents is open to **quantitative structure–activity studies** (QSARs), as described in Chapter 13.

10.3.2 Extension of the structure

The strategy of extension involves the addition of another functional group to the lead compound in order to probe for extra binding interactions with the target. Lead compounds are capable of fitting the binding site and have the necessary functional

meta (inductive electron-withdrawing effect)

para (electron-withdrawing effect due to resonance *and* inductive effects)

Figure 10.39 Electronic effects of aromatic substitutions.

groups to interact with some of the important binding regions present. It is possible, however, that they do not interact with all the binding regions available. For example, a lead compound may bind to three binding regions in the binding site but fail to use a fourth (Fig. 10.40). Therefore, why not add extra functional groups to probe for that fourth region?

Extension tactics are often used to find extra hydrophobic regions in a binding site, by adding various alkyl or arylalkyl groups (see Box 10.1). Such groups could be added to functional groups such as alcohols, phenols, amines and carboxylic acids should they be present in the drug. Other functional groups could be added to probe for extra hydrogen bonding interactions or ionic interactions.

Frequently, extension strategies are used to convert an agonist into an antagonist. This may happen because an extra binding interaction is found which is not used by the natural agonist or substrate. Since the

---

BOX 10.1 THE USE OF EXTENSION TACTICS

Extension tactics were used in the development of antihypertensive agents which inhibit an enzyme known as **angiotensin-converting enzyme** (ACE). Adding a phenylalkyl group to the lead compound (I) increased activity (*extension strategy*). Varying the length of the alkyl chain connecting the aromatic ring to (I) demonstrated that the phenylethyl group was the best substituent (*chain extension strategy*). This structure showed a 1000-fold improvement in inhibition, demonstrating that the extra aromatic ring was binding to a hydrophobic pocket in the enzyme's active site (see also section 10.3.11.2).

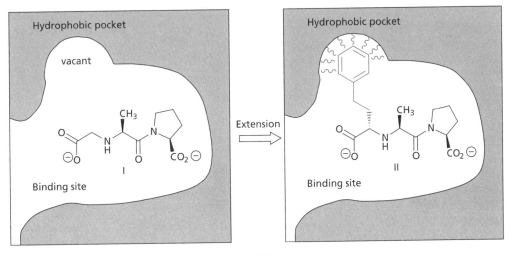

ACE inhibitors.

overall binding interaction is different, antagonist activity results rather than agonist activity.

The extension tactic has been employed successfully to produce more active analogues of morphine (sections 21.3.2 and 21.3.4) and more active adrenergic agents (sections 20.9–20.11). It was also used to improve the activity and selectivity of the protein kinase inhibitor, imatinib (section 18.6.2.2).

10.3.3 Chain extension/contraction

Some drugs have two important binding groups linked together by a chain, in which case it is possible that the chain length is not ideal for the best interaction. Therefore, shortening or lengthening the chain length is a useful tactic to try (Fig. 10.41, see also Box 10.1).

10.3.4 Ring expansion/contraction

If a drug has one or more rings, it is generally worth synthesizing analogues where one of these rings is expanded or contracted. The principle behind this approach is much the same as varying the substitution pattern of an aromatic ring. Expanding or contracting the ring puts the binding groups in different positions relative to each other and may lead

to better interactions with specific regions in the binding site (Fig. 10.42).

Varying the size of rings can also bring substituents into a good position for binding. For example, during the development of the antihypertensive agent **cilazaprilat** (another ACE inhibitor), the bicyclic structure I showed promising activity (Fig. 10.43). The important binding groups were the two carboxylate groups and the amide group. By carrying out various ring contractions and expansions, cilazaprilat was identified as the structure having the best interaction with the binding site.

10.3.5 Ring variations

A popular strategy used for compounds containing an aromatic or heteroaromatic ring is to replace the original ring with a range of other heteroaromatic rings of different ring size and heteroatom positions. For example, several non-steroidal anti-inflammatory agents (NSAIDs) have been reported, all consisting of a central ring with 1,2-biaryl substitution. Different pharmaceutical companies have varied the central ring to produce a range of active compounds (Fig. 10.44).

Admittedly, a lot of these changes are merely ways of avoiding patent restrictions (**'me too' drugs**),

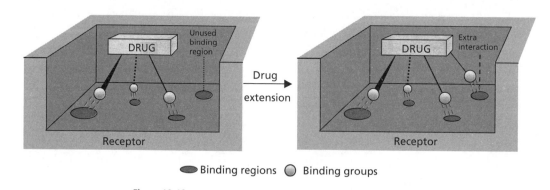

Figure 10.40 Extension of a drug to provide a fourth binding group.

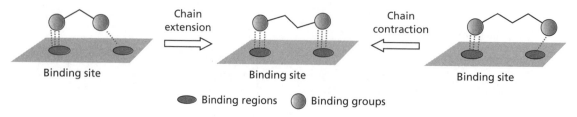

Figure 10.41 Chain contraction/extension.

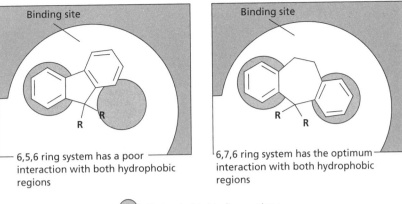

Figure 10.42 Ring expansion.

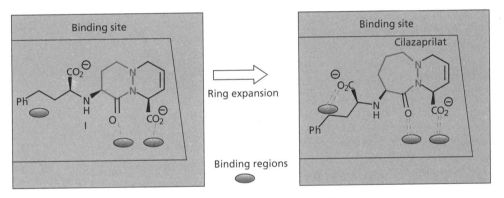

Figure 10.43 Development of cilazaprilat.

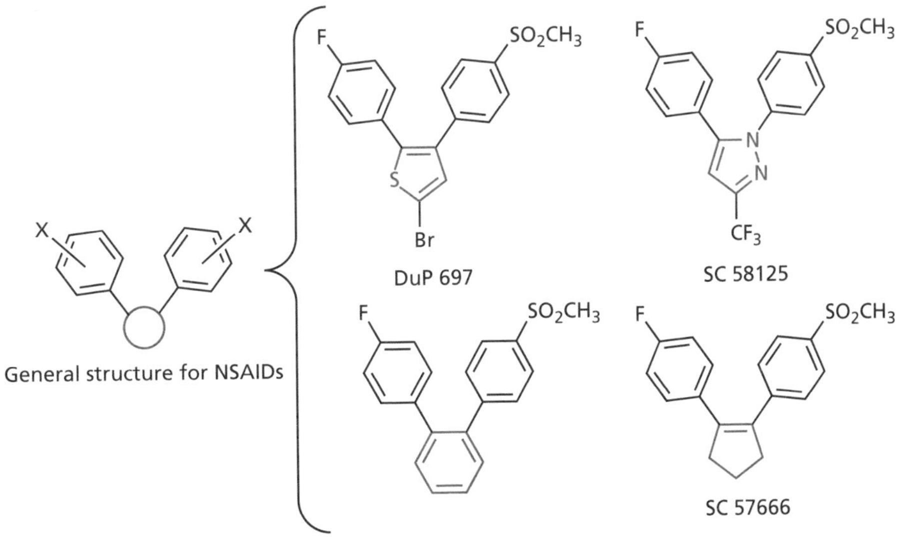

Figure 10.44 Non-steroidal anti-inflammatory drugs (NSAIDS).

but there can often be significant improvements in activity, as well as increased selectivity and reduced side effects. For example, the antifungal agent (I) (Fig. 10.45) acts against a fungal enzyme which is also present in humans. Replacing the imidazole ring of structure (I) with a 1,2,4-triazole ring to give UK 46245 resulted in better selectivity against the fungal form of the enzyme.

One advantage of altering an aromatic ring to a heteroaromatic ring is that it introduces the possibility of an extra hydrogen bonding interaction with the binding site, should a suitable binding region be available. For example, structure I (Fig. 10.46) was the lead compound for a project looking into novel antiviral agents. Replacing the aromatic ring with a pyridine ring resulted in an additional binding interaction with the target enzyme. Further development led eventually to the antiviral agent **nevirapine** (Fig. 10.46).

10.3.6 Ring fusions

Extending a ring by ring fusion can sometimes result in increased interactions or increased selectivity.

Structure I   UK 46245

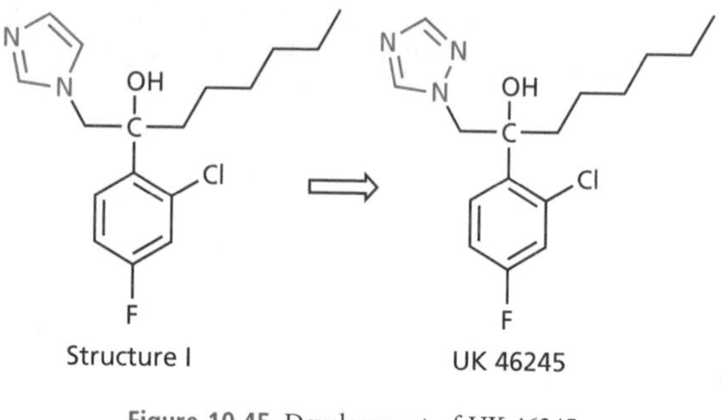

Figure 10.45 Development of UK 46245.

One of the major advances in the development of the selective β-blockers was the replacement of the aromatic ring in **adrenaline** with a naphthalene ring system (**pronethalol**) (Fig. 10.47). This resulted in a compound that was able to distinguish between two very similar receptors, the α- and β-receptors for adrenaline. One possible explanation for this could be that the β-receptor has a larger van der Waals binding area for the aromatic system than the α-receptor and can interact more strongly with pronethalol than with adrenaline. Another possible explanation is that the naphthalene ring system is sterically too big for the α-receptor, but is just right for the β-receptor.

10.3.7 Isosteres and bioisosteres

Isosteres (section 10.1.15) have often been used in drug design to vary the character of the molecule in a rational way with respect to features such as size, polarity, electronic distribution, and bonding. Some isosteres can be used to determine the importance of size towards activity, whereas others can be used to determine the importance of electronic factors. For example, fluorine is often used as an isostere of hydrogen since it is virtually the same size. However, it is more electronegative and can be used to vary the electronic properties of the drug without having any steric effect.

The presence of fluorine in place of an enzymatically labile hydrogen can also disrupt an enzymatic reaction, as C–F bonds are not easily broken. For example, the antitumour drug **5-fluorouracil** described in section 18.3.2 is accepted by its target enzyme because it appears little different from the normal substrate—**uracil**. However, the mechanism of the

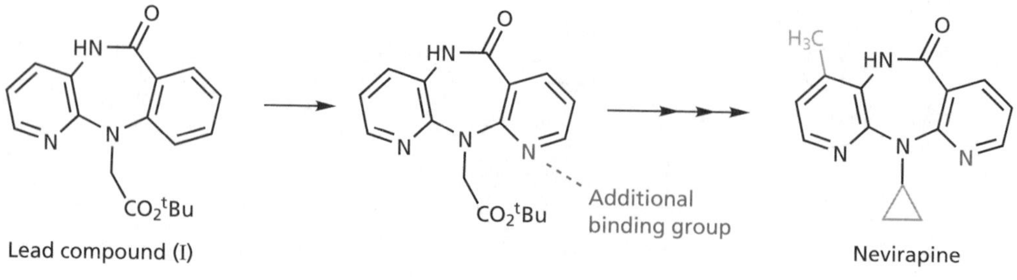

Lead compound (I)

Nevirapine

Figure 10.46 Development of nevirapine.

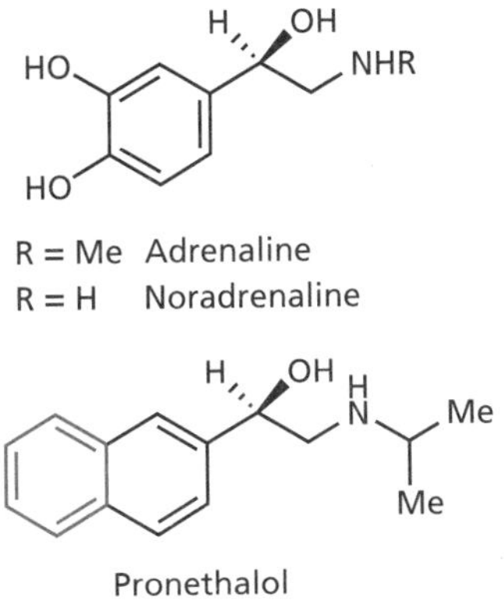

R = Me Adrenaline
R = H   Noradrenaline

Pronethalol

Figure 10.47 Ring variation of adrenaline.

enzyme-catalysed reaction is totally disrupted, as the fluorine has replaced a hydrogen which is normally lost during the enzyme mechanism.

Several non-classical isosteres have been used in drug design as replacements for particular functional groups. Non-classical isosteres are groups which do not obey the steric and electronic rules used to define classical isosteres, but which have similar physical and chemical properties. For example, the structures shown in Fig. 10.48 are non-classical isosteres for a thiourea group. They are all planar groups of similar size and basicity.

The term **bioisostere** is used in drug design and includes both classical and non-classical isosteres. A bioisostere is a group that can be used to replace another group while retaining the desired biological activity. Bioisosteres are often used to replace a functional group that is important for target binding but is problematic in one way or another. For example, the thiourea group was present as an important binding group in early histamine antagonists, but was responsible for toxic side effects. Replacing it with bioisosteres allowed the important binding interactions to be retained for histamine antagonism but avoided the toxicity problems (section 22.2.6). Further examples of the use of bioisosteres are given in sections 11.1.5, 11.2.2, and 17.7.4. It is important to realize that bioisosteres are specific for a particular group of compounds and their target. Replacing a functional group with a bioisostere is not guaranteed to retain activity for every drug at every target (see for example section 11.2.2).

As stated above, bioisosteres are commonly used in drug design to replace a problematic group while retaining activity. In some situations, the use of a bioisostere can actually increase target interactions and selectivity. For example, a pyrrole ring has frequently been used as a non-classical isostere for an amide. Carrying out this replacement on the dopamine antagonist **sultopride** led to increased activity and selectivity towards the dopamine D_3 receptor over the dopamine D_2 receptor (Fig. 10.49). Such agents show promise as antipsychotic agents which lack the side effects associated with the D_2 receptor.

Transition-state isosteres are a special type of isostere used in the design of transition-state analogues. These are drugs that are used to inhibit enzymes (section 4.8.4). During an enzymatic reaction, a substrate goes through a transition state before it becomes product. It is proposed that the transition state is bound more strongly than either the substrate or the product, so it makes sense to design drugs based on the structure of the transition state rather than the structure of the substrate or the product. However, the transition state is inherently unstable

Figure 10.49 Isosteric change for an amide group.

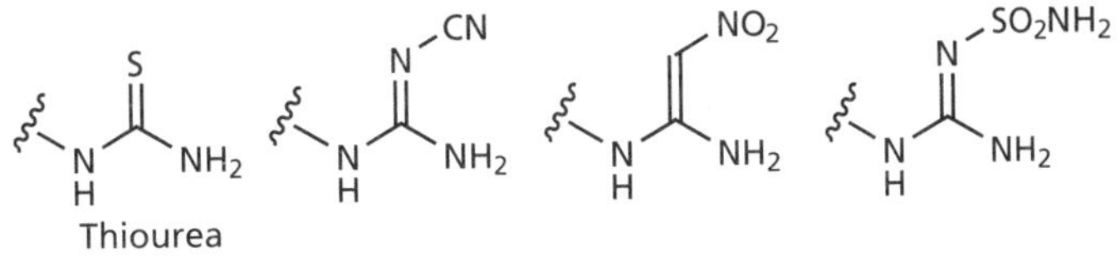

Figure 10.48 Non-classical isosteres for a thiourea group.

and so transition-state isosteres are moieties that are used to mimic the crucial features of the transition state but which are stable. For example, the transition state of an amide hydrolysis resembles the tetrahedral reaction intermediate shown in Fig. 10.50. This is a geminal diol, which is inherently unstable. The hydroxy-ethylene moiety shown is a transition-state isostere since it shares the same tetrahedral geometry, retains one of the hydroxyl groups, and is stable to hydrolysis. Further examples of the use of transition-state iso-steres are given in sections 17.7.4, 17.8.3, and 18.3.4.

10.3.8 Simplification of the structure

Simplification is a strategy which is commonly used on the often complex lead compounds arising from natural sources (see Box 10.2). Once the essential groups of such a drug have been identified by SAR, it is often possible to discard the non-essential parts of the structure without losing activity. Consideration is given to removing functional groups which are not part of the pharmacophore, simplifying the carbon skeleton, and removing asymmetric centres.

This strategy is best carried out in small stages. For example, consider our hypothetical natural product glipine (Fig. 10.51). The essential groups have been marked, and we might aim to synthesize simplified compounds in the order shown. These still retain the groups that we have identified as being essential—the pharmacophore.

Chiral drugs pose a particular problem. The easiest and cheapest method of synthesizing a chiral drug is to make the racemate. However, both enantiomers then have to be tested for their activity and side effects, doubling the number of tests that have to be carried out. This is because different enantiomers can have different activities. For example, compound **UH-301** (Fig. 10.52) is inactive as a racemate, whereas its enantiomers have opposing agonist and antagonist activity at the serotonin receptor (5-HT_{1A}).

The use of racemates is discouraged, and it is best to separate the enantiomers of the racemic drug, or to carry out an asymmetric synthesis. Both options inevitably add to the cost of the synthesis, and so designing a structure that lacks some or all of the asymmetric centres, can be advantageous and represents a simplification of the structure. For example, the cholesterol-lowering agent **mevinolin** has eight asymmetric centres, but a second generation of cholesterol-lowering agents has been developed which contain far fewer (e.g. **HR 780**; Fig. 10.52).

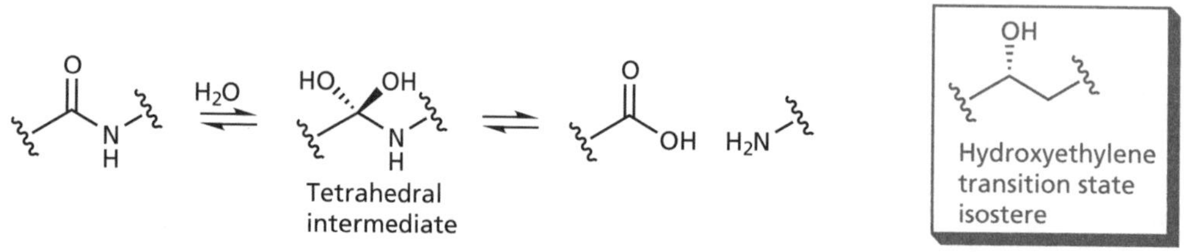

Figure 10.50 Transition state isosteres.

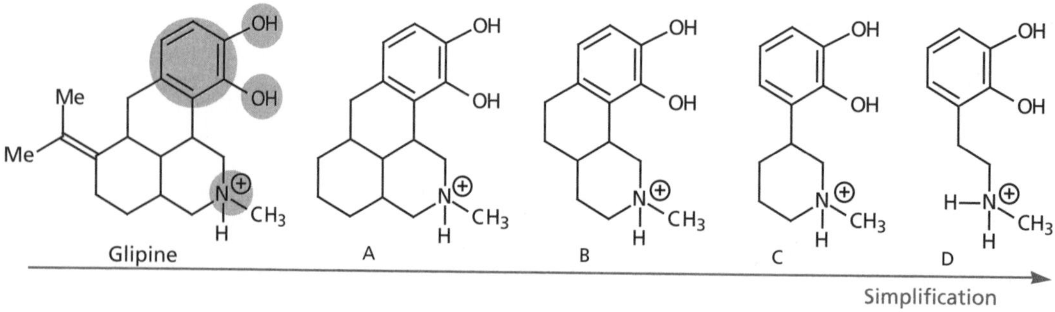

Figure 10.51 Glipine analogues.

BOX 10.2 SIMPLIFICATION

Simplification tactics have been used successfully with the alkaloid **cocaine**. Cocaine has local anaesthetic properties, and its simplification led to the development of local anaesthetics which could be easily synthesized in the laboratory. One of the earliest was **procaine** (**Novocaine**), discovered in 1909 (Fig. 1). Simplification tactics have also proved effective in the design of simpler morphine analogues (section 21.3.3).

More recently, the microbial metabolite **asperlicin** was simplified to **devazepide**, retaining the benzodiazepine and indole skeletons inherent in the structure (Fig. 2). Both asperlicin and devazepide act as antagonists of a neuropeptide chemical messenger called **cholecystokinin** (CCK) which has been implicated in causing panic attacks. Therefore, antagonists may be of use in treating such attacks.

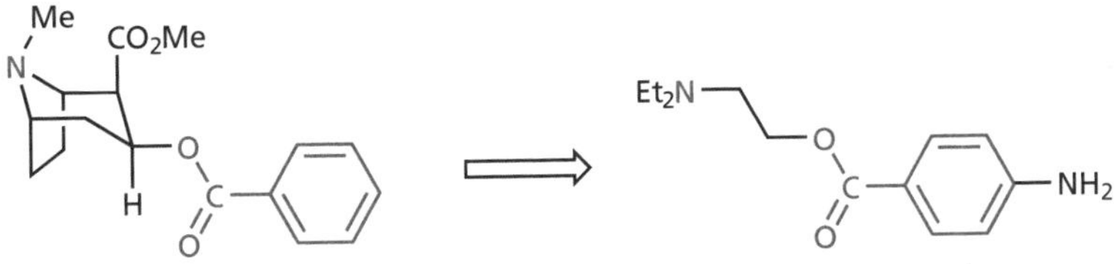

Figure 1 Simplification of cocaine (pharmacophore shown in colour).

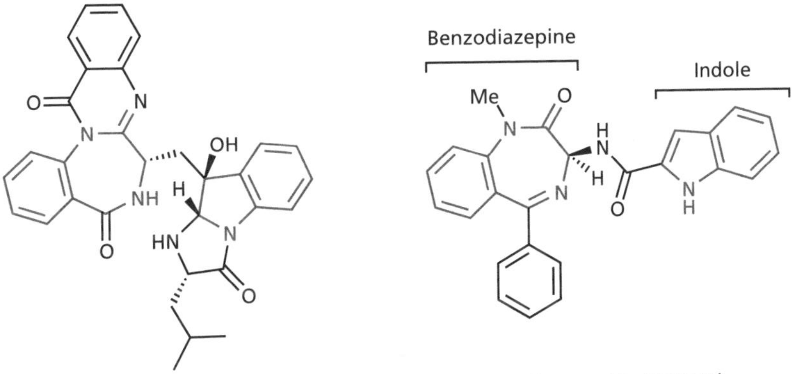

Figure 2 Simplification of asperlicin.

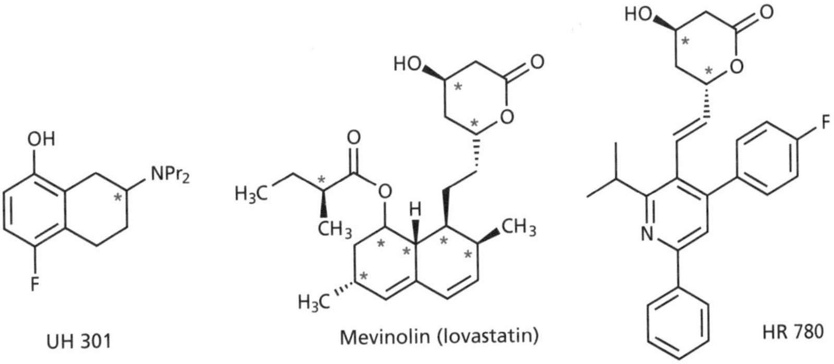

Figure 10.52 UH 301, mevinolin and HR 780.

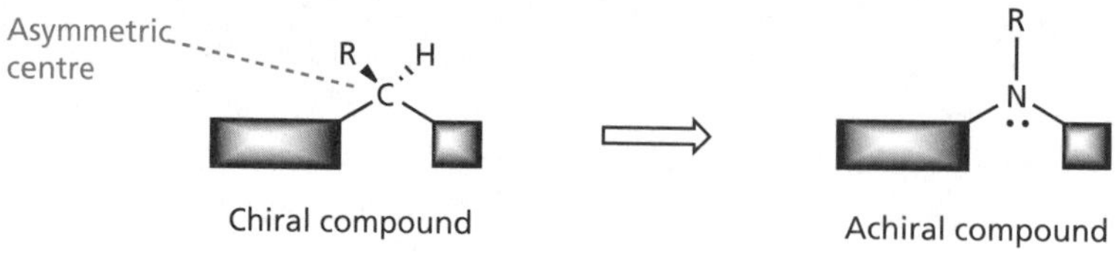

Figure 10.53 Replacing an asymmetric carbon with nitrogen.

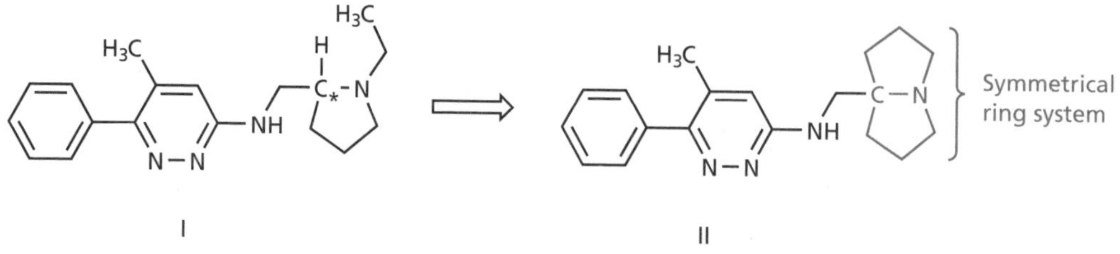

Figure 10.54 Introducing symmetry.

Various tactics can be used to remove asymmetric carbon centres. For example, replacing the carbon centre with nitrogen has been effective in many cases (Fig. 10.53). An illustration of this can be seen in the design of thymidylate synthase inhibitors described in section 15.15.1. (It should be noted that the introduction of an amine in this way may well have significant effects on the pharmacokinetics of the drug in terms of log P, basicity, polarity, etc.; see Chapters 8 and 11.) Another tactic is to introduce symmetry where originally there was none. For example, the muscarinic agonist (II) was developed from (I) in order to remove asymmetry, and has the same activity (Fig. 10.54).

The advantage of simpler structures is that they are easier, quicker, and cheaper to synthesize in the laboratory. Usually the complex lead compounds obtained from natural sources are impractical to synthesize and have to be extracted from the source material—a slow, tedious, and expensive business. There are, however, potential disadvantages in over-simplifying molecules. Simpler molecules are often more flexible and can sometimes bind differently to their targets compared to the original lead compound, resulting in different effects. It is best to simplify in small stages, checking that the desired activity is retained at each stage. Oversimplification may also result in reduced activity, reduced selectivity, and increased side effects. We shall see why in the next section (section 10.3.9).

10.3.9 Rigidification of the structure

Rigidification has often been used to increase the activity of a drug or to reduce its side effects. In order to understand why this tactic can work, let us consider again our hypothetical neurotransmitter from Chapter 5 (Fig. 10.55). This is quite a simple molecule and highly flexible. Bond rotation can lead to a large number of conformations or shapes. As seen in Fig. 10.55, conformation I is the conformation recognized by the receptor—the **active conformation**. Other conformations such as II have the ionized amino group too far away from the anionic centre to interact efficiently, so this is an inactive conformation for our model receptor binding site. It is quite possible, however, that a different receptor exists which *is* capable of binding conformation II. If this is the case, then our model neurotransmitter could switch on two different receptors and give two different biological responses.

The body's own neurotransmitters are highly flexible molecules (section 5.2), but fortunately the body is quite efficient at releasing them close to their target receptors, then quickly inactivating them so that they do not make the journey to other receptors. This is not the case for drugs. They have to be sturdy enough to travel through the body and consequently will interact with all the receptors that are prepared to accept them. The more flexible a drug molecule is, the more likely it will interact with more than one receptor and produce other biological responses (side effects).

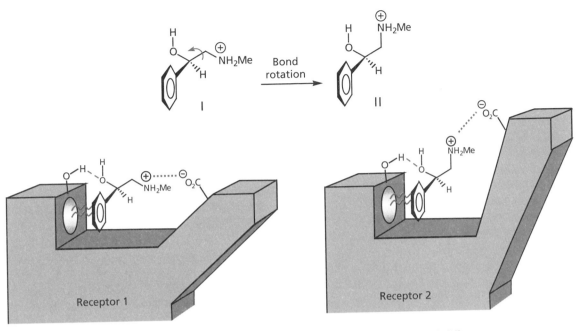

Figure 10.55 Two conformations of a neurotransmitter which are capable of binding with different receptors.

The strategy of rigidification is to lock the molecule into a more rigid conformation so that it cannot take up these other shapes or conformations. Consequently, other receptor interactions and side effects are eliminated. This same strategy should also increase activity since, by locking the drug into the active conformation, the drug is ready to fit its target receptor site and does not need to adjust to the correct conformation. There is an important consequence when it comes to the thermodynamics of binding as well. A flexible molecule has to adopt a single active conformation in order to bind to its target, which means that it has to become more ordered. This results in a decrease in entropy, and since the free energy of binding is related to entropy by the equation $\Delta G = \Delta H - T\Delta S$, any decrease in entropy will adversely affect ΔG. This in turn lowers the binding affinity (K_i), which is related to ΔG by the equation $\Delta G = -RT \ln K_i$. A totally rigid molecule, on the other hand, is already in its active conformation, and there is no loss of entropy involved in binding to the target. If the binding interactions (ΔH) are exactly the same as for the more flexible molecule, the rigid molecule will have the better overall binding affinity.

Incorporating the skeleton of a flexible drug into a ring is the usual way of locking a conformation,

and for our model compound the analogue shown in Fig. 10.56 would be suitably rigid.

An example of rigidification concerns the acyclic pentapeptide shown in Fig. 10.57. This is a highly flexible molecule that acts as an inhibitor of a proteolytic enzyme. It was decided to rigidify the structure by linking the asparagine residue with the aromatic ring of the phenylalanine residue to form a macrocyclic ring. The resulting structure showed a 400-fold increase in inhibition.

Similar rigidification tactics have been useful in the development of the antihypertensive agent **cilazapril** (section 15.10.2) and the sedative **etorphine** (section 21.3.4).

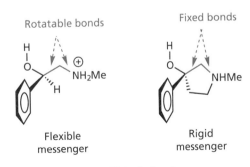

Figure 10.56 Locked analogue.

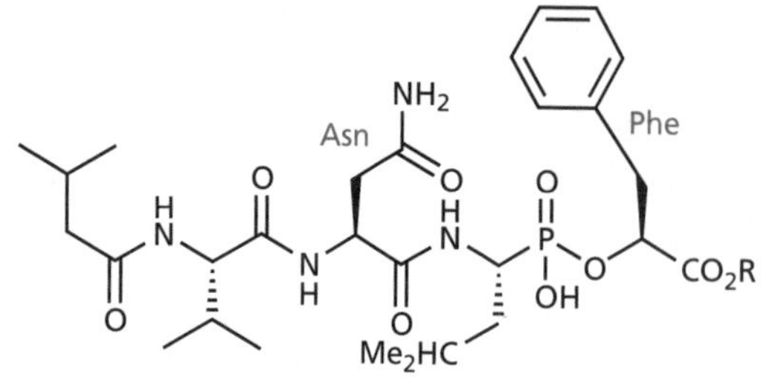

Acyclic pentapeptide (K_i 42 nM)

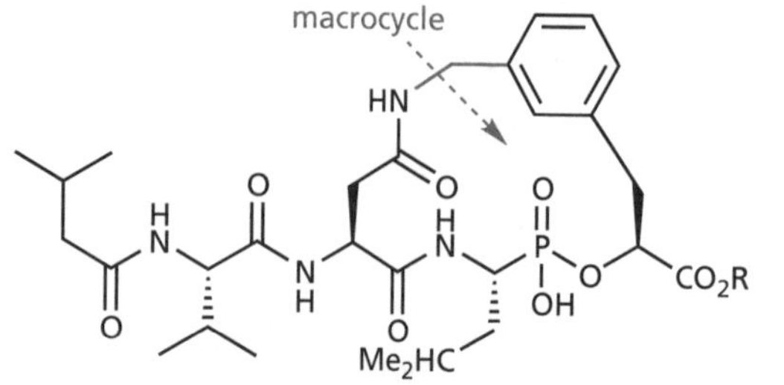

Rigidified pentapeptide (K_i 0.1 nM)

Figure 10.57 Rigidification of an acyclic pentapeptide.

Locking a rotatable bond into a ring is not the only way a structure can be rigidified. A flexible side chain can be partially rigidified by incorporating a rigid functional group such as a double bond, alkyne, amide, or aromatic ring (see Box 10.3).

Rigidification also has potential disadvantages. Rigidified structures may be more complicated to synthesize. There is also no guarantee that rigidification will retain the active conformation; it is perfectly possible that rigidification will lock the compound into an inactive conformation. Another disadvantage involves drugs acting on targets which are prone to mutation. If a mutation alters the shape of the binding site, then the drug may no longer be able to bind, whereas a more flexible drug could adopt a different conformation that *could* bind.

10.3.10 Conformational blockers

We have seen how rigidification tactics can restrict the number of possible conformations for a compound. Another tactic that has the same effect is the use of conformational blockers. In certain situations, a quite simple substituent can hinder the free rotation of a single bond. For example, introducing a methyl substituent to the dopamine (D_3) antagonist (I in

---

BOX 10.3 RIGIDIFICATION TACTICS IN DRUG DESIGN

The diazepine (I) is an inhibitor of platelet aggregation and binds to its target receptor by means of a guanidine functional group and a diazepine ring system. These binding groups are linked together by a highly flexible chain and it was considered desirable to make this chain less flexible. Structures II and III are examples of active compounds where the connecting chain between the guanidine group and the bicyclic system has been partially rigidified by the introduction of rigid functional groups.

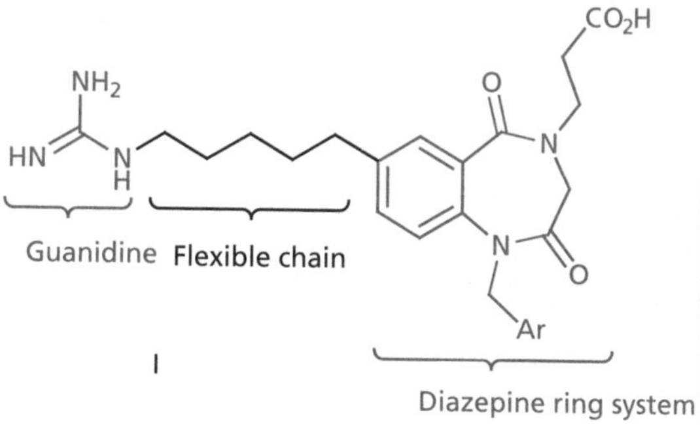

I

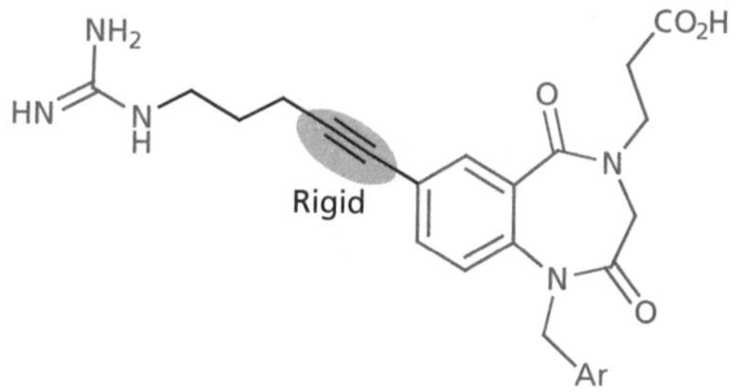

II

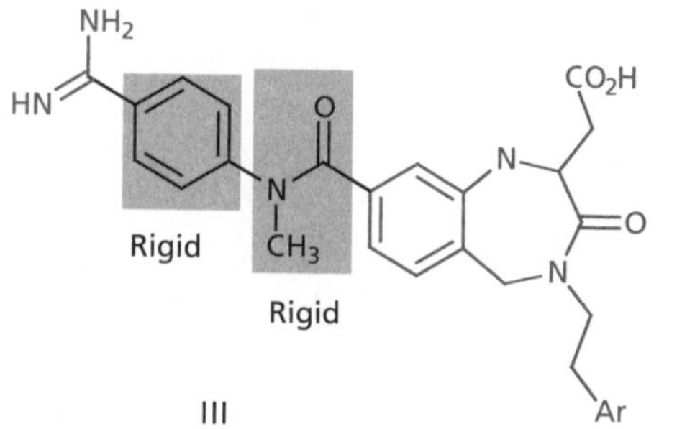

III

Rigidification of flexible chains.

Fig. 10.58) gives structure II and results in a dramatic reduction in affinity. The explanation lies in a bad steric clash between the new methyl group and an *ortho* proton on the neighbouring ring which prevents both rings being in the same plane. Free rotation around the bond between the two rings is no longer possible and so the structure adopts a conformation where the two rings are at an angle to each other. In structure I, free rotation around the connecting bond allows the molecule to adopt a conformation where the aromatic rings are co-planar—the active conformation for the receptor. In this case, a conformational blocker 'rejects' the active conformation. An example of a conformational blocker favouring the active conformation can be seen in the case study described in section 15.18.10, and also in the development of the anticancer agent imatinib (section 18.6.2.2). In the latter case, conformational restraint not only increased activity, but also introduced selectivity between two similar target binding sites.

10.3.11 Structure-based drug design and molecular modelling

10.3.11.1 The tools: X-ray crystallography and molecular modelling

So far we have discussed the traditional strategies of drug design. These were frequently carried out with no knowledge of the target structure, and the results

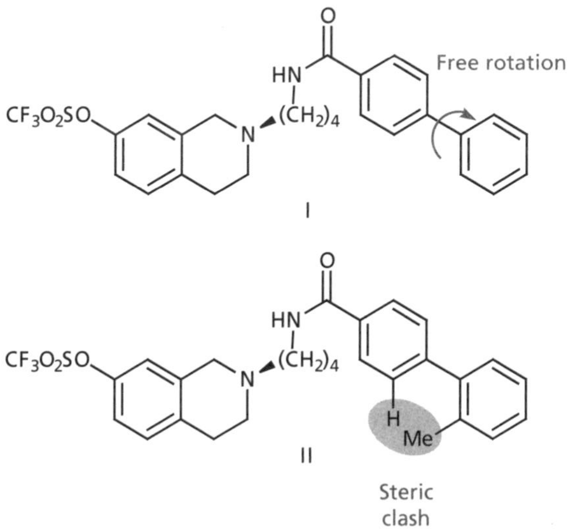

obtained were useful in providing information about the target binding site. Clearly, if a drug has an important binding group, there must be a complementary binding group present in the binding site of the receptor or enzyme.

If the macromolecular target can be isolated and crystallized, then it may be possible to determine the structure by using X-ray crystallography. Unfortunately, this does not reveal where the binding site is, and so it is better to crystallize the protein with a known inhibitor or antagonist (ligand) bound to the binding site. X-ray crystallography can then be used to determine the structure of the complex and this can be downloaded to a computer. Molecular modelling software is then used to identify where the ligand is, and thus identify the binding site. Moreover, by measuring the distances between the atoms of the ligand and neighbouring atoms in the binding site, it is possible to identify important binding interactions such that binding groups in the ligand and binding regions in the binding site can be identified. Once this has been done, the ligand can be removed from the binding site *in silico*, and novel lead compounds can be inserted *in silico* to see how well they fit. (The term *in silico* indicates that the process concerned is being carried out virtually, on a computer using molecular modelling software.) Regions in the binding site which are not occupied by the lead compound can be identified and used to guide the medicinal chemist as to what modifications and additions can be made to design a new drug that occupies more of the available space and binds more strongly as a result. The drug can then be synthesized and tested for activity. If it proves active, the target protein can be crystallized with the new drug bound to the binding site, and then X-ray crystallography and molecular modelling can be used again to identify the structure of the complex and to see if binding took place as expected. This approach is known as structure-based drug design and is covered in more detail in sections 15.10–15.18.

A related process is known as *de novo* drug design. This involves the design of a novel drug structure, based on a knowledge of the binding site alone. This is quite a demanding exercise, but there are examples where *de novo* design has successfully led to a novel lead compound which can then be the start point for structure-based drug design (section 15.15).

Structure-based drug design cannot be used in all cases. Sometimes the target for a lead compound may

Figure 10.58 Conformational blocking.

not have been identified, and even if it has been identified, it may not be possible to crystallize it. This is particularly true for membrane-bound proteins. One way round this is to identify a protein which is thought to be similar to the target protein and which *has* been crystallized and studied by X-ray crystallography. The structural and mechanistic information obtained from that analogous protein can then be used to design drugs for the target protein (sections 10.3.11.2 and 15.14).

Molecular modelling can also be used to study different compounds which are thought to interact with the same target. The structures can be compared and the important pharmacophore identified (section 15.11), allowing the design of novel structures containing the same pharmacophore. Compound databanks can be searched for those pharmacophores to identify novel lead compounds (section 15.13).

There are many other applications of molecular modelling in medicinal chemistry, some of which are described in Chapter 15. However, a warning! It is important to appreciate that molecular modelling studies usually tackle only one part of a much bigger problem—the design of an effective drug. True, one might design a compound that binds perfectly to a particular enzyme or receptor, but if the compound cannot be synthesized or never reaches the target protein in the body, it is a useless drug.

10.3.11.2 Case study: the design of ACE inhibitors

The design of ACE inhibitors demonstrates how it is possible to design drugs for a protein target in a rational manner even if the structure of the target has not been determined. The **angiotensin-converting enzyme** (ACE) is a membrane-bound enzyme which has been difficult to isolate and study. It is a member of a group of enzymes called the **zinc metalloproteinases** and catalyses the hydrolysis of a dipeptide fragment from the end of a decapeptide called **angiotensin I** to give the octapeptide **angiotensin II** (Fig. 10.59).

Angiotensin II is an important hormone that causes blood vessels to constrict, resulting in a rise in blood pressure. Therefore, ACE inhibitors are potential antihypertensive agents because they inhibit the production of angiotensin II. Although the enzyme ACE could not be isolated, the design of ACE inhibitors was helped by studying the structure and mechanism of another zinc metalloproteinase that could—an enzyme called **carboxypeptidase**. This enzyme splits the terminal amino acid from a peptide chain as shown in Fig. 10.60 and is inhibited by L-**benzylsuccinic acid**.

The active site of carboxypeptidase (Fig. 10.61) contains a charged arginine unit (Arg-145) and a zinc ion, which are both crucial in binding the substrate peptide. The peptide binds such that the terminal carboxylic acid is ionically bound to the arginine unit,

Asp-Arg-Val-Tyr-Ile-His-Pro-Phe-His-Leu $\xrightarrow{\text{ACE}}$ Asp-Arg-Val-Tyr-Ile-His-Pro-Phe + His-Leu

Angiotensin I   Angiotensin II

Figure 10.59 Reaction catalysed by ACE.

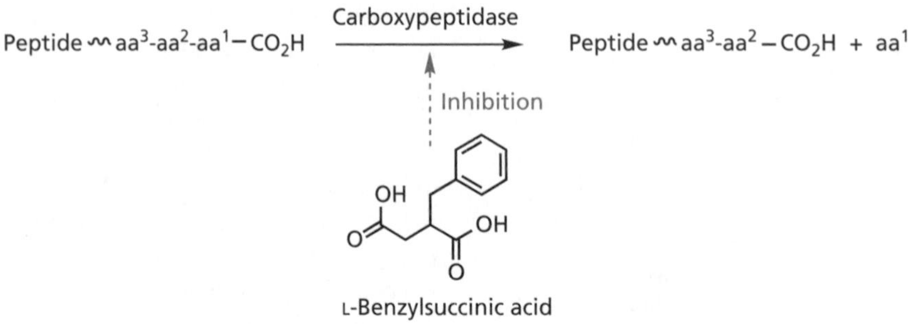

Peptide $\sim$ aa^3-aa^2-aa^1$-CO_2H$ $\xrightarrow{\text{Carboxypeptidase}}$ Peptide $\sim$ aa^3-aa^2$-CO_2H$ + aa^1

Inhibition

L-Benzylsuccinic acid

Figure 10.60 Hydrolysis by carboxypeptidase.

while the carbonyl group of the terminal peptide bond is bound to the zinc ion. There is also a pocket called the S1′ pocket which can accept the side chain of the terminal amino acid (Phe in the example shown). Hydrolysis of the terminal peptide then takes place.

The design of the carboxypeptidase inhibitor L-benzylsuccinic acid was based on the hydrolysis products arising from this enzymatic reaction. The benzyl group was included to occupy the S1′ pocket, while the adjacent carboxylate anion was present to

form an ionic interaction with Arg-145. The second carboxylate was present to act as a ligand to the zinc ion, mimicking the carboxylate ion of the hydrolysis product. L-Benzylsuccinic acid binds as shown in Fig. 10.62. However, it cannot be hydrolysed as there is no peptide bond present, and so the enzyme is inhibited for as long as the compound stays attached.

An understanding of the above mechanism and inhibition helped in the design of ACE inhibitors. First

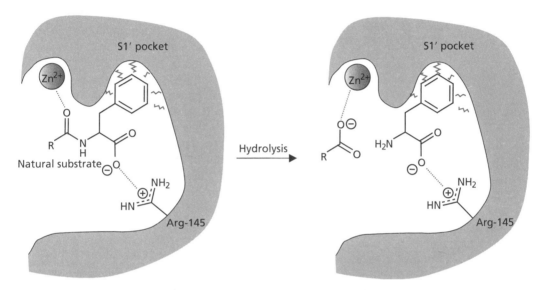

Figure 10.61 Binding site interactions for carboxypeptidase.

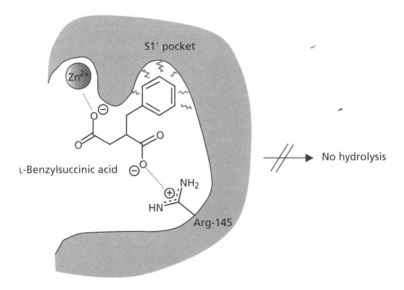

Figure 10.62 Inhibition by L-benzylsuccinic acid.

of all, it was assumed that the active site contained the same zinc ion and arginine group. However, as ACE splits a dipeptide unit from the peptide chain rather than an amino acid, these groups are likely to be further apart and so an analogous inhibitor to benzylsuccinic acid would be a succinyl-substituted amino acid. **Succinyl proline** was chosen, since proline is

Succinyl proline

Glu-Trp-Pro-Arg-Pro-Gln-Ile-Pro

Teprotide

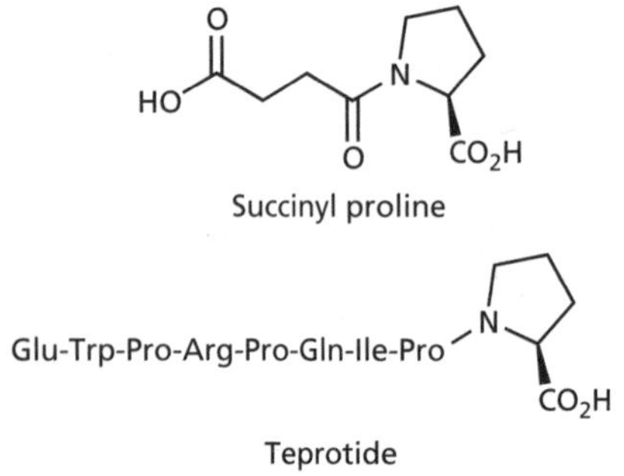

Figure 10.63 ACE inhibitors.

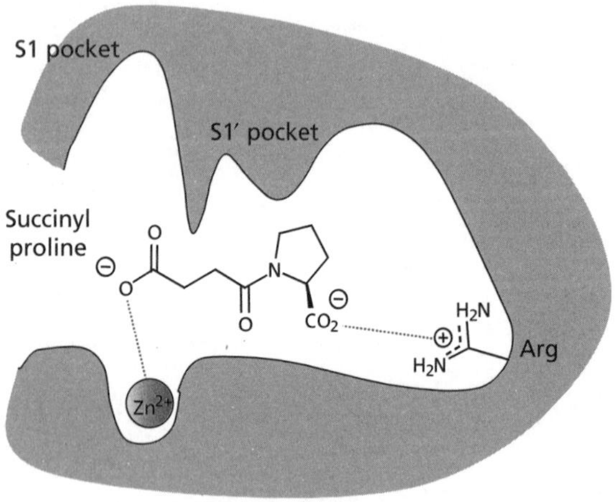

present on the terminus of **teprotide** (a known inhibitor of ACE; Fig. 10.63).

Succinyl proline did indeed inhibit ACE and it was proposed that both carboxylate groups were ionized, one interacting with the arginine group and one with the zinc ion (Fig. 10.64). It was now argued that there must be pockets available to accommodate amino acid side chains (pockets S1 and S1'). The strategy of extension was now employed to find a group that would fit the S1' pocket and increase the binding affinity. A methyl group fitted the bill and resulted in an increase in activity (Fig. 10.65). The next step was to see whether there was a better group than the carboxylate ion to interact with zinc, and it was discovered that a thiol group led to increased activity. This resulted in **captopril**, which was the first non-peptide ACE inhibitor to become commercially available.

The next advance involved extension strategies aimed at finding a group that would fit the S1 pocket— normally occupied by the phenylalanine residue in angiotensin I. This time **glutarylproline** was used as the lead compound instead of succinyl proline, resulting in the ACE inhibitor **enalaprilate** (Fig. 10.66).

Lisinopril (Fig. 9.10) is another example of an ACE inhibitor which is similar to enalaprilate, but where the methyl substituent has been extended to an aminobutyl substituent. In 2003, a crystal structure of ACE complexed with lisinopril was finally determined by X-ray crystallography. This provided a detailed picture of the 3D structure of ACE and how lisinopril binds to the active site. In fact, there is a marked difference in structure between ACE and carboxypeptidase A, which means that ACE inhibitors do not bind in the way that was originally thought. For example, the ionic interaction originally thought to involve arginine involves a lysine residue instead. Now that an accurate picture has been obtained of the active site and the manner in which lisinopril binds, it is possible that a new generation of ACE inhibitors will be designed with improved

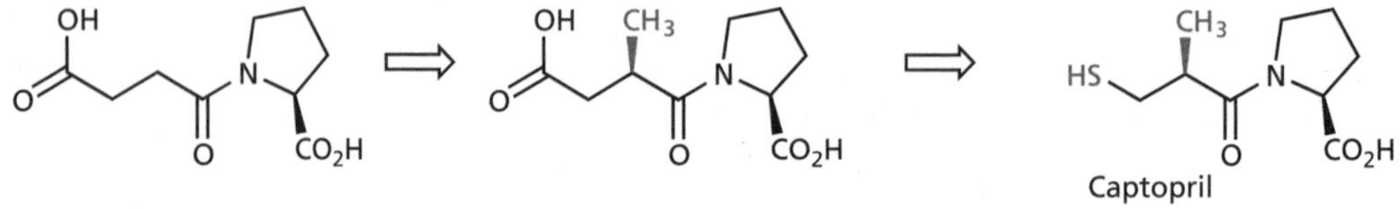

Figure 10.64 Binding site interaction for ACE.

Figure 10.65 Development of captopril.

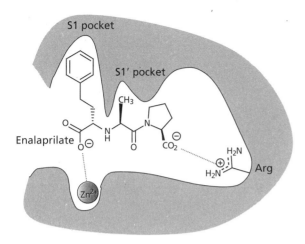

Figure 10.66 Enalaprilate.

binding characteristics using structure-based drug design.

10.3.12 Drug design by NMR

The use of NMR in designing lead compounds has already been discussed in section 9.4.10. This can also be seen as a method of drug design, since the focus is not only on designing a lead compound, but in designing a *potent* lead compound. Usually, drug design aims to optimize a lead compound once it has been discovered. In the NMR method, optimization of the component parts (epitopes) is carried out first to maximize binding interactions; then they are linked together to produce the final compound.

NMR is also being increasingly used to identify the structure of target proteins that cannot be crystallized and studied by X-ray crystallography. Once the structure has been identified, molecular modelling techniques can be used for drug design as described in section 10.3.11.

10.3.13 The elements of luck and inspiration

It is true to say that drug design has become more rational, but the role of chance or the need for hard-working, mentally alert medicinal chemists has not yet been eliminated. Most of the drugs still on the market were developed by a mixture of rational design, trial and error, hard graft, and pure luck. There

is a growing number of drugs that were developed by rational design such as the ACE inhibitors (section 10.3.11.2), thymidylate synthase inhibitors (section 15.15.1), HIV protease inhibitors (section 17.7.4), neuraminidase inhibitors (section 17.8.3), **pralidoxime** (section 19.18), and **cimetidine** (section 22.2), but they are still in the minority.

Frequently, the development of drugs is helped by watching the literature to see what works on related compounds and what doesn't, then trying out similar alterations to one's own work. It is often a case of groping in the dark, with the chemist asking whether the addition of a group at a certain position will have a steric effect, an electronic effect, or a bonding effect. Even when drug design is carried out on rational lines, good fortune often has a role to play (see Box 10.4).

KEY POINTS

- Drug optimization aims to maximize the interactions of a drug with its target binding site in order to improve activity and selectivity, and to minimize side effects. Designing a drug that can be synthesized efficiently and cheaply is another priority.

- The length and size of alkyl substituents can be modified to fill up hydrophobic pockets in the binding site or to introduce selectivity for one target over another. Alkyl groups attached to heteroatoms are most easily modified.

- Aromatic substituents can be varied in character and/or ring position.

- Extension is a strategy where extra functional groups are added to the lead compound in order to interact with extra binding regions in the binding site.

- Chains connecting two important binding groups can be modified in length in order to maximize the interactions of each group with the corresponding binding regions.

- Rings linking important binding groups can be expanded or contracted such that the binding groups bind efficiently with relevant binding regions.

- Rings acting as scaffolds for important binding groups can be varied in order to give novel classes of drugs which may have improved properties.

- Rings can be fused to existing rings in order to maximize binding interactions or to increase selectivity for one target over another.

BOX 10.4 A SLICE OF LUCK

The development of the β-blocker **propranolol** was aided by a slice of good fortune. Chemists at ICI were trying to improve on a drug called **pronethalol**, the first β-blocker to reach the market. It was known that the naphthalene ring and the ethanolamine segment were important to activity and selectivity, so these groups had to be retained. Therefore, it was decided to study what would happen if the distance between these two groups was extended (*chain extension*). Perhaps by doing so, the two groups would interact with their respective binding regions more efficiently.

Various segments were to be inserted, one of which was the OCH$_2$ moiety (Fig. 1a). The analogue that would

have been obtained is shown as structure I. β-Naphthol was the starting material, but was not immediately available. Rather than waste the day, α-naphthol was used instead (Fig. 1b). The result was propranolol, which has proved a successful drug in the treatment of angina for many years. When the original target structure was eventually made, it showed little improvement over the original lead compound, pronethalol.

A further interesting point concerning this work is that propranolol had been synthesized some years earlier. However, the workers involved had not been searching for β-blocking activity and had not recognized the potential of the compound.

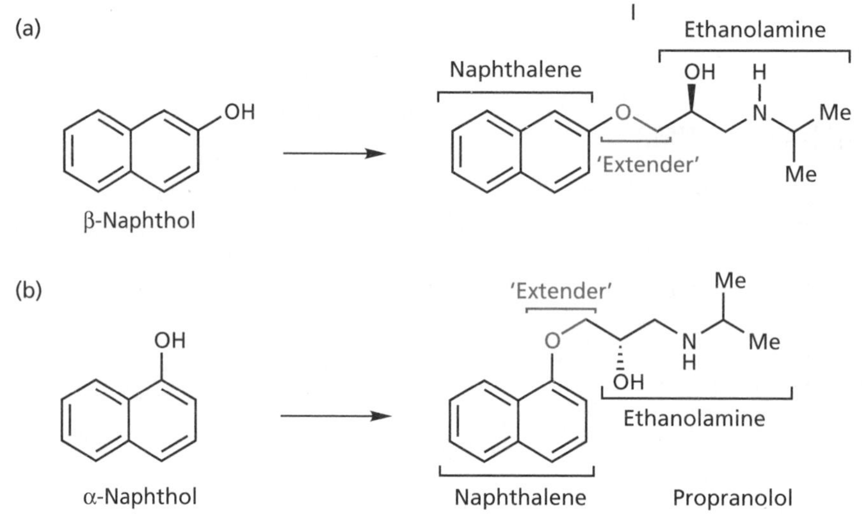

Figure 1 The development of the β-blocker propranolol.

- Classical and non-classical isosteres are frequently used in drug optimization.

- Simplification involves removing functional groups from the lead compound that are not part of the pharmacophore. Unnecessary parts of the carbon skeleton or asymmetric centres can also be removed in order to design drugs that are easier and cheaper to synthesize. Oversimplification can result in molecules that are too flexible, resulting in decreased activity and selectivity.

- Rigidification is used on flexible lead compounds. The aim is to reduce the number of conformations available while

retaining the active conformation. Locking rotatable rings into ring structures or introducing rigid functional groups are common methods of rigidification.

- Conformational blockers are groups which are introduced into a lead compound to reduce the number of conformations that the molecule can adopt through steric interactions.

- Structure-based drug design makes use of X-ray crystallography and computer-based molecular modelling to study how a lead compound and its analogues bind to a target binding site.

- NMR studies can be used to determine protein structure and to design novel drugs.

- Serendipity plays a role in drug design and optimization.

10.4 A case study: oxamniquine

The development of **oxamniquine** (Fig. 10.67) is a nice example of how traditional strategies can be used in the development of a drug where the molecular target is unknown. It also demonstrates that strategies can be used in any order and may be used more than once.

Oxamniquine is an important drug in developing counties, used in the treatment of **schistosomiasis** (**bilharzia**). This disease affects an estimated 200 million people and is contracted by swimming or wading in infected water. The disease is carried by a snail whose flukes can penetrate human skin and enter the blood supply. In the body, eggs are produced which become trapped in organs and tissues, and this in turn leads to the symptoms of the disease.

The first stage in the development of oxamniquine was to find a lead compound, and so a study was made of compounds that were active against the parasite. The tricyclic structure **lucanthone** (Fig. 10.67) was chosen. It was known to be effective against some forms of the disease, but it was also toxic and had to be injected at regular intervals to remain effective. The goal was to increase the activity of the drug, broaden its activity, reduce side effects, and make it orally active.

Having found a lead compound, it was decided to try *simplifying* the structure to see whether the tricyclic

system was really necessary. Several compounds were made, and the most interesting structure was one where the two rings seen on the left in Fig. 10.67 had been removed. This gave a compound called **mirasan** (Fig. 10.67), which retained the right-hand aromatic ring containing the methyl and β-aminoethylamino side chains *para* to each other. *Varying substituents* showed that an electronegative chloro substituent, positioned where the sulfur atom had been, was beneficial to activity. Mirasan was active against the bilharzia parasite in mice, but not in humans.

It was now reasoned that the β-aminoethylamino side chain was important to receptor binding and would adopt a particular conformation in order to bind efficiently. This conformation would be only one of many conformations available to a flexible molecule such as mirasan, and so there would only be a limited chance of it being adopted at any one time. Therefore, it was decided to restrict the number of possible conformations by incorporating the side chain into a ring (*rigidification*). This would increase the chances of the molecule having the correct conformation when it approached its target binding site. There was the risk that the active conformation itself would be disallowed by this tactic and so, rather than incorporate the whole side chain into a ring, compounds were first designed such that only portions of the chain were included.

The bicyclic structure (I in Fig. 10.68) contains one of the side chain bonds fixed in a ring to prevent rotation round that bond. It was found that this gave a dramatic improvement in activity. The compound was still not active in humans, but, unlike mirasan, it was active in monkeys. This gave hope that the chemists were on the right track. Further *rigidification* led to structure II in Fig. 10.68, where two of the side

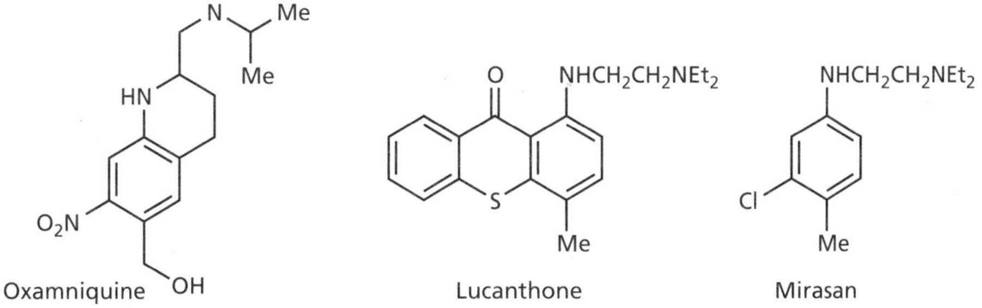

Figure 10.67 Oxamniquine, lucanthone and mirasan.

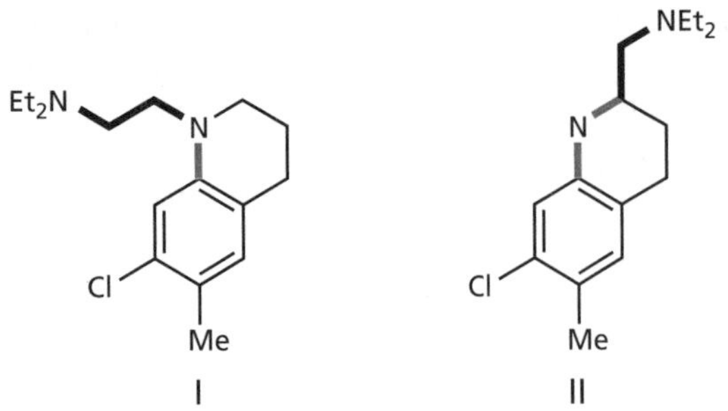

Figure 10.68 Bicyclic structures I and II (restricted bonds in colour).

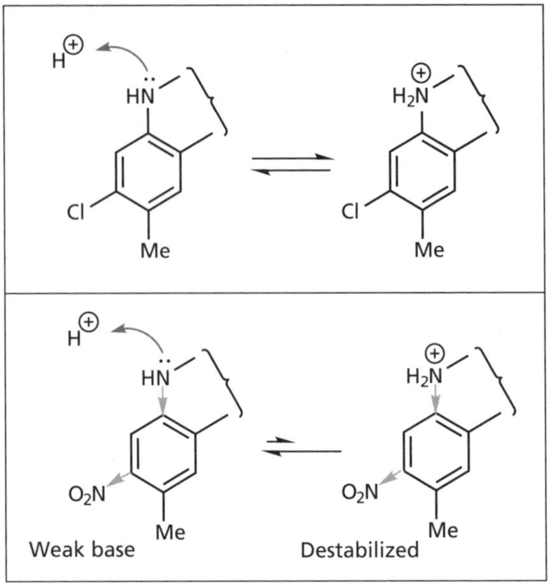

Figure 10.69 Effect of aromatic substituents on pK_a.

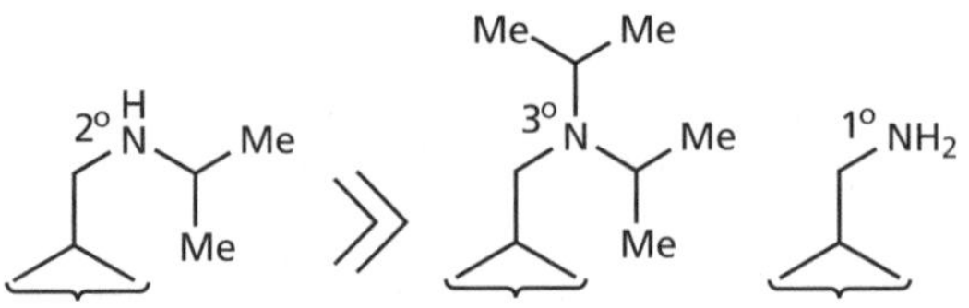

chain bonds were constrained. This compound showed even more activity in mouse studies and it was decided to concentrate on this compound.

By now, the structure of the compound had been altered significantly from mirasan. When this is the case, it is advisable to check whether past results still hold true. For example, does the chloro group still have to be *ortho* to the methyl group? Can the chloro group be changed for something else? Novel structures may fit the binding site slightly differently from the lead compound such that the binding groups are no longer in the optimum positions for binding.

Therefore, structure II was modified by *varying substituents* and *substitution patterns* on the aromatic ring, and by *varying alkyl substituents* on the amino groups. Chains were also *extended* to search for other possible binding regions.

The results and possible conclusions were as follows.

- The substitution pattern on the aromatic ring could not be altered and was essential for activity. Altering the substitution pattern presumably places the essential binding groups out of position with respect to their binding regions.

- Replacing the chloro substituent with more electronegative substituents improved activity, with the nitro group being the best substituent. Therefore, an electron-deficient aromatic ring is beneficial to activity. One possible explanation for this could be the effect of the neighbouring aromatic ring on the basicity of the cyclic nitrogen atom. A strongly electron deficient aromatic ring would pull the cyclic nitrogen's lone pair of electrons into the ring, thus

Figure 10.70 Relative activity of amino side chains.

reducing its basicity (Fig. 10.69). This in turn might improve the pK_a of the drug such that it is less easily ionized and is able to pass through cell membranes more easily (section 11.1.4).

- The best activities were found if the amino group on the side chain was secondary rather than primary or tertiary (Fig. 10.70).

- The alkyl group on this nitrogen could be increased up to four carbon units with a corresponding increase in activity. Longer chains led to a reduction in activity. The latter result might imply that large substituents are too bulky and prevent the drug from binding to the binding site. Acyl groups eliminated activity altogether, emphasizing the importance of this nitrogen atom. Most likely, it is ionized and interacts with the receptor through an ionic bond (Fig. 10.71).

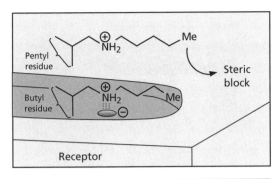

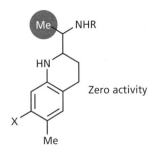

Figure 10.73 Addition of a methyl group.

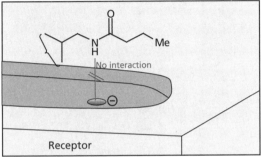

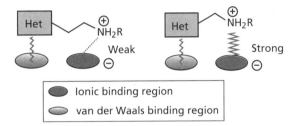

Figure 10.74 Effect of extension of the side chain.

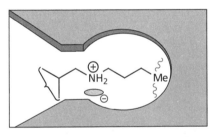

Figure 10.71 Proposed ionic binding interaction.

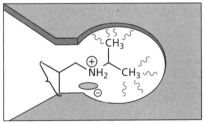

Figure 10.72 Branching of the alkyl chain.

- Branching of the alkyl chain increased activity. A possible explanation is that branching increases van der Waals interactions to a hydrophobic region of the binding site (Fig. 10.72). Alternatively, the lipophilicity of the drug might be increased, allowing easier passage through cell membranes.

- Putting a methyl group on the side chain eliminated activity (Fig. 10.73). A methyl group is a bulky group compared with a proton and it is possible that it prevents the side chain taking up the correct binding conformation—*conformational blocking*.

- Extending the length of the side chain by an extra methylene group eliminated activity (Fig. 10.74). This tactic was tried in case the binding groups were not far enough apart for optimum binding. This result suggests the opposite.

The optimum structure based on these results was structure III (Fig. 10.75). It has one asymmetric centre and, as one might expect, the activity was much greater in one enantiomer than it was in the other.

The tricyclic structure IV (Fig. 10.75) was also constructed. In this compound, the side chain is fully incorporated into a ring structure, drastically restricting the number of possible conformations (*rigidification*). As mentioned earlier, there was a risk that the active

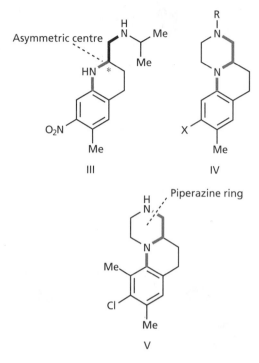

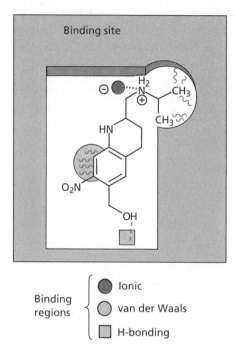

Figure 10.75 The optimum structure (III) and the tricyclic structures (IV) and (V) (restricted bonds in colour).

Figure 10.76 Bonding interactions of oxamniquine.

conformation would no longer be allowed, but in this case good activity was still obtained. The same *variations* as above were carried out to show that a secondary amine was essential and that an electronegative group on the aromatic ring was required. However, some conflicting results were obtained compared with the previous results for structure III. A chloro substituent on the aromatic ring was better than a nitro group, and it could be in either of the two possible *ortho* positions relative to the methyl group. These results demonstrate that optimizing substituents in one structure does not necessarily mean that they will be optimum in another. One possible explanation for the chloro substituent being better than the nitro is that a less electronegative substituent is required to produce the optimum pK_a or basicity for membrane permeability.

Adding a further methyl group to the aromatic ring to give structure (V) (Fig. 10.75) increased activity. It was proposed that the bulky methyl group could

interact with the piperazine ring, causing it to twist out of the plane of the other two rings (*conformational blocker*). The resulting increase in activity suggests that a better fitting conformation is obtained for the binding site.

Compound V was three times more active than structure III. However, structure III was chosen for further development. The decision to choose III rather than V was based on preliminary toxicity results, as well as the fact that it was cheaper to synthesize. Structure III is a simpler molecule and in general simpler molecules are easier and cheaper to synthesize.

Further studies on the metabolism of related compounds then revealed that the aromatic methyl group on these compounds is oxidized to a hydroxymethylene group, and that the resulting metabolites were more active compounds. This suggested that the new hydroxyl group was involved in an extra hydrogen bonding interaction with the binding site. Therefore, the methyl group on III was replaced by a hydroxymethylene group to give oxamniquine (Fig. 10.76). The drug was put on the market in 1975, 11 years after the start of the project. It is now believed that compound III itself is inactive because of the lack of the extra hydrogen bonding interaction.

- Lucanthone was the lead compound, which was simplified to give mirasan.

- The strategies of rigidification, substituent variation, extension, chain variation, and conformational blocking were all involved in the design of oxamniquine.

- Metabolic studies contributed to the final design of oxamniquine.

KEY POINTS

- Oxamniquine is an example of a drug designed by classical methods.

QUESTIONS

1. **DU 122290** was developed from **sultopride** (Fig. 10.49) and shows improved activity and selectivity. Suggest possible reasons for this.

2. Methotrexate inhibits the enzyme dihydrofolate reductase. The pteridine ring system of methotrexate binds to the binding site as shown in Fig. 10.27. Suggest how dihydrofolate (the natural substrate for the enzyme) might bind.

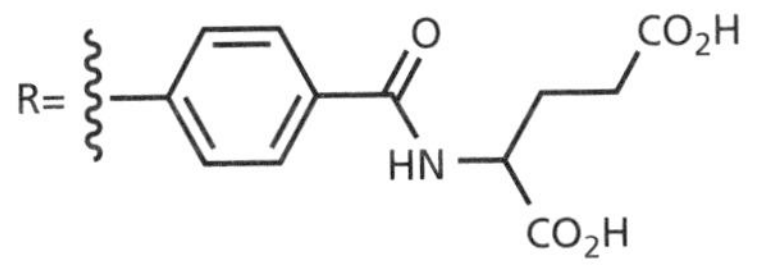

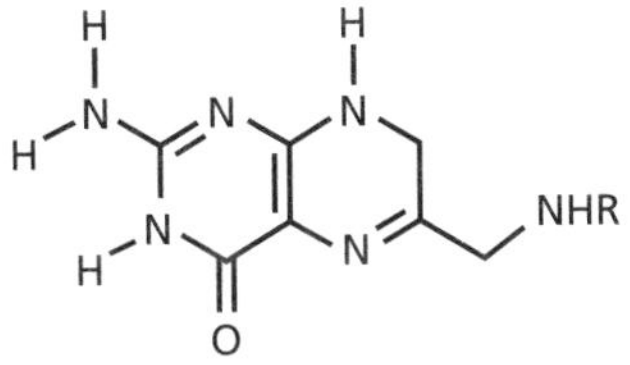

Dihydrofolate

3. A lead compound containing a methyl ester was hydrolysed to give a carboxylic acid. An *in vivo* bioassay suggested that the ester was active and that the acid was

inactive. However, an *in vitro* bioassay suggested that the ester was inactive and that the acid was active. Explain these contradictory results.

4. A lead compound contains an aromatic ring. The following structures were made as analogues. Structures I and II were similar in activity to the lead compound, whereas structure III showed a marked increase in activity. Explain these results and describe the strategies involved.

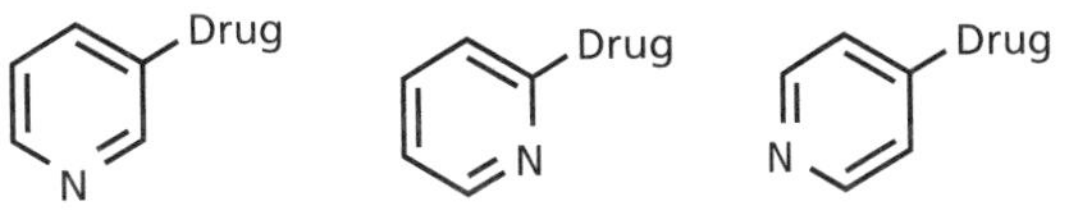

5. The pharmacophore of cocaine is shown in Box 10.2. Identify possible cyclic analogues which are simpler than cocaine and which would be expected to retain activity.

6. Procaine (Box 10.2) has been a highly successful local anaesthetic and yet there are three bonds between the important ester and amine binding groups, compared to four in cocaine. This might suggest that these groups are too close together in procaine. In fact, this is not the case. Suggest why not.

7. The aromatic amine on procaine is not present in cocaine. Comment on its possible role.

8. Explain how you would apply the principles of rigidification to structure IV below in order to improve its pharmacological properties. Give two specific examples of rigidified structures.

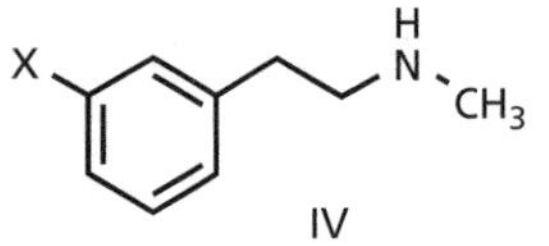

IV

9. Combretastatin is an anticancer agent discovered from an African plant. Analogue V is more active than combretastatin whereas analogue VI is less active. What strategy was used in designing analogues V and VI? Why is analogue V more active than combretastatin, and analogue VI less active?

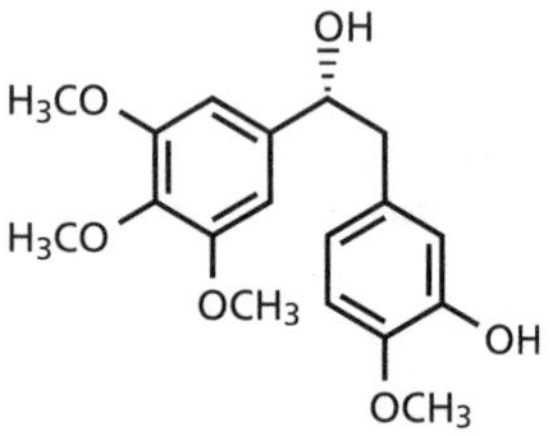

Combretastatin

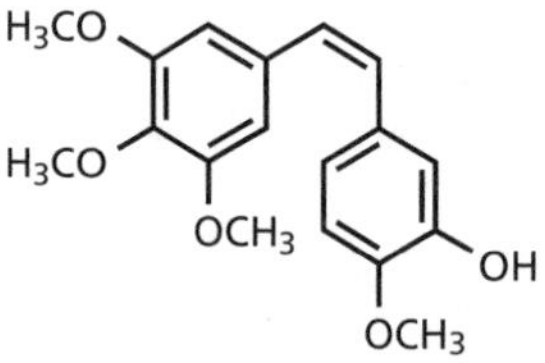

V

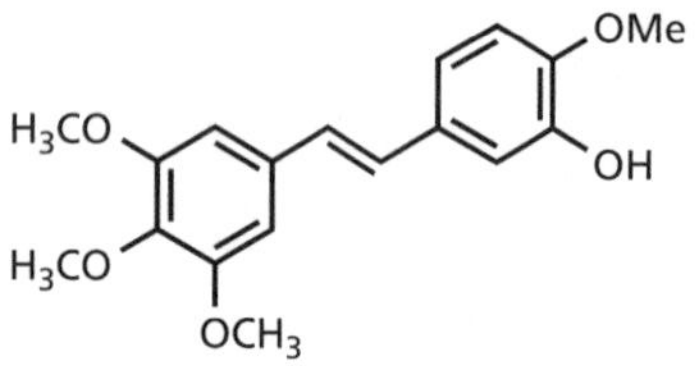

VI

10. Structure VII is a serotonin antagonist. A methyl group has been introduced into analogue VIII, resulting in increased activity. What role does the methyl group play and what is the term used for such a group? Explain why increased activity arises.

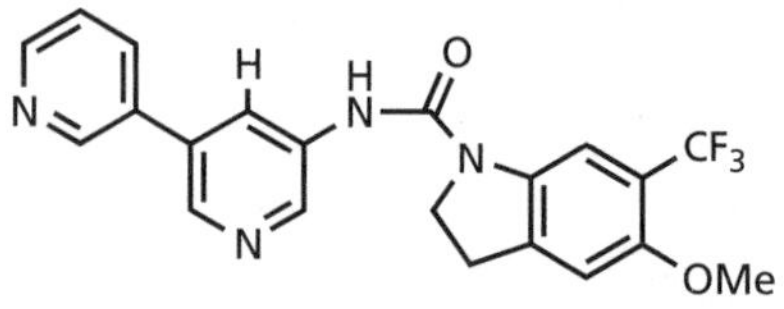

VII

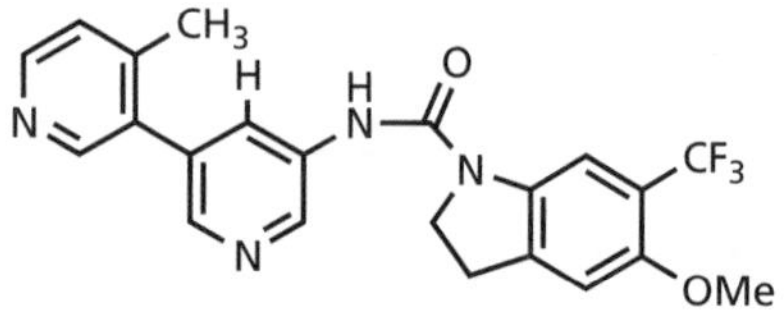

VIII

FURTHER READING

Acharya, K. R. *et al.* (2003) ACE revisited: a new target for structure-based drug design. *Nature Reviews Drug Discovery*, 2, 891–902.

Bolin, J. T., Filman, D. J., Matthews, D. A., Hamlin, R. C., and Kraut, J. (1982) Crystal structures of *Escherichia coli* and *Lactobacillus casei* dihydrofolate reductase refined at 1.7 Å resolution. I. General features and binding of methotrexate. *Journal of Biological Chemistry*, 257, 13650–13662 (methotrexate binding).

Ganellin, C. R. and Roberts, S. M. (eds.) (1994) Angiotensin-converting enzyme (ACE) inhibitors and the design of cilazapril. Chapter 9 in: *Medicinal chemistry—the role of organic research in drug research*, 2nd edn. Academic Press, London.

Hruby, V. J. (2002) Designing peptide receptor agonists and antagonists. *Nature Reviews Drug Discovery*, 1, 847–858.

Jeffrey, G. A. (1991) *Hydrogen bonding in biological structures.* Springer-Verlag, New York.

Khan, A. R. *et al.* (1998) Lowering the entropic barrier for binding conformationally flexible inhibitors to enzymes. *Biochemistry*, **37**, 16839–16845.

Luca, S. *et al.* (2003) The conformation of neurotensin bound to its G protein-coupled receptor. *Proceedings of the National Academy of Sciences of the USA*, **100**, 10706–10711 (active conformation by NMR).

Meyer, E. G. *et al.* (1995) Backward binding and other structural surprises. *Perspectives in Drug Discovery and Design*, **3**, 168–195.

Roberts, S. M. and Price, B. J. (eds.) (1985) Oxamniquine: a drug for the tropics. Chapter 14 in: *Medicinal chemistry—the role of organic research in drug research*. Academic Press, London.

Saunders, J. (2000) Inhibitors of angiotensin converting enzyme as effective antihypertensive agents. Chapter 1 in: *Top drugs: top synthetic routes*. Oxford University Press, Oxford.

Zaman, M. A., Oparil, S., and Calhoun, D. A. (2002) Drugs targeting the renin-angiotensin-aldosterone system. *Nature Reviews Drug Discovery*, **1**, 621–636.

Titles for general further reading are listed on p. 711.

11 Drug design: optimizing access to the target

In Chapter 10, we looked at drug design strategies aimed at optimizing the binding interactions of a drug with its target. However, the compound with the best binding interactions is not necessarily the best drug to use in medicine. The drug needs to overcome many barriers if it is to reach its target in the body (Chapter 8). In this chapter we shall study design strategies which can be used to counter such barriers and which involve modification of the drug itself. There are other methods of aiding a drug in reaching its target, which include linking the drug to polymers or antibodies or encapsulating it within a polymeric carrier. These topics are discussed in sections 8.9 and 18.9. In general, the aim is to design drugs that will be absorbed into the blood supply, will reach their target efficiently and be stable enough to survive the journey, and will be eliminated in a reasonable period of time. This all comes under the banner of a drug's pharmacokinetics.

11.1 Improving absorption

The ease with which a drug is absorbed is determined by its hydrophilic/hydrophobic properties (section 8.2). Polarity and ionization are both important in this respect. Drugs which are too polar or strongly ionized do not easily cross the cell membranes of the gut wall. One solution is to inject them, but they are also quickly excreted (section 8.5). Non-polar drugs, on the other hand, are poorly soluble in aqueous solution, resulting in poor absorption. If they are injected, they are taken up by fat tissue. In general, the polarity and ionization of compounds can be altered by changing easily accessible substituents. Such changes are particularly open to a quantitative approach known as QSAR (quantitative structure–activity relationships), discussed in Chapter 13.

11.1.1 Variation of alkyl or acyl substituents to vary polarity

Molecules can be made less polar by masking a polar functional group with an alkyl or acyl group. For example, an alcohol or a phenol can be converted to an ether or ester, a carboxylic acid can be converted to an ester or amide, and primary and secondary amines can be converted to amides or to secondary and tertiary amines. Polarity is decreased not only by masking the polar group, but by the addition of an extra hydrophobic alkyl group—larger alkyl groups having a greater hydrophobic effect. One has to be careful in masking polar groups, though, as they may be important in binding the drug to its target, and masking them may prevent binding. If this turns out to be the case, it is often useful to mask the polar group temporarily such that the mask is removed once the drug is absorbed (section 11.6). Alternatively, extra alkyl groups could be added to the carbon skeleton of the molecule, but this usually involves a more involved synthesis.

If the molecule is not sufficiently polar, then the opposite strategy can be used—i.e. replacing large alkyl groups with smaller alkyl groups, or removing them entirely.

Sometimes there is a benefit in increasing the size of one alkyl group and decreasing the size of another. This is called a **methylene shuffle** and has been used to modify the hydrophobicity of compounds. An example of this is in the design of second-generation anti-impotence drugs based on **sildenafil (Viagra)**. It was found that adding extra bulk on the right-hand side of the molecule (as drawn in the figure) increased the drug's selectivity to its target. However, this also made the drug too lipophilic. Therefore, a methylene shuffle was carried out to alter a propyl and a methyl group to two ethyl groups. This resulted in reduced lipophilicity and better *in vivo* activity. The compound (**UK 343664**) entered clinical trials (Fig. 11.1).

11.1.2 Varying polar functional groups to vary polarity

A polar functional group could be added to a drug to increase its polarity. For example, the antifungal agent **tioconazole** is only used for skin infections because it is non-polar and is poorly soluble in blood. Introducing a polar hydroxyl group and more polar

heterocyclic rings led to the orally active antifungal agent **fluconazole** with improved solubility and enhanced activity against systemic infection (i.e. in the blood supply) (Fig. 11.2).

In contrast, the polarity of an excessively polar drug could be lowered by removing polar functional groups. This strategy has been particularly successful with lead compounds derived from natural sources (e.g. alkaloids or endogenous peptides). It is important, though, not to remove functional groups which are important to the drug's binding interactions with its target. In some cases, a drug may have too many essential polar groups. For example, the antibacterial agent in Fig. 11.3 has good *in vitro* activity but poor *in vivo* activity because of the large number of polar groups. Some of these groups can be removed or masked, but most of them are required for activity. As a result, the drug cannot be used clinically.

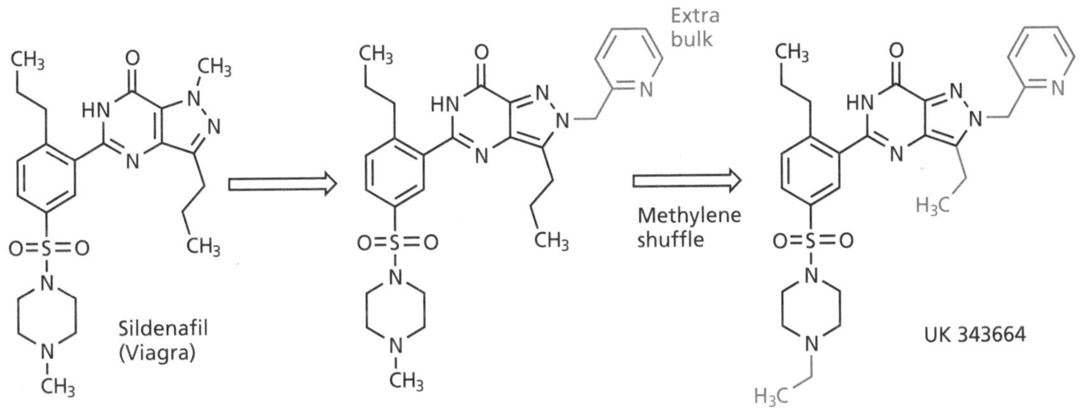

Figure 11.1 Methylene shuffle.

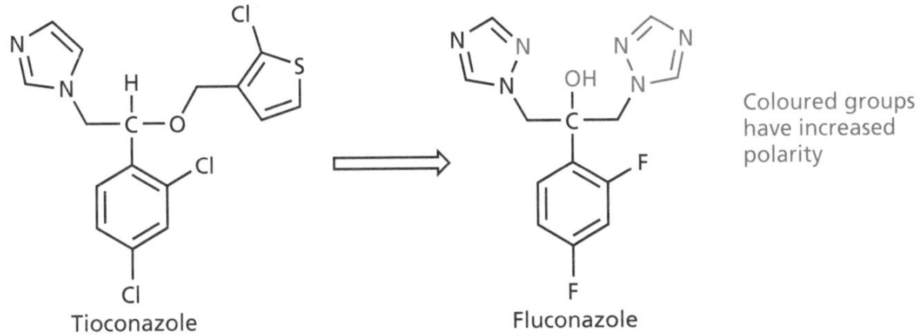

Figure 11.2 Increasing polarity in antifungal agents.

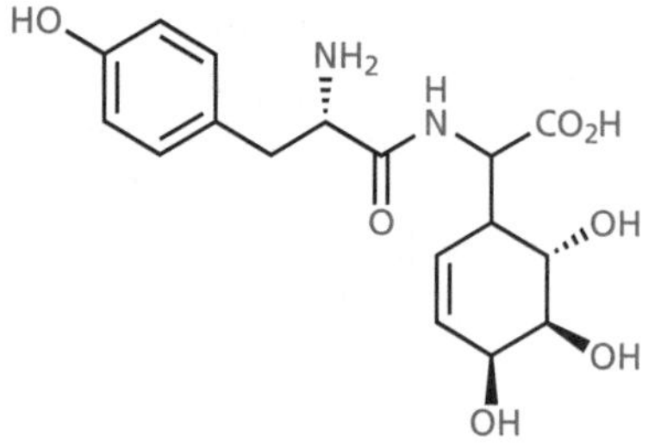

Figure 11.3 Excess polarity (coloured) in a drug.

Structure I

Amidine

11.1.3 Variation of *N*-alkyl substituents to vary pK_a

Drugs with a pK_a outside the range 6–9 tend to be too strongly ionized and are poorly absorbed through cell membranes. The pK_a can often be altered to bring it into the preferred range, however. For example, the pK_a of an amine can be altered by varying the alkyl substituents. It is sometimes difficult to predict how such variations will affect the pK_a. Extra *N*-alkyl groups or larger *N*-alkyl groups have an increased electron-donating effect which should increase basicity, but increasing the size or number of alkyl groups increases the steric bulk around the nitrogen atom. This hinders water molecules from solvating the ionized form of the base and prevents stabilization of the ion. This in turn decreases the basicity of the amine. Therefore, there are two different effects acting against each other. Nevertheless, varying alkyl substituents is a useful tactic to try.

A variation of this tactic is to 'wrap up' a basic nitrogen within a ring. For example, the benzamidine structure (I in Fig. 11.4) has antithrombotic activity, but the amidine group present is too basic for effective absorption. Incorporating the group into an iso-quinoline ring system (**PRO 3112**) reduced basicity and increased absorption.

11.1.4 Variation of aromatic substituents to vary pK_a

The pK_a of an aromatic amine or carboxylic acid can be varied by adding electron-donating or electron-withdrawing substituents to the ring. The position of the substituent relative to the amine or carboxylic acid is important if the substituent interacts with the ring through resonance (section 13.2.2). An illustration of this can be seen in the development of **oxamniquine** (section 10.4).

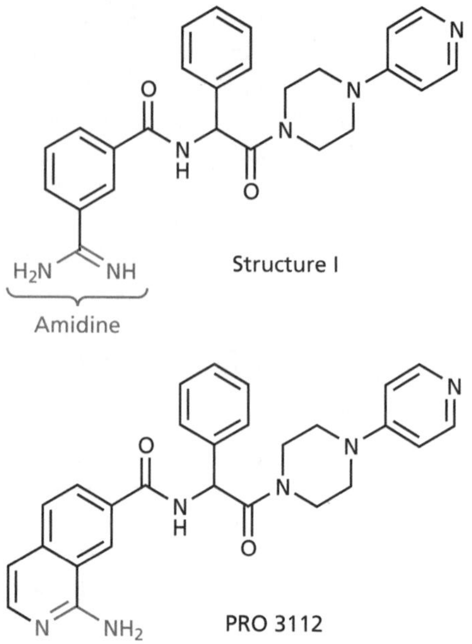

PRO 3112

Figure 11.4 Varying basicity in antithrombotic agents.

11.1.5 Bioisosteres for polar groups

The use of bioisosteres has already been described in section 10.3.7 in the design of compounds with improved target interactions. Bioisosteres have also been used as substitutes for important functional groups which are required for target interactions but pose pharmacokinetic problems. For example, a carboxylic acid is a highly polar group which can ionize and hinder absorption of any drug containing it. One way of getting round this problem is to mask it as an ester prodrug (section 11.6.1.1). Another strategy is to replace it with a bioisostere which has similar physicochemical properties but which offers some advantage over the original carboxylic acid. Several bioisosteres have been used for carboxylic acids, but among the most popular are 5-substituted tetrazoles (Fig. 11.5). Like carboxylic acids,

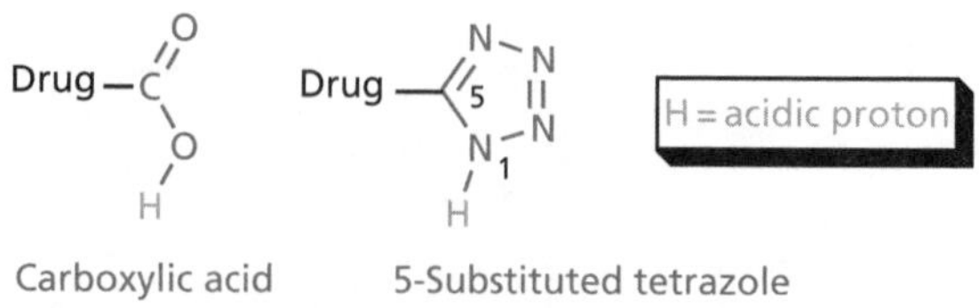

Carboxylic acid   5-Substituted tetrazole

H = acidic proton

Figure 11.5 5-Substituted tetrazole as a bioisostere for a carboxylic acid.

tetrazoles contain an acidic proton and are ionized at pH 7.4. They are also planar in structure. However, they have an advantage in that the tetrazole anion is 10 times more lipophilic than a carboxylate anion and drug absorption is enhanced as a result. They are also resistant to many of the metabolic reactions that occur on carboxylic acids (see Box 11.1).

BOX 11.1 THE USE OF BIOISOSTERES TO INCREASE ABSORPTION

The biphenyl structure (Fig. 1a) was shown by Du Pont to inhibit the receptor for angiotensin II, and had potential as an antihypertensive agent. However, the drug had to be injected as it showed poor absorption through the gut wall. Replacing the carboxylic acid with a tetrazole ring led to **losartan** (Fig. 1b), which was launched in 1994.

(a)

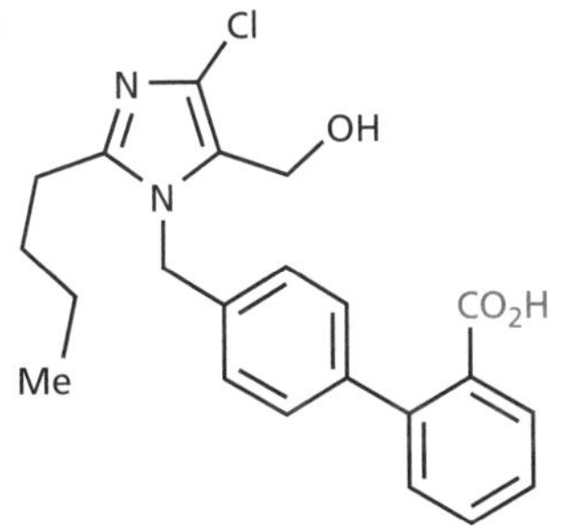

Structure I

(b)

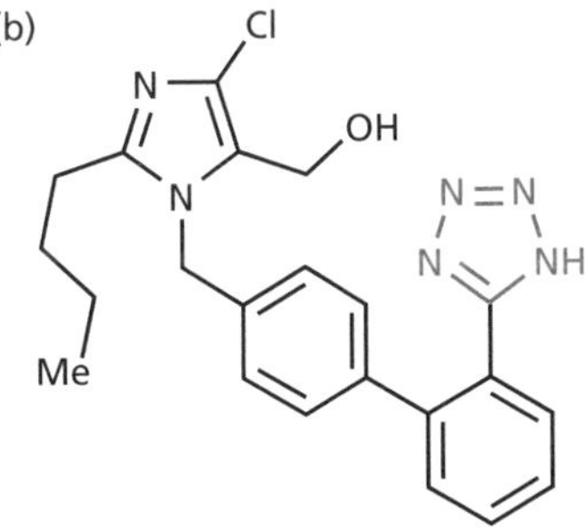

Losartan

Figure 1 Development of losartan.

11.2 Making drugs more resistant to chemical and enzymatic degradation

There are various strategies that can be used to make drugs more resistant to hydrolysis and drug metabolism, and thus prolong their activity.

11.2.1 Steric shields

Some functional groups are more susceptible to chemical and enzymatic degradation than others. For example, esters and amides are particularly prone to hydrolysis. A common strategy that is used to protect such groups is to add **steric shields**, designed to hinder the approach of a nucleophile or an enzyme to the susceptible group. These usually involve the addition of a bulky alkyl group close to the functional group. For example, the t-butyl group in the antirheumatic agent **D1927** serves as a steric shield and blocks hydrolysis of the terminal peptide bond (Fig. 11.6). Steric shields have also been used to protect penicillins from lactamases (section 16.5.1.8) and to prevent drugs interacting with cytochrome P450 enzymes (section 15.18.8).

11.2.2 Electronic effects of bioisosteres

Another popular tactic used to protect a labile functional group is to stabilize the group electronically using a bioisostere. A bioisostere is a chemical group used to replace another chemical group within the drug, without affecting the important biological activity. Other features such as the drug's stability may also be improved. Isosteres and non-classical isosteres

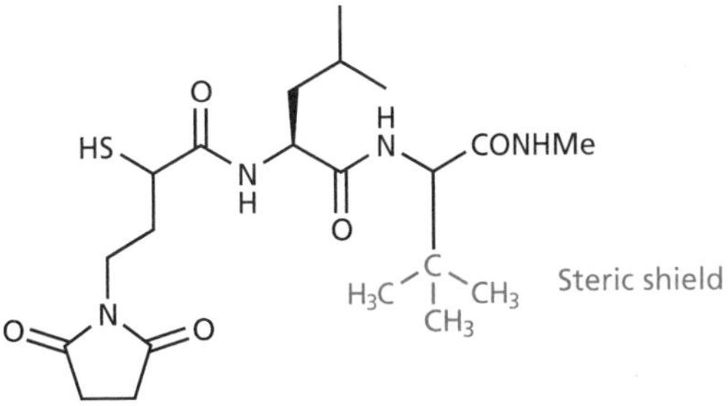

Figure 11.6 The use of a steric shield to protect the antirheumatic agent D1927.

are frequently used as bioisosteres (see also sections 10.1.15, 10.3.7, and 11.1.5). For example, replacing the methyl group of an ethanoate ester with NH_2 results in a urethane functional group which is more stable than the original ester (Fig. 11.7). The NH_2 group is the same valency and size as the methyl group and therefore has no steric effect, but it has totally different electronic properties since it can feed electrons into the carboxyl group and stabilize it from hydrolysis. The cholinergic agonist **carbachol** is stabilized in this way (section 19.9.2), as is the cephalosporin **cefoxitin** (section 16.5.2.4).

Alternatively, a labile ester group could be replaced by an amide group (NH replacing O). Amides are more resistant to chemical hydrolysis, due again to the lone pair of the nitrogen feeding its electrons into the carbonyl group and making it less electrophilic.

It is important to realize that bioisosteres are not general and are often specific to a particular field. Replacing an ester with a urethane or an amide may work in one category of drugs but not another. One must also appreciate that bioisosteres are different from isosteres. It is the retention of important biological activity that determines whether a group is a bioisostere, not the valency. Therefore, non-isosteric groups can be used as bioisosteres. For example, a pyrrole ring was used as a bioisostere for an amide bond in the development of the dopamine antagonist **Du 122290** from **sultopride** (section 10.3.7). Similarly, thiazolyl rings were used as bioisosteres for pyridine rings in the development of ritonavir (section 17.7.4.4).

One is not confined to the use of bioisosteres to increase stability. Groups or substituents having an inductive electronic effect have frequently been incorporated into molecules to increase the stability of a labile functional group. For example, electron-withdrawing groups were incorporated into the side chain of penicillins to increase their resistance to acid

hydrolysis (section 16.5.1.8). The inductive effects of groups can also determine the ease with which ester prodrugs are hydrolysed (Box 11.3).

11.2.3 Stereoelectronic modifications

Steric hindrance and electronic stabilization have often been used together to stabilize labile groups. For example, **procaine** is a good local anaesthetic, but it is short-lasting because its ester group is quickly hydrolysed. Changing the ester group to the less reactive amide group reduces chemical hydrolysis. Furthermore, the presence of two *ortho*-methyl groups on the aromatic ring helps to shield the carbonyl group from attack by nucleophiles or enzymes. This results in the longer-acting local anaesthetic **lidocaine** (Fig. 11.8). Since steric and electronic influences are both involved, the modifications are defined as stereoelectronic. Further successful examples of stereoelectronic modification are demonstrated by **oxacillin** (Chapter 16) and **bethanechol** (section 19.9.3).

11.2.4 Metabolic blockers

Some drugs are metabolized by the introduction of polar groups at particular positions in their skeleton. For example, the oral contraceptive **megestrol acetate** is oxidized at position 6 to give a hydroxyl group at that position (Fig. 11.9). The introduction of a polar hydroxyl group allows the formation of polar conjugates which can be quickly eliminated from the system. By introducing a methyl group at position 6, metabolism is blocked and the activity of the drug is prolonged.

On the same lines, a popular method of protecting aromatic rings from metabolism at the *para* position is to introduce a fluorine substituent. For example, **CGP 52411** (Fig. 11.10) is an enzyme inhibitor which acts on the kinase active site of the epidermal growth factor receptor (section 6.7.2). It went forward for

Figure 11.7 Isosteric replacement of a methyl with an amino group.

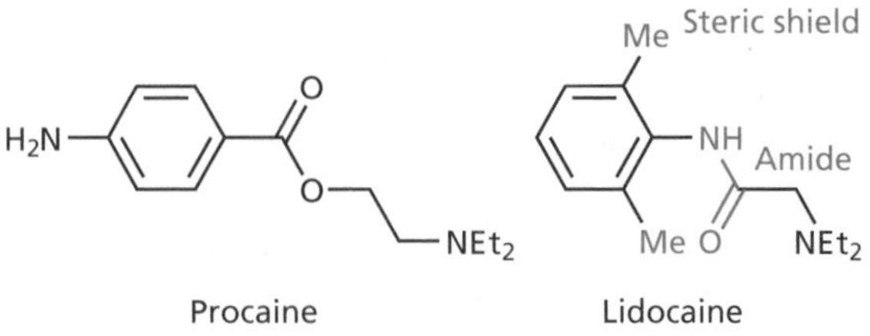

Figure 11.8 Stereoelectronic modification.

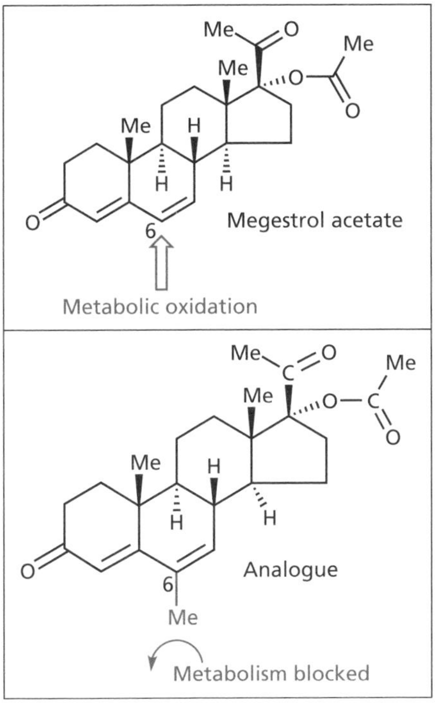

Figure 11.9 Metabolic blocking.

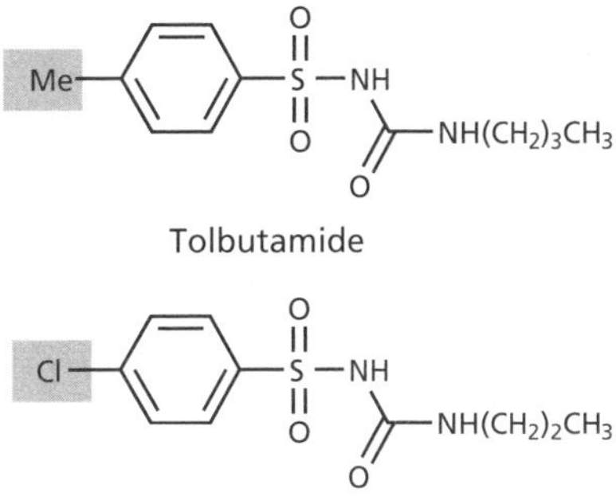

Figure 11.11 Replacing metabolically labile groups.

clinical trials as an anticancer agent and was found to undergo oxidative metabolism at the *para* position of the aromatic rings. Fluorine substituents were successfully added to block this metabolism, and the analogue **CGP 53353** was also put forward for clinical trials.

11.2.5 Removal of susceptible metabolic groups

Certain chemical groups are particularly susceptible to metabolic enzymes. For example, methyl groups on aromatic rings are often oxidized to carboxylic acids (section 8.4.2). These acids can then be quickly eliminated from the body. Other common

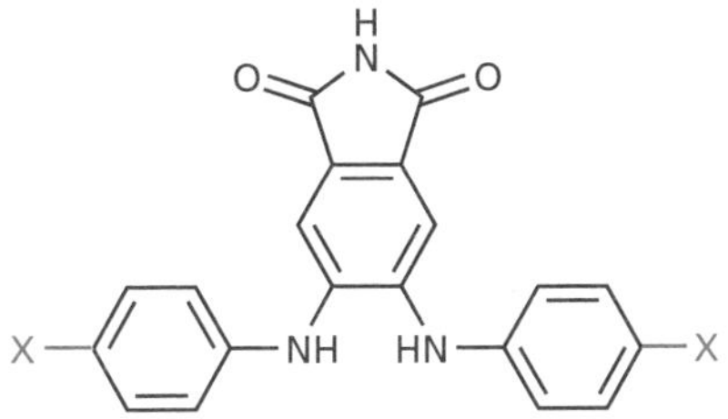

Figure 11.10 Fluorine as a metabolic blocker. X = H, CGP 52411; X = OH, metabolite; X = F, CGP 53353.

metabolic reactions include aliphatic and aromatic C-hydroxylations, N- and S-oxidations, O- and S-dealkylations, and deamination (section 8.4).

Susceptible groups can sometimes be removed or replaced by groups that are stable to oxidation, in order to prolong the lifetime of the drug. For example, the methyl group of the antidiabetic **tolbutamide** was replaced by a chlorine atom to give **chlorpropamide,** which is much longer lasting (Fig. 11.11). Another example is the replacement of a susceptible ester in cephalosporins to give more stable cephalosporins such as **cephaloridine** and **cefalexin** (section 16.5.2.3).

11.2.6 Group shifts

Removing or replacing a metabolically vulnerable group is feasible if the group concerned is not involved in important binding interactions with the binding site. If the group *is* important, then we have to use a different strategy.

There are two possible solutions. We can either mask the vulnerable group on a temporary basis by using a prodrug (section 11.6) or we can try shifting the vulnerable group within the molecular skeleton. The latter tactic was used in the development of **salbutamol** (Fig. 11.12). Salbutamol was introduced in 1969 for the treatment of asthma and is an analogue of the neurotransmitter **noradrenaline**—a catechol structure containing two *ortho*-phenolic groups.

One of the problems faced by catechol compounds is metabolic methylation of one of the phenolic groups. Since both phenol groups are involved in hydrogen bonds to the receptor, methylation of one of the phenol groups disrupts the hydrogen bonding and makes the compound inactive. For example, the noradrenaline analogue (I in Fig. 11.13) has useful

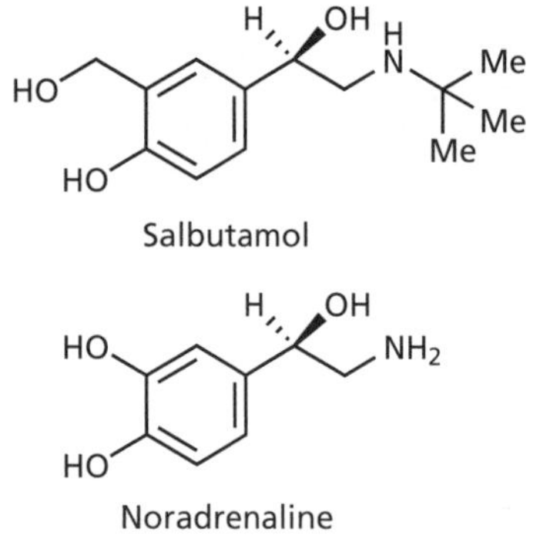

Salbutamol

Noradrenaline

Figure 11.12 Salbutamol and noradrenaline.

antiasthmatic activity, but the effect is of short duration because the compound is rapid metabolized to the inactive methyl ether (II).

Removing the OH or replacing it with a methyl group prevents metabolism, but also prevents the important hydrogen bonding interactions with the binding site. So how can this problem be solved? The answer was to move the vulnerable hydroxyl group out from the ring by one carbon unit. This was enough to make the compound unrecognizable to the metabolic enzyme, but not to the receptor binding site.

Fortunately, the receptor appears to be quite lenient over the position of this hydrogen bonding group and it is interesting to note that a hydroxyethyl group is also acceptable (Fig. 11.14). Beyond that, activity is lost because the OH group is out of range, or too large to fit. These results demonstrate that it is better to consider a binding region within the receptor binding

site as an available volume, rather than imagining it as being fixed at one spot. A drug can then be designed such that the relevant binding group is positioned into any part of that available volume.

Shifting an important binding but metabolically susceptible group worked for salbutamol, but one cannot guarantee that the same tactic will always be successful. Shifting the group may make the molecule unrecognizable both to its target and to the metabolic enzyme.

11.2.7 Ring variation

Certain ring systems are often found to be susceptible to metabolism, and so varying the ring can often improve metabolic stability. For example, the imidazole ring of the antifungal agent **tioconazole** mentioned previously is susceptible to metabolism, but replacement with a 1,2,4-triazole ring as in **fluconazole** results in improved stability (Fig. 11.2).

11.3 Making drugs less resistant to drug metabolism

So far, we have looked at how the activity of drugs can be prolonged by inhibiting their metabolism. However, a drug that is extremely stable to metabolism and is very slowly excreted can pose just as many problems as one that is susceptible to metabolism. It is usually

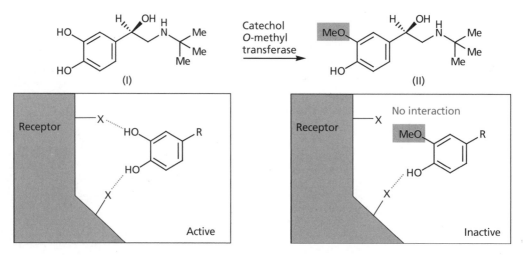

Figure 11.13 Metabolic methylation of a noradrenaline analogue. X denotes an electronegative atom.

desirable to have a drug that does what it is meant to do, then stops doing it within a reasonable time. If not, the effects of the drug could last too long and cause toxicity and lingering side effects. Therefore, designing drugs with decreased chemical and metabolic stability can sometimes be useful.

11.3.1 Introducing metabolically susceptible groups

Introducing groups that are susceptible to metabolism is a good way of shortening the lifetime of a drug (see

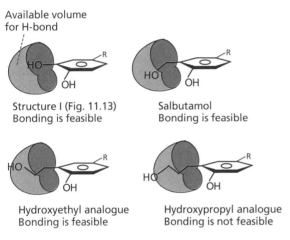

Figure 11.14 Viewing a binding region as an available volume.

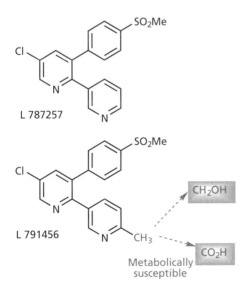

Figure 11.15 Adding a metabolically labile methyl group.

Box 11.2). For example, a methyl group was introduced to the antiarthritic agent **L 787257** to shorten its lifetime. The methyl group of **L 791456** was metabolically oxidized to a polar alcohol as well as to a carboxylic acid (Fig. 11.15).

BOX 11.2 SHORTENING THE LIFETIME OF A DRUG

Antiasthmatic drugs are usually taken by inhalation, to reduce the chances of side effects elsewhere in the body. However, a significant amount is swallowed and can be absorbed into the blood supply from the gastrointestinal tract. Therefore, it is desirable to have an antiasthmatic drug which is potent and stable in the lungs but which is rapidly metabolized in the blood supply. **Cromakalim** has useful antiasthmatic properties but has cardiovascular side effects if it gets into the blood supply. Structures **UK 143220** and **UK 157147** were developed from cromakalim so that they would be quickly metabolized in the blood supply. UK 143220 contains an ester which is quickly hydrolysed by esterases to produce an inactive carboxylic acid, while UK 157147 contains a phenol group which is quickly conjugated by metabolic conjugation enzymes and eliminated. Both these compounds were considered as clinical candidates.

Metabolically labile analogues of cromakalim.

11.3.2 Self-destruct drugs

A self-destruct drug is one which is chemically stable under one set of conditions but becomes unstable and spontaneously degrades under another set of conditions. The advantage of a self-destruct drug is that inactivation does not depend on the activity of metabolic enzymes, which could vary from patient to patient. The best example of a self-destruct drug is the neuromuscular blocking agent **atracurium,** which is stable at acid pH but self-destructs when it meets the slightly alkaline conditions of the blood (section 19.12.2). This means that the drug has a short duration of action, allowing anaesthetists to control its blood levels during surgery by providing it as a continuous intravenous drip.

KEY POINTS

- The polarity or pK_a of a lead compound can be altered by varying alkyl substituents or functional groups, allowing the drug to be absorbed more easily.

- Drugs can be made more resistant to metabolism by introducing steric shields to protect susceptible functional groups. It may also be possible to modify the functional group itself to make it more stable. When both tactics are used together, this is termed a stereoelectronic modification.

- Metabolically stable groups can be added to block metabolism at certain positions.

- Groups which are susceptible to metabolism may be modified or removed to prolong activity, as long as the group is not required for drug–target interactions.

- Metabolically susceptible groups which are necessary for drug–target interactions can be shifted in order to make them unrecognizable by metabolic enzymes, as long as they are still recognizable to the target.

- Varying a heterocyclic ring in the lead compound can sometimes improve metabolic stability.

- Drugs which are slowly metabolized may linger too long in the body and cause side effects.

- Groups which are susceptible to metabolic or chemical change can be incorporated to reduce a drug's lifetime.

11.4 Targeting drugs

One of the major goals in drug design is to find ways of targeting drugs to the exact locations in the body where they are most needed. The principle of targeting drugs can be traced back to Paul Ehrlich, who developed antimicrobial drugs that were selectively toxic for microbial cells over human cells. Drugs can also be made more selective to distinguish between different targets within the body, as discussed in Chapter 10. Here, we discuss other tactics related to the targeting of drugs.

11.4.1 Targeting tumour cells—'search and destroy' drugs

A major goal in cancer chemotherapy is to target drugs efficiently against tumour cells rather than normal cells. One method of achieving this is to design drugs which make use of specific molecular transport systems. The idea is to attach the active drug to an important 'building block' molecule that is needed in large amounts by the rapidly dividing tumour cells. This could be an amino acid or a nucleic acid base (e.g. uracil mustard; section 18.2.3). Of course, normal cells require these building blocks as well, but tumour cells often grow more quickly than normal cells and require the building blocks more urgently. Therefore, the uptake is greater in tumour cells.

A more recent idea has been to attach the active drug (or a poison such as **ricin**) to **monoclonal antibodies** which can recognize antigens unique to the tumour cell. Once the antibody binds to the antigen, the drug or poison is released to kill the cell. The difficulties in this approach include the identification of suitable antigens and the production of antibodies in significant quantity. Nevertheless, the approach has great promise for the future and is covered in more detail in section 18.9.2. Another tactic which has been used to target anticancer drugs is to administer an enzyme–antibody conjugate where a suitable enzyme is chosen to activate an anticancer prodrug and the antibody is chosen to direct the enzyme to the tumour. This is a strategy known as ADEPT and is covered in more detail in section 18.9.3. Other targeting strategies include ADAPT and GDEPT, covered in sections 18.9.4 and 18.9.5 respectively. Antibodies are also being studied as a means of targeting drugs to viruses (section 17.10.5).

11.4.2 Targeting gastrointestinal tract infections

If a drug is to be targeted against an infection of the gastrointestinal tract, it must be prevented from being absorbed into the blood supply. This can easily be done by using a fully ionized drug which is incapable of crossing cell membranes. For example, highly ionized sulfonamides are used against gastrointestinal infections because they are incapable of crossing the gut wall (section 16.4.1).

11.4.3 Targeting peripheral regions rather than the central nervous system

It is often possible to target drugs such that they act peripherally and not in the central nervous system. By increasing the polarity of drugs, they are less likely to cross the blood–brain barrier (section 8.3.5) and this means they are less likely to have central nervous system side effects. Achieving selectivity for the central nervous system over the peripheral regions of the body is not so straightforward.

11.5 Reducing toxicity

It is often found that a drug fails clinical trials because of toxic side effects. This may be due to toxic metabolites, in which case the drug should be made more resistant to metabolism as described earlier (section 11.2). It is also worth checking to see whether there are any functional groups present which are particularly prone to producing toxic metabolites. For example, it is known that functional groups such as aromatic nitro groups, aromatic amines, bromoarenes, hydrazines, hydroxylamines, or polyhalogenated groups are often metabolized to toxic products (see section 8.4 for typical metabolic reactions).

Side effects might also be reduced or eliminated by varying apparently harmless substituents. For example, the halogen substituents of the antifungal agent **UK 47265** were varied in order to find a compound that was less toxic to the liver. This led to the successful antifungal agent **fluconazole** (Fig. 11.16).

Varying the position of substituents can often reduce or eliminate side effects. For example, the dopamine antagonist **SB 269652** inhibits cytochrome P450 enzymes as a side effect. Placing the cyano group at a different position prevented this inhibition (Fig. 11.17).

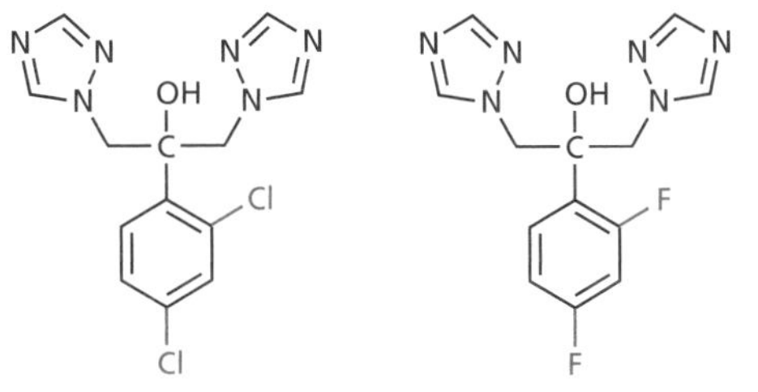

KEY POINTS

- Strategies designed to target drugs to particular cells or tissues are likely to lead to safer drugs with fewer side effects.

- Drugs can be linked to amino acids or nucleic acid bases to target them against fast-growing and rapidly dividing cells.

- Drugs can be targeted to the gastrointestinal tract by making them ionized or highly polar such that they cannot cross the gut wall.

- The central nervous system side effects of peripherally acting drugs can be eliminated by making the drugs more polar so that they do not cross the blood–brain barrier.

- Drugs with toxic side effects can sometimes be made less toxic by varying the nature or position of substituents, or by preventing their metabolism to a toxic metabolite.

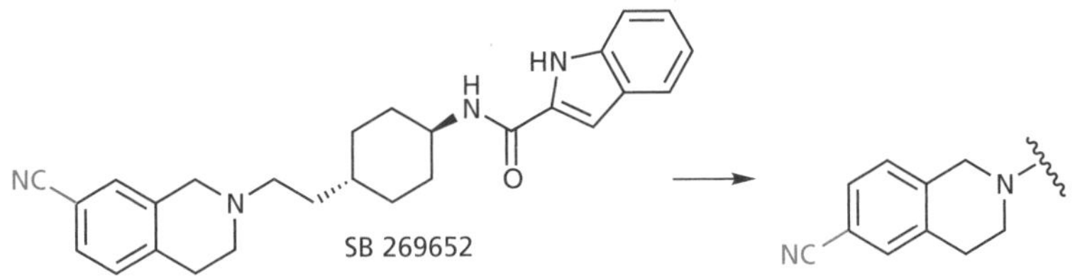

UK 47265   Fluconazole

Figure 11.16 Varying aromatic substituents to reduce toxicity.

SB 269652

Figure 11.17 Varying substituents to reduce side effects.

11.6 Prodrugs

Prodrugs are compounds which are inactive in themselves, but which are converted in the body to the active drug. They have been useful in tackling problems such as acid sensitivity, poor membrane permeability, drug toxicity, bad taste, and short duration of action. Usually, a metabolic enzyme is involved in converting the prodrug to the active drug and so a good knowledge of drug metabolism and the enzymes involved allows the medicinal chemist to design a suitable prodrug which turns drug metabolism into an advantage rather than a problem. Prodrugs have been designed to be activated by a variety of metabolic enzymes. Ester prodrugs which are hydrolysed by esterase enzymes are particularly common, but prodrugs have also been designed which are activated by N-demethylation, decarboxylation, and the hydrolysis of amides and phosphates. Not all prodrugs are activated by metabolic enzymes, however. For example, photodynamic therapy involves the use of an external light source to activate prodrugs. When designing prodrugs, it is important to ensure that the prodrug is effectively converted to the active drug once it has been absorbed into the blood supply, but it is also important to ensure that any groups that are cleaved from the molecule are non-toxic.

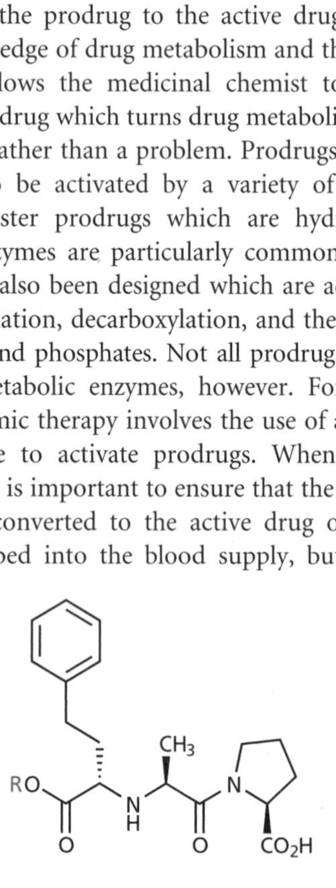

11.6.1 Prodrugs to improve membrane permeability

11.6.1.1 Esters as prodrugs

Prodrugs have proved very useful in temporarily masking an 'awkward' functional group which is important to target binding, but which hinders the drug from crossing the cell membranes of the gut wall. For example, a carboxylic acid functional group may have an important role to play in binding a drug to its binding site via ionic or hydrogen bonding. However, the very fact that it is an ionizable group may prevent it from crossing a fatty cell membrane. The answer is to protect the acid function as an ester. The less polar ester can cross fatty cell membranes, and once it is in the bloodstream it is hydrolysed back to the free acid by esterases in the blood. Examples of ester prodrugs used to aid membrane permeability include **enalapril**, which is the prodrug for the antihypertensive agent **enalaprilate** (Fig. 11.18), and **pivampicillin**, which is a penicillin prodrug (section 16.5.1).

Not all esters are hydrolysed equally efficiently, and a range of esters may need to be tried to find the best one (Box 11.3). It is possible to make esters more susceptible to hydrolysis by introducing electron-withdrawing groups to the alcohol moiety (e.g. OCH_2CF_3, OCH_2CO_2R, $OCONR_2$, OAr). The inductive effect of these groups aids the hydrolytic mechanism by making the alcohol a better leaving group (Fig. 11.19). Care has to be taken, however, not to make the ester too reactive in case it becomes chemically unstable and is hydrolysed before it reaches the blood supply.

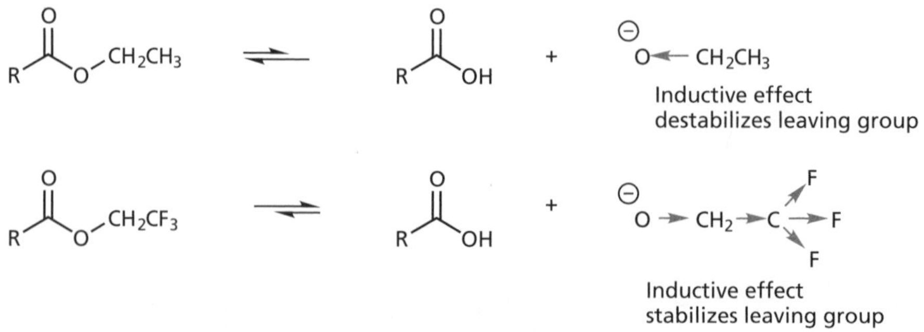

Figure 11.18 Enalapril (R = Et); Enalaprilate (R = H).

Figure 11.19 Inductive effects on the stability of leaving groups.

The protease inhibitor **candoxatrilat** has to be given intra-venously because it is too polar to be absorbed from the gastrointestinal tract. Different esters were tried as prodrugs to get round this problem. It was found that an ethyl ester was absorbed but was inefficiently hydrolysed. A more ac-tivated ester was required, and a 5-indanyl ester proved to be the best. The 5-indanol released on hydrolysis is non-toxic.

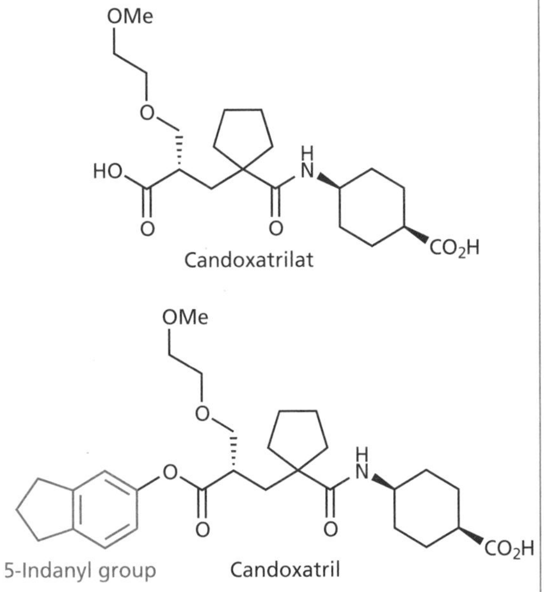

Candoxatrilat

5-Indanyl group   Candoxatril

Protease inhibitors.

11.6.1.2 *N*-Methylation

N-Demethylation is a common metabolic reaction in the liver, so polar amines can be *N*-methylated to reduce polarity and improve membrane permeability. Several hypnotics and antiepileptics take advantage of this reaction, for example **hexobarbitone** (Fig. 11.20).

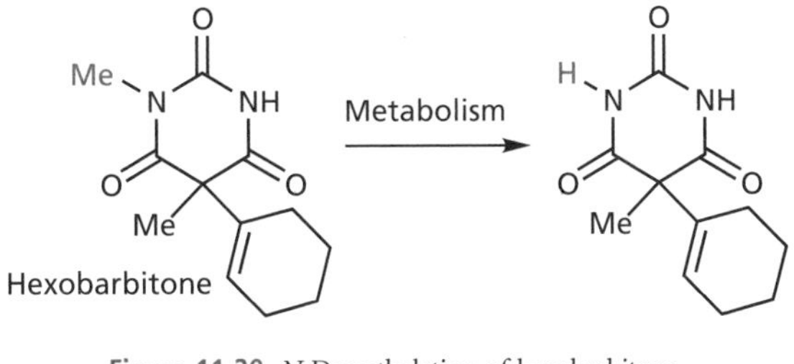

Hexobarbitone

Figure 11.20 *N*-Demethylation of hexobarbitone.

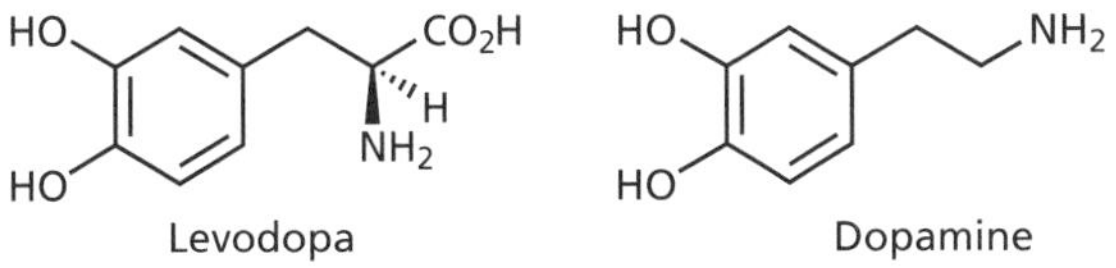

Levodopa   Dopamine

Figure 11.21 Levodopa and dopamine.

11.6.1.3 Trojan horse approach for carrier proteins

Another way round the problem of membrane per-meability is to design a prodrug which can take ad-vantage of carrier proteins (section 3.7.1) in the cell membrane, such as the ones responsible for carrying amino acids into a cell. A well-known example of such a prodrug is **levodopa** (Fig. 11.21). Levodopa is a prodrug for the neurotransmitter **dopamine** and has been used in the treatment of Parkinson's disease—a condition due primarily to a deficiency of that neurotransmitter in the brain. Dopamine itself cannot be used, since it is too polar to cross the blood–brain barrier. Levodopa is even more polar and seems an unlikely prodrug, but it is also an amino acid and so it is recognized by the carrier proteins for amino acids which carry it across the cell membrane. Once in the brain, a decarboxylase enzyme removes the acid group and generates dopamine (Fig. 11.22).

11.6.2 Prodrugs to prolong drug activity

Sometimes prodrugs are designed to be converted slowly to the active drug, thus prolonging a drug's activity. For example, **6-mercaptopurine** (Fig. 11.23) suppresses the body's immune response and is there-fore useful in protecting donor grafts. Unfortunately, the drug tends to be eliminated from the body too quickly. The prodrug **azathioprine** has the advantage that it is slowly converted to 6-mercaptopurine by being attacked by glutathione, allowing a more sus-tained activity. The rate of conversion can be altered, depending on the electron-withdrawing ability of the heterocyclic group. The greater the electron-withdrawing power, the faster the breakdown. The NO_2 group is therefore present to ensure an efficient conversion to 6-mercaptopurine, since it is strongly electron-withdrawing on the heterocyclic ring.

There is a belief that the well-known sedatives **Valium** (Fig. 11.24) and **Librium** might be prodrugs and are active because they are metabolized by *N*-demethylation to **nordazepams**. Nordazepam itself

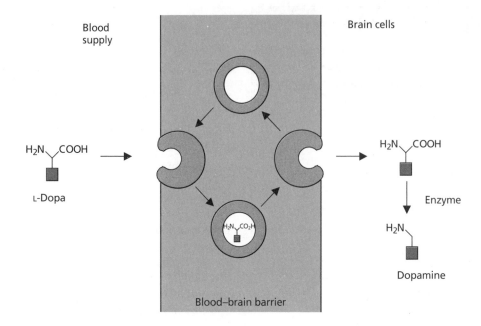

Figure 11.22 Transport of levodopa across the blood–brain barrier.

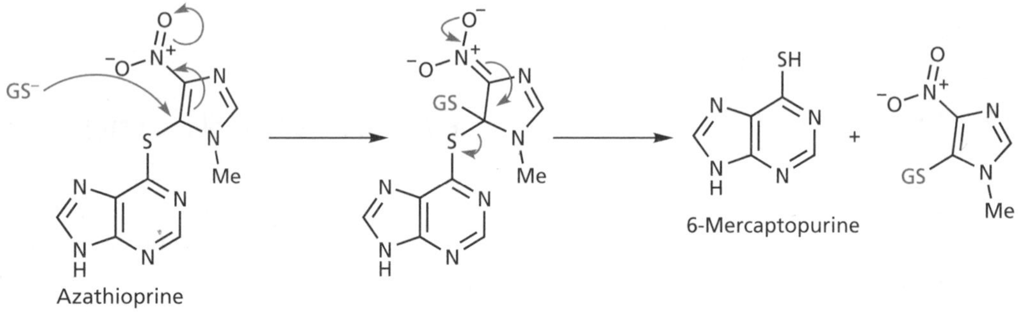

Figure 11.23 Azathioprine acts as a prodrug for 6-mercaptopurine (GS = glutathione).

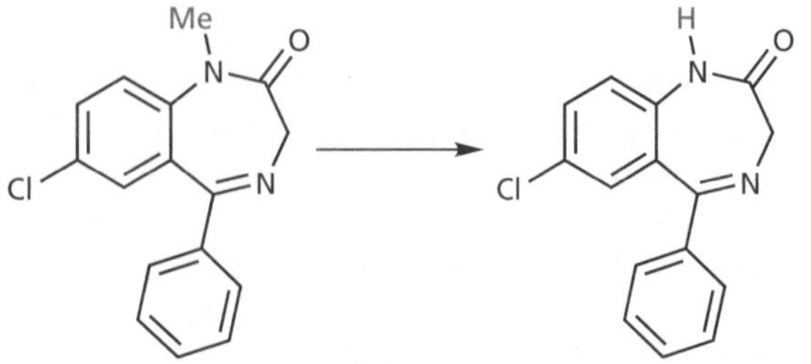

Figure 11.24 Valium (diazepam) as a possible prodrug for nordazepam.

has been used as a sedative, but loses activity quite quickly as a result of metabolism and excretion. **Valium,** if it is a prodrug for nordazepam, demonstrates again how a prodrug can be used to lead to a more sustained action.

Another approach to maintaining a sustained level of drug over long periods is to deliberately associate a very lipophilic group to the drug. This means that most of the drug is stored in fat tissue, and if the lipophilic group is only slowly removed, the drug is steadily and slowly released into the bloodstream. The antimalarial agent cycloguanil pamoate (Fig. 11.25) is one such agent. The active drug is bound ionically to an anion with a large lipophilic group.

Similarly, lipophilic esters of the antipsychotic drug **fluphenazine** are used to prolong its action. The prodrug is given by intramuscular injection and slowly diffuses from fat tissue into the blood supply where it is rapidly hydrolysed (Fig. 11.26).

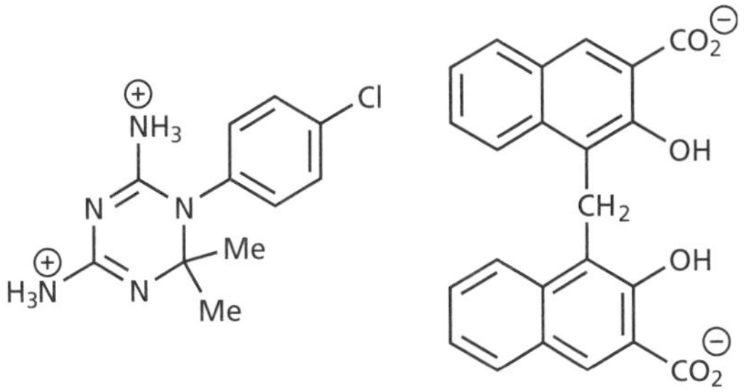

11.6.3 Prodrugs masking drug toxicity and side effects

Prodrugs can be used to mask the side effects and toxicity of drugs (Box 11.4). For example, **salicylic acid** is a good painkiller, but causes gastric bleeding due to the free phenolic group. This is overcome by masking the phenol as an ester (**aspirin**) (Fig. 11.27). The ester is later hydrolysed to free the active drug. (Note that aspirin is a drug in its own right, with an anti-inflammatory action due to its acetylation of cyclooxygenases.)

Figure 11.25 Cycloguanil pamoate.

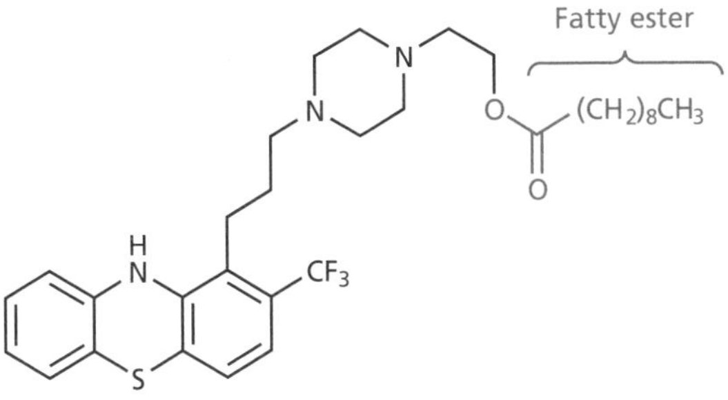

Figure 11.26 Fluphenazine decanoate.

BOX 11.4 PRODRUGS MASKING TOXICITY AND SIDE EFFECTS

LDZ is an example of a diazepam prodrug which avoids the drowsiness side effects associated with **diazepam**. These side effects are associated with the high initial plasma levels of diazepam following administration. The use of a prodrug avoids this problem. An aminopeptidase enzyme hydrolyses the prodrug to release a non-toxic lysine moiety, and the resulting amine spontaneously cyclizes to the diazepam (as shown below).

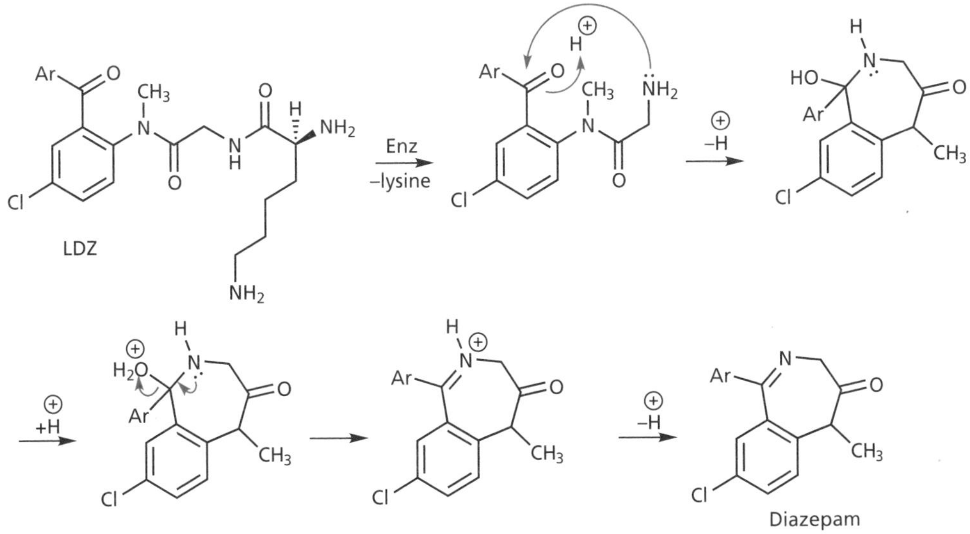

LDZ as a diazepam prodrug.

Prodrugs can be used to give a slow release of drugs that would be too toxic to give directly. **Propiolalde-hyde** is useful in the aversion therapy of alcohol, but is not used itself because it is an irritant. The prodrug **pargylene** can be converted to propiolaldehyde by enzymes in the liver (Fig. 11.28).

Cyclophosphamide is a successful, non-toxic prodrug which can be safely taken orally. Once absorbed, it is metabolized in the liver to a toxic alkylating agent which is useful in the treatment of cancer (section 18.2.3).

Many important antiviral drugs such as **aciclovir** and **penciclovir** are non-toxic prodrugs which show selective toxicity towards virally infected cells. This is

because they are converted to toxic triphosphates by a viral enzyme which is only present in infected cells (section 17.6.1).

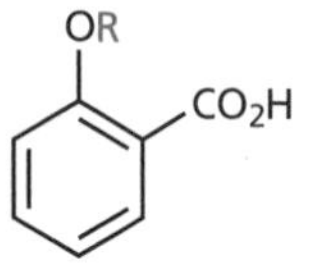

11.6.4 Prodrugs to lower water solubility

Some drugs have a revolting taste! One way to avoid this problem is to reduce their water solubility so they do not dissolve on the tongue. For example, the bitter taste of the antibiotic **chloramphenicol** can be avoided by using the palmitate ester (Fig. 11.29). This is more hydrophobic because of the masked alcohol and the long chain fatty group that forms the ester. It does not dissolve easily on the tongue, and is quickly hydrolysed once swallowed.

11.6.5 Prodrugs to improve water solubility

Prodrugs have been used to increase the water solubility of drugs (Box 11.5). This is particularly useful for drugs which are given intravenously, as it means that higher concentrations and smaller volumes can be used. For example, the succinate ester of **chloramphenicol** (Fig. 11.29) increases the latter's water solubility due to the extra carboxylic acid that is present. Once the ester is hydrolysed, chloramphenicol is released along with succinic acid which is naturally present in the body.

Figure 11.27 Aspirin (R = CH₃CO) and salicylic acid (R = H).

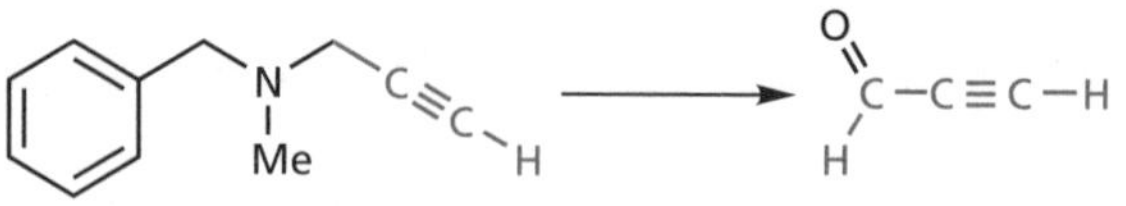

Figure 11.28 Pargylene as a prodrug for propiolaldehyde.

BOX 11.5 PRODRUGS TO IMPROVE WATER SOLUBILITY

Polar prodrugs have been used to improve the absorption of non-polar drugs from the gut. Drugs have to have some water solubility if they are to be absorbed, otherwise they dissolve in fatty globules and fail to interact effectively with the gut wall.

The steroid **oestrone** is one such drug. By using a lysine ester prodrug, water solubility and absorption is increased. Hydrolysis of the prodrug releases the active drug, and the amino acid lysine as a non-toxic by-product.

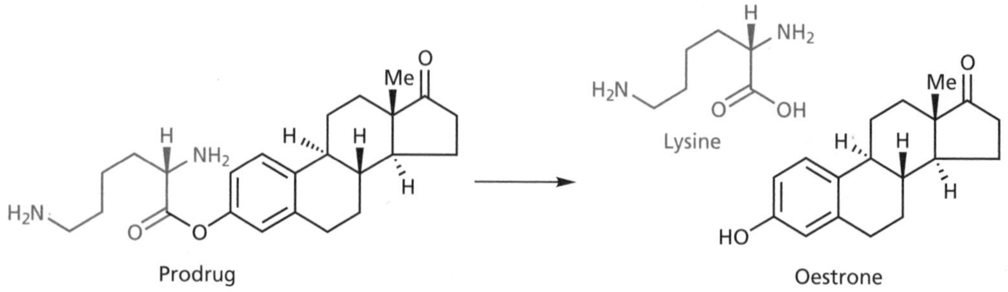

Lysine ester of oestrone to improve water solubility and absorption.

Prodrugs designed to increase water solubility have proved useful in preventing the pain associated with some injections, which is caused by the poor solubility of the drug at the site of injection.

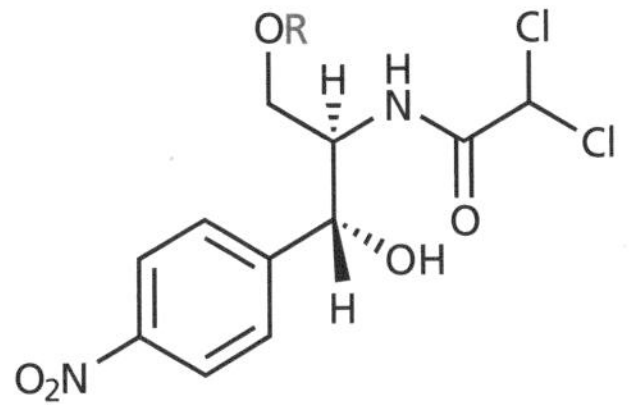

Figure 11.29 Chloramphenicol prodrugs. R = H, chloramphenicol; R = CO(CH$_2$)$_{14}$CH$_3$, chloramphenicol palmitate; R = CO(CH$_2$)$_2$-CO$_2$H, chloramphenicol succinate.

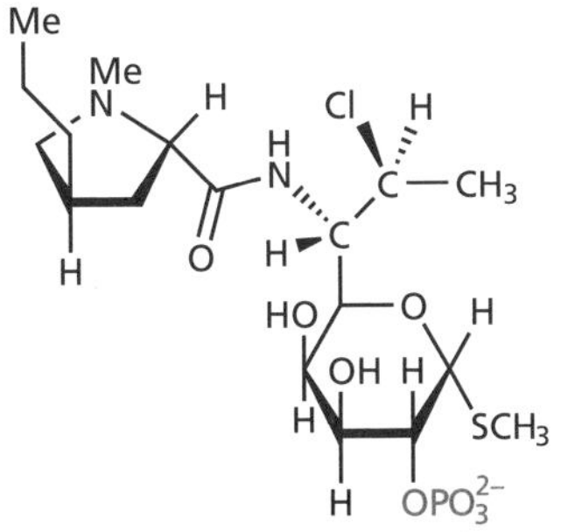

Figure 11.30 Clindamycin phosphate.

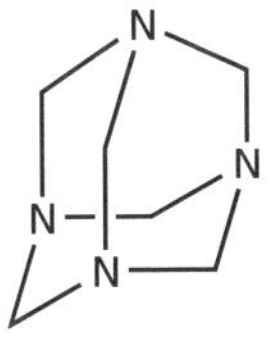

Figure 11.31 Methenamine.

For example, the antibacterial agent **clindamycin** is painful when injected, but using a phosphate ester prodrug improves solubility because of the ionic phosphate group, and thus prevents the pain (Fig. 11.30).

11.6.6 Prodrugs used in the targeting of drugs

Methenamine (Fig. 11.31) is a stable, inactive compound when the pH is more than 5. At a more acidic pH, however, the compound spontaneously degrades to generate formaldehyde, which has antibacterial properties. This is useful in the treatment of urinary tract infections. The normal pH of blood is slightly alkaline (7.4) and so methenamine passes round the body unchanged. However, once it is excreted into the infected urinary tract, it encounters urine which is acidic as a result of the bacterial infection. Consequently, methenamine degrades to generate formaldehyde just where it is needed.

Prodrugs have also been used to target sulfonamides against intestinal infections (section 16.4.1).

The targeting of prodrugs to tumour cells by antibody-related strategies was mentioned in section 11.4.1 and is described in more detail in section 18.9. Antibody–drug conjugates can also be viewed as prodrugs and are also described in that section.

11.6.7 Prodrugs to increase chemical stability

The antibacterial agent **ampicillin** decomposes in concentrated aqueous solution as a result of intramolecular attack of the side chain amino group on the lactam ring. **Hetacillin** (Fig. 11.32) is a prodrug which locks up the offending nitrogen in a ring and prevents this reaction. Once the prodrug has been

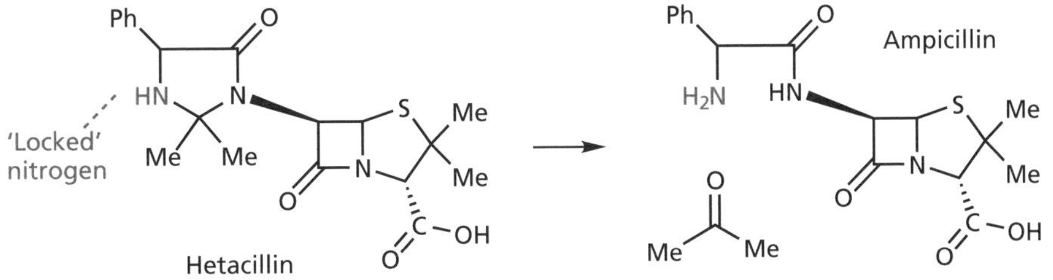

Figure 11.32 Hetacillin and ampicillin.

administered, hetacillin slowly decomposes to release ampicillin and acetone.

11.6.8 Prodrugs activated by external influence (sleeping agents)

Conventional prodrugs are inactive compounds which are normally metabolized in the body to the active form. A variation of the prodrug approach is the concept of a 'sleeping agent'. This is an inactive compound which is only converted to the active drug by some form of external influence. The best example of this approach is the use of photosensitizing agents such as porphyrins or chlorins in cancer treatment—a strategy known as **photodynamic therapy**. Given intravenously, these agents accumulate within cells and have some selectivity for tumour cells. By themselves, the agents have little effect, but if the cancer cells are irradiated with light, the porphyrins are converted to an excited state and react with molecular oxygen to produce highly toxic singlet oxygen. This is covered in section 18.10.

KEY POINTS

- Prodrugs are inactive compounds which are converted to active drugs in the body—usually by drug metabolism.

- Esters are commonly used as prodrugs to make a drug less polar, allowing it to cross cell membranes more easily. The nature of the ester can be altered to vary the rate of hydrolysis.

- Introducing a metabolically susceptible N-methyl group can sometimes be advantageous in reducing polarity.

- Prodrugs with a similarity to important biosynthetic building blocks may be capable of crossing cell membranes with the aid of carrier proteins.

- The activity of a drug can be prolonged by using a prodrug which is converted slowly to the active drug.

- The toxic nature of a drug can be reduced by using a prodrug which is slowly converted to the active compound, preferably at the site of action.

- Prodrugs which contain metabolically susceptible polar groups are useful in improving water solubility. They are particularly useful for drugs which have to be injected, or for drugs which are too hydrophobic for effective absorption from the gut.

- Prodrugs which are susceptible to pH or chemical degradation can be effective in targeting drugs or increasing stability in solution prior to injection.

- Prodrugs which are activated by light are the basis for photodynamic therapy.

11.7 Drug alliances

Some drugs are found to affect the activity or pharmacokinetic properties of other drugs, and this can be put to good use. The following are some examples.

11.7.1 'Sentry' drugs

In this approach, a second drug is administered alongside the principal drug. The role of the second drug is to guard or assist the principal drug. Usually, the second drug inhibits an enzyme that metabolizes the principal drug. For example, **clavulanic acid** inhibits the enzyme **β-lactamase** and is therefore able to protect penicillins from that particular enzyme (section 16.5.4.1).

The antiviral drug **Kaletra** used in the treatment of AIDS is a combination of two drugs called **ritonavir** and **lopinavir**. Although the former has antiviral activity, it is principally present to protect lopinavir. Lopinavir is metabolized by the metabolic cytochrome P450 enzyme (CYP3A4). Ritonavir is a strong inhibitor of this enzyme and so the metabolism of lopinavir is decreased, allowing lower doses to be used for therapeutic plasma levels (section 17.7.4.4).

Another example is to be found in the drug therapy of Parkinson's disease. The use of **levodopa** as a prodrug for **dopamine** has already been described (section 11.6.1.3). To be effective, however, large doses of levodopa (3–8 g per day) are required, and over a period of time these dose levels lead to side effects such as nausea and vomiting. Levodopa is susceptible to the enzyme **dopa decarboxylase** and as a result, much of the levodopa administered is decarboxylated to dopamine before it reaches the central nervous system (Fig. 11.33). This build-up of dopamine in the peripheral blood supply leads to the observed nausea and vomiting.

The drug **carbidopa** has been used successfully as an inhibitor of dopa decarboxylase and allows smaller doses of levodopa to be used. Furthermore, since it is

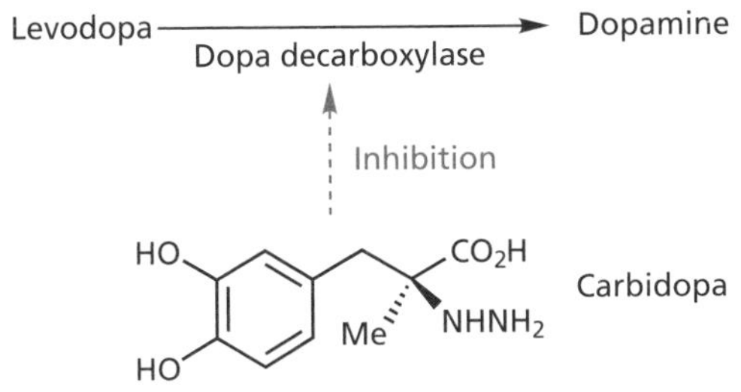

Figure 11.33 Inhibition of levodopa decarboxylation.

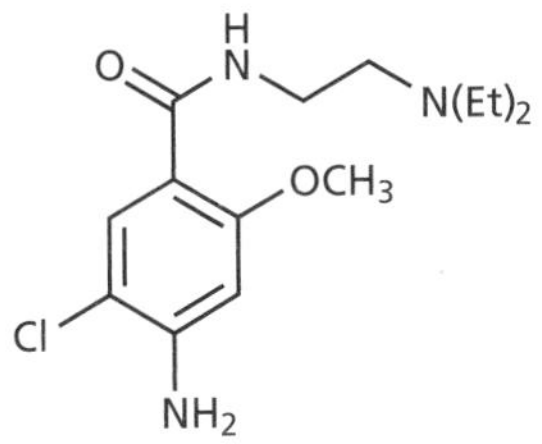

Figure 11.34 Metoclopramide.

a highly polar compound containing two phenolic groups, a hydrazine moiety, and an acidic group, it is unable to cross the blood–brain barrier and so cannot prevent the conversion of levodopa to dopamine in the brain. Carbidopa is marketed as a mixture with levodopa and is called co-careldopa.

Several important peptides and proteins could be used as drugs if it were not for the fact that they are quickly broken down by **protease** enzymes. One way round this problem is to inhibit the protease enzymes. **Candoxatril** (Box 11.3) is a protease inhibitor which has some potential in this respect and is under clinical evaluation.

Finally, the action of penicillins can be prolonged if they are administered alongside probenecid (Box 16.6).

11.7.2 Localizing a drug's area of activity

Adrenaline is an example of a drug which has been used to localize the area of activity for another drug. When injected with the local anaesthetic **procaine**, adrenaline constricts the blood vessels in the vicinity of the injection and so prevents procaine being rapidly removed from the area by the blood supply.

11.7.3 Increasing absorption

Metoclopramide (Fig. 11.34) is administered alongside analgesics in the treatment of migraine. Its function is to increase gastric motility, leading to faster absorption of the analgesic and quicker pain relief.

KEY POINTS

- A sentry drug is a drug which is administered alongside another drug to enhance the latter's activity.

- Many sentry drugs protect their partner drug by inhibiting an enzyme which acts on the latter.

- Sentry drugs have also been used to localize the site of action of local anaesthetics and to increase the absorption of drugs from the gastrointestinal tract.

11.8 Endogenous compounds as drugs

Endogenous compounds are molecules which occur naturally in the body. Many of these could be extremely useful in medicine. For example, the body's hormones are natural chemical messengers, so why not use them instead of synthetic drugs that are foreign to the body? In this section we look at important molecules such as neurotransmitters, hormones, peptides, and oligonucleotides and see how feasible it is to use them as drugs.

11.8.1 Neurotransmitters

Many non-peptide neurotransmitters are simple molecules which can be easily prepared in the laboratory, so why are these not used commonly as drugs? For example, if there is a shortage of dopamine in the brain, why not administer more dopamine to make up the balance?

Unfortunately, this is not possible for a number of reasons. Many neurotransmitters are not stable enough to survive the acid of the stomach, and would have to be injected. Even if they were injected, there is little chance that they would survive to reach their target receptors. The body has efficient mechanisms which inactivate neurotransmitters as soon as they have passed on their message from nerve to target cell. Therefore, any neurotransmitter injected into the blood supply would be swiftly inactivated by enzymes or by cellular uptake. Even if they were not inactivated, they would be poor drugs indeed, leading to many undesirable side effects. For example, the shortage of neurotransmitter may only be at one small area in the brain; the situation may be normal elsewhere. If we gave the natural neurotransmitter, how would we stop it producing an overdose of transmitter at these other sites? Of course this is a problem with all drugs, but it has been discovered that the receptors for a specific neurotransmitter are not all identical. There are slightly different subtypes of a particular receptor, and their distribution around the body is not uniform. One subtype of receptor may be common in one type of tissue, whereas a different subtype is common in another tissue. The medicinal chemist can design synthetic drugs which take advantage of that difference, ignoring receptor subtypes which the natural neurotransmitter would not. In this respect, the medicinal chemist has actually improved on nature.

We cannot even assume that the body's own neurotransmitters are perfectly safe, and free from the horrors of tolerance and addiction associated with drugs such as **heroin**. It is quite possible to be addicted to one's own neurotransmitters and hormones. Some people are addicted to exercise, and are compelled to exercise long hours each day in order to feel good. The very process of exercise leads to the release of hormones and neurotransmitters which can produce a 'high', and this drives susceptible people to exercise more and more. If they stop exercising, they suffer withdrawal symptoms such as deep depression.

The same phenomenon probably drives mountaineers into attempting feats which they know quite well might lead to their death. The thrill of danger produces hormones and neurotransmitters which in turn produce a 'high'. This may also explain why some individuals choose to become mercenaries and risk their lives travelling the globe in search of wars to fight.

To conclude, many of the body's own neurotransmitters are known and can be easily synthesized, but they cannot be effectively used as medicines.

11.8.2 Natural hormones as drugs

Unlike neurotransmitters, natural hormones have potential in drug therapy as they normally circulate round the body and behave like drugs. Indeed, **adrenaline** is commonly used in medicine to treat (amongst other things) severe allergic reactions. Most hormones are peptides and proteins, and some naturally occurring peptide and protein hormones are already used in medicine. These include **insulin, calcitonin, human growth factor, interferons**, and **colony stimulating factors**. Other hormones have proved ineffective, though. This is because peptides and proteins suffer serious drawbacks because of their susceptibility to digestive and metabolic breakdown. Furthermore, proteins are large molecules which could possibly induce an adverse immunological response.

11.8.3 Peptides and proteins as drugs

Because of the difficulties already mentioned, there is often a reluctance to use peptides and proteins as drugs, but this does not mean that peptide drugs have no role to play in medicinal chemistry. For example, the immunosuppressant **ciclosporin** can be administered orally (section 8.2). Another important peptide drug is **goserelin** (Fig. 11.35), which is

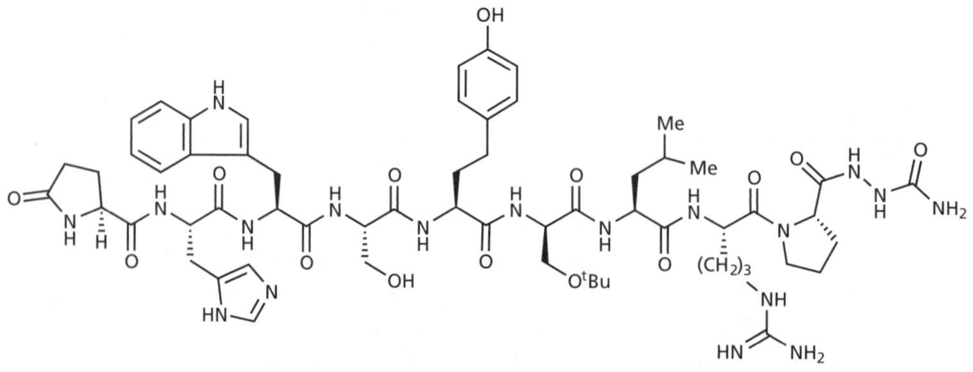

Figure 11.35 Goserelin (Zoladex).

administered as a subcutaneous implant and is used against breast and prostate cancers, earning $700 million dollars a year for its maker. In 2003, **enfuvirtide** was approved as the first of a new class of anti-HIV drugs (section 17.7.5). It is a polypeptide of 36 amino acids which is injected subcutaneously and offers another weapon in the combination therapies used against HIV. **Teriparatide** is a polypeptide consisting of 34 amino acids, which has been approved for the treatment of osteoporosis and is administered by subcutaneous injection. Peptide drugs can be useful if one chooses the right disease and method of administration.

Proteins also have an important role as drugs. For example, a variety of antibodies have been approved for the treatment of cancer (section 18.9.1). Biotechnology companies are now producing an ever-increasing number of peptide, protein, and antibody-based drugs with the aid of recombinant DNA and monoclonal antibody technology. Unfortunately, proteins are large molecules and this can pose problems. They cannot easily be administered orally and so have to be injected. They can also induce an immune response, which involves the body producing antibodies against the proteins, resulting in serious side effects.

Solutions to these problems are appearing, though. It has been found that linking the polymer **polyethylene glycol** (PEG) to a protein can increase the latter's solubility and stability, as well as decreasing the likelihood of an immune response (Fig. 11.36). PEGylation, as it is called, also prevents the removal of small proteins from the blood supply by the kidneys or the reticuloendothelial system. The increased size of the PEGylated protein means that it is not filtered into the kidney nephrons and remains in the blood supply.

The PEG molecules surrounding the protein can be viewed as a kind of hydrophilic polymeric shield which both protects and disguises the protein. The PEG polymer has the added advantage that it shows little toxicity. The enzymes L-**asparaginase** and

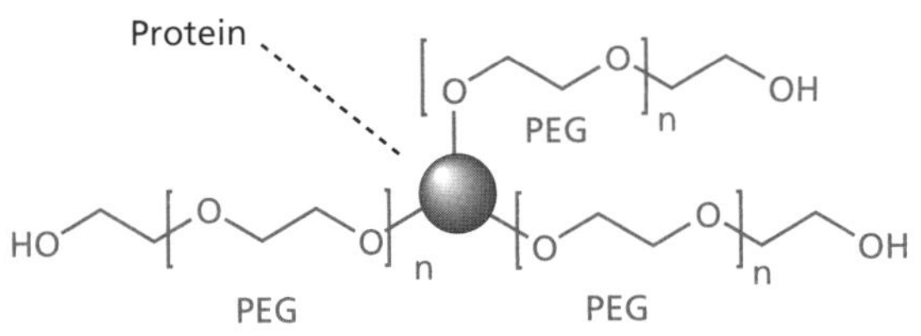

Figure 11.36 PEGylated protein.

adenosine deaminase have been treated in this way to give protein–PEG conjugates called **pegaspargase** and **pegademase**, which have been used for the treatment of leukaemia and SCID syndrome respectively. (SCID, or severe combined immunodeficiency disease, is an immunological defect associated with a lack of adenosine deaminase.) The conjugates have longer plasma half-lives than the enzymes alone and are less likely to produce an immune response. **Interferon** has similarly been PEGylated to give a preparation called **peginterferon** α2b which is used for the treatment of hepatitis C. **Pegvisomant** is the PEGylated form of **human growth hormone antagonist** and is used for the treatment of a condition known as acromegaly which results in abnormal enlargement of the skull, jaw, hands, and feet due to the excessive production of growth hormone. **Pegfilgrastim** is the PEGylated form of the drug **filgrastim** and is used as an anticancer agent. PEGylation has also been used to protect liposomes for drug delivery (section 8.9).

Unfortunately, the treatment of antibodies with PEG tends to be counterproductive, as it prevents the antibody acting out its role as a targeting system. However, controlling the PEGylation such that it only occurs on the thiol group of cysteine residues could be beneficial as it would limit the number of PEG molecules attached and make it more likely that the antibody is still functional.

11.8.4 Peptidomimetics

Although there are problems associated with the use of peptides and proteins as drugs, there is certainly no doubt that they serve as highly important lead compounds for the design of novel drugs. Current examples include renin inhibitors (section 4.8.4), protease inhibitors (section 17.7.4), LHRH agonists (section 18.4.5), matrix metalloproteinase inhibitors (section 18.7.1), and enkephalin analogues (section 21.6). Peptides will continue to be important lead compounds, as many of the new targets in medicinal chemistry, such as the protein kinases, involve proteins as substrates. Various strategies have now been developed to disguise the peptide nature of the lead compound and to design an orally active molecule that is more stable to digestive and metabolic enzymes and is more easily absorbed, so that it attains an acceptable level in the blood supply. Such analogues are known as **peptidomimetics**, and the fraction of an orally

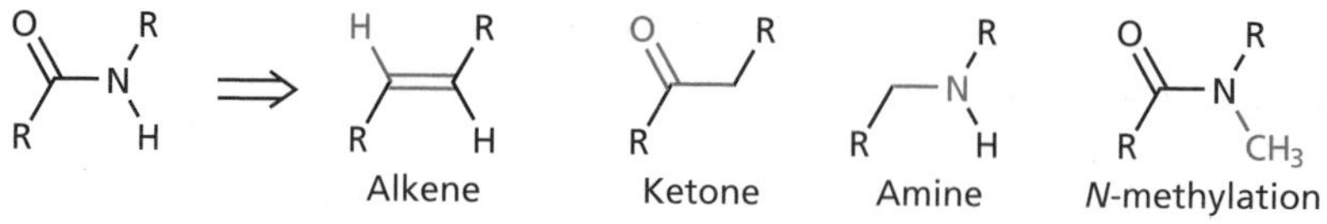

Figure 11.37 Examples of modifications carried out on a peptide bond.

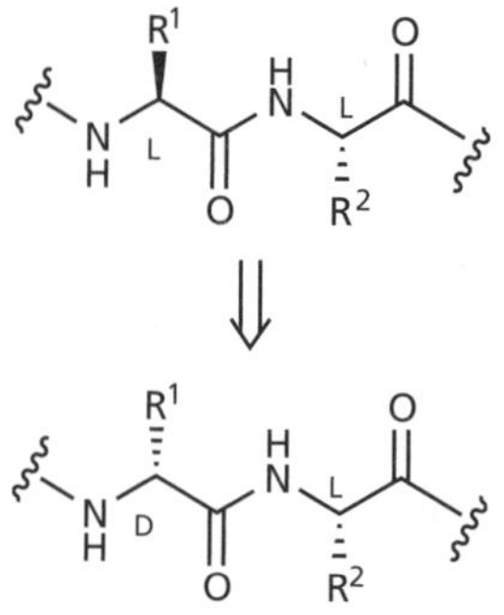

administered peptidomimetic that reaches the blood supply is known as its **bioavailability**.

One approach that is used to increase bioavailability is to replace a chemically or enzymatically susceptible peptide bond with a functional group that is more stable to hydrolytic attack. For example, a peptide bond might be replaced by an alkene (Fig. 11.37). If the compound retains activity, then the alkene represents a **bioisostere** for the peptide link. An alkene has the advantage that it mimics the double bond nature of a peptide bond. However, the peptide bonds in lead compounds are often involved in hydrogen bond interactions with the binding site, where the NH acts as a hydrogen bond donor and the carbonyl C=O acts as a hydrogen bond acceptor. Replacing both of these groups may result in a significant drop in binding strength. Alternatively, one might replace the amide with a ketone or an amine such that only one possible interaction is lost. The problem now is that the double bond nature of the original amide group is lost, resulting in greater chain flexibility and a possible drop in binding affinity (see section 10.3.9).

A different approach is to retain the amide but to protect it or disguise it. One strategy that has been used successfully is to methylate the nitrogen of the amide group. The methyl group may help to protect the amide from hydrolysis by acting as a steric shield, or prevent an important hydrogen bonding interaction taking place between the NH of the original amide and the active site of the peptidase enzyme which would normally hydrolyse it.

A second strategy is to replace an L-amino acid with the corresponding D-enantiomer (Fig. 11.38). Such a move alters the relative orientation of the side chain with respect to the rest of the molecule and can make the molecule unrecognizable to digestive or metabolic enzymes, especially if the side chain is involved in binding interactions. The drawback to this strategy is that the resulting peptidomimetic may become unrecognizable to the desired target as well.

A third strategy is to replace natural amino acid residues with unnatural ones. This is a tactic that has

Figure 11.38 Replacing an L-amino acid with a D-amino acid.

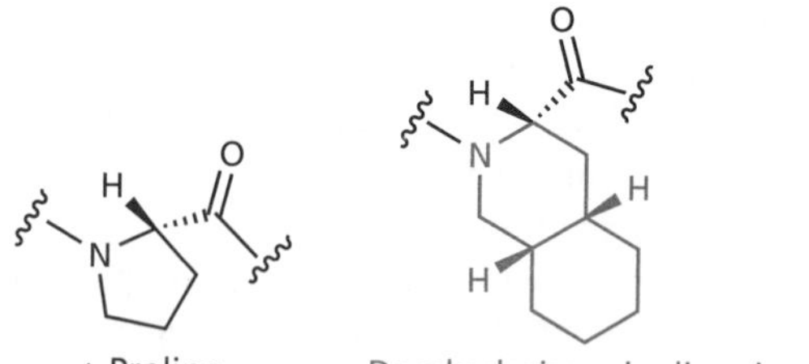

Figure 11.39 Replacing a natural residue with an unnatural one.

worked successfully in structure-based drug design where the binding interactions of the peptidomimetic and a protein target are studied by X-ray crystallography and molecular modelling. The idea is to identify subsites in the protein target into which various amino acid residues fit and bind. The residues are then replaced by groups which are designed to fit the subsites better, but which are not found on natural amino acids. This increases the binding affinity of the peptidomimetic to the target binding site, and at the same time makes it unrecognizable to digestive and metabolic enzymes. For example, the lead compound for the antiviral drug **saquinavir** contained an L-proline residue that occupied a hydrophobic subsite of a viral protease enzyme. The proline residue was replaced by a decahydroisoquinoline ring which filled the hydrophobic subsite more fully, resulting in better binding interactions (Fig. 11.39).

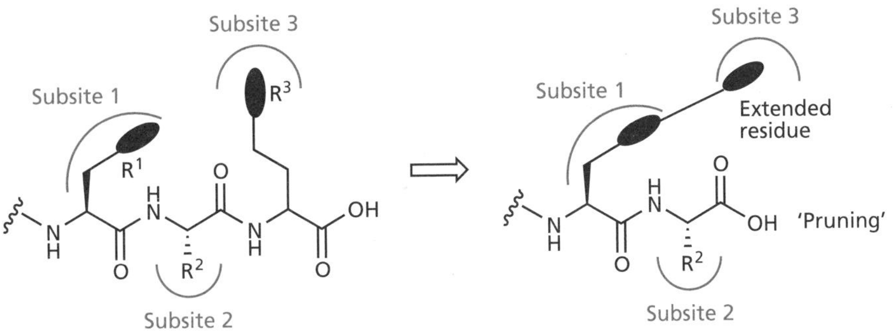

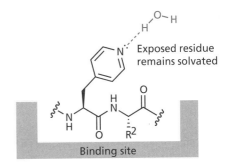

Figure 11.40 Extended residues.

It is even possible to design extended groups which fill two different subsites (Fig. 11.40). This means that the peptidomimetic can be pruned to a smaller molecule. The resulting decrease in molecular weight often leads to better absorption.

Peptidomimetics are often lipophilic in nature, and this can pose a problem because poor water solubility may result in poor oral absorption. Water solubility can be increased by increasing the polarity of residues. For example, an aromatic ring could be replaced by a pyridine ring. However, it is important that this group is not involved in any binding interactions with the target and remains exposed to the surrounding water medium when the peptidomimetic is bound (Fig. 11.41). Otherwise it would have to be desolvated and this would carry an energy penalty which would result in a decreased binding affinity.

The structure-based design of protease inhibitors and matrix metalloproteinase inhibitors is described in sections 17.7.4 and 18.7.1 respectively, and illustrates many of the principles described above.

11.8.5 Oligonucleotides as drugs—antisense drugs

The therapeutic potential of antisense drugs was mentioned in section 7.4. However, there are disadvantages to the use of oligonucleotides as drugs, as they are rapidly degraded by enzymes called **nucleases**. They are also large and highly charged, and are not easily absorbed through cell membranes. Attempts to stabilize these molecules and to reduce their polarity have involved modifying the phosphate linkages in the sugar phosphate backbone. For example, phosphorothioates and methylphosphonates have been extensively studied, and oligonucleotides containing

Figure 11.41 Altering exposed residues to increase water solubility.

these linkages show promise as therapeutic agents (Fig. 11.42). An antisense oligonucleotide with such a modified backbone has been accepted as an antiviral drug (section 17.6.3). Alterations to the sugar moiety have also been tried. For example placing a methyl group at position 2', or using the α-anomer of a deoxyribose sugar, increases resistance to nucleases. Bases have also been modified to improve and increase the number of hydrogen bonding interactions with target nucleic acids.

KEY POINTS

- Neurotransmitters are not effective as drugs as they have a short lifetime in the body, and have poor selectivity for the various types and subtypes of a particular target.

- Hormones are more suitable as drugs, and several are used clinically. Others are susceptible to digestive or metabolic enzymes, and show poor absorption when taken orally. Adverse immune reactions are possible.

- Peptides and proteins generally suffer from poor absorption or metabolic susceptibility. Peptidomimetics are compounds

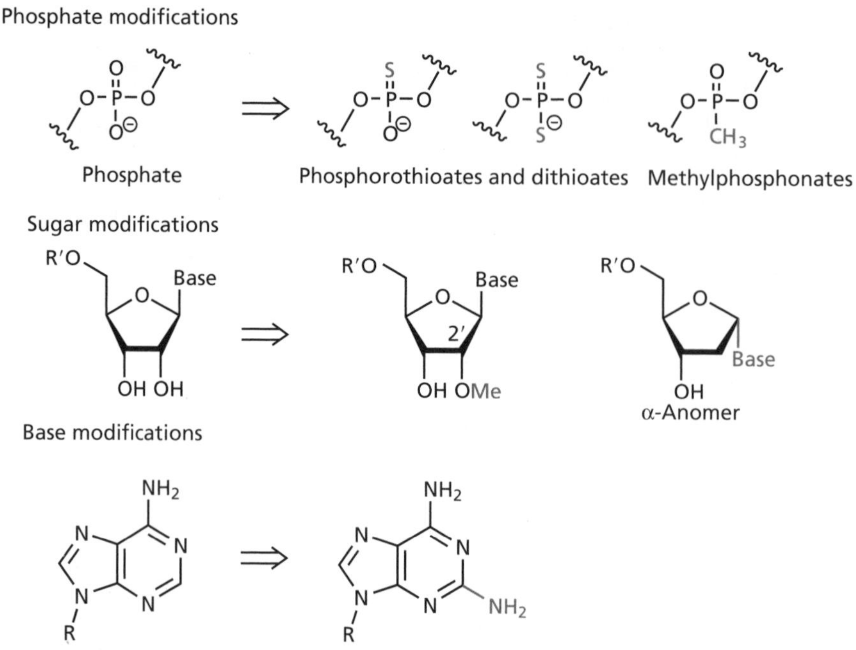

Phosphate modifications

Phosphate   Phosphorothioates and dithioates   Methylphosphonates

Sugar modifications

α-Anomer

Base modifications

Figure 11.42 Modifications on oligonucleotides.

that are derived from peptide lead compounds, but have been altered to disguise their peptide character.

- Oligonucleotides are susceptible to metabolic degradation, but can be stabilized by modifying the sugar phosphate backbone so that it is no longer recognized by relevant enzymes.

QUESTIONS

1. Suggest a mechanism by which methenamine (Fig. 11.31) is converted to formaldehyde under acid conditions.

2. Suggest a mechanism by which ampicillin (Fig. 11.32) decomposes in concentrated solution.

3. Carbidopa (Fig. 11.33) protects levodopa from decarboxylation in the peripheral blood supply, but is too polar to cross the blood–brain barrier into the central nervous system. Carbidopa is reasonably similar in structure to levodopa, so why can it not mimic levodopa and cross the blood–brain barrier by means of a carrier protein?

4. Acetylcholine (Fig. 5.3) is a neurotransmitter that is susceptible to chemical and enzymatic hydrolysis. Suggest strategies that could be used to stabilize the ester group of acetylcholine and show the sort of analogues which might have better stability.

5. Decamethonium is a neuromuscular blocking agent which requires both positively charged nitrogen groups to be active. Unfortunately, it is slowly metabolized and lasts too long in the body. Suggest analogues which might be expected to be metabolized more quickly and lead to inactive metabolites.

$$\overset{\oplus}{Me_3N}(CH_2)_{10}\overset{\oplus}{NMe_3}$$

Decamethonium

6. Miotine has been used in the treatment of a muscle-wasting disease, but there are side effects because a certain amount of the drug enters the brain. Suggest how one might modify the structure of miotine to eliminate this side effect.

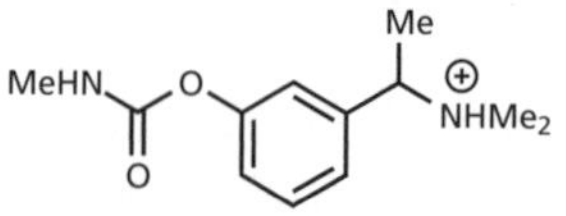

Miotine

7. The oral bioavailability of the antiviral drug aciclovir is only 15–30%. Suggest why this may be the case, and how one might increase the bioavailability of this drug.

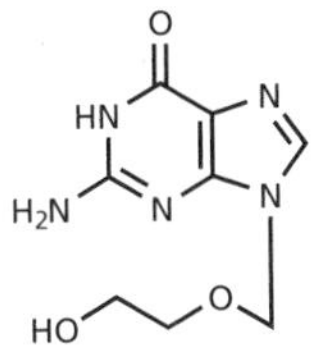

Aciclovir

8. The tetrapeptide H-Cys-Val-Phe-Met-OH is susceptible to metabolism by an aminopeptidase enzyme which hydrolyses

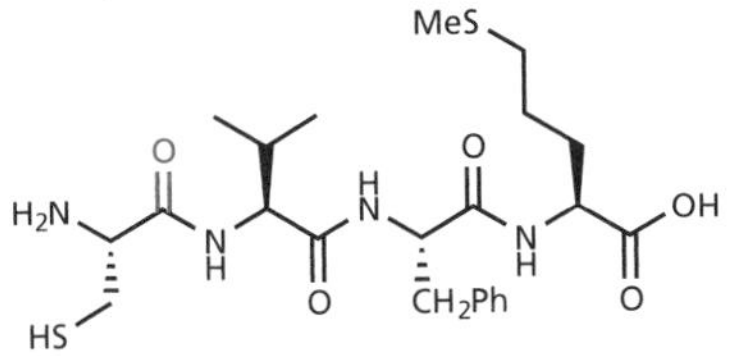

H-Cys-Val-Phe-Met-OH

the peptide bond between cysteine and valine. In order to prevent this, an analogue lacking the carbonyl group in blue

was prepared. A solution of this compound was found to be chemically unstable and the amino acid methionine was identified as being present in the solution. What was the rationale for preparing this analogue, and why did it prove unstable?

9. CGP 52411 is a useful inhibitor of a protein kinase enzyme. Studies on structure–activity relationships demonstrate that substituents on the aromatic rings such as Cl, Me, or OH are bad for activity. Drug metabolism studies show that *para*-hydroxylation occurs to produce inactive metabolites. How would you modify the structure to protect it from metabolism?

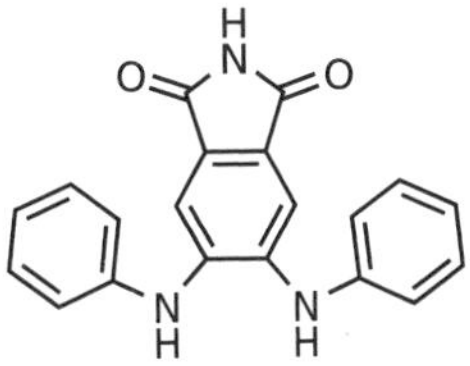

CGP 52411

FURTHER READING

Berg, C., Neumeyer, K., and Kirkpatrick, P. (2003) Teriparatide. *Nature Reviews Drug Discovery*, **2**, 257–258.

Duncan, R. (2003) The dawning era of polymer therapeutics. *Nature Reviews Drug Discovery*, **2**, 347–360.

Ganellin, C. R. and Roberts, S. M. (eds.) (1994) Fluconazole, an orally active antifungal agent. Chapter 13 in: *Medicinal chemistry—the role of organic research in drug research*, 2nd edn. Academic Press, London.

Harris, J. M. and Chess, R. B. (2003) Effect of pegylation on pharmaceuticals. *Nature Reviews Drug Discovery*, **2**, 214–221.

Herr, R. J. (2002) 5-Substituted-1H-tetrazoles as carboxylic acid isosteres: medicinal chemistry and synthetic methods. *Bioorganic and Medicinal Chemistry*, **10**, 3379–3393.

Matthews, T. *et al.* (2004) Enfuvirtide: the first therapy to inhibit the entry of HIV-1 into host CD4 lymphocytes. *Nature Reviews Drug Discovery*, **3**, 215–225.

Opalinska, J. B. and Gewirtz, A. M. (2002) Nucleic-acid therapeutics: basic principles and recent applications. *Nature Reviews Drug Discovery*, **1**, 503–514.

Pardridge, W. M. (2002) Drug and gene targeting to the brain with molecular Trojan horses. *Nature Reviews Drug Discovery*, **1**, 131–139.

Rotella, D. P. (2002) Phosphodiesterase 5 inhibitors: current status and potential applications, *Nature Reviews Drug Discovery*, **1**, 674–682.

Titles for general further reading are listed on p. 711.

12 Drug development

The methods by which lead compounds are discovered were discussed in Chapter 9. In Chapters 10 and 11 we looked at how lead compounds can be optimized to improve their target interactions and pharmacokinetic properties. In this chapter, we look at the various issues that need to be tackled before a promising-looking drug candidate reaches the clinic and goes into full-scale production. The drug development phase is significantly more expensive in terms of time and money than either lead discovery or drug design, and many drugs will fall by the wayside. On average, for every 10 000 structures synthesized during drug design, 500 will reach animal testing, 10 will reach phase I clinical trials and only 1 will reach the market place. The average overall development cost of a new drug was recently estimated as $800 million or £444 million.

Three main issues are involved in drug development. First, the drug has to be tested to ensure that it is not only safe and effective, but can be administered in a suitable fashion. This involves preclinical and clinical trials covering toxicity, drug metabolism, stability, formulation, and pharmacological tests. Second, there are the various patenting and legal issues. Third, the drug has to be synthesized in ever-increasing quantities for testing and eventual manufacture. This is a field known as chemical and process development. Many of these issues have to be tackled in parallel.

12.1 Preclinical and clinical trials

12.1.1 Toxicity testing

One of the first priorities for a new drug is to test if it has any toxicity. This often starts with *in vitro* tests on genetically engineered cell cultures and/or *in vivo* testing on transgenic mice to examine any effects on cell reproduction, and to identify potential carcinogens. Any signs of carcinogenicity would prevent the drug being taken any further.

The drug is also tested for acute toxicity by administering sufficiently large doses *in vivo* to produce a toxic effect or death over a short period of time. Different animal species are used in the study and the animals are dissected to test whether particular organs are affected. Further studies on acute toxicity then take place over a period of months, where the drug is administered to laboratory animals at a dose level expected to cause toxicity but not death. Blood and urine samples are analysed over that period, and then the animals are killed, and the tissues are analysed by pathologists for any sign of cell damage or cancer.

Finally, long-term toxicology tests are carried out over a period of years at lower dose levels to test the drug for chronic toxic effects, carcinogenicity, special toxicology, mutagenicity, and reproduction abnormalities.

The toxicity of a drug used to be measured by its LD_{50} value (the lethal dose required to kill 50% of a group of animals). The ratio of LD_{50} to ED_{50} (the dose required to produce the desired effect in 50% of test animals) is known as the **therapeutic ratio** or **therapeutic index**. A therapeutic ratio of 10 indicates an $LD_{50} : ED_{50}$ ratio of 10:1. This means that a tenfold increase in the ED_{50} dose would result in a 50% death rate. The dose–response curves for a drug's therapeutic and lethal effects can be compared to determine whether the therapeutic ratio is safe or not (Fig. 12.1). Ideally, the curves should not overlap on the x-axis, which means that the more gradual

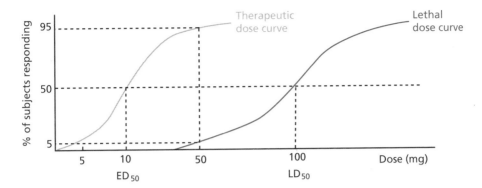

Figure 12.1 Comparison of therapeutic and lethal dose curves.

the two slopes, the riskier the drug will be. The example, Fig. 12.1, shows the therapeutic and lethal dose–response curves for a sedative. Here, a 50-mg dose of the drug will act as a sedative for 95% of the test animals, but will be lethal for 5%. Such a drug would be unacceptable, even though it is effective in 95% of cases treated.

A better measure of a drug's safety is to measure the ratio of the lethal dose for 1% of the population to the effective dose for 99% of the population. A sedative drug with the ratio $LD_1 : ED_{99}$ of 1 would be safer than the one shown in Fig. 12.1.

However, LD values and therapeutic ratios are not the best indicators of a drug's toxicity, as they fail to register any non-lethal or long-term toxic effects. Therefore, toxicity testing should include a large variety of different *in vitro* and *in vivo* tests designed to reveal different types of toxicity. This is not foolproof, however, and a new and unexpected toxic effect may appear during later clinical trials which will require the development of a new test. For example, when **thalidomide** was developed, nobody appreciated that drugs could cause fetal deformities and so there was no test for this. Moreover, even if there had been such tests available, only *in vivo* tests on rabbits would have detected the potential risk.

Many promising drugs fail toxicity testing— a frustrating experience indeed for the drug design teams. For example, the antifungal agent **UK 47265** was an extremely promising antifungal agent, but *in vivo* tests on mice, dogs, and rats showed that it had liver toxicity and was potentially **teratogenic**. The design team had to synthesize more analogues and finally discovered the clinically useful drug **fluconazole** (section 11.5).

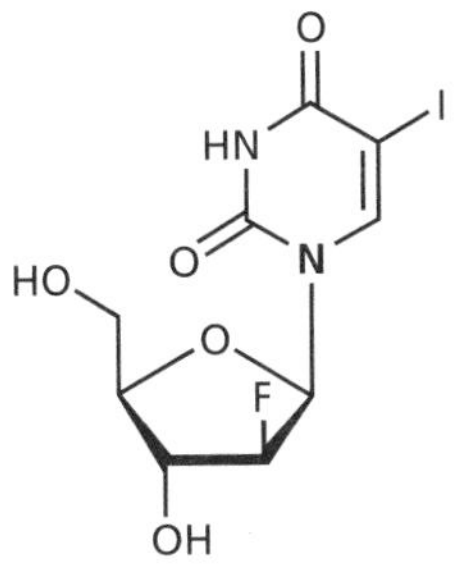

Figure 12.2 Fialuridine.

It should also be borne in mind that it is rare for a drug to be 100% pure. There are bound to be minor impurities present arising from the synthetic route used, and these may well have an influence on the toxicity of the drug. The toxicity results of a drug prepared by one synthetic route may not be the same for the same drug synthesized by a different route, and so it is important to establish the manufacturing synthesis as quickly as possible (section 12.3).

Another aim of toxicity testing is to discover what dose levels are likely to be safe for future clinical trials. Animal toxicity tests do not, however, always highlight potential problems and the toxic properties in test animals may differ from those ultimately observed in humans. For example, the antiviral agent **fialuridine** (Fig. 12.2) underwent clinical trials for the treatment of hepatitis B after toxicity testing on animals. Severe unexpected delayed liver and kidney toxicity suddenly occurred resulting in half the patients (15 in total) suffering liver failure. Five of them died and two of the survivors required liver transplants. It was later found that the drug was incorporated into mitochondrial DNA—something that was not observed in the animal toxicity tests.

Finally, it is unlikely that the thorny problem of animal testing will disappear for a long time. There are so many variables involved in a drug's interaction with the body that it is impossible to anticipate them all. One has also to take into account that the drug will be metabolized to other compounds, all with their own range of biological properties. It appears impossible, therefore, to predict whether a potential drug will be safe by *in vitro* tests alone. Therein lies the importance of animal experiments. Only animal tests can test for the unexpected. Unless we are prepared to volunteer ourselves as guinea-pigs, animal experiments will remain an essential feature of preclinical trails for many years to come.

12.1.2 Drug metabolism studies

The body has an arsenal of metabolic enzymes that can modify foreign chemicals in such a way that they are rapidly excreted (section 8.4). The structures formed from these reactions are called drug metabolites, and it is important to find out what metabolites are formed from any new drug. The structure and stereochemistry of each metabolite has to be determined and the metabolite tested to see what sort of biological activity it might have. This is a safety issue, since some metabolites might prove toxic and others may have side effects that will affect the dose levels that can be used in clinical trials. Ideally, any metabolites that are formed should be inactive and quickly excreted. However, it is quite likely that they will have some form of biological activity (see Box 12.1).

In order to carry out such studies, it is necessary to synthesize the drug labelled with an isotope such as deuterium (^{2}H or D), carbon-13 (^{13}C), tritium (^{3}H or T), or carbon-14 (^{14}C). This makes it easier to detect any metabolites that might be formed. Metabolites containing radioisotopes such as ^{3}H and ^{14}C can be detected at small levels by measuring their β-radiation. Metabolites containing stable heavy isotopes such as deuterium can be detected by mass spectrometry or, in the case of ^{13}C, NMR spectroscopy.

Normally, a synthesis is carried out to include the isotopic label at a specific position in the molecule.

It may be possible to use the established synthetic route for the drug, but in many cases a different route may have to be developed in order to incorporate the label in an efficient manner. Usually, it is preferable to include the label at the latest possible stage of the synthesis. It is not necessary to label every single molecule of the drug, as detection methods are sensitive enough to detect the label even if only a small proportion of the molecules are labelled.

Deuterium or tritium can be very easily incorporated into any molecule containing an exchangeable proton such as those of an alcohol, carboxylic acid, or phenol. This is done simply by shaking a solution of the drug with D_2O or T_2O. Unfortunately, the label is just as easily lost as a result of proton exchange with water in the test animal. Therefore, it is best to carry out a synthesis that places the label on the carbon skeleton of the drug. Nevertheless, there is always the possibility that deuterium or tritium could be lost through a metabolic reaction, in such a way that the metabolite is not detected.

Introducing a carbon isotope often means devising a different synthetic route from the normal one. The effort is often worthwhile, though, as there is less chance of the isotope being lost as a result of a metabolic reaction. Having said that, it is not impossible that an isotope could be lost in this way. Labelling an *N*-methyl group is asking for problems, as *N*-demethylation is a well-known metabolic reaction.

Once a labelled drug has been synthesized, a variety of *in vitro* and *in vivo* tests can be carried out. *In vivo* tests are carried out by administering the labelled drug to a test animal in the normal way, then taking blood and urine samples for analysis to see if any metabolites have been formed. For radiolabelled drugs, this can be done by using high-performance liquid chromatography (HPLC) with a radioactivity detector. It is important to choose the correct animal for these studies, since there are significant metabolic differences across different species. *In vivo* drug metabolism tests are also carried out as part of phase I clinical trials to see whether the drug is metabolized differently in humans from any of the test animals.

In vitro drug metabolism studies can also be carried out using perfused liver systems, liver microsomal fractions, or pure enzymes. Many of the individual cytochrome P450 enzymes that are so important in drug metabolism are now commercially available.

Drug metabolism studies can sometimes be useful in drug design. On several occasions it has been found that an active drug *in vivo* is inactive *in vitro*. This is often a sign that the structure is not really active at all, but is being converted to the active drug by metabolism. The story of **oxamniquine** (section 10.4) illustrates this. Another example was the discovery that the antihypertensive compound I was less active *in vitro* than it was *in vivo*, implying that it was being converted into an active metabolite. Further studies led to the discovery that the active metabolite was **cromakalim**, which proved superior to Structure I as an antihypertensive agent.

Discovery of cromakalim.

12.1.3 Pharmacology, formulation, and stability tests

Although the pharmacology of the drug may have been studied during the drug discovery and drug design stages, it is usually necessary to carry out more tests to see whether the drug has activity at targets other than the intended one, and to gain a better insight into the drug's mechanism of action. These studies also determine a dose–response relationship and define the drug's duration of action.

Formulation studies involve developing a preparation of the drug which is both stable and acceptable to the patient. For orally taken drugs, this usually involves incorporating the drug into a tablet or a capsule. It is important to appreciate that a tablet contains a variety of other substances apart from the drug itself, and studies have to be carried out to ensure that the drug is compatible with these other substances. Preformulation involves the characterization of a drug's physical, chemical, and mechanical properties in order to choose what other ingredients should

be used in the preparation. Formulation studies then consider such factors as particle size, polymorphism, pH, and solubility, as all of these can influence bioavailability and hence the activity of a drug. The drug must be combined with inactive additives by a method which ensures that the quantity of drug present is consistent in each dosage unit. The dosage should have a uniform appearance, with an acceptable taste, tablet hardness, or capsule disintegration.

It is unlikely that these studies will be complete by the time clinical trials commence. This means that simple preparations are developed initially for use in phase I clinical trials (section 12.1.4.1). These typically consist of hand-filled capsules containing a small amount of the drug and a diluent. Proof of the long-term stability of these formulations is not required, as they will be used in a matter of days. Consideration has to be given to what is called the **drug load**—the ratio of the active drug to the total contents of the dose. A low drug load may cause homogeneity problems. A high drug load may pose flow problems or require large capsules if the compound has a low bulk density.

By the time phase III clinical trials are reached (section 12.1.4.3), the formulation of the drug should have been developed to be close to the preparation that will ultimately be used in the market. A knowledge of stability is essential by this stage, and conditions must have been developed to ensure that the drug is stable in the preparation. If the drug proves unstable, it will invalidate the results from clinical trials since it would be impossible to know what the administered dose actually was. Stability studies are carried out to test whether temperature, humidity, ultraviolet light, or visible light have any effect, and the preparation is analysed to see if any degradation products have been formed. It is also important to check whether there are any unwanted interactions between the preparation and the container. If a plastic container is used, tests are carried out to see whether any of the ingredients become adsorbed on to the plastic, and whether any plasticizers, lubricants, pigments, or stabilizers leach out of the plastic into the preparation. Even the adhesives for the container label need to be tested, to ensure they do not leach through the plastic container into the preparation.

12.1.4 Clinical trials

Once the preclinical studies described above have been completed, the company decides whether to proceed to clinical trials. Usually, this will happen if the drug has the desired effect in animal tests, demonstrates a distinct advantage over established therapies, and has acceptable pharmacokinetics, few metabolites, a reasonable half-life, and no serious side effects. Clinical trials are the province of the clinician rather than the scientist, but this does not mean that the research team can wash its hands of the candidate drug and concentrate on other things. Many promising drug candidates fail this final hurdle, and further analogues may need to be prepared before a clinically acceptable drug is achieved. Clinical trials involve testing the drug on volunteers and patients, so the procedures involved must be ethical and beyond reproach. These trials can take 5–7 years to carry out, involve hundreds to thousands of patients, and be extremely expensive. There are four phases of clinical trials.

12.1.4.1 Phase I studies

Phase I studies take about a year and involve 100–200 volunteers. They are carried out on healthy human volunteers to provide a preliminary evaluation of the drug's safety, its pharmacokinetics and the dose levels that can be administered, but they are not intended to demonstrate whether the drug is effective or not.

The drug is tested at different dose levels to see what levels can be tolerated. For each dose level, 6–12 subjects are given the active drug and 2–4 subjects are given a placebo. Normally, the initial dose is a tenth of the highest safest dose used in animal testing. Pharmacokinetic studies are then carried out in order to follow the drug and its metabolites. After a full safety assessment has been made, a higher dose is given, and this is continued until mild adverse effects are observed. This indicates the maximum tolerated dose, and further studies will then concentrate on smaller doses.

During the study, volunteers do not take medication, caffeine, alcohol, or cigarettes. This is to avoid any complications that might arise due to drug–drug interactions (section 8.4.6). As a result, these effects may appear in later phase studies. Studies are carried out early on, however, to determine whether there are any interactions between the drug and food. This is

essential in order to establish when the dose should be taken relative to meals.

Another study involves 4–8 healthy volunteers being given a radiolabelled drug in order to follow the absorption, distribution, and excretion of the drug. These studies also determine how the drug is metabolized in humans.

Studies may be carried out on special age groups of volunteers. For example, drugs intended for Alzheimer's disease are tested on healthy elderly volunteers to test the drug's pharmacokinetics in that particular population. Studies may also be carried out to test whether there are interactions with any other drugs likely to be taken by such a cohort. For example, a drug for Alzheimer's disease will be used mostly on elderly patients who are likely to be taking drugs such as diuretics or anticoagulants such as aspirin. Special studies may be carried out on volunteers with medical conditions that will affect the pharmacokinetics of the drug. These include patients with abnormal rates of metabolism, liver or kidney problems, inflammatory bowel disease or other gastrointestinal diseases.

Bioavailability refers to the fraction of administered drug that reaches the blood supply in a set period of time, and this can vary depending on a variety of factors such as the crystal form of the drug, whether the drug is administered as a tablet or a capsule, or a variation in the constituents of a tablet or capsule. As a result, it is important to check that bioavailability remains the same should there be any alteration to the manufacturing, formulation, or storage processes. Such checks are called **bioequivalence** studies. For example, bioequivalence studies are required when different dosage forms are used in the early and late phases of clinical trials. Powder-filled capsules are frequently used in phase I, whereas tablets are used in phases II and III. Therefore, it is necessary to establish that these formulations show bioequivalence in healthy volunteers. In addition, it has to be demonstrated that dissolution of both formulations is similar.

In situations where the drug is potentially toxic and is to be used on a life-threatening disease such as AIDS or cancer, volunteer patients are used for phase I studies rather than healthy volunteers.

The decision whether to proceed to phase II can be difficult, as only a limited amount of safety data is available. Any adverse effects that are observed may or may not be due to the drug. For example, abnormal liver function in a healthy patient may be due to the drug or to alcohol. Nevertheless, evidence of a serious adverse effect will usually result in clinical trials being terminated.

12.1.4.2 Phase II studies

Phase II studies generally last about 2 years and may start before phase I studies are complete. They are carried out on patients to establish whether the drug has the therapeutic property claimed, to study the pharmacokinetics and short-term safety of the drug, and to define the best dosing regimes. Phase II trials can be divided into early and late studies (IIa and IIb respectively).

Initial trials (phase IIa) involve a limited number of patients to see if the drug has any therapeutic value at all and to see if there are any obvious side effects. If the results are disappointing, clinical trials may be terminated at this stage.

Later studies (IIb) involve a larger numbers of patients. They are usually carried out as double-blind placebo-controlled studies. This means that the patients are split into two groups where one group receives the drug, and the other group receives a placebo. In a double-blind study neither the doctor nor the patient knows whether a placebo or drug is administered. In the past, it has been found that investigators can unwittingly 'give the game away' if they know which patient is getting the actual drug. The studies demonstrate whether the patients receiving the drug show an improvement relative to the patients receiving the placebo. Different dosing levels and regimes are also determined to find the most effective. Most phase II trials require 20–80 patients per dose group to demonstrate efficacy.

Some form of rescue medication may be necessary for those patients taking a placebo. For example, it would be unethical to continue asthmatic patients on a placebo if they suffer a severe asthmatic attack. A conventional drug would be given and its use documented. The study would then compare how frequently the placebo group needed to use the rescue medicine compared to those taking the new drug. With life-threatening diseases such as AIDS or cancer, the use of a placebo is not ethical and an established drug is used as a standard comparison.

The **endpoint** is the measure that is used to determine whether a drug is successful or not. It can be any parameter that is relevant, measurable, sensitive, and ethically acceptable. Examples of endpoints include blood assays, blood pressure, tumour regression, and the disappearance of an invading pathogen from tissues or blood. Less defined endpoints include perception of pain, use of rescue medications, and level of joint stiffness.

12.1.4.3 Phase III studies

Phase III studies normally take about 3 years and can be divided into phases IIIa and IIIb. These studies may begin before phase II studies are completed. The drug is tested in the same way as in phase II, using double-blind procedures, but on a much larger sample of patients. Patients taking the drug are compared with patients taking a placebo or another available treatment. Comparative studies of this sort must be carried out without bias and this is achieved by randomly selecting the patients—those who will receive the new drug and those who will receive the alternative treatment or placebo. Nevertheless, there is always the possibility of a mismatch between the two groups with respect to factors such as age, race, sex, and disease severity, and so the greater the number of patients in the trial the better.

Phase IIIa studies establish whether the drug is really effective or whether any beneficial effects are psychological. They also allow further 'tweaking' of dose levels to achieve the optimum dose. Any side effects not previously detected may be picked up with this larger sample of patients. If the drug succeeds in passing phase IIIa, it can be registered. Phase IIIb studies are carried out after registration, but before approval. They involve a comparison of the drug with those drugs that are already established in the field.

In certain circumstances where the drug shows a clear beneficial effect early on, the phase III trials may be terminated earlier than planned. Some patients in the phase III studies will be permitted to continue taking the drug if it has proved effective, and will be monitored to assess the long-term safety of the drug.

12.1.4.4 Phase IV studies

The drug is now placed on the market and can be prescribed, but it is still monitored for effectiveness and for any rare or unexpected side effects. In a sense, this phase is a never-ending process as unexpected side effects may crop up many years after the introduction

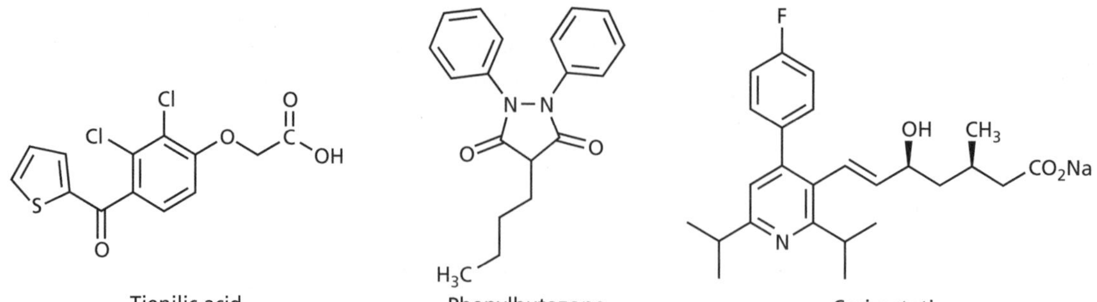

Figure 12.3 Drugs which have been removed from the market due to rare toxic side effects.

of the drug. In the UK, the medicines committee runs a voluntary yellow card scheme where doctors and pharmacists report suspected adverse reactions to drugs. This system has revealed serious side effects for a number of drugs after they had been put on the market. For example, the β-blocker **practolol** had to be withdrawn after several years of use because some patients suffered blindness and even death. The toxic effects were unpredictable and are still not understood, and so it has not been possible to develop a test for this effect.

The diuretic agent **tienilic acid** (Fig. 12.3) had to be withdrawn from the market because it damaged liver cells in 1 out of every 10 000 patients. The anti-inflammatory agent **phenylbutazone** (Fig. 12.3) can cause a rare but fatal side effect in 22 patients out of every million treated with the drug! Such a rare toxic effect would clearly not be detected during phase III trials. A more recent example is **cerivastatin** (Fig. 12.3), which was marketed as a potent anti-cholesterol drug. Unfortunately, it had to be withdrawn in 2001 as a result of adverse drug–drug interactions which resulted in muscle damage and the deaths of 40 people worldwide. The withdrawal of drugs due to such rare events could potentially be avoided if the genomes of individual patients were 'fingerprinted' to establish who might be at risk from such rare effects.

12.1.4.5 Ethical issues

In phases I–III of clinical trials, the permission of the patient is mandatory. However, ethical problems can still arise. For example, unconscious patients and mentally ill patients cannot give consent, but might benefit from the improved therapy. Should one include them or not? The ethical problem of including children in clinical trials is also a thorny issue, and so most clinical trials exclude them. However, this means that most licensed drugs have been licensed for adults. When it comes to prescribing for children, clinicians are left with the problem of deciding what dose levels to use, and simple arithmetic mistakes can have tragic consequences.

KEY POINTS

- Toxicity tests are carried out *in vitro* and *in vivo* on drug candidates to assess acute and chronic toxicity. During animal studies, blood and urine samples are taken for analysis. Individual organs are analysed for tissue damage or abnormalities. Toxicity testing is important in defining what the initial dose level should be for phase I clinical trials.

- Drug metabolism studies are carried out on animals and humans to identify drug metabolites. The drug candidate is labelled with an isotope in order to aid the detection of metabolites.

- Pharmacology tests are carried out to determine a drug's mechanism of action and to determine whether it acts at targets other than the intended one.

- Formulation studies aim to develop a preparation of the drug which can be administered during clinical trials and beyond. The drug must remain stable in the preparation under a variety of environmental conditions.

- Clinical trials involve four phases. In phase I, healthy volunteers are normally used to evaluate the drug's safety, its pharmacokinetics, and the dose levels that can safely be administered. Phase II studies are carried out on patients to assess whether the drug is effective, to give further information on the most effective dosing regime and to identify side effects. Phase III studies are carried out on larger numbers of patients to ensure that results are statistically sound, and to detect less common side effects. Phase IV studies are ongoing and monitor the long-term use of the drug in specific patients, as well as the occurrence of rare side effects.

12.2 Patenting and regulatory affairs

12.2.1 Patents

Having spent enormous amounts of time and money on research and development, a pharmaceutical company quite rightly wants to reap the benefit of all its hard work. To do so, it needs to have the exclusive rights to sell and manufacture its products for a reasonable period of time, and at a price which will not only recoup its costs, but will generate sufficient profits for further research and development. Without such rights, a competitor could synthesize the same product without suffering the expense involved in designing and developing it.

Patents allow companies the exclusive right to the use and profits of a novel pharmaceutical for a limited term. In order to gain a patent, the company has to first submit or **file** the patent. This should reveal what the new pharmaceutical is, what use it is intended for, and how it can be synthesized. This is no straightforward task. Each country has its own patents, so the company has to first decide in which countries it is going to market its new drug and then file the relevant patents. Patent law is also very precise, and varies from country to country. Therefore, submitting a patent is best left to the patent attorneys and lawyers who are specialists in the field. The cost and effort involved in obtaining patents from different countries can be reduced in two ways. First, a patent application can be made to the **European Patent Office** (EPO). If it is approved, a European patent is granted which can then be converted to country-specific patents relevant to the individual countries belonging to the **European Patent Convention** (EPC). There are 27 such countries and the applicant can decide how many of the 27 individual patents should be taken out. A second approach is to file an international application which designates one or more of the 122 countries who have signed up to the **Patent Cooperation Treaty** (PCT).

An **International Search Report** (ISR) and **International Preliminary Examination Report** (IPER) can be obtained, which can then be used when applying for patents from individual countries. No PCT or international patents are awarded, but the reports received help the applicant to decide which patent applications to individual countries are likely to succeed.

Once a patent has been filed, the patent authorities decide whether the claims are novel and whether they satisfy the necessary requirements for that patent body. One universal golden rule is that the information supplied has not previously been revealed, either in print or by word of mouth. As a result, pharmaceutical companies only reveal their work after the structures involved have been safely patented.

It is important that a patent is filed as soon as possible. Such is the competition between the pharmaceutical companies that it is highly likely that a novel agent discovered by one company may be discovered by a rival company only weeks or months later. This means that patents are filed as soon as a novel agent or series of agents is found to have significant activity. Usually, the patent is filed before the research team has had the chance to start all the extensive preclinical tests that need to be carried out on novel drugs. It may not even have synthesized all the possible structures it is intending to make. Therefore, the team is in no position to identify which specific compound in a series of structures is likely to be the best drug candidate. As a result, most patents are designed to cover a series of compounds belonging to a particular structural class, rather than one specific structure. Even if a specific structure has been identified as the best drug candidate, it is best to write the patent to cover a series of analogues. This prevents a rival company making a close analogue of the specified structure and selling it in competition. All the structures that are to be protected by the patent should be specified in the patent, but only a representative few need to be described in detail.

Patents in most countries run for 20 years after the date of filing. This sounds a reasonable time span, but it has to be remembered that the protection period starts from the time of filing, not from when the drug comes onto the market. A significant period of patent protection is lost due to the time required for preclinical tests, clinical trials, and regulatory approval. This often involves a period of 6–10 years. In some cases, this period may threaten to be even longer, in which case the company may decide to abandon the project as the duration of patent protection would be deemed too short to make sufficient profits. This illustrates the point that not all patents lead to a commercially successful product.

Patents can be taken out to cover specific products, the medicinal use of the products, the synthesis of the products, or preferably all three of these aspects. Taking out a patent which only covers the synthesis of a novel product offers poor patent protection. A rival company could quite feasibly develop a different synthesis to the same structure and then sell it legally.

One of the current issues in the patent area is **chiral switches**. In the period 1983–87, 30% of approved drugs were pure enantiomers, 29% were racemates, and 41% were achiral. Nowadays, most of the drugs reaching the market are either achiral or pure enantiomers. The problem with racemates is that each enantiomer usually has a different level of activity. Moreover, the enantiomers often differ in the way they are metabolized and in their side effects. Consequently, it is better to market the pure enantiomer rather than the racemate. The issue of chiral switching relates mostly to racemic drugs that have been on the market for several years and are approaching the end of their patent life. By switching to the pure enantiomer, companies can argue that it is a new invention and take out a new patent. However, they have to prove that the pure enantiomer is an improvement on the original racemate and that they could not reasonably have been expected to know that when the racemate was originally patented. A full appreciation of how different enantiomers can have different biological properties was realized in the 1980s, and so chiral switches have normally been carried out on racemic drugs that reached the market before that. Timing is important, and the company ideally wants to have the pure enantiomer reaching the market just as the patent on the original racemate is expiring. It is not possible to patent a new drug in its racemic form today, then expect to market it later as the pure enantiomer, as the issue of stereoisomerism is now an established fact. **Bupivacaine** is an example of an established chiral drug that has undergone chiral switching. It is a long-lasting local anaesthetic used as a spinal and epidural anaesthetic for childbirth and hip replacements, and acts by blocking sodium ion

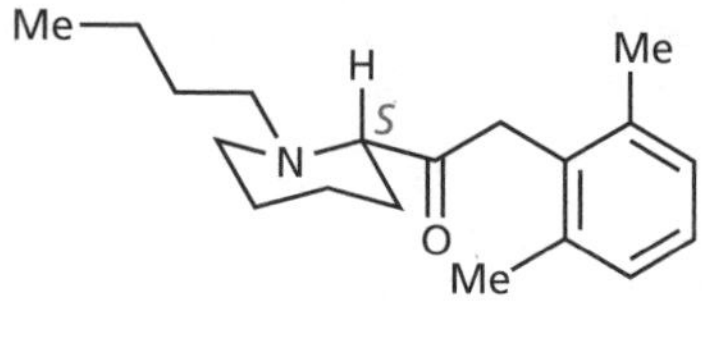

Figure 12.4 Levobupivacaine.

channels in nerve axons. Unfortunately, it also affects the heart, which prevents it being used for intravenous injections. The S-enantiomer of bupivacaine is called **levobupivacaine** (Fig. 12.4). It has less severe side effects and is a safer local anaesthetic used for the same purposes as bupivacaine. Other examples of drugs that have undergone chiral switches include **salbutamol** (section 20.10.3) and **omeprazole** (section 22.3).

12.2.2 Regulatory affairs

12.2.2.1 The regulatory process

Regulatory bodies such as the **Food and Drug Administration** (FDA) in the USA and the **European Agency for the Evaluation of Medicinal Products** (EMEA) in Europe come into play as soon as a pharmaceutical company believes it has a useful drug. Before clinical trials can begin, the company has to submit the results of its scientific and preclinical studies to the relevant regulatory authority. In the USA, this takes the form of an **Investigational Exemption to a New Drug Application** (IND), which is a confidential document submitted to the FDA. The IND should contain information regarding the chemistry, manufacture, and quality control of the drug, as well as information on its pharmacology, pharmacokinetics, and toxicology. The FDA assesses this information and then decides whether clinical trials can begin. Dialogue then continues between the FDA and the company throughout the clinical trials. Any adverse results must be reported to the FDA, who will discuss with the company whether the trials should be stopped.

If the clinical trials proceed smoothly, the company applies to the regulatory authority for marketing approval. In the USA, this involves the submission of a **New Drug Application** (NDA) to the FDA; in Europe, the equivalent submission is called a **Marketing Authorization Application** (MAA). An NDA or MAA is typically 400–700 volumes in size, with each volume containing 400 pages! The application has to state what the drug is intended to do, along with scientific and clinical evidence for its efficacy and safety. It should also give details of the chemistry and manufacture of the drug, as well as the controls and analysis which will be in place to ensure that the drug has a consistent quality. Any advertising and marketing material must be submitted, to ensure that it makes accurate claims and that the drug is being promoted for its intended use. The labelling of a drug preparation must also be approved, to ensure that it instructs physicians about the mechanism of action of the drug, the medical situations for which it should be used, and the correct dosing levels and frequency. Possible side effects, toxicity, or addictive effects should be detailed, as well as special precautions which might need to be taken (e.g. avoiding drugs that interact with the preparation).

The FDA has inspectors who will visit clinical investigators to ensure that their records are consistent with those provided in the NDA, and that patients have been adequately protected. An approval letter is finally given to the company and the product can be launched, but the FDA will continue to monitor the promotion of the product as well as further information regarding any unusual side effects.

Once an NDA is approved, any modifications to a drug's manufacturing synthesis or analysis must be approved. In practice, this means that the manufacturer will stick with the route described in the NDA and perfect that, rather than consider alternative routes.

An abbreviated NDA can be filed by manufacturers who wish to market a generic variation of an approved drug whose patent life has expired. The manufacturer is only required to submit chemistry and manufacturing information, and demonstrate that the product is comparable with the product already approved.

The term **new chemical entity** (NCE) or **new molecular entity** (NME) refers to a novel drug structure. In the 1960s about 70 NCEs were reaching the market each year, but this had dropped to less than 30 per year by 1971. In part, this was due to more stringent testing regulations that were brought in after the thalidomide disaster. Another factor, however, was the fact that the number of lead compounds was

starting to run out at the same time. In an attempt to address this, there has been an emphasis since the 1970s on understanding the mechanism of the disease and designing new drugs in a scientific fashion. Although the approach has certainly been more scientific than the previous trial and error approach, the number of NCEs reaching the market is still low. In 2002, only 18 NCEs were approved by the FDA and 13 by the EMEA.

12.2.2.2 Fast tracking and orphan drugs

The regulations of many regulatory bodies include the possibility of **fast tracking** certain types of drug, so that they reach the market as quickly as possible. Fast tracking is made possible by demanding a smaller number of phase II and phase III clinical trials before the drug is put forward for approval. Fast tracking is carried out for drugs that show promise for diseases where no current therapy exists, and for drugs that show distinct advantages over existing ones in the treatment of life-threatening diseases such as cancer.

Orphan drugs are drugs that are effective against relatively rare medical problems. Because there is not a large market for such drugs, pharmaceutical companies are unlikely to reap much benefit and may decide not to develop and market the drug. In the USA an orphan drug is defined as one that is used for less than 200 000 people. Financial and commercial incentives are given to firms in order to encourage the development and marketing of such drugs.

12.2.2.3 Good laboratory, manufacturing, and clinical practice

Good Laboratory Practice (GLP) and Good Manufacturing Practice (GMP) are scientific codes of practice for a pharmaceutical company's laboratories and production plants. They detail the scientific standards that are necessary, and the company must prove to regulatory bodies that it is adhering to these standards.

GLP regulations apply to the various research laboratories involved in pharmacology, drug metabolism, and toxicology studies. GMP regulations apply to the production plant and chemical development laboratories. They encompass the various manufacturing procedures used in the production of the drug, as well as the procedures used to ensure that the product is of a consistently high quality.

As part of GMP regulations, the pharmaceutical company is required to set up an independent quality control unit which monitors a wide range of factors including employee training, the working environment, operational procedures, instrument calibration, batch storage, labelling, and the quality control of all solvents, intermediates, and reagents used in the process. The analytical procedures which are used to test the final product must be defined, as well as the specifications that have to be met. Each batch of drug which is produced must be sampled to ensure that it passes those specifications. Written operational instructions must be in place for all special equipment (e.g. freeze dryers), and standard operating procedures (SOPs) must be written for the use, calibration, and maintenance of equipment.

Detailed and accurate paperwork on the above procedures must be available for inspection by the regulatory bodies. This includes calibration and maintenance records, production reviews, batch records, master production records, inventories, analytical reports, equipment cleaning logs, batch recalls, and customer complaints. Although record-keeping is crucial, it is possible the extra paperwork involved can stifle innovations in the production process.

Investigators involved in clinical research must demonstrate that they can carry out the work according to Good Clinical Practice (GCP) regulations. The regulations require proper staffing, facilities, and equipment for the required work, and each test site involved must be approved. There must also be evidence that a patient's rights and well-being are properly protected. In the USA, approval is given by the **Institutional Review Board** (IRB). While the work is in progress, regulatory authorities may carry out data audits to ensure that no research misconduct is taking place (e.g. plagiarism, falsification of data, poor research procedures, etc.). In the UK, the **General Medical Council** or the **Association of British Pharmaceutical Industry** can discipline unethical researchers. Problems can arise due to the pressures which are often placed on researchers to obtain their results as speedily as possible. This can lead to hasty decisions, resulting in mistakes and poorly thought-out procedures. There have also been individuals who have

deliberately falsified results or have cut corners. Sometimes personal relationships can prove a problem. The investigator can be faced with a difficult dilemma between doing the best for the patient, and maintaining good research procedures. Patients may also mislead clinicians if they are desperate for a new cure, and falsify their actual condition in order to take part in the trial. Other patients have been known to get their drugs analysed to see if they are getting a placebo or drug.

KEY POINTS

- Patents are taken out as soon as a useful drug has been identified. They cover a structural class of compounds rather than a single structure.

- A significant period of the patent is lost as a result of the time taken to get a drug to the market place.

- Patents can cover structures, their medicinal use, and their method of synthesis.

- Regulatory bodies are responsible for approving the start of clinical trials and the licensing of new drugs for the market place.

- Drugs that show promise in a field which is devoid of a current therapy may be fast tracked.

- Special incentives are given to companies to develop orphan drugs—drugs that are effective in rare diseases.

- Pharmaceutical companies are required to abide by professional codes of practice known as Good Laboratory Practice, Good Manufacturing Practice, and Good Clinical Practice.

12.3 Chemical and process development

12.3.1 Chemical development

Once a compound goes forward for preclinical tests, it is necessary to start the development of a large-scale synthesis as soon as possible. This is known as chemical development, and is carried out in specialist laboratories. To begin with, a quantity of the drug may be obtained by scaling up the synthetic route used by the research laboratories. In the longer term, however, such routes often prove unsuitable for large-scale manufacture. There are several reasons for this. During the drug discovery/design phase, the emphasis is on producing as many different compounds as possible in as short a period of time as possible. The yield is unimportant as long as sufficient material is obtained for testing. The reactions are also done on a small scale, which means that the cost is trivial, even if expensive reagents or starting materials are used. Hazardous reagents, solvents, or starting materials can also be used because of the small quantities involved.

The priorities in chemical development are quite different. A synthetic route has to be devised which is straightforward, safe, cheap, efficient, and high yielding, has the minimum number of synthetic steps, and will provide a consistently high-quality product which meets predetermined specifications of purity.

During chemical development, the conditions for each reaction in the synthetic route are closely studied and modified in order to get the best yields and purity. Different solvents, reagents, and catalysts may be tried. The effects of temperature, pressure, reaction time, excess reagent or reactant, concentration, and method of addition are studied. Consideration is also given to the priorities required for scale up. For example, the original synthesis of aspirin from salicylic acid involved acetylation with acetyl chloride (Fig. 12.5). Unfortunately, the by-product from this is hydrochloric acid, which is corrosive and environmentally hazardous. A better synthesis involves acetic anhydride as the acylating agent. The by-product formed here is acetic acid, which does not have the unwanted properties of hydrochloric acid, and can also be recycled.

Therefore, the final reaction conditions for each stage of the synthesis may be radically different from the original conditions, and it may even be necessary to abandon the original synthesis and devise a completely different route.

Once the reaction conditions for each stage have been optimized, the process needs to be scaled up. The priorities here are cost, safety, purity, and yield. Expensive or hazardous solvents or chemicals should be avoided and, if possible, replaced by cheaper, safer alternatives. Experimental procedures may have to be modified. Several operations carried out on a research scale are impractical on large scale. These include the use of drying agents, rotary evaporators, and separating funnels. Alternative procedures for these operations which can be used on large scale are, respectively, removing water as an azeotrope, distillation, and stirring the different phases.

There are several stages in chemical development. In the first stage about a kilogram of drug is required for short-term toxicology and stability tests, analytical research, and pharmaceutical development. Often, the original synthetic route will be developed quickly and scaled up in order to produce this quantity of material, as time is of the essence. The next stage is to produce about 10 kg for long-term toxicology tests, as well as for formulation studies. Some of the material may also be used for phase I clinical trials. The third stage involves a further scale up to the pilot plant, where about 100 kg is prepared for phase II and phase III clinical trials.

Because of the time scales involved, the chemical process used to synthesize the drug during stage 1 may differ markedly from that used in stage 3. However, it is important that the quality and purity of the drug remains as constant as possible for all the studies carried out. Therefore, an early priority in

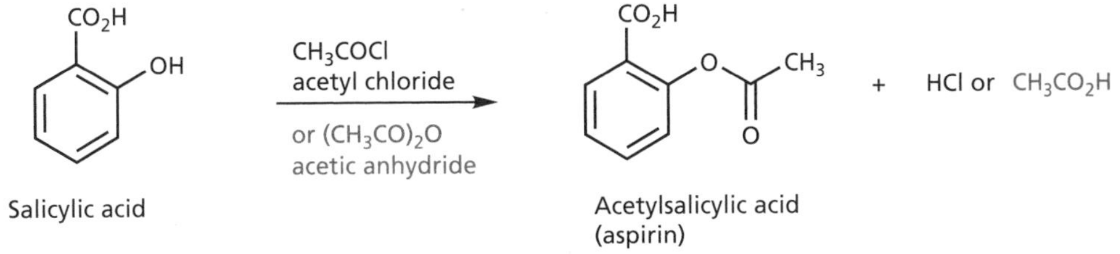

Figure 12.5 Synthesis of aspirin.

chemical development is to optimize the final step of the synthesis and to develop a purification procedure which will consistently give a high-quality product. The **specifications** of the final product are defined and determine the various analytical tests and purity standards required. These define predetermined limits for a range of properties such as melting point, colour of solution, particle size, polymorphism, and pH. The product's chemical and stereochemical purities must also be defined, and the presence of any impurities or solvent should be identified and quantified. Acceptable limits for different compounds are proportional to their toxicity. For example, the specifications for ethanol, methanol, mercury, sodium, and lead are 2%, 0.05%, 1 ppm, 300 ppm, and 2 ppm respectively. Carcinogenic compounds such as benzene or chloroform should be completely absent, which means in practice that they must not be used as solvents or reagents in the final stages of the synthesis.

All future batches of the drug must meet these specifications. Once the final stages have been optimized, future development work can then look to optimize or alter the earlier stages of the synthesis.

In some development programmes, the structure originally identified as the most promising clinical prospect may be supplanted by another structure that demonstrates better properties. The new structure may be a close analogue of the original compound, but such a change can have radical effects on chemical development and require totally different conditions to maximize the yields for each synthetic step (see Box 12.2).

BOX 12.2 SYNTHESIS OF EBALZOTAN

Ebalzotan is an antidepressant drug produced by Astra which works as a selective serotonin (5-HT$_{1A}$) antagonist. The original synthesis from structure I involved six steps (Fig. 1) and included several expensive and hazardous reagents, resulting in a paltry overall yield of 3.7%. Development of the route involved the replacement of 'problem' reagents and optimization of the reaction conditions, leading to an increase in the overall yield to 15%. Thus, the expensive and potentially toxic reducing agent sodium cyanoborohydride was replaced by hydrogen gas over a palladium catalyst. In the demethylation step, BBr$_3$—which is corrosive, toxic, and expensive—was replaced by HBr, which is cheaper and less toxic.

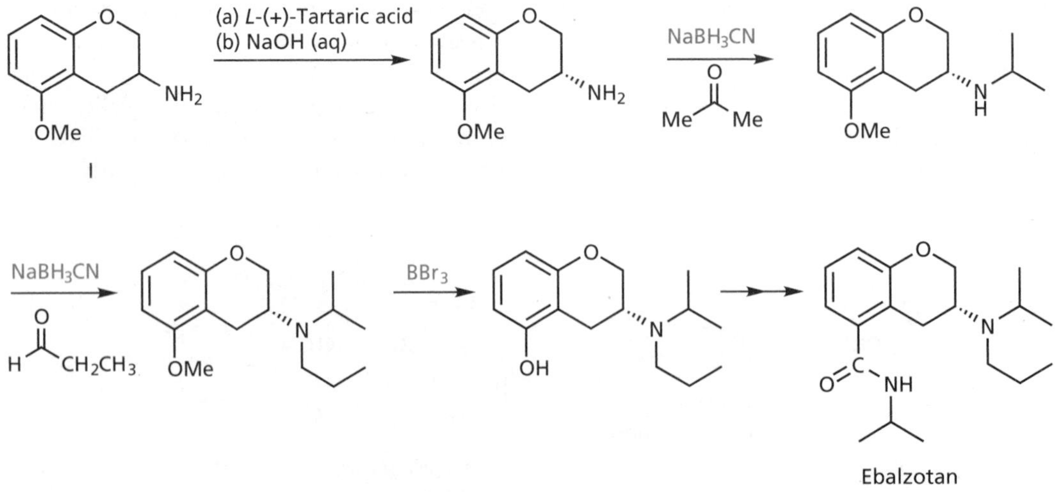

Figure 1 Synthesis of ebalzotan.

12.3.2 Process development

Process development aims to ensure that the number of reactions in the synthetic route is as small as possible and that all the individual stages in the process are integrated with each other such that the full synthesis runs smoothly and efficiently on a production scale. The aim is to reduce the number of operations to the minimum. For example, rather than isolating each intermediate in the synthetic sequence, it is better to move it directly from one reaction vessel to the next for the subsequent step. Ideally, the only purification step carried out is on the final product.

Environmental and safety issues are extremely important. Care is taken to minimize the risk of chemicals escaping into the surrounding environment, and chemical recycling is carried out as much as possible. The use of 'green technology' such as electrochemistry, photochemistry, ultrasound, or microwaves may solve potential environmental problems.

Cost is a high priority, and it is more economic to run the process such that a small number of large batches are produced rather than a large number of small batches. Extreme care must be taken over safety. Any accident in a production plant has the potential to be a major disaster, so there must be strict adherence to safety procedures, and close monitoring of the process when it is running. However, the overriding priority is that the final product should still be produced with a consistently high purity in order to meet the required specifications.

Process development is very much aimed at optimizing the process for a specific compound (Box 12.3). If the original structure is abandoned in favour of a different analogue, the process may have to be rethought completely.

The regulatory authorities require that every batch of a drug product is analysed to ensure that it meets the required specifications, and that all impurities are characterized, identified, and quantified. The identification of impurities is known as **impurity profiling** and typically involves their isolation by preparative HPLC, followed by NMR spectroscopy or mass spectrometry to identify their structure.

BOX 12.3 SYNTHESIS OF ICI D7114

ICI D7114 is an agonist at adrenergic β_3-receptors and was developed for the treatment of obesity and non-insulin-dependent diabetes, The original synthetic route used in the research laboratory is shown in Fig. 1.

The overall yield was only 1.1%, and there were various problems in applying the route to the production scale. The first reaction involves hydroquinone and ethylene dibromide, both of which could react twice to produce side products. Moreover, ethylene bromide is a carcinogen, and toxic vinyl bromide is generated as a volatile side product during the reaction. A chromatographic separation was required after the second stage to remove a side product—a process which is best avoided on large scale. The use of high-pressure hydrogenation at 20 times atmospheric pressure to give structure III was not possible on the plant scale, as the equipment available could only achieve 500 kP (about 5 times atmospheric pressure). Finally, the product has an asymmetric centre, and so it was necessary to carry out a resolution. This involved forming a salt with a chiral acid and carrying out eight crystallizations—a process that would be totally unsuitable at production scale.

The revised synthetic route shown in Fig. 2 avoided these problems and improved the overall yield to 33%. To avoid the possibility of any dialkylated side product being formed, para-benzyloxyphenol was used as starting material. Ethylene dimesylate was used in place of the carcinogenic ethylene dibromide. As a result, vinyl bromide was no longer generated as a side product. The alkylated product (IV) was not isolated, but was treated in situ with benzylamine, thus cutting down number of operations involved. Hydrogenolysis of the benzyl ether group was carried out in the presence of methanesulfonic acid, the latter helping to prevent hydrogenolysis of the N-benzyl group. This gave structure II, which was one of the intermediates in the original synthesis. Alkylation gave structure V and an asymmetric reaction was carried out with the epoxide to avoid the problem of resolution. The product from this reaction (VI) could be hydrogenated to the final product in situ without the need to isolate VI, again cutting down the number of operations.

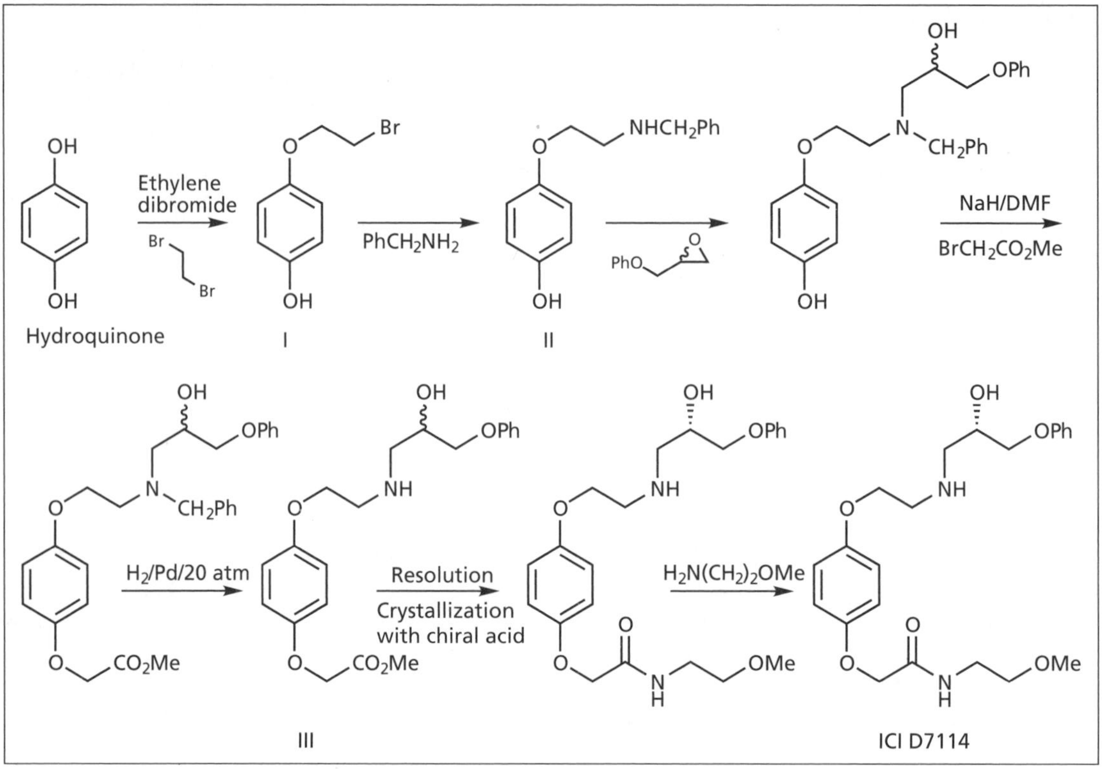

Figure 1 Research synthesis of ICI D7114.

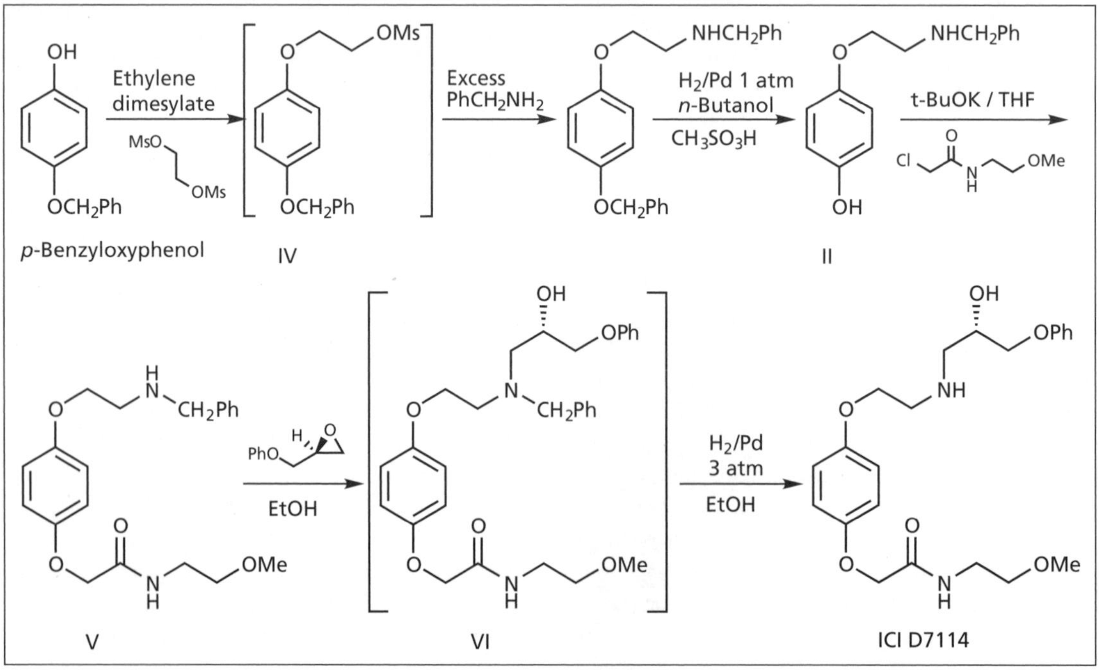

Figure 2 Revised synthetic route to ICI D7114.

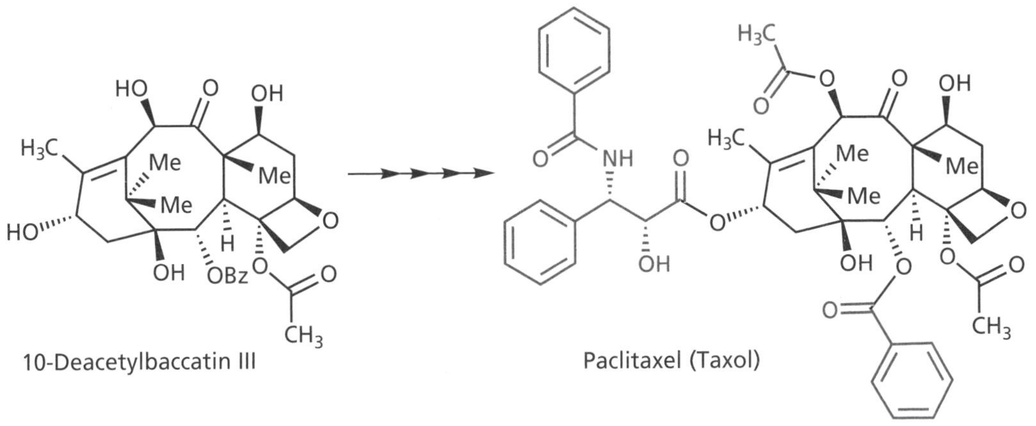

Figure 12.6 Semisynthetic synthesis of paclitaxel (Taxol).

12.3.3 Choice of drug candidate

The issues surrounding chemical and process development can affect the choice of which drug candidate is taken forward into drug development. If it is obvious that a particular structure is going to pose problems for large-scale production, an alternative structure may be chosen which poses fewer problems, even if it is less active.

12.3.4 Natural products

Not all drugs can be fully synthesized. Many natural products have quite complex structures that are too difficult and expensive to synthesize on an industrial scale. These include drugs such as **penicillin, morphine**, and **paclitaxel (Taxol)**. Such compounds can only be harvested from their natural source—a process which can be tedious, time consuming, and expensive, as well as being wasteful on the natural resource. For example, four mature yew trees have to be cut down to obtain enough paclitaxel to treat one patient! Furthermore, the number of structural analogues that can be obtained from harvesting is severely limited.

Semi-synthetic procedures can sometimes get round these problems. This often involves harvesting a biosynthetic intermediate from the natural source, rather than the final compound itself. The intermediate could then be converted to the final product by conventional synthesis. This approach can have two advantages. First, the intermediate may be more easily extracted in higher yield than the final product itself. Second, it may allow the possibility of synthesizing analogues of the final product. The semisynthetic penicillins are an illustration of this approach (section 16.5.1.6).

Another more recent example is that of paclitaxel. It is manufactured by extracting **10-deacetylbaccatin III** from the needles of the yew tree, then carrying out a four-stage synthesis (Fig. 12.6).

KEY POINTS

- Chemical development involves the development of a synthetic route which is suitable for the large-scale synthesis of a drug.

- The priorities in chemical development are to develop a synthetic route which is straightforward, safe, cheap, efficient, has the minimum number of synthetic steps, and will provide a consistently good yield of high-quality product that meets predetermined purity specifications.

- An early priority in chemical development is to define the purity specifications of the drug and to devise a purification procedure which will satisfy these requirements.

- Process development aims to develop a production process which is safe, efficient, economic, environmentally friendly, and produces product of a consistent yield and quality to satisfy purity specifications.

- Drugs derived from natural sources are usually produced by harvesting the natural source or through semi-synthetic methods.

QUESTIONS

1. Discuss whether the doubly labelled atropine molecule shown below is suitable for drug metabolism studies.

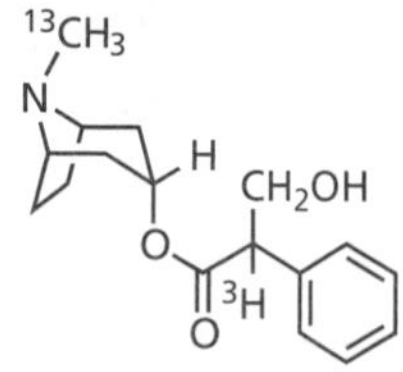

Atropine

2. What is meant by a placebo, and what sort of issues need to be considered in designing a suitable placebo?

3. Usually, a 'balancing act' of priorities is required during chemical development. Explain what this means.

4. Discuss whether chemical development is simply a scale-up exercise.

5. The following synthetic route was used for the initial synthesis of fexofenadine (R=CO_2H)—an analogue of terfenadine (R=CH_3). The synthesis was suitable for the large-scale synthesis of terfenadine, but not for fexofenadine. Suggest why not. (Hint: consider the electronic effects of R.)

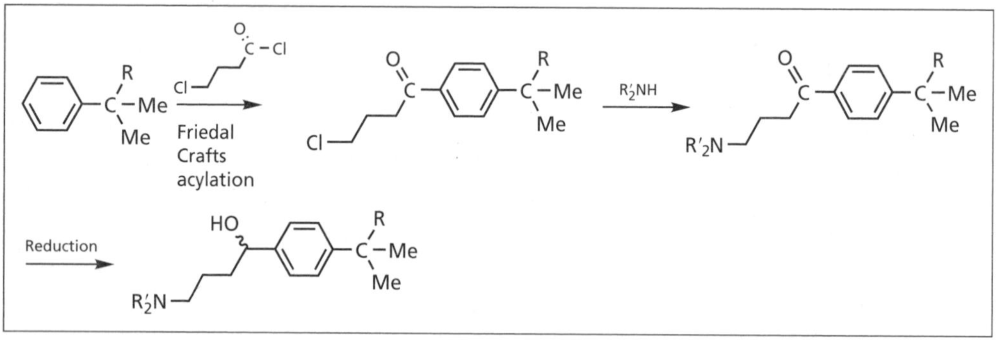

6. The following reaction was carried out using ethanol or water as solvents but gave poor yields in both cases. Suggest why this might be the case, and how these problems could be overcome.

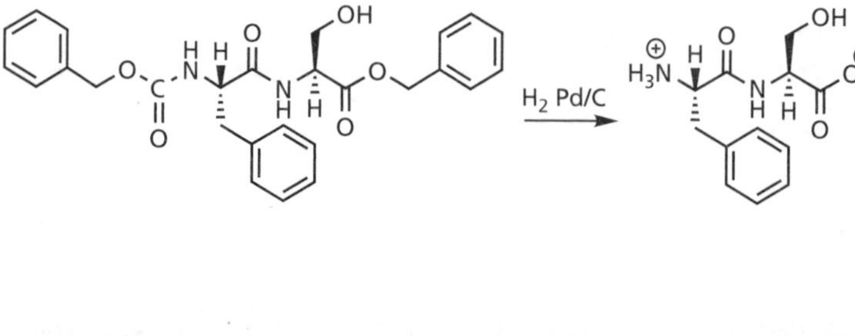

7. The following reaction was carried out with heating under reflux at 110°C. However, the yield was higher when the condenser was set for distillation. Explain.

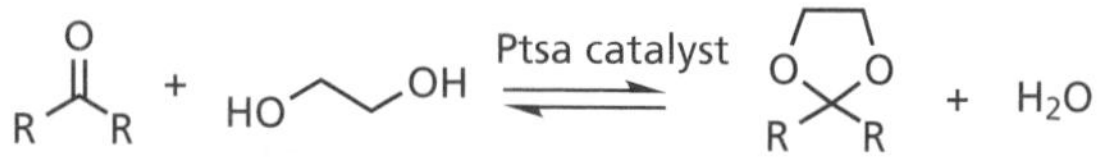

8. What considerations do you think have to be taken into account when choosing a solvent for scale up? Would you consider diethyl ether or benzene as a suitable solvent?

9. Phosphorus tribromide was added to an alcohol to give an alkyl bromide, but the product was contaminated with an ether impurity. Explain how this impurity might arise and how the reaction conditions could be altered to avoid the problem.

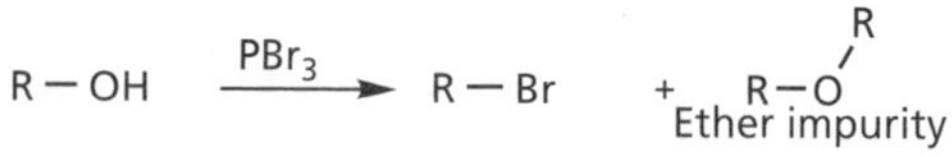

FURTHER READING

Preclinical trials

Cavagnaro, J. A. (2002) Preclinical safety evaluation of biotechnology-derived pharmaceuticals. *Nature Reviews Drug Discovery*, 1, 469–475.

Lindpaintner, K. (2002) The impact of pharmacogenetics and pharmacogenomics on drug discovery. *Nature Reviews Drug Discovery*, 1, 463–469.

Matfield, M. (2002) Animal experimentation: the continuing debate. *Nature Reviews Drug Discovery*, 1, 149–152.

Nicholson, J. K. *et al.* (2002) Metabonomics: a platform for studying drug toxicity and gene function. *Nature Reviews Drug Discovery*, 1, 153–161.

Pritchard, J. F. (2003) Making better drugs: decision gates in non-clinical drug development. *Nature Reviews Drug Discovery*, 2, 542–553.

Ulrich, R. and Friend, S. H. (2002) Toxicogenomics and drug discovery: will new technologies help us produce better drugs. *Nature Reviews Drug Discovery*, 1, 84–88.

Chemical and process development

Collins, A. N., Sheldrake, G. N., and Crosby, J. (eds.) (1997) *Chirality in industry I and II.* John Wiley and Sons, Chichester.

Lee, S. and Robinson, G. (1995) *Process development: fine chemicals from grams to kilograms.* Oxford University Press, Oxford.

Repic, O. (1998) *Principles of process research and chemical development in the pharmaceutical industry.* John Wiley and Sons, Chichester.

Saunders, J. (2000) *Top drugs: top synthetic routes.* Oxford University Press, Oxford.

Patenting

Agranat, I., Caner, H., and Caldwell, J. (2002) Putting chirality to work: the strategy of chiral switches. *Nature Reviews Drug Discovery*, 1, 753–768.

Webber, P. M. (2003) Protecting your inventions: the patent system. *Nature Reviews Drug Discovery*, 2, 823–830.

Regulatory affairs

Engel, L. W. and Straus, S. E. (2002) Development of therapeutics: opportunities within complementary and alternative medicine. *Nature Reviews Drug Discovery*, 1, 229–237.

Haffner, M. E., Whitley, J., and Moses, M. (2002) Two decades of orphan product development. *Nature Reviews Drug Discovery*, 1, 821–825.

Maeder, T. (2003) The orphan drug backlash. *Scientific American*, May, 71–77.

Reichert, J. M. (2003) Trends in development and approval times for new therapeutics in the United States. *Nature Reviews Drug Discovery*, 2, 695–702.

Clinical trials

Issa, A. M. (2002) Ethical perspectives on pharmacogenomic profiling in the drug development process. *Nature Reviews Drug Discovery*, 1, 300–308.

Schreiner, M. (2003) Paediatric clinical trials: redressing the balance. *Nature Reviews Drug Discovery*, 2, 949–961.

Titles for general further reading are listed on p. 711.

PART C

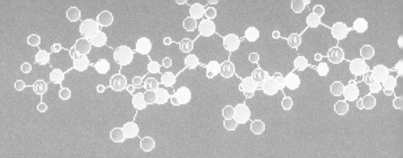

Tools of the trade

In Part C, we study three topics which are invaluable tools in the discovery and design of drugs. The topics covered are quantitative structure–activity relationships (QSAR), combinatorial chemistry, and the use of computers in medicinal chemistry. It should be emphasized that these are not the only topics that could be considered as 'tools of the trade'. For example, one could easily argue that organic synthesis should be included, as a knowledge of synthesis is crucial for drug design. There is no point designing drugs that are impossible to synthesize. However, organic synthesis is a major area in itself and a single chapter would do no justice to it, especially when there are many excellent undergraduate texts covering the subject in depth.

In Chapter 13, we look at QSAR. This topic has been around for many years and it is a well-established tool in medicinal chemistry. QSAR attempts to relate the physicochemical properties of compounds to their biological activity in a quantitative fashion by the use of equations. In traditional QSAR, this typically involves studying a series of analogues with different substituents and studying how the physicochemical properties of the substituents affect the biological activities of the analogues. Typically, the hydrophobic, steric, and electronic properties of each substituent are considered when setting up a QSAR equation. With the advent of computers and suitable software programmes, traditional QSAR studies have been largely superseded by three-dimensional quantitative structure–activity relationships (3D QSAR), where the physicochemical properties of the complete molecule are calculated and then related to biological activity.

Combinatorial chemistry is the subject of Chapter 14. This is a method of rapidly preparing large numbers of compounds in an automated or semi-automated fashion, usually by solid phase synthetic methods. The technique was developed to meet the urgent need for new lead compounds for the ever increasing number of novel targets discovered by genomic and proteomic projects. It is now an effective method of producing large numbers of analogues for drug development and for studies into structure–activity relationships.

In Chapter 15, we look at some of the operations that can be carried out using computers, and which aid the drug design process. Computers and molecular modelling software packages have now become an integral part of the drug design process and have been instrumental in a more scientific approach to medicinal chemistry.

All of these tools are extremely important in medicinal chemistry research, but they are meant to be used in combination with each other and should also be used when they are appropriate. Critics have sometimes argued that none of these tools has ever resulted in a clinically useful drug *per se*. That is not the point. Just as one cannot build a house with just a hammer, so it is unrealistic to suggest that a drug can be discovered by using a single scientific tool.

13 Quantitative structure–activity relationships (QSAR)

In Chapters 10 and 11 we studied the various strategies which can be used in the design of drugs. Several of these strategies involved a change in shape such that the new drug had a better 'fit' for its target binding site. Other strategies involved a change in functional groups or substituents such that the drug's pharmacokinetics or binding site interactions were improved. These latter strategies often involved the synthesis of analogues containing a range of substituents on aromatic or heteroaromatic rings or accessible functional groups. The number of possible analogues that could be made is infinite, if we were to try and synthesize analogues with every substituent and combination of substituents possible. Therefore, it is clearly advantageous if a rational approach can be followed in deciding which substituents to use. The quantitative structure–activity relationship (QSAR) approach has proved extremely useful in tackling this problem.

The QSAR approach attempts to identify and quantify the physicochemical properties of a drug and to see whether any of these properties has an effect on the drug's biological activity. If such a relationship holds true, an equation can be drawn up which quantifies the relationship and allows the medicinal chemist to say with some confidence that the property (or properties) has an important role in the pharmacokinetics or mechanism of action of the drug. It also allows the medicinal chemist some level of prediction. By quantifying physicochemical properties, it should be possible to calculate in advance what the biological activity of a novel analogue might be. There are two advantages to this. First, it allows the medicinal chemist to target efforts on analogues which should have improved activity and thus cut down the number of analogues that have to be made. Second, if an analogue is discovered which does not fit the equation, it implies that some other feature is important and provides a lead for further development.

What are these physicochemical features that we have mentioned? Essentially, they could be any structural, physical, or chemical property of a drug. Clearly, any drug will have a large number of such properties and it would be a Herculean task to quantify and relate them all to biological activity at the same time. A simple, more practical approach is to consider one or two physicochemical properties of the drug and to vary these while attempting to keep other properties constant. This is not as simple as it sounds, since it is not always possible to vary one property without affecting another. Nevertheless, there have been numerous examples where the approach has worked.

13.1 Graphs and equations

In the simplest situation, a range of compounds is synthesized in order to vary one physicochemical property (e.g. $\log P$) and to test how this affects the biological activity ($\log 1/C$) (we will come to the meaning of $\log 1/C$ and $\log P$ in due course). A graph is then drawn to plot the biological activity on the y-axis versus the physicochemical feature on the x-axis (Fig. 13.1). It is then necessary to draw the best possible line through the data points on the graph. This is done by a procedure known as '**linear regression analysis by the least squares method**'. This is quite a mouthful and can produce a glazed expression on any chemist who is not mathematically orientated. In fact,

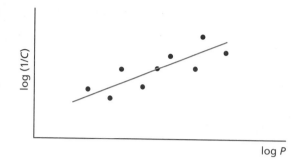

Figure 13.1 Biological activity versus log P.

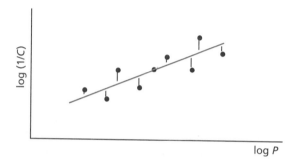

Figure 13.2 Proximity of data points to line of best fit.

the principle is quite straightforward. If we draw a line through a set of data points, most of the points will be scattered on either side of the line. The best line will be the one closest to the data points. To measure how close the data points are, vertical lines are drawn from each point (Fig. 13.2). These verticals are measured and then squared in order to eliminate the negative values. The squares are then added up to give a total (the sum of the squares). The best line through the points will be the line where this total is a minimum. The equation of the straight line will be $y = k_1 x + k_2$ where k_1 and k_2 are constants. By varying k_1 and k_2, different equations are obtained until the best line is obtained. This whole process can be speedily done using relevant software.

The next stage in the process is to see whether the relationship is meaningful. As any good politician knows, you can make figures mean anything and so statistical evidence has to be obtained to support the QSAR equation and to quantify the goodness of fit. Otherwise, we may have obtained a straight line through points which are so random that it means nothing. The **regression or correlation coefficient** (r) is a measure of how well the equation explains

the variance in activity observed in terms of the physicochemical parameters present in the equation. An explanation of how r is derived is given in Appendix 3. For a perfect fit, $r = 1$, in which case the observed activities would be the same as those calculated by the equation. Such perfection is impossible with biological data and so r values greater than 0.9 are considered acceptable. The regression coefficient is often quoted as r^2, in which case values over 0.8 are considered a good fit. If r^2 is multiplied by 100 it indicates the percentage variation in biological activity that is accounted for by the physicochemical parameters used in the equation. Thus, an r^2 value of 0.85 signifies that 85% of the variation in biological activity is accounted for by the parameters used. There are dangers in putting too much reliance on r, as the value obtained takes no account of the number of compounds (n) involved in the study and it is possible to obtain higher values of r by increasing the number of compounds tested.

Therefore, another statistical measure for the goodness of fit should be quoted alongside r. This is the standard error of estimate or the standard deviation (s). Ideally, s should be zero, but this would assume there were no experimental errors in the experimental data or the physicochemical parameters. In reality, s should be small, but not smaller than the standard deviation of the experimental data. It is therefore necessary to know the latter to assess whether the value of s is acceptably low. Appendix 3 shows how s is obtained and demonstrates that the number of compounds (n) in the study influences the value of s.

Statistical tests called Fisher's F-tests are often quoted. These tests are used to assess the significance of the coefficients k for each parameter in the QSAR equation. Normally p values (derived from the F-test) should be less than or equal to 0.05 if the parameter is significant. If this is not the case, the parameter should not be included in the QSAR equation.

13.2 Physicochemical properties

Many physical, structural, and chemical properties have been studied by the QSAR approach, but the most common are hydrophobic, electronic, and steric properties. This is because it is possible to quantify these effects. Hydrophobic properties can be easily quantified for complete molecules or for individual substituents. On the other hand, it is more difficult to quantify electronic and steric properties for complete molecules, and this is only really feasible for individual substituents.

Consequently, QSAR studies on a variety of totally different structures are relatively rare and are limited to studies on hydrophobicity. It is more common to find QSAR studies being carried out on compounds of the same general structure, where substituents on aromatic rings or accessible functional groups are varied. The QSAR study then considers how the hydrophobic, electronic, and steric properties of the substituents affect biological activity. The three most studied physico-chemical properties are now considered in some detail.

13.2.1 Hydrophobicity

The hydrophobic character of a drug is crucial to how easily it crosses cell membranes (section 8.2) and may also be important in receptor interactions. Changing substituents on a drug may well have significant effects on its hydrophobic character and hence its biological activity. Therefore, it is important to have a means of predicting this quantitatively.

The partition coefficient (*p*)

The hydrophobic character of a drug can be measured experimentally by testing the drug's relative distribution in an *n*-octanol/water mixture. Hydrophobic molecules will prefer to dissolve in the *n*-octanol layer of this two-phase system, whereas hydrophilic molecules will prefer the aqueous layer. The relative distribution is known as the partition coefficient (*P*) and is obtained from the following equation:

$$P = \frac{\text{Concentration of drug in octanol}}{\text{Concentration of drug in aqueous solution}}$$

Hydrophobic compounds have a high *P* value, whereas hydrophilic compounds have a low *P* value.

Varying substituents on the lead compound will produce a series of analogues having different hydrophobicities and therefore different *P* values. By plotting these *P* values against the biological activity of these drugs, it is possible to see if there is any relationship between the two properties. The biological activity is normally expressed as $1/C$, where C is the concentration of drug required to achieve a defined level of biological activity. The reciprocal of the

concentration $(1/C)$ is used, since more active drugs will achieve a defined biological activity at lower concentration.

The graph is drawn by plotting $\log(1/C)$ versus $\log P$. In studies where the range of the $\log P$ values is restricted to a small range (e.g. $\log P = 1–4$), a straight-line graph is obtained (Fig. 13.1) showing that there is a relationship between hydrophobicity and biological activity. Such a line would have the following equation:

$$\log(1/C) = k_1 \log P + k_2$$

For example, the binding of drugs to serum albumin is determined by their hydrophobicity and a study of 42 compounds resulted in the following equation:

$$\log(1/C) = 0.75 \log P + 2.30 \quad (n = 42, r = 0.960, \\ s = 0.159)$$

The equation shows that serum albumin binding increases as $\log P$ increases. In other words, hydrophobic drugs bind more strongly to serum albumin than hydrophilic drugs. Knowing how strongly a drug binds to serum albumin can be important in estimating effective dose levels for that drug. When bound to serum albumin, the drug cannot bind to its receptor and so the dose levels for the drug should be based on the amount of unbound drug present in the circulation. The equation above allows us to calculate how strongly drugs of similar structure will bind to serum albumin and gives an indication of how 'available' they will be for receptor interactions. The r value of 0.96 is close to 1, which shows that the line resulting from the equation is a good fit. The value of r^2 is 92%, which indicates that 92% of the variation in serum albumin binding can be accounted for by the different hydrophobicities of the drugs tested. This means that 8% of the variation is unaccounted for, partly as a result of the experimental errors involved in the measurements.

Despite such factors as serum albumin binding, it is generally found that increasing the hydrophobicity of a lead compound results in an increase in biological activity. This reflects the fact that drugs have to cross hydrophobic barriers such as cell membranes in order to reach their target. Even if no barriers are to be crossed (e.g. in *in vitro* studies), the drug has to interact with a target system such as an enzyme or receptor where the binding site is usually hydrophobic.

Therefore, increasing hydrophobicity aids the drug in crossing hydrophobic barriers or in binding to its target site.

This might imply that increasing $\log P$ should increase the biological activity *ad infinitum*. In fact, this does not happen. There are several reasons for this. For example, the drug may become so hydrophobic that it is poorly soluble in the aqueous phase. Alternatively, it may be 'trapped' in fat depots and never reach the intended site. Finally, hydrophobic drugs are often more susceptible to metabolism and subsequent elimination.

A straight-line relationship between $\log P$ and biological activity is observed in many QSAR studies because the range of $\log P$ values studied is often relatively narrow. For example, the study carried out on serum albumin binding was restricted to compounds having $\log P$ values in the range 0.78 to 3.82. If these studies were to be extended to include compounds with very high $\log P$ values then we would see a different picture. The graph would be parabolic, as shown in Fig. 13.3. Here, the biological activity increases as $\log P$ increases until a maximum value is obtained. The value of $\log P$ at the maximum ($\log P^0$) represents the optimum partition coefficient for biological activity. Beyond that point, an increase in $\log P$ results in a decrease in biological activity.

If the partition coefficient is the only factor influencing biological activity, the parabolic curve can be expressed by the equation

$$\log(1/C) = -k_1(\log P)^2 + k_2 \log P + k_3$$

Note that the $(\log P)^2$ term has a minus sign in front of it. When P is small, the $(\log P)^2$ term is very small and the equation is dominated by the $\log P$ term. This represents the first part of the graph, where activity increases with increasing P. When P is large, the $(\log P)^2$ term is more significant and eventually 'overwhelms' the $\log P$ term. This represents the last part of the graph where activity drops with increasing P. k_1, k_2, and k_3 are constants and can be determined by a suitable software program.

There are relatively few drugs where activity is related to the $\log P$ factor alone. Such drugs tend to operate in cell membranes where hydrophobicity is the dominant feature controlling their action. The best example of drugs which operate in cell membranes are the general anaesthetics. Although they also bind to GABA-A receptors, general anaesthetics are thought to function by entering the central nervous system and 'dissolving' into cell membranes where they affect membrane structure and nerve function. In such a scenario, there are no specific drug–receptor interactions and the mechanism of the drug is controlled purely by its ability to enter cell membranes (i.e. its hydrophobic character). The general anaesthetic activity of a range of ethers was found to fit the following parabolic equation:

$$\log(1/C) = -0.22(\log P)^2 + 1.04 \log P + 2.16$$

According to this equation, anaesthetic activity increases with increasing hydrophobicity (P), as determined by the $\log P$ factor. The negative $(\log P)^2$ factor shows that the relationship is parabolic and that there is an optimum value for $\log P$ ($\log P^0$) beyond which increasing hydrophobicity causes a decrease in anaesthetic activity. With this equation, it is now possible to predict the anaesthetic activity of other compounds, given their partition coefficients. However, there are limitations. The equation is derived purely for anaesthetic ethers and is not applicable to other structural types of anaesthetics. This is generally true in QSAR studies. The procedure works best if it is applied to a series of compounds which have the same general structure.

However, QSAR studies have been carried out on other structural types of general anaesthetics, and in each case a parabolic curve has been obtained. Although, the constants for each equation are different, it is significant that the optimum hydrophobicity (represented by $\log P^0$) for anaesthetic activity is close to 2.3, regardless of the class of anaesthetic being studied. This finding suggests that all general

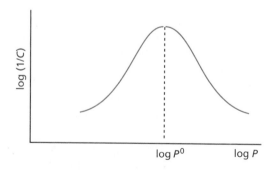

Figure 13.3 Parabolic curve of $\log(1/C)$ vs. $\log P$.

anaesthetics are operating in a similar fashion, controlled by the hydrophobicity of the structure.

Since different anaesthetics have similar log P^0 values, the log P value of any compound can give some idea of its potential potency as an anaesthetic. For example, the log P values of the gaseous anaesthetics **ether**, **chloroform**, and **halothane** are 0.98, 1.97, and 2.3 respectively. Their anaesthetic activity increases in the same order.

Since general anaesthetics have a simple mechanism of action based on the efficiency with which they enter the central nervous system, it implies that log P values should give an indication of how easily any compound can enter the central nervous system. In other words, compounds having a log P value close to 2 should be capable of entering the central nervous system efficiently. This is generally found to be true. For example, the most potent barbiturates for sedative and hypnotic activity are found to have log P values close to 2.

As a rule of thumb, drugs which are to be targeted for the central nervous system should have a log P value of approximately 2. Conversely, drugs which are designed to act elsewhere in the body should have log P values significantly different from 2 in order to

avoid possible central nervous system side effects (e.g. drowsiness) see Box 13.1.

The substituent hydrophobicity constant (π)

We have seen how the hydrophobicity of a compound can be quantified using the partition coefficient P. In order to get P, we have to measure it experimentally and that means that we have to synthesize the compounds. It would be much better if we could calculate P theoretically and decide in advance whether the compound is worth synthesizing. QSAR would then allow us to target the most promising-looking structures. For example, if we were planning to synthesize a range of barbiturate structures, we could calculate log P values for them all and concentrate on the structures which had log P values closest to the optimum log P^0 value for barbiturates.

Partition coefficients can be calculated by knowing the contribution that various substituents make to hydrophobicity. This contribution is known as the substituent hydrophobicity constant (π) and is a measure of how hydrophobic a substituent is, relative to hydrogen. The value can be obtained as follows. Partition coefficients are measured experimentally for a standard compound such as benzene with and without a substituent (X). The hydrophobicity constant (π_X) for the substituent (X) is then obtained using the following equation:

$$\pi_X = \log P_X - \log P_H$$

where P_H is the partition coefficient for the standard compound and P_X is the partition coefficient for the standard compound with the substituent.

A positive value of π indicates that the substituent is more hydrophobic than hydrogen; a negative value indicates that the substituent is less hydrophobic. The π values for a range of substituents are shown in Table 13.1. These π values are characteristic for the substituent and can be used to calculate

---

BOX 13.1 ALTERING LOG P TO REMOVE CENTRAL NERVOUS SYSTEM SIDE EFFECTS

The cardiotonic agent (I) was found to produce 'bright visions' in some patients, which implied that it was entering the central nervous system. This was supported by the fact that the log P value of the drug was 2.59. In order to prevent the drug entering the central nervous system, the 4-OMe group was replaced by a 4-S(O)Me group. This particular group is approximately the same size as the methoxy group, but more hydrophilic. The log P value of the new drug (**sulmazole**) was found to be 1.17. The drug was now too hydrophilic to enter the central nervous system and was free of central nervous system side effects.

Cardiotonic agents.

Table 13.1 Values of π for various substituents

Group	CH$_3$	t-Bu	OH	OCH$_3$	CF$_3$	Cl	Br	F
π (aliphatic substituents)	0.50	1.68	−1.16	0.47	1.07	0.39	0.60	−0.17
π (aromatic substituents)	0.52	1.68	−0.67	−0.02	1.16	0.71	0.86	0.14

how the partition coefficient of a drug would be affected if these substituents were present. The P value for the lead compound would have to be measured experimentally, but once that is known, the P value for analogues can be calculated quite simply.

As an example, consider the log P values for benzene (log $P = 2.13$), chlorobenzene (log $P = 2.84$), and benzamide (log $P = 0.64$) (Fig. 13.4). Benzene is the parent compound, and the substituent constants for Cl and $CONH_2$ are 0.71 and -1.49 respectively. Having obtained these values, it is now possible to calculate the theoretical log P value for *meta*-chlorobenzamide:

$$\log P_{\text{(chlorobenzamide)}} = \log P_{\text{(benzene)}} + \pi_{Cl} + \pi_{CONH_2}$$
$$= 2.13 + 0.71 + (-1.49)$$
$$= 1.35$$

The observed log P value for this compound is 1.51.

It should be noted that π values for aromatic substituents are different from those used for aliphatic substituents. Furthermore, neither of these sets of π values are in fact true constants and they are accurate only for the structures from which they were derived. They can be used as good approximations when studying other structures, but it is possible that the values will have to be adjusted in order to get accurate results.

In order to distinguish calculated log P values from experimental ones, the former are referred to as **Clog P**

values. There are software programs which will calculate Clog P values for a given structure.

P vs. π

QSAR equations relating biological activity to the partition coefficient P have already been described, but there is no reason why the substituent hydrophobicity constant π cannot be used in place of P if only the substituents are being varied. The equation obtained would be just as relevant as a study of how hydrophobicity affects biological activity. That is not to say that P and π are exactly equivalent—different equations would be obtained with different constants. Apart from that, the two factors have different emphases. The partition coefficient P is a measure of the drug's overall hydrophobicity, and is therefore an important measure of how efficiently a drug is transported to its target and bound to its binding site. The π factor measures the hydrophobicity of a specific region on the drug's skeleton and if it is present in the QSAR equation, it could emphasize any important hydrophobic interactions involving that region of the molecule with the binding site.

Most QSAR equations have a contribution from P or from π, but there are examples of drugs for which they have only a slight contribution. For example, a study on antimalarial drugs showed very little relationship between antimalarial activity and hydrophobic character. This finding supports the theory that these drugs act in red blood cells, since previous research has shown that the ease with which drugs enter red blood cells is not related to their hydrophobicity.

13.2.2 Electronic effects

The electronic effects of various substituents will clearly have an effect on a drug's ionization or polarity. This in turn may have an effect on how easily a drug can pass through cell membranes or how strongly it can interact with a binding site. It is therefore useful to measure the electronic effect of a substituent.

As far as substituents on an aromatic ring are concerned, the measure used is known as the **Hammett substituent constant** (σ). This is a measure of the electron-withdrawing or electron-donating ability of a substituent, and has been determined by measuring the dissociation of a series of substituted

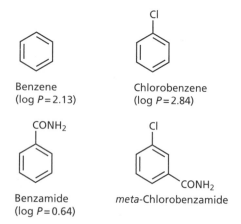

Benzene
(log $P = 2.13$)

Chlorobenzene
(log $P = 2.84$)

Benzamide
(log $P = 0.64$)

meta-Chlorobenzamide

Figure 13.4 Values for log P.

benzoic acids compared to the dissociation of benzoic acid itself.

Benzoic acid is a weak acid and only partially ionizes in water (Fig. 13.5). An equilibrium is set up between the ionized and non-ionized forms, where the relative proportions of these species are known as the **equilibrium** or **dissociation constant** K_H (the subscript H signifies that there are no substituents on the aromatic ring).

$$K_H = \frac{[PhCO_2^-]}{[PhCO_2H]}$$

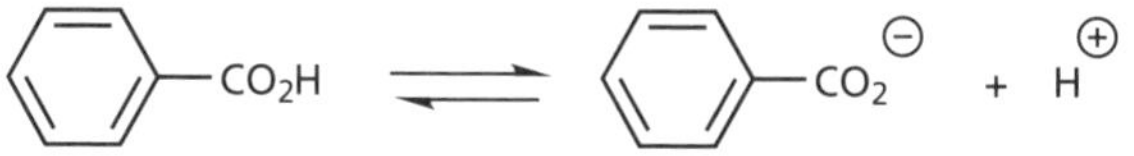

Figure 13.5 Ionization of benzoic acid.

When a substituent is present on the aromatic ring, this equilibrium is affected. Electron-withdrawing groups, such as a nitro group, result in the aromatic ring having a stronger electron-withdrawing and stabilizing influence on the carboxylate anion. The equilibrium will therefore shift more to the ionized form such that the substituted benzoic acid is a stronger acid and has a larger K_X value (X represents the substituent on the aromatic ring) (Fig. 13.6).

If the substituent X is an electron-donating group such as an alkyl group, then the aromatic ring is less able to stabilize the carboxylate ion. The equilibrium shifts to the left and a weaker acid is obtained with a smaller K_X value (Fig. 13.7).

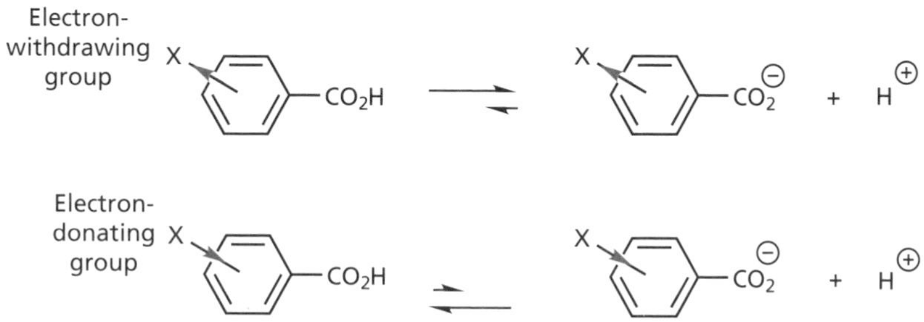

Figure 13.6 Position of equilibrium dependent on substituent group X.

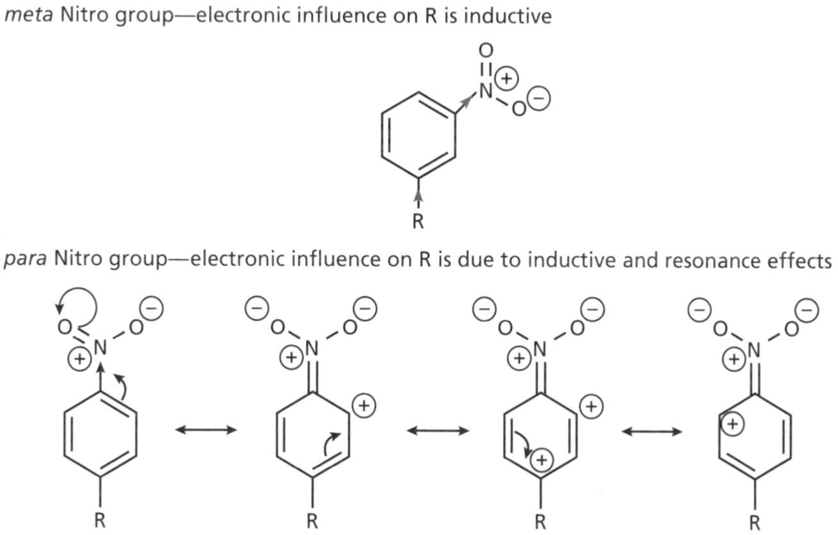

Figure 13.7 Substituent effects of a nitro group.

The Hammett substituent constant (σ_X) for a particular substituent (X) is defined by the following equation:

$$\sigma_X = \log(K_X/K_H) = \log K_X - \log K_H$$

Benzoic acids containing electron-withdrawing substituents will have larger K_X values than benzoic acid itself (K_H) and therefore the value of σ_X for an electron-withdrawing substituent will be positive. Substituents such as Cl, CN, or CF_3 have positive σ values.

Benzoic acids containing electron-donating substituents will have smaller K_X values than benzoic acid itself and hence the value of σ_X for an electron-donating substituent will be negative. Substituents such as Me, Et, and *t*-Bu have negative values of σ. The Hammett substituent constant for H is zero.

The Hammett constant takes into account both resonance and inductive effects. Therefore, the value of σ for a particular substituent will depend on whether the substituent is *meta* or *para*. This is indicated by the subscript *m* or *p* after the σ symbol. For example, the nitro substituent has $\sigma_p = 0.78$ and $\sigma_m = 0.71$. In the *meta* position, the electron-withdrawing power is due to the inductive influence of the substituent, whereas at the *para* position inductive and resonance both play a part and so the σ_p value is greater (Fig. 13.7).

For the hydroxyl group, $\sigma_m = 0.12$ and $\sigma_p = -0.37$. At the *meta* position, the influence is inductive and electron-withdrawing. At the *para* position, the electron-donating influence due to resonance is more significant than the electron-withdrawing influence due to induction (Fig. 13.8).

Most QSAR studies start off by considering σ and if there is more than one substituent, the σ values are summed ($\sum \sigma$). However, as more compounds are synthesized, it is possible to fine-tune the QSAR equation. As mentioned above, σ is a measure of a substituent's inductive and resonance electronic effects. With more detailed studies, the inductive and resonance effects can be considered separately. Tables of constants are available which quantify a substituent's inductive effect (F) and its resonance effect (R). In some cases, it might be found that a substituent's effect on activity is due to F rather than R, and vice versa. It might also be found that a substituent has a more significant effect at a particular position on the ring and this can also be included in the equation.

There are limitations to the electronic constants described so far. For example, Hammett substituent constants cannot be measured for *ortho* substituents since such substituents have an important steric, as well as electronic, effect.

There are very few drugs whose activities are solely influenced by a substituent's electronic effect, as hydrophobicity usually has to be considered as well. Those that do are generally operating by a mechanism whereby they do not have to cross any cell membranes (see Box 13.2). Alternatively, *in vitro* studies on isolated enzymes may result in QSAR equations lacking

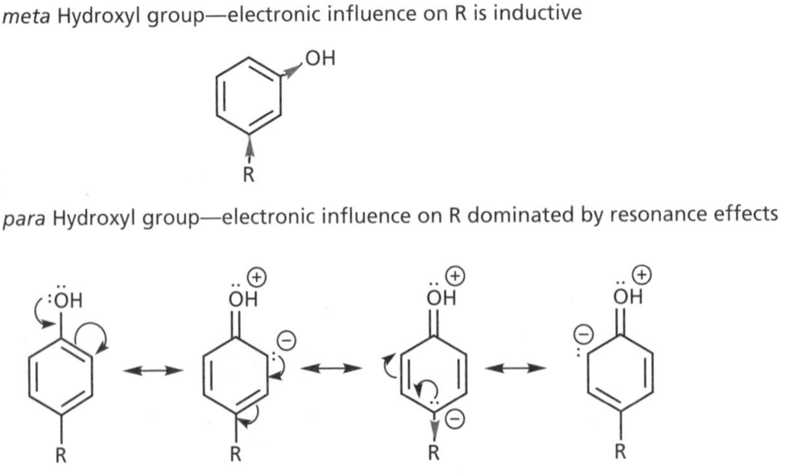

meta Hydroxyl group—electronic influence on R is inductive

para Hydroxyl group—electronic influence on R dominated by resonance effects

Figure 13.8 Substituent effects of a phenol.

The insecticidal activity of diethyl phenyl phosphates is one of the few examples where activity is related to electronic factors alone:

$$\log(1/C) = 2.282\sigma - 0.348$$
$$(r^2 = 0.952, r = 0.976, s = 0.286)$$

The equation reveals that substituents with a positive value for σ (i.e. electron-withdrawing groups) will increase activity. The fact that the a hydrophobic parameter is not present is a good indication that the drugs do not have to pass into or through a cell membrane to have activity. In fact, these drugs are known to act on an enzyme called **acetylcholinesterase** which is situated on the outside of cell membranes (section 19.16).

The value of r is close to 1, which demonstrates that the line is a good fit, and the value of r^2 demonstrates that 95% of the data are accounted for by the σ parameter.

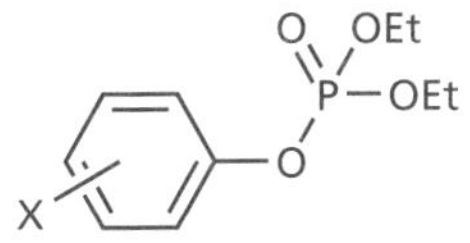

Diethyl phenyl phosphates

the hydrophobicity factor, since there are no cell membranes to be considered.

The constants σ, R, and F can only be used for aromatic substituents and are therefore only suitable for drugs containing aromatic rings. However, a series of aliphatic electronic substituent constants are available. These were obtained by measuring the rates of hydrolysis for a series of aliphatic esters (Fig. 13.9). Methyl ethanoate is the parent ester and it is found that the rate of hydrolysis is affected by the substituent X. The extent to which the rate of hydrolysis is affected

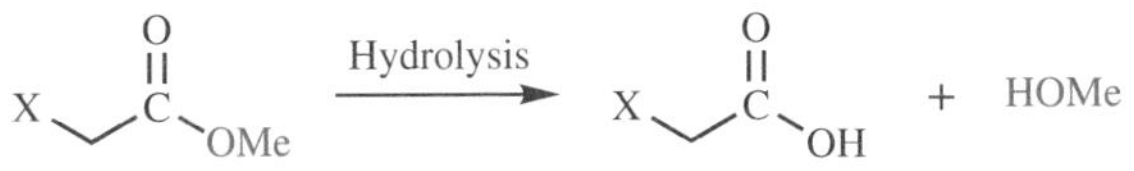

Figure 13.9 Hydrolysis of an aliphatic ester.

is a measure of the substituent's electronic effect at the site of reaction (i.e. the ester group). The electronic effect is purely inductive and is given the symbol σ_I. Electron-donating groups reduce the rate of hydrolysis and therefore have negative values. For example, σ_I values for methyl, ethyl, and propyl are -0.04, -0.07, and -0.36 respectively. Electron-withdrawing groups increase the rate of hydrolysis and have positive values. The σ_I values for NMe_3^+ and CN are 0.93 and 0.53 respectively.

It should be noted that the inductive effect is not the only factor affecting the rate of hydrolysis. The substituent may also have a steric effect. For example, a bulky substituent may shield the ester from attack and lower the rate of hydrolysis. It is therefore necessary to separate out these two effects. This can be done by measuring hydrolysis rates under basic conditions and also under acidic conditions. Under basic conditions, steric and electronic factors are important, whereas under acidic conditions only steric factors are important. By comparing the rates, values for the electronic effect (σ_I), and the steric effect (E_s) (see below) can be determined.

13.2.3 Steric factors

The bulk, size, and shape of a drug will influence how easily it can approach and interact with a binding site. A bulky substituent may act like a shield and hinder the ideal interaction between a drug and its binding site. Alternatively, a bulky substituent may help to orientate a drug properly for maximum binding and increase activity. Steric properties are more difficult to quantify than hydrophobic or electronic properties. Several methods have been tried, of which three are described here. It is highly unlikely that a drug's biological activity will be affected by steric factors alone, but these factors are frequently found in Hansch equations (section 13.3).

13.2.3.1 Taft's steric factor (E_s)

Attempts have been made to quantify the steric features of substituents by using Taft's steric factor (E_s). The value for E_s can be obtained by comparing the rates of hydrolysis of substituted aliphatic esters against a standard ester under acidic conditions. Thus,

$$E_s = \log k_x - \log k_o$$

where k_x represents the rate of hydrolysis of an aliphatic ester bearing the substituent X and k_o represents the rate of hydrolysis of the reference ester.

The substituents that can be studied by this method are restricted to those which interact sterically with the tetrahedral transition state of the reaction and not by resonance or internal hydrogen bonding. For example, unsaturated substituents which are conjugated to the ester cannot be measured by this procedure. Examples of E_s values are shown in Table 13.2. Note that the reference ester is X = Me. Substituents such as H and F which are smaller than a methyl group result in a faster rate of hydrolysis ($k_x > k_o$), making E_s positive. Substituents which are larger than methyl reduce the rate of hydrolysis ($k_x < k_o$), making E_s negative. A disadvantage of E_s values is that they are a measure of an *intramolecular* steric effect, whereas drugs interact with target binding sites in an *intermolecular* manner. For example, consider the E_s values for *i*-Pr, *n*-Pr, and *n*-Bu. The E_s value for the branched isopropyl group is significantly greater than that for the linear *n*-propyl group, since the bulk of the substituent is closer to the reaction centre. Extending the alkyl chain from *n*-propyl to *n*-butyl has little effect on E_s. The larger *n*-butyl group is extended away from the reaction centre and so has little additional steric effect on the rate of hydrolysis. As a result, the E_s value for the *n*-butyl group undervalues the steric effect which this group might have if it was present on a drug that was approaching a binding site.

13.2.3.2 Molar refractivity

Another measure of the steric factor is provided by a parameter known as molar refractivity (*MR*). This is a measure of the volume occupied by an atom or a group of atoms. The *MR* is obtained from the following equation:

$$MR = \frac{(n^2 - 1)}{(n^2 + 2)} \times \frac{MW}{d}$$

where *n* is the index of refraction, *MW* is the molecular weight, and *d* is the density. The term *MW/d* defines a volume, and the $(n^2 - 1)/(n^2 + 2)$ term provides a correction factor by defining how easily the substituent can be polarized. This is particularly significant if the substituent has π electrons or lone pairs of electrons.

13.2.3.3 Verloop steric parameter

Another approach to measuring the steric factor involves a computer program called **Sterimol** which calculates steric substituent values (Verloop steric parameters) from standard bond angles, van der Waals radii, bond lengths, and possible conformations for the substituent. Unlike E_s, the Verloop steric parameters can be measured for any substituent. For example, the Verloop steric parameters for a carboxylic acid group are demonstrated in Fig. 13.10. *L* is the length of the substituent and B_1–B_4 are the radii of the group in different dimensions.

13.2.4 Other physicochemical parameters

The physicochemical properties most commonly studied by the QSAR approach have been described above, but other properties have been studied including dipole moments, hydrogen bonding, conformations, and interatomic distances. Difficulties in quantifying these properties limit the use of these parameters, however. Several QSAR formulae have been developed based on the highest occupied and/or the lowest unoccupied molecular orbitals of the test compounds. The calculation of these orbitals can be carried out using semi-empirical quantum mechanical methods (section 15.7.3). Indicator variables for different substituents can also be used. These are described in section 13.7.

Table 13.2 Values of E_s for various substituents

Substituent	H	F	Me	Et	*n*-Pr	*n*-Bu	*i*-Pr	*i*-Bu	Cyclopentyl
E_s	1.24	0.78	0	−0.07	−0.36	−0.39	−0.47	−0.93	−0.51

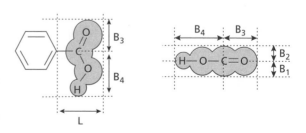

Figure 13.10 Verloop parameters for a carboxylic acid group.

13.3 Hansch equation

In section 13.2 we looked at the physicochemical properties commonly used in QSAR studies and how it is possible to quantify them. In a simple situation where biological activity is related to only one such property, a simple equation can be drawn up. The

biological activity of most drugs, however, is related to a combination of physicochemical properties. In such cases, simple equations involving only one parameter are relevant only if the other parameters are kept constant. In reality, this is not easy to achieve and equations which relate biological activity to a number of different parameters are more common. These equations are known as Hansch equations and they

---

BOX 13.3 HANSCH EQUATION FOR A SERIES OF ANTIMALARIAL COMPOUNDS

A series of 102 phenanthrene aminocarbinols was tested for antimalarial activity. In the structure shown, X represents up to four substituents on the left-hand ring and Y represents up to four substituents on the right hand ring. Experimental log P values for the structures were not available and equations were derived which compared the activity with some or all of the following terms:

π_{sum} the π constants for *all* the substituents in the molecule (i.e. all the X and Y substituents, as well as the amino substituents R and R″). This term was used in place of log P to represent the overall hydrophobicity for the molecule

σ_{sum} the σ constants for *all* the substituents in the molecule

$\sum \pi_X$ the sum of the π constants for all the substituents X in the left-hand ring

$\sum \pi_Y$ the sum of the π constants for all the substituents Y in the right-hand ring

$\sum \pi_{X+Y}$ the sum of the π constants for all the substituents X and Y in both the left- and right-hand rings

$\sum \sigma_{X+Y}$ the sum of the σ constants for all the substituents X and Y in both the left- and right-hand rings

$\sum \sigma_X$ the sum of the σ constants for all the substituents X in the left-hand ring

$\sum \sigma_Y$ the sum of the σ constants for all the substituents Y in the right-hand ring

Equations such as equations 1–3 were derived which matched activity against one of the above terms, but none of them had an acceptable value of r^2.

A variety of other equations were derived which included two of the above terms but these were not satisfactory either. Finally, an equation was derived which contained six terms and proved satisfactory:

$$\log(1/C) = -0.015(\pi_{sum})^2 + 0.14\pi_{sum} + 0.27\sum \pi_X$$
$$+ 0.40\sum \pi_Y + 0.65\sum \sigma_X + 0.88\sum \sigma_X$$
$$+ 2.34 \quad (n = 102, r = 0.913, r^2 = 0.834,$$
$$s = 0.258)$$

The equation shows that antimalarial activity increases very slightly as the overall hydrophobicity of the molecule (π_{sum}) increases (the constant 0.14 is low). The $(\pi_{sum})^2$ term shows that there is an optimum overall hydrophobicity for activity and this is found to be 4.44. Activity increases if hydrophobic substituents are present on ring X and in particular on ring Y. This could be taken to imply that some form of hydrophobic interaction is involved near both rings. Electron-withdrawing substituents on both rings are also beneficial to activity, more so on ring Y than ring X. The r^2 value is 0.834, which is above the minimum acceptable value of 0.8.

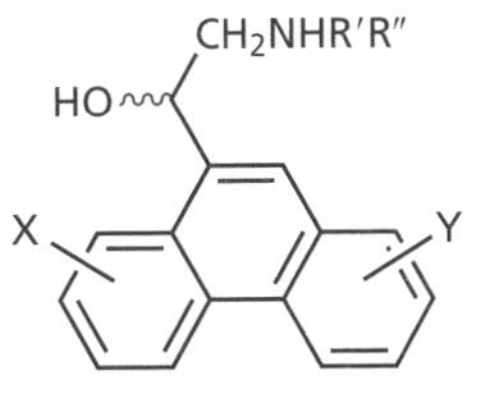

Phenanthrene aminocarbinols

(1) $\log\left(\dfrac{1}{C}\right) = 0.557\Sigma\pi_{X+Y} + 2.699$ $(n = 102, r = 0.768, r^2 = 0.590, s = 0.395)$

(2) $\log\left(\dfrac{1}{C}\right) = 0.017\pi_{sum} + 3.324$ $(n = 102, r = 0.069, r^2 = 0.005, s = 0.616)$

(3) $\log\left(\dfrac{1}{C}\right) = 1.218\sigma_{sum} + 2.721$ $(n = 102, r = 0.814, r^2 = 0.663, s = 0.359)$

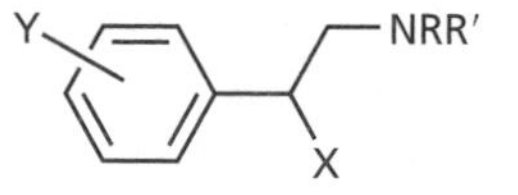

$$\log\left(\frac{1}{C}\right) = 1.22\,\pi - 1.59\,\sigma + 7.89$$

$$(n=22, r^2=0.841, s=0.238)$$

β-Halo-arylamines

Figure 13.11 QSAR equation for β-halo-arylamines.

Table 13.3 Values of π and MR for various substituents

Substituent	H	Me	Et	n-Pr	n-Bu	OMe	NHCONH$_2$	I	CN
π		0.00	0.56	1.02	1.50	2.13	−0.02	−1.30	1.12	−0.57
MR		0.10	0.56	1.03	1.55	1.96	0.79	1.37	1.39	0.63

usually relate biological activity to the most commonly used physicochemical properties (log P or π, σ, and a steric factor). If the range of hydrophobicity values is limited to a small range then the equation will be linear, as follows:

$$\log(1/C) = k_1 \log P + k_2\sigma + k_3 E_s + k_4$$

If the log P values are spread over a large range, then the equation will be parabolic for the reasons described in section 13.2.1:

$$\log(1/C) = -k_1 (\log P)^2 + k_2 \log P + k_3\sigma + k_4 E_s + k_5$$

The constants k_1–k_5 are determined by computer software in order to get the best fitting equation. Not all the parameters will necessarily be significant. For example, the adrenergic blocking activity of β-halo-arylamines (Fig. 13.11) was related to π and σ and did not include a steric factor. This equation tells us that biological activity increases if the substituents have a positive π value and a negative σ value. In other words, the substituents should be hydrophobic and electron donating.

When carrying out a Hansch analysis, it is important to choose the substituents carefully to ensure that the change in biological activity can be attributed to a particular parameter. There are plenty of traps for the unwary. Take, for example, drugs which contain an amine group. One of the studies most frequently carried out on amines is to synthesize analogues containing a homologous series of alkyl substituents on the nitrogen atom (i.e. Me, Et, n-Pr, n-Bu). If activity increases with the chain length of the substituent, is it due to increasing hydrophobicity, increasing size, or both? If we look at the π and MR values of these substituents, we find that both sets of values increase in a similar fashion across the series and we would not be able to distinguish between them (Table 13.3). In this example, a series of substituents would have to be chosen where π and MR are not correlated. The

substituents H, Me, OMe, NHCOCH$_2$, I, and CN would be more suitable.

13.4 Craig plot

Although tables of π and σ factors are readily available for a large range of substituents, it is often easier to visualize the relative properties of different substituents by considering a plot where the y-axis is the value of the σ factor and the x-axis is the value of the π factor. Such a plot is known as a Craig plot. The example shown in Fig. 13.12 is the Craig plot for the σ and π factors of *para*-aromatic substituents. There are several advantages to the use of such a Craig plot.

- The plot shows clearly that there is no overall relationship between π and σ. The various substituents are scattered around all four quadrants of the plot.

- It is possible to tell at a glance which substituents have positive π and σ parameters, which substituents have negative π and σ parameters, and which substituents have one positive and one negative parameter.

- It is easy to see which substituents have similar π values. For example, the ethyl, bromo, trifluoromethyl, and trifluoromethylsulfonyl groups are all approximately on the same vertical line on the plot. In theory, these groups could be interchangeable on drugs where the principal factor affecting biological activity is the π factor. Similarly, groups which form a horizontal line can be identified as being isoelectronic or having similar σ values (e.g. CO$_2$H, Cl, Br, I).

- The Craig plot is useful in planning which substituents should be used in a QSAR study. In order to derive the most accurate equation involving π and σ,

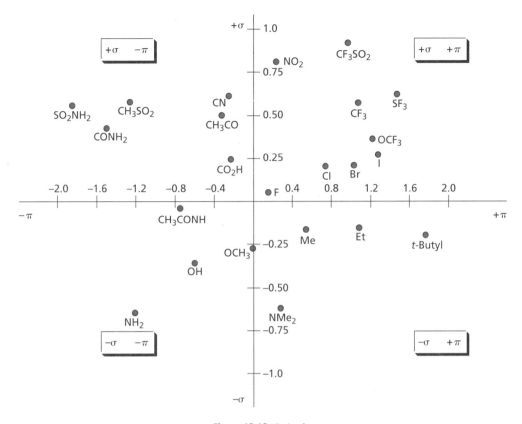

Figure 13.12 Craig plot.

analogues should be synthesized with substituents from each quadrant. For example, halogen substituents are useful representatives of substituents with increased hydrophobicity and electron-withdrawing properties (positive π and positive σ), whereas an OH substituent has more hydrophilic and electron-donating properties (negative π and negative σ). Alkyl groups are examples of substituents with positive π and negative σ values, whereas acyl groups have negative π and positive σ values.

- Once the Hansch equation has been derived, it will show whether π or σ should be negative or positive in order to get good biological activity. Further developments would then concentrate on substituents from the relevant quadrant. For example, if the equation shows that positive π and positive σ values are necessary, then further substituents should only be taken from the top right quadrant.

Craig plots can also be drawn up to compare other sets of physicochemical parameters, such as hydrophobicity and *MR*.

13.5 Topliss scheme

In certain situations, it might not be feasible to make the large range of structures required for a Hansch equation. For example, the synthetic route involved might be so difficult that only a few structures can be made in a limited time. In these circumstances, it would be useful to test compounds for biological activity as they are synthesized and to use these results to determine the next analogue to be synthesized.

A Topliss scheme is a flow diagram which allows such a procedure to be followed. There are two Topliss schemes, one for aromatic substituents (Fig. 13.13) and

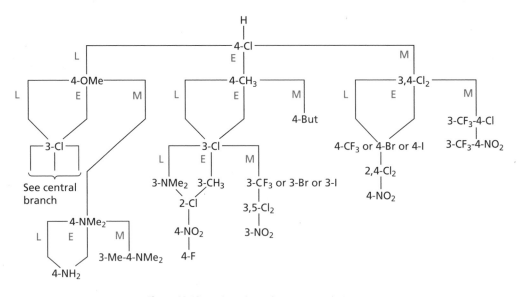

Figure 13.13 Topliss scheme for aromatic substituents.

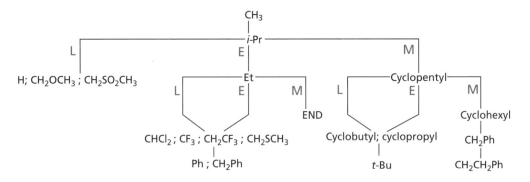

Figure 13.14 Topliss scheme for aliphatic side chain substituents.

one for aliphatic side chain substituents (Fig. 13.14). The schemes were drawn up by considering the hydrophobicity and electronic factors of various substituents and are designed such that the optimum substituent can be found as efficiently as possible. They are not meant to be a replacement for a full Hansch analysis, however. Such an analysis would be carried out in due course, once a suitable number of structures have been synthesized.

The Topliss scheme for aromatic substituents (Fig. 13.13) assumes that the lead compound has been tested for biological activity and contains a monosubstituted aromatic ring. The first analogue in the scheme is the 4-chloro derivative, as this derivative is usually easy to synthesize. The chloro substituent is more hydrophobic and electron-withdrawing than hydrogen, and therefore π and σ are positive.

Once the chloro analogue has been synthesized, the biological activity is measured. There are three possibilities. The analogue will have less activity (L), equal activity (E), or more activity (M). The type of activity observed will determine which branch of the Topliss scheme is followed next.

If the biological activity increases, the (M) branch is followed and the next analogue to be synthesized is the 3,4-dichloro-substituted analogue. If, on the other hand, the activity stays the same, then the (E) branch is followed and the 4-methyl analogue is synthesized. Finally, if activity drops, the (L) branch is followed and the next analogue is the 4-methoxy analogue.

Biological results from the second analogue now determine the next branch to be followed in the scheme.

What is the rationale behind this?

Let us consider the situation where the 4-chloro derivative increases in biological activity. The chloro substituent has positive π and σ values, which implies that one or both of these properties are important to biological activity. If both are important, then adding a second chloro group should increase biological activity yet further. If it does, substituents are varied to increase the π and σ values even further. If it does not, then an unfavourable steric interaction or excessive hydrophobicity is indicated. Further modifications then test the relative importance of π and steric factors.

Now consider the situation where the 4-chloro analogue drops in activity. This suggests either that negative π and/or σ values are important to activity or that a *para* substituent is sterically unfavourable. It is assumed that an unfavourable σ effect is the most likely reason for the reduced activity and so the next substituent is one with a negative σ factor (i.e. 4-OMe). If activity improves, further changes are suggested to test the relative importance of the σ and π factors. On

the other hand, if the 4-OMe group does not improve activity, it is assumed that an unfavourable steric factor is at work and the next substituent is a 3-chloro group. Modifications of this group would then be carried out in the same way as shown in the centre branch of Fig. 13.13.

The last scenario is where the activity of the 4-chloro analogue is little changed from the lead compound. This could arise from the drug requiring a positive π value and a negative σ value. As both values for the chloro group are positive, the beneficial effect of the positive π value might be cancelled out by the detrimental effects of a positive σ value. The next substituent to try in that case is the 4-methyl group, which has the necessary positive π value and negative σ value. If this still has no beneficial effect, then it is assumed that there is an unfavourable steric interaction at the *para* position and the 3-chloro substituent is chosen next. Further changes continue to vary the relative values of the π and σ factors.

The validity of the Topliss scheme was tested by looking at structure–activity results for various drugs which had been reported in the literature. For example, the biological activities of 19 substituted

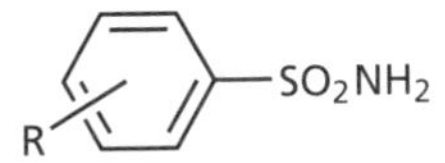

Order of synthesis	R	Biological activity	High potency
1	H	–	
2	4-Cl	More	
3	3,4-Cl$_2$	Less	
4	4-Br	Equal	
5	4-NO$_2$	More	*

Figure 13.15 Biological activity of substituted benzenesulfonamides.

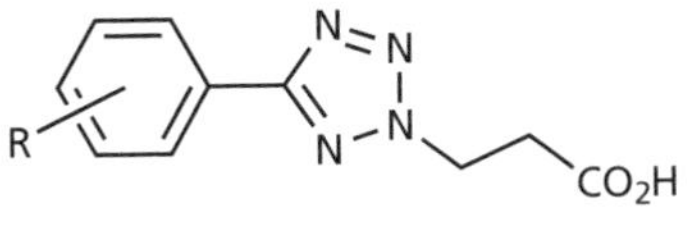

Order of synthesis	R	Biological activity	High potency
1	H	–	
2	4-Cl	Less	
3	4-OMe	Less	
4	3-Cl	More	*
5	3-CF$_3$	Less	
6	3-Br	More	*
7	3-I	Less	
8	3,5-Cl$_2$	More	*

Figure 13.16 Anti-inflammatory activities of substituted aryltetrazolylalkonoic acids.

benzenesulfonamides (Fig. 13.15) have been reported. The second most active compound was the nitro-substituted analogue, which would have been the fifth compound synthesized if the Topliss scheme had been followed.

Another example comes from the anti-inflammatory activities of substituted aryltetrazolylalkanoic acids (Fig. 13.16), of which 28 were synthesized. Using the Topliss scheme, 3 out of the 4 most active structures would have been prepared from the first 8 compounds synthesized.

The Topliss scheme for aliphatic side chains (Fig. 13.14) was set up following a similar rationale to the aromatic scheme, and is used in the same way for side groups attached to a carbonyl, amino, amide, or similar functional group. The scheme attempts to differentiate only between the hydrophobic and electronic effects of substituents and not their steric properties. Thus, the substituents involved have been chosen to try to minimize any steric differences. It is assumed that the lead compound has a methyl group. The first analogue suggested is the isopropyl analogue. This has an increased π value and in most cases would be expected to increase activity. It has been found from experience that the hydrophobicity of most lead compounds is less than optimum.

Let us concentrate first of all on the situation where activity increases. Following this branch, a cyclopentyl group is now used. A cyclic structure is used since it has a larger π value, but keeps any increase in steric factor to a minimum. If activity rises again, more hydrophobic substituents are tried. If activity does not rise, then there could be two explanations. Either the optimum hydrophobicity has been passed or there is an electronic effect (σ_I) at work. Further substituents are then used to determine which is the correct explanation.

Let us now look at the situation where the activity of the isopropyl analogue stays much the same. The most likely explanation is that the methyl group and the isopropyl group are on either side of the hydrophobic optimum. Therefore, an ethyl group is used next, since it has an intermediate π value. If this does not lead to an improvement, it is possible that there is an unfavourable electronic effect. The groups used have been electron-donating, and so electron-withdrawing groups with similar π values are now suggested.

Finally, we shall look at the case where activity drops for the isopropyl group. In this case, hydrophobic and/or electron-donating groups could be bad for activity and the groups suggested are suitable choices for further development.

13.6 Bioisosteres

Tables of substituent constants are available for various physicochemical properties. A knowledge of these constants allows the medicinal chemist to identify substituents which may be potential bioisosteres. Thus, the substituents CN, NO_2, and COMe have similar hydrophobic, electronic, and steric factors, and might be interchangeable. Such interchangeability was observed in the development of **cimetidine** (section 22.2). The important thing to note is that groups can be bioisosteric in some situations, but not others. Consider for example the table shown in Fig. 13.17. This table shows physicochemical parameters for six different substituents. If the most important physicochemical parameter for biological activity is σ_p, then the $COCH_3$ group (0.50) would be a reasonable bioisostere for the $SOCH_3$ group (0.49). If, on the other hand, the dominant parameter is π,

Substituent	$\underset{CH_3}{\overset{\displaystyle O}{\overset{\|}{C}}}$	$\underset{CH_3}{\overset{\displaystyle NC\diagdown\diagup CN}{\overset{C}{\overset{\|}{C}}}}$	$\underset{CH_3}{\overset{\displaystyle O}{\overset{\|}{S}}}$	$\overset{\displaystyle O}{\underset{\displaystyle O}{\overset{\|}{S}}}-CH_3$	$\overset{\displaystyle O}{\underset{\displaystyle O}{\overset{\|}{S}}}-NHCH_3$	$\underset{NMe_2}{\overset{\displaystyle O}{\overset{\|}{C}}}$
π	−0.55	0.40	−1.58	−1.63	−1.82	−1.51
σ_p	0.50	0.84	0.49	0.72	0.57	0.36
σ_m	0.38	0.66	0.52	0.60	0.46	0.35
MR	11.2	21.5	13.7	13.5	16.9	19.2

Figure 13.17 Physicochemical parameters for six substituents.

then a more suitable bioisostere for $SOCH_3$ (-1.58) would be SO_2CH_3 (-1.63).

13.7 Free–Wilson approach

In the Free–Wilson approach to QSAR, the biological activity of a parent structure is measured then compared with the activities of a range of substituted analogues. An equation is then derived which relates biological activity to the presence or otherwise of particular substituents (X_1-X_n):

$$\text{activity} = k_1X_1 + k_2X_2 + k_3X_3 + \cdots + k_nX_n + Z$$

In this equation, X_n is defined as an **indicator variable** and is given the value 1 or 0, depending on whether the substituent (n) is present or not. The contribution that each substituent makes to the activity is determined by the value of k_n. Z is a constant representing the overall average activity of the structures studied.

Since the approach considers the overall effect of a substituent to biological activity rather than its various physicochemical properties, there is no need for physicochemical constants and tables, and the method only requires experimental measurements of biological activity. This is particularly useful when trying to quantify the effect of unusual substituents that are not listed in the tables, or when quantifying specific molecular features which cannot be tabulated.

The disadvantage in the approach is the large number of analogues which have to be synthesized and tested to make the equation meaningful. For example, each of the terms k_nX_n refers to a specific substituent at a specific position in the parent structure. Therefore, analogues would not only have to have different substituents, but have them at different positions of the skeleton as well.

Another disadvantage is the difficulty in rationalizing the results and explaining why a substituent at a particular position is good or bad for activity. Finally, the effects of different substituents may not be additive. There may be intramolecular interactions which affect activity.

Nevertheless the idea of using indicator variables can be useful in certain situations and they can also be used as part of a Hansch equation. An example of this can be seen in the following study (section 13.9) case.

13.8 Planning a QSAR study

When starting a QSAR study it is important to decide which physicochemical parameters are going to be studied and to plan the analogues such that the parameters under study are suitably varied. For example, it would be pointless to synthesize analogues where the hydrophobicity and steric volume of the substituents are correlated, if these two parameters are to go into the equation.

It is also important to make enough structures to make the results statistically meaningful. As a rule of thumb, five structures should be made for every parameter studied. Typically, the initial QSAR study would involve the two parameters π and σ, and possibly E_s. Craig plots could be used in order to choose suitable substituents.

Certain substituents are worth avoiding in the initial study, as they may have properties other than those being studied. For example, it is best to avoid substituents that might ionize (CO_2H, NH_2, SO_3H), and groups that might easily be metabolized (e.g. esters or nitro groups).

If there are two or more substituents, the initial equation usually considers the total π and σ contribution. As more analogues are made, it is often possible to consider the hydrophobic and electronic effect of substituents at specific positions of the molecule. Furthermore, the electronic parameter σ can be split into its inductive and resonance components (F and R). Such detailed equations may show up a particular localized requirement for activity. For example, a hydrophobic substituent may be favoured in one part of the skeleton, while an electron-withdrawing substituent is favoured at another. This in turn gives clues about the binding interactions involved between drug and receptor.

13.9 Case study

An example of how a QSAR equation can become more specific as a study develops is demonstrated from work carried out on the antiallergic activity of a series of pyranenamines (Fig. 13.18). In this study, substituents were varied on the aromatic ring, and the

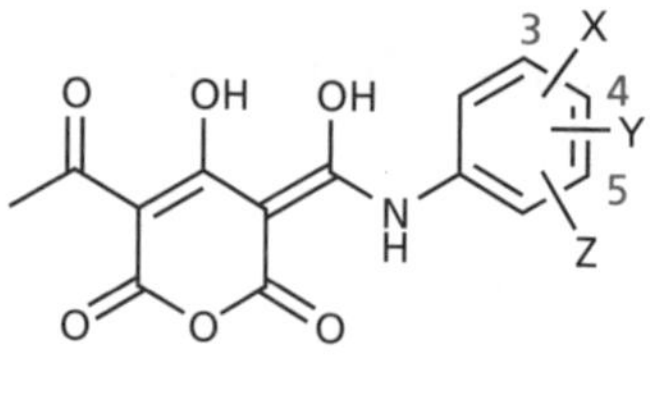

Figure 13.18 Structure of pyranenamines.

remainder of the molecule was kept constant. Nineteen compounds were synthesized and the first QSAR equation was obtained by considering π and σ:

$$\log(1/C) = -0.14\sum\pi - 1.35\left(\sum\sigma\right)^2 - 0.72$$
$$(n = 19, r^2 = 0.48, s = 0.47, F_{2,16} = 7.3)$$

where $\sum\pi$ and $\sum\sigma$ are the total π and σ values for all the substituents present.

The negative coefficient for the π term shows that activity is inversely proportional to hydrophobicity, which is quite unusual. The $(\sum\sigma)^2$ term is also quite unusual. It was chosen because there was no simple relationship between activity and σ. In fact, it was observed that activity decreased if the substituent was electron-withdrawing *or* electron-donating. Activity was best with neutral substituents. To take account of this, the $(\sum\sigma)^2$ term was introduced. As the coefficient in the equation is negative, activity is lowered if σ is anything other than zero.

A further range of compounds was synthesized with hydrophilic substituents to test this equation, making a total of 61 structures. This resulted in the following inconsistencies.

- The activities for the substituents 3-NHCOMe, 3-NHCOEt, and 3-NHCOPr were all similar, but according to the equation these activities should have dropped as the alkyl group got larger, due to increasing hydrophobicity.

- Activity was greater than expected if there was a substituent such as OH, SH, NH$_2$, or NHCOR at position 3, 4, or 5.

- The substituent NHSO$_2$R was bad for activity.

- The substituents 3,5-(CF$_3$)$_2$ and 3,5-(NHCOMe)$_2$ had much greater activity than expected.

- An acyloxy group at the 4-position resulted in an activity five times greater than predicted by the equation.

These results implied that the initial equation was too simple and that properties other than π and σ were important to activity. At this stage, the following theories were proposed to explain the above results.

- The similar activities for 3-NHCOMe, 3-NHCOEt, and 3-NHCOPr could be due to a steric factor. The substituents had increasing hydrophobicity, which is bad for activity, but they were also increasing in size and it was proposed that this was good for activity. The most likely explanation is that the size of the substituent is forcing the drug into the correct orientation for optimum receptor interaction.

- The substituents which unexpectedly increased activity when they were at positions 3, 4, or 5 are all capable of hydrogen bonding. This suggests an important hydrogen bonding interaction with the receptor. For some reason, the NHSO$_2$R group is an exception, which implies there is some other unfavourable steric or electronic factor peculiar to this group.

- The increased activity for 4-acyloxy groups was explained by suggesting that these analogues are acting as prodrugs. The acyloxy group is less polar than the hydroxyl group and so these analogues would be expected to cross cell membranes and reach the receptor more efficiently than analogues bearing a free hydroxyl group. At the receptor, the ester group could be hydrolysed to reveal the hydroxyl group which would then take part in hydrogen bonding with the receptor.

- The structures having substituents 3,5-(CF$_3$)$_2$ and 3,5-(NHCOMe)$_2$ are the only di-substituted structures where a substituent at position 5 has an electron-withdrawing effect, so this feature was also introduced into the next equation.

The revised QSAR equation was as follows:

$$\log(1/C) = -0.30\sum\pi - 1.5\left(\sum\sigma\right)^2 + 2.0(F\text{-}5)$$
$$+ 0.39(345\text{-}HBD) - 0.63(NHSO_2)$$
$$+ 0.78(M\text{-}V) + 0.72(4\text{-}OCO) - 0.75$$
$$(n = 61, r^2 = 0.77, s = 0.40, F_{7,53} = 25.1)$$

The π and σ parameters are still present, but a number of new parameters have now been introduced:

- The *F-5* term represents the inductive effect of a substituent at position 5. The coefficient is positive and large, showing that an electron-withdrawing group substantially increases activity. However, only 2 compounds of the 61 synthesized had a 5-substituent, so there might be quite an error in this result.

- The *M-V* term represents the volume of any *meta* substituent. The coefficient is positive, indicating that substituents with a large volume at the *meta* position increase activity.

- The advantage of having hydrogen bonding substituents at position 3, 4, or 5 is accounted for by including a hydrogen bonding term (*345-HBD*). The value of this term depends on the number of hydrogen bonding substituents present. If one such group is present, the *345-HBD* term is 1. If two such groups are present, the parameter is 2. Therefore, for each hydrogen bonding substituent present at positions 3, 4, or 5, $\log(1/C)$ increases by 0.39. This sort of term is known as an **indicator variable**, which is the basis of the **Free–Wilson** approach described earlier. There is no tabulated value one can use for a hydrogen bonding substituent and so the contribution that this term makes to the biological activity is determined by the value of k, and whether the relevant group is present or not. Indicator variables were also used for the following terms.

- The *NHSO₂* term was introduced because this group was bad for activity despite being capable of hydrogen bonding. The negative coefficient indicates the drop in activity. A figure of 1 is used for any $NHSO_2R$ substituent present, resulting in a drop of activity by 0.63.

- The *4-OCO* term is 1 if an acyloxy group is present at position 4, and so $\log(1/C)$ is increased by 0.72 if this is the case.

A further 37 structures were synthesized to test steric and *F-5* parameters, as well as exploring further groups capable of hydrogen bonding. Since hydrophilic substituents were good for activity, a range of very hydrophilic substituents were also tested to see if there was an optimum value for hydrophilicity. The results obtained highlighted one more anomaly, in that two hydrogen bonding groups *ortho* to each other were

bad for activity. This was attributed to the groups hydrogen bonding with each other rather than to the receptor. A revised equation was obtained as follows:

$$\log(1/C) = -0.034\left(\sum\pi\right)^2 - 0.33\left(\sum\pi\right) + 4.3(F\text{-}5)$$
$$+ 1.3(R\text{-}5) - 1.7\left(\sum\sigma\right)^2 + 0.73(345\text{-}HBD)$$
$$- 0.86(HB\text{-}INTRA) - 0.69(NHSO_2)$$
$$+ 0.72(4\text{-}OCO) - 0.59$$
$$(n=98, r^2=0.75, s=0.48, F_{9,88}=28.7)$$

The main points of interest from this equation are as follows:

- Increasing the hydrophilicity of substituents allowed the identification of an optimum value for hydrophobicity ($\sum\pi = -5$) and introduced the $(\sum\pi)^2$ parameter into the equation. The value of -5 is remarkably low and indicates that the receptor site is hydrophilic.

- As far as electronic effects are concerned, it is revealed that the resonance effects of substituents at the 5-position also have an influence on activity.

- The unfavourable situation where two hydrogen bonding groups are *ortho* to each other is represented by the *HB-INTRA* parameter. This parameter is given the value 1 if such an interaction is possible and the negative constant (-0.86) shows that such interactions decrease activity.

- It is interesting to note that the steric parameter is no longer significant and has disappeared from the equation.

The compound having the greatest activity has two $NHCOCH(OH)CH_2OH$ substituents at the 3- and 5-positions and is 1000 times more active than the original lead compound. The substituents are very polar and are not ones that would normally be used. They satisfy all the requirements determined by the QSAR study. They are highly polar groups which can take part in hydrogen bonding. They are *meta* with respect to each other, rather than *ortho*, to avoid undesirable intramolecular hydrogen bonding. One of the groups is at the 5-position and has a favourable *F-5* parameter. Together the two groups have a negligible $(\sum\sigma)^2$ value. Such an analogue

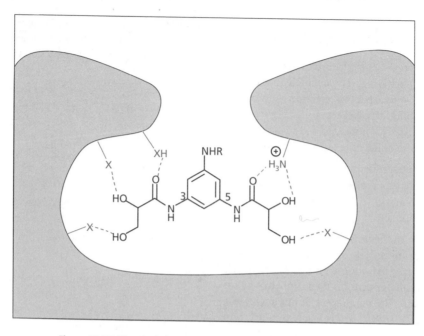

Figure 13.19 Hypothetical receptor binding interactions of a pyranenamine.

would certainly not have been obtained by trial and error, and this example demonstrates the strengths of the QSAR approach.

All the evidence from this study suggests that the aromatic ring of this series of compounds is fitting into a hydrophilic pocket in the receptor which contains polar groups capable of hydrogen bonding. It is further proposed that a positively charged residue such as arginine, lysine, or histidine might be present in the pocket which could interact with an electronegative substituent at position 5 of the aromatic ring (Fig. 13.19).

This example demonstrates that QSAR studies and computers are powerful tools in medicinal chemistry. However, it also shows that the QSAR approach is a long way from replacing the human factor. One cannot put a series of facts and figures into a computer and expect it to magically produce an instant explanation of how a drug works. The medicinal chemist still has to interpret results, propose theories, and test those theories by incorporating the correct parameters into the QSAR equation. Imagination and experience still count for a great deal.

KEY POINTS

- QSAR relates the physicochemical properties of a series of drugs to their biological activity by means of a mathematical equation.

- The commonly studied physicochemical properties are hydrophobicity, electronic factors, and steric factors.

- The partition coefficient is a measure of a drug's overall hydrophobicity. Values of log P are used in QSAR equations, with larger values indicating greater hydrophobicity.

- The substituent hydrophobicity constant is a measure of the hydrophobic character of individual substituents. The value is different for aliphatic and aromatic substituents and is only directly relevant to the class of structures from which the values were derived. Positive values represent substituents more hydrophobic than hydrogen; negative values represent substituents more hydrophilic than hydrogen.

- The Hammett substituent constant is a measure of how electron-withdrawing or electron-donating an aromatic substituent is. It is measured experimentally and is dependent on the relative position of the substituent on the ring. The value takes into account both inductive and resonance effects.

- The parameters *F* and *R* are constants quantifying the inductive and resonance effects of an aromatic substituent.
- The inductive effect of aliphatic substituents can be measured experimentally and tabulated.
- Steric factors can be measured experimentally or calculated using physical parameters or computer software.
- The Hansch equation is a mathematical equation which relates a variety of physicochemical parameters to biological activity for a series of related structures.
- The Craig plot is a visual comparison of two physicochemical properties for a variety of substituents. It facilitates the choice of substituents for a QSAR study such that the values of each property are not correlated.
- The Topliss scheme is used when structures can only be synthesized and tested one at a time. The scheme is a guide to which analogue should be synthesized next in order to get good activity. There are different schemes for aromatic and aliphatic substituents.
- Indicator variables are used when there are no tabulated or experimental values for a particular property or substituent. The Free–Wilson approach to QSAR only uses indicator variables, whereas the Hansch approach can use a mixture of indicator variables and physicochemical parameters.

13.10 3D QSAR

In recent years, a method known as 3D QSAR has been developed in which the 3D properties of a molecule are considered as a whole rather than by considering individual substituents or moieties. This has proved remarkably useful in the design of new drugs. Moreover, the necessary software and hardware are readily affordable and relatively easy to use. The philosophy of 3D QSAR revolves around the assumption that the most important features about a molecule are its overall size and shape, and its electronic properties (electrostatic fields).

If these features can be defined, then it is possible to study how they affect biological properties. There are several approaches to 3D QSAR, but the method which has gained ascendancy was developed by the company Tripos and is known as **CoMFA** (Comparative Molecular Field Analysis). CoMFA methodology is based on the assumption that drug–receptor interactions are non-covalent and that changes in biological activity correlate with the changes in the steric and/or electrostatic fields of the drug molecules.

13.10.1 Defining steric and electrostatic fields

In order to define the necessary steric and electrostatic fields, a molecule is constructed on the computer using molecular modelling software (section 15.3). If several stable conformations are possible then the **active conformation** (sections 10.2 and 15.10) has to be identified and energy minimized (section 15.4). The **pharmacophore** (section 10.2) is then identified (Fig. 13.20).

The next stage is to place the pharmacophore into a lattice or grid. The intersections of this lattice are

Figure 13.20 Identification of active conformation and pharmacophore.

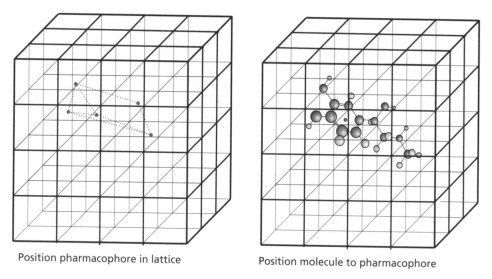

Position pharmacophore in lattice   Position molecule to pharmacophore

Figure 13.21 Positioning pharmacophore and molecules into lattice.

called lattice (or grid) points and these define the 3D space around the pharmacophore. The position of the pharmacophore is kept constant and acts as a reference point when positioning different molecules into the lattice. For each molecule studied, the active conformation and pharmacophore is identified and then the molecule is placed into the lattice such that its pharmacophore matches the reference pharmacophore (Fig. 13.21).

Once a molecule has been placed into the lattice, the steric and electrostatic fields around it are measured. This is done by placing a probe atom such as a proton or sp^3 hybridized carbocation at each of the grid points in turn and using the software to calculate the steric and electrostatic interactions between the probe and the molecule (Fig. 13.22). As far as the steric field is concerned, this will increase as the probe atom gets closer to the molecule. As far as the electrostatic field is concerned, there will be an attraction between the positively charged probe and electron-rich regions of the molecule, and a repulsion between the probe and electron-deficient regions of the molecule.

The steric and electrostatic fields at each grid point are then tabulated, and this is repeated for each molecule in the study. A particular value for the steric energy is then chosen which will define the shape of the molecule, and the grid points having that value are then connected by contour lines to define the steric field. This is done for each molecule. A similar process is carried out to measure the electrostatic interactions between

the positively charged probe atom and the test molecule. Electron-rich and electron-deficient regions for each molecule are then defined by suitable contour lines.

13.10.2 Relating shape and electronic distribution to biological activity

Defining the size, shape, and electronic distribution of a series of molecules is relatively straightforward and is carried out automatically by the software program. The next stage is to relate these properties to the biological activity of the molecules. This is less straightforward, and differs significantly from traditional QSAR. In traditional QSAR, there are relatively few variables involved. For example, if we consider hydrophobicity, π, σ, and a size factor for each molecule, then we have four variables per molecule to compare against biological activity. With 100 molecules in the study, there are far more molecules than variables and it is possible to come up with an equation relating variables to biological activity as previously described.

In 3D QSAR, the variables for each molecule are the calculated steric and electronic interactions at a couple of thousand lattice points. With 100 molecules in the study, the number of variables now far outweighs the number of structures, and it is not possible to relate these to biological potency by the standard multiple linear regression analysis described in section 13.1. A different statistical procedure has to

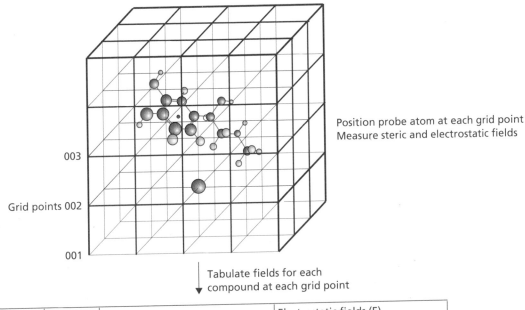

Position probe atom at each grid point
Measure steric and electrostatic fields

Tabulate fields for each
compound at each grid point

Compound	Biological activity	Steric fields (S) at grid points (001–998)					Electrostatic fields (E) at grid points (001–098)				
		S001	S002	S003	S004	S005 etc	E001	E002	E003	E004	E005 etc
1	5.1										
2	6.8										
3	5.3										
4	6.4										
5	6.1										

Partial least squares
analysis (PLS)

QSAR equation   Activity = aS001 + bS002 + · · · + mS998 + nE001 + · · · + yE998 + z

Figure 13.22 Measuring steric and electronic fields.

be followed, using a technique called **partial least squares** (PLS). Essentially, it is an analytical computing process which is repeated over and over again (iterated) to try to find the best formula relating biological property against the various variables. As part of the process, the number of variables is reduced as the software filters out those which are clearly unrelated to biological activity.

An important feature of the analysis is that a structure is deliberately left out as the computer strives to form some form of relationship. Once a formula has been defined, the formula is tested against the structure which was left out. This is called **cross-validation** and tests how well the formula predicts the biological property for the molecule which

was left out. The results of this are fed back into another round of calculations, but now the structure which was left out is included in the calculations and a different structure is left out. This leads to a new improved formula which is once again tested against the compound that was left out, and so the process continues until cross-validation has been carried out against all the structures.

At the end of the process, the final formula is obtained (Fig. 13.22). The predictability of this final equation is quantified by the **cross-validated correlation coefficient** r^2, which is usually referred to as q^2. In contrast to normal QSAR, where r^2 should be greater than 0.8, values of q^2 greater than 0.3 are considered significant. It is more useful, though, to

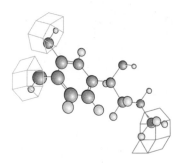

Figure 13.23 Definition of favourable and unfavourable interactions round representative molecule.

give a graphical representation showing which regions around the molecule are important to biological activity on steric or electronic grounds. Therefore, a steric map shows a series of coloured contours indicating beneficial and detrimental steric interactions around a representative molecule from the set of molecules tested (Figure 13.23). A similar contour map is created to illustrate beneficial or detrimental electrostatic interactions.

An example of a 3D QSAR study is described in section 15.18.9 and in the case study (section 13.10.6).

13.10.3 Hydrophobic potential

In traditional QSAR, the hydrophobic factor is very important and most equations include a log P value. Surprisingly, the calculation of a hydrophobic factor in 3D QSAR is not so crucial and most successful studies have been based on steric and electrostatic factors alone. Nevertheless, 3D QSAR studies can be carried out using a hydrophobic factor if the probe atom is replaced by a water molecule in order to calculate the hydrophobic potential at each lattice point.

13.10.4 Advantages of 3D QSAR over traditional QSAR

Some of the problems involved with a traditional QSAR study include the following:

- Only molecules of similar structure can be studied.
- The validity of the numerical descriptors is open to doubt. These descriptors are obtained by measuring reaction rates and equilibria constants in model reactions and are listed in tables. However, separating one property from another is not always possible in experimental measurement. For example, the Taft steric factor is not purely a measure of the steric factor because the measured reaction rates used to define it are also affected by electronic factors. Also, the n-octanol/water partition coefficients which are used to measure log P are known to be affected by the hydrogen bonding character of molecules.

- The tabulated descriptors may not include entries for unusual substituents.

- It is necessary to synthesize a range of molecules where substituents are varied in order to test a particular property (e.g. hydrophobicity). However, synthesizing such a range of compounds may not be straightforward or feasible.

- Traditional QSAR equations do not directly suggest new compounds to synthesize.

These problems are avoided with 3D QSAR, which has the following advantages:

- Favourable and unfavourable interactions are represented graphically by 3D contours around a representative molecule. A graphical picture such as this is easier to visualize than a mathematical formula.

- In 3D QSAR the properties of the test molecules are calculated individually by computer program. There is no reliance on experimental or tabulated factors. There is no need to confine the study to molecules of similar structure. As long as one is confident that all the compounds in the study share the same pharmacophore and interact in the same way with the target, they can all be analysed in a 3D QSAR study.

- The graphical representation of beneficial and non-beneficial interactions allows medicinal chemists to design new structures. For example, if a contour map shows a favourable steric effect at one particular location, this implies that the target binding site has space for further extension at that location. This may lead to further favourable receptor–drug interactions.

- Both traditional and 3D QSAR can be used without needing to know the structure of the biological target.

13.10.5 Potential problems of 3D QSAR

There are several pitfalls which have to be avoided when carrying out 3D QSAR. The main ones are:

- Care has to be taken to ensure that each molecule is in the active conformation when it is built on the computer.

- Each molecule must be properly aligned on the screen with respect to the others, so that their pharmacophores match up.

Knowing the active conformation is possible in rigid structures such as steroids, but it is more difficult with flexible molecules that are capable of several bond rotations. Therefore, it is useful to have a conformationally restrained analogue which is biologically active and which can act as a guide to the likely active conformation. More flexible molecules can then be constructed on the computer with the conformation most closely matching that of the more rigid analogue. If the structure of the target binding site is known, this can be useful in deciding the likely active conformations of molecules and how they should be aligned before the 3D QSAR analysis.

It is also crucial to identify the likely pharmacophore so that important atoms or groups of atoms are positioned in the same area of space for each molecule. It may be difficult to identify the pharmacophore in some molecules, however. In that case, a pharmacophore mapping exercise can be carried out by computer (section 15.11). This is likely to be successful if there are some rigid active compounds available, allowing a restriction on the number of possible conformations.

One has to be careful to ensure that all the compounds in the study interact with the target in similar ways. For example a QSAR study on all possible **acetylcholinesterase** inhibitors is doomed to failure. In the first place, the great diversity of structures involved makes it impossible to align these structures in an unbiased way or to generate a 3D pharmacophore. Second, the various inhibitors do not interact with the target enzyme in the same way. X-ray crystallographic studies of enzyme–inhibitor complexes show that the inhibitors **tacrine**, **edrophonium**, and **decamethonium** all have different binding orientations in the active site.

3D QSAR provides a summary of how structural changes in a drug affect biological activity, but it is dangerous to assume too much. For example, a 3D QSAR model may show that increasing the bulk of the molecule at a particular location increases activity. This might suggest that there is an accessible hydrophobic pocket allowing extra binding interactions. On the other hand, it is possible that the extra steric bulk causes the molecule to bind in a different orientation from the other molecules in the analysis, and that this is the reason for the increased activity.

KEY POINTS

- 3D QSAR is a computer based system which involves the use of software to measure steric and electrostatic fields round a series of structures.

- A comparison of the steric and electrostatic fields for different molecules against their biological activity allows the definition of steric and electrostatic interactions which are favourable and unfavourable for activity. These can be displayed visually as contour lines.

- It is necessary to define the active conformation and pharmacophore for each molecule in the study.

- Unlike conventional QSAR studies, molecules of different structural classes can be compared if they share the same pharmacophore.

- 3D QSAR does not depend on experimentally measured parameters.

13.10.6 Case study: inhibitors of tubulin polymerization

Colchicine (Fig. 13.24) is a lead compound for agents which act as inhibitors of tubulin polymerization (section 3.7.2) and which might be useful in the treatment of arthritis. Other lead compounds have been discovered which bind to tubulin at the same binding site and so a study was carried out to compare the various structural classes interacting in this way.

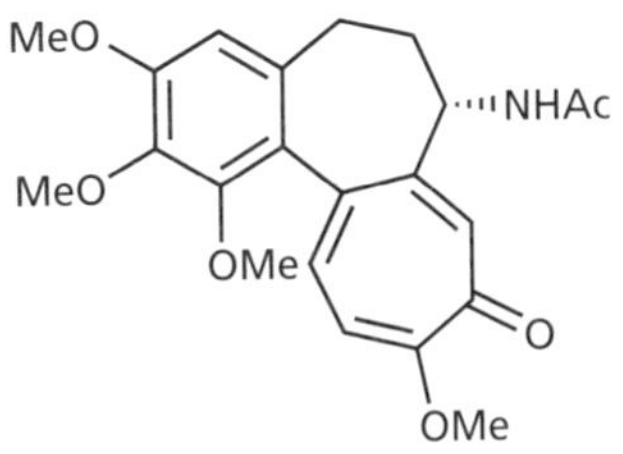

Figure 13.24 Colchicine.

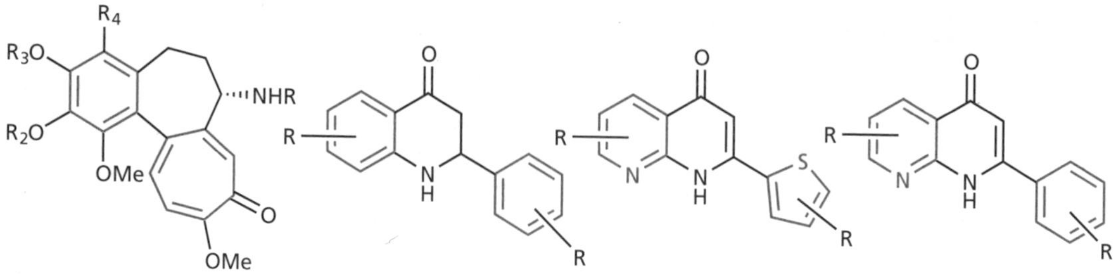

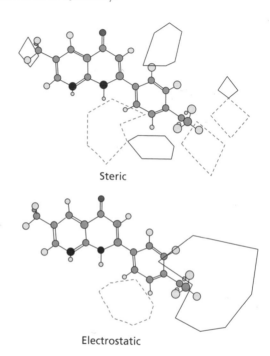

Figure 13.25 Structural classes used in the 3D QSAR study.

In this 3D QSAR study, 104 such agents were tested, belonging to 4 distinct families of compounds (Fig. 13.25); 51 compounds were used as a 'training set' for the analysis itself and 53 were used as a 'testing set' to test the predictive value of the results. Both sets contained a mixture of structural classes having both low and high activity.

The first task was to work out how to align these different classes of molecule. Colchicine is the most rigid of the four, and also has a high affinity for tubulin. Therefore, it was chosen as the template on to which the other structures would be aligned. The relevant pharmacophore in colchicine was identified as the two aromatic rings. Molecular modelling was now carried out on each of the remaining structures to generate various conformations. Each conformation was compared with colchicine to find the one that would allow the pharmacophores in each structure to be aligned. This was then identified as the active conformation.

Once the active conformations for each structure had been identified, they were fitted into the lattice of grids previously described such that each structure was properly aligned. The steric and electrostatic fields round each molecule were calculated using a probe atom, then the 3D QSAR analysis carried out to relate the fields with the measured biological activity.

The results of the 3D QSAR analysis are summarized as contour lines round a representative molecule (Fig. 13.26). For the steric interactions, solid contours represent fields which are favourable for activity, and the dashed lines show fields which are unfavourable. For the electrostatic interactions, solid lines are regions where positive charged species improve affinity, and dashed lines indicate regions where negatively charged groups are favourable.

The results showed that introducing steric bulk round the aromatic ring is more crucial to activity

Steric

Electrostatic

Figure 13.26 Results of the 3D QSAR analysis ($q^2 = 0.637$).

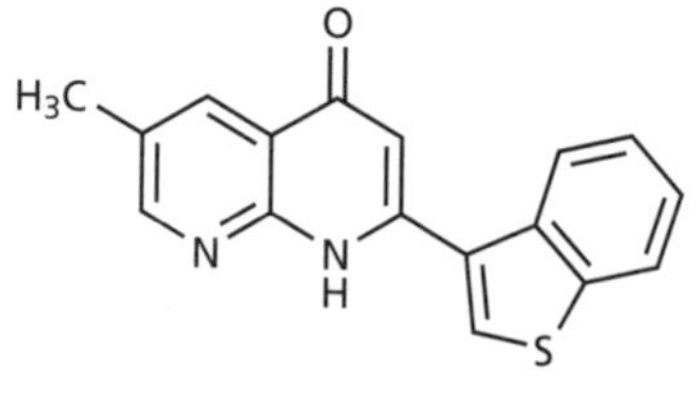

Figure 13.27 Novel agent designed on the basis of the 3D QSAR study.

than introducing steric bulk round the bicyclic system. Based on this evidence, the structure shown in Fig. 13.27 was synthesized. The predicted value of pIC_{50} for the compound was 5.62. The actual value was in close agreement at 6.04. ($pIC_{50} = -\log[IC_{50}]$, where IC_{50} is the concentration of inhibitor required to produce 50% enzyme inhibition.)

The steric fields of the Tripos' CoMFA analysis (Fig. 13.26) were subsequently placed into a model of the binding site. It was found that the bad steric regions were in the same regions as the peptide backbone, whereas the favourable steric areas were in empty spaces.

QUESTIONS

1. Using values from Table 13.1, calculate the log P value for structure (I) (log P for benzene $= 2.13$).

2. Several analogues of a drug are to be prepared for a QSAR study which will consider the effect of various aromatic substituents on biological activity. You are asked whether the substituents (SO_2NH_2, CF_3, CN, CH_3SO_2, SF_5, $CONH_2$, OCF_3, CO_2H, Br, I) are relevant to the study. What are your thoughts?

3. A lead compound has a monosubstituted aromatic ring present as part of its structure. An analogue was synthesized containing a *para*-chloro substituent which had approximately the same activity. It was decided to synthesize an analogue bearing a methyl group at the *para*-position. This showed increased activity. What analogue would you prepare next, and why?

4. The following QSAR equation was derived for the pesticide activity of structure (II). Explain what the various terms mean and whether the equation is a valid one. Identify what kind of substituents would be best for activity.

$$log(1/C) = 1.08\pi_x + 2.41F_x + 1.40R_x - 0.072MR_x + 5.25$$
$$(n = 16, r^2 = 0.840, s = 0.59)$$

5. A QSAR equation for the anticonvulsant (III) was derived as follows:

$$log(1/C) = 0.92\pi_x - 0.34\pi_x^2 + 3.18$$
$$(n = 15, r^2 = 0.902, s = 0.09, \pi_o = 1.35)$$

What conclusions can you draw from this equation? Would you expect activity to be greater if X $=$ CF_3 rather than H or CH_3?

6. The following QSAR equation is related to the mutagenic activity of a series of nitrosoamines: $log(1/C) = 0.92\pi + 2.08\sigma - 3.26$ ($n = 12$, $r^2 = 0.794$, $s = 0.314$). What sort of substituent is likely to result in high mutagenic activity?

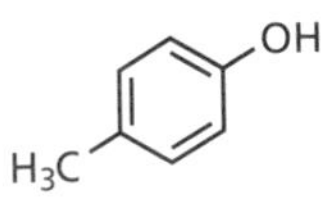

I 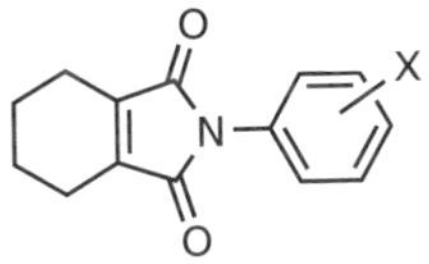   II 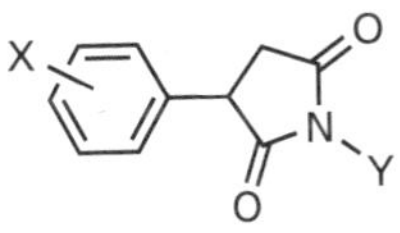   III

FURTHER READING

Craig, P. N. (1971) Interdependence between physical parameters and selection of substituent groups for correlation studies. *Journal of Medicinal Chemistry*, **14**, 680–684.

Cramer, R. D. et al. (1979) Application of quantitative structure–activity relationships in the development of the antiallergenic pyranenamines. *Journal of Medicinal Chemistry*, **22**, 714–725.

Cramer, R. D., Patterson, D. E., and Bunce, J. D. (1988) Comparative field analysis (CoMFA). *Journal of the American Chemical Society*, **110**, 5959–5967.

Hansch, C. and Leo, A. (1995) *Exploring QSAR*. American Chemical Society, Washington, DC.

Kubini, H., Folkers, G., and Martin, Y. C. (eds.) (1998) *3D QSAR in drug design*. Kluwer/Escom, Dordrecht.

Martin, Y. C. and Dunn, W. J. (1973) Examination of the utility of the Topliss schemes by analog synthesis. *Journal of Medicinal Chemistry*, **16**, 578–579.

Taft, R. W. (1956) Separation of polar, steric and resonance effects in reactivity. Chapter 13 in: *Steric effects in organic chemistry*, ed. M. S. Newman. John Wiley and Sons, New York.

Topliss, J. G. (1972) Utilization of operational schemes for analog synthesis in drug design. *Journal of Medicinal Chemistry*, **15**, 1006–1011.

Verloop, A., Hoogenstraaten, W., and Tipker, J. (1976) Development and application of new steric substituent parameters in drug design. *Medicinal Chemistry*, **11**, 165–207.

van de Waterbeemd, H., Testa, B., and Folkers, G. (eds.) (1997) *Computer-assisted lead finding and optimization.* Wiley-VCH, New York.

Zhang, S.-X. *et al.* (2000) Antitumor agents. 199. Three-dimensional quantitative structure–activity relationship study of the colchicine binding site ligands using comparative molecular field analysis. *Journal of Medicinal Chemistry*, **43**, 167–176.

Titles for general further reading are listed on p. 711.

14 Combinatorial synthesis

Combinatorial synthesis has been one of the most rapidly developing fields in the pharmaceutical industry in recent years and is now seen as an essential tool in both the discovery and the development of new drugs. So what is combinatorial synthesis, and why is it important?

Put at its simplest, combinatorial synthesis is a means of producing a large number of compounds in a short period of time, using a defined reaction route and a large variety of starting materials and reagents. Usually, this is done on a very small scale using solid phase synthesis, so that the process can be automated or semi-automated. This allows each reaction of the synthetic route to be carried out in several reaction vessels at the same time and under identical conditions, but using different reagents for each vessel.

Combinatorial synthesis can be carried out such that a single product is obtained in each different reaction flask—a process known as **parallel synthesis**. Alternatively, the process can be designed such that **mixtures** of compounds are produced in each reaction vessel.

14.1 Combinatorial synthesis in medicinal chemistry

Medicinal chemistry requires the rapid synthesis of a large number of compounds for a variety of reasons (Fig. 14.1).

The need to find new lead compounds in drug discovery has been the major driving force in the development of combinatorial synthesis. There has been a rapid explosion in the number of new drug targets discovered by genomic and proteomic projects across the world, many of which provide the opportunity of developing new treatments for old diseases. With so many new targets being discovered, pharmaceutical companies are faced with the problem of identifying the function of each target and finding a lead compound to interact with it. Before the advent of combinatorial chemistry, the need to find a lead compound was becoming the limiting factor in the whole process. Herein lies the real need for combinatorial synthesis. Whereas in the past the driving force was the discovery of a lead compound, the driving force now is the discovery of new drug targets. It has been stated that a pharmaceutical company might expect to set up and carry out lead discovery programmes against about 100 targets per year and will need to screen over a million compounds if it is to find a lead compound quickly and efficiently. Combinatorial chemistry provides a means of producing that many compounds.

Both parallel and mixed combinatorial syntheses can be used to generate large quantities of structures in order to find that lead compound, but the latter method can generate significantly more structures in a set period of time and increase the chances of finding a lead compound. In the procedure, mixtures of compounds are deliberately produced in each reaction flask, allowing chemists to produce thousands and even millions of novel structures in the time that they would take to synthesize a few dozen by conventional means. This method of synthesis goes against the grain of conventional organic synthesis—traditionally, chemists set out to produce a single identifiable structure which can be purified and characterized. The structures in each reaction vessel of a mixed combinatorial synthesis are not separated and purified, but are

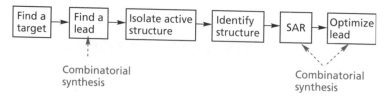

Figure 14.1 The role of combinatorial synthesis in drug discovery and drug optimization.

tested for biological activity as a whole. If there is no activity, then there is no need to study that mixture any more and it is stored. If activity *is* observed, the challenge is to identify which component of the mixture is the active compound. We shall discuss this in more detail later on. Overall, there is an economy of effort, as a negative result for a mixture of 100 compounds saves the effort of synthesizing, purifying, and identifying each component of that mixture. On the other hand, identifying the active component of an active mixture is not straightforward.

In a sense, a mixed combinatorial synthesis can be looked upon as the synthetic equivalent of nature's chemical pool. Through evolution, nature has produced a huge number and variety of chemical structures, some of which are biologically active. Traditional medicinal chemistry dips into that pool to pick out the **active principles** and develop them. A mixed combinatorial synthesis produces pools of purely synthetic structures that we can dip into for active compounds. The diversity of structures from the natural pool is far greater than that likely to be achieved by combinatorial synthesis, but isolating, purifying, and identifying new agents from natural sources is a relatively slow process and there is no guarantee that a lead compound will be discovered against a specific drug target. The advantage of combinatorial chemistry (both mixed and parallel) is the fact that it produces new compounds faster than those derived from natural sources and can produce a diversity not found in the traditional banks of synthetic compounds held by pharmaceutical companies.

The other two areas of medicinal chemistry where a large number of compounds have to be synthesized are in structure–activity relationship studies and in drug optimization. Parallel rather than mixed syntheses are used here, as each compound has to be tested individually. Although far more compounds could be produced by mixed combinatorial synthesis, the time saved in producing them would be lost in separating and identifying them all. Indeed, the use of mixed

combinatorial synthesis is declining and parallel synthesis is used far more frequently in the pharmaceutical industry.

Combinatorial synthesis is not without its critics, and cynics have described its use in drug discovery as nothing more than a technologically advanced form of trial and error. Admittedly, there is some truth in that, but it would be wrong to say that combinatorial synthesis removes the intellectual challenge from drug discovery and development. There is more to combinatorial chemistry than devising a reaction sequence and programming a machine to churn out thousands of compounds. A certain amount of thought has to go into the design of the process to ensure that it will be as efficient as possible in producing a 'hit'. Careful thought also has to go into what types of novel structure are likely to be pharmaceutically active before synthesizing them. The future will see an increasing use of robots and machines to synthesize new structures, but that does not mean we have to think like them.

14.2 Solid phase techniques

Although some combinatorial experiments have been performed in solution, the majority have been achieved using solid phase techniques where the reaction is carried out on a solid support such as a resin bead. There are several advantages to this.

- A range of different starting materials can be bound to separate beads. The beads can then be mixed together such that all the starting materials can be treated with another reagent in a single experiment. The starting materials and products are still physically distinct, as they are bound to separate beads. In most cases, mixing all the starting materials together in solution chemistry is a recipe for disaster, with polymerizations and side reactions producing a tarry mess.

- Since the starting materials and products are bound to a solid support, excess reagents or unbound by-products can easily be removed by washing the resin.

- Large excesses of reagents can be used to drive the reactions to completion (greater than 99%) because of the ease with which excess reagent can be removed.

- If one uses low loadings (less than 0.8 mmol/g support), undesired side reactions such as cross-linking can be suppressed.

- Intermediates in a reaction sequence are bound to the bead and do not need to be purified.

- The individual beads can be separated at the end of the experiment to give individual products.

- The polymeric support can be regenerated and reutilized if appropriate cleavage conditions and suitable anchor/linker groups are chosen (see later).

- Automation is possible.

Solid phase synthesis was pioneered by Merrifield for the synthesis of peptides, and most of the early work carried out on combinatorial synthesis was performed on peptides. Peptides have serious disadvantages as drugs (section 11.8.3), however, and so

a large amount of research has been carried out to extend solid phase synthetic methods to the synthesis of small non-peptide molecules. The essential requirements for solid phase synthesis are:

- a cross-linked insoluble polymeric support which is inert to the synthetic conditions (e.g. a resin bead)

- an anchor or linker covalently linked to the resin, having a reactive functional group such that substrates can be attached to it

- a bond linking the substrate to the linker which will be stable to the reaction conditions used in the synthesis

- a means of cleaving the product or the intermediates from the linker

- chemical protecting groups for functional groups not involved in the synthetic route.

14.2.1 The solid support

The earliest form of resin used by Merrifield was polystyrene beads where the styrene is partially cross-linked with 1% divinylbenzene. The beads are derivatized with a chloromethyl group (the anchor/linker) to which amino acids can be coupled via an ester group (Fig. 14.2). This ester group is stable to the

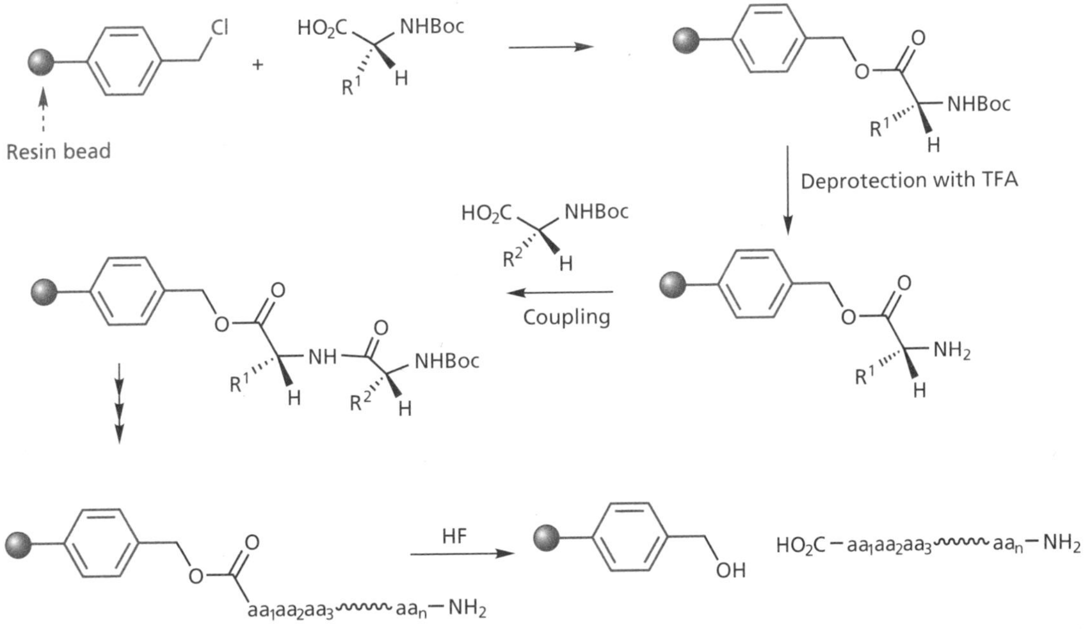

Figure 14.2 Peptide synthesis (Boc = *tert*-butyloxycarbonyl = *t*-BuO-CO).

reaction conditions used in peptide synthesis, but can be cleaved at the end of the synthesis using vigorous acidic conditions (hydrofluoric acid).

One disadvantage of polystyrene beads is the fact that they are hydrophobic and the growing peptide chain is hydrophilic. As a result, the growing peptide chain is not solvated and often folds in on itself, forming internal hydrogen bonds. This in turn hinders access of further amino acids to the exposed end of the growing chain. To address this, more polar solid phases were developed such as **Sheppard's polyamide resin**. Other resins have been developed to be more suitable for the combinatorial synthesis of non-peptides. For example, **Tentagel resin** is 80% poly-ethylene glycol grafted to cross-linked polystyrene and provides an environment similar to ether or tetra-hydrofuran. Regardless of the polymer that is used, the bead should be capable of swelling in solvent, yet remain stable. Swelling is important because most of the reactions involved in solid phase synthesis take place in the interior of the bead rather than on the surface. It is wrong to think of resin beads as being like miniature marbles with an impenetrable surface. Each bead is a polymer and swelling involves unfolding of the polymer chains such that solvent and reagents can

move between the chains into the heart of the polymer (Fig. 14.3).

Although beads are the common shape for the solid support, a range of other shapes such as pins have been designed to maximize the surface area available for reaction and hence maximize the amount of compound linked to the solid support. Functionalized glass surfaces have also been used and are suitable for oligo-nucleotide synthesis.

14.2.2 The anchor/linker

The anchor/linker is a molecular unit covalently attached to the polymer chain making up the solid support. It contains a reactive functional group with which the starting material in the proposed synthesis can react and hence become attached to the resin. The resulting link must be stable to the reaction conditions used throughout the synthesis, but be easily cleaved to release the final compound once the synthesis is complete (Fig. 14.4). Since the linkers are distributed along the length of the polymer chain, most of them will be in the interior of the polymer bead, emphasizing the importance of the bead swelling if the starting material is to reach them.

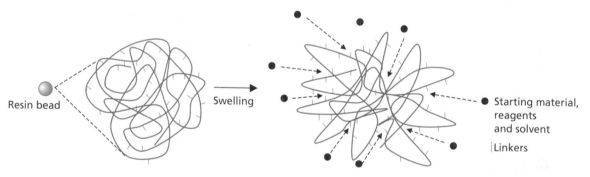

Figure 14.3 Swelling of a resin bead.

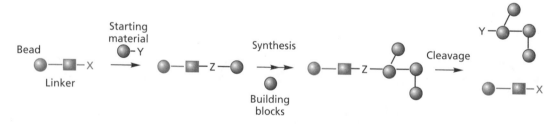

Figure 14.4 Anchor/linker. X, Y, Z are functional groups.

Different linkers are used depending on

- the functional group which will be present on the starting material
- the functional group which is desired on the final product once it is released.

Resins having different linkers are given different names. For example, the **Wang resin** has a linker which is suitable for the attachment and release of carboxylic acids, whereas the **Rink resin** is suitable for the attachment of carboxylic acids and the release of carboxamides (the linkage point is circled in Fig. 14.5). The dihydropyran-derivatized resin is suitable for the attachment and release of alcohols.

The Wang resin can be used in peptide synthesis whereby an *N*-protected amino acid is linked to the resin by means of an ester link. This ester link remains stable to coupling and deprotection steps in the peptide synthesis and can then be cleaved using trifluoroacetic acid (TFA) to release the final peptide from the bead (Fig. 14.6).

Starting materials with a carboxylic acid (RCO_2H) can be linked to the Rink resin via an amide link. Once the reaction sequence is complete, treatment with TFA releases the product with a primary amide group rather than the original carboxylic acid ($\mathbf{R'CONH_2}$; Fig. 14.7).

Primary and secondary alcohols (ROH) can be linked to a dihydropyran-functionalized resin. Linking the alcohol is done in the presence of pyridinium 4-toluenesulfonate (PPts) in dichloromethane. Once the reaction sequence has been completed, cleavage can be carried out using TFA (Fig. 14.8).

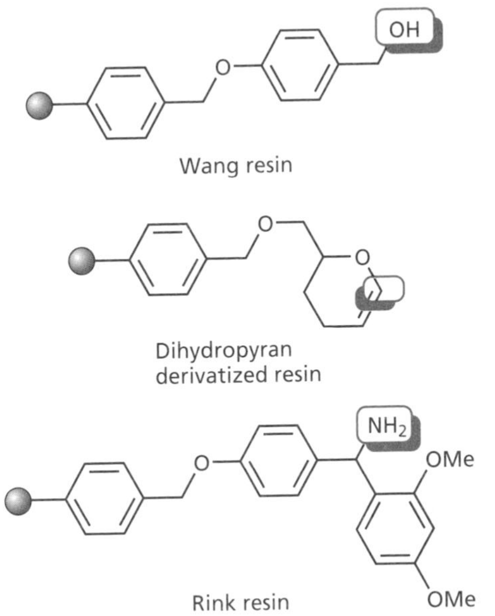

Figure 14.5 Types of resin.

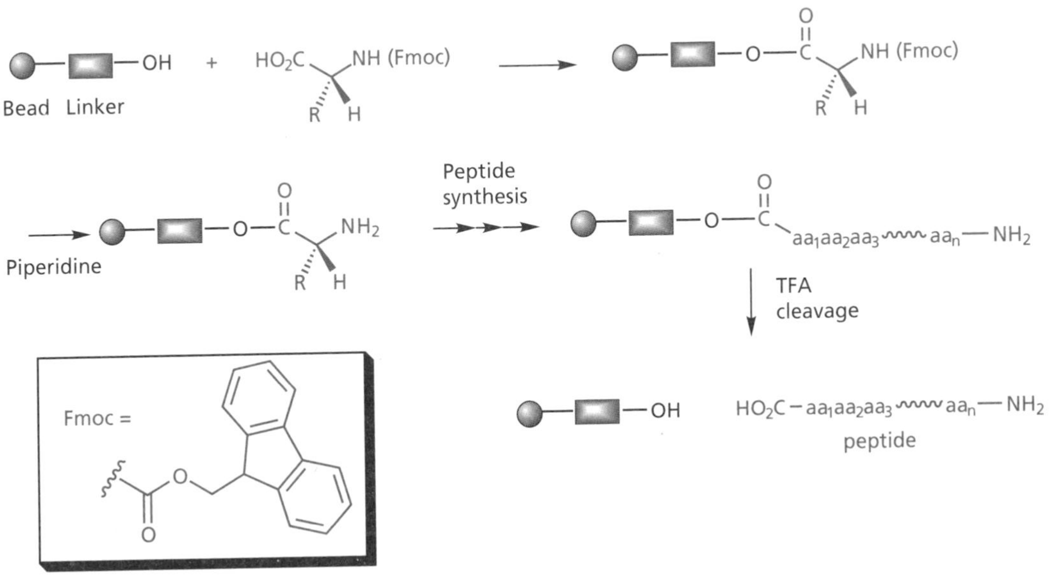

Figure 14.6 Peptide synthesis with Wang resin.

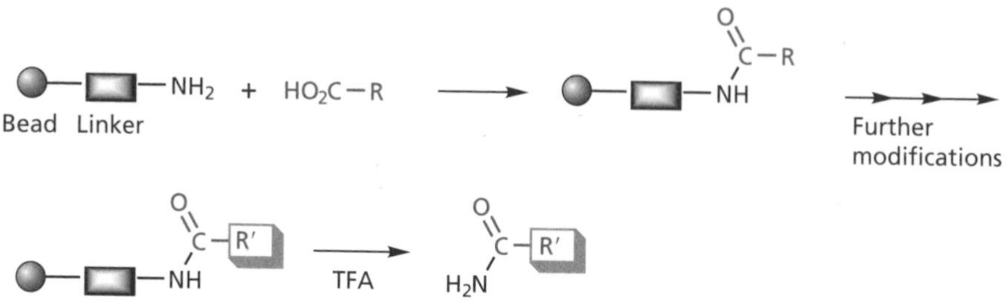

Figure 14.7 Combinatorial synthesis with a Rink resin.

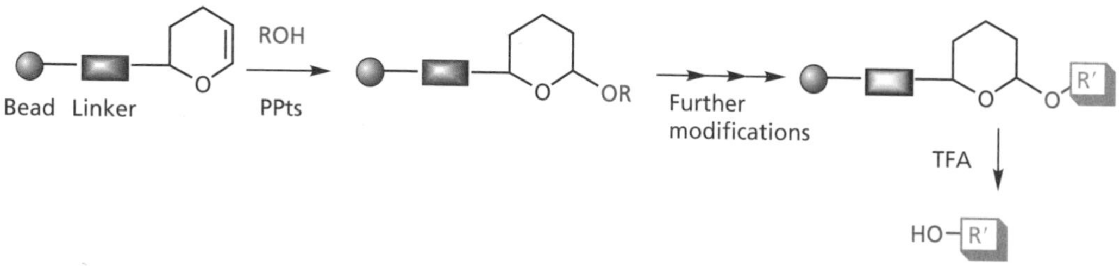

Figure 14.8 Combinatorial synthesis with a dihydropyran-functionalized resin.

14.2.3 Protecting groups and synthetic strategy

When a molecule is being constructed by solid phase synthesis, it is important to protect important functional groups which are not involved in the synthetic route. The selection of suitable protecting groups is extremely important. They should be stable to the reaction conditions involved in the synthesis, but be capable of being removed in high yield under mild conditions once the synthesis is complete. As far as peptide synthesis is concerned, two main protecting group strategies are used.

14.2.3.1 Boc /benzyl protection strategy

This strategy is suitable for the Merrifield resin as illustrated in Fig. 14.2. The *N*-terminus of each amino acid used in the synthesis is protected by a *tert*- or *t*-butyloxycarbonyl (Boc) group. Once each amino acid has been added to the growing peptide chain, its Boc group is removed with TFA to free up the amino group and the next protected amino acid can be coupled on to the chain. The bond which connects the growing peptide chain to the linker is stable to TFA and remains unaffected by the synthesis. However, this bond is susceptible to strong acid and once the synthesis is complete, hydrofluoric acid is used to release the

peptide. Functional groups on the amino acid residues also have to be protected during the synthesis, which means that the protecting groups have to be stable to TFA. Benzyl-type protecting groups fit the bill. They are stable to TFA but susceptible to hydrofluoric acid, so treatment with hydrofluoric acid releases the final peptide and deprotects the residues at the same time. One major disadvantage with this procedure is the need to use hydrofluoric acid. This is a particularly nasty chemical which dissolves glass, so expensive Teflon equipment has to be used. The harsh conditions can also result in peptide decomposition in some cases. Finally, there is a serious health risk and it is important to ensure that no hydrofluoric acid gets on the skin. It is reputed that a garage worker once decided to take advantage of the glass-dissolving properties of hydrofluoric acid, by soaking a rag with it and removing scratches from a car windscreen. Unfortunately, his hands had to be surgically removed as well.

14.2.3.2 Fmoc / t-Bu strategy

The alternative Fmoc/*t*-Bu strategy involves milder conditions and is partly illustrated in Fig. 14.5 with a Wang resin. The 9-fluorenylmethoxycarbonyl (Fmoc) group is an alternative protecting group for the terminal amino group and can be removed using a mild base such as piperidine. Functional groups on amino acid residues can be protected with a *t*-butyl group which can be removed by TFA. Since totally different reaction conditions are involved in removing the Fmoc group and the *t*-butyl group, one base and one acid, the protecting group strategy is defined as being orthogonal. The link to the Wang resin is also susceptible to TFA and avoids the need to use hydrofluoric acid. Thus, the peptide can be cleaved from the resin, and at the same time the functional groups on the residues are deprotected.

KEY POINTS

- Combinatorial chemistry normally involves the automated solid phase synthesis of large numbers of compounds in a much shorter period of time than conventional synthesis.

- Combinatorial synthesis has proved valuable in lead discovery, structure–activity relationships, and drug optimization.

- Parallel combinatorial synthesis involves the synthesis of a different compound in each reaction vial and is useful in all aspects of medicinal chemistry where synthesis is required.

- Mixed combinatorial synthesis involves the synthesis of mixtures of compounds in each reaction vial and is useful in discovering lead compounds.

- Solid phase synthesis has several advantages. Intermediates do not need to be isolated or purified. Reactants and reagents can be used in excess. Impurities and excess reagents or reactants are easily removed. Different compounds are linked to different solid phase surfaces such that they are physically separated, allowing them to undergo reactions and work-up procedures in the same reaction vessel.

- The solid support consists of a polymeric surface and a linker molecule which allows a starting material to be covalently linked to the support.

- Different linkers are used depending on the functional group which is available on the starting material and the functional group which is desired on the product.

14.3 Methods of parallel synthesis

Having looked at the basics of solid phase synthesis, we shall now look at some methods of combinatorial synthesis. The simplest examples involve parallel syntheses where a single reaction product is produced in each reaction vessel.

14.3.1 Houghton's teabag procedure

The teabag procedure is a manual approach to parallel synthesis and has been used for the parallel synthesis of more than 150 peptides at a time. The polymeric support resin (typically 100 mg) is sealed in polypropylene meshed containers (3 × 4 cm) (known as teabags) and each teabag is labelled (Fig. 14.9).

The teabags are then placed in polyethylene bottles which act as the reaction vessels. In the case of a peptide synthesis, the first amino acid is added to the resin—a different amino acid to each bottle used. All the teabags in one specific bottle now have the same amino acid linked to the resin. The teabags from every bottle are now combined in one vessel for deprotection and washing. This allows all the amino acids to be deprotected at one time, avoiding the need to carry out deprotection separately on each amino acid. The teabags can then be redistributed between the bottles for the addition of a second amino acid, recombined for deprotection and washing, redistributed for addition of the next amino acid, and so on. The communal deprotection and washing procedure greatly speeds up the synthetic process. The advantage of this approach is that it is cheap and can be carried out in a laboratory without the need for expensive equipment. The major problem is the fact that it is manual, and this limits the quantity and speed with which new structures can be synthesized. Thus, pharmaceutical industries now use automation or semi-automation for the parallel synthesis of structures.

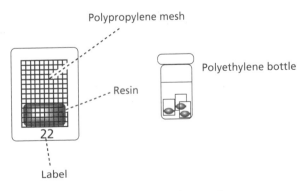

Figure 14.9 Houghton's teabag procedure.

14.3.2 Automated parallel synthesis

Automated or semi-automated synthesizers can cope with the parallel synthesis of 6, 12, 42, 96, or 144 structures depending on the instrument and the size of the reaction tubes used. The addition of solvent, starting materials, and reagents can be carried out automatically using syringes. Automated work-up procedures such as the removal of solvent, washing, and liquid–liquid separations are also possible. Reactions can be stirred and carried out under inert atmospheres, and the reactions can be heated or cooled as required. A good description of automated and semi-automated synthesizers is given in the textbook by Beck-Sickenger and Weber (see Further reading).

14.4 Methods in mixed combinatorial synthesis

14.4.1 General principles

Combinatorial synthesis is often designed to produce a mixture of products in each reaction vessel, starting with a wide range of starting materials and reagents. This does not mean that all possible starting materials are thrown together in the one reaction flask. Planning has to go into designing a combinatorial synthesis to minimize the effort involved and to maximize the number of different structures obtained.

As an example, suppose we wish to synthesize all the possible dipeptides of five different amino acids. Using orthodox chemistry, we would synthesize these one at a time. There are 25 possible dipeptides and so we would have to carry out 25 separate experiments (Fig. 14.10).

Using combinatorial synthesis, the same products could be obtained with far less effort. If all five different amino acids are separately bound to resin beads, the beads can then be mixed together and treated with a second amino acid to produce all possible dipeptides in five experiments. For example, in one experiment the five different amino acids could be combined with glycine to produce 5 of the 25 possible dipeptides (Fig. 14.11). This mixture could then be tested for activity. If the results were positive, the emphasis would be on identifying which of the dipeptides was active. If there was no activity present, then the mixture could be ignored and stored.

In studies such as these, one can generate large numbers of mixtures, many of which are inactive. However, these mixtures are not discarded. Although they may not contain a lead compound on this particular occasion, they may provide the necessary lead compound for a different target in medicinal chemistry. Therefore, all the mixtures (both active and inactive) resulting from a combinatorial synthesis are stored and are referred to as **combinatorial or compound libraries.**

The example above produced 25 compounds in 5 mixtures. However, combinatorial synthesis can

Gly	25 separate	Gly-Gly	Ala-Gly	Phe-Gly	Val-Gly	Ser-Gly
Ala	experiments	Gly-Ala	Ala-Ala	Phe-Ala	Val-Ala	Ser-Ala
Phe	⟶	Gly-Phe	Ala-Phe	Phe-Phe	Val-Phe	Ser-Phe
Val		Gly-Val	Ala-Val	Phe-Val	Val-Val	Ser-Val
Ser		Gly-Ser	Ala-Ser	Phe-Ser	Val-Ser	Ser-Ser

Figure 14.10 Traditional synthesis of dipeptides.

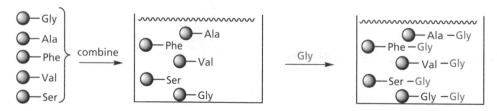

Figure 14.11 Synthesis of five different dipeptides.

be used to produce different structures numbering in the order of thousands or millions. Since the quantities involved are extremely small (perhaps only 5 beads containing 1 individual structure), even huge numbers like these can be stored and used for later studies. Because of the vast number and the small quantities of compounds present in these mixtures, the exact structures of each component in a mixture is not known, although the chemist would have a general idea of the type of structure likely to be present based on the type of synthesis carried out and reagents used. The library acts as a source of potential new leads in the same way as a plant or herb would.

14.4.2 The mix and split method

When generating large quantities of different structures, it is important to minimize the effort involved, and the mix and split method is a popular way of doing this. An example best illustrates the principle. Let us assume that we want to make all the possible tripeptides of three different amino acids (e.g. glycine, valine, alanine). The mix and split method would work like this:

- **Stage 1:** Link each amino acid to a solid support (Fig. 14.12). (To clarify the diagrams, each sphere represents a resin bead with suitable linker unit.)

- **Stage 2:** Mix the beads together and separate into three equal portions (Fig. 14.13).

- **Stage 3:** React each portion with a different amino acid (Fig. 14.14).

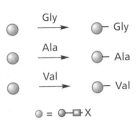

Figure 14.12

All nine possible dipeptides have now been synthesized in three separate experiments. Samples of each portion could be retained for **recursive deconvolution** (see section 14.6.2).

- **Stage 4:** Isolate all the beads, mix them together and split into three equal portions. Each portion will now have all nine possible dipeptides (Fig. 14.15).

- **Stage 5:** React each portion with one of the three amino acids (Fig. 14.16). All 27 possible tripeptides have now been synthesized in another 3 experiments.

In this example, amino acids were linked together, but any monomer unit or combination of chemical

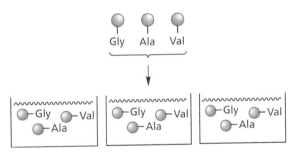

Figure 14.13

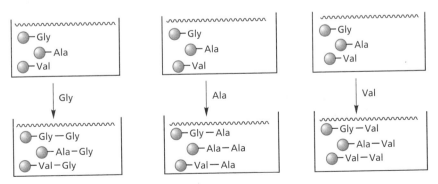

Figure 14.14

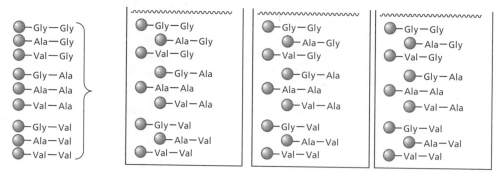

Figure 14.15

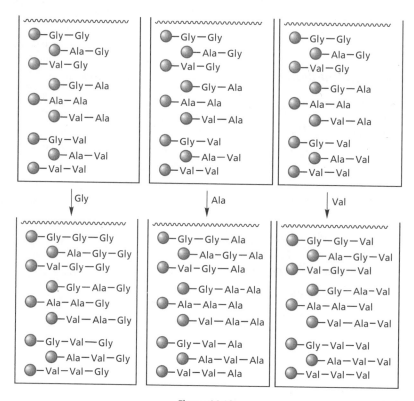

Figure 14.16

structures could be linked together using the same strategy, and so the process is not limited to peptide synthesis.

14.4.3 Mix and split in the production of positional scanning libraries

A variation of the mix and split method allows the creation of positional scanning libraries. In this method, the same library compounds are prepared several times and in each library a different residue in the sequence is held constant. For example a series of hexapeptide libraries totalling 34 million compounds was produced, each library consisting of 6 sets of mixtures, and with each mixture containing 1 889 568 peptides. In each of the mixtures one of the amino acid positions was held constant.

The first library consisted of six mixtures where the first amino acid was constant *within* each individual mixture but was different *between* mixtures. The most active mixture was then identified and since the amino acid at position 1 was constant in that mixture, it could be identified. The second library consisted of a series of six mixtures where the second amino acid was constant within each mixture but different between the mixtures. Thus it was known which amino acid was present at position 2 in the most active of these mixtures. Testing all the mixtures in all the libraries revealed the preferred residues at each of the six residue positions.

However, one has to be careful here. Although the most active amino acid at each position can be identified, it does not mean that the most active structure is the one linking each of these amino acids. After all, the activity being measured is the combined effect of several different hexapeptide structures in the active mixtures. Therefore, it is quite possible that the most active hexapeptide is in a mixture which has a lower overall activity than another mixture, since the latter might contain a large number of compounds with moderate activity.

One way around this is to identify the top three to four amino acids at each position, rather than just the most active one. Once these have been identified, all the possible variations of these amino acids could be synthesized. For example, several libraries of a hexapeptide were prepared in a search for structures which would bind to the μ opiate receptor. The most active amino acids at each position were found to be tyrosine (Y), glycine (G), phenylalanine (F), phenylalanine (F), leucine (L), and arginine (R). Linking these together gave a hexapeptide (YGFFLR) which was only weakly active. However, the results showed that mixtures containing glycine or phenylalanine at position 3; phenylalanine, tyrosine, methionine (M), or leucine at position 5; and phenylalanine, tyrosine, or arginine at position 6 were also active.

Therefore, all 24 hexapeptides having the sequence YG(G/F)(F)(F/M/L)(F/Y/R) were made, and the most active was found to be YGGFMY. It is interesting to note that the endogenous analgesic Met-enkephalin is the pentapeptide YGGFM (section 21.6).

14.5 Isolating the active component in a mixture: deconvolution

Once a compound mixture proves to be biologically active, the tricky job of identifying the active component (or components) now needs to be carried out. Isolating and identifying the most active compound in a mixture is known as **deconvoluting** the mixture. There are several methods of doing this.

14.5.1 Micromanipulation

Each bead in a mixture contains only one type of structural product. Therefore, the individual beads can be separated and the product cleaved and then tested. This procedure can be aided by colorimetric analysis, where products are tested for activity when they are still bound to the beads. The active beads would be distinguished by a colour reaction and can then be picked out by micromanipulation.

14.5.2 Recursive deconvolution

Micromanipulation is tedious, and has serious drawbacks when large quantities of beads have to be handled. A method known as recursive deconvolution can be useful in cutting down the amount of work involved and can be illustrated by considering the library of tripeptides described in section 14.4.2. Here we synthesized three mixtures. Let us assume that one of these three mixtures shows activity. How do we find out which of the nine possible tripeptides is the active component? We could synthesize all nine possible tripeptides separately and test each one. However, we could cut down the work if samples of the dimer mixtures that were produced during the combinatorial synthesis had been retained (Fig. 14.14).

Let us assume that the third tripeptide mixture in section 14.4.2 showed activity (Fig. 14.16). This means that the active tripeptide has valine at the N-terminus. The next stage is to take the three dipeptide mixtures which were retained and link valine to each mixture (Fig. 14.17). This gives us the nine tripeptides we need in three separate mixtures. Now the second and third amino acids are the same in each mixture. The three mixtures can be tested. If one of these mixtures is active, we can identify the second and third amino acid. Let us assume that the mixture containing alanine and valine is the active mixture. The three component tripeptides in this mixture can now be individually synthesized and tested.

In this example, we looked at tripeptides involving three different amino acids, but typical combinatorial syntheses involve a larger variety of monomeric

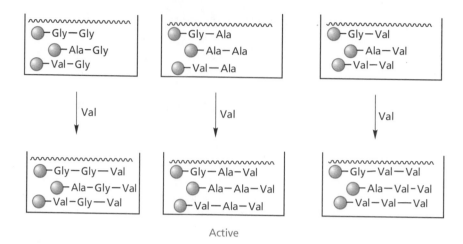

Figure 14.17 Recursive deconvolution.

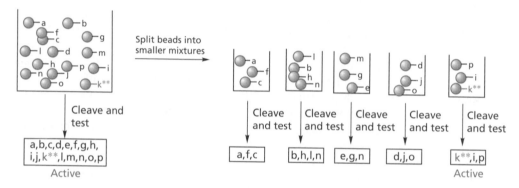

Figure 14.18 Sequential release.

units, and the larger the number of monomers the larger the number of compounds produced. For example, a series of 34 million hexapeptides was synthesized by this method using 18 amino acids and 324 mixtures.

Appropriate use of the mix and split method, and the retention of intermediate mixtures for recursive deconvolution is crucial in economizing the effort involved in identifying active components.

14.5.3 Sequential release

Linkers have been devised which allow release of a certain percentage of the product from the bead. The process can be repeated, releasing the product sequentially rather than all at once. Therefore, a mixture of beads can be treated to release some of the bound product for testing. If the mixture is active, the same beads are split into smaller mixtures and further product is released and tested (Fig. 14.18). This process can be repeated several times until the active bead is identified (Fig. 14.19).

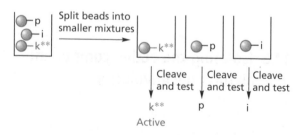

Figure 14.19 Identification of active bead.

BOX 14.1 DOUBLE CLEAVABLE LINKER

With a double cleavable linker, it is possible to release products in two stages. This makes it possible to release a set of compounds for biological testing, then to release the second set of compounds for structural analysis. An example of a double cleavable linker for peptides is shown below. The first cleavage is initiated by addition of neutral buffer, the second by addition of a base.

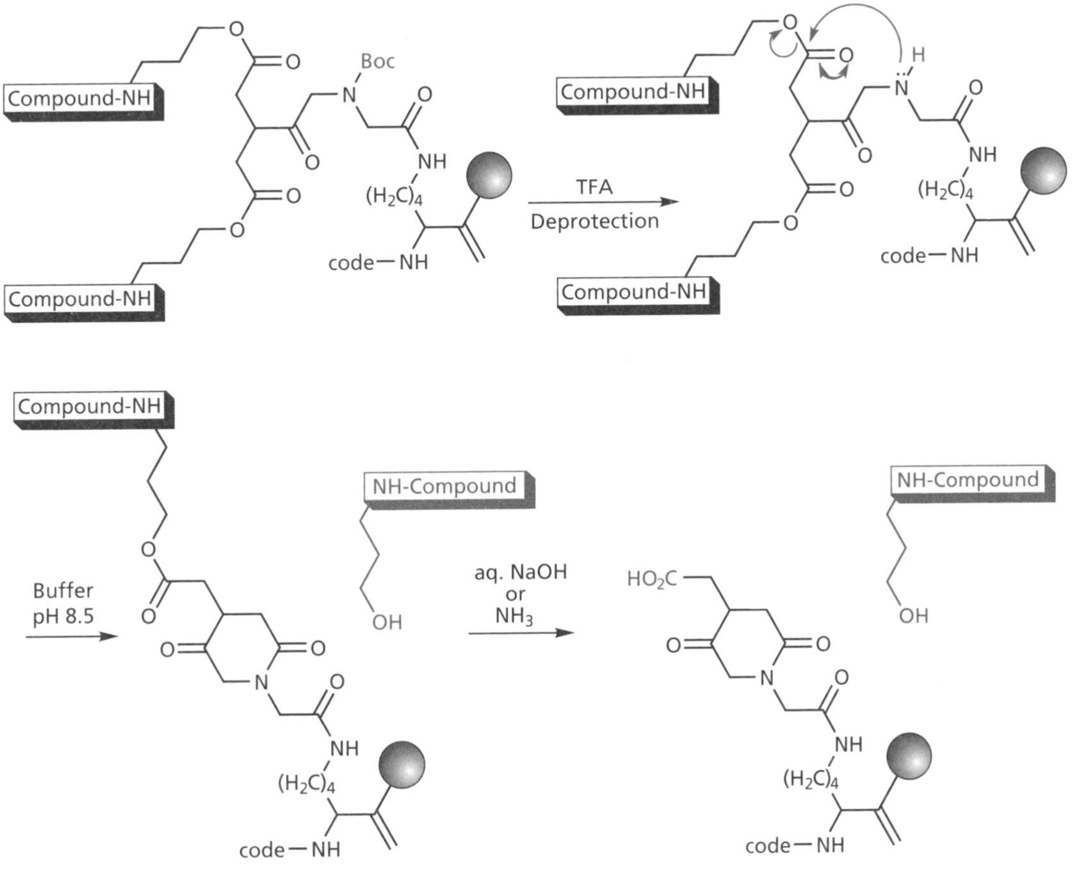

Double cleavable linker.

KEY POINTS

- Houghton's teabag procedure is a manual approach to combinatorial synthesis.

- Most combinatorial syntheses are carried out using automated or semi-automated synthesizers.

- The mix and split method allows the efficient synthesis of large numbers of compounds with a minimum number of operations.

- The compounds synthesized in a combinatorial synthesis are stored as combinatorial libraries.

- Positional scanning libraries involve the synthesis of sets of libraries where a component is kept constant in each subset of each library. This allows the identification of components which are good for activity.

- Deconvolution is the process by which the active component in a combinatorial synthetic can be identified.

- Recursive deconvolution involves the use of product mixtures obtained at different stages of a mix and split combinatorial synthesis in order to narrow down the likely structure of the active component in a final mixture.

- Linkers are available which permit the sequential release of a product bound to a particular resin.

14.6 Structure determination of the active compound(s)

The direct structural determination of components in a compound mixture is no easy task, but advances have been made in obtaining interpretable mass, NMR, Raman, infrared and ultraviolet spectra on products attached to a single resin bead. Peptides can be sequenced while still attached to the bead. Each 100-μm bead contains about 100 pmole of peptide, which is enough for microsequencing. With non-peptides, the structural determination of an active compound can be achieved by a systematic iterative resynthesis as described in section 14.5, although this can be tedious. Alternatively, **tagging** procedures can be used during the synthesis.

14.6.1 Tagging

In this process, two molecules are built up on the same bead. One of these is the intended structure; the other is a molecular tag (usually a peptide or oligonucleotide) which will act as a code for each step of the synthesis.

For this to work, the bead must have a multiple linker capable of linking both the target structure and the molecular tag. A starting material is added to one part of the linker, and an encoding amino acid or nucleotide to another part. After each subsequent stage of the combinatorial synthesis, an amino acid or nucleotide is added to the growing tag to indicate what reagent was used. One example of a multiple linker is called the **Safety CAtch Linker** (**SCAL**) (Fig. 14.20), which includes lysine and tryptophan. Both these amino acids have a free amino group.

The target structure is constructed on the amino group of the tryptophan moiety and after each stage of the synthesis a tagging amino acid is built on to the amino groups of the lysine moiety. Figure 14.21 illustrates the procedure for a synthesis involving three reagents, so that by the end of the process there is a tripeptide tag where each amino acid defines the identity of the variable groups R, R', and R″ in the target structure.

The non-peptide target structure can be cleaved by reducing the two sulfoxide groups in the safety catch linker, then treating with acid. Under these conditions, the tripeptide sequence remains attached to the bead and can be sequenced on the bead to identify the structure of the compound which was released. The same strategy can be used with an oligonucleotide as the tagging molecule instead. The oligonucleotide can be amplified by replication and the code read by DNA sequencing.

There are drawbacks to tagging processes; they are time consuming and require elaborate instrumentation. Building the coding structure itself also adds

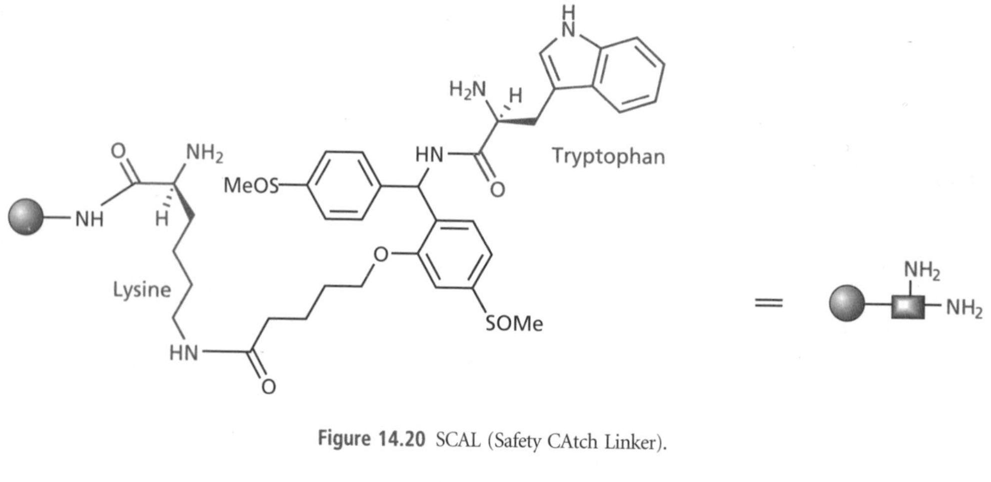

Figure 14.20 SCAL (Safety CAtch Linker).

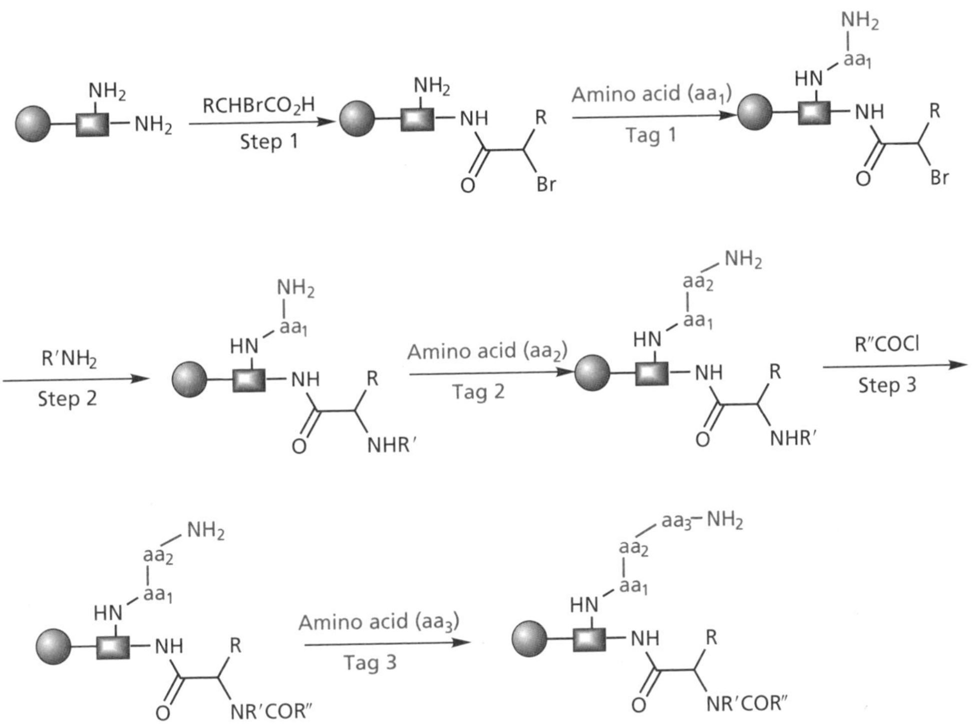

Figure 14.21 Tagging.

extra restraints on the protection strategies that can be employed, and may impose limitations on the reactions that can be used. In the case of oligonucleotides, their inherent instability can prove a problem. Another possible problem with tagging is the possibility of an unexpected reaction taking place, resulting in a different structure from that expected. Nevertheless, the tagging procedure is still valid since it identifies the starting materials and the reaction conditions, and when these are repeated on larger scale any unusual reactions would be discovered.

These tagging methods require the use of a specific molecular tag to represent each reagent used in the synthesis. Moreover, the resultant molecular tag has to be sequenced at the end of the synthesis. A more efficient method of tagging and identifying the final product is to use some form of encryption or 'bar code'. For example, it is possible to identify which one of seven possible reagents has been used in the first stage of a synthesis with the use of only three molecular labels (A–C). This is achieved by adding different combinations of the three tags to set up a triplet code on the bead. Thus, adding just one of the tags (A, B, or C) will allow the identification of three of the reagents.

Adding two of the tags at the same time allows the identification of another three reagents, and adding all three tags at the same time allows the identification of a seventh reagent. The presence (1) or absence (0) of the tag forms a triplet code: the presence of a single molecular tag (A, B, or C) gives the triplet codes 100, 010, and 001, the presence of two different tags is indicated by another three triplet codes (110, 101, 011), and the presence of all three tags is represented by 111. The tags are linked to the bead by means of a photocleavable bond, so irradiating the bead releases all the tags. These can then be passed through a gas chromatogram and identified by their retention time.

Three different molecular tags could now be used to represent seven reagents in the second stage, and so on. All the tags used to represent the second reagent would have longer retention times than the tags used to represent the first reagent. Similarly, all subsequent tags would have longer retention times. Once the synthesis is complete, all the tags are released simultaneously and passed through the gas chromatogram as before. The 'bar code' is then read from the chromatogram in one go, not only identifying the reagents used but the order in which they were used (Fig. 14.22).

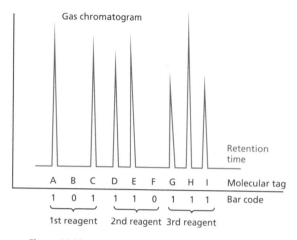

Figure 14.22 Identification of reagents and order of use by bar coding.

14.6.2 Photolithography

Photolithography is a technique which permits miniaturization and spatial resolution such that specific products are synthesized on a plate of immobilized solid support. In the synthesis of peptides, the solid support surface contains an amino group protected by the photolabile protecting group **nitroveratryloxy-carbonyl** (**NVOC**) (Fig. 14.23). Using a mask, part of the surface is exposed to light resulting in deprotection of the exposed region. The plate is then treated with a protected amino acid and the coupling reaction takes place only on the region of the plate which has been deprotected. The plate is then washed to remove excess amino acid. The process can be repeated on a different region using a different mask and so different peptide chains can be built on different parts of the plate; the sequences are known from the record of masks used.

Incubation of the plate with a protein receptor can then be carried out to detect active compounds which bind to the binding site of the receptor. A convenient method to assess such interactions is to incubate the plate with a fluorescently tagged receptor. Only those regions of the plate which contain active compounds will bind to the receptor and fluoresce. The fluorescence intensity can be measured using fluorescence microscopy and is a measure of the affinity of the compound for the receptor. Alternatively, testing can be carried out such that active compounds are detected by radioactivity or chemiluminescence.

The photodeprotection described above can be achieved in high resolution. At a 20-μm resolution,

plates can be prepared with 250 000 separate compounds per square centimetre.

KEY POINTS

- Tagging involves the construction of a tagging molecule on the same solid support as the target molecule. Tagging molecules are normally peptides or oligonucleotides.

- After each stage of the target synthesis, the peptide or oligonucleotide is extended and the amino acid or nucleotide used defines the reactant or reagent used in that stage.

- Photolithography is a technique involving a solid support surface containing functional groups protected by photo-labile groups. Masks are used to reveal defined areas of the plate to light, thus removing the protecting groups and allowing a reactant to be linked to the solid support. A record of the masks used determines what reactions have been carried out at different regions of the plate.

14.7 Limitations of combinatorial synthesis

The 20 natural amino acids could combine to form 10 240 billion possible decapeptides. In principle, combinatorial chemistry could be used to synthesize them all! There are limitations, however, since one has to consider the practical details of weight and volume. First, how many beads will be needed for a combinatorial synthesis? For statistical reasons, the number of beads should exceed the number of target molecules by a factor of 10. Otherwise, one might not sample all the possible structures present. For example, if there are only 5 beads representing each of the 3.2 million components of a pentapeptide library, and one fifth of the whole mixture is removed as a sample, then the probability of finding all the peptides present in that sample is only 63%.

Assuming that one uses the required excess of beads, the weight of beads required to make a complete library of dipeptides would be 8.4 mg. In order to make a complete library of tetrapeptides you would need 3.4 g, which is still practical. In order to make a complete library of decapeptides, however, you would need 215.3 tonnes!

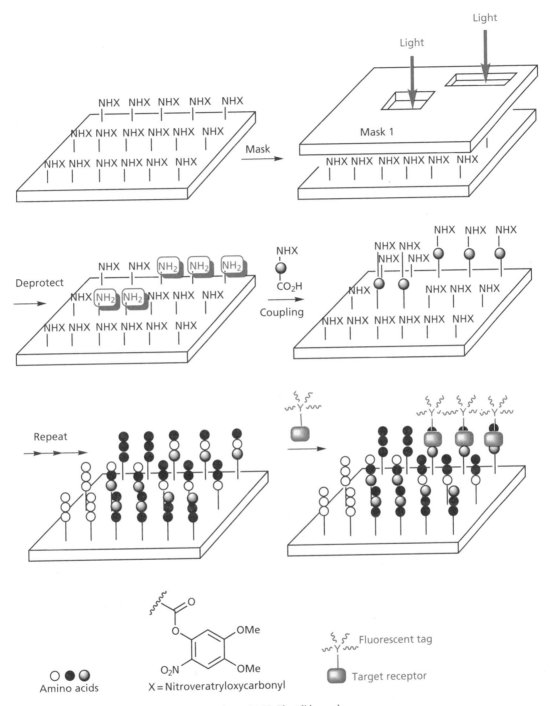

Figure 14.23 Photolithography.

14.8 Examples of combinatorial syntheses

Combinatorial chemistry has proved its worth in throwing up new lead compounds in a variety of fields. Much of the early work in combinatorial chemistry was carried out on peptides since the solid phase procedures had already been developed for that field. This resulted in the discovery of new HIV protease inhibitors, antimicrobial agents, opiate receptor ligands, and aspartic acid protease inhibitors. However peptides are not ideal drug candidates, as they usually have poor oral bioavailability (section 11.8.3).

The first move away from peptides was to use the same peptide coupling procedures but with non-natural amino acids. Peptides could also be modified once they were built, by reactions such as N-methylation. Peptides have also been built linking N-substituted glycine units to produce structures

known as **peptoids** where the side chain is attached to the nitrogen rather than the α-carbon. Some of these have been shown to be ligands for various important receptors and show increased metabolic stability.

A disadvantage with all the above structures is the fact that they are linear, flexible molecules linked together by a regular molecular backbone. The real interest in combinatorial chemistry began when it became possible to produce heterocyclic combinatorial libraries. Heterocycles are less susceptible to metabolism, and have better pharmacokinetic properties. They are more rigid and diversity is possible by varying the substituents around the heterocyclic 'core'. One of the earliest examples of a heterocyclic solid phase synthesis was that of 1,4-benzodiazepines (Fig. 14.24). This was a good synthesis, as three distinct units were involved which could be varied. The final product has five variable substituents, two of which can be positionally varied on the aromatic ring.

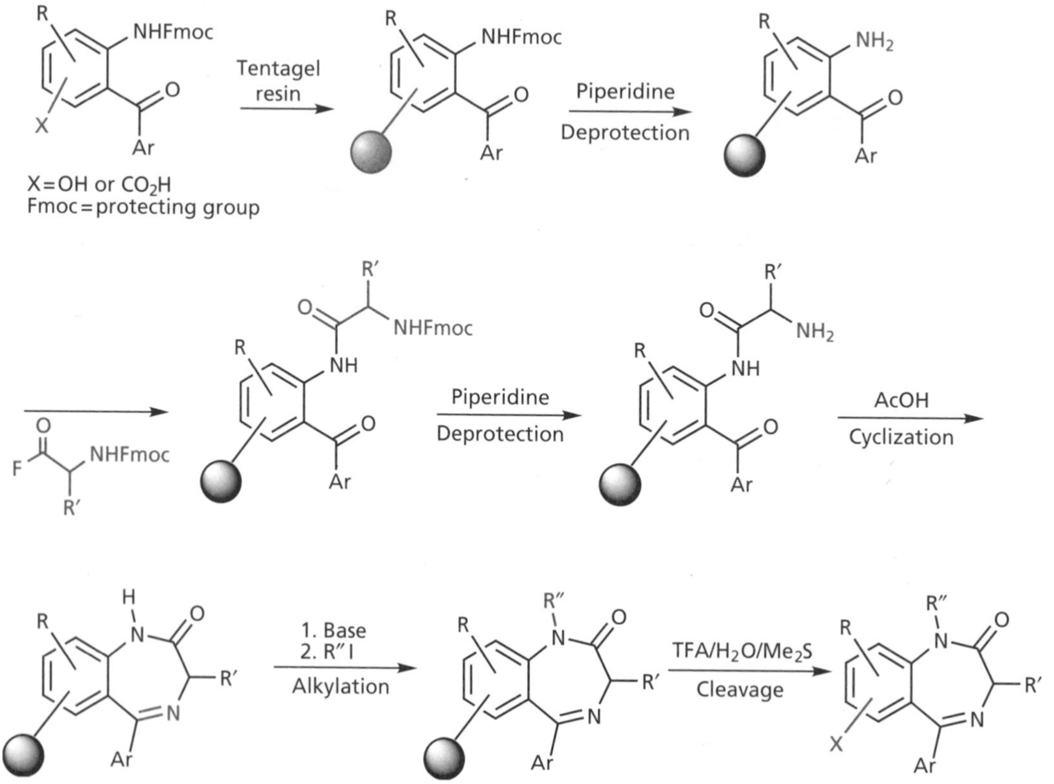

Figure 14.24 Synthesis of 1,4-benzodiazepines.

A limitation with this synthesis was the need to have a phenol or carboxylic acid group present to allow attachment to the solid support. This meant that the final product retained this functional group when released. Later, an alterative heterocyclic synthesis was devised which avoided this problem (Fig. 14.25). A selection of amino acids was linked to resin beads through the carboxylic acid group. Reaction with a variety of imines gave the adducts shown. Treatment

with TFA released the adducts, which then cyclized to give the final products. The advantage of this synthesis lies in the fact that the functional group released from the resin takes part in the final cyclization and does not remain as an extra, and possibly redundant, group.

A similar strategy was employed for the synthesis of hydantoins (Fig. 14.26) and a large variety of heterocyclic compounds have now been synthesized by combinatorial methods (Fig. 14.27).

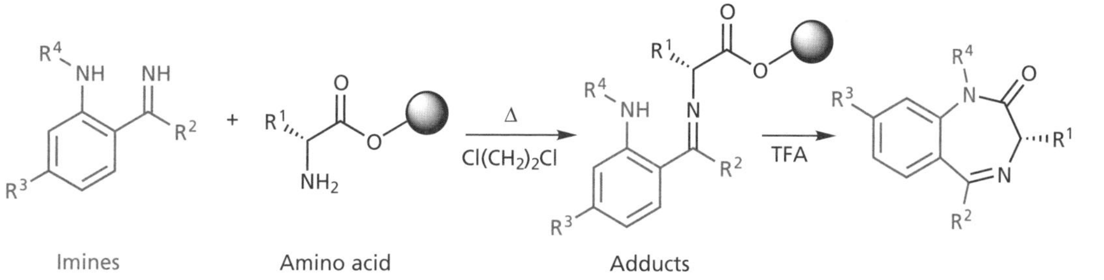

Figure 14.25 Alternative benzodiazepine synthesis.

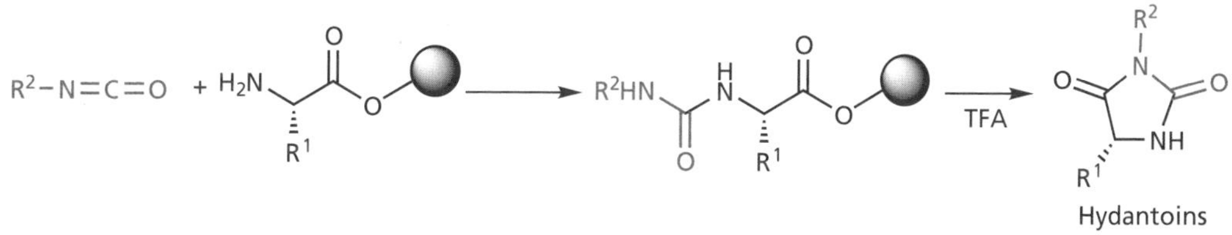

Figure 14.26 Synthesis of hydantoins.

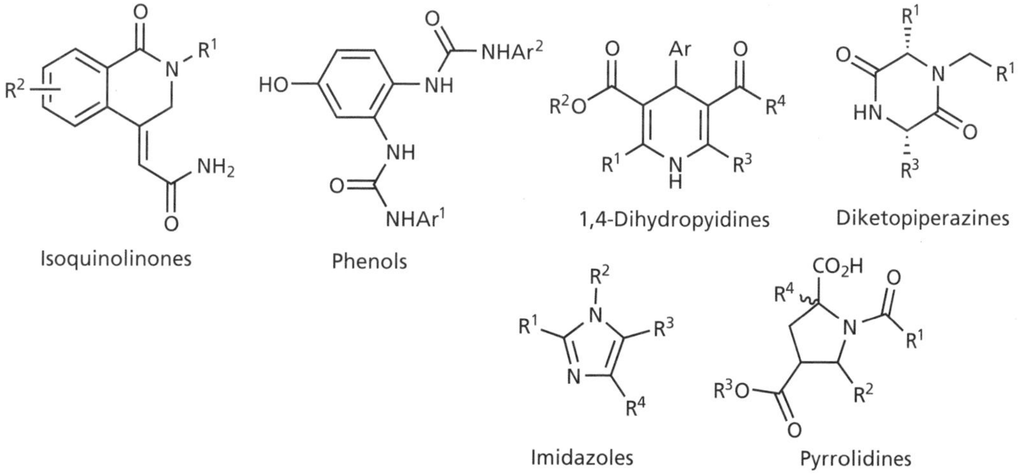

Figure 14.27 Examples of heterocycles prepared by combinatorial synthesis.

The range of reactions which can be carried out on solid phase has also been extended: most common reactions are now feasible, including moisture-sensitive and organometallic reactions such as aldol condensations, DIBAL reductions, Wittig reactions, LDA reductions, Heck couplings, Stille couplings, and Mitsunobu reactions.

With heterocycles, it is important to spend some time optimizing the reaction conditions before launching into a full-scale library synthesis. Otherwise, the reactions might proceed in a totally different way from expected or might not take place at all. As there is no easy way of analysing what is happening on the beads, this could lead to millions of useless compounds being synthesized before the problem is identified. Therefore, it is a good idea to carry out some model syntheses using a couple of 'worst case' scenarios—structures that might not be expected to react well because of steric or electronic factors. These model studies can be used to optimize conditions or to avoid using monomers which are particularly unreactive. When the library synthesis is being carried out, a parallel synthesis of a single compound by solid phase should be carried out as a further check.

14.9 Dynamic combinatorial chemistry

Dynamic combinatorial chemistry is an exciting development which has been used in the search for new lead compounds as an alternative to the classic mix and split combinatorial syntheses. In the latter process, stable products are built up stepwise following a particular synthetic route and using a variety of building blocks for each synthetic step involved. The different building blocks are added to different reaction flasks as described earlier. Once all the products have been prepared, they are screened to find the most active compound.

The aim of dynamic combinatorial chemistry is to synthesize all the different compounds in one flask at the same time, screen them *in situ* as they are being formed, and thus identify the most active compound in a much shorter period of time (Box 14.2). How can

this be achieved? There are several important principles which are followed.

- The best way of screening the compounds is to have the desired target present in the reaction flask along with the building blocks. This means that any active compounds can bind to the target as soon as they are formed. The trick is then to identify which of the products are binding.

- The reactions involved should be reversible. If this is the case, a huge variety of products are constantly being formed in the flask then breaking back down into their constituent building blocks. The advantage of this may not seem obvious, but it allows the possibility of 'amplification' where the active compound is present to a greater extent than the other possible products. By having the target present, active compounds become bound and are effectively removed from the equilibrium mixture. The equilibrium is now disturbed such that more of the active product is formed. Thus the target serves not only to screen for active compounds but to amplify them as well.

- In order to identify the active compounds, it is necessary to 'freeze' the equilibrium reaction such that it no longer takes place. This can be done by carrying out a further reaction which converts all the equilibrium products into stable compounds that cannot revert back to starting materials.

A simple example of dynamic combinatorial synthesis involved the reversible formation of imines from aldehydes and primary amines (Fig. 14.28). A total of three aldehydes and four amines were used in the study (Fig. 14.29), allowing the possibility of 12 different imines in the equilibrium mixture. The building blocks were mixed together with the target enzyme **carbonic anhydrase** and allowed to interact. After a suitable period of time, sodium cyanoborohydride was

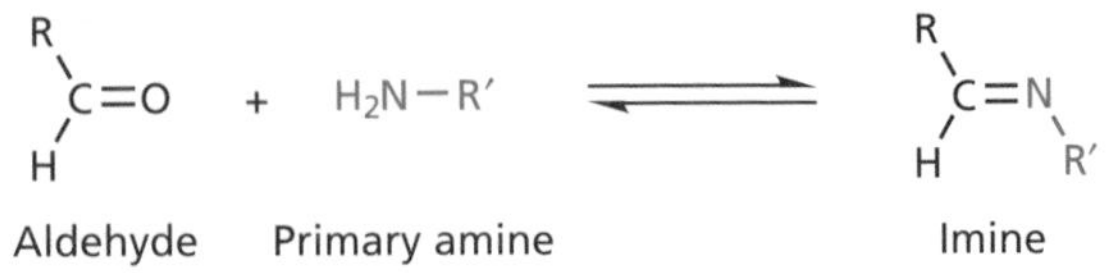

Figure 14.28 Imine formation.

added to reduce all the imines present to secondary amines so that they could be identified (Fig. 14.30). The mixture was separated by reverse-phase HPLC, allowing each product to be quantified and identified. These results were compared with those obtained when the experiment was carried out in the absence of carbonic anhydrase, making it possible to identify which products had been amplified. In this experiment, the sulfonamide shown in Fig. 14.31 was significantly amplified, which demonstrated that the corresponding imine was an active compound.

The above example illustrates a simple case involving one reaction and two sets of building blocks, but it is feasible to have more complex situations. For example, a molecule with two or more functional groups could be present to act as a scaffold on to which various substituents could be added from the building blocks available (Fig. 14.32). The use of a central scaffold has another benefit: it helps the amplification process. If the number of scaffold molecules present is equal to the number of target molecules, then the number of products formed cannot be greater than the number of targets available. If any of these products binds to the target, the effect on the equilibrium will be greater than if there were far more products than targets available.

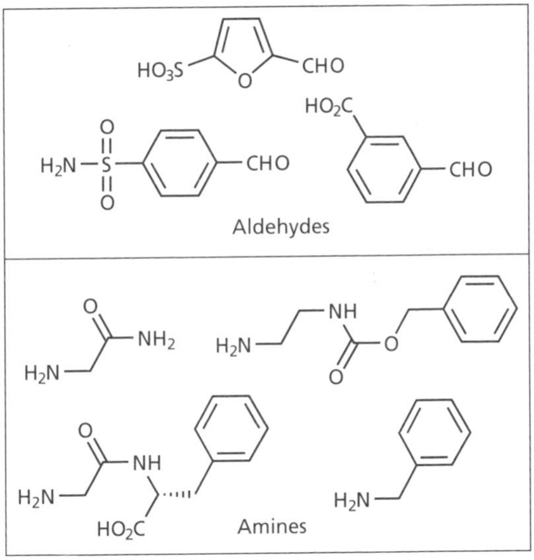

Figure 14.29 Aldehyde and amine building blocks.

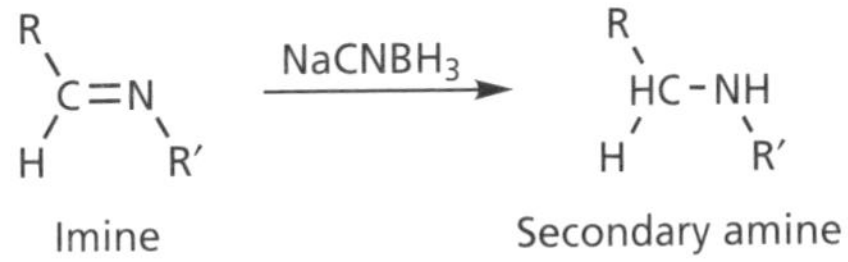

Figure 14.30 Reduction of imines.

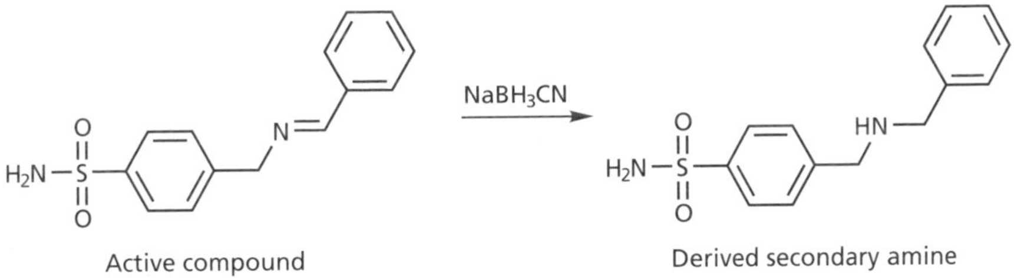

Figure 14.31 Amplified product.

Figure 14.32 Use of a scaffold molecule.

BOX 14.2 DYNAMIC COMBINATORIAL SYNTHESIS OF VANCOMYCIN DIMERS

Vancomycin is an antibiotic that works because it masks the building blocks required for bacterial cell wall synthesis (section 16.5.5.2). Binding takes place specifically between the antibiotic and a peptide sequence (L-Lys-D-Ala-D-Ala) which is present in the building block. It is also known that this binding promotes dimerization of the vancomycin–target complex, which suggests that covalently linked vancomycin dimers might be more effective antibacterial agents than vancomycin itself. A dynamic combinatorial synthesis was carried out to synthesize a variety of different vancomycin dimers covalently linked by bridges of different lengths. The vancomycin monomers used had been modified such that they contained long-chain alkyl substituents with double bonds at the end. Reaction between the double bonds in the presence of a catalyst then lead to bridge formation through a reaction known as olefin metathesis (Fig. 1).

The tripeptide target was present to accelerate the rate of bridge formation and to promote formation of vancomycin dimers having the ideal length of bridge. As shown in Fig. 2, the vancomycin monomers bind the tripeptide which encourages the self-assembly of non-covalently linked dimers. Once formed, those dimers having the correct length of substituent are more likely to react together to form the covalent bridge.

Having established the optimum length of bridge, another experiment was carried out on eight vancomycin monomers which had the correct length of 'tether' but varied slightly in their structure. The mixture of 36 possible products was analysed by mass spectrometry to indicate the relative proportion of each dimer formed. Eleven of the 36 compounds were then synthesized separately and it was found that their antibacterial activity matched their level of amplification, i.e. the compounds present in greater quantities had the greater activity.

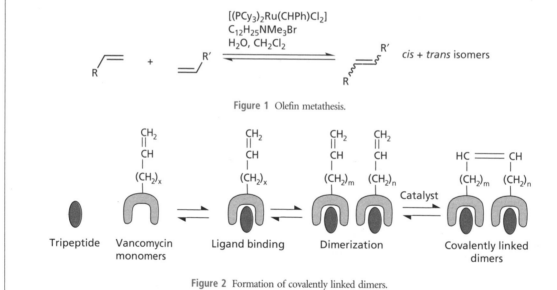

Figure 1 Olefin metathesis.

Figure 2 Formation of covalently linked dimers.

There are certain limitations to dynamic combinatorial chemistry:

- Conditions must be chosen such that the target does not react with one of the building blocks, or is unstable under the reaction conditions used.

- The target is normally in an aqueous environment, so the reactions have to be carried out in aqueous solution.

- The reactions themselves have to undergo fast equilibration rates to allow the possibility of amplification.

- It is important to avoid using some building blocks that are more likely to react than others, as this would bias the equilibrium towards particular products and confuse the identification of the amplified product.

14.10 Planning and designing a combinatorial synthesis

14.10.1 'Spider like' scaffolds

In order to find a new lead compound from combinatorial synthesis, we need to generate a large number of different structures, but we also need to ensure that these structures are as diverse as possible. This may not seem very likely if we are restricted to using a single reaction sequence, but if we are careful in the type of molecule we synthesize and the method by which we synthesize it, then such diversity *is* possible. In general, it is best to synthesize 'spider-like' molecules, so called because they consist of a central body (called the **centroid** or **scaffold**) from which various 'arms' (substituents) radiate (Fig. 14.33). These arms contain different functional groups which are used to probe a

binding site for binding regions once the spider-like molecule has entered (Fig. 14.34). The chances of success are greater if the 'arms' are evenly spread around the scaffold, as this allows a more thorough exploration of the 3D space (**conformational space**) around the molecule. The molecules made in the synthesis are planned in advance to ensure that they contain different functional groups on their arms and at different distances from the central scaffold.

14.10.2 Designing 'drug-like' molecules

The 'spider-like' approach increases the chances of finding a lead compound which will interact with a target binding site, but it is also worth remembering that compounds with good binding interactions do not necessarily make good medicines. There are also the pharmacokinetic issues to be taken into account (Chapter 8) and so it is worthwhile introducing certain restrictions to the types of molecule that will be produced in order to increase the chance that the lead compound will be orally active. In general, the chances of oral activity are increased if the structure obeys Lipinski's rule of five (section 8.2), i.e.:

- a molecular weight less than 500
- a calculated log P value less than $+5$
- no more than 5 hydrogen bond donating groups
- no more than 10 hydrogen bond accepting groups

Groups such as esters which are liable to be easily metabolized should be avoided. In addition, scaffolds or substituents likely to result in toxic compounds should be avoided; for example alkylating groups or aromatic nitro groups.

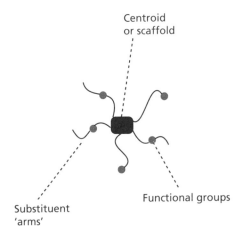

Figure 14.33 'Spider like' molecule.

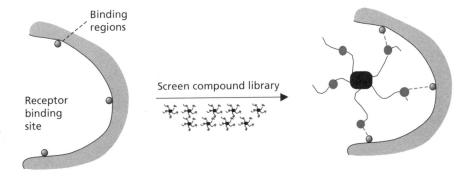

Figure 14.34 Probing for an interaction.

14.10.3 Scaffolds

Most scaffolds are synthesized by the synthetic route used for the combinatorial synthesis, and the synthesis used determines the number and variety of substituents which can be attached to the scaffold. The ideal scaffold should be small, in order to allow a wide variation of substituents (see Box 14.3). It should also have its substituents widely dispersed round its structure (spider-like) rather than restricted to one part of the structure (tadpole-like) if the conformational space around it is to be fully explored (Fig. 14.35). Finally, the synthesis should allow each of the substituents to be varied independently of each other.

Scaffolds can be flexible (e.g. a peptide backbone) or rigid (a cyclic system). They may contain groups that are capable of forming useful bonding interactions with the binding site, or they may not. Some scaffolds are already common in medicinal chemistry (e.g. benzodiazepine, hydantoin, tetrahydroisoquinoline, and benzenesulfonamide) and are associated with a diverse range of activities. Such scaffolds are termed '**privileged**' scaffolds.

14.10.4 Substituent variation

The variety of substituents chosen in a combinatorial synthesis depends on their availability and the diversity required. This would include such considerations as structure, size, shape, lipophilicity, dipole moment, electrostatic charge, and functional groups present. It is usually best to identify which of these factors should be diversified before commencing the synthesis.

14.10.5 Designing compound libraries for lead optimization

When using combinatorial synthesis to optimize a known lead structure, the variations planned should

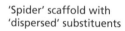
'Spider' scaffold with 'dispersed' substituents   'Tadpole' scaffold with 'restricted' substituents

Figure 14.35 Dispersed and restricted substituents.

take into account several factors such as the biological and physical properties of the compound, its binding interactions, and the potential problems of particular substituents. For example, if the binding interactions of a target receptor with its usual ligand are known, this knowledge can be used to determine what size of compounds would be best synthesizing, the types of functional groups that ought to be present and their relative positions. For example, if the target was a zinc-containing protease (e.g. angiotensin converting enzyme), a library of compounds containing a carboxylic acid or thiol group would be relevant.

14.10.6 Computer-designed libraries

It has been claimed that half of all known drugs involve only 32 scaffolds. Furthermore, it has been stated that a relatively small number of moieties account for the large majority of side chains in known drugs. This may imply that it is possible to define 'drug-like molecules' and use computer software programs to design more focused combinatorial compound libraries. Descriptors used in this approach include log P, molecular weight, number of hydrogen bond donors, number of hydrogen bond acceptors, number of rotatable bonds, aromatic density, the degree of branching in the structure, and the presence or absence of specific functional groups. One can also choose to filter out compounds that do not obey the rules mentioned in section 14.10.2. Computer programs can also be used to identify the structures which should be synthesized in order to maximize the number of different pharmacophores produced (section 15.16).

14.11 Testing for activity

We shall now look in more detail at how the products from combinatorial synthesis are tested for biological activity.

14.11.1 High-throughput screening

Because combinatorial synthesis produces a large quantity of structures in a very short period of time, biological testing has to be carried out quickly and automatically. The process is known as high-throughput screening (HTS) and was developed

BOX 14.3 EXAMPLES OF SCAFFOLDS

Benzodiazepines, hydantoins, β-lactams, and pyridines are examples of extremely good scaffolds. They all have small molecular weights and there are various synthetic routes available which produce the substitution patterns required to fully explore the conformational space about them. For example, it is possible to synthesize benzodiazepines such that there are variable substituents round the whole structure.

Peptide scaffolds are flexible scaffolds which have the capacity to form hydrogen bonds with target binding sites. They are easy to synthesize and a large variety of different substituents is possible by using the amino acid building blocks. Further substitution is possible on the terminal amino and carboxylic acid groups. The substituents are widely distributed along the peptide chain, allowing a good exploration of conformational space. If we consider Lipinski's rule of five, the peptide scaffold should ideally be restricted to di- and tripeptides in order to keep the molecular weight below 500. It is interesting to note that the antihypertensive agents **captopril** and **enalapril** are dipeptide-like and are orally active, whereas larger peptides such as the **enkephalins** are not orally active. Oral activity has also been a problem with

those HIV protease inhibitors having molecular weights over 500 (section 17.7.4).

Some of the scaffolds shown below have various disadvantages. Although **glucose** has a small molecular weight and the possibility of five variable substituents, it contains multiple hydroxyl groups. Attaching different substituents to similar groups usually requires complex protection and deprotection strategies. Nevertheless, the potential of sugar-based drugs is so great that a lot of progress has been made in developing combinatorial syntheses based on sugar scaffolds.

Steroids might appear attractive as scaffolds. However, the molecular weight of the steroid skeleton itself (314) limits the size of the substituents which can be added if we wish to keep the overall molecular weight below 500. Furthermore, there are relatively few positions where substituents can be easily attached. This limits the conformational space which can be explored around the steroid scaffold.

The indole scaffold shown suffers the disadvantage in having its variable substituents located in the same region of the molecule, preventing a full exploration of conformational space (i.e. it is a 'tadpole-like' scaffold).

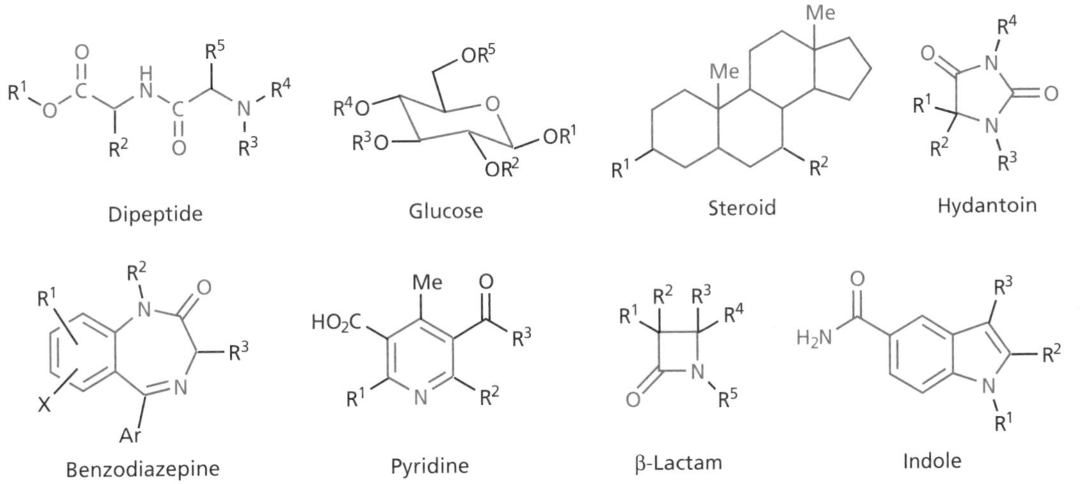

Examples of scaffolds.

before combinatorial synthesis. Indeed, the existence of HTS was one of the pressures to develop combinatorial synthesis. Since biological testing was so rapid and efficient, the pharmaceutical companies soon ran out of novel structures to test, and the synthesis of new structures became the limiting factor in the whole process of drug discovery. Combinatorial synthesis solved that problem, and the number of new compounds synthesized each year has increased dramatically. In fact, there are now so many compounds being produced that the focus is on making HTS even more efficient. Traditionally, compounds are automatically tested and analysed on a plate containing 96 small wells each with a capacity of 0.1 ml. There is now a move to use test plates of similar size but containing 1536 wells, where the test volumes are only 1–10 μl. Moreover, methods such as fluorescence and chemi-luminescence are being developed which will allow the simultaneous identification of active wells. Further miniaturization of open systems is unlikely, because of the problems of evaporation involving small volumes less than 1 μl. However, miniaturization using closed systems is on the horizon. The next major advance will involve the science of **microfluidics**, which involves the manipulation of tiny volumes of liquids in con-fined space. Microfluidic circuits on a chip can be used to control fluids electronically, allowing separation of an analytical sample using capillary electrophoresis. Companies are now developing machines that com-bine ultra-small-scale synthesis and miniaturized analysis. A single 10×10 cm silicon wafer can be microfabricated to support 10^5 separate syntheses/bioassays on a nanolitre scale!

14.11.2 Screening 'on bead' or 'off bead'

Sometimes structures can be tested for biological activity when they are still attached to the solid phase. 'On bead' screening assays involve interactions with targets which are tagged with an enzyme, fluorescent probe, radionuclide, or chromophore. A positive interaction results in a recognizable effect such as fluorescence or a colour change. These screening assays are rapid, and 10^8 beads can be readily screened. Active beads can then be picked out by micromanipulation and the structure of the active compound determined.

A false negative might be obtained if the solid phase sterically interferes with the assay. If such interference is suspected, it is better to release the drug from the solid phase before testing. This avoids the uncertainty of false negatives. On the other hand, there are cases where the compounds released prove to be insoluble in the test assay and give a negative result, whereas they give a positive result when attached to the bead.

KEY POINTS

- A sufficient excess of beads should be present relative to the number of structures planned, such that all possible products are represented when samples are taken for analysis.
- Combinatorial synthesis has been used for the synthesis of peptides, peptoids, and heterocyclic structures. Most organic reactions are feasible.
- Dynamic combinatorial chemistry involves the equilibrium formation of a mixture of compounds in the presence of a target. Binding of a product with the target amplifies that product in the equilibrium mixture.
- A scaffold is the core structure of a molecule round which variations are possible through the use of different substituents.
- Spider-like scaffolds allow substituent variation around the whole molecule, making it possible to explore all the conformational space around the scaffold. This increases the possibility of finding a lead compound which will bind to a target binding site.
- Lipinski's rule of five can be used when planning combinat-orial libraries to increase the chances of identifying an orally active lead compound.
- A privileged scaffold is one which is commonly present in known drugs.
- Computer software is available to assist in the planning of combinatorial libraries.
- High-throughput screening allows the automated analysis of large numbers of samples for their biological activity against defined targets. The analysis requires only small quantities of each sample.
- Screening can be carried out on compounds attached to resin beads, or on compounds which have been released into solution.

QUESTIONS

1. Identify three stages of the drug discovery, design, and development process where combinatorial chemistry is of importance.

2. A pharmaceutical laboratory wishes to synthesize all the possible dipeptides containing the amino acids tyrosine, lysine, phenylalanine, and leucine. Identify the number of

possible dipeptides and explain how the lab would carry this out using combinatorial techniques.

3. What particular precautions have to be taken with the amino acids tyrosine and lysine in the above synthesis?

4. Identify the advantages and disadvantages of the following structures as scaffolds.

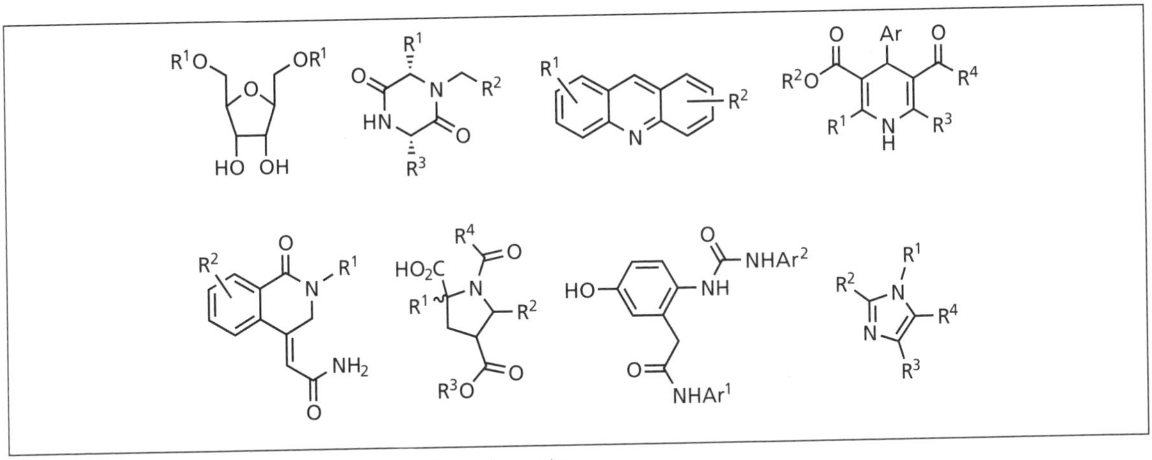

molecules suitable for tagging purposes (A–I), seven bromo acids (B1–B7), seven amines (A1–A7), and seven acid chlorides (C1–C7). Construct a suitable coding system for the synthesis.

6. Based on your coding scheme from Question 5, what product is present on the bead if the released tags resulted in the gas chromatogram shown in Fig. 14.22?

5. You wish to carry out the combinatorial synthesis shown in Fig. 14.21 using bar coding techniques rather than the conventional tagging scheme shown in the figure. You have nine

FURTHER READING

Beck-Sickinger, A. and Weber, P. (2002) *Combinatorial strategies in biology and chemistry*. John Wiley and Sons, New York.

Braeckmans, K. *et al.* (2002) Encoding microcarriers: present and future technologies. *Nature Reviews Drug Discovery*, 1, 447–456.

DeWitt, S. H. *et al.* (1993) 'Diversomers': An approach to nonpeptide, nonoligomeric chemical diversity. *Proceedings of the National Academy of Sciences of the USA*, 90, 6909–6913.

Dolle, R. E. (2003) Comprehensive survey of combinatorial library synthesis: 2002. *Journal of Combinatorial Chemistry*, 5, 693–753.

Geysen, H. M. *et al.* (2003) Combinatorial compound libraries for drug discovery: an ongoing challenge. *Nature Reviews Drug Discovery*, 2, 222–230.

Houlton, S. (2002) Sweet synthesis. *Chemistry in Britain*, April, 46–49.

Jung, G. (ed.) (1996) *Combinatorial peptide and nonpeptide libraries*. VCH, Weinheim.

Le, G. T. *et al.* (2003) Molecular diversity through sugar scaffolds. *Drug Discovery Today*, 8, 701–709.

Ley, S. V. and Baxendale, I. R. (2002) New tools and concepts for modern organic synthesis. *Nature Reviews Drug Discovery*, 1, 573–586.

Nicolaou, K. C. *et al.* (2000) Target-accelerated combinatorial synthesis and discovery of highly potent antibiotics effective against vancomycin-resistant bacteria. *Angewandte Chemie, International Edition*, 39, 3823–3828.

Ramstrom, O. and Leh, J.-M. (2002) Drug discovery by dynamic combinatorial libraries. *Nature Reviews Drug Discovery*, 1, 26–36.

Terret, N. K. (1998) *Combinatorial Chemistry*. Oxford University Press, Oxford.

Titles for general further reading are listed on p. 711.

15 Computers in medicinal chemistry

Computers are an essential tool in modern medicinal chemistry and are important in both drug discovery and drug development. Rapid advances in computer hardware and software have meant that many of the operations which were once the exclusive province of the expert can now be carried out on ordinary laboratory computers with little specialist expertise in the molecular or quantum mechanics involved. In the next few sections, we shall look at examples of how computers are used in medicinal chemistry. However, it has to be appreciated that it is not possible to do full justice to this subject in a single chapter and the author has been fairly selective in what has been included. Moreover, the pace of change is such that the material reported here could well be out of date within a few months of publication, if not before!

15.1 Molecular and quantum mechanics

The various operations which are carried out in molecular modelling involve the use of programs or algorithms which calculate the structure and property data for the molecule in question. For example, it is possible to calculate the energy of a particular arrangement of atoms (conformation), modify the structure to create an energy minimum, and calculate properties such as charge, dipole moment, and heat of formation. The mathematical details of these operations are too involved to be included in an introductory text, but it is important to appreciate a few general principles about how these processes are carried out. The computational methods that are used to calculate structure and property data can be split into two categories—molecular mechanics and quantum mechanics.

15.1.1 Molecular mechanics

In molecular mechanics, equations are used which follow the laws of classical physics and apply them to nuclei without consideration of the electrons. In essence, the molecule is treated as a series of spheres (the atoms) connected by springs (the bonds). Equations derived from classical mechanics are used to calculate the different interactions and energies (**force fields**) resulting from bond stretching, angle bending, non-bonded interactions, and torsional energies. (Torsional energies are associated with atoms that are separated from each other by three bonds. The relative orientation of these atoms is defined by the dihedral or torsion angle—see for example Fig. 15.18.) These calculations require data or parameters which are stored in tables within the program and which describe interactions between different sets of atoms. The energies calculated by molecular mechanics have no meaning as absolute quantities, but are useful when comparing different conformations of the same molecule. Molecular mechanics is fast and is less intensive on computer time than quantum mechanics. However, it cannot calculate electronic properties because electrons are not included in the calculations.

15.1.2 Quantum mechanics

Quantum mechanics uses quantum physics to calculate the property of a molecule by considering the interactions between the electrons and nuclei of the molecule. Unlike molecular mechanics, atoms are not treated as solid spheres. In order to make the calculations feasible, various approximations have to be made.

- Nuclei are regarded as motionless. This is reasonable, since the motion of the electrons is much faster in comparison. As electrons are considered to be moving around fixed nuclei, it is possible to describe electronic energy separately from nuclear energy.

- It is assumed that the electrons move independently of each other, so the influence of other electrons and nuclei is taken as an average.

Quantum mechanical methods can be subdivided into two broad categories—*ab initio* and semi-empirical. The former is more rigorous and does not require any stored parameters or data. However it is expensive on computer time and is restricted to small molecules. Semi-empirical methods compute for valence electrons only. They are quicker though less accurate, and can be carried out on larger molecules. There are various forms of semi-empirical software (i.e. programs such as MINDO/3, MNDO, MNDO-d, AM1, and PM3). These methods are quicker, because they use further approximations and make use of stored parameters.

15.1.3 Choice of method

The method of calculation chosen depends on what calculation needs to be done, as well as the size of the molecule. As far as size of molecule is concerned, *ab initio* calculations are limited to molecules containing tens of atoms, semi-empirical calculations on molecules containing hundreds of atoms, and molecular mechanics on molecules containing thousands of atoms.

Molecular mechanics is useful for the following operations or calculations:

- energy minimization
- identifying stable conformations
- energy calculations for specific conformations
- generating different conformations
- studying molecular motion.

Quantum mechanical methods are suitable for calculating the following:

- molecular orbital energies and coefficients
- heat of formation for specific conformations

- partial atomic charges calculated from molecular orbital coefficients
- electrostatic potentials
- dipole moments
- transition-state geometries and energies
- bond dissociation energies.

15.2 Drawing chemical structures

Chemical drawing packages do not require the calculations described in section 15.1, but they are often integrated into molecular modelling programs. In the not so distant past, drawing chemical structures for a report or a scientific paper was a rather tedious business which involved tracing the skeleton of the molecule with templates, then using a typewriter to add elements and substituents. Positioning the paper in the typewriter to get the substituent at the correct position was quite an art! Various software packages such as ChemDraw, ChemWindow, and Isis/Draw are now available which can be used to construct diagrams quickly and to a professional standard. For example, the diagrams in this book have all been prepared using the ChemDraw package.

Some drawing packages are linked to other items of software which allow quick calculations of various molecular properties. For example, the following properties for **adrenaline** were obtained using ChemDraw Ultra; the structure's correct IUPAC chemical name, molecular formula, molecular weight, exact mass, and theoretical elemental analysis. It was also possible to get calculated predictions of the compound's ^{1}H and ^{13}C NMR chemical shifts, melting point, freezing point, log P value, molar refractivity, and heat of formation (Fig. 15.1).

15.3 3D structures

Molecular modelling software allows the chemist to construct a 3D molecular structure on the computer. There are several software packages available such as Chem3D, Alchemy, Sybyl, Hyperchem, ChemX, Discovery Studio Pro, and CAChe. The 3D model can be made by constructing the molecule atom by atom, and

Calculated properties
C$_9$H$_{13}$NO$_3$
Exact mass: 183.09
Mol. wt.: 183.20
C, 59.00; H, 7.15; N, 7.65; O, 26.20

4-(1-Hydroxy-2-methylamino-ethyl)-benzene-1,2-diol

Predicted properties
log *P* = –0.61 – 0.63
Molar refractivity 48.66–49.08 [cm.cm.cm/mol]
Boiling point 618.55 K; Freezing point 539.03 K
Heat of formation –451.22 kJ/mol

Predicted ^{13}C NMR

Predicted ^{1}H NMR

Figure 15.1 Drawing chemical structures.

Stereocentre

Prepared in ChemDraw

Converted to 3D in Chem3D

Carbon
Oxygen
Nitrogen
Hydrogen

Figure 15.2 Conversion of a 2D drawing to a 3D model.

bond by bond. It is also possible to automatically convert a 2D drawing into a 3D structure, and most molecular modelling packages have this facility. For example, the 2D structure of adrenaline in Fig. 15.2 was drawn in ChemDraw, then copied and pasted into Chem3D, resulting in the automatic construction of the 3D model shown. The 3D structures of a large number of small molecules can also be accessed from the Cambridge Structural Database (CSD) and downloaded so that they can be studied. This database contains over 200 000 molecules which have been crystallized and their structure determined by X-ray crystallography.

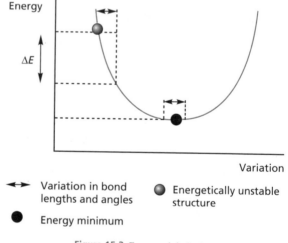

Energy

ΔE

Variation

Variation in bond lengths and angles

Energetically unstable structure

Energy minimum

Figure 15.3 Energy minimization.

15.4 Energy minimization

Whichever software program is used to create a 3D structure, a process called energy minimization should be carried out once the structure is built. This is because the construction process may have resulted in unfavourable bond lengths, bond angles, or torsion angles. Unfavourable non-bonded interactions may also be present (i.e. atoms from different parts of the molecule occupying the same region of space). The energy minimization process is usually carried out by a molecular mechanics program which calculates the energy of the starting molecule, then varies the bond lengths, bond angles, and torsion angles to create a new structure. The energy of the new structure is calculated to see whether it is energetically more stable or not. If the starting

BOX 15.1 ENERGY MINIMIZING APORPHINE

A 2D structure of **aporphine** was converted to a 3D structure using Chem3D. However, the catechol ring was found to be non-planar with different lengths of C–C bond. Energy minimization corrected the deformed aromatic ring, resulting in the desired planarity and the correct length of bonds.

Energy minimization carried out on aporphine.

structure is inherently unstable, a slight alteration in bond angle or bond length will have a large effect on the overall energy of the molecule resulting in a large energy difference (ΔE; Fig. 15.3). The program will recognize this and carry out more changes, recognizing those which lead to stabilization and those which do not. Eventually, a structure will be found where structural variations result in only slight changes in energy—an energy minimum. The program will interpret this as the most stable structure and will stop at that stage (Box 15.1).

possible to display the structure in different formats (i.e. cylindrical bonds, wire frame, ball and stick, space filling; Fig. 15.5).

There is another format known as the ribbon format which is suitable for portraying regions of protein secondary structure such as α-helices. This often simplifies the highly complex-looking structure of a protein, allowing easier visualization of its secondary and tertiary structure. The ball and stick model of an α-helical decapeptide consisting of 10 alanine units is shown in Fig. 15.6, along with the same molecule displayed as a ribbon.

15.5 Viewing 3D molecules

Once a structure has been energy minimized, it can be rotated in various axes to study its shape from different angles. For example, the 3D structure of **adrenaline** is shown from different aspects in Fig. 15.4. It is also

15.6 Molecular dimensions

Once a 3D model of a structure has been constructed, it is a straightforward procedure to measure all its bond lengths, bond angles, and torsion (or dihedral)

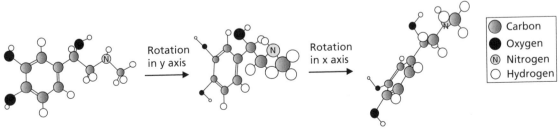

Figure 15.4 Viewing a 3D model in different axes.

angles. These values can be read from relevant tables or by highlighting the relevant atoms and bonds on the structure itself. The various bond lengths, bond angles and torsion angles measured for **adrenaline** are illustrated in Fig. 15.7. It is also a straightforward process to measure the separation between any two atoms in a molecule.

15.7 Molecular properties

Various properties of the 3D structure can be calculated once it has been built and minimized. For example, the steric energy is automatically measured

as part of the minimization process and takes into account the various strain energies within the molecule, such as bond stretching or bond compression, deformed bond angles, deformed torsion angles, nonbonded interactions arising from atoms too close to each other in space, and unfavourable dipole–dipole interactions. The steric energy is useful when comparing different conformations of the same structure, but the steric energies of different molecules should not be compared.

Other properties for the structure can be calculated, such as the predicted heat of formation, dipole moment, charge density, electrostatic potential, electron spin density, hyperfine coupling constants, partial charges, polarizability, and infrared vibrational frequencies. Some of these are described below.

15.7.1 Partial charges

It is important to realize that the valence electrons in molecules are not fixed to any one particular atom and can move around the molecule as a whole. As the electrons are likely to spend more of their time nearer electronegative atoms than electropositive atoms, this distribution is not uniform and results in some parts of the molecule being slightly positive and other parts being slightly negative. For example, the partial charges for **histamine** are shown in Fig. 15.8.

The calculation of partial charges has important consequences in the way we view ions. Conventionally,

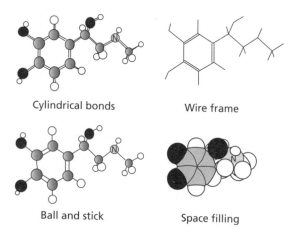

Figure 15.5 Different methods of visualizing molecules.

Cylindrical bonds

Wire frame

Ball and stick

Space filling

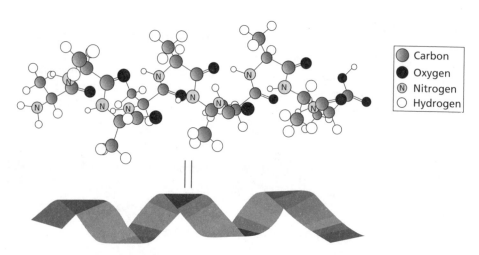

| Carbon |
| Oxygen |
| Nitrogen |
| Hydrogen |

Figure 15.6 Ribbon representation of a helical decapeptide (Chem3D).

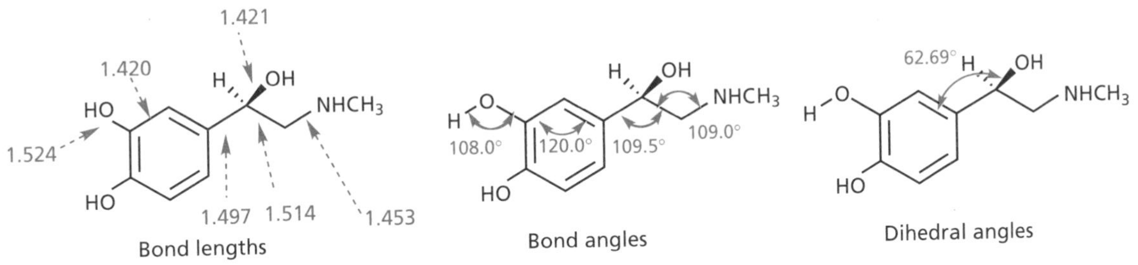

Figure 15.7 Molecular dimensions for adrenaline (Chem3D).

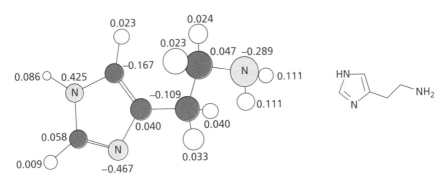

Figure 15.8 Partial charges for histamine.

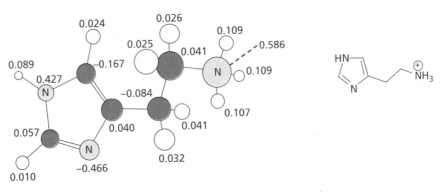

Figure 15.9 Charge distribution on the histamine ion.

we consider charges to be fixed on a particular atom (unless delocalization is possible). For example, the histamine ion is normally drawn showing the positive charge on the terminal nitrogen atom (Fig. 15.9). In fact, calculation of partial charges shows that some of the positive charge is localized on the hydrogens attached to the terminal nitrogen. This has

important consequences in the way we think of ionic interactions between a drug and its binding site. It implies that charged areas in the binding site and the drug are more diffuse than one might think. This in turn suggests that we have wider scope in designing novel drugs. For example, in the classical viewpoint of charge distribution, a certain molecule might be considered to have its charged centre too far away from the corresponding 'centre' in the binding site. If these charged areas are actually more diffuse, then this is not necessarily true (Fig. 15.10).

It is worth pointing out, however, that such calculations are carried out on structures in isolation from their environment. In the body, histamine is in an aqueous environment and would be surrounded by water molecules which would solvate the charge and consequently have an effect on charge distribution. Furthermore, water has a high dielectric constant, which means that electrostatic interactions are more effectively masked than in a hydrophobic environment.

Partial charges can also be represented by dot clouds. The size of each cloud represents the amount of charge, and the clouds can be coloured to show what sort of charge it is. The dot clouds shown in Fig. 15.11 represent the relative sizes of partial charges on the histamine ion.

15.7.2 Molecular electrostatic potentials

Another way to consider charge distribution is to view the molecule as a whole rather than as individual atoms and bonds, such that one can identify areas of the molecule which are electron rich or electron poor. This is particularly important in the **3D QSAR** technique of **CoMFA** described in section 13.10. It can also be useful in identifying how compounds with different structures might line up to interact with corresponding electron rich and electron poor areas in a binding site.

Molecular electrostatic potentials (MEPs) can be calculated by placing a proton 'probe' at different positions in space around the molecule. The interaction of the probe with the partial charges of the molecule is then measured. Alternatively, the MEP can be calculated using quantum mechanics by considering the molecular orbitals. The MEP for histamine shown in Fig. 15.12 was calculated using the semi-empirical method AM1.

An example of how electrostatic potentials have been used in drug design can be seen in the design of the **cromakalim** analogue (II; Fig. 15.13), where the cyanoaromatic ring was replaced by a pyridine ring.

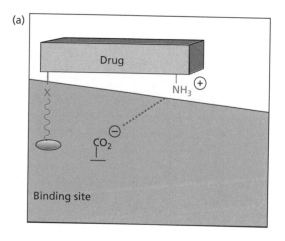

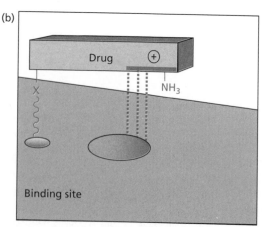

Figure 15.10 Ionic interactions. (a) Classical view of an ionic interaction and (b) diffuse ionic interaction.

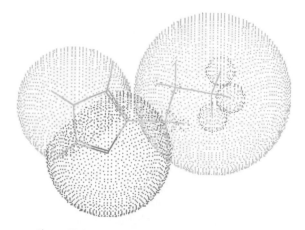

Figure 15.11 Dot cloud representation of partial charge.

This was part of a study looking into analogues of cromakalim which would have similar antihypertensive properties, but which might have different pharmacokinetics. In order to retain activity, it was important that any replacement heteroaromatic ring was as similar in character to the original aromatic ring as possible. Consequently, the MEPs of various bicyclic systems were calculated and compared with the parent bicyclic system (III; Fig. 15.14). In order to simplify the analysis, the study was carried out in 2D within the plane of the bicyclic systems, and maps were created showing areas of negative potential (Fig. 15.15). The contours represent the various levels of the MEP, and can be taken to indicate possible hydrogen bonding regions around

each molecule. The analysis demonstrated that the bicyclic system (IV) had similar electrostatic properties to (III), resulting in the choice of structure (II) as an analogue.

15.7.3 Molecular orbitals

The molecular orbitals of a compound can be calculated using quantum mechanics. For example, **ethene** can be shown to have 12 molecular orbitals. The **highest occupied molecular orbital** (**HOMO**) and **lowest unoccupied molecular orbital** (**LUMO**) are shown in Fig. 15.16 (see Box 15.2).

15.7.4 Spectroscopic transitions

It is possible to calculate the infrared or ultraviolet transitions for a molecule. Although a theoretical infrared spectrum can be generated, it is highly unlikely that it will accurately match the actual infrared spectrum. Nevertheless, the position and

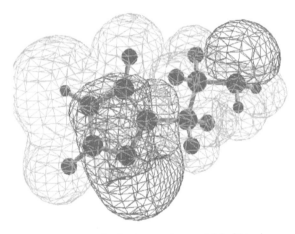

Figure 15.12 Molecular electrostatic potential for histamine.

Figure 15.13 Ring variation on cromakalim.

Figure 15.14 Bicyclic models in cromakalim study.

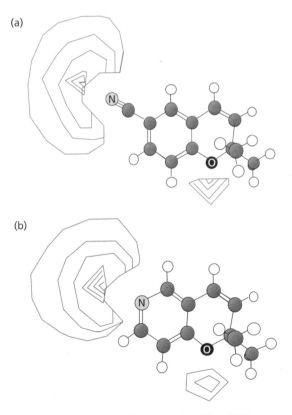

(a)

(b)

Figure 15.15 MEPs of bicyclic models (III) and (IV).

BOX 15.2 STUDY OF HOMO AND LUMO ORBITALS

A study of a molecule's HOMO and LUMO orbitals is useful because frontier molecular orbital theory states that these orbitals are the most important in terms of a molecule's reactivity. HOMO and LUMO orbitals can also help to explain drug–receptor interactions. For example, **ketanserin** (Fig. 1) is an antagonist at serotonin receptors, but has a greater binding affinity than would be expected from normal bonding interactions.

In order to explain this greater binding affinity, it was proposed that a charge transfer interaction was taking place between the electron-deficient fluorobenzoyl ring system of

ketanserin and an electron-rich tryptophan residue which was known to be nearby in the binding site. To check this, HOMO and LUMO energies were calculated for a model complex between the indole system of tryptophan and the fluorobenzoyl system of ketanserin (Fig. 2). This showed that the HOMO for the indole/fluorobenzoyl complex resided on the indole structure whereas the LUMO was on the fluorobenzoyl moiety, indicating that charge transfer was possible. With other antagonists, there was not this same clear-cut separation between the HOMO and LUMO orbitals, with the indole system being involved in both orbitals.

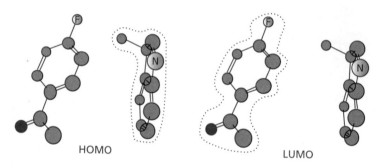

Figure 1 Ketanserin.

Electron-deficient fluorobenzoyl system

Figure 2 HOMO and LUMO molecular orbitals (dot surfaces) for the indole/fluorobenzoyl complex.

HOMO

LUMO

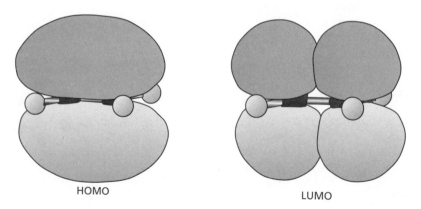

HOMO

LUMO

Figure 15.16 HOMO and LUMO molecular orbitals for ethene.

identification of specific absorptions can be identified and can be useful in the design of drugs. For example, it is found that the activity of penicillins is related to the position of the β-lactam carbonyl stretching vibration in the infrared spectrum. Calculating the theoretical wavenumber for a range of β-lactam structures can be useful in identifying which ones are likely to have useful activity before synthesizing them.

KEY POINTS

- Several chemical drawing packages include software allowing the calculation of various physical properties.

- Molecular modelling software makes use of programs based on molecular mechanics and quantum mechanics.

- Molecular mechanics programs use equations based on classical physics to calculate force fields. Atoms are treated as spheres, bonds as springs. Electrons are ignored. This method is suitable for energy minimization and conformational analysis.

- Quantum mechanical methods are *ab initio* or semi-empirical. The former is more rigorous but is restricted to small molecules. These methods are suitable for measuring molecular properties such as molecular orbital energies and coefficients.

- Energy minimization has to be carried out on any molecule constructed with molecular modelling software. The process involves alteration of bond lengths, bond angles, torsion angles, and non-bonded interactions until a stable conformation is obtained.

- Molecular modelling software allows the dimensions of a molecule to be accurately measured as well as its partial charges, molecular electrostatic potentials, and molecular orbitals.

15.8 Conformational analysis

15.8.1 Local and global energy minima

In section 15.4, we saw how energy minimization is carried out on a 3D structure to produce a stable conformation. The structure obtained, however, is not necessarily the most stable conformation. This is because energy minimization stops as soon as it reaches the first stable conformation it finds, and that will be the one closest to the starting structure. This is illustrated in Fig. 15.17, where the most stable conformation is separated from a less stable conformation by an energy saddle. If the 3D structure initially created is on the energy curve at the position shown, energy minimization will stop when it reaches the first stable conformation it encounters—**a local energy minimum**. At this point, variations in structure result in low energy changes and so the minimization will stop. In order to cross the saddle to the more stable conformation, structural variations would have to be carried out which increase the strain energy of the structure and these will be rejected by the program. The minimization program has no way of knowing that there is a more stable conformation (**a global energy minimum**) beyond the energy saddle. Therefore, in order to identify the most stable conformation, it is necessary to generate different conformations of the molecule and to compare their steric energies. There are two methods of doing this—molecular dynamics or stepwise rotation of bonds.

15.8.2 Molecular dynamics

Molecular dynamics is a molecular mechanics program designed to mimic the movement of atoms within a molecule. The software program works by treating the atoms in the structure as moving spheres. After one femtosecond (1×10^{-15} s) of movement, the

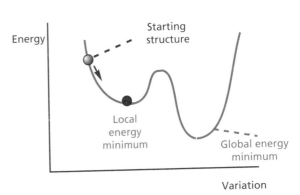

Figure 15.17 Local and global energy minima.

position and velocity of each atom in the structure is determined. The forces acting on each atom are then calculated by considering bond lengths, bond angles, torsional terms, and non-bonded interactions with surrounding atoms. The potential energy of each atom is calculated and Newton's laws of motion are then used to determine the acceleration and direction of movement of each atom (the kinetic energy). This allows the program to predict the velocity and position of each atom a femtosecond later. The procedure is then repeated for every femtosecond of the process. The femtosecond duration is important, as it is an order of magnitude less than the rate of a bond stretching vibration, and so an atom is only allowed to move a fraction of a bond length between each calculation. If this was not the case, and atoms were allowed to move greater distances, one might get the situation where two atoms occupy the same area of space. The calculated forces and potential energies would then be huge leading to atoms moving with excessive velocities and accelerations and the system would fail.

Molecular dynamics can be used to generate a variety of different conformations by 'heating' the molecule to 900 K. Of course, this does not mean that the inside of your computer is about to melt. It means that the program allows the structure to undergo bond stretching and bond rotation as if it *were* being heated. As a result, energy barriers between different conformations are overcome, allowing the

crossing of energy saddles. In the process, the molecule is 'heated' at a high temperature (900 K) for a certain period (e.g. 5 ps), then 'cooled' to 300 K for another period (e.g. 10 ps) to give a final structure.

The process can be repeated automatically as many times as desired, to give as many different structures as is practical. Each of these structures can then be recovered and energy minimized, and its steric energy measured. By carrying out this procedure, it is usually possible to identify some distinct conformations that might be more stable than the initial conformation. For example, the two-dimensional drawing of **butane** shown in Fig. 15.18 was imported into Chem-3D and energy minimized. Because of the way the molecule was represented, energy minimization stopped at the first local energy minimum it found—the gauche conformation having a steric energy of 12.076 kJ/mole. The molecular dynamics program was run to generate other conformations and successfully produced the fully staggered *trans* conformation which, after energy minimization, had a steric energy of 9.135 kJ/mole, i.e. it was more stable by about 3 kJ/mole. In fact, this particular problem could be solved more efficiently by the stepwise rotation of bonds described below (section 15.8.3). Molecular dynamics is more useful for creating different conformations of molecules which are not conducive to stepwise bond rotation (e.g. cyclic systems—see Box 15.3), or which would take too long to analyse by that process (large molecules).

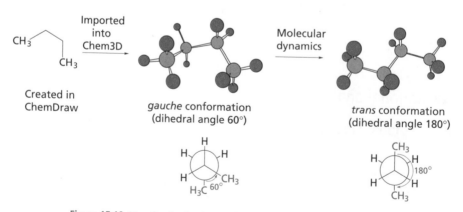

Figure 15.18 Use of molecular dynamics to find the most stable conformation.

The twist boat conformation of **cyclohexane** is not the
most stable conformation of cyclohexane, but remains as
the twist boat when energy minimization is carried out.
'Heating' the molecule by molecular dynamics produces a
variety of different conformations, including the more
stable chair conformation.

Molecular
dynamics
→

Twist boat
11.917 kcal/mol

Chair
6.558 kcal/mol

Generation of the cyclohexane chair conformation by
molecular dynamics in Chem3D.

Finally, it has to be remembered that in the real
world biomolecules are surrounded by water and that
this can affect the relative stability of different conformations. Therefore, it is advisable to include water
molecules in the modelling system before carrying out
molecular dynamics experiments.

15.8.3 Stepwise bond rotation

Although molecular dynamics can be used to generate
different conformations, there is no guarantee that it
will identify all the conformations which are possible
for a structure. A more systematic process is to generate
different conformations by automatically rotating
every single bond by a set number of degrees. For
example, 12 different conformations of butane were
generated by automatically rotating the central bond in
30° steps. The steric energy of each conformation was
calculated and graphed (Fig. 15.19), revealing that the
most stable conformation was the fully staggered one,
and the least stable conformation was the eclipsed one.
In this operation, energy minimization is not carried
out on each structure, because the aim is to identify
both stable and unstable conformations.

Some modelling software packages can automatically identify all the rotatable single bonds in a structure. Bonds to hydrogen or to methyl groups
are excluded in this analysis, as rotations of
these bonds do not generate significantly different

conformations. Once the rotatable bonds have been
identified, the program generates all the possible
conformations which can arise from rotating these
bonds by a set amount determined by the operator.
The number of conformations generated will depend
on the number of rotatable bonds present and the set
amount of rotation. For example, a structure with
three rotatable bonds could be analysed for conformations resulting from 10° increments at each bond
to generate 46 656 conformations. With four rotatable
bonds, 30° increments would generate 20 736 conformations.

In general, about 1000 conformations per second
can be processed on a standard benchtop computer.
However, it is important to be as efficient as possible
and care should be taken in deciding how much each
bond should be rotated at a time to ensure that a
representative but manageable number of conformations is created.

It is possible to make the process more efficient,
depending on the information desired. For example, if
we were only interested in identifying stable conformations, the program can automatically filter
out conformations which are eclipsed or near eclipsed.
It is also possible to filter out 'nonsense' conformations, i.e. conformations where some atoms occupy
the same position in space. Such conformations can
arise because bond rotations are being carried out by
the program without analysing what is happening
elsewhere in the molecule.

Once a series of conformations have been
generated, they can be tabulated and sorted into their
order of stability. The most stable conformations can
then be energy minimized and their structures
compared.

15.8.4 Monte Carlo methods in conformational searching

The Monte Carlo method is commonly used to search
for different conformations. In molecular dynamics,
a search for the most stable conformation involves the
generation of random conformations which are all
analysed separately. This means that the same amount
of processing time is taken up by high-energy and
low-energy conformations. The Monte Carlo method
introduces a bias towards stable conformations such
that more processing time is spent on those—a process known as **importance sampling**. It works by

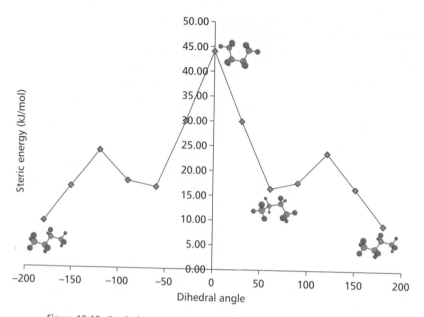

Figure 15.19 Graph showing relative stabilities of various butane conformations.

comparing the energies of randomly created conformations as they are formed. These random conformations are generated by altering the positions of atoms relative to each other by carrying out a bond rotation. This is different from molecular dynamics because it is purely a space move and does not involve an analysis of the movement, velocity, or acceleration of the atoms. If a more stable conformation is generated, the program identifies this and carries out smaller conformational changes to explore the local minima related to that conformation. On its own, this could mean the program only identifies a local minimum and fails to find the global minimum, so at regular intervals, larger changes in the conformation are carried out such that the energy barriers can be crossed in order to access other stable conformations. The program can also accept a certain number of high-energy conformations, depending on the frequency with which they are formed.

15.9 Structure comparisons and overlays

Using molecular modelling, it is possible to compare the 3D structures of two or more molecules. For example, suppose we wish to compare the structures

of **cocaine** and **procaine**. Both of these compounds have a local anaesthetic property, and structure–activity relationships indicate that the important pharmacophore for local anaesthesia is the presence of an amine, an ester, and an aromatic ring. These functional groups are present in cocaine and procaine, but the pharmacophore also requires the functional groups to be in the same relative positions in space with respect to each other. Looking at the 2D structures of procaine and cocaine, it would be tempting to match up corresponding bonds as shown in Fig. 15.20, but this would place the nitrogen atoms one bond length apart in the overlay. With molecular modelling, the important atoms of the structures can be matched up, in this case the nitrogens and the aromatic rings of both structures. The software then strives to find the best fit, resulting in the overlay shown in Fig. 15.21. Here the procaine molecule has been laid across the centre of the bicyclic system in cocaine, so that both the aromatic rings and nitrogen atoms overlap.

How does a software program know when a best fit has been achieved? This is done by calculating the root mean square distance (RMSD) between all the atom pairs which have to be matched up, and to find the relative orientation of the molecules where this value is a minimum. For example, in the overlay between cocaine and procaine, the pairs of atoms to be matched

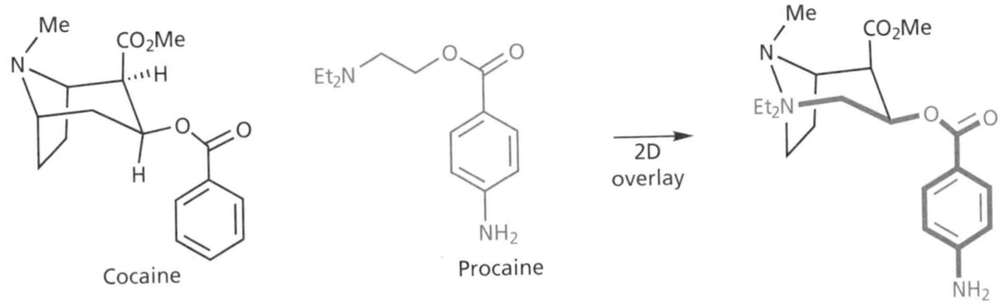

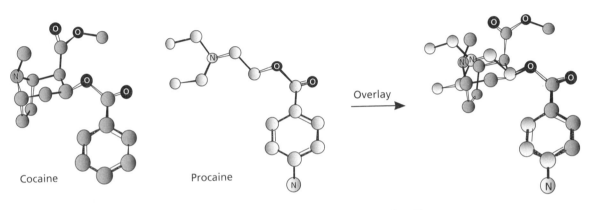

Figure 15.20 2D-Overlay of cocaine and procaine.

Figure 15.21 Overlay of cocaine and procaine using Chem3D.

up are defined as the two nitrogens in each molecule and the corresponding aromatic carbons. The distance between the nitrogens in each molecule is measured, as are the distances between each of the corresponding aromatic carbons. The RMSD for all the atom pairs is then calculated. One of the structures is then moved in stages with respect to the other and the calculations repeated until a minimum value of RMSD is obtained, corresponding to the best fit.

It is important to appreciate that the fitting process described above is carried out on a rigid basis; that is, the molecules are locked in the one conformation and no bond rotations are permitted. Therefore, it is important that each molecule is in the active conformation before carrying out the fitting process. If the active conformations are not known, it is possible to carry out overlays where one or both molecules

can change conformations in order to get the best fit, but this would be more expensive on computer time.

Some modelling software programs have the capacity to automatically overlay two molecules without the operator having to define the centres or how they should be matched up. The program searches each molecule for what it considers to be important centres. These are centres which are normally involved in binding interactions (i.e. aromatic rings, hydrogen bond donors, hydrogen bond acceptors, positively charge centres, acidic centres, and basic centres). As far as an aromatic ring is concerned, the centre of the ring represents the whole ring. For hydrogen bond donors (X) or hydrogen bond acceptors (X–H), the heteroatom (X) is defined as the centre. To be precise, the centre for a hydrogen bond donor should be the hydrogen atom, but there is a large uncertainty as to where this atom is located because of bond rotation, and so the heteroatom is defined as the centre of an 'available volume' within which the hydrogen atom is located. Certain functional groups can be defined as being more than one type of centre. For example, the hydroxyl group is considered both as a hydrogen bond donor and a hydrogen bond acceptor centred on the oxygen. A primary amine is considered a hydrogen bond donor, hydrogen bond acceptor, basic centre, and positively charged centre (because it could be protonated). Once the centres for each molecule have been identified, the program then strives to overlay them such that equivalent centres are matched up.

KEY POINTS

- Energy minimization produces the nearest stable conformation to the structure presented, and not necessarily the global conformation.

- Molecular dynamics can be carried out on a molecule to generate different conformations which, on energy minimization, give a range of stable conformations. Alternatively, bonds can be rotated in a stepwise process to generate different conformations.

- Molecular modelling can be used to overlay two molecules in order to assess their similarity.

15.10 Identifying the active conformation

A problem frequently encountered in drug design is trying to decide what shape or conformation a molecule is in when it fits its target binding site—the active conformation. This is particularly true for simple flexible molecules, which can adopt a large number of conformations. One might suggest that the most stable conformation will be the active conformation, as the molecule is most likely to be in that conformation. However, it is possible that a less stable conformation could be the active conformation. This is because the binding interactions with the target result in an energy gain which may compensate for the energy required to adopt that conformation.

15.10.1 X-ray crystallography

The easiest way of identifying an active conformation is to study the X-ray crystal structure of a target protein with its ligand (the drug) attached. The crystal structure of the ligand itself can be obtained from the Cambridge Structural Database (CSD), and the crystal structure of protein–ligand complexes can be obtained from the Brookhaven National Laboratory Protein Data Bank (PDB) The protein–ligand complex can be downloaded and studied using molecular modelling software and the active conformation of the ligand identified. Not all proteins can be easily crystallized, however, and other methods of identifying active conformations may have to be used.

15.10.2 Comparison of rigid and non-rigid ligands

Identification of active conformations is made much easier if one of the active compounds is a rigid molecule which has only one possible conformation. The geometry of the pharmacophore (the important binding centres) can thus be determined for the rigid molecule. More flexible molecules can then be analysed to find a conformation which will place the important binding groups in the same relative geometry (see Box 15.4).

If a fully rigid molecule is not available to act as a template, it may be possible to match up different structures which have an element of rigidity somewhere in their skeleton. For example, structures I–III in Fig. 15.22 are all antagonists of the $5HT_{2A}$ (serotonin) receptor. An active conformation for structure I could be proposed which matches the rigid moieties of structures II and III. Structure II gives the conformation for the top half of the molecule, and structure III gives the conformation of the bottom half.

Another method of finding active conformations is to consider all the reasonable conformations for a range of active compounds, and then to determine the common volume or space into which the various important binding groups can be placed in order to interact with the binding site. A study such as this was carried out in order to determine the active conformation of the antihypertensive agent **captopril** (Fig. 15.23). Because captopril is flexible, the exact 3D

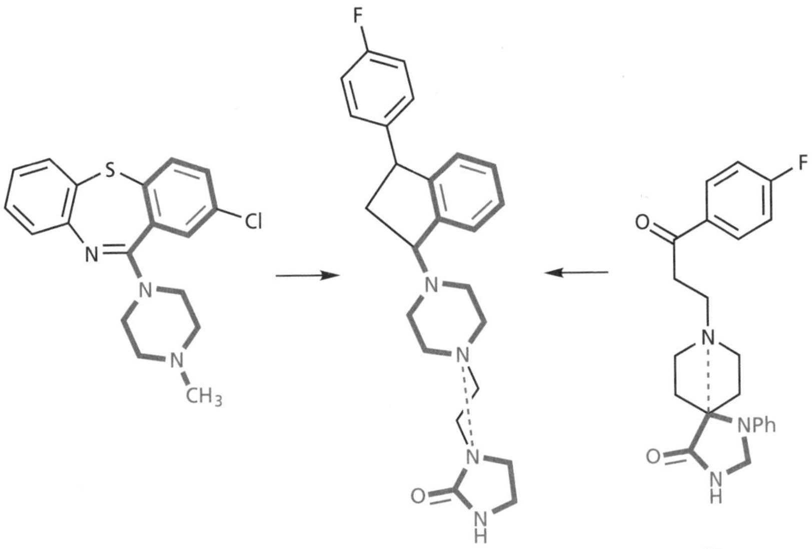

Figure 15.22 Defining the active conformation of serotonin antagonists. The dashed lines represent dummy bonds defining the relative positions of connected atoms.

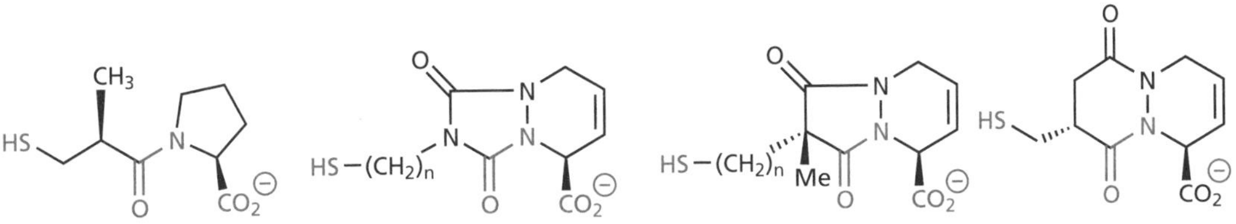

Figure 15.23 Captopril and rigid analogues (binding groups coloured).

relationship of the important binding groups (i.e. the carboxyl, amide, and thiol groups) in the active conformation is not known. There was also no X-ray crystallographic data available to reveal how captopril binds to its target binding site. To address this problem, a variety of rigid analogues (I–III) was synthesized, where the amide and the carboxyl group were fixed in space with respect to each other, but because of bond rotation the thiol group could access an area of space. The biological activity of these compounds was then measured to study how the relative orientation of the thiol group with respect to the other two groups affected biological activity. The possible conformations for captopril arising from bond rotation around the two bonds shown in Fig. 15.24 were determined using molecular modelling, and a spatial

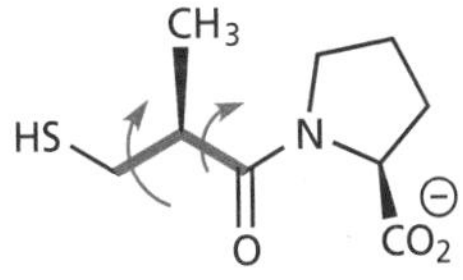

Figure 15.24 Bond rotations in captopril.

map (A) was generated to show the possible regions in space that were accessible to the thiol group (Fig. 15.25).

Some of the conformations involved in this analysis are high-energy eclipsed conformations, and these were filtered out of the analysis by programming the software to reject conformations with steric energy greater than 200 kJ/mol. When this was done, the spatial map (B) showed that the thiol group was

BOX 15.4 IDENTIFICATION OF AN ACTIVE CONFORMATION

The neuromuscular blocking agent **tubocurarine** is a fairly rigid structure where the two quaternary nitrogen atoms represent the pharmacophore. Molecular modelling allows the distance between these atoms to be measured as 11.527 Å (Fig. 1). **Decamethonium** also acts as a neuromuscular blocking agent, but it is an extremely flexible molecule, which means that a large number of conformations are possible.

The most stable conformation is the extended one where the quaternary nitrogens are 14.004 Å apart. Using molecular dynamics, a variety of different conformations for decamethonium can be generated as described in section 15.8.2. One of these conformations has the quaternary nitrogens 11.375 Å apart—a possible candidate for the active conformation (Fig. 2).

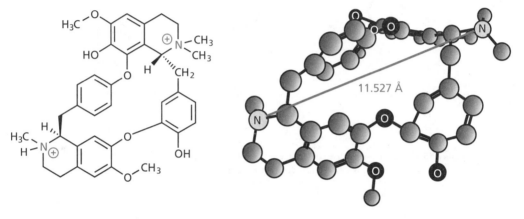

Figure 1 Computer generated model of tubocurarine (Chem3D).

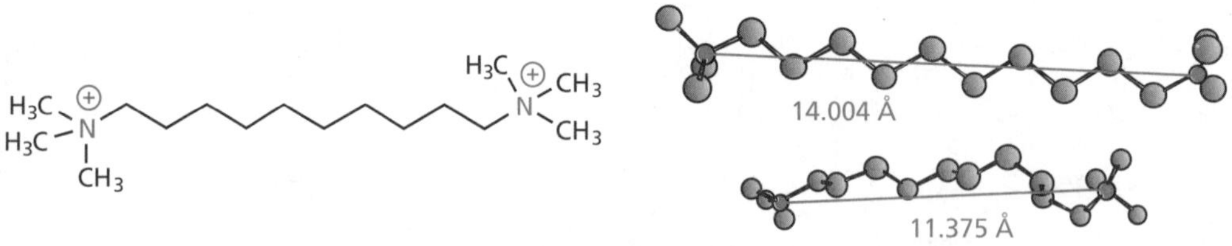

Figure 2 Computer generated conformations of decamethonium (Chem3D).

restricted to two main regions in space with respect to the other two binding groups.

A spatial map for one of the active rigid analogues was now generated in the same manner and compared with the one generated by captopril. The overlap between the maps was considered to be the most likely location for the thiol group. The process was then repeated for the other rigid analogues, further narrowing down the possible area that would be occupied by the thiol group. The study identified two 'hot spots' (C) for the thiol group. Conformations of captopril which placed the thiol group in those

'hot spots' were then considered to be likely active conformations.

15.11 3D pharmacophore identification

A 3D pharmacophore represents the relative position of important binding groups in space and disregards the molecular skeleton that holds them there. Thus,

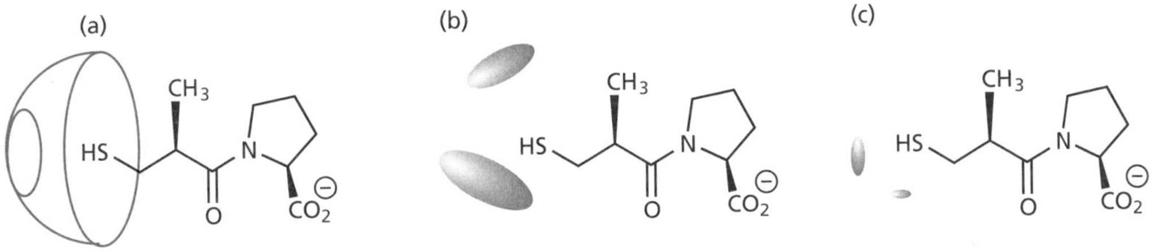

Figure 15.25 Generating spatial maps in conformational analysis. (a) All possible conformations, (b) stable conformations, and (c) after overlaps.

the 3D pharmacophore for a particular binding site should be common to all the various ligands which bind to it. Once the 3D pharmacophore has been identified, structures can be analysed to see whether they can adopt a stable conformation which will contain the required pharmacophore. If they do, and there are no steric clashes with the binding site, the structure should be active. There are several 3D chemical databases which can be searched for relevant structures, such as the CSD.

15.11.1 X-ray crystallography

The crystal structure of a target protein with its ligand bound to the binding site can be used to identify the 3D pharmacophore. The protein–ligand complex can be downloaded on to the computer and studied to identify the bonding interactions that hold the ligand in the binding site. This is done by measuring the distances between likely binding groups in the drug and complementary binding groups in the binding site, to see whether they are within bonding distance. Once the binding groups on the ligand have been identified, their positions can be mapped to produce the pharmacophore. The Brookhaven Protein Databank stores the crystal coordinates of proteins and other large macromolecules with and without bound ligands.

15.11.2 Structural comparison of active compounds

If the structure of the target is unknown, a 3D pharmacophore can be identified from the structures of a range of active compounds. Ideally, the active conformations and the important binding groups of the various compounds should be known. The molecules can then be overlaid as previously described in

section 15.9 to ensure that the important binding groups are matched up as closely as possible. It would be rare for the binding groups to match up exactly, so an allowed region in space for each important binding group can be identified for the 3D pharmacophore.

15.11.3 Automatic identification of pharmacophores

It is possible to identify possible 3D pharmacophores for a range of active compounds using some software programs, even if the important binding groups are unknown or uncertain. First of all, the program identifies potential binding centres in a particular molecule. These are the hydrogen bond donors, hydrogen bond acceptors, aromatic rings, acidic groups, and basic groups. It is also possible to search for hydrophobic centres involving hydrocarbon skeletons of three or more carbon atoms. Here, the hydrophobic centre is calculated as the midpoint of the carbon atoms in question.

If dopamine were the structure being analysed, four important binding centres would be identified—the aromatic ring, both phenolic groups (hydrogen bond donors and acceptors) and the amine nitrogen (hydrogen bond donor, hydrogen bond acceptor, base, and a positively charged centre if protonated) (Fig. 15.26). The program now identifies the various triangles which connect up the important centres. In the case of dopamine, there are apparently four such triangles. Each one is defined by the length of each side and the type of binding centres present, resulting in a set of pharmacophore triangles. Some of the points specified represent more than one type of binding centre, so this means that there will actually be more than four pharmacophore triangles. For example, if one of the points is a phenol, then it represents a hydrogen bond donor or a hydrogen bond acceptor. Therefore, any triangle including this point

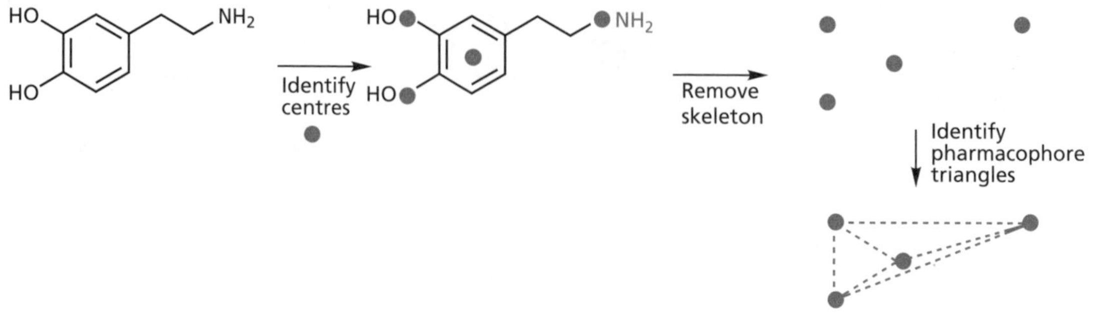

Figure 15.26 Pharmacophore ID in dopamine.

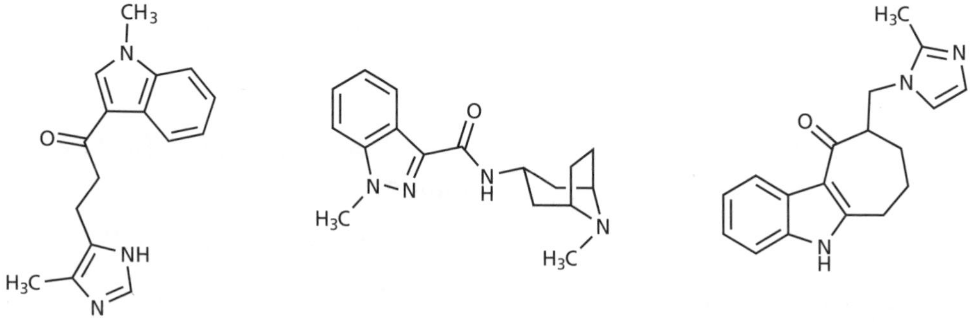

Figure 15.27 Test structures.

must result in two pharmacophore triangles—one for the hydrogen bond donor and one for the hydrogen bond acceptor.

Of course, this analysis has only been carried out on one conformation of dopamine. The program is now used to generate a range of different conformations as described in section 15.8.3, and for each conformation another set of pharmacophore triangles is defined. Adding all these together gives the total number of possible pharmacophore triangles for dopamine in all the conformations created.

Another structure with dopamine-like activity is now analysed. Once all its pharmacophore triangles have been determined, they are compared with those for dopamine and the pharmacophores that are common to both structures are identified. The process is then repeated for all the active compounds, until pharmacophore triangles common to all the structures have been identified. These are then plotted on a 3D plot where the x-, y-, and z-axes correspond to the lengths of the three sides of each triangle. This produces a visual display which allows easy identification of distinct pharmacophores. Closely

similar pharmacophores can be quickly spotted as they are clustered close together in specific regions of the plot.

For example, the three structures in Fig. 15.27 were analysed and found to have 38 common pharmacophores. When these were plotted, seven distinct groups of pharmacophore were identified (Fig. 15.28). Each pharmacophore present in the grid can be highlighted and revealed. For example, one of the possible active pharmacophores consists of two hydrogen bond acceptors and an aromatic centre. Note that it is advisable to begin this exercise with the most active compound and then proceed through the structures in order of activity.

The above analysis can be simplified enormously if certain groups are known to be essential for binding. The program can then be run such that only triangles containing these centres are included. For example, if the nitrogen atom of dopamine is known to be an essential bonding centre, then the triangles connecting the phenolic oxygens and the aromatic ring for each conformation can be omitted.

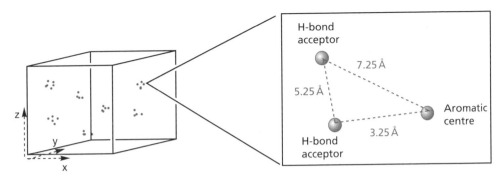

Figure 15.28 Pharmacophore plot.

15.12 Docking procedures

15.12.1 Manual docking

Molecular modelling can be used to dock or fit a molecule into a model of its binding site. If the binding groups on the ligand and the binding site are known, they can be defined by the operator such that each binding group in the ligand is paired with its complementary group in the binding site. The ideal bonding distance for each potential interaction is then defined and the docking procedure is started. The program then moves the molecule around within the binding site to try and get the best fit as defined by the operator. In essence, the procedure is similar to the overlay or fitting process described in section 15.9, only this time the paired groups are not directly overlaid but fitted such that the groups are within preferred bonding distances of each other. Both the ligand and the protein remain in the same conformation throughout the process and so this is a rigid fit. Once a molecule has been successfully docked, fit optimization is carried out. This is essentially the same as energy minimization, but carried out on the ligand–target protein complex. Different conformations of the molecule can be docked in the same way and the interaction energies measured to identify which conformation fits the best.

15.12.2 Automatic docking

Automatic docking can be carried out where the software itself decides how it will dock the ligand. First, an X-ray crystal structure of the target protein

is defined and downloaded on to the computer. The amino acids in the binding pocket are then left displayed while the remainder of the protein is hidden. (The structure is still present, but hiding the atoms makes analysis and visualization easier.)

The task for the docking program is twofold. First, it has to place the ligand within the active site in different orientations or binding modes. Second, it has to 'score' the different binding modes to identify the best ones. The most simple approach to automatic docking is to treat the ligand and the macromolecular target as rigid bodies. This is acceptable if the active conformation of the ligand is known or if the ligand is a rigid cyclic structure. At the next level of complexity, the target is still considered as a rigid body but the ligand is allowed to be flexible and to adopt different conformations. The most complex situation is where both the target and ligand are considered to be flexible. This last situation is extremely expensive in terms of computer time, and most docking studies are carried out by assuming a rigid target.

The first problem with any program is how to position the ligand within the active site. If you or I were handed real models of the target and the ligand, we would consider the space available in the binding site, eye up the ligand, and have a pretty good idea how we could place the ligand into the binding site before we actually do it. In other words, humans have a spatial awareness which includes the ability to judge the shape of an empty space. This does not come naturally to computers, and the empty space of a binding site has to be defined in a way that a computer program can understand before ligands can be inserted.

The DOCK program was one of the earliest programs to tackle this problem. It defined the empty space of the binding site by identifying a collection of different sized spheres which would fill up the binding site (Fig. 15.29) according to the following guidelines:

- Each sphere must touch the molecular surface of the binding site twice, but no more than twice.

- The said molecular surface is defined as the van der Waals radii of the atoms making up the binding site. Each point of the molecular surface that is contacted by a sphere represents the van der Waals radius of a particular atom in the binding site.

- The sphere of smallest radius is chosen for each pair of points on the molecular surface.

- Each atom of the binding site must have only one sphere in contact with its van der Waals radius. The procedure above may produce several spheres that make contact with a specific atom and different other atoms in the binding site. If that is the case, the sphere with the largest radius is chosen and the others are rejected.

- Spheres are allowed to overlap.

- Spheres with radii over 5 Å are rejected, as such spheres tend to project out of the binding site surface.

Once the analysis and filtering process has been carried out, there will be the same number of spheres as atoms in the binding site. The centre of each sphere then accurately defines a unique position of 3D space within the binding site.

A similar exercise is carried out for the ligand molecule. This time the spheres are chosen to touch the inner surface of the ligand's molecular surface such that the set of spheres selected represent the shape and size of the molecule. Again, there will be the same number of spheres as atoms in the ligand.

A matching operation is now carried out in a systematic fashion. A ligand sphere is matched to one of the receptor spheres. The distances between the centre of the ligand sphere and the centres of all the other ligand spheres are measured. The same operation is carried out for the receptor sphere and all the other receptor spheres. The program now strives to pair up two further spheres, one from the ligand and one from the receptor. It succeeds if it finds a pairing where the measured distances for the new pairing match the measured distances for the first pairing. Further pairings are then sought until no more can be found. The minimum number of pairing required for an acceptable docking is four.

The whole procedure is now repeated systematically with different ligand and receptor spheres used for the initial pairing to find a variety of docking modes.

A restriction can be included to reduce the number of initial pairings and subsequent calculation time. A centroid for the ligand spheres is defined which

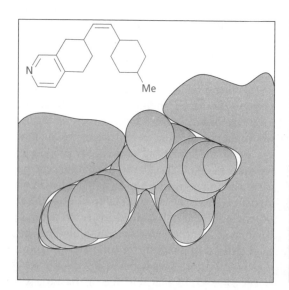
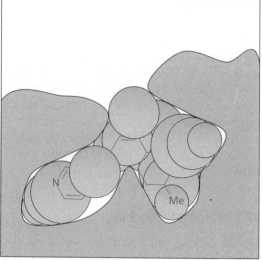

Figure 15.29 The DOCK program.

will represent the heart of the molecule. Similarly, a centroid for the receptor spheres is defined which represents the heart of the binding site. An initial pairing will only be permitted if the distance from the ligand centroid to the centre of a ligand sphere, matches the corresponding distance for the receptor centroid and sphere.

Once a set of pairings and a potential docking has been identified, the program fits the relevant ligand spheres onto their paired receptor spheres in much the same manner as an overlay (section 15.9). This is repeated for all the other possible docking or binding modes.

For each binding mode, a check is carried out to see whether there are any steric clashes between the atoms of the ligand and the atoms of the binding site. If there are any such clashes, the binding mode is disregarded. If not, then an interaction energy between the ligand and the binding site is calculated to give a score for that binding mode. The binding modes with the highest scores are then stored so that they can be analysed further by the operator. The calculation of the binding energy should obviously consider the binding interactions between the ligand and the binding site, but it should also take into account issues such as desolvation, the difference in energy between a ligand's different conformations, and the decrease in entropy resulting from a flexible molecule being bound in a fixed conformation. As far as binding interactions are concerned, calculations can be speeded up by pre-calculating what the electrostatic and van der Waals interactions will be at different positions of the binding site. This is done by creating a grid within the binding site and placing a probe at each point of the grid to measure the relevant interactions between the probe and the binding site (compare section 13.10). These values are then stored and automatically accessed whenever an atom of the ligand is placed at a particular grid point.

The original DOCK program considered the ligand and the target as rigid bodies. By incorporating the types of software required to generate different conformations (section 15.8), docking programs can be used to assess the possible binding modes of a flexible ligand with a rigid binding site. This increases the complexity and time taken for the docking study in line with the number of different conformations considered.

A different approach to docking was used by the ChemX program, and was based on the use of pharmacophores. It helps to know which amino acids in

the binding site are involved in binding, but if this information is not available, an analysis of potentially important binding centres in the binding site can be carried out (i.e. hydrogen bond donors and acceptors, acidic centres, basic centres, positive and negative centres, and aromatic rings). Once all the possible centres have been identified, a location is then defined in the binding site where a complementary group should be positioned for a bonding interaction (Fig. 15.30).

In this process, it is assumed that aromatic rings interact face on, but face-to-edge interactions could be included if necessary. The complementary centre for a hydrogen bond donor or acceptor is placed along the line of the carbon heteroatom bond. This is an unlikely position for a hydrogen bond, as the interaction is more likely to be at an angle to allow interaction with the lone pair of the heteroatom. However, the defined position is midway between the most likely positions for hydrogen bonding and a suitable tolerance value is chosen to ensure that the ideal positions are included (i.e. the complementary centres are treated as volumes of space rather than distinct points, with the size of the volume being determined by the operator). Having identified the potential complementary centres, all the possible pharmacophore triangles involving these centres are calculated (Fig. 15.31). A large number of triangles may be generated, and it is a good idea to simplify the situation before carrying out this step. For example, some centres can be rejected if they are on the margins of the binding site. On the other hand, other binding centres which are known to be important could be defined such that they must be included in pharmacophore triangles.

Structures can now be docked into the binding site. The program will do this by searching for pharmacophores in the structure which match pharmacophores in the binding site. Once a match is found, the pharmacophore triangles are overlaid, resulting in docking. Successful dockings are then studied individually to see what conformation or conformations fit.

This docking analysis can be carried out automatically on a range of compounds to distinguish those that are capable of fitting the binding site from those that are not. In this sort of analysis, the process can be speeded up by aborting the full search for all a structure's conformations as soon as one fit is found.

A 'bump filter' can also be included to quickly assess if a conformation has a bad steric interaction with the

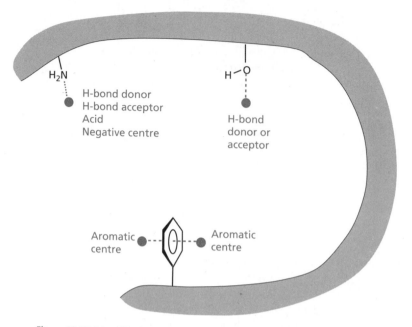

Figure 15.30 Identifying binding centres and locations for complementary groups.

Figure 15.31 Pharmacophore triangles from Fig. 15.30.

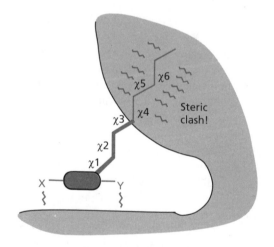

Figure 15.32 'Bump' filter.

binding site and then reject it. Such 'bump filter rejections' may also reject a range of other possible conformations if they too contain the undesired feature. For example in Fig. 15.32, the structure has been docked using groups X and Y. There is also a long alkyl chain which can take up a large number of conformations. The program systematically varies the torsion angles of this chain, working outwards from the heart of the molecule. If a conformation having specific torsion angles χ_1, χ_2, and χ_3 is rejected because of a bad steric interaction with the receptor, the program will not bother varying χ_4–χ_6.

Another way to identify the location of important binding centres in the binding site is to place a potential energy grid into the binding site and place probe atoms at every grid point, in order to calculate interaction energies between the probe and the receptor (compare section 13.10). The grid points with

the highest interaction energies are then retained as the centres for the binding site. The type of centre identified (i.e. hydrogen bonding, hydrophobic, etc.) depends on the type of probe atom used. When it comes to docking, positive scores are generated when the correct centre on the ligand is matched with the corresponding centre in the binding site. The molecule is moved around the binding site until a maximum score is achieved, and this is taken as the best fit.

A 'goodness of fit' is then measured, taking into account the shape fit (a measure of how well the ligand makes contact with the surface of the binding site) and the chemical fit (checking whether hydrogen bonding, polar, electrostatic or hydrophobic interactions are good or bad). Good interactions are scored positively and bad ones negatively.

A potential energy grid can be used for the calculation of ligand–receptor interaction energies. The idea is to pre-calculate interaction energies with the receptor for various atom types placed on the grid points. A 3D table of these interaction energies is then stored and when a ligand is docked, the atoms closest to each grid point are identified and the interaction energies automatically read off the table and summed to give a score for the goodness of fit.

15.13 Automated screening of databases for lead compounds

The automated docking procedures described in section 15.12.2 can be used to screen a variety of different 3D structures to see whether they fit the binding site of a particular target (**electronic screening** or **database mining**). This is useful for a pharmaceutical company wishing to screen its own or other chemical stocks (libraries) for suitable lead compounds.

Screening of databases can also be done purely by searching for suitable pharmacophores. The process is speeded up by a quick filter which eliminates any structure that does not contain the necessary centres. The operator has the ability to vary the tolerances involved in the search in order to find pharmacophores which nearly match the desired structures.

15.14 Protein mapping

Drug design is made easier if the structure of the target protein and its binding site are known. The best way of obtaining this information is from X-crystallography of protein crystals, preferably with a ligand bound to the binding site. Unfortunately, not all proteins are easily crystallized (e.g. membrane

proteins). In cases like this, model proteins and binding sites may be constructed to aid the drug design process.

15.14.1 Constructing a model protein

A model of a protein can be created using molecular modelling if the primary amino acid sequence is known and the X-ray structure of a related protein has been determined. Of particular interest in this respect was the protein **bacteriorhodopsin** which has been crystallized and its structure determined by X-ray crystallography (Fig. 15.33). Bacteriorhodopsin is structurally similar to **G-protein-coupled receptors** which contain seven transmembrane helices (section 6.3). Many of the important receptors in medicinal chemistry belong to this family of proteins and so the structure of bacteriorhodopsin has been a vital template in constructing what is termed as **homology models** of these membrane-bound receptors. By identifying the primary amino acid sequence of the target receptor and looking for suitable stretches

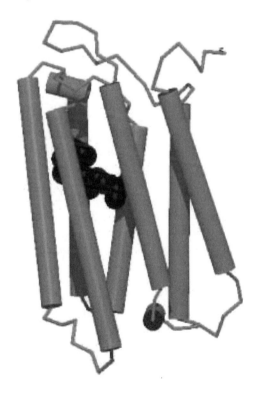

Figure 15.33 Bacteriorhodopsin with bound retinal.

of hydrophobic amino acids, it is possible to identify the seven transmembrane helices and then to use bacteriorhodopsin as a template in order to construct the helices in a similar position relative to each other. The linking loops can then be modelled in to give the total 3D structure. Unfortunately, bacteriorhodopsin is not a G-protein-coupled receptor and so it is not the ideal template for constructing these model receptors. More recently, in 2000, the crystal structure of the bovine receptor **rhodopsin** was successfully determined by X-ray crystallography. This *is* a G-protein-coupled receptor and provides a better template for the construction of more accurate receptor models.

If a new protein has been discovered, its primary structure is first determined. Suitable software is then used to compare its primary sequence with the primary sequences of other proteins in order to find a closely related protein. This involves comparing the sequences with respect to conserved amino acids, hydrophobic regions, and secondary structure. Once a reference protein of similar structure has been identified, it is used as a template in order to build the peptide backbone of the new protein. First of all, regions which are similar in the new protein and the template protein are identified. The backbone for the new protein is constructed to match the corresponding region in the template protein. This leaves connecting regions whose structure cannot be determined from the template. A suitable conformation for these intervening sequences might be found by searching the protein databases for a similar sequence in another protein. Alternatively, a loop may be generated to connect two known regions. Once the backbone has been constructed, the side chains are added in energetically favourable conformations. Energy minimization is carried out and the structure is refined with molecular dynamics in the absence and presence of ligand. Once the model has been constructed, it is tested experimentally. For example, the model would indicate that certain amino acids might be important in the binding site. These could then be mutated to see if this has an effect on ligand binding. Studies such as this have identified amino acids which are important in binding neurotransmitters in a range of G-protein-coupled receptors (Fig. 15.34).

This study shows interesting similarities and differences between the four receptors studied. For example, all four binding sites interact with ligands having a charged nitrogen group and contain a hydrophobic pocket lined with aromatic residues to receive it. There are several conserved aromatic residues in this pocket at positions 307, 613, and 616. An aspartate residue at position 311 is also present in all cases and is capable of forming an ionic interaction with the charged nitrogen on the ligand.

There are also significant differences between the binding sites, which account for the different ligand selectivities. For example, the amino acids at positions 505 and 508 in the catecholamine receptors are serine, whereas the corresponding amino acids in the cholinergic receptors are alanine. The amino acid at position 617 in the catecholamine receptors is phenylalanine, allowing an interaction with the aromatic portion of the catecholamines, whereas in the cholinergic receptor this amino acid is arginine, allowing a hydrogen bonding interaction with the ester group of acetylcholine.

15.14.2 Constructing a binding site

Rather than construct a complete model protein, it is possible to use molecular modelling to design a model binding site based on the structures of the compounds which bind to it. In order to do this effectively, a range of structurally different compounds with a range of activities is chosen. The active conformations are identified as far as possible and a 3D pharmacophore identified as described previously. The molecules are then aligned with each other such that their pharmacophores are matched. Each molecule is then placed in a potential energy grid and different probes are placed at each grid point in turn to measure interaction energies between the molecule and the probe atom (compare section 13.10). An aromatic CH probe is used to measure hydrophobic interactions, and an aliphatic OH probe is used to measure polar interactions. The interactions are then displayed by isoenergy contours (typically -6.0 kJ mol^{-1} for hydrophobic interactions and -17.0 kJ mol^{-1} for polar interactions). Figure 15.35 shows **altanserin** and the moieties of the molecule that can participate in hydrophobic interactions (e.g. the aromatic and heteroaromatic rings) as well as the moieties that can participate in hydrogen bonding (e.g. carbonyl oxygens or nitrogen).

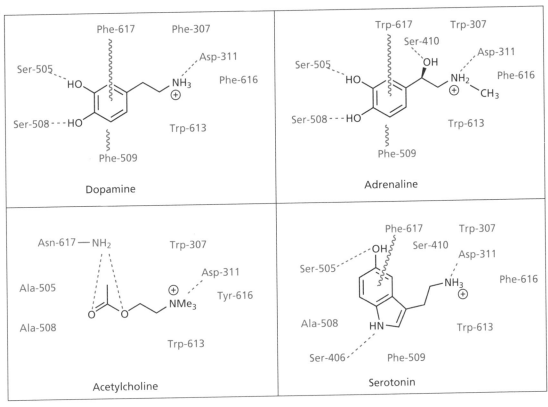

Figure 15.34 Important amino acids in various binding sites. (The numbering indicates the position of each amino acid on the seven possible helices of a G-protein-coupled receptor. For example, 311 indicates position 11 on helix 3.)

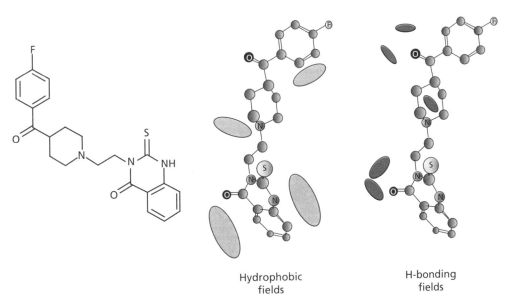

Figure 15.35 Analysis of hydrophobic and hydrogen bonding fields around altanserin.

The fields for all the molecules in the study can then be compared to identify common fields. Once these have been identified, suitable amino acids can be positioned to allow the required interaction. For example, an aspartate residue could be used to allow an ionic interaction, and amino acids such as phenylalanine, tryptophan, isoleucine, leucine, or valine could be used for a hydrophobic interaction.

Once built, known compounds can be docked to the model receptor binding site, the complex minimized, and binding energies calculated. These can then be compared with experimental binding affinities to see how well the model agrees with experiment. If the results make sense, the model can then be used for the design and synthesis of new agents (Box 15.5).

BOX 15.5 CONSTRUCTING A RECEPTOR MAP

A range of structures including altanserin and **ketanserin** was used to construct a model receptor binding site for the 5-HT$_{2a}$ receptor. Taking ketanserin as the representative structure for these compounds, various hydrogen bonding, ionic bonding and hydrophobic bonding interactions were identified. Structure–activity relationships (SAR) were then used to identify whether any of these proposed interactions were important or not. In this case, SAR indicated that the two carbonyl groups were not important, and so the hydrogen bonding regions derived from these groups probably do not exist in the receptor binding site. Suitable amino acids can now be placed in the relevant positions. The choice of which amino acids should be used is helped by knowing the amino acid sequence of the target protein and the structure of a comparable protein. The 5-HT$_{2a}$ receptor belongs to a superfamily of proteins that includes bacteriorhodopsin, whose structure is known. Allying this information with the primary amino acid sequence of the receptor led to the choice of amino acids shown in Fig. 1.

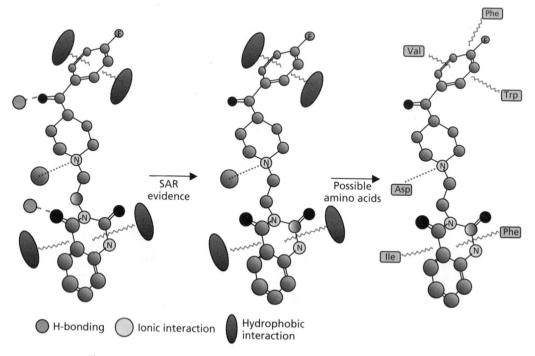

Figure 1 Receptor map for the 5HT$_{2a}$-receptor using ketanserin as representative ligand.

15.15 *De novo* design

De novo design involves the design of novel structures based on the structure of the binding site with which they are meant to interact.

15.15.1 Thymidylate synthase inhibitors

In theory, it should be possible to design a drug for a particular target if one knows the structure of the binding site. The process sounds quite straightforward. A molecular skeleton could be designed which would fit the binding site and fill the available space. Suitable functional groups could then be incorporated into the structure to ensure binding interactions with nearby amino acid residues.

In reality, *de novo* design is not as straightforward as it seems and it is rare to obtain an ideal compound purely by this method. In practice, *de novo* design has been better at designing structures which act as good lead compounds for further development. A good illustration is the design of inhibitors for the enzyme **thymidylate synthase**. This enzyme catalyses the methylation of **deoxyuridylate monophosphate** (**dUMP**) to **deoxythymidylate monophosphate** (**dTMP**) using **5,10-methylenetetrahydrofolate** as a coenzyme (Fig. 15.36). Inhibitors of this enzyme have been shown to be antitumour agents which prevent the biosynthesis of one of the required building blocks for DNA. Traditional inhibitors have been modelled on dUMP or the enzyme cofactor 5,10-methylenetetrahydrofolate (Fig. 15.37), which means that these inhibitors are structurally related to the natural substrate and cofactor. Unfortunately, this increases the possibility of side effects due to inhibition of other enzymes and receptors which use these molecules as natural ligands. Therefore, it was decided that a novel structure should be designed which was unrelated to either of the natural substrates.

Before starting the *de novo* design, a good supply of the enzyme was required. Although human thymidylate synthase was not readily available in large

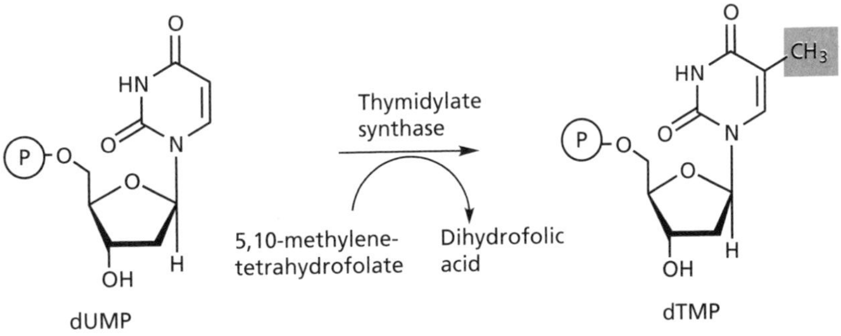

Figure 15.36 Reaction catalysed by thymidylate synthase.

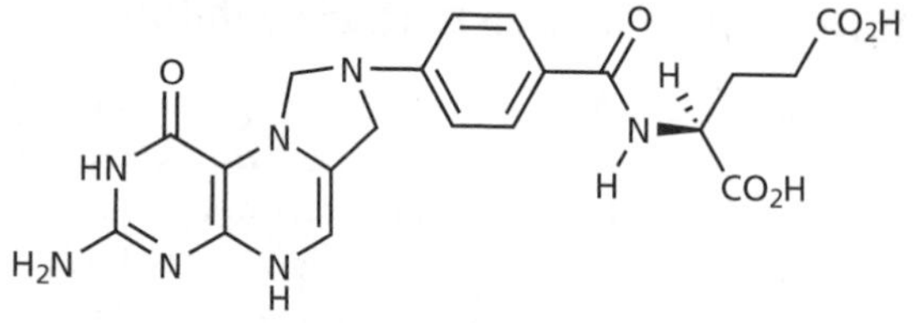

Figure 15.37 5,10-Methylenetetrahydrofolate.

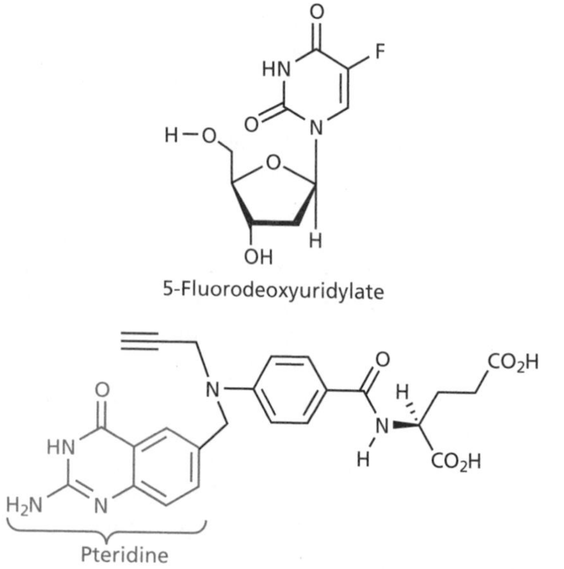

5-Fluorodeoxyuridylate

CB3717

Figure 15.38 Inhibitors of thymidylate synthase.

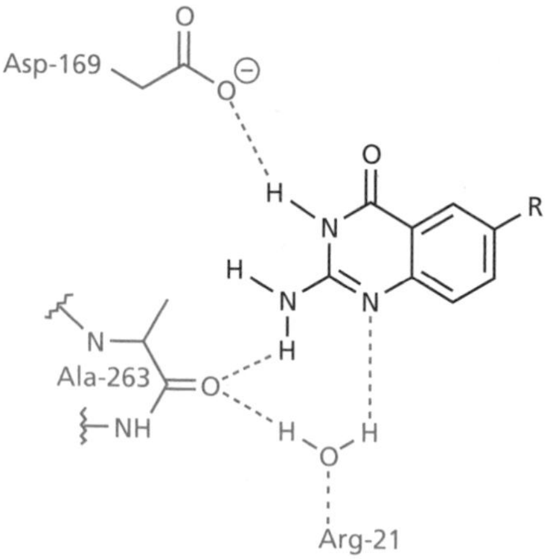

Figure 15.39 Binding interactions in active site.

quantities, it was possible to obtain good quantities of the bacterial version from *E. coli*, by using recombinant DNA technology to clone the gene and then expressing it in fast-growing cells. The bacterial enzyme is not identical to the human version, but it is very similar and so it was considered a reasonable analogue. The enzyme was crystallized along with the known inhibitors **5-fluorodeoxyuridylate** and **CB3717** (Fig. 15.38). These structures mimic the substrate and the coenzyme respectively and bind to the sites normally occupied by these structures. The structure of the enzyme–inhibitor complex was then determined by X-ray crystallography, and downloaded on to the computer.

A study of the enzyme–inhibitor complex revealed where the inhibitors were bound and also the binding interactions involved. For CB3717, the binding interactions around the pteridine portion of the inhibitor

were identified as involving hydrogen bonding interactions to two amino acids (the carboxylate ion of Asp-169 and the main chain peptide link next to Ala-263). There was also a hydrogen bonding interaction to a water molecule which acted as a hydrogen bonding bridge to Arg-21 (Fig. 15.39). Using molecular modelling, the inhibitor was deleted from the binding site to allow further analysis of the empty binding site. Generating the empty binding site from the enzyme–ligand complex is better than studying the empty binding site from the pure enzyme, because the latter does not take into account the induced fit that occurs on ligand binding.

A grid was set up within the binding site, and an aromatic CH probe was placed at each grid point to measure hydrophobic interactions and thus identify hydrophobic regions (see section 13.10). From this analysis, it was discovered that the pteridine portion of CB3717 was positioned in a hydrophobic pocket, despite the presence of the hydrogen bonding interactions which held it there. The boundaries of this hydrophobic region were determined and a naphthalene ring was found to be a suitable hydrophobic molecule to fit the pocket, yet still leave room for the addition of a functional group which would be capable of forming the important hydrogen bonds.

The functional group chosen was a cyclic amide which was fused to the naphthalene scaffold to create a naphthostyryl scaffold (Fig. 15.40). Modelling suggested that the NH portion of the amide would bind to Asp-169 while the carbonyl group would bind to the water molecule identified above. A substituent was now added to the naphthostyryl scaffold in order to

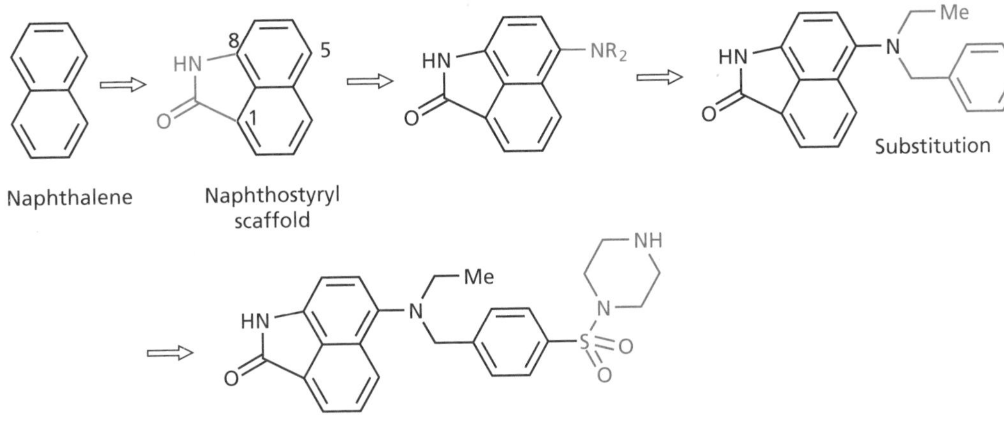

Figure 15.40 Design of an inhibitor.

gain access to the space normally occupied by the benzene ring of the cofactor. A dialkylated amine was chosen as the linking unit and was placed at position 5 of the structure. There were several reasons for this. First of all, adding an amine at this position was easy to carry out synthetically. Second, the two substituents on the amine could be easily varied, which would allow fine tuning of the compound. Last, by using an amine, it would be possible to have a branching point which could have different substituents without adding an asymmetric centre. If a carbon atom had been added instead, two different substituents would have inevitably led to an asymmetric centre.

Modelling demonstrated that a benzyl group was a suitable substituent for the amine in order to access the space normally occupied by the benzene ring of the cofactor. A (phenylsulfonyl)piperazine group was then added to the *para* position of the aromatic ring, in order to make the molecule more water soluble—a necessary property if the synthesized structures were to be bound to the enzyme and crystallized for further X-ray crystallographic studies.

This structure was now synthesized and tested for inhibition against both the bacterial and human versions of the enzyme and was found to be active, with higher activity for the human enzyme. A crystal structure of the novel inhibitor bound to the bacterial enzyme was successfully obtained and studied to see whether the inhibitor had fitted the binding site as expected. In fact, it was found that the naphthalene ring of the inhibitor was wedged deeper into the pocket than expected, because of more favourable hydrophobic interactions. As a result, the cyclic

amide failed to form the direct hydrogen bond interaction to Asp-169 which had been planned, and was hydrogen bonding to a bridging water molecule instead. The lactam carbonyl oxygen was also too close to Ala-263, and this had caused this residue to shift 1 Å from its usual position. This in turn had displaced the water molecule which had been the intended target for hydrogen bonding (Fig. 15.41).

By studying the actual position of the structure in the binding site, it was possible to identify four areas where extra substituents could fill up empty space and perhaps improve binding. These are shown in Fig. 15.42.

Various structures were proposed then overlaid on the lead compound (still docked within the binding site) to see whether they fitted the binding site. Only those which passed this test and were in stable conformations were synthesized and tested for activity (41 in total). The optimum substituent at each position was then identified.

- In region 1 (R^1), modelling showed that this substituent fitted into a hydrophobic pocket which became hydrophilic the deeper one got. This suggested that a hydrogen bonding substituent at the end of an alkyl chain might be worth trying (*extension*) and, indeed, a CH_2CH_2OH group led to an improvement in binding affinity. It was also found that a methyl group was better than the original ethyl group.

- In region 2 (R^2), the carbonyl oxygen was replaced by an amidine group which would be capable of hydrogen bonding to the carbonyl oxygen of

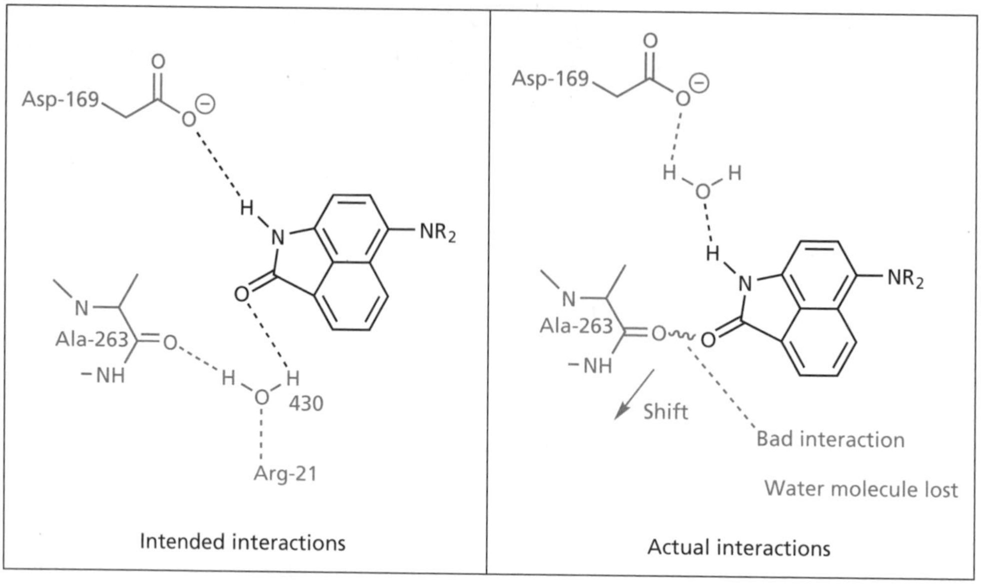

Figure 15.41 Intended versus actual interactions.

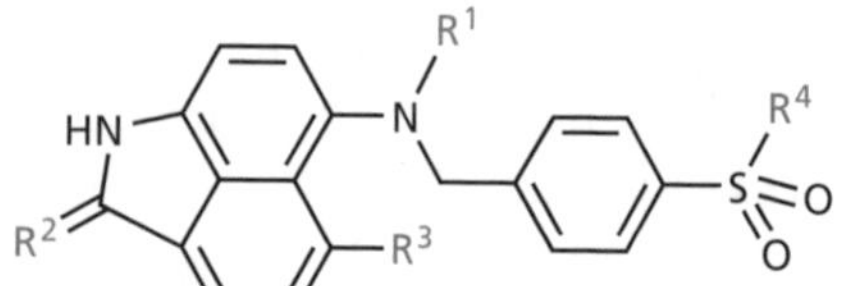

Figure 15.42 Variable positions (in colour).

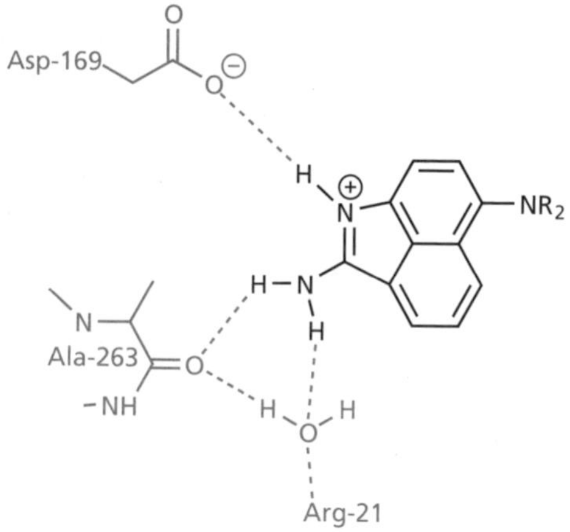

Ala-263, rather than repelling it. An added advantage in using a basic amidine group was the fact that there was a good chance that it would become protonated, allowing a stronger ionic interaction with Asp-169 as well as a better hydrogen bonding interaction with Ala-263. When this structure was synthesized, it was found to have improved inhibition, and a crystal structure of the enzyme–inhibitor complex showed that the expected interactions were taking place (Fig. 15.43). Moreover, Ala-263 had returned to its original position, allowing the return of the bridging water molecule.

• In region 3 (R³), there was room for a small group such as a chlorine atom or a methyl group and both of these substituents led to an increase in activity.

• Region 4 (R⁴) was relatively unimportant for inhibitory activity, since groups at this position protrude out of the active site into the surrounding solvent and have only minimal contact with the enzyme. Nevertheless, the piperazine ring was replaced by a morpholine group because the latter

Figure 15.43 Binding interactions.

had some advantages with respect to selectivity and pharmacological properties.

Having identified the optimum groups at each position, structures were synthesized combining some or all of these groups. The presence of the amidine resulted in the best improvement in activity and so the presence of this group was mandatory. Interestingly, adding all the optimum groups is not as beneficial as adding some of them.

The modified structure (Fig. 15.44) was synthesized and was found to be a potent inhibitor which was 500 times more active than the original amide. A crystal

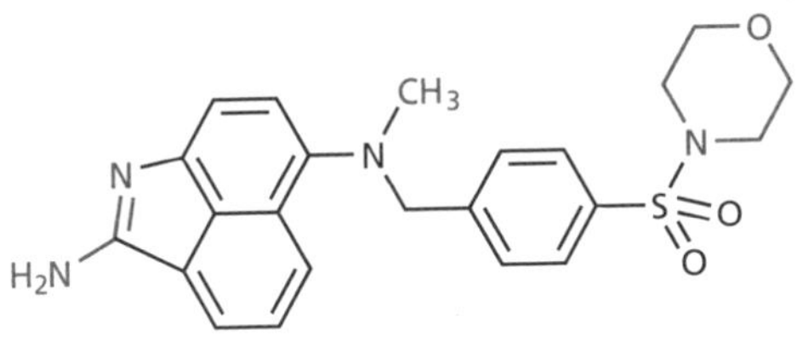

Figure 15.44 Modified inhibitor.

structure of the enzyme–inhibitor complex showed a much better fit, and the compound was put forward for clinical trials as an antitumour agent.

A couple of general points are worth mentioning about the strategies involved in *de novo* design. First, it may be tempting to design a molecule which completely fills the available space in the binding site. However, this would not be a good idea for the following reasons:

- The position of atoms in the crystal structure is accurate only to 0.2–0.4 Å and allowance should be made for that.

- It is possible that the designed molecule may not bind to the binding site exactly as predicted. If the intended fit is too tight, a slight alteration in the binding mode may prevent the molecule binding at all. It would be better to have a loose-fitting structure in the first instance and to check whether it binds as intended. If it does not, the loose fit gives the molecule a chance to bind in an alternative fashion.

- It is worth leaving scope for variation and elaboration in the molecule. This allows fine tuning of the molecule's binding affinity and pharmacokinetics.

Other points to take into consideration in *de novo* design are the following:

- Flexible molecules are better than rigid molecules, because the former are more likely to find an alternative binding conformation should they fail to bind as expected. This allows modifications to be carried out based on the actual binding mode. If a rigid molecule fails to bind as predicted, it may not bind at all.

- It is pointless designing molecules which are difficult or impossible to synthesize.

- Similarly, it is pointless designing molecules which need to adopt an unstable conformation in order to bind.

- Consideration of the energy losses involved in water desolvation should be taken into account.

- There may be subtle differences in structure between receptors and enzymes from different species.

For example, the computer modelling studies described above were carried out on bacterial thymidylate synthase enzyme rather than the human version. Fortunately, the activities of the designed inhibitors were actually greater for the human enzyme than for the bacterial enzyme and this was put down to the fact that the hydrophobic space available for the naphthalene ring was larger in the human enzyme than in the bacterial one. Fortunately most changes carried out had beneficial effects for both enzymes, with one exception: adding a methyl group at R^3 to the amidine led to an increase in activity for the bacterial enzyme but not for the human enzyme.

15.15.2 Automated *de novo* design

Computer software programs have been written which automatically design novel structures to fit known binding sites. One of the best known of these is called **LUDI**, which works by fitting molecular fragments to different regions of the binding site then linking the fragments together (Fig. 15.45). There are three stages to the process.

Stage 1: Identification of interaction sites

First of all, the atoms present in the binding site are analysed to identify those that can take part in hydrogen bonding interactions and those that can take part in van der Waals interactions. Oxygen atoms and tertiary nitrogen atoms are identified as hydrogen bond acceptors. A hydrogen attached to oxygen or nitrogen is identified as a hydrogen bond donor. Aromatic and aliphatic carbons are identified as such, and are capable of taking part in van der Waals interactions.

Interaction sites can then be defined. These are positions in the binding site where a ligand atom could be placed to interact with any of the above atoms. For example, suppose the binding site contains a methyl group (Fig. 15.46). The program would identify the carbon of that group as an aliphatic carbon capable of taking part in van der Waals interactions. This is a non-directional interaction, so a sphere is constructed around the carbon atom with a radius corresponding to the ideal distance for such an interaction (4 Å).

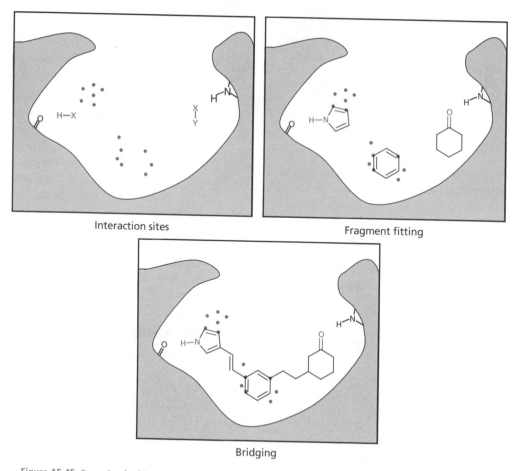

Figure 15.45 Stages involved in automated *de novo* design using LUDI. H–X, hydrogen bond donor interaction site; X–Y, hydrogen bond acceptor interaction site. The dots indicate aromatic interaction sites.

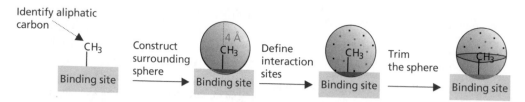

Figure 15.46 Identification of aliphatic interaction sites round a methyl group (LUDI).

A number of points (normally 14) are then placed evenly over the surface of the sphere to define aliphatic interaction sites. Regions of the sphere which overlap or come too close to the atoms making up the binding site (i.e. less than 3 Å separation) are rejected, along with any of the 14 points that were on that part of the surface. The remaining points are then used as the aliphatic interaction sites. A similar process is involved in identifying aromatic interaction sites surrounding an aromatic carbon atom.

Identifying interaction sites for hydrogen bonds is carried out in a different fashion. As hydrogen bonds are directional, it is important to define not only the distance between the ligand and the binding region, but the relevant orientation of the atoms as well. This can be done by defining the hydrogen bond interaction site as

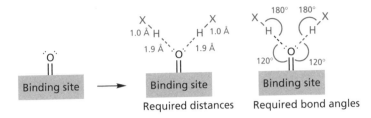

Figure 15.47 The interaction sites for a hydrogen bond donor, represented by H–X (LUDI).

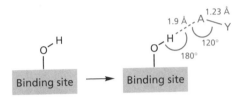

Figure 15.48 The interaction site for a hydrogen bond acceptor represented by A–Y (LUDI). A denotes the hydrogen bond acceptor.

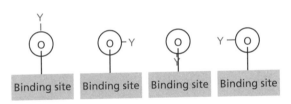

Figure 15.49 Four possible interaction sites for A–Y (A is hidden).

a vector involving two atoms. The position of these atoms is determined by the ideal bond lengths and bond angles for a hydrogen bond. For example, if the binding site has a carbonyl group present, then there are two possible hydrogen bond interaction sites (X–H) which can be determined (Fig. 15.47).

If the binding site has a hydrogen bond donor present, then interaction sites for a hydrogen bond acceptor would be determined in a similar fashion. For example, Fig. 15.48 shows how the interaction site for a hydrogen bond acceptor is determined when the binding site has a hydroxyl group present.

There are in fact four interaction sites that are normally calculated in this situation. The other three can be visualized if we take a viewpoint view along the line of atoms O–H–A and vary the relative position of Y as shown (Fig. 15.49).

As with the van der Waals interaction sites, hydrogen bond interaction sites are checked to ensure that

they are no closer than 1.5 Å from any atom present in the binding site. If they are, they are rejected.

Stage 2: Fitting molecular fragments

Once interaction sites have been determined, the LUDI program accesses a library of several hundred molecular fragments such as those shown in Fig. 15.50. The molecules chosen are typically 5–30 atoms in size and are usually rigid in structure because the fitting procedure assumes rigid fragments. Some fragments are included which can adopt different conformations. For these fragments, a selection of different conformations has to be present in the library if they are to be represented fairly in the fitting process. Each conformation is treated as a separate entity during the fitting process.

The atoms which are going to be used in the fitting process have to be predetermined for each fragment. Similarly, the interaction sites on to which each atom can be fitted has to be predetermined. For example, the methyl carbons of an acetone fragment are defined as aliphatic and can only be fitted onto aliphatic interaction sites. The carbonyl group is defined as a hydrogen bond acceptor and can only be fitted on to that interaction site (Fig. 15.51). The best fit will be the one that matches up the fragment with the maximum number of interaction sites.

The program can 'try out' the various fragments in its library and identify those that can be matched up or fitted to the available interaction sites in the binding site.

Stage 3: Fragment bridging

Once fragments have been identified and fitted to the binding site, the final stage is to link them up. The program first identifies the molecular fragments that are closest to each other in the binding site, then identifies the closest hydrogen atoms (Fig 15.52). These now define the link sites for the bridge. The program now tries out various molecular bridges from

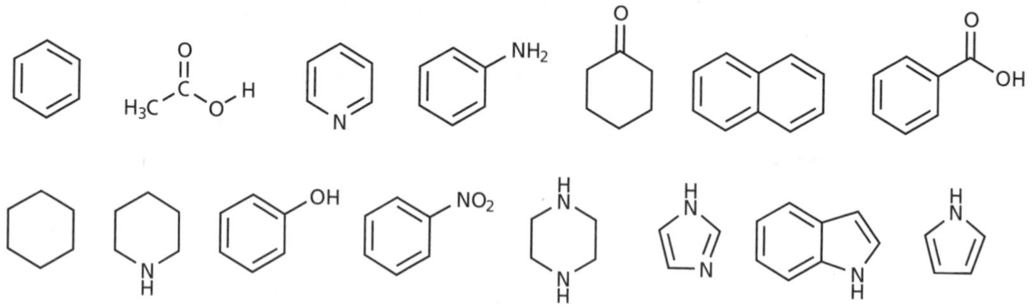

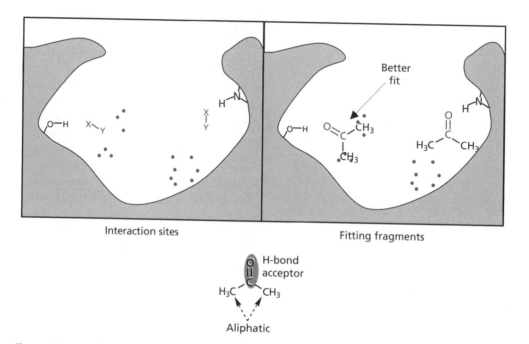

Figure 15.50 Examples of molecular fragments used by LUDI.

Interaction sites

Better fit

Fitting fragments

H-bond acceptor

Aliphatic

Figure 15.51 Fitting fragments. X–Y, hydrogen bond acceptor interaction site. The dots indicate aliphatic interaction sites.

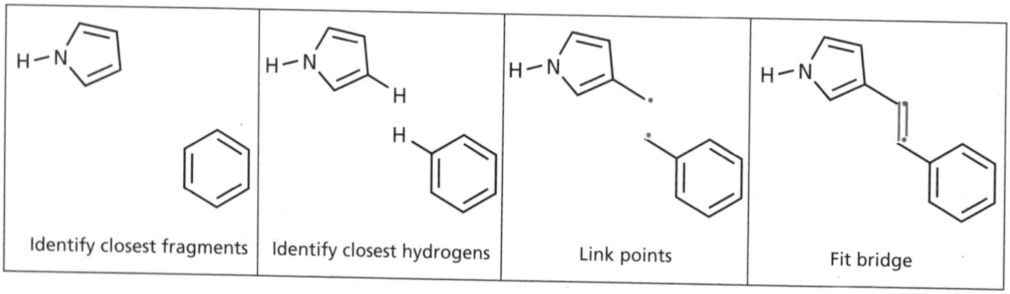

Identify closest fragments | Identify closest hydrogens | Link points | Fit bridge

Figure 15.52 The bridging process (LUDI).

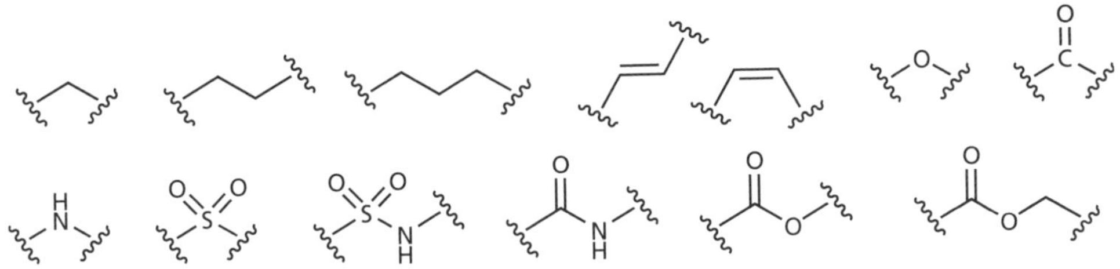

Figure 15.53 Examples of molecular bridges (LUDI).

a stored library to find out which one fits best. Examples of the types of molecular bridges that are stored are shown in Fig. 15.53. Once a suitable bridge has been found, a final molecule is created.

KEY POINTS

• *De novo* design involves the design of a novel ligand, based on the structure of the binding site.

• An X-ray crystal structure of the target protein complexed with a ligand allows identification of the binding site and the binding mode of the ligand.

• A new ligand should initially be loose fitting and flexible to allow for any alterations in the way binding takes place compared to what is predicted.

• The X-ray structure of the protein complexed with the new ligand will give valuable information on the actual binding mode of the new ligand and allow modifications to be made which will maximize bonding.

• The new ligand should be capable of synthesis and be capable of interacting with the binding site using a stable conformation.

• Energy losses resulting from desolvation of the ligand should be taken into account when calculating stabilization energies arising from ligand–protein binding.

• Automated programs for *de novo* design identify interaction sites in the binding site, then match molecular fragments to these sites. Bridges are then designed to link the fragments.

15.16 Planning combinatorial syntheses

Combinatorial synthesis (Chapter 14) is a method of creating a large number of compounds on a small scale and in a short time. The combinatorial synthesis

could be carried out to synthesize all the possible compounds from the available starting materials and reagents. However, molecular modelling can help to focus the study so that a smaller number of structures are made but the chances of success are retained.

One method of doing this is based on the identification of **pharmacophore triangles** (section 15.11.3). Let us assume that the combinatorial synthesis is being carried out to generate 1000 compounds with as diverse a range of structures as possible. The number of different pharmacophores generated from the 1000 compounds would be an indication of the structural diversity. Therefore, a library of compounds which generates 100 000 different pharmacophores would be superior to a library of similar size which produces only 100 different pharmacophores. Pharmacophore searching can be done on all the possible synthetic targets to select those structures that will give the widest diversity. These compounds would then be synthesized.

First, all the possible synthetic targets are automatically ranked on their level of rigidity. This can be achieved by identifying the number of rotatable bonds. Pharmacophore searching then starts with the most rigid structure. All the possible pharmacophore triangles are identified for the structure. If different conformations are possible, these are generated and the various pharmacophore triangles arising from these are added to the total. The next structure is then analysed for all of its pharmacophore triangles. Again, triangles are identified for all the possible conformations. The pharmacophores from the first and second structures are then compared. If more than 10% of the pharmacophores from the second structure are different from those of the first, both structures are added to the list for the intended library. Both sets of pharmacophores are combined, and the next structure is analysed for all of its pharmacophores. These are compared with the total number of pharmacophores from structures 1 and 2 and if there are 10% new pharmacophores represented, the third structure is added to the list and the pharmacophores for all three structures are added together for comparison with the next structure. This process is repeated throughout all the target structures, eliminating all compounds which generate less than 10% of new pharmacophores. In this way, it is possible to cut the number of structures which need to be synthesized by 80–90% with only a 10% drop in the number of pharmacophores generated.

There is a good reason for starting this analysis with a rigid structure. A rigid structure has only a few conformations, and there is a good chance that most of these will be represented when the structure interacts with its target. Therefore, one can be confident that the associated pharmacophores are also represented. If the analysis started with a highly flexible molecule having a large number of conformations, there is less chance that all the conformations and their associated pharmacophores will be fairly represented when the structure meets its target binding site. Rigid structures which express some of these conformations more clearly would not be included in the library, as they would be rejected during the analysis. As a result, some pharmacophores which should be present are actually left untested.

It is possible to use modelling software to carry out a substituent search when planning a combinatorial synthetic library. Here, one defines the common scaffold created in the synthesis, as well as the number of substituents which are attached and their point of attachment. Next, the general structures of the starting materials used to introduce these substituents are defined. The substituents which can be added to the structure can then be identified by having the computer search databases for commercially available starting materials. The program then generates all the possible structures which can be included in the library, based on the available starting materials. Once these have been identified, they can be analysed for pharmacophore diversity as described above.

Alternatively, the various possible substituents can be clustered into similar groups on the basis of their structural similarity. This allows starting materials to be pre-selected, choosing a representative compound from each group. The structural similarity of different substituents would be based on a number of criteria such as the distance between important binding centres, the types of centre present, particular bonding patterns, and functional groups.

15.17 Database handling

The development of a drug requires the analysis of a large amount of data. For example, activity against a range of targets has to be measured to ensure that the compounds have not only good activity against their intended target, but also good selectivity with respect to a range of other targets. When it comes to rationalizing results, many other parameters have to be considered such as molecular weight, log P, and pK_a. The handling of such large amounts of data requires dedicated software.

Several software programs are available for the handling of data which allow medicinal chemists to assess biological activity versus physical properties, or to compare the activities of a series of compounds at two different targets. Such programs permit results to be presented in a visual qualitative fashion, allowing a quick identification of any likely correlations between different sets of data. For example, if one wanted to see whether the log P value of a series of compounds was related to their α- or β-adrenergic activity, then a 2D plot could be drawn up comparing α-adrenergic activity to β-adrenergic activity. The log P value of each compound could then be indicated by a colour code for the various points on the plot. In this way, it would be easy to see whether these three properties were related. Such an analysis might show, for example, that a high log P is associated with compounds having low α-adrenergic activity and high β-adrenergic activity.

Some programs can be used to assess the biological results from a combinatorial synthetic study. First the scaffold used in the synthesis is defined, then the substituents are defined. Once the biological test results are obtained, a tree diagram can be drawn up to assess which substitution point is most important for activity. For example, supposing there were three substitution points on the scaffold, the program could analyse the data to identify which of the substitution points was the most important in controlling the activity. The data relevant for this particular substituent could then be split into three groups corresponding to good, average, and poor activity. For each of these groups, the program could be used to identify the next most important substitution point, and so on.

- Molecular modelling can be used to plan intended combinatorial libraries such that the maximum number of pharmacophores are generated for the minimum number of structures.

- Structures are analysed for their various conformations and resulting pharmacophores, starting with the most rigid structures.

- Each structure is compared with a growing bank of pharmacophores to assess whether it presents a significantly different number of pharmacophores compared to the structures that went before.

Serotonin (5-HT)   *meta*-Chlorophenylpiperazine (*m*CPP)

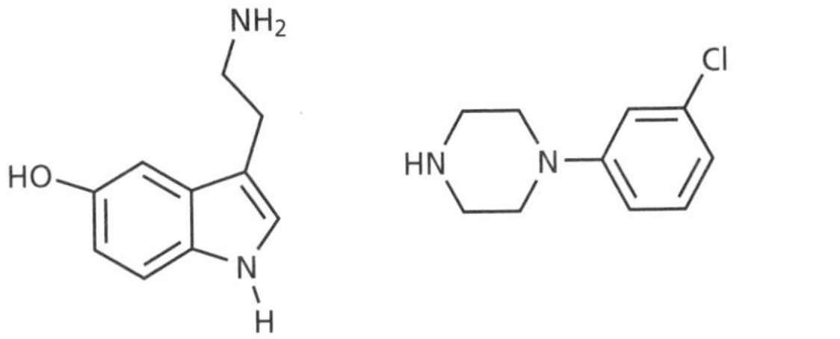

Figure 15.54 Serotonin and *m*CPP.

15.18 Case study

It was once claimed that drugs could be designed using molecular modelling alone. However, to date, no drug has been designed purely by molecular modelling. Even the thymidylate synthase inhibitors were modified in the light of experimental results. Molecular modelling should be seen as one of several powerful tools which the medicinal chemist has to hand. The following case study illustrates how computer modelling has been used as an aid in drug design alongside many of the tools and strategies described in previous chapters. The project in question was carried out by SmithKline Beecham and was aimed at finding novel anxiolytic and antidepressant agents.

15.18.1 The target

First, a suitable target had to be identified and it was decided to design an antagonist for a serotonin receptor. **Serotonin** (or 5-hydroxytryptamine) (Fig. 15.54) is an important neurotransmitter in the central nervous system, and abnormalities in serotonin levels are thought to be involved in a variety of disorders such as anxiety, depression, and migraine. Antagonists are therefore useful in treating these diseases. Unfortunately, these agents have various problems and side effects associated with them. The discovery of various types of serotonin receptor allows the possibility of designing antagonists which might be more selective in their action and have less side effects as a result.

There are seven main types of serotonin receptor (5-HT$_1$–5-HT$_7$) and several subtypes of these. For example, the 5-HT$_2$ receptor consists of three subtypes, (5-HT$_{2A}$, 5-HT$_{2B}$, and 5-HT$_{2C}$). The 5-HT$_{2C}$ receptor is of particular interest, as there is some evidence that this receptor might be involved in anxiety. For example, **meta-chlorophenylpiperazine** (*m*CPP) is an agonist which shows some selectivity for the 5-HT$_{2B}$ and 5-HT$_{2C}$ receptors and has been shown to cause anxiety in animal and human studies. As the 5-HT$_{2B}$ receptor is mainly in the peripheral nervous system, it is likely that these central nervous system effects are caused by the agonist acting on the 5-HT$_{2C}$ receptor which is only present in the central nervous system. The serotonin antagonists which were on the market at the start of this project did not show any selectivity for the 5-HT$_{2C}$ over the 5-HT$_{2A}$ receptor, and so it was argued that a selective antagonist might have improved properties over established drugs.

Having identified a likely target, the aim was now to identify a suitable profile of activity for the intended drug. Clearly, the antagonist should be as selective as possible for the 5-HT$_{2C}$ receptor in order to reduce the possibility of side effects. In particular, the challenge was to obtain selectivity with respect to the 5-HT$_{2A}$ receptor. Furthermore, the new drug should not affect metabolic enzymes (e.g. cytochrome P450 enzymes), again to reduce possible side effects and to avoid the possibility of drug–drug interactions. The ideal drug should also be non-sedating, have no interaction with alcohol, have a fast onset of action and a high response rate, and show no withdrawal effects.

15.18.2 Testing procedures

Both *in vitro* and *in vivo* tests were required for the study. The *in vitro* tests involved radioligand binding studies carried out on human 5-HT$_{2A}$, 5-HT$_{2B}$,

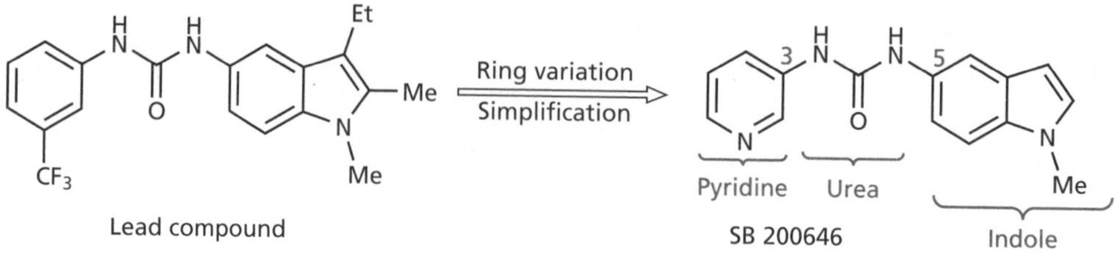

Figure 15.55 Simplification and ring variation.

and 5-HT$_{2C}$ receptors which had been cloned and expressed in fast growing cells. Testing for any inhibitory activity against cytochrome P450 enzymes was also done *in vitro*.

In vivo activity was measured by the ability of compounds to block hypoactivity in rats, brought on by administering *m*CPP.

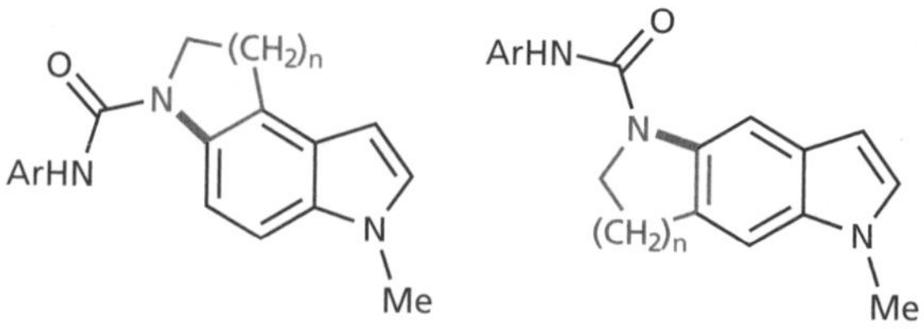

15.18.3 Lead compound to SB 200646

The lead compound for the project was a drug that had been produced by Lilly Pharmaceuticals. One of the problems with this compound was its insolubility in water, and so it was decided to replace the phenyl ring with a more polar pyridine ring (*ring variation*) (Fig. 15.55).[1] Various analogues were synthesized to find the best substitution positions for the urea group linking the ring systems, for both the pyridine ring and the indole ring. This revealed that 3- and 5-substitution respectively were ideal (*substituent variation*). Finally, the methyl and ethyl substituents on the indole ring were not essential and could be absent (*simplification*). The final compound (**SB 200646**) proved to be the first selective 5HT$_{2B/2C}$ antagonist, having modest 5HT$_{2C}$ affinity *in vitro*, some oral activity *in vivo*, and a 50-fold selectivity over the closely related 5HT$_{2A}$ receptor.

15.18.4 SB 200646 to SB 206553

The urea functional group is the most flexible region of SB 200646 and so the next tactic was to reduce this flexibility by locking one of the bonds into a ring (*rigidification*). The aim here was to reduce the total number of conformations, while retaining the active conformation. Four conformationally restrained structures were prepared (Fig. 15.56; n = 1, 2). In both

Figure 15.56 Rigidification (locked bond shown in bold).

cases, the introduction of a six-membered ring decreased affinity, whereas the five-membered ring increased affinity.

The difference in affinity between the six- and five-membered rings was rationalized by a molecular modelling study, which showed a marked difference in geometry between the two ring systems (Fig. 15.57). With a five-membered ring (A), the overall structure is roughly planar. This is not the case with the six-membered ring (B) where the overall structure is more twisted. Clearly, the planar molecule interacts with the receptor more strongly than the twisted molecule.

Therefore, rigidification was carried out using the five-membered ring, leading to **SB 206553** (Fig. 15.58) and resulting in a 10-fold increase in *in vitro* affinity and a 160-fold selectivity over the 5-HT$_{2A}$ receptor. The compound was also shown to relieve anxiety *in vivo*, with a 4-fold increase in potency in the rat hypolocomotion assay. Unfortunately, this compound was found to undergo metabolism in the body resulting in *N*-demethylation (section 8.4.2) and the production of a non-selective, active metabolite.

15.18.5 Analogues of SB 206553

Various strategies were tried to replace or stabilize the metabolically labile *N*-methyl group of SB 206553. First, a number of tricyclic analogues of SB 206553 were synthesized to explore the effects of fusing the

---

[1] The italicized terms in brackets throughout this case study refer to strategies described in Chapters 10 and 11.

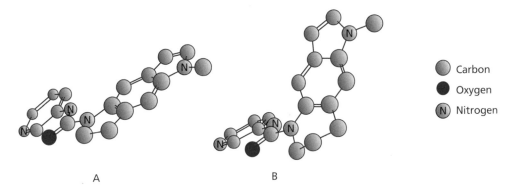

Figure 15.57 Relative geometries of the rigidified analogues.

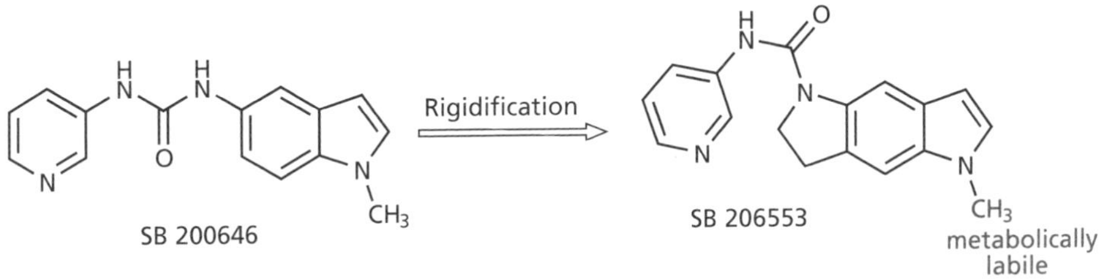

A   B

Carbon
Oxygen
N Nitrogen

Rigidification

SB 200646

SB 206553
metabolically
labile

Figure 15.58 Development of SB 206553.

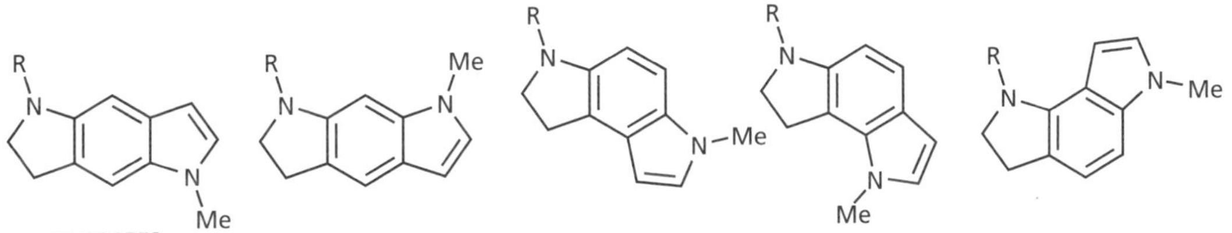

R   Me   R   R   R

Me

N—Me   N—Me

SB 206553
Me   Me

Figure 15.59 Tricyclic analogues of SB 206553 (R = CONH(3-pyridyl)).

five-membered ring at different positions of the aromatic ring (*group shift*) (Fig. 15.59). Such a variation shifts the methyl group relative to the benzene ring, which might make it unrecognizable to metabolic enzymes, yet still allow the molecule to interact with its target binding site. Unfortunately, all of these compounds showed decreased activity and selectivity, demonstrating that this strategy failed to make the required distinction.

Removing the *N*-methyl group from the indole ring to replace it with NH led to a loss of affinity, suggesting that the methyl group was important and might bind to a hydrophobic pocket in the binding site. In order to explore the size of this proposed pocket, the methyl group was replaced by larger alkyl

groups (*varying substituents*) to see whether this would lead to increased hydrophobic interactions and an increased affinity. It was found that groups such as Et, Pr, and ^{i}Pr resulted in similar 5-HT$_{2C}$ affinity and slightly increased selectivity, whereas an *N*-benzyl group was bad for both affinity and selectivity. Substitution elsewhere in the pyrrole ring was also bad for affinity (Fig 15.60).

An isostere for the pyrrole ring was now sought, and various five-membered heterocycles were tried (Fig. 15.61) (*ring variation and isosteres*). A thiophene ring led to a loss in selectivity and this was attributed to the loss of the methyl group which could no longer fit the proposed hydrophobic pocket. A substituted furan ring was used instead, with a methyl substituent positioned at an analogous position. As this group is attached to carbon rather than nitrogen it was expected that it would be more stable to metabolism. The resulting compound retained good *in vitro* affinity and selectivity, but had poor oral activity *in vivo*. However, *in vivo* activity was better if the compound was administered intravenously, suggesting that the compound was poorly absorbed from the gut or was rapidly metabolized in the liver.

15.18.6 Molecular modelling studies on SB 206553 and analogues

Having synthesized several analogues, molecular modelling was carried out to overlay both active and inactive compounds, using the urea group as the common feature. The space occupied by each molecule was then analysed and compared with its affinity for both the 5-HT$_{2A}$ and 5-HT$_{2C}$ receptors. It was then assumed that compounds which showed poor activity must have substituents in sterically disallowed areas of the relevant binding site. This allowed a map to be created (Fig. 15.62) showing a volume in space (coloured dark blue) which could be accessed by molecules having 5-HT$_{2C}$ affinity, and which included a volume (coloured light blue) that was disallowed for 5-HT$_{2A}$ affinity. This crucial light blue area was in the region of the *N*-methyl group of SB 206553. (The pyridine ring

was not included in the calculations because of the large number of conformations it can adopt.)

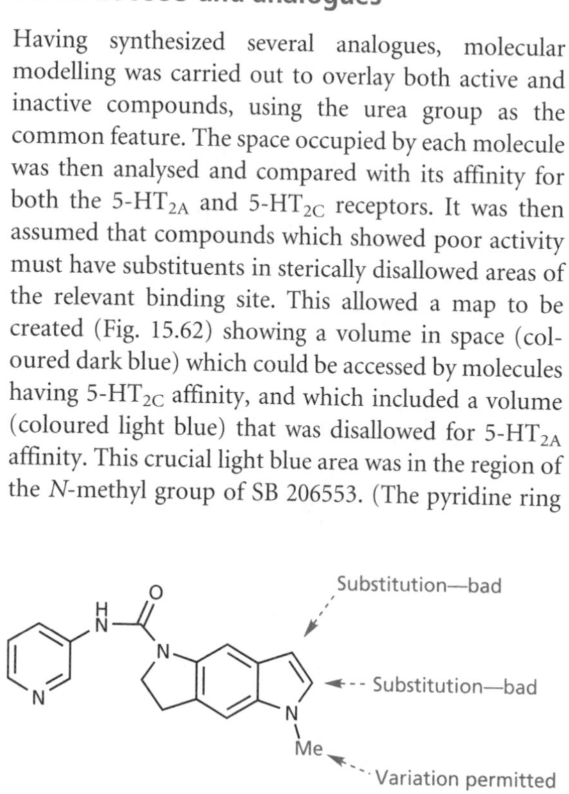

A model 5-HT$_{2C}$ receptor was now built using molecular modelling software. The amino acid sequence for the 5-HT$_{2C}$ receptor was obtained and compared with other related serotonin receptors to identify the likely transmembrane regions. The receptor is a member of the 7-TM receptor family (section 6.3) and so a model was built by analogy with the crystal structure of **bacteriorhodopsin** to create a model binding site for the receptor (section 15.14).

The next stage was to dock SB 206553 into the binding site. Standard agonists and antagonists for serotonin receptors have a strongly basic nitrogen which binds to an aspartate residue. However, SB 206553 does not have this feature. Moreover, the molecule is quite rigid and does not fit into the normal agonist cavity. Therefore, other binding modes were investigated and because SAR studies emphasized the importance of the carbonyl oxygen in the urea group, the receptor model was studied to see whether there were any amino acid residues which could act as possible hydrogen bonding donors. Two serine residues were identified fairly close to the important aspartate residue, one of which (Ser-312) is unique to the 5-HT$_2$ class of receptors. Since these compounds do not bind to 5-HT$_1$ receptors, it was presumed that this was an important binding group.

SB 206553 was positioned into the pocket such that it interacted with Ser-312, and it was found that a second hydrogen bonding interaction involving the urea carbonyl group was possible to another serine residue (Ser-315). This binding interaction also allowed the two aromatic regions (the pyridine and tricyclic ring) to fit into hydrophobic pockets, one of which was dominated by aromatic amino acids, and the other made up of both aliphatic and aromatic amino acids. It could not be determined, however, which ring fitted which pocket and so both

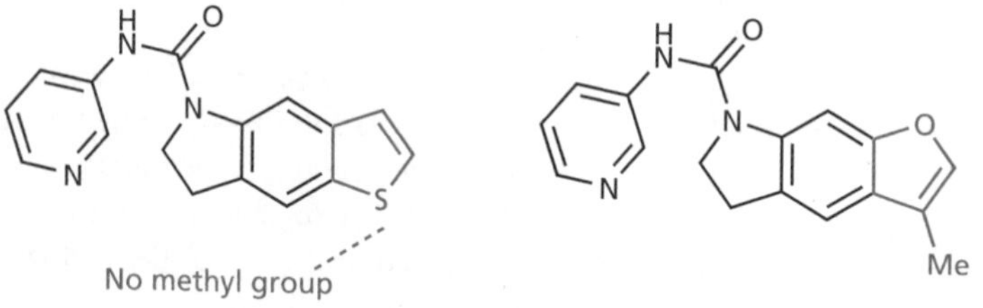

Figure 15.60 Varying substituents on SB 206553.

Figure 15.61 Ring variation of SB 206553.

possibilities were considered. Once docked, energy minimizations were carried out for both binding modes, and interaction energies were calculated. The more stable binding mode was the one where the pyridine ring was placed in the more aromatic of the hydrophobic pockets (Fig. 15.63).

In this binding mode, the carbonyl oxygen of the urea interacts through hydrogen bonds with two serine

residues, while the indole *N*-methyl group is placed in a hydrophobic pocket adjacent to two valine residues (Val-608 and Val-212) (Fig. 15.64). The receptor–ligand model was now studied to identify methods of achieving better selectivity for the 5-HT$_{2C}$ receptor over the 5-HT$_{2A}$ receptor. These studies revealed a difference between the two receptors in the pocket occupied by the *N*-methyl group. In the 5-HT$_{2C}$ receptor there are two valine residues in this pocket. In the 5-HT$_{2A}$ receptor, these amino acids are replaced by leucine. Leucine is bulkier than valine, so it was proposed that the hydrophobic pocket is smaller in the 5-HT$_{2A}$ receptor, making it more difficult for the *N*-methyl group of SB 206553 to fit and thus accounting for the observed selectivity.

15.18.7 SB 206553 to SB 221284

The next stage was to try and take advantage of the above findings. The metabolically labile *N*-methyl group had to be removed, but it was important to replace it with something that could fit the identified hydrophobic pocket in order to retain selectivity. It was decided to remove the pyrrole ring containing the

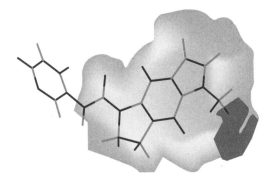

Figure 15.62 5-HT$_{2C}$ allowed volume (light blue) includes a 5-HT$_{2A}$ disallowed volume (dark blue) which is crucial for selectivity (shown around the structure of SB 206553).

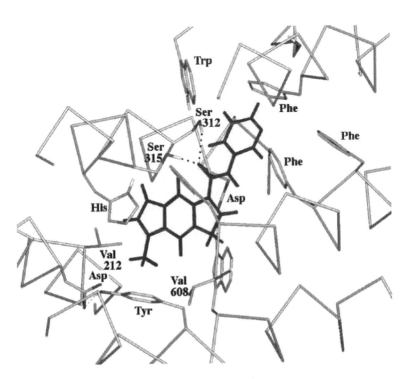

Figure 15.63 Stable binding mode for SB 206553.

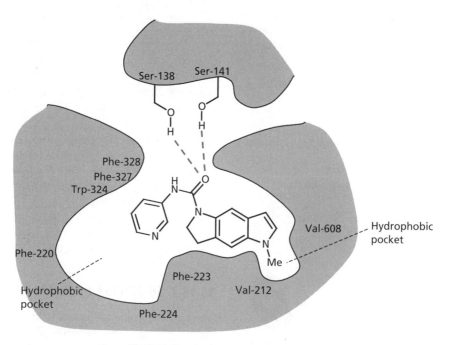

Figure 15.64 Binding site showing two hydrophobic pockets.

N-methyl group and add substituents to the remaining bicyclic indoline system (A) to act as isosteres for the lost ring (Fig. 15.65). These structures were fairly difficult to synthesize, so an initial study was carried out on the simpler phenyl urea system (B). A large number of analogues (86) were prepared by **combinatorial parallel synthesis** (Chapter 14) in order to find the best substituents for that system.

Mono-, di-, and tri-substituted structures were prepared at various positions and it was found that the preferred substitution pattern consisted of two substituents, *ortho* to each other at positions 3 and 4. Substitution at position 2 was bad for affinity and it was rationalized that this was due to a steric effect between the substituent and the urea group which caused the urea group to move out of the plane relative to the aromatic ring. This would result in a twist in the molecule and poorer binding as a consequence.

Having identified positions 3 and 4 as the ideal substitution points, the nature of the substituents was studied to see whether good affinity was related to particular substituents. At position 3, small hydrophobic, electron-withdrawing substituents were preferred (e.g. Cl or CF$_3$). In contrast, small hydrophobic electron-donating groups were preferred at position 4

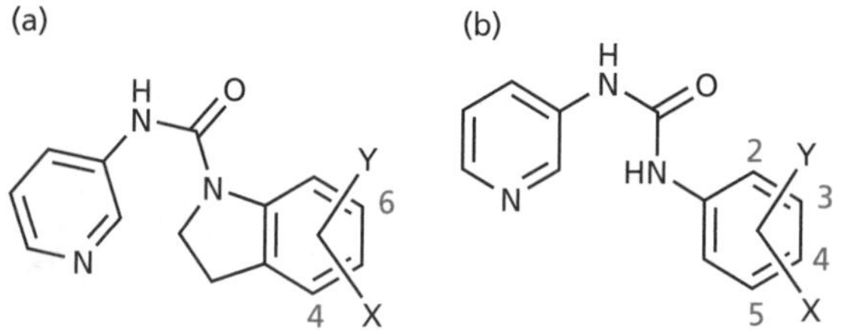

Figure 15.65 Search for an isostere of the pyrrole ring. (a) Indolines and (b) phenylureas.

(e.g. SMe or OMe). **QSAR studies** (Chapter 13) showed a good correlation between the lipophilicity (π) of small substituents at position 4 and 5-HT$_{2C}$ binding affinity (pK_i; see section 5.14) (Fig. 15.66).

Having established the preferred substituents in the phenyl urea series, it was important to check that the results were also valid for the indoline series and so a selection of indolines were prepared and their affinities checked against the corresponding phenylureas. The graph (Fig. 15.67) showed the same substituent effect for the indolines as for the phenylureas. As expected, the indolines were more potent than the corresponding phenylureas. The dotted line on the graph is what would be expected if the indolines and phenylureas had equal activity.

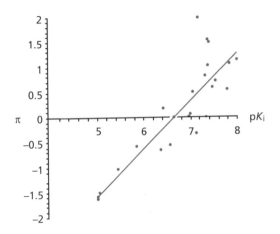

Figure 15.66 QSAR plot relating lipophilicity of 4-substituents vs binding affinity.

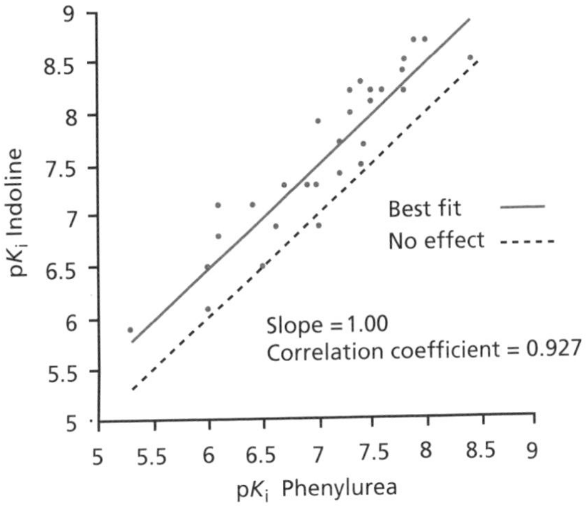

Figure 15.67 Comparison of pK_i values for indolines and phenylureas.

In the indoline series (Fig. 15.68), an electron-withdrawing group at position 6 was preferred (e.g. Cl or CF$_3$). The substituent at position 5 is the one expected to fit the hydrophobic pocket which is so crucial to selectivity. Increasing the bulk of this substituent using different alkyl groups (*varying substituents*) increased the selectivity but also led to a fall in affinity. The best balance of affinity versus selectivity was achieved with a thioether or an ether group. The best selectivity was found with groups such as SEt, S(^{n}Pr), or O(^{i}Pr). However, *in vivo* activity was better with smaller groups such as SMe or OMe. Consequently, **SB 221284** (Fig. 15.68) was chosen for further study and proved to be a potent inhibitor of 5-HT$_{2B/2C}$ receptors with good selectivity over the 5-HT$_{2A}$ receptor. Unfortunately, the compound had a major drawback in that it inhibited cytochrome P450 enzymes.

15.18.8 Modelling studies on SB 221284

Ligand docking studies were now carried out with SB 221284 on the model 5-HT$_{2C}$ receptor (Figs. 15.69 and 15.70). The resulting complex showed that the pyridine and indoline rings were in the same hydrophobic pockets as before. The hydrogen bonding interactions to the two serine residues were present, and the *S*-methyl group fitted into the hydrophobic pocket previously occupied by the *N*-methyl group. Moreover, it was noted that the CF$_3$ group helped to orientate the thioether correctly for this interaction. The CF$_3$ group, which was *ortho* to the thiomethyl group, restricted the latter's rotation (*conformational blocker*), thus favouring conformations where the thiomethyl group was directed towards its binding pocket. It was also found that both the pyridine ring and the SMe group were twisted out of the plane of the indoline ring and were almost perpendicular to it.

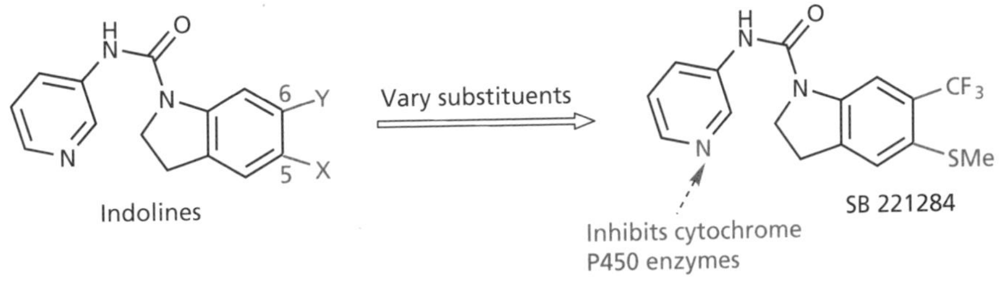

Figure 15.68 Development of SB 221284.

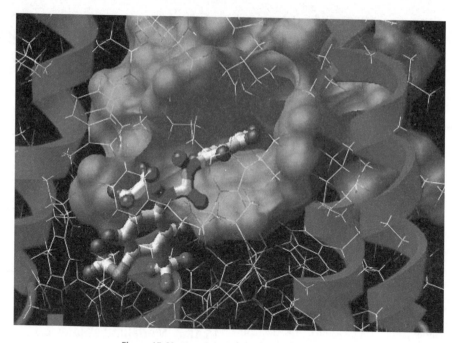

Figure 15.69 Ligand docking studies on SB 221284.

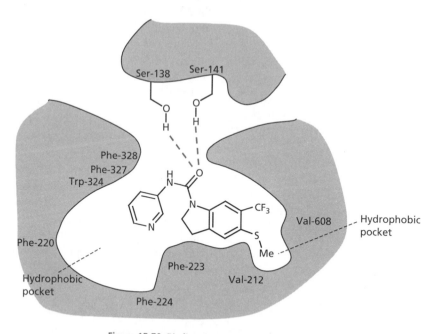

Figure 15.70 Binding site interactions for SB 221284.

Attention now turned to the hydrophobic pocket occupied by the pyridine ring and it was observed that this pocket was quite deep, suggesting that further aromatic substituents could be added to increase the binding interactions. Adding an extra substituent to the pyridine ring might also be expected to have a separate beneficial effect. Previous work had shown that the pyridine nitrogen in this class of compounds was responsible for the cytochrome P450 inhibitory activity and that this activity could be prevented if substituents were placed close to the pyridine nitrogen to act as steric shields (section 11.2.1).

15.18.9 3D QSAR studies on analogues of SB 221284

As a check on the authenticity of the receptor model, 55 substituted analogues of SB 221284 were synthesized and a 3D QSAR study (section 13.10) was carried out. Each structure was docked into the model receptor and the receptor–ligand complex minimized. The ligands were then removed from the receptor and subjected to CoMFA analysis, leaving out 8 structures to serve as a test set. Therefore, 47 compounds were analysed and a QSAR equation was derived which was checked against the test set of 8 structures. Affinity values for the 47 structures used in the analysis were calculated and a graph of predicted affinity versus actual affinity showed a good relationship (Fig. 15.71). Further analysis showed that the steric fields derived for the CoMFA analysis were of more importance to the equation than the electrostatic fields.

A graphical representation of the steric fields was produced which showed beneficial areas for affinity in dark blue and detrimental areas in light blue (Fig. 15.72). The large number of detrimental areas which were present suggested that the indoline ring and its substituents were in a tight pocket allowing little scope for further variation. These results were in agreement with what had been expected from the proposed receptor binding model.

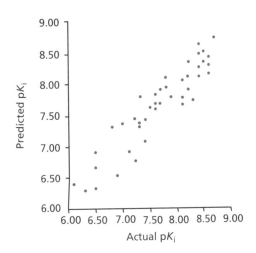

Figure 15.71 3D QSAR study on SB 221284 analogues—predicted vs actual pKi (q^2 = 0.656).

15.18.10 SB 221284 to SB 228357

Work was now carried out to determine whether adding further aromatic groups to the pyridine ring would have the expected beneficial effects on affinity, and also remove cytochrome P450 inhibition. For this work, compounds were prepared with a methoxy group at position 5 instead of a methylthio group (Fig. 15.73). It had previously been shown that the methoxy and methylthio groups were essentially equivalent as far as affinity and selectivity were concerned.

Substitutions at various positions of the pyridine ring were carried out, with position 5 being the most promising. An aromatic substituent (structure I)

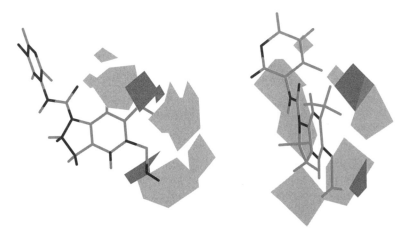

Figure 15.72 3D QSAR study on SB 221284—Steric fields (detrimental in light blue, beneficial in dark blue). Diagram shows the molecule and the steric fields from two different viewpoints.

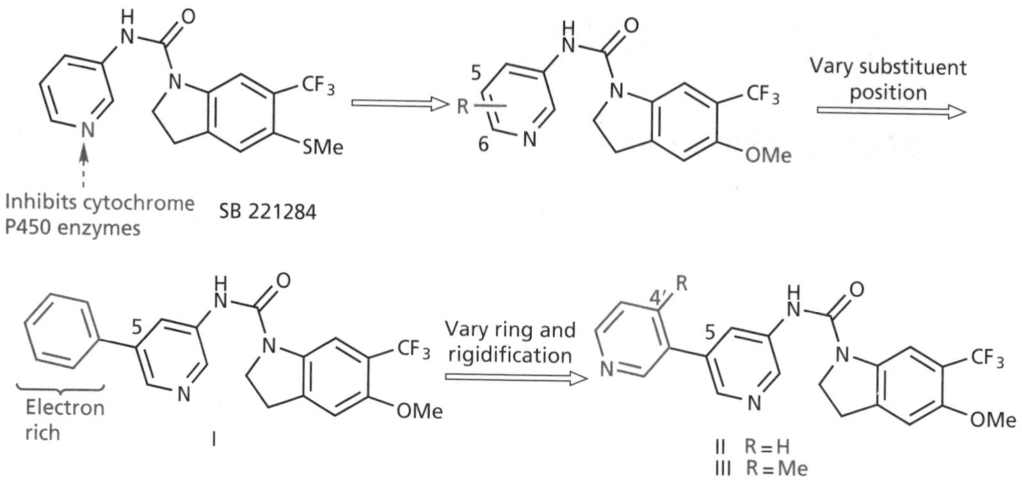

Figure 15.73 Substitution on the pyridine ring.

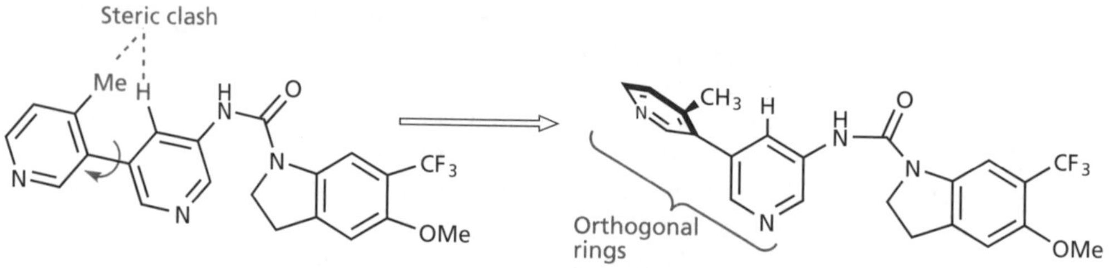

Figure 15.74 Conformation constraint due to the *ortho* methyl group.

resulted in slightly increased affinity, increased selectivity and also led to a 100-fold drop in cytochrome P450 inhibition. However, the level of cytochrome P450 inhibition remaining was still unacceptable and the structures also had poor oral activity. It was proposed that the aromatic substituent might be susceptible to metabolism since it is relatively electron rich, and so fluorine substituents were added to make the ring less electron rich. Unfortunately, this was detrimental to both affinity and selectivity.

A pyridine ring was then introduced in place of the aromatic ring in order to increase water solubility and metabolic stability (structure II). This resulted in a 10-fold increase in affinity, which suggested that an additional binding interaction might be taking place involving the pyridine nitrogen. Regrettably, affinity for the 5-HT$_{2A}$ receptor increased even more significantly, resulting in a drop in selectivity.

Fortunately, selectivity was recovered by placing a methyl group at position 4' (structure III). This group

forces the two heteroaromatic rings out of plane with each other, because of a steric clash with the *ortho* proton of the neighbouring ring (Fig. 15.74). The methyl group is therefore acting as a conformational blocker and has forced the molecule into a conformation which is favoured by the 5-HT$_{2C}$ receptor rather than the 5-HT$_{2A}$ receptor.

Structure III had slightly increased affinity for the 5-HT$_{2C}$ receptor and moderate oral activity. However, it still inhibited P450 enzymes although the level of inhibition had fallen.

Placing the heteroaromatic substituent at position 6 of the pyridine ring was also tried, as this should have a greater shielding effect and cause a greater reduction in cytochrome P450 inhibition. This was indeed observed. Selectivity between the 5-HT$_{2C}$ and 5-HT$_{2A}$ receptors was also increased 2000-fold, but the compound was orally inactive.

The pyridine nitrogen was identified as the cause of cytochrome P450 inhibition, so it was decided to

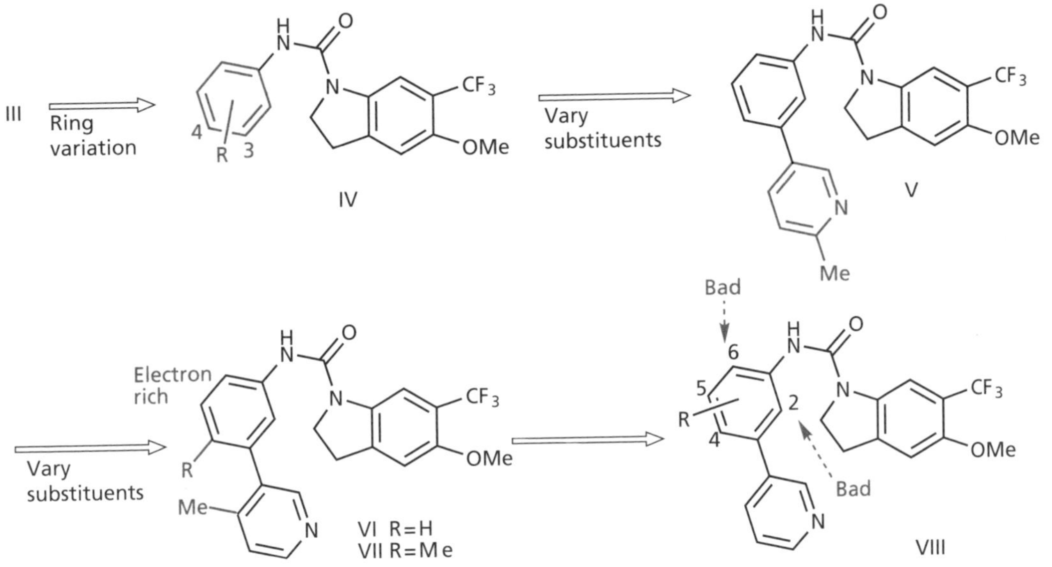

Figure 15.75 Development of an agent with 5HT$_{2C}$ affinity and no 5-HT$_{2A}$ affinity (structure VII).

remove it altogether and to replace the pyridine ring with an aromatic ring (structure IV) (Fig. 15.75). This led to a fall in water solubility, but this could be restored by adding a pyridine ring as a substituent (structure V). This structure had improved selectivity, higher 5-HT$_{2C}$ affinity, and potent oral activity, but was still observed to inhibit cytochrome P450 enzymes.

Moving the methyl group in the pyridine substituent to the *ortho* position (structure VI) resulted in restricted rotation between the two rings and forced them into a non-planar conformation with respect to each other (*conformational restraint*). This resulted in increased selectivity. Placing an *ortho* methyl in the aromatic ring as well (structure VII) increased 5-HT$_{2C}$ affinity yet further and effectively abolished affinity for the 5-HT$_{2A}$ receptor. Molecular modelling showed that the aromatic and pyridine rings were at right angles to each other, implying that this is a good conformation for binding to the 5-HT$_{2C}$ receptor but not the 5-HT$_{2A}$ receptor. Structure VII also had good oral activity, although it still retained some cytochrome P450 activity. Nevertheless, structure VI had the better overall *in vitro* profile with high *in vivo* potency and was chosen for further modification.

One problem with structure VI was its short duration of action—an indication that it was being metabolized, presumably at the electron-rich phenyl ring. A series of analogues was therefore prepared with electron-withdrawing substituents on the aromatic ring (structure VIII). Substituents at positions 2 and 6 were found to be bad for affinity and selectivity. Substitutions at this position are likely to twist the ring out of plane with the urea group. Substituents were better tolerated at positions 4 and 5, resulting in good affinity and selectivity. As mentioned previously, substitution at position 4 is particularly good for selectivity as it forces the neighbouring pyridine ring out of the same plane. However, the crucial feature at this stage was to increase the duration of activity, and SB 228357 with fluorine at position 5 was found to have activity lasting 6 hours Fig. 15.76).

Further receptor–ligand modelling studies suggested that there was still space in the hydrophobic pocket to allow further extension. An oxygen linker was therefore placed between the pyridine and phenyl rings to allow a deeper occupation of the pocket (structure IX; Fig. 15.72). This structure had poor oral activity though, probably due to reduced solubility. As a result, the phenyl ring was replaced by the original pyridine ring (structure X). An *ortho* methyl group was added as R^2 (structure XI) to increase the torsional angle between the two rings, resulting in increased selectivity and affinity for the 5-HT$_{2C}$ receptor. Moreover, selectivity over the 5-HT$_{2B}$ receptor was found for the first time (80-fold). Structure XI had low cytochrome P450 activity and good *in vivo* activity, but there were worries that the 5-methoxy group on the indoline ring would be metabolically labile and so this was replaced by a methyl group. This compound

(**SB 243213**) had a good profile, with negligible cytochrome P450 activity. The receptor–ligand model also showed that the pyridine ring substituent was well inserted into the hydrophobic pocket (Figs. 15.77 and 15.78).

SB 243213 and SB 228357 were both tested to make sure that they were inactive against a range of other

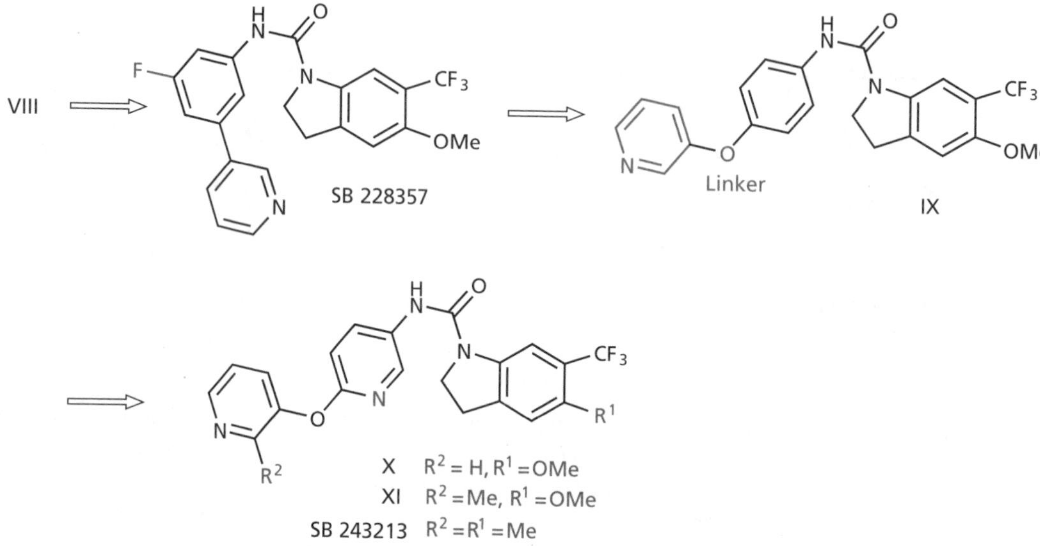

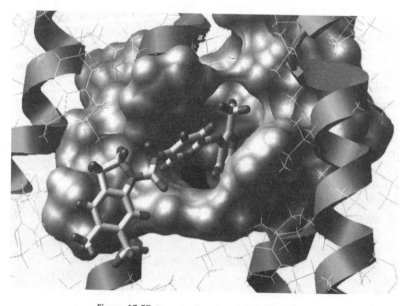

Figure 15.76 Development of SB 228357 and SB 243213.

Figure 15.77 Receptor–ligand model for SB 243213.

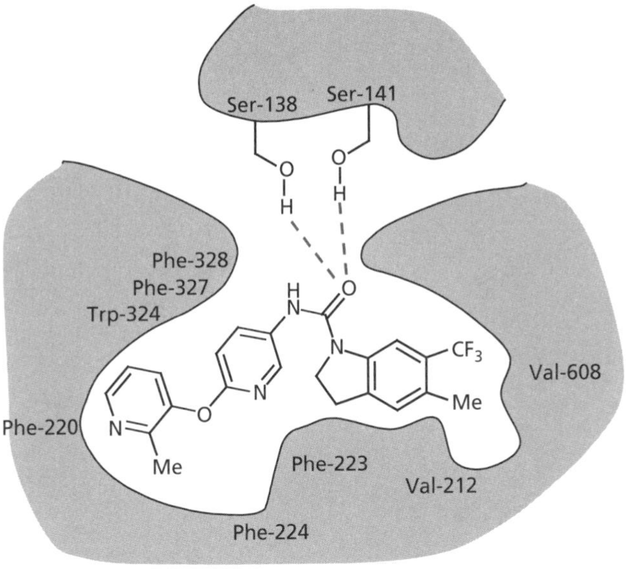

Figure 15.78 Binding of SB 243213.

receptors, ion channels and enzymes. Long-term studies were also carried out to see whether chronic dosing would affect receptor density or sensitivity. One of the key advantages of SB 243213 over SB 228357 is its much improved solubility, so the former was put forward for phase I clinical trials as a non-sedating antidepressant/anxiolytic.

QUESTIONS

1. What is meant by energy minimization and how is it carried out?

2. What is meant by the terms local and global energy minima, and what relevance have they to conformational analysis?

3. What two properties should be known about two drugs if they are to be overlaid as a comparison?

4. Is it reasonable to assume that the most stable conformation of a drug is the active conformation?

5. You are carrying out a *de novo* design of a ligand for a binding site that contains a hydrogen bonding region and two hydrophobic pockets. Structures I and II below are both suitable candidates. Compare the relevant merits of these structures and decide which one you would synthesize first to test your binding theory.

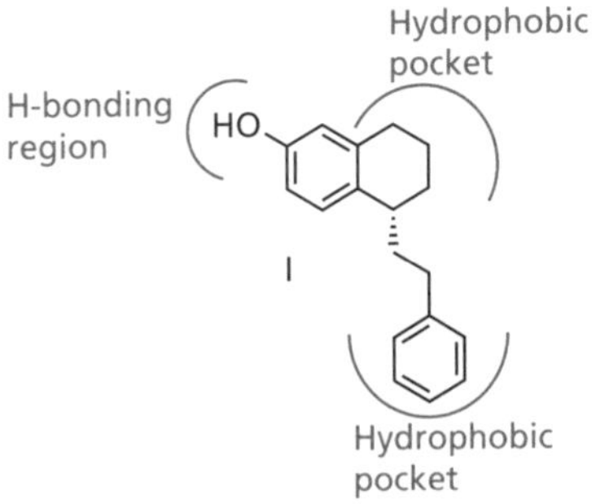

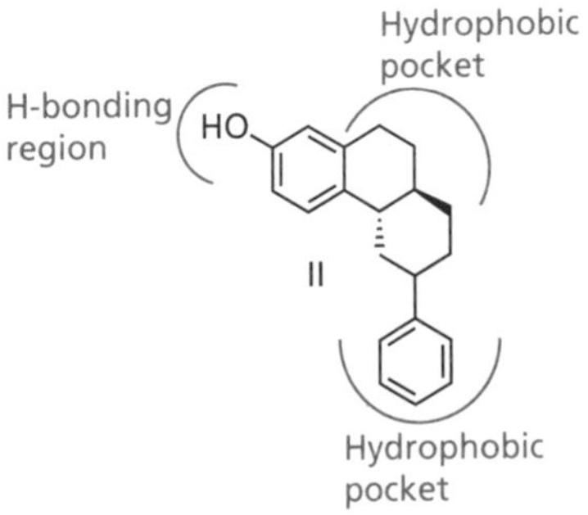

6. Both structures I and II above show poor water solubility. It is suggested that the phenyl group be replaced by a pyridine ring. What would be the advantages and disadvantages of this idea? Have you any alternative ideas?

7. Assuming that structures I and II both bound to the binding site as predicted, what further modifications might you make to increase binding interactions?

8. Why were such modifications not carried out earlier?

9. The following 8 structures have been tested for activity as receptor agonists. Five are active and 3 are inactive. Assess the structures and discuss what the pharmacophore might be for agonist activity.

10. How would you go about carrying out overlays of the active structures in Question 9?

FURTHER READING

Agrafiotis, D. K., Lobanov, V. S., and Salemme, F. R. (2002) Combinatorial informatics in the post-genomics era. *Nature Reviews Drug Discovery*, **1**, 337–346.

Bikker, J. A., Trumpp-Kallmeyer, S., and Humblet, C. (1998) G-protein coupled receptors: models, mutagenesis, and drug design. *Journal of Medicinal Chemistry*, **41**, 2911–2927.

Bohm, H.-J. (1992) The computer program LUDI: A new method for the *de novo* design of enzyme inhibitors. *Journal of Computer-Aided Molecular Design*, **6**, 61–78.

Bourne, P. E. and Wessig, H. (eds.) (2003) *Structural bioinformatics.* John Wiley and Sons, New York.

Bromidge, S. M., Dabbs, S., Davies, D. T. *et al.* (2000) Biarylcarbamoylindolines are novel and selective 5-HT$_{2C}$ receptor inverse agonists: identification of 5-methyl-1-[[2-[(2-methyl-3-pyridyl)oxy]-5-pyridyl]carbamoyl]-6-trifluoromethylindoline (SB 243213) as a potential antidepressant/anxiolytic agent. *Journal of Medicinal Chemistry*, **43**, 1123–1134.

Ganellin, C. R. and Roberts, S. M. (eds.) (1994) Discovery and development of cromokalim and related potassium channel activators. Chapter 8 in *Medicinal chemistry—the role of organic research in drug research*, 2nd edn. Academic Press, New York.

Greer, J., Erickson, J. W., Baldwin, J. J., and Varney, M. D. (1994) Application of the three-dimensional structures of protein target molecules in structure-based design. *Journal of Medicinal Chemistry*, **37**, 1035–1054.

Leach, A. R. (2001) *Molecular modelling: principles and applications*, 2nd edn. Pearson Education, London.

Miller, M. A. (2002) Chemical database techniques. *Nature Reviews Drug Discovery*, **1**, 220–227.

Navia, M. A. and Murcko, M. A. (1992) Use of structural information in drug design. *Current Opinion in Structural Biology*, **2**, 202–216.

Richards, G. (2002) Virtual screening using grid computing: the screensaver project. *Nature Reviews Drug Discovery*, **1**, 551–555.

van de Waterbeemd, H., Testa, B., and Folkers, G. (eds.) (1997) *Computer-assisted lead finding and optimization.* Wiley-VCH, New York.

Titles for general further reading are listed on p. 711.

PART D

Selected topics in medicinal chemistry

In Part D, we look at seven particular topics in medicinal chemistry. The topics are of the authors choosing, and demonstrate a personal preference or interest. There are many other fascinating topics that could have been included, but sadly there is a limit to what can appear in a textbook of this kind.

The different chapters illustrate different methods of classifying drugs, and demonstrate the advantages and disadvantages of such classifications. For example, there are four chapters where drugs are classified by their pharmacological effect, mainly those with antibacterial, antiviral, anticancer, and antiulcer activities. The advantage of this classification is that it gives an overall view of the many different types of drug that can be used to treat these diseases. The disadvantage is the volume of information that has to be imparted. Since these diseases have many causes or mechanisms, there are many different possible targets for drugs. This means that the types of drugs which can be used in each of these fields are extremely varied in their structure and mechanism of action.

In contrast, the chapters on cholinergic and adrenergic agents concentrate on drugs which interact with specific biological systems, mainly the cholinergic and adrenergic nervous system. Since the target systems are more focused, there are fewer targets to consider. This has the advantage that it is easier to rationalize the drug structures involved and to study their mechanisms of action. It is also possible to understand why drugs having an effect on these systems can be used in particular fields of medicine, such as in the treatment of asthma or in cardiovascular medicine. The disadvantage in concentrating on a particular biological system is that it ignores drugs that could be antiasthmatic or cardiovascular by acting on a different biological system.

A study of opiate analgesics is included, where the drugs have been classed by their chemical structure. This is a useful classification for medicinal chemists since the drugs involved have the same pharmacological activity and targets, making their study well focused. The disadvantage with this classification is that analgesics with different structures and mechanisms of action are not included.

Topics have also been chosen to include several traditional fields of medicinal chemistry as well as those which are more recent. For example, the opiate analgesics were discovered more than 100 years ago, while the majority of antibacterial agents were discovered in the mid twentieth century. In contrast, progress in antiviral therapy has been relatively recent. A comparison of these chapters illustrates the changing face of medicinal chemistry, from one of trial and error to a more scientific approach where diseases are understood at the molecular level and drugs are then designed accordingly.

16 Antibacterial agents

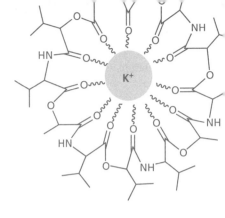

The fight against bacterial infections over the last 70 years has been one of the great success stories of medicinal chemistry, yet it remains to be seen whether it will last. Bacteria such as *Staphylococcus aureus* have the worrying ability to gain resistance to known drugs, so the search for new drugs is never ending. Although deaths from bacterial infection have dropped in the developed world, bacterial infection is still a major cause of death in the developing world. For example, the World Health Organization estimated that tuberculosis was responsible for about 2 million deaths in 2002 and that one in three of the world's population was infected. The same organization estimated that, in the year 2000, 1.9 million children died worldwide of respiratory infections with 70% of these deaths occurring in Africa and Asia. They also estimated that each year 1.4 million children died from gut infections and the diarrhoea resulting from these infections. In the developed world, deaths from food poisoning due to virulent strains of *Escherichia coli* have attracted widespread publicity, and tuberculosis has returned as a result of the AIDS epidemic.

The topic of antibacterial agents is a large one and terms are used in this chapter which are unique to this particular field. Rather than clutter the text with explanations and definitions, Appendix 5 contains explanations of such terms as aerobic and anaerobic organisms, antibacterial and antibiotic substances, cocci, bacilli, streptococci, and staphylococci. Appendix 5 also explains briefly the difference between bacteria, algae, protozoa, and fungi. The emphasis in this chapter is on agents that act against bacteria, but some of those described also act against protozoal infections and this may be mentioned in the text.

16.1 The history of antibacterial agents

There is evidence of antibacterial herbs or potions being used for many centuries. For example, the Chinese used mouldy soybean curd to treat carbuncles, boils, and other infections. Greek physicians used wine, myrrh, and inorganic salts. In the middle ages, certain types of honey were used to prevent infections following arrow wounds. Of course in those days, there was no way of knowing that bacteria were the cause of these infections.

Bacteria are single-cell microorganisms which were first identified in the 1670s by van Leeuwenhoek, following his invention of the microscope. It was not until the nineteenth century, however, that their link with disease was appreciated. This followed the elegant experiments carried out by the French scientist Pasteur, who demonstrated that specific bacterial strains were crucial to fermentation, and that these and other microorganisms were more widespread than was previously thought. The possibility that these microorganisms might be responsible for disease began to take hold.

An early advocate of a 'germ theory of disease' was the Edinburgh surgeon Lister. Despite the protests of several colleagues who took offence at the suggestion that they might be infecting their own patients, Lister introduced **carbolic acid** as an antiseptic and sterilizing agent for operating theatres and wards. The improvement in surgical survival rates was significant.

During the latter half of the nineteenth century, scientists such as Koch were able to identify the microorganisms responsible for diseases such as tuberculosis, cholera, and typhoid. Methods of vaccination were studied, and research was carried out to try and find effective antibacterial agents or antibiotics. The scientist who can lay claim to be the father of chemotherapy—the use of chemicals against infection—was Paul Ehrlich. Ehrlich spent much of his career studying histology, then immunochemistry, and won a Nobel prize for his contributions to immunology. In 1904, however, he switched direction and entered a field which he defined as chemotherapy. Ehrlich's '**principle of chemotherapy**' was that a chemical could directly interfere with the proliferation of microorganisms, at concentrations tolerated by the host. This concept was popularly known as the '**magic bullet**', where the chemical was seen as a bullet which could search out and destroy the invading microorganism without adversely affecting the host. The process is one of **selective toxicity**, where the chemical shows greater toxicity to the target microorganism than to the host cells. Such selectivity can be represented by a '**chemotherapeutic index**', which compares the minimum effective dose of a drug versus the maximum dose that can be tolerated by the host. This measure of selectivity was eventually replaced by the currently used **therapeutic index**.

By 1910, Ehrlich had successfully developed the first example of a purely synthetic antimicrobial drug. This was the arsenic-containing compound **salvarsan** (Fig. 16.1). Although it was not effective against a wide range of bacterial infections, it did prove effective against the protozoal disease of sleeping sickness (trypanosomiasis), and the spirochete disease of syphilis. The drug was used until 1945 when it was replaced by penicillin (see also Box 16.9).

Over the next 20 years, progress was made against a variety of protozoal diseases, but little progress was made in finding antibacterial agents until the introduction of **proflavine** in 1934 (Fig. 16.1), a drug which was used during the Second World War against bacterial infections in deep surface wounds.

Unfortunately, it was too toxic to be used against systemic bacterial infections (i.e. those carried in the bloodstream) and there was still an urgent need for agents which would fight these infections.

This need was answered in 1935 when it was discovered that a red dye called **prontosil** was effective against streptococcal infections *in vivo*. As discussed later, prontosil was eventually recognized as a prodrug for a new class of antibacterial agents—the **sulfa drugs** or **sulfonamides**. The discovery of these drugs was a real breakthrough, as they represented the first drugs to be effective against systemic bacterial infections. In fact, they were the only effective drugs until penicillin became available in the early 1940s.

Although **penicillin** was discovered in 1928, it was not until 1940 that effective means of isolating it were developed by Florey and Chain. Society was then rewarded with a drug which revolutionized the fight against bacterial infection and proved even more effective than the sulfonamides. Despite penicillin's success, it was not effective against all types of infection and the need for new antibacterial agents still remained. Penicillin is an example of a toxic fungal metabolite that kills bacteria and allows the fungus to compete for nutrients. The realization that fungi might be a source for novel antibiotics spurred scientists into a huge investigation of microbial cultures from all round the globe.

In 1944, the antibiotic **streptomycin** was discovered from a systematic search of soil organisms. It extended the range of chemotherapy to the tubercle bacillus and a variety of Gram-negative bacteria. This compound was the first example of a series of antibiotics known as the **aminoglycoside** antibiotics. After the Second World War, the search continued leading to the discovery of **chloramphenicol** (1947), the peptide antibiotics (e.g. **bacitracin;** 1945), the tetracycline antibiotics (e.g. **chlortetracycline;** 1948), the macrolide antibiotics (e.g. **erythromycin;** 1952), the

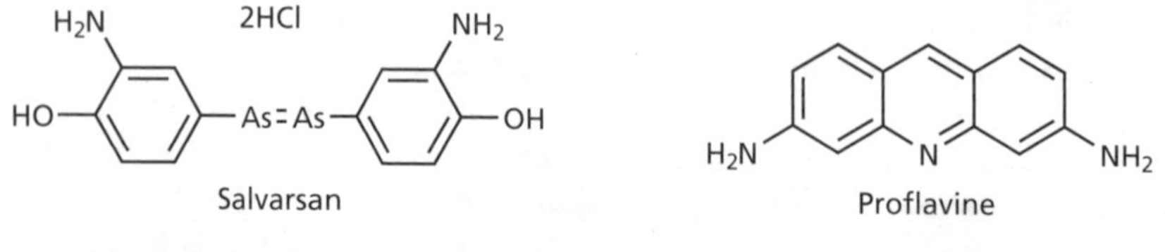

Figure 16.1 Salvarsan and proflavine. (The structure of salvarsan shown here is a simplification; it is in fact a cyclic trimer with no As=As bonds.)

cyclic peptide antibiotics (e.g. **valinomycin**), and in 1955 the first example of a second major group of β-lactam antibiotics, **cephalosporin C**.

As far as synthetic agents were concerned, **isoniazid** was found to be effective against human tuberculosis in 1952, and in 1962 **nalidixic acid** (the first of the quinolone antibacterial agents) was discovered. A second generation of this class of drugs was introduced in 1987 with **ciprofloxacin**.

Many antibacterial agents are now available and the vast majority of bacterial diseases have been brought under control (e.g. syphilis, tuberculosis, typhoid, bubonic plague, leprosy, diphtheria, gas gangrene, tetanus, and gonorrhoea). This represents a great achievement for medicinal chemistry and it is perhaps sobering to consider the hazards society faced in the days before penicillin. Septicaemia was a risk faced by mothers during childbirth and could lead to death. Ear infections were common, especially in children, and could lead to deafness. Pneumonia was a frequent cause of death in hospital wards. Tuberculosis was a major problem, requiring special isolation hospitals built away from populated centres. A simple cut or a wound could lead to severe infection requiring the amputation of a limb, and the threat of peritonitis lowered the success rates of surgical operations. This was in the 1930s—still within living memory for many. Perhaps those of us born since the Second World War take the success of antibacterial agents too much for granted.

16.2 The bacterial cell

The success of antibacterial agents owes much to the fact that they can act selectively against bacterial cells rather than animal cells. This is largely because bacterial and animal cells differ both in their structure and in their biosynthetic pathways. Let us consider some of the differences between the bacterial cell (defined as prokaryotic) (Fig. 16.2) and the animal cell (defined as eukaryotic).

Differences between bacterial and animal cells:

- The bacterial cell does not have a defined nucleus, whereas the animal cell does.

- Animal cells contain a variety of structures called organelles (mitochondria, endoplasmic reticulum, etc.), whereas the bacterial cell is relatively simple.

- The biochemistry of a bacterial cell differs significantly from that of an animal cell. For example, bacteria may have to synthesize essential vitamins which animal cells can acquire intact from food. The bacterial cells must have the enzymes to catalyse these reactions. Animal cells do not, because the reactions are not required.

- The bacterial cell has a cell membrane and a cell wall, whereas the animal cell has only a cell membrane. The cell wall is crucial to the bacterial cell's survival. Bacteria have to survive a wide range of environments and osmotic pressures, whereas animal cells do not. If a bacterial cell lacking a cell wall was placed in an

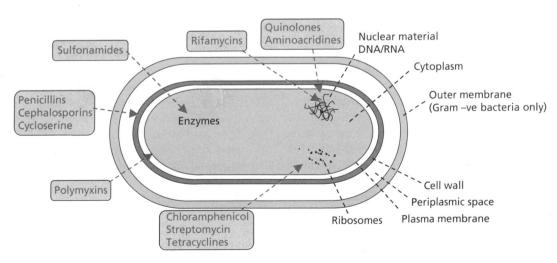

Figure 16.2 The bacterial cell and drug targets.

aqueous environment containing a low concentration of salts, water would freely enter the cell due to osmotic pressure. This would cause the cell to swell and eventually burst. The scientific term for this is **lysis.** The cell wall does not stop water flowing into the cell directly, but it does prevent the cell from swelling and so indirectly prevents water entering the cell. Bacteria can be characterized by a staining technique which allows them to be defined as Gram positive or Gram negative (Appendix 5). The staining technique involves the addition of a purple dye followed by washing with acetone. Bacteria with a thick cell wall (20–40 nm) absorb the dye and are stained purple, and are defined as Gram positive. Bacteria with a thin cell wall (2–7 nm) absorb only a small amount of dye which is washed out by acetone. These bacteria are stained pink with a second dye and are said to be Gram negative. Although Gram-negative bacteria have a thin cell wall, they have an additional outer membrane not present in Gram positive bacteria. This outer membrane is made up of lipopolysaccharides—similar in character to the cell membrane. These differences in cell walls and membranes have important consequences for the different vulnerabilities of Gram-positive and Gram-negative bacteria to antibacterial drugs.

16.3 Mechanisms of antibacterial action

There are five main mechanisms by which antibacterial agents act (Fig. 16.2).

- **Inhibition of cell metabolism:** Antibacterial agents which inhibit cell metabolism are called **antimetabolites**. These compounds inhibit the metabolism of a microorganism, but not the metabolism of the host. They can do this by inhibiting an enzyme-catalysed reaction which is present in the bacterial cell, but not in animal cells. The best-known examples of antibacterial agents acting in this way are the sulfonamides. It is also possible for antibacterial agents to show selectivity against enzymes which are present in both the bacterial and mammalian cell, as long as there are significant differences in structure between the two.

- **Inhibition of bacterial cell wall synthesis:** Inhibition of cell wall synthesis leads to bacterial cell lysis and death. Agents operating in this way include penicillins, cephalosporins, and vancomycin. As animal cells do not have a cell wall, they are unaffected by such agents.

- **Interactions with the plasma membrane:** Some antibacterial agents interact with the plasma membrane of bacterial cells to affect membrane permeability. This has fatal results for the cell. Polymyxins and tyrothricin operate in this way.

- **Disruption of protein synthesis:** Disruption of protein synthesis means that essential proteins and enzymes required for the cell's survival can no longer be made. Agents which disrupt protein synthesis include the rifamycins, aminoglycosides, tetracyclines, and chloramphenicol.

- **Inhibition of nucleic acid transcription and replication:** Inhibition of nucleic acid function prevents cell division and/or the synthesis of essential proteins. Agents acting in this way include nalidixic acid and proflavine.

We now consider these mechanisms in more detail.

16.4 Antibacterial agents which act against cell metabolism (antimetabolites)

16.4.1 Sulfonamides

16.4.1.1 The history of sulfonamides

The best example of antibacterial agents acting as antimetabolites are the sulfonamides (sometimes called the sulfa drugs). The sulfonamide story began in 1935 when it was discovered that a red dye called **prontosil** (Fig. 16.3) had antibacterial properties *in vivo* (i.e. when given to laboratory animals). Strangely enough, no antibacterial effect was observed *in vitro*. In other words, prontosil could not kill bacteria grown in the test tube. This remained a mystery until it was discovered that prontosil was metabolized by bacteria present in the small intestine of the test animal to give a product called **sulfanilamide** (Fig. 16.3). It was this compound which was the true antibacterial agent.

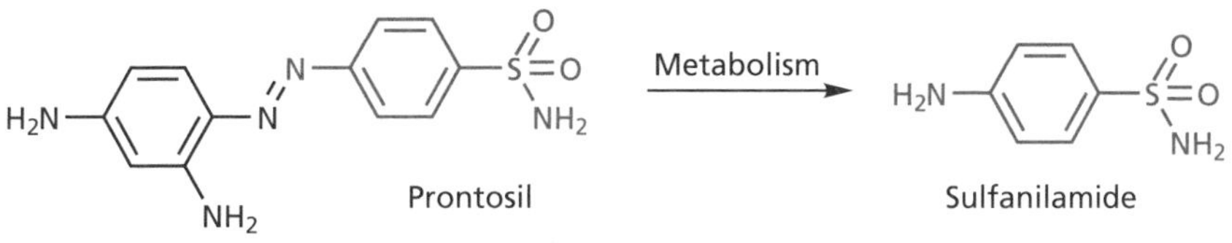

Figure 16.3 Metabolism of prontosil.

Thus, prontosil was an early example of a **prodrug** (section 11.6). Sulfanilamide was synthesized in the laboratory and became the first synthetic antibacterial agent found to be active against a wide range of infections. Further developments led to a range of sulfonamides which proved effective against Gram-positive organisms, especially pneumococci and meningococci.

Despite their undoubted benefits, sulfa drugs have proved ineffective against infections such as *Salmonella*—the organism responsible for typhoid. Other problems have resulted from the way these drugs are metabolized, as toxic products are frequently obtained. This led to the sulfonamides being superseded by penicillin.

16.4.1.2 Structure–activity relationships

The synthesis of a large number of sulfonamide analogues (Fig. 16.4) led to the following conclusions:

- The *para*-amino group is essential for activity and must be unsubstituted (i.e. $R^1 = H$). The only exception is when $R^1 =$ acyl (i.e. amides). The amides themselves are inactive but can be metabolized in the body to regenerate the active compound (Fig. 16.5). Thus amides can be used as sulfonamide prodrugs.

- The aromatic ring and the sulfonamide functional group are both required.

- The aromatic ring must be *para*-substituted only.

- The sulfonamide nitrogen must be primary or secondary.

- R^2 is the only possible site that can be varied in sulfonamides.

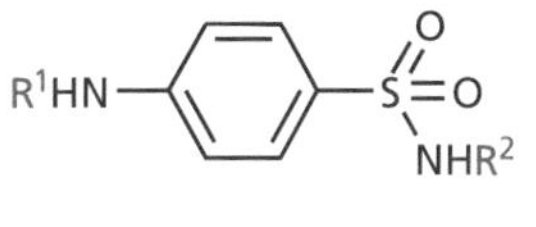

Figure 16.4 Sulfonamide analogues.

16.4.1.3 Sulfanilamide analogues

R^2 is often varied by incorporating a large range of heterocyclic or aromatic structures, which affects the extent to which the drug binds to plasma protein. This in turn controls the blood levels and lifetime of the drug. Thus, a drug that binds strongly to plasma protein will be slowly released into the blood circulation and will be longer lasting. Varying R^2 can also affect the solubility of sulfonamides. Therefore, variations of R^2 affect the pharmacokinetics of the drug, rather than its mechanism of action (Box 16.1).

16.4.1.4 Applications of sulfonamides

Before the appearance of penicillin, the sulfa drugs were the drugs of choice in the treatment of infectious diseases. Indeed, they played a significant part in world history by saving Winston Churchill's life during the Second World War. While visiting North Africa, Churchill became ill with a serious infection and was bedridden for several weeks. At one point, his condition was deemed so serious that his daughter was flown out from Britain to be at his side. Fortunately, he responded to the novel sulfonamide drugs of the day.

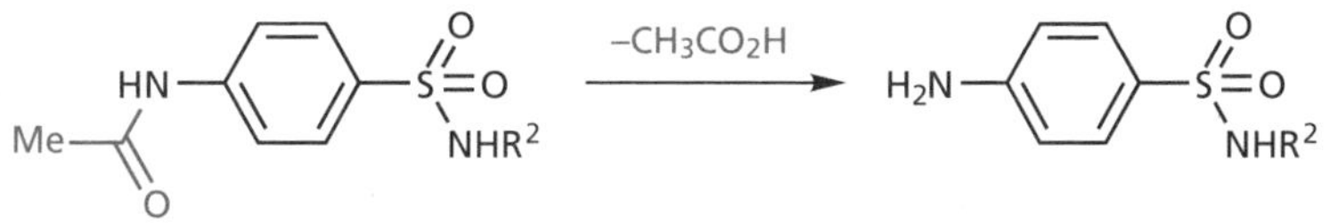

Figure 16.5 Metabolism of acyl group to regenerate active compound.

The primary amino group of sulfonamides is acetylated in the body and the resulting amides have reduced solubility which can lead to toxic effects. For example, the metabolite formed from **sulfathiazole** (an early sulfonamide) is poorly soluble and can prove fatal if it blocks the kidney tubules (Fig. 1). It is interesting to note that certain populations are more susceptible to this than others. For example, Japanese and Chinese people metabolize sulfathiazole more quickly than the average American, and are therefore more susceptible to its toxic effects.

It was discovered that the solubility problem could be overcome by replacing the thiazole ring in sulfathiazole with a pyrimidine ring to give **sulfadiazine** (Fig. 2). The reason for the improved solubility lies in the acidity of the sulfonamide NH proton. In sulfathiazole, this proton is not very acidic (high pK_a). Therefore, sulfathiazole and its metabolite are mostly un-ionized at blood pH. Replacing the thiazole ring with a more electron-withdrawing pyrimidine ring increases the acidity of the NH proton by stabilizing the resulting anion. Therefore, sulfadiazine and its metabolite are significantly ionized at blood pH. As a consequence, they are more soluble and less toxic. Sulfadiazine was also found to be more active than sulfathiazole and soon replaced it in therapy. Silver sulfadiazine cream is still used topically to prevent infection of burns, although it is really the silver ions that provide the antibacterial effect.

Figure 1 Metabolism of sulfathiazole.

Figure 2 Sulfadiazine.

Penicillins largely superseded sulfonamides, and for a long time sulfonamides took a back seat. There has been a revival of interest, however, with the discovery of a new 'breed' of longer-lasting sulfonamides. One example of this new generation is **sulfadoxine** (Fig. 16.6), which is so stable in the body that it need only be taken once a week. The combination of sulfadoxine and **pyrimethamine** is called **Fanisdar** and has been used for the treatment of malaria.

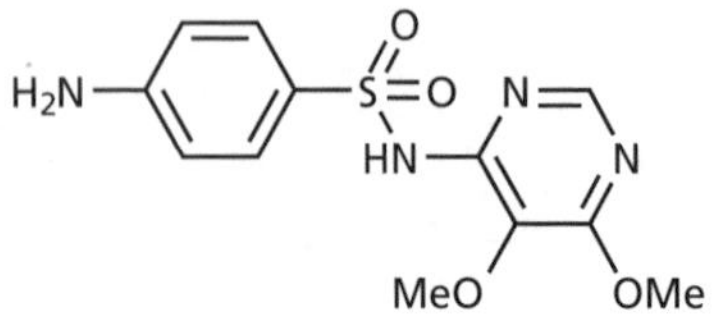

The sulfa drugs presently have the following applications in medicine:

- treatment of urinary tract infections
- eye lotions
- treatment of infections of mucous membranes
- treatment of gut infections.

It is also worth noting that sulfonamides have occasionally found uses in other areas of medicine (section 9.4.4.2 and Box 16.2).

Figure 16.6 Sulfadoxine.

BOX 16.2 TREATMENT OF INTESTINAL INFECTIONS

Sulfonamides have been particularly useful against intestinal infections, and can be targeted against these by the use of prodrugs. For example, **succinyl sulfathiazole** is a prodrug of sulfathiazole (Fig. 1). The succinyl moiety contains an acidic group, which means that the prodrug is ionized in the slightly alkaline conditions of the intestine. As a result, it is not absorbed into the bloodstream and is retained in the intestine. Slow enzymatic hydrolysis of the succinyl group then releases the active sulfathiazole where it is needed.

Benzoyl substitution (Fig. 2) on the aniline nitrogen has also given useful prodrugs which are poorly absorbed through the gut wall because they are too hydrophobic (section 8.2). They can be used in the same way.

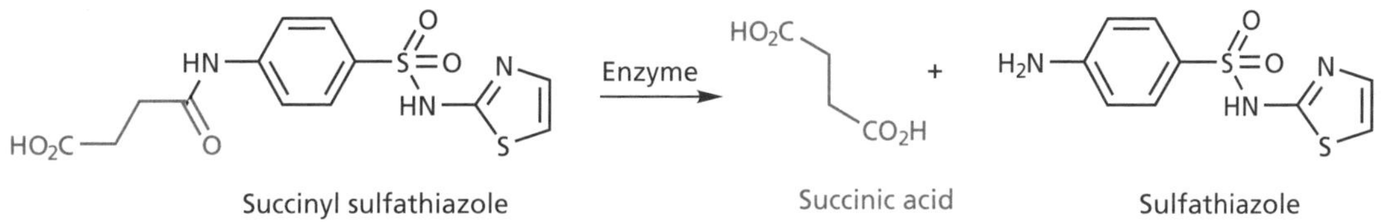

Figure 1 Succinyl sulfathiazole is a prodrug of sulfathiazole.

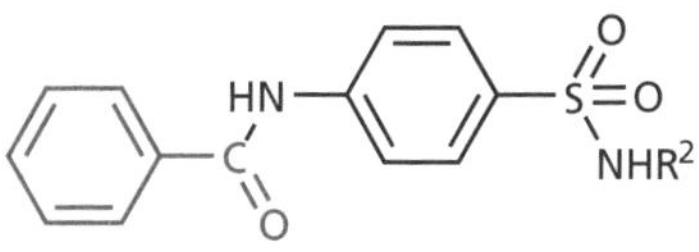

Figure 2 Substitution on the aniline nitrogen with benzoyl groups.

16.4.1.5 Mechanism of action

The sulfonamides act as competitive enzyme inhibitors of dihydropteroate synthetase and block the biosynthesis of **tetrahydrofolate** in bacterial cells (Fig. 16.7). Tetrahydrofolate is important in both human and bacterial cells, as it is an enzyme cofactor that provides one carbon units for the synthesis of the pyrimidine nucleic acid bases required for DNA synthesis (section 18.3.1). If pyrimidine and DNA synthesis is blocked, then the cell can no longer grow and divide.

Note that sulfonamides do not actively kill bacterial cells. They do, however, prevent the cells growing and multiplying. This gives the body's own defence systems enough time to gather their resources and wipe out the invader. Antibacterial agents which inhibit cell growth are classed as **bacteriostatic**, whereas agents such as penicillin which actively kill bacterial cells are classed as **bactericidal**. Because sulfonamides rely on a healthy immune system to complete the job they have started, they are not recommended for patients with a weakened immune system. This includes people with AIDS, and patients who are undergoing cancer chemotherapy or have had an organ transplant and are taking immunosuppressant drugs.

Sulfonamides act as inhibitors by mimicking **p-aminobenzoic acid** (PABA)—one of the normal substrates for dihydropteroate synthetase. The sulfonamide molecule is similar enough in structure to PABA that the enzyme is fooled into accepting it into its active site (Fig. 16.8). Once it is bound, the sulfonamide prevents PABA from binding. As a result, dihydropteroate is no longer synthesized. One might ask why the enzyme does not join the sulfonamide to the other component of dihydropteroate to give a dihydropteroate analogue containing the sulfonamide skeleton. This can in fact occur, but it does the cell no good at all because the analogue is not accepted by the next enzyme in the biosynthetic pathway.

Sulfonamides are competitive enzyme inhibitors, so inhibition is reversible. This is demonstrated by certain organisms such as staphylococci, pneumococci, and gonococci which can acquire resistance by synthesizing more PABA. The more PABA there is in the cell, the more effectively it can compete with the sulfonamide

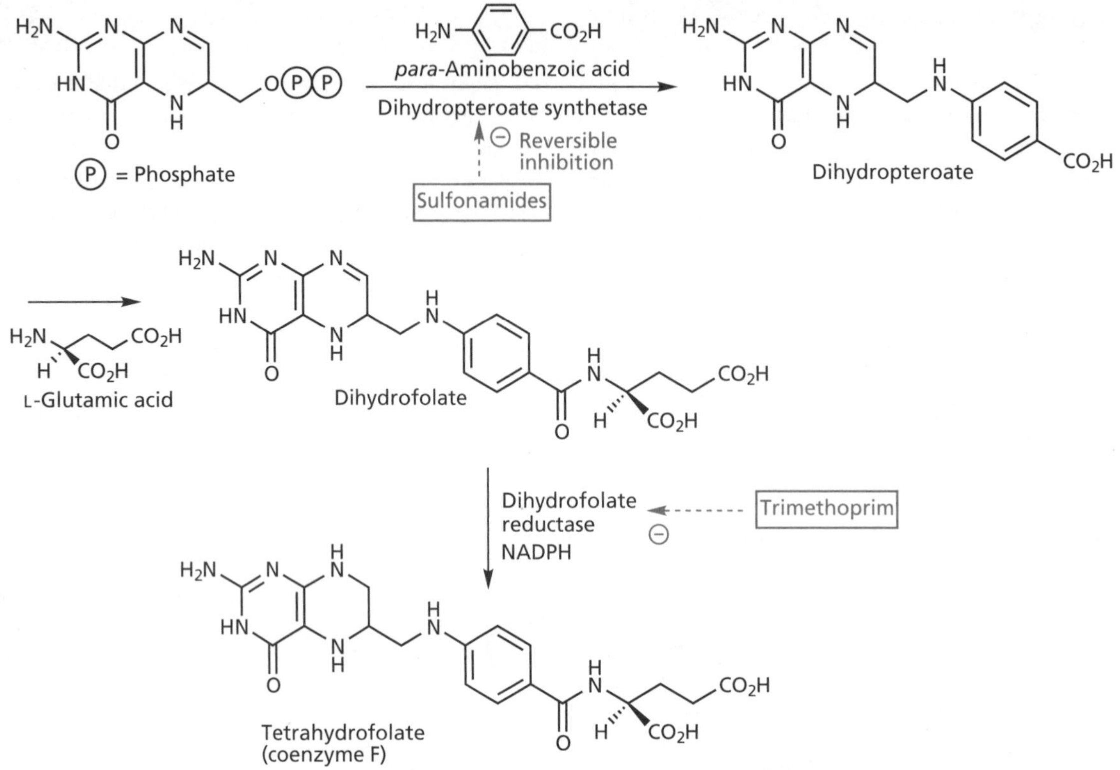

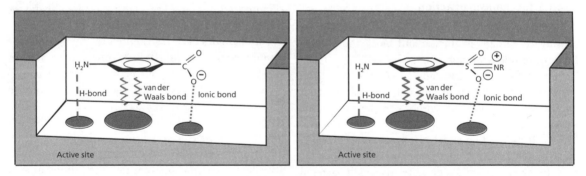

Figure 16.7 Mechanism of action of sulfonamides.

Figure 16.8 Sulfonamide prevents PABA from binding by mimicking PABA.

inhibitor to reach the enzyme's active site. In such cases, the dose levels of sulfonamide have to be increased to bring back the same level of inhibition. Resistance to sulfonamides can also arise by mutations which modify the target enzyme such that it has less affinity for sulfonamides, or by decreased permeability of the cell membrane to the sulfonamide.

Tetrahydrofolate is clearly necessary for the survival of bacterial cells, but it is also vital for the survival of

human cells, so why are the sulfa drugs not toxic to humans? The answer lies in the fact that human cells synthesize tetrahydrofolate in a different manner and do not contain the enzyme dihydropteroate synthetase. In human cells, tetrahydrofolate is synthesized from folic acid which is obtained from the diet as a vitamin. Folic acid is brought across cell membranes by a transport protein. We could now ask, 'If human cells can acquire folic acid from the diet, why can't bacterial cells infecting the human body do the same, then convert it to tetrahydrofolate?' In fact, bacterial cells are unable to acquire folic acid since they lack the necessary transport protein required to carry it across the cell membrane.

To sum up, the success of sulfonamides is due to two metabolic differences between mammalian and bacterial cells:

- Bacteria have a susceptible enzyme which is not present in mammalian cells.

- Bacteria lack the transport protein which would allow them to acquire folic acid from outside the cell.

16.4.2 Examples of other antimetabolites

Other antimetabolites in medical use include **trimethoprim** and a group of compounds known as **sulfones** (Fig. 16.9).

16.4.2.1 Trimethoprim

Trimethoprim is an orally active diaminopyrimidine structure which has proved to be a highly selective antibacterial and antimalarial agent. It acts against **dihydrofolate reductase**—the enzyme which carries out the conversion of dihydrofolate to tetrahydrofolate—leading to the inhibition of DNA synthesis and cell growth.

Dihydrofolate reductase is present in mammalian cells as well as bacterial cells, but mutations over millions of years have resulted in a significant difference in structure between the two enzymes such that trimethoprim recognizes and inhibits the bacterial enzyme more strongly. In fact, trimethoprim is 100 000 times more active against the bacterial enzyme.

Trimethoprim is often given in conjunction with the sulfonamide **sulfamethoxazole** (Fig. 16.9) in a preparation called **co-trimoxazole**. The sulfonamide inhibits the incorporation of PABA into dihydropteroate, while trimethoprim inhibits dihydrofolate reductase. Therefore, two enzymes in the one biosynthetic route are inhibited (Fig. 16.7). This is a very effective method of inhibiting a biosynthetic route and has the advantage that the doses of both drugs can be kept down to a safe level. To get the same level of inhibition using a single drug, the dose level would have to be much higher, leading to possible side effects. This approach has been described as '**sequential blocking**'.

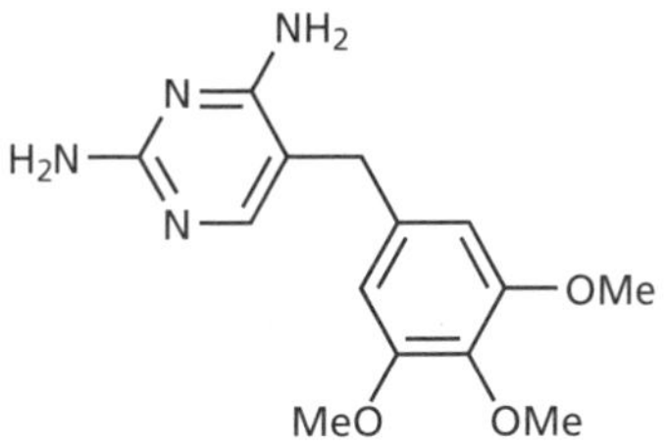

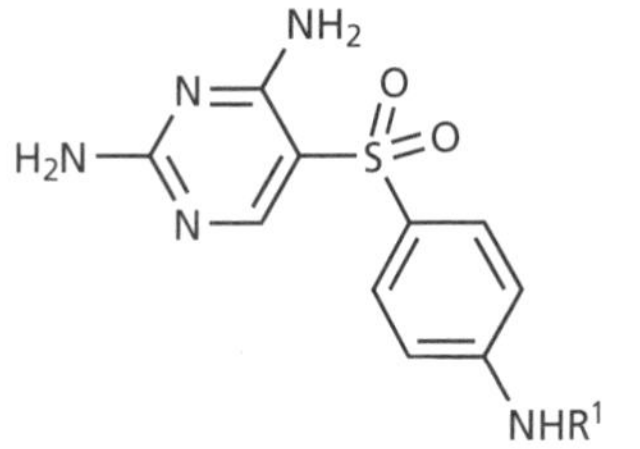

Figure 16.9 Examples of antimetabolites in medical use.

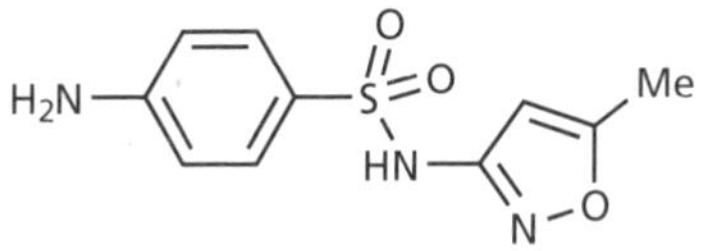

Resistance to trimethoprim has been observed in strains of *E. coli* which produce a new form of the target enzyme that has less affinity for the drug.

16.4.2.2 Sulfones

The sulfones (Fig. 16.9) are the most important drugs used in the treatment of leprosy. It is believed that they inhibit the same bacterial enzyme inhibited by the sulfonamides, i.e. dihydropteroate synthetase.

KEY POINTS

- The principle of chemotherapy or the magic bullet involves the design of chemicals which show selective toxicity against bacterial cells rather than mammalian cells.

- Early antibacterial agents were salvarsan, prontosil, and the sulfonamides. Following the discovery of penicillin, several classes of antibiotics were isolated from fungal strains.

- The bacterial cell differs in various respects from mammalian cells allowing the identification of drug targets which are unique to bacterial cells, or which differ significantly from equivalent targets in mammalian cells.

- Antibacterial agents act on five main targets—cell metabolism, the cell wall, the plasma membrane, protein synthesis, and nucleic acid function.

- Sulfonamides require a primary aromatic amine group and a secondary sulfonamide group for good activity.

- Adding an aromatic or heteroaromatic group to the sulfonamide nitrogen provides a variety of sulfonamides with different pharmacokinetic properties.

- N-Acetylation of sulfonamides is a common metabolic reaction.

- Sulfonamides are used to treat infections of the urinary tract, gastrointestinal tract, and mucous membranes. They are also used in eye lotions.

- Sulfonamides are similar in structure to para-aminobenzoic acid—a component of dihydropteroate. As a result, they can bind to the bacterial enzyme responsible for dihydropteroate synthesis and act as an inhibitor.

- Mammals synthesize tetrahydrofolate from folic acid acquired from the diet. They lack the enzyme targeted by sulfonamides. Bacteria lack the transport mechanisms required to transport folic acid into the cell.

- Trimethoprim inhibits dihydrofolate reductase—an enzyme which converts folic acid to tetrahydrofolate. It has been used in combination with sulfamethoxazole in a strategy known as sequential blocking.

- Sulfones are used in the treatment of leprosy.

16.5 Antibacterial agents which inhibit cell wall synthesis

16.5.1 Penicillins

16.5.1.1 History of penicillins

In 1877, Pasteur and Joubert discovered that certain moulds could produce toxic substances which killed bacteria. Unfortunately, these substances were also toxic to humans and were of no clinical value. They did demonstrate, however, that moulds could be a potential source of antibacterial agents.

In 1928, Fleming noted that a bacterial culture that had been left several weeks open to the air had become infected by a fungal colony. Of more interest was the fact that there was an area surrounding the fungal colony where the bacterial colonies were dying.

He correctly concluded that the fungal colony was producing an antibacterial agent which was spreading into the surrounding area. Recognizing the significance of this, he set out to culture and identify the fungus and showed it to be a relatively rare species of *Penicillium*. It has since been suggested that the *Penicillium* spore responsible for the fungal colony originated from another laboratory in the building, and that the spore was carried by air currents to be blown through the window of Fleming's laboratory. This in itself appears to be a remarkable stroke of good fortune. However, a series of other chance events were involved in the story—not least the weather! A period of early cold weather had encouraged the fungus to grow while the bacterial colonies had remained static. A period of warm weather then followed, which encouraged the bacteria to grow. These weather conditions were the ideal experimental conditions required for (a) the fungus to produce penicillin during the cold spell and (b) the antibacterial properties of penicillin to be revealed during the hot spell. If the weather had been consistently cold, the bacteria would not have grown significantly and the death of cell colonies close to the fungus would not have been seen. Alternatively, if the weather had been consistently warm, the bacteria would have outgrown the fungus and little penicillin would have been produced. As a final twist to the story, the crucial agar plate had been stacked in a bowl of disinfectant ready for washing up, but was actually placed above the surface of the disinfectant. It says much for Fleming's observational powers that he bothered to take any notice of a discarded culture plate, and that he spotted the crucial area of inhibition.

Fleming spent several years investigating the novel antibacterial substance and showed it to have significant antibacterial properties and to be remarkably non-toxic to mammals. Unfortunately, the substance was also unstable and Fleming was unable to isolate and purify the compound. He therefore came to the conclusion that penicillin was too unstable to be used clinically.

The problem of isolating penicillin was eventually solved in 1938 by Florey and Chain by using processes such as freeze-drying and chromatography which allowed isolation of the antibiotic under much milder conditions than had previously been available. By 1941, Florey and Chain were able to carry out the first clinical trials on crude extracts of penicillin and achieved spectacular success. Further developments aimed at producing the new agent in large quantities were developed in the USA, and by 1944 there was enough penicillin to treat casualties arising from the D-Day landings.

Although the use of penicillin was now widespread, the structure of the compound was still not settled and the unusual structures being proposed proved a source of furious debate. The issue was finally settled in 1945 when Dorothy Hodgkin established the structure by X-ray crystallographic analysis. The structure was quite surprising at the time, as penicillin was clearly a highly strained molecule—which explained why Fleming had been unsuccessful in purifying it.

The full synthesis of such a highly strained molecule presented a huge challenge—one that was met successfully by Sheehan in 1957. Unfortunately, the full synthesis was too involved to be of commercial use, but in the following year Beechams isolated a biosynthetic intermediate of penicillin called **6-aminopenicillanic acid (6-APA)**. This revolutionized the field of penicillins by providing the starting material for a huge range of semi-synthetic penicillins.

Penicillins have been used widely, and often carelessly. As a result, penicillin-resistant bacteria have evolved and have become an increasing problem. The fight against penicillin-resistant bacteria was helped in 1976, when Beechams discovered a natural product called **clavulanic acid** which proved highly effective in protecting penicillins from the bacterial enzymes which attack them (section 16.5.4).

16.5.1.2 Structure of benzylpenicillin and phenoxymethylpenicillin

Penicillin (Fig. 16.10) contains a highly unstable-looking bicyclic system consisting of a four-membered β-lactam ring fused to a five-membered thiazolidine ring. The skeleton of the molecule suggests that it is derived from the amino acids cysteine and valine, and this has been established (Fig. 16.11). The overall shape of the molecule is like a half-open book, as shown in Fig. 16.12.

The acyl side chain (R) varies, depending on the components of the fermentation medium. For example, corn steep liquor (the fermentation medium first used for mass production of penicillin) contains high levels of phenylacetic acid ($PhCH_2CO_2H$) and gives **benzylpenicillin** (**penicillin G**; R = benzyl). A fermentation medium containing phenoxyacetic acid ($PhOCH_2CO_2H$) gives **phenoxymethylpenicillin** (**penicillin V**; R = $PhOCH_2$) (Fig. 16.10).

16.5.1.3 Properties of benzylpenicillin

The properties of benzylpenicillin (penicillin G) are summarized below.

- Active versus non-β-lactamase producing Gram-positive bacilli (e.g. *Meningitis*, *Gonorrhoea*, and

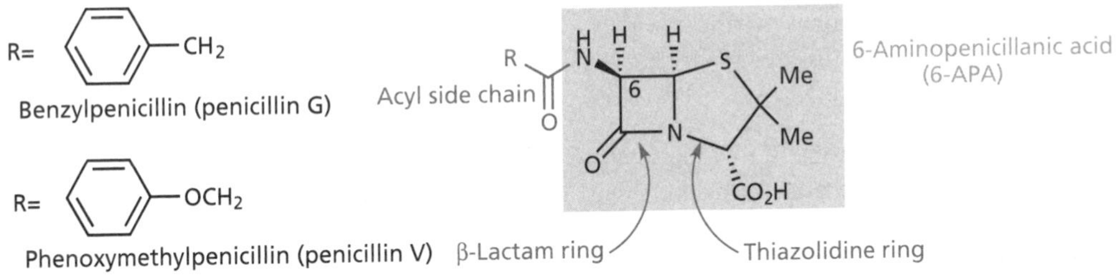

Figure 16.10 The structure of penicillin.

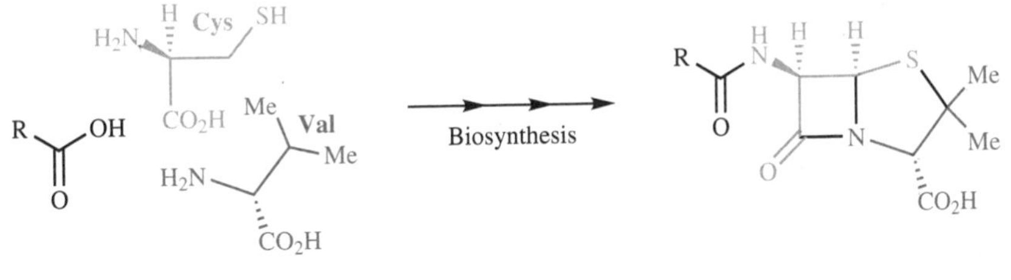

Figure 16.11 Biosynthetic precursors of penicillin.

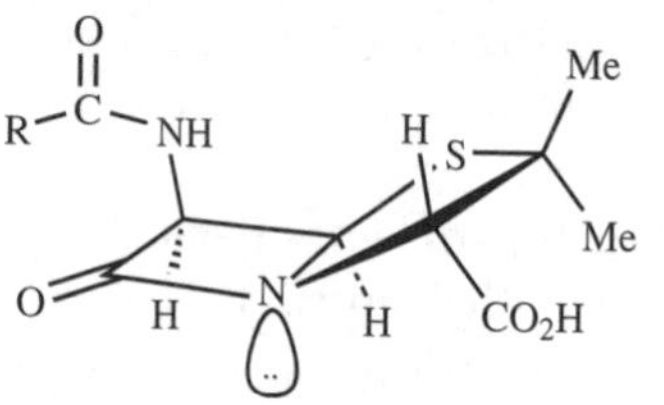

Figure 16.12 Shape of penicillin.

early strains of staphylococci) and several Gram-negative cocci (e.g. *Neisseria*). β-Lactamases are enzymes produced by penicillin-resistant bacteria which catalyse the degradation of penicillins.

- Active versus anaerobic microorganisms.

- Currently the agent of choice for treatment of infections such as those caused by *Streptococcus pyogenes*, and susceptible strains of *Streptococcus pneumoniae* and enterococci.

- Active only against rapidly dividing bacteria.

- Bactericidal rather than bacteriostatic.

- Non-toxic! This point is worth emphasizing. The penicillins are amongst the safest drugs known to medicine. They are not totally safe, however, and some people are allergic to them (see below).

- Not active over a wide range (or spectrum) of bacterial species. Generally shows poor activity against Gram-negative bacteria.

- Ineffective when taken orally, since it is broken down in the acid conditions of the stomach. Penicillin G can only be administered by injection.

- Sensitive to all known β-lactamases.

- Causes allergic reactions in some individuals, varying from a rash to immediate anaphylactic shock. Penicillins are small molecules and should not produce this effect. However, it is possible for them to react with nucleophilic groups on body proteins such that the β-lactam ring is opened and the penicillin is covalently linked to the protein. This results in the penicillin skeleton being 'recognized' by the immune response and the generation of antibodies against the penicillin structure. Anaphylactic reactions occur in 0.2% of patients with a fatality rate of 0.001%. Less serious allergic reactions are more common (1–4%).

Clearly, there are several problems associated with the use of penicillin G, the most serious being

sensitivity to acid, sensitivity to β-lactamases, and a narrow spectrum of activity. Therefore, there is scope for producing analogues with improved properties. Before looking at penicillin analogues, we shall look at penicillin's mechanism of action.

16.5.1.4 Mechanism of action of penicillin

Structure of the cell wall

In order to understand penicillin's mechanism of action, we have to first look at the structure of the bacterial cell wall and the mechanism by which it is formed. Bacteria have cell walls in order to survive a large range of environmental conditions such as varying pH, temperature, and osmotic pressure. Without a cell wall, water would continually enter the cell as a result of osmotic pressure, causing the cell to swell and burst (lysis). The cell wall is very porous and does not block the entry of water, but it does prevent the cell swelling. Animal cells do not have a cell wall, making it the perfect target for antibacterial agents such as penicillins.

The wall is a peptidoglycan structure (Fig. 16.13). In other words, it is made up of peptide and sugar units. The structure of the wall consists of a parallel series of sugar backbones containing two types of sugar—**N-acetylmuramic acid** (NAM) and **N-acetylglucosamine** (NAG) (Fig. 16.14). Peptide chains are bound to the NAM sugars, and it is interesting to note the presence of D-amino acids in these chains. In human biochemistry there are only L-amino acids, whereas bacteria have racemase enzymes that can convert L-amino acids into D-amino acids. In the final stage of cell wall biosynthesis, the peptide chains are linked together by the displacement of D-alanine from one chain by glycine in another.

About 30 enzymes are involved in the overall biosynthesis of the cell wall, but it is the final cross-linking reaction which is inhibited by penicillin. This leads to a cell wall framework that is no longer interlinked (Fig. 16.15). As a result, the wall becomes fragile and can no longer prevent the cell from swelling and bursting. The enzyme responsible for the cross-linking reaction is known as the **transpeptidase enzyme**. There are several types of the enzyme, which vary in character from one bacterial species to another, but they are all inhibited to various degrees by penicillins.

There are significant differences in the thickness of the cell wall between Gram-positive and Gram-negative bacteria. The cell wall in Gram-positive

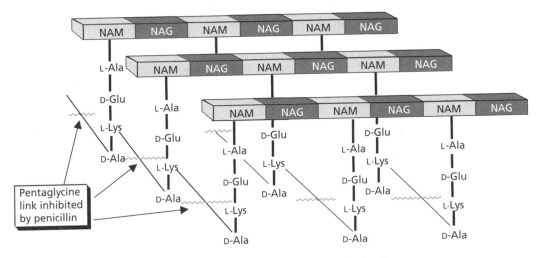

Figure 16.13 Peptidoglycan structure of bacterial cell walls.

Figure 16.14 Sugars contained in the cell wall structure of bacteria. R = H, *N*-acetylglucosamine (NAG); R = CHMeCO$_2$H, *N*-acetylmuramic acid (NAM).

bacteria consists of 50–100 peptidoglycan layers, whereas in Gram-negative bacteria it consists of only 2 layers.

The transpeptidase enzyme and its inhibition

The transpeptidase enzyme is bound to the outer surface of the cell membrane and is similar to a class of enzymes called the serine proteases, so called because they contain a serine residue in the active site and catalyse the hydrolysis of peptide bonds. In the normal mechanism (see Fig. 16.16a below), serine acts as a nucleophile to split the peptide bond between the two unusual D-alanine units on a peptide chain. The terminal alanine departs the active site, leaving the peptide chain bound to the active site. The pentaglycyl moiety of another peptide chain now enters the active site and the terminal glycine forms a peptide bond to the alanine group, displacing it from serine and linking the two chains together.

It has been proposed that penicillin has a conformation which is similar to the transition-state conformation taken up by the D-Ala-D-Ala moiety during the cross-linking reaction (Fig. 16.16), and that the enzyme mistakes penicillin for D-Ala-D-Ala and binds it to the active site. Once bound, penicillin is subjected to nucleophilic attack by serine. The enzyme can attack the β-lactam ring of penicillin and open it in the same way as it did with the peptide bond. However, penicillin is cyclic and as a result the molecule is not split in two and nothing leaves the active site. Subsequent hydrolysis of the ester group linking the penicillin to the active site does not take place either, since the penicillin structure blocks access to the pentaglycine chain or water.

If penicillin *is* acting as a mimic for a D-Ala-D-Ala moiety, this provides another explanation for its lack of toxicity, as there are no D-amino acids or D-Ala-D-Ala segments in any human protein. It is unlikely that any of the body's serine protease enzymes would recognize either the segment or penicillin itself. As a result, penicillin is selective for the bacterial transpeptidase enzyme and is ignored by the body's own serine proteases.

This theory has one or two anomalies, though. For example, **6-methylpenicillin** (Fig. 16.17) is a closer analogue to D-Ala-D-Ala. It should fit the active site better and have higher activity. In fact, it is found to have lower activity.

An alternative theory is that penicillin does not bind to the active site itself, but to a site nearby. By doing so, the penicillin structure overlaps the active site and prevents access to the normal reagents—the umbrella effect (section 5.7.2). If a nucleophilic group (not necessarily in the active site) attacks the β-lactam ring,

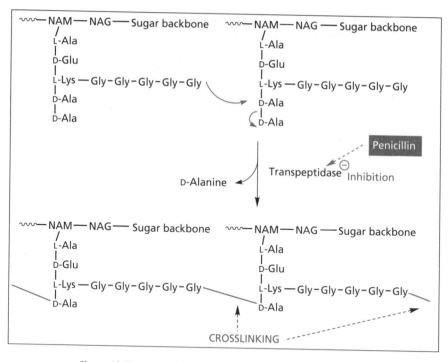

Figure 16.15 Cross-linking of bacterial cell walls inhibited by penicillin.

(a) Transpeptidase cross-linking

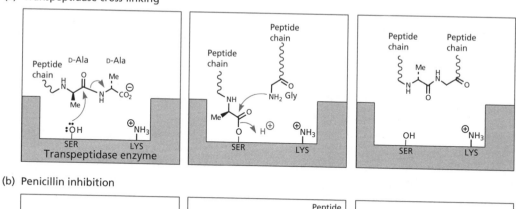

(b) Penicillin inhibition

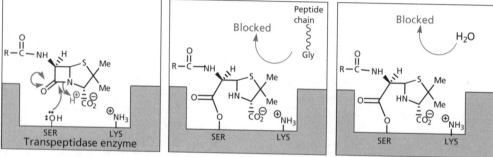

Figure 16.16 Mechanisms of transpeptidase cross-linking and penicillin inhibition.

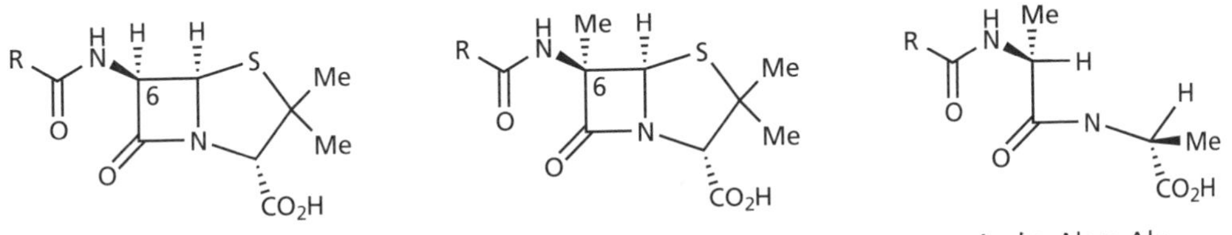

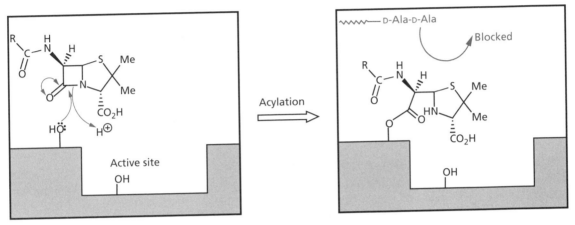

Figure 16.17 Comparison of penicillin, 6-methylpenicillin, and acyl-D-Ala-D-Ala.

Figure 16.18 Alternative 'umbrella' mechanism of inhibition.

the penicillin becomes bound irreversibly, permanently blocking the active site (Fig. 16.18).

16.5.1.5 Resistance to penicillin

Bacterial strains vary in their susceptibility to penicillin. Some species, such as streptococci, are quite vulnerable, whereas a bacterium like *Pseudomonas aeruginosa* is particularly resistant (see Box 16.3). Other species, such as *Staphylococcus aureus*, are initially vulnerable, but acquire resistance when they are exposed to penicillin over a period of time. There are several reasons for this varied susceptibility.

Physical barriers

If penicillin is to inhibit the transpeptidase enzyme, it has to reach the outer surface of the bacterial cell membrane where the enzyme is located. Thus, penicillin has to pass through the cell walls of both Gram-positive and Gram-negative bacteria. The cell wall is much thicker in Gram-positive bacteria than in Gram-negative bacteria, so one might think that penicillin would be more effective against Gram-negative bacteria. However, this is not the case. Although the cell wall is a strong, rigid structure, it is also highly porous and small molecules like penicillin can move through it without difficulty. One can imagine the cell wall as several layers of chicken wire, and the penicillin molecules as small pebbles able to pass through the gaps.

If the cell wall does not prevent penicillin reaching the cell membrane, what does? As far as Gram-positive bacteria are concerned there is no barrier, and that is why penicillin G has good activity against these organisms. As far as Gram-negative bacteria are concerned, they have an outer lipopolysaccharide membrane surrounding the cell wall which is impervious to water and polar molecules such as penicillin (Fig. 16.19). That can explain why Gram-negative bacteria are generally resistant, but not why some Gram-negative bacteria are susceptible and some are not. Should they not all be resistant?

The answer lies in protein structures called **porins**, which are located in the outer membrane. These act as pores through which water and essential nutrients can pass to reach the cell. Small drugs such as penicillin can also pass this way, but whether they do or not depends on the structure of the porin as well as the characteristics of the penicillin (i.e. its size, structure,

BOX 16.3 *PSEUDOMONAS AERUGINOSA*

Pseudomonas aeruginosa is an example of an **opportunistic pathogen**. Such organisms are not normally harmful to healthy individuals. Many people carry the organism without being aware of it, as their immune system keeps the organism under control. Once that immune system is weakened though, the organism can start multiplying and lead to serious illness. Hospital inpatients are particularly at risk, especially those suffering from shock, or those undergoing AIDS or cancer chemotherapy. Burn victims are particularly prone to *P. aeruginosa* skin infections and this can lead to septicaemia which can prove fatal. The organism is also responsible for serious lung infections in patients undergoing mechanical ventilation.

The cells of *P. aeruginosa* are rod shaped and can appear blue or green in colour, which is why it was given the name aeruginosa. It prefers to grow in moist environments and has

been isolated from soil, water, plants, animals, and humans. It can even grow in distilled water and contact lens solutions. In hospitals, there are several possible sources of infection including respiratory equipment, sinks, uncooked vegetables, and flowers brought by visitors.

P. aeruginosa is a difficult organism to treat because it has an intrinsic resistance to a wide variety of antibacterial agents including many penicillins, cephalosporins, tetracyclines, quinolones, and chloramphenicol. There are two reasons for this. The outer membrane of the cell has a low permeability to drugs, and even if a drug does enter the cell, there is an efflux system which can pump it back out again. Nevertheless, there are drugs which have proved effective against the organism, in particular aminoglycosides such as tobramycin or gentamicin, and penicillins such as ticarcillin. These are often given in combination with each other.

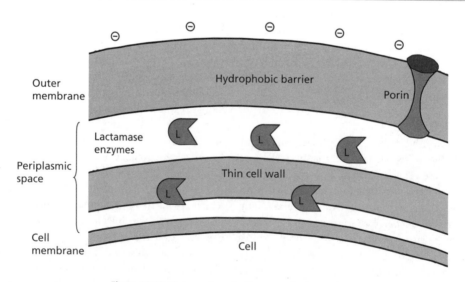

Tobramycin.

Figure 16.19 Outer surface of a Gram-negative bacterial cell.

and charge). In general, drugs have less chance of passing through the porins if they are large, have a negative charge, and are hydrophobic. In contrast, a small hydrophilic drug that can exist as a zwitterion will move through quickly. Therefore, porins play a crucial role in controlling the amount of penicillin capable of reaching the periplasmic space between the outer membrane and cell membranes. If access is slow, the concentration of penicillin at the transpeptidase enzyme may not be sufficient to inhibit it effectively.

Presence of β-lactamase enzymes

The presence of β-lactamase enzymes is the most important mechanism by which bacteria gain resistance to penicillin. β-Lactamases are enzymes which have mutated from transpeptidases and so they are quite similar in nature. For example, they have a serine residue in the active site and can open up the β-lactam ring of penicillin to form an ester link to the structure. Unlike the transpeptidase enzyme, β-lactamases are able to hydrolyse the ester link and shed the ring-opened penicillin. They do this so effectively that 1000 penicillin molecules are hydrolysed per second (Fig. 16.20).

Some Gram-positive bacterial strains are resistant to penicillin because they can release β-lactamase into the surrounding environment such that penicillin is intercepted before it reaches the cell membrane. The enzyme eventually dissipates through the cell wall and is lost, so the bacterium has to keep generating the enzyme to maintain its protection. *Staphylococcus aureus* is a Gram-positive bacterium that used to be susceptible to penicillin, but 95% of *S. aureus* strains now release a β-lactamase which hydrolyses penicillin G.

Most if not all Gram-negative bacteria produce β-lactamases, which makes them more resistant to penicillins. Moreover, the β-lactamase released is trapped in the periplasmic space between the cell membrane and the outer membrane because it cannot pass through the latter. As a result, any penicillin managing to penetrate the outer membrane encounters a higher concentration of β-lactamase than it would with Gram-positive bacteria. This might suggest again that all Gram negative bacteria should be resistant to penicillin. However, there are various types of β-lactamase enzyme produced by both Gram-positive and Gram-negative bacteria which vary in their substrate selectivity. Some are selective for penicillins (penicillinases), some for cephalosporins (section 16.5.2) (cephalosporinases) and some for both penicillins and cephalosporins. The differing levels of enzyme and their differing affinities for different β-lactams account for the varying susceptibilities of Gram-negative bacteria to different β-lactams.

High levels of transpeptidase enzyme produced

In some Gram-negative bacteria, excess quantities of transpeptidase are produced, and penicillin is incapable of inactivating all the enzyme molecules present.

Affinity of the transpeptidase enzyme to penicillin

There are several forms of the transpeptidase enzyme present within any bacterial cell and these vary in their affinity for the different β-lactams. Differences in the relative proportions of these enzymes across bacterial species account in part for the variable susceptibility of these bacteria to different penicillins. For example, early strains of *Staphylococcus aureus* contained transpeptidase enzymes which had a high affinity for penicillin and were effectively inhibited. Penicillin-resistant strains of *Staphylococcus aureus* acquired a transpeptidase enzyme called penicillin binding protein (PBP) 2a, which has a much lower affinity to penicillins. The presence of low-affinity transpeptidases is also a problem with enterococci and pneumococci.

Transport back across the outer membrane of gram-negative bacteria

There are proteins in the outer membrane of some Gram-negative bacteria which are capable of pumping penicillin out of the periplasmic space, thus lowering

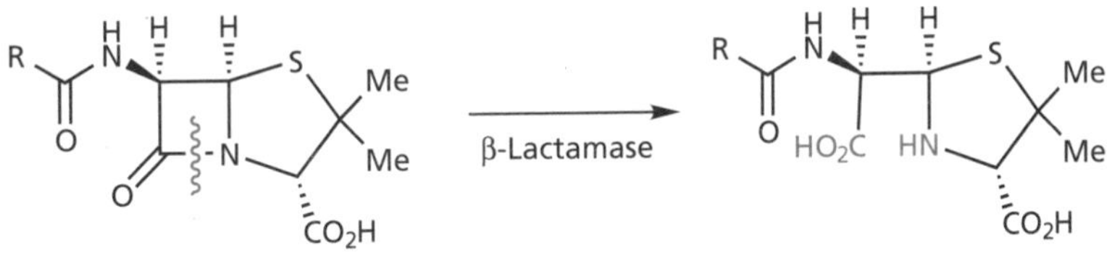

Figure 16.20 β-Lactamase deactivation of penicillin.

its concentration and effectiveness. The extent to which this happens varies from species to species and also depends on the structure of the penicillin. This is known as an **efflux** process.

Mutations and genetic transfers

Mutations can occur which will affect any or all of the above mechanisms such that they are more effective in resisting the effects of β-lactams. Small portions of DNA carrying the genes required for resistance can also be transferred from one cell to another by means of genetic vehicles called **plasmids**. These are small pieces of circular bacterial extrachromosomal DNA. If the transferred DNA contains a gene coding for a β-lactamase enzyme or some other method of improved resistance, then the recipient cell acquires immunity. Genetic material can also be transferred between bacterial cells by viruses and by the uptake of free DNA released by dead bacteria.

16.5.1.6 Methods of synthesizing penicillin analogues

Having studied the mechanism of action of penicillin G and the various problems surrounding resistance, we now look at how analogues of penicillin G can be synthesized which might have improved stability and activity. A method of preparing analogues is required which is cheap, efficient, and flexible. Sheehan's full synthesis of penicillin is too long and low yielding (1%) to be practical, so that limits the options to fermentation methods or semi-synthetic procedures.

Fermentation

Originally, the only way to prepare different penicillins was to vary the fermentation conditions. Adding different carboxylic acids to the fermentation medium resulted in penicillins with different acyl side chains (e.g. **phenoxymethylpenicillin;** Fig. 16.10). Unfortunately, there was a limitation to the sort of carboxylic acid which was accepted by the biosynthetic route (i.e. only acids of general formula RCH_2CO_2H). This in turn restricted the variety of analogues which could be obtained. The other major disadvantage was the tedious and time-consuming nature of the method.

Semi-synthetic procedure

In 1959, Beechams isolated a biosynthetic intermediate of penicillin from *Penicillium chrysogenum* grown in a fermentation medium which was deficient in a carboxylic acid. The intermediate (**6-aminopenicillanic acid; 6-APA**) proved to be one of Sheehan's synthetic intermediates, and so it was possible to use this to synthesize a huge number of analogues by a semi-synthetic method. Thus, fermentation yielded 6-APA, which could then be treated with a range of acid chlorides (Fig. 16.21).

6-APA is now produced more efficiently by hydrolysing penicillin G or penicillin V with an enzyme (**penicillin acylase**) (Fig. 16.22) or by a chemical method that allows the hydrolysis of the side chain in the presence of the highly strained β-lactam ring. The latter procedure is described in more detail in

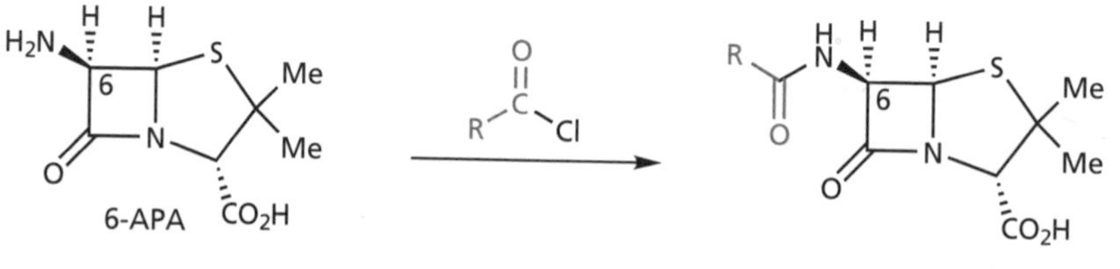

Figure 16.21 Penicillin analogues synthesized by acylating 6-APA.

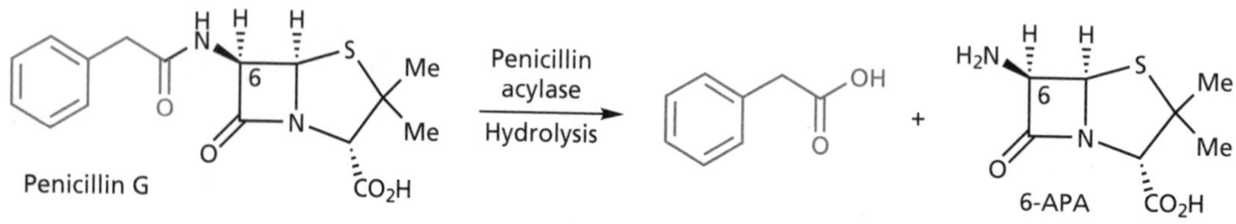

Figure 16.22 Production of 6-APA.

section 16.5.2.2 where it is used to hydrolyse the side chain from cephalosporins.

We have emphasized the drive to make penicillin analogues with varying acyl side chains, but what is so special about the acyl side chain? Could changes not be made elsewhere in the molecule? In order to answer these questions we need to look at the structure–activity relationships of penicillins.

16.5.1.7 Structure–activity relationships of penicillins

A large number of penicillin analogues have been synthesized and studied. The results of these studies led to the following SAR conclusions (Fig. 16.23).

- The strained β-lactam ring is essential.

- The free carboxylic acid is essential. This is usually ionized and penicillins are administered as sodium or potassium salts. The carboxylate ion binds to the charged ammonium ion of a lysine residue in the binding site.

- The bicyclic system is important. (This confers further strain on the β-lactam ring—the greater the strain, the greater the activity, but the greater the instability of the molecule to other factors.)

- The acylamino side chain is essential.

- Sulfur is usual but not essential.

- The stereochemistry of the bicyclic ring with respect to the acylamino side chain is important.

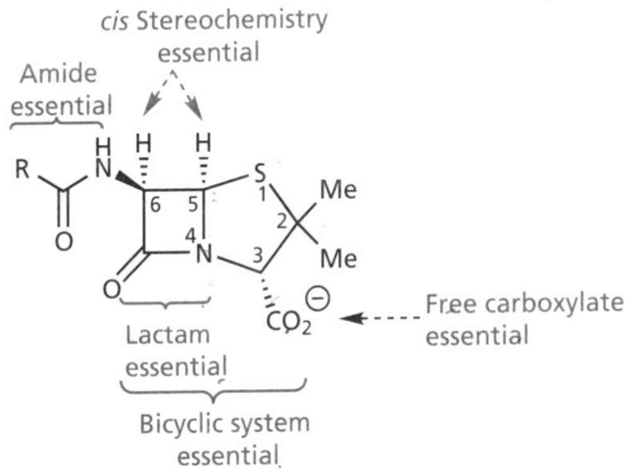

The results of this analysis led to the inevitable conclusion that very little variation is tolerated by the penicillin nucleus and that any variations are restricted to the acylamino side chain.

16.5.1.8 Penicillin analogues

In this section we consider the penicillin analogues which proved successful in tackling the problems of acid sensitivity, β-lactamase sensitivity, and limited breadth of activity.

Acid sensitivity of penicillins

There are three reasons for the acid sensitivity of penicillin G.

- **Ring strain:** The bicyclic system in penicillin consists of a four-membered ring fused to a five-membered ring. As a result, penicillin suffers large angle and torsional strains. Acid-catalysed ring opening relieves these strains by breaking open the more highly strained β-lactam ring (Fig. 16.24).

- **A highly reactive β-lactam carbonyl group:** The carbonyl group in the β-lactam ring is highly susceptible to nucleophiles and does not behave like a normal tertiary amide. The latter is resistant to nucleophilic attack, because the carbonyl group is stabilized by the neighbouring nitrogen atom as shown in Fig. 16.25; the nitrogen can feed its lone pair of electrons into the carbonyl group to form a dipolar resonance structure with bond angles of 120°. This resonance stabilization is impossible for the β-lactam ring, because of the increase in angle

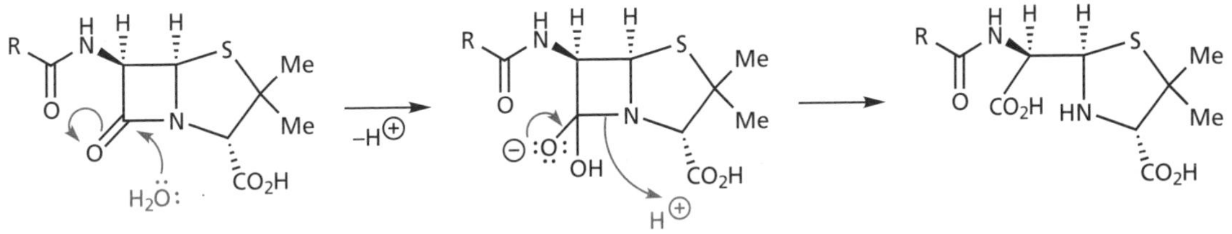

Figure 16.23 Structure–activity relationships of penicillins.

Figure 16.24 Ring opening of the β-lactam ring.

strain that would result in having a double bond within a β-lactam ring. The preferred bond angles for a double bond are 120°, but the bond angles of the β-lactam ring are constrained to 90°. As a result, the lone pair is localized on the nitrogen atom, and the carbonyl group is more electrophilic than one would expect for a tertiary amide.

• **Influence of the acyl side chain (neighbouring group participation)**: Figure 16.26 demonstrates how the neighbouring acyl group can actively participate in a mechanism to open up the lactam

ring. Thus, penicillin G has a self-destruct mechanism built into its structure.

Acid-resistant penicillins

It can be seen that countering acid sensitivity is a difficult task. Nothing can be done about the first two factors, as the β-lactam ring is vital for antibacterial activity. Therefore, only the third factor can be tackled. The task then becomes one of reducing the amount of neighbouring group participation taking place. This was achieved by placing an electron-withdrawing

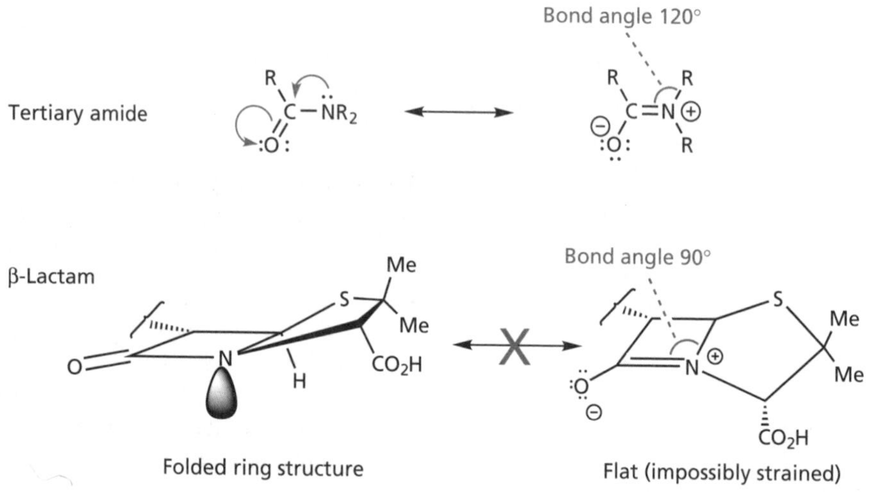

Figure 16.25 Comparison of tertiary amide and β-lactam carbonyl groups.

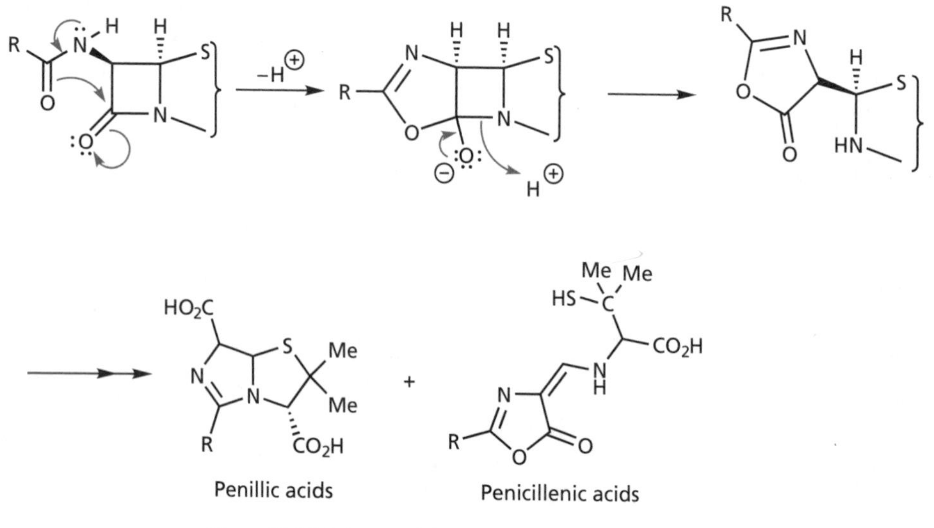

Figure 16.26 Influence of the acyl side chain on acid sensitivity.

group in the side chain which could draw electrons away from the carbonyl oxygen and reduce its tendency to act as a nucleophile (Fig. 16.27).

Phenoxymethylpenicillin (penicillin V) has an electronegative oxygen on the acyl side chain with the electron-withdrawing effect required. The molecule has better acid stability than penicillin G and is stable enough to survive the acid in the stomach, so it can be given orally.

A range of penicillin analogues with an electron-withdrawing substituent (X) on the α-carbon of the side chain (Fig. 16.27) have also proved resistant to acid hydrolysis and can be given orally (e.g. **ampicillin**; see Fig. 16.30).

To conclude, the problem of acid sensitivity is fairly easily solved by having an electron-withdrawing group on the acyl side chain.

Penicillinase-resistant penicillins

The problem of β-lactamases became critical in 1960, when the widespread use of penicillin G led to an alarming increase of penicillin-resistant *S. aureus* infections. At one point, 80% of all *S. aureus* infections in hospitals were due to virulent, penicillin-resistant strains. Alarmingly, these strains were also resistant to all other available antibiotics. Fortunately, a solution to the problem was just around the corner—the design of penicillinase-resistant penicillins.

The strategy of steric shields (section 11.2.1) was successfully used to block penicillin from accessing the penicillinase or β-lactamase active site by placing a bulky group on the side chain (Fig. 16.28). However, there was a problem. If the steric shield was too bulky, then it also prevented the penicillin from attacking the transpeptidase target enzyme. Therefore, a great deal of work had to be done to find an ideal shield which would be large enough to ward off the lactamase enzyme, but would be small enough to allow the penicillin to bind to the target enzyme. The fact that the β-lactam ring interacts with both enzymes in the same way highlights the difficulty in finding the ideal shield.

Fortunately, shields were found which could make that discrimination. **Methicillin** (Fig. 16.28) was the first effective semi-synthetic penicillin with resistance to the *S. aureus* β-lactamase and reached the clinic just in time to treat the growing *S. aureus* problem. The steric shields are the two *ortho*-methoxy groups on the aromatic ring.

Methicillin is by no means an ideal drug, however. With no electron-withdrawing group on the side chain, it is acid sensitive, and has to be injected. It has only 1/50 the activity of penicillin G against organisms that are penicillin G sensitive, it shows poor activity against some streptococcal strains, and it is inactive against Gram-negative bacteria. Better penicillinase resistant agents have since been developed (see Box 16.4),

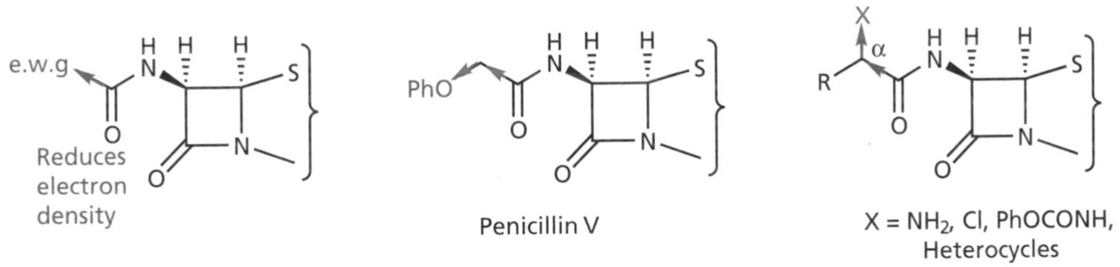

Figure 16.27 Reduction of neighbouring group participation with an electron-withdrawing group (e.w.g.).

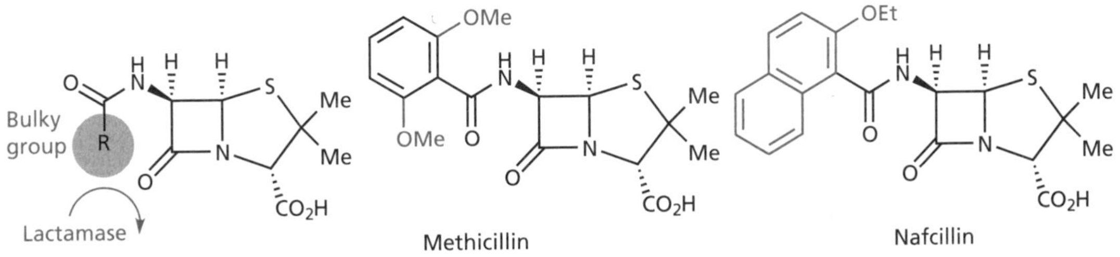

Figure 16.28 Blocking penicillin from reaching the penicillinase active site.

and methicillin is no longer used clinically. **Nafcillin** contains a naphthalene ring as its steric shield. It has more intrinsic activity than methicillin against staphylococci and streptococci, and is also administered by injection.

To sum up, acid-resistant penicillins would be the drugs of first choice against an infection. If the bacteria

---

BOX 16.4 THE ISOXAZOLYL PENICILLINS

The incorporation of an isoxazolyl ring into the penicillin side chain led to orally active compounds which were stable to the β-lactamase enzyme of *S. aureus*. The isoxazolyl ring acts as the steric shield but it is also electron-withdrawing, giving the structure acid stability.

Oxacillin, cloxacillin, flucloxacillin and dicloxacillin are all useful against *S. aureus* infections. The only difference between them is the type of halogen substitution on the aromatic ring. The influence of these groups is found to be pharmacokinetic; that is, they influence such factors as absorption of the drug and plasma protein binding. For example, cloxacillin is better absorbed through the gut wall than oxacillin, whereas flucloxacillin is less bound to plasma protein, resulting in higher levels of the free drug in the blood supply.

Having pointed out the advantages of these drugs over methicillin, it is worth putting things into context by pointing out that these four penicillins have inferior activity to the original penicillins when they are used against bacteria that lack the lactamase enzyme. They are also inactive against Gram-negative bacteria.

Oxacillin	$R^1 = R^2 = H$
Cloxacillin	$R^1 = Cl, R^2 = H$
Flucloxacillin	$R^1 = Cl, R^2 = F$
Dicloxacillin	$R^1 = Cl, R^2 = Cl$

Incorporation of a five-membered heterocycle.

---

proved resistant due to the presence of a penicillinase enzyme (e.g. penicillin-resistant *S. aureus* and *S. epidermidis*), then the therapy would be changed to penicillinase-resistant penicillins.

Unfortunately, 95% of *S. aureus* strains detected in hospitals have become resistant to methicillin and the other penicillinase-resistant penicillins as a result of mutations to the transpeptidase enzyme. These bacteria are referred to as MRSA. The abbreviation stands for <u>m</u>ethicillin-<u>r</u>esistant *S. <u>a</u>ureus*, but the term applies to all the penicillinase-resistant penicillins and not just methicillin.

Broad-spectrum penicillins

There are a variety of factors affecting whether a particular bacterial strain will be susceptible to a penicillin. The spectrum of activity shown by any penicillin depends on its structure, its ability to cross the cell membrane of Gram-negative bacteria, its susceptibility to β-lactamases, its affinity for the transpeptidase target enzyme, and the rate at which it is pumped back out of cells by Gram-negative organisms. All these factors vary in importance across different bacterial species and so there are no clear-cut tactics which can be used to improve the spectrum of activity. Consequently, the search for broad-spectrum antibiotics was one of trial and error which involved making a huge variety of analogues. These changes were again confined to variations in the side chain and gave the following results:

- Hydrophobic groups on the side chain (e.g. penicillin G) favour activity against Gram-positive bacteria, but result in poor activity against Gram-negative bacteria.

- If the hydrophobic character is increased, there is little effect on Gram-positive activity, but what activity there is against Gram-negative bacteria drops even more.

- Hydrophilic groups on the side chain have little effect on Gram-positive activity (e.g. **penicillin T**) or cause a reduction of activity (e.g. **penicillin N**) (Fig. 16.29, Table 16.1). However, they lead to an increase in activity against Gram-negative bacteria.

- Enhancement of Gram-negative activity is found to be greatest if the hydrophilic group (e.g. NH_2, OH, CO_2H) is attached to the carbon, α to the carbonyl group on the side chain.

Those penicillins having useful activity against both Gram-positive and Gram-negative bacteria are known as **broad-spectrum antibiotics.** There are three classes of broad-spectrum antibiotics. All have an α-hydrophilic group which aids the passage of these penicillins through the porins of the Gram negative bacterial outer cell membrane.

Broad-spectrum penicillins—the aminopenicillins

Ampicillin (Fig. 16.30; Beechams 1964) and **amoxicillin** are very similar in structure, the only difference being an extra phenol group on amoxicillin. Amoxicillin has similar properties to ampicillin, but is better absorbed through the gut wall.

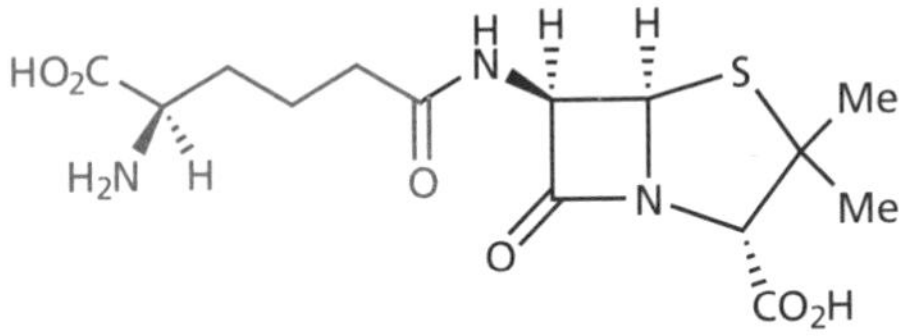

Penicillin N

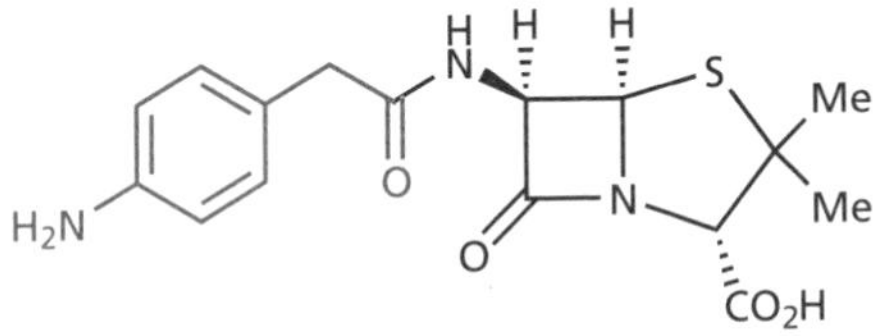

Penicillin T

Figure 16.29 Effect of sidechain hydrophilic groups on antibacterial activity.

Table 16.1 Effect of side chain hydrophilic groups on antibacterial activity (with respect to penicillin G)

Penicillin N		Penicillin T	
Gram positive	Gram negative	Gram positive	Gram negative
1%	Greater	Similar	2–4 × greater

The properties of ampicillin and amoxicillin are as follows:

- Similar spectrum of activity to penicillin G but more active against Gram-negative cocci and enterobacteria.
- Acid-resistant due to the NH_2 group, and therefore orally active.
- Non-toxic.
- Sensitive to β-lactamases (no shield).
- Inactive against *Pseudomonas aeruginosa* (a particularly resistant species).
- Can cause diarrhoea due to poor absorption through the gut wall. All penicillins used at high doses for prolonged periods will abolish the normal bacterial flora in the intestines. This allows the colonization of resistant Gram-negative bacilli or fungi leading to intestinal problems.
- Amoxicillin has been used in the treatment of bronchitis, pneumonia, typhoid, gonorrhoea, and urinary tract infections. Its spectrum of activity is increased when administered with clavulanic acid (section 16.5.4.1).

The problem of poor absorption through the gut wall is due to the dipolar nature of the molecule—it has both a free amino group and a free carboxylic acid function. This problem can be alleviated by using a prodrug where one of the polar groups is masked with a protecting group which can be removed metabolically once the prodrug has been absorbed (Box 16.5).

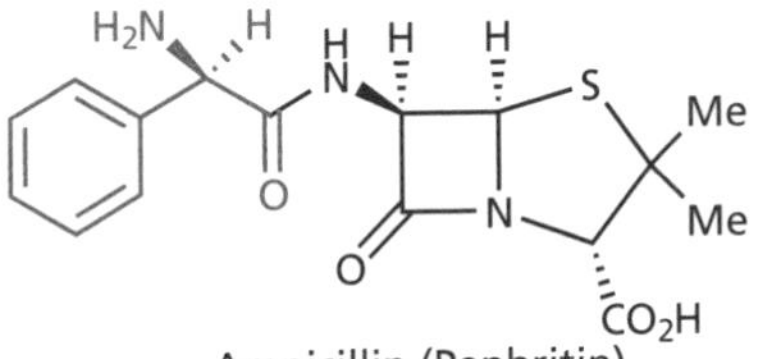

Ampicillin (Penbritin)

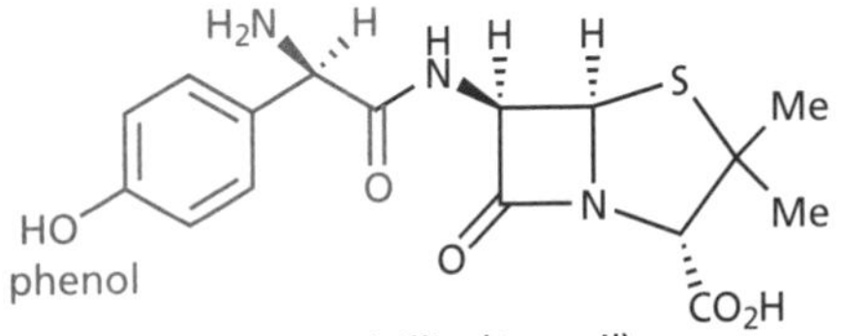

Amoxicillin (Amoxil)

Figure 16.30 Broad spectrum antibiotics—the aminopenicillins.

BOX 16.5 AMPICILLIN PRODRUGS

Pivampicillin, talampicillin, and bacampicillin are prodrugs of ampicillin (Fig. 1). In all three cases, the esters used to mask the carboxylic acid group seem rather elaborate and one may ask why a simple methyl ester is not used. The answer is that methyl esters of penicillins are not metabolized in humans. The bulky penicillin skeleton is so close to the ester that it acts as a steric shield, and prevents the esterase enzymes which catalyse this reaction from accepting the penicillin ester as a substrate.

Fortunately, acyloxymethyl esters *are* susceptible to esterases. These 'extended' esters contain a second ester group further away from the penicillin nucleus, which is more exposed to attack. The hydrolysis products are inherently unstable and decompose spontaneously to release formaldehyde

and reveal the free carboxylic acid (Fig. 2). The release of formaldehyde is not ideal, as it is a toxic chemical. However, it is formed naturally in the body through enzymatic demethylation of various compounds found in the diet, and the levels produced from the prodrugs described above cause little problem because the drugs are only taken for a short duration of time.

Such extended esters can be used to prepare prodrugs of other penicillins, but one has to be careful that one does not go to the other extreme and make the penicillin too lipophilic. For example, the 1-acyloxyalkyl ester of penicillin G is too lipophilic and has poor solubility in water. Fortunately, the problem can easily be avoided by making the extended ester more polar (e.g. by attaching valine as in Fig. 3).

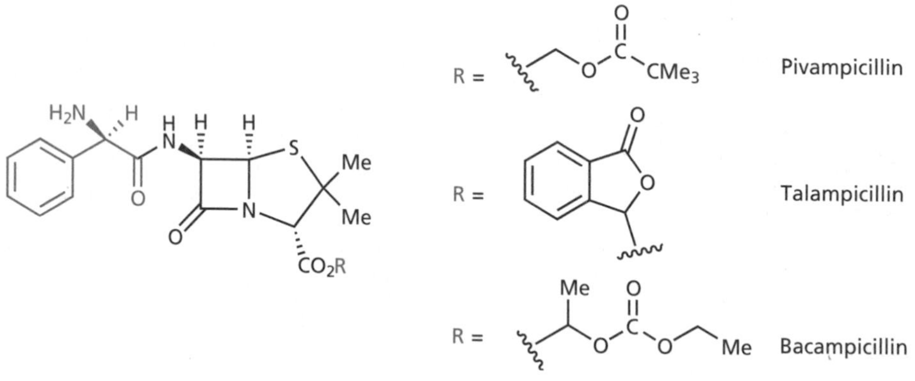

Figure 1 Prodrugs used to aid absorption of antibiotic through gut wall.

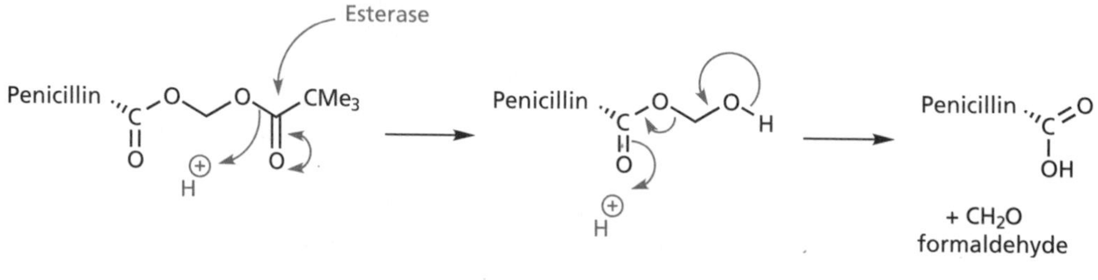

Figure 2 Decomposition of acyloxymethyl esters.

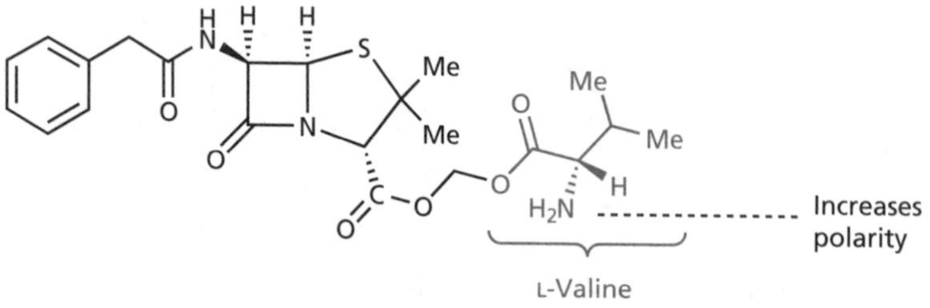

Figure 3 Polar extended ester for Penicillin G.

Broad-spectrum penicillins—the carboxypenicillins

Carbenicillin (Fig. 16.31) was the first example of this class of compounds and the first to show activity against *Pseudomonas aeruginosa*. Carbenicillin has activity against a wider range of Gram-negative bacteria than ampicillin, and was used particularly against penicillin-resistant strains. The broad activity against Gram-negative bacteria is due to the hydrophilic acid group (ionized at pH 7) on the side chain. It is particularly interesting to note that the stereochemistry of this group is important. The α-carbon is asymmetric and only one of the two enantiomers is active.

Unfortunately, carbenicillin has several disadvantages. It is less active than ampicillin against various other bacterial strains and requires high dose levels. It has toxic side effects, and a marked reduction in activity against Gram-positive bacteria (note the hydrophilic acid group). It is also acid sensitive and has to be injected. Better penicillins such as the ureidopenicillins have since been developed and so the use of carbenicillin is now discouraged.

Carfecillin and indanyl carbenicillin (Fig. 16.31) are prodrugs for carbenicillin and show an improved absorption through the gut wall. Aryl esters are used rather than alkyl esters since the former are chemically more susceptible to hydrolysis because of the electron-withdrawing inductive effect of the aryl ring. An extended ester is not required in this case since the aryl ester is further from the β-lactam ring and is not shielded. These agents proved useful for the treatment of urinary tract infections but have generally been superseded by fluoroquinolone antibacterial agents (section 16.8.1).

Ticarcillin is similar in structure to carbenicillin but has a thiophene ring in place of the phenyl group. Like carbenicillin, it is administered by injection, and has an identical antibacterial spectrum, but it has several advantages over carbenicillin. Smaller doses can be used, it is 2–4 times more effective against *P. aeruginosa* and it has fewer side effects. It can also be administered with clavulanic acid to broaden its spectrum of activity (section 16.5.4).

Broad-spectrum penicillins—the ureidopenicillins

Ureidopenicillins (Fig. 16.32) have a urea functional group at the α-position. In general, they are more active than the carboxypenicillins against streptococci

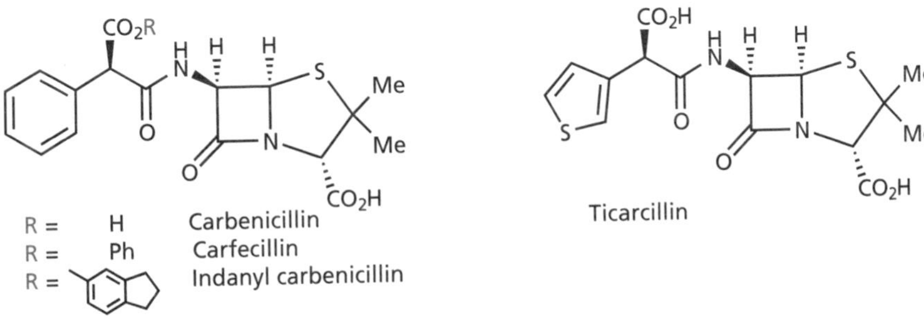

Figure 16.31 Carboxypenicillins.

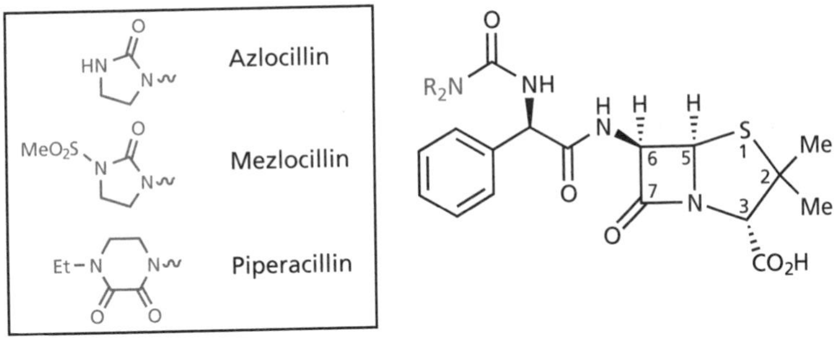

Figure 16.32 Ureidopenicillins.

and *Haemophilus* species. They show similar activity against Gram-negative aerobic rods such as *P. aeruginosa* but are generally more active against other Gram negative bacteria. Unfortunately, they have to be injected. Examples include **azlocillin,** which is 8–16 times more active than carbenicillin against *P. aeruginosa,* and is used primarily for the treatment of infections caused by that organism. It is susceptible to β-lactamases. **Mezlocillin** has a similar spectrum of activity to carbenicillin but is more active since it has a higher affinity for transpeptidases and can cross the outer membrane of Gram-negative bacteria more effectively. **Piperacillin** is similar to ampicillin in its activity against Gram-positive species. It also has good activity against anaerobic species of both cocci and bacilli and can be used against a variety of infections. Piperacillin can be administered alongside **tazobactam** as **Tazocin** to widen its spectrum of activity (section 16.5.4.2 and Box 16.6).

BOX 16.6 SYNERGISM OF PENICILLINS WITH OTHER DRUGS

There are several examples in medicinal chemistry where the presence of one drug enhances the activity of another. In many cases this can be dangerous, leading to an effective overdose of the enhanced drug. In some cases, though, it can be useful. There are two interesting examples where the activity of penicillin has been enhanced by the presence of another drug.

One of these is the effect of clavulanic acid, described in section 16.5.4.1. The other is the administration of penicillins with a compound called **probenecid**. Probenecid is a moderately lipophilic carboxylic acid and as such is similar to penicillin. It is found that probenecid can block facilitated transport of penicillin through the kidney tubules. In other words, probenecid slows down the rate at which penicillin is excreted by competing with it in the excretion mechanism. Probenecid also competes with penicillin for binding sites on albumin. As a result, penicillin levels in the bloodstream are enhanced and the antibacterial activity increases—a useful tactic if faced with a particularly resistant bacterium.

Probenecid.

KEY POINTS

- Penicillins have a bicyclic structure consisting of a β-lactam ring fused to a thiazolidine ring. The strained β-lactam ring reacts irreversibly with the transpeptidase enzyme responsible for the final cross-linking of the bacterial cell wall.
- Penicillin analogues can be prepared by fermentation or by a semi-synthetic synthesis from 6-aminopenicillanic acid. Variation of the penicillin structure is limited to the acyl side chain.
- Penicillins can be made more resistant to acid conditions by incorporating an electron-withdrawing group into the acyl side chain.
- Steric shields can be added to penicillins to protect them from bacterial β-lactamase enzymes.
- Broad spectrum activity is associated with the presence of an α-hydrophilic group on the acyl side chain of penicillin.
- Prodrugs of penicillins are useful in masking polar groups and improving absorption from the gastrointestinal tract. Extended esters are used which undergo enzyme-catalysed hydrolysis to produce a product which spontaneously degrades to release the penicillin.
- Probenecid can be administered with penicillins to hinder the excretion of penicillins.

16.5.2 Cephalosporins

16.5.2.1 Cephalosporin C

Discovery and structure of cephalosporin C

The second major group of β-lactam antibiotics to be discovered were the cephalosporins. The first cephalosporin (**cephalosporin C**) was derived from a fungus obtained in the mid 1940s from sewer waters on the island of Sardinia. This was the work of an Italian professor, who noted that the waters surrounding the sewage outlet periodically cleared of microorganisms. He reasoned that an organism might be producing an antibacterial substance and so he collected samples and managed to isolate a fungus called *Cephalosporium acremonium* (now called *Acremonium chrysogenum*). The crude extract from this organism was shown to have antibacterial properties, and in 1948, workers at Oxford University isolated cephalosporin C, but it was not until 1961 that the structure was established by X-ray crystallography.

The structure of cephalosporin C (Fig. 16.33) has similarities to that of penicillin in that it has a bicyclic system containing a four-membered β-lactam ring. This time the β-lactam ring is fused to a six-membered dihydrothiazine ring. The larger ring relieves the strain in the bicyclic system to some extent, but it is still a reactive system. A study of the cephalosporin skeleton reveals that cephalosporins can be derived from the same biosynthetic precursors as penicillin, i.e. cysteine and valine (Fig. 16.34).

Properties of cephalosporin C

Cephalosporin C is not particularly potent compared to penicillins (1/1000 the activity of penicillin G), but the antibacterial activity it does have is more evenly directed against Gram-negative and Gram-positive bacteria. Another in-built advantage of cephalosporin C is its greater resistance to acid hydrolysis and to β-lactamase enzymes. It is also less likely to cause allergic reactions. Therefore, cephalosporin C was seen as a useful lead compound for the development of broad-spectrum antibiotics. The aim was to produce analogues with increased potency, while retaining the breadth of activity. Cephalosporin C itself has been used in the treatment of urinary tract infections, since it is found to concentrate in the urine and survive the body's hydrolytic enzymes.

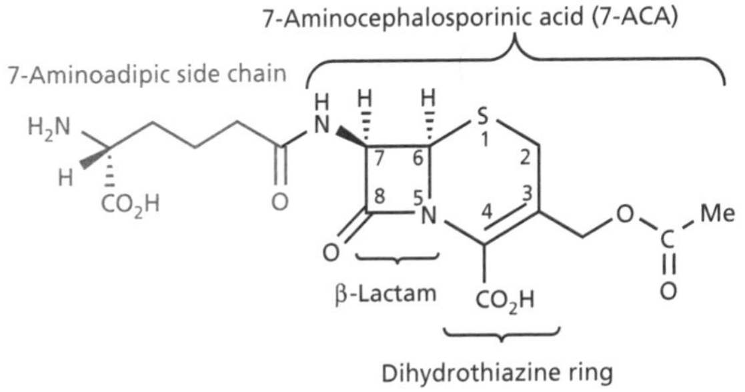

Structure–activity relationships of cephalosporin C

Many analogues of cephalosporin C have been made which demonstrate the importance of the β-lactam ring within the bicyclic system, the ionized carboxylate group at position 4, and the acylamino side chain at position 7. These results tally closely with those obtained for the penicillins. The strain effect of a six-membered ring fused to a four-membered ring is less than for penicillin, but this is partially offset by the effect of the acetyloxy group at position 3. This can act as a good leaving group in the inhibition mechanism (Fig. 16.35).

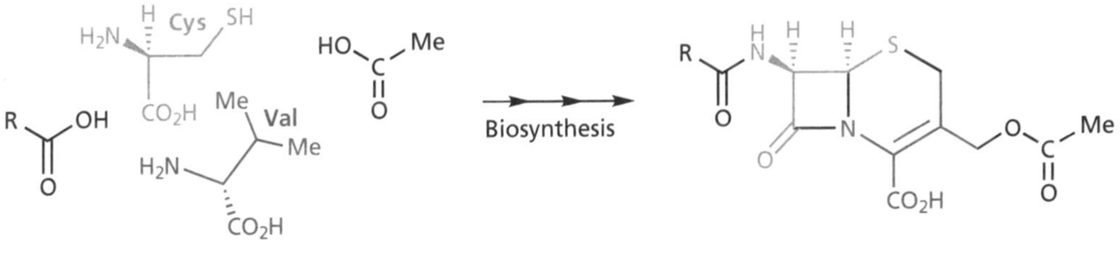

Figure 16.33 Cephalosporin C.

Figure 16.34 Biosynthetic precursors of cephalosporin C.

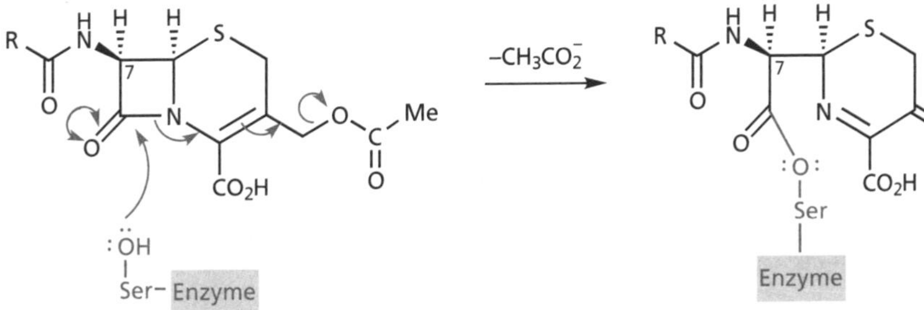

Figure 16.35 Mechanism of action by which cephalosporins inhibit the transpeptidase enzyme.

There is a limited number of places where modifications can be made (Fig. 16.36), but there are more possibilities than with penicillins:

- variations of the 7-acylamino side chain
- variations of the 3-acetoxymethyl side chain
- extra substitution at carbon 7.

16.5.2.2 Synthesis of cephalosporin analogues at the 7-position

Access to analogues with varied side chains at the 7-position initially posed a problem. Unlike

penicillins, it proved impossible to obtain cephalosporin analogues by fermentation. Similarly, it was not possible to obtain **7-ACA (7-aminocephalosporinic acid)** either by fermentation or by enzymatic hydrolysis of cephalosporin C, thus preventing the semi-synthetic approach analogous to the preparation of penicillins from 6-APA (section 16.5.1.6).

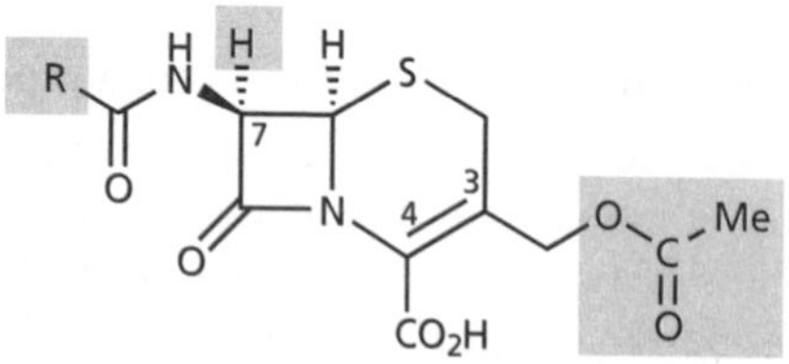

Therefore, a way had to be found of obtaining 7-ACA from cephalosporin C by chemical hydrolysis. This is no easy task, as a secondary amide has to be hydrolysed in the presence of a highly reactive β-lactam ring. Normal hydrolytic procedures are not suitable, and so a special method had to be worked out (Fig. 16.37).

The strategy used takes advantage of the fact that the β-lactam nitrogen is unable to share its lone pair of electrons with its neighbouring carbonyl group. The first step of the procedure requires the formation of a double bond between the nitrogen on the side chain and its neighbouring carbonyl group (Fig. 16.38). This is only possible for the secondary amide group, since

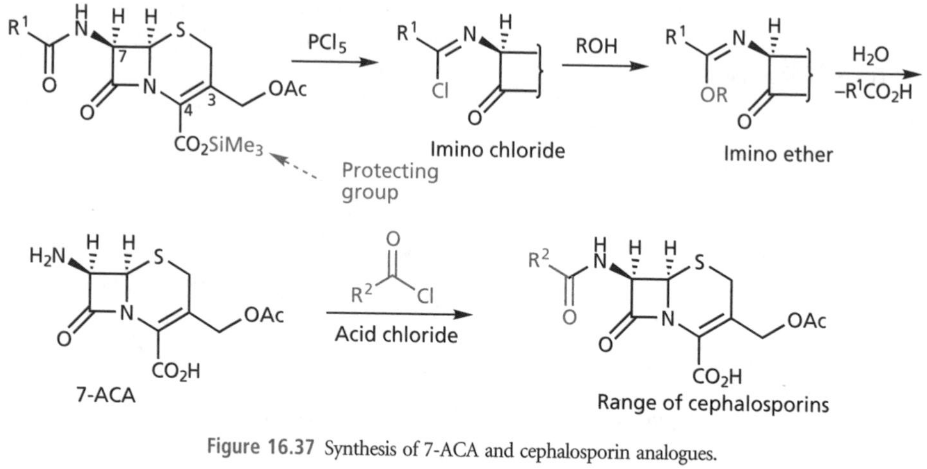

Figure 16.36 Positions for possible modification of cephalosporin C. The shading indicates positions which can be varied.

Figure 16.37 Synthesis of 7-ACA and cephalosporin analogues.

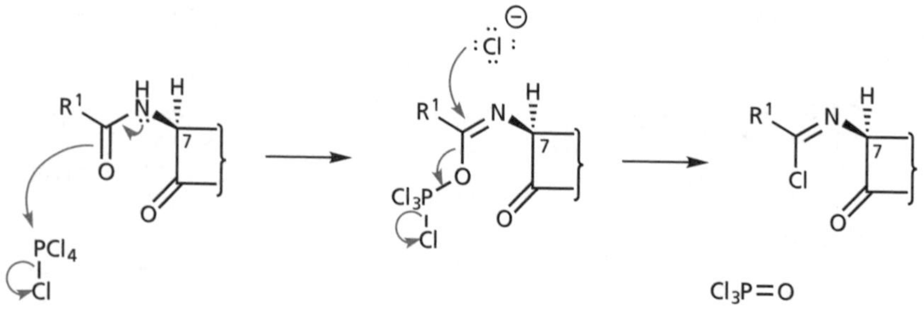

Figure 16.38 Mechanism for imino chloride formation.

ring constraints prevent the β-lactam nitrogen forming a double bond within the β-lactam ring. A chlorine atom is now introduced to form an imino chloride which can then be treated with an alcohol to give an imino ether. This functional group is now more susceptible to hydrolysis than the β-lactam ring and so treatment with aqueous acid successfully gives the desired 7-ACA which can then be acylated to give a range of analogues.

16.5.2.3 First-generation cephalosporins

In general, the first-generation cephalosporins have lower activity than comparable penicillins, but a better range. Their best activity is against Gram-positive cocci, but they can be used to treat some community-derived Gram-negative infections (i.e. infections not caught in a hospital). They can also be used against *S. aureus* and streptococcal infections when penicillins have to be avoided. Most are poorly absorbed through the gut wall and have to be injected. As with penicillins, the appearance of resistant organisms has posed a problem, particularly with Gram-negative organisms. These contain β-lactamases which are more effective than the β-lactamases of Gram-positive organisms. Steric shields are successful in protecting cephalosporins from these β-lactamases, but also prevent them from inhibiting the transpeptidase target enzymes.

One of the most commonly used first-generation cephalosporins is **cephalothin** (Fig. 16.39), which is more active than penicillin G against some Gram-negative bacteria, and is less likely to cause allergic reactions. It can also be used against penicillinase-producing *S. aureus* strains.

A disadvantage with cephalothin is the fact that the acetyloxy group at position 3 is readily hydrolysed by esterase enzymes to give a less active alcohol (Fig. 16.40). The acetyloxy group is important to the mechanism of inhibition and acts as a good leaving group, whereas the alcohol is a much poorer leaving group. Therefore, it would be useful if this metabolism could be blocked to prolong activity. Replacing the ester with a metabolically stable pyridinium group gives **cephaloridine** (Fig. 16.41). The pyridine can still act as a good leaving group for the inhibition mechanism, but is not cleaved by esterases. Cephaloridine exists as a zwitterion and is soluble in water, but like most first-generation

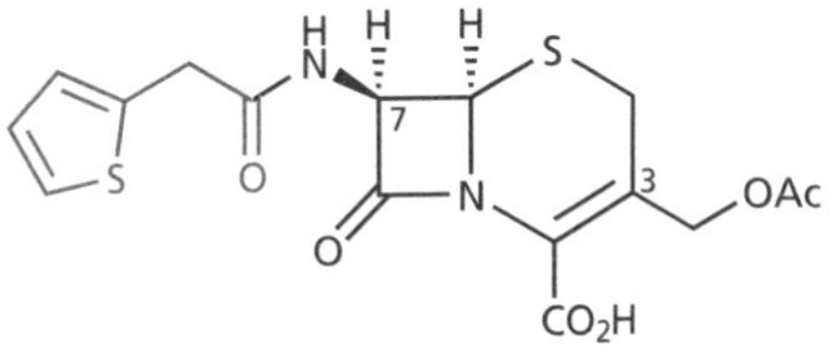

Figure 16.39 Cephalothin.

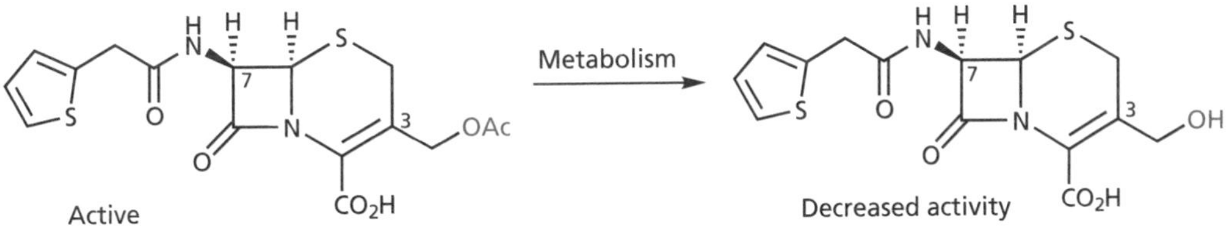

Figure 16.40 Metabolic hydrolysis of cephalothin.

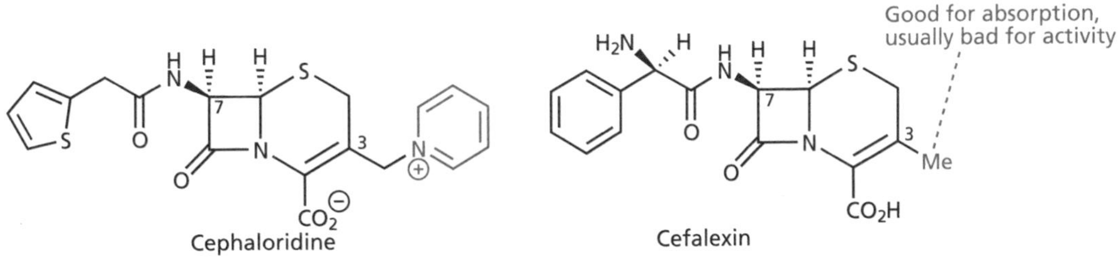

Figure 16.41 Cephaloridine and Cefalexin.

BOX 16.7 SYNTHESIS OF 3-METHYLATED CEPHALOSPORINS

The synthesis of 3-methylated cephalosporins involves the use of a penicillin starting material as shown below. The synthesis, which was first demonstrated by Eli Lilly Pharmaceuticals, involves oxidation of sulfur followed by an acid catalysed ring expansion, where the five-membered thiazolidine ring in penicillin is converted to the six-membered dihydrothiazine ring in cephalosporin.

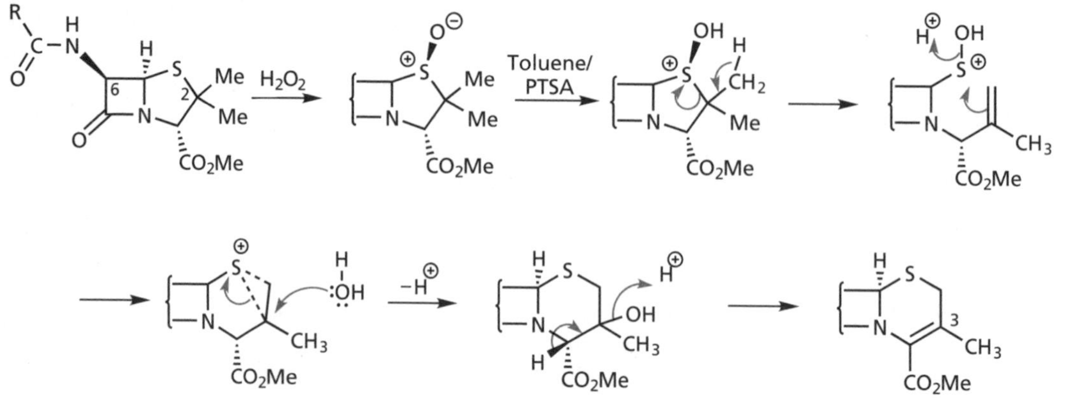

Synthesis of 3-methylated cephalosporins.

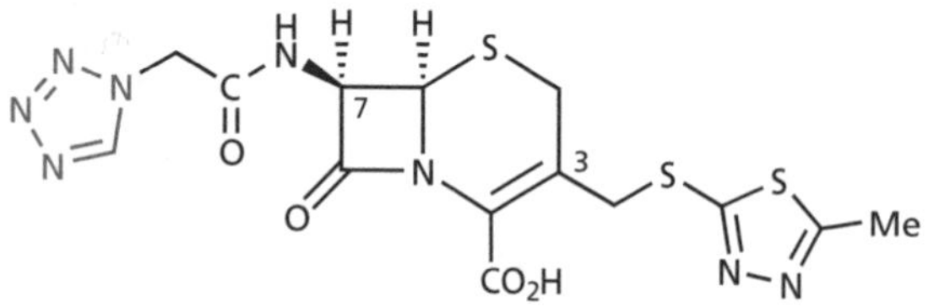

Figure 16.42 Cefozolin.

cephalosporins, it is poorly absorbed through the gut wall and has to be injected.

Cefalexin (Fig. 16.41) has a methyl substituent at position 3. This is bad for activity, since it is not a good leaving group, but it appears to help oral absorption, and cefalexin is one of the few cephalosporins which can be taken orally. The mechanism of absorption through the gut wall is poorly understood and it is not clear why the 3-methyl group is so advantageous. Although the methyl group at position 3 is bad for activity, the presence of a hydrophilic amino group at the α-carbon of the 7-acylamino side chain helps to compensate. The synthesis of 3-methylated cephalosporins is described in Box 16.7.

Cefazolin (Fig. 16.42) is another example of a first-generation cephalosporin. It is recommended for use as a prophylactic to prevent infection when surgical procedures are used to implant foreign bodies.

16.5.2.4 Second-generation cephalosporins

Cephamycins

A methoxy substituent at position 7 of the cephalosporin skeleton has proved advantageous. Compounds such as these are known as cephamycins. The parent compound **cephamycin C** (Fig. 16.43) was isolated from a culture of *Streptomyces clavuligerus* and was the first β-lactam to be isolated from a bacterial source. Modification of the side chain gave **cefoxitin** (Fig. 16.43) which showed a broader spectrum of activity than most first-generation cephalosporins. This is due to greater resistance to β-lactamase enzymes which may be due to the steric hindrance provided by the extra methoxy group. It is interesting to note that introduction of the methoxy group at the corresponding 6-α-position of penicillins results in loss of activity. Cefoxitin shows good metabolic stability to esterases due to the presence of a urethane group rather than an ester at position 3 (section 11.2.2).

Oximinocephalosporins

A major advance in cephalosporin research has been the development of the oximinocephalosporins. These contain an iminomethoxy group at the α-position of the acyl side chain, which significantly increases the stability of cephalosporins against the β-lactamases produced by

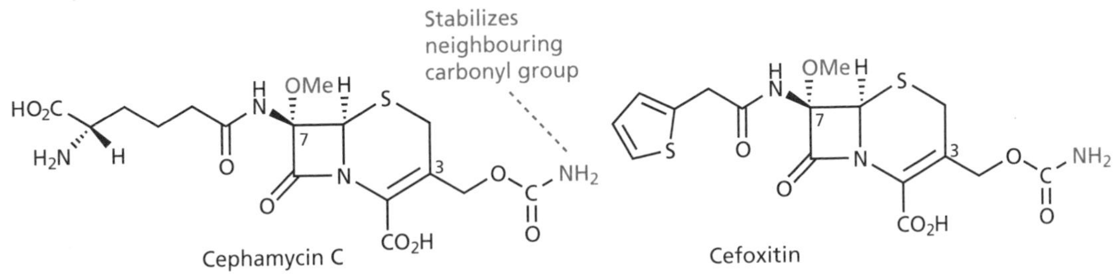

Figure 16.43 Cephamycin C and cefoxitin.

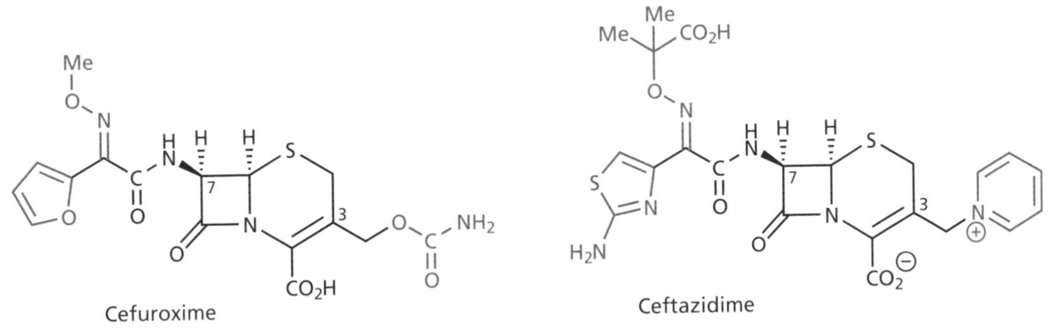

Figure 16.44 Cefuroxime and ceftazidime.

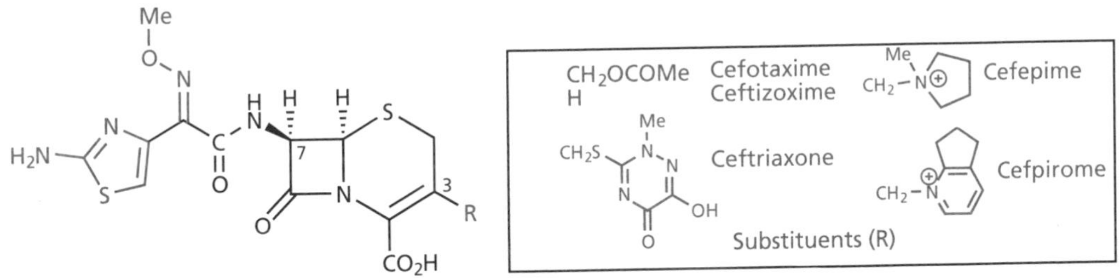

Figure 16.45 Third- and fourth-generation oximinocephalosporins.

some organisms (e.g. *Haemophilus influenza*). The first useful agent in this class of compounds was **cefuroxime** (Fig. 16.44) which, like cefoxitin, has an increased resistance to β-lactamases and mammalian esterases. Unlike cefoxitin it retains activity against streptococci and to a lesser extent staphylococci. The drug has a wide spectrum of activity, and is useful against organisms which have become resistant to penicillin. However, it is not active against 'difficult' bacteria such as *P. aeruginosa*. It is used clinically against respiratory infections caused by *H. influenza*, *Moraxella catarrhalis*, and susceptible strains of *S. pneumoniae*.

In general, the second-generation cephalosporins have variable activity against Gram-positive cocci, but increased activity against Gram-negative bacteria.

16.5.2.5 Third-generation cephalosporins

Replacing the furan ring of the aforesaid oximino-cephalosporins with an aminothiazole ring enhances the penetration of cephalosporins through the outer membrane of Gram-negative bacteria and may also increase affinity for the transpeptidase enzyme. As a result, third-generation cephalosporins containing this ring have a marked increase in activity against these bacteria. A variety of such structures have been prepared (Figs. 16.44–45) with different substituents at the 3 position to vary the pharmacokinetic properties of the analogues. They all have good activity against Gram-negative bacteria, but vary in their activity against Gram-positive cocci. The ability to attack *P. aeruginosa* also varies from structure to structure, and they lack activity against the MRSA organisms and *Enterobacter* species. Nevertheless, they play a major role in antimicrobial therapy due to their activity against Gram-negative bacteria, many of which are resistant to other β-lactams. Because such infections are uncommon outside hospitals, physicians are discouraged from prescribing these drugs routinely, and they are viewed as 'reserve troops' to be used for troublesome infections which do not respond to the more commonly prescribed β-lactams.

Ceftazidime (Fig. 16.44) is an injectable cephalosporin which is unusual among third-generation cephalosporins in that it has excellent activity against *P. aeruginosa* as well as other Gram-negative bacteria. As the drug can cross the blood–brain barrier, it can be used to treat meningitis. Compared to the other aminothiazole structures, ceftazidime has good activity against streptococci, but loses activity against strains of methicillin-susceptible *S. aureus*. This is due to a decreased binding affinity for the transpeptidase enzyme present in *S. aureus*.

16.5.2.6 Fourth-generation cephalosporins

Cefepime and **cefpirome** (Fig. 16.45) are oximino-cephalosporins which have been classed as fourth-generation cephalosporins. They are zwitterionic compounds having a positively charged substituent at the 3-position and a negatively charged carboxylate group at the 4-position. This property appears to radically enhance the ability of these compounds to penetrate the outer membrane of Gram-negative bacteria. They are also found to have a good affinity for the transpeptidase enzyme and a low affinity for a variety of β-lactamases. Fourth-generation cephalosporins have activity against Gram-positive cocci and a broad array of Gram-negative bacteria including *P. aeruginosa* and many of the enterobacterial species.

16.5.2.7 Resistance to cephalosporins

The activity of a specific cephalosporin against a particular bacterial cell is dependent on the same factors as those for penicillins; that is, ability to reach the transpeptidase enzyme, stability to any β-lactamases which might be present, and the affinity of the antibiotic for the target. For example, most cephalosporins (with the exception of cephaloridine) are stable to the β-lactamase produced by *S. aureus*, and as this is a Gram-positive organism, the β-lactams can reach the transpeptidase enzyme without difficulty. Therefore, the relative ability of cephalosporins to inhibit *S. aureus* comes down to their affinity for the target transpeptidase enzyme. Agents such as the cephamycins and ceftazidime have poor affinity whereas other cephalosporins have a higher affinity. The MRSA organism contains a modified transpeptidase enzyme (PBP2a) for which both penicillins and cephalosporins have a poor affinity.

KEY POINTS

- Cephalosporins contain a strained β-lactam ring fused to a dihydrothiazine ring.

- In general, first-generation cephalosporins offer advantages over penicillins in that they have greater stability to acid and β-lactamases, and have a good ratio of activity against Gram-positive and Gram-negative bacteria. However, they have poor oral availability and are generally lower in activity.

- Variation of the 7-acylamino side chain alters antimicrobial activity whereas variation of the side chain at position 3 predominantly alters the metabolic and pharmacokinetic properties of the compound. Introduction of a methoxy substitution at C-7 is possible.

- Semi-synthetic cephalosporins can be prepared from 7-aminocephalosporanic acid (7-ACA).

- 7-ACA is obtained from the chemical hydrolysis of cephalosporins. This requires prior activation of the side chain to make it more reactive than the β-lactam ring.

- Deacetylation of cephalosporins occurs metabolically to produce inactive metabolites. Metabolism can be blocked by replacing the susceptible acetoxy group with metabolically stable groups.

- Methyl substitution at the 3-position of cephalosporins is good for oral absorption but bad for activity unless a hydrophilic group is present at the α-position of the acyl side chain.

- 3-Methylated cephalosporins can be synthesized from penicillins.

- Cephamycins are cephalosporins containing a methoxy group at the 7-position.

- Oximinocephalosporins have resulted in second, third, and fourth generations of cephalosporins with increased potency and broader spectra of activity, particularly against Gram-negative bacteria.

16.5.3 Other β-lactam antibiotics

Although penicillins and cephalosporins are the best known and most researched β-lactams, there are other β-lactam structures which are of great interest in the antibacterial field.

Carbapenems

Thienamycin (Fig. 16.46) was the first example of this class of compounds, and was isolated from *Streptomyces cattleya* in 1976. It is potent, with an extraordinarily broad range of activity against Gram-positive and Gram-negative bacteria, including *P. aeruginosa*. It has low toxicity and shows a high resistance to β-lactamases. This resistance has been ascribed to the presence of the hydroxyethyl side chain. Unfortunately, it shows poor metabolic and chemical stability, and is not absorbed from the gastrointestinal tract. The unexpected features of the structure of thienamycin are the missing sulfur atom and acylamino side chain, both of which were thought to be essential to antibacterial activity. Furthermore, the stereochemistry of

the side chain at substituent 6 is opposite from the usual stereochemistry in penicillins, another factor in the resistance of this agent to β-lactamases.

Imipenem and **meropenem** are analogues of thienamycin which are used clinically. Imipenem is useful in treating some infections which do not respond to cephalosporins, or in treating infections which have become resistant to more conventional β-lactams. The drug is metabolized by a dehydropeptidase enzyme to produce metabolites that are toxic to the kidney, but this can be alleviated by administrating the drug alongside **cilastatin**—a dehydropeptidase inhibitor which protects imipenem from metabolism. This preparation is known as **Primaxin**. Meropenem is slightly less active than imipenem against Gram-positive bacteria but is more active against Gram-negative bacteria. Unlike imipenem, it is active against *P. aeruginosa*. It can be administered on its own, as the different substituent at the 2-position makes it more resistant to dehydropeptidases. Both meropenem and imipenem penetrate the outer membrane of Gram-negative bacteria through porins, but meropenem enters more efficiently, so it has a higher activity against these bacteria. The drug has been used to treat pneumonia, meningitis, abdominal infections, and urinary tract infections. **Ertapenem** was approved in 2002 and is similar in structure to meropenem. It has an extra methyl substituent on the carbapenem ring which provides further stability against dehydropeptidases,

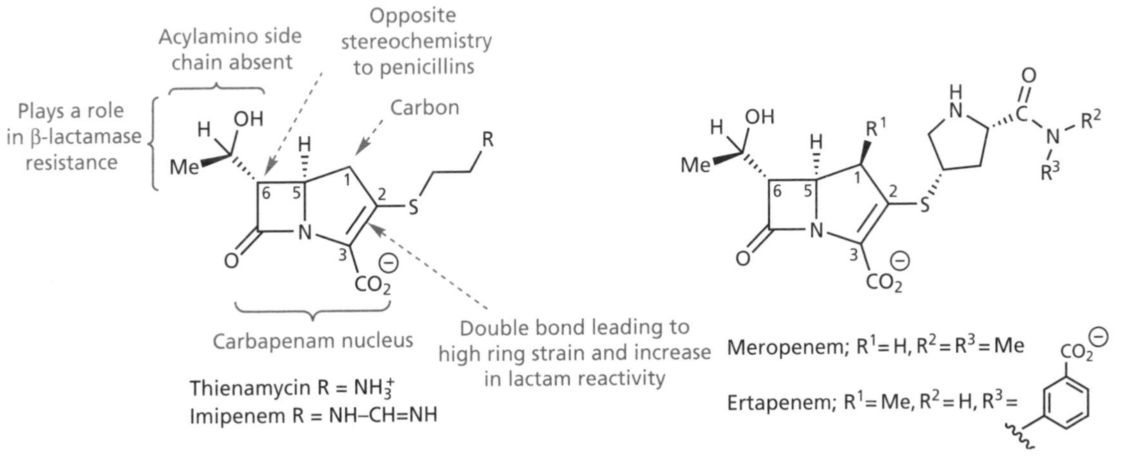

Figure 16.46 Carbapenems.

while the ionized benzoic acid contributes to high protein binding and prolongs the half-life of the drug such that once daily dosing is allowed.

In general, the carbapenems have the broadest spectrum of activity of all the β-lactam antibiotics.

Monobactams

Monocyclic β-lactams such as the nocardicins (Fig. 16.47) have been isolated from natural sources. At least seven nocardicins have been isolated by the Japanese company Fujisawa. They show moderate activity *in vitro* against a narrow group of Gram-negative bacteria including *P. aeruginosa*. However, it is surprising that they should show any activity at all, since they contain a single β-lactam ring unfused to any other ring system. The presence of a fused second ring has always been thought to be essential in order to strain the β-lactam ring sufficiently for antibacterial activity.

One explanation for the surprising activity of the nocardicins is that they operate via a different mechanism from penicillins and cephalosporins. There is some evidence supporting this in that the nocardicins are inactive against Gram-positive

bacteria and generally show a different spectrum of activity from the other β-lactam antibiotics. It is possible that these compounds act on cell wall synthesis by inhibiting a different enzyme. They also show low levels of toxicity.

Aztreonam is an example of a monobactam which has reached the clinic as an intravenous antibacterial agent. It can be safely used for patients with allergies to penicillin or cephalosporins. Developed from a naturally occurring monobactam isolated from *Chromobacterium violaceum*, it has no activity against Gram-positive organisms or anaerobic bacteria, as it does not bind to the transpeptidases produced by these organisms. However, it can bind to and inhibit the transpeptidases produced by Gram-negative aerobic organisms.

16.5.4 β-Lactamase inhibitors

16.5.4.1 Clavulanic acid

Clavulanic acid (Fig. 16.48) was isolated from *Streptomyces clavuligerus* by Beechams in 1976. It has weak and unimportant antibiotic activity, but it is a powerful and irreversible inhibitor of most

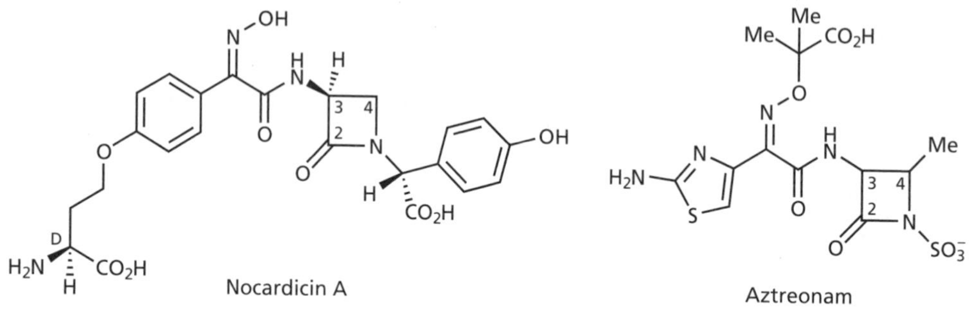

Figure 16.47 Monobactams.

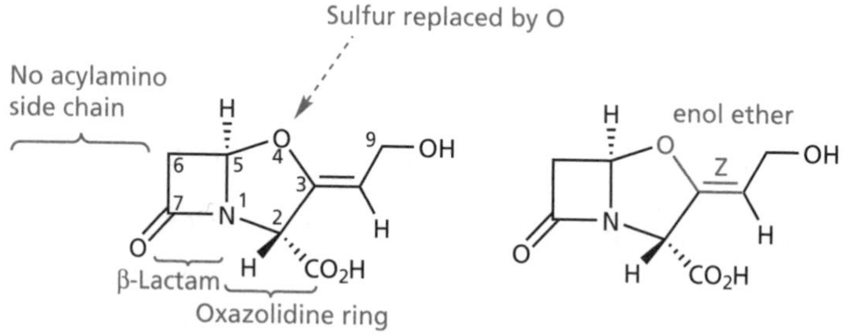

Figure 16.48 Clavulanic acid.

β-lactamases and as such is now used in combination with traditional penicillins such as amoxicillin (**Augmentin**). This allows the dose levels of amoxicillin to be decreased and also increases the spectrum of activity. It should be noted that there are various types of β-lactamases; clavulanic acid is effective against most but not all.

The structure of clavulanic acid was the first example of a naturally occurring β-lactam ring that was not fused to a sulfur-containing ring: it is fused instead to an oxazolidine ring structure. It is also unusual in that it does not have an acylamino side chain.

Many analogues have now been made and the essential requirements for β-lactamase inhibition are:

- Strained β-lactam ring.
- Enol ether.
- The double bond of the enol ether has the *Z* configuration. (Activity is reduced but not eliminated if the double bond is *E*.)
- No substitution at C-6.
- (*R*)-stereochemistry at positions 2 and 5.
- Carboxylic acid group.

The variability allowed is therefore strictly limited to the 9-hydroxyl group. Small hydrophilic groups appear to be ideal, suggesting that the original hydroxyl group is involved in a hydrogen bonding interaction with the active site of the β-lactamase enzyme.

Clavulanic acid is a mechanism-based irreversible inhibitor and could be classed as a suicide substrate (section 4.8.5). The drug fits the active site of β-lactamase, and the β-lactam ring is opened by a serine residue in the same manner as penicillin. However, the acyl-enzyme intermediate then reacts further with another enzymatic nucleophilic group (possibly NH_2) to bind the drug irreversibly to the enzyme (Fig. 16.49). The mechanism requires the loss or gain of protons at various stages, and an amino acid such as histidine in the active site would be capable of acting as a proton donor/acceptor (compare sections 4.5.3 and 19.16.3).

Clavulanic acid is also administered intravenously with ticarcillin as **Timentin**.

16.5.4.2 Penicillanic acid sulfone derivatives

The agents **sulbactam** and **tazobactam** have also been developed as β-lactamase inhibitors and are used

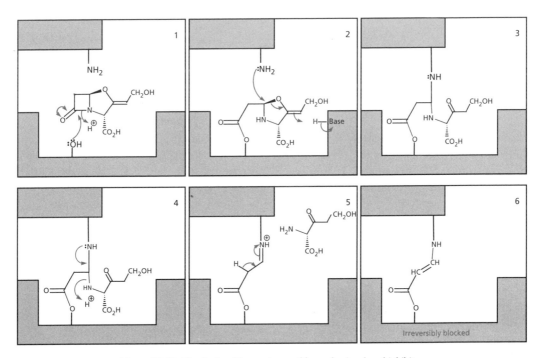

Figure 16.49 Clavulanic acid as an irreversible mechanism-based inhibitor.

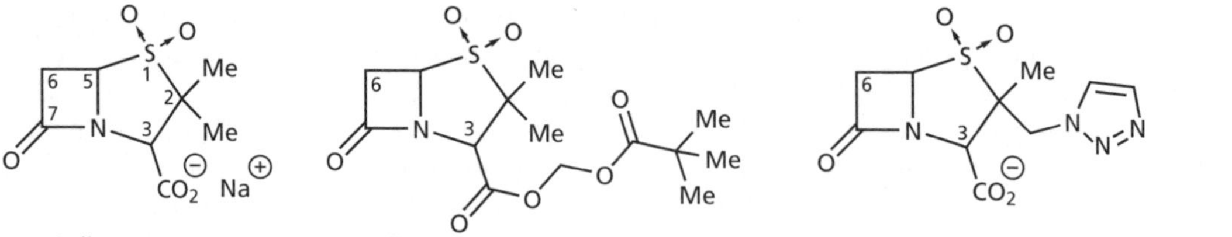

Figure 16.50 Penicillanic acid sulfones.

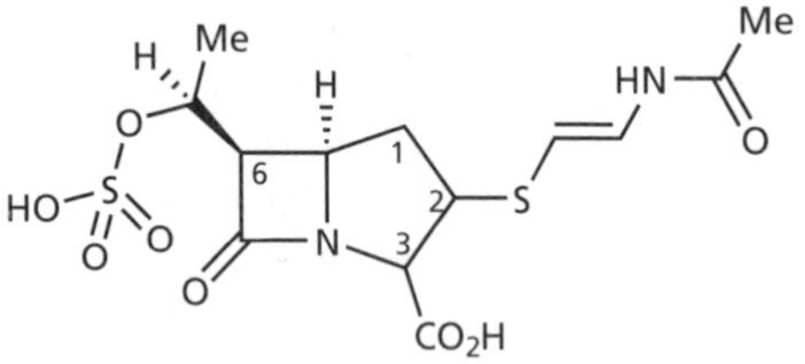

Figure 16.51 MM 13902.

clinically (Fig. 16.50). They too act as suicide substrates for β-lactamase enzymes and have similar properties. Sulbactam has a broader spectrum of activity against β-lactamases than clavulanic acid, but is less potent. It is combined with ampicillin for intravenous administration in a preparation called **Unasyn.**

Tazobactam is similar to sulbactam and has a similar spectrum of activity against β-lactamases. However, its potency is more like clavulanic acid. It is administered intravenously with piperacillin in a preparation called **Tazocin or Zosyn.** This combination has the broadest spectrum of activity of the three combinations described.

16.5.4.3 Olivanic acids

The olivanic acids (e.g. **MM 13902**) (Fig. 16.51) were isolated from strains of *Streptomyces olivaceus* and are carbapenem structures like thienamycin. They are very strong inhibitors of β-lactamase, in some cases 1000 times more potent than clavulanic acid. They are also effective against the β-lactamases which can break down cephalosporins and which are unaffected by clavulanic acid. Unfortunately, olivanic acids lack chemical stability.

16.5.5 Other drugs which act on bacterial cell wall biosynthesis

β-Lactams are not the only antibacterial agents which inhibit cell wall biosynthesis. The antibacterial agents

vancomycin, D-**cycloserine**, and **bacitracin** also inhibit biosynthesis, though at different stages. In order to synthesise the cell wall, NAM is linked to three amino acids, then to the dipeptide D-Ala-D-Ala (Fig. 16.52). This dipeptide is derived from two L-alanine units which are first racemized, then linked together.

NAM with its pentapeptide chain is then linked to a C_{55} carrier lipid with the aid of a translocase enzyme and carried to the outer surface of the cell membrane, where the lipid carrier acts as an anchor holding the glycopeptide in place for the subsequent steps. These steps involve the addition of NAG and a pentaglycine chain to give the complete 'building block'. A transglycosidase enzyme catalyses the attachment of the disaccharide building block to the growing cell wall and at the same time a carrier lipid is released to pick up another molecule of NAM/pentapeptide. Cross-linking between the various chains of the cell wall finally takes place catalysed by the transpeptidase enzyme as described previously (section 16.5.1.4).

16.5.5.1 D-cycloserine and bacitracin

D-Cycloserine (Fig. 16.53) is a simple molecule produced by *Streptomyces garyphalus* which has broad-spectrum activity and acts within the cytoplasm to prevent the formation of D-Ala-D-Ala. It does this by mimicking the structure of D-alanine and inhibiting the enzymes L-alanine racemase (responsible for racemizing L-alanine to D-alanine) and D-Ala-D-Ala ligase (responsible for linking the two D-alanine units together).

Bacitracin is a polypeptide complex produced by *Bacillus subtilis* which binds to the lipid carrier responsible for transporting the NAM/pentapeptide unit across the cell membrane and prevents it from carrying out that role. The drug is used topically for skin infections.

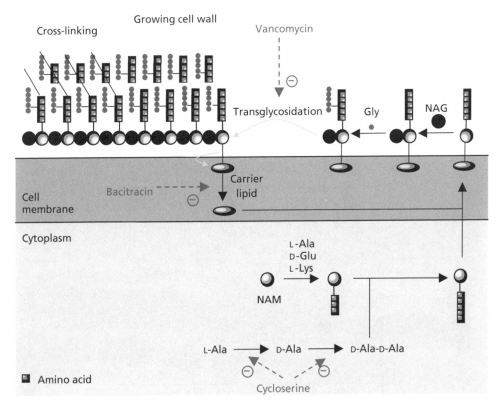

Figure 16.52 Cell wall biosynthesis.

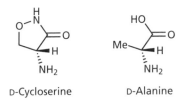

Figure 16.53 D-Cycloserine as a mimic for D-alanine.

16.5.5.2 The glycopeptides—vancomycin and vancomycin analogues

Vancomycin (Fig. 16.54) is a narrow-spectrum bactericidal glycopeptide produced by a microorganism called *Streptomyces orientalis* which was found in Borneo and India. Its name is aptly derived from the verb 'to vanquish'. Vancomycin was introduced in 1956 for the treatment of infections due to penicillin-resistant *S. aureus*, but was discontinued when methicillin became available. Since then, it has been reintroduced and it is now the main stand-by drug for MRSA. Since it is not absorbed orally, it is administered by injection but it is given orally to treat gut infections due to a microorganism called *Clostridium difficile*. This organism may appear following the use of broad-spectrum antibiotics, and produces harmful toxins. Vancomycin and related glycopeptides are often the last resort in treating patients with drug-resistant infections. As such, they have become extremely important and a great deal of research is currently being carried out in this area.

Vancomycin is derived biosynthetically from a linear heptapeptide containing five aromatic residues. These undergo oxidative coupling with each other to produce three cyclic moieties within the structure. Chlorination, hydroxylation, and the final addition of two sugar units then complete the structure (Fig. 16.55).

The cyclizations described transform a highly flexible heptapeptide molecule into a rigid structure which holds the peptide backbone in a fixed conformation. Moreover, there is an extra element of rigidity to the structure which may not be apparent at first sight. The aromatic rings (A–E) cannot rotate and are fixed in

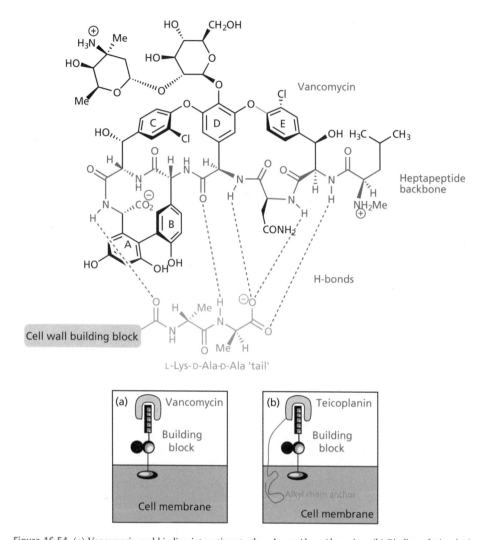

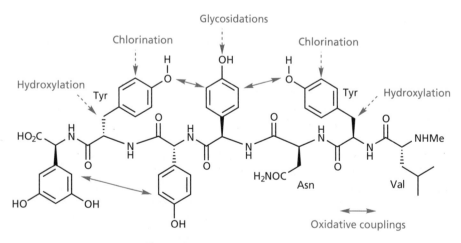

Figure 16.54 (a) Vancomycin and binding interactions to the L-Lys-D-Ala-D-Ala moiety. (b) Binding of teicoplanin.

Figure 16.55 Biosynthesis of vancomycin.

space because of hindered single bond rotation. For example, the aromatic rings C and E have a chloro substituent which prevents these rings becoming co-planar with ring D. Similarly, rings A and B have phenol substituents which prevent them becoming coplanar.

The fixed conformation of the hexapeptide chain is important to vancomycin's unique mechanism of action, which involves targeting the cell wall's building blocks rather than a protein or a nucleic acid. To be specific, there is a pocket in the vancomycin structure into which the tail of the building block's pentapeptide moiety can fit. It is then held there by the formation of five hydrogen bonds between it and the hexapeptide chain of vancomycin (Fig. 16.54). Dimerization can now occur where a highly stable vancomycin dimer is bound to two tails. Because vancomycin is a large molecule, it covers the tails and acts as a steric shield, blocking access to the transglycosidase and transpeptidase enzymes (Fig. 16.56).

Dimerization occurs head to tail such that the heptapeptide chains of each vancomycin molecule interact through four hydrogen bonds (Fig. 16.57). The sugar and chloro groups also play an important role in this dimerization and activity drops if either of these groups is absent.

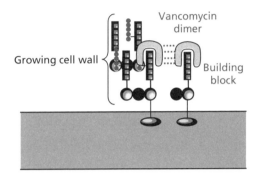

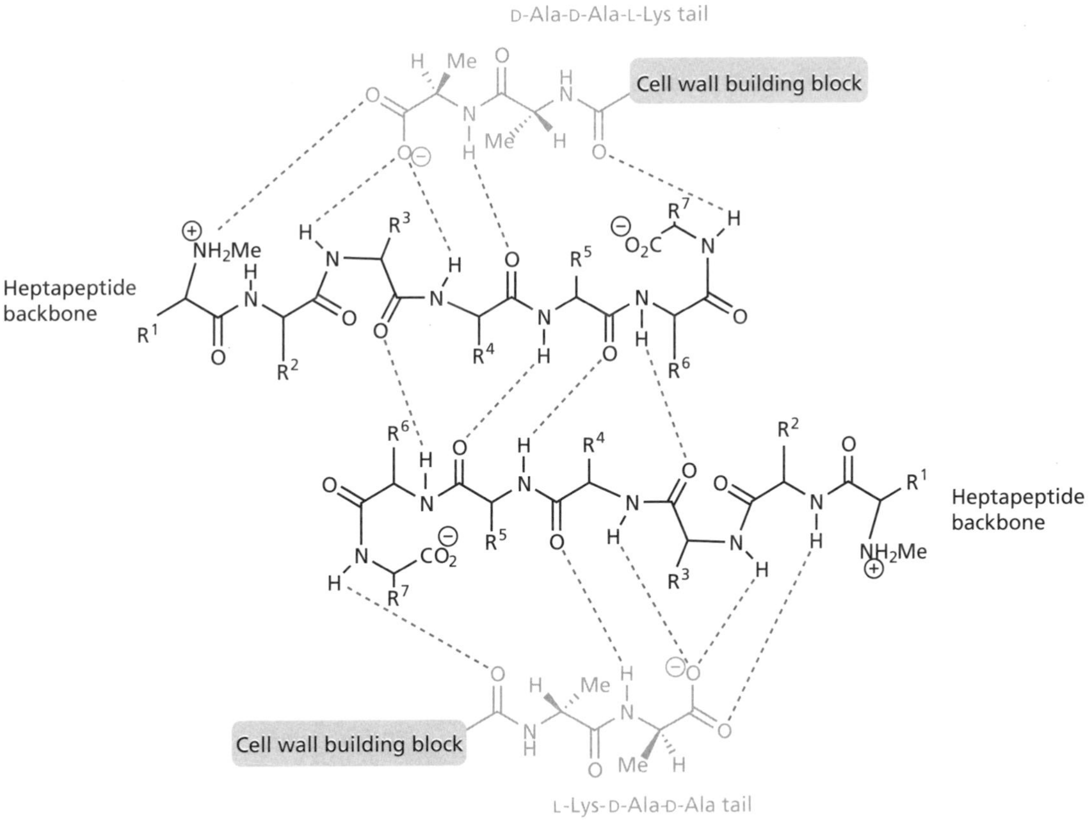

Figure 16.56 'Capping' of pentapeptide 'tails' by vancomycin.

D-Ala-D-Ala-L-Lys tail

Cell wall building block

Heptapeptide backbone

Heptapeptide backbone

Cell wall building block

L-Lys-D-Ala-D-Ala tail

Figure 16.57 Dimerization of vancomycin. The dashed lines represent hydrogen bonds.

Because vancomycin is such a large molecule, it is unable to cross the outer cell membrane of Gram-negative bacteria and consequently lacks activity against those organisms. It is also unable to cross the inner cell membrane of Gram-positive bacteria, but this is not required since the construction of the cell wall takes place outside the cell membrane.

Bacterial resistance to vancomycin has been slow to develop, although some hospital strains of *S. aureus* were identified in 1996 which do show resistance (**VRSA**). Of particular concern is the appearance of **vancomycin-resistant enterococci** (VRE) in 1989. These are organisms that can cause life-threatening gut infections in patients whose immune system is weakened. Resistance in the latter organisms has arisen from a modification of the cell wall precursors where the terminal D-alanine group in the pentapeptide

chain has been replaced by D-lactic acid, resulting in a terminal ester link rather than an amide link (Fig. 16.58). This removes one of the NH groups involved in the hydrogen bonding interaction with vancomycin. It may not sound like much, but it is sufficient to weaken the binding affinity and make the antibiotic ineffective. The building block is still acceptable to the glycosylase and transpeptidase enzyme. In the latter case, lactate acts as the leaving group rather than D-alanine.

Teicoplanin (Fig. 16.59) is another member of the vancomycin family which has been isolated from a soil microorganism called *Actinoplanes teichomyceticus*. Unlike vancomycin, it fails to dimerize, but the long alkyl chain anchors the antibiotic to the outer surface of the cell membrane where it is perfectly placed to interact with the building blocks of the cell wall

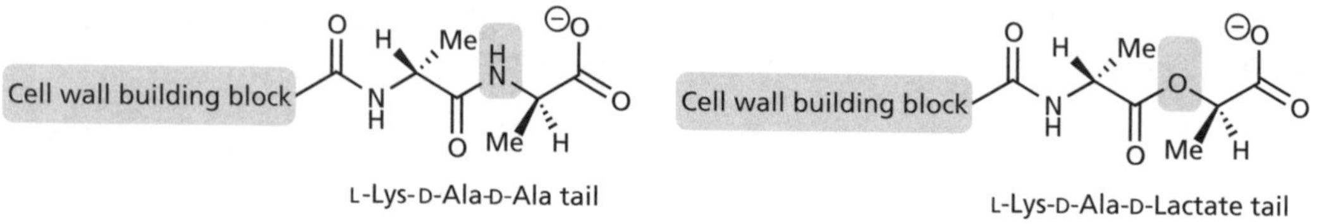

L-Lys-D-Ala-D-Ala tail

L-Lys-D-Ala-D-Lactate tail

Figure 16.58 Modification of pentapeptide chain leading to resistance.

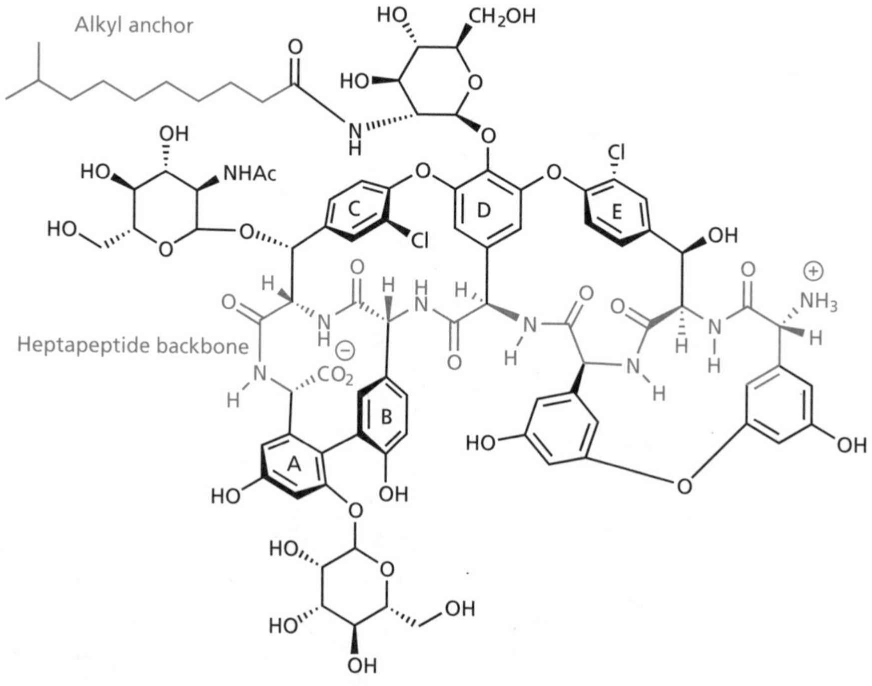

Figure 16.59 Teicoplanin.

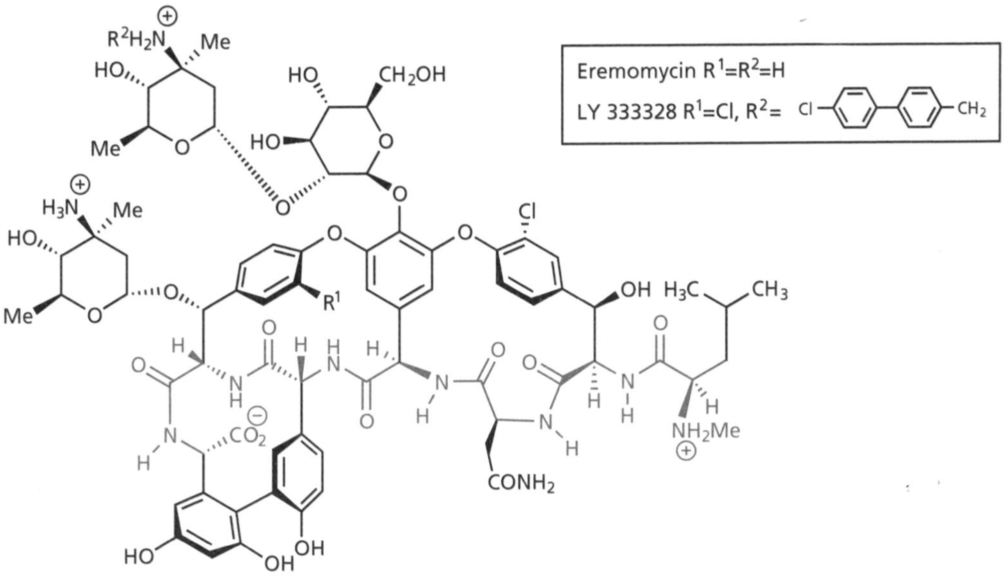

Figure 16.60 Eremomycin and LY 333328.

synthesis (Fig. 16.54). It is used clinically for the treatment of Gram-positive infections and is less toxic than vancomycin.

Another naturally occurring member of the vancomycin family is **eremomycin** (Fig. 16.60). A biphenyl hydrophobic 'tail' was added to act as an anchor, resulting in a compound (**LY 333328**) which is 1000 times more active than vancomycin.

Although the complexity of the glycopeptides is an advantage in their targeting and selectivity, it is a problem when it comes to synthesizing analogues. Therefore, work is being carried out in an attempt to prepare simplified analogues of vancomycin which are easier to synthesize, yet retain the desired selectivity. Structures such as those shown in Fig. 16.61 have been prepared which are capable of binding to D-Ala-D-Ala and D-Ala-D-Lac. These now represent lead compounds for the development of effective antibacterial agents.

There are another two mechanisms by which glycopeptide antibiotics may have an antibacterial activity. First, it is possible that glycopeptide dimers disrupt the cell membrane structure. This is supported by the fact that glycopeptide antibiotics enhance the activity of aminoglycosides by increasing their absorption through the cell membrane. Second, RNA synthesis is known to be disrupted in the presence of

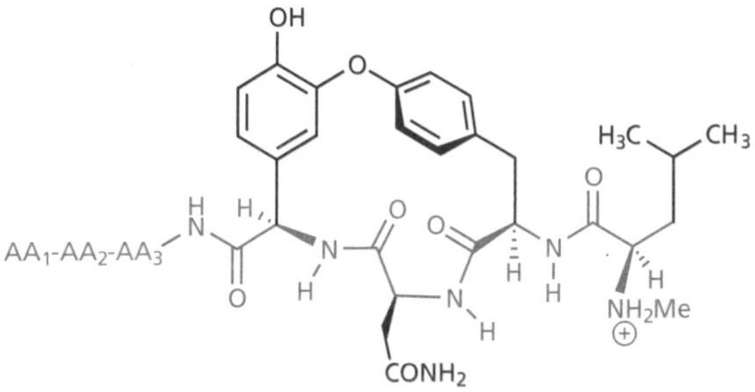

Figure 16.61 Simplified analogues of the glycopeptides.

glycopeptides. The possibility of three different mechanisms of action explains why bacteria are slow to acquire resistance to the glycopeptides.

KEY POINTS

- β-Lactamase inhibitors are β-lactam structures that have negligible antibacterial activity but inhibit β-lactamases. They can be administered alongside penicillins to protect them from β-lactamases and to broaden their spectrum of activity.

- Carbapenems and monobactams are examples of other β-lactam structures with clinically useful antibacterial activity.

- Glycopeptides such as vancomycin bind to the building blocks for cell wall synthesis, preventing their incorporation into the cell wall. They also block the cross-linking reaction for those units already incorporated in the wall. The glycopeptides are the drugs of last resort against drug resistant strains of bacteria.

- Bacitracin binds to and inhibits the carrier lipid responsible for carrying the cell wall components across the cell membrane.

- Cycloserine inhibits the synthesis of D-Ala-D-Ala.

16.6 Antibacterial agents which act on the plasma membrane structure

16.6.1 Valinomycin and gramicidin A

The peptides **valinomycin** (Fig. 16.62) and **gramicidin A** (Fig. 16.63) both act as ion-conducting antibiotics (ionophores) and allow the uncontrolled movement of ions across the cell membrane. Unfortunately, these agents show no selective toxicity for bacterial cells over mammalian cells and are therefore useless as

therapeutic agents. Their mechanism of action is interesting nevertheless.

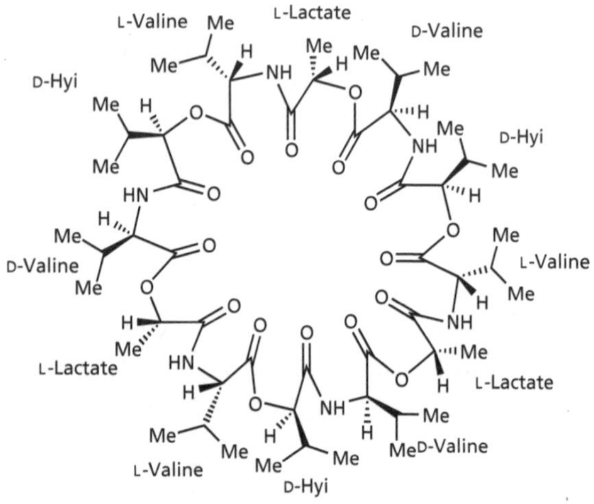

Valinomycin is a cyclic structure obtained from *Streptomyces* fermentation. It contains three molecules of L-valine, three molecules of D-valine, three molecules of L-lactic acid, and three molecules of D-hydroxyisovalerate. These four components are linked in an ordered fashion such that there is an alternating sequence of ester and amide linking bonds around the cyclic structure. This is achieved by the presence of a lactic or hydroxyisovaleric acid unit between each of the six valine units. Further ordering can be observed by noting that the L and D portions of valine alternate around the cycle, as do the lactate and hydroxyisovalerate units.

Valinomycin acts as an ion carrier and could be looked upon as an inverted detergent. Since it is cyclic, it forms a doughnut-type structure where the polar carbonyl oxygens of the ester and amide groups face inwards while the hydrophobic side chains of the valine and hydroxyisovalerate units point outwards. This is clearly favoured because the hydrophobic side chains can interact via van der Waals interactions with the fatty lipid interior of the cell membrane, while the polar hydrophilic groups are clustered together in the centre of the doughnut to produce a hydrophilic environment. This hydrophilic centre is large enough to accommodate an ion and it is found that a 'naked' potassium ion (i.e. one with no surrounding water molecules) fits the space and is complexed by the amide carboxyl groups (Fig. 16.64).

Valinomycin can therefore collect a potassium ion from the inner surface of the membrane, carry it across the membrane and deposit it outside the cell, thus disrupting the ionic equilibrium of the cell (Fig. 16.65). Normally, cells contain a high concentration of potassium ions and a low concentration of sodium ions. The fatty cell membrane prevents passage of ions between the cell and its environment, and ions can only pass through the cell membrane aided by specialized and controlled ion transport systems. Valinomycin introduces an uncontrolled ion transport system, which proves fatal.

D-Hyi = D-Hydroxyisovaleric acid

Figure 16.62 Valinomycin.

Val-Gly-Ala-Leu-Ala-Val-Val-Val-Trp-Leu-Trp-Leu-Trp-Leu-Trp-NH-CH$_2$-CH$_2$-OH

Figure 16.63 Gramicidin A.

Valinomycin is specific for potassium ions over sodium ions, and one might be tempted to think that sodium ions would be too small to be properly complexed. The real reason is that sodium ions do not lose their surrounding water molecules very easily and would have to be transported as the hydrated ion. As such, they are too big for the central cavity of valinomycin.

The ionophores **nigericin, monensin A**, and **lasalocid A** (Fig. 16.66) function in much the same way as valinomycin and are used in veterinary medicine to control the levels of bacteria in the rumen of cattle and the intestines of poultry.

Gramicidin A (Fig. 16.63) is a peptide containing 15 amino acids, which is thought to coil into a helix such that the outside of the helix is hydrophobic and interacts with the membrane lipids, while the inside of the helix contains hydrophilic groups, thus allowing the passage of ions. Therefore, gramicidin A could be viewed as an escape tunnel through the cell membrane. In fact, one molecule of gramicidin would not be long enough to traverse the membrane and it has been proposed that two gramicidin helices align themselves end-to-end in order to achieve the length required (Fig. 16.67).

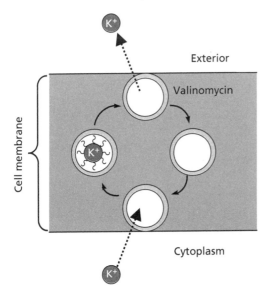

Figure 16.64 Potassium ion in the hydrophilic centre of valinomycin.

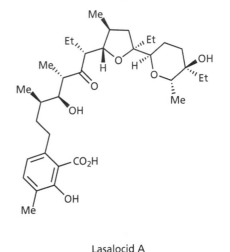

Figure 16.66 Ionophores used in veterinary medicine.

16.6.2 Polymyxin B

The polypeptide antibiotic **polymyxin B** (Fig. 16.68) derives from a soil bacterium called *Bacillus polymyxa*. It also operates within the cell membrane and shows a selective toxicity for bacterial cells over animal cells. This appears to be related to the ability of the compound to bind selectively to the different plasma membranes. The mechanism of this selectivity is not fully understood. Polymyxin B acts like valinomycin, but it causes the leakage of small molecules such as nucleosides from the cell. The drug is injected intramuscularly and is useful against *Pseudomonas* strains which are resistant to other antibacterial agents. It can be used topically for the treatment of minor skin infections and has good activity against Gram-negative bacteria. It is less effective against Gram-positive bacteria, as it is difficult for such a big molecule to pass through the thicker cell wall.

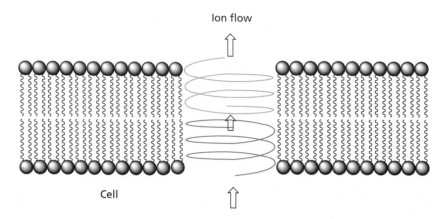

Figure 16.67 Gramicidin helices aligned end-to-end traversing membrane.

Figure 16.68 Polymyxin B (DAB = αγ-diaminobutyric acid with peptide link through the α-amino group).

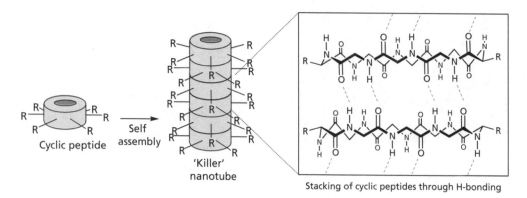

Figure 16.69 Self assembly of 'killer nanotubes'.

16.6.3 Killer nanotubes

Work is currently in progress to design cyclic peptides which will self-assemble in the cell membranes of bacteria to form tubules that have been labelled as 'killer nanotubes' (Fig. 16.69). Once formed, the nanotubes would allow molecules to leach out from the cell and result in cell death. The cyclic peptides concerned are designed to have 6–8 alternating D and L amino acids such that the amide groups are perpendicular to the plane of the cyclic structure, and the residues are pointing outwards in the same plane. This means that the residues do not interfere with the stacking process and the amide groups in each cyclic peptide form hydrogen bonds to the cyclic peptides above and below it, thus promoting the stacking process. Modifying the types of residues present has been successful in introducing selectivity *in vitro* for bacterial cells versus red blood cells. For example, the inclusion of a basic amino acid such as lysine is useful for selectivity. Lysine has a primary amino group which can become protonated and gain a positive charge. This encourages the structures to target bacterial membranes, since the latter tend to have a negative charge on their surface. *In vivo* studies have also been carried out successfully on mice.

16.6.4 Cyclic lipopeptides

Daptomycin (Fig. 16.70) was approved in 2003 for the treatment of Gram-positive infections and is a member of a new class of structures called the **cyclic lipopeptides**. It is a natural product derived from a bacterial strain *Streptomyces roseosporus*. It works by disrupting multiple functions of the bacterial cell membrane and is used in the treatment of skin infections including those caused by MRSA. The lipid portion of the molecule is derived from decanoic acid and the yield of product obtained is increased if decanoic acid is added to the fermentation medium.

KEY POINTS

- Ionophores act on the plasma membrane and result in the uncontrolled movement of ions across the cell membrane leading to cell death.
- Polymyxin B operates selectively on the plasma membrane of bacteria and causes the uncontrolled movement of small molecules across the membrane.
- Cyclic peptides are being designed which will self-assemble to form nanotubes in the cell membranes of bacteria.
- Cyclic lipopeptides are a new class of antibiotic.

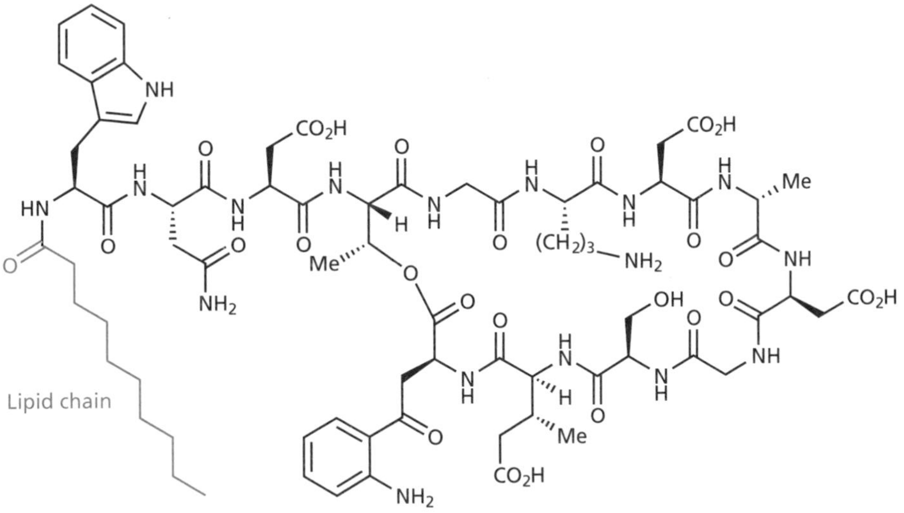

Figure 16.70 Daptomycin.

16.7 Antibacterial agents which impair protein synthesis—translation

The agents described in this section all inhibit protein synthesis by binding to ribosomal RNA and inhibiting different stages of the translation process (Fig. 16.71). Selective toxicity is due to either different diffusion rates through the cell barriers of bacterial versus mammalian cells or to a difference between the ribosomal target structures. The bacterial ribosome is a 70S particle (see section 7.2.1) made up of a 30S subunit and a 50S subunit. The 30S subunit binds mRNA and initiates protein synthesis. The 50S subunit combines with the 30S subunit–mRNA complex to form a ribosome, then binds aminoacyl tRNA and catalyses the building of the protein chain. There are two main binding sites for the tRNA molecules. The peptidyl site (P-site) binds the tRNA bearing the peptide chain. The acceptor aminoacyl site (A-site) binds the tRNA bearing the next amino acid, to which the peptide chain will be transferred (see also section 7.2.2). The ribosomes of eukaryotic cells are bigger (80S), consisting of a 60S large subunit and a 40S small subunit. They are sufficiently different in structure from prokaryotic ribosomes that it is possible for drugs to make a distinction between them.

16.7.1 Aminoglycosides

Streptomycin (Fig. 16.72) was isolated from the soil microorganism *Streptomyces griseus* in 1944, and is an

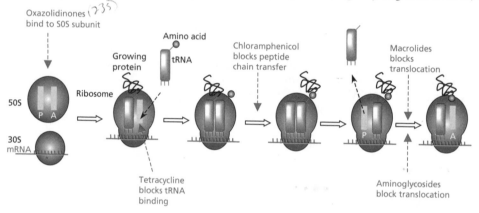

Figure 16.71 Stages at which antibiotics inhibit translation.

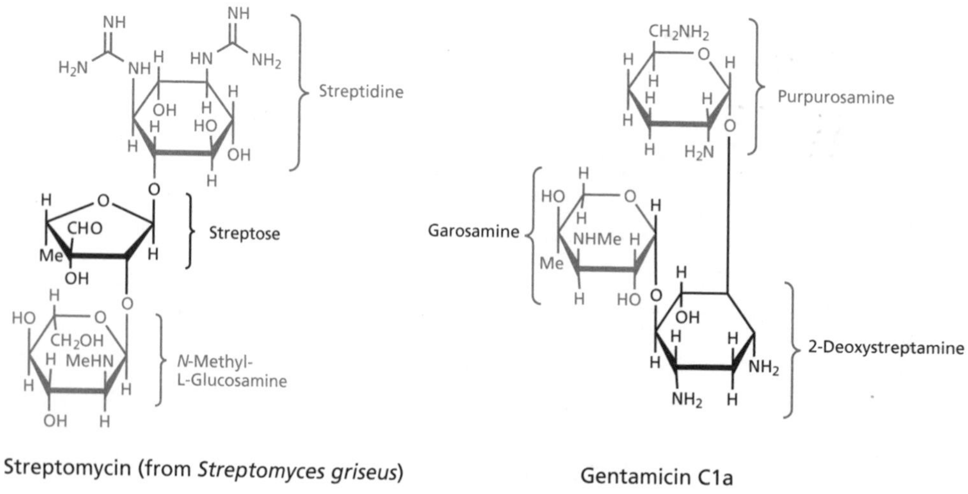

Streptomycin (from *Streptomyces griseus*)

Gentamicin C1a

Figure 16.72 Aminoglycosides.

example of an aminoglycoside—a carbohydrate structure which includes basic amine groups. Streptomycin was the next most important antibiotic to be discovered after penicillin, and proved to be the first effective agent used against tuberculosis. However, resistance soon developed and a multidrug therapy involving streptomycin, isoniazid, and *para*-aminosalicylic acid was used until the early 1970s. At that point, rifampicin became available allowing different multidrug therapies to be used—one of which still includes streptomycin.

A variety of other aminoglycosides such as **gentamicin C1a** (Fig. 16.72) have been isolated from various organisms. At pH 7.4 they are positively charged, which is beneficial to activity and also means that they work best in slightly alkaline conditions. The ionic charge is important to how these agents are absorbed through the outer membrane of Gram-negative bacteria. An ionic interaction takes place with various negatively charged groups in the membrane (lipopolysaccharides, phospholipids, and proteins) which displaces magnesium and calcium ions. These ions normally act as bridges between adjacent lipopolysaccharide molecules, and their displacement results in rearrangement of these molecules to produce pores through which the drug can pass. The drug still has to cross the cell membrane to enter the cell and this is an energy-dependent process. Once across the membrane, it is trapped inside the cell and so it is possible for high concentrations of the drug to build up within.

Binding to bacterial ribosomes now takes place to inhibit protein synthesis. The binding is specifically to the 30S ribosomal subunit and prevents the movement of the ribosome along mRNA so that the triplet code on mRNA can no longer be read. In some cases, protein synthesis is terminated and the shortened proteins end up in the cell membrane. This can lead to a further increase in cell permeability, resulting in an even greater uptake of the drug. Aminoglycosides are bactericidal rather than bacteriostatic and it is thought that their activity may be due to their effects, both on the ribosomes and the outer cell membrane.

Because the ribosomes in human cells are different in structure from those in bacterial cells they have a much lower binding affinity for the aminoglycosides, which explains their selectivity.

Aminoglycosides are fast acting, but they can also cause ear and kidney problems if the dose levels are not carefully controlled. They are effective in the treatment of infections caused by aerobic Gram-negative bacteria, including *P. aeruginosa*. Indeed, they used to be the only compounds effective against that organism.

Some Gram-negative bacteria are resistant to aminoglycosides, due mainly to enzymes which catalyse reactions such as *O*-phosphorylations, *O*-adenylations (addition of an adenine group), and *N*-acylations. Resistance can also occur from alterations of the ribosomes such that they bind aminoglycosides less strongly, or by less efficient uptake mechanisms.

Because the aminoglycosides are polar in nature, they have to be injected. They are also unable to cross the blood–brain barrier efficiently and so they cannot be used for the treatment of meningitis, unless they are injected directly into the central nervous system. The activity of aminoglycosides is increased if they are administered with agents which disrupt cell wall synthesis, since this increases uptake of the former. However, bacteriostatic agents should not be taken with aminoglycosides, as these inhibit the energy-dependent uptake process by which the aminoglycosides cross the cell membrane.

16.7.2 Tetracyclines

The tetracyclines are bacteriostatic antibiotics which have a broad spectrum of activity and are the most widely prescribed form of antibiotic after penicillins. They are also capable of attacking the malarial parasite. One of the best known tetracyclines is **chlortetracyclin** (**aureomycin**) (Fig. 16.73), which was isolated in 1948 from a mud-growing microorganism in Missouri called *Streptomyces aureofaciens*—so called because it of its golden colour. It is a broad-spectrum antibiotic, active against both Gram-positive and Gram-negative bacteria, with increased activity under acidic conditions. Unfortunately, it has side effects because it kills the intestinal flora that make vitamin K, which is needed as part of the blood clotting process. As a result, it is now restricted to topical use. Further tetracyclines such as **tetracycline** and **doxycycline** (Fig. 16.73) have been synthesized or discovered.

The tetracyclines inhibit protein synthesis by binding to the 30S subunit of ribosomes and preventing aminoacyl-tRNA from binding. This stops the further addition of amino acids to the growing protein chain. Protein release is also inhibited.

In general the tetracyclines can be divided into short-lasting compounds such as chlortetracycline, an intermediate group of compounds such as **demeclocycline**, and longer-acting compounds such as doxycycline. Tetracycline and doxycycline are the most useful agents in the family, the latter being quickly absorbed when taken orally.

In the case of Gram-negative bacteria, tetracyclines cross the outer membrane by passive diffusion through the porins. Passage across the inner membrane is dependent on a pH gradient, which suggests that a proton-driven carrier is involved. Selectivity is due to the ability of bacterial cells to concentrate these agents faster than human cells. This is fortunate, because tetracyclines are capable of inhibiting protein synthesis in mammalian cells—particularly in mitochondria.

Widespread resistance to tetracyclines has occurred, caused partly by the use of tetracyclines to cure animal infections, and as a food additive to promote the growth of newborn animals. Resistance can arise through several mechanisms. Some organisms have effective efflux mechanisms which pump the drug back out the cell. Resistance can also arise from alterations in the bacterial ribosomes such that they have lower affinity for the agents.

The tetracyclines were originally used for many types of respiratory infections but have been largely replaced by β-lactams because of the problems of resistance. They are still useful, however, in the treatment of sexually transmitted diseases, Lyme disease, acne, and a variety of different infections.

Tetracyclines should be avoided for young children and pregnant mothers as they can bind to developing teeth and bone, leading to tooth discolouration.

16.7.3 Chloramphenicol

Chloramphenicol (Fig. 16.73) was originally isolated from a microorganism called *Streptomyces venezuela* found in a field near Caracas, Venezuela. It is now prepared synthetically and has two asymmetric centres. Only the *R,R*-isomer is active.

Chloramphenicol binds to the 50S subunit of ribosomes and appears to act by inhibiting the movement of ribosomes along mRNA, probably by inhibiting the peptidyl transferase reaction by which the peptide chain is extended. Since it binds to the same region as macrolides and lincosamides, these drugs cannot be used in combination. The nitro group and both alcohol groups are involved in binding interactions. The dichloroacetamide group is also important, but can be replaced by other electronegative groups.

In some regions of the world, chloramphenicol is the drug of choice for the treatment of typhoid when more expensive drugs cannot be afforded. It can also be used in severe bacterial infections which are insensitive to other antibacterial agents, and is widely used against eye infections. However, the drug should only be used in these restricted scenarios as it is quite toxic, especially to bone marrow. The nitro group is suspected to be responsible for this, although intestinal bacteria are capable of reducing this group to an amino group. The drug is inadequately metabolized in babies, leading to a combination of symptoms described as the **grey baby syndrome**, which can be fatal. In adults, the drug undergoes a phase II conjugation reaction to form a glucuronic acid conjugate (section 8.4.5) which is excreted. This reaction fails to take place efficiently in newborn babies and so the drug levels increase to toxic levels.

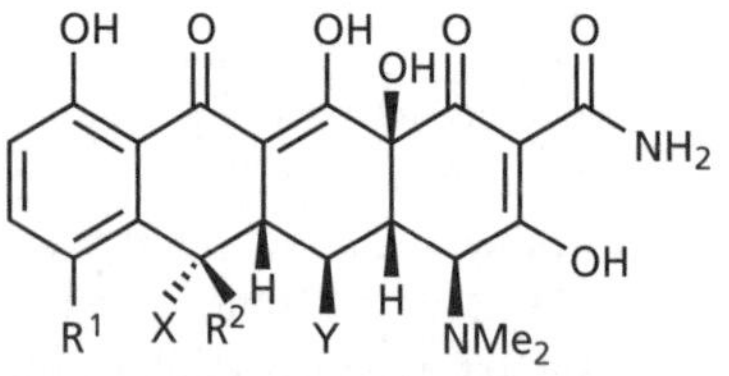

Chlortetracyclin (Aureomycin) (R^1 = Cl, R^2 = MeX = OH, Y = H)
Tetracycline (R^1 = H, R^2 = MeX = OH, Y = H)
Doxycycline (Vibramycin) (R^1 = H, R^2 = MeX = H, Y = OH)
Demeclocycline (R^1 = Cl, R^2 = H, X = OH, Y = H)

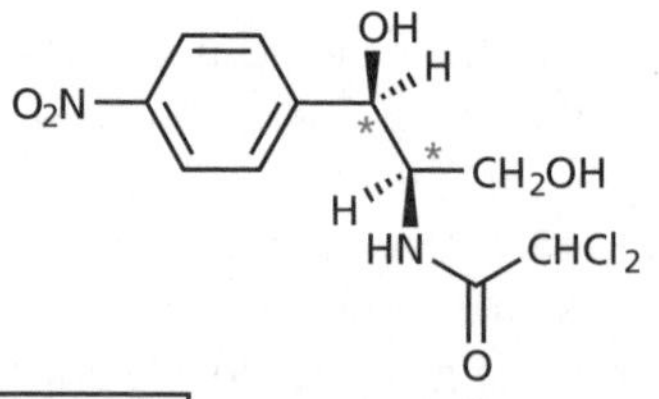

Chloramphenicol

Figure 16.73 Tetracyclines and chloramphenicol. The asterisks indicate asymmetric centres.

Bacteria with resistance to the drug contain an enzyme called chloramphenicol acetyltransferase, which catalyses the acylation of the hydroxyl groups.

16.7.4 Macrolides

Macrolides are bacteriostatic agents. The best-known example of this class of compounds is **erythromycin**—a metabolite isolated in 1952 from the soil microorganism *Streptomyces erythreus* in the Philippines, and one of the safest antibiotics in clinical use. The structure (Fig. 16.74) consists of a 14-membered macrocyclic lactone ring with a sugar and an aminosugar attached. The sugar residues are important for activity.

Erythromycin acts by binding to the 50S subunit of bacterial ribosomes. It works by inhibiting translocation, but other mechanisms of action also appear likely. Since erythromycin and chloramphenicol bind to the same region of the ribosome, they should not be administered together as they will compete with each other and be less effective.

Erythromycin was used against penicillin-resistant staphylococci, but newer penicillins are now used for these infections. It is, however, the drug of choice against a variety of diseases including legionnaires' disease and diphtheria, and it is sometimes used as an alternative to penicillin G. Topically, it can be used for the treatment of acne. Although it is not stable to stomach acids, it can be taken orally in a tablet form. The formulation of the tablet involves a coating that is designed to protect the tablet during its passage through the stomach, but which is soluble once it reaches the intestines (enterosoluble).

The acid sensitivity of erythromycin is due to the presence of a ketone and two alcohol groups which are set up for the acid-catalysed intramolecular formation of a ketal (Fig. 16.75). One way of preventing this is to protect the hydroxy groups. For example, **clarithromycin** is a methoxy analogue of erythromycin which is more stable to gastric juices and has improved oral absorption. It is one of the drugs used in the treatment of ulcers caused by the presence of *Helicobacter pylori* (section 22.4). Another method of increasing acid stability is to increase the size of the macrocycle to a 16-membered ring.

Resistance to macrolides is due to effective efflux mechanisms which pump the drug back out the cell. The ribosomal target site may also change in character such that binding is weakened. Enzyme catalysed modifications can also occur. Recently there has been research into finding novel macrolides which can be effective against respiratory infections due to resistant strains of *Streptococcus pneumoniae*, as well as to *H. influenza*.

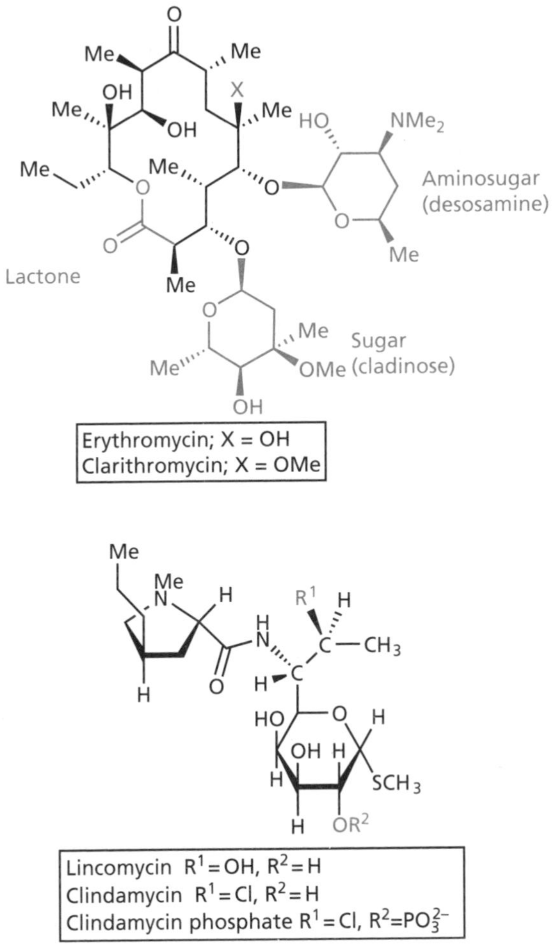

Erythromycin; X = OH
Clarithromycin; X = OMe

Lincomycin R^1 = OH, R^2 = H
Clindamycin R^1 = Cl, R^2 = H
Clindamycin phosphate R^1 = Cl, R^2 = PO$_3^{2-}$

Figure 16.74 Macrolides and lincosamides.

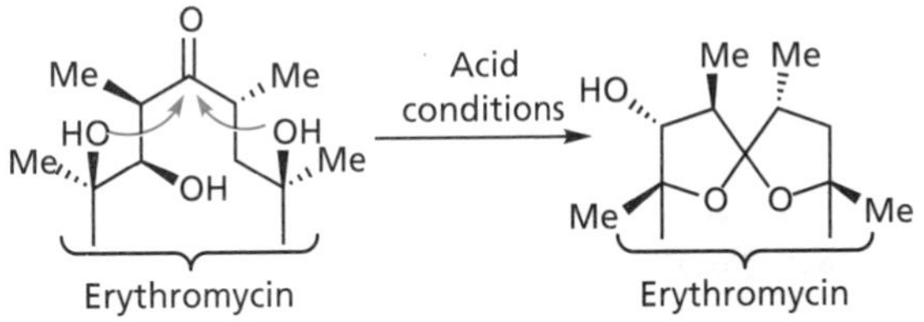

Figure 16.75 Intramolecular ketal formation in erythromycin.

16.7.5 Lincosamides

The lincosamide antibiotics (Fig. 16.74) have similar antibacterial properties to the macrolides and act in the same fashion. **Lincomycin** was the first of these agents and was isolated in 1962 from a soil organism called *Streptomyces lincolnensis* found near Lincoln, Nebraska, USA. Chemical modification led to the clinically useful **clindamycin**, which has increased activity and oral absorption. Clindamycin is used primarily for peripheral infections involving *B. fragilis* or other penicillin-resistant anaerobic bacteria. It is also used topically for the treatment of acne.

16.7.6 Streptogramins

Pritinamycin is a mixture of macrolactone structures obtained from *Streptomyces pristinaespiralis* which has been used orally in the treatment of Gram-positive cocci infections, including MRSA. Two of the components (**quinupristin** and **dalfopristin**) have been isolated and are used intravenously in combination (**Synercid**). These agents bind to different regions of the 50S subunit of the bacterial ribosome and form a complex with it. It is found that the binding of dalfopristin increases the binding affinity for quinupristin and so the two agents act in synergy with each other. Quinupristin inhibits peptide chain elongation, while dalfopristin interferes with the transfer of the peptide chain from one tRNA to the next. At present these agents are reserved for life-threatening infections for which there are no alternative therapies.

16.7.7 Oxazolidinones

The oxazolidinones are a new class of clinically useful synthetic antibacterial agents which have been discovered in recent years. They have a broad spectrum of activity and are active against bacterial strains which have acquired resistance to other antibacterial agents acting against protein synthesis. This is because the oxazolidinones inhibit protein synthesis at a much earlier stage than previous agents. Before protein synthesis can start, a 70S ribosome has to be formed by the combination of a 30S ribosome with a 50S ribosome. The oxazolidinones bind to the 50S ribosome and prevent this from happening. As a result, translation cannot start. Other agents that inhibit protein synthesis do so during the translation process itself (Fig. 16.71).

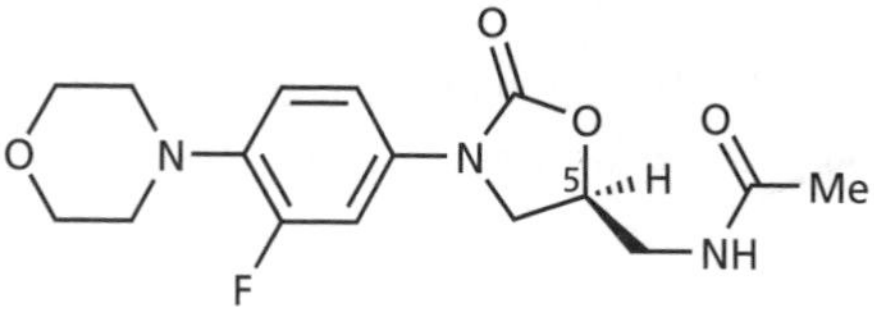

Figure 16.76 Linezolid.

Linezolid (Fig. 16.76) was the first of this class of compounds to reach the market, in 2000. It has good activity against most clinically important Gram-positive bacteria, including MRSA. It can also be taken orally with 100% uptake from the gastrointestinal tract. Unfortunately, there is a high level of side effects related to its use and since it is a bacteriostatic agent, there is a greater risk of bacterial resistance developing.

16.8 Agents which act on nucleic acid transcription and replication

16.8.1 Quinolones and fluoroquinolones

The quinolone and fluoroquinolone antibacterial agents are particularly useful in the treatment of urinary tract infections and also for the treatment of infections which prove resistant to the more established antibacterial agents.

Nalidixic acid (Fig. 16.77) was the first therapeutically useful agent in this class of compounds, and was produced in 1962. It is active against Gram-negative bacteria and was useful in the short-term therapy of urinary tract infections. It can be taken orally, but unfortunately, bacteria can develop a rapid resistance to it. Various analogues have been synthesized which have similar properties to nalidixic acid, but provide no great advantage.

A big breakthrough was made in the 1980s, with the development of **enoxacin** (Fig. 16.77) which has a greatly increased spectrum of activity against Gram-negative and Gram-positive bacteria. It was also active against the highly resistant *P. aeruginosa*. The development of enoxacin was based on the discovery that a single fluorine atom at position 6 greatly increased activity, as well as increasing uptake into the bacterial cell. A basic substituent such as a piperazinyl ring at position 7 was beneficial for a variety of reasons including improved oral absorption, tissue distribution,

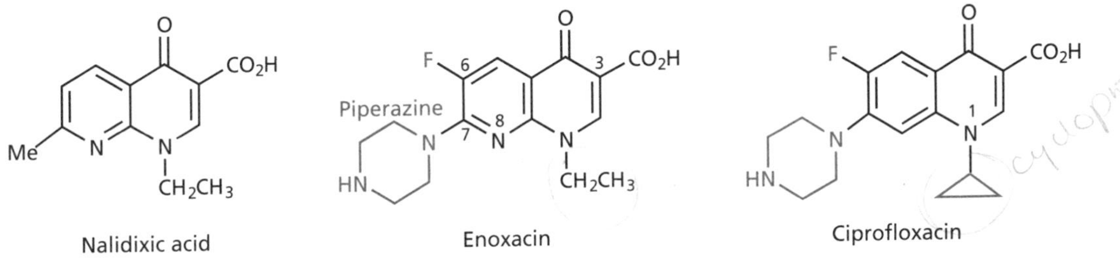

Figure 16.77 Quinolones and fluoroquinolones.

and metabolic stability, as well as an improvement in the level and spectrum of activity, particularly against Gram-negative bacteria such as *P. aeruginosa*. Many of these benefits are due to the ability of the basic substituent at position 7 to form a zwitterion with the carboxylic acid group at position 3.

The introduction of a cyclopropyl substituent at position 1 increased broad-spectrum activity, while replacement of the nitrogen at position 8 with carbon reduced adverse reactions and increased activity against *S. aureus*. This led to **ciprofloxacin** (Fig. 16.77), the most active of the fluoroquinolones against Gram-negative bacteria (Box 16.8). It has been used in the treatment of a large range of infections involving the urinary, respiratory, and gastrointestinal tracts (e.g. travellers' diarrhoea), as well as infections of skin, bone, and joints. It has been claimed that ciprofloxacin may be the most active broad-spectrum antibacterial agent on the market. Furthermore, bacteria are slow to acquire resistance to it, unlike nalidixic acid.

The quinolones and fluoroquinolones inhibit the replication and transcription of bacterial DNA. During these processes, the DNA double helix is unwound by enzymes called **helicases**. The uncoiling process creates tension due to excess supercoiling of the remaining DNA double helix. (You can demonstrate this by pulling apart the strands of rope or string.) This tension needs to be relieved if the process is to continue. The bacterial enzyme **topoisomerase IV** carries out this role by breaking both strands of the DNA chain, crossing them over, then resealing them. The fluoroquinolones act by inhibiting topoisomerase IV in Gram-positive bacteria and show a 1000-fold selectivity for the bacterial enzyme over the corresponding enzyme in human cells.

Inhibition arises by the formation of a ternary complex involving the drug, the enzyme, and bound DNA (Fig. 16.78). This occurs because the binding site

for fluoroquinolones only appears once the enzyme bound DNA has been split and the strands are ready to be crossed over. At that point, four fluoroquinolone molecules become bound in a stacking arrangement such that their aromatic rings are coplanar. Bonding interactions also take place between the substituents at N1. Meanwhile, the carbonyl and carboxylate groups of the fluoroquinolones interact with DNA by hydrogen bonding, while the fluoro substituent, the substituent at C-7, and the carboxylate ion are involved in binding interactions with the enzyme. The resulting complex is stable and so the cut strands of the DNA are not resealed.

In Gram-negative bacteria, the main target for fluoroquinolones is a topoisomerase II enzyme called DNA gyrase. It has the same role as topoisomerase IV in reverse and is required when the DNA double helix is being supercoiled after replication and transcription.

Resistance to the fluoroquinolones is slow to appear, but when it appears it is mainly due to efflux mechanisms which pump the drug back out of the cell. Less common resistance mechanisms include mutations to the topoisomerase enzymes which reduce their affinity to the agents, and alteration of porins in the outer membrane of Gram-negative organisms to limit access.

A large number of fluoroquinolones have now been synthesized. The importance of the binding interactions shown in Fig. 16.78 is emphasized by the fact that agents with good activity all have a similar bicyclic ring system which includes a pyridone ring and a carboxylic acid at position 3.

A problem with first- and second-generation fluoroquinolones is that they generally show only moderate activity against *S. aureus*, with resistance being quick to arise. Furthermore, only marginal activity is shown against anaerobes and *Streptococcus pneumoniae*. Third-generation fluoroquinolones such as **grepafloxacin**,

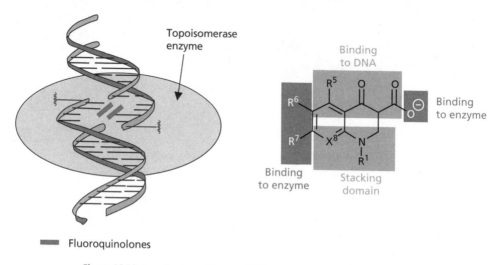

Figure 16.78 Complex formed between DNA, topoisomerase and fluroquinolones.

BOX 16.8 SYNTHESIS OF CIPROFLOXACIN

The synthesis of ciprofloxacin is a seven-stage route and is applicable to a wide range of fluoroquinolones. It involves the construction of the right-hand pyridone ring on to an aromatic ring which already contains the fluoro substituent.

The cyclopropyl substituent is incorporated just before ring closure and the piperazinyl substituent is added at the final stage of the synthesis.

Synthesis of ciprofloxacin.

trovafloxacin, and **clinafloxacin** (Fig. 16.79) began to be developed in the early 1990s to tackle these issues. All of these agents show improved activity against *Streptococcus pneumoniae*, while maintaining activity against enterobacteria.

16.8.2 Aminoacridines

Aminoacridine agents such as the yellow-coloured **proflavine** (Fig. 16.80) are topical antibacterial agents which were used particularly in the Second World War to treat deep surface wounds. The best agents are completely ionized at pH 7 and they interact directly with bacterial DNA. The flat tricyclic ring intercalates between the DNA base pairs and interacts with them by van der Waals forces, while the ammonium cations form ionic bonds with the negatively charged phosphate groups on the sugar phosphate backbone. Once inserted, it deforms the DNA double helix and

prevents the normal functions of replication and transcription. Despite the success of this drug, it is not suitable for the treatment of systemic bacterial infections as it is toxic to host cells.

16.8.3 Rifamycins

Rifampicin (Fig. 16.81) is a semi-synthetic rifamycin made from **rifamycin B**—an antibiotic isolated from *Streptomyces mediterranei* in 1957. It inhibits Gram-positive bacteria and works by binding non-covalently to DNA-dependent RNA polymerase and inhibiting the start of RNA synthesis. The DNA-dependent RNA polymerases in eukaryotic cells are unaffected, since the drug binds to a peptide chain not present in the mammalian RNA polymerase. It is therefore highly selective.

The drug is bactericidal and is mainly used in the treatment of tuberculosis and staphylococci infections that resist penicillin. It is also used in combination

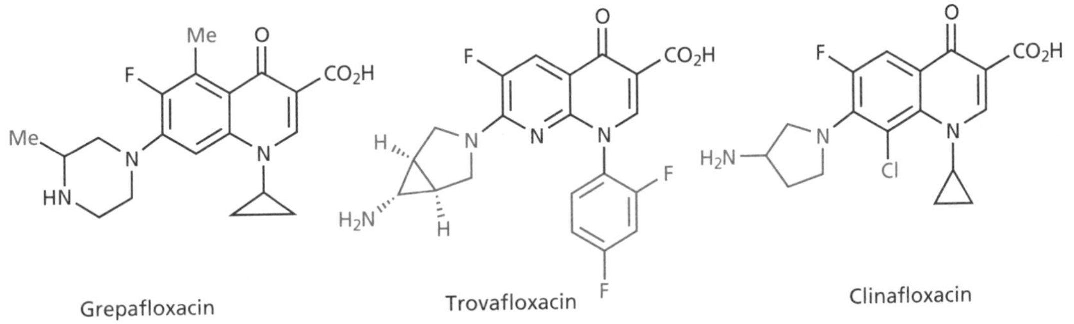

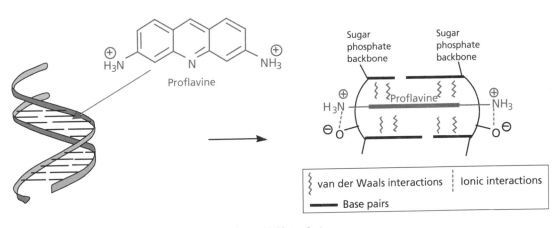

Grepafloxacin   Trovafloxacin   Clinafloxacin

Figure 16.79 Third-generation fluoroquinolones.

Proflavine

Figure 16.80 Proflavine.

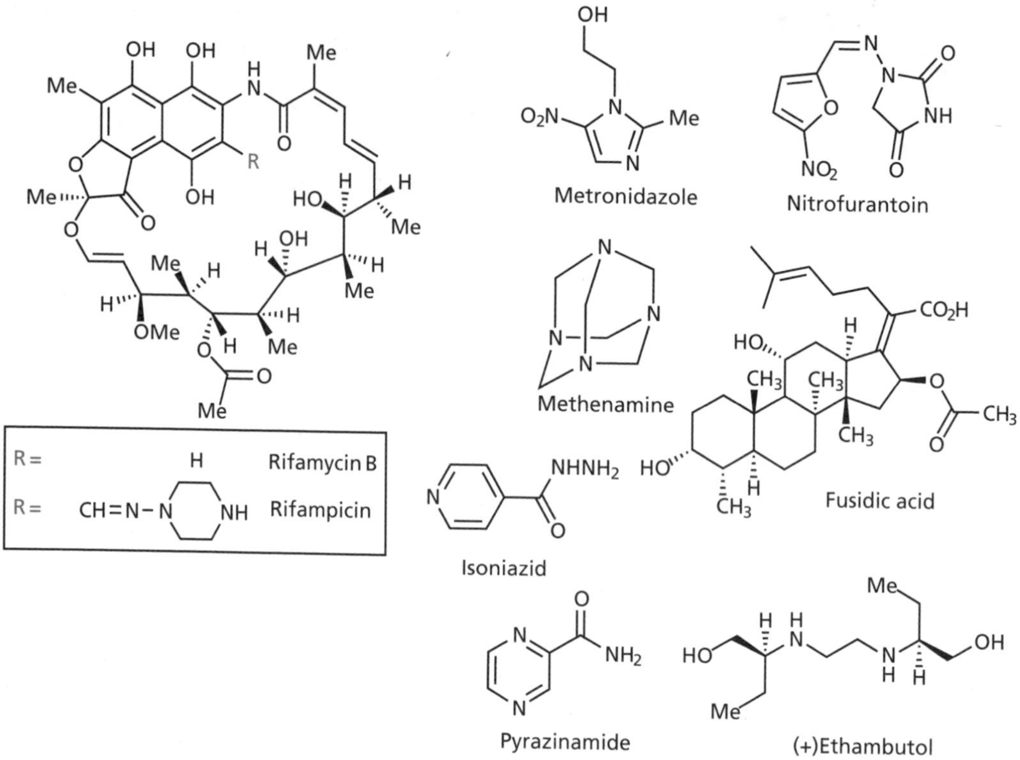

Figure 16.81 Miscellaneous agents.

with dapsone in treating leprosy. It is a very useful antibiotic, showing a high degree of selectivity against bacterial cells over mammalian cells. Unfortunately, it is also expensive, which discourages its use against a wider range of infections. The flat naphthalene ring and several of the hydroxyl groups are essential for activity. It is actually a zwitterion, and has good solubility both in lipids and aqueous acid.

Rifampicin poses a special problem when treating tuberculosis in AIDS patients, since it enhances the activity of the cytochrome P450 enzyme family (CYP3A). These enzymes metabolize the HIV protease inhibitors used in HIV therapy, thus lowering their effectiveness. Increased cytochrome P450 activity also decreases the effect of oral anticoagulants, oral contraceptives, and barbiturates.

16.8.4 Nitroimidazoles and nitrofurantoin

Metronidazole (Fig. 16.81) is a nitroimidazole structure which was introduced in 1959 and was originally used against protozoa. In the 1970s, it was discovered that it had the best activity in treating infections caused by anaerobic bacteria, including difficult-to-treat organisms such as *Bacteroides fragilis* and *Clostridium difficile.*

The mechanism of action involves the drug entering the bacterial cell where the nitro group is reduced. This lowers the concentration of metronidazole within the cell setting up a concentration gradient down which more drug can flow. The reduction mechanism also proves toxic to the cell since free radicals are formed which act on DNA.

The drug is effective against *Guardia* infections derived from polluted water supplies. It is well distributed round the body and crosses the blood–brain barrier, so it can be used for the treatment of brain abscesses and other central nervous system infections involving anaerobic bacteria. It is administered with amoxicillin (or with tetracycline and bismuth) in the treatment of gastric ulcers involving *H. pylori* (section 22.4). Finally, nitroimidozoles are commonly combined with cephalosporins or aminoglycosides to treat infections involving both aerobic and anaerobic organisms. Resistance is rare, although it has been observed.

Nitrofurantoin is used to treat urinary tract infections. It too undergoes reduction within bacterial cells to form radical species which act on DNA.

16.9 Miscellaneous agents

Methenamine is used to treat urinary tract infections. It degrades in acid conditions to give formaldehyde which is the active agent (section 11.6.6).

Fusidic acid is a steroid structure derived from the fungus *Fusidium coccineum* and is used as a topical agent. It can penetrate intact and damaged skin, so it is useful for the treatment of boils. It has also been used to eradicate MRSA colonies carried in the nasal passages of hospital patients and health workers.

Isoniazid (Fig. 16.81) is the most widely used drug for the treatment of tuberculosis. It acts by inhibiting the synthetic pathways leading to mycolic acid, an important constituent of mycobacterial cell walls. It is activated in bacterial cells by a catalase-peroxidase enzyme. Resistant strains of tuberculosis (TB) block the action of this enzyme. **Ethambutol** and **pyrazinamide** (Fig. 16.81) are synthetic compounds which are both front-line drugs in the treatment of tuberculosis. Ethambutol inhibits arabinosyl transferase enzymes that are involved in the biosynthesis of the mycobacterial cell wall.

KEY POINTS

- Aminoglycosides, tetracyclines, chloramphenicol, streptogramins, lincosamides, and macrolides inhibit protein synthesis by binding to the bacterial ribosomes involved in the translation process.

- Resistance can arise from a variety of mechanisms such as drug efflux, altered binding affinity of the ribosome, altered membrane permeability, and metabolic reactions.

- Oxazolidinones prevent the formation of the 70S ribosome by binding to the 50S subunit.

- Quinolones and fluoroquinolones inhibit topoisomerase enzymes, resulting in inhibition of replication and transcription.

- Aminoacridines are useful topical antibacterial agents which can intercalate with bacterial DNA and hinder replication and transcription.

- Rifamycins inhibit the enzyme RNA polymerase and prevent RNA synthesis. This in turn prevents protein synthesis. Rifampicin is used to treat tuberculosis and staphylococcus infections.

- Nitroimidazoles are used against infections caused by protozoa and anaerobic bacteria.

16.10 Drug resistance

Medicinal chemists are still actively seeking new and improved antibacterial agents to combat the worrying ability of bacteria to acquire resistance to current drugs. For example, 60% of *Streptococcus pneumoniae* strains are resistant to β-lactams and 60% of *Staphylococcus aureus* strains are resistant to **methicillin**. The last resort in treating *S. aureus* infections is **vancomycin**, but resistance is beginning to appear to that antibiotic as well. Some strains of *Enterococcus faecalis* appearing in urinary and wound infections are resistant to all known antibiotics and are untreatable. If antibiotic resistance continues to grow, medicine could be plunged back to the 1930s. Indeed many of today's advanced surgical procedures would become too risky to carry out due to the risks of infection. Old diseases are already making a comeback. For example, a new antibiotic resistant strain of tuberculosis (**multidrug-resistant TB—MDRTB**) appeared in New York and took 4 years and $10 million to bring under control. These strains were resistant to two of the front-line drugs used against TB (isoniazid and rifamycin), and had various levels of resistance against another two (streptomycin and ethambutol). Other examples of bacterial strains acquiring resistance include penicillin-resistant meningococci and pneumococci in South Africa, penicillin-resistant gonococci in Asia and Africa, ampicillin-resistant *H. influenza* in the USA and Europe, and chloramphenicol-resistant meningococci in France and South-east Asia. Resistance to trimethoprim in some of the developing nations has meant that the drug has become ineffective as a treatment of dysentery within 10 years.

Drug resistance can arise due to a variety of factors described in section 16.5.1.5, but the cell must have the necessary genetic information. This information can be obtained by mutation or by the transfer of genes between cells.

16.10.1 Drug resistance by mutation

Bacteria multiply at such a rapid rate that there is always a chance that a mutation will render a bacterial cell resistant to a particular agent. This feature has been known for a long time and is the reason why patients should fully complete a course of antibacterial treatment even though their symptoms may have disappeared well before the end of the course. If this rule is adhered to, the vast majority of the invading bacterial cells will be wiped out, leaving the body's own defence system to mop up any isolated survivors or resistant cells. If the treatment is stopped too soon, however, then the body's defences struggle to cope with the survivors. Any isolated resistant cell is then given the chance to multiply, resulting in a new infection which will, of course, be completely resistant to the original drug. Patients failing to complete a course of therapy was a major factor in the appearance of MDRTB.

These mutations occur naturally and randomly and do not require the presence of the drug. Indeed, it is likely that a drug-resistant cell is present in a bacterial population even before the drug is encountered. This was demonstrated with the identification of **streptomycin**-resistant cells from old cultures of *E. coli* which had been freeze-dried to prevent multiplication before the introduction of streptomycin into medicine.

16.10.2 Drug resistance by genetic transfer

A second way in which bacterial cells can acquire drug resistance is by gaining that resistance from another bacterial cell. This occurs because it is possible for genetic information to be passed on directly from one bacterial cell to another. There are two main methods by which this can take place—**transduction** and **conjugation**.

In transduction, small segments of genetic information known as **plasmids** are transferred by means of bacterial viruses (**bacteriophages**) which leave the resistant cell and infect a non-resistant cell. If the plasmid contains the gene required for drug resistance, then the recipient cell will be able to use that information and gain resistance. For example, the genetic information required to synthesize β-**lactamases** can be passed on in this way, rendering bacteria resistant to penicillins. The problem is particularly prevalent in hospitals where currently over 90% of staphylococcal infections are resistant to antibiotics such as penicillin, erythromycin, and tetracycline. It may seem odd that hospitals should be a source of drug-resistant strains of bacteria. In fact, they are the perfect breeding ground. Drugs commonly used in hospitals are present in the air in trace amounts. It has been shown that breathing in these trace amounts kills sensitive bacteria in the nose and allows the nostrils to act as a breeding ground for resistant strains.

In conjugation, bacterial cells pass genetic material directly to each other. This is a method used mainly by Gram-negative, rod-shaped bacteria in the colon, and involves two cells building a connecting bridge of sex pili through which the genetic information can pass.

16.10.3 Other factors affecting drug resistance

The more useful a drug is, the more it will be used and the greater the possibilities of resistant bacterial strains emerging. The original penicillins were used widely in human medicine, but were also commonly used in veterinary medicine. Antibacterial agents have also been used in animal feeding to increase animal weight and this, more than anything else, has resulted in drug-resistant bacterial strains. It is sobering to think that many of the original bacterial strains which were treated so dramatically with penicillin V or penicillin G are now resistant to those early penicillins. In contrast, these two drugs are still highly effective antibacterial agents in poorer, developing African nations, where the use (and abuse) of the drugs have been far less widespread.

The ease with which different bacteria acquire resistance varies. For example, *S. aureus* is notorious for its ability to acquire drug resistance due to the ease with which it can undergo transduction. On the other hand, the microorganism responsible for syphilis seems incapable of acquiring resistance and is still susceptible to the original drugs used against it.

16.10.4 The way ahead

The ability of bacteria to gain resistance to drugs is an ever-present challenge to the medicinal chemist and it is important to continue designing new antibacterial agents. Identifying potential new targets is essential in this never-ending battle. The sequencing of genomes and a study of the proteins present in bacterial cells promises to give more detailed understanding of the molecular details of infectious agents leading to the identification of new drug targets. For example, *Mycobacterium tuberculosis*—the causative agent of tuberculosis—has a complex cell wall where three types of polymers are attached to peptidoglycan. The detailed mechanisms by which these polymers are synthesized and incorporated into the cell wall are being investigated to identify new targets for antibacterial drugs which will disrupt the cell wall structure.

It is also beginning to be appreciated that the drugs with the least susceptibility to resistance are those with several different modes of action. Therefore, designing drugs which act on a number of different targets rather than one specific target is more likely to be successful.

Examples of new targets include enzymes known as **aminoacyl tRNA synthetases**. These enzymes are an ancient group of enzymes responsible for attaching amino acids to tRNA. As they are ancient, there is a considerable sequence divergence between the bacterial and human enzymes, making selective inhibition possible. Isoleucyl tRNA synthetase is one such enzyme which is known to be inhibited by **mupirocin** (Fig. 16.82)—a clinically useful antibiotic isolated from *Pseudomonas fluorescens* with activity against MRSA. Mupirocin is used as a topical agent for skin infections, and has also been used to combat the transmission of

S. aureus within hospitals by treating the nasal passages of patients and hospital staff. Unfortunately, the widespread use of the agent for this purpose has led to strains of *S. aureus* with increasing resistance to the drug. Research is now being carried out to find novel inhibitors for a different aminoacyl tRNA synthetase present in *S. aureus*, mainly tyrosine tRNA synthetase.

Another potential approach in countering resistance is to modify the antibiotics such that they gain resistance to the mechanisms of resistance used against them! For example, **kanamycin** is an aminoglycoside which is no longer used since resistant bacteria can phosphorylate one of the hydroxyl groups present (Fig. 16.83). An active analogue has been synthesized which replaces the susceptible alcohol with a ketone (Fig. 16.84). This ketone is in equilibrium with the hydrated gem-diol. When phosphorylation occurs on the diol, the phosphate group thus formed acts as a good leaving group and the ketone is regenerated. *In vitro* tests showed that this agent was active against strains of bacteria which are resistant to kanamycin.

Another approach is to design molecules with an inbuilt self-destruct mechanism. One of the problems with antibiotics in medicine or veterinary practice is that much of the active antibiotic is excreted, giving bacteria in the environment the opportunity to gain resistance. This problem should be reduced by incorporating a self-destruct mechanism which kicks in once the antibiotic is excreted. For example work has been carried out on a cephalosporin containing a protected hydrazine group (Fig. 16.85). The protecting group concerned is *ortho*-nitrobenzylcarbamate which is susceptible to light. Once the antibiotic is excreted and exposed to light, the protecting group is lost allowing the nucleophilic hydrazine moiety to react with the β-lactam ring and deactivate the molecule. This works *in vitro* but has still be tested *in vivo*.

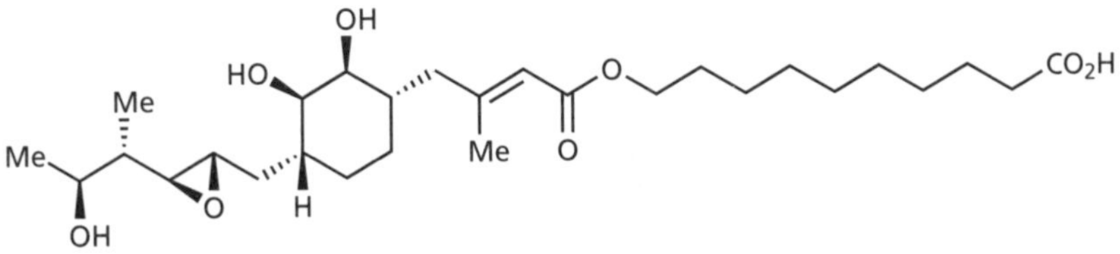

Figure 16.82 Mupirocin.

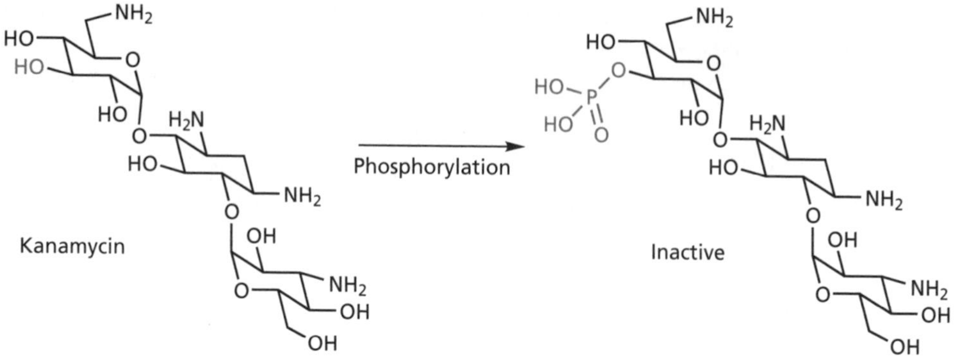

Figure 16.83 Resistance to kanamycin.

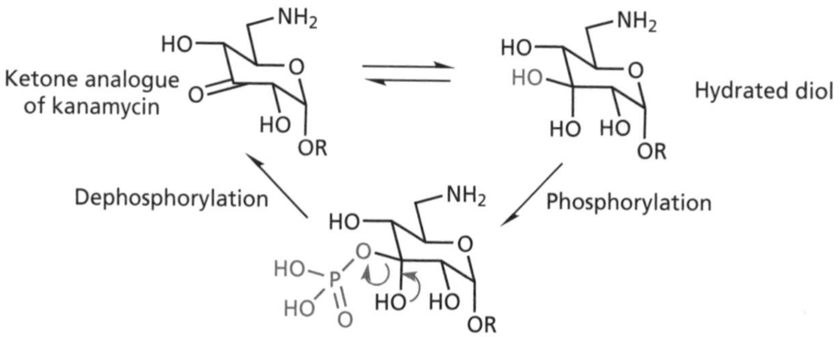

Figure 16.84 Analogue of kanamycin resistant to a resistance mechanism.

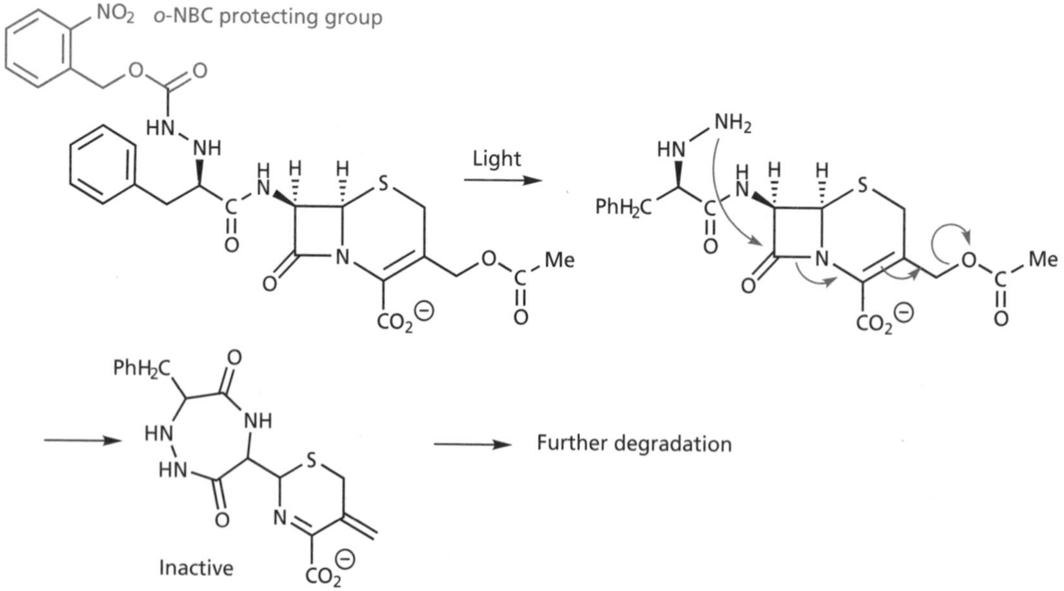

Figure 16.85 Self-destruct mechanism.

BOX 16.9 ORGANOARESENICALS AS ANTIPARASITIC DRUGS

The first effective antimicrobial drug to be synthesized was the organoaresenical, salvarsan (section 16.1). In the late 1940s another organoaresenical called **melarsoprol** was introduced into medicine and is the first-choice drug for the treatment of trypanosomiasis and sleeping sickness (Fig. 1).

One of the mechanisms by which melarsoprol might act is through a reaction with the cysteine residues of enzymes involved in glycolysis (Fig. 2). As a result, glycolysis is blocked leading to a loss of cell motility and eventual cell death. Other mechanisms of action have been proposed.

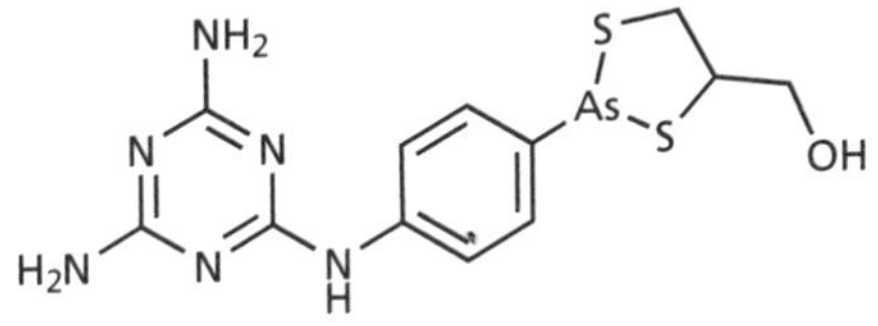

Figure 1 Melarsoprol.

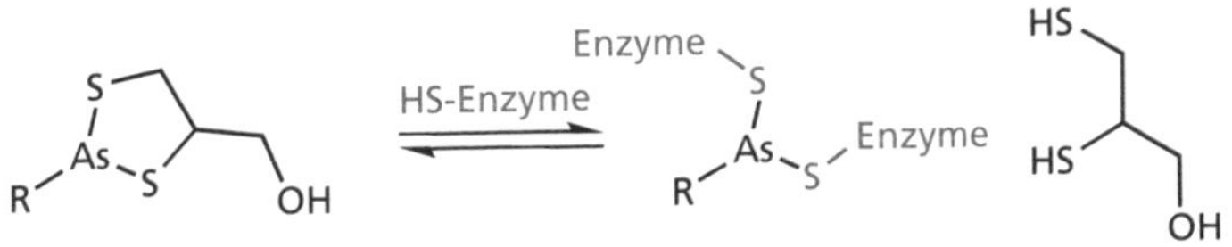

Figure 2 Mechanism of action of melarsoprol.

KEY POINTS

- Bacteria can gain resistance to antibacterial drugs. Different strains gain resistance more easily than others. *Staphylococcus aureus* strains are quick to gain antibacterial resistance. The MRSA strain is a *S. aureus* strain that is resistant to most antibacterials, including methicillin.

- Vancomycin is the antibacterial agent of last resort in the treatment of resistant bacterial strains.

- There are many mechanisms by which bacteria can acquire resistance against antibacterial agents, but they all result from a change in the cell's genetic make-up.

- Drug resistance can result from mutation of a cell's genetic information, or from transfer of genetic information from one cell to another. Genetic information can be transferred from one cell to another by transduction or conjugation.

- Care has to be take to use antibacterial agents in a responsible manner to reduce the chances of resistance developing.

- It is important to identify new targets which can be used for the design of novel antibacterial agents.

QUESTIONS

1. How would you convert penicillin G to 6-aminopenicillanic acid (6-APA) using chemical reagents? Suggest how you would make ampicillin from 6-APA.

2. Penicillin is produced biosynthetically from cysteine and valine. If the biosynthetic pathway could accept different amino acids, what sort of penicillin analogues might be formed if valine was replaced by alanine, phenylalanine, glycine, or lysine? What sort of penicillin analogue might be formed if cysteine was replaced by serine? (see Appendix 1 for amino acid structures).

3. Referring to Question 2, why do you think penicillin analogues like this are not formed during the fermentation process?

4. The activity of sulfonamides is decreased if they are taken at the same time as procaine. Suggest why this might be the case.

5. Discuss whether you think the following penicillin analogue would be a useful antibacterial agent.

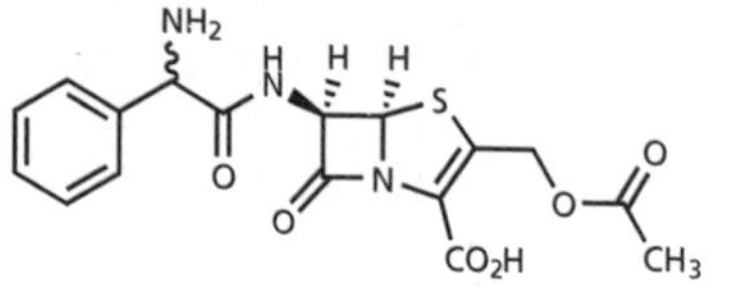

Penicillin analogue

6. Explain what effect replacing the methoxy groups on methicillin with ethoxy groups might have on the properties of the agent.

7. The following structure is an analogue of cefoxitin. What sort of properties do you think it might have compared to cefoxitin itself?

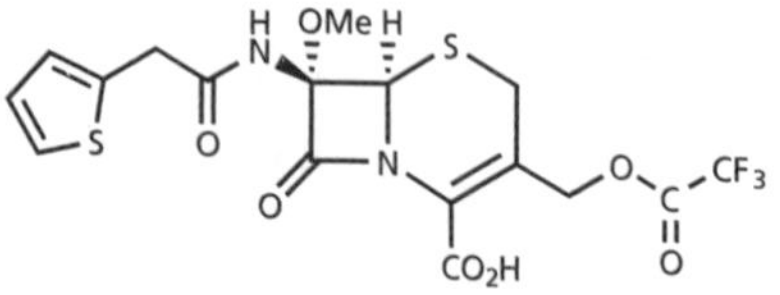

Cefoxitin analogue

8. Show the mechanism by which the prodrug bacampicillin is converted to ampicillin. What are the by-products?

9. Which of the following structures would you expect to have the best antibacterial activity?

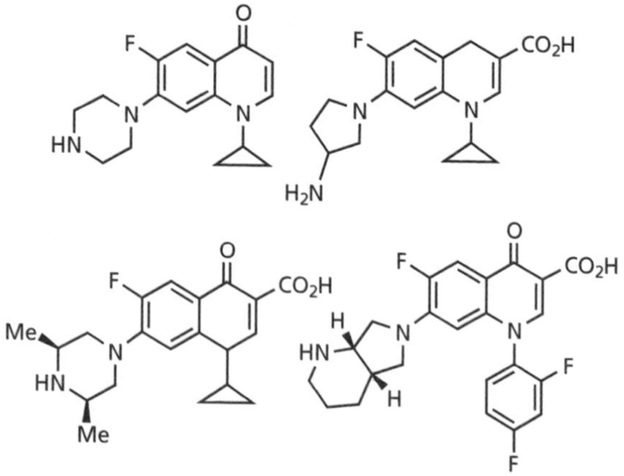

10. Devise a synthesis for the structure chosen in Question 9.

FURTHER READING

Armstrong, D. and Cohen, J. (eds.) (1999) *Infectious diseases*, Section 7, Mosby, London.

Coates, A. *et al.* (2002) The future challenges facing the development of new antimicrobial drugs. *Nature Reviews Drug Discovery*, 1, 895–910.

Evans, J. (1998) TB: Know your enemy. *Chemistry in Britain*, November, 38–42

Hook, V. (1997) Superbugs step up the pace. *Chemistry in Britain*, May, 34–35.

Mendell, G., Bennett, J. E., and Dolin, R. (eds.), (2000) *Mendell, Douglas and Bennett's principles and practice of infectious diseases*, 5th edn, Vols. 1 and 2. Churchill Livingstone, Edinburgh.

Raja, A. *et al.* (2003) Daptomycin. *Nature Reviews Drug Discovery*, 2, 943–944.

β-Lactams and other agents acting on cell walls

Axelsen, P. H. and Li, D. (1998) A rational strategy for enhancing the affinity of vancomycin towards depsipeptide ligands. *Bioorganic and Medicinal Chemistry*, 6, 877–881.

Nicolaou, K. C. *et al.* (1999) Chemistry, biology, and medicine of the glycopeptide antibiotics. *Angewandte Chemie, International Edition*, 38, 2096–2152.

Agents acting on the cell membrane

Mann, J. (2001) Killer nanotubes. *Chemistry in Britain*, November, 22.

Linezolid

Ford, C. (2001) First of a kind. *Chemistry in Britain*, March, 22–24.

Genin, M. J. *et al.* (1998) N-C-linked (azolylphenyl)oxazolidinones. *Journal of Medicinal Chemistry*, 41, 5144–5147.

Quinolones and other agents acting on nucleic acids

Andriole, V. T. (ed.) (1998) *The quinolones*, 2nd edn. Academic Press, New York.

Saunders, J. (2000) Quinolones as anti-bacterial DNA gyrase inhibitors. Chapter 10 in: *Top drugs: top synthetic routes.* Oxford University Press, Oxford.

Agents acting against protein synthesis

Agouridas, C. *et al.* (1998) Synthesis and antibacterial activity of ketolides. *Journal of Medicinal Chemistry*, **41**, 4080–4100.

Titles for general further reading are listed on p. 711.

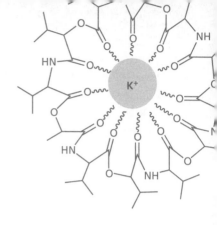

17 Antiviral agents

17.1 Viruses and viral diseases

Viruses are non-cellular infectious agents which take over a host cell in order to survive and multiply. There are a large variety of different viruses which are capable of infecting bacterial, plant, and animal cells, with more than 400 different viruses known to infect humans.

Viruses can be transmitted in a variety of ways. Those responsible for diseases such as influenza (flu), chickenpox, measles, mumps, viral pneumonia, rubella, and smallpox can be transmitted through the air by an infected host sneezing or coughing. Other viruses can be transmitted by means of arthropods or ticks, leading to diseases such as Colorado tick fever and yellow fever. Some viruses are unable to survive long outside the host and are transmitted through physical contact. The viruses responsible for AIDS, cold sores, the common cold, genital herpes, certain leukaemias, and rabies are examples of this kind. Finally, food-borne or water-borne viruses can lead to hepatitis A and E, poliomyelitis, and viral gastroenteritis.

Historically, viral infections have proved devastating to human populations. It has been suggested that smallpox was responsible for the major epidemics which weakened the Roman Empire during the periods AD 165–180 and AD 251–266. Smallpox was also responsible for the decimation of indigenous tribes in both North and South America during European colonization. In some areas, it is estimated that 90% of the population died from the disease. Various flu epidemics and pandemics have proved devastating.

The number of deaths worldwide due to the flu pandemic of 1918–1919 is estimated to be over 20 million, much greater than the number killed by military action in the First World War.

The African continent has its fair share of lethal viruses including Ebola and the virus responsible for Lassa fever. In the past, viral diseases such as these occurred in isolated communities and were easily contained. Nowadays, with cheap and readily available air travel, tourists are able to visit remote areas, thus increasing the chances of rare or new viral diseases spreading round the world. Therefore, it is important that world health authorities monitor potential risks and take appropriate action when required. The outbreak of severe acute respiratory syndrome (SARS) in the Far East during 2003 could have had a devastating effect worldwide if it had been ignored. Fortunately, the world community acted swiftly and the disease was brought under control relatively quickly. Nevertheless, the SARS outbreak serves as a timely warning of how dangerous viral infections can be. Scientists have warned of a nightmare scenario involving the possible evolution of a 'supervirus'. Such an agent would have a transmission mode and infection rate equivalent to flu, but a much higher mortality rate. There are already lethal viruses which can be spread rapidly and have a high mortality rate. Fortunately, the latency period between infection and detectable symptoms is short and so it is possible to contain the outbreak, especially if it is in isolated communities. If such viral infections evolved such that the latency period increased to that of AIDS, they could result in devastating pandemics equivalent to the plagues of the Middle Ages.

Considering the potential devastation that viruses can wreak on society, there are fears that terrorists might one day try to release lethal viral strains on civilian populations. This has been termed **bioterrorism**. To date, no terrorist group has carried out such an action, but it would be wrong to ignore the risk.

It is clear that research into effective antiviral drugs is a major priority in medicinal chemistry.

17.2 Structure of viruses

At their simplest, viruses can be viewed as protein packages transmitting foreign nucleic acid between host cells. The type of nucleic acid present depends on the virus concerned. All viruses contain one or more molecules of either RNA or DNA, but not both. They can therefore be defined as RNA or DNA viruses. Most RNA viruses contain single-stranded RNA (ssRNA), but some viruses contain double-stranded RNA. If the base sequence of the RNA strand is identical to viral mRNA, it is called the positive (+) strand. If it is complementary, it is called the negative (−) strand. Most DNA viruses contain double-stranded DNA, but a small number contain single-stranded DNA. The size of the nucleic acid varies widely, with the smallest viral genomes coding for 3–4 proteins and the largest coding for over 100 proteins.

The viral nucleic acid is contained and protected within a protein coat called the **capsid**, Capsids are usually made up of protein subunits called **protomers** which are generated in the host cell and can interact spontaneously to form the capsid in a process called **self-assembly**. Once the capsid contains the viral nucleic acid, the whole assembly is known as the **nucleocapsid**. In some viruses, the nucleocapsid may contain viral enzymes which are crucial to its replication in the host cell. For example the flu virus contains an enzyme called **RNA-dependent RNA polymerase** within its nucleocapsid (Fig. 17.1).

Additional membranous layers of carbohydrates and lipids may be present surrounding the nucleocapsid, depending on the virus concerned. These are usually derived from the host cell, but they may also contain viral proteins which have been coded by viral genes.

The complete structure is known as a **virion** and this is the form that the virus takes when it is outside

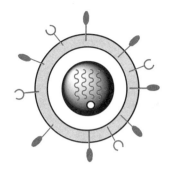

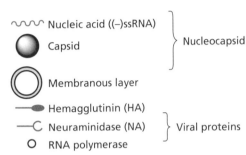

Figure 17.1 Diagrammatic representation of the flu virus.

the host cell. The size of a virion can vary from 10 to 400 nm. As a result, most viruses are too small to be seen by a light microscope and require the use of an electron microscope.

17.3 Life cycle of viruses

The various stages involved in the life cycle of a virus are as follows (Fig. 17.2):

- **Adsorption:** A virion has to first bind to the outer surface of a host cell. This involves a specific molecule on the outer surface of the virion binding to a specific protein or carbohydrate present in the host cell membrane. The relevant molecule on the host cell can thus be viewed as a 'receptor' for the virion. Of course, the host cell has not produced this molecule to be a viral receptor. The molecules concerned are usually glycoproteins which have crucial cellular functions such as the binding of hormones. The virion takes advantage of these, however, and once the virion is bound, the next stage can take place—introduction of the viral nucleic acid into the host cell.

- **Penetration and uncoating:** Different viruses introduce their nucleic acid into the host cell by

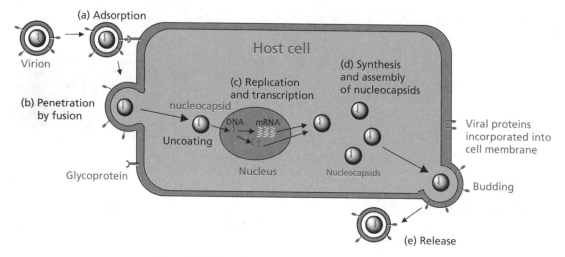

Figure 17.2 Life cycle of a DNA virus such as herpes simplex.

different methods. Some inject their nucleic acid through the cell membrane; others enter the cell intact and are then uncoated. This can also happen in a variety of ways. The viral envelope of some virions fuses with the plasma membrane and the nucleocapsid is then introduced into the cell (Fig. 17.2). Other virions are taken into the cell by endocytosis where the cell membrane wraps itself round the virion and is then pinched off to produce a vesicle called an **endosome** (see for example Fig. 17.39). These vesicles then fuse with **lysosomes**, and host cell enzymes aid the virus in the uncoating process. Low endosomal pH also triggers uncoating. In some cases, the virus envelope fuses with the lysosome membrane and the nucleocapsid is released into the cell. Whatever the process, the end result is the release of viral nucleic acid into the cell.

- **Replication and transcription:** Viral genes can be defined as *early* or *late*. Early genes take over the host cell such that viral DNA and/or RNA is synthesized. The mechanism involved varies from virus to virus. For example, viruses containing negative single-strand RNA use a viral enzyme called RNA-dependent RNA polymerase (or transcriptase) to synthesize mRNA which then codes for viral proteins.

- **Synthesis and assembly of nucleocapsids:** Late genes direct the synthesis of capsid proteins and these self-assemble to form the capsid. Viral nucleic acid is then taken into the capsid to form the nucleocapsid.

- **Virion release:** Naked virions (those with no outer layers round the nucleocapsid) are released by cell lysis where the cell is destroyed. In contrast, viruses with envelopes are usually released by a process known as **budding** (Fig. 17.2). Viral proteins are first incorporated into the host cell's plasma membrane. The nucleocapsid then binds to the inner surface of the cell membrane and at the same time viral proteins collect at the site and host cell proteins are excluded. The plasma membrane containing the viral proteins then wraps itself round the nucleocapsid and is pinched off from the cell to release the mature virion.

The life cycle stages of herpes simplex, flu virus, and HIV are illustrated in Figs 17.2, 17.12, and 17.39 respectively.

17.4 Vaccination

Vaccination is the preferred method of protection against viral disease and has proved extremely successful against childhood diseases such as polio, measles, and mumps, as well as historically serious diseases such as smallpox and yellow fever. The first successful vaccination was carried out by Edward

Jenner in the eighteenth century. Having observed that a milkmaid had contracted the less virulent cowpox and had subsequently become immune to smallpox, he inoculated people with material from cowpox lesions and discovered that they too gained immunity from smallpox.

Vaccination works by introducing the body to foreign material which bears molecular similarity to some component of the virus, but which lacks its infectious nature or toxic effects. The body then has the opportunity to recognize the molecular fingerprint of the virus (i.e. specific antigens), and the immune system is primed to attack the virus should it infect the body. Usually a killed or weakened version of the virus is administered so that it does not lead to infection itself. Alternatively, fragments of the virus (subunit vaccines) can be used if they display a characteristic antigen. Vaccination is a preventive approach and is not usually effective on patients who have already become infected.

Vaccines are currently under investigation for the prevention or treatment of HIV, dengue fever, genital herpes, and haemorrhagic fever caused by the Ebola virus. There are difficulties surrounding the HIV and flu viruses, however, because rapid gene mutation in these viruses results in constant changes to the amino acid composition of glycoproteins normally present on the viral surface. Since these glycoproteins are the important antigens that trigger the immune response, any changes in their structure 'disguise' the virus and the body's primed immune system fails to recognize it.

Another problem concerning vaccination relates to patients with a weakened immune response. The main categories of patients in this situation are cancer patients undergoing chemotherapy, patients undergoing organ transplants (where the immune system has been deliberately suppressed to prevent organ rejection), and AIDS patients. Vaccination in these patients is less likely to be effective and the weakened immune response also leads to increased chances of infections such as pneumonia.

In situations where infection has occurred and the immune system is unable to counter the invasion, antiviral drugs can help to bring the disease under control and allow the immune system to regain ascendancy.

17.5 Antiviral drugs: general principles

Antiviral drugs are useful in tackling viral diseases where there is a lack of an effective vaccine, or where infection has already taken place. The life cycle of a virus means that for most of its time in the body it is within a host cell and is effectively disguised both from the immune system and from circulating drugs. Since it also uses the host cell's own biochemical mechanisms to multiply, the number of potential drug targets that are unique to the virus is more limited than those that can be identified for invading microorganisms. Thus, the search for effective antiviral drugs has proved more challenging than that for antibacterial drugs. Indeed, the first antiviral agents appeared relatively late on in the 1960s, and only three clinically useful antiviral drugs were in use during the early 1980s. Early antiviral drugs included **idoxuridine** and **vidarabine** for herpes infections, and **amantadine** for influenza A.

Since then, progress has accelerated for two principle reasons—the need to tackle the AIDS pandemic, and the increased understanding of viral infectious mechanisms resulting from viral genomic research.

In 1981, it was noticed that gay men was unusually susceptible to diseases such as pneumonia and fungal infections—ailments which were previously only associated with patients whose immune response had been weakened. The problem soon reached epidemic proportions and it was discovered that a virus (the **human immunodeficiency virus**—HIV) was responsible. It was found that this virus infected T-cells—cells which are crucial to the immune response—and was therefore directly attacking the immune response. With a weakened immune system, infected patients proved susceptible to a whole range of opportunistic secondary diseases resulting in the term **acquired immune deficiency syndrome** (AIDS). This discovery led to a major research effort into understanding the disease and counteracting it—an effort which kick-started more general research into antiviral chemotherapy. Fortunately, the tools needed to carry out effective research appeared on the scene at about the same time, with the **advent of viral genomics**. The full genome of any virus can now be quickly determined and compared with those of other viruses, allowing the identification of how the genetic sequence is split into genes. Although the genetic sequence is unlikely to be identical from one virus to another, it is possible to identify similar genes coding for similar proteins with similar functions. These proteins can then be studied as potential drug targets. Standard genetic engineering methods allow the production of pure copies of the target protein by inserting the viral gene into a bacterial cell, thus allowing sufficient quantities of the protein to be synthesized and isolated (section 7.6). The protein can be used for screening as well as for studying drug–protein interactions.

Good drug targets are proteins which are likely to have the following characteristics:

- They are important to the life cycle of the virus, such that their inhibition or disruption has a major effect on infection.

- They bear little resemblance to human proteins, thus increasing the chances of good selectivity and minimal side effects.

- They are common to a variety of different viruses and have a specific region which is identical in its amino acid composition. This makes the chances of developing a drug with broad antiviral activity more likely.

- They are important to the early stages of the virus life cycle, so that the virus has less chance of spreading through the body and producing symptoms.

Most antiviral drugs in use today act against HIV, herpesviruses (responsible for a variety of ailments including cold sores and encephalitis), hepatitis B, and hepatitis C. Diseases such as herpes and HIV are chronic in developed countries, and intensive research has been carried out to develop drugs to combat them. In contrast, less research has been carried out on viral diseases prevalent in developing countries, such as tropical (dengue) and haemorrhagic (Ebola) fevers.

Most antiviral drugs in use today disrupt critical stages of the virus life cycle or the synthesis of virus-specific nucleic acids. Excluding drugs developed for the treatment of HIV, more drugs are available for the treatment of DNA viruses than for RNA viruses. Few drugs show a broad activity against both DNA and RNA viruses.

Studies of the human genome are also likely to be useful for future research. The identification of human proteins which stimulate the body's immune response or the production of antibodies would provide useful leads for the development of drugs that would have an antiviral effect by acting as immunomodulators.

KEY POINTS

- Viruses pose a serious health threat, and there is a need for new antiviral agents.

- Viruses consist of a protein coat surrounding nucleic acid which is either RNA or DNA. Some viruses have an outer membranous coat which is derived from the host cell.

- Viruses are unable to self-multiply and require to enter a host cell in order to do so.

- Vaccination is effective against many viruses, but is less effective against viruses which readily mutate.

- Research into antiviral drugs has increased in recent years as a result of the AIDS epidemic and the need to find drugs to combat it.

- Antiviral research has been aided by advances in viral genomics and genetic engineering, as well as the use of X-ray crystallography and molecular modelling.

17.6 Antiviral drugs used against DNA viruses

Most of the drugs which are active against DNA viruses have been developed against herpesviruses to combat diseases such as cold sores, genital herpes, chickenpox, shingles, eye diseases, mononucleosis, Burkitt's lymphoma, and Kaposi's sarcoma. Nucleoside analogues have been particularly effective.

17.6.1 Inhibitors of viral DNA polymerase

Aciclovir was discovered by compound screening and was introduced into the market in 1981. It represented a revolution in the treatment of **herpes** infections, being the first relatively safe, non-toxic drug to be used systemically. It is used for the treatment of infections due to herpes simplex 1 and 2 (i.e. herpes simplex encephalitis and genital herpes), as well as **varicella-zoster viruses** (VZV) (i.e. chickenpox and shingles). Aciclovir has a nucleoside-like structure and contains the same nucleic acid base as deoxyguanosine, but lacks the complete sugar ring. In virally infected cells, it is phosphorylated in three stages to form a triphosphate which is the active agent, and so aciclovir itself is a prodrug (Fig. 17.3).

Nucleotide triphosphates are the building blocks for DNA replication where a new DNA strand is constructed using a DNA template—a process catalysed by the enzyme **DNA polymerase** (Fig. 17.4). Aciclovir triphosphate prevents DNA replication in two ways. First, it is sufficiently similar to the normal deoxyguanosine triphosphate building block (Fig. 17.5) that it can bind to DNA polymerase and inhibit it. Second, DNA polymerase can catalyse the attachment of the aciclovir nucleotide to the growing DNA chain. Since the sugar unit is incomplete and lacks the required hydroxyl group normally present at position 3′ of the sugar ring, the nucleic acid chain cannot be extended any further. Thus, the drug acts as a chain terminator.

However, what is to stop aciclovir triphosphate inhibiting DNA polymerase in normal, uninfected cells? The answer lies in the fact that aciclovir is only converted to the active triphosphate in infected cells. The explanation for this lies in the first phosphorylation reaction catalysed by the enzyme thymidine kinase. Although this enzyme is present in host cells, the herpesvirus carries its own version. It turns out that aciclovir is more readily converted to its monophosphate by viral thymidine kinase (100-fold) than by host cell thymidine kinase. Once formed, the monophosphate is converted to the active triphosphate by cellular enzymes. In normal uninfected cells, therefore, aciclovir is a poor substrate for cellular thymidine kinase and remains as the prodrug. This, along with the fact that there is a selective uptake of aciclovir by infected cells, explains its excellent activity and much reduced toxicity relative to previous drugs. Another feature which enhances its safety is that aciclovir triphosphate shows a 50-fold selective action against viral DNA polymerases relative to cellular polymerases.

The oral bioavailability of aciclovir is quite low (15–30%) and to overcome this, various prodrugs were developed to increase water solubility. **Valaciclovir** (Fig. 17.6) is an L-valyl ester prodrug and is hydrolysed to aciclovir in the liver and gut wall. When this prodrug is given orally, blood levels of aciclovir are obtained which are equivalent to those obtained by intravenous administration. Valaciclovir is particularly useful in the treatment of VZV infections. **Desciclovir** (Fig.17.6) is an analogue of aciclovir which lacks the

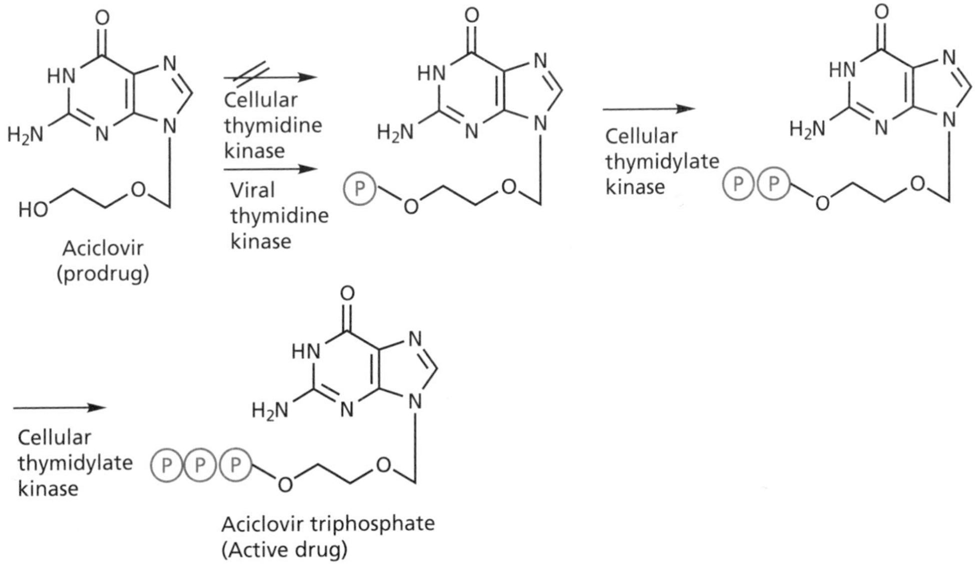

Figure 17.3 Activation of aciclovir. Ⓟ represents phosphate groups.

Normal replication

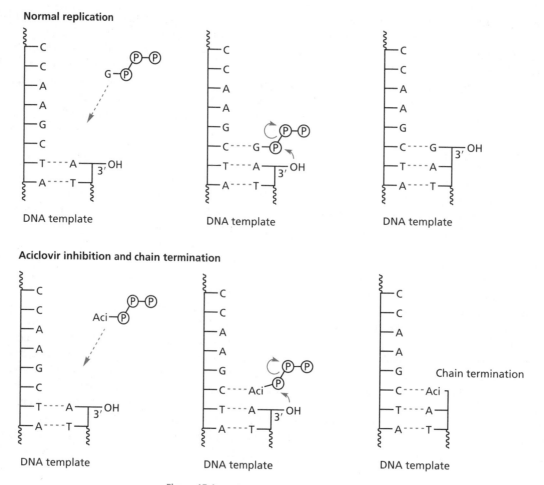

Aciclovir inhibition and chain termination

Figure 17.4 Aciclovir acting as a chain terminator.

carbonyl group at position 6 of the purine ring and is more water soluble. Once in the blood supply, metabolism by cellular xanthine oxidase oxidizes the 6-position to give aciclovir. Desciclovir is somewhat more toxic than aciclovir itself, however, and this limits its potential.

Unfortunately, strains of herpes are appearing which are resistant to aciclovir. This can arise due to mutations, either of the viral thymidine kinase enzyme such that it no longer phosphorylates aciclovir, or of viral DNA polymerase such that it no longer recognizes the activated drug.

Aciclovir is not effective against all types of herpesvirus. There are eight herpesviruses which are divided into three subfamilies. Aciclovir is effective against the α-subfamily but not the β-subfamily, because the latter produces a different thymidine

kinase that fails to phosphorylate aciclovir. The analogue **ganciclovir** (Fig. 17.6), however, is phosphorylated by thymidine kinases produced by both the α- and β-subfamilies and can be used against both viruses. Ganciclovir contains an extra hydroxymethylene group which increases its similarity to deoxyguanosine. Unfortunately, the drug is not as safe as aciclovir as it can be incorporated into cellular DNA. Nevertheless, it can be used for the treatment of **cytomegalovirus** (CMV) infections. This is a virus which causes eye infections and can lead to blindness. Aciclovir is not effective in this infection, because CMV does not encode a viral thymidine kinase. Ganciclovir, on the other hand, can be converted to its monophosphate by kinases other than thymidine kinase. Since ganciclovir has a low oral bioavailability, the valine prodrug **valganciclovir** (Fig. 17.6) has been introduced for the treatment of CMV infections.

Penciclovir (Fig. 17.7) is an analogue of ganciclovir where a methylene group has replaced the oxygen in the acyclic 'sugar' moiety. Its biological properties are closer to aciclovir than to ganciclovir, however. It is

metabolized to the active triphosphate in the same way as aciclovir and essentially has the same spectrum of activity, but it has better potency, a faster onset and a longer duration of action. It is used topically for the treatment of cold sores (HSV-1) and intravenously for the treatment of HSV in immunocompromised patients. Like aciclovir, penciclovir has poor oral bioavailability and is poorly absorbed from the gut due to its polarity. Therefore, **famciclovir** (Fig. 17.7) is used as a prodrug. The two alcohol groups are masked as esters making the structure less polar, and leading to better absorption. The acetyl groups are then hydrolysed by esterases, and the purine ring is oxidized by aldehyde oxidase in the liver to generate penciclovir. Phosphorylation reactions then take place in virally infected cells as described previously.

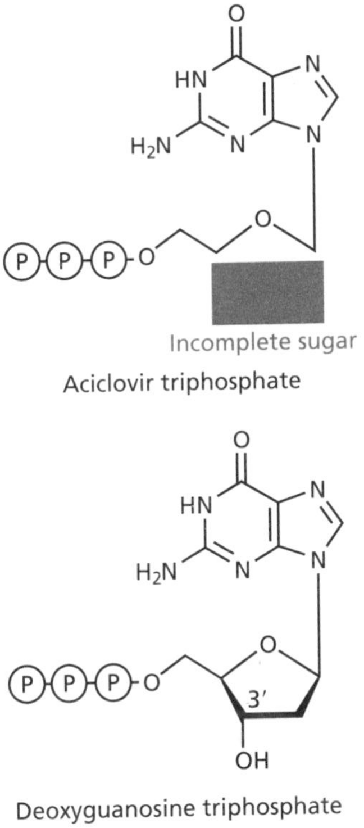

Some viruses are immune from the action of the above antiviral agents because they lack the enzyme thymidine kinase. As a result, phosphorylation fails to take place. **Cidofovir** was designed to combat this problem (Fig. 17.8). It is an analogue of deoxycytidine 5-monophosphate where the sugar and phosphate groups have been replaced by an acyclic group and a phosphonomethylene group respectively. The latter group acts as a bioisostere for the phosphate group and is used because the phosphate group itself would be more susceptible to enzymatic hydrolysis. Since a phosphate equivalent is already present, the drug does not require thymidine kinase to become activated. Two more phosphorylations can now take place catalysed by cellular kinases to convert cidofovir to the active 'triphosphate'.

Cidofovir is a broad-spectrum antiviral agent which shows selectivity for viral DNA polymerase, and

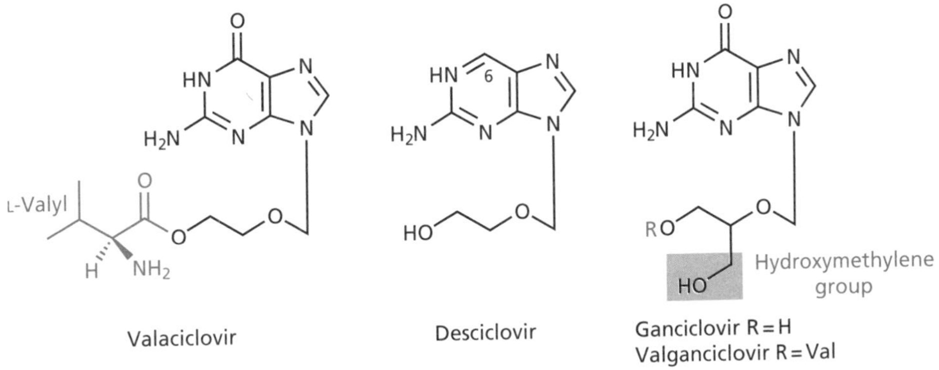

Figure 17.5 Comparison of aciclovir triphosphate and deoxyguanosine triphosphate.

Figure 17.6 Prodrugs and analogues of aciclovir.

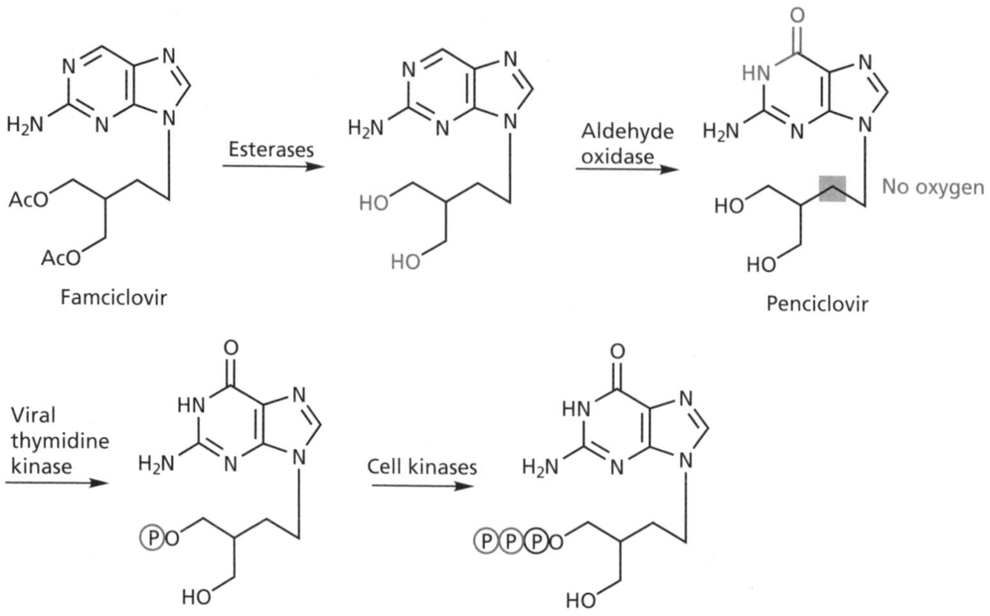

Figure 17.7 Penciclovir and famciclovir. Ⓟ represents a phosphate group.

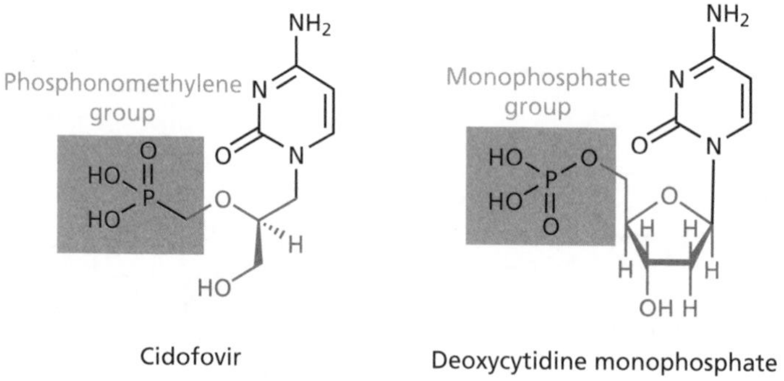

Figure 17.8 Comparison of cidofovir and deoxycytidine monophosphate.

is used to treat retinal inflammation caused by CMV. Unfortunately the drug is extremely polar and has a poor oral bioavailability (5%). It is also toxic to the kidneys, but this can be reduced by co-administering probenecid (Box 16.6).

In contrast to aciclovir, **idoxuridine, trifluridine**, and **vidarabine** (Fig. 17.9) are phosphorylated equally well by viral and cellular thymidine kinase and so there is less selectivity for virally infected cells. As a result, these drugs have more toxic side effects. Idoxuridine, like trifluridine, is an analogue of deoxythymidine and was the first nucleoside-based antiviral agent licensed in the USA. It can be used for the topical treatment of herpes keratitis, but trifluridine is the drug of choice for this disease since it is effective at lower dose-frequencies. The triphosphate inhibits viral DNA polymerase as well as thymidylate synthetase.

Vidarabine (Fig. 17.9) is the purine counterpart of the pyrimidine nucleoside cytarabine (ara-C) and was an early antiviral drug with clinical applications. Aciclovir is now used in preference because of its lower toxicity.

Foscarnet (Fig. 17.9) inhibits viral DNA polymerase, but is non-selective and toxic. Since it is highly charged, it has difficulty crossing cell membranes.

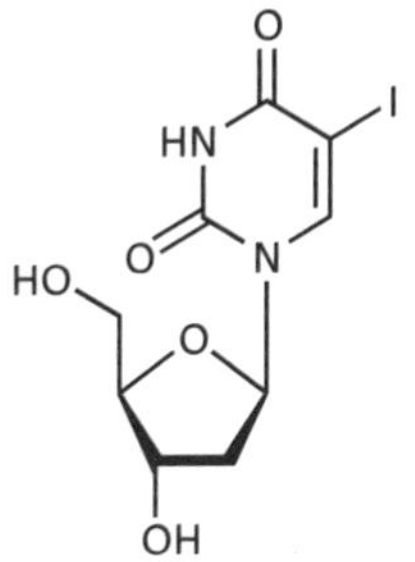

Figure 17.9 Miscellaneous antiviral agents.

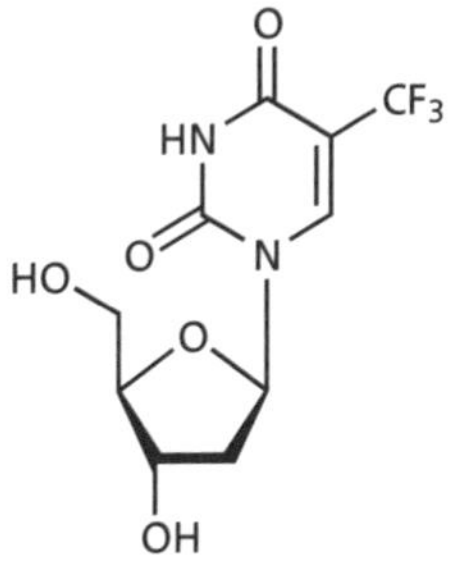

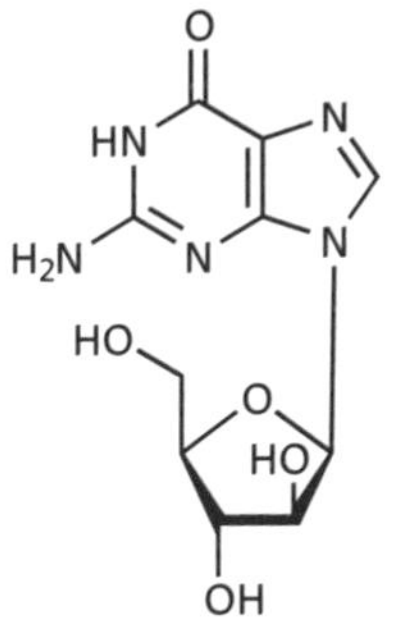

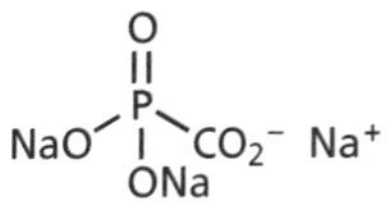

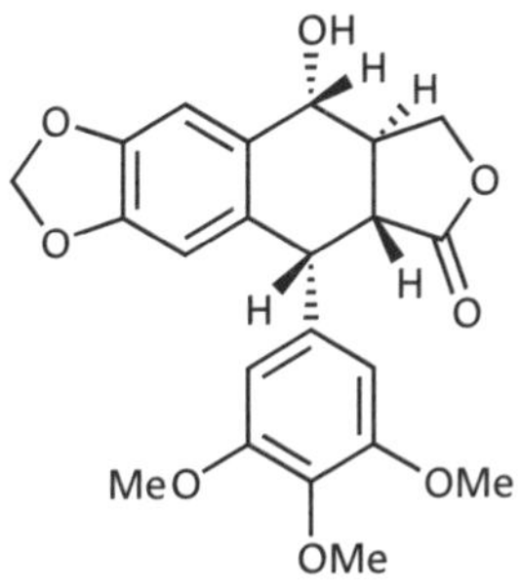

Figure 17.10 Podophyllotoxin.

d(*P*-thio)(G-C-G-T-T-T-G-C-T-C-T-T-C-T-T-C-T-T-G-C-G)

Figure 17.11 Fomivirsen.

It was discovered in the 1960s and is used in the treatment of CMV retinitis where it is approximately equal in activity to ganciclovir. It can also be used in immunocompromised patients for the treatment of HSV and VZV strains which prove resistant to aciclovir. It does not undergo metabolic activation.

17.6.2 Inhibitors of tubulin polymerization

The plant product **podophyllotoxin** (Fig. 17.10) has been used clinically to treat genital warts (caused by the DNA virus **papillomavirus**), but it is not as effective as **imiquimod** (section 17.10.4). It is a powerful inhibitor of tubulin polymerization (sections 3.7.2 and 18.5.1).

17.6.3 Antisense therapy

Fomivirsen (Fig. 17.11) is the first, and so far the only, DNA antisense molecule that has been approved as an antiviral agent. It consists of 21 nucleotides and a phosphonothioate backbone rather than a phosphate backbone to increase the metabolic stability of the molecule (section 11.8.5). The drug blocks the translation of viral RNA and is used against retinal inflammation caused by CMV in AIDS patients. Because of its high polarity it is administered as an ocular injection (intravitreal).

KEY POINTS

- Nucleoside analogues have been effective antiviral agents used against DNA viruses, mainly herpesviruses.

- Nucleoside analogues are prodrugs which require to be phosphorylated to a triphosphate in order to be active. They have a dual mechanism of action whereby they inhibit viral DNA polymerase and also act as DNA chain terminators.

- Nucleoside analogues show selectivity for virally infected cells over normal cells if viral thymidine kinase is required to catalyse the first of three phosphorylation steps. They are also taken up more effectively into virally infected cells and their triphosphates inhibit viral DNA polymerases more effectively than cellular DNA polymerases.

- Agents containing a bioisostere for a phosphate group can be used against DNA viruses lacking thymidine kinase.

- Inhibitors of tubulin polymerization have been used against DNA viruses.

- A DNA antisense molecule has been designed as an antiviral agent.

17.7 Antiviral drugs acting against RNA viruses: HIV

17.7.1 Structure and life cycle of HIV

HIV (Fig. 17.12) is an example of a group of viruses known as the retroviruses. There are two variants of HIV. HIV-1 is responsible for AIDS in America, Europe, and Asia, whereas HIV-2 occurs mainly in western Africa. HIV has been studied extensively over the last 20 years and a vast research effort has resulted in a variety of antiviral drugs which have proved successful in slowing down the disease, but not eradicating it. At present, clinically useful antiviral drugs act against two targets—the viral enzymes **reverse transcriptase** and **protease.** There is a need to develop effective drugs against a third target, and a good knowledge of the life cycle of HIV is essential in identifying suitable targets (Fig. 17.12).

HIV is an RNA virus which contains two identical strands of (+)ssRNA within its capsid. Also present are the viral enzymes reverse transcriptase and **integrase**, as well as other proteins called p6 and p7. The capsid is made up of protein known as p24, and surrounding the capsid there is a layer of matrix protein (p17), then a membranous envelope which originates from host cells and which contains the viral glycoproteins **gp120 and gp41**. Both of these proteins are crucial to the processes of adsorption and penetration. Gp41 traverses the envelope and is bound non-covalently to gp120, which projects from the surface. When the virus approaches the host cell, gp120

(a)

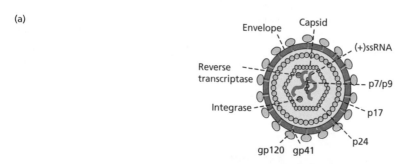

(b)

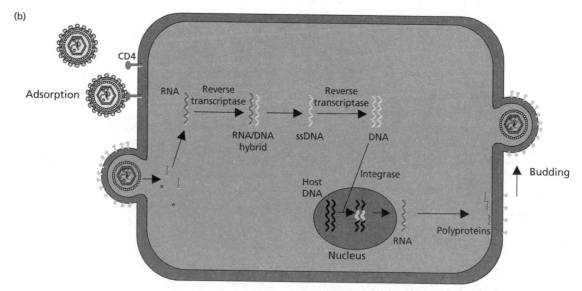

Figure 17.12 (a) Structure of virus particle and (b) life cycle of the human immunodeficiency virus (HIV).

interacts and binds with a transmembrane protein called **CD4** which is present on host T-cells. The gp120 proteins then undergo a conformational change which allows them to bind simultaneously to chemokine receptors (CCR5 and CXCR4) on the host cell (not shown). Further conformational changes peel away the gp120 protein so that viral protein gp41 can reach the surface of the host cell and anchor the virus to the surface. The gp41 then undergoes a conformational change and pulls the virus and the cell together so that their membranes can fuse.

Once fusion has taken place, the HIV nucleocapsid enters the cell. Disintegration of the protein capsid then takes place, probably aided by the action of a viral enzyme called protease. Viral RNA and viral enzymes are then released into the cell cytoplasm. The released viral RNA is not capable of coding directly for viral proteins, or of self-replication. Instead, it is converted into DNA and incorporated into the host cell DNA. The conversion of RNA into DNA is not a process that occurs in human cells, so there are no host enzymes to catalyse the process. Therefore, HIV carries its own enzyme—reverse transcriptase—to do this. This enzyme is a member of a family of enzymes known as the DNA polymerases, but is unusual in that it can use an RNA strand as a template. The enzyme first catalyses the synthesis of a DNA strand using viral RNA as a template. This leads to a (+)RNA-(-)DNA hybrid. Reverse transcriptase catalyses the degradation of the RNA strand, then uses the remaining DNA strand as a template to catalyse the synthesis of double-stranded DNA (proviral DNA). Proviral DNA is now spliced into the host cell's DNA, a process catalysed by integrase—an enzyme also carried by the virion. Once the proviral DNA has been incorporated into host DNA, it is called the provirus and can remain dormant in host cell DNA until activated by cellular processes. When that occurs, transcription of the viral genes *env*, *gag*, and *pol* takes place to produce viral RNA, some of which will be incorporated into new virions, and the rest of which is used in translation to produce three large non-functional polyproteins, one derived from the *env* gene, one from the *gag* gene, and the other from the *gag-pol* genes. The first of these polyproteins is cleaved by cellular proteinases and produces the viral glycoproteins (gp120 and gp41) which are incorporated into the cell membrane. The remaining two polypeptides

(Pr55 and Pr160) are not split by cellular proteinases. Instead, they move to the inner membrane surface. The viral glycoproteins in the cell membrane also concentrate in this area and cellular proteins are excluded. Budding then takes place to produce an immature membrane-bound virus particle. During the budding process a viral enzyme called protease is released from the gag–pol polypeptide. This is achieved by the protease enzyme autocatalysing the cleavage of susceptible peptide bonds linking it to the rest of the polypeptide. Once released, the protease enzyme dimerizes and cleaves the remaining polypeptide chains to release reverse transcriptase, integrase, and viral structural proteins. The capsid proteins now self-assemble to form new nucleocapsids containing viral RNA, reverse transcriptase, and integrase.

It has also been observed that a viral protein called Vpu has an important part to play in the budding process. Vpu binds to the host membrane protein CD4 and triggers a host enzyme to tag the CD4 protein with a protein called ubiquitin. Proteins that are tagged with ubiquitin are marked out for destruction by the host cell and so the CD4 proteins in the host cell are removed. This is important, as the CD4 proteins could complex with the newly synthesized viral proteins gp120 and prevent the assembly of the new viruses.

17.7.2 Antiviral therapy against HIV

Until 1987, no anti-HIV drug was available, but an understanding of the life cycle of HIV has led to the identification of several possible drug targets. At present, most drugs that have been developed act against the viral enzymes reverse transcriptase and protease. However, a serious problem with the treatment of HIV is the fact that the virus undergoes mutations extremely easily. This results in rapid resistance to antiviral drugs. Experience has shown that treatment of HIV with a single drug has a short-term benefit, but in the long term the drug serves only to select mutated viruses which are resistant. As a result, current therapy involves combinations of different drugs acting on both reverse transcriptase and protease. This has been successful in delaying the progression to AIDS and increasing survival rates, but there is a need to develop effective drugs against a third target.

The available drugs for highly active antiretroviral therapy (**HAART**) include:

- nucleoside reverse transcriptase inhibitors (NRTIs)—zidovudine, didanosine, zalcitabine, stavudine, lamivudine, and abacavir
- non-nucleoside reverse transcriptase inhibitors (NNRTIs)—nevirapine, delavirdine, efavirenz
- protease inhibitors (PIs)—saquinavir, ritonavir, indinavir, nelfinavir.

Currently, PIs are used with reverse transcriptase inhibitors (divergent therapy) or with another PI (convergent therapy). A combination of two NRTIs plus a PI is recommended, but one can also use two PIs with a NRTI, or an NNRTI with two NRTIs.

The demands on any HIV drug are immense, especially since it is likely to be taken over long periods of time. It must have a high affinity for its target (in the picomolar range) and be effective in preventing the virus multiplying and spreading. It should show low activity for any similar host targets in the cell, and be safe and well tolerated. It must be active against as large a variety of viral isolates as possible, or else it only serves to select resistant variants. It needs to be synergistic with other drugs used to fight the disease and be compatible with other drugs used to treat opportunistic diseases and infections arising from the weakened immune response. The drug must stay above therapeutic levels within the infected cell and in the circulation. It must be capable of being taken orally and with a minimum frequency of doses, and preferably should be able to cross the blood–brain barrier in case the virus lurks in the brain. Finally, it must be inexpensive as it is likely to be used for the lifetime of the patient.

17.7.3 Inhibitors of viral reverse transcriptase

17.7.3.1 Nucleoside reverse transcriptase inhibitors

Since the enzyme reverse transcriptase is unique to HIV, it serves as an ideal drug target. Nevertheless, the enzyme is still a DNA polymerase and care has to be taken that inhibitors do not have a significant inhibitory effect on cellular DNA polymerases. Various nucleoside-like structures have proved useful as antiviral agents. The vast majority of these are not active themselves but are phosphorylated by three cellular enzymes to form an active nucleotide triphosphate. This is the same process previously described in section 17.6.1, but one important difference is the requirement for all three phosphorylations to be carried out by cellular enzymes as HIV does not produce a viral kinase.

Zidovudine (Fig. 17.13) was originally developed as an anticancer agent but was the first drug to be approved for use in the treatment of AIDS. It is an analogue of deoxythymidine, where the sugar 3'-hydroxyl group has been replaced by an azido group. It inhibits **reverse transcriptase** as the triphosphate. Furthermore, the triphosphate is attached to the growing DNA chain. Since the sugar unit has an azide substituent at the 3' position of the sugar ring, the nucleic acid chain cannot be extended any further. Unfortunately, zidovudine can cause severe side effects

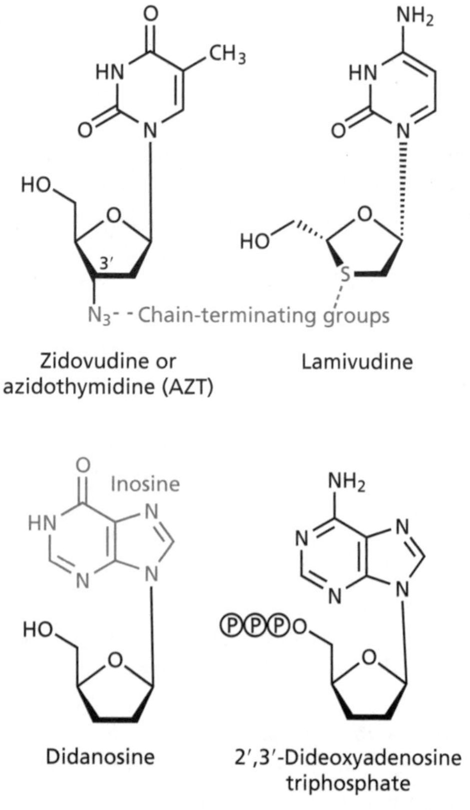

Figure 17.13 Inhibitors of viral reverse transcriptase.

such as anaemia. Studies of the target enzyme have allowed the development of less toxic nucleoside analogues such as **lamivudine** (an analogue of deoxycytidine where the 3′ carbon has been replaced by sulfur). This drug has also been approved for treatment of hepatitis B.

Didanosine (Fig. 17.13) was the second anti-HIV drug approved for use in the USA (1988). Its activity was unexpected, since the nucleic acid base present is inosine—a base which is not naturally incorporated into DNA. However, a series of enzyme reactions converts this compound into 2′,3′-dideoxyadenosine triphosphate which is the active drug.

Other clinically useful NRTIs include **abacavir** (the only guanosine analogue), **stavudine**, and **zalcitabine** (Fig. 17.14). Abacavir was approved in 1998 and has been used successfully in children, in combination with the PIs nelfinavir and saquinavir. Zalcitabine also acts against hepatitis B, but long-term toxicity means that it is unacceptable for the treatment of chronic

viral diseases which are not life threatening. **Tenofovir disoproxil** and **adefovir dipivoxil** are prodrugs of modified nucleosides. Both structures contain a monophosphate group protected by two extended esters. Hydrolysis *in vivo* reveals the phosphate group which can then be phosphorylated to the triphosphate as described previously. Tenofovir disoproxil was approved for HIV-1 treatment in 2001. It remains in infected cells longer than many other antiretroviral drugs, allowing once-daily dosing. Adefovir dipivoxil was approved by the US FDA in 2002 for the treatment of chronic hepatitis B. It is also active on viruses such as CMV and herpes.

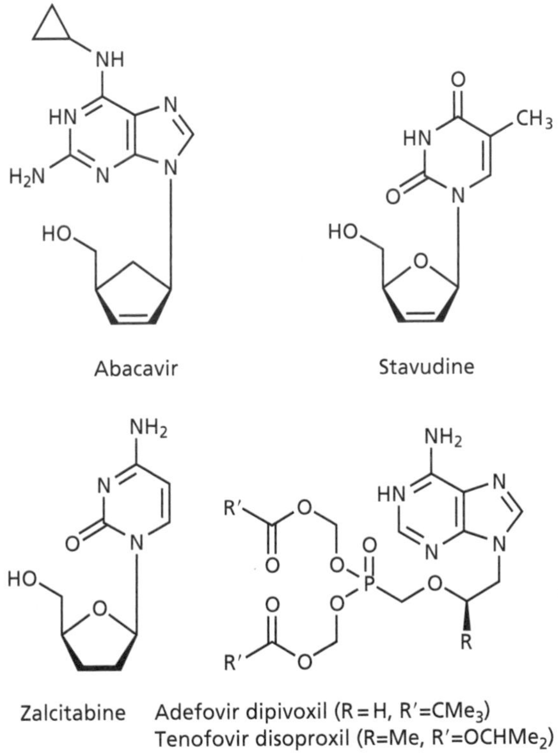

17.7.3.2 Non-nucleoside reverse transcriptase inhibitors

The NNRTIs (Fig. 17.15) are generally hydrophobic molecules that bind to an allosteric binding site which is hydrophobic in nature. Since the allosteric binding site is separate from the substrate binding site, the NNRTIs are non-competitive, reversible inhibitors. They include first-generation NNRTIs such as **nevirapine** and **delavirdine** as well as second-generation drugs such as **efavirenz**. X-ray crystallographic studies on enzyme–inhibitor complexes show that the allosteric binding site is adjacent to the substrate binding site. Binding of an NNRTI to the allosteric site results in an induced fit which locks the neighbouring substrate-binding site into an inactive conformation. NNRTIs show a higher selectivity for HIV-1 reverse transcriptase over host DNA polymerases than do the NRTIs. As a result, NNRTIs are les toxic and have fewer side effects. Unfortunately, rapid resistance emerges due to mutations in the NNRTI binding site, the most common being the replacement of Lys-103 with asparagine. This mutation is called K103N and is defined as a **pan-class resistance mutation**. It can be prevented if the NNRTI is combined with an NRTI from the start of treatment. The two types of drugs can be used together as the binding sites are distinct.

Nevirapine was developed from a lead compound discovered through a random screening programme, and has a rigid butterfly-like conformation that makes it chiral. One 'wing' interacts through hydrophobic and van der Waals interactions with aromatic residues in the binding site while the other wing interacts with aliphatic residues. The other NNRTI inhibitors bind to the same pocket and appear to function as π electron donors to aromatic side chain residues.

Abacavir   Stavudine

Zalcitabine   Adefovir dipivoxil (R = H, R′=CMe₃)
Tenofovir disoproxil (R=Me, R′=OCHMe₂)

Figure 17.14 Further inhibitors of viral reverse transcriptase.

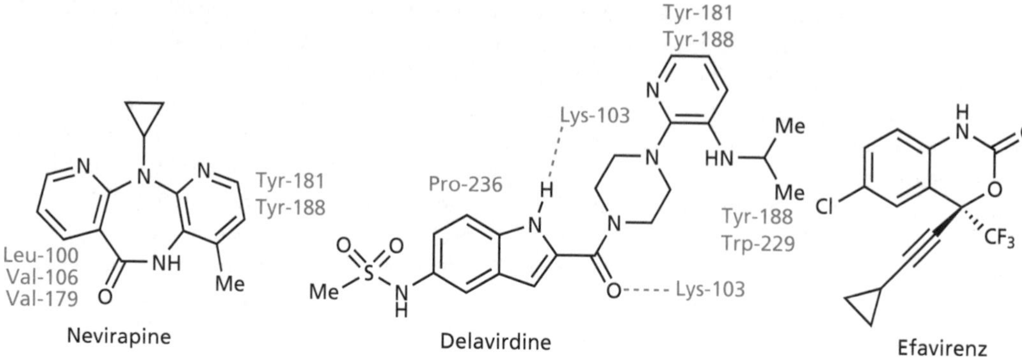

Figure 17.15 Non-nucleoside reverse transcriptase inhibitors in clinical use (interactions with amino acids in the binding site shown in blue).

Delavirdine was developed from a lead compound discovered by a screening programme of 1500 structurally diverse compounds. It is larger than other NNRTIs and extends beyond the normal pocket such that it projects into surrounding solvent. The pyridine region and isopropylamine groups are the most deeply buried parts of the molecule and interact with tyrosine and tryptophan residues. There are also extensive hydrophobic contacts. Unlike other first generation NNRTIs, there is hydrogen bonding to the main peptide chain next to Lys-103. The indole ring of delavirdine interacts with Pro-236, and mutations involving Pro-236 lead to resistance. Analogues having a pyrrole ring in place of indole might avoid this problem.

Second- and third-generation NNRTIs were developed specifically to find agents that were active against resistant variants as well as wild-type virus. This development has been helped by X-ray crystallographic studies, which show how the structures bind to the binding site. It has been shown from sequencing studies that in most of the mutations that cause resistance to first-generation NNRTIs a large amino acid is replaced by a smaller one, implying that an important binding interaction has been lost. Interestingly, mutations that replace an amino acid with a larger amino acid appear to be detrimental to the activity of the enzyme, and no mutations have been found which block NNRTIs sterically from entering the binding site.

Efavirenz is a benzoxazinone structure and is the only second-generation NNRTI on the market in 2004. It has activity against many mutated variants but has less activity against the mutated variant K103N. Nevertheless, activity drops less than for nevirapine, and a study of X-ray structures of each complex revealed that the cyclopropyl group of efavirenz has fewer interactions with Tyr-181 and Tyr-188 than nevirapine does. Consequently, mutations of these amino acids have less effect on efavirenz than they do on nevirapine. Efavirenz is also a smaller structure and can shift its binding position when K103N mutation occurs, allowing it to form hydrogen bonds to the main peptide chain of the binding site.

X-ray crystallographic studies of enzyme complexes with second generation NNRTIs reveal that these agents contain a non-aromatic moiety which interacts with the aromatic residues Tyr-181, Tyr-188, and Trp-229 at the top of the binding pocket. The ability to form hydrogen bonds to the main peptide chain, and a relatively small bulk, are important since they allow compounds to change their binding mode when mutations occur.

A large number of third-generation NNRTIs are being studied, including those shown in Fig. 17.16. **Emivirine** was developed from a lead compound found by screening structures similar to aciclovir. Resistance is slow to occur and requires two mutations to occur in the binding site. The isopropyl group at C-5 forces the aromatic residue of Tyr-181 into an orientation which allows an enhanced interaction with the aromatic substituent at C-6.

SJ3366 inhibits HIV-1 replication at a concentration below 1 nM with a therapeutic index greater than 4 million and, unlike other NNRTIs, inhibits HIV-2. **DPC083** was developed from efavirenz.

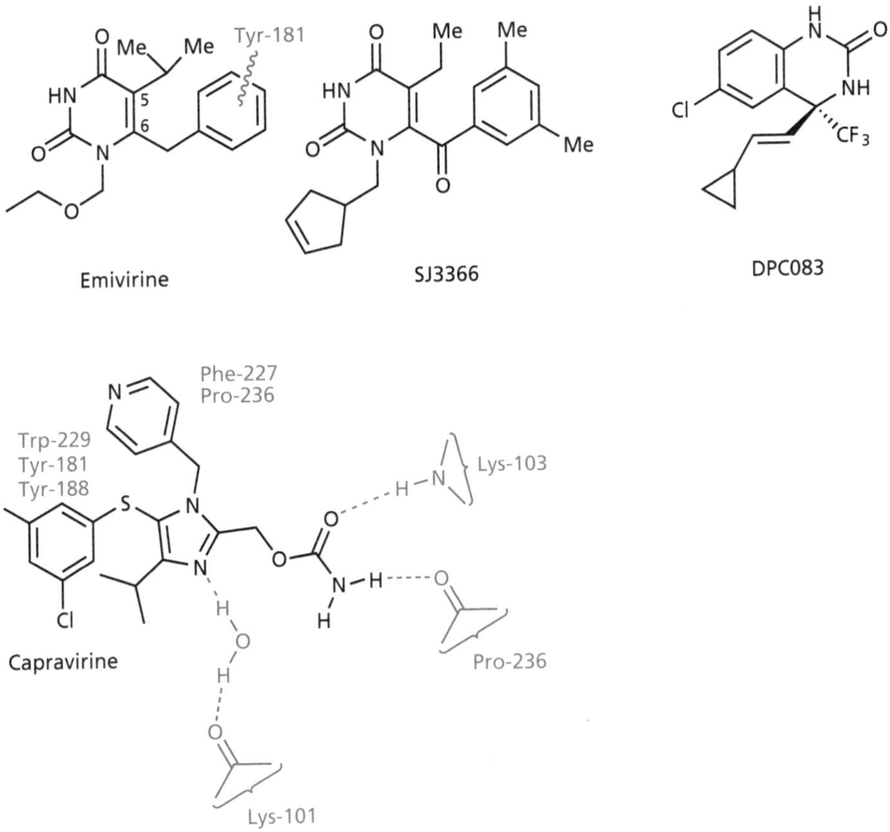

Emivirine

SJ3366

DPC083

Capravirine

Figure 17.16 Examples of third-generation NNRTIs.

Capravirine retains activity against mutants, including those with the K103N mutation. It has three hydrogen bonding interactions with the main chain of the protein active site, including one to Pro-236 which is not present in other inhibitors. It is suggested that hydrogen bonding to the main chain rather than to residues makes the molecule less vulnerable to mutations because binding to the polypeptide chain remains constant. The molecule is also quite flexible and this may allow it to adopt different orientations in mutated binding sites such that it can still bind. It is also larger than most other inhibitors and has potentially more interactions. This means that the loss of one interaction due to a mutation is less critical to the overall binding strength. So far, resistance to capravirine is found only after two mutations in the binding site. The chloro-substituted aromatic ring occupies the top of the binding pocket and makes more contact with the highly conserved Trp-229 indole side chain

than the aromatic rings of first generation compounds. Hydrophobic interactions also take place to the aromatic rings of Tyr-181 and Tyr-188. The pyridine ring interacts with the side chains of Phe-227 and Pro-236.

NRTIs generally have good oral bioavailability, are only minimally bound to plasma proteins, and are excreted through the kidneys. They act against both HIV-1 and HIV-2. NNRTIs are restricted to HIV-1 activity and are generally metabolized by the liver. They can interact with other drugs and bind more strongly to plasma proteins.

17.7.4 Protease inhibitors

In the mid 1990s, the use of X-ray crystallography and molecular modelling led to the structure-based design of a series of inhibitors which act on the viral enzyme **HIV protease**. Like the reverse transcriptase inhibitors, PIs have a short-term benefit when they are used alone, but resistance soon develops. Consequently, combination therapy is now the accepted method of treating HIV infections. When protease and reverse transcriptase inhibitors are used together, the antiviral activity is enhanced and viral resistance is slower to develop.

Unlike the reverse transcriptase inhibitors, the PIs are not prodrugs and do not need to be activated. Therefore, it is possible to use *in vitro* assays involving virally infected cells in order to test their antiviral activity. The protease enzyme can also be isolated, allowing enzyme assays to be carried out. In general, the latter are used to measure IC_{50} levels as a measure of how effective novel drugs are in inhibiting the protease enzyme. The IC_{50} is the concentration of drug required to inhibit the enzyme by 50%. Thus, the lower the IC_{50} value, the more potent the inhibitor. A good PI does not necessarily mean a good antiviral drug, however. In order to be effective, the drug has to cross the cell membrane of infected cells, and so *in vitro* whole-cell assays are often used alongside enzyme studies to check cell absorption. EC_{50} values are a measure of antiviral activity and represent the concentration of compounds required to inhibit 50% of the cytopathic effect of the virus in isolated lymphocytes. Another complication is the requirement for anti-HIV drugs to have a good oral bioavailability (i.e. to be orally active). This is a particular problem with the PIs. As we shall see, most PIs are designed from peptide lead compounds. Peptides are well known to have poor pharmacokinetic properties (i.e. poor absorption, metabolic susceptibility, rapid excretion, limited access to the central nervous system, and high plasma protein binding). This is due mainly to high molecular weight, poor water solubility, and susceptible peptide linkages. In the following examples, we will find that potent PIs were discovered relatively quickly, but that these had a high peptide character. Subsequent work was then needed to reduce the peptide character of these compounds in order to achieve high antiviral activity, alongside acceptable levels of oral bioavailability and half-life.

Clinically useful PIs are generally less well orally absorbed than reverse transcriptase inhibitors and are also susceptible to first pass metabolic reactions involving the cytochrome P450 isozyme (CYP3A4). This metabolism can result in drug–drug interactions with many of the other drugs given to AIDS patients to combat opportunistic diseases (e.g. rifabutin, ketoconazole, rifampin, terfenadine, astemizole, cisapride).

17.7.4.1 The HIV protease enzyme

The HIV protease enzyme (Fig. 17.17) is an example of an enzyme family called the **aspartyl proteases**— enzymes which catalyse the cleavage of peptide bonds and which contain an aspartic acid in the active site that is crucial to the catalytic mechanism. The enzyme is relatively small and can be obtained by synthesis. Alternatively, it can be cloned and expressed in fast-growing cells, then purified in large quantities. The enzyme is easily crystallized with or without an inhibitor bound to the active site, and this has meant that it has proved an ideal candidate for structure-based drug design. This involves the X-ray crystallographic study of enzyme–inhibitor complexes and the design of novel inhibitors based on those studies.

The HIV protease enzyme is a dimer made up of two identical protein units, each consisting of 99 amino acids. The active site is at the interface between the protein units and, like the overall dimer, it is symmetrical, with twofold rotational (C2) symmetry. The amino acids Asp-25, Thr-26, and Gly-27 from each monomer are located on the floor of the active site, and each monomer provides a flap to act as the ceiling. The enzyme has a broad substrate specificity and can cleave a variety of peptide bonds in viral polypeptides, but crucially it can cleave bonds between a proline residue and an aromatic residue (phenylalanine or tyrosine) (Fig. 17.18). The cleavage of a peptide bond next to proline is unusual and does not occur with mammalian proteases such as renin, pepsin, or cathepsin D, and so the chances are good of achieving selectivity against HIV protease over mammalian proteases. Moreover, the viral enzyme and its active site are symmetrical. This is not the case with mammalian proteases, again suggesting the possibility of drug selectivity.

There are eight binding subsites in the enzyme, four on each protein unit, located on either side of the catalytic region (Fig. 17.18). These subsites accept the amino acid residues of the substrate and are numbered S1–S4 on one side and S1′–S4′ on the other side. The relevant residues on the substrate are numbered P1–P4 and P1′–P4′ (Fig. 17.19). The nitrogen and oxygen of each peptide bond in the substrate backbone is involved in a hydrogen bonding interaction with the enzyme, as shown in Fig. 17.19. This includes hydrogen bonding interactions between two of the carbonyl oxygens on the

'Flaps'

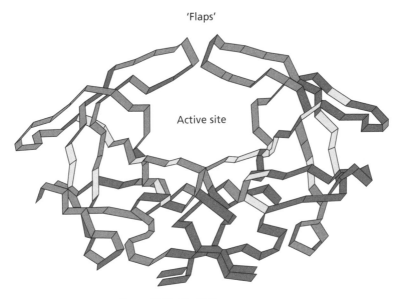

Figure 17.17 The HIV protease enzyme.

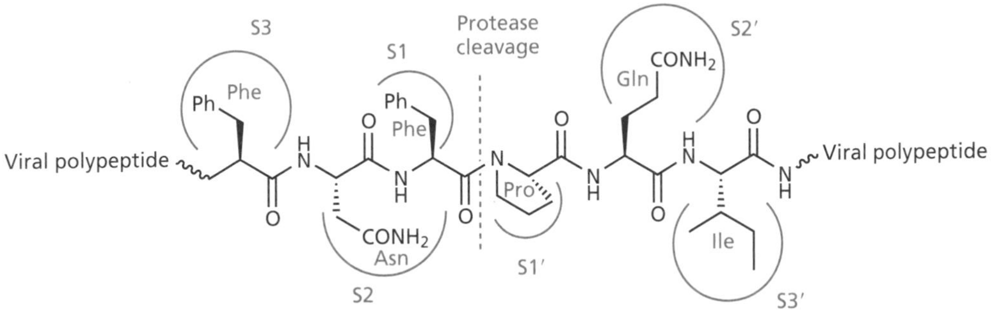

Figure 17.18 Protease catalysed cleavage of an aromatic-proline peptide bond
(six of the eight binding subsites are shown).

substrate and a water molecule which acts as hydrogen bonding bridge to two isoleucine NH groups on the enzyme flaps. This hydrogen bonding network has the effect of closing the flaps over the active site once the substrate is bound.

There are two variants of HIV protease. The protease enzyme for HIV-2 shares 50% sequence identity with HIV-1. The greatest variation occurs outwith the active site, and so inhibitors are found to bind similarly to both enzymes.

The aspartic acids Asp-25 and Asp-25' are involved in the catalytic mechanism and are on the floor of the active site, each contributed from one of the protein subunits. The carboxylate residues of these aspartates and a bridging water molecule are involved in the mechanism by which the substrate's peptide bond is hydrolysed (Fig. 17.20).

17.7.4.2 Design of HIV protease inhibitors

A similar hydrolytic mechanism to that shown in Fig. 17.20 takes place for a mammalian aspartyl protease called **renin**. This enzyme was extensively studied before the discovery of HIV protease, and a variety of renin inhibitors were designed as antihypertensive agents. These agents act as **transition-state inhibitors**, and many of the discoveries and strategies resulting

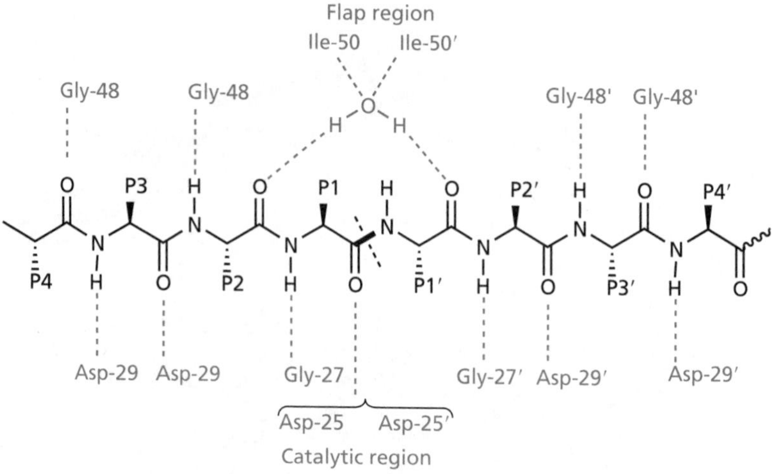

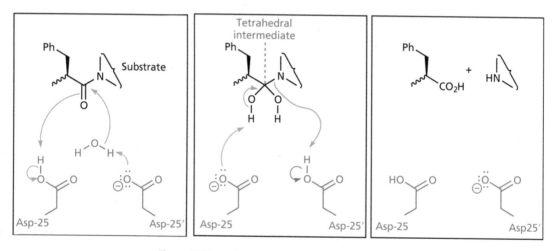

Figure 17.19 Interactions involving the substrate peptide backbone.

Figure 17.20 Mechanism of the protease-catalysed reaction.

from the development of renin inhibitors were adapted to the design of HIV PIs.

Transition-state inhibitors are designed to mimic the transition state of an enzyme catalysed reaction. The advantage of this approach is that the transition state is likely to be bound to the active site more strongly than either the substrate or product. Therefore, inhibitors resembling the transition state are also likely to be bound more strongly. In the case of the protease-catalysed reaction, the transition state resembles the tetrahedral intermediate shown in Fig. 17.20. Since such structures are inherently

unstable, it is necessary to design an inhibitor which contains a **transition-state isostere**. Such an isostere would have a tetrahedral centre to mimic the tetrahedral centre of the transition state, yet be stable to hydrolysis. Fortunately, several such isosteres had already been developed in the design of renin inhibitors (Fig. 17.21). Thus, a large number of structures were synthesized incorporating these isosteres, with the hydroxyethylamine isostere proving particularly effective. This isostere has a hydroxyl group which mimics one of the hydroxyl groups of the tetrahedral intermediate and binds to the aspartate residues in the

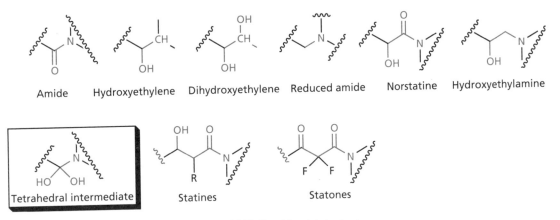

Figure 17.21 Transition state isosteres.

active site. It is also found that the stereochemistry of this group is important to activity, with the *R*-configuration generally being preferred. The nature of the P1′ group affects the stereochemical preference of the hydroxyl group.

Having identified suitable transition-state isosteres, inhibitors were designed based on the enzyme's natural peptide substrates, since these contain amino acid residues which fit the eight subsites and allow a good binding interaction between the substrate and the enzyme. In theory, it might make sense to design inhibitors such that all eight subsites are filled, to allow stronger interactions. However, this leads to structures with a high molecular weight and consequently poor oral bioavailability.

Therefore, most of the PIs were designed to have a core unit spanning the S1 to S1′ subsites. Further substituents were then added at either end to fit into the S2/S3 and S2′/S3′ subsites. Early inhibitors such as saquinavir (see below) have amino acid residues at P2 and P2′. Unfortunately, these compounds have a high molecular weight and a high peptide character leading to poor pharmacokinetic properties. More recent inhibitors contain a variety of novel P2 and P2′ groups which were designed to reduce the molecular weight of the compound as well as its peptide character, in order to increase aqueous solubility and oral bioavailability. The S2 and S2′ subsites of the protease enzyme appear to contain both polar (Asp-29, Asp-30) and hydrophobic (Val-32, Ile-50, Ile-84) amino acids, allowing the design of drugs containing hydrophobic P2 groups capable of hydrogen bonding. It has also been possible

to design a P1 group that can span both the S1 and S3 subsites, allowing the removal of the P3 moiety thus lowering the molecular weight. The P2 group is usually attached to P1 by an acyl link, because the carbonyl oxygen concerned acts as an important hydrogen bond acceptor to the bridging water molecule described previously (Fig. 17.19).

We shall now look at how individual PIs were discovered and developed.

17.7.4.3 Saquinavir

Saquinavir was developed by Roche, and as the first PI to reach the market it serves as the benchmark for all other PIs. The design of saquinavir started by considering a viral polypeptide substrate (pol, see section 17.7.1) and identifying a region of the polypeptide which contains a phenylalanine–proline peptide link. A pentapeptide sequence Leu[165]-Asn-Phe-Pro-Ile[169] was identified and served as the basis for inhibitor design. The peptide link normally hydrolysed is between Phe-167 and Pro-168, and so this link was replaced by a hydroxyethylamine transition-state isostere to give a structure which successfully inhibited the enzyme (Fig. 17.22). The amino acid residues for Leu-Asn-Phe-Pro-Ile are retained in this structure and bind to the five subsites S3–S2′. Despite that, enzyme inhibition is relatively weak. The compound also has a high molecular weight and high peptide-like character, both of which are detrimental to oral bioavailability.

Consequently, the Roche team set out to identify a smaller inhibitor, starting from the simplest possible substrate for the enzyme—the dipeptide Phe-Pro

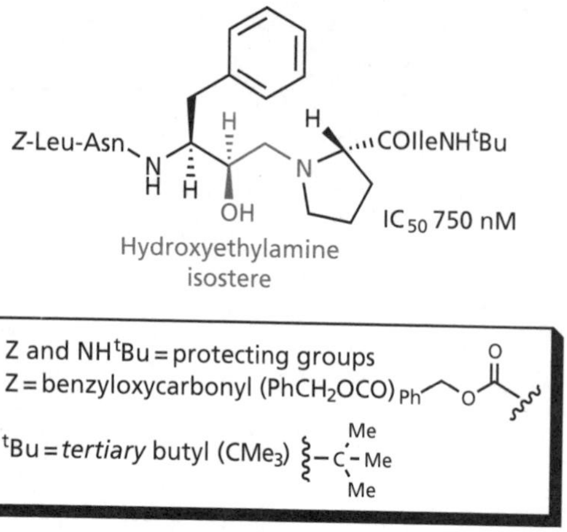

Figure 17.22 Pentapeptide analogue incorporating a hydroxyethylamine transition state isostere.

(Fig. 17.23). The peptide link was replaced by the hydroxylamine transition-state isostere and the resulting N- and C-protected structure (I) was tested and found to have weak inhibitory activity. The inclusion of an asparagine group (structure II) to occupy the S2 subsite resulted in a 40-fold increase in activity, and a level of activity greater than the pentapeptide analogue (Fig. 17.22). This might seem an unexpected result, as the latter occupies more binding subsites. However, it has been found that the crucial interaction of inhibitors is in the core region S2–S2'. If the addition of extra groups designed to bind to other subsites weakens the interaction to the core subsites, it can lead to an overall drop in activity. This was supported by the fact that the inclusion of leucine in structure II

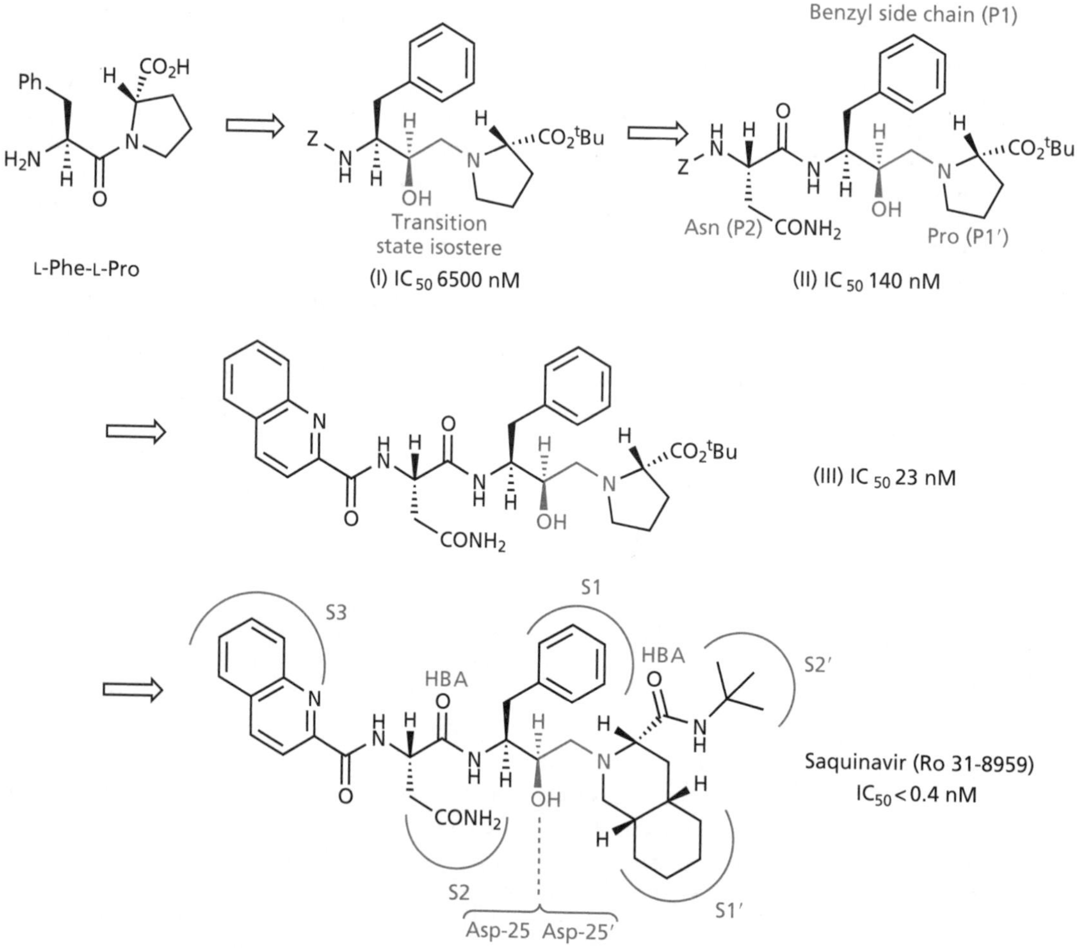

Figure 17.23 Development of saquinavir (Z = PhCH₂OCO).

resulted in a drop in activity, despite the fact that leucine can occupy the S3 subsite.

Structure II was adopted as the new lead compound and the residues P1 and P2 were varied to find the optimum groups for the S1 and S2 subsites. As it turned out, the benzyl group and the asparagine side chain were already the optimum groups. An X-ray crystallographic study of the enzyme–inhibitor complex was carried out and revealed that the protecting group (Z) occupied the S3 subsite, which proved to be a large hydrophobic pocket. As a result, it was possible to replace this protecting group with a larger quinoline ring system to occupy the subsite more fully and lead to a sixfold increase in activity (structure III). Variations were also carried out on the carboxyl half of the molecule. Proline fits into the S1′ pocket but it was found that it could be replaced by a bulkier decahydroisoquinoline ring system. The t-butyl ester protecting group was found to occupy the S2′ subsite and could be replaced by a t-butylamide group which proved more stable in animal studies. The resulting structure was saquinavir having a further 60-fold increase in activity. The R-stereochemistry of the transition-state hydroxyl group is essential. If the configuration is S, all activity is lost.

X-ray crystallography of the enzyme–saquinavir complex (Figs. 17.23 and 17.24) confirmed the following:

- The substituents on the drug occupy the five subsites S3–S2′.

- The position of the t-butylamine nitrogen is such that further substituents on the nitrogen would be incapable of reaching the S3′ subsite.

- There are hydrogen bonding interactions between the hydroxyl group of the hydroxyethylamine moiety and the catalytic aspartates (Asp-25 and Asp-25′).

- The carbonyl groups on either side of the transition-state isostere act as hydrogen bond acceptors to a bridging water molecule. The latter forms hydrogen bonds to the isoleucine groups in the enzyme's flap region in a similar manner to that shown in Fig. 17.19.

Saquinivir shows a 100-fold selectivity for both HIV-1 and HIV-2 proteases over human proteases. Clinical trials were carried out in early 1991 and the drug reached the market in 1995. Approximately 45% of patients develop clinical resistance to the drug over a 1-year period, but resistance can be delayed if it is given in combination with reverse transcriptase inhibitors. The oral bioavailability of saquinavir is only 4% in animal studies, although this is improved if the drug is taken with meals. The compound is also highly bound to plasma proteins (98%). As a result, the drug has to be taken in high doses to maintain therapeutically high plasma levels. Various efforts have been made to design simpler analogues of saquinavir which have lower molecular weight, less peptide character, and consequently better oral bioavailability.

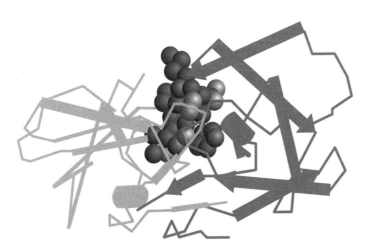

Figure 17.24 Saquinavir bound to the active site of HIV protease.

17.7.4.4 Ritonavir and lopinavir

Ritonavir was developed by Abbott Pharmaceuticals to take advantage of the symmetrical properties of the protease enzyme and its active site. Since the active site has C2 symmetry, a substrate is capable of binding 'left to right' or 'right to left' as the binding subsites S1–S4 are identical to subsites S1′–S4′. This implies that it should be possible to design inhibitors having C2 symmetry. This should have several advantages. First, symmetrical inhibitors should show greater selectivity for the viral protease over mammalian aspartyl proteases, since the active sites of the latter are not symmetrical. Second, symmetrical molecules might be less recognizable to peptidases, resulting in improved oral bioavailability. Third, the development of saquinavir showed that a benzyl residue was the optimum binding group for the S1 subsite. Since the S1′ subsite is identical to S1, a symmetrical inhibitor having benzyl groups fitting both S1 and S1′ subsites should bind more strongly and have improved activity. This argument could also be extended for the binding groups fitting the S2/S2′ subsites and so on.

Since there was no lead compound having C2 symmetry, one had to be designed which not only had the necessary C2 symmetry, but which also matched up with the C2 symmetry of the active site once it was bound. The first lead compound was designed by considering the tetrahedral reaction intermediate derived from the natural substrate. It was assumed that the axis of C2 symmetry for the active site passed through the reaction centre of this intermediate (Fig. 17.25). Since the benzyl group was known to be optimum for binding to the S1 subsite, the left-hand portion of the molecule was retained and the

right-hand portion was deleted. The left-hand moiety was then rotated such that two benzyl residues were present in the correct orientation for C2 symmetry. The resulting geminal diol is inherently unstable, so one of the alcohols was removed leading to the simplest target alcohol (I; R=H). In order to check whether this target molecule would match the C2 symmetry of the active site when bound, a molecular modelling experiment was carried out whereby the inhibitor was constructed in the active site. The results of this analysis were favourable and so the target alcohol was synthesized. Although it only had weak activity as an enzyme inhibitor, and was inactive against the virus *in vitro*, it still served as a lead compound designed by *de novo* techniques.

The next stage was to extend the molecule to take advantage of the S2 and S2′ subsites. A variety of structures was synthesized and tested revealing vastly improved activity when valine was added, and further improvement when the valines had N-protecting groups (A 74704; Fig. 17.26). A 74704 also showed *in vitro* activity against the HIV virus itself, and was resistant to proteolytic degradation. The structure was co-crystallized with recombinant protease enzyme and studied by X-ray crystallography to reveal a symmetrical pattern of hydrogen bonding between the inhibitor and the enzyme (Fig. 17.27). It was also found that a water molecule (Wat-301) still acted as a hydrogen bonding bridge between the carbonyl groups of P2 and P2′, and the NH groups of Ile-50 and Ile-50′ on the flaps of the enzyme. The C2 symmetry axes of the inhibitor and the active site passed within 0.2 Å of each other and deviated by an angle of only 6°, demonstrating that the design philosophy had been valid.

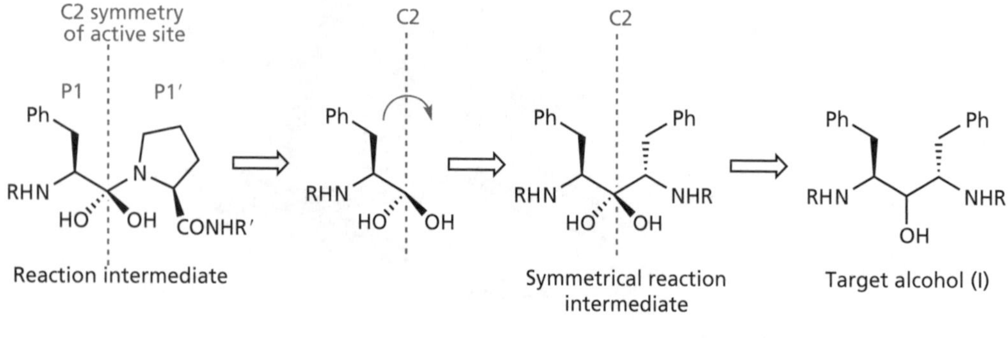

Figure 17.25 Design of a symmetrical lead compound acting as an inhibitor.

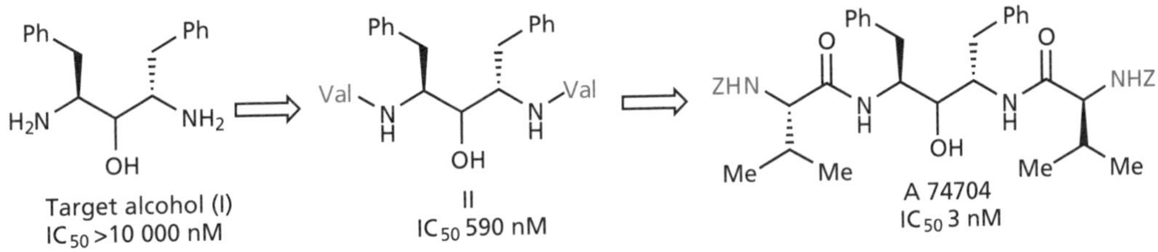

Figure 17.26 Development of A 74704 (Z = PhCH$_2$OCO).

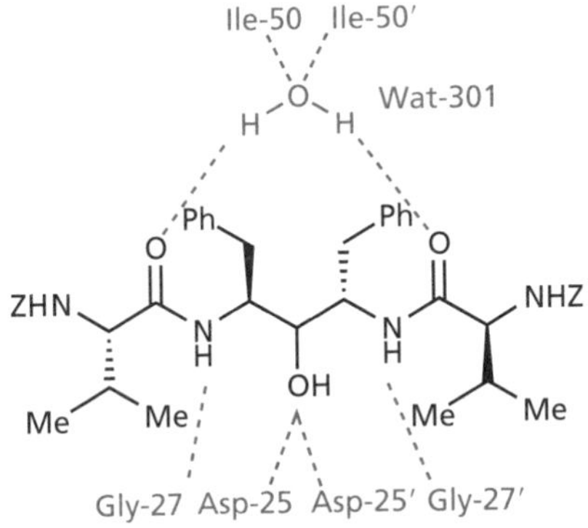

Figure 17.27 Binding interactions between the active site and the backbone of the A 74704 (Z = PhCH$_2$OCO).

Further analysis of the crystal structure suggested that the NH groups on the inhibitor were binding to Gly-27 and Gly-27′ but were too close to each other to allow optimum hydrogen bonding. To address this, it was decided to design symmetrical inhibitors where the relevant NH groups would be separated by an extra bond. In order to achieve this, the axis of C2 symmetry was placed through the centre of the susceptible bond. The same exercise described above was then carried out, leading to a target diol (Fig. 17.28).

Diol structures analogous to the alcohols previously described were synthesized. Curiously, it was found that the absolute configuration of the diol centres had little effect on activity, and in general, the activity of the diols was better than the corresponding alcohols. For example, the diol equivalent of A 74704 (Fig. 17.29) had a 10-fold better level of activity. Unfortunately, this compound had poor water solubility. A crystal structure of the enzyme–inhibitor complex was studied, which revealed that the terminal portions of the molecule were exposed to solvation. This meant that more polar groups could be added at those positions without affecting binding. Consequently, the terminal

phenyl groups were replaced by more polar pyridine rings. The urethane groups near the terminals were also replaced by urea groups, leading to A 77003 which had improved water solubility. Unfortunately, the oral bioavailability was still unsatisfactory and so the structure entered clinical trials as an intravenous antiviral agent rather than an oral one.

Modelling studies of how A 77003 might bind to the active site suggested two possible binding modes, one where each of the diol hydroxyl groups was symmetrically hydrogen bonding to each of the aspartate residues, and an asymmetric binding where only one of the hydroxyl groups hydrogen bonded to both aspartate groups. To investigate this further, X-ray crystallography was carried out on the enzyme–inhibitor complex revealing that asymmetric binding was taking place whereby the (R)-OH took part in hydrogen bonding with both aspartate residues, and the (S)-OH was only able to form a single hydrogen bonding interaction. This analysis also showed that the increased separation of the amide NHs failed to improve the geometry of the hydrogen bonding interactions with Gly-27 and Gly-27′. Thus, the improved activity of the diols over the alcohols must have arisen for reasons other than those expected. Results such as this are not totally unexpected when carrying out *de novo* design, since molecules can be flexible enough to bind differently from the manner predicted. The better activity for the diols may in fact be due to better binding of the P′ groups to the S′ subsites.

The fact that the (S)-hydroxyl group makes only one hydrogen bonding interaction suggested that it might be worth removing it, as the energy gained from only one hydrogen bonding interaction might be less than the energy required to desolvate the hydroxyl group of water before binding. This led to A 78791, which had improved activity and was shown by X-ray crystallography to bind in the same manner as A 77003.

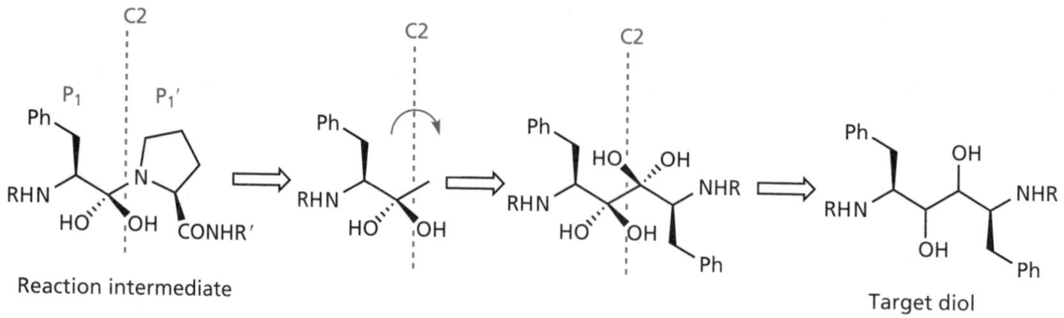

Figure 17.28 Design of a symmetrical diol inhibitor.

A study was then carried out to investigate what effect variations of molecular size, aqueous solubility, and hydrogen bonding would have on the pharmacokinetics and activity of these agents. This led to A 80987, where the P2′ valine was removed and the urea groups near the terminal were replaced by urethane groups. In general, it was found that the presence of N-methylureas was good for water solubility and bioavailability, whereas the presence of carbamates was good for plasma half-life and overall potency. Thus, it was possible to fine-tune these properties by a suitable choice of group at either end of the molecule.

Despite being smaller, A 80987 retained activity and had improved oral bioavailability. However, it had a relative short plasma lifetime, bound strongly to plasma proteins, and it was difficult to maintain therapeutically high levels. Metabolic studies then showed that A 80987 was N-oxidized at either or both pyridine rings, and that the resulting metabolites were excreted mainly in bile. In an attempt to counter this, both steric and electronic strategies were tried. First, alkyl groups were placed on the pyridine ring at the vacant position *ortho* to the nitrogen. These were intended to act as a steric shield, but proved ineffective in preventing metabolism. It was then proposed that metabolism might be reduced if the pyridine rings were less electron rich and so methoxy or amino substituents were added as electron-withdrawing groups. However, this too failed to prevent metabolism. Finally, the pyridine ring at P3 was replaced by a variety of heterocycles in an attempt to find a different ring system which would act as a bioisostere but would be less susceptible to metabolism. The best results were obtained using the more electron-deficient 4-thiazolyl ring. Although water solubility decreased, it could be restored by reintroducing an N-methylurea group in place of one of the urethanes. Further improvements in activity were obtained by placing hydrophobic alkyl groups at the 2 position of the thiazole ring (P3), and by subsequently altering the position of the hydroxyl group in the transition-state isostere. This led to A 83962, which showed an 8-fold increase in potency over A 80987.

Attention now turned to the pyridine group at P2′. Replacement with a 5-thiazolyl group proved beneficial, indicating that a hydrogen bonding interaction was taking place between the thiazolyl N and Asp-30 (specifically the NH of the peptide backbone). This matched a similar hydrogen bonding interaction involving the pyridine N in A 80987.

The improved bioavailability is due principally to better metabolic stability (20 times more stable than A 80987) and it was possible to get therapeutic plasma levels of the drug lasting 24 hours following oral administration.

Ritonavir reached the market in 1996. It is active against both HIV-1 and HIV-2 proteases and shows selectivity for HIV proteases over mammalian proteases. Despite a high molecular weight and the presence of amide bonds, it has better bioavailability than many other PIs. This is because the compound is a potent inhibitor of the cytochrome enzyme CYP3A4 and thus shuts down its own metabolism. Care has to be taken when drugs affected by this metabolism are taken alongside ritonavir, and doses should be adjusted accordingly. On the other hand, this property can be useful in combination therapy with other PIs which are normally metabolized by this enzyme (e.g. saquinavir, indinavir, nelfinavir, amprenavir), as their lifetime and plasma levels may be increased. Ritonavir is highly protein bound (99%).

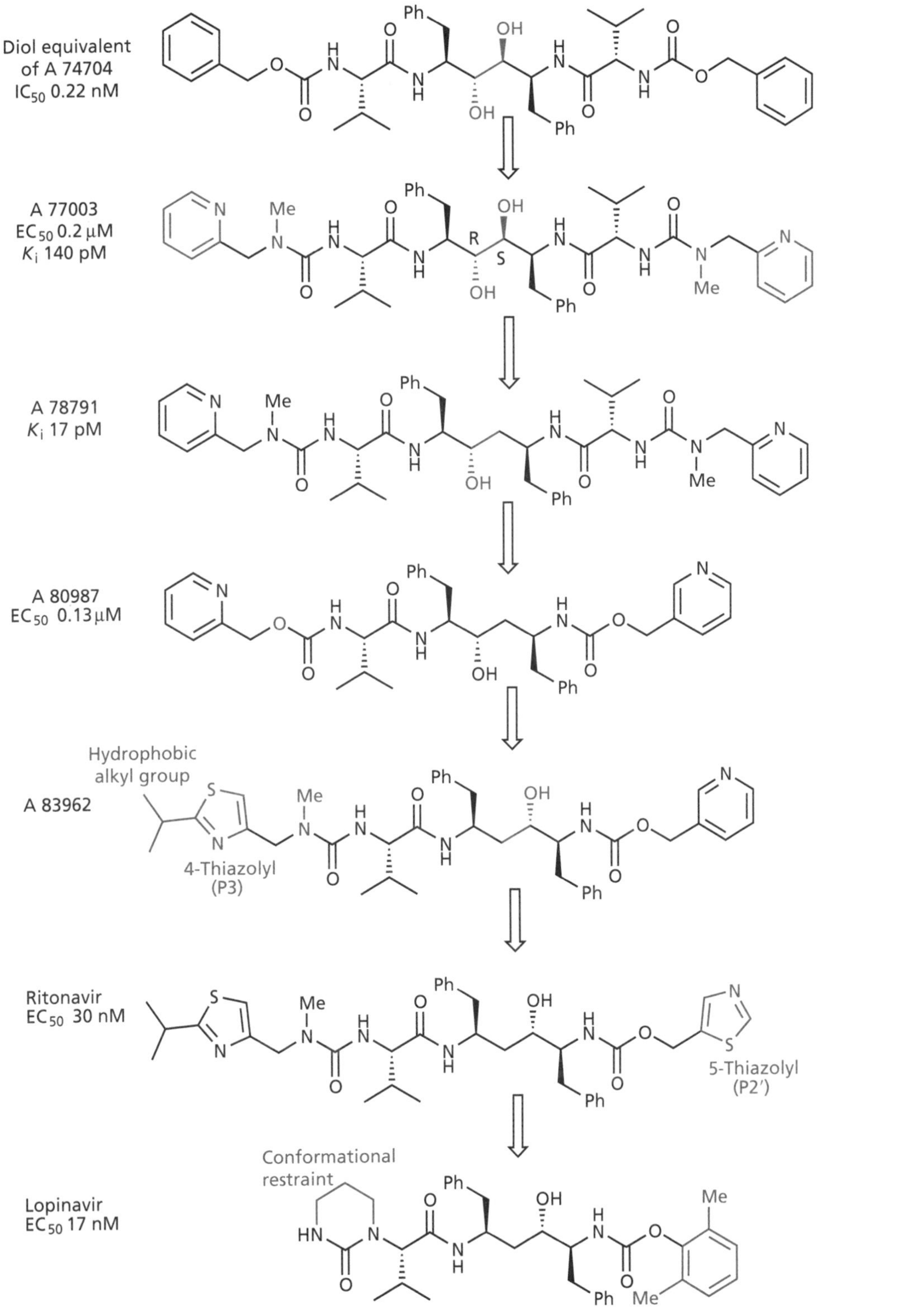

Figure 17.29 Development of ritonavir (ABT 538) and lopinavir (ABT 378).

Resistant strains of the virus have developed when ritonavir is used on its own. These arise from a mutation of valine at position 82 of the enzyme to either alanine, threonine or phenylalanine. X-ray crystallography shows that there is an important hydrophobic interaction between the isopropyl substituent on the P3 thiazolyl group of ritonavir and the isopropyl side chain of Val-82 which is lost as a result of this mutation. Further drug development led to lopinavir (Fig. 17.29) where the P3 thiazolyl group was removed and a cyclic urea group was incorporated to introduce conformational constraint. This allowed enhanced hydrogen bonding interactions with the S2 subsite, which balanced out the loss of binding due to the loss of the thiazolyl group. As this structure does not have any interactions with Val-82, it is active against the ritonavir resistant strain.

Lopinavir and ritonavir are marketed together as a single capsule combination called Kaletra. Each capsule contains 133 mg of lopinavir and 33 mg of ritonavir, with the latter serving as a cytochrome P450 inhibitor to increase the level of lopinavir present in the blood supply.

17.7.4.5 Indinavir

The design of indinavir included an interesting hybridization strategy (Fig. 17.30). Merck had designed a potent PI that included a hydroxyethylene transition-state isostere (L 685434). Unfortunately, it suffered from poor bioavailability and liver toxicity. At this point, the Merck workers concluded that it might be possible to take advantage of the symmetrical nature of the active site. Since the S and S' subsites are equivalent, it should be possible to combine half of one PI with half of another to give a structurally distinct hybrid inhibitor. A modelling study was carried out to check the hypothesis and the Merck team decided to combine the P' half of L 685434 with the P' half

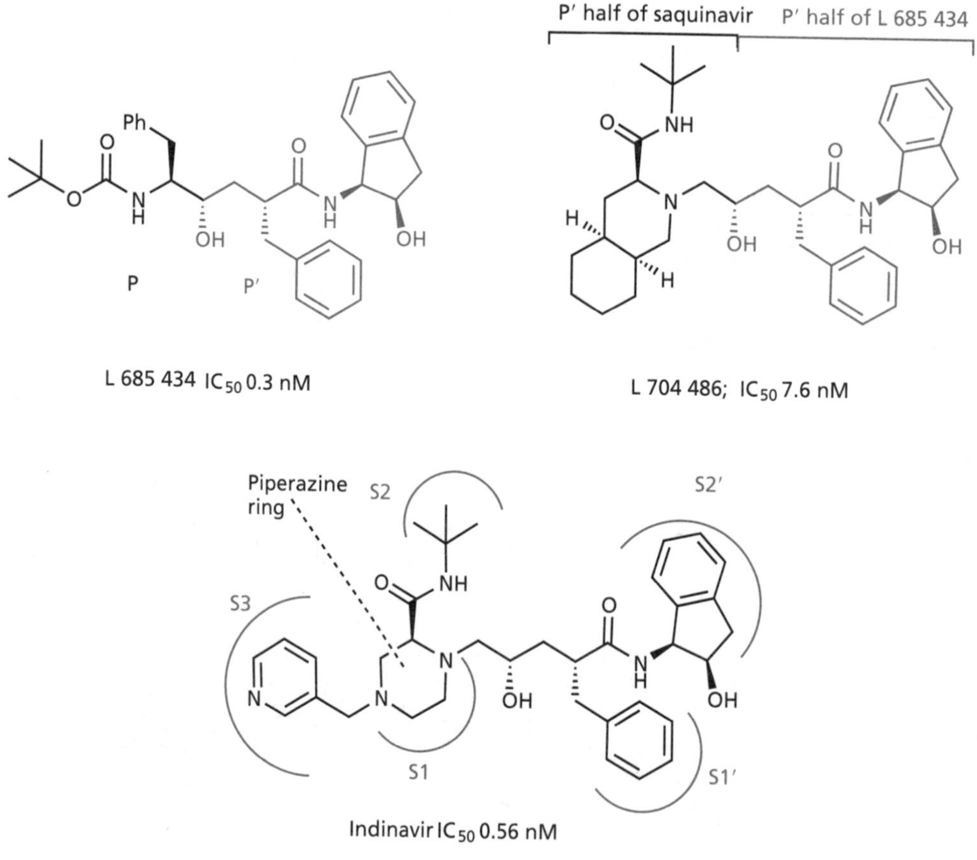

L 685 434 IC$_{50}$ 0.3 nM

L 704 486; IC$_{50}$ 7.6 nM

Indinavir IC$_{50}$ 0.56 nM

Figure 17.30 Development of indinavir.

of saquinavir. The P′ moiety of saquinavir was chosen for its solubility-enhancing potential, and the P′ moiety of L 685434 is attractive for its lack of peptide character. The resulting hybrid structure (L 704486) was less active as an inhibitor but was still potent. Moreover, the presence of the decahydroisoquinoline ring system resulted in better water solubility and oral bioavailability (15%), as intended.

Further modifications were aimed at improving binding interactions, aqueous solubility, and oral bioavailability. The decahydroisoquinoline ring was replaced by a piperazine ring, the additional nitrogen helping to improve aqueous solubility and oral bioavailability. It also made it possible to add a pyridine substituent to access the S3 subsite and improve binding. This resulted in indinavir, which reached the market in 1996. It has better oral bioavailability than saquinavir and is less highly bound to plasma proteins (60%).

17.7.4.6 Nelfinavir

The development of nelfinavir was based on work carried out by the Lilly company, aimed at reducing the molecular weight and peptide character of PIs. Structure-based drug design had led them to develop AG1254 (Fig. 17.31), which contains an extended substituent at P1 capable of spanning the S1 and S3 subsites of the enzyme and binding to both subsites. This did away with the need for a separate P3 group and allowed the design of compounds with a lower molecular weight. They also designed a new P2 group to replace an asparagine residue which had been present in their lead compound. This group was designed to bind effectively to the S2 subsite and, since it was different from any amino acid residue, the peptide character of the compound was reduced. Unfortunately, the antiviral activity of AG1254 was not sufficiently high and the compound had poor aqueous solubility.

The company decided to switch direction and see what effect their newly designed substituents would have if they were incorporated into saquinavir, and this led ultimately to nelfinavir. A crystal structure of nelfinavir bound to the enzyme showed that the molecule is bound in an extended conformation where the binding interactions involving the molecular backbone are similar to saquinavir. A tightly bound water molecule serves as a hydrogen bonding bridge between the two amide carbonyls of the inhibitor and

the flap region of the enzyme in a similar manner to other enzyme–inhibitor complexes. The crystal structure also showed that the S-phenyl group resides mainly in the S1 site and partially extends into the S3 site. The substituted benzamide occupies the S2 pocket with the methyl substituent interacting with valine and isoleucine through van der Waals interactions, and the phenol interacting with Asp-30 through hydrogen bonding.

Nelfinavir was marketed in 1997 and is used as part of a four-drug combination therapy. Like indinavir and ritonavir, nelfinavir is more potent than saquinavir because of its better pharmacokinetic profile. Compared to saquinavir, it has a lower molecular

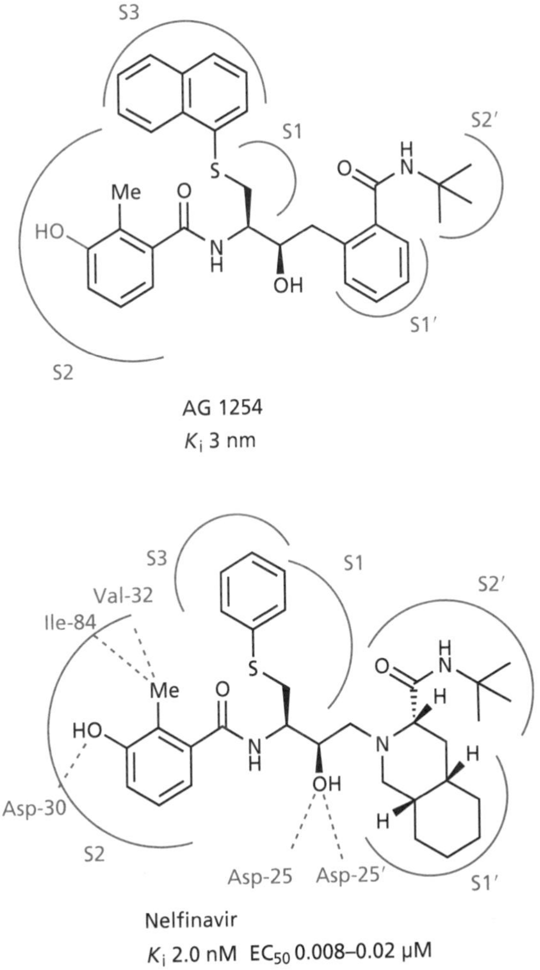

AG 1254
K_i 3 nm

Nelfinavir
K_i 2.0 nM   EC$_{50}$ 0.008–0.02 µM

Figure 17.31 AG 1254 and nelfinavir.

weight and log *P* and an enhanced aqueous solubility, resulting in enhanced oral bioavailability. It can inhibit the metabolic enzyme CYP3A4 and thus affects the plasma levels of other drugs metabolized by this enzyme. It is 98% bound to plasma proteins.

17.7.4.7 Palinavir

Palinavir (Fig. 17.32) is a highly potent and specific inhibitor of HIV-1 and HIV-2 proteases. The left-hand or P half of the molecule is the same as saquinavir and the molecule contains the same hydroxyethylamine transition-state mimic. The right-hand (P′) side is different, and was designed using the same kind of extension strategy used in nelfinavir. In this case, the P1′ substituent was extended to occupy the S1′ and S3′ subsites. This was achieved by replacing the original proline group at P1′ with 4-hydroxypipecolinic acid and adding a pyridine-containing substituent to access the S3′ subsite.

The crystal structure of the enzyme–inhibitor complex shows that the binding pockets S3–S3′ are all occupied. Two carbonyl groups interact via the bridging water molecule to isoleucines in the enzyme flaps. The hydroxyl group interacts with both catalytic aspartate residues. Finally, the oxygen atoms and nitrogen atoms of all the amides are capable of hydrogen bonding to complementary groups in the active site. Work is currently in progress to simplify palinavir by introducing a single group that will span two binding subsites, thus allowing the removal of the P3 binding group.

17.7.4.8 Amprenavir

Amprenavir (Fig. 17.33) was designed by Vertex Pharmaceuticals as a non-peptide-like PI using saquinavir as the lead compound. Saquinavir suffers from having a high molecular weight and a high peptide character, both of which are detrimental to oral bioavailability. Therefore, it was decided to design a simpler analogue with a lower molecular weight and less peptide character, but retaining good activity. First, the decahydroisoquinoline group in saquinavir was replaced by an isobutyl sulfonamide group to give structure I. This also had the advantage of reducing the number of asymmetric centres from six to three, allowing easier synthesis of analogues. Further simplification and reduction of peptide character was carried out by replacing the P2 and P3 groups with a tetrahydrofuran (THF) carbamate which had been previously found by Merck to be a good binding group for the S2 subsite. Finally, an amino group was introduced on the phenylsulfonamide group to increase water solubility and to enhance oral absorption.

Amprenavir was licensed to GlaxoWellcome and was approved in 1999. It is reasonably specific for the viral protease relative to mammalian proteases, and is about 90% protein bound. It has good oral bioavailability (40–70% in animal studies). Further work has shown that a fused bis-tetrahydrofuryl ring system is a better binding group for the S2 pocket than a single THF ring, because of extra hydrogen bonding interactions involving the THF oxygens.

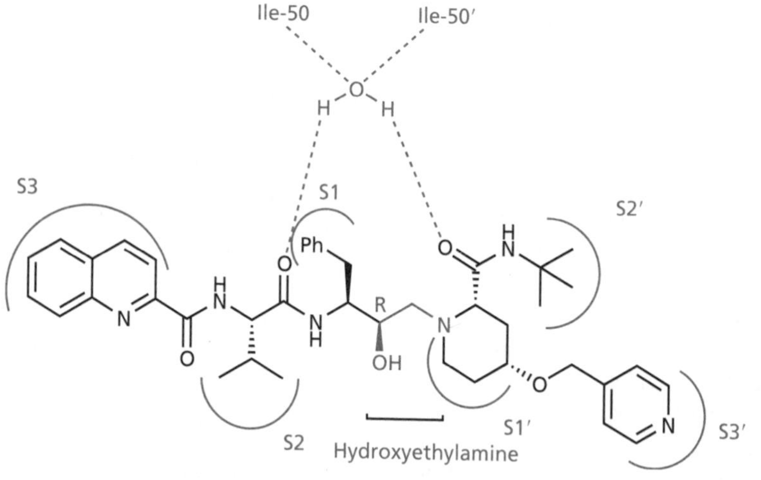

Figure 17.32 Palinavir and binding interactions.

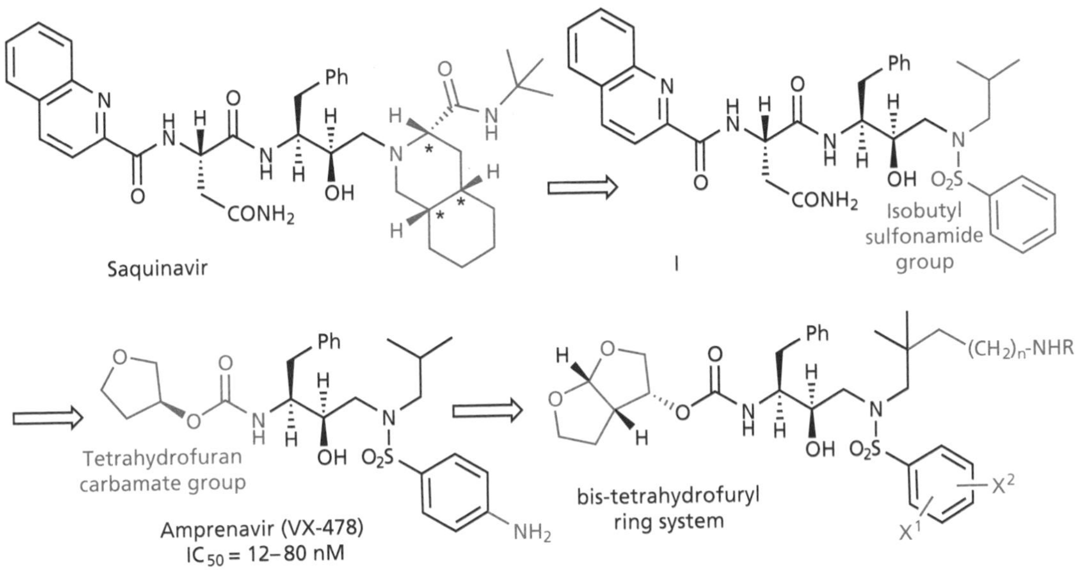

Figure 17.33 Development of amprenavir and further developments.

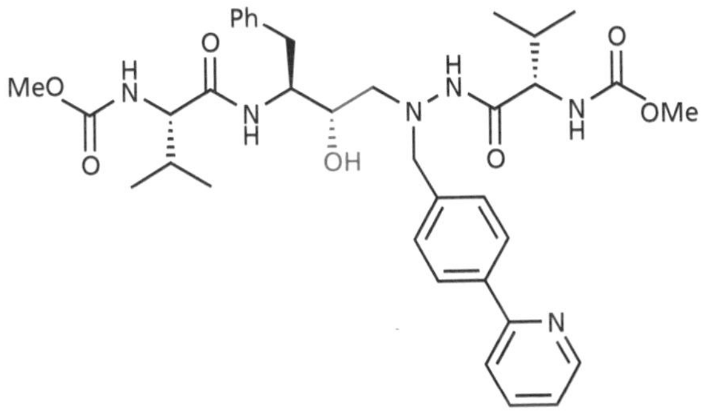

Figure 17.34 Atazanavir.

d(P-thio)(T-C-T-T-C-C-T-C-T-C-T-C-T-A-C-C-C-A-C-G-C-T-C-T-C)

Figure 17.35 Trecovirsen.

17.7.4.9 Atazanavir

Atazanavir (Fig. 17.34) was approved in June 2003 as the first once-daily HIV-1 PI to be used as part of a combination therapy. It is similar to the early compounds leading towards ritonavir.

17.7.5 Inhibitors of other targets

Antisense agents are being developed to block the production of the HIV protein Tat, which is needed for the transcription of other HIV genes. **Trecovirsen** (Fig. 17.35) is a phosphorothioate oligonucleotide containing 25 nucleotides and has been designed to hybridize with the mRNA derived from the HIV gene *gag*, to prevent its translation into HIV proteins. It was withdrawn from clinical trials due to toxicity, but a similar oligonucleotide (**GEM92**) with increased stability is currently in clinical trials.

Other agents under study for the treatment of HIV include integrase inhibitors and cell entry inhibitors. Blocking entry of a virus into a host cell is particularly desirable, as it is so is early in the life cycle. **Enfuvirtide** was approved in March 2003 as the first member of a new class of fusion inhibitors. It is a polypeptide consisting of 36 amino acids which matches the C-terminal end of the viral protein gp41. It works by forming an α-helix and binding to a group of three similar α-helices belonging to the gp41 protein. This association prevents the process by which the virus enters the host cell. In order to bring about fusion, the gp41 protein anchors the virus to the cell membrane of the host cell. It then undergoes a conformational change where it builds a grouping of six helices using the three already present as the focus for that grouping (Fig. 17.36). This pulls the membranes of the virion and the host cell together so that they can fuse. By binding to the group of three helices, enfuvirtide

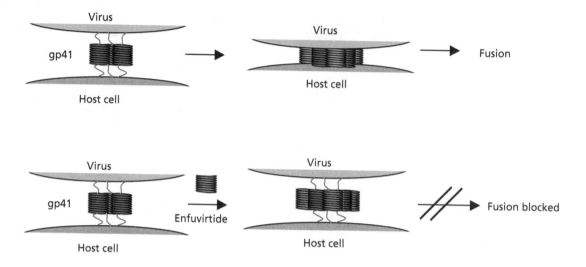

Figure 17.36 Enfuvirtide as a fusion inhibitor.

JM3100

N-Butyldeoxynojirimycin

Figure 17.37 Agents that inhibit cell entry.

blocks formation of the required hexamer and prevents fusion.

The manufacture of enfuvirtide involves 106 steps, which makes it expensive and may limit its use. A smaller compound (**BMS 378806**) is being investigated which binds to gp120 and prevents the initial binding of the virus to CD4 on the cell surface.

***N*-Butyldeoxynojirimycin** is a carbohydrate that inhibits **glycosidases**—enzymes that catalyse the trimming of carbohydrate moieties which are linked to viral proteins. If this process is inhibited, too many

carbohydrate groups end up attached to a protein, resulting in the protein adopting a different conformation. It is thought that the gp120 protein is affected in this way and cannot be peeled away as described in section 17.7.1 to reveal the gp41 protein.

Bicyclams such as **JM 3100** (Fig. 17.37) block the CCR5 chemokine receptor and are under investigation.

KEY POINTS

- HIV is a retrovirus containing RNA as its genetic material, and is responsible for AIDS.

- The two main viral targets for anti-HIV drugs are the enzymes reverse transcriptase and protease. Combination therapy is the favoured treatment, but there is a need to develop drugs which are effective against a third target.

- The potency and safety demands for anti-HIV drugs are high, as they are likely to be used for the lifetime of the patient.

- Reverse transcriptase is a DNA polymerase which catalyses the conversion of single-stranded RNA to double-stranded DNA. No such biochemical process occurs in normal cells.

- Nucleoside reverse transcriptase inhibitors are prodrugs that are converted by cellular enzymes to active triphosphates which act as enzyme inhibitors and chain terminators.

- Non-nucleoside reverse transcriptase inhibitors act as enzyme inhibitors by binding to an allosteric binding site.

- The protease enzyme is a symmetrical dimeric structure consisting of two identical protein subunits. An aspartic acid residue from each subunit is involved in the catalytic mechanism.

- The protease enzyme is distinct from mammalian proteases in being symmetrical and being able to catalyse the cleavage of peptide bonds between proline and aromatic amino acids.

- Protease inhibitors are designed to act as transition-state inhibitors. They contain a transition-state isostere which is tetrahedral but stable to hydrolysis. Suitable substituents are added to fill various binding pockets usually occupied by the amino acid residues of polypeptide substrates.

- To obtain an orally active PI, it is important to maximize the binding interactions with the enzyme, while minimizing the molecular weight and peptide character of the molecule.

- Cell fusion inhibitors have been developed, one of which has reached the market.

17.8 Antiviral drugs acting against RNA viruses: flu virus

17.8.1 Structure and life cycle of the influenza virus

Influenza (or flu) is an airborne respiratory disease caused by an RNA virus which infects the epithelial cells of the upper respiratory tract. It is a major cause of mortality, especially amongst the elderly, or amongst patients with weak immune systems. The most serious pandemic occurred in 1918 with the death of at least 20 million people worldwide caused by the Spanish flu virus. Epidemics then occurred in 1957 (Asian flu), 1968 (Hong Kong flu), and 1977 (Russian flu). Despite the names given to these flus, it is likely that they all derived from China where families are in close proximity to poultry and pigs, increasing the chances of viral infections crossing from one species to another.[1] In 1997, there was an outbreak of flu

---

[1] On the other hand, there has been a recent theory that the 1918 pandemic originated in army transit camps in France. The living conditions in these camps were similar to communities in China in the sense that large numbers of soldiers were camping in close proximity to pigs and poultry used as food stocks. The return of the forces to all parts of the globe after the First World War could explain the rapid spread of the virus.

in Hong Kong which killed 6 out of 18 people. This was contained by slaughtering infected chickens, duck, and geese which had been the source of the problem. If action had not been swift, it is possible that this flu variant could have become a pandemic and wiped out 30% of the world's population. This emphasizes the need for effective antiviral therapies to combat flu.

The nucleocapsid of the flu virus contains (−)single-stranded RNA and a viral enzyme called **RNA polymerase** (see Fig. 17.1). Surrounding the nucleocapsid, there is a membranous envelope derived from host cells which contains two viral glycoproteins called **neuraminidase** (NA) and **haemagglutinin** (HA) (which acquired its name because it can bind virions to red blood cells and cause haemagglutination). These glycoproteins are spike-like objects which project about 10 nm from the surface and are crucial to the infectious process.

In order to reach the epithelial host cells of the upper respiratory tract the virus has to negotiate a layer of protective mucus, and it is thought that the viral protein NA is instrumental in achieving this. The mucosal secretions are rich in glycoproteins and glycolipids which bear a terminal sugar substituent called **sialic acid** (also called N-acetylneuraminic acid). Neuraminidase (also called sialidase) is an enzyme which is capable of cleaving the sialic acid sugar moiety from these glycoproteins and glycolipids (Fig. 17.38), thus degrading the mucus layer and allowing the virus to reach the surface of epithelial cells.

Once the virus reaches the epithelial cell, adsorption takes place whereby the virus binds to cellular glyco-conjugates which are present in the host cell membrane, and which have a terminal sialic acid moiety. The viral protein HA is crucial to this process. Like NA, it recognizes sialic acid but instead of catalysing the cleavage of the sialic acid from the glycoconjugate, HA binds to it (Fig. 17.39). Once the virion has been adsorbed, the cell membrane bulges inwards taking the virion with it to form a vesicle called an endosome—a process called **receptor-mediated endocytosis**. The pH in the endosome then decreases, causing HA in the virus envelope to undergo a dramatic conformational change whereby the hydrophobic ends of the protein spring outward and extend towards the endosomal membrane. After contact, fusion occurs and the RNA nucleocapsid is released into the cytoplasm of the host cell. Disintegration of the nucleocapsid releases viral

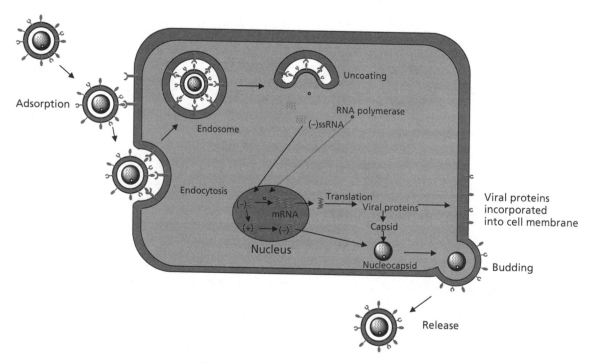

Figure 17.38 Action of neuraminidase (sialidase).

Figure 17.39 Life cycle of the influenza virus.

RNA and viral RNA polymerase, which both enter the cell nucleus.

Viral RNA polymerase now catalyses the copying of (−)viral RNA to produce (+) viral RNA which departs the nucleus and acts as the mRNA required for the translation of viral proteins. Copies of (−) viral RNA are also produced in the nucleus and exported out of the nucleus.

Capsid proteins spontaneously self-assemble in the cytoplasm with incorporation of (−) RNA and newly produced RNA polymerase to form new nucleocapsids. Meanwhile, the freshly synthesized viral proteins HA and NA are incorporated into the membrane of the host cell. Newly formed nucleocapsids then move to the cell membrane and attach to the inner

surface. HA and NA move through the cell membrane to these areas, and at the same time host cell proteins are excluded. Budding then takes place and a new virion is released. NA aids this release by hydrolysing any interactions that take place between HA on the virus and sialic acid conjugates on the host membrane.

There is an important balance between the rate of desialylation by NA (to aid the virion's departure from the host cell) and the rate of attachment by HA to sialylated glycoconjugates (to allow access to the cell). If NA was too active, it would hinder infection of the cell by destroying the receptors recognized by HA. On the other hand, if the enzyme activity of NA was too weak, the newly formed virions would remain adsorbed to the host cell after budding, preventing them from infecting other cells. It is noticeable that the amino acids present in the active site of NA are highly conserved, unlike amino acids elsewhere in the protein. This demonstrates the importance of the enzyme's activity level.

Since HA and NA are on the outer surface of the virion, they can act as antigens (i.e. molecules which can potentially be recognized by antibodies and the body's defence systems). In theory, it should also be possible to prepare vaccines which will allow the body to gain immunity from the flu virus. Such vaccinations are available, but they are not totally protective and they lose what protective effect they have with time. This is because the flu virus is adept at varying the amino acids present in HA and NA, thus making these antigens unrecognizable to the antibodies which originally recognized them—a process called **antigenic variation**. The reason it takes place can be traced back to the RNA polymerase enzyme which is a relatively error-prone enzyme and means that the viral RNA which codes for HA and NA is not consistent. Variations in the code lead to changes in the amino acids present in NA and HA which results in different types of flu virus, based on the antigenic properties of their NA and HA. For example, there are nine antigenic variants of NA.

There are three groups of flu virus, classified as A, B, and C. Antigenic variation does not appear to take place with influenza C, and occurs slowly with influenza B. With influenza A, however, variation occurs almost yearly. If the variation is small, it is called **antigenic drift**. If it is large, it is called **antigenic shift** and it is this that can lead to the more serious epidemics and pandemics. There are two influenza A virus subtypes which are epidemic in humans—those with H1N1 and H3N2 antigens (where H and N stand for HA and NA respectively). A major aim in designing effective antiviral drugs is to find a drug which will be effective against the influenza A virus and remain effective despite antigenic variations. In general, vaccination is the preferred method of preventing flu, but antiviral drugs also have their place, for both the prevention and treatment of flu when vaccination proves unsuccessful.

17.8.2 Ion channel disrupters: adamantanes

The adamantanes were discovered by random screening and are the earliest antiviral drugs used

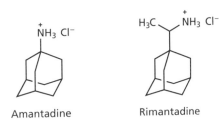

Figure 17.40 The adamantanes.

clinically against flu, decreasing the incidence of the disease by 50–70%. **Amantadine** and **rimantadine** (Fig 17.40) are related adamantanes with similar mechanisms of action and can inhibit viral infection in two ways. At low concentration (<1 μg/ml), they inhibit the replication of influenza A viruses by blocking a viral ion channel protein called matrix (M2) protein. At higher concentration (>50 μg/ml)) the basic nature of the compounds becomes important and they buffer the pH of endosomes and prevent the acidic environment needed for HA to fuse the viral membrane with that of the endosome. These mechanisms inhibit penetration and uncoating of the virus.

Unfortunately, the virus can mutate in the presence of amantadine to form resistant variants. Amantadine binds to a specific region of the M2 ion channel, and resistant variants have mutations which alter the width of the channel. Research carried out to find analogues which might still bind to these mutants proved unsuccessful. Work has also been carried out in an attempt to find an analogue which might affect the ion channel and pH levels at comparable concentrations. This has focused on secondary and tertiary amines with increased basicity, as well as alteration of the structure to reduce activity for the ion channel. The rationale is that resistant flu variants are less likely to be produced if the drug acts on two different targets at the same time. Rimantadine was approved in 1993 as a less toxic alternative to amantadine for the treatment of influenza A. Unfortunately, neither agent is effective against influenza B, since this virus does not contain the matrix (M2) protein. Side effects are also a problem, possibly due to effects on host cell ion channels.

17.8.3 Neuraminidase inhibitors

17.8.3.1 Structure and mechanism of neuraminidase

Since NA plays two crucial roles in the infectious process (section 17.8.1), it is a promising target for potential antiviral agents. Indeed, a screening program for NA inhibitors was carried out as early as 1966 although without success. Following on from this, researchers set out to design a mechanism-based transition-state inhibitor. This work progressed slowly until the enzyme was isolated and its crystal structure studied by X-ray crystallography and molecular modelling.

Neuraminidase is a mushroom-shaped tetrameric glycoprotein anchored to the viral membrane by a single hydrophobic sequence of some 29 amino acids. As a result, the enzyme can be split enzymatically from the surface and studied without loss of antigenic or enzymic activity. X-ray crystallographic studies have shown that the active site is a deep pocket located centrally on each protein subunit. There are two main types of the enzyme (corresponding to the influenza viruses A and B) and various subtypes. Due to the ease with which mutations occur, there is a wide diversity of amino acids making up the various types and subtypes of the enzyme. However, the 18 amino acids making up the active site itself are constant. As mentioned previously, the absolute activity of the enzyme is crucial to the infectious process and any variation that affects the active site is likely to affect the activity of the enzyme. This in turn will adversely affect the infectious process. Since the active site remains

constant, any inhibitor designed to fit it has a good chance of inhibiting all strains of the flu virus. Moreover, it has been observed that the active site is quite different in structure from the active sites of comparable bacterial or mammalian enzymes, so there is a strong possibility that inhibitors can be designed that are selective antiviral drugs.

The enzyme has been crystallized with sialic acid (the product of the enzyme-catalysed reaction) bound to the active site, and the structure determined by X-ray crystallography. A molecular model of the complex was created based on these results, which included 425 added water molecules. It was then energy minimized in such a way that it resembled the observed crystal structure as closely as possible. From this it was calculated that sialic acid was bound to the active site through a network of hydrogen bonds and ionic interactions as shown in Fig. 17.41.

The most important interactions involve the carboxylate ion of sialic acid, which is involved in ionic interactions and hydrogen bonds with three arginine residues, particularly with Arg-371. In order to achieve these interactions, the sialic acid has to be distorted from its most stable chair conformation (where the carboxylate ion is in the axial position) to a pseudo-boat conformation where the carboxylate ion is equatorial.

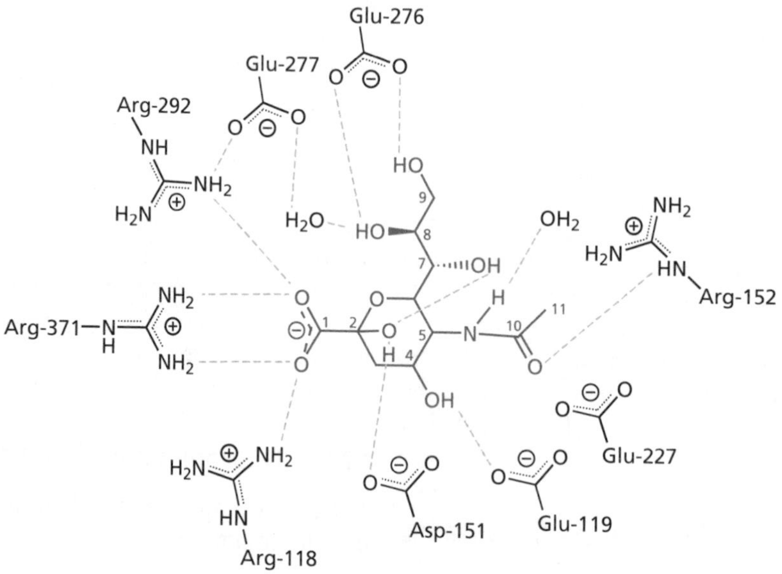

Figure 17.41 Hydrogen bonding interactions between sialic acid and the active site.

There are three other important binding regions or pockets within the active site. The glycerol side chain of sialic acid fills one of these pockets, interacting with glutamate residues and a water molecule by hydrogen bonding. The hydroxyl group at C-4 of sialic acid is situated in another binding pocket interacting with a glutamate residue. Finally, the acetamido substituent of sialic acid fits into a hydrophobic pocket which is important for molecular recognition. This pocket includes the hydrophobic residues Trp-178 and Ile-222 which lie close to the methyl carbon (C-11) of sialic acid as well as the hydrocarbon backbone of the glycerol side chain.

It was further established that the distorted pyranose ring binds to the floor of the active site cavity through its hydrophobic face. The glycosidic OH at C-2 is also shifted from its normal equatorial position to an axial position where it points out of the active site and can form a hydrogen bond to Asp-151, as well as an intramolecular hydrogen bond to the hydroxyl group at C-7.

Based on these results, a mechanism of hydrolysis was proposed which consists of four major steps (Fig. 17.42). The first step involves the binding of the substrate (sialoside) as described above. The second step involves proton donation from an activated water facilitated by the negatively charged Asp-151, and formation of an endocyclic sialosyl cation transition-state intermediate. Glu-277 is proposed to stabilize the developing positive charge on the glycosidic oxygen as the mechanism proceeds.

The final two steps of the mechanism are formation and release of sialic acid. Support for the proposed mechanism comes from kinetic isotope studies which indicate it is an S_N1 nucleophilic substitution. NMR studies have also been carried out which indicate that sialic acid is released as the α-anomer. This is consistent with an S_N1 mechanism having a high degree of stereofacial selectivity. Possibly expulsion of the product from the active site is favoured by mutarotation to the more stable β-anomer.

Finally, site directed mutagenesis studies have shown that the activity of the enzyme is lost if Arg-152 is replaced by lysine and Glu-277 by aspartate. These replacement amino acids contain similarly charged residues but have a shorter residue chain. As a result, the charged residues are unable to reach the required area of space in order to stabilize the intermediate.

17.8.3.2 Transition-state inhibitors: development of zanamivir (Relenza)

The transition state shown in Fig. 17.42 has a planar trigonal centre at C-2 and so sialic acid analogues containing a double bond between positions C-2 and C-3 were synthesized to achieve that same trigonal geometry at C-2. This resulted in the discovery of the inhibitor 2-deoxy-2,3-dehydro-N-acetylneuraminic acid (Neu5Ac2en) in 1969 (Fig. 17.43). In order to achieve the required double bond, the hydroxyl group originally present at C-2 of sialic acid had to be omitted, which resulted in lower hydrogen bonding interactions with the active site. On the other hand, the inhibitor does not need to distort from a favourable chair shape in order to bind, and the energy saved by this more than compensates for the loss of one hydrogen bonding interaction. The inhibitor was crystallized with the enzyme and studied by X-ray crystallography and molecular modelling to show that the same binding interactions were taking place with the exception of the missing hydroxyl group at C-2. Unfortunately, this compound also inhibited bacterial and mammalian sialidases and could not be used therapeutically. Moreover, it was inactive in vivo.

Following the development of a model active site, the search for new inhibitors centred around the use of molecular modelling software to evaluate likely binding regions within the model active site. This involved setting up a series of grid points within the active site and placing probe atoms at each point to measure interactions between the probe and the active site (compare section 13.10). Different atomic probes were used to represent various functional groups. These included the oxygen of a carboxylate group, the nitrogen of an ammonium cation, the oxygen of a hydroxyl group, and the carbon of a methyl group. Multiatom probes were also used. A multiatom probe is positioned such that one atom of the probe is placed at the grid point and then energy calculations are performed for all the atoms within the probe and a total probe energy is assigned to that grid point. The probe is then rotated such that each possible hydrogen bonding orientation is considered and the most favourable interaction energy accepted.

The most important result from these studies was the discovery that the region around the 4-OH of sialic acid could interact with an ammonium or guanidinium ion. As a result, sialic acid analogues, having an

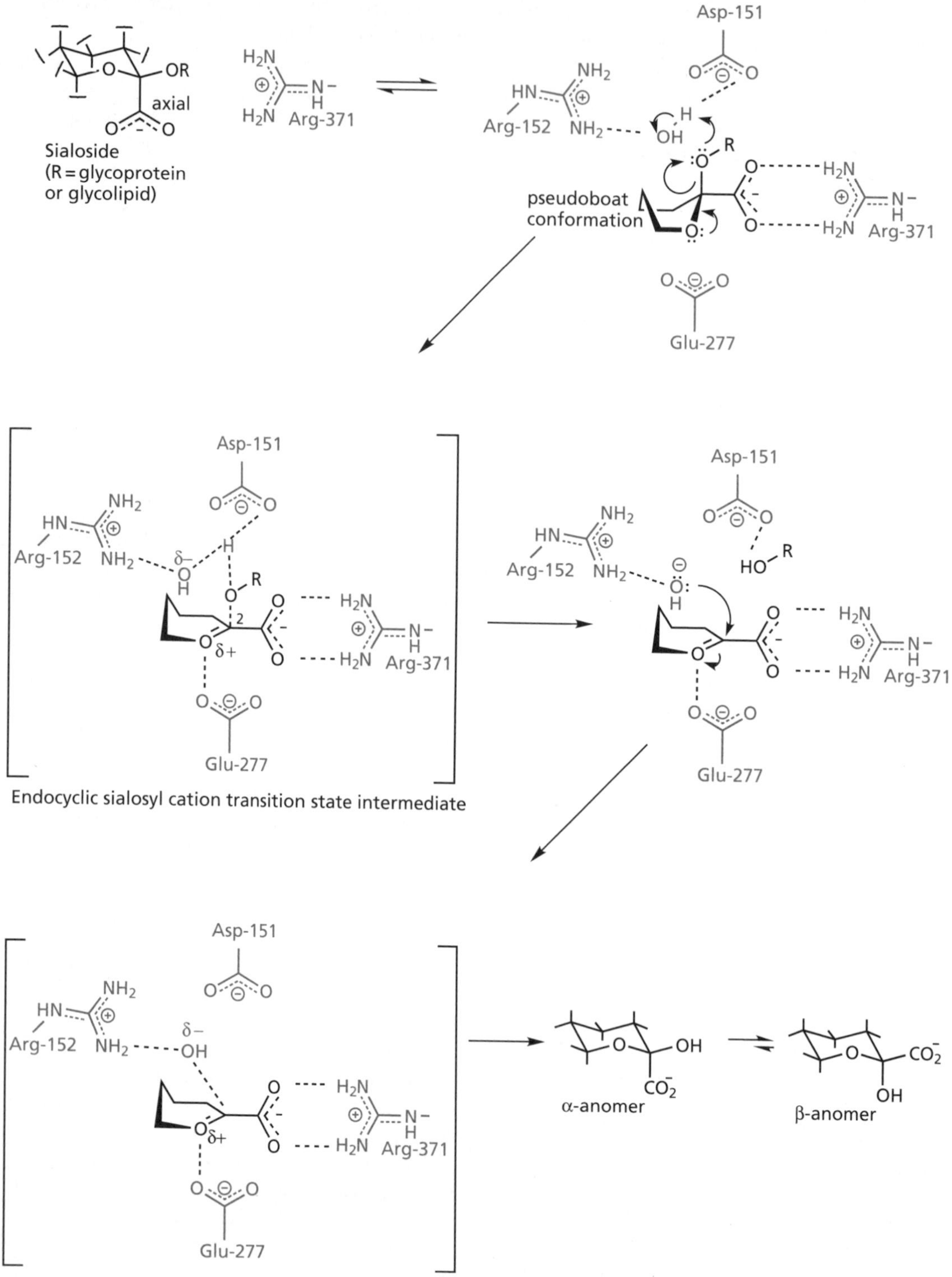

Figure 17.42 Proposed mechanism of hydrolysis.

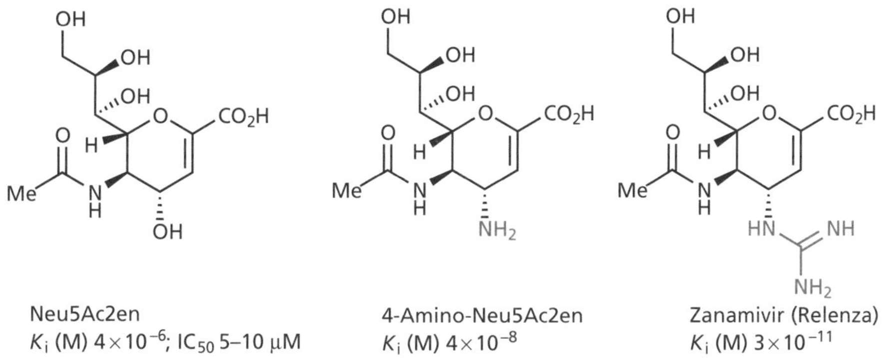

Figure 17.43 Transition state inhibitors.

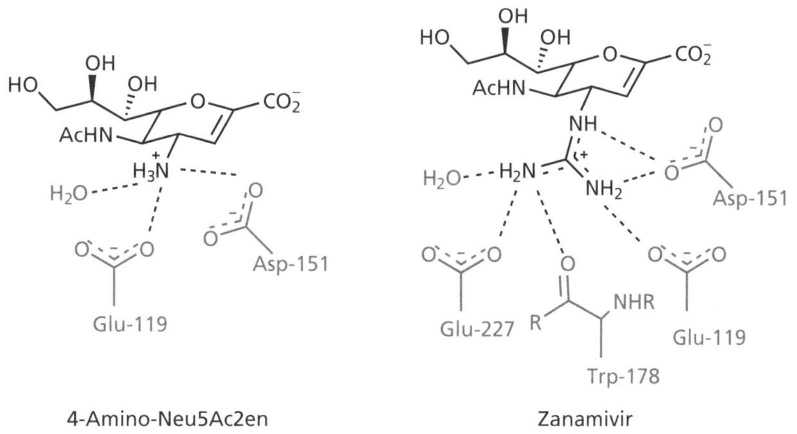

Figure 17.44 Binding interactions of the ammonium and guanidinium moieties at C4.

amino or guanidinyl group at C-4 instead of hydroxyl, were modelled in the active site to study the binding interactions and to check whether there was room for the groups to fit.

These results were favourable and so the relevant structures were synthesized and tested for activity. 4-Amino-Neu5Ac2en (Fig. 17.43) was found to be more potent than Neu5Ac2en. Moreover, it was active in animal studies and showed selectivity against the viral enzyme, implying that the region of the active site which normally binds the 4-hydroxyl group of the substrate is different in the viral enzyme from comparable bacterial or mammalian enzymes. A crystal structure of the inhibitor bound to the enzyme confirmed the binding pattern predicted by the molecular modelling (Fig. 17.44).

Molecular modelling studies had suggested that the larger guanidinium group would be capable of

even greater hydrogen bonding interactions as well as favourable van der Waals interactions. The relevant structure (zanamivir; Fig. 17.43) was indeed found to be a more potent inhibitor having a 100-fold increase in activity. X-ray crystallographic studies of the enzyme–inhibitor complex demonstrated the expected binding interactions (Fig. 17.44). Moreover, the larger guanidino group was found to expel a water molecule from this binding pocket, which is thought to contribute a beneficial entropic effect. Zanamivir is a slow-binding inhibitor with a high binding affinity to influenza A neuraminidase. It was approved by the US FDA in 1999 for the treatment of influenza A and B, and was marketed by Glaxo Wellcome and Biota. Unfortunately, the polar nature of the molecule means it has poor oral bioavailability (<5%), and it is administered by inhalation.

Following on from the success of these studies, 4-epi-amino-Neu5Ac2en (Fig. 17.45) was synthesized to place the amino group in another binding region predicted by the GRID analysis. This structure proved to be a better inhibitor than Neu5Ac2en but not as good as zanamivir. The pocket into which this amino group fits is small and there is no room for larger groups.

17.8.3.3 Transition-state inhibitors: 6-carboxamides

A problem with the inhibitors described in section 17.8.3.2 is their polar nature. The glycerol side chain is particularly polar and has important binding interactions with the active site. However, it was found that it could be replaced by a carboxamide side chain with retention of activity (Fig. 17.45).

A series of 6-carboxamide analogues was prepared to explore their structure–activity relationships. Secondary carboxamides where $R_{cis} = H$ showed similar weak inhibition against both A and B forms of the neuraminidase enzyme. Tertiary amides having an alkyl substituent at the *cis* position resulted in a pronounced improvement against the A form of the enzyme, with relatively little effect on the activity against the B form. Thus, tertiary amides showed a marked selectivity of 30–1000-fold for the A form of the enzyme. Good activity was related to a variety of different-sized R_{trans} substituents larger than methyl, but the size of the R_{cis} group was more restricted and optimum activity was achieved when R_{cis} was ethyl or *n*-propyl.

The 4-guanidino analogues are more active than corresponding 4-amino analogues but the improvement is slightly less than that observed for the glycerol series, especially where the 4-amino analogue is already highly active.

Crystal structures of the carboxamide (I in Fig. 17.45) bound to both enzymes A and B were determined by X-ray crystallography. The dihydropyran portion of the carboxamide (I) binds to both the A and B forms of the enzyme in essentially the same manner as observed for zanamivir. The important binding interactions involve the carboxylate ion, the 4-amino group and the 5-acetamido group—the latter occupying a hydrophobic pocket lined by Trp-178 and Ile-222 (Fig. 17.46).

A significant difference, though, is in the region occupied by the carboxamide side chain. In the sialic acid analogues, the glycerol side chain forms intermolecular hydrogen bonds to Glu-276. These interactions are not possible for the carboxamide side chain. Instead, the Glu-276 side chain changes conformation and forms a salt bridge with the guanidino side chain of Arg-224 and reveals a lipophilic pocket into which the R_{cis} *n*-propyl substituent can fit. The size of this pocket is optimal for an ethyl or propyl group which matches the structure–activity (SAR) results. The R_{trans} phenethyl group lies in an extended lipophilic cleft on the enzyme surface formed between Ile-222 and Ala-246. This region can accept a variety of substituents, again consistent with SAR results.

Comparison of the X-ray crystal structures of the native A and B enzymes shows close similarity of position and orientation of the conserved active site residues except in the region occupied normally by the glycerol side chain, particularly as regards Glu-276. Zanamivir can bind to both A and B forms with little or no distortion of the native structures. Binding of the carboxamide (I) to the A form is associated with a

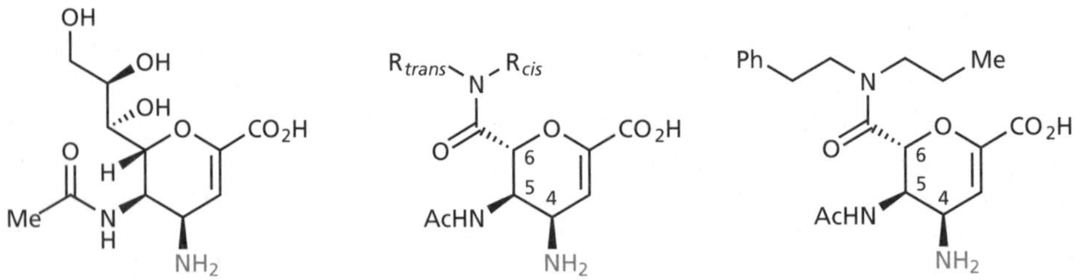

4-Epi-amino-Neu5Ac2en
K_i (M) 3×10^{-7}

6-Carboxamides

I

Figure 17.45 4-Epi-amino-Neu5Ac2en and carboxamides.

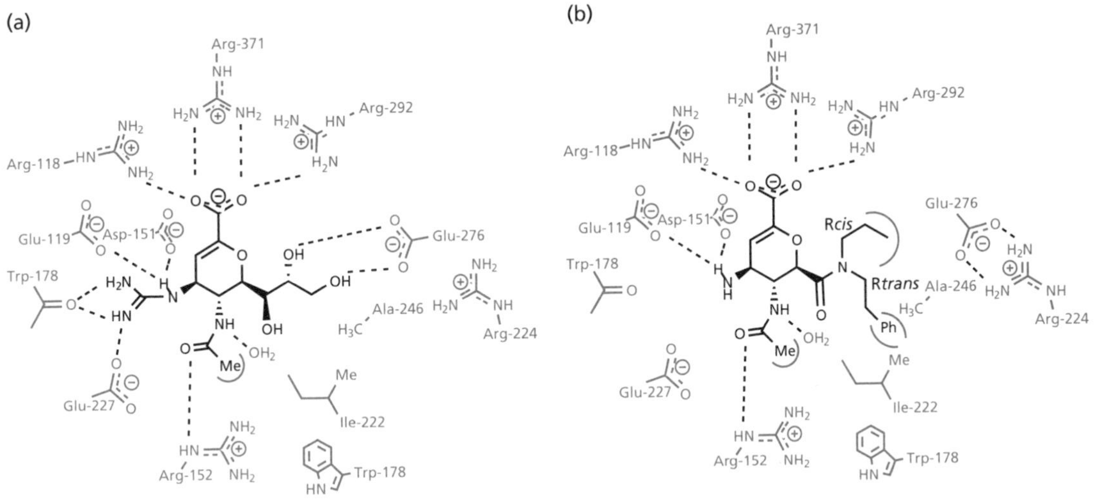

(a)

(b)

Figure 17.46 Binding interactions of zanamivir and carboxamides: (a) binding of zanamivir to the active site; (b) binding of carboxamide (I) to the active site.

change in torsion angles of the Glu-276 side chain such that the residue can form the salt bridge to Arg-224, but there is little distortion of the protein backbone in order to achieve this. In contrast, when the carbox-amide binds to the B form of the enzyme, there is a significant distortion of the protein backbone required before the salt bridge is formed. Distortion in the B enzyme structure also arises around the phe-nethyl substituent. This implies that binding of the carboxamide to the B form involves more energy ex-penditure than to A and this can explain the observed specificity.

17.8.3.4 Carbocyclic analogues: development of oseltamivir (Tamiflu)

The dihydropyran oxygen of Neu5Ac2en and related inhibitors has no important role to play in binding these structures to the active site of neuraminidase. Therefore, it should be possible to replace it with a methylene isostere to form carbocyclic analogues such as structure I in Fig. 17.47. This would have the ad-vantage of removing a polar oxygen atom which would increase hydrophobicity and potentially increase oral bioavailability. Moreover, it would be possible to synthesize cyclohexene analogues such as structure II which more closely match the stereochemistry of the reaction's transition state than previous inhibitors—compare the reaction intermediate in Fig. 17.47 which

can be viewed as a transition-state mimic. Such agents might be expected to bind more strongly and be more potent inhibitors.

Structures I and II were synthesized to test this theory, and it was discovered that structure II was 40 times more potent than structure I as an inhibitor. Since the substituents are the same, this indicates that the conformation of the ring is crucial for inhibitory activity. Both structures have half chair conformations but these are different due to the position of the double bond.

It was now planned to replace the hydroxyl group on the ring with an amino group to improve binding interactions (compare section 17.8.3.2), and to remove the glycerol side chain to reduce polarity. In its place a hydroxyl group was introduced for two reasons. First, the oxonium double bond in the transition state is highly polarized and electron deficient whereas the double bond in the carbocyclic structures is electron rich. Introducing the hydroxyl substituent in place of the glycerol side chain means that the oxygen will have an inductive electron-withdrawing effect on the car-bocyclic double bond and reduce its electron density. The second reason for adding the hydroxyl group was that it would be possible to synthesize ether analogues which would allow the addition of hydrophobic groups to fill the binding pocket previously occupied by the glycerol side chain (compare section 17.8.3.3).

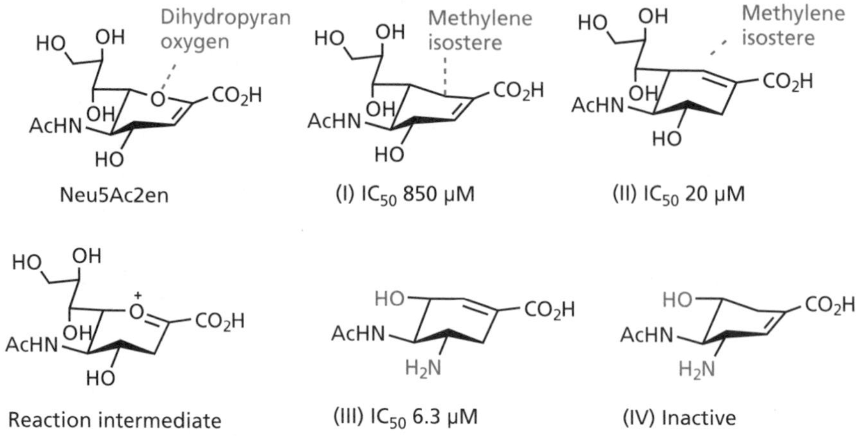

Figure 17.47 Comparison of Neu5Ac2en, reaction intermediate and carbocyclic structures.

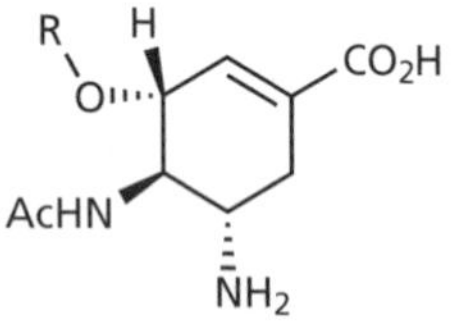

Figure 17.48 Alkoxy analogues.

The resultant structure III was synthesized and proved to be a potent inhibitor. In contrast, the isomer IV failed to show any inhibitory activity.

A series of alkoxy analogues of structure III was now synthesized in order to maximize hydrophobic interactions in the region of the active site previously occupied by the glycerol side chain (Fig. 17.48). For linear alkyl chains, potency increased as the carbon chain length increased from methyl to *n*-propyl. Beyond that, activity was relatively constant (150–300 nM) up to and including *n*-nonyl, after which activity dropped. Although longer chains than propyl increase hydrophobic interactions, there is a downside in that there is partial exposure of the side chain to water outside of the active site.

Branching of the optimal propyl group was investigated. There was no increase in activity when methyl branching was at the β-position but the addition of a methyl group at the α-position increased activity by 20-fold. Introduction of an α-methyl group introduces an asymmetric centre, but both isomers were found to have similar activity indicating two separate hydrophobic pockets. The optimal side chain proved to be a pentyloxy side chain (R = CH(Et)$_2$).

The *N*-acetyl group is required for activity and there is a large drop in activity without it. The binding region for the *N*-acetyl group has limitations on the functionality and size of groups which it can accept. Any variations tend to reduce activity. This was also observed with sialic acid analogues.

Replacing the amino group with a guanidine group improves activity, as with the sialic acid series. However, the improvement in activity depends on the type of alkyl group present on the side chain, indicating that individual substituent contributions may not be purely additive.

The most potent of the above analogues was the pentyloxy derivative (GS4071) (Fig. 17.49). This was co-crystallized with the enzyme and the complex was studied by X-ray crystallography revealing that the alkoxy side chain makes several hydrophobic contacts in the region of the active site normally occupied by the glycerol side chain. In order to achieve this, the carboxylate group of Glu-276 is forced to orientate outwards from the hydrophobic pocket as observed with the carboxamides. The overall gain in binding energy from these interactions appears to be substantial, as a guanidino group is not required to achieve

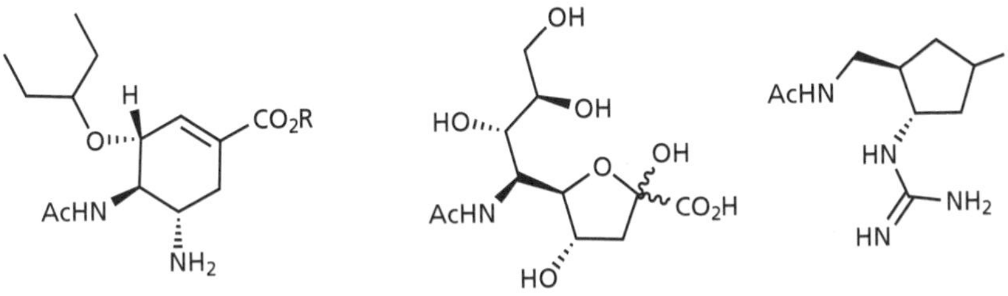

Figure 17.49 Oseltamivir and other ring systems.

low nanomolar inhibition. Interactions elsewhere are similar to those observed with previous inhibitors.

Oseltamivir (Tamiflu) (Fig. 17.49) is the ethyl ester prodrug of GS4071 and was approved in 1999 for the treatment of influenza A and B. The drug is marketed by Hoffman La Roche and Gilead Sciences. It is taken orally and is converted to GS4071 by esterases in the gastrointestinal tract.

17.8.3.5 Other ring systems

Work has been carried out to develop new NA inhibitors where different ring systems act as scaffolds for the important binding groups (Fig. 17.49).

The five-membered THF (I) is known to inhibit neuraminidase with a potency similar to Neu5Ac2en. It has the same substituents as Neu5Ac2en, although their arrangement on the ring is very different. Nevertheless, a crystal structure of (I) bound to the enzyme shows that the important binding groups (carboxylate, glycerol, acetamido and C-4-OH) can fit into the required pockets. The central ring or scaffold is significantly displaced from the position occupied by the pyranose ring of Neu5Ac2en in order to allow this. This indicates that the position of the central ring is not crucial to activity, and that the relative position of the four important binding groups is more important.

Five-membered carbocyclic rings have also been studied as suitable scaffolds. Structure II (Fig. 17.49) was designed such that the guanidine group would fit the negatively charged binding pocket previously described. A crystal structure of the inhibitor with the enzyme showed that the guanidine group occupies the desired pocket and displaces the water molecule originally present. It is involved in charge-based interactions with Asp-151, Glu-119, and Glu-227, analogous to zanamivir.

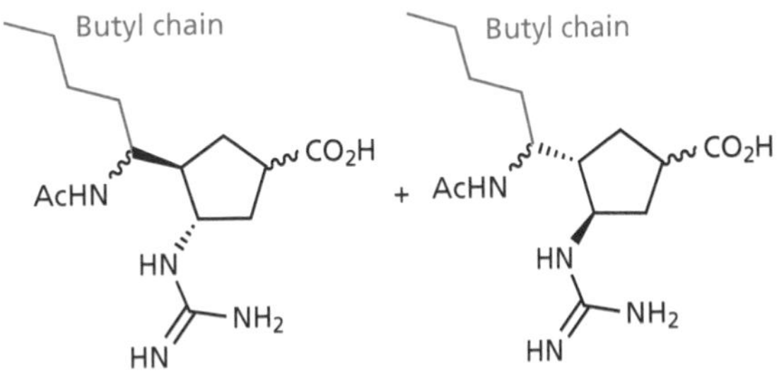

Figure 17.50 Mixture of isomers.

Modelling studies suggested that the addition of a butyl chain to the structure would allow van der Waals interactions with a small hydrophobic surface in the binding site. The target structure now has four asymmetric centres, and a synthetic route was used which controlled the configuration of two of these. As a result, four racemates or eight isomers were prepared as a mixture (Fig. 17.50). Neuraminidase crystals were used to select the most active isomer of the mixture by soaking a crystal of the enzyme in the solution of isomers for a day and then collecting X-ray diffraction data from the crystal. This showed the active isomer to be structure I in Fig. 17.51. The structure binds to the active sites of both influenza A and B neuraminidases with the *n*-butyl side chain adopting two different binding modes. In the influenza B neuraminidase, the side chain is positioned against a hydrophobic surface formed by Ala-246, Ile-222, and Arg-224. In the A version, the chain is in a region formed by the reorientation of the side chain of Glu-276.

BCX-1812 (Fig. 17.51) was designed to take advantage of both hydrophobic pockets in the active site. It was prepared as a racemic mixture, and a crystal of the neuraminidase enzyme was used to bind

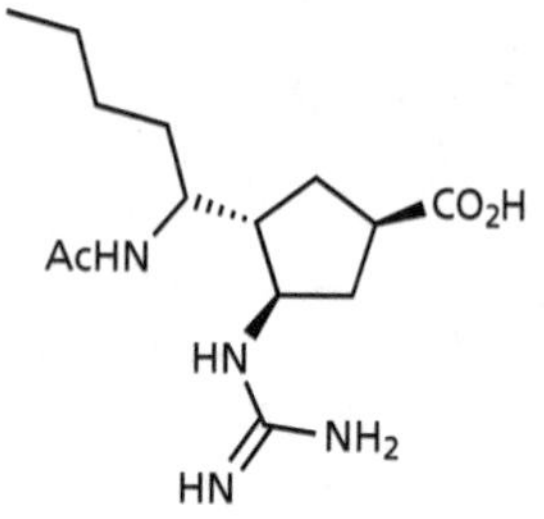

Figure 17.51 Development of BCX 1812.

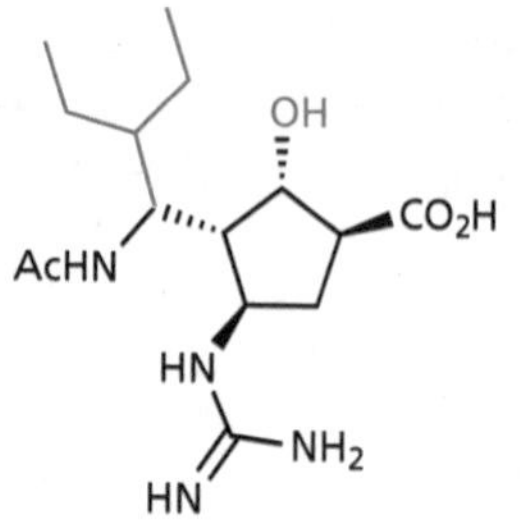

to the active isomer. Once identified, this was then prepared by a stereospecific synthesis. The relative stereochemistry of the substituents was the same as in (I).

In vitro tests of BCX-1812 versus strains of influenza A and B show it to be as active as zanamivir and GS4071. It is also four orders of magnitude less active against bacterial and mammalian neuraminidases, making it a potent and highly specific inhibitor of flu virus neuraminidase. *In vivo* tests carried out on mice showed it to be orally active and the compound is undergoing clinical trials.

17.8.3.6 Resistance studies

Studies have been carried out to investigate the likelihood of viruses acquiring resistance to the drugs mentioned above. This is done by culturing the viruses in the presence of the antiviral agents to see if mutation leads to a resistant strain.

Zanamivir has a broad-spectrum efficacy against all type A and B strains tested, and interacts only with conserved residues in the active site of NA. Thus, in order to gain resistance, one of these important amino acids has to mutate. A variant has been observed where Glu-119 has been mutated to glycine. This has reduced affinity for zanamivir, and the virus can replicate in the presence of the drug. Removing Glu-119 affects the binding interactions with the 4-guanidinium group of zanamivir without affecting interactions with sialic acid. Zanamivir-resistant mutations were also found where a mutation occurred in HA around the sialic acid binding site. This mutation weakened affinity for sialic acid and so lowered binding. Thus, mutant viruses were able to escape more easily from the infected cell after budding. No such mutations have appeared during clinical trials, however.

Another mutation has been observed where Arg-292 is replaced by lysine. In wild-type NA, Arg-292 binds to the carboxylate group of the inhibitor and is partly responsible for distorting the pyranose ring from the chair to the boat conformation. In the mutant structure, the amino group of Lys-292 forms an ionic interaction with Glu-276 which normally binds the 8 and 9 hydroxyl groups of the glycerol side chain. This results in a weaker interaction with both inhibitors and substrate leading to a weaker enzyme.

One conclusion that has arisen from studies on easily mutatable targets is the desirability to find an inhibitor which is least modified from the normal substrate and which uses the same interactions for binding as much as possible.

KEY POINTS

- The flu virus contains (−)ssRNA and has two glycoproteins called HA and neuraminidase in its outer membrane.
- HA binds to the sialic acid moiety of glycoconjugates on the outer surface of host cells, leading to adsorption and cell uptake.
- NA catalyses the cleavage of sialic acid from glycoconjugates. It aids the movement of the virus through mucus and releases the virus from infected cells after budding.
- HA and NA act as antigens for flu vaccines. However, the influenza A virus readily mutates these proteins, requiring new flu vaccines each year.
- The adamantanes are antiviral agents which inhibit influenza A by blocking a viral ion channel called the matrix (M2) protein. At high concentration they buffer the pH of endosomes. They are ineffective against influenza B, which lacks the matrix (M2) protein.
- Neuraminidase has an active site which remains constant for the various types and subtypes of the enzyme and which is different from the active sites of comparable mammalian enzymes.
- There are four important binding pockets in the active site. The sialic acid moiety is distorted from its normal chair conformation when it is bound.
- The mechanism of reaction is proposed to go through an endocyclic sialosyl cation transition state. Inhibitors were designed to mimic this state by introducing an endocyclic double bond.

- Successful antiviral agents have been developed using structure-based drug design.

- Different scaffolds can be used to hold the four important binding groups.

- There is an advantage in designing drugs which use the same binding interactions as the natural ligand when the target undergoes facile mutations.

17.9 Antiviral drugs acting against RNA viruses: cold virus

The agents used against flu are ineffective against colds, as these infections are caused by a different kind of virus called a rhinovirus. Colds are less serious than flus. Nevertheless, research has taken place to find drugs which can combat them.

There are at least 89 serotypes of **human rhinoviruses** (HRV) and they belong to a group of viruses called the **picornaviruses** which include the polio, hepatitis A, and foot and mouth disease viruses. They are amongst the smallest of the animal RNA viruses, containing a positive strand of RNA coated by an icosahedral shell made up of 60 copies of 4 distinct proteins, VP1–VP4 (Fig. 17.52). The proteins VP1–VP3 make up the surface of the virion. The smaller VP4 protein lies underneath to form the inner surface

and is in contact with the viral RNA. At the junction between each VP1 and VP3 protein, there is a broad canyon 25 Å deep and this is where attachment takes place between the virus and the host cell. On the canyon floor, there is a pore which opens into a hydrophobic pocket within the VP1 protein. This pocket is either empty or occupied by a small molecule called a **pocket factor**. So far the identity of the pocket factor has not been determined but it is known from X-ray crystallographic studies that it is a fatty acid containing seven carbon atoms.

When the virus becomes attached to the host cell, a receptor molecule on the host cell fits into the canyon and induces conformational changes which see the VP4 protein and the N-terminus of VP1 move to the exterior of the virus—a process called externalization. This is thought to be important to the process by which the virus is uncoated and releases its RNA into the host cell. It is thought that the pocket factor regulates the stability of the virion. When it is bound to the pocket, it stabilizes the capsid and prevents the conformational changes that are needed to cause infection.

A variety of drugs having antiviral activity are thought to mimic the pocket factor by displacing it and binding to the same hydrophobic pocket. The drugs concerned are called capsid binding agents and are characteristically long-chain hydrophobic molecules. Like the pocket factor, they stabilize the capsid by locking it into a stable conformation and prevent the conformational changes required for uncoating.

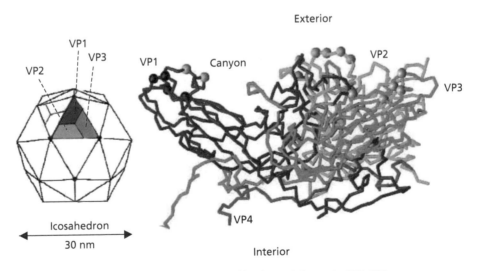

Figure 17.52 Structure of human rhinovirus and the proteins VP1–VP4.

They also raise the canyon floor and prevent the receptor on the host cell from fitting the canyon and inducing uncoating (Fig. 17.53).

Pleconaril (Fig. 17.54) is one such drug that has undergone phase III clinical trials which demonstrate that it has an effect on the common cold. It is an orally active, broad-spectrum agent which can cross the blood–brain barrier. The drug may also be useful against the enteroviruses which cause diarrhoea, viral meningitis, conjunctivitis, and encephalitis since they are similar in structure to the rhinoviruses.

The development of pleconaril started when a series of isoxazoles was found to have antiviral activity. This led to the discovery of **disoxaril** (Fig. 17.54) which entered phase I clinical trials but proved to be too toxic. X-ray crystallographic studies of VP1–drug complexes involving disoxaril and its analogues showed that the oxazoline and phenyl rings were roughly coplanar and located in a hydrophilic region of the pocket near the pore leading into the centre of the virion (Fig. 17.53). The hydrophobic isoxazole ring binds into the heart of the hydrophobic pocket

and the chain provides sufficient flexibility for the molecule to bend round a corner in the pocket. Binding moves Met-221 which normally seals off the pocket, and this also causes conformational changes in the canyon floor. Structure-based drug design was carried out to find safer and more effective antiviral agents. For example, the chain cannot be too short or too long or else there are steric interactions. Placing additional hydrophobic groups on to the phenyl ring improves activity against the HRV2 strain, since increased interactions are possible with a phenylalanine residue at position 116 rather than leucine. The structure **WIN 54954** was developed and entered clinical trials, but results were disappointing because extensive metabolism resulted in 18 different metabolic products due mainly to hydrolysis of the oxazoline ring. Further structure-based drug design studies led to modifications of the phenyl and oxazoline moieties. This included the introduction of a trifluoromethyl group to block metabolism, resulting in pleconaril, which has 70% oral bioavailability.

Figure 17.53 Binding of disoxaril (possible hydrogen bonds shown as dashed lines).

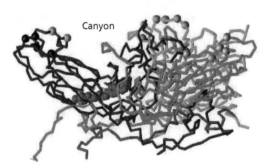

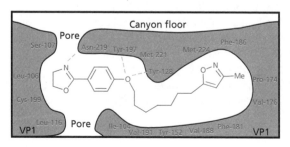

Figure 17.54 Caspid binding agents.

17.10 Broad-spectrum antiviral agents

Very few broad-spectrum antiviral agents that act on specific targets have reached the clinic. The following are some examples.

17.10.1 Agents acting against cytidine triphosphate synthetase

Cytidine triphosphate is an important building block for RNA synthesis and so blocking its synthesis inhibits the synthesis of viral mRNA. The final stage in the biosynthesis of cytidine triphosphate is the amination of uridine triphosphate—a process that is catalysed by the enzyme cytidine triphosphate synthetase. **Cyclopentenyl cytosine** (Fig. 17.55) is a carbocyclic nucleoside that is converted in the cell to the triphosphate which then inhibits this final enzyme in the biosynthetic pathway. The drug has broad antiviral activity against more than 20 RNA and DNA viruses, and has also been studied as an anticancer drug.

17.10.2 Agents acting against S-adenosylhomocysteine hydrolase

The 5'-end of a newly transcribed mRNA is capped with a methyl group in order to stabilize it against phosphatases and nucleases, as well as enhancing its translation. S-Adenosylhomocysteine hydrolase is an intracellular enzyme that catalyses this reaction, and many viruses need it to cap their own viral mRNA. **3-Deazaneplanocin A** (Fig. 17.55) is an analogue of cyclopentenyl cytosine, and acts against a range of RNA and DNA viruses by inhibiting S-adenosylhomocysteine hydrolase.

17.10.3 Ribavirin (or virazole)

Ribavarin (Fig. 17.55) is a synthetic nucleoside that induces mutations in viral genes and is used against hepatitis C infection. It was the first synthetic, non-interferon-inducing broad-spectrum antiviral nucleoside and can inhibit both RNA and DNA viruses by a variety of mechanisms, although it is only licensed for hepatitis C and respiratory syncytial virus. Nevertheless, it has also been used in developing countries for the treatment of tropical and haemorrhagic fevers such as Lassa fever when there is no alternative effective treatment. Tests show that it is useful in combination with other drugs such as rimantadine. Its dominant mechanism of action appears to be depletion of intracellular pools of GTP by inhibiting inosine-5'-monophosphate dehydrogenase. Phosphorylation of ribavarin results in a triphosphate which inhibits guanyl transferase and prevents the 5' capping of mRNAs. The triphosphate can also inhibit viral RNA-dependent RNA polymerase. Due to these multiple mechanisms of action, resistance is rare. The drug's main side effect is anaemia, and it is a suspected teratogen.

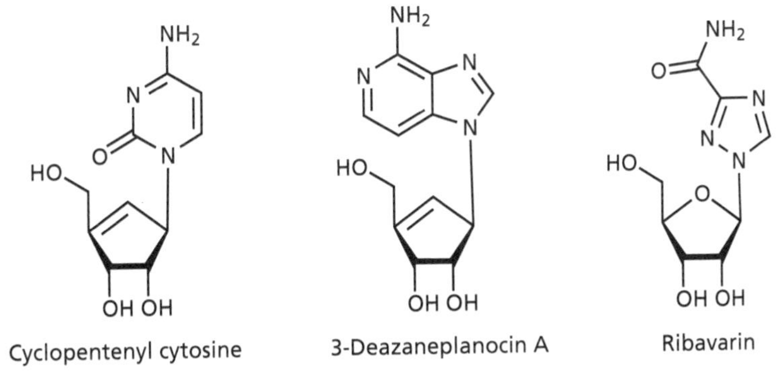

Figure 17.55 Broad-spectrum antiviral agents.

17.10.4 Interferons

Interferons are small natural proteins which were discovered in 1957, and which are produced by host cells as a response to 'foreign invaders'. Once produced, interferons inhibit protein synthesis and other aspects of viral replication in infected cells. In other words, they shut the cell down. This can be described as an intracellular immune response. Administering interferons to patients has been seen as a possible approach to treating flu, hepatitis, herpes, and colds.

There are several interferons, which are named according to their source: α-interferons from lymphocytes, β-interferons from fibroblasts, and γ-interferons from T-cells. α-Interferon (also called **alferon** or IFN-alpha) is the most widely used of the three types. In the past, isolating interferons from their natural cells was difficult and expensive, but recombinant DNA techniques allow the production of genetically engineered interferons in larger quantities (section 7.6). Recombinant α-interferon is produced in three main forms. The α-2a and α-2b are natural forms, and **alfacon-1** is the unnatural form. They have proved successful therapeutically, but can have serious toxic side effects. At present, α-interferon is used clinically against hepatitis B infections. It is also used with ribavarin against hepatitis C infections.

Interferon production in the body can also be induced by agents known as **immunomodulators**. One such example is **avridine** (Fig. 17.56) which is used as a vaccine adjuvant for the treatment of animal diseases such as foot and mouth. **Imiquimod** (Fig. 17.56) also induces the production of α-interferon, as well as other cytokines that stimulate the immune system. It is effective against genital warts.

17.10.5 Antibodies and ribozymes

Producing an antibody capable of recognizing an antigen that is unique to a virion would allow the virion to be marked for destruction by the body's immune system. **Palivizumab** is a humanized monoclonal antibody which was approved in 1998. It is used for the treatment of respiratory syncytial infection in babies and blocks viral spread from cell to cell by targeting a specific protein of the virus. Another monoclonal antibody is being tested for the treatment of hepatitis B.

It has been possible to identify sites in viral RNA that are susceptible to cutting by ribozymes—enzymatic forms of RNA. One such ribozyme is being tested in patients with hepatitis C and HIV. Ribozymes could be generated in the cell by introducing genes into infected cells—a form of gene therapy. Other gene therapy projects are looking at genes that would code for specialized antibodies capable of seeking out targets inside infected cells, or that would code for proteins which would latch on to viral gene sequences within the cell.

17.11 Bioterrorism and smallpox

The first effective antiviral drug to be used clinically was an agent called **methisazone** (Fig. 17.57) which was used in the 1960s against smallpox. The drug was no longer required once the disease was eradicated through worldwide vaccination. In recent years, however, there have been growing worries that terrorists might acquire smallpox and unleash it on a world no longer immunized against the disease. As a result, there has been a regeneration of research into

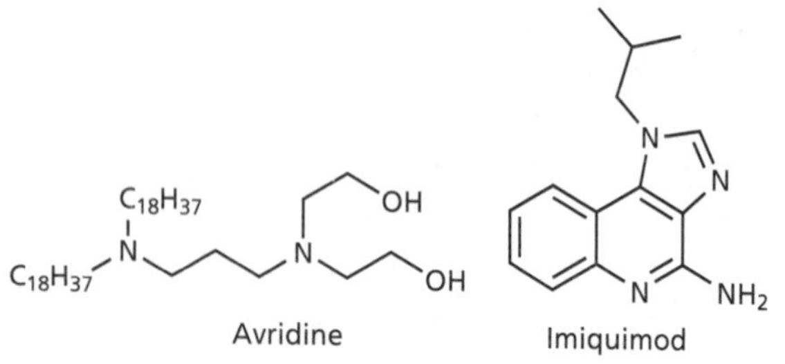

Figure 17.56 Immunomodulators.

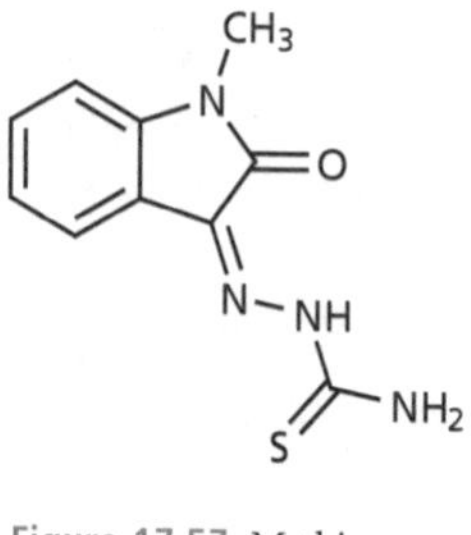

Figure 17.57 Methisazone.

finding novel antiviral agents which are effective against this disease.

- There are few broad-spectrum antiviral agents currently available.

- The best broad-spectrum antiviral agents appear to work on a variety of targets, reducing the chances of resistance.

- Interferons are chemicals produced in the body which shut down infected host cells and limit the spread of virus.

- Antibodies and ribozymes are under investigation as antiviral agents.

QUESTIONS

1. Consider the structures of the PIs given in section 17.7.4 and suggest a hybrid structure that might also act as a PI.

2. Consider the structure of the PIs in section 17.7.4 and suggest a novel structure with an extended subsite ligand.

3. What disadvantage might the following structure have as an antiviral agent compared to cidofovir?

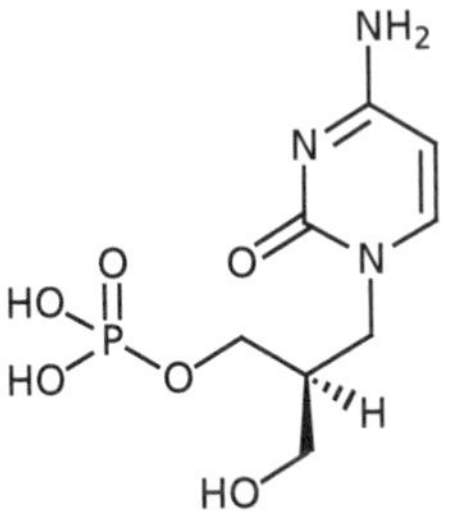

4. Zanamivir has a polar glycerol side chain which has good interactions with a binding pocket through hydrogen bonding, yet carboxamides and oseltamivir have hydrophobic substituents which bind more strongly to this pocket. How is this possible?

5. Show the mechanism by which the prodrugs tenofovir disoproxil and adefovir dipivoxil are converted to their active forms. Why are extended esters used as prodrugs for these compounds?

6. Capravirine has a side chain which takes part in important hydrogen bonding to Lys-103 and Pro-236 in the allosteric binding site of reverse transcriptase, yet the side chain has a carbonyl group. Discuss whether this is prone to enzymatic hydrolysis and inactivation.

7. All the PIs bind to the active site, with a water molecule acting as a hydrogen bonding bridge to the enzyme flaps. Suggest what relevance this information might have in the design of novel PIs.

8. The following structures were synthesized during the development of L 685434. Identify the differences between the two structures and suggest why one is more active than the other.

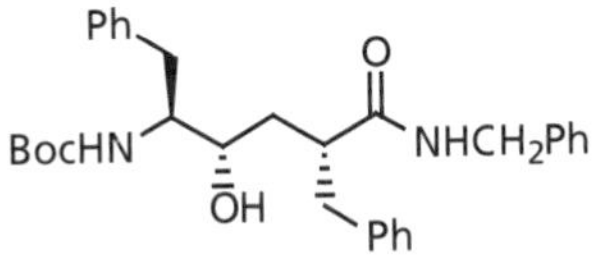

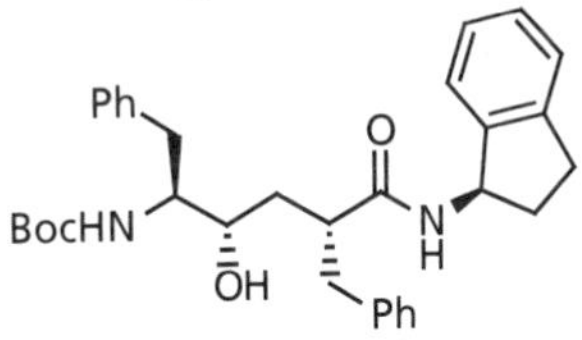

FURTHER READING

Carr, A. (2003) Toxicity of antiretroviral therapy and implications for drug development. *Nature Reviews Drug Discovery*, **2**, 624–634.

Coen, D. M. and Schaffer, P. A. (2003) Antiherpesvirus drugs: a promising spectrum of new drugs and drug targets. *Nature Reviews Drug Discovery*, **2**, 278–288.

De Clercq, E. (2002) Strategies in the design of antiviral drugs. *Nature Reviews Drug Discovery*, 1, 13–25.

Driscoll, J. S. (2002) *Antiviral agents.* Ashgate, Aldershot.

Greer, J., Erickson, J. W., Baldwin, J. J., and Varney, M. D. (1994) Application of the three-dimensional structures of protein target molecules in structure-based design. *Journal of Medicinal Chemistry*, 37, 1035–1054.

Milroy, D and Featherstone, J. (2002) Antiviral market overview. *Nature Reviews Drug Discovery*, 1, 11–12.

Tan, S-L. *et al.* (2002) Hepatitis C therapeutics: Current status and emerging strategies. *Nature Reviews Drug Discovery*, 1, 867–881.

HIV

Beaulieu, P. R. *et al.* (2000) *Journal of Medicinal Chemistry*, 43, 1094–1108 (palinavir).

Campiani, G. *et al.* (2002) Non-nucleoside HIV-1 reverse transcriptase (RT) inhibitors: past, present and future perspectives. *Current Pharmaceutical Design*, 8, 615–657.

De Clercq, E. (2003) The bicyclam AMD3100 story. *Nature Reviews Drug Discovery*, 2, 581–587.

HIV databases: http://www.hiv.lanl.gov/content/index

Kempf, D. J. and Sham, H. L. (1996) Protease inhibitors. *Current Pharmaceutical Design*, 2, 225–246.

Matthews, T. *et al.* (2004) Enfuvirtide: the first therapy to inhibit the entry of HIV-1 into host CD4 lymphocytes. *Nature Reviews Drug Discovery*, 3, 215–225.

Miller, J. F., Furfine, E. S., Hanlon, M. H. *et al.* (2004) Novel arylsulfonamides possessing sub-picomolar HIV protease activities and potent anti-HIV activity against wild-type and drug-resistant viral strains. *Bioorganic and Medical Chemistry Letters*, 14, 959–963 (amprenavir).

Raja, A., Lebbos, J., and Kirkpatrick, P. (2003) Atazanavir sulphate. *Nature Reviews Drug Discovery*, 2, 857–858.

Tomasselli, A. G. and Heinrikson, R. L. (2000) Targeting the HIV-protease in AIDS therapy. *Biochimica et Biophysica Acta*, 1477, 189–214.

Tomasselli, A. G., Thaisrivongs, S., and Heinrikson, R. L. (1996) Discovery and design of HIV protease inhibitors as drugs for AIDS. *Advances in Antiviral Drug Design*, 2, 173–228.

Werber, Y. (2003) HIV drug market. *Nature Reviews Drug Discovery*, 2, 513–514.

Flu

Babu, Y. S., Chand, P., Bantia, S. *et al.* (2000) Discovery of a novel, highly potent, orally active, and selective influenza neuraminidase inhibitor through structure-based drug design. *Journal of Medicinal Chemistry*, 43, 3482–3486.

Ezzell, C. (2001) Magic bullets fly again. *Scientific American*, October, 28–35 (antibodies).

Kim, C. U., Lew, W., Williams. M. A. *et al.* (1998) Structure–activity relationship studies of novel carbocyclic influenza neuraminidase inhibitors. *Journal of Medicinal Chemistry*, 41, 2451–2460 (oseltamivir).

Laver, W. G., Bischofberger, N., and Webster, R. G. (1999) Disarming flu viruses. *Scientific American*, January, 56–65.

Taylor, N. R., Cleasby, A., Singh, O. *et al.* (1998) Dihydropyrancarboxamides related to zanamivir: a new series of inhibitors of influenza virus sialidases. 2. crystallographic and molecular modeling study of complexes of 4-amino-4H-pyran-6-carboxamides and sialidase from influenza virus types A and B. *Journal of Medicinal Chemistry*, 41, 798–807 (carboxamides).

von Itzstein, M., Dyason, J. C., Oliver, S. W. *et al.* (1996) A study of the active site of influenza virus sialidase: an approach to the rational design of novel anti-influenza drugs. *Journal of Medicinal Chemistry*, 39, 388–391 (zanamivir).

Titles for general further reading are listed on p. 711.

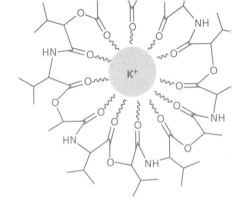

18 Anticancer agents

18.1 Cancer: an introduction

18.1.1 Definitions

Cancer still remains one of the most feared diseases in the modern world. According to the World Health Organization it affected one person in three and caused a quarter of all deaths in the developed world during the year 2000. After heart disease, it is the largest cause of death. Cancer cells are formed when normal cells lose the normal regulatory mechanisms that control growth and multiplication. They become 'rogue cells' and often lose the specialized characteristics that distinguish one type of cell from another (for example a liver cell from a blood cell). This is called a loss of **differentiation**. The term **neoplasm** means new growth and is a more accurate terminology for the disease. The terms **cancer** and **tumour**, however, are more commonly accepted and will be used throughout this chapter. (The word tumour actually means a local swelling.) If the cancer is localized it is said to be **benign**. If the cancer cells invade other parts of the body and set up secondary tumours—a process known as **metastasis**—the cancer is defined as **malignant**. It is the latter form of cancer which is life threatening. A major problem in treating cancer is the fact that it is not a single disease. There are more than 200 different cancers resulting from different cellular defects, and so a treatment that is effective in controlling one type of cancer may be ineffective on another.

18.1.2 Causes of cancer

Possibly as many as 30% of cancers are caused by smoking, while another 30% are diet related.

Carcinogenic chemicals in smoke, food, and the environment may cause cancer by inducing gene mutations or interfering with normal cell differentiation. The birth of a cancer (**carcinogenesis**) can be initiated by a chemical—usually a **mutagen**—but other triggering events such as exposure to further mutagens are usually required before a cancer develops.

Viruses have been implicated in at least six human cancers and are the cause of about 15% of the world's cancer deaths. For example, the Epstein–Barr virus is the cause of Burkitt's lymphoma and nasopharyngeal carcinoma. Human papillomaviruses are sexually transmitted and can lead to cancer of the cervix. Hepatitis B may cause 80% of all liver cancers, and HIV can cause Kaposi's sarcoma and lymphoma. Viruses can bring about cancer in several ways. They may bring oncogenes (see below) into the cell and insert them into the genome. For example, Rous sarcoma virus carries a gene for abnormal tyrosine kinase. Some viruses carry one or more promoters or enhancers. If these are integrated next to a cellular oncogene, the promoter stimulates its transcription leading to cancer. The bacterium *Helicobacter pylori* is responsible for many stomach ulcers (section 22.4) and is also implicated in stomach cancer.

The treatments used to combat cancer (radiotherapy and chemotherapy) can actually induce a different cancer in surviving patients. For example, 5% of patients cured of Hodgkin's disease developed acute leukaemia. Nevertheless, the risk of a second cancer is outweighed by the benefit of defeating the original one.

Some patients are prone to certain cancers for genetic reasons. Damaged genes can be passed from one generation to another, increasing the risk of cancer in subsequent generations.

18.1.3 Genetic faults leading to cancer: proto-oncogenes and oncogenes

18.1.3.1 Activation of proto-oncogenes

Proto-oncogenes are genes which normally code for proteins involved in the control of cell division and differentiation. If they are mutated, this disrupts the normal function and the cell can become cancerous. The proto-oncogene is then defined as an **oncogene**. The *ras* gene is one example. Normally, it codes for a protein called Ras which is involved in the signalling pathway leading to cell division (section 6.7.4). In normal cells, this protein has a self-regulating ability and can switch itself off. If the gene becomes mutated, an abnormal Ras protein is produced which loses this ability and is continually active, leading to continuous cell division. It has been shown that mutation of the *ras* gene is present in 20–30% of human cancers. Oncogenes may also be introduced to the cell by viruses.

18.1.3.2 Inactivation of tumour suppression genes (anti-oncogenes)

If DNA is damaged in a normal cell, there are cellular 'policemen' that can detect the damage and block DNA replication. This gives the cell time to repair the damaged DNA before the next cell division. If repair does not prove possible, the cell commits suicide (**apoptosis**). Tumour suppression genes are genes which code for proteins that are involved in these processes of checking, repair and suicide. *TP53* is an important example of such a gene and codes for the protein of the same name (p53). If the *TP53* gene is damaged, the repair mechanisms become less efficient, defects are carried forward from one cell generation to another and as the damage increases, the chances of the cell becoming cancerous increase.

Genetic defects can lead to the following cellular defects, all of which are associated with cancer:

- abnormal signalling pathways
- insensitivity to growth-inhibitory signals
- abnormalities in cell cycle regulation
- evasion of programmed cell death (apoptosis)
- limitless cell division (immortality)
- ability to develop new blood vessels (angiogenesis)
- tissue invasion and metastasis.

It is thought that most if not all of these conditions have to be met before a defective cell can spawn a life-threatening malignant growth. Thus, a single defect can be kept under control by a series of safeguards. This can explain why cancers may take many years to develop after exposure to a damaging mutagen such as asbestos or coal dust. That first exposure may have caused mutations in some cells, but cellular chemistry has the control systems in place to cope and to keep the cells in check. A lifetime's exposure to other damaging mutagens such as tobacco smoke, however, results in further genetic damage which overwhelm the safeguards one by one until the abnormal cell finally breaks free of its shackles and becomes cancerous.

The various hurdles and safeguards that a potential cancer cell has to overcome explains why cancers are relatively rare early on in life and are more common in later years. This also helps to explain why cancer is so difficult to treat once it does appear. Since so many cellular safeguards have already been overcome, it is unlikely that tackling one specific cellular defect is going to be totally effective. As a result, traditional anticancer drugs have tended to be highly toxic agents and act against a variety of different cellular targets by different mechanisms. Unfortunately, since they are potent cellular poisons, they also affect normal cells and produce serious side effects. Such agents are said to be **cytotoxic**, and dose levels have to be chosen which are high enough to affect the tumour but are bearable to the patient. In recent years, anticancer drugs have been developed which target specific abnormalities in a cancer cell, allowing them to be more selective and have less serious side effects. However, bearing in mind the number of defects in a cancer cell, it is unlikely that a single agent of this kind will be totally effective and it is more likely that these new agents will be most effective when they are used in combination with other drugs having different mechanisms of action, or with surgery and radiotherapy.

We now look at the various defects that are common in cancer cells.

18.1.4 Abnormal signalling pathways

Whether a normal cell grows and divides depends on the various signals it receives from surrounding cells. The most important of these signals come from hormones called **growth factors**. These are extracellular chemical messengers which activate protein kinase receptors in the cell membrane (section 6.7). The receptors concerned trigger a signal transduction pathway which eventually reaches the nucleus and instructs the transcription of the proteins and enzymes required for cell growth and division. Most if not all cancers suffer from some defect in this signalling process, such that the cell is constantly instructed to multiply. The signalling process is complex, so there are various points at which it can go wrong.

Many cancer cells are capable of growing and dividing in the absence of external growth factors. They can do this by producing the growth factor themselves, then releasing it such that it stimulates its own receptors. Examples include growth factors called **platelet-derived growth factor** (PDGF) and **transforming growth factor α** (TGF-α). Other cancer cells can produce abnormal receptors which are constantly switched on despite the lack of growth factors (e.g. Erb-B2 receptors in breast cancer cells). It is also possible for receptors to be **over-expressed**. This means that an oncogene is too active, and codes for excessive protein receptor. Once this is in the cell membrane, the cell becomes supersensitive to low levels of growth factor.

There are many points where things could go wrong in the signal transduction pathways. For example, the Ras protein is a crucial feature in the signal transduction pathways leading to cell growth and division. Abnormal Ras protein is locked in the 'on' position and is constantly active despite the lack of an initial signal from a growth factor.

18.1.5 Insensitivity to growth-inhibitory signals

Several external hormones such as **transforming growth factor β** (TGF-β) counteract the effects of stimulatory growth factors, and signal the inhibition of cell growth and division. Insensitivity to these signals raises the risk of a cell becoming cancerous. This can arise from damage to the genes coding for the receptors for these inhibitory hormones—the tumour suppression genes.

18.1.6 Abnormalities in cell cycle regulation

A cycle of events takes place during cell growth and multiplication which involves four phases known as G_1, S, G_2, and M (Fig. 18.1). As part of this process,

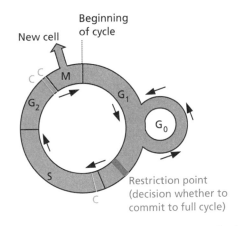

Figure 18.1 The cell cycle. G_1, gap 1—cell enlarges and makes new proteins; S, synthesis of DNA; G_2, gap 2—cell prepares to divide; M, mitosis—cell divides; G_0, resting stage—no growth; C, checkpoints.

decisions have to be made by the cell whether to move from one stage to another, depending on the balance of those chemical signals promoting growth and those inhibiting it.

The G_1 phase (gap 1) is where a cell is actively growing in size and preparing to copy its DNA in response to various growth factors or internal signals. The next phase is the S phase (synthesis) where replication of DNA takes place. Once the cell's chromosomes are copied, there is another interval called the G_2 phase (gap 2) during which the cell readies itself for cell division. This gap or interval is crucial, as it gives the cell time to check the copied DNA and to repair any damaged copies. Finally, there is the M phase (mitosis) where cell division takes place to produce two daughter cells, each containing a full set of chromosomes. The daughter cells can then enter the cell cycle again (G_1). Alternatively, they may move into a dormant or resting state (G_0).

Within the cell cycle, there are various decision points which determine whether the cell should continue to the next phase. For example, there is a decision point called the restriction point (R) during the G_1 phase which frequently becomes abnormal in tumour cells. There are also various surveillance mechanisms known as checkpoints which assess the integrity of the process. For example, a delay will take place during the G_2 phase if DNA damage is detected. This gives sufficient time for damaged DNA to be repaired or for the cell to commit suicide (apoptosis). These checkpoints can also be defective in tumour cells.

Control of the cell cycle involves a variety of proteins called **cyclins**, and enzymes called **cyclin-dependent kinases** (CDKs) (Fig. 18.2). There are at least 15 types of cyclin and 9 types of CDK, and each has a role to play at different stages of the cell cycle. Examples are shown in Fig. 18.2. Binding of a cyclin with its associated kinase activates the enzyme and serves to move the cell from one phase of the cell cycle to another. For example, when a cell is in the G_1 phase,

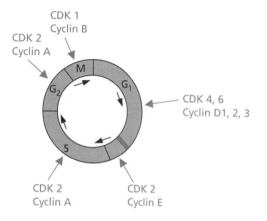

Figure 18.2 Control of the cell cycle.

a decision has to be made whether to move into the S phase and start copying DNA. This decision is taken depending on the balance of stimulatory versus inhibitory signals being received through signal transduction. If the balance is towards cell growth and division, there is an increase in cyclin D. This binds to CDK4 and CDK6. The resulting complexes phosphorylate a powerful growth-inhibitory molecule known as **pRB** which normally binds and inactivates a transcription factor. Phosphorylation alters pRB such that it can no longer bind to the transcription factor and the latter is free to bind to specific regions of DNA. This results in the transcription of specific genes which lead to the production of proteins capable of moving the cell towards the S phase (e.g. cyclin E and thymidine kinase). Once cyclin E has been produced, it combines with CDK2 and this complex is responsible for progression from the G_1 phase to the S phase. Other activated cyclin–CDK complexes are important in different phases of the cell cycle. For example, cyclin A–CDK2 is required for progression through the S phase, and cyclin B–CDK1 is necessary for mitosis.

Restraining proteins are present which can modify the effect of cyclins (Fig. 18.3). These include **p15** and

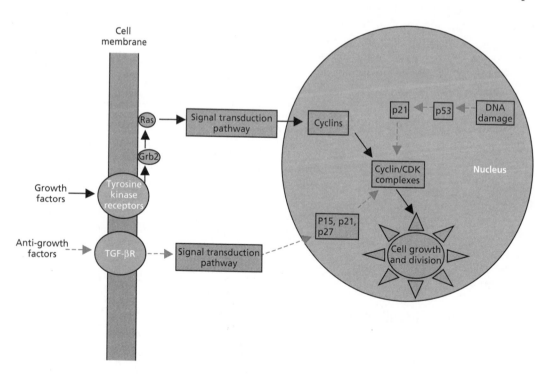

Figure 18.3 Signalling pathways that activate and inhibit cell growth and division.

p16 which block the activity of the cyclin D–CDK complex. Another is the inhibitory protein **p21** which is controlled by **p53**—an important protein that monitors the health of the cell and the integrity of DNA.

To sum up, progression through the cell cycle is regulated by sequential activation of cyclins and CDKs—a process which can be down-regulated by the CDK inhibitors. The whole process is normally tightly controlled, such that there is an accumulation of a relevant cyclin–CDK complex followed by rapid degradation of the cyclin once its task is complete.

Overactive cyclins or CDKs have been associated with several cancers. For example, breast cancer cells often produce excess cyclin D and E, and skin melanoma has lost the gene that codes for the inhibitory protein p16. Half of all human tumours lack a proper functioning p53 protein, which means that the level of the inhibitory protein p21 falls. In viral-related cervical cancers, both the pRB and p53 proteins are often disabled.

Oncogenic alteration of cyclins, CDKs, cyclin-dependent kinase inhibitors (CKIs), and other components of the pRB pathway have been reported in 90% of human cancers, especially in the G_1 phase. Thus, excessive production of cyclins or CDKs, or insufficient production of CKIs, can lead to a disruption of the normal regulation controls and lead to cancer. Efforts have been made to identify how one can restore the control of the cancer cell cycle by targeting molecular abnormalities. These can include CDK inhibition, down-regulation of cyclins, up-regulation of CDK inhibitors, degradation of cyclins, or inhibition of tyrosine kinases that trigger the cell cycle activation in the first place.

18.1.7 Apoptosis and the p53 protein

There is a built-in cellular destruction process called **apoptosis**, which is the normal way in which the body protects itself against abnormal or faulty cells. Each cell essentially monitors itself for a series of different chemical signals. Should any of these be absent, a self-destruct mechanism is automatically initiated (Fig. 18.4). Apoptosis is also important in destroying cells that escape from their normal tissue environment. Cancer cells which metastasize have undergone genetic changes that allow them to avoid this process.

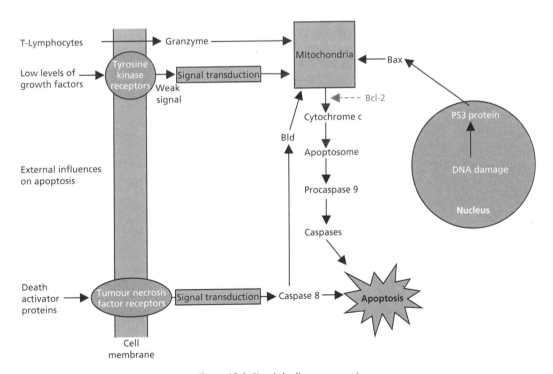

Figure 18.4 Signals leading to apoptosis.

Two distinct pathways for apoptosis have been characterized:

- An **extrinsic route** where apoptosis results from external factors, which can take three forms. First, there could be a sustained lack of growth factors or hormones. Second, there are proteins called **death activator proteins** which can bind to cell membrane proteins called **tumour necrosis factor receptors** (TNF-R). This triggers a signalling process initiating apoptosis. Finally, the immune system produces T-lymphocytes which circulate the body searching for damaged cells. Once found, the lymphocyte perforates the cell membrane of the damaged cell and injects an enzyme called **granzyme** which initiates apoptosis.

- An **intrinsic pathway** can arise from factors such as DNA damage arising from exposure to chemicals, drugs, or oxidative stress. The cell has monitoring systems which can detect damage and lead to the increased production of the tumour suppressor protein p53. At sufficient levels, this protein will trigger apoptosis.

The various signals described above converge on the mitochondria which contain proteins capable of promoting apoptosis, in particular **cytochrome c**. Release of cytochrome c from mitochondria results in the assembly of a large oligomeric protein complex known as an **apoptosome** which is made up of a scaffolding protein called **Apaf-1**. The apoptosome then recruits and activates an enzyme known as **procaspase 9** which in turn activates **caspases**. Caspases are protease enzymes containing a cysteine residue in the active site which is important to the catalytic mechanism. They are proteases, so they set about destroying the cell's proteins and this leads to destruction of the cell.

Considering the fatal effect caspases have on the cell, it is not surprising that there are various checks and balances to ensure that apoptosis does not occur too readily. A family of proteins regulate the process, some members of the family such as **Bad** or **Bax** promoting it and others such as **Bcl-2** and **Bcl-X** suppressing it. The relative levels of these proteins is dependent on the various monitoring procedures within the cell. For example, genetic damage leading to increased levels of p53 induces apoptosis by up-regulating the expression of Bax.

The survival of each cell in the body is therefore dependent on the balance of internal and external signals regulating cell growth, as well as the balance of regulatory chemicals promoting or inhibiting apoptosis. A defect in the complex systems leading to apoptosis could inhibit apoptosis, increasing the likelihood of carcinogenesis. For example, it has been found that the gene coding for p53 is the most frequently mutated gene in cancer (30–70%). Damage to this gene means a lack of the apoptosis-inducing p53 protein, and an increased chance that the defective cell will survive to become cancerous. The genes coding for the apoptosis suppressors Bcl-2 and Bcl-X$_L$ are also known to be over-expressed in several tumour types.

Defects in the apoptosis mechanisms also has serious consequences for radiotherapy and many chemotherapeutic drugs, since both these procedures act by triggering apoptosis. For example, many traditional drugs damage DNA. This in itself may not be fatal to the cell, but the cell's monitoring systems detects the damage and goes into self-destruct mode. If the mechanisms involved are defective then apoptosis does not occur and the drugs are not as effective.

18.1.8 Telomeres

Cancer cells are often described as becoming 'immortal'. This is because there is no apparent limit to the number of times they can divide. The lifetime of normal cells is predetermined by the possible number of times their DNA can be replicated (about 50–60 cell divisions).

Structures called **telomeres** (Fig. 18.5) play a key role in this immortalization process. A telomere consists of a polynucleotide region at the 3' ends of chromosomes, which contains several thousand repeats of a short (6 bp) sequence. The purpose of the telomere is to act as a 'splice' for the end of the chromosome and to stabilize and protect the DNA. After each replication process, about 50–100 base pairs are lost from the telomere because DNA polymerase is unable to completely replicate the 3' ends of chromosomal DNA. Eventually, the telomere becomes too short to be effective and the DNA becomes unstable, either unravelling or linking up with another DNA end to end. This proves fatal to the cell and apoptosis is triggered.

It is observed that in the early stages of cancer many cancer cells are also restricted to the number of times

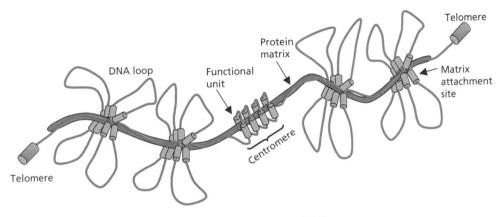

Figure 18.5 Chromatin, telomeres, and DNA.

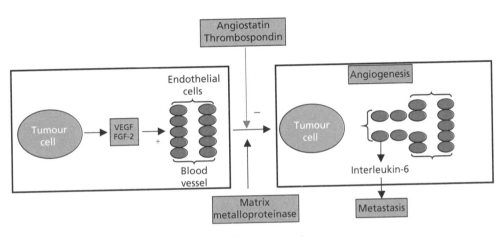

Figure 18.6 Angiogenesis.

they can divide, but eventually a cancer cell develops which breaks free of this restriction and becomes immortal, i.e. it is capable of dividing an indefinite number of times. These cells maintain the length of their telomere by expressing an enzyme called **telomerase**—a member of a group of enzymes called the RNA-dependent DNA polymerases. Telomerase has the ability to add hexanucleotide repeats on to the end of telomeric DNA and thus maintain its length. This is an important process during the development of an embryo when telomerase is responsible for creating the telomeres in the first place, but after birth the gene encoding the enzyme is suppressed. Immortal cells have found a way of removing that suppression such that the enzyme is expressed once more. The telomerase enzyme is expressed in over 85% of cancers.

Several efforts have been made to design drugs which will inhibit telomerase, but no such drugs have reached the clinic to date.

18.1.9 Angiogenesis

As a tumour grows, its cancerous cells require a steady supply of amino acids, nucleic acid bases, carbohydrates, oxygen, and growth factors if they are to continue multiplying. This means that the tumour has to have a good blood supply. As a tumour grows in size, however, its cells become increasingly remote from the blood supply and become starved of these resources. This is particularly true for the cells in the centre of the tumour. In order to counter this, tumour cells release growth factors such as **vascular endothelial growth factor** (VEGF) and **fibroblast growth factor** (FGF-2) which interact with receptors on the endothelial cells of nearby blood vessels and stimulate these cells to divide, leading to the branching and extension of existing capillaries—a process known as **angiogenesis** (Fig. 18.6). Vascular growth factors are present in normal cells and are usually released when tissues have been damaged. The resulting angiogenesis helps in the repair of the injured tissues and is normally controlled by angiogenesis inhibitors such as **angiostatin** and **thrombospondin**. Unfortunately, this balance is disturbed in tumour growth. As a result, tumours are able to receive the increased blood supply required for their survival. Moreover, the chances of cancer cells escaping from the primary source and metastasizing are increased, not only because of the increased availability of blood vessels but also because the newly developing endothelial cells can release proteins such as **interleukin-6** that stimulate metastasis. The blood vessels arising from angiogenesis are abnormal in that they are disorganized in structure, dilated, and leaky. The cells also display molecules called **integrins** on their surface which are absent from mature vessels and which protect the new cells from apoptosis. Before angiogenesis can begin, the basement membrane round the blood vessels has to be broken down and this is carried out by enzymes known as **matrix metalloproteinases (MMPs)**. This then allows the endothelial cells to migrate towards the tumour. Dissolution of the matrix also allows angiogenesis factors to be released to encourage angiogenesis.

Inhibiting angiogenesis is a tactic which can help to tackle cancer. Drugs have been developed that inhibit angiogenesis and break down the abnormal blood vessels. Angiogenesis inhibitors are generally safer and less toxic than traditional chemotherapeutic agents, but are unlikely to be used on their own. Instead they will be probably be used alongside standard cancer treatments such as surgery, chemotherapy, and radiation. Angiogenesis inhibitors appear to 'normalize' the abnormal blood vessels of tumours before they kill them. This normalization can help anticancer agents reach tumours more effectively. In the longer term, it serves to stall tumour growth, then shrink it by breaking up abnormal capillaries. As a result, the tumour becomes starved of nutrients and growth should decrease.

Some anticancer treatments take advantage of the leaky blood vessels which result from angiogenesis. Anticancer drugs can be encapsulated into liposomes, nanospheres, and other drug delivery systems which are too big to escape from normal blood vessels, but can escape through the walls of the leakier blood vessels supplying the tumour. As a result, the anticancer drug is concentrated at the tumour. Since tumours generally do not develop an effective lymphatic system, the polymeric drug delivery systems tend to be trapped at the tumour site.

Despite angiogenesis, there are regions of a well-developed tumour which fail to receive an adequate blood supply. As a result, cells in the centre of the tumour are starved of oxygen and nutrients, and may well stop growing and become dormant. This can pose a serious problem, as most anticancer drugs act best on actively dividing cells. Anticancer therapy may well be successful in halting a cancer and eliminating most of it. Once the treatment is stopped, though, the dormant cells start multiplying and the tumour reappears. Worryingly, it has been observed that such cells are more likely to metastasize.

Another consequence of an insufficient blood supply and lack of oxygen is that cells in the centre of the tumour are forced to revert to glycolysis in order to produce energy. This leads to a build-up of acidic by-products within the cell. The cells address this problem by exporting acidic protons into the extracellular space. As a result, the environment around tumours tends to be more acidic than in normal tissues. Several anticancer therapies have attempted to take advantage of this difference in acidities—for example, the selective localization of porphyrins in photodynamic therapy.

Angiogenesis inhibitors are relatively safe, so it may be possible to use these drugs as a prophylactic to prevent the appearance of cancer in susceptible individuals.

Several of the drugs described in the following sections inhibit angiogenesis. These include combretastatin (section 18.5.1), VEGF receptor kinase inhibitors (section 18.6.2.4), matrix metalloproteinase inhibitors (section 18.7.1), TNP-470 (section 18.7.5), thalidomide (section 18.8.1), endostatin and angiostatin (section 18.8.3), and bevacizumab (section 18.9.1).

18.1.10 Tissue invasion and metastasis

Not all cancers are life threatening. Benign tumours are growths which remain localized in a particular part of the body and can grow to the size of a football without a fatal result. Malignant cancers, on the other hand, are life threatening because the cells involved have the ability to break away from the primary tumour, invade a blood vessel or a lymphatic vessel, travel through the circulation, and set up tumours elsewhere in the body. In order to do this, these cells have to overcome a series of controls that are designed to keep cells in their place.

Cells have a molecular signature on their surface which identifies whether they are in the correct part of the body or not. These are cell adhesion molecules (e.g. E-cadherin) which ensure that cells adhere to cells of similar character, and to an insoluble mesh-work of protein filling the space between them—the extracellular matrix. This is particularly true of epi-thelial cells—the cell layers forming the outer surface of skin and the lining of the gut, lungs, and other organs.

Adhesion to the extracellular matrix is particularly important, as it is necessary if cells to survive—an anchoring requirement which involves molecules called **integrins**. If a normal cell becomes detached, it stops growing and apoptosis is triggered. This prevents cells from one part of the body straying to other parts of the body. Moreover, normal cells can only survive if their adhesion molecules match the relevant extra-cellular matrix.

Cell adhesion molecules are missing in metastasized cancer cells, allowing them to break away from the primary tumour. Such cells also appear to be anchorage independent: they do not self-destruct once they have become free, and can latch on to extracellular matrix in other parts of the body to set up secondary tumours. It is thought that oncogenes in these cells code for proteins which send false messages back to the nucleus implying that the cell is still attached.

It is noticeable that most cancers derive from epithelial cells. Once an epithelial cell has gained the ability to split away from its neighbours, it needs to gain access to the blood supply if it is to spread round the body. However, epithelial cells grow on a basement membrane—a thin layer of extracellular matrix which acts as a physical barrier to the movement of the cells. Cancer cells and white blood cells are the only cells capable of breaching this barrier. White blood cells need to do this in order to reach areas of infection, whereas cancer cells breach the barrier to spread the disease. Both types of cell contain the **matrix metalloproteinase** enzyme that hydrolyses the proteins composing the barrier. Once a cancer cell breaks through the basement barrier, it has to break down a similar barrier surrounding the blood vessel in order to enter the blood supply. It then spreads round the body carried by the blood supply until it finally adheres to the blood vessel and breaks out by the opposite process in order to reach new tissue. It is estimated that fewer than one in 10 000 such cells succeed in setting up a secondary tumour, but it only needs one such cell to start up a secondary tumour, and once metastasis has occur-red, the prospects of survival are slim. Circulating tumour cells usually get trapped in the first network of capillaries they meet, and this is where they are most likely to set up secondary tumours. For most tissues, the focus for secondary tumours will be the lungs. In the case of cells originating from the intestines, it is the liver. Some cancer cells produce factors that cause platelets to initiate blood clotting around them such that they increase in size, become stickier, and stick to the blood vessel wall, allowing them to escape.

18.1.11 Treatment of cancer

There are three traditional approaches to the treatment of cancer—surgery, radiotherapy, and chemotherapy. This chapter is devoted to cancer chemotherapy, but it is important to appreciate that it is usually used alongside surgery and radiotherapy. Moreover, it is often the case that combination therapy (the simul-taneous use of various anticancer drugs with different mechanisms of action) is more effective than using a single drug. The advantages include increased effici-ency of action and decreased toxicity.

Since cancer cells are derived from normal cells, identifying targets that are unique to cancer cells is not easy. As a result, most traditional anticancer drugs act against targets which are present in both types of cell. Therefore, the effectiveness and selectivity of such drugs is dependent on them becoming more concen-trated in cancer cells than normal cells. This often turns out to be the case, since cancer cells are generally growing faster than normal cells, and so they accu-mulate nutrients, synthetic building blocks, and drugs more quickly. Unfortunately, not all cancer cells grow rapidly; cells in the centre of a tumour may be dormant and evade the effects of the drug. Conversely, there are normal cells in the body which grow rapidly, such as bone marrow cells. As a result, they too accumulate anticancer drugs, resulting in bone mar-row toxicity—a common side effect of cancer chemotherapy which results in a weakening of the immune response and a decreased resistance to infection. Indeed, many cancer patients are prone to pathogens which would not normally be infectious. Such secondary infections can be difficult to treat, and care has to be taken over which antibacterial drugs are used. For example, bacteriostatic antibacterial agents may not be effective, as they rely on the normal functioning of the immune system. Other typical side effects of traditional anticancer drugs are impaired wound healing, loss of hair, damage to the epithelium of the gastrointestinal tract, depression of growth in children, sterility, teratogenicity, nausea, and kidney damage.

Most traditional anticancer drugs work by disrupting the function of DNA and are classed as cytotoxic. Some act on DNA directly; others (antimetabolites) act indirectly by inhibiting the enzymes involved in DNA synthesis. Having said that, cancer chemotherapy is now entering a new era which can be described as molecular targeted therapeutics—highly selective agents which target specific molecular targets that are abnormal or over-expressed in the cancer cell. Progress in this area has arisen from a better understanding of the cellular chemistry involved in particular cancer cells. The development of kinase inhibitors such as **imatinib (Glivec)** is a much-heralded illustration of this approach (section 18.6.2). The use of antibodies and gene therapy is another area of research which shows huge potential (section 18.9).

Knowledge of the cell cycle is important in chemotherapy. Some drugs are more effective during one part of the cell cycle than another. For example, drugs which affect microtubules are effective when cells are actively dividing (the M phase), whereas drugs acting on DNA are more effective if the cells are in the S phase. Some drugs are effective regardless of the phase; for example, alkylating agents such as cisplatin. For this reason, anticancer drugs are most effective against cancers which are rapidly proliferating, since they are more likely to become susceptible when they reach the relevant part of the cell cycle. Conversely, slower-growing cancers are less effectively treated.

A better understanding of the molecular mechanisms behind specific cancers is yielding better and more specific treatments. The genetic analysis of tumours in individual patients is also likely to become an important feature in the early detection and identification of cancer, as well as allowing the best treatment to be used for a particular individual. Genetic fingerprinting should also identify individuals who are at risk to particular cancers so that they can be regularly screened. The importance in detecting cancer early on cannot be overemphasized. Unfortunately, the physical symptoms of most tumours do not become apparent until they are well established. By that time, it may be too late.

Although cancer is difficult to treat, there have been notable successes in treating rapidly growing cancers such as Hodgkin's disease, Burkitt's lymphoma, testicular cancer, and several childhood malignancies. Early diagnosis also improves the chances of successful treatment in other cancers. At present, four cancers account for over half of all new cases (lung, breast, colon, and prostate).

Finally, one of the best ways of reducing cancers is to reduce the risk. Public education campaigns are important in highlighting the dangers of smoking, excessive drinking, and hazardous solvents, as well as promoting healthy diets and lifestyles. The benefits of eating high-fibre foods, fruit, and vegetables are clear. Indeed, there have been various research projects aimed at identifying the specific chemicals in these foods which are responsible for this protective property. For example, dithiolthiones are a group of chemicals in broccoli, cauliflower, and cabbage which appear to have protective properties, one of which involves the activation of enzymes in the liver to detoxify carcinogens. **Genistein** (Fig. 18.7) is a protective compound found in soy products used commonly in Asian diets. It is notable that Asian populations have a low incidence of breast, prostate, and colon cancers. **Epigallocatechin gallate**, an antioxidant present in green tea, is another potential protective agent. Synthetic drugs are also being investigated as possible cancer preventives (e.g. finasteride, aspirin, ibuprofen, and difluoromethylornithine).

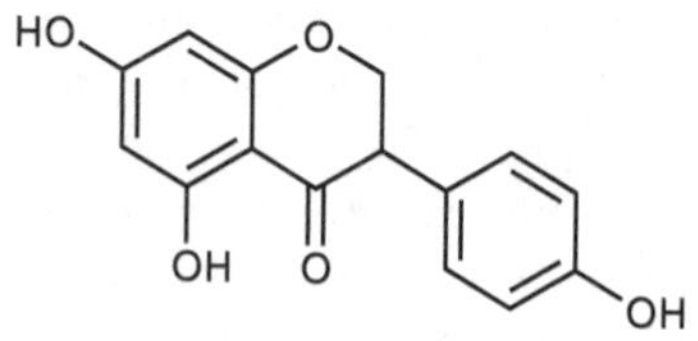

Figure 18.7 Genistein.

18.1.12 Resistance

Resistance to anticancer drugs is a serious problem. Resistance can be intrinsic or acquired.

- **Intrinsic resistance** means that the tumour shows little response to an anticancer agent from the very start. This can be due to a variety of possible mechanisms such as slow growth rate, poor uptake of the drug, or the biochemical/genetic properties of the cell. Tumour cells in the centre of the tumour may be in the resting state and be intrinsically resistant as a result.

- When a tumour is initially susceptible to a drug but becomes resistant, it is said to show **acquired resistance**. This is due to the presence of a mixture of drug-sensitive and drug-resistant cells within the tumour. The drug wipes out the drug-sensitive cells, but this only serves to select out and enrich the drug-resistant cells. The survival of even one such cell can lead to failure of the treatment, as that one survivor can spawn a newer drug resistant tumour. One might ask why a single tumour should contain drug-sensitive and drug-resistant cells since it is likely to have developed from a single cell in the first place. The reason is that cancer cells by their very nature are genetically unstable and so mutations are bound to have occurred during tumour growth which will result in resistant cells.

There are several molecular mechanisms by which resistance can take place as a result of mutation. For example, resistance can be due to decreased uptake of drug by the cell or increased synthesis of the target against which the drug is directed. Some drugs need to be activated in the cell, and the cancer cell may adapt such that these reactions no longer take place. Alternative metabolic pathways may be found to avoid the effects of antimetabolites. Drugs may be actively expelled from the cell in a process known as **efflux**. A cell membrane carrier protein called **P-glycoprotein** is particularly important in this last mechanism. This protein normally expels toxins from normal cells, but mutations in cancer cells can result in an increased expression of the protein such that anticancer drugs are efficiently removed as soon as they enter the cell. Unfortunately, the P-glycoprotein can eject a wide diversity of molecules. As a result, cells with excess P-glycoprotein are resistant to a variety of different anticancer drugs, even if they have not been exposed to them before. This is known as **multidrug resistance** (**MDR**). For example, cells acquiring resistance to the vinca alkaloids are also resistant to dactinomycin (actinomycin D) and anthracyclines.

Efforts have been made to counter this form of resistance by developing drugs which compete for the P-glycoprotein or inhibit it. The calcium ion channel blocker **verapamil** effectively competes for the P-glycoprotein and allows the build-up of an anticancer drug within the cancer cell. Unfortunately, verapamil cannot be used clinically because of its own inherent activity, but there is potential in this approach.

Since it is likely that a drug-resistant cell may be present in a cancer, it makes sense to use combinations of anticancer drugs with different targets, to increase the chances of finding a weakness in every cell.

KEY POINTS

- Cancer cells have defects in the normal regulatory controls governing cell growth and division. Such defects arise from mutations resulting in the activation of oncogenes and the inactivation of tumour suppression genes.

- Defects in signalling pathways are commonly found in cancer cells. The pathways stimulating cell growth and division are overactive as a result of the overproduction of a crucial protein in the pathway or the production of an abnormal protein. The proteins involved include growth factors, receptors, signal proteins, and kinases.

- The production of regulatory proteins which suppress cell growth and division is suppressed in many cancers.

- The cell cycle consists of four phases. Progression through the cell cycle is controlled by cyclins and cyclin-dependent kinases, moderated by restraining proteins. Defects in this system have been detected in 90% of cancers.

- Apoptosis is a destructive process leading to cell death. Cells have monitoring systems which check the general health of the cell and trigger the process of apoptosis if there are too many defects. Regulatory proteins have a moderating influence on apoptosis. Defects in apoptosis increase the chances of defective cells developing into cancerous cells and reduce the effectiveness of several drugs.

- Telomeres act as splices to stabilize the ends of DNA. Normally, they decrease in size at each replication until they are too short to be effective, resulting in cell death. Cancer cells activate the expression of an enzyme called telomerase to maintain the telomere and become immortal.

- Angiogenesis is the process by which tumours stimulate the growth of new blood vessels to provide the nutrients required for continued growth. Agents which inhibit angiogenesis are useful in anticancer therapy to inhibit tumour growth and to enhance the effectiveness of other drugs.

- Metastasis is the process by which cancer cells break free of the primary tumour, enter the blood supply, and set up secondary tumours in other tissues. To do this, the regulatory controls which fix cells to a specific environment, and which destroy cells that become detached, are overruled.

- Surgery, radiotherapy, and chemotherapy are used to treat cancer. Chemotherapy usually involves combinations of drugs having different targets or mechanisms of action. Traditional anticancer drugs are generally cytotoxic; more modern drugs are selective in their action.

- Cancer cells can have intrinsic or acquired resistance to anticancer drugs. Resistance may be due to poor uptake of the drug, increased production of the target protein, mutations which prevent the drug binding to its target, alternative metabolic pathways, or efflux systems which expel drugs from the cell.

18.2 Drugs acting directly on nucleic acids

18.2.1 Intercalating agents

Intercalating drugs contain a planar aromatic or heteroaromatic ring system which can slip into the double helix of DNA (section 7.1.2) and distort its structure. Once bound, the drug can inhibit the enzymes involved in the replication and transcription processes.

Dactinomycin (Fig. 18.8) (previously called actinomycin D) is a naturally occurring antibiotic isolated from a *Streptomyces* species. It contains two cyclic pentapeptides, but the important feature is the flat tricyclic heteroaromatic structure which slides into the double helix via the minor groove. It appears to favour interactions with guanine–cytosine base pairs and, in particular, between two adjacent guanine bases on alternate strands of the helix. The molecule is further held in position by hydrogen bond interactions between the nucleic acid bases of DNA and the cyclic pentapeptides positioned on the outside of the helix. The 2-amino group of guanine plays a particularly important role in this interaction. The resulting bound complex is very stable and prevents the unwinding of the double helix. This in turn prevents DNA-dependent RNA polymerase from catalysing the synthesis of mRNA and thus prevents transcription. It is also possible that intercalation-mediated topoisomerase inhibition takes place (see below). Eventually, apoptosis is triggered leading to cell death. The drug is given intravenously and is used to treat paediatric solid tumours including Wilms' tumour and Ewing's tumour.

Doxorubicin (previously called adriamycin) (Fig. 18.8) belongs to a group of naturally occurring antibiotics called the **anthracyclines** and was isolated

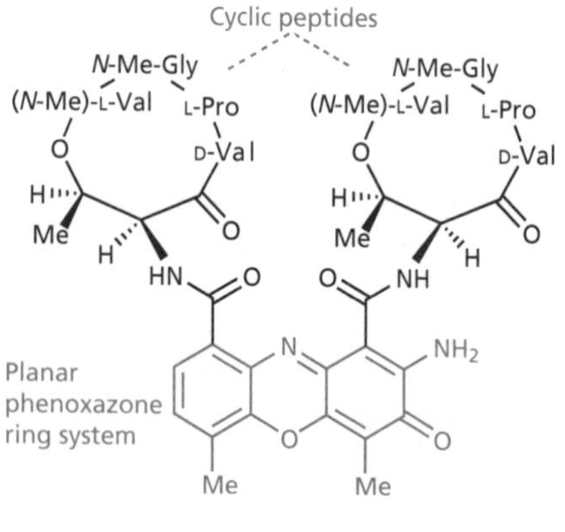

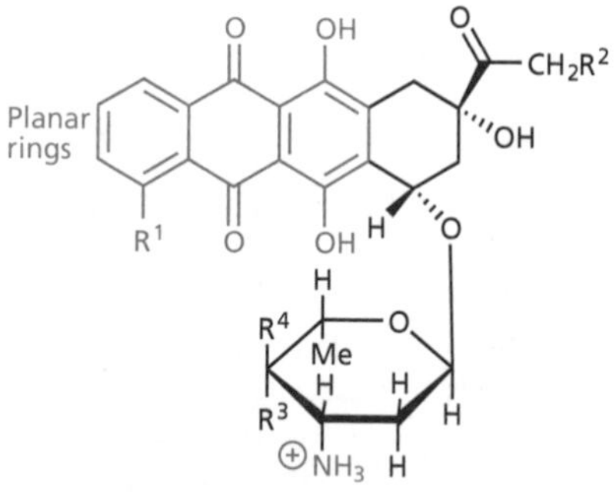

Anthracyclines	R^1	R^2	R^3	R^4
Doxorubicin	OMe	OH	OH	H
Epirubicin	OMe	OH	H	OH
Daunorubicin	OMe	H	H	OH
Idarubicin	H	H	H	OH

Figure 18.8 Dactinomycin and the anthracyclines.

from *Streptomyces peucetius*. It is used to treat a broad spectrum of solid tumours and has a tetracyclic system where three of the rings are planar. The drug approaches DNA via the major groove of the double helix and intercalates using the planar tricyclic system. The charged amino group attached to the sugar is also important, as it forms an ionic bond with the negatively charged phosphate groups on the DNA backbone. This is supported by the fact that the tetracyclic ring system without the aminosugar has poor activity. Intercalation prevents the normal action

of an enzyme called **topoisomerase II**. This enzyme is crucial to effective replication of DNA and to the mitosis process. During replication, the DNA double helix has to be unravelled, but this can lead to increased tension and entanglement further down the helix. In order to relieve this tension, the enzyme binds to regions where two regions of double helix are in near contact (Fig. 18.9). The enzyme binds to one of these DNA double helices and a tyrosine residue is used to nick both strands of the DNA (Fig. 18.10). This results in a temporary covalent bond between the

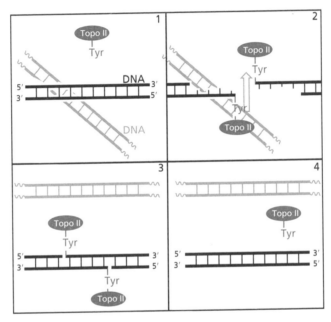

Figure 18.9 Action of topoisomerase II.

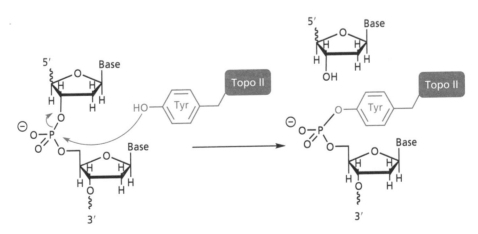

Figure 18.10 'Nicking' of the DNA chain by topoisomerase II.

enzyme and the resulting 5′ end of each strand, thus stabilizing the DNA. The strands are now pulled in opposite directions to form a gap through which the intact DNA region can be passed. The enzyme then reseals the strands and departs. When doxorubicin is intercalated in DNA it stabilizes the DNA–enzyme complex at the point where the enzyme is covalently linked to the 'nicked' DNA—the cleavable or cleavage complex. Agents which do this are referred to as topoisomerase II poisons rather than inhibitors, as they do not prevent the enzyme functioning but stall it. It is thought that an excessive number of these stabilized DNA–enzyme complexes results in permanent DNA breaks when they are encountered by the DNA polymerases and helicases involved in replication, and that this triggers apoptosis. Since these enzymes are most active during active growth and division, the topoisomerase II poisons are most effective against rapidly proliferating cells.

A second mechanism by which doxorubicin can prove harmful to DNA involves the hydroxyquinone moiety which can chelate iron to form a doxorubicin–DNA–iron complex. Reactive oxygen species are then generated, leading to single-strand breaks in the DNA chain. This mechanism is considered less important than the interaction with topoisomerase II, but it has been implicated in the cardiotoxicity of doxorubicin.

A third proposed mechanism involves intercalated doxorubicin inhibiting the helicases which unravel DNA into single DNA strands.

A variety of other anthracyclines are also used in cancer chemotherapy, mainly **epirubicin, daunorubicin** (also called cerubidine, daunomycin, or rubidomycin), and **idarubicin** (idamycin) (Fig. 18.8). Idarubicin is a second-generation anthracycline. It lacks the methoxy group at position 4, so it is more polar and has an altered metabolism which prolongs its half-life.

The anthracyclines are important drugs, but they are orally inactive and have to be administered by intravenous injection. Another drawback is that they have cardiotoxic side effects. Multidrug resistance can develop due to amplification of the gene coding for the P-glycoprotein (section 18.1.12). The use of liposomes as carriers for the delivery of doxorubicin has proved therapeutically useful (section 8.9).

Mitoxantrone (Fig. 18.11) is a synthetic anthracenedione structure having a similar mechanism of action to the anthracyclines. It is used for the treatment

of leukaemia and lymphomas as well as in combination therapies for advanced breast and ovarian cancers. It was designed as a simplified, easily synthesized analogue of the anthracyclines, and lacks the sugar moiety of the anthracyclines, as this was thought to be responsible for cardiotoxic side effects. It was recognized that the amino group of the aminosugar moiety was important for binding and so a substituent was added which would place the required nitrogen in the same relative position. The tetracyclic ring system was truncated to a tricyclic system by removing the non-planar ring, and two identical substituents were added to form a symmetrical molecule which would be easier to synthesize. The resulting tricyclic system is planar and can still intercalate with DNA. Structure–activity relationships (SAR) studies demonstrated the importance of a pharmacophore involving one of the phenol groups, a carbonyl group, and the amino group in the side chain. Because the molecule is symmetrical there are two such pharmacophores, but activity remains much the same for analogues containing only one. It was also demonstrated that the amino group linking the side

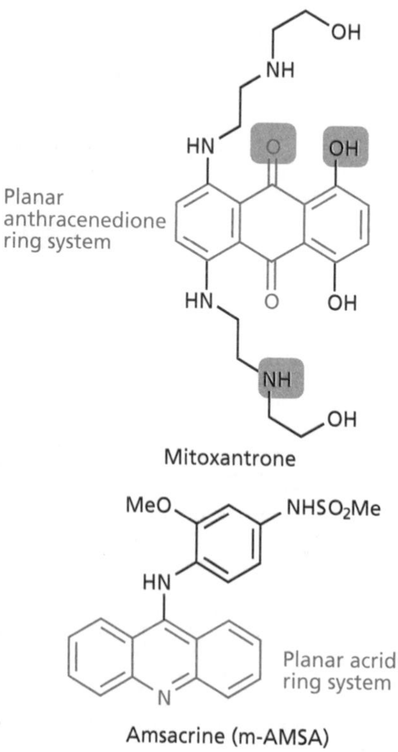

Figure 18.11 Mitoxantrone (pharmacophoric groups highlighted) and amsacrine.

chain to the tricyclic ring system was important to activity.

Mitoxantrone does not have the same level of cardiotoxicity associated with the anthracyclines. It intercalates DNA preferentially at guanine–cytosine base pairs such that the side chains lie in the minor groove of DNA and it is thought to interact with topoisomerase II in a similar fashion to doxorubicin. Other mechanisms of action have been proposed including inhibition of microtubule assembly and inhibition of protein kinase C.

Amsacrine (Fig. 18.11) contains an acridine tricyclic system capable of intercalating into DNA. It also stabilizes topoisomerase-cleavable complexes. The drug is used intravenously for the treatment of acute non-lymphocytic leukaemia.

Bleomycins (Fig. 18.12) are large, water-soluble glycoproteins derived from *Streptomyces verticillus*. Each bleomycin molecule has a complex structure including a bithiazole ring system which intercalates with DNA. Once the structure has become intercalated, the nitrogen atoms of the primary amines, pyrimidine ring, and imidazole ring chelate a ferrous ion which then interacts with oxygen and is oxidized to a ferric ion, leading to the generation of superoxide or

hydroxyl radicals which abstract hydrogen atoms from DNA. This leads to cutting of the DNA strands, particularly between purine and pyrimidine nucleotides. The mechanism involves free radical abstraction of the C-4 proton of deoxyribose leading to oxidative cleavage of the C-3'–C-4' bond. Bleomycin also appears to prevent the enzyme **DNA ligase** from repairing the damage caused, and it is found that DNA synthesis is inhibited more than RNA or protein synthesis. The clinical drug 'Bleomycin' is a mixture of bleomycin A_2 and bleomycin B_2, and is used intravenously or intramuscularly in combination therapies against certain types of skin cancer, testicular carcinoma, and lymphomas. Unlike most anticancer agents, it produces very little bone marrow depression, but it is quite toxic, particularly to the skin and mucous membranes where the drug is found to accumulate and survive metabolic inactivation. This inactivation is related to an enzyme in the body which hydrolyses the primary amide to a carboxylic acid. The pK_a of the nearby amino group is affected and the carboxylic acid group displaces it as a ligand for the ferrous ion. The resulting complex is less able to form the activated iron–oxygen–bleomycin complex required for free radical generation and so the drug is inactivated. It is found

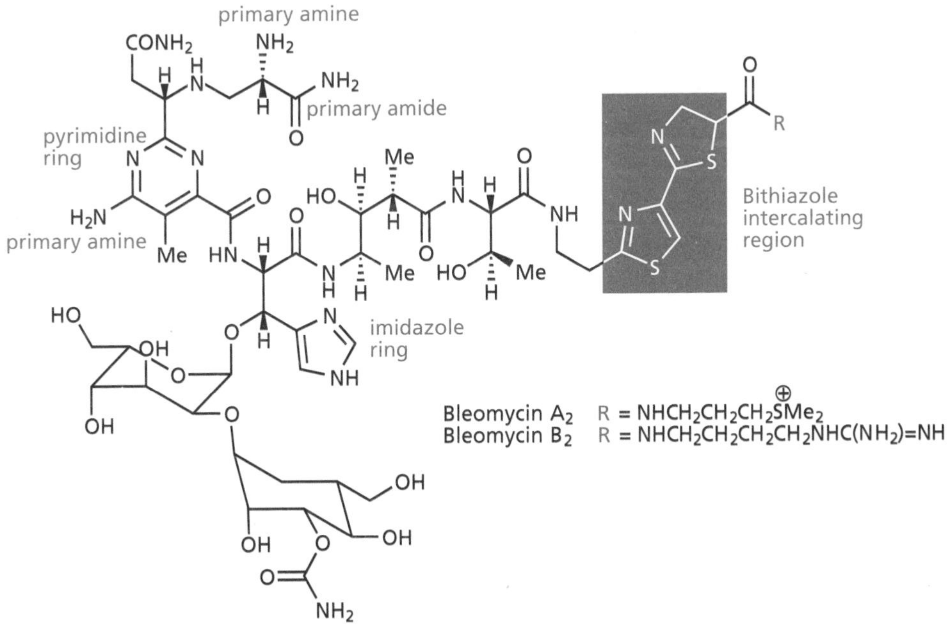

Bleomycin A_2 R = $NHCH_2CH_2CH_2\overset{\oplus}{S}Me_2$
Bleomycin B_2 R = $NHCH_2CH_2CH_2CH_2NHC(NH_2)=NH$

Figure 18.12 Bleomycins.

that skin has very low levels of this enzyme and so levels of the active drug accumulate here to toxic levels.

18.2.2 Non-intercalating agents which inhibit the action of topoisomerase enzymes on DNA

18.2.2.1 Podophyllotoxins

Etoposide and **teniposide** (Fig. 18.13) are potent anti-cancer agents which are used clinically for a variety of conditions such as testicular cancer and small-cell lung cancer. They are semi-synthetic derivatives of **epipodophyllotoxin**—an isomer of a naturally occurring agent called **podophyllotoxin** (Fig. 18.13) where the substituent at position 4 has the opposite configuration.

Both agents stabilize the normally transient covalent intermediate formed between the 5′ end of both DNA strands and topoisomerase II, thus preventing the resealing of the DNA strands which have been cleaved by the enzyme. DNA strand breakage is also thought to occur by a free radical process since oxidation of the 4′-phenolic group can produce a semiquinone free radical. The 4′-OH group is crucial to the mechanism of action resulting in breaks in single-stranded DNA. Evidence supporting this comes from the fact that the 4′-methoxy structures are inactive. The presence of the glucoside moiety also increases the ability to induce breaks.

The drugs show selectivity for cancer cells despite the fact that topoisomerase II is present in both cancer cells and normal cells. This is thought to be due to elevated enzyme levels or enzyme activity in the cancer cells. It has also been found that teniposide is more readily taken up by cells than etoposide and has a greater cytotoxic effect. This is thought to be because teniposide is less polar and can cross cell membranes more easily.

Etoposide suffers from poor water solubility but this can be improved by using a phosphate ester prodrug. Resistance can arise due to over-expression of the P-glycoprotein involved in the efflux mechanism, or to mutations in the topoisomerase enzyme which weaken interactions with the drug. A range of etoposide analogues has been synthesized in an effort to find agents which have better aqueous solubility, improved activity against drug-resistant cancer cells and less susceptibility to metabolic inactivation.

18.2.2.2 Camptothecins

Camptothecin (Fig. 18.14) is a naturally occurring cytotoxic alkaloid which was extracted from a Chinese bush (*Camptotheca acuminata*) in 1966. It targets the enzyme **topoisomerase I**. This enzyme is similar to topoisomerase II in that it relieves the torsional stress of supercoiled DNA during replication, transcription, and repair of DNA. The difference is that it cleaves only one strand of DNA, whereas topoisomerase II cleaves both strands. The enzyme catalyses a reversible transesterification reaction similar to that shown in Fig. 18.10, but where the tyrosine residue of the enzyme is linked to the 3′ phosphate end of the DNA strand rather than the 5′ end. This creates a cleavable complex with a single-strand break which allows relaxation of torsional strain, either by allowing passage of the intact strand through the nick or by free

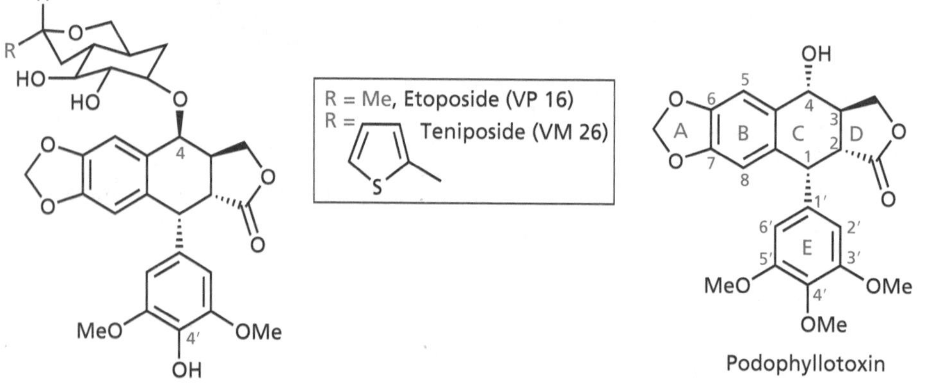

Figure 18.13 Podophyllotoxins.

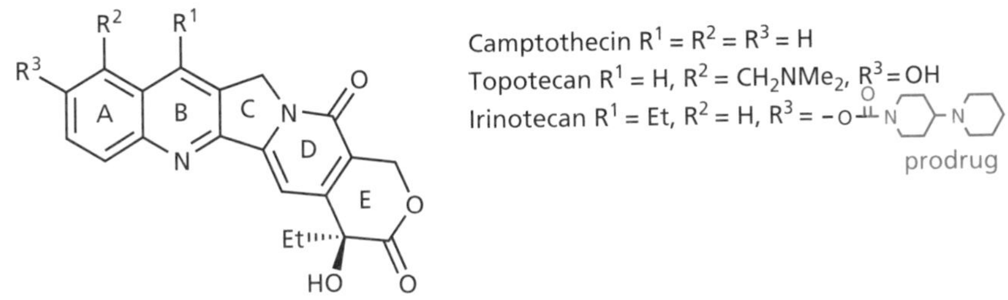

Figure 18.14 Camptothecins.

rotation of the DNA about the uncleaved strand. Once the torsional strain has been relieved, the enzyme rejoins the cleaved strand of DNA and dissociates from the relaxed double helix.

The camptothecins act as topoisomerase I poisons rather than inhibitors, by stabilizing the normally transient DNA–enzyme cleavable complex and inhibiting the rejoining of the strand. As a result, single-strand breaks accumulate in the DNA. These can be repaired if the drug departs, but if replication is taking place when the drug–enzyme–DNA complex is present, an irreversible double-strand break takes place which leads to cell death. This mechanism of drug action requires DNA synthesis to be in progress, but it has been observed that these agents are also toxic to cancer cells which are not synthesizing new DNA. This is due to an alternative mechanism of action—possibly the induction of destructive enzymes such as serine proteases and endonucleases.

The camptothecins show selectivity for cancer cells over normal cells when the cancer cells in question show higher levels of topoisomerase I than normal cells. Topoisomerase I can also be more active in certain cancer cells, which may also account for the antitumour selectivity observed.

The lactone group is important for activity, but at blood pH it is in equilibrium with the less active ring-opened carboxylate structure. Introducing substituents into the A and B rings can alter the relative binding affinities of these structures to serum albumin such that the level of the lactone present is altered accordingly. Unfortunately, camptothecin itself shows poor aqueous solubility and has unacceptable toxic side effects.

Irinotecan and **topotecan** (Fig. 18.14) are semisynthetic analogues of camptothecin which retain the important lactone group and which were developed to improve aqueous solubility by adding suitable polar functional groups such as alcohols and amines.

Topotecan is used in the treatment of ovarian cancer and small-cell lung cancer. Irinotecan is a urethane prodrug that is converted to the active alcohol by carboxylesterases, predominantly in the liver. It is used in combination therapy with fluoropyrimidines for the treatment of advanced colorectal cancer, and has a potential role in a variety of other cancers. Unfortunately, carboxylesterases are not very efficient at hydrolysing the urethane group in irinotecan, and only 2–5% of an injected dose is actually converted. Gene therapy and ADEPT strategies are being explored to try to improve this process (section 18.9).

Resistance to these drugs arises from mutations to the topoisomerase I enzyme.

18.2.3 Alkylating agents

Alkylating agents are highly electrophilic compounds that react with nucleophiles to form strong covalent bonds. There are several nucleophilic groups in DNA such as N-1 and N-3 of adenine bases, N-3 of cytosine, and in particular N-7 of guanine. Drugs with two alkylating groups can react with a guanine on each chain and cross-link the strands such that they disrupt replication or transcription. Alternatively, the drug could link two guanine groups on the same chain such that the drug is attached like a limpet to the side of the DNA helix. That portion of DNA becomes masked and the necessary enzymes required for DNA function cannot gain access.

Miscoding due to alkylated guanine units is also possible. The guanine base usually exists as the keto tautomer and base-pairs with cytosine. Once alkylated, however, guanine prefers the enol tautomer and is more likely to base-pair with thymine. Such miscoding ultimately leads to an alteration in the amino acid sequence of proteins, which in turn leads to disruption of protein structure and function.

Unfortunately, alkylating agents can alkylate nucleophilic groups on proteins as well as DNA, which means they have poor selectivity. Nevertheless, alkylating drugs have been useful in the treatment of cancer. Tumour cells often divide more rapidly than normal cells and so disruption of DNA function affects these cells more drastically than normal cells. It should

also be noted that these drugs can be mutagenic and carcinogenic in their own right as a result of the damage they wreak on DNA.

18.2.3.1 Nitrogen mustards

The nitrogen mustards get their name because they are related to the sulfur-containing mustard gases used during the First World War. The nitrogen mustard compound **chlormethine** (Fig 18.15) was the first alkylating agent to be used medicinally, in 1942. The nitrogen atom is able to displace a chloride ion

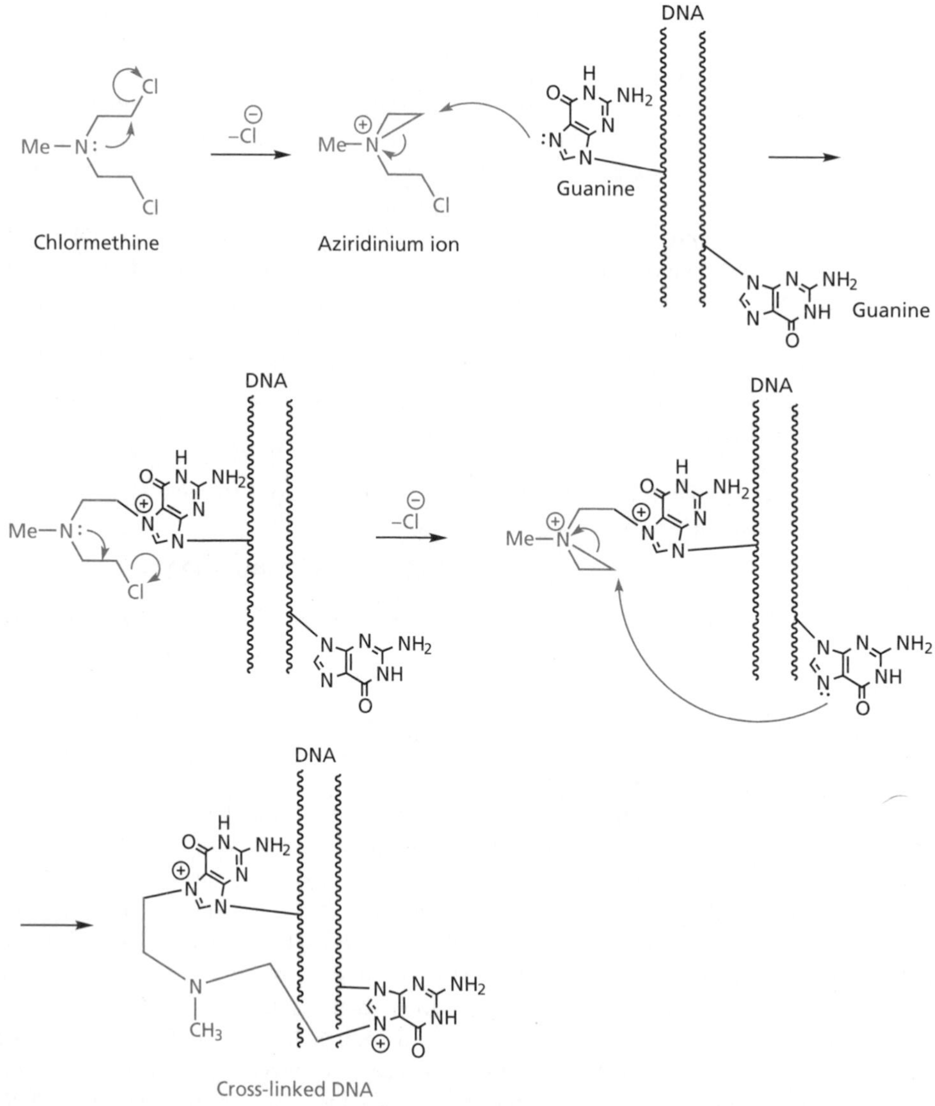

Figure 18.15 Alkylation of DNA by the nitrogen mustard compound chlormethine.

intramolecularly to form the highly electrophilic aziridinium ion. Alkylation of DNA can then take place. Since the process can be repeated, cross-linking between chains or within the one chain will occur. Monoalkylation of DNA guanine units is also possible if the second alkyl halide reacts with water, but cross-linking is the major factor by which these drugs inhibit replication and act as anticancer agents. Chlormethine is highly reactive and can react with water, blood, and tissues. It is too reactive to survive the oral route and has to be administered intravenously. It is mainly used for the treatment of Hodgkin's lymphoma, as part of a multidrug regime.

The side reactions mentioned above can be reduced by lowering the reactivity of the alkylating agent. For example, putting an aromatic ring on the nitrogen atom instead of a methyl group (I in Fig. 18.16) has such an effect. The lone pair of the nitrogen interacts with the π system of the ring and is less available to displace the chloride ion. As a result, the intermediate aziridinium ion is less easily formed and only strong nucleophiles such as guanine will react with it. The alkylating agent **melphalan** (Fig. 18.16) takes advantage of this property and has the added advantage

of having a moiety which mimics the amino acid phenylalanine. As a result, the drug is more likely to be recognized as an amino acid and be taken into cells by carrier proteins. The increased stability also means that the drug can be given orally. Phenylalanine is a biosynthetic precursor for melanin and it was hoped that this would help to target the drug to skin melanomas. Unfortunately, such targeting has not been particularly significant. The drug is used in the treatment of multiple myeloma, as well as ovarian and breast cancers.

A similar approach has been to attach a nucleic acid building block. For example, **uracil mustard** (Fig. 18.16) contains the nucleic acid base uracil. This drug has been used successfully in the treatment of chronic lymphatic leukaemia and shows a certain amount of selectivity for tumour cells over normal cells. Since tumour cells generally divide faster than normal cells, nucleic acid synthesis is faster and so tumour cells need more of the nucleic acid building blocks. The tumour cells scavenge more than their fair share of the building blocks and accumulate the cytotoxic drug more effectively. Unfortunately, this approach has not achieved the high levels of

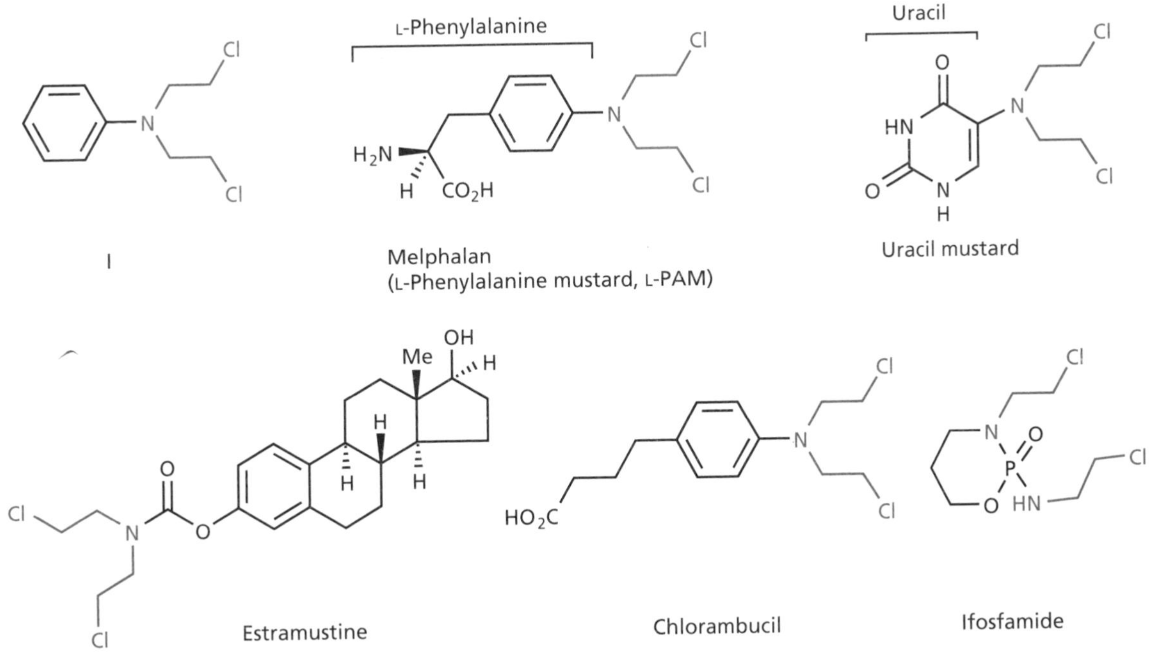

Figure 18.16 Mustard-like alkylating agents.

selectivity desired for effective eradication of all relevant tumour cells.

Other examples of alkylating agents include **estramustine** (Fig. 18.16), where the alkylating group has been linked to the hormone **oestradiol**. Since oestradiol normally crosses cell membranes into cells, it carries the alkylating agent with it. The link to the steroid is through a urethane functional group which lowers the nitrogen's nucleophilicity. **Chlorambucil** (Fig. 18.16) is an orally active drug used primarily in the treatment of chronic lymphocytic leukaemia.

Resistance to alkylating agents can arise through reaction with cellular thiols and decreased cellular uptake.

Cyclophosphamide (Fig. 18.17) is the most commonly used alkylating agent in cancer chemotherapy and has a broad application. It acts as a prodrug and is not toxic itself, being converted into the cytotoxic alkylating agent in the body by the process shown in Fig. 18.17. Metabolism in the liver by cytochrome P450 enzymes oxidizes the ring. Ring opening then takes place and non-enzymatic hydrolysis splits acrolein from the molecule to generate the cytotoxic alkylating agent. Here, the nucleophilicity of the nitrogen has been reduced because it is part of a phosphoramide group, and so the active agent is more selective for stronger nucleophiles such as guanine.

Cyclophosphamide itself is relatively non-toxic, and can be taken orally without causing damage to the gut wall. It was also hoped that the high level of phosphoramidase enzyme present in some tumour cells would lead to a greater concentration of alkylating agent in these cells and result in some selectivity of action. Unfortunately, the acrolein released is responsible for toxicity to the kidneys and bladder (haemorrhagic cystitis) and results in inflammation, oedema, bleeding, ulceration, and cell death. One possible explanation for the toxicity of acrolein is that it alkylates cysteine residues in cell proteins. Certainly, toxicity can be reduced by co-administrating sulfhydryl donors such as **N-acetylcysteine** or **sodium-2-mercaptoethane sulfonate (mesna)** $(HSCH_2CH_2SO_3^-)$ which interact with the acrolein. **Ifosfamide** (Fig. 18.16) is a related drug with a similar mechanism and similar problems.

18.2.3.2 Nitrosoureas

Lomustine and **carmustine** (Fig. 18.18) are examples of chloroethylnitrosoureas which are lipid soluble and can cross the blood–brain barrier. As a result, they have been used in the treatment of brain tumours and meningeal leukaemia. The drug decomposes spontaneously in the body to form two active compounds—an alkylating agent and a carbamoylating agent (Fig. 18.19). The organic isocyanate which is formed carbamoylates lysine residues in proteins and may inactivate DNA repair enzymes. The alkylating agent reacts initially with the O-6 position of a guanine moiety in one strand of DNA, then with the O-6

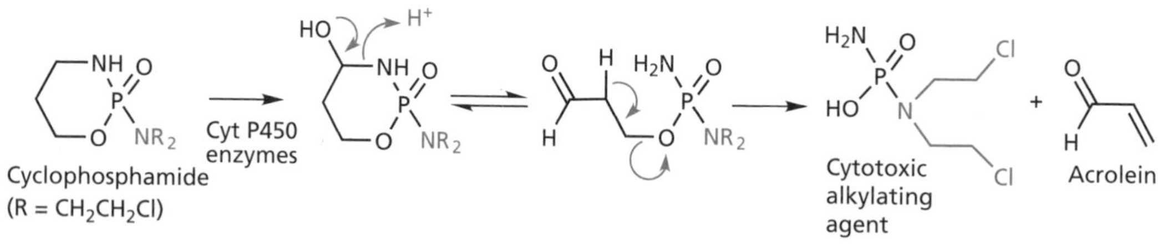

Figure 18.17 Phosphoramide mustard from cyclophosphamide.

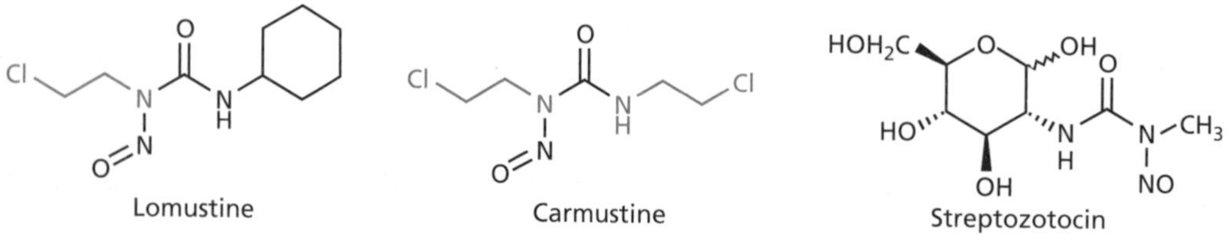

Figure 18.18 Nitrosourea alkylating agents.

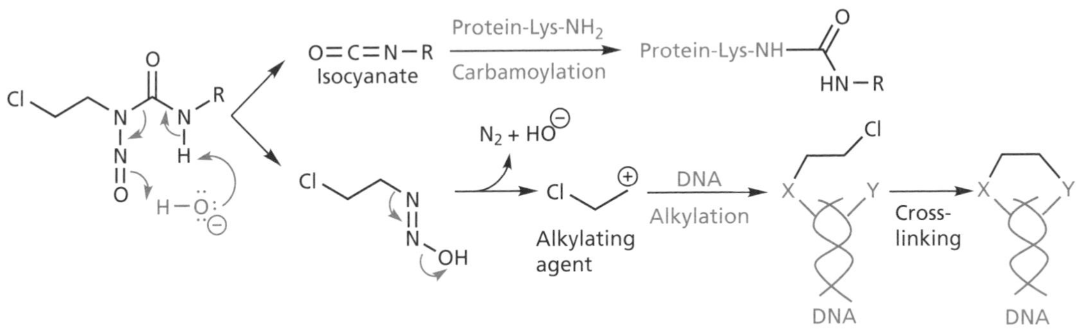

Figure 18.19 Mechanisms of action for nitrosoureas.

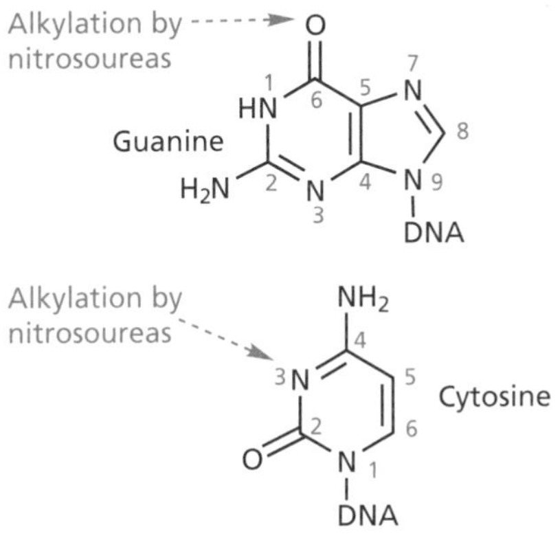

Figure 18.20 Alkylation sites on guanine and cytosine for nitrosoureas.

position of a guanine unit or the N-3 position of cytosine in the other strand to produce interstrand cross-linking (Figs. 18.19 and 18.20). Lomustine can be given orally, but carmustine is given intravenously because it is rapid metabolized.

Streptozotocin (Fig. 18.18) is a naturally occurring nitrosourea isolated from *Streptomyces achromogenes*, used for the treatment of pancreatic islet-cell carcinoma. There is a specific uptake of the drug into the pancreas where it carbamoylates proteins.

18.2.3.3 Busulfan

Busulfan (Fig. 18.21) is used in the treatment of chronic granulocytic leukaemia and may increase the life expectancy of patients by about a year. It acts selectively on the bone marrow and has little effect on lymphoid tissue or the gastrointestinal tract. The sulfonate groups are good leaving groups and play a similar role to the chlorines in the nitrogen mustards. The mechanism involves a direct S_N2

nucleophilic substitution of the sulfonate groups and does not involve an intermediate similar to the aziridinium ion. The N-7 position of guanine residues is the most susceptible site for this reaction to form interstrand cross-linking. Resistance to busulfan is related to the rapid removal and repair of the DNA cross-links.

18.2.3.4 Cisplatin and cisplatin analogues

Cisplatin (Fig. 18.22) is a very useful antitumour agent for the intravenous treatment of testicular and ovarian tumours. It is also used in various combination therapies to treat other forms of cancer and is one of the most frequently used anticancer drugs. Its discovery was fortuitous in the extreme, arising from research carried out in the 1960s to investigate the effects of an electric current on bacterial growth. During these experiments, it was discovered that bacterial cell division was inhibited. Further research led to the discovery that an electrolysis product from the platinum electrodes was responsible for the inhibition and the agent was eventually identified as *cis*-diammonia dichloroplatinum (II), now known as cisplatin.

In the body, the chloride groups of cisplatin are replaced by neutral water ligands to give reactive positively charged species (Fig 18.23). These bind strongly to DNA in regions containing several guanine units, forming Pt–DNA links within strands (intrastrand binding). It is likely that this takes place to the N-7 and O-6 of adjacent guanine molecules. The hydrogen bonds involved in base-pairing guanine to cytosine are disrupted by the cross-links, leading to localized unwinding of the DNA helix and inhibition of transcription. Unfortunately, cisplatin is associated with very severe nausea and vomiting, but the

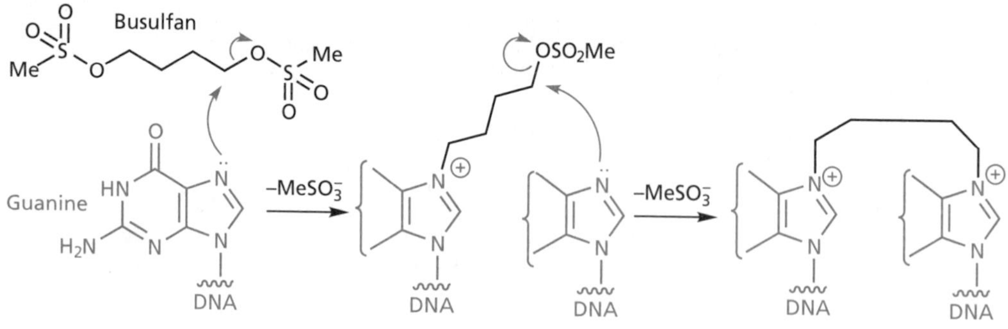

Figure 18.21 Cross-linking with busulfan.

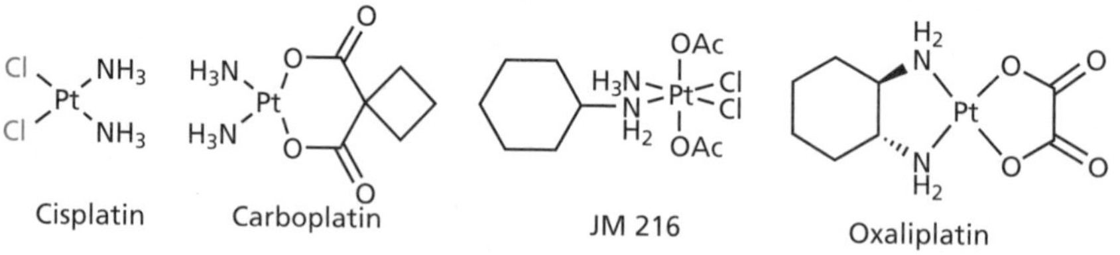

Figure 18.22 Platinum-based anticancer drugs.

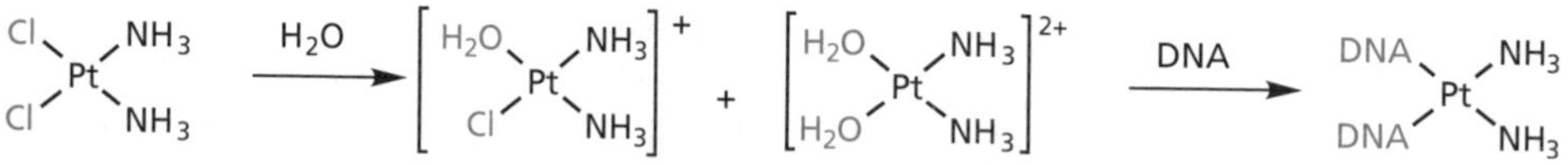

Figure 18.23 Activation of cisplatin.

administration of the 5-HT$_3$ receptor antagonist **ondansetron** (Box 9.2) is effective in combating this problem.

Carboplatin (Fig. 18.22) is a derivative of cisplatin with reduced side effects. A range of other platinum drugs (Fig. 18.22), such as the first orally active compound **JM216**, have also been developed in an attempt to overcome tumour resistance. Most of these compounds are still undergoing clinical trials, but **oxaliplatin** was approved in 1999 for the treatment of colorectal cancer and shows a better safety profile than cisplatin or carboplatin. It also shows a lack of cross-resistance to cisplatin and carboplatin. Tumour resistance to cisplatin and similar agents has been attributed to enhanced DNA repair mechanisms. The diaminocyclohexane ring is responsible for the lack of cross-resistance but is bad for water solubility. This can be counteracted by introducing an oxalato ligand as the leaving group, since this group improves water solubility.

18.2.3.5 Dacarbazine and procarbazine

Dacarbazine (Fig. 18.24) is a prodrug which is activated by N-demethylation in the liver—a reaction catalysed by cytochrome P450 enzymes (section 8.4.2). Formaldehyde is then lost spontaneously to form a product which spontaneously degrades to form 5-aminoimidazole-4-carboxamide (AIC) and a methyl-diazonium ion, the latter being the alkylating agent. Reaction of this ion with RNA or DNA results in methylation mainly at the 7-position of guanine. DNA fragmentation can also occur. AIC has no cytotoxic effect and is present naturally as an intermediate in purine synthesis. Dacarbazine is used clinically in combination therapies for the treatment of melanoma, Hodgkin's lymphoma, and soft tissue sarcomas.

Procarbazine (Fig. 18.25) is a related compound which is used in combination therapy for the treatment of advanced Hodgkin's disease, small-cell carcinomas of the lung, non-Hodgkin's lymphomas, and malignant melanoma. It is also a prodrug and is thought to

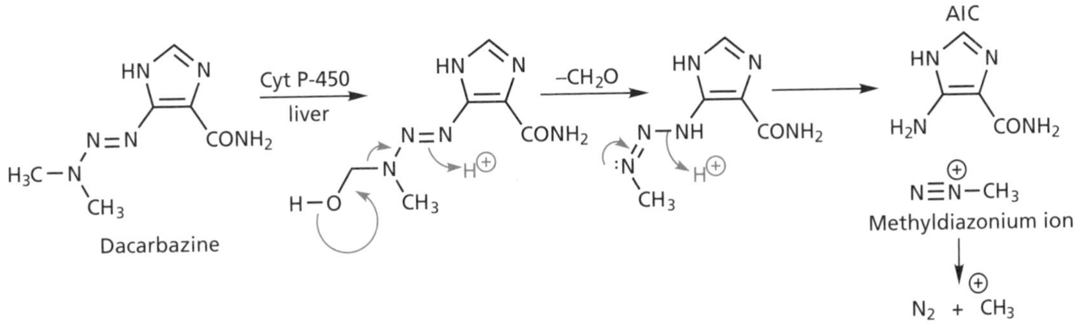

Figure 18.24 Mechanism of action of dacarbazine.

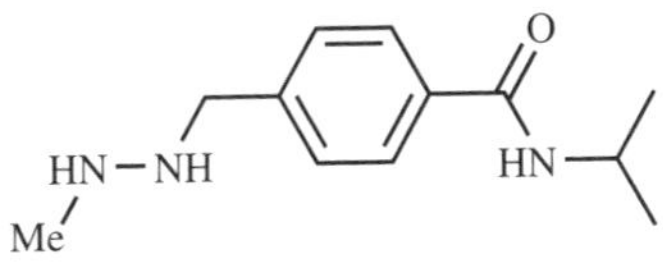

Figure 18.25 Procarbazine.

undergo oxidation by cytochrome P450 enzymes to produce the methyldiazonium ion.

18.2.3.6 Mitomycin C

Mitomycin C (Fig. 18.26) is an anticancer drug which is administered intravenously and is converted to an alkylating agent in the body. This is initiated by an enzyme catalysed reduction of the quinone ring system to a hydroquinone. Loss of methanol and opening of the three-membered aziridine ring then takes place to generate the alkylating agent. Guanine residues on different DNA strands are then alkylated, leading to interstrand cross-linking, and the inhibition of DNA replication and cell division (Fig. 18.26). The drug is used for the treatment of several types of cancer including breast, stomach, gullet, and bladder cancers. Since a reduction step is involved in the mechanism, it has been proposed that this drug should be more effective against tumours in an oxygen-starved environment such as the centre of solid tumour masses. Mitomycin C has many side effects, and is one of the most toxic anticancer drugs in clinical use.

18.2.3.7 CC 1065 analogues

CC 1065 (Fig. 18.27) is a naturally occurring anti-cancer agent which binds to the minor groove of DNA, then alkylates an adenine base. It is 1000 times more active *in vitro* than doxorubicin and cisplatin.

Adozelesin is a simplified synthetic analogue and is being considered for use in antibody–drug conjugates (section 18.9.2).

18.2.4 Chain cutters

Calicheamicin γ¹ (Fig. 18.28) is an antitumour agent which was isolated from a bacterium. It binds to the minor groove of DNA and cuts the DNA chain by producing highly reactive radical species. The driving force behind this reaction is the formation of an aromatic ring from the unusual enediyne system. The reaction starts with a nucleophile attacking the trisulfide group (Fig. 18.29). The sulfur which is freed then undergoes an intramolecular Michael addition with a reactive α,β-unsaturated ketone. The resulting intermediate then cycloaromatizes to produce an aromatic diradical species which snatches two hydrogens from DNA. As a result, the DNA becomes a diradical. Reaction with oxygen then leads to chain cutting.

18.2.5 Antisense therapy

The biopharmaceutical company Genta has developed an antisense drug (section 7.4) called **oblimersen** which consists of 18 deoxynucleotides linked by a phosphorothioate backbone (section 11.8.5). It binds to the initiation codon of the mRNA molecule carrying the genetic instructions for Bcl-2. Bcl-2 is a protein which suppresses cell death (apoptosis), and so suppressing its synthesis will increase the chances of apoptosis taking place when chemotherapy or radiotherapy is being employed. This is currently being tested in phase III clinical trials in combination with the anticancer drugs docetaxel and irenotecan.

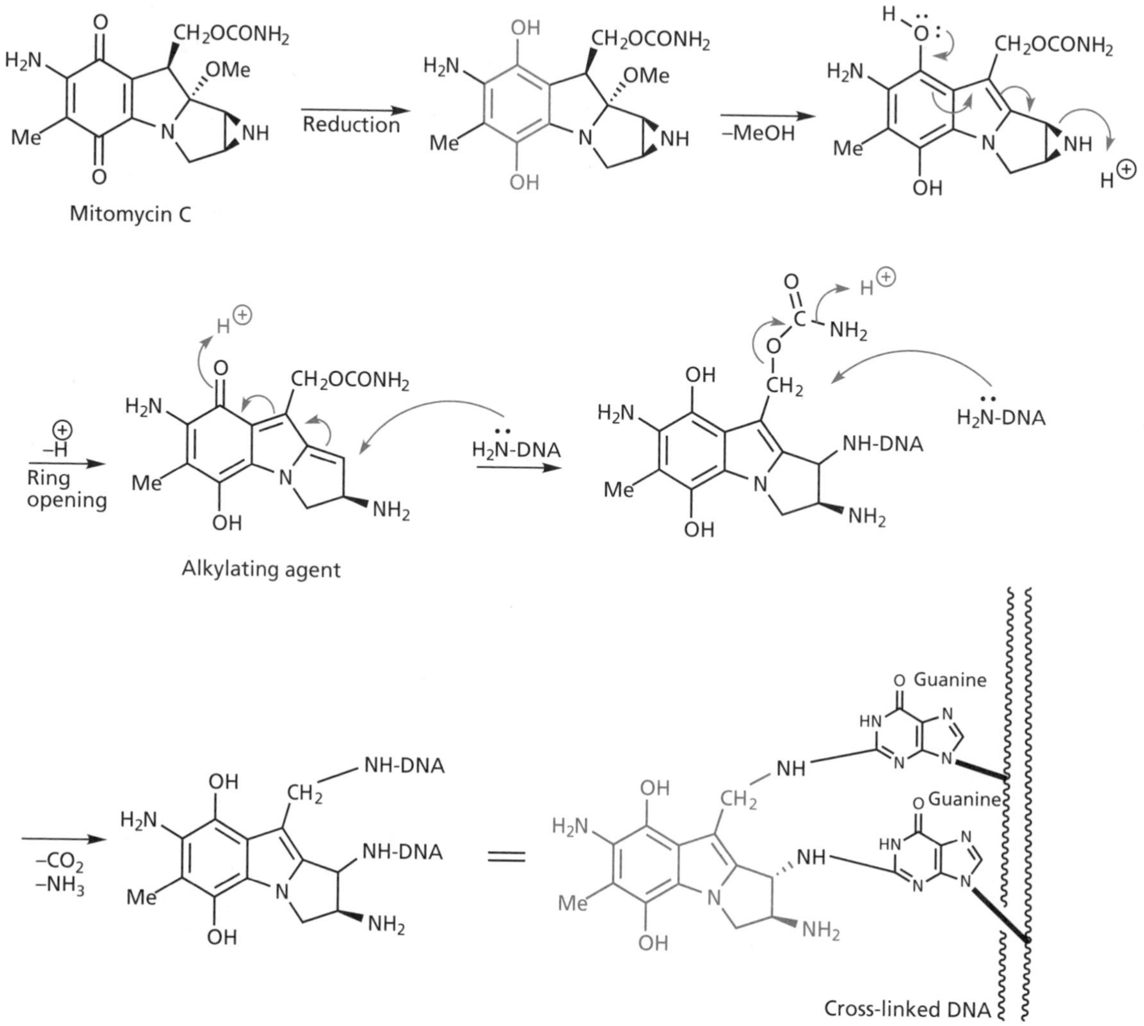

Figure 18.26 DNA cross-linking by mitomycin C.

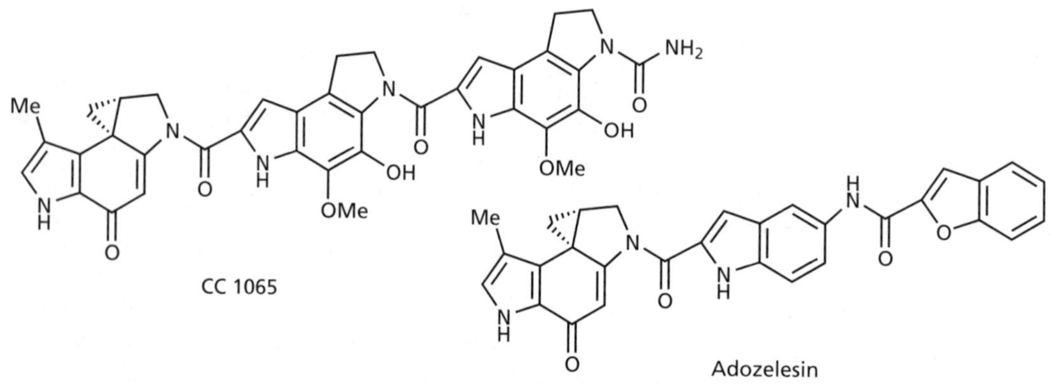

Figure 18.27 CC 1065 and adozelesin.

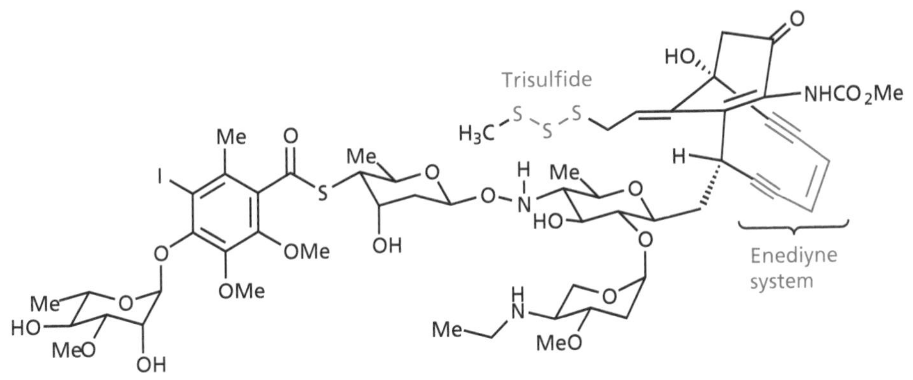

Figure 18.28 Calicheamicin γ^1.

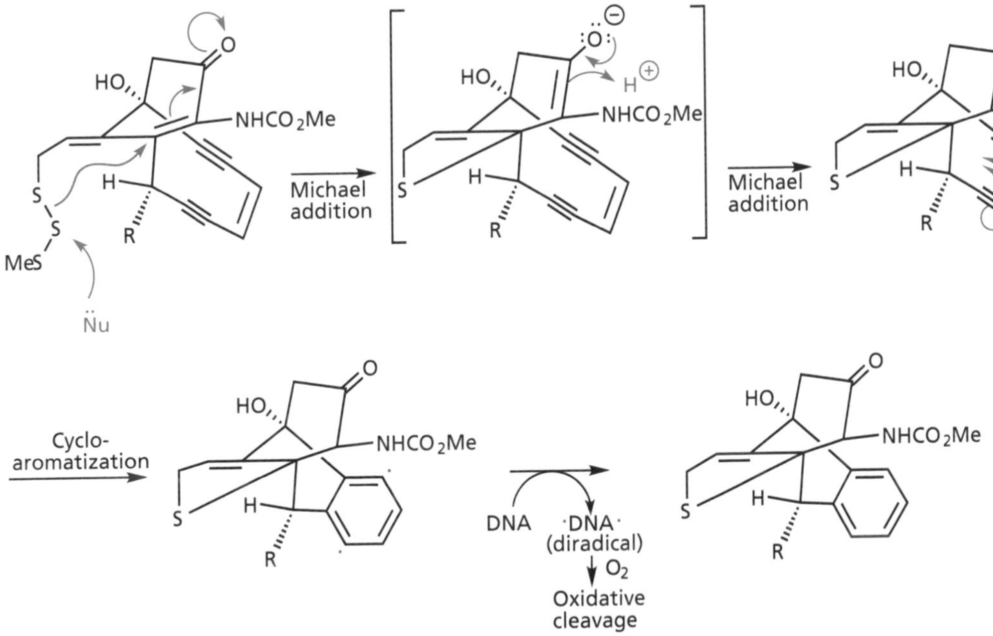

Figure 18.29 Mechanism of action of calicheamicin γ^1.

Phosphorothioate oligonucleotides are also being investigated that will target the genetic instructions for Raf and PKCγ, two proteins which are involved in signal transduction pathways.

KEY POINTS

- Intercalating drugs contain planar aromatic or heteroaromatic ring systems which can slide between the base pairs of the DNA double helix.

- The anthracyclines are intercalating drugs that act as topoisomerase II poisons, stabilizing the cleavage complex formed between the enzyme and DNA.

- Bleomycins are intercalating drugs which form complexes with iron ions. These complexes generate reactive oxygen species that cleave the strands of DNA.

- Etoposide and teniposide are non-intercalating drugs that act as topoisomerase II poisons.

- Camptothecins are non-intercalating drugs that act as topoisomerase I poisons. They stabilize an enzyme–DNA complex where a single strand of DNA has been cleaved.

- Alkylating agents contain electrophilic groups that react with nucleophilic centres on DNA. If two electrophilic groups are present, interstrand, and/or intrastrand cross-linking of the DNA is possible.

- Nitrogen mustards react with guanine groups on DNA to produce cross-linking. The reactivity of the agents can be lowered by attaching electron-withdrawing groups to the nitrogen to increase selectivity against DNA over proteins. Incorporation of important biosynthetic building blocks aids the uptake into rapidly dividing cells.

- Nitrosoureas have a dual mechanisms of action whereby they alkylate DNA and carbamoylate proteins.

- Cisplatin and its analogues are alkylating agent which cause intrastrand cross-linking. They are commonly used for the treatment of testicular and ovarian cancers.

- Dacarbazine and procarbazine are prodrugs which are activated by enzymes to produce a methyldiazonium ion that acts as an alkylating agent.

- Mitomycin C is a natural product that is converted to an alkylating agent by enzymatic reduction. Interstrand cross-linking takes place between guanine groups.

- CC1065 analogues are highly potent alkylating agents which are being considered for use in antibody–drug conjugates.

- Calicheamicin is a natural product which reacts with nucleophiles to produce a diradical species. Reaction with DNA ultimately leads to cutting of the DNA chains.

- Antisense molecules have been designed to inhibit the mRNA molecules that code for proteins which suppress apoptosis.

18.3 Drugs acting on enzymes: antimetabolites

The drugs described in section 18.2 interact directly with DNA to inhibit its various functions. Another method of disrupting DNA function is to inhibit the enzymes involved in the synthesis of DNA or its nucleotide building blocks. The inhibitors involved are described as antimetabolites. The action of antimetabolites leads to the inhibition of DNA function, or the synthesis of abnormal DNA which may trigger the processes leading to apoptosis.

18.3.1 Dihydrofolate reductase inhibitors

Dihydrofolate reductase (DHFR) is an enzyme which is crucial in maintaining levels of the enzyme cofactor **tetrahydrofolate** (FH_4) (Figs. 18.30 and 18.31).

Without this cofactor, the synthesis of the DNA building block (dTMP) would grind to a halt, which in turn would slow down DNA synthesis and cell division. The enzyme catalyses the reduction of the vitamin folic acid to FH_4 in two steps via dihydrofolate (FH_2). Once formed, FH_4 picks up a single carbon unit to form N^5,N^{10}-methylene FH_4, which then acts as a source of one-carbon units for various biosynthetic pathways, including the methylation of deoxyuridine monophosphate (dUMP) to form deoxythymidine monophosphate (dTMP). N^5,N^{10}-methylene FH_4 is converted back to FH_2 in the process and dihydrofolate reductase is vital in restoring the N^5,N^{10}-methylene FH_4 for further reaction.

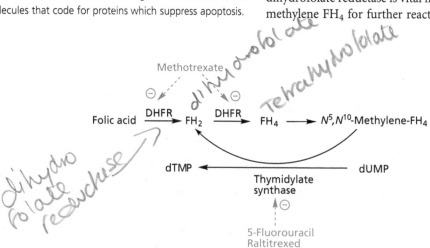

Figure 18.30 Reactions catalysed by dihydrofolate reductase and thymidylate synthase.

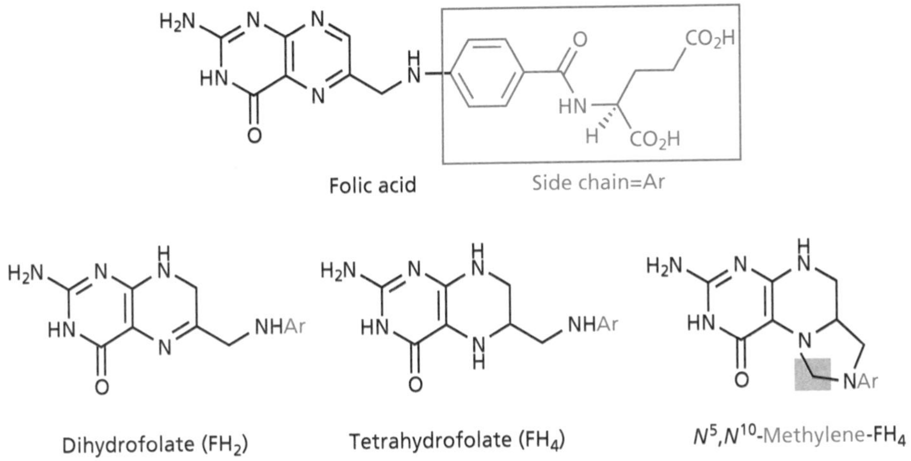

Folic acid   Side chain=Ar

Dihydrofolate (FH$_2$)   Tetrahydrofolate (FH$_4$) N^5,N^{10}-Methylene-FH$_4$

Figure 18.31 Structures of folic acid and related cofactors.

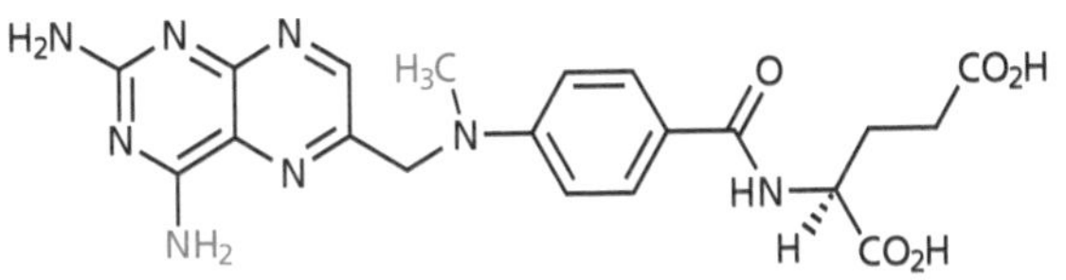

Figure 18.32 Methotrexate.

Methotrexate (Fig. 18.32) is one of the most widely used antimetabolites in cancer chemotherapy. It is very similar in structure to the natural folates, differing only in additional amino and methyl groups. It has a stronger binding affinity for the enzyme, due to an additional hydrogen bond or ionic bond which is not present when FH$_2$ binds. As a result, methotrexate prevents the binding of FH$_2$ and its conversion to N^5,N^{10}-methylene FH$_4$. Depletion of the cofactor has its greatest effect on the enzyme thymidylate synthase resulting in the lowered synthesis of dTMP.

Methotrexate tends to accumulate in cells as a result of polyglutamylation. This is an enzyme-catalysed process which involves the addition of glutamate groups to the glutamate moiety already present in the molecule. This also happens to natural folates, and the reaction serves to increase the charge and size of the folates such that they are trapped within the cell.

The drug can be administered orally and by various other methods. It is used to treat a wide variety of cancers either alone or in combination with other drugs. Resistance to methotrexate can arise from enhanced expression of DHFR, or diminished uptake of methotrexate by the reduced folate carrier (RFC)—a membrane transport protein responsible for the cellular uptake of both folates and antifolates.

18.3.2 Inhibitors of thymidylate synthase

Methotrexate has an indirect effect on thymidylate synthase by lowering the amount of N^5,N^{10}-methylene FH$_4$ cofactor required. **5-Fluorouracil** (Fig. 18.33) is an anticancer drug used for the treatment of breast, liver, and skin cancers, which inhibits this enzyme directly.

It does so by acting as a prodrug for a suicide substrate. 5-Fluorouracil is converted in the body to the fluorinated analogue of 2'-deoxyuridylic acid monophosphate (FdUMP) (Fig. 18.33) which then combines with the enzyme and the cofactor to form a suicide substrate *in situ* (Fig. 18.34). Up until this point, nothing unusual has happened and the reaction mechanism has been proceeding normally. The tetrahydrofolate has formed a covalent bond to the uracil skeleton via the methylene unit, which is usually transferred to uracil. Now things start to go wrong. At this stage, a proton is usually lost from the 5-position of uracil. However, 5-fluorouracil has a fluorine atom at that position instead of a hydrogen. Further reaction is impossible, since it would require fluorine to leave as a positive ion. As a result, the fluorouracil skeleton remains covalently and irreversibly bound to the active site. The synthesis of thymidine is now terminated, which in turn stops the synthesis of DNA. Consequently, replication and cell division are blocked. 5-Fluorouracil is a particularly useful drug for the

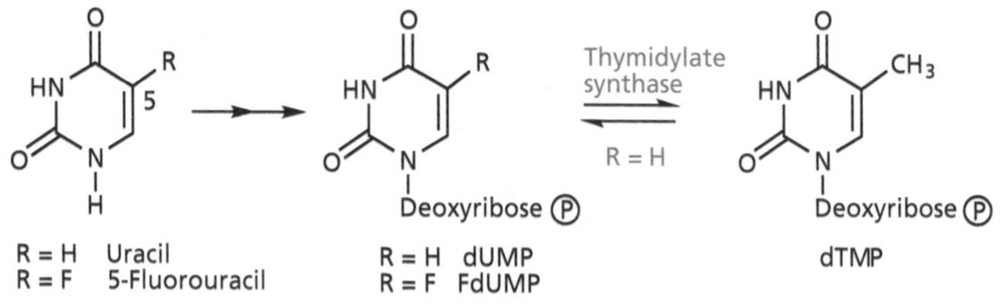

Figure 18.33 Biosynthesis of dTMP.

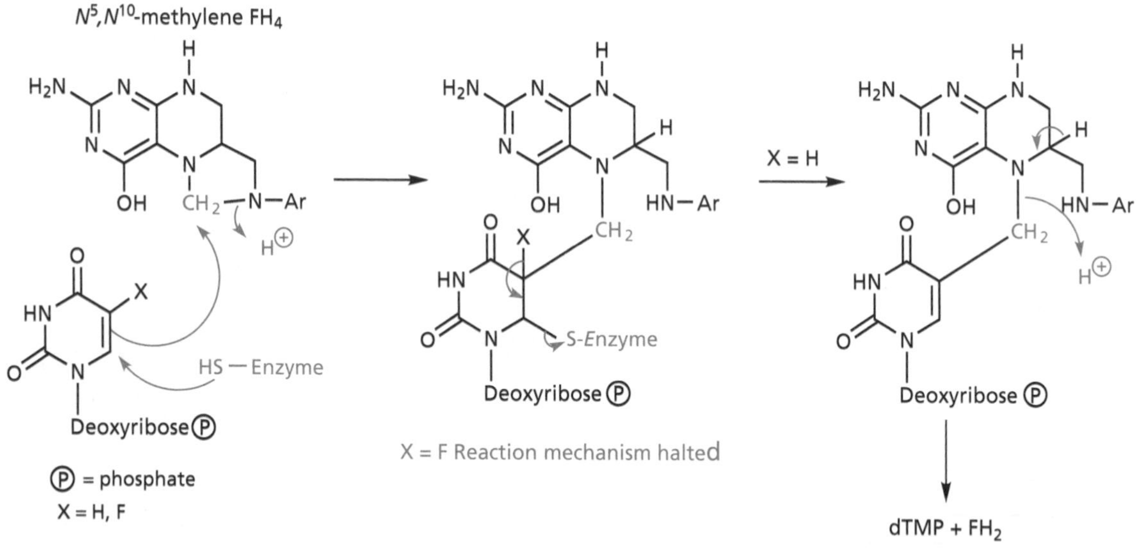

R = H Uracil
R = F 5-Fluorouracil

R = H dUMP
R = F FdUMP

dTMP

N^5,N^{10}-methylene FH$_4$

P = phosphate
X = H, F

X = H

X = F Reaction mechanism halted

dTMP + FH$_2$

Figure 18.34 Use of 5-fluorouracil as a prodrug for a suicide substrate.

treatment of skin cancer because it shows a high level of selectivity for cancer cells over normal skin cells. It is also used for the treatment of several commonly occurring solid tumours including breast, head, neck, gastric, and pancreatic cancers. Unfortunately, it also has neurotoxic and cardiotoxic side effects. Resistance can occur if the cell produces excess quantities of dUMP to compete with the drug for the active site.

5-Fluorouracil binds to the same region of the active site as uracil. Inhibitors which bind to the cofactor binding region have also been developed. **Ralitrexed** (Fig. 18.35) is an injectable cytotoxic drug which is used in the treatment of advanced colorectal cancer. It is the first of a new generation of highly specific folate-based thymidylate synthase inhibitors. Acquired resistance includes impaired

cellular uptake, decreased polyglutamation by folyl-polyglutamate synthetase (FPGS), or increased thymidylate synthase expression. Another agent under study is ZD9331, which is not a substrate for FPGS and can overcome cell resistance where cells have decreased FPGS expression.

18.3.3 Inhibitors of ribonucleotide reductase

Ribonucleotide reductase is responsible for the conversion of ribonucleotide diphosphates to deoxyribonucleotide diphosphates (Fig. 18.36). The enzyme contains an iron cofactor which is crucial to the reaction mechanism. This involves the iron cofactor reacting with a tyrosine residue to generate and stabilize a tyrosine free radical which then abstracts

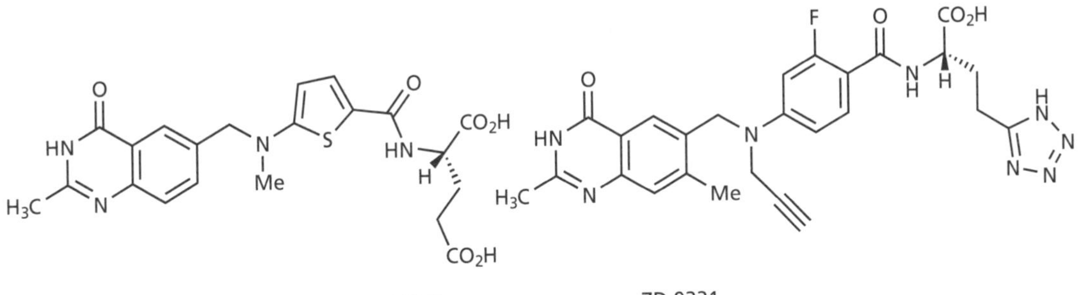

Raltitrexed (Tomudex; ZD 1694)

ZD 9331

Figure 18.35 Ralititrexed and ZD 9331.

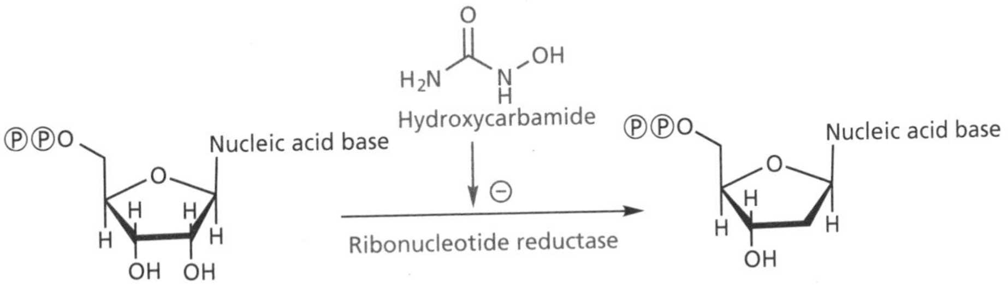

Hydroxycarbamide

Figure 18.36 Reaction catalysed by ribonucleotide reductase.

Tetrahedral intermediate

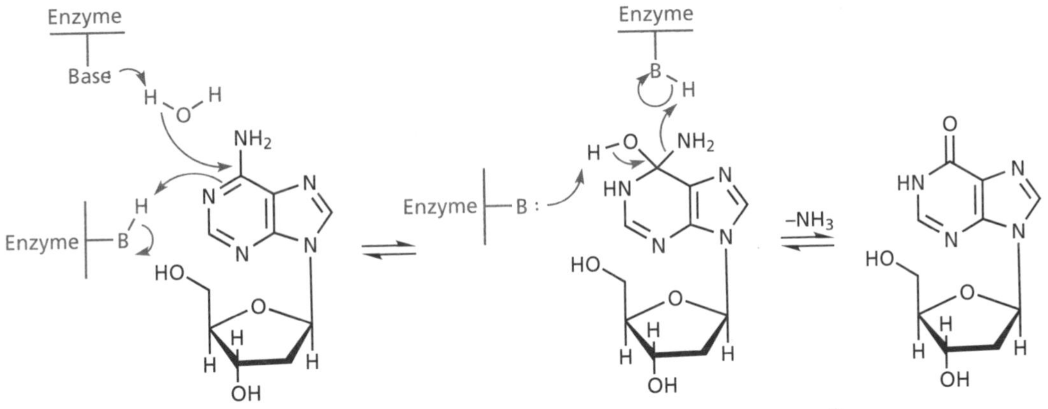

Figure 18.37 Mechanism of adenosine deaminase.

a proton from the substrate, initiating the mechanism by which reaction takes place. **Hydroxycarbamide** (Fig. 18.36) is a clinically useful agent which inhibits the enzyme by destabilizing the iron centre. It is orally administered for the treatment of busulfan-resistant chronic granulocytic leukaemia, and has been used in combination therapy for the treatment of head, neck,

and cervical cancers. Resistance can arise due to increased expression of the enzyme.

18.3.4 Inhibitors of adenosine deaminase

Ribonucleotide reductase is inhibited directly by hydroxycarbamide, but it can also be inhibited indirectly by increasing the level of natural allosteric inhibitors such as dATP (allosteric inhibitors are described in section 4.8.3). The enzyme adenosine deaminase catalyses the deamination of adenosine to inosine (Fig. 18.37) and it is found that inhibition of the enzyme leads to a build-up of dATP in the cell, which in turn inhibits ribonucleotide reductase.

The antileukaemia drug **pentostatin** (Fig. 18.38) is a natural product isolated from *Streptomyces antibioti-cus*, and is a powerful inhibitor of adenosine deam-inase ($K_i = 2.5$ pM). It acts as a transition-state inhibitor, mimicking the tetrahedral intermediate in Fig. 18.37.

18.3.5 Inhibitors of DNA polymerases

DNA polymerases catalyse the synthesis of DNA using the four deoxyribonucleotide building blocks dATP, dGTP, dCTP, and dTTP (Chapter 7). The anticancer drug **cytarabine** (Fig. 18.39) is an analogue of 2' deoxycytidine and acts as a prodrug. It is phos-phorylated in cells to the corresponding triphosphate

(ara-CTP) which acts as a competitive inhibitor. In addition, ara-CTP can act as a substrate for DNA polymerases and become incorporated into the growing DNA chain. This can lead to chain termina-tion or prevent replication of the modified DNA. All of these effects result in the inhibition of DNA synthesis and repair. Cytarabine is used intravenously for the treatment of a wide variety of leukaemias. **Gemcitabine** is an analogue of cytarabine with fewer side effects and is used intravenously to treat pancre-atic cancer, non-small-cell lung cancer, and breast cancer. The purine analogue **fludarabine** is also metabolized to a triphosphate and has the same mechanism of action as cytarabine. It also inhibits transcription and can be incorporated into RNA. The drug is administered intravenously for the treatment of chronic lymphatic leukaemia and is available as the 5' monophosphate prodrug (**Fludara**) to improve solubility.

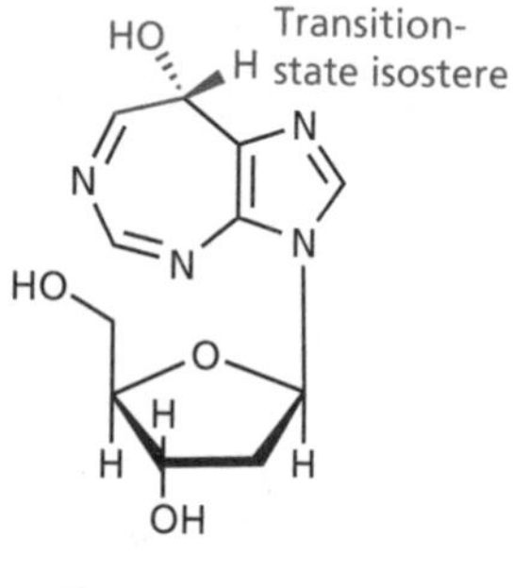

18.3.6 Purine antagonists

The thiopurines **6-mercaptopurine** and **6-tioguanine** (Fig. 18.40) are prodrugs which are converted to their corresponding nucleoside monophosphates by cellular enzymes. The monophosphates then inhibit purine synthesis at a number of points. They are also incor-porated into RNA and DNA, leading to complex effects which end in cell death. Both agents are con-verted to a common product (thio-GMP) which is subsequently converted to thio-GTP and thio-dGTP, before incorporation into RNA and DNA respectively. The drugs are used primarily for the treatment of acute leukaemias and are more effective in children than in adults.

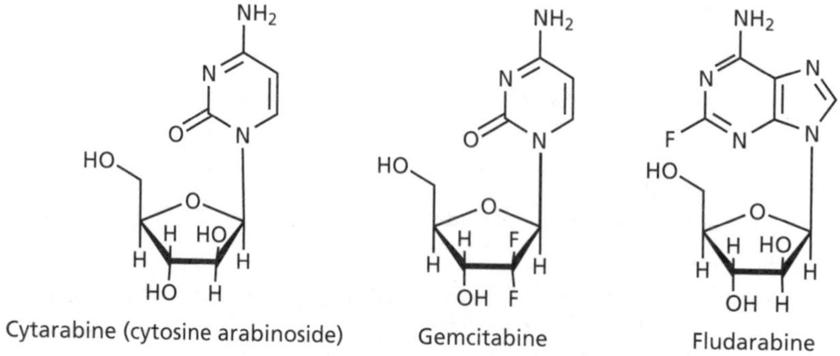

Figure 18.38 Pentostatin.

Figure 18.39 Inhibitors of DNA polymerase.

KEY POINTS

- Antimetabolites are agents which inhibit the enzymes involved in the synthesis of DNA or its building blocks.

- Thymidylate synthase catalyses the synthesis of dTMP from dUMP. The cofactor required for this reaction is regenerated by the enzyme dihydrofolate reductase. Inhibition of either enzyme is useful in anticancer therapy.

- Ribonucleotide reductase catalyses the conversion of ribonucleotide diphosphates to deoxyribonucleotide diphosphates. It can be inhibited directly by drugs, or indirectly by inhibiting adenosine deaminase. In the latter case, a build up of dATP results in allosteric inhibition.

- Various nucleosides and purines act as prodrugs and are converted in the cell to agents that inhibit DNA polymerases.

6-Mercaptopurine   6-Tioguanine

ThioGTP   ThiodGTP

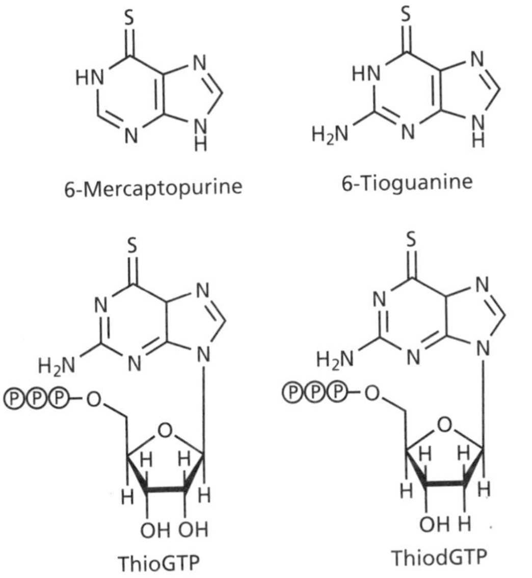

Figure 18.40 Purine antagonists ⓟ represents phosphate.

The active agents also act as substrates and are incorporated into growing DNA, leading to chain termination or the inhibition of replication.

18.4 Hormone-based therapies

Hormone-based therapies are used against cancers which are hormone dependent. If the cancer cell requires a specific hormone, then a hormone can be administered which has an opposing effect. Alternatively hormone antagonists can be used to block the action of the required hormone. Steroid hormones combine with intracellular receptors to form complexes that act as **nuclear transcription factors**. In other words, they control whether transcription takes place or not. This is described in more detail in section 6.8. The types of hormones used in anticancer therapy include glucocorticoids (so classed because they promote gluconeogenesis—the biosynthesis of glucose from non-carbohydrate precursors), oestrogens, and progestins. Analogues of the luteinizing hormone-releasing hormones (LHRH, also called gonadotropin-releasing hormone) can be used to treat breast cancer and prostate cancer. **Octreotide** is an analogue of **somatostatin** and is used to treat hormone secreting tumours of the gastrointestinal tract.

18.4.1 Glucocorticoids

The glucocorticoids **prednisolone** and **prednisone** (Fig. 18.41) are used orally for the treatment of leukaemias and lymphomas. Prednisone acts as a prodrug and is enzymatically converted to prednisolone in the body.

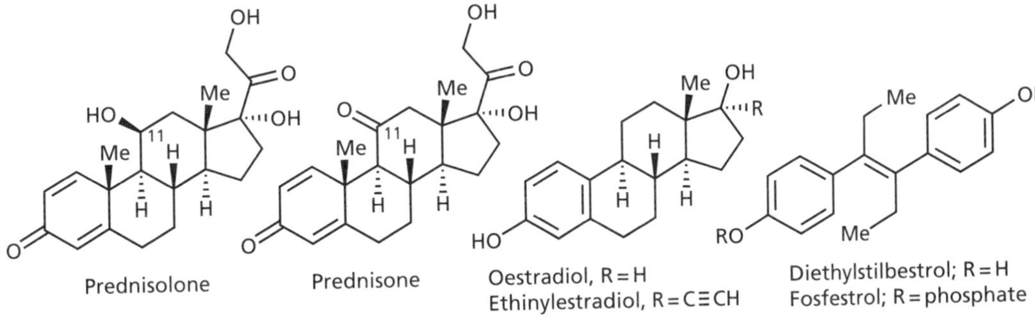

Prednisolone   Prednisone   Oestradiol, R = H   Diethylstilbestrol; R = H
                                          Ethinylestradiol, R = C≡CH   Fosfestrol; R = phosphate

Figure 18.41 Glucocorticoids and oestrogens.

18.4.2 Oestrogens

Oestrogens (spelt estrogens in the USA) are used primarily to treat prostate cancer. They inhibit the production of luteinizing hormone (LH) and by doing so decrease the synthesis of testosterone. The most commonly used agents are **ethinylestradiol** (a derivative of **oestradiol**) and **diethylstilbestrol** (a non-steroidal oestrogen) (Fig. 18.41). **Fosfestrol** is the diphosphate prodrug of diethylstilbestrol and is used for the treatment of hormone-resistant metastatic prostate cancer. It is only activated in target cells where it can reach higher concentrations than if the drug itself was administered.

18.4.3 Progestins

Progestins are used primarily to treat advanced endometrial carcinoma that cannot be treated by surgery or radiation. They have also been used as a second line drug for the treatment of metastatic breast cancer. The most commonly used agents are **medroxyprogesterone acetate** and **megestrol acetate** (Fig. 18.42), which can both be administered orally.

18.4.4 Androgens

Androgens are sometimes used to treat metastatic breast cancer. They are thought to suppress production of LH, resulting in a decrease in oestrogen synthesis. Unfortunately, they have a masculizing effect and so they are only used in a minority of cases. The most commonly used agents are **fluoxymesterone** and **testosterone propionate** (Fig. 18.42). The latter is a prodrug which is converted to **dihydrotestosterone**.

18.4.5 LHRH agonists

LHRH is a decapeptide hormone which binds to receptors on anterior pituitary cells and stimulates the release of LH. On long-term exposure to LHRH, the receptor becomes desensitized leading to a drop in LH levels. Since LH stimulates the synthesis of testosterone, this results in lowered testosterone levels. LHRH agonists are used to treat advanced prostate cancer. The two agents most commonly used are **leuprolide** and **goserelin** (Fig. 18.43) which are both decapeptide analogues of LHRH designed to be more resistant to peptidase degradation. This normally takes place next to glycine at position 6, and replacing this amino acid with an unnatural D-amino acid makes this region unrecognizable to the enzyme. Substitution of the glycine residue at position 10 with a suitable group also increases receptor affinity.

Both agents are administered as their acetates. Leuprolide acetate can be administered daily or once monthly when it is contained in microspheres. In the latter case, the drug is released slowly from the

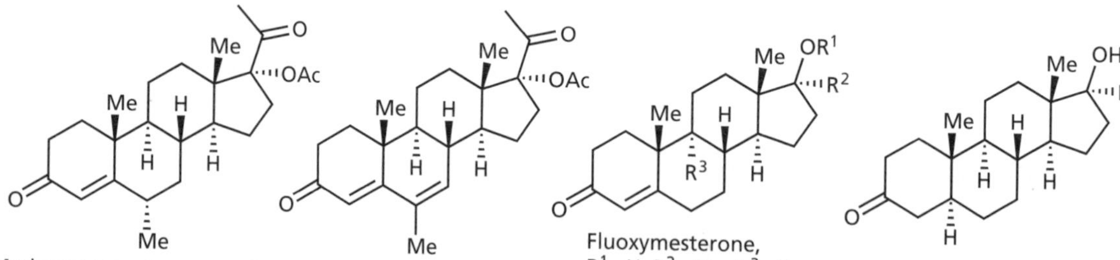

Medroxyprogesterone acetate

Megestrol acetate

Fluoxymesterone, $R^1 = H$, $R^2 = Me$, $R^3 = F$
Testosterone propionate, $R^1 = COEt$, $R^2 = R^3 = H$

Dihydrotestosterone

Figure 18.42 Progestins and androgens.

1 2 3 4 5 6 7 8 9 10
pyroGlu-His-Trp-Ser-Tyr-Gly-Leu-Arg-Pro-Gly-NH_2 — LHRH
pyroGlu-His-Trp-Ser-Tyr-(D-Leu)-Leu-Arg-Pro-ethylamide — Leuprolide
pyroGlu-His-Trp-Ser-Tyr-(D-(t-Bu)Ser)-Leu-Arg-Pro-Azgly-NH_2 — Goserelin

Figure 18.43 LHRH agonists.

microspheres over several weeks. Goserelin acetate can be provided as a slow-release implant where the drug is contained within a biodegradable cylindrical polymer rod. This can be implanted into subcutaneous fat every 28 days.

18.4.6 Antioestrogens

Tamoxifen and **raloxifene** (Fig. 18.44) are synthetic agents which antagonize oestrogen receptors and prevent oestradiol from binding. They are used for the treatment of hormone-dependent breast cancer.

The mechanism by which these agents work has been extensively studied and is described in section 6.8. More recent antioestrogens include **toremifene** and **fulvestrant**, which was approved in 2002.

18.4.7 Antiandrogens

Flutamide and **cyproterone acetate** (Fig. 18.45) are used to block the action of androgens at their receptors and are used in the treatment of prostate cancer.

At present, prostate cancer is treated with a combined therapy of an LHRH agonist and an

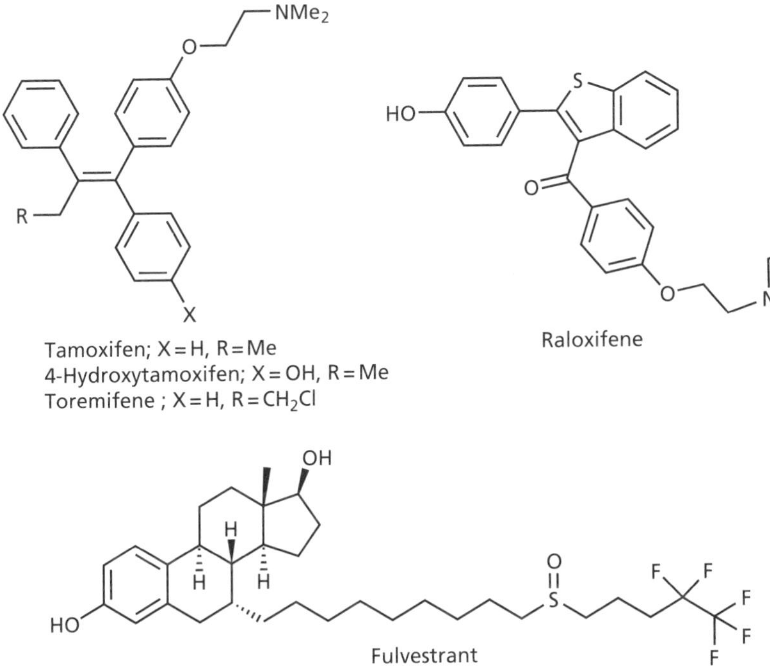

Tamoxifen; X = H, R = Me
4-Hydroxytamoxifen; X = OH, R = Me
Toremifene ; X = H, R = CH₂Cl

Raloxifene

Fulvestrant

Figure 18.44 Antioestrogens.

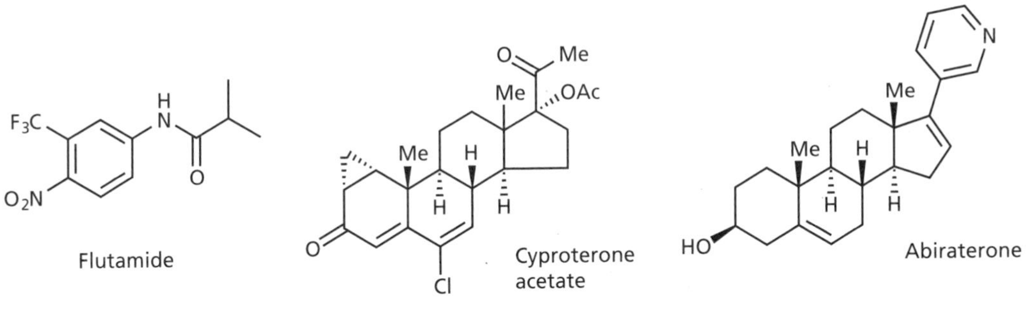

Flutamide

Cyproterone acetate

Abiraterone

Figure 18.45 Antiandrogens.

antiandrogen. It may be possible, however, to use a single agent to inhibit androgen formation by inhibiting a metabolic enzyme called cytochrome $P450_{17a}$, which is involved in the formation of androgens. **Abiraterone** (Fig. 18.45) is a potent inhibitor of this enzyme and has entered clinical trials.

18.4.8 Aromatase inhibitors

Aromatase inhibitors tend to be used as second line drugs for the treatment of oestrogen-dependent breast cancers that prove resistant to tamoxifen. Aromatase is a membrane-bound enzyme complex consisting of two proteins—one is a cytochrome P450 enzyme containing haem (CYP19), and the other is a reductase enzyme using NADPH as cofactor. Aromatase catalyses the last stage in the biosynthesis of oestrogens from androgens where an aromatic ring is formed (Fig. 18.46). The cytochrome enzyme contains haem which serves to bind the steroid substrate and oxygen, then catalyse the oxidation. Since the enzyme catalyses the last step of this synthesis, it has been seen as an important target for the design of antioestrogenic drugs. There are two types of inhibitor which are used clinically—reversible competitive inhibitors and irreversible inhibitors acting as suicide substrates.

Aminoglutethimide (Fig. 18.47) is an early example of a reversible competitive inhibitor, but has disadvantages in that binds to various cytochrome P450 enzymes and inhibits a range of steroid hydroxylations. This results in undesirable side effects. Drug design based on aminoglutethimide as the lead compound has resulted in more selective inhibitors such as **anastrazole** and **letrozole**, which are used to treat breast cancer. The N-4 nitrogen of the triazole ring interacts with the haem iron of aromatase and prevents binding of the steroid substrate. The anilino nitrogen of aminoglutethimide serves the same purpose.

4-Hydroxyandrostenedione (Fig. 18.48) acts as a suicide substrate that permanently inactivates aromatase and is more selective in its action than aminoglutethimide.

18.4.9 Adrenocortical suppressors

Mitotane (Fig. 18.48) interferes with the synthesis of adrenocortical steroids and is used in the treatment of adrenal cortical tumours.

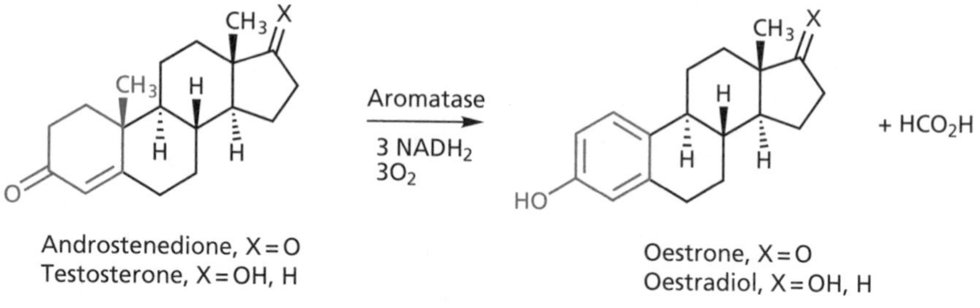

Figure 18.46 Reaction catalysed by aromatase.

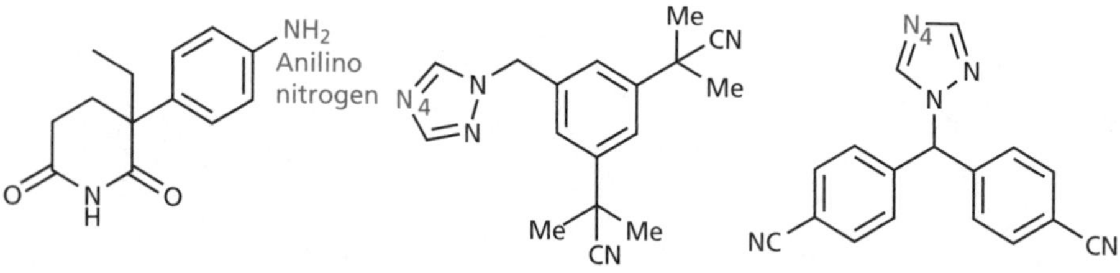

Figure 18.47 Reversible competitive inhibitors of aromatase.

- Hormone-based therapy is used against cancers which are hormone dependent. Hormones can be administered which counteract the offending hormone. Alternatively, antihormonal compounds are administered to prevent the offending hormone from binding to its receptor.

- Glucocorticoids, oestrogens, progestins, androgens, and LHRH are used in hormone-based therapy.

- Agents which act as receptor antagonists are used to block oestrogens and androgens.

- Enzyme inhibitors are used to block the synthesis of hormones. An important target is the enzyme aromatase which catalyses the last step leading to oestrogens.

18.5 Drugs acting on structural proteins

Tubulin is a structural protein which is crucial to cell division (section 3.7.2)—a process which involves the polymerization and depolymerization of microtubules using tubulin proteins as building blocks. A variety of drugs interfere with this process by either binding to tubulin and inhibiting the polymerization process, or binding to the microtubules to stabilize them and inhibit depolymerization. Either way, the balance between polymerization and depolymerization is disrupted leading to a toxic effect and the inability of the cell to divide.

Agents which bind to tubulin prevent polymerization but do not prevent depolymerization, and so this eventually leads to dissolution of the microtubules and destruction of the mitotic spindle required for cell division.

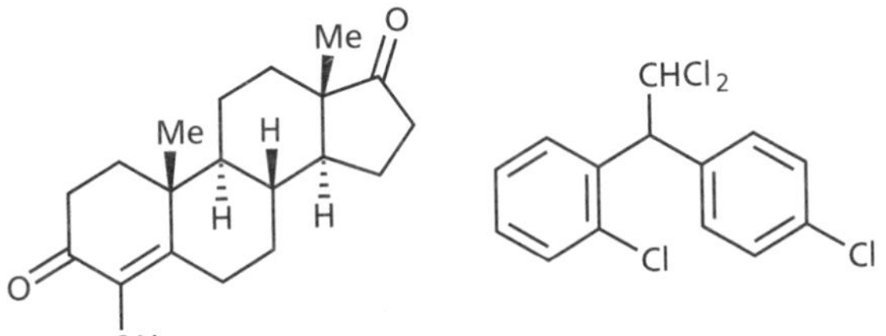

18.5.1 Agents which inhibit tubulin polymerization

Vincristine, **vinblastine**, **vindesine**, and **vinorelbine** (Fig. 18.49) are alkaloids derived from the Madagascar periwinkle plant (*Catharanthus roseus*, formerly known as *Vinca rosea*), which bind to tubulin to prevent polymerization. Vincristine is used in combination therapy to treat acute leukaemias, Hodgkin's lymphoma, small-cell lung carcinoma, and a variety of other tumours. Vinblastine has been used in combination therapies for the treatment of lymphomas, testicular cancer, and ovarian cancer. Vindesine has been used in the treatment of various leukaemias and lymphomas. Finally, vinorelbine is widely used for the treatment of breast cancer and non-small-cell lung carcinomas. Resistance can arise from over-expression of the P-glycoprotein involved in transporting drugs out of the cell.

Phyllanthoside (Fig. 18.50) is another natural product which has entered clinical trials, and is thought

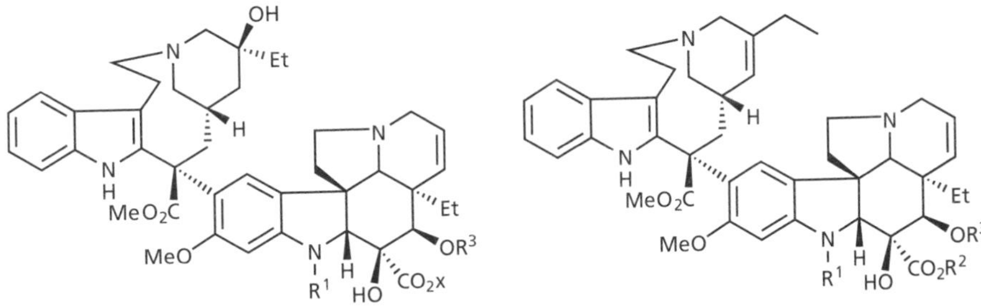

Figure 18.48 4-Hydroxyandrostenedione and mitotane.

Vinblastine (R^1 = Me; x = OMe; R^3 = COMe)
Vincristine (R^1 = CHO; x = OMe; R^3 = COMe)
Vindesine (R^1 = Me; x = NH_2; R^3 = H)

Vinorelbine

Figure 18.49 The *Vinca* alkaloids.

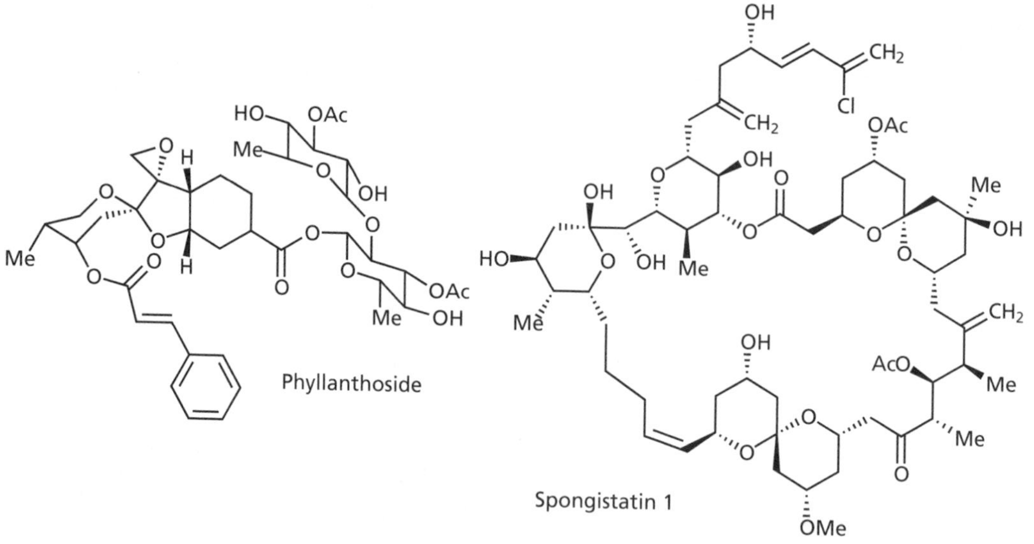

Figure 18.50 Natural products inhibiting microtubule formation.

to bind to tubulin and prevent polymerization. It was obtained from the roots of a Costa Rican tree in the early 1970s.

A variety of other naturally occurring agents have been extracted from marine sources and shown to inhibit microtubule formation. For example, **spongistatin 1** (Fig. 18.50) was extracted from a marine sponge in the Maldives and shows potential as an anticancer agent.

Analogues of the naturally occurring compound **podophyllotoxin** (Fig. 18.13) have already been mentioned in section 18.2.2.1 for their effect on topoisomerase II. Curiously, podophyllotoxin itself has a completely different mechanism of action where it forms a complex with tubulin and prevents the synthesis of microtubules. Podophyllotoxins belong to a group of compounds called lignans and have been isolated from plant sources such the American mandrake or May apple (*Podophyllum peltatum*), and from the Himalayan plant *Podophyllum emodi*. Extracts of these plants have been used for over 1000 years to treat a variety of diseases including cancers. For example, it has been recorded that the roots of the wild chervil (*Anthriscus sylvestris*) were used as a treatment for cancer, and it has been shown that these roots contain deoxypodophyllotoxin. The crude extracts from plants such as those described above is known as **podophyllin** and was shown in 1942 to be effective in the treatment of venereal warts. Podophyllotoxin was

eventually isolated from this extract and was used for a while as an anticancer agent. However, its use was restricted because of severe side effects. It is still the agent of choice for the treatment of genital warts, but it must be handled with care because of its toxicity. A structural similarity has been noted between podophyllotoxin and **colchicine**—another compound which interacts with tubulin (section 3.7.2).

It is interesting to note that the activity of **epipodophyllotoxin** (section 18.2.2.1) is an order of magnitude lower than podophyllotoxin in inhibiting tubulin polymerization, and when bulky sugar molecules are present as in etoposide, activity is removed altogether. This implies that the sugar moieties in etoposide form a bad steric interaction with tubulin which prevents binding.

Cryptophycins (Fig. 18.51) have been isolated from blue-green algae and shown to have an anticancer mechanism which involves the inhibition of microtubule formation. They also inhibit the mechanisms by which microtubules and mitotic spindles function. **Cryptophycin 52** is being considered for clinical trials.

Maytansine 1 (Fig. 18.51) belongs to a group of natural products called the maytansinoids which were extracted from an Ethiopian shrub. It has some similarities in structure to the cryptophycins and also inhibits tubulin polymerization, having an activity 1000 times greater than vincristine. Clinical trials had to be abandoned due to its toxic effects and poor therapeutic

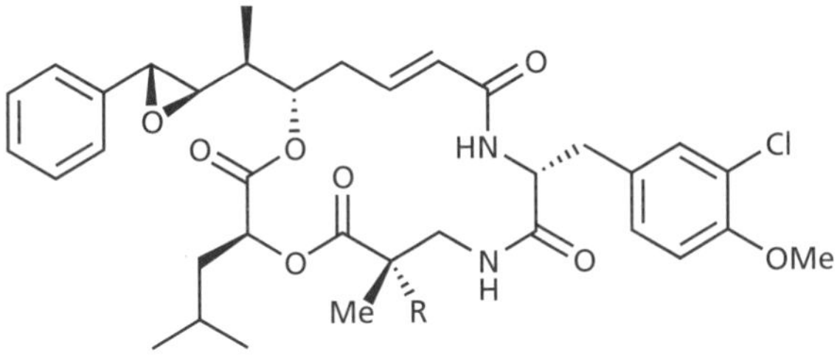

Figure 18.51 Cryptophycins and maytansine 1.

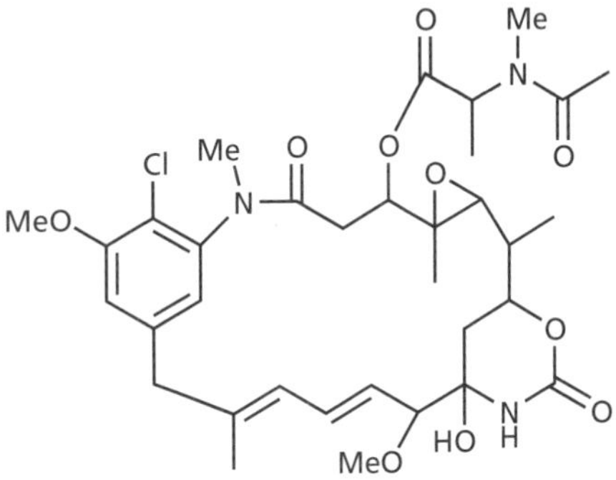

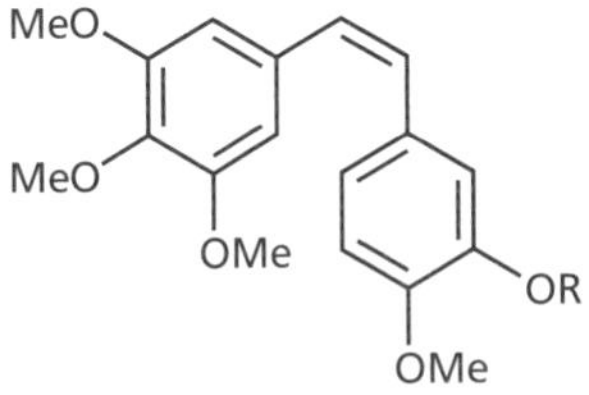

Figure 18.52 Combretastatins.

window, but it is now being considered as a suitable drug for antibody–drug conjugates (section 18.9).

Combretastatins (Fig. 18.52) are natural products derived from the African bush willow (*Combretum caffrum*), a plant which was used by the Zulus as a medicine and as a charm to ward off enemies. **Combretastatin A-4** is the most active structure in this family and has reached clinical trials as its more water-soluble phosphate prodrug. It shares many of the structural features of other tubulin binding drugs such as colchicine and podophyllotoxin, and binds to tubulin at the same binding region as colchicine. The relative orientation of the two aromatic rings is important and therefore the *cis*-geometry of the double bond is crucial to activity. The drug has been shown to selectively inhibit the blood supply to tumours and prevent angiogenesis.

18.5.2 Agents which inhibit tubulin depolymerization

Paclitaxel (Taxol) (Fig. 18.53) is derived from the bark of yew trees (*Taxus* spp.). The term **taxoids** is used as

a general term for paclitaxel and its derivatives. Paclitaxel was first identified in 1971 following a screening programme for new anticancer agents carried out by the US National Cancer Institute. It shows outstanding therapeutic activity against solid tumours and was approved for clinical use in 1992 for the treatment of breast and ovarian cancers. **Docetaxel** (Fig. 18.53) is a semi-synthetic taxoid which was approved for the treatment of advanced breast cancer in 1996. Both drugs are in clinical trials for the treatment of a variety of other cancers.

The agents bind to the β-subunit of tubulin (one of the two separate proteins that make up tubulin). In contrast to the drugs described in section 18.5.1, the binding of paclitaxel accelerates polymerization and stabilizes the resultant microtubules, thus inhibiting depolymerization. As a result, the cell division cycle is halted mainly at the G_2/M stage. Apoptosis then takes place. The benzoyl and acetyl substituents, at positions 2 and 4 respectively, play an important role in this binding interaction, as do the side chain and the oxetane ring.

Obtaining sufficient paclitaxel was initially a problem, as the bark from two yew trees was required to supply sufficient paclitaxel for one patient! A full synthesis of paclitaxel was achieved in 1994 but was impractical for large-scale production because it involved 30 steps and gave a low overall yield. Fortunately, it has been possible to carry out a semi-synthetic synthesis (section 12.3.4) using a related natural product which can be harvested from the yew needles without damaging the tree. The semi-synthetic route involves docetaxel as an intermediate.

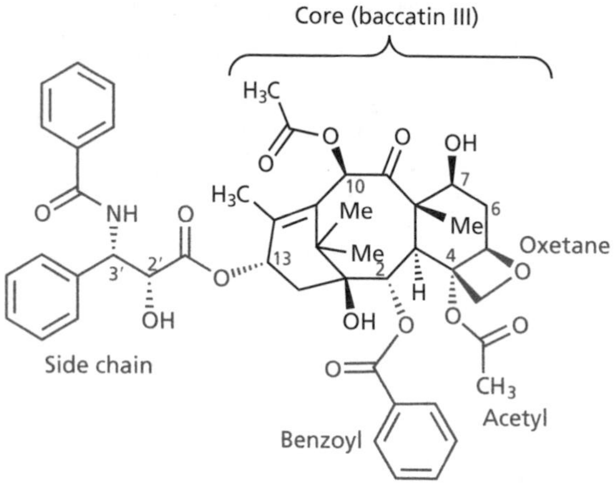

Core (baccatin III)

Paclitaxel

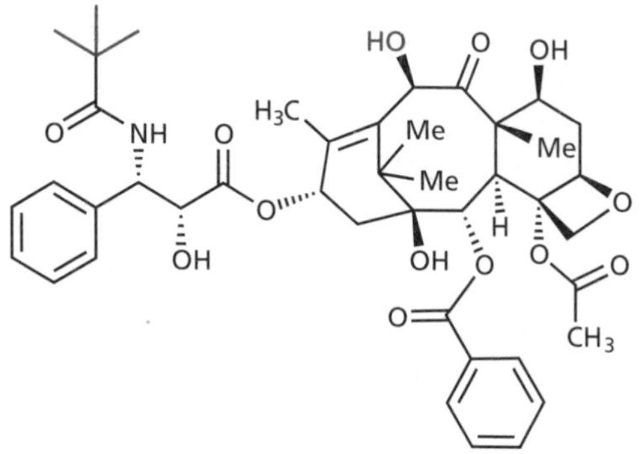

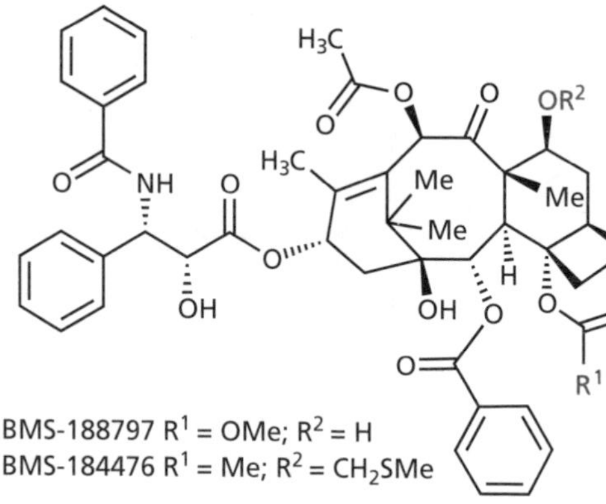

BMS-188797 R^1 = OMe; R^2 = H
BMS-184476 R^1 = Me; R^2 = CH$_2$SMe

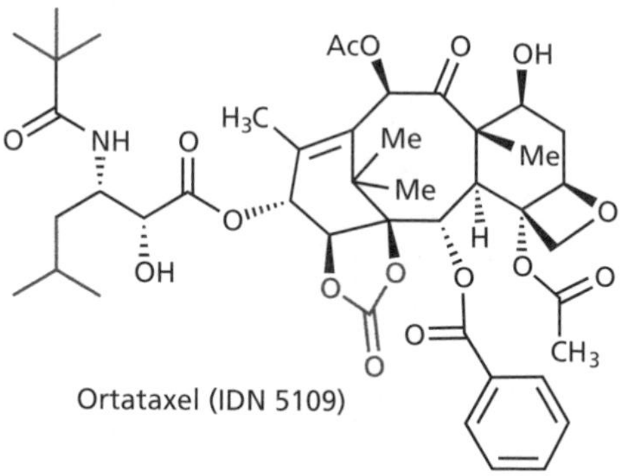

Docetaxel

Ortataxel (IDN 5109)

Figure 18.53 Paclitaxel (Taxol), with important binding groups in colour, and docetaxel (Taxotere).

Figure 18.54 Analogues of paclitaxel.

A problem with the use of taxoids such as paclitaxel and docetaxel is the fact that they cannot be taken orally, and they also have various undesirable side effects. Moreover, therapy often leads to the development of multidrug resistance. This involves several mechanisms including tubulin mutation leading to weaker binding interactions, and over-expression of the P-glycoprotein transport protein leading to faster efflux from the cell. Semi-synthetic taxoids are currently being investigated to find compounds with better oral bioavailability, improved pharmacological properties, and activity against drug-resistant cancers. The binding groups mentioned above dominate the 'lower' or 'southern' half of the molecule (as the structure is normally presented), and so the variations that are possible in this region are restricted. In contrast, it is possible to carry out more variations in the 'northern' half of the molecule. This can affect the *in vivo* efficacy of the molecule allowing modification of aqueous solubility and pharmacokinetic properties. **BMS 188797** and **BMS 184476** (Fig. 18.54) are two taxoids which have recently been developed and have reached clinical trials.

More substantial variations led to a second generation of taxoids where potency was increased by 2–3 orders of magnitude. For example, it was possible to replace the aromatic rings of paclitaxel with other hydrophobic groups. Having a suitable acyl group at position 10 has also been found to increase activity against drug resistant strains. Such compounds have the ability, not only to bind to tubulin, but to inhibit the P-glycoprotein efflux pump. The first orally active taxoid structure **ortataxel** (Fig. 18.54) has also been developed and has entered clinical trials. Finally, the addition of a methyl substituent at C-2' has been found to increase activity by inhibiting rotation of the C-2'–C-3' bond.

Since the discovery of paclitaxel a variety of other natural products have been found to have a similar

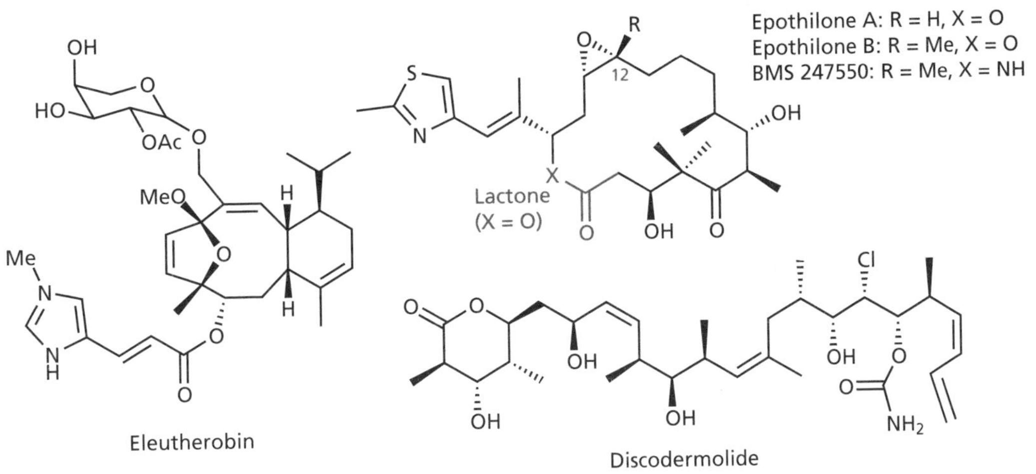

Figure 18.55 Recently discovered natural products which inhibit microtubule depolymerization.

mechanism of action, and are currently being studied as potential anticancer agents (Fig. 18.55). These include bacterial metabolites called **epothilones**, and marine natural products such as **eleutherobin** from coral, and **discodermolide** from a marine sponge. These compounds show several advantages over paclitaxel. First, the epothilones and discodermolide do not appear to be substrates for the P-glycoprotein efflux system, and are therefore potentially effective against drug-resistant cancer cells. Second, it has been found that discodermolide enhances the activity of paclitaxel and it may be possible to use these drugs in combination therapy. Third, the epothilones have better aqueous solubility than paclitaxel, which may allow the development of better formulations.

A drawback with the epothilones is their metabolic lability, which results from the cleavage of the lactone ring by esterases. They have also been shown to be highly toxic in animal studies. Therefore, research is being carried out to find analogues with improved properties. This has led to **BMS 247550** (Fig. 18.55) which has entered clinical trials. The lactone in this structure has been replaced by a more stable amide group which stabilizes the molecule to metabolism and also reduces its toxic side effects.

These novel agents bind to the same region of tubulin as paclitaxel and a 3D pharmacophore has been developed which encompasses the different structures and which is being used as the basis for the design of hybrid molecules that may lead to a third generation of taxoids.

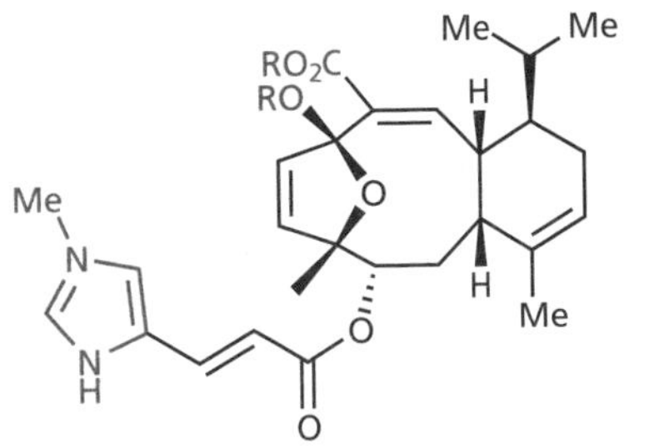

Figure 18.56 Sarcodictyins.

Sarcodictyins (Fig. 18.56) are similar in structure to eleutherobin and are also active against drug-resistant cancers. SAR studies of these compounds have demonstrated the importance of the coloured groups in Fig. 18.56.

KEY POINTS

- Agents which inhibit the polymerization or depolymerization of microtubules are important anticancer agents.

- The vinca alkaloids, podophyllotoxin, the combretastatins, and a variety of other natural products bind to tubulin and inhibit the polymerization process.

- Paclitaxel and its derivatives bind to tubulin and accelerate polymerization by stabilizing the resulting microtubules. Newer analogues are being investigated which show better oral bioavailability, improved pharmacological properties and activity against drug-resistant cancers.

- A variety of natural products have been discovered which have a similar mechanism of action to paclitaxel.

18.6 Inhibitors of signalling pathways

Most traditional anticancer drugs are cytotoxic both to cancer and normal cells, and any selectivity relies on a greater concentration of the agents within cancer cells. Now, cancer chemotherapy is on the verge of a revolution. Advances in genetics and molecular biology have led to an ever-increasing understanding of the molecular processes behind specific cancers, and the identification of a variety of molecular targets which are either unique to a cancer cell or are over-expressed compared to normal cells. The design of agents that will act on these targets promises the development of more selective anticancer agents with less toxic side effects. Identifying defects in a cell's signalling pathways and identifying suitable targets has already resulted in clinically useful drugs. Suitable targets include the receptors for growth hormones, and the various signal proteins and kinases in the signal transduction pathways. The following sections illustrate some of the most promising lines of re-search, but it should be appreciated that there is a vast amount of research being carried out in this area and it is not possible to give a comprehensive coverage of it all.

18.6.1 Inhibition of farnesyl transferase and the Ras protein

It has been observed that an abnormal form of the signalling protein **Ras** (sections 6.7.3–6.7.4) is present in 30% of human cancers, and is particularly preval-ent in colonic and pancreatic cancers. Abnormal Ras derives from a mutation of the *ras* gene to form a *ras* oncogene. Ras proteins are an inherent component of the cellular signalling pathways which control cell growth and multiplication. They are small G-proteins which bind GDP when they are in the resting state, and bind GTP when they are in the active state. Binding to GTP is temporary, as the protein can autocatalyse its hydrolysis back to GDP and return

to the resting state. Mutant Ras proteins, however, persistently bind GTP and fail to hydrolyse it, such that they are constantly active. Since Ras is an integral part of the signalling pathways that control cell growth and division, it is believed that this contri-butes to the development of cancer. Finding methods of 'neutralizing' Ras, therefore, could be useful in combating cancer.

One of these approaches centres round a zinc metalloenzyme called farnesyl transferase (FT). This enzyme is responsible for attaching a 15-carbon farnesyl group to the Ras protein when it is in the cytoplasm of the cell. The farnesyl group is hydro-phobic and acts as a hook and an anchor to hold the Ras protein at the inner surface of the cell membrane. This is necessary if the Ras protein is to interact with other elements of the signal transduction process. Inhibitors of the FT enzyme have been shown *in vitro* to reverse malignancy in cancer cells containing the *ras* oncogene without affecting normal cells.

The enzyme mechanism (Fig. 18.57) involves the binding of farnesyl diphosphate (FPP) to the active site, followed by the Ras substrate. This order of binding is important, as FPP is actually involved in binding the Ras protein. Farnesylation can then take place where a cysteine residue on the Ras protein displaces a pyrophosphate leaving group from FPP. Magnesium and iron ions are present in the active site as cofactors. The former is involved in com-plexing the negatively charged pyrophosphate group to make it a good leaving group, while the latter interacts with the thiol group of cysteine, enhancing its nucleophilicity. Following farnesylation, the ter-minal tripeptide of Ras is cleaved and the resulting carboxylic acid is methylated to a methyl ester. Methylation is important or the charged carboxylic acid would hinder the binding to the cell membrane by the farnesyl chain.

Inhibitors have been developed which mimic the terminal tetrapeptide moiety of the Ras protein. This region, which is common to different types of Ras protein, is known as the **CaaX peptide** (C stands for cysteine, a is valine, isoleucine, or leucine, and X is methionine, glutamine, or serine). Studies on a variety of tetrapeptides were carried out to see what effect modification of these positions would have. These studies showed that placing an aromatic amino acid such as phenylalanine next to X transformed the tetrapeptide from a substrate into an inhibitor

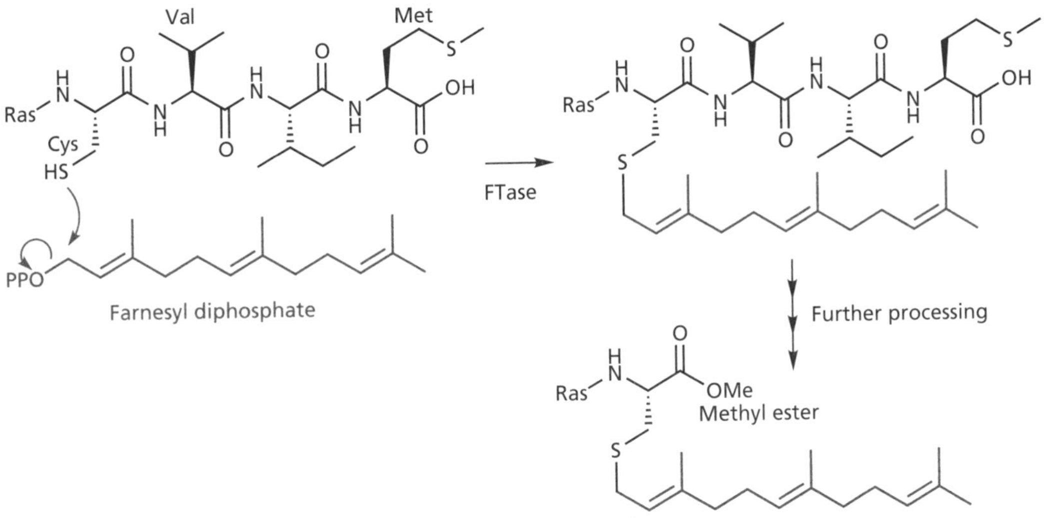

Figure 18.57 Mechanism of farnesyl transferase.

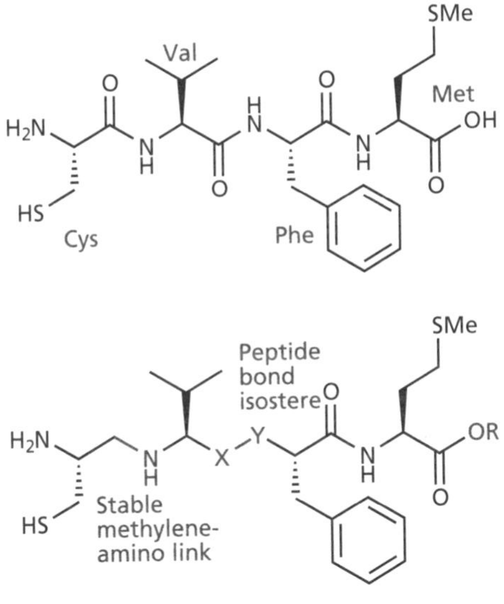

Figure 18.58 Design of farnesyl transferase inhibitors.

(Fig. 18.58). This then served as a lead compound for further development which involved the replacement of two of the peptide links with stable isosteric groups and the masking of the carboxylic acid. We shall now look more closely at why these modifications were carried out.

As the target enzyme is within the cell, inhibitors have to cross cell membranes, be metabolically stable, and have suitable pharmacokinetic properties. However, the lead compound is a tetrapeptide and suffers from various disadvantages.

First, it has a polar carboxylic acid group which is bad for absorption. This problem can be overcome by masking the acid group as an ester prodrug such that the molecule can cross cell membranes more easily. Once inside the cell, the ester can be hydrolysed to give the required carboxylic acid.

Second, the lead compound has peptide bonds that are susceptible to metabolism, particulary to

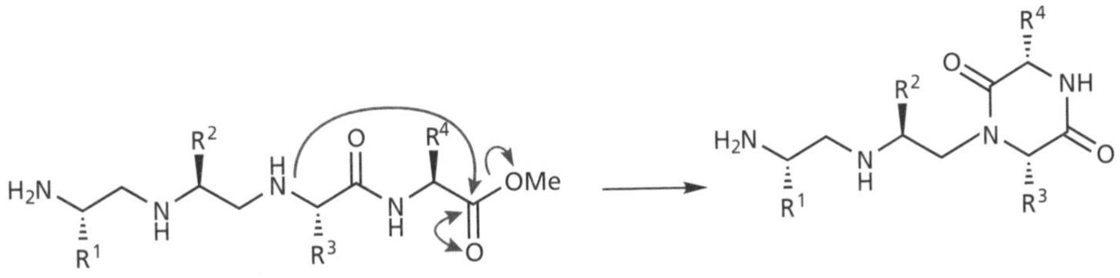

Figure 18.59 Dioxopiperazine formation.

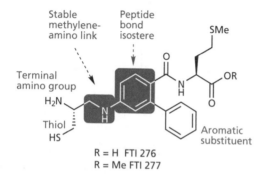

R = H L 739750
R = ⁱPr L 744832

R = H FTI 276
R = Me FTI 277

Figure 18.60 FTase inhibitors.

Figure 18.61 AZD-3409.

and their respective ester prodrugs (Fig. 18.60). These have shown promising *in vivo* results on cancers in transgenic mice, without obvious toxicity. Both structures contain the important thiol group as a ligand for the zinc ion cofactor, the stable methyleneamino moiety, and the aromatic substituent which is beneficial to inhibitory activity. The isostere for the middle peptide link is a methyleneoxy group in L 739750 and an aromatic ring in FTI 276. The methylthio substituent has been replaced by a sulfone group in L 739750, as this was found to increase activity.

The terminal amino group in both drugs is important for binding interactions. It is ionized and forms an ionic bond with an ionized phosphate group of FPP. The carboxylic acid group is also important for binding.

Although the thiol group is important as a ligand for zinc, there are problems associated with it, since it is potentially toxic and can be oxidized by metabolic enzymes. The AstraZeneca compound **AZD-3409** (Fig. 18.61) has the thiol group and the carboxylic acid groups masked in a double prodrug strategy to try to alleviate this problem. A pyrrolidine ring is also present and introduces conformational rigidity to this portion of the molecule. The drug is a potent inhibitor of FTase ($K_i < 1$ nM) and it also inhibits a related prenylating[1] enzyme called **geranylgeranyltransferase** (GGTase) ($K_i = 8$ nM). This enzyme can also catalyse prenylations but uses a 20-carbon structure called **geranylgeranyl diphosphate** as the prenylating agent. Normally, GGTase prenylates proteins having the CaaX motif where X = Leu, but it is possible that it may prenylate proteins which are normally prenylated by FT if the latter enzyme is inhibited. Such a reaction would allow these proteins to bind to cell membranes and still be functional, thus bypassing the inhibition of FTase. Therefore, an agent capable of inhibiting both enzymes may be beneficial.

Non-peptide inhibitors have also been designed and are undergoing clinical trials. Structure I in Fig. 18.62 has an imidazole ring acting as the zinc ligand rather than a thiol group, since the latter group has been associated with undesirable side effects. The imidazole

aminopeptidases. Replacing two of these bonds with stable methyleneamino groups avoided this problem but introduced a different one. The resulting amines are more nucleophilic than the original peptide groups and one of these was able to carry out an intramolecular cyclisation with the terminal ester to form an inactive dioxopiperazine structure (Fig. 18.59). This could be avoided by replacing the offending amine with an ether.

These features are seen in some of the most notable inhibitors studied so far, mainly **L 739750**, **FTI 276**,

---

[1] Prenylation is the term used to describe the formation of a covalent bond between a molecule such as a protein and a prenyl moiety such as a farnesyl or a geranylgeranyl group.

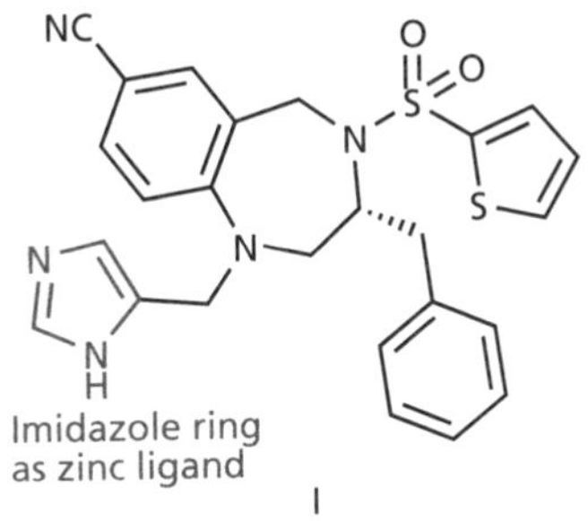

I
IC$_{50}$ 1.4 nM

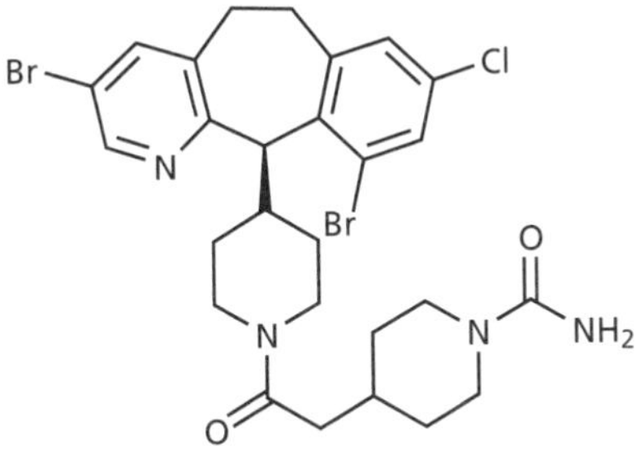

Imidazole ring
as zinc ligand

Lonafarnib
IC$_{50}$ 1.9 nM

Figure 18.62 Non-peptide-like inhibitors of FTase undergoing clinical trails.

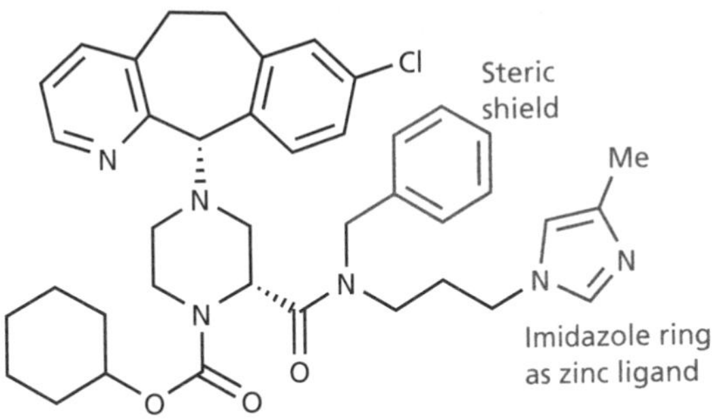

Sch 226374; IC$_{50}$ 0.36 nM

Steric shield

Imidazole ring
as zinc ligand

Figure 18.63 Development of a lonafarnib analogue.

a variety of different protein substrates other than Ras, it is possible that inhibition affects other cellular processes resulting in the observed anticancer activity. These proteins could include several nuclear proteins such as centromere associated proteins and protein phosphatases. The former are associated with chromosome alignment and the mitotic checkpoint. Inhibition of these proteins could prevent the cell entering the mitosis phase.

18.6.2 Protein kinase inhibitors

Protein kinases are enzymes which phosphorylate specific amino acids in protein substrates. It is estimated that there are over 2000 different types of protein kinase. Many are enzymes within the cytoplasm of the cell (sections 6.5–6.7) while others (protein kinase receptors) traverse the cell membrane and play a dual role as receptor and enzyme (section 6.7). The latter structures have an extracellular binding site to receive an external molecular messenger, and an intracellular kinase active site which is activated when the messenger binds to the receptor's binding site. The chemical messengers involved are a wide variety of growth hormones and growth factors which trigger the start of a signalling cascade that involves the various cytoplasmic protein kinases. This process ultimately controls transcription of specific genes in DNA leading to cell growth and cell division. In many cancers, it has been observed that there is an excess of a particular growth hormone or growth factor, or an excessive quantity of a particular protein kinase or protein kinase receptor. Since these structures are intimately involved

ring had previously been shown to act as a zinc ligand in other structures.

Lonafarnib (Fig. 18.62) was developed from a compound that was discovered by screening compound libraries, and is 10 000 times more active than the original lead compound. Remarkably, it has no ligand for the zinc cofactor and so structure-based drug design was carried out to introduce a suitable group at the correct position, resulting in **Sch 226374** (Fig 18.63). The imidazole ring acts as the zinc ligand while an aromatic ring acts as a steric shield to protect it from metabolism (see also Box 18.1).

Although farnesyl transferase inhibitors (FTIs) show promise as anticancer agents, it is questionable whether their observed activity is due solely to their action in inhibiting the farnesylation of the Ras protein. For example, there are three human Ras proteins (H-Ras, N-Ras, and K-Ras). FTIs inhibit the farnesylation of all three but can inhibit the *cellular* functions of H-Ras only, since the other two Ras proteins can be prenylated by GGTase and become linked to the cell membrane. Nevertheless, anticancer effects are still observed in cancers where it is the K-Ras protein that is being expressed. Since farnesyl transferase can accept

BOX 18.1 DEVELOPMENT OF A NON-PEPTIDE FARNESYL TRANSFERASE INHIBITOR

The screening of compound libraries produced the lead compound (I) for **tipifarnib**—an agent undergoing clinical trials as a farnesyl transferase inhibitor. SAR studies were carried out on structure I and established the importance of both aromatic rings to activity. Alterations were then carried out to improve activity, mainly the introduction of a *meta*-chloro substituent (structure II), *N*-methylation of the quinolone (structure III), altering the position of the nitrogens on the imidazole ring (structure IV), and finally the introduction of a primary amino group. The imidazole ring acts as a ligand for the zinc cofactor in the enzyme's active site.

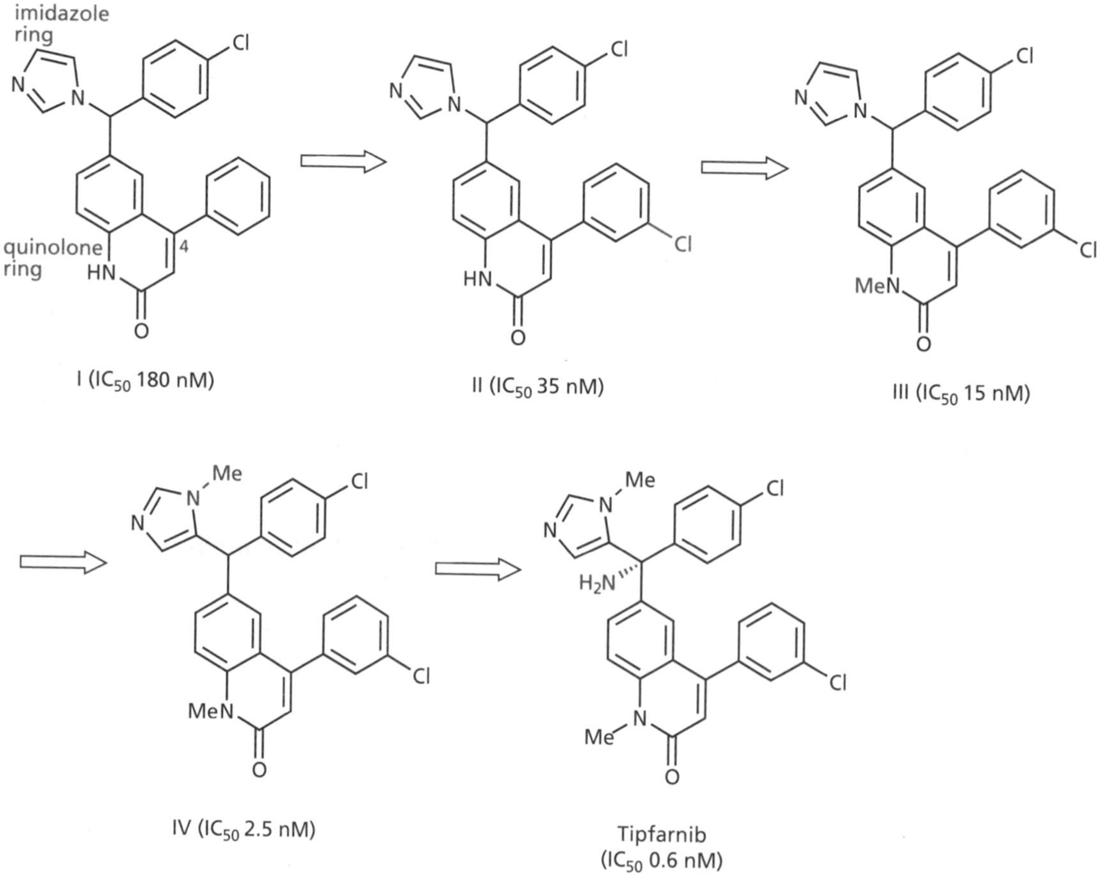

Development of tipifarnib.

in the signal transduction processes which drive cell growth and cell division, it is reasonable to assume that protein kinase inhibitors would be useful as anticancer agents.

Protein kinases can be divided into two main categories—the tyrosine kinases and the serine-threonine kinases. (More recently, histidine kinases have been discovered which phosphorylate the nitrogen of a histidine residue.) The tyrosine kinases phosphorylate the phenol group of tyrosine residues, whereas the serine-threonine kinases phosphorylate the alcohol group of serine and threonine residues (section 6.5). All the kinases use ATP as the phosphorylating agent, and so there is a region within the active site that binds ATP, as well as a region that binds the substrate. Considering the fact that there are so many kinases, and that they all use ATP as the phosphorylating agent, many scientists thought that the protein kinases would be poor targets for drugs. It was believed that the ATP binding region would be very similar for all protein kinases, so that any inhibitor would show poor selectivity and end up inhibiting all protein kinases. This has not turned out to be the case. Crystal structures of protein kinases bound to ATP have been studied and it has been shown that ATP is bound quite loosely to the binding region, and that there are areas of the binding region which remain unoccupied. There are also significant differences in the amino acids present in the ATP binding region from one active site to another. As a result, it is quite possible to design selective inhibitors. Indeed, all the protein kinase inhibitors of clinical interest have been designed to bind to this region of the active site, rather than the substrate binding site.

The binding interactions of ATP with the kinase active site of one protein kinase receptor (the epidermal growth factor receptor, EGF-R) are shown in Fig. 18.64 and are representative of all the kinase binding sites. The purine base is buried deep in the binding site and makes two important hydrogen bonding interactions with the protein backbone. The ribose sugar is bound into a ribose binding pocket and the triphosphate chain lies along a cleft leading to the surface of the enzyme. The ionized triphosphate interacts with two metal ions and with several amino acids through hydrogen bonding. There are also various areas of unoccupied space—in particular, a hydrophobic pocket opposite the ribose binding pocket. These unoccupied regions show a structural variability between different kinases, allowing the possibility of selective inhibitors designed to access these regions. A knowledge of how ATP and the various protein kinase inhibitors are bound to the kinase active site has helped enormously in the design of potent and selective agents.

18.6.2.1 Kinase inhibitors of the epidermal growth factor receptor

EGF-R is a membrane-bound tyrosine kinase receptor having an extracellular binding site for epidermal growth factor (EGF), and an intracellular kinase active site (section 6.7.2). The over-expression or alteration of EGF-R arises from a mutation to the normal gene such that it becomes an oncogene. Abnormal or over-expressed EGF-R protein has been associated with a variety of cancers in the breast, lung, brain, prostate, gastrointestinal tract, and ovaries, and several agents have been studied as kinase inhibitors. The most

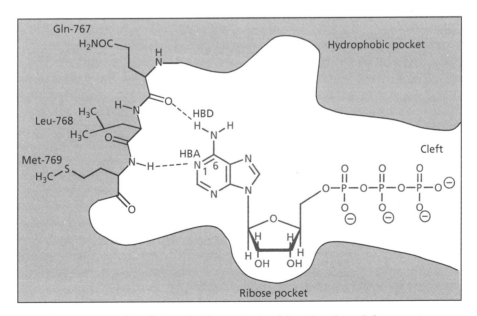

Figure 18.64 Binding of ATP to the kinase active site of the epidermal growth factor receptor.

advanced of these is **gefitinib** (**Iressa**) (Fig. 18.65), which has been approved for the treatment of refractory lung cancers.

Gefitinib was developed by AstraZeneca and belongs to a group of structures known as the 4-anilinoquinazolines (see Box 18.2). It was developed from a potent inhibitor (I in Fig. 18.66) which had various important features previously identified by SAR studies, mainly a secondary amine, electron-donating substituents at positions 6 and 7, and a small lipophilic substituent on the aromatic ring. The structure had useful *in vitro* activity, but its *in vivo* activity was hampered by the fact that it was rapidly metabolized by cytochrome P450 enzymes to give two metabolites. Oxidation of the aromatic methyl group resulted in metabolite II, and oxidation of the

aromatic *para*-position resulted in metabolite III. Both these types of positions are well known to be vulnerable to oxidative metabolism (section 8.4). Therefore, it was decided to modify the structure such that both metabolic routes were blocked. In structure IV in Fig. 18.66, the methyl group was replaced by a chloro substituent. This has a similar size and lipophilicity to a methyl group, but is resistant to oxidation. A fluoro substituent was chosen to block oxidation of the aromatic *para*-position. Fluorine is essentially the same size as hydrogen and so there is little risk of any adverse steric effects arising from its introduction. Although the resulting compound was less active *in vitro* as an enzyme inhibitor, it showed better *in vivo* activity since it proved resistant to metabolism. Further modifications were then carried out to optimize the pharmacokinetic properties of the drug. A variety of alkoxy substituents at the 6-position were tried, culminating in the discovery of gefitinib. This contains a morpholine ring, which is well known to enhance water solubility because of its basic nitrogen atom.

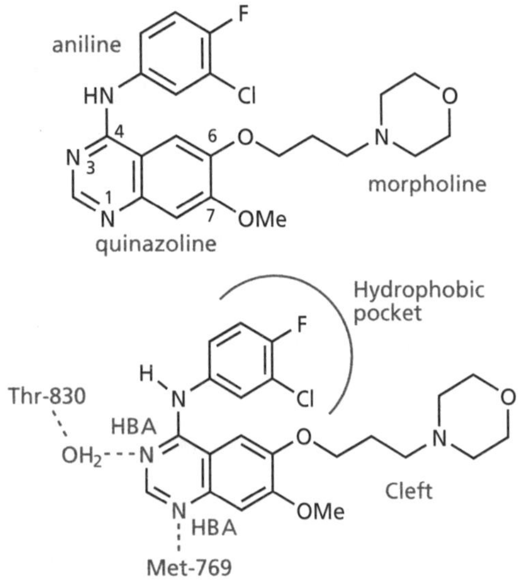

A molecular modelling study was carried out to investigate how gefitinib might bind to a model of the kinase binding site, based on a knowledge of the structure of similar active sites. Gefitinib was found to mimic ATP and bind to the ATP-binding region of the kinase active site. The binding of ATP itself involves two important hydrogen bonding interactions between the purine base of ATP and the protein backbone between amino acids Gln-767 and Met-769 (Fig. 18.64). One of these interactions involves the purine group acting as a hydrogen bond donor, while the other involves purine acting as a hydrogen bond acceptor. Gefitinib has two nitrogen atoms in the quinazoline ring which both act as hydrogen bond acceptors (Fig. 18.65). One of these

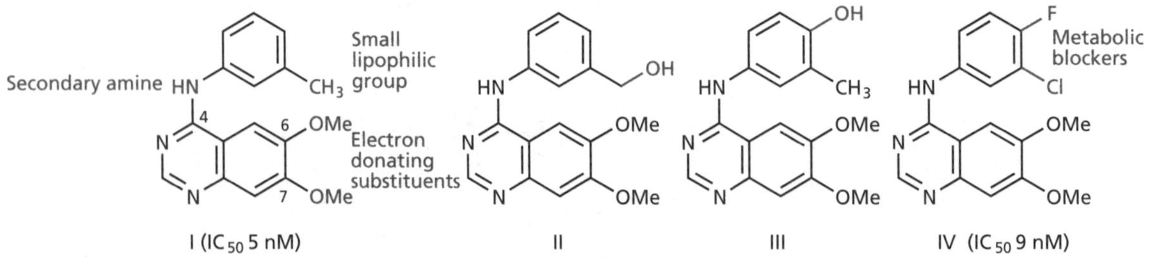

Figure 18.65 Structure and binding interactions of gefitinib (Iressa).

Figure 18.66 Design of a metabolically stable analogue of structure (I).

BOX 18.2 GENERAL SYNTHESIS OF GEFITINIB AND RELATED ANALOGUES

A general synthesis for gefitinib and its analogues starts from a quinazolinone starting material which acts as the central scaffold for the molecule. The synthesis is then a case of introducing the two important substituents. Selective demethylation reveals a phenol which is protected by an acetate group. Chlorination is now carried out on the carbonyl group and the resulting chloro substituent is substituted by an aniline to introduce the first important substituent. Deprotection of the phenol group and reaction with an alkyl halide introduces the second important substituent.

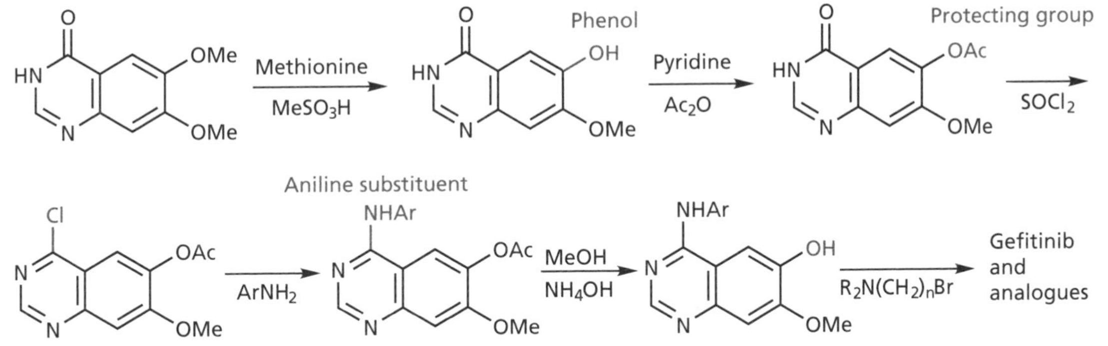

Synthesis of gefitinib (Iressa) and related analogues.

interacts directly with Met-769, while the other interacts with a bridging water molecule which interacts in turn with the hydroxyl group of Thr-830. The aniline ring occupies the normally vacant hydrophobic pocket opposite the ribose binding pocket (Fig. 18.64).

Other EGF-R kinase inhibitors include **PKI-166** and **Tarceva** (Fig. 18.67), both of which are undergoing clinical trials. PKI-166 is a pyrrolopyrimidine structure which binds differently from the quinazoline structures above. Here the important hydrogen bonding interactions to the adenine binding region involve an N on the pyrimidine ring and an NH on the pyrrole ring.

The binding interactions of ATP, quinazolines, and pyrrolopyrimidines illustrate an important point. All three classes of compound contain a pyrimidine ring with an NH substituent at position 4. Having seen how this group binds for ATP (Fig. 18.64), it would be tempting to assume that it binds in the same manner for the other structures containing it. The fact that it doesn't illustrates the importance of analysing crystal structures of enzyme–inhibitor complexes and not making assumptions. The binding site for ATP is quite spacious, so it is perfectly feasible for molecules to bind in different modes. Indeed, it is possible for different molecules

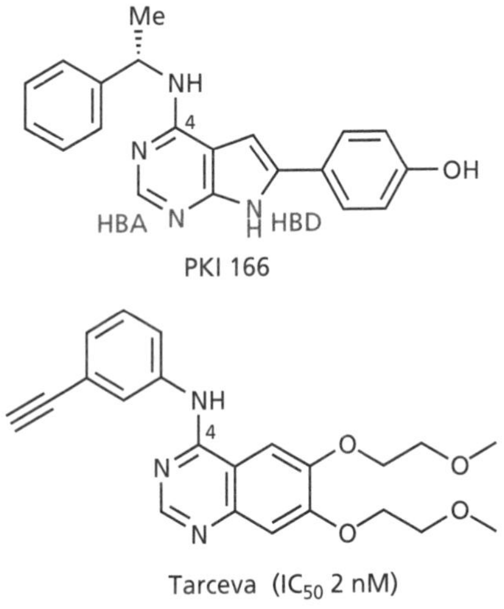

Figure 18.67 Inhibitors of the EGF-R kinase.

within the same structural class to bind in different modes depending on the substituents that are present.

18.6.2.2 Inhibitors of the Abelson tyrosine kinase

As the first protein kinase inhibitor to reach the market, **Imatinib** (**Glivec** or **Gleevec**; Fig. 18.68)

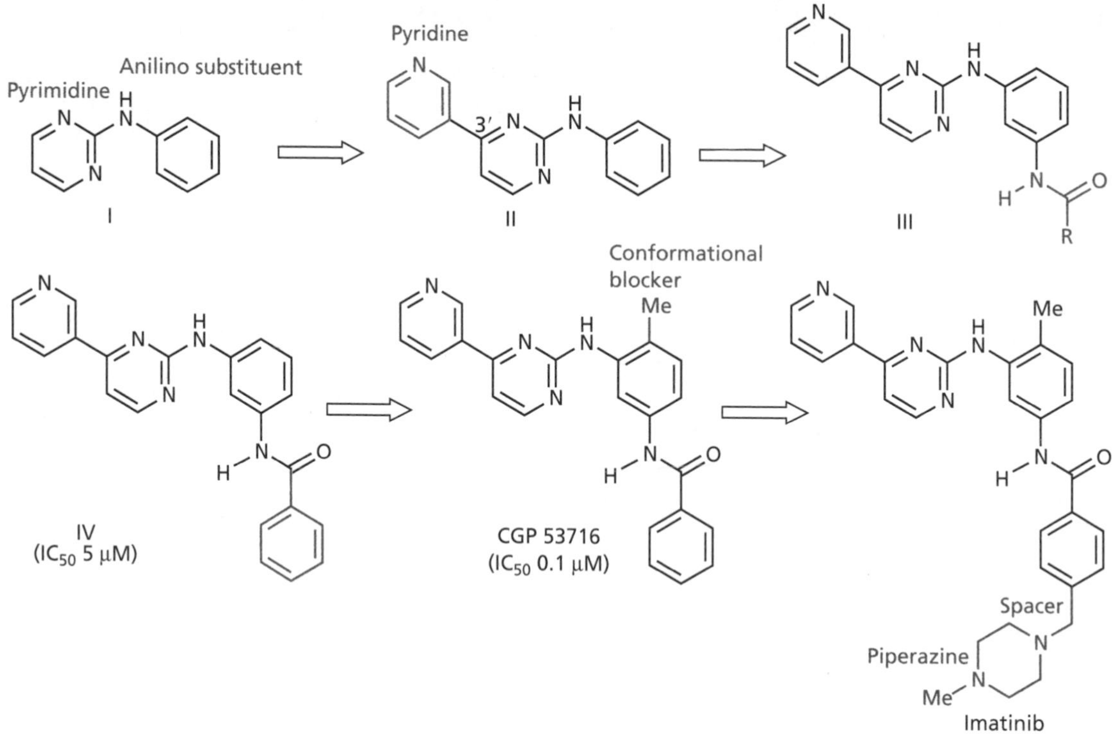

Figure 18.68 Development of imatinib.

represents a milestone in anticancer therapy. It was introduced for the treatment of a rare blood cancer called chronic myeloid leukaemia (CML). The cancer cells involved contain an abnormal protein kinase which is not found in normal cells. The protein kinase concerned is a member of the tyrosine kinase family and has been named **Bcr-Abl**. This name is derived from the genes which code for the protein (i.e. the c-abl and bcr genes). In normal cells these genes are distinct and on different chromosomes, so they code for separate proteins. In the cancer cells associated with CML, part of one chromosome has been transferred to another resulting in a shortened chromosome called the Philadelphia chromosome—a characteristic feature of this type of cancer. The result of this genetic transfer is the formation of a hybrid gene (bcr–abl) which is not properly regulated and which codes for excessive levels of the hybrid protein kinase. This in turn leads to excessive quantities of white blood cells (leukocytes). The tyrosine kinase active site resides on the Abl portion of the hybrid protein.

Imatinib is a selective inhibitor of this protein kinase and has been successful in 90% of patients. As such, it marks a new era in cancer chemotherapy since it is the first drug to target a molecular structure which is unique to a cancer cell. Its synthesis is described in Box 18.3.

The lead compound (I in Fig 18.68) used for the development of imatinib was a phenylamino-pyrimidine structure identified by random screening of large compound libraries. The original aim of this search was to find inhibitors of a different protein kinase known as protein kinase C (PKC)—a serine-threonine kinase. Strong inhibition of PKC was achieved by adding a pyridyl substituent at the 3' position of the pyrimidine (II). Adding an amide group to the aromatic ring then led to structures which showed inhibitory activity against tyrosine kinases as well. For example, structure IV inhibited serine-threonine protein kinases such as PKC-α and was also a relatively weak inhibitor of tyrosine kinases. A series of chemically related structures was then synthesized to test SAR against a variety of protein kinases and to optimize activity against tyrosine kinases. Introduction of an *ortho* methyl group as a conformational blocker (section 10.3.10) resulted in CGP 53716, which had enhanced activity against tyrosine kinases and no activity against serine-threonine kinases, demonstrating that the molecule had been forced to adopt a conformation which suited binding to tyrosine kinases but not to serine-threonine kinases. Further modifications were then carried out to maximize activity and selectivity with the addition of a piperazine ring. This ring is also important for aqueous solubility, as it contains a basic nitrogen which allows the formation of water soluble salts. A C_1 spacer was introduced between the aromatic ring and the piperazine ring, as aniline moieties are known to have mutagenic properties.

Molecular modelling studies have been carried out where imatinib has been docked into a model of the ATP binding site of Abl kinase. The model was created by studying the X-ray crystal structure of a related protein kinase called fibroblast growth factor receptor (FGF-R), complexed with a similar aminopyrimidine inhibitor. The active site was then modified to act as a model for the Abl kinase by altering the amino acids to those present in Abl kinase. Imatinib was docked into the active site such that it bound in a similar fashion to the related inhibitor bound to FGF-R.

The binding interactions observed for imatinib (Fig. 18.69) matched many of the SAR results which had been obtained. For example, an amino NH group in imatinib is involved as a hydrogen bond donor to Met-318. If this group is alkylated, all activity is lost. It is also noteworthy that the *ortho*-methyl group which acts as a conformational blocker makes a van der Waals interaction with the side chain of Leu-370. The equivalent residue in the serine-threonine kinase (PKC) is a bulkier methionine residue which cannot accommodate this group, and this helps to explain the lack of activity against this protein kinase. The model also demonstrates the importance of the piperazinyl group, since it forms an ionic interaction with Glu-258. Glu-258 is conserved in three protein kinases (Abl, c-Kit, and PDGF-R) and imatinib is an inhibitor of all three. In contrast, Glu-258 is absent from the tyrosine kinases (EGFR and c-SRC) that are not inhibited by imatinib. Therefore, this ionic interaction is important to the selectivity of the agent.

As mentioned above, imatinib inhibits the tyrosine kinases PDGF-R and c-Kit as well as v-Abl. c-Kit is a receptor kinase which is activated by stem cell factor (SCF), and is implicated in certain types of stomach cancer. As a result, imatinib has also been approved for the treatment of relevant stomach cancers.

The platelet-derived growth factor receptor (PDGF-R) is responsive to the platelet-derived growth factor (PDGF). It has been suggested that excessive levels of PGDF contribute to the appearance of certain types of cancer, so imatinib may be useful in this respect as well.

The fact that imatinib is not totally selective and inhibits various different kinases led to the concern that it would have serious side effects. Fortunately, this

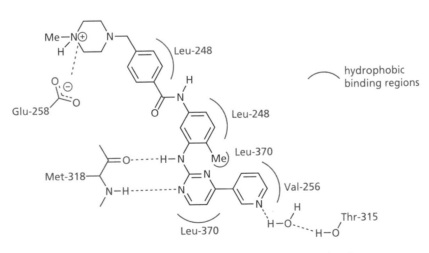

Figure 18.69 Binding interactions of imatinib in a model binding site.

is not the case. It appears that normal cells are able to survive inhibition of these kinases, whereas the survival of CML cells relies crucially on Bcr-Abl. Therefore, reliance of a cancer cell on an abnormally functioning protein sensitizes it to agents which target that protein.

Acquired resistance to imatinib has been observed due to mutations in the Abl kinase domain that prevent the drug from binding. Specifically, a substitution by isoleucine for threonine has been observed at position 315. Thr-315 forms an important hydrogen bond to imatinib, which is lost by this mutation. Other point mutations have been observed as well. Alternative signalling pathways may be adopted in some resistant cells, and an increased expression of the target receptor may occur in others. One way of tackling the problem of resistance is to design drugs which have a dual target. For example, structures are being developed that inhibit the kinase active sites of Abl and Src.

18.6.2.3 Inhibitors of cyclin-dependent kinases

CDKs are involved in the control of the cell cycle (section 18.1.6) but are over-expressed or overactive in many cancer cells. Since these enzymes are inactive in normal resting cells, drugs that target them should have fewer and less toxic side effects than conventional cytotoxic drugs. CDKs are serine-threonine kinases which are activated by cyclins and inhibited by cyclin-dependent kinase inhibitors (CKIs). There are at least nine CDKs, and typically they are small proteins of about 300 amino acids. At present, a variety of inhibitors have been identified which compete with ATP for the kinase active site. **Flavopiridol** (Fig. 18.70) is one such structure which is undergoing clinical trials and looks promising as part of a combination therapy. It is a semi-synthetic flavone derived from **rohitukine**—a natural product extracted from an Indian plant. One possible problem with flavopiridol is its lack of selectivity between different CDKs, and so analogues which show such selectivity may prove beneficial. As far as the binding interactions are concerned, flavopiridol binds to the same region of the active site as ATP. The benzopyran ring lies in the adenine binding region such that the ketone acts as a hydrogen bond acceptor and the OH acts as a hydrogen bond donor. The piperidine ring lies in the region normally occupied by the first phosphate

---

BOX 18.3 GENERAL SYNTHESIS OF IMATINIB AND ANALOGUES

The synthesis of imatinib and its analogues involves the pyridine structure I as starting material. The central pyrimidine ring is then constructed in two stages to give structure II. The remaining two steps involve reduction of an aromatic nitro group to an amine, and acylation to give the final product. The route described allows the synthesis of a large variety of amides from intermediate III.

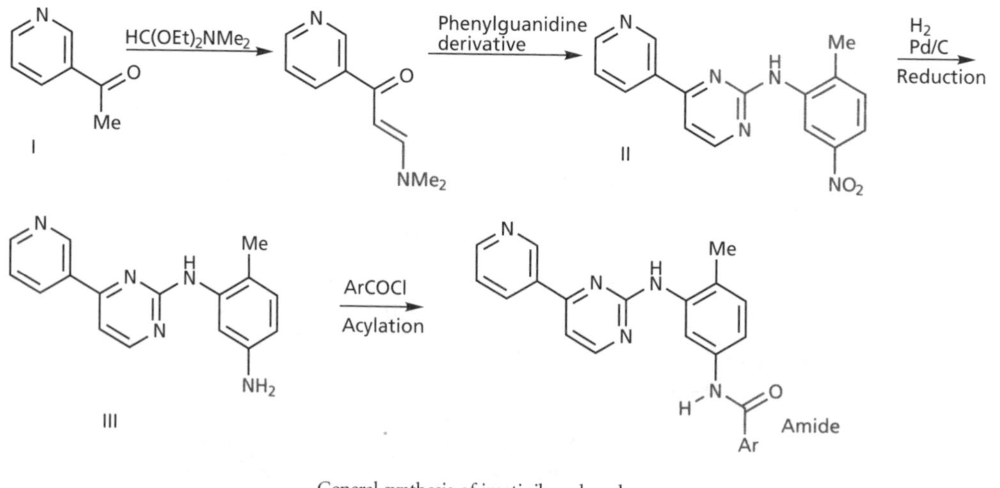

General synthesis of imatinib and analogues.

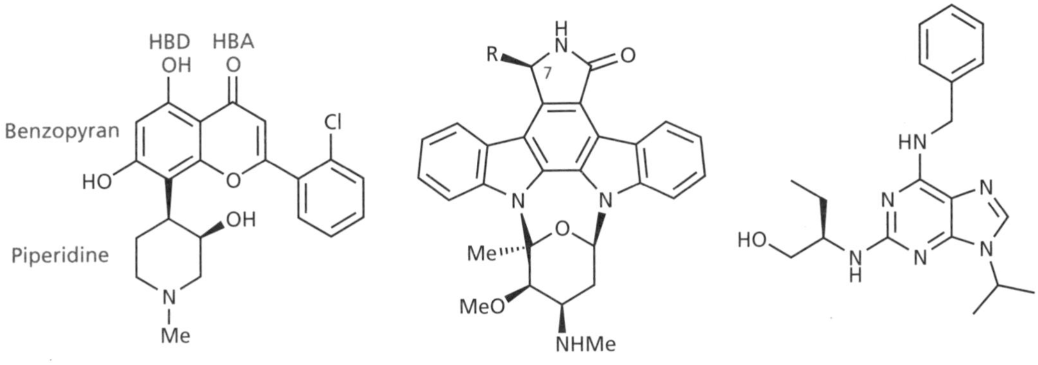

Figure 18.70 Inhibitors of cyclin-dependent kinases.

moiety of ATP, and makes several hydrogen bonding interactions with water and nearby amino acid residues. The phenyl group lies over the ribose binding pocket.

Flavopiridol is also thought to inhibit the expression of cyclins D1 and D3. It has been found to have an antiangiogenic effect and can induce apoptosis.

Another CDK inhibitor which has entered clinical trials is **7-hydroxystaurosporin** (Fig. 18.70). This is a derivative of a natural compound called **staurosporine**, which is a non-selective inhibitor of protein kinases. Staurosporine has been an extremely important lead compound for a variety of projects aimed at developing selective kinase inhibitors. The 7-hydroxy derivative shows greater selectivity than staurosporine itself but still inhibits a variety of kinases including CDKs.

Roscovitine (Fig. 18.70) shows selectivity for CDK2 and competes with ATP for the binding site. It induces apoptosis and is currently in clinical trials.

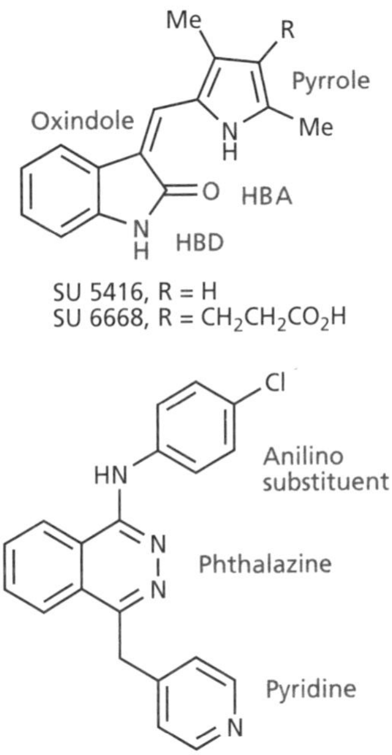

18.6.2.4 Kinase inhibitors of FGF-R and VEGF-R

Elevated expression levels of fibroblast growth factor receptor (FGF-R) and the vascular endothelial growth factor receptor (VEGF-R) and their corresponding ligands have been associated with angiogenesis, and kinase inhibitors of these receptors may disrupt the angiogenesis process. This will starve tumours of nutrients in the long term while improving access of conventional drugs in the short term. The receptor for the vascular endothelial growth factor (VEGF) is a protein kinase receptor which dimerizes when it binds

Figure 18.71 Kinase inhibitors of VEGF-R and FGF-R.

VEGF and triggers the start of a signal transduction pathway leading to angiogenesis. Inhibitors designed to bind to the ATP binding region of the kinase active site have entered clinical trials as potential anticancer agents (Fig. 18.71). For example, **SU-5416** is an oxindole structure that reached phase III clinical trials for the treatment of colorectal cancer. The oxindole ring can provide a hydrogen bond acceptor and a hydrogen bond donor to bind to the same region of the peptide backbone as the adenine ring of ATP. **PTK-787** is a phthalazine structure produced by Novartis which is also in phase III clinical trials.

Despite the potential of these agents, inhibition of the VEGF receptor is not guaranteed to inhibit angiogenesis completely since there are other growth factors which can promote angiogenesis; for example PDGF, TGF-β, TNF-α, and interleukins IL-1 and IL-8. It is therefore likely that a variety of agents targeting angiogenesis in different ways may be the best approach.

18.6.2.5 Other kinase targets

There are other types of protein kinases and protein kinase receptors which have been found to be over-expressed in certain types of cancer cells, and research is being carried out to find selective inhibitors of these targets as well. The main targets other than those already described are PDGF-R, PKC, mitogen-activated protein kinases (MAPK), insulin growth factor 1 receptor (IGF-1R), protein kinase B (PKB), and c-Src tyrosine kinase. IP$_3$ kinase inhibitors are being studied as antiangiogenesis agents.

It is worth noting that most research has been carried out on ATP mimics, but there is potential in designing inhibitors that bind to the substrate binding region. Many researchers feel that such inhibitors could be more selective and safer to use.

KEY POINTS

- Many cancers produce abnormal or over-expressed proteins that are involved in the signalling pathways that stimulate cell growth and division. Agents which act selectively against these targets are less likely to have the serious side effects associated with traditional cytotoxic agents.

- An abnormal form of the Ras protein which is permanently active is associated with many cancers. Inhibiting the farnesyl transferase enzyme prevents the Ras protein becoming attached to the cell membrane and prevents it interacting with other elements of the signal transduction process. The observed anticancer effects of FT inhibitors may be due to the effect on a variety of proteins other than just Ras.

- FT inhibitors contain a group which acts a ligand for the zinc cofactor in the enzyme. Early inhibitors were modelled on the end tetrapeptide moiety of Ras. Newer agents are less

peptide like and are smaller molecules with improved pharmacokinetic properties.

- Protein kinases are enzymes which use ATP to phosphorylate hydroxyl or phenol groups in protein substrates. Protein kinase receptors are proteins in the cell membrane which play a dual role of receptor and enzyme, and which are activated by growth factors. Both the messengers and the receptors have been implicated in various cancers by being over-expressed or abnormal in nature. Anticancer agents have been designed to bind to the ATP binding region of the kinase active site, and to act as inhibitors. Since there is a variation in the active site from kinase to kinase, it is possible to design agents which are selective between different kinases.

18.7 Miscellaneous enzyme inhibitors

In previous sections we looked at enzyme inhibitors associated with DNA synthesis and function, as well as enzymes involved in signal transduction. In this section, we look at the inhibition of enzymes which have not been covered so far, but which have important roles to play in angiogenesis, metastasis, and apoptosis.

18.7.1 Matrix metalloproteinase inhibitors

MMPs are zinc-dependent enzymes which play an important role in the invasiveness and metastasis of cancer cells—processes that have few anticancer agents acting against them. The MMPs are extremely destructive enzymes involved in the normal turnover and remodelling of the extracellular matrix or connective tissue (section 18.1.9). This process is usually tightly controlled by natural protein inhibitors. However, excessive activity can result in various problems including chronic degenerative diseases, inflammation, and tumour invasiveness. There are four main groups of MMPs—collagenases, gelatinases, stromelysins, and the membrane type (MT)—and a number of these are implicated in tumour growth, invasion, metastasis, and angiogenesis.

Matrix metalloproteinase inhibitors (MMPIs) can be used to inhibit the breakdown of the extracellular matrix, to make it more difficult for cancer cells to escape and metastasize. They can also be used to inhibit angiogenesis by blocking the release of VEGF from storage depots in the extracellular matrix.

A variety of first-generation inhibitors have been developed which are based on the natural protein substrates for the collagenase enzymes of this family. These enzymes catalyse the cleavage of peptide bonds between glycine and isoleucine (or leucine) in protein substrates (Fig. 18.72). The substrate is thought to bind such that the carbonyl oxygen of glycine co-ordinates to the zinc cofactor, while the neighbouring NH acts as a hydrogen bond donor to an alanine residue. A water molecule is held between zinc and a glutamate residue, and acts as the nucleophile to hydrolyse the peptide bond, assisted by the negatively charged glutamate residue and the zinc cofactor.

A variety of peptide-based inhibitors have been designed. In general, they have the following features:

- Replacement of the susceptible peptide bond with a moiety which is stable to hydrolysis.

- One or more substituents that can fit into the enzyme subsites and form van der Waals interactions.

The subsites normally accept the amino acid residues of the substrate.

- At least one functional group capable of forming a hydrogen bond to the enzyme backbone.

- A group capable of strong interactions with the zinc ion cofactor, such as a thiol, carboxylate, or hydroxamic acid.

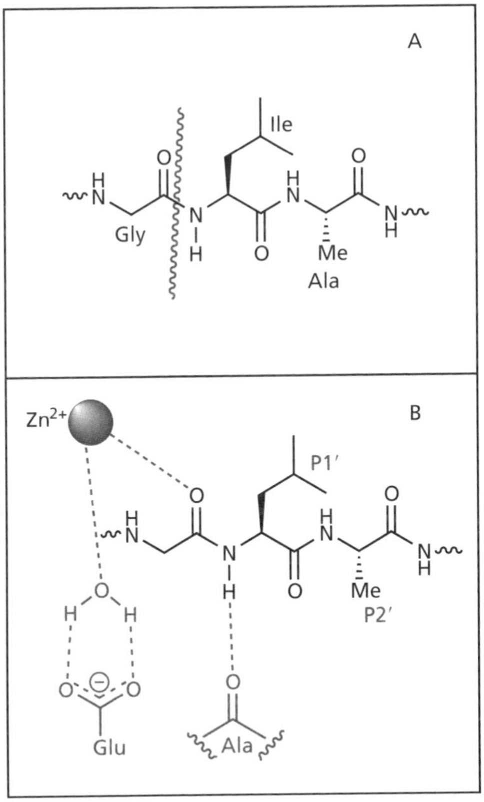

Early work showed that inhibitors mimicking the substrate groups P1' and P2' to the right of the scissile peptide link (right-hand-side inhibitors) were more effective than those mimicking the left-hand side. It was also discovered that a hydroxamate group was a particularly good ligand for the zinc cofactor, as it can act as a bidentate ligand using both oxygen atoms, while the NH group forms a hydrogen bond interaction with a carbonyl oxygen on the enzyme backbone.

These features can be seen in **marimastat** (Fig. 18.73) which is an orally active, synthetic compound in phase III clinical trials for breast and prostate cancer. A hydroxamic acid group is present to form a strong bidentate interaction with zinc. Substituents (P1'–P3') are also present to fit three binding pockets in the active site. The NH moiety of the amide bond normally between glycine and iso-leucine has been replaced by a hydroxymethylene group which prevents the normal hydrolysis reaction taking place. The hydroxyl group is also beneficial for

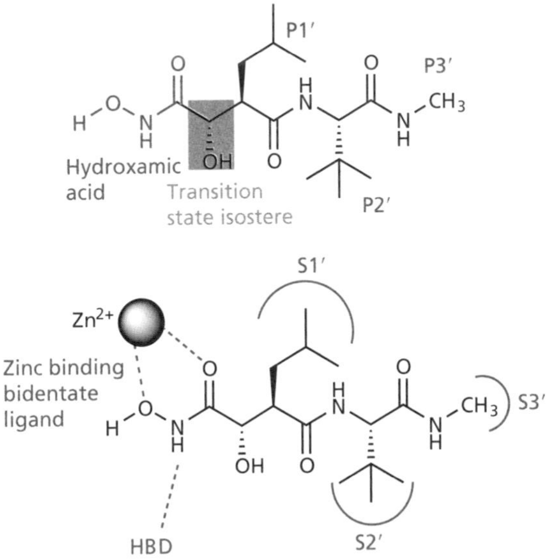

Figure 18.72 Peptide bond cleaved by matrix metalloproteinases.

Figure 18.73 Structure and binding interactions of marimastat.

inhibitory activity and aqueous solubility. Binding studies suggest that the hydroxyl group is directed away from the protein surface and is hydrogen bonded to water. The *t*-butyl substituent (P2′) also serves as a steric shield to protect the terminal amide from hydrolysis.

The nature of the P1′ group can be varied, and determines the activity and selectivity of the inhibitors against the various metalloproteinases. The nature of the P2′ group can also be varied, but it is beneficial to have a bulky group to act as a steric shield to protect the peptide bonds. It also serves to desolvate the peptide bonds such that energy is not expended on desolvation prior to binding.

Another peptide-like structure to reach clinical trials is **BMS 275291** (Fig. 18.74), which has a thiol group as the ligand for zinc. The hydantoin ring is thought to access the S1 subsite. The right-hand half of the structure is identical to marimastat.

Unfortunately, inhibitors such as marimastat lack selectivity and produce side effects such as tendinitis. They also have poor pharmacokinetic properties, due

to their peptide nature. For example, they have poor aqueous solubility and show susceptibility to peptidases in the gastrointestinal tract. Therefore, work is now under way to develop a second generation of nonpeptide like metalloproteinase inhibitors which are more selective in their action (Fig. 18.75). Examples include **CGS 27023A** and **prinomastat**, which are both in clinical trials. These structures have decreased peptide character, but still suffer from a lack of selectivity. CGS 27023A contains an isopropyl group which appears to protect the hydroxamic acid group from metabolism. Activity increased if the isopropyl group was incorporated into a ring in order to restrain the number of possible conformations. Further extension of the P1′ substituent then led to prinomastat. Second-generation inhibitors such as **BAY 12-9655** have a carboxylate group as the zinc ligand.

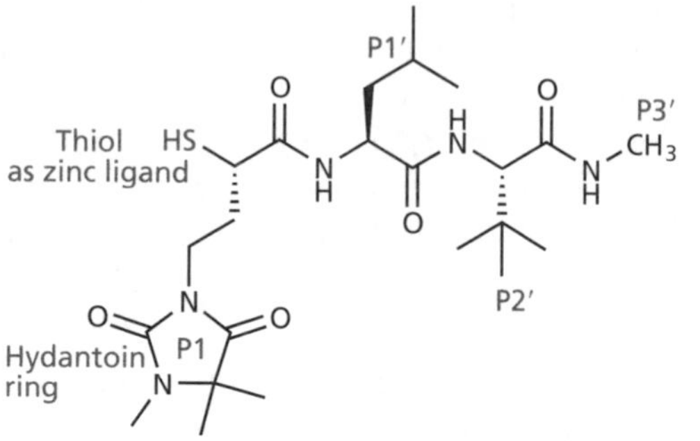

18.7.2 Cyclooxygenase-2 inhibitors

Cyclooxygenase (COX; see section 4.8.6) is an enzyme involved in prostaglandin synthesis and is overexpressed in several human cancers. It has two isozymes: COX-1, which is present in normal cells and plays an important role in many processes, and COX-2 which is not present in normal tissues, but is induced by mitogenic and inflammatory stimuli resulting in enhanced prostaglandin synthesis in cancerous or inflamed tissue respectively. COX-2 is thought to be important to various important mechanisms of carcinogenesis including angiogenesis, apoptosis, tumour invasiveness, and multidrug resistance via efflux pumps.

Therefore, it has been proposed that selective inhibitors of COX-2 could help prevent further

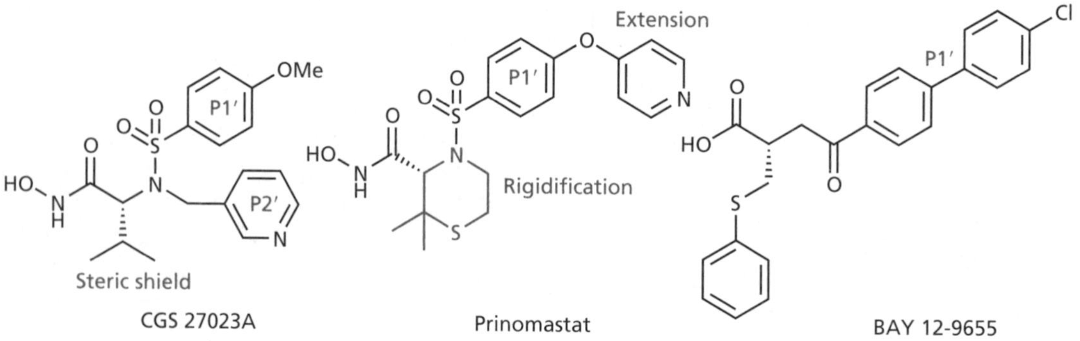

Figure 18.74 BMS 275291 (D 2163).

Figure 18.75 Second-generation MMIs.

growth of tumours and/or augment the effect of currently used anticancer drugs. It is possible to design COX-2 selective inhibitors, that distinguish between an isoleucine group in the binding site of COX-1 and a valine group at the equivalent position in COX-2. A clinical trial involving the COX-2 inhibitor **celecoxib** has shown that there may be potential in using COX-2 inhibitors for both the prevention and treatment of cancers, especially of the colon and prostate.

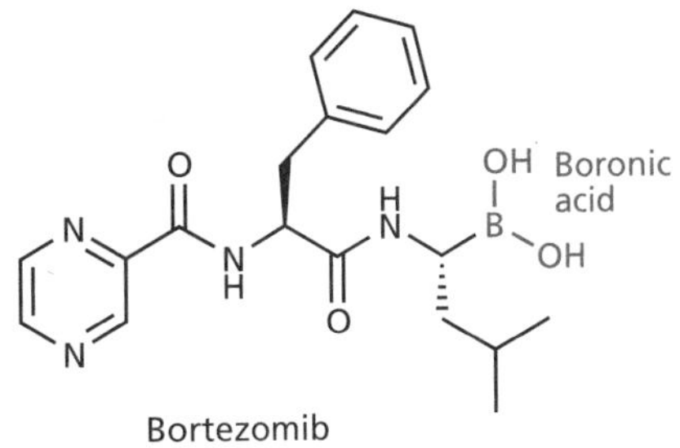

18.7.3 Proteasome inhibitors

The proteasome is a complex structure, which can be viewed as the cell's rubbish disposal unit. It is an ATP-dependent multicatalytic protease that destroys proteins. Its prime role is to eliminate damaged or misfolded proteins and to degrade key regulatory proteins. Considering the destructive power of this structure, it is important that it destroys only defective proteins and not normal ones. Therefore, cells mark their defective proteins with a molecular label so that they can be recognized by the protein killing machine. This molecular label is a protein called **ubiquitin**.

Since the proteasome is so destructive, one might think it would be best to boost its activity in tumour cells. In fact, the opposite strategy is adopted and research is looking at agents that inhibit its action. The rationale lies in the fact that the proteasome removes regulatory proteins which have done their job. Blocking the proteasome will result in an accumulation of various regulatory proteins, which leads to a cellular crisis and triggers apoptosis. One of the proteins that accumulates as a result of proteasome inhibition is the apoptosis promoter Bax (section 18.1.7).

Bortezomib (Fig. 18.76) is a boronic acid dipeptide that inhibits the proteasome, possibly by forming a boron–threonine bond at active sites. It became the first proteasome inhibitor to be approved for the treatment of multiple myeloma. Unlike most anti-cancer drugs, bortezomib is not prone to multidrug resistance. **Aclarubicin** (**aclacinomycin A**) is an anthraquinone which affects proteasomes by inhibiting the chymotrypsin activity of the structure. The tetracyclic moiety and the three sugar rings are all necessary for activity.

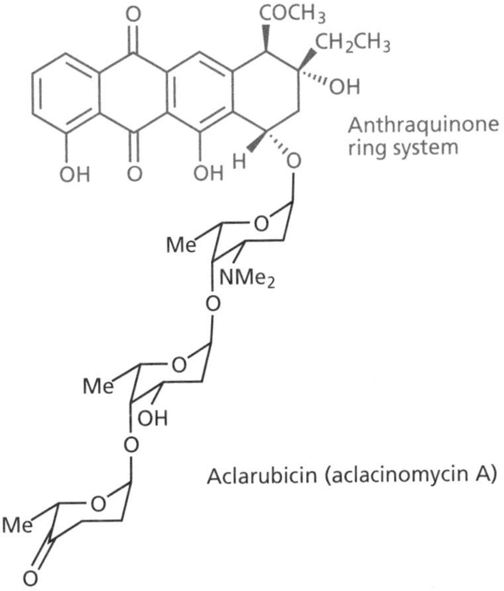

Figure 18.76 Proteasome inhibitors.

18.7.4 Histone deacetylase inhibitors

In structures called **chromatins** (Fig. 18.5), DNA is wrapped around proteins, most of which are histones. The histones assist in DNA packaging and also have a regulatory role. There is a repeating pattern of 8 histone proteins along the length of the chromatin structure, with each octet associated with about 200 base pairs of DNA. Each of these repeating units is known as a **nucleosome**.

Histone acetylase is an enzyme that adds acetyl groups to the lysine residues of histone tails which stick out from the chromatin structure. Histone deacetylase is an enzyme that removes them. Acetylation facilitates the binding of transcription factors to specific DNA sequences by destabilizing the nucleosomes which are bound to the promoter region of the target genes.

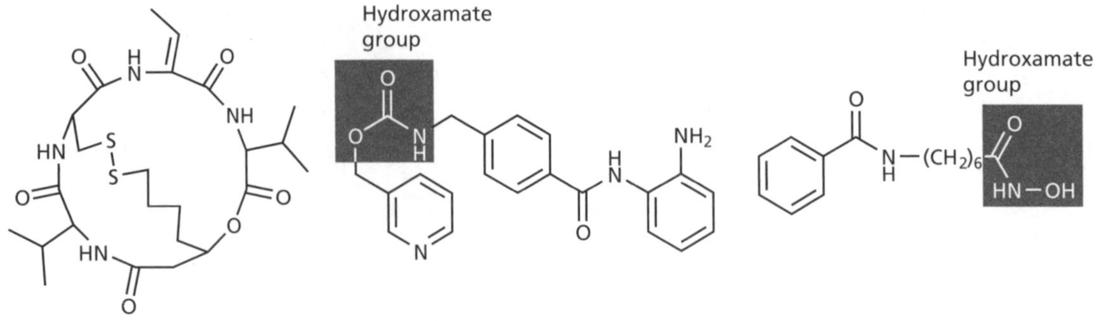

Figure 18.77 Histone deacetylase inhibitors.

Inhibitors for these enzymes are currently being studied, and several inhibitors of histone deacetylase are in clinical trials (Fig. 18.77). **Depsipeptide** is a natural product derived from a bacterial strain. It is reduced inside cells to an active compound which promotes apoptosis and inhibits cell proliferation and angiogenesis. Synthetic agents such as **MS-275** are also being investigated. Since the enzyme contains a zinc cofactor, many of these agents contain a hydroxamate group to act as a ligand; for example, **suberoylanilide hydroxamic acid**.

18.7.5 Other enzyme targets

There are many other enzymes which are being studied as potential targets for anticancer agents. For example, inhibiting the **telomerase** enzyme should prevent cells becoming immortal and be useful in anticancer therapy. Several powerful inhibitors have been developed, although no telomerase inhibitor has reached the clinic to date.

The inhibition of regulatory enzymes may be useful in shutting down a biosynthetic pathway that is too active. One example is inhibition of **tyrosine hydroxylase** (section 20.12.1).

Methionine aminopeptidase is an enzyme that plays a key role in endothelial cell proliferation and blocking it should inhibit angiogenesis. **TNP-470** (Fig. 18.78) is an analogue of a fungal product called **fumagillin** which acts as an inhibitor of this enzyme and is being studied as an antiangiogenesis agent.

The activation of **caspases** to induce apoptosis is another possible approach to novel anticancer agents.

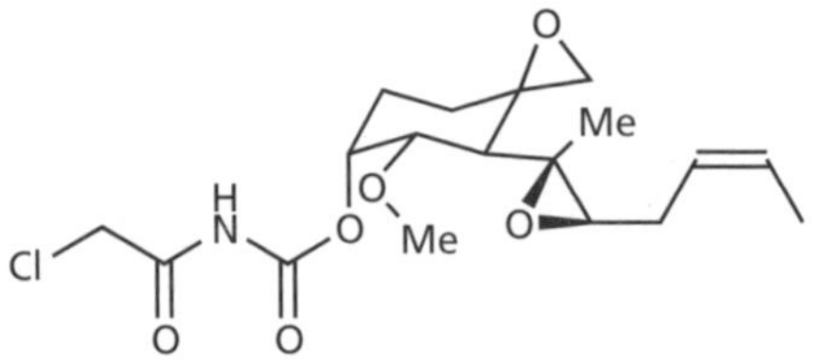

Figure 18.78 TNP 470.

18.8 Miscellaneous anticancer agents

The field of anticancer research is a vast one, with a wide diversity of novel structures being investigated. The following are examples of various structures which act at different targets or whose targets have not been identified.

18.8.1 Synthetic agents

Thalidomide (Fig. 18.79) was originally marketed as a safe, non-toxic sedative and antiemetic in the 1950s and rapidly became popular to counter the effects of morning sickness during pregnancy. Unfortunately, it was instrumental in one of the major medical disasters of modern times when it produced teratogenic effects in developing fetuses and led to babies being born with stunted limbs and other developmental deformities—the so called 'thalidomide babies'. Since thalidomide was considered such a safe drug at the time, few suspected that it was the cause of the problem and by the time it was linked to the deformities and withdrawn in 1961, 8000–12 000 babies had been affected. Consequently, thalidomide gained a lasting notoriety

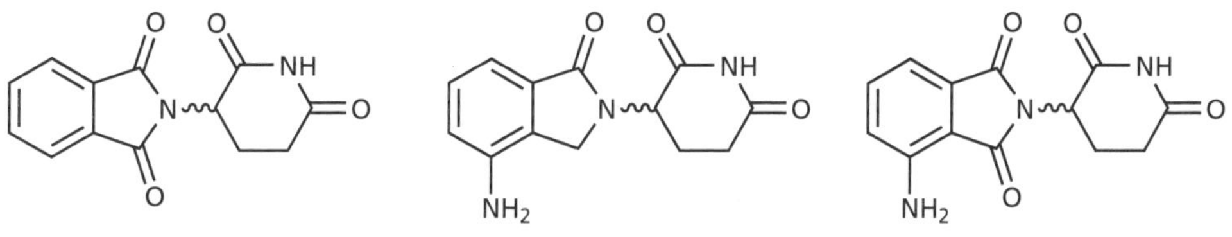

Figure 18.79 Thalidomide and thalidomide analogues.

which was instrumental in a significant tightening of the regulations surrounding the testing of drugs. Despite its notorious past, there has been continuing interest in thalidomide since it has some remarkable properties which indicate a wide variety of clinical uses. Early on, it was recognized that thalidomide has an anti-inflammatory property which was useful in treating leprosy. This activity was eventually linked to thalidomide's ability to inhibit the synthesis of the pro-inflammatory endogenous cytokine TNF-α, which is produced by monocytes. In 1998, the drug was approved for the treatment of leprosy. Thalidomide has a raft of other properties, though. For example, it is an immunosuppressant and could possibly be used for the treatment of autoimmune diseases or in countering the immune response to allow organ transplants to be accepted by the host. Interest in thalidomide as an anticancer agent started when it was found that it inhibited angiogenesis by an unknown mechanism. Tests showed that thalidomide did indeed have anticancer activity and it entered phase III clinical trials for the treatment of renal cancer and multiple myeloma on that basis. Since then, it has been discovered that the anticancer properties of thalidomide are more complex than its effect on angiogenesis alone. In some patients, thalidomide can boost the immune system by a variety of mechanisms rather than suppress it, and this too may account for its anticancer activity. Since thalidomide can suppress or boost the immune system depending on individual circumstances, it is known as an **immunomodulator**. Thalidomide also appears to directly arrest the growth of cells and promote apoptosis.

Analogues of thalidomide have been synthesized, with the aim of removing its teratogenic properties. **Revimid** and **actimid** (Fig. 18.79) are two such examples which are showing great promise and are in clinical trials. Both contain an amino substituent on the aromatic ring which is found to be crucial in producing a safer drug. Revamid entered clinical trials in the year 2000 and was given orphan drug status in 2001 for the treatment of the currently incurable disease of multiple myeloma. In 2003, it entered phase III clinical trials and was given fast-track status. Such is the pace of progress with these drugs that they may well be approved before their mechanism of action is fully understood.

Arsenic trioxide is an orphan drug used for a variety of leukaemias. It is thought to promote cell suicide by targeting the cell's mitochondria.

18.8.2 Natural products

Pancratistatin (Fig. 18.80) is a natural product isolated from a plant called *Pancratium littoralis* which belongs to the genus *Narcissus*. Records show that extracts from plants in this genus were used by Hippocrates in 200 BC to treat breast cancer. The drug also inhibits angiogenesis, and its phosphate prodrug shows potential as an anticancer drug. The exact mechanism of action is still to be determined. One or more of the hydroxyl groups at positions C-2, C-3, and C-4 are thought to be important.

Bryostatin 1 (Fig. 18.80) is a natural product which was isolated from a marine invertebrate off the coast of California in 1981. It was shown to boost the immune system and make it more effective against cancers. It is undergoing clinical trials and promises to be effective, either on its own or in combination with other established anticancer drugs such as taxol, vincristine, fludarabine, or cisplatin.

Dolostatins are natural products which were isolated from the marine sea hare off the island of Mauritius in the Indian Ocean. A full synthesis has been developed to produce **dolostatin 10**, **auristatin PE**, and **dolostatin 15** (Fig. 18.81), which are all undergoing clinical trials.

Cephalostatin 1 (Fig. 18.81) is a very potent anticancer agent which was isolated from a marine worm. Its mechanism of action has not been established,

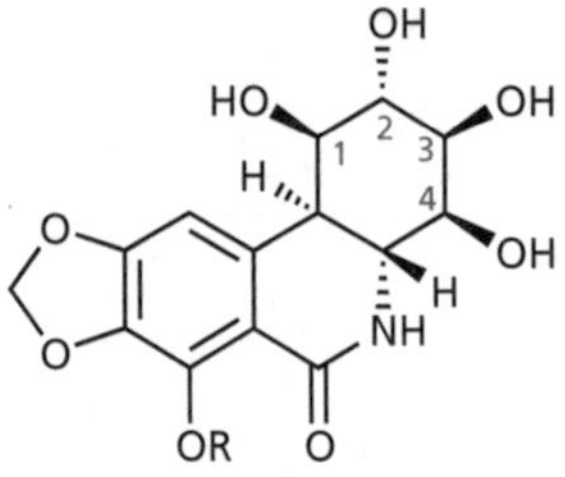

Pancratistatin; R = H
Pancratistatin prodrug; R = phosphate

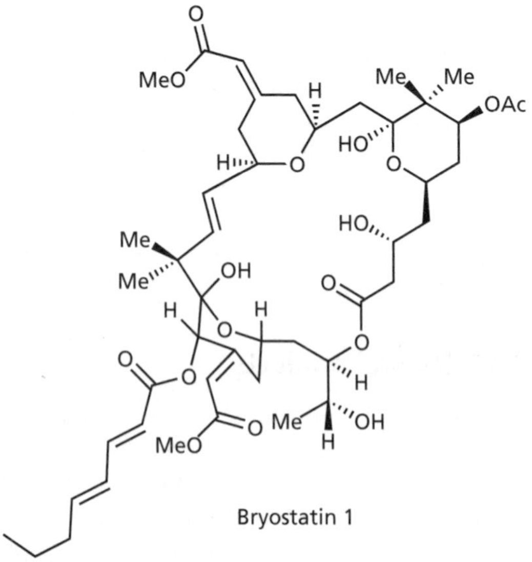

Bryostatin 1

Figure 18.80 Miscellaneous natural products having anticancer activity.

although it has been suggested that it spans the lipid bilayer of cells and disrupts membrane structure. It has not yet entered clinical trials, and there is a need to develop an efficient synthesis of the compound to obtain sufficient quantities.

18.8.3 Protein therapy

A variety of proteins are being considered as antiangiogenesis agents. For example, **angiostatin** and **endostatin** are two naturally occurring proteins in the body which inhibit the formation of new blood vessels and are being studied in cancer therapy. **α-Interferon** inhibits the release of growth factors such as VEGF and is in phase III clinical trials for various cancers.

Cancer cells are more sensitive than normal cells to a natural death-inducing protein with the catchy title of TNF-related apoptosis-inducing ligand (**TRAIL**), which stimulates apoptosis. Injecting purified TRAIL

in vivo might selectively stimulate increased death rates in cancer cells. This is currently being studied in animals.

A variety of proteins are being considered as immunostimulants including **γ-interferon, aldesleukin** (a preparation of IL-2), and **tretinoin**.

Some cancer cells have lost the ability to carry out normal synthetic routes, as a result of gene mutation and the production of inactive enzymes. For example, some leukaemia cells lose the capacity to synthesize the amino acid asparagine and have to obtain it from the blood supply. The enzyme asparaginase can catalyse the degradation of asparagine, so providing that enzyme should break down asparagine in the blood supply and starve the cancer cells of this amino acid. A preparation of the asparaginase enzyme (**crisantaspase**) is used to treat certain cancers of this type (see also section 11.8.3).

KEY POINTS

- MMPs are zinc-dependent enzymes which degrade the extracellular matrix and encourage the processes of angiogenesis, tumour propagation, and metastasis.

- Inhibitors of MMPs have been based on small peptide sequences of the protein substrates. In general, the susceptible peptide bond has been replaced by a stable bond, substituents have been incorporated to fit binding subsites and a ligand for the zinc cofactor has been included.

- Second-generation inhibitors of MMPs are non-peptide in nature and are more selective in their action.

- Inhibitors of COX-2 are being studied as potential anticancer agents.

- The proteasome is a destructive enzyme complex which breaks down proteins. Inhibition leads to a build-up of conflicting regulatory proteins, which triggers apoptosis.

- Histone acetylases and deacetylases are involved in the regulation of transcription. Inhibitors of histone deacetylase are in clinical trials.

- A variety of other enzymes are potential targets for novel anticancer agents.

- A large number of synthetic structures and natural products have anticancer properties by unknown mechanisms. Others appear to work by having several different mechanisms.

- Protein therapy has proved useful in the treatment of certain cancers.

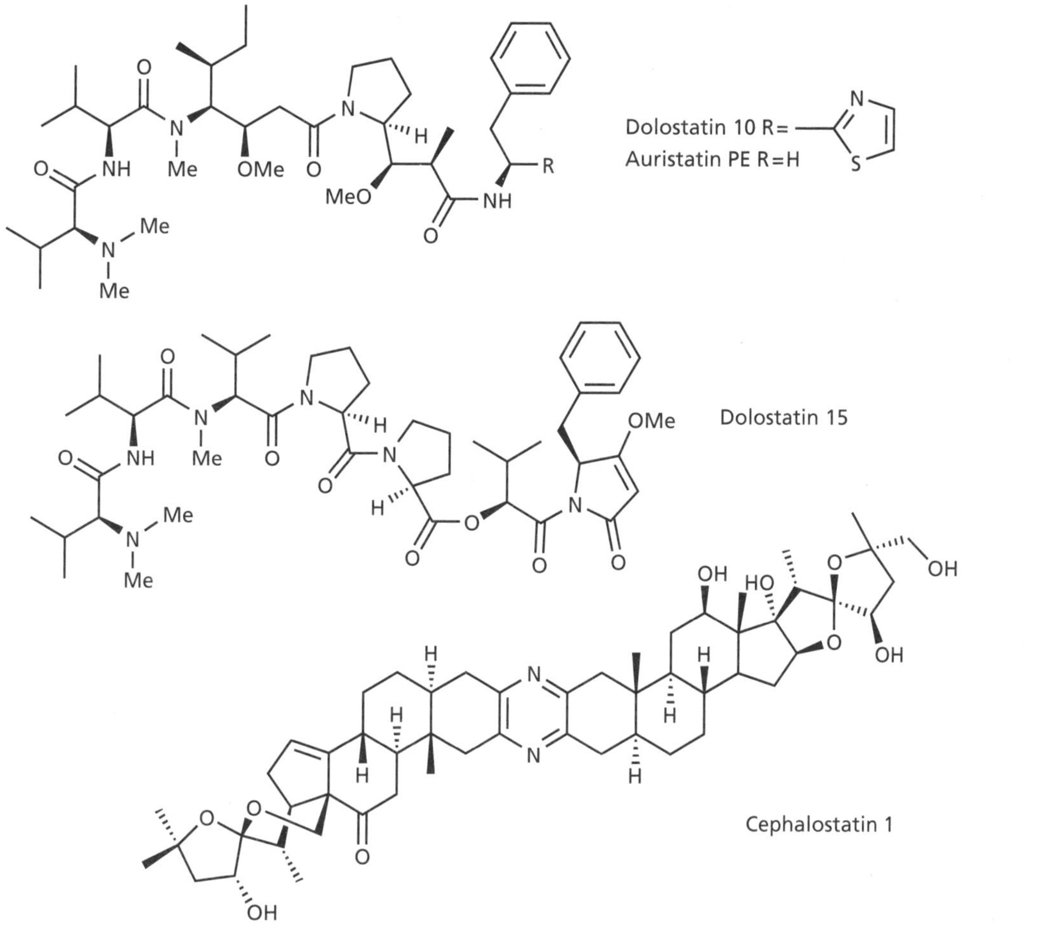

Figure 18.81 Natural products with anticancer activity.

18.9 Antibodies, antibody conjugates, and gene therapy

18.9.1 Monoclonal antibodies

Cancer cells have unusual shapes and altered plasma membranes that contain distinctive antigens which have been over-expressed. This allows the possibility of using antibodies against the disease. Although the antigens concerned are likely to be present in some normal cells, they are likely to be present to a greater extent on the cancer cells making the latter more vulnerable. Monoclonal antibodies (section 3.9) have been produced for numerous tumour-associated antigens, and a few have reached the clinic as anticancer agents. These serve to activate the body's immune response to direct killer cells against the tumour. Alternatively, if the antigen is an over-expressed receptor, the antibody may bind to it, such that the chemical messenger can no longer bind. In this case, the antibody acts as a receptor antagonist.

- **Trastuzumab** is a humanized monoclonal antibody which targets the extracellular region of the HER-2 growth factor receptor and was approved in 1998 for the treatment of metastatic breast cancer in combination with paclitaxel. HER-2 is a member of the EGF-R family of tyrosine kinase receptors, which is over-expressed in 25% of breast cancers. When the antibody binds to the receptor, it induces the immune response to attack the specified cell.

It also promotes internalization and degradation of the receptor.

- **Alemtuzumab** is humanized antibody that is used for B-cell chronic lymphocytic leukaemia where other therapies have not worked. The antibody binds to a receptor (the CD52 antigen) which is found both on normal and cancerous immune cells (B- and T-lymphocytes). Although the agent shows no selectivity for cancer cells over normal cells, normal cells recover quicker after treatment.

- **Rituximab** is a chimeric antibody (section 3.9) targeting the CD20 receptor on B-lymphocytes and was approved in 1997 for the treatment of B-cell lymphoma.

- **Cetuximab** is a chimeric monoclonal antibody that was rejected by the US FDA but has subsequently been accepted by the Swiss health authorities as a treatment for advanced colorectal cancer. It targets the extracellular domain of EGF-R and blocks access to EGF.

- **Bevacixumab** is a humanized monoclonal antibody that disables the growth factor VEGF required for angiogenesis and is in phase III clinical trials for the treatment of breast and colorectal cancers.

18.9.2 Antibody–drug conjugates

The monoclonal antibodies described above have an anticancer activity in the 'naked' form (i.e. without a drug attached), but the level of activity is usually too low to be effective and so a better strategy is to attach an anticancer drug to the antibody (an antibody–drug conjugate) such that the drug is delivered selectively to the cancer cell.

One of the original aims in designing antibody–drug conjugates was to deliver anticancer agents to tumour cells in greater concentrations than was possible by conventional therapy. There is often a narrow therapeutic window between the levels of drug that are effective and the levels leading to unacceptable toxicity. Antibody–drug conjugates were seen as a means of avoiding this problem, as it was anticipated that targeting would lead to higher concentrations of the drug at tumour cells. The first generation of such conjugates involved antibodies linked to anticancer agents such as methotrexate, the vinca alkaloids, and doxorubicin, but the results were disappointing. The

anticancer activity achieved was less than using the drug itself, and yet the toxicity problem remained the same.

It was later realized that the lifetime of the antibody–drug conjugate was substantially greater than that of the drug itself, which contributed to the toxicity problem. Furthermore, delivery to the tumour and penetration into the tumour was limited due to the size of the conjugate. This meant that the concentration of antibody–drug reaching the tumour was actually less than when the drug itself was used. Since the rate of delivery and penetration is determined by the antibody, there is little that can be done to improve it. Therefore, it was realized that highly potent anticancer agents would have to be attached to the antibodies in order to be effective at the levels attained at the tumour cell. This rules out anticancer drugs such as doxorubicin, etoposide, 5-fluorouracil, and cisplatin. More potent anticancer drugs are available, but if they should become detached from the antibody during circulation, they are likely to cause severe toxicity. Therefore, it is important that any such drug is attached to the antibody by a stable bond and remains bound until it enters the cancer cell.

The requirements for antibody–drug conjugates now include the following:

- The antibody has to be humanized to avoid an immune response.

- It has to show selectivity for an antigen which is overexpressed in cancer cells rather than normal cells.

- It then needs to be internalized into the cell by receptor-mediated endocytosis such that the antibody–drug can be delivered into the cell.

- The link between the antibody and the drug should be stable until cell entry has taken place and then cleaved to release and activate the drug.

There are various ways in which a drug can be linked to an antibody. For example, there are lysine residues present throughout the whole molecule which contain a nucleophilic primary amino group. A number of drug molecules could be added to the one antibody by acylating or alkylating these groups. There is a problem with this approach, however, since it is quite possible for a molecule to be attached to the region responsible for 'recognizing' the antigen. This would prevent antibody–antigen binding. Moreover, the masking of polar amino groups may lead to precipitation of the antibody–drug complex.

A better approach is to reduce the four intrastrand disulfide links at the hinge region of the antibody (Fig. 18.82) to produce eight thiol groups, and to attach drugs to these by alkylation or via a disulfide linkage. This has the advantage that it can be carried out in a controlled fashion and the drugs do not mask the antigen recognition site. The disadvantage is that a maximum of only eight drugs can be added to any one antibody molecule. One way round this may be to add a linker molecule to the antibody which could itself bear several drug molecules.

Another method of attaching the drug to the antibody is to take advantage of the carbohydrate region between the two heavy chains. Mild oxidation of vicinal diols in the sugar rings produces aldehyde groups to which drugs can be linked through an imine functional group (Fig. 18.83). Further reduction can then be carried out to form more stable amine linkages.

It is important that the linker is cleaved once the antibody–drug complex enters the cancer cell. Various linkers have been tried such as acid-labile linkers, peptidase-labile linkers, and disulfide linkers. The disulfide linker can be cleaved by disulfide exchange with an intracellular thiol such as glutathione which has a higher concentration within cells than in plasma.

The drug itself needs to be highly potent ($IC_{50} < 10^{-10}$ M), which involves the use of drugs that are 100–1000 times more cytotoxic than conventional cytotoxic drugs. A variety of agents are being investigated including radioactive isotopes, ricin, diphtheria toxin, *Pseudomonas aeruginosa* exotoxin A, maytansinoids, adozelesin, calicheamicin γ_1, highly potent taxoids, and highly potent doxorubicin analogues (see Box 18.4).

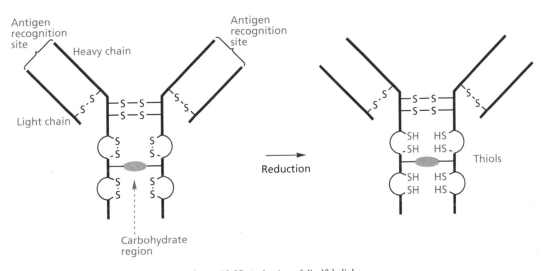

Figure 18.82 Reduction of disulfide links.

Figure 18.83 Linking drugs to carbohydrates.

BOX 18.4 GEMTUZUMAB: AN ANTIBODY–DRUG CONJUGATE

A humanized antibody has been linked to the highly potent anticancer drug calicheamicin. The trisulfide group normally present in calicheamicin was first modified to a disulfide with a hydrazide linker attached, while the antibody was treated with periodate to generate aldehyde groups at the carbohydrate region. The two molecules were then linked up by reacting the hydrazine group on the drug with the aldehyde groups on the antibody (Fig. 1). The resulting conjugate is called gemtuzumab and has been approved for the treatment of acute myeloid leukaemia (AML) in patients older than 60 years.

When the antibody–drug complex reaches the target leukaemic cell, the antibody attaches itself to a CD33 antigen and the antibody–drug complex is then taken into the cell by endocytosis. It is thought that the drug is then released from the antibody in lysosomes or endosomes by acidic hydrolysis of the hydrazone, and that reduction of the disulfide group occurs later in the cell nucleus to produce the active thiol (Fig. 2).

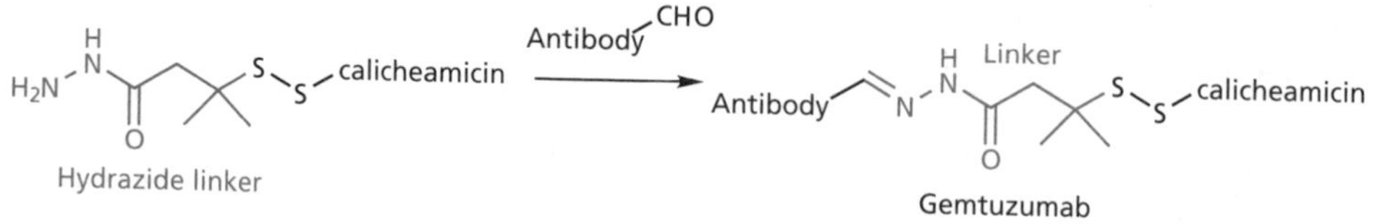

Figure 1 Linking calicheamicin to an antibody.

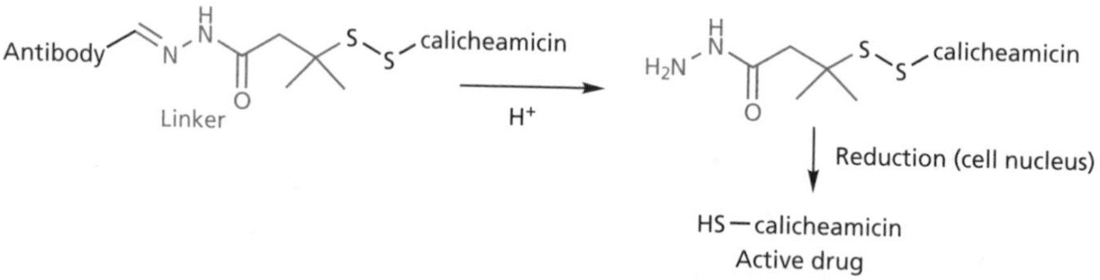

Figure 2 Release and activation of calicheamicin.

Ibritumomab and **tositumomab** are conjugated antibodies which carry radioactive isotopes to an antigen called CD20 on the surface of B-lymphocytes. These cells grow uncontrollably in non-Hodgkin's lymphoma. Ibritumomab carries ^{90}Y and tositumomab carries ^{131}I. Ibritumomab is the first approved drug involving radioimmunotherapy for the treatment of non-Hodgkin's lymphoma.

18.9.3 Antibody-directed enzyme prodrug therapy (ADEPT)

Antibody-directed enzyme prodrug therapy (ADEPT) involves two steps. The first is the administration of an antibody–enzyme complex. The antibody is raised against tumour-selective antigens and is linked to an enzyme such as bacterial carboxypeptidase. This complex then gets bound to the tumour. Unlike antibody–drug conjugates, the antibody–enzyme complex needs to remain attached to the surface of the cell and should not be internalized. A certain period of time needs to elapse to give the complex time to bind to target cells, and for unbound complex to be cleared from the blood supply. A prodrug of a cytotoxic drug is then administered. The prodrug is designed such that it will be stable in the blood supply and can only be cleaved and activated by the enzyme complexed to the antibody. This means that the toxic drug is only produced at the tumour and can be administered in higher doses than the parent drug. **CJS149** is an example of one such prodrug which is activated by a bacterial carboxypeptidase (Fig. 18.84). An advantage of ADEPT over antibody–drug conjugates is that the enzyme is catalytic and can generate a large number of active drug molecules at the site of the tumour. These can then diffuse into the tumour and affect cells which might not have any antibody attached to them.

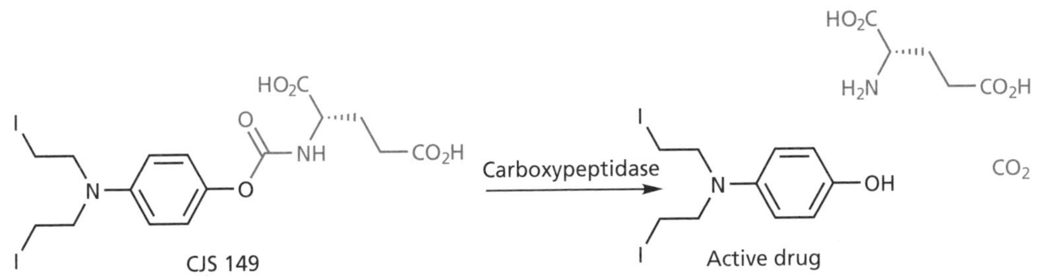

Figure 18.84 Activation of a prodrug by carboxypeptidase.

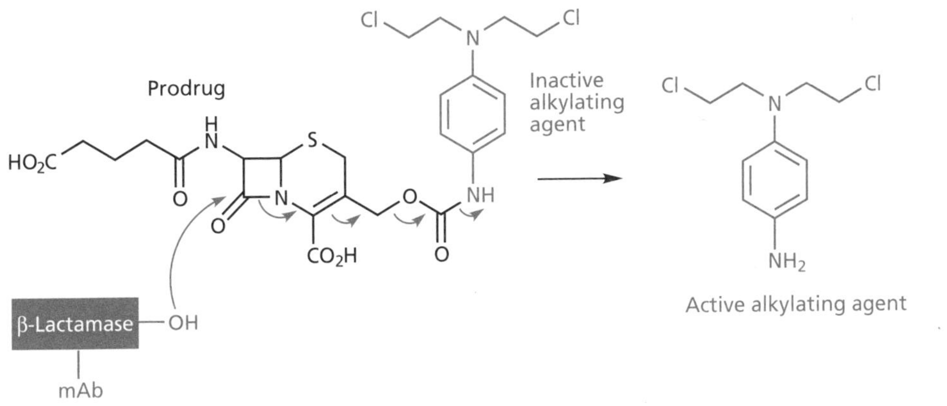

Figure 18.85 ADEPT strategy to release an alkylating agent.

A lot of research has been carried out on ADEPT using enzymes derived from bacteria such as carboxypeptidase G2, penicillin G acylase, and β-lactamase. The advantage of using a foreign enzyme is that enzymes can be chosen that are not present in the mammalian cell and so there is no chance of the prodrug being activated by mammalian enzymes during its circulation round the body. It is also possible to use foreign enzymes even if a counterpart exists in the body, as long as the latter are only present in low levels in the blood and/or they are structurally distinct. Prodrugs can be designed that selectively react with the foreign enzyme rather than the mammalian version. Examples of enzymes in this category include β-glucuronidase and nitroreductase.

Many studies have been carried out on ADEPT. One example is an antibody–β-lactamase complex capable of reacting with the cephalosporin prodrug of an alkylating agent (Fig. 18.85). This takes advantage of the mechanism by which cephalosporins react

with β-lactamase to eliminate a leaving group (section 16.5.2.1).

One of the problems associated with ADEPT is the possibility of an immune response to the antibody–enzyme complex since the enzyme is a foreign protein. For this reason, it may be preferable to use human enzymes along with prodrugs that are already approved for anticancer use. Research has been carried out on human enzymes such as alkaline phosphatase, carboxypeptidase A, and β-glucuronidase. The advantage of using a human enzyme is the decreased chance of an immune response, but the disadvantage is the increased risk of prodrug activation occurring during circulation in the blood supply.

Another problem may be insufficient enzyme activity. For example, the activation of irinotecan has been achieved using a particularly active isozyme of human carboxylesterase enzyme isolated from the liver. The isozyme concerned (hCE-2) was 26 times more active than another isozyme, hCE-1, but was still too low to be effective for ADEPT. Nevertheless, the isozyme may be suitable for gene therapy (section 18.9.5) where greater concentrations of the isozyme could be achieved within the cell than could be brought to the cell by antibodies.

The time gap between the administration of the antibody–enzyme complex and the prodrug is critical. Enough time must be provided to ensure that unbound complex has dropped to low levels, otherwise the prodrug will be activated in the blood supply. However, the longer the time gap, the more chance the levels of the antibody–enzyme complex will drop at the tumour. One way to tackle this problem is a three-stage ADEPT strategy. The antibody–enzyme complex is administered as before. Sufficient time is given for the complex to concentrate at the tumour, then a second antibody is administered which targets the conjugate and speeds up its clearance from the blood supply. The second antibody can be galactosylated to speed up its clearance rate such that it only has time to target circulating conjugate and does not survive long enough to penetrate the tumour. Finally, the prodrug is added as before.

18.9.4 Antibody-directed abzyme prodrug therapy (ADAPT)

Abzymes are antibodies which have a catalytic property. It is possible that prodrugs could be designed that act as antigens for these antibodies and are activated by the catalytic properties of the antibody. This can be done by immunizing mice with a transition-state analogue of the reaction that is desired, followed by isolation of the monoclonal antibodies by hybridization techniques. Since the antibody targets the prodrug rather than antigens on the cancer cell, this fails to target drugs to cancer cells. However, it should be possible to construct hybrid antibodies where one arm recognizes antigens on cancer cells while the other arm recognizes the prodrug and activates it. This approach is still in its early stages, but it has several potential advantages over ADEPT. For example, it should be possible to design catalytic mechanisms that do not occur naturally; allowing highly selective activation of prodrugs at tumours. It also removes the risk of an immune response due to foreign enzymes. At present the catalytic activity of abzymes is too low to be useful and much more research has to be carried out.

18.9.5 Gene-directed enzyme prodrug therapy (GDEPT)

Gene-directed enzyme prodrug therapy (GDEPT) involves the delivery of a gene to the cancer cell. Once delivered, the gene codes for an enzyme capable of transforming a prodrug into an active drug. As the enzyme will be produced inside the cell, the prodrug is required to enter the cell.

The main challenge in GDEPT is delivering the gene selectively to tumour cells. In one method the gene is packaged inside a virus such as a retrovirus or adenovirus. In the case of adenoviruses, the desired genes could be spliced into the viral DNA such that the virus inserts into host cell DNA on infection. The virus is also genetically modified such that it is no longer virulent and can do no harm to normal cells. Non-viral vectors have also been tried, such as cationic lipids and peptides. So far, it has not been possible to achieve the required selectivity for cancer cells over normal cells, and so the delivery vector has to be administered directly to the tumour.

The enzymes which are ultimately produced by the introduced genes should not be present in normal cells, so that prodrug activation only occurs in tumour cells. One advantage of GDEPT over ADEPT is the fact that foreign enzymes could be generated inside cancer cells and hidden from the immune response. The thymidine kinase enzyme produced by herpes simplex virus has been studied intensively in GDEPT. This enzyme activates the antiviral drugs aciclovir and ganciclovir (section 17.6.1). Since these drugs are poor substrates for mammalian thymidine kinase, activation will only be significant in the tumour cells containing the viral form of the enzyme. Several clinical trials have been carried out using this approach.

One problem associated with GDEPT is that it is unlikely for all tumour cells to receive the necessary gene to activate the prodrug. It is therefore important that the anticancer drug is somehow transferred between cells in the tumour—a so-called 'bystander' effect. This may occur by a variety of means such as release of the activated drug from the infected cell, direct transfer through intercellular gap junctions or by the release of drug-carrying vesicles following cell death.

GDEPT has been used to introduce the genes for the bacterial enzymes nitroreductase and carboxypeptidase G2 into cancer cells. Prodrugs were then administered which were converted to alkylating agents by the resulting enzymes. One of the problems with carboxypeptidase G2 is the difficulty some of the prodrugs have in crossing cell membranes. In order to overcome this problem the gene was modified such that the resulting enzyme was incorporated into the cell membrane with the active site revealed on the outer surface of the cell.

Gene therapy aimed at activating the prodrug **irinotecan** is being explored to try and improve the process by which the urethane is hydrolysed to the active drug (section 18.2.2.2). This could involve the introduction of a gene encoding a more active carboxypeptidase enzyme into tumour cells. For example, rabbit liver carboxypeptidase is 100–1000 times more efficient than the human form of the enzyme.

18.9.6 Other forms of gene therapy

Gene therapy could also be used to introduce the genes coding for regulatory proteins which have been suppressed in cancer cells. For example, attempts have been made to introduce the gene for the p53 protein via a virus vector.

KEY POINTS

- Monoclonal antibodies have been targeted against antigens which are over-expressed in certain cancer cells. They are useful in the treatment of breast cancers, colorectal cancer, and lymphomas.

- Antibody–drug conjugates involve the linking of a highly potent drug or radioisotope to an antibody. The conjugate is designed to target specific cancer cells and then be enveloped by the cell such that the drug can be released inside the cell.

- Antibodies should be humanized to avoid the immune response.

- Drugs can be attached to antibodies via lysine residues. Alternatively, the antibody can be modified to produce thiol or aldehyde groups to which drugs can be attached.

- ADEPT involves an antibody–enzyme conjugate which is targeted to specific cancer cells. Once the antibody has become attached to the outer surface of cancer cells, a prodrug is administered which is activated by the enzyme at the tumour site.

- ADAPT involves an antibody which has catalytic activity designed to activate a prodrug. At present, the activity of such abzymes is too low to be useful.

- GDEPT involves the delivery of a gene into a cancer cell. The gene codes for an enzyme capable of activating an anticancer prodrug.

18.10 Photodynamic therapy

Conventional prodrugs are inactive compounds which are normally metabolized in the body to their active form. A variation of the prodrug approach is the concept of a **sleeping agent**. This is an inactive compound which is converted to the active drug by some form of external influence. The best example of this approach is the use of photosensitizing agents such as porphyrins or chlorins in cancer treatment—**photodynamic therapy** (**PDT**). Porphyrins occur naturally in chlorophyll in plants and haemoglobin in red blood cells. They usually complex a metal ion in the centre of the molecule (magnesium in chlorophyll and iron in haemoglobin), and in this form they are non-toxic. However, if they lack the central ion, they have the potential to do great damage. Given intravenously, these agents accumulate within cells and have some selectivity for tumour cells. By themselves, the agents have little effect, but if the cancer cells are irradiated with red light or a red laser, the porphyrins are converted to an excited state and react with molecular oxygen to produce highly toxic singlet oxygen. Singlet oxygen can then attack proteins and unsaturated lipids in the cell membrane leading to the formation of hydroxyl radicals which further react with DNA leading to cell destruction. **Temoporfin (Foscan)** (Fig. 18.86) is an example of a chlorin photosensitizing agent undergoing clinical trials for the treatment of a variety of cancers.

Unfortunately, the porphyrin structures used for PDT are inherently hydrophobic, which makes them difficult to formulate. Encapsulation using liposomes, oils, or polymeric micelles is one method of avoiding this problem, and has the advantage that tumours engulf and retain macromolecules more readily than would be the case with normal tissue. This is because

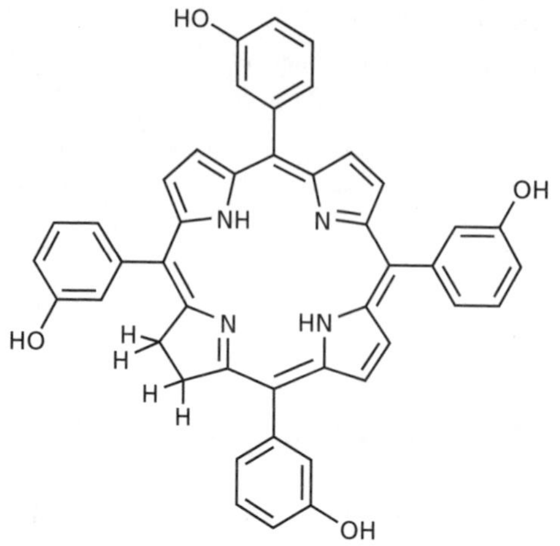

Figure 18.86 Temoporfin.

the blood vessels nourishing tumours are leaky (see section 18.1.9) and release larger molecules than would be released from normal blood vessels.

Despite this, problems still remain. For example, the liposomes which carry the agent can be engulfed and destroyed by cells of the reticuloendothelial system. The most serious disadvantage with PDT, however, is photosensitivity. Once the drug has been released from liposomes and activated, it is free to circulate round the body and accumulate in the eyes and skin, leading to phototoxic side effects which

render the patient highly sensitive to light. Indeed, it is this property that first highlighted the possibility of using porphyrins in PDT. Porphyria is a disease where porphyrins accumulate in the skin and result in photosensitization and disfigurement. Victims are unable to tolerate sunlight, and disfigurements can include erosion of the gums revealing red, fanglike teeth. It is likely that sufferers from this disease may have inspired the medieval vampire legends. Indeed, it is interesting to note that victims would have been averse to garlic, as components of garlic exacerbate the symptoms and cause an agonizing reaction. It was the observation that porphyrins could break down cells that led to the idea that these agents could be used to break down cancer cells.

Problems such as photosensitivity have limited the application of PDT, but research is underway to find improved methods of delivering the agent.

KEY POINTS

- PDT involves the irradiation of tumours containing porphyrin photosensitizers. This produces reactive oxygen species which are fatal to the cell. Photosensitivity is a serious problem, as the porphyrins can accumulate in the eyes and skin where they become activated by daylight.

QUESTIONS

1. Sch 226374 (Fig. 18.63) contains an aromatic ring which protects an imidazole ring from metabolism. Why do you think this imidazole ring is so susceptible to metabolism?

2. Do you think sulfonamides would be suitable antibacterial agents to treat opportunistic infections in cancer patients?

3. Since esters are commonly used as prodrugs, esterases would be suitable enzymes to use in ADEPT. What are your thoughts on this statement?

4. Staurosporin (Fig 18.70) is a kinase inhibitor that shows no selectivity, but is a useful lead compound for potential antitumour agents. A simplification strategy resulted in arcyiaflavin (shown below) which is selective for PKC. There are

three reasons why this molecule is simpler. Explain what they are and why simplification is desirable.

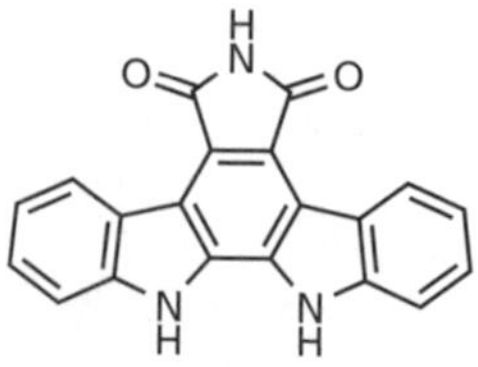

Arcyriflavin

5. CGP 52411 (shown below) is a further simplification of arcyriaflavin A. Remarkably, this compound is inactive against PKC and is selective for the kinase active site of the epidermal growth factor receptor. Suggest why there might be such a drastic change in selectivity.

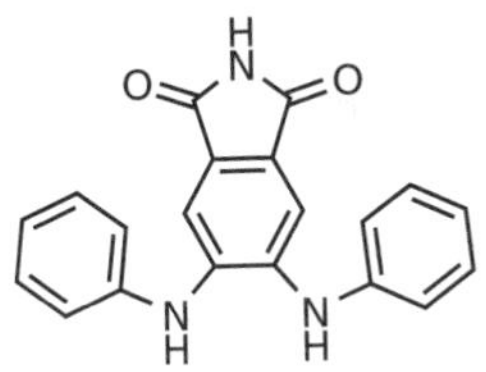

CGP 52411

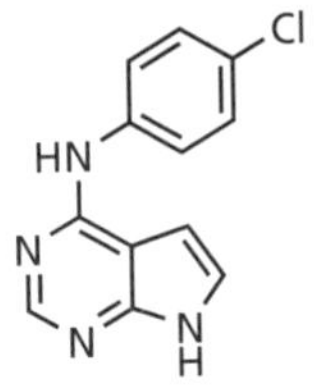

CGP 59326

Pyrazolopyrimidines

6. Further studies showed that CGP 52411 was bound to the ATP binding site of the kinase active site (Fig. 18.64). Suggest how this structure might bind, showing the binding interactions that are possible. SAR studies show that substitution on any of the NH groups or the aromatic rings is bad for activity.

7. CGP 59326 and the pyrazolopyrimidine structures shown are also useful inhibitors of the kinase active site of the epidermal growth factor receptor. Suggest how they might be bound to the active site.

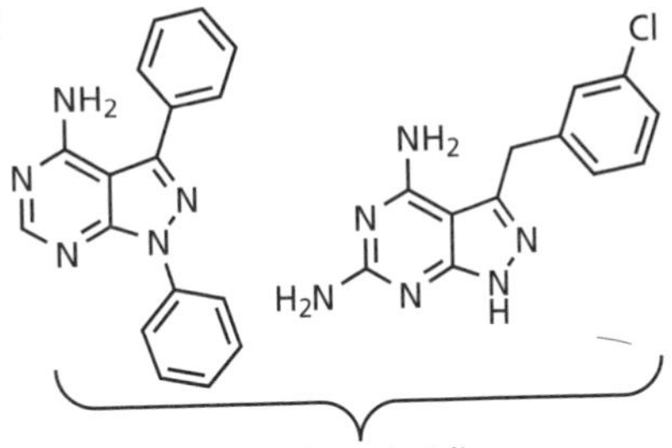

8. In the development of imatinib, a conformational blocker was introduced (Fig. 18.68). Suggest a conformation which would be feasible in the lead compound that would be prevented by the conformational blocker.

9. Imatinib has a pyrimidine ring where one of the nitrogens is involved in an important hydrogen bond interaction. It has been suggested that it should be possible to produce an analogue where the pyrimidine ring is replaced by a pyridine ring. What are your thoughts on this suggestion?

10. Alkylating agents have been observed to cause breaks in the DNA chain as shown below. Suggest a mechanism.

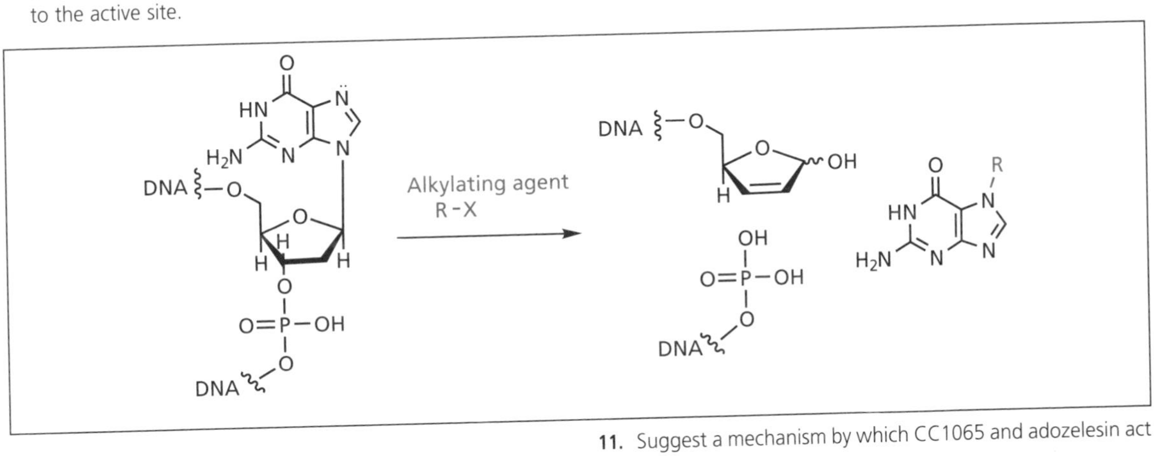

11. Suggest a mechanism by which CC1065 and adozelesin act as alkylating agents.

FURTHER READING

Atkins, J. H. and Gershell, L. J. (2002) Selective anticancer drugs. *Nature Reviews Drug Discovery*, 1, 491–492.

Elsayed, Y. A. and Sausville, E. A. (2001) Selected novel anticancer treatments targeting cell signalling proteins. *The Oncologist*, 6, 517–537.

Featherstone, J. and Griffiths, S. (2002) Drugs that target angiogenesis. *Nature Reviews Drug Discovery*, 1, 413–414.

Goldberg, A. L., Elledge, S. J., and Harper, J. W. (2001) The cellular chamber of doom. *Scientific American*, January, 68–73.

Hanahan, D. and Weinberg, R. A. (2000) The hallmarks of cancer. *Cell*, **100**, 57–70.

Jain, R. K. and Carmeliet, P. F. (2001) Vessels of death. *Scientific American*, December, 27–33 (angiogenesis).

Jordan, V. C. (2003) Tamoxifen: a most unlikely pioneering medicine. *Nature Reviews Drug Discovery*, 2, 205–213.

Neidle, S. and Parkinson, G. (2002) Telomere maintenance as a target for anticancer drug discovery. *Nature Reviews Drug Discovery*, 1, 383–393.

Ojima, I., Vite, G. D., and Altmann, K.-H. (eds.) (2001) *Anticancer agents*. ACS Symposium Series 796, American Chemical Society, Washington, DC.

Opalinska, J. B. and Gewirtz, A. M. (2002) Nucleic-acid therapeutics: basic principles and recent applications. *Nature Reviews Drug Discovery*, 1, 503–514.

Reed, J. C. (2002) Apoptosis-based therapies. *Nature Reviews Drug Discovery*, 1, 111–121.

Wayt Gibbs, W. (2003) Untangling the roots of cancer. *Scientific American*, July, 48–57.

Weissman, K. (2003) Life and cell death. *Chemistry in Britain*, August, 19–22.

Zhang, J. Y. (2002) Apoptosis-based anticancer drugs. *Nature Reviews Drug Discovery*, 1, 101–102.

Topoisomerase poisons

Fortune, J. M. and Osheroff, N. (2000) Topoisomerase II as a target for anticancer drugs. *Progress in Nucleic Acid Research*, **64**, 221–253.

Pommier, Y. *et al.* (1998) Mechanism of action of eukaryotic DNA topoisomerase I and drugs targeted to the enzyme. *Biochimica et Biophysica Acta*, **1400**, 83–106.

Agents acting on tubulin and microtubules

Farina, V. (ed.) (1995) The chemistry and pharmacology of taxol and its derivatives. Elsevier, Amsterdam.

Hormonal therapy

Brzozowski, A. M. *et al.* (1997) Molecular basis of agonism and antagonism in the oestrogen receptor. *Nature*, **389**, 753–758.

Jordan, C. (2003) Antiestrogens and selective estrogen receptor modulators as multifunctional medicines. *Journal of Medicinal Chemistry*, **46**, 1081–1111.

Photodynamic therapy

Lane, N. (2003) New light on medicine. *Scientific American*, January, 26–33.

Farnesyltransferase inhibitors

Bell, I. M. (2004) Inhibitors of farnesyltransferase: a rational approach to cancer chemotherapy? *Journal of Medicinal Chemistry*, **47**, 1–10.

Protein kinase inhibitors

Barker, A. J. *et al.* (2001) Studies leading to the identification of ZD1839 (Iressa). *Bioorganic and Medical Chemistry Letters*, 11, 1911–1914.

Capdeville, R. *et al.* (2002) Glivec (STI571, imatinib), a rationally developed, targeted anticancer drug. *Nature Reviews Drug Discovery*, 1, 493–502.

Dancey, J. and Sausville, E. A. (2003) Issues and progress with protein kinase inhibitors for cancer treatment. *Nature Reviews Drug Discovery*, 2, 296–313.

Zaiac, M. (2002) Taking aim at cancer. *Chemistry in Britain*, November, 44–46.

Miscellaneous enzyme inhibitors and other agents

Bartlett, J. B., Dredge, K., and Dalgleish, A. G. (2004) The evolution of thalidomide and its IMiD derivatives as anti-cancer agents. *Nature Reviews Cancer*, 4, 314–322.

Johnstone, R. W. (2002) Histone-deacetylase inhibitors: novel drugs for the treatment of cancer. *Nature Reviews Drug Discovery*, 1, 287–299.

McLaughlin, F., Finn, P., and La Thangue, N. B. (2003) The cell cycle, chromatin and cancer. *Drug Discovery Today*, 8, 793–802.

Whittaker, M., *et al.* (1999) Design and therapeutic application of matrix metalloproteinase inhibitors. *Chemical Reviews*, 99, 2735–2776.

Antibodies and gene therapy

Dubowchik, G. M. and Walker, M. A. (1999) Receptor-mediated and enzyme-dependent targeting of cytotoxic anticancer drugs. *Pharmacology and Therapeutics*, **83**, 67–123.

Ezzell, C. (2001) Magic bullets fly again. *Scientific American,* October, 28–35 (antibodies).

Senter, P. D. and Springer, C. J. (2001) Selective activation of anticancer prodrugs by monoclonal antibody–enzyme conjugates. *Advanced Drug Delivery Reviews,* 53, 247–264.

Titles for general further reading are listed on p. 711.

19 Cholinergics, anticholinergics, and anticholinesterases

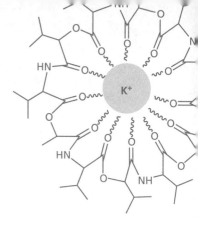

In Chapter 16, we discussed the medicinal chemistry of antibacterial agents and noted the success of these agents in combating many of the diseases that have afflicted humanity over the years. This success was aided in no small way by the fact that the 'enemy' could be identified, isolated, and conquered—first in the Petri dish, then in the many places in the body where it could hide. After this success, medicinal chemists set out to tackle the many other human ailments that are not infection-based; problems such as heart disorders, depression, schizophrenia, ulcers, autoimmune disease, and cancer. In all these ailments, the body itself has ceased to function properly in some way or other. There is no 'enemy' as such.

So what can medicinal chemistry do if there is no enemy to fight, save for the human body's failings? The first logical step is to understand what exactly has gone wrong. However, the mechanisms and reaction systems of the human body can be extremely complex. A vast array of human functions proceed each day with the greatest efficiency and with the minimum of outside interference. Breathing, digestion, temperature control, excretion, posture—these are all day-to-day operations that we take for granted—until they go wrong of course! Considering the complexity of the human body, it is perhaps surprising that its workings don't go wrong more often than they do.

Even if the problem is identified, what can a mere chemical do in a body filled with complex enzymes and interrelated chemical reactions? If it is even possible for a single chemical to have a beneficial effect, which of the infinite number of organic compounds would we use? The problem might be equated to finding the computer virus which has invaded your home computer software, or perhaps trying to trace where a missing letter went in the mail.

All is not doom and gloom, though. The ancient herbal remedies of the past helped to raise the curtain on some of the body's jealously guarded secrets. Even the toxins of snakes, spiders, and plants gave important clues to the workings of the body and provided lead compounds to possible cures. Over the last 100 years or so, many biologically active compounds have been extracted from their natural sources, then purified and identified. Chemists subsequently rang the changes on these lead compounds until an effective drug was identified. The process depended on trial and effort, chance and serendipity, but with this effort came a better understanding of how the body works and how drugs interact with the body. In more recent years, rapid advances in the biological sciences and in molecular modelling have resulted in medicinal chemistry moving from being a game of chance to being a science, where the design of new drugs is based on logical theories.

In this chapter, we concentrate on one particular field of medicinal chemistry—cholinergic and anticholinergic drugs. These are drugs which act on the peripheral and central nervous systems. We shall concentrate on the cholinergics to start with.

19.1 The peripheral nervous system

The peripheral nervous system (Fig. 19.1) is the part of the nervous system which is peripheral to the central nervous system (CNS—the brain and spinal

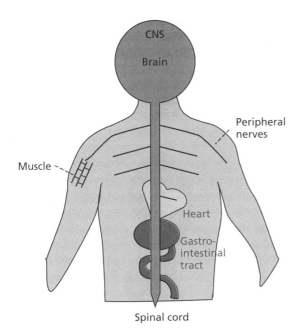

Figure 19.1 The peripheral nervous system.

column). There are many divisions and subdivisions of the peripheral system, which can lead to confusion. The first distinction we can make is between the following:

- sensory nerves (which take messages from the body to the CNS)
- motor nerves (which carry messages from the CNS to the rest of the body).

Here we are only concerned with the motor nerves.

19.2 Motor nerves of the peripheral nervous system

These nerves take messages from the CNS to various parts of the body such as skeletal muscle, smooth muscle, cardiac muscle, and glands (Fig. 19.1). The messages can be considered as similar to electrical pulses, but the analogy with electricity should not be taken too far since the pulse is a result of ion flow across the membranes of nerves and not a flow of electrons (see Appendix 4).

It should be evident that the workings of the human body depend crucially on an effective motor nervous system. Without it, we would not be able to operate our muscles and we would end up as flabby blobs, unable to move or breathe. We would not be able to eat, digest, or excrete our food, because the smooth muscle of the gastrointestinal tract (GIT) and the urinary tract are innervated by motor nerves. We would not be able to control body temperature, as the smooth muscle controlling the diameter of our peripheral blood vessels would cease to function. Finally, our heart would resemble a wobbly jelly rather than a powerful pump. In short, if the motor nerves failed to function, we would be in a mess! Let us now look at the motor nerves in more detail.

The motor nerves of the peripheral nervous system have been divided into three subsystems (Fig. 19.2): the somatic motor nervous system, the autonomic motor nervous system and the enteric nervous system. These are considered in the following sections.

19.2.1 The somatic motor nervous system

The somatic motor nerves carry messages from the CNS to the skeletal muscles. There are no synapses (junctions) *en route*, and the neurotransmitter at the neuromuscular junction is **acetylcholine.** The final result of such messages is contraction of skeletal muscle.

19.2.2 The autonomic motor nervous system

The autonomic motor nerves carry messages from the CNS to smooth muscle, cardiac muscle, and the adrenal medulla. This system can be divided into two subgroups:

- **Parasympathetic nerves:** These leave the CNS, travel some distance, then synapse with a second nerve which then proceeds to the final synapse with smooth muscle. The neurotransmitter at both synapses is acetylcholine.
- **Sympathetic nerves:** These leave the CNS, but almost immediately synapse with a second nerve (the neurotransmitter here is acetylcholine) which then proceeds to the same target organs as the parasympathetic nerves. Here, they synapse with different receptors on the target organs and use a different

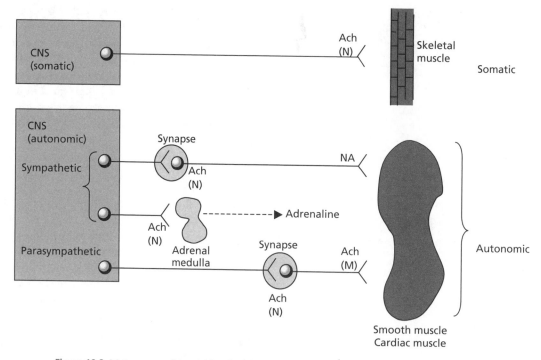

Figure 19.2 Motor nerves of the peripheral nervous system. N, Nicotinic receptor; M, Muscarinic receptor; Ach, Acetylcholine; N, Noradrenaline.

neurotransmitter—**noradrenaline** (for their actions, see section 19.4 and Chapter 20). The only exception to this are the nerves which go directly to the **adrenal medulla**. The neurotransmitter released here is noradrenaline and this stimulates the adrenal medulla to release the hormone **adrenaline**. This hormone then circulates in the blood system and interacts with noradrenaline receptors as well as other adrenaline receptors not directly supplied with nerves.

Note that the nerve messages are not sent along continuous channels analogous to telephone lines. Gaps (**synapses**) occur between different nerves and also between nerves and their target organs (Fig. 19.3). If a nerve wishes to communicate its message to another nerve or a target organ, it can only do so by releasing a chemical which has to cross the synaptic gap and bind to receptors on the target cell. This interaction between neurotransmitter and receptor can then stimulate other processes which, in the case of a second nerve, continues the message. Since these chemicals effectively carry a message from a nerve, they have become known as chemical messengers or **neurotransmitters**. The very fact that they are chemicals

and that they carry out a crucial role in nerve transmission allows the medicinal chemist to design and synthesize organic compounds which can mimic (**agonists**) or block (**antagonists**) the natural neurotransmitters.

19.2.3 The enteric system

The third subgroup of the peripheral nervous system is the enteric system, which is located in the walls of the intestine. It receives messages from sympathetic and parasympathetic nerves, but it also responds to local effects to provide local reflex pathways which are important in the control of GIT function. A large variety of neurotransmitters are involved including **serotonin, neuropeptides**, and **ATP. Nitric oxide (NO)** is also involved as a chemical messenger.

19.3 The neurotransmitters

There are a large variety of neurotransmitters in the CNS and the enteric system, but as far as the peripheral nervous system is concerned, we need

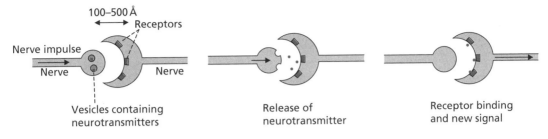

Figure 19.3 Signal transmission at a synapse.

Acetylcholine

R = H   Noradrenaline
R = Me   Adrenaline

Figure 19.4 Acetylcholine, noradrenaline and adrenaline.

only consider two—**acetylcholine** and **noradrenaline** (Fig. 19.4).

19.4 Actions of the peripheral nervous system

The actions of the peripheral nervous system can be divided into two systems—**somatic** and **autonomic**. Stimulation of the somatic peripheral system leads to the contraction of skeletal muscle. The autonomic system can be further classified as **sympathetic** or **parasympathetic**:

- **Sympathetic:** Noradrenaline is released at target organs and leads to the contraction of cardiac muscle and an increase in heart rate. It relaxes smooth muscle and reduces the contractions of the gastrointestinal and urinary tracts. It also reduces salivation and reduces dilatation of the peripheral blood vessels. In general, the sympathetic nervous system promotes the '**fight or flight**' response by shutting down the body's housekeeping

roles (digestion, defecation, urination, etc.), and stimulating the heart. The stimulation of the adrenal medulla releases the hormone **adrenaline**, which reinforces the action of noradrenaline.

- **Parasympathetic:** The stimulation of the parasympathetic system leads to the opposite effects from those of the sympathetic system. Acetylcholine is released at the target organs and reacts with receptors specific to it and not to noradrenaline.

Note that the sympathetic and parasympathetic nervous systems oppose each other in their actions and could be looked upon as a brake and an accelerator. The analogy is not quite apt, because both systems are always operating and the overall result depends on which effect is the stronger.

Failure in either of these systems would clearly lead to a large variety of ailments involving heart, skeletal muscle, digestion, etc. Such failure might be the result of either a deficit or an excess of neurotransmitter. Therefore, treatment involves the administration of drugs which can act as agonists or antagonists depending on the problem.

There is a difficulty with this approach, however. Usually, the problem we wish to tackle occurs at a certain location where there might, for example, be a lack of neurotransmitter. Application of an agonist to make up for low levels of neurotransmitter at the heart might solve the problem there, but would lead to problems elsewhere in the body. At these other locations, the levels of neurotransmitter would be normal and applying an agonist would lead to an 'overdose' and cause unwanted side effects. Therefore, drugs showing selectivity to certain parts of the body over others are clearly preferred.

This selectivity has been achieved to a great extent with both the cholinergic agonists/antagonists and the noradrenaline agonists/antagonists. In this chapter we concentrate on the former.

19.5 The cholinergic system

19.5.1 The cholinergic signalling system

Let us look first at what happens at synapses involving acetylcholine as the neurotransmitter. Figure 19.5 shows the synapse between two nerves and the events involved when a message is transmitted from one nerve cell to another. The same general process takes place when a message is passed from a nerve cell to a muscle cell.

1. The first stage involves the biosynthesis of acetylcholine (Fig. 19.6). Acetylcholine is synthesized from **choline** and **acetyl coenzyme A** in the nerve ending of the presynaptic nerve. The reaction is catalysed by the enzyme **choline acetyltransferase**.

2. Acetylcholine is incorporated into membrane-bound vesicles by means of a specific carrier protein.

3. The arrival of a nerve signal leads to an opening of calcium ion channels and an increase in intracellular calcium concentration. This induces the vesicles to fuse with the cell membrane and in doing so release the transmitter into the synaptic gap.

4. Acetylcholine crosses the synaptic gap and binds to the cholinergic receptor leading to stimulation of the second nerve.

5. Acetylcholine moves to an enzyme called **acetylcholinesterase** which is situated on the postsynaptic nerve and which catalyses the hydrolysis of acetylcholine to produce choline and ethanoic acid.

6. Choline is taken up into the presynaptic nerve by a carrier protein to continue the cycle.

The most important thing to note is that there are several stages where it is possible to use drugs to either promote or inhibit the overall process. The greatest success so far has been with drugs targeted at stages 4

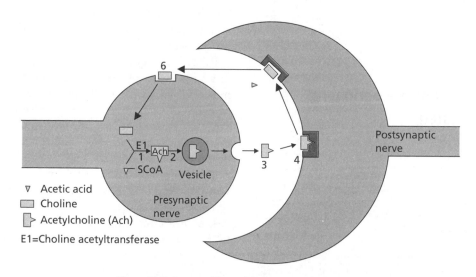

Figure 19.5 Synapse with acetylcholine as neurotransmitter.

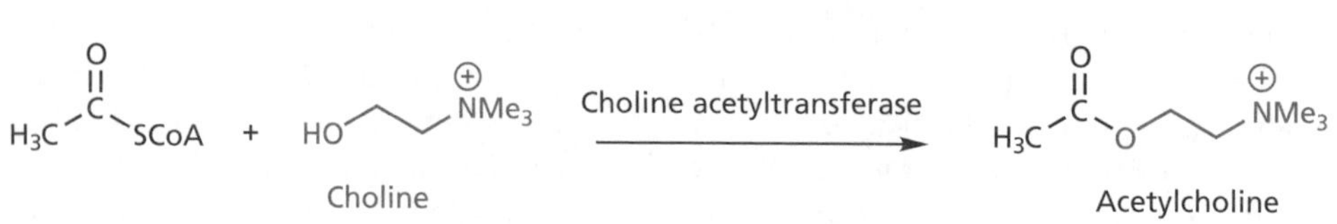

Figure 19.6 Biosynthesis of acetylcholine.

and 5 (i.e. the cholinergic receptor and the acetyl-cholinesterase enzyme). These are considered in more detail in subsequent sections.

19.5.2 Presynaptic control systems

Cholinergic receptors (called autoreceptors) are present at the terminal of the presynaptic nerve (Fig. 19.7). The purpose of these receptors is to provide a means of local control over nerve transmission. When acetylcholine is released from the nerve, some of it will find its way to these autoreceptors and switch them on. This has the effect of inhibiting further release of acetylcholine.

The presynaptic nerve also contains receptors for **noradrenaline** which act as another control system for acetylcholine release. Branches from the sympathetic nervous system lead to the cholinergic synapses and when the sympathetic nervous system is active, noradrenaline is released and binds to these receptors. Once again, the effect is to inhibit acetylcholine release. This indirectly enhances the activity of noradrenaline at its target organs by lowering cholinergic activity.

The chemical messenger **nitric oxide (NO)** can also influence acetylcholine release, but in this case it promotes release. A large variety of other chemical messengers including co-transmitters (see below) are also implicated in presynaptic control. The important thing to appreciate is that presynaptic receptors offer another possible drug target to influence the cholinergic nervous system.

19.5.3 Co-transmitters

Co-transmitters are messenger molecules released along with acetylcholine. The particular co-transmitter released depends on the location and target cell of the nerves. Each co-transmitter interacts with its own receptor on the postsynaptic cell. Co-transmitters have a variety of structures and include peptides such as **vasoactive intestinal peptide** (VIP), **gonadotrophin-releasing hormone** (GnRH), and **substance P**. The role of these agents appear to be as follows:

- They are longer lasting and reach more distant targets than acetylcholine, leading to longer-lasting effects.
- The balance of co-transmitters released varies under different circumstances (e.g. presynaptic control) and so can produce different effects.

19.6 Agonists at the cholinergic receptor

One point might have occurred to the reader. If there is a lack of acetylcholine acting at a certain part of the body, why not just administer more acetylcholine? After all, it is easy enough to make in the laboratory (Fig. 19.8).

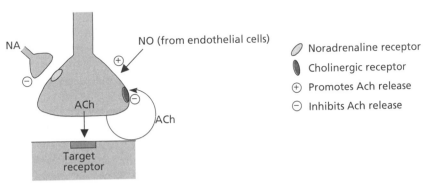

Figure 19.7 Presynaptic control systems.

Figure 19.8 Synthesis of acetylcholine.

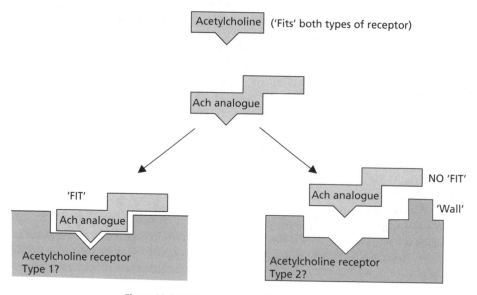

Figure 19.9 Binding sites for two cholinergic receptors.

There are three reasons why this is not feasible:

- Acetylcholine is easily hydrolysed in the stomach by acid catalysis and cannot be given orally.
- Acetylcholine is easily hydrolysed in the blood, both chemically and by enzymes (esterases).
- There is no selectivity of action. Additional acetylcholine will switch on all acetylcholine receptors in the body.

Therefore, we need analogues of acetylcholine which are more stable to hydrolysis and which are more selective with respect to where they act in the body. We shall look at selectivity first.

There are two ways in which selectivity can be achieved. First, some drugs might be distributed more efficiently to one part of the body than another. Second, cholinergic receptors in various parts of the body might be slightly different. This difference would have to be quite subtle—not enough to affect the interaction with the natural neurotransmitter acetylcholine, but enough to distinguish between two different synthetic analogues. We could, for example, imagine that the binding site for the cholinergic receptor is a hollow into which the acetylcholine molecule can fit (Fig. 19.9). We might then imagine that some cholinergic receptors in the body have a 'wall' bordering this hollow, while other cholinergic receptors do not. Thus, a synthetic analogue of acetylcholine

Figure 19.10 Nicotine and muscarine.

which is slightly bigger than acetylcholine itself would bind to the latter receptor, but not the former. This theory might appear to be wishful thinking, but it is now established that cholinergic receptors in different parts of the body are indeed subtly different.

This is not just a peculiarity of acetylcholine receptors. Subtle differences have been observed for other types of receptors such as those for dopamine, noradrenaline, and serotonin, and there are many types and subtypes of receptor for each chemical messenger (see Chapters 5 and 6).

The first indications that different types of cholinergic receptor existed came from the action of natural compounds. It was discovered that the compounds **nicotine** (present in tobacco) and **muscarine** (the active principle of a poisonous mushroom) (Fig. 19.10) were both acetylcholine agonists, but that they had different physiological effects.

Nicotine was found to be active at the synapses between different nerves and also at the synapses between nerves and skeletal muscle, but had poor activity elsewhere. Muscarine was active at the synapses of nerves with smooth muscle and cardiac muscle, but showed poor activity at the sites where nicotine was active. From these results, it was concluded that there was one type of acetylcholine receptor on skeletal muscles and at nerve synapses (the **nicotinic receptor**), and a different type of acetylcholine receptor on smooth muscle and cardiac muscle (the **muscarinic receptor**).

Muscarine and nicotine were the first compounds to indicate that receptor selectivity was possible, but they are unsuitable as medicines themselves because they have undesirable side effects resulting from their interactions with other receptors. In the search for a good drug, it is important to gain selectivity for one class of receptor over another (e.g. the acetylcholine receptor in preference to a noradrenaline receptor), and selectivity between receptor types (e.g. the muscarinic receptor in preference to a nicotinic receptor). It is also preferable to gain selectivity for particular subtypes of a receptor. For example, not every muscarinic receptor is the same throughout the body. At present, five subtypes of the muscarinic receptor have been discovered by cloning and have been numbered m_1–m_5. Four of these have been identified in tissues and are numbered M_1–M_4.

The principle of selectivity was proven with nicotine and muscarine, and so the race was on to design novel drugs which had the selectivity of nicotine or muscarine, but not the side effects.

KEY POINTS

- The cholinergic nervous system involves nerves which use the neurotransmitter acetylcholine as a chemical messenger. These include the motor nerves which innervate skeletal muscle, nerves which synapse with other nerves in the peripheral nervous system, and the parasympathetic nerves innervating cardiac and smooth muscle.

- There are two types of cholinergic receptor. Muscarinic receptors are present in smooth and cardiac muscle. Nicotinic receptors are present in skeletal muscle and in nerve-nerve synapses.

- Acetylcholine is hydrolysed by the enzyme acetylcholinesterase when it departs the cholinergic receptor. The hydrolytic product choline is taken up into presynaptic nerves and acetylated back to acetylcholine. The cholinergic

receptor and the enzyme acetylcholinesterase are useful drug targets.

- Acetylcholine cannot be used as a drug, because it is rapidly hydrolysed by acid and enzymes. It shows no selectivity for different types and subtypes of cholinergic receptor.

19.7 Acetylcholine: structure, SAR, and receptor binding

The first stage in any drug development is to study the lead compound and to find out which parts of the molecule are important to activity so that they can be retained in future analogues (i.e. structure–activity relationships—SAR). These results also provide information about what the binding site of the cholinergic receptor looks like and help decide what changes are worth making in new analogues.

In this case, the lead compound is acetylcholine itself. The results described below are valid for both the nicotinic and muscarinic receptors and were obtained by the synthesis of a large range of analogues.

- The positively charged nitrogen atom is essential to activity. Replacing it with a neutral carbon atom eliminates activity.

- The distance from the nitrogen to the ester group is important.

- The ester functional group is important.

- The overall size of the molecule cannot be altered greatly. Bigger molecules have poorer activity.

- The ethylene bridge between the ester and the nitrogen atom cannot be extended (Fig. 19.11).

- There must be two methyl groups on the nitrogen. A larger, third alkyl group is tolerated, but more than one large alkyl group leads to loss of activity.

- Bigger ester groups lead to a loss of activity.

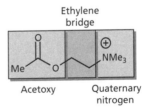

Figure 19.11 Acetylcholine.

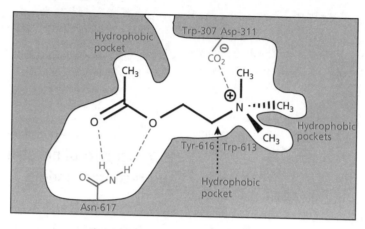

Figure 19.12 Muscarinic receptor binding site.

Conclusions: clearly, there is a tight fit between acetylcholine and its binding site, which leaves little scope for variation. The above findings tally with a receptor binding site as shown in Fig. 19.12.

It is proposed that important hydrogen bonding interactions exist between the ester group of acetylcholine and an asparagine residue. It is also thought that a small hydrophobic pocket exists which can accommodate the methyl group of the ester, but nothing larger. This interaction is thought to be more important in the muscarinic receptor than the nicotinic receptor.

The evidence suggests that the NMe_3^+ group is placed in a hydrophobic pocket lined with three aromatic amino acids. It is also thought that the pocket contains two smaller hydrophobic pockets, which are large enough to accommodate two of the three methyl substituents on the NMe_3^+ group. The third methyl substituent on the nitrogen is positioned in an open region of the binding site and so it is possible to replace it with other groups. A strong ionic interaction has been proposed between the charged nitrogen atom and the anionic side group of an aspartic acid residue. The existence of this ionic interaction represents the classical view of the cholinergic receptor, but there is an alternative suggestion which states that there may be an induced dipole interaction between the NMe_3^+ group and the aromatic residues in the hydrophobic pocket.

There are several reasons for this. First of all, the positive charge on the NMe_3^+ group is not localized on the nitrogen atom, but is spread over the three methyl groups (compare section 15.7.1). Such a diffuse charge is less likely to be involved in a localized ionic

Figure 19.13 Bond rotations in acetylcholine leading to different conformations.

interaction, and it has been shown by model studies that NMe_3^+ groups can be stabilized by binding to aromatic rings. It might seem strange that a hydrophobic aromatic ring should be capable of stabilizing a positively charged group, but it has to be remembered that aromatic rings are electron-rich, as shown by the fact they can undergo reaction with electrophiles. It is thought that the diffuse positive charge on the NMe_3^+ group is capable of distorting the π electron cloud of aromatic rings to induce a dipole moment (section 2.3.4). Dipole interactions between the NMe_3^+ group and an aromatic residue such as tyrosine would then account for the binding. The fact that three aromatic amino acids are present in the pocket adds weight to the argument.

Of course, it is possible that both types of binding interactions are taking place, which will please both parties!

A large amount of effort has been expended trying to identify the active conformation of acetylcholine; that is, the shape adopted by the neurotransmitter when it binds to the cholinergic receptor. This has been no easy task, as acetylcholine is a highly flexible molecule (Fig. 19.13), in which bond rotation along the length of its chain can lead to at least nine possible stable conformations (or shapes).

In the past, it was assumed that a flexible neurotransmitter such as acetylcholine would adopt its most stable conformation when binding. In the case of acetylcholine, that would be the conformation represented by the sawhorse and Newman projections shown in Fig. 19.14. This assumption is invalid, though, since there is not a massive energy difference between alternative stable conformations such as the gauche conformation shown in Fig. 19.15. The stabilization energy gained from binding interactions within the binding site could more than compensate for any energy penalties involved in adopting a less stable conformation.

In order to try and establish the active conformation of acetylcholine, rigid cyclic molecules have been studied which contain the skeleton of acetylcholine within their structure; for example, muscarine and the analogues shown in Fig. 19.16. In these structures, the portion of the acetylcholine skeleton which is included in a ring is locked into a particular conformation because bonds within rings cannot rotate freely. If such molecules bind to the cholinergic receptor, this indicates that this particular conformation is 'allowed' for activity.

Many such structures have been prepared, but it has not been possible to identify one specific active conformation for acetylcholine. This probably indicates that the cholinergic receptor has a certain amount of latitude and can recognize the acetylcholine skeleton within the rigid analogues, even when it is not in the ideal active conformation. Nevertheless, such studies have shown that the separation between the ester group and the quaternary nitrogen is important for binding and that this distance differs for the muscarinic and the nicotinic receptor (Fig. 19.17).

Having identified the binding interactions and pharmacophore of acetylcholine, we shall now look at how acetylcholine analogues were designed with improved stability.

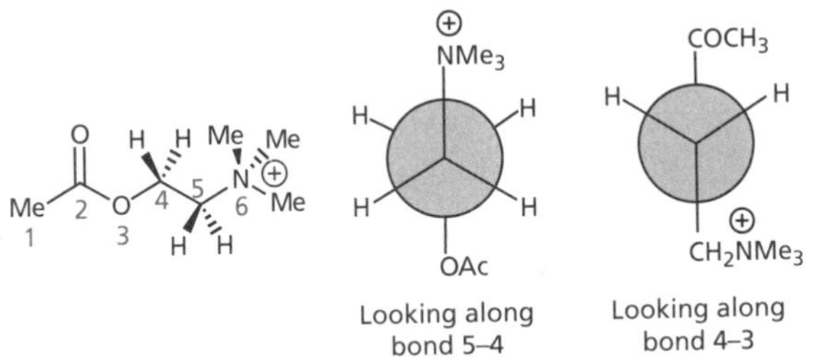

Figure 19.14 The sawhorse and Newman projections of acetylcholine.

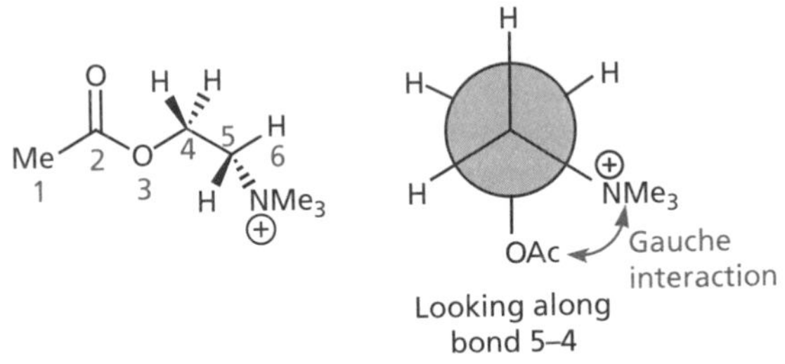

19.8 The instability of acetylcholine

As described previously, acetylcholine is prone to hydrolysis. This is explained by considering one of the conformations that the molecule can adopt (Fig. 19.18). In this conformation, the positively charged nitrogen interacts with the carbonyl oxygen and has an electron-withdrawing effect. To compensate, the oxygen atom pulls electrons from the neighbouring carbon atom and makes that carbon atom electron-deficient and more prone to nucleophilic attack. Water is a poor nucleophile, but because the carbonyl group is more electrophilic, hydrolysis takes place relatively easily. This influence of the nitrogen ion is known as

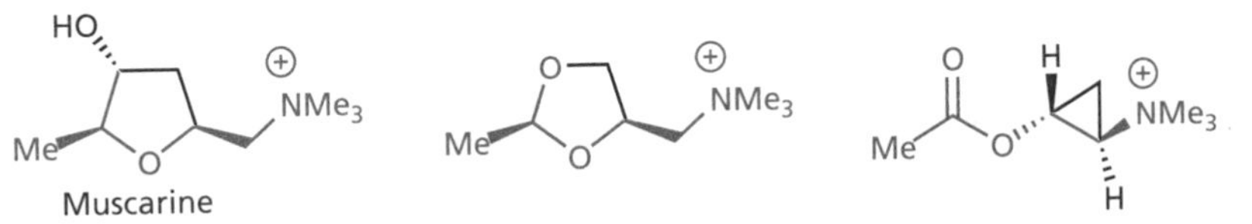

Figure 19.15 A gauche conformation for acetylcholine.

Figure 19.16 Rigid molecules incorporating the acetylcholine skeleton (C–C–O–C–C–N).

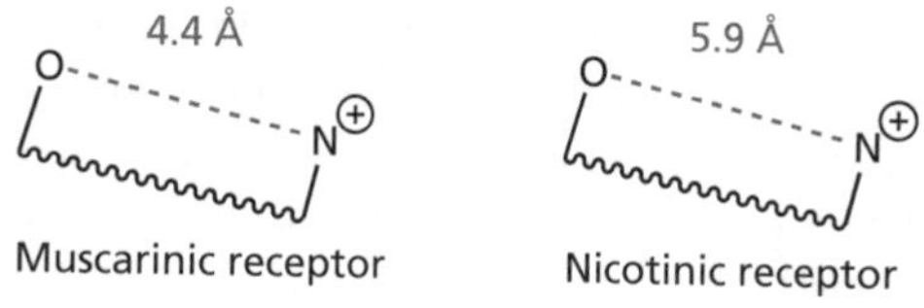

Figure 19.17 Pharmacophore of acetylcholine.

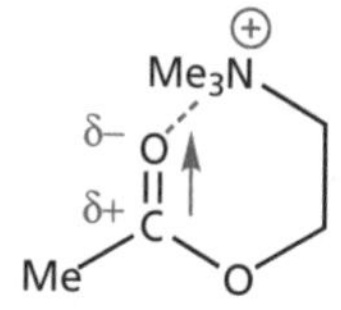

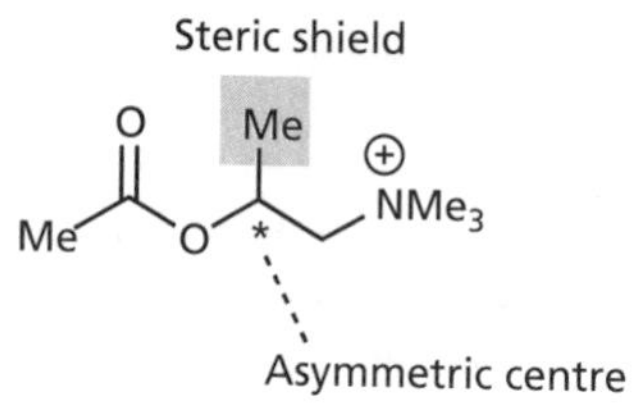

Figure 19.19 Methacholine.

Figure 19.18 Neighbouring group participation. The arrow indicates the inductive pull of oxygen which increases the electrophilicity of the carbonyl carbon.

neighbouring group participation or **anchimeric assistance**.

We shall now look at how the problem of hydrolysis was overcome, but it should be appreciated that we are doing this with the benefit of hindsight. At the time the problem was tackled, the SAR studies were incomplete and the format of the cholinergic receptor binding site was unknown.

19.9 Design of acetylcholine analogues

There are two possible approaches to tackling the inherent instability of acetylcholine: steric shields and electronic stabilization.

19.9.1 Steric shields

The principle of steric shields was described in section 11.2.1 and can be demonstrated with **methacholine** (Fig. 19.19) Here, an extra methyl group has been placed on the ethylene bridge as a steric shield to protect the carbonyl group. The shield hinders the approach of any potential nucleophile and also hinders binding to esterase enzymes, thus slowing down chemical and enzymatic hydrolysis. As a result, methacholine is three times more stable to hydrolysis than acetylcholine.

The obvious question to ask now is, why not put on a bigger alkyl group like an ethyl group or a propyl group? Alternatively, why not put a bulky group on the acyl half of the molecule, since this would be closer to the carbonyl centre and have a greater shielding effect?

In fact, these approaches were tried. They certainly increased stability but they lowered cholinergic activity. We should already know why—the fit between acetylcholine and its receptor is so tight that there is little scope for enlarging the molecule. The extra methyl group is acceptable, but larger substituents hinder the molecule binding to the cholinergic receptor and decrease its activity.

Introducing a methyl steric shield has another useful effect. It was discovered that methacholine has significant muscarinic activity, but very little nicotinic activity. Therefore, methacholine shows good selectivity for the muscarinic receptor. This is perhaps more important than the gain in stability.

Selectivity for the muscarinic receptor can be explained if we compare the active conformation of methacholine with muscarine (Fig. 19.20), since the methyl group of methacholine occupies the same position as a methylene group in muscarine. This is only possible for the S-enantiomer of methacholine, and when the two enantiomers of methacholine were separated and their activities compared, it was found that the S-enantiomer was indeed the more active enantiomer. It is not used therapeutically, however.

19.9.2 Electronic effects

The use of electronic factors to stabilize functional groups has been described in sections 11.2.2–11.2.3, and was used in the design of **carbachol** (Fig. 19.21), a long-acting cholinergic agent which is resistant to hydrolysis. Here, the acyl methyl group has been replaced by NH_2 such that the ester is replaced by a urethane or carbamate group. This functional group is more resistant to hydrolysis, since the lone pair of nitrogen can interact with the neighbouring carbonyl group and lower its electrophilic character (Fig. 19.22).

The tactic worked, but it was by no means a foregone conclusion that it would. Although the NH_2

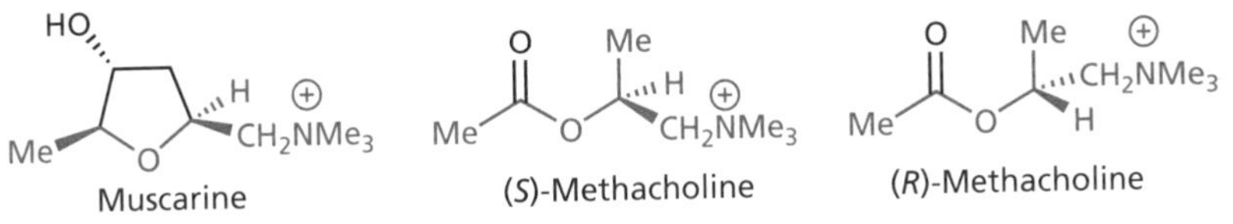

Figure 19.20 Comparison of muscarine and the R- and S-enantiomers of methacholine.

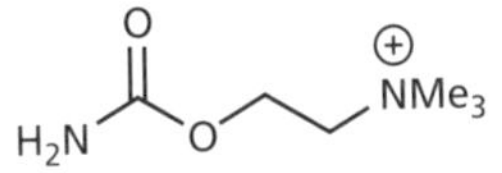

Figure 19.21 Carbachol.

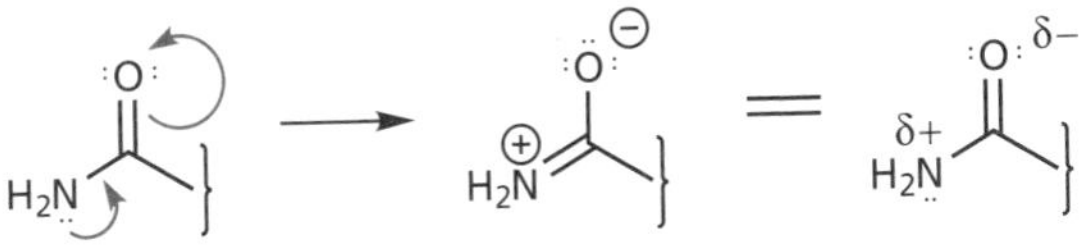

Figure 19.22 Resonance structures of carbachol.

* Asymmetric centre

Figure 19.23 Bethanechol.

group is equivalent in size to the methyl group, the former is polar and the latter is hydrophobic, and this implies that the polar NH_2 group has to fit into a hydrophobic pocket in the receptor. Fortunately, it does and activity is retained. This means that the amino group acts as a **bioisostere** for the methyl group. A bioisostere is a group which can replace another group without affecting the pharmacological activity of interest (section 10.3.7). Thus, the amino group is a bioisostere for the methyl group as far as the cholinergic receptor is concerned, but not as far as the esterase enzymes are concerned.

The inclusion of the electron-donating amino group greatly increases chemical and enzymatic stability. Unfortunately, carbachol shows very little selectivity between the muscarinic and nicotinic receptors. Nevertheless, it was used clinically for the treatment of glaucoma where it can be applied locally, thus avoiding the problems of receptor selectivity. Glaucoma arises when the aqueous contents of the eye cannot be drained. This raises the pressure on the eye and can lead to blindness. Agonists cause eye muscles to

contract, thus relieving the blockage and allowing drainage.

19.9.3 Combining steric and electronic effects

We have seen that the β-methyl group of methacholine increases stability and introduces receptor selectivity. Therefore, it made sense to add a β-methyl group to carbachol. The resulting compound is **bethanechol** (Fig. 19.23) which, as expected, is both stable to hydrolysis and selective in its action. It is occasionally used therapeutically in stimulating the GIT and urinary bladder after surgery. Both these organs are 'shut down' with drugs during surgery.

19.10 Clinical uses for cholinergic agonists

19.10.1 Muscarinic agonists

The clinical uses for muscarinic agonists are:

- treatment of glaucoma
- 'switching on' the GIT and urinary tract after surgery
- treatment of certain heart defects by decreasing heart muscle activity and heart rate.

Pilocarpine (Fig. 19.24) is an example of a muscarinic agonist which is used in the treatment of glaucoma. It is an alkaloid obtained from the leaves of shrubs belonging to the genus *Pilocarpus*. Although there is no quaternary ammonium group present in pilocarpine, it is assumed that the drug is protonated before it interacts with the muscarinic receptor. Molecular modelling shows that pilocarpine can adopt a conformation having the correct pharmacophore for the muscarine receptor; that is a separation between nitrogen and oxygen of 4.4 Å.

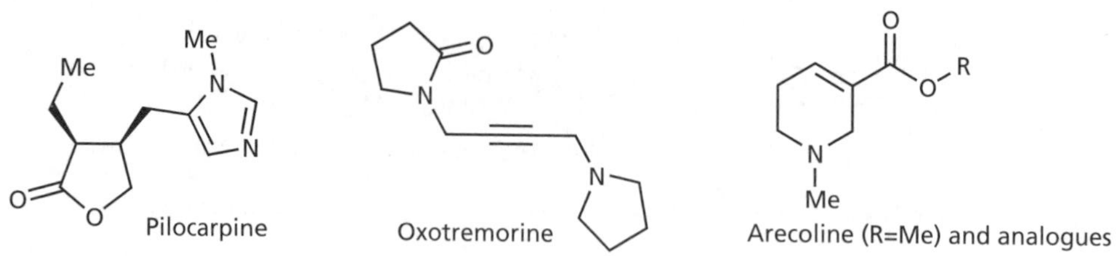

Figure 19.24 Examples of muscarinic agonists.

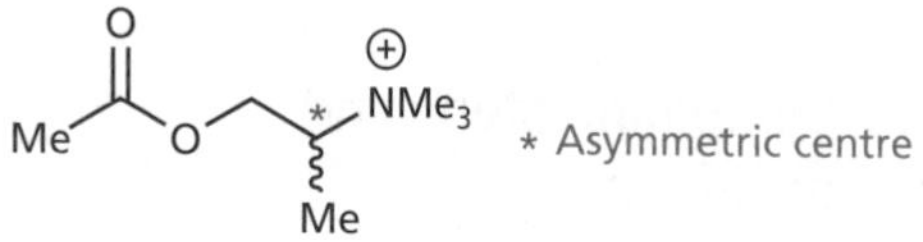

Figure 19.25 Example of a selective nicotinic agonist.

Pilocarpine is also being considered for the treatment of Alzheimer's disease, as are other muscarinic agonists such as **oxotremorine** and various **arecoline** analogues (Fig. 19.24). At present anticholinesterases are used clinically for the treatment of this disease (section 19.19).

19.10.2 Nicotinic agonists

These drugs are used in the treatment of myasthenia gravis, an autoimmune disease where the body has produced antibodies against its own cholinergic receptors. This leads to a reduction in the number of available receptors and so fewer messages reach the muscle cells. This in turn leads to severe muscle weakness and fatigue. Administering an agonist increases the chance of activating what few receptors remain. An example of a selective nicotinic agonist is shown in Fig. 19.25. This agent is very similar in structure to methacholine and differs only in the position of the methyl substituent. This is sufficient, however, to completely alter receptor selectivity. Despite that, this particular compound is not used clinically and anticholinesterases (section 19.16) are the preferred treatment.

KEY POINTS

- Acetylcholine fits snugly into the binding site of cholinergic receptors and there is little scope for variation. Two of the N-methyl groups and the acyl methyl group fit into hydrophobic pockets. The ester is involved in hydrogen bonding and the quaternary nitrogen is involved in ionic interactions and/or induced dipole interactions.

- Rigid analogues of acetylcholine have been used to try and identify the active conformation.

- Acetylcholine is unstable to acid due to neighbouring group participation. Stable analogues have been designed using steric shields and/or electronic effects.

19.11 Antagonists of the muscarinic cholinergic receptor

19.11.1 Actions and uses of muscarinic antagonists

Antagonists of the cholinergic receptor are drugs which bind to the receptor but do not switch it on. By binding to the receptor, an antagonist acts like a plug at the receptor binding site and prevents acetylcholine from binding (Fig. 19.26). The overall effect on the body is the same as if there was a lack of acetylcholine. Therefore, antagonists have the opposite clinical effect from agonists.

The antagonists described in this section act only at the muscarinic receptor and therefore affect nerve transmissions to glands, the CNS, and the smooth muscle of the GIT and urinary tract. The clinical effects and uses of these antagonists reflect this.

The clinical effects of muscarinic antagonists are:

- reduction of saliva and gastric secretions
- reduction of the motility of the GIT and urinary tract by relaxing smooth muscle
- dilatation of eye pupils
- CNS effects.

The clinical uses are:

- shutting down the GIT and urinary tract during surgery

- ophthalmic examinations
- relief of peptic ulcers
- treatment of Parkinson's disease
- treatment of anticholinesterase poisoning
- treatment of motion sickness.

19.11.2 Muscarinic antagonists

The first antagonists to be discovered were natural products—in particular alkaloids; nitrogen-containing compounds derived from plants.

19.11.2.1 Atropine and hyoscine

Atropine (Fig. 19.27) is present in the roots of *Atropa belladonna* (deadly nightshade) and is included in a root extract which was once used by Italian women to dilate their eye pupils so that they would appear more beautiful—hence the name belladonna. Clinically, atropine has been used to decrease gastrointestinal motility and to counteract anticholinesterase poisoning.

Atropine has an asymmetric centre but exists as a racemate. Usually, natural products exist exclusively as one enantiomer. This is also true for atropine, which is present in the plants of the genus Solanaceae as a single enantiomer called **hyoscyamine**. As soon as the natural product is extracted into solution, however,

racemization takes place. The asymmetric centre in atropine is easily racemized, as it is next to a carbonyl group. This makes the proton attached to the asymmetric centre acidic and easily removed.

Hyoscine (or scopolamine) (Fig. 19.27) is obtained from the thorn apple (*Datura stramonium*) and is very similar in structure to atropine. It has been used in treating motion sickness.

These two compounds bind to the cholinergic receptor, but at first sight, they do not look anything like acetylcholine. If we look more closely, though, we can see that a basic nitrogen and an ester group are present, and if we superimpose the acetylcholine skeleton on to the atropine skeleton, the distance between the ester and the nitrogen groups are similar in both molecules (Fig. 19.28). There is, of course, the problem that the nitrogen in atropine is uncharged, whereas the nitrogen in acetylcholine has a full positive charge. This implies that the nitrogen atom in atropine is protonated when it binds to the cholinergic receptor.

Therefore, atropine has two important binding features shared with acetylcholine—a charged nitrogen when protonated and an ester group. It is able to bind to the receptor, but why is it unable to switch it on? Because atropine is a larger molecule than acetylcholine, it is capable of binding to other binding regions outside the acetylcholine binding site. As a result, it interacts

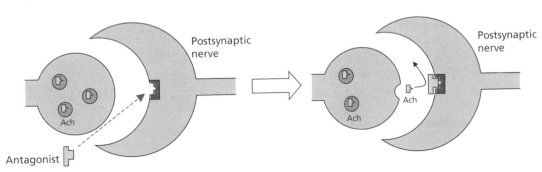

Figure 19.26 Action of an antagonist.

Figure 19.27 Atropine and hyoscine.

differently with the receptor, and does not induce the same conformational changes as acetylcholine. This means that the receptor is not activated.

Since both atropine and hyoscine are tertiary amines rather than quaternary salts, they are able to cross the blood–brain barrier as the free base. Once they are in the brain they can become protonated and antagonize muscarinic receptors in the CNS. This leads to CNS effects. For example, hallucinogenic activity is brought on with high doses, and both hyoscine and atropine were used by witches centuries ago to produce that very effect. Other CNS effects observed in atropine poisoning are restlessness, agitation, and hyperactivity.

In recent times, the disorientating effect of scopolamine has seen it being used as a truth drug for the interrogation of spies, and so it is no surprise to find it cropping up in various novels. An interesting application for scopolamine was described in Jack Higgins' novel *Day of Judgement* where it was used in association with **succinyl choline** to torture one hapless victim. Succinyl choline was applied to the conscious victim in order to create initial convulsive muscle spasms, followed by paralysis, inability to breathe, agonizing pain, and a living impression of death. Scopolamine was then used to erase the

memory of this horror, so that the impact would be just as bad when the process was repeated!

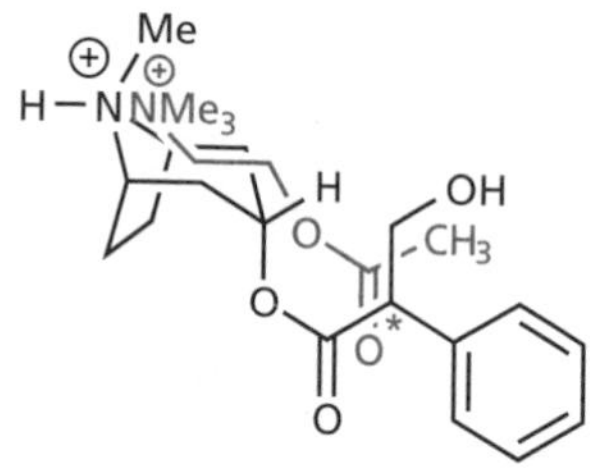

19.11.2.2 Structural analogues based on atropine

In order to reduce CNS side effects, quaternary salts of atropine and atropine analogues are used clinically (Fig. 19.29). For example, **ipratropium** is used as a bronchodilator and **atropine methonitrate** is used to lower motility in the GIT.

A large number of different analogues of atropine were synthesized to investigate the SAR of atropine, revealing the importance of the aromatic ring, the ester group, and the basic nitrogen (which is ionized). It was further discovered that the complex ring system was not necessary for antagonist activity, so simplification could be carried out. For example, **amprotropine** (Fig. 19.29) is active and has an ester group separated from an amine by three carbon atoms. Chain contraction to two carbon atoms can be carried out without loss of activity and a large variety of active antagonists have been prepared having the general formula shown in Fig. 19.30; for example, **tridihexethyl bromide** and **propantheline bromide**. These studies came up with the following generalizations:

- The alkyl groups (R) on nitrogen can be larger than methyl (in contrast to agonists).

- The nitrogen can be tertiary or quaternary, whereas agonists must have a quaternary nitrogen. Note, however, that the tertiary nitrogen is probably charged when it interacts with the receptor.

- Very large acyl groups are allowed (R^1 and R^2 = aromatic or heteroaromatic rings). This is in contrast to agonists, where only the acetyl group is permitted.

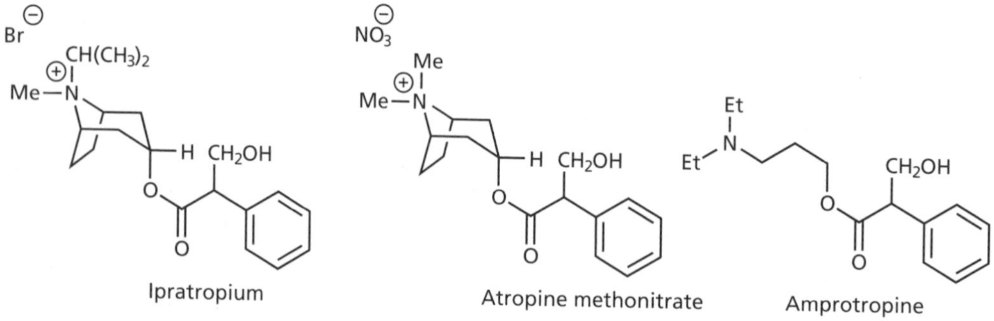

Figure 19.28 Acetylcholine skeleton superimposed on to the atropine skeleton.

Figure 19.29 Structural analogues of atropine.

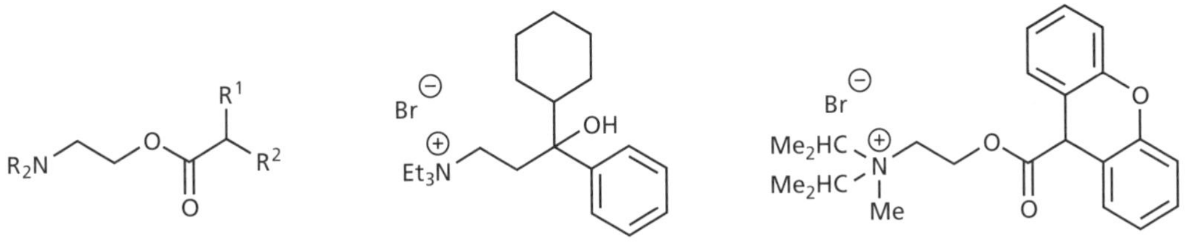

R¹ and R² = Aromatic
or heteroaromatic

Tridihexethyl bromide

Propanthreline bromide

Figure 19.30 Simplified analogues of atropine.

This last point appears to be the most crucial in determining whether a compound will act as an antagonist or not. The acyl group has to be bulky, but it also has to have that bulk arranged in a certain manner: there must be some sort of branching in the acyl group. For example, the molecule shown in Fig. 19.31 has a large unbranched acyl group but is not an antagonist.

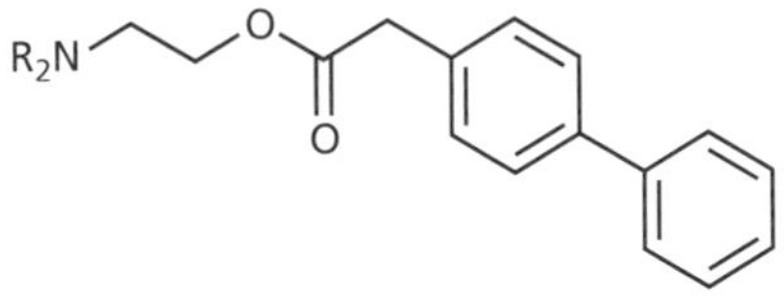

The conclusion that can be drawn from these results is that there must be hydrophobic binding regions next to the normal acetylcholine binding site. The overall shape of the acetylcholine binding site plus the extra binding regions would have to be T- or Y-shaped in order to explain the importance of branching in antagonists (Fig. 19.32). A structure such as **propantheline**, which contains the complete acetylcholine skeleton as well as the hydrophobic acyl side chain, not surprisingly binds more strongly to the receptor than acetylcholine itself (Fig. 19.32). The extra bonding interaction means that the conformational changes induced in the receptor (if any are induced at all) will be different from those induced by acetylcholine and will fail to induce the secondary biological response. As long as the antagonist is bound, acetylcholine is unable to bind and pass on its message (see also Box 19.1).

A large variety of antagonists have proved to be useful medicines, with many showing selectivity for specific organs. For example, atropine methonitrate acts at the intestine to relieve spasm, ipratropium is useful as an antiasthmatic, **tropicamide** and **cyclopentolate** (Fig. 19.33) are used in eye drops to dilate pupils for ophthalmic examination, and **trihexyphenidyl** and **benzatropine** (Fig. 19.33) are used centrally to counteract movement disorders due to Parkinson's disease. Some agents act selectively to decrease gastric secretion; others are useful in ulcer therapy. The selectivity of action for these drugs owes more to their distribution properties than to receptor

Figure 19.31 Analogue with no branching on the acyl group.

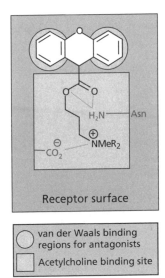

van der Waals binding
regions for antagonists

Acetylcholine binding site

Figure 19.32 Propantheline, which binds strongly to the receptor.

selectivity. In other words, the compounds can reach some parts of the body more easily than others. Having said that, the antagonist **pirenzepine** (Fig. 19.33), which was used in the treatment of peptic ulcers, is a selective M_1 antagonist with no activity against M_2 receptors.

BOX 19.1 PHOTOAFFINITY LABELLING

Since antagonists bind more strongly than agonists, they are better compounds to use for the labelling and identification of receptors in tissue preparations. An antagonist labelled with a radioactive isotope of H or C is used, and the radioactivity reveals where the receptor is located. Ideally, the antagonist should bind irreversibly by forming a covalent bond to the receptor. One useful tactic is to take an established antagonist and to incorporate a reactive chemical centre into the molecule. This reactive centre is usually electrophilic so that it will react with any suitably placed nucleophile close to the binding site; for example, the OH of a serine residue or the SH of a cysteine residue. In theory, the antagonist should bind to the receptor in the usual way and the electrophilic group will react with any nucleophilic amino acid within range. In practice, the procedure is not always as simple as this, as the highly reactive electrophilic centre might react with a nucleophilic group on another protein before it reaches the receptor and its binding site. One way to avoid this problem is to include a latent reactive centre which can only be activated once the antagonist has bound. One favourite method is **photoaffinity labelling**, where the reactive centre is activated by light. Chemical groups such as diazoketones or azides can be converted to highly reactive carbenes and nitrenes respectively, when irradiated.

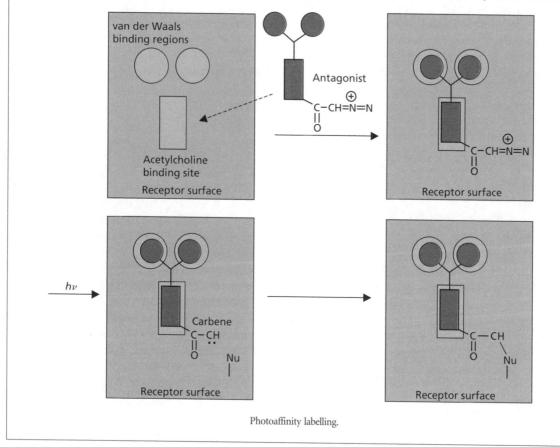

Photoaffinity labelling.

KEY POINTS

- Cholinergic antagonists bind to cholinergic receptors but fail to activate them. They block binding of acetylcholine and have a variety of clinical uses.

- Muscarinic antagonists normally contain a tertiary or quaternary nitrogen, a functional group involving oxygen and a branch point containing two hydrophobic ring substituents.

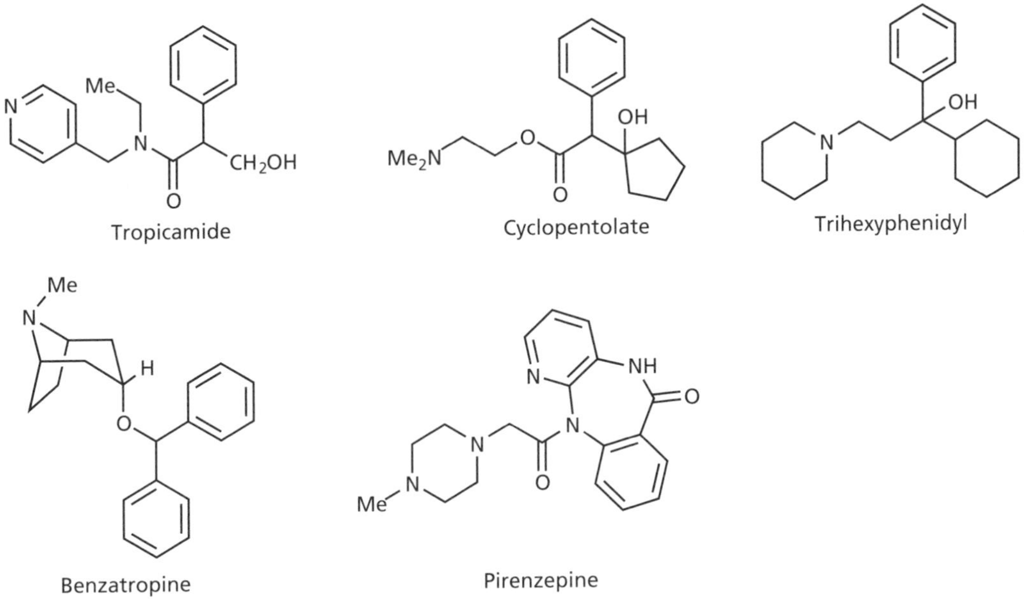

Figure 19.33 Some examples of clinically useful cholinergic antagonists.

19.12 Antagonists of the nicotinic cholinergic receptor

19.12.1 Applications of nicotinic antagonists

Nicotinic receptors are present in nerve synapses at ganglia, as well as at the neuromuscular synapse. However, drugs are able to show a level of selectivity between these two sites, mainly because of the distinctive routes which have to be taken to reach them. Antagonists of ganglionic nicotinic receptor sites are not therapeutically useful, because they cannot distinguish between the ganglia of the sympathetic nervous system and the ganglia of the parasympathetic nervous system (both use nicotinic receptors). Consequently, they have many side effects. However, antagonists of the neuromuscular junction are therapeutically useful and are known as neuromuscular blocking agents.

19.12.2 Nicotinic antagonists

19.12.2.1 Curare and tubocurarine

Curare was first identified in the sixteenth century when Spanish soldiers in South America found themselves under attack by indigenous people using poisoned arrows. It was discovered that the Indians were using a crude, dried extract from a plant called *Chondrodendron tomentosum*, which stopped the heart and also caused paralysis. We now know that curare is a mixture of compounds. The active principle, however, is an antagonist of acetylcholine which blocks nerve transmissions from nerve to muscle.

It might seem strange to consider such a compound for medicinal use, but at the right dose levels and under proper control, there are useful applications for this sort of action. The main application is in the relaxation of abdominal muscles in preparation for surgery. This allows the surgeon to use lower levels of general anaesthetic than would otherwise be required, and therefore increase the safety margin for operations.

As mentioned above, curare is actually a mixture of compounds, and it was not until 1935 that the active principle (**tubocurarine**) was isolated. The determination of the structure took even longer, and it was not established until 1970 (Fig. 19.34). Tubocurarine was used clinically as a neuromuscular blocker, but it had undesirable side effects since it also acted as an antagonist at the nicotinic receptors of the autonomic nervous system (Fig. 19.2). Better agents are now available.

The structure of tubocurarine presents a problem to our theory of receptor binding. Although it has a couple of charged nitrogen centres, there is no ester to interact with the acetyl binding region. Studies on the compounds discussed so far show that the positively charged nitrogen on its own is not sufficient for good binding, so why should tubocurarine bind to the cholinergic nicotinic receptor?

The answer lies in the fact that the molecule has two positively charged nitrogen atoms (one tertiary, which is protonated, and one quaternary). Originally, it was believed that the distance between the two centres (1.15 nm) might be equivalent to the distance between two separate cholinergic receptors and that the large tubocurarine molecule could act as a bridge between the two receptor sites, thus spreading a blanket over the two receptors and blocking access to acetylcholine. However pleasing that theory may be, the dimensions of the nicotinic receptor make this unlikely. The nicotinic receptor is a protein dimer made up of two identical protein complexes separated by 9–10 nm—far

too large to be bridged by the tubocurarine molecule (section 19.14 and Fig. 19.35).

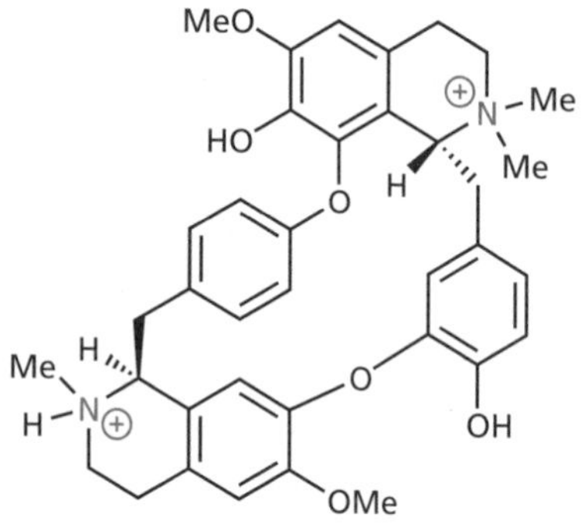

Another possibility is that the tubocurarine molecule bridges two acetylcholine binding sites within the one protein complex. As there are two such sites within the complex, this appears an attractive theory. However, the two sites are more than 1.15 nm apart and so this too seems unlikely. It has now been proposed that one of the positively charged nitrogens on tubocurarine binds to the anionic binding region of the acetylcholine binding site, while the other binds to a nearby cysteine residue 0.9–1.2 nm away (Fig. 19.35).

Despite the uncertainty surrounding the binding interactions of tubocurarine, it seems highly probable that two ionic binding regions are involved. Such an interaction is extremely strong and would more than make up for the lack of the ester binding interaction. It is also clear that the distance between the two positively charged nitrogen atoms is crucial to activity. Therefore, analogues that retain this distance should also be good antagonists. Strong evidence for this comes from the fact that the simple molecule decamethonium is a good antagonist.

19.12.2.2 Decamethonium and suxamethonium

Decamethonium (Fig. 19.36) is as simple an analogue of tubocurarine as one could imagine. It is a straight-chain molecule and as such is capable of a large number of conformations. The fully extended conformation places the nitrogen atoms 1.4 nm apart, but bond rotations can result in other conformations that position the nitrogen centres 1.14 nm apart, which compares well with the equivalent distance in tubocurarine (1.15 nm) (see also section 15.10.2).

Figure 19.34 Tubocurarine.

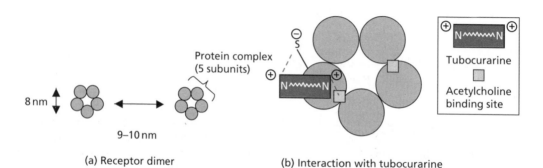

(a) Receptor dimer (b) Interaction with tubocurarine

Figure 19.35 Tubocurarine binding to and blocking the cholinergic receptor.

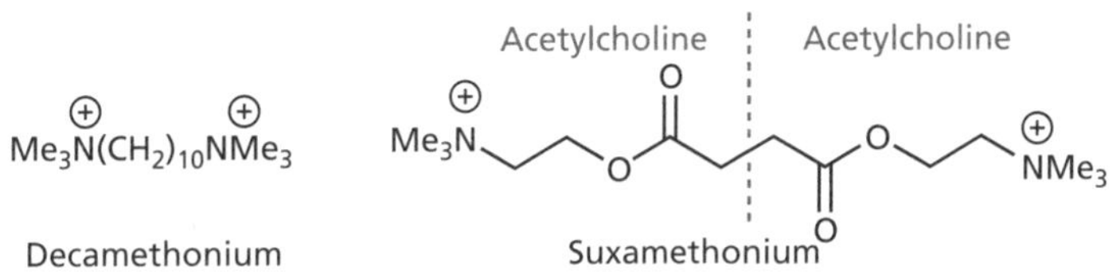

Figure 19.36 Decamethonium and suxamethonium.

The drug binds strongly to cholinergic receptors and has proved a useful clinical agent. It suffers from several disadvantages, however. For example, when it binds initially to the nicotinic receptor, it acts as an agonist rather than an antagonist. In other words, it switches on the receptor and this leads to a brief contraction of the muscle. Once this effect has passed, the drug remains bound to the receptor, blocking access to acetylcholine, and acting as an antagonist. (A theory of how such an effect might take place is described in section 5.10.) Another disadvantage is that it binds too strongly, so patients take a long time to recover from its effects. Both decamethonium and suxamethonium have effects on the autonomic ganglia, which explains some of their side effects. Decamethonium also lacks total selectivity for the neuromuscular junction and has an effect on cholinergic receptors in the heart. This leads to an increased heart rate and a fall in blood pressure.

We now face the opposite problem from the one faced when designing cholinergic agonists. Instead of stabilizing a molecule, we need to introduce some instability—a sort of timer control whereby the molecule can be inactivated more quickly. Success was first achieved with **suxamethonium** (Fig. 19.36) where two ester groups were incorporated into the chain in such a way that the distance between the charged nitrogen remained the same. The ester groups are susceptible to chemical and enzymatic hydrolysis and once this takes place, the molecule can no longer bridge the two binding regions on the receptor and is inactivated. The ester groups were also introduced such that suxamethonium mimics two acetylcholine molecules linked end on. Suxamethonium has a fast onset and short duration of action (5–10 minutes), but suffers from various side effects. Furthermore, about 1 person in every 2000 lacks the enzyme which hydrolyses suxamethonium. Nevertheless, it is still used clinically in short surgical procedures such as the insertion of tracheal tubes.

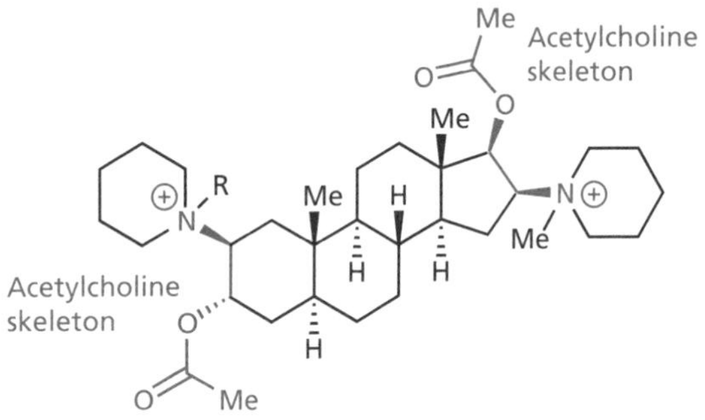

Figure 19.37 Pancuronium (R=Me) and vecuronium (R=H).

19.12.2.3 Pancuronium and vecuronium

The design of pancuronium and vecuronium (Fig. 19.37) was based on tubocurarine, but involved a steroid nucleus acting as a spacer between the two nitrogen groups. The distance between the quaternary nitrogens is 1.09 nm, compared to 1.15 nm in tubocurarine. Acyl groups were also added to introduce two acetylcholine skeletons into the molecule in order to improve affinity for the receptor sites. These compounds have a faster onset of action than tubocurarine and do not affect blood pressure. They are not as rapid in onset as suxamethonium and have a longer duration of action (45 minutes). Their main advantage is that they have fewer side effects and so they are widely used clinically.

19.12.2.4 Atracurium and mivacurium

The design of atracurium (Fig. 19.38) was based on the structures of tubocurarine and suxamethonium. As a drug it is superior to both, since it lacks cardiac side effects and is rapidly broken down in blood. This rapid breakdown allows the drug to be administered as an intravenous drip.

The rapid breakdown was designed into the molecule by incorporating a self-destruct mechanism. At blood pH, the molecule can undergo a **Hofmann elimination** (Fig. 19.39). Once this happens, the compound is inactivated because the positive charge

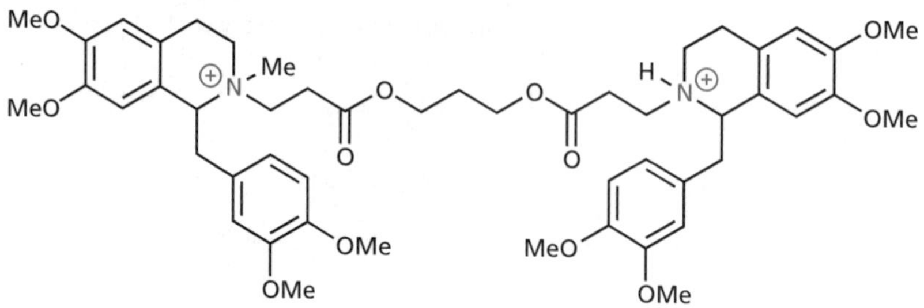

Figure 19.38 Atracurium.

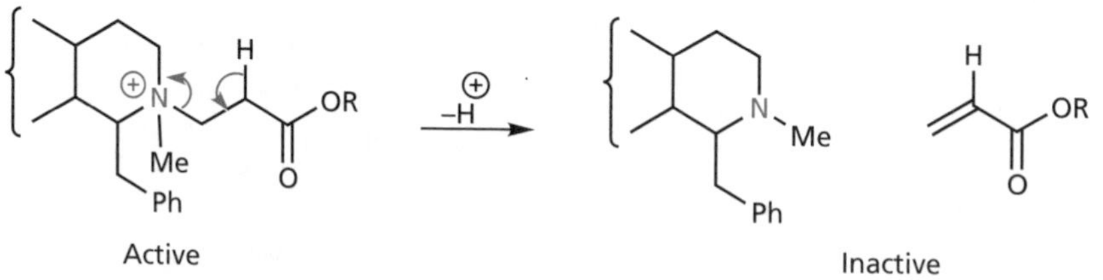

Figure 19.39 Hofmann elimination of atracurium.

on the nitrogen is lost and the molecule is split in two. It is a particularly clever example of drug design, in that the very element responsible for the molecule's biological activity promotes its deactivation.

The important features of atracurium are:

- **The spacer:** the 13-atom chain which connects the two quaternary centres.

- **The blocking units:** the cyclic structures at either end of the molecule which block the binding site from acetylcholine.

- **The quaternary centres:** essential for receptor binding. If one is lost through Hofmann elimination, the binding interaction is too weak and the antagonist leaves the binding site.

- **The Hofmann elimination:** The ester groups within the spacer chain are crucial to the rapid deactivation process. Hofmann eliminations normally require strong alkaline conditions and high temperatures—hardly normal physiological conditions. However, if a good electron-withdrawing group is present on the carbon, β to the quaternary nitrogen centre, it allows the reaction to proceed under much milder conditions (at pH 7.4, blood is mildly alkaline). The electron-withdrawing group

increases the acidity of the hydrogen on the β-carbon such that it is easily lost. The Hofmann elimination does not occur at acid pH, and so the drug is stable in solution at a pH of 3–4 and can be stored safely in a refrigerator.

Since the drug only acts very briefly (approx 30 minutes), it is added intravenously for as long as it is needed. As soon as surgery is over, the intravenous drip is stopped and antagonism ceases almost instantaneously. Another major advantage is that the drug does not require enzymes to become deactivated, and so deactivation occurs at a constant rate between patients. With previous neuromuscular blockers, deactivation depended on metabolic mechanisms involving enzymic deactivation and/or excretion. The efficiency of these processes varies from patient to patient and is particularly poor for patients with kidney failure or with low levels of plasma esterases.

Mivacurium (Fig. 19.40) is a newer drug which is similar to atracurium and is rapidly inactivated by plasma enzymes as well as by the Hofmann elimination. It has a faster onset (about 2 minutes) and shorter duration of action (about 15 minutes), although the duration is longer if the patients have liver disease or enzyme deficiencies.

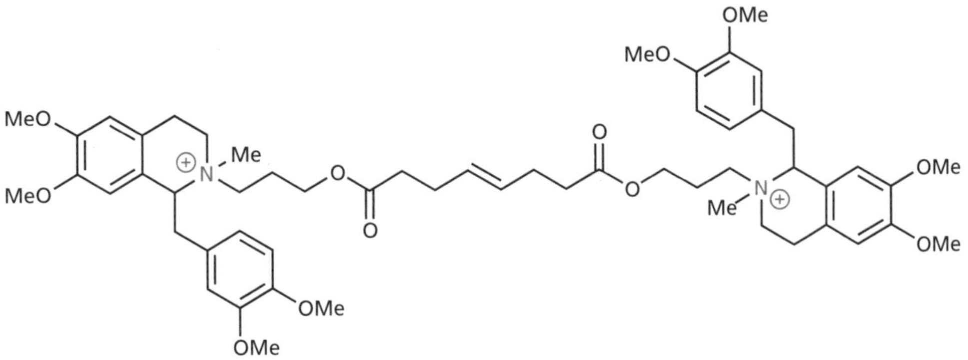

Figure 19.40 Mivacurium.

19.13 Other cholinergic antagonists

Local anaesthetics and barbiturates appear to prevent the changes in ion permeability which would normally result from the interaction of acetylcholine with the nicotinic receptor. They do not, however, bind to the acetylcholine binding site. It is believed that they bind instead to the part of the receptor which is on the inside of the cell membrane, perhaps binding to the ion channel itself and blocking it.

Certain snake toxins have been found to bind irreversibly to the nicotinic receptor, thus blocking cholinergic transmissions. These include toxins such as **α-bungarotoxin** from the Indian cobra. The toxin is a polypeptide containing 70 amino acids which crosslinks the α- and β-subunits of the cholinergic receptor (section 19.14).

KEY POINTS

- Nicotinic antagonists are useful as neuromuscular blockers in surgery.

- The pharmacophore for a nicotinic antagonist consists of two charged nitrogen atoms separated by a spacer molecule such that the centres are a specific distance apart.

- One of the charged nitrogens binds to the cholinergic binding site; the other interacts with a nucleophilic group neighbouring the binding site.

- Neuromuscular blockers should have a fast onset of action, minimal side effects, and a short duration of action to allow fast recovery. The lifetime of neuromuscular blockers can be decreased by introducing ester groups which are susceptible to enzymatic hydrolysis.

- Neuromuscular blockers which chemically degrade by means of the Hofmann elimination are not dependent on metabolic reactions and are more consistent from patient to patient.

19.14 Structure of the nicotinic receptor

The nicotinic receptor has been successfully isolated from the electric ray (*Torpedo marmorata*), a fish found in the Atlantic and the Mediterranean, allowing the receptor to be carefully studied. As a result, a great deal is known about its structure and operation. It is a protein complex made up of five subunits, two of which are the same. The five subunits (two α, one β, one γ, and one δ) form a cylindrical or barrel shape which traverses the cell membrane as shown in Fig. 19.41 (see also section 6.2.1).

The centre of the cylinder acts as an ion channel for sodium, and a gating or lock system is controlled by the interaction of the nicotinic receptor with acetylcholine. When acetylcholine is unbound, the gate is shut. When acetylcholine binds, the gate is opened.

The amino acid sequence for each subunit has been established, and it is known that there is extensive secondary structure. The binding site for acetylcholine is situated mainly on the α-subunit and there are two binding sites per receptor protein. It is usually found that nicotinic receptors occur in pairs, linked together by a disulfide bridge between the delta subunits (Fig. 19.42).

This is the make up of the nicotinic receptor at neuromuscular junctions. The nicotinic receptors at ganglia and in the CNS are more diverse in nature, involving different α- and β-subunits. This allows drugs to act selectively on neuromuscular rather than neuronal receptors. For example, decamethonium is only a weak antagonist at autonomic ganglia, whereas **epibatidine** (extracted from a South American frog) is a selective agonist for neuronal receptors. The snake toxin **α-bungarotoxin** is specific for receptors at neuromuscular junctions.

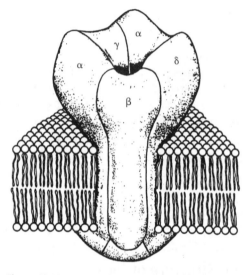

Figure 19.41 Schematic diagram of the nicotinic receptor. Taken from C. M. Smith and A. M. Reynard, *Textbook of pharmacology*, W. B. Saunders and Co. (1992).

19.15 Structure of the muscarinic receptor

Muscarinic receptors belong to the superfamily of G-protein-coupled receptors (section 6.3) which operate by activation of a signal transduction process (section 6.4–6.6). As mentioned above, four subtypes of muscarinic receptors that have been identified in the body (M_1–M_4) and these tend to be concentrated in specific tissues. For example, M_2 receptors occur mainly in the heart whereas M_4 receptors are found mainly in the CNS. M_2 receptors are also used as the autoreceptors on presynaptic cholinergic nerves (section 19.5.2).

The m_1, m_3, and m_5 receptors are associated with a signal transduction process involving the secondary messenger inosine triphosphate (IP_3). The m_2 and m_4 receptors involve a process which inhibits the production of the secondary messenger cAMP. Lack of M_1 activity is thought to be associated with dementia.

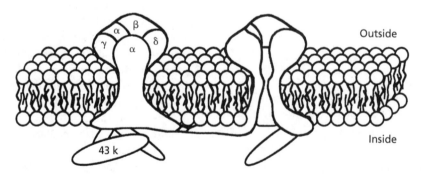

Figure 19.42 Nicotinic receptor pair. Taken from T. Nogrady, *Medicinal chemistry, a biochemical approach*, 2nd edn, Oxford University Press (1988).

- The nicotinic receptor is an ion channel consisting of five protein subunits. There are two binding sites for each ion channel.

- The muscarinic receptor is a G-protein-coupled receptor. Various subtypes of muscarinic receptor predominant in different tissues.

19.16 Anticholinesterases and acetylcholinesterase

19.16.1 Effect of anticholinesterases

Anticholinesterases are inhibitors of acetylcholinesterase—the enzyme that hydrolyses acetylcholine. If acetylcholine is not destroyed, it can return to reactivate the cholinergic receptor and increase cholinergic effects (Fig. 19.43). Therefore, an acetylcholinesterase inhibitor will have the same biological effect as a cholinergic agonist.

19.16.2 Structure of the acetylcholinesterase enzyme

The acetylcholinesterase enzyme has a fascinating tree-like structure (Fig. 19.44). The trunk of the tree is a collagen molecule which is anchored to the cell membrane. There are three branches (disulfide bridges) leading off from the trunk, each of which holds the acetylcholinesterase enzyme above

the surface of the membrane. The enzyme itself is made up of four protein subunits, each of which has an active site. Therefore, each enzyme tree has twelve active sites. The trees are rooted immediately next to the cholinergic receptors so that they efficiently capture acetylcholine as it departs the receptor. In fact, the acetylcholinesterase enzyme is one of the most efficient enzymes known. A soluble cholinesterase enzyme called butyrylcholinesterase is also present in various tissues and plasma. This enzyme has broader substrate specificity than acetylcholinesterase and can hydrolyse a variety of esters.

19.16.3 The active site of acetylcholinesterase

The design of anticholinesterases depends on the shape of the enzyme active site, the binding interactions involved with acetylcholine, and the mechanism of hydrolysis.

19.16.3.1 Binding interactions at the active site

There are two important areas to be considered—the anionic binding region and the ester binding region (Fig. 19.45).

Note that:

- Acetylcholine binds to the acetylcholinesterase enzyme by ionic bonding to an aspartate residue (compare section 19.7), and by hydrogen bonding to a tyrosine residue.

- Aspartate, histidine, and serine residues in the active site are involved in the mechanism of hydrolysis.

The anionic binding region in acetylcholinesterase is very similar to the anionic binding region in the

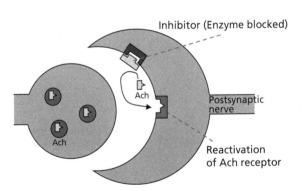

Figure 19.43 Effect of anticholinesterases.

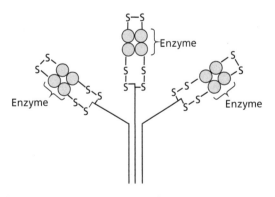

Figure 19.44 The acetylcholinesterase enzyme.

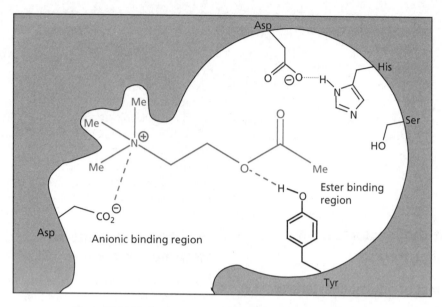

Figure 19.45 Binding interactions at the active site of acetylcholinesterase.

cholinergic receptor, and may be identical. There are thought to be two hydrophobic pockets large enough to accommodate methyl residues but nothing larger. The positively charged nitrogen is thought to be bound to a negatively charged aspartate residue, and may also interact with aromatic amino acids by induced dipole interactions (compare section 19.7).

19.16.3.2 Mechanism of hydrolysis

The histidine residue acts as an acid–base catalyst throughout the mechanism, while serine plays the part of a nucleophile. This is not a particularly good role for serine, as an aliphatic alcohol is a poor nucleophile and is unable to hydrolyse an ester, but the acid–base catalysis provided by histidine overcomes that disadvantage. The aspartate residue interacts with the histidine residue and serves to orientate and activate the ring (compare chymotrypsin—section 4.5.3). There are several stages to the mechanism (Fig. 19.46):

1. Acetylcholine approaches and binds to the active site. Serine acts as a nucleophile and uses a lone pair of electrons to form a bond to the ester of acetylcholine. Nucleophilic addition to the ester takes place and opens up the carbonyl group.

2. The histidine residue catalyses this reaction by acting as a base and removing a proton, thus making serine more nucleophilic.

3. Histidine now acts as an acid catalyst and protonates the 'O–R' portion of the intermediate, turning it into a much better leaving group.

4. The carbonyl group reforms and expels the alcohol portion of the ester (i.e. choline).

5. The acyl portion of acetylcholine is now covalently bound to the active site. Choline leaves the active site and is replaced by water.

6. Water acts as a nucleophile and uses a lone pair of electrons on oxygen to attack the acyl group.

7. Water is normally a poor nucleophile, but once again histidine aids the process by acting as a basic catalyst and removing a proton.

8. Histidine acts as an acid catalyst by protonating the intermediate.

9. The carbonyl group is reformed and the serine residue is released. As it is now protonated, it is a much better leaving group.

10. Ethanoic acid leaves the active site and the cycle can be repeated.

The enzymatic process is remarkably efficient due to the close proximity of the aspartate residue (not shown), the serine nucleophile and the histidine acid–base catalyst. As a result, enzymatic hydrolysis by acetylcholinesterase is 10^8 (one hundred million)

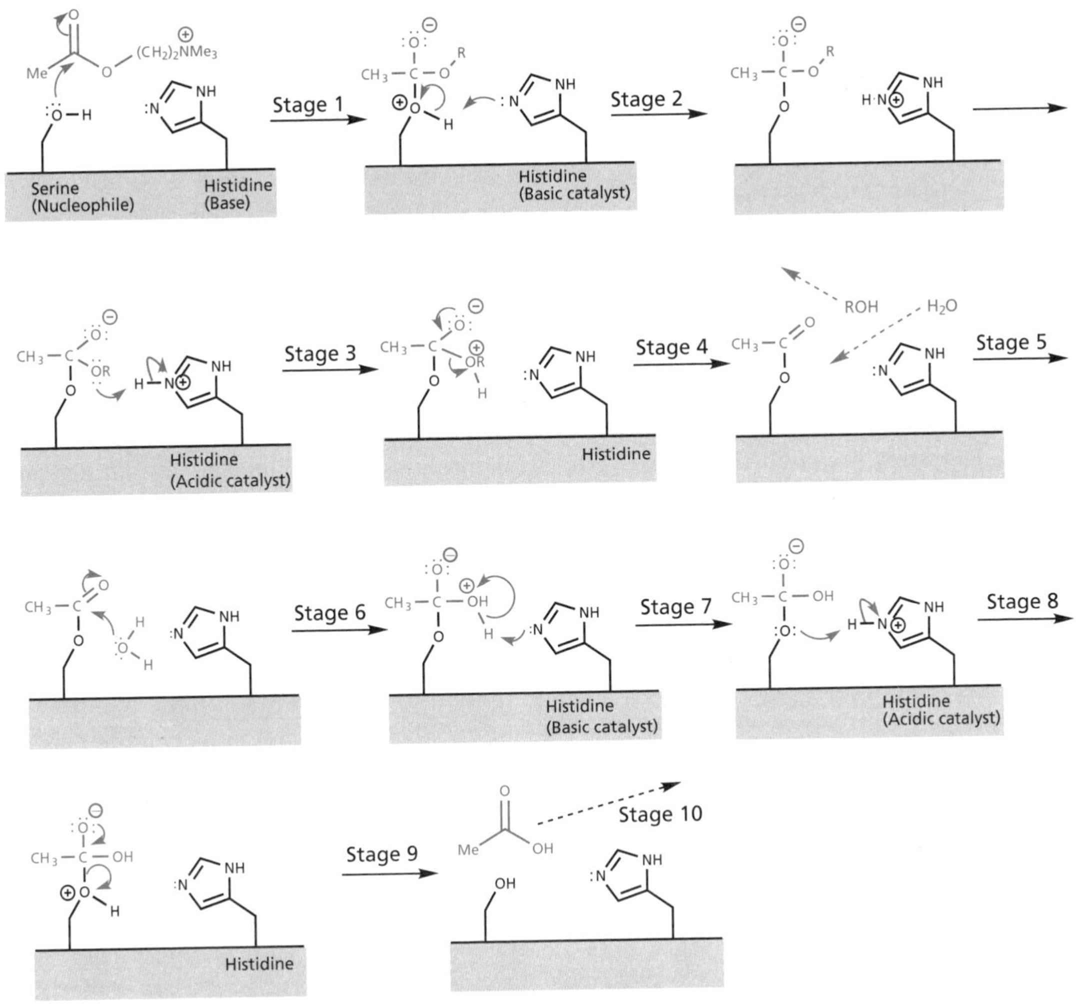

Figure 19.46 Mechanism of hydrolysis (the aspartate component of the catalytic triad is not shown).

times faster than chemical hydrolysis. The process is so efficient that acetylcholine is hydrolysed within 100 μs of reaching the enzyme.

19.17 Anticholinesterase drugs

Anticholinesterase drugs act as inhibitors of the enzyme acetylcholinesterase. This inhibition can be either reversible or irreversible depending on how the drug interacts with the active site. Two main groups of acetylcholinesterases are considered here—carbamates and organophosphorus agents.

19.17.1 Carbamates

19.17.1.1 Physostigmine

As in so many fields of medicinal chemistry, it was a natural product that provided the lead for the carbamate inhibitors. The natural product was **physostigmine** (Fig. 19.47) (also called eserine) which was discovered in 1864 as a product of the poisonous calabar bean (the ordeal bean, *Physostigma venenosum*) from West Africa. Extracts of these beans were fed to criminals to assess whether they were guilty or innocent. Death indicated a guilty verdict. The structure was established in 1925, and physostigmine is still used clinically to treat glaucoma.

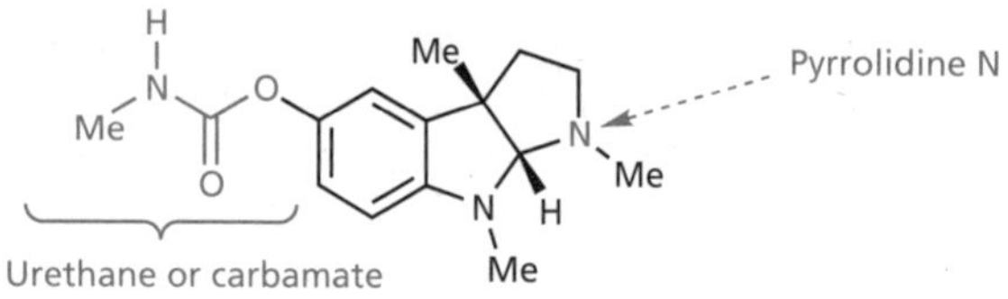

Figure 19.47 Physostigmine.

SAR studies of physostigmine demonstrate that:

- the carbamate group is essential to activity
- the benzene ring is important
- the pyrrolidine nitrogen is important and is ionized at blood pH.

Working backwards, the positively charged pyrrolidine nitrogen is important because it binds to the anionic binding region of the enzyme. The benzene ring may be involved in some extra hydrophobic bonding with the active site. Alternatively, it may be important in the mechanism of inhibition as it provides a good leaving group. The carbamate group is the crucial group responsible for physostigmine's inhibitory properties. To understand why, we must look at what happens when physostigmine acts as the substrate for acetylcholinesterase (Fig. 19.48).

The first four stages proceed as normal, with histidine catalysing the nucleophilic attack of the serine residue on physostigmine (stages 1 and 2). The leaving group (this time a phenol) is expelled with the aid of acid catalysis from histidine (stages 3 and 4), and departs the active site to be replaced by a water molecule.

The next stage turns out to be extremely slow. Despite the fact that histidine can still act as a basic catalyst, water finds it difficult to attack the carbamoyl intermediate. This step becomes the rate-determining step for the whole process and the overall rate of hydrolysis of physostigmine is 40×10^6 times slower than that of acetylcholine. As a result, the cholinesterase active site becomes blocked and is unable to react with acetylcholine.

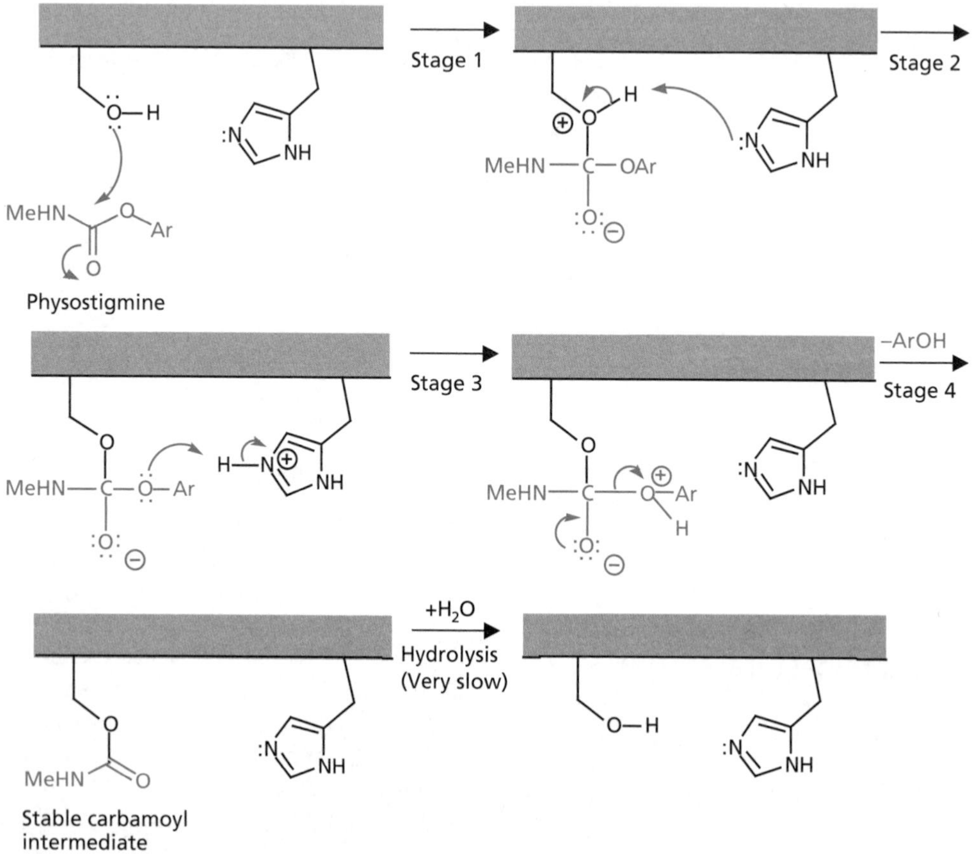

Figure 19.48 Mechanism of inhibition by physostigmine.

The reason why this final stage is so slow is that the carbamoyl–enzyme complex is stabilized because the nitrogen can feed a lone pair of electrons into the carbonyl group and drastically reduce its electrophilic character (Fig. 19.49) (compare section 19.9.2).

19.17.1.2 Analogues of physostigmine

Physostigmine has limited medicinal use because of serious side effects, and it has only been used in the treatment of glaucoma or as an antidote for atropine poisoning. Simpler analogues, however, have been used in the treatment of myasthenia gravis and as an antidote to curare poisoning.

Miotine (Fig. 19.50) still has the necessary carbamate, aromatic, and tertiary aliphatic nitrogen groups. It is active as an antagonist but has disadvantages: it is susceptible to chemical hydrolysis, and it can cross the blood–brain barrier (see section 8.3.5) as the free base, resulting in side effects due to its action in the CNS.

Neostigmine and **pyridostigmine** (Fig. 19.50) were designed to deal with both the problems described above. First, a quaternary nitrogen atom is present so that there is no chance of the free base being formed. Since the molecule is permanently charged, it cannot cross the blood–brain barrier and so the drug is free of CNS side effects (section 8.3.5). Increased stability is achieved by using a dimethylcarbamate group rather than a methylcarbamate group. There are two possible explanations for this, based on two possible hydrolysis mechanisms.

Mechanism 1 (Fig. 19.51) involves nucleophilic substitution by water. The rate of the reaction depends on the electrophilic character of the carbonyl group and if this is decreased, the rate of hydrolysis is decreased. We have already seen how the lone pair of the neighbouring nitrogen can reduce the electrophilic character of the carbonyl group. The presence of a second methyl group on the nitrogen has an inductive donating effect, which increases electron density on the nitrogen and further encourages the nitrogen lone pair to interact with the carbonyl group.

Mechanism 2 (Fig. 19.52) is a fragmentation involving loss of the phenolic group before addition of the nucleophile. This mechanism requires the loss of a proton from the nitrogen. Replacing this hydrogen with a methyl group would severely inhibit the reaction, since the mechanism would require the loss of a methyl cation—a highly unfavoured process.

Whichever mechanism is involved, the presence of the second methyl group acts to discourage the process. Two further points to note about neostigmine are the following:

- The quaternary nitrogen is 4.7 Å away from the ester group.
- The direct bonding of the quaternary centre to the aromatic ring reduces the number of conformations

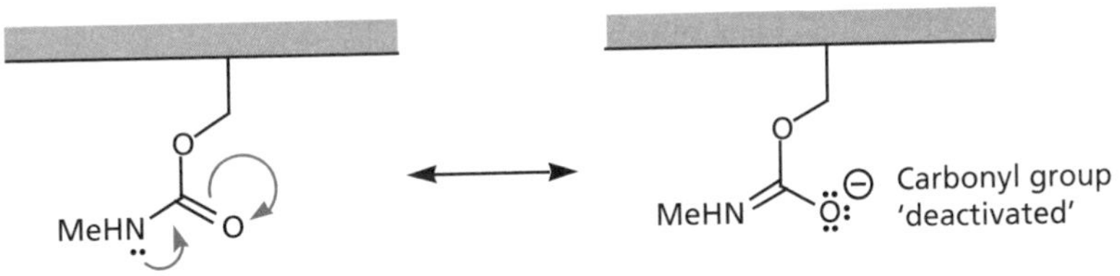

Figure 19.49 Stabilization of the carbamoyl–enzyme intermediate.

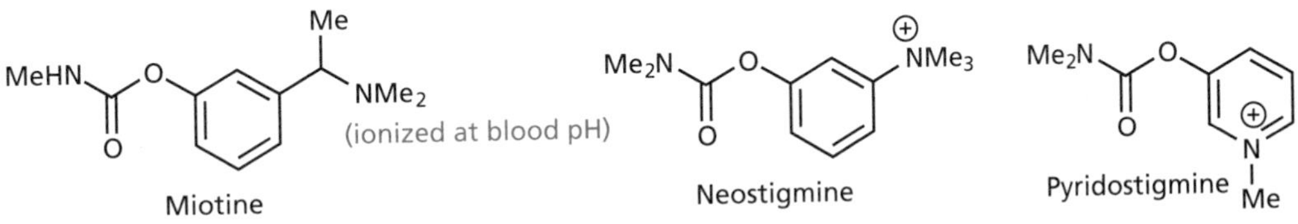

Figure 19.50 Analogues of physostigmine.

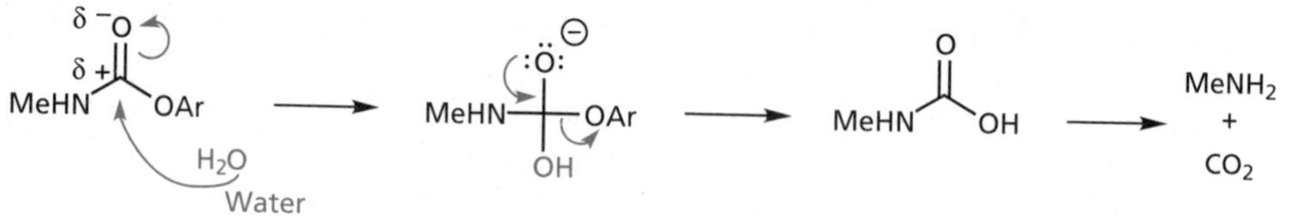

Figure 19.51 Mechanism 1.

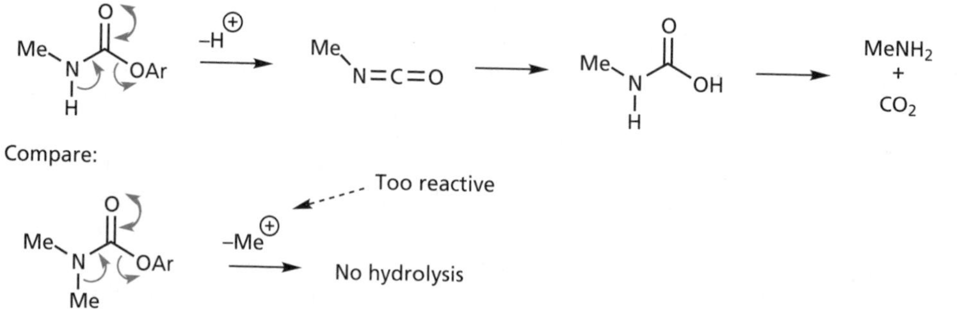

Figure 19.52 Mechanism 2.

that the molecule can take up. This is an advantage if the active conformation is retained, since the molecule is more likely to be in the active conformation when it approaches the active site.

Both neostigmine and pyridostigmine are in use today. They are given intravenously to reverse the actions of neuromuscular blockers, or used orally in the treatment of myasthenia gravis. Pyridostigmine was one of the drugs used in the chemical cocktail provided to allied troops in Iraq during Operation Desert Shield. The agent was present to help protect against possible exposure to organophosphate nerve gases.

19.17.2 Organophosphorus compounds

The potential of organophosphorus agents as nerve gases was first recognized by German scientists in the 1920s and 30s, and research was carried out to investigate their potential as weapons of war. When the Second World War broke out, governments in the UK, USA, Sweden, and Russia recognized the danger of Germany perfecting these weapons and started their own research efforts during the 1940s. In the UK, this was carried out at the Porton Down Defence Centre. Fortunately these agents were never used, but researchers in different countries continued work to find suitable antidotes which could protect troops from a possible attack. It has not been proved whether the organophosphate nerve gases have ever been used in combat, but many believe that they were part of the chemical weapons arsenal that was used against the Kurds by the Iraqi government. It has also been proposed that **sarin** may have been released when Iraqi chemical plants and ammunition dumps were bombed during 1990–91, and that this might be a possible cause of the mystery illness that afflicted many of the veterans of that war—Gulf War syndrome. Bosnians, Serbs, and Croats have also been accused of using nerve gases during the break up of Yugoslavia in the 1990s. Certainly, nerve gases have been used by terrorist groups: the most notorious example was the release of sarin in the Tokyo subway during 1996.

The organophosphate nerve gases are examples of the weapons of mass destruction which several Western countries feared might be used by Iraq on its neighbours, or supplied to extremist groups. The invasion of Iraq in 2003 was designed to combat this threat, but subsequent searches failed to reveal any such weapons.

It would be wrong to give the impression that the only use for organophosphates is as weapons of war and terror. They are also extremely important insecticides, used in agriculture and animal husbandry, and have found a variety of uses in medicine. We shall consider these aspects in the following sections.

19.17.2.1 Nerve gases

The nerve gases **dyflos** and **sarin** (Fig. 19.53) were discovered and perfected long before their mode of action was known. Both agents inhibit acetyl-cholinesterase by irreversibly phosphorylating the serine residue at the active site (Fig. 19.54).

The early part of the mechanism is similar to the normal mechanism, but the phosphorylated adduct which is formed is extremely resistant to hydrolysis. Consequently, the enzyme is permanently inactivated. As acetylcholine cannot be hydrolysed, the cholinergic system is continually stimulated. This results in permanent contraction of skeletal muscle, leading to death.

19.17.2.2 Medicines

Once the mechanism of action of nerve agents was discovered, compounds such as **ecothiopate**

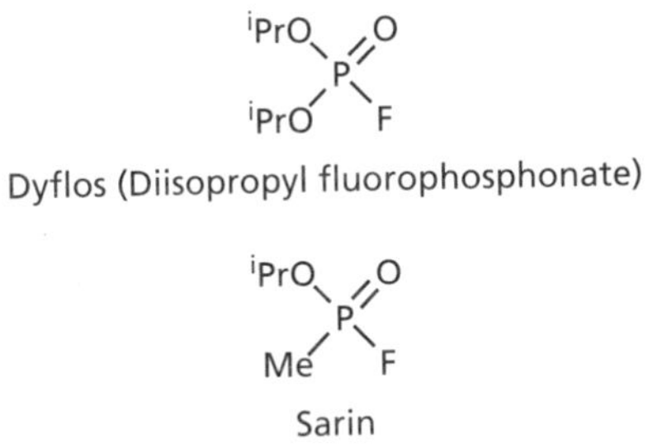

Figure 19.53 Nerve gases.

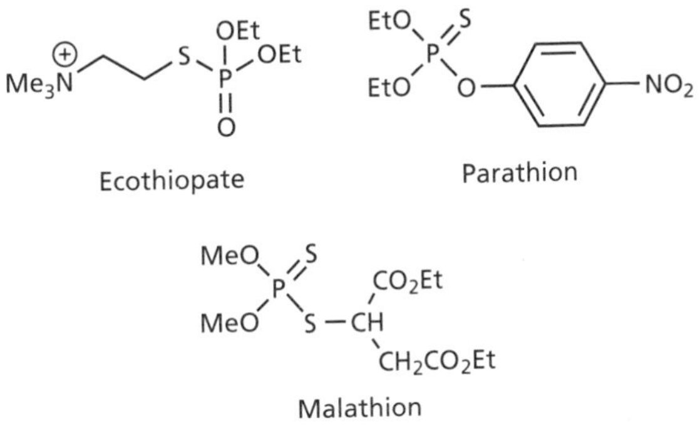

Figure 19.55 Medicines and insecticides.

(Fig. 19.55) were designed to fit the active site more effectively by including a quaternary amine to bind with the anionic region. This meant that lower doses would be more effective. Ecothiopate is used medicinally in the form of eye drops for the treatment of glaucoma, and has advantages over dyflos which has also been used in this way. Unlike dyflos, ecothiopate slowly hydrolyses from the enzyme over a matter of days.

19.17.2.3 Insecticides

The insecticides parathion and malathion (Fig. 19.56) are good examples of how a detailed knowledge of biosynthetic pathways can be put to good use. Parathion and malathion are relatively non-toxic compared to nerve gases, since the P=S double bond prevents these molecules from inhibiting acetyl-cholinesterase enzymes. The equivalent compounds containing a P=O double bond, on the other hand, are lethal compounds.

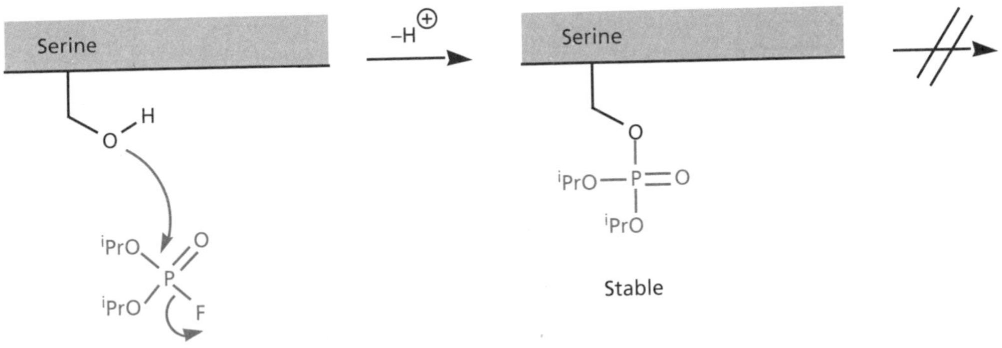

Figure 19.54 Action of dyflos.

Fortunately, there are no metabolic pathways in mammals which can convert the P=S double bond to a P=O double bond. Such a pathway does exist in insects, however, and in these species parathion and malathion act as prodrugs. They are metabolized by oxidative desulfurization to give the active anticholinesterases, which irreversibly bind to the insects' acetylcholinesterase enzymes and lead to death. In mammals, the same compounds are metabolized in a different way to give inactive compounds which are then excreted (Fig. 19.56). Despite this, organophosphate insecticides are not totally safe, and prolonged exposure to them causes serious side effects if they are not handled with care. Parathion has high lipid solubility and is easily absorbed through mucous membranes, and can also be absorbed through the skin.

Preparations of malathion are used medicinally for the treatment of head lice, crab lice, and scabies, but should not be used too frequently or over prolonged periods.

19.18 Pralidoxime: an organophosphate antidote

Pralidoxime (Fig. 19.57) represents one of the early examples of rational drug design. It is an antidote to organophosphate poisoning if given quickly enough

and was designed as such. A drug acting as an antidote to organophosphate poisoning has to displace the organophosphate moiety from serine. This requires hydrolysis of the phosphate-serine bond, but this is a strong bond and not easily broken. Therefore, a stronger nucleophile than water is required.

The literature revealed that phosphates can be hydrolysed with hydroxylamine (Fig. 19.58). This proved too toxic a compound to be used on humans, so the next stage was to design an equally reactive nucleophilic group which would specifically target the acetylcholinesterase enzyme. If such a compound could be designed, then there was less chance of the antidote taking part in toxic side-reactions.

The designers' job was made easier by the knowledge that the organophosphate group does not fill the active site, and the anionic binding site is vacant. The obvious thing to do was to find a suitable group to bind to this anionic centre and attach a hydroxylamine moiety to it. Once positioned in the active site, the hydroxylamine group could react with the phosphate ester (Fig. 19.57).

Pralidoxime was the result. The positive charge is provided by a methylated pyridine ring and the nucleophilic side-group is attached to the *ortho* position, since it was calculated that this would place the nucleophilic hydroxyl group in exactly the correct position to react with the phosphate ester. The results were spectacular, with pralidoxime showing a potency as an antidote 10^6 times greater than hydroxylamine.

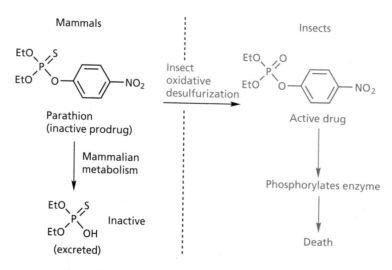

Figure 19.56 Metabolism of insecticides in mammals and insects.

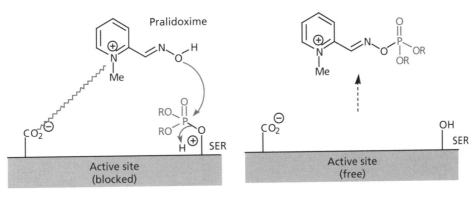

Figure 19.57 Pralidoxime as an antidote for organophosphate poisoning.

Figure 19.58 Hydrolysis of phosphates.

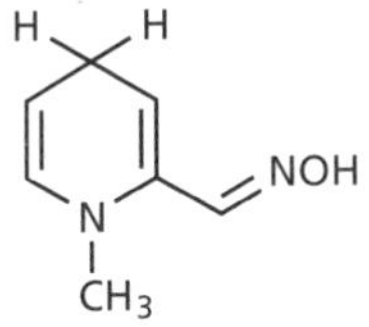

Figure 19.59 ProPAM.

Because pralidoxime has a quaternary nitrogen, it is fully charged and cannot pass through the blood–brain barrier into the CNS. This means that the antidote cannot work on any enzymes that have been inhibited in the brain. Pro-2-PAM (Fig. 19.59) is a prodrug of pralidoxime which avoids this problem. As a tertiary amine it can pass through the blood–brain barrier, and is oxidized to pralidoxime once it has entered the CNS.

19.19 Anticholinesterases as 'smart drugs'

Acetylcholine is an important neurotransmitter in the CNS as well as in the peripheral nervous system. In recent years, it has been proposed that the memory loss, intellectual deterioration, and personality changes associated with Alzheimer's disease may in part be due to loss of cholinergic nerves in the brain. Although Alzheimer's disease is primarily a disease of the elderly, it can strike victims as young as 30. The disease destroys neurons in the brain and is associated with the appearance of plaques and tangles of nerve fibres.

Research has been carried out into the use of anticholinesterases for the treatment of Alzheimer's disease—the so-called 'smart drugs'. There is no evidence that these compounds assist general memory improvement and so they are not a student's answer to exam cramming! The treatment does not offer a cure for Alzheimer's disease either, but it can alleviate the symptoms by allowing the brain to make more use of the cholinergic receptors still surviving. Unlike anticholinesterases acting in the periphery, 'smart drugs' have to cross the blood–brain barrier, and so structures containing quaternary nitrogen atoms are not suitable. Tests with physostigmine were carried out in 1979 but the compound was not ideal as it does not enter the brain sufficiently well and shows short-lived, non-selective inhibition. The first drug to be approved for the treatment of Alzheimer's was **tacrine** (Fig. 19.60) in 1993. However, this is an extremely toxic drug. Other agents which have been introduced since include **donepezil** in 1997, **rivastigmine** in 2000,

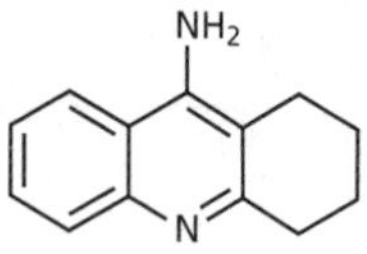

Tacrine (Cognex, Parke-Davis)

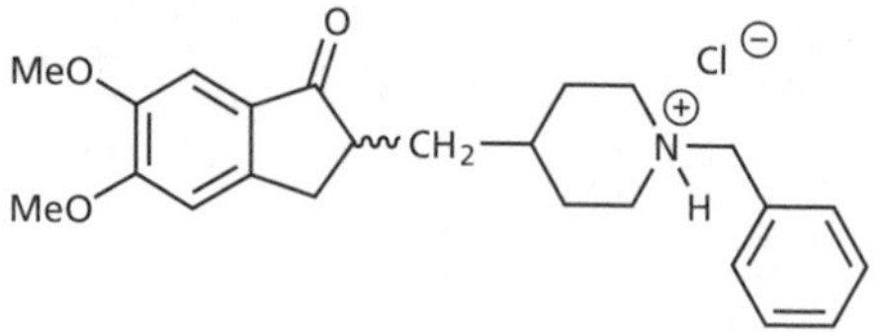

Donepezil (Aricept, Eisai)

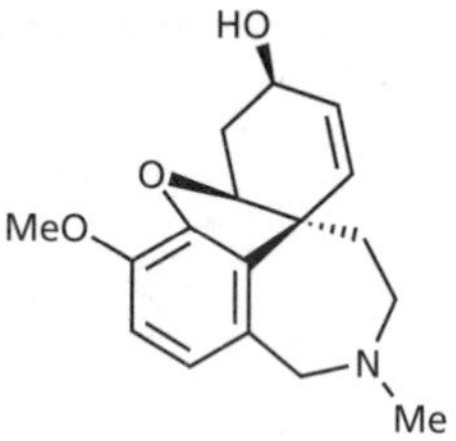

Galantamine (Reminyl, Shire)

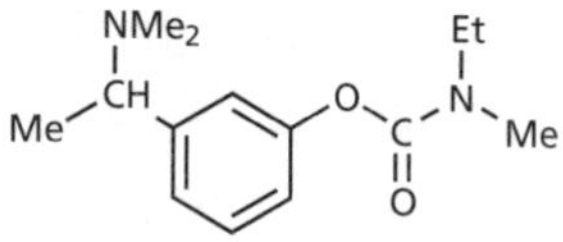

Rivastigmine (Exelon, Novartis)

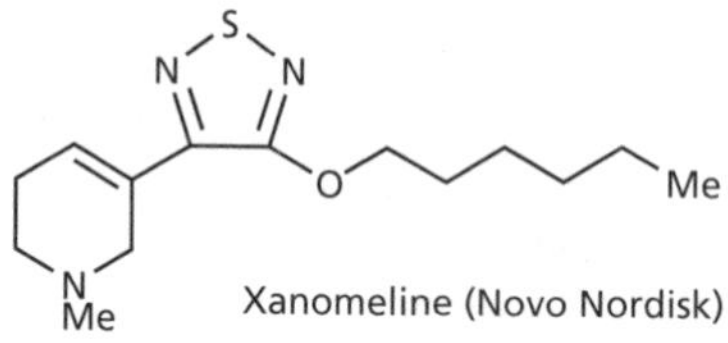

Xanomeline (Novo Nordisk)

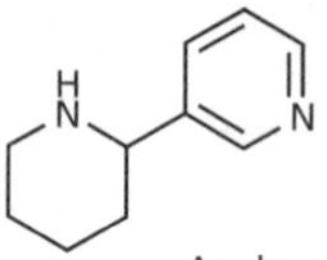

Anabaseine

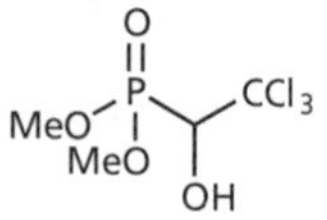

Metrifonate (Bayer)

Figure 19.60 'Smart drugs'.

and **galantamine** (obtained from daffodils or snow-drop bulbs) in 2001 (see Box 19.2). Rivastigmine (an analogue of physostigmine) was the first drug to be approved in all countries of the European Union. It shows selectivity for the brain and has beneficial effects on cognition, memory, concentration, and functional abilities such as day-to-day tasks or hobbies. The drug has a short half-life, reducing the risk of accumulation or drug–drug interactions. **Metrifonate** (an organophosphate) and **anabaseine** (from ants and marine worms) have also been tested for the treatment of Alzheimer's.

The anticholinesterase drugs have been shown to be beneficial in the early stages of Alzheimer's disease, but are of less benefit when the disease has become advanced. One disadvantage with the long-term use of these agents is the fact that they increase acetylcholine levels all round the body and not just in the brain, and this leads to gastrointestinal side effects. Another problem is that the increased acetylcholine levels result in an increased activation of presynaptic cholinergic receptors which act as a feedback control to lower the amounts of acetylcholine released. As a result, there has been research into finding selective cholinergic agonists that could be used to treat the symptoms of the disease.

BOX 19.2 MOSSES PLAY IT SMART

An extract from the club moss *Huperzia serrata* has been used for centuries in Chinese herbal medicine to treat ailments varying from confusion in Alzheimer's disease to schizophrenia. The extract contains a novel alkaloid called **huperzine A**, which acts as an anticholinesterase. Binding is very specific and so the drug can be used in small doses, thus minimizing the risk of side effects. Huperzine A has been undergoing clinical trials in China and has been shown to have memory-enhancing effects.

A synthetic route to the natural product has been worked out, which has allowed the synthesis of different analogues. None of these is as active as the natural product. The tricyclic ring system seems to be necessary for good activity, ruling out the possibility of significant simplification. All the functional groups in the molecule are also required for good activity.

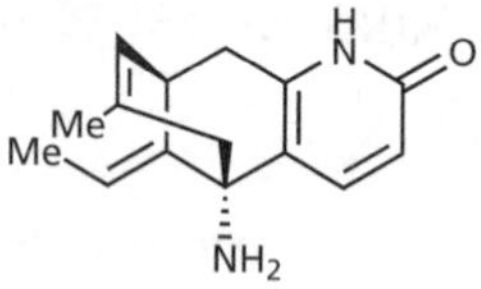

Huperzine A.

KEY POINTS

- Anticholinesterases inhibit the enzyme acetylcholinesterase and have the same clinical effects as cholinergic agonists.

- The active site for acetylcholinesterase is similar to the binding site for the cholinergic receptor, but also includes the amino acids histidine and serine.

- Histidine acts as an acid–base catalyst while serine acts as a nucleophile during the hydrolytic mechanism.

- The carbamate inhibitors are derived from the lead compound physostigmine. They react with acetylcholinesterase to produce a carbamoyl-bound intermediate which is stable and slow to hydrolyse.

- Organophosphorus agents have been used as nerve gases, medicines, and insecticides. They irreversibly phosphorylate serine in the active site.

- Pralidoxime was designed as an antidote for organophosphate poisoning. It can bind to the active site of phosphorylated enzymes and displace the phosphate group from serine.

- Anticholinesterases have been used as smart drugs in the treatment of Alzheimer's disease. They have to cross the blood–brain barrier and cannot be permanently charged.

QUESTIONS

1. Based on the binding site described in Fig. 19.12, suggest whether the following structures are likely to act as agonists or not.

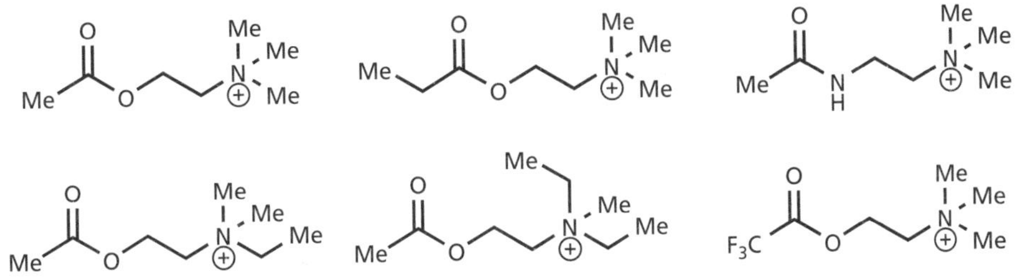

2. Suggest a mechanism by which atropine is racemized.

3. A fine balance of binding interactions is required of a neurotransmitter. What do you think is meant by this and what consequences does it have for drug design?

4. Suggest how the binding interactions holding acetylcholine to the active site of acetylcholinesterase might aid in the hydrolysis of acetylcholine.

5. Explain how the following diester could act as a prodrug for pilocarpine.

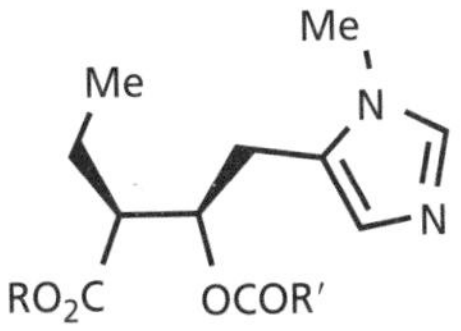

Diester prodrug for pilocarpine

6. What advantage do you think the pilocarpine analogue shown below might have over pilocarpine itself, and why?

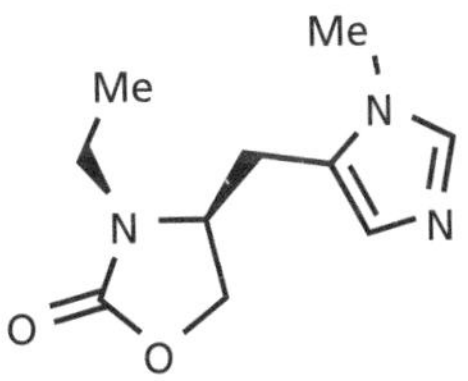

Pilocarpine analogue

7. Arecoline (shown below) has been described as a cyclic 'reverse ester' bioisostere of acetylcholine. What is meant

by this, and what similarity is there if any between arecoline and acetylcholine?

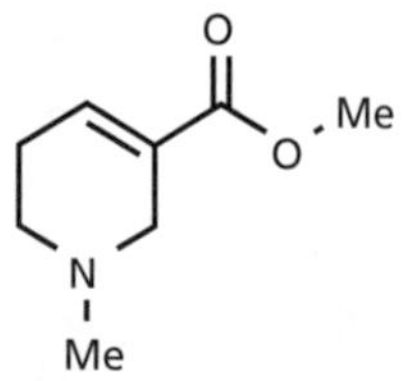

Arecoline

8. Arecoline has a very short duration of action. Why do you think this is?

9. Suggest analogues of arecoline that might have better properties, such as a longer duration of action.

FURTHER READING

Hardman, J. G. *et al.* (eds.) (1996) Anticholinesterase agents. In: *The pharmacological basis of therapeutics.* McGraw-Hill, New York, pp. 161–176.

Quinn, D. M. (1987) Acetylcholinesterase. *Chemical Reviews,* **87**, 955–975.

Roberts, S. M. and Price, B. J. (eds.) (1985) Atracurium design and function. Chapter 8 in: *Medicinal chemistry—the role of organic research in drug research.* Academic Press, London.

Teague, S. J. (2003) Implications of protein flexibility for drug discovery. *Nature Reviews Drug Discovery,* **2**, 527–541.

Titles for general further reading are listed on p. 711.

20

The adrenergic nervous system

20.1 The adrenergic system

20.1.1 Peripheral nervous system

In Chapter 19, we studied the cholinergic system and the important role it plays in the peripheral nervous system. **Acetylcholine** is the crucial neurotransmitter in the cholinergic system and has specific actions at various synapses and tissues. The other important player in the peripheral nervous system (sections 19.1–19.2) is the adrenergic system, which makes use of the chemical messengers **adrenaline** and **noradrenaline**. Noradrenaline (also called norepinephrine) is the neurotransmitter released by the sympathetic nerves which feed smooth muscle and cardiac muscle, whereas adrenaline (epinephrine) is a hormone released along with noradrenaline from the adrenal medulla, and circulates in the blood supply in order to reach adrenergic receptors.

The action of noradrenaline at various tissues is the opposite to that of acetylcholine, which means that tissues are under a dual control. For example, if noradrenaline has a stimulant activity at a specific tissue, acetylcholine has an inhibitory activity at that same tissue. Both the cholinergic and adrenergic systems have a 'background' activity, so the situation is analogous to driving a car with one foot on the brake and one foot on the accelerator. The overall effect on the tissue depends on which effect is predominant.

The adrenergic nervous system has a component that the cholinergic system does not have—the facility to release adrenaline during times of danger or stress. This is known as the '**fight or flight**' response.

Adrenaline activates adrenergic receptors around the body, preparing the body for immediate physical action, whether that be to fight the perceived danger or to flee from it. This means that the organs required for physical activity are activated, while those that are not important are suppressed. For example, adrenaline stimulates the heart and dilates the blood vessels to muscles so that the latter are supplied with sufficient blood for physical activity. At the same time, smooth muscle activity in the gastrointestinal tract is suppressed, as digestion is not an immediate priority. This 'fight or flight' response is clearly an evolutionary advantage and stood early humans in good stead when faced with an unexpected encounter with a grumpy old bear. Nowadays, it is unlikely that you will meet a grizzly bear on your way to the supermarket, but the 'fight or flight' response is still functional when you are faced with modern dangers such as crazy drivers. It also functions in any situation of stress such as an imminent exam, important football game, or public performance. In general, the effects of noradrenaline are the same as those of adrenaline although noradrenaline constricts blood vessels to skeletal muscle rather than dilates them.

20.1.2 The central nervous system

There are also adrenergic receptors in the central nervous system (CNS), and noradrenaline is important in many functions of the CNS including sleep, emotion, temperature regulation, and appetite. However, the emphasis in this chapter is on the peripheral role of adrenergic agents.

20.2 Adrenergic receptors

20.2.1 Types of adrenergic receptor

In Chapter 19, we saw that there are two types of cholinergic receptor, with subtypes of each. The same holds true for adrenergic receptors. The two main types of adrenergic receptor are called the α- and β-adrenoceptors. Both the α- and the β-adrenoceptors are G-protein-coupled receptors (section 6.3), but differ in the type of G-protein with which they couple (G_o for α-adrenoceptors; G_s for β-adrenoceptors).

For each type of receptor, there are various receptor subtypes with slightly different structures. The α-adrenoceptor consists of α_1 and α_2 subtypes, which differ in structure and also differ in the type of secondary message which is produced. The α_1-receptors produce **inositol triphosphate** (IP_3) and **diacylglycerol** (DG) as secondary messengers (sections 6.4 and 6.6), whereas the α_2-receptors inhibit the production of the secondary messenger **cAMP** (sections 6.4–6.5). The β-adrenoceptor consists of β_1, β_2, and β_3 subtypes, all of which activate the formation of cAMP. To complicate matters slightly further, both the α_1- and α_2-adrenoceptors have further sub-categories (α_{1A}, α_{1B}, α_{1C}, α_{2A}, α_{2B}, α_{2C}).

All of these adrenergic receptor types and subtypes are 'switched on' by adrenaline and noradrenaline, but the fact that they have slightly different structures means that it should be possible to design drugs which will be selective agonists and switch on only a few or even just one of them. This is crucial in designing drugs that have minimal side effects and act at specific organs in the body, for, as we shall see, the various adreno-ceptors are not evenly distributed around the body. By the same token, it should be possible to design selective antagonists with minimal side effects that switch off particular types and subtypes of adrenoceptor.

20.2.2 Distribution of receptors

The various adrenoceptor types and subtypes are not uniformly distributed around the body and certain tissues contain more of one type of adrenoceptor than another. Table 20.1 describes various tissues, the types of adrenoceptor which predominate in these tissues, and the effect of activating these receptors.

A few points are worth highlighting here:

- Activation of α-receptors generally contracts smooth muscle (except in the gut), whereas activation of β-receptors generally relaxes smooth muscle. This latter effect is mediated through the most common of the β-adrenoceptors—the β_2-receptor. In the heart, the β_1-adrenoceptors predominate and activation results in contraction of muscle.

Table 20.1 Distribution and effects of adrenoceptors in different parts of the body

Organ or tissue	Predominant adrenoceptors	Effect of activation	Physiological effect
Heart muscle	β_1	Muscle contraction	Increased heart rate and force
Bronchial smooth muscle	α_1	Smooth muscle contraction	Closes airways
	β_2	Smooth muscle relaxation	Dilates and opens airways
Arteriole smooth muscle (not supplying muscles)	α	Smooth muscle contraction	Constricts arterioles and increases blood pressure (hypertension)
Arteriole smooth muscle (supplying muscle)	β_2	Smooth muscle relaxation	Dilates arterioles and increases blood supply to muscles
Veins	α	Smooth muscle contraction	Constricts veins and increases blood pressure (hypertension)
	β_2	Smooth muscle relaxation	Dilates veins and decreases blood pressure (hypotension)
Liver	α_1 and β_2	Activates enzymes which metabolize glycogen and deactivates enzymes which synthesize glycogen	Breakdown of glycogen to produce glucose
GI tract smooth muscle	α_1, α_2, and β_2	Relaxation	Shuts down digestion
Kidney	β_2	Increases renin secretion	Increases blood pressure
Fat cells	β_3	Activates enzymes	Fat breakdown

- Different types of adrenoceptor explain why adrenaline can have different effects at different parts of the body. For example, the blood vessels supplying skeletal muscle have mainly β_2-adrenoceptors and are dilated by adrenaline, whereas the blood vessels elsewhere have mainly α-adrenoceptors and are constricted by adrenaline. Since more blood vessels are constricted than are dilated in the system, the overall effect of adrenaline is to increase the blood pressure, yet at the same time provide sufficient blood for the muscles in the 'fight or flight' response.

20.2.3 Clinical effects

The main clinical use for adrenergic agonists is in the treatment of asthma. Activation of β_2-adrenoceptors causes the smooth muscles of the bronchi to relax, thus widening the airways. Agonists acting selectively on α_1-adrenoceptors cause vasoconstriction and can be used alongside local anaesthetics in dentistry to localize and prolong the effect of the anaesthetic at the site of injection. They are also used as nasal decongestants. Selective α_2-agonists are used in the treatment of glaucoma, hypertension, and pain.

The main uses for adrenergic antagonists are in treating angina and hypertension. Agents which block the α-receptors act on the α-receptors of blood vessels, causing relaxation of smooth muscle, dilatation of the blood vessels, and a drop in blood pressure. Selective α_1-antagonists are now preferred for the treatment of hypertension and are also being investigated as potential agents for the treatment of benign prostatic hyperplasia. Selective α_2-antagonists are being studied for the treatment of depression. Agents which block β_1 receptors act on β_1 receptors in the heart, slowing down the heart rate and reducing the force of contractions. β-Blockers also have a range of other effects in other parts of the body which combine to lower blood pressure.

20.3 Endogenous agonists for the adrenergic receptors

The term endogenous refers to any chemical which is naturally present in the body. As far as the adrenergic system is concerned, the body's endogenous chemical messengers are the neurotransmitter noradrenaline

and the hormone adrenaline. Both act as agonists and switch on adrenoceptors. They belong to a group of compounds called the **catecholamines**—so called because they have an alkylamine chain linked to a **catechol** ring (the 1,2-benzenediol ring) (Fig. 20.1).

20.4 Biosynthesis of catecholamines

The biosynthesis of noradrenaline and adrenaline starts from the amino acid L-**tyrosine** (Fig. 20.2). The enzyme **tyrosine hydroxylase** catalyses the introduction of a second phenol group to form **levodopa**, which is then decarboxylated to give **dopamine**—an important neurotransmitter in its own right. Dopamine is then hydroxylated to **noradrenaline**, which is the end product in adrenergic nerves. In the adrenal medulla, however, noradrenaline is N-methylated to form **adrenaline**. The biosynthesis of the catecholamines is controlled by regulation of tyrosine hydroxylase—the first enzyme in the pathway. This enzyme is inhibited by noradrenaline, the end product of biosynthesis, thus allowing self-regulation of catecholamine synthesis and control of catecholamine levels.

20.5 Metabolism of catecholamines

Metabolism of catecholamines in the periphery takes place within cells and involves two enzymes—**monoamine oxidase** (MAO) and **catechol O-methyltransferase**

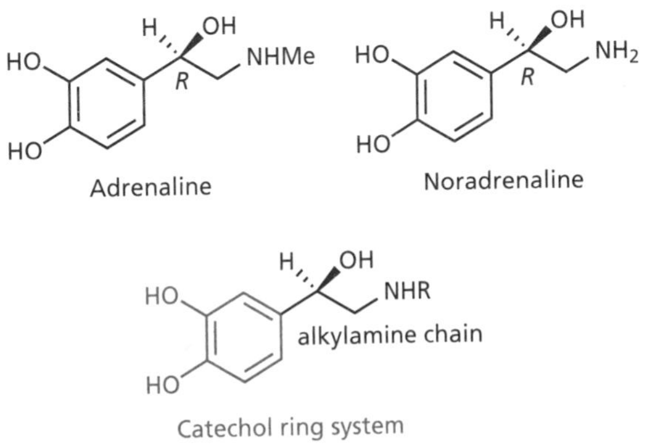

Adrenaline

Noradrenaline

Catechol ring system

Figure 20.1 Adrenergic transmitters.

(COMT). MAO converts catecholamines to their corresponding aldehydes. These compounds are inactive as adrenergic agents and undergo further metabolism as shown in Fig. 20.3 for noradrenaline. The final carboxylic acid is polar and excreted in the urine.

An alternative metabolic route is possible, which results in the same product. This time the enzyme COMT catalyses the methylation of one of the phenolic groups of the catecholamine. The methylated product is oxidized by MAO, then converted to the final carboxylic acid and excreted (Fig. 20.4).

Metabolism in the CNS is slightly different, but still involves MAO and COMT as the initial enzymes.

20.6 Neurotransmission

20.6.1 The neurotransmission process

The mechanism of neurotransmission is shown in Fig. 20.5 and applies to adrenergic nerves innervating smooth or cardiac muscle, as well as synaptic connections within the CNS.

Noradrenaline is biosynthesized in a presynaptic nerve, then stored in membrane-bound vesicles. When a nerve impulse arrives at the nerve terminal it stimulates the opening of calcium ion channels, and this promotes the fusion of the vesicles with the cell membrane to release noradrenaline. The neurotransmitter then diffuses to adrenergic receptors

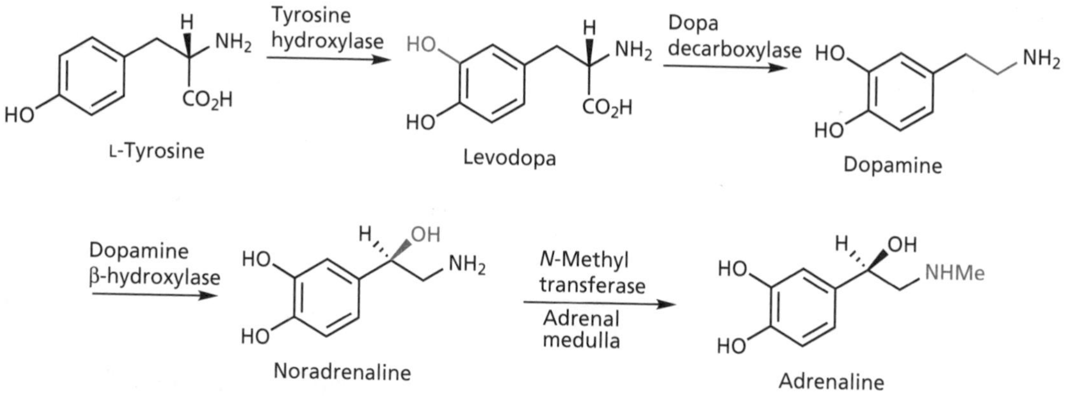

Figure 20.2 Biosynthesis of noradrenaline and adrenaline.

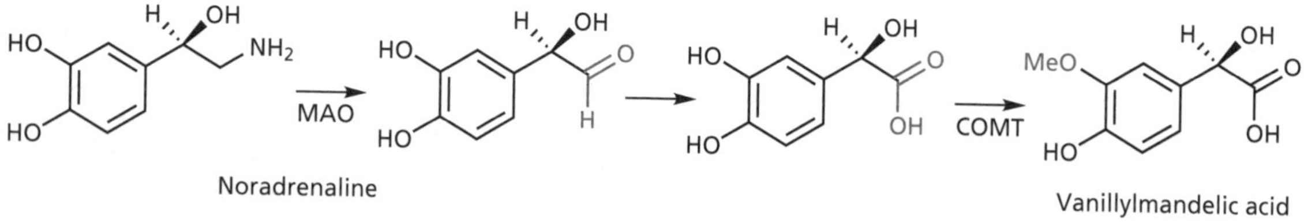

Figure 20.3 Metabolism with MAO and COMT.

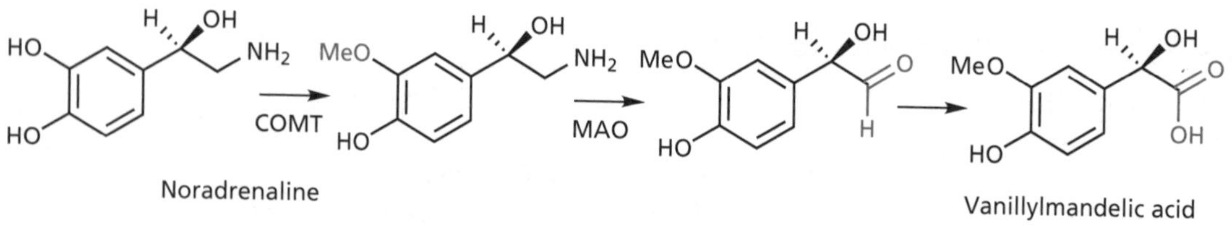

Figure 20.4 Metabolism with COMT and MAO.

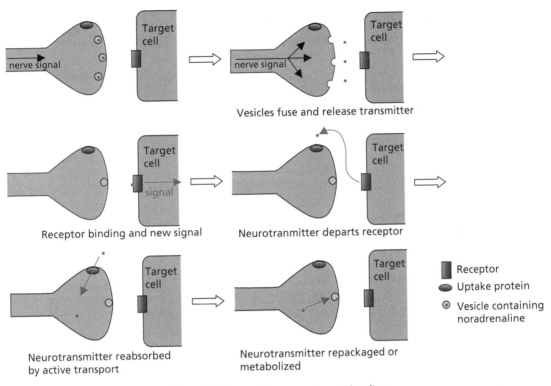

Figure 20.5 Transmission process for noradrenaline.

on the target cell, where it binds and activates the receptor—leading to the signalling process which will eventually result in a cellular response. Once the message has been received, noradrenaline departs and is taken back into the nerve terminal through active transport by means of a carrier protein which is specific for noradrenaline. Once in the cell, noradrenaline is repackaged into the vesicles. Some of the noradrenaline is metabolized before it is repackaged, but this is balanced out by noradrenaline biosynthesis.

20.6.2 Co-transmitters

The process of adrenergic neurotransmission is actually more complex than that illustrated in Fig. 20.5. For example, noradrenaline is not the only neurotransmitter released during the process. **ATP** and a protein called **chromogranin A** are released from the vesicles along with noradrenaline, and act as co-transmitters. They interact with their own specific receptors on the target cell and allow a certain

variation in the speed and type of message which the target cell receives. For example, ATP leads to a fast response in smooth muscle contraction.

20.6.3 Presynaptic receptors and control

A further feature of the neurotransmission process not shown in Fig. 20.5 is the existence of presynaptic receptors which have a controlling effect on noradrenaline release (Fig. 20.6). There are a variety of these receptors, each of which responds to a specific chemical messenger. For example, there is an adrenergic receptor (the α_2-adrenoceptor) which interacts with released noradrenaline and has an inhibitory effect on further release of noradrenaline. Thus, noradrenaline acts to control its own release by a negative feedback system.

There are receptors specific for **prostaglandins** released from the target cell. For example, the prostaglandin **PGE_2** appears to inhibit transmission, whereas **$PGF_{2\alpha}$** appears to facilitate it. Thus, the target cell itself can have some influence on the adrenergic signals coming to it.

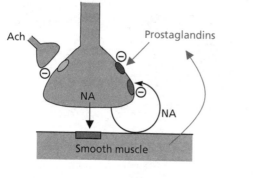

Cholinergic receptor
Presynaptic adrenergic receptor
Prostaglandin receptor
Postsynaptic adrenergic receptor
⊖ Activation of receptor reduces noradrenaline release
NA Noradrenaline

Figure 20.6 Presynaptic receptors.

There are presynaptic muscarinic receptors that are specific for **acetylcholine** and serve to inhibit release of noradrenaline. These receptors respond to side branches of the cholinergic nervous system which synapse on to the adrenergic nerve. This means that when the cholinergic system is active, it sends signals along its side branches to inhibit adrenergic transmission. Therefore, as the cholinergic activity to a particular tissue increases, the adrenergic activity decreases, both of which enhance the overall cholinergic effect (cf. section 19.5.2).

20.7 Drug targets

Having studied the nerve transmission process, it is now possible to identify several potential drug targets which will affect the process (Fig. 20.7):

1. biosynthetic enzymes
2. vesicle carriers
3. exocytosis and release of transmitter
4. adrenergic receptors
5. uptake by carrier protein
6. metabolic enzymes
7. presynaptic adrenergic receptors.

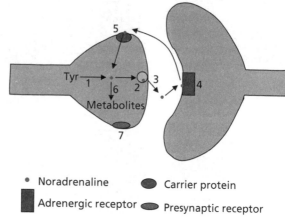

• Noradrenaline   ● Carrier protein
■ Adrenergic receptor   ◗ Presynaptic receptor

Figure 20.7 Drug targets affecting noradrenaline transmission.

In the next section we concentrate on the adrenergic receptor. In later sections, we will consider some of the other possible drug targets.

KEY POINTS

• The neurotransmitter involved in the adrenergic nervous system is noradrenaline. Adrenaline is a hormone which is released by the adrenal medulla at times of stress and activates adrenergic receptors.

• The sympathetic nerves innervating smooth muscle and cardiac muscle release noradrenaline.

• Adrenergic receptors are G-protein-coupled receptors. There are two main types: the α- and β-adrenoceptors. There are various subtypes of each.

• The different types and subtypes of adrenoceptor predominate in different tissues. Drugs which show receptor selectivity also show tissue selectivity.

• The major use of adrenergic agonists is in the treatment of asthma. The major use of adrenergic antagonists is in cardiovascular medicine.

• Adrenaline, noradrenaline, and dopamine are catecholamines.

• The biosynthesis of catecholamines starts from tyrosine and involves levodopa as an intermediate.

• Catecholamines are metabolized by monoamine oxidase and catechol O-methyltransferase.

• Noradrenaline is synthesized in presynaptic nerves, and packaged in vesicles prior to release. Once released, it activates receptors on target cells. It is then is taken up into presynaptic nerves by a carrier protein and repacked into vesicles. A certain percentage of noradrenaline is metabolized.

- Adrenergic receptors are the main targets for adrenergic drugs.

20.8 The adrenergic binding site

The adrenergic receptors are G-protein linked receptors which consist of seven transmembrane (TM) helices (section 6.3). Three of these helices (TM3, 5, and 6) are involved in the binding site, illustrated for the β-adrenoceptor in Fig. 20.8. Mutagenesis studies have indicated the importance of an aspartic acid residue (Asp-113), a phenylalanine residue (Phe-290) and two serine residues (Ser-207 and Ser-204). These groups can bind to adrenaline or noradrenaline as shown in the figure. The serine residues interact with the phenolic groups of the catecholamine via hydrogen bonding. The aromatic ring of Phe-290 interacts with the catechol ring by van der Waals interactions, while Asp-113 interacts with the protonated nitrogen of the catecholamine by ionic bonding. There is also a hydrogen bonding interaction between a hydrogen bonding group of the receptor and the alcohol function of the catecholamine.

20.9 Structure–activity relationships

20.9.1 Important binding groups on catecholamines

Support for the above binding site interactions is provided by studies of structure–activity relationships (SAR) on catecholamines. These emphasize the importance of having the alcohol group, the intact catechol ring system with both phenolic groups unsubstituted, and the ionized amine (Fig. 20.9).

Some of the evidence supporting these conclusions is as follows:

- **The alcohol group:** The *R*-enantiomer of noradrenaline is more active than the *S*-enantiomer, indicating that the secondary alcohol is involved in a hydrogen bonding interaction. Compounds lacking the hydroxyl group (e.g. dopamine) have a greatly reduced interaction. Some of the activity is retained, indicating that the alcohol group is important but not essential.

- **The amine:** The amine is normally protonated and ionized at physiological pH. This is important, since replacing nitrogen with carbon results in a large drop in activity. Activity is also affected by the number of substituents on the nitrogen. Primary and secondary amines have good adrenergic activity, whereas tertiary amines and quaternary ammonium salts do not.

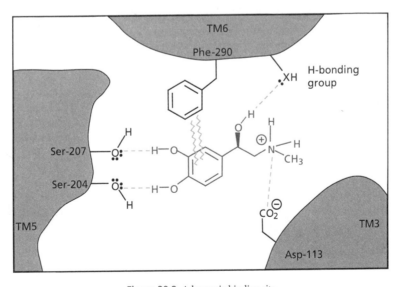

Figure 20.8 Adrenergic binding site.

- **The phenol substituents:** The phenol groups are important. For example, **tyramine** and **amphetamine** (Fig. 20.10) have no affinity for adrenoceptors. Having said that, the phenol groups can be replaced by other groups capable of interacting with the binding site by hydrogen bonding. This is particularly true for the *meta* phenol group which can be replaced by groups such as CH_2OH, CH_2CH_2OH, NH_2, NHMe, NHCOR, NMe_2, and $NHSO_2R$.

- **Alkyl substitution:** Alkyl substitution on the side chain linking the aromatic ring to the amine decreases activity at both α- and β-adrenergic receptors. This may be a steric effect which blocks hydrogen bonding to the alcohol, or which prevents the molecule adopting the active conformation.

powerful β-stimulant devoid of α-agonist activity. The presence of a bulky *N*-alkyl group, such as isopropyl or *t*-butyl, is particularly good for β-adrenoceptor activity. These results indicate that the β-adrenoceptor has a hydrophobic pocket into which a bulky alkyl group can fit, whereas the α-adrenoceptor does not (Fig. 20.12).

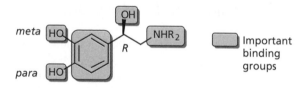

Figure 20.9 Important binding groups.

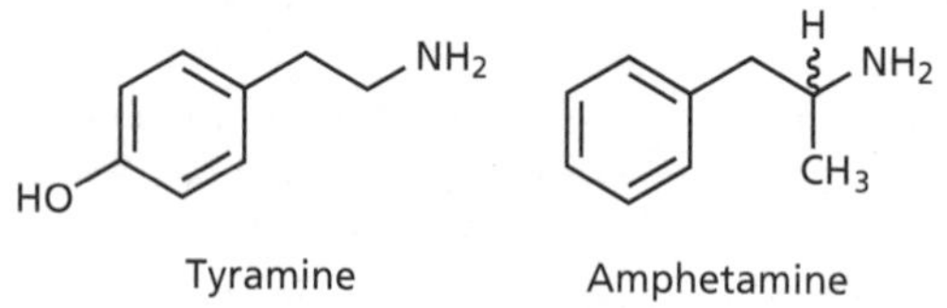

20.9.2 Selectivity for α- versus β-adrenoceptors

SAR studies demonstrate certain features which introduce a level of receptor selectivity between the α- and β-adrenoceptors:

- **N-Alkyl substitution:** It was discovered that adrenaline has the same potency for both types of adrenoceptor, whereas noradrenaline has a greater potency for α-adrenoceptors than for β-adrenoceptors. This indicates that an *N*-alkyl substituent has a role to play in receptor selectivity. Further work demonstrated that increasing the size of the *N*-alkyl substituent resulted in loss of potency at the α-receptor but an increase in potency at β-receptors. For example, the synthetic analogue **isoprenaline** (Fig. 20.11) is a

Tyramine   Amphetamine

Figure 20.10 Agents which have no affinity for the adrenergic receptor.

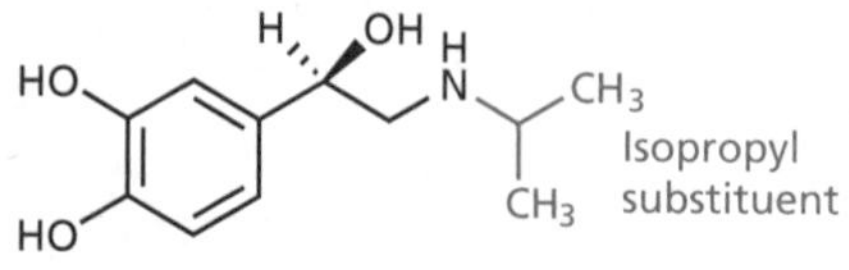

Figure 20.11 Isoprenaline.

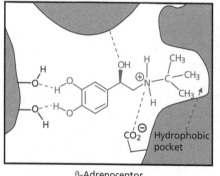

β-Adrenoceptor

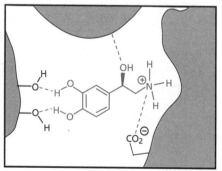

α-Adrenoceptor

Figure 20.12 Comparison of β- and α-adrenoceptor binding sites.

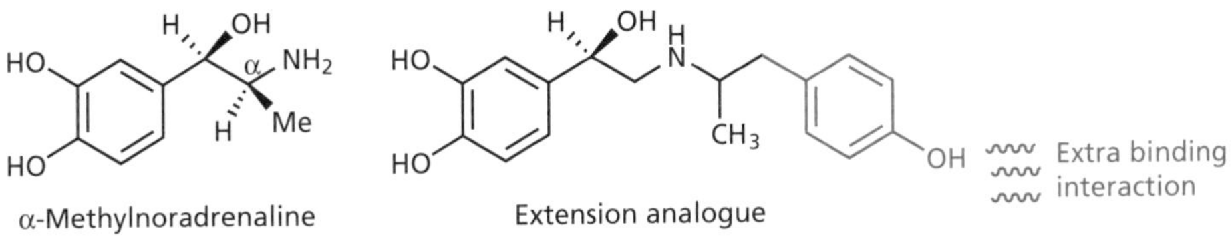

Figure 20.13 α-Methylnoradrenaline and extension analogue of noradrenaline.

- **Phenol groups:** The phenol groups seem particularly important for the β-receptors. If they are absent, activity drops more significantly for the β-receptors than for the α-receptor.

- **α-Methyl substitution:** Addition of an α-methyl group (e.g. α-methylnoradrenaline; Fig. 20.13) increases α_2-receptor selectivity.

- **Extension:** As mentioned above, isopropyl or *t*-butyl substituents on the amine nitrogen are particularly good for β-selectivity. Increasing the length of the alkyl chain offers no advantage, but if a polar functional group is placed at the end of the alkyl group, the situation changes. In particular, adding a phenol group to the end of a C_2 alkyl chain results in a dramatic rise in activity, demonstrating that an extra polar binding region has been accessed which can take part in hydrogen bonding. For example, the activity of the extension analogue shown in Fig. 20.13 is increased by a factor of 800.

Figure 20.14 Ephedrine.

medicine for many years. There are two asymmetric centres, and ephedrine exists as a racemate of the R, S and S, R stereoisomers. It activates both α- and β-adrenoceptors and has been used extensively in non-prescription preparations as a bronchodilator. It has also been used as a vasopressor and cardiac stimulant. Since it lacks the phenolic groups of adrenaline, it is not susceptible to metabolism by catechol O-methyltransferase. Furthermore, it can enter the brain more easily as it is more lipophilic and can act there as a stimulant. Ephedrine is the active constituent of herbal remedies that contain the dried plant material *ma-huang*.

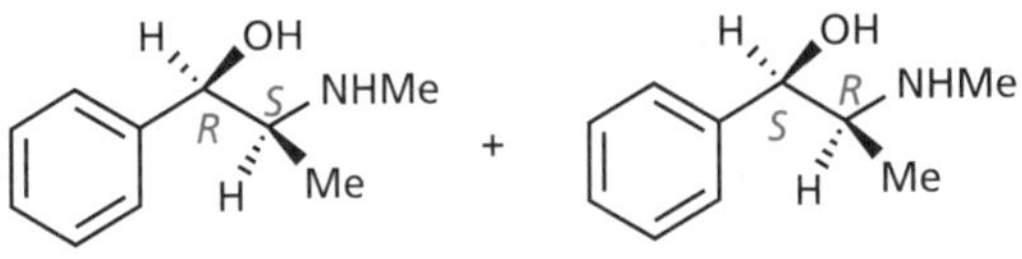

20.10 Adrenergic agonists

20.10.1 General adrenergic agonists

Adrenaline itself is an obvious agonist for the overall adrenergic system, and it is frequently used in emergency situations such as cardiac arrest or anaphylactic reactions caused by hypersensitivity to some foreign chemical such as a bee sting or penicillin. Adrenaline is also administered with local anaesthetics to constrict blood vessels and to prolong the local anaesthetic activity at the site of injection. Apart from these conditions, however, it is preferable to have agonists which are selective for specific adrenoceptors.

Ephedrine (Fig 20.14) is a natural product present in various plants which have been used in folk

20.10.2 α_1, α_2, β_1, and β_3-agonists

In general, there is limited scope for agonists at these receptors, although there is potential for antiobesity drugs which act on the β_3 receptor. The β_1-agonist **dobutamine** (Fig. 20.15) is used to treat cardiogenic shock. Agonists acting on the α-adrenoceptors are less useful, as these agents constrict blood vessels, raise blood pressure, and can cause cardiovascular problems. However, selective α_1 and α_2-agonists have found a number of uses as described in section 20.2.3. **Clonidine** is a selective α_2-agonist which is used for the treatment of hypertension. There is also strong evidence that it acts as an analgesic, especially if it is injected directly into the spinal cord. Selective α_1-agonists such as **oxymetazoline** and **xylometazoline** act as vasoconstrictors, and are widely used as topical medicines for the treatment of nasal congestion and bloodshot eyes.

20.10.3 β₂-Agonists and the treatment of asthma

The most useful adrenergic agonists in medicine today are the β₂-agonists. These can be used to relax smooth muscle in the uterus to delay premature labour, but they are more commonly used for the treatment of asthma. Activation of the β₂-adrenoceptor results in smooth muscle relaxation and since β₂-receptors predominate in bronchial smooth muscle, this leads to dilatation of the airways.

Adrenaline was one of the early compounds used to dilate the airways and it is still used today in cases of emergency. Unfortunately, adrenaline has a short duration of action and since it has no selectivity for β₂-receptors, it switches on all possible adrenergic receptors, leading to a whole range of side effects, particularly cardiovascular effects.

In order to increase the selectivity of action, adrenaline was replaced by **isoprenaline** (Fig. 20.11) as an antiasthmatic agent. Isoprenaline was taken by inhalation and was selective for β-receptors over

α-receptors because of its bulky *N*-alkyl substituent, but showed no selectivity between the different subtypes of β-receptors. Therefore, isoprenaline not only activated the β₂-receptors in the airways, it also activated the β₁ receptors of the heart, again leading to unwanted cardiovascular effects. The search was now on to find an agonist selective to β₂-receptors which could be inhaled, and which would also have a long duration of action. Further research demonstrated that selectivity between different types of β-receptors could be obtained by introducing alkyl substituents to the side chain linking the aromatic ring and the amine, and/or varying the alkyl substituents on the nitrogen. For example, **isoetharine** (Fig. 20.16) was shown to be selective for β₂-receptors, but was short lasting.

This short duration of action is because drugs such as isoetharine and adrenaline are taken up by tissues and methylated by the metabolic enzyme COMT, to form an inactive ether. In order to prevent this, attempts were made to modify the *meta* phenol group and make it more resistant to metabolism (Fig. 20.17). This was no easy task, as the phenolic group is

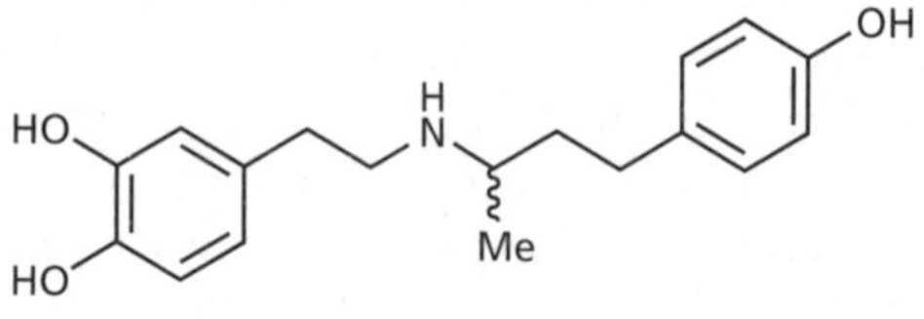

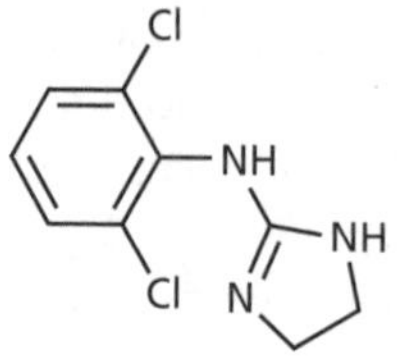

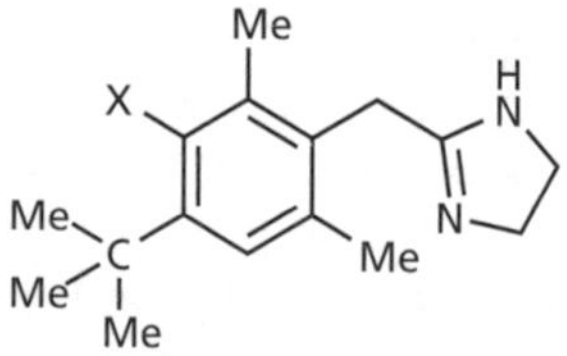

Figure 20.15 Adrenergic agonists.

Figure 20.16 Metabolism of isoetharine.

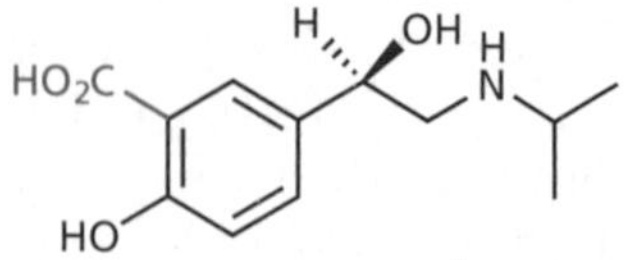

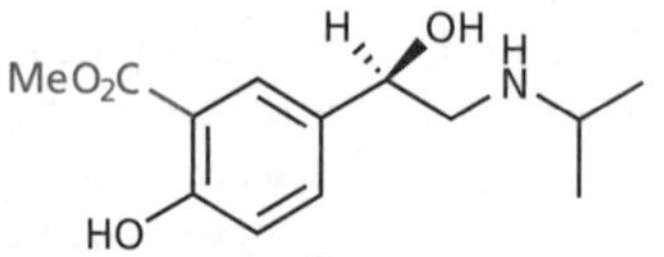

Figure 20.17 Variation of the *meta* substituent.

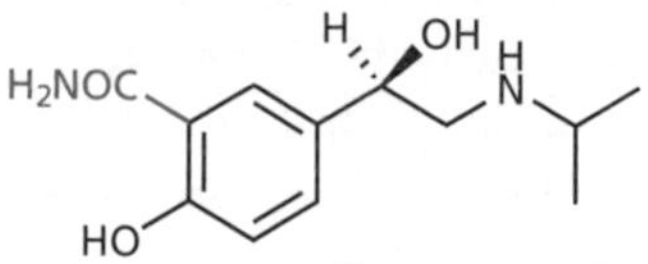

important to activity, so it was necessary to replace it with a group which could still bind to the receptor and retain biological activity, but would not be recognized by the metabolic enzyme. Various factors had to be taken into account in finding a suitable group to replace the phenol, and at the time this work was carried out it was not clear exactly why this moiety was important. Was it taking part in hydrogen bonding, or was it ionized and taking part in ionic bonding? Did the phenol have an important electronic influence on the aromatic ring which affected binding? Was the size of the phenol group important? In order to test these various possibilities, the *meta* phenol group was replaced by a variety of different substituents.

A carboxylic acid group (A in Fig. 20.17) was tried first, but this compound had no activity. An ester and an amide were then tried (B and C), but these compounds proved to be β-antagonists rather than agonists. The introduction of a sulfonamide group (MeSO$_2$NH) was more successful, resulting in a long-lasting selective β$_2$-agonist called **soterenol** (Fig. 20.18).

However, this compound was never used clinically because a better compound was obtained in **salbutamol** (known as albuterol in the USA)–Box 20.1. Here, the *meta* phenol group of the catecholamine skeleton was replaced by a hydroxymethylene group. Salbutamol

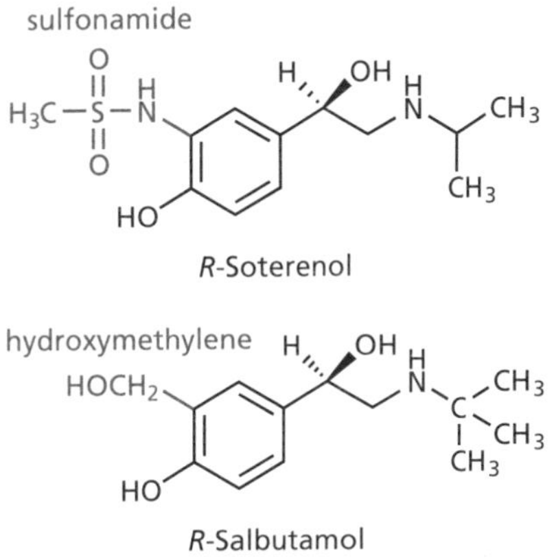

Figure 20.18 Selective β$_2$ agonists.

BOX 20.1 SYNTHESIS OF SALBUTAMOL

Salbutamol is an important antiasthmatic drug that can be synthesized from aspirin. A Fries rearrangement of aspirin produces a ketoacid which is then esterified. A bromoketone is then prepared which allows the introduction

of an amino group by nucleophilic substitution. The methyl ester and ketone are then reduced, and finally the N-benzyl protecting group is removed by hydrogenolysis.

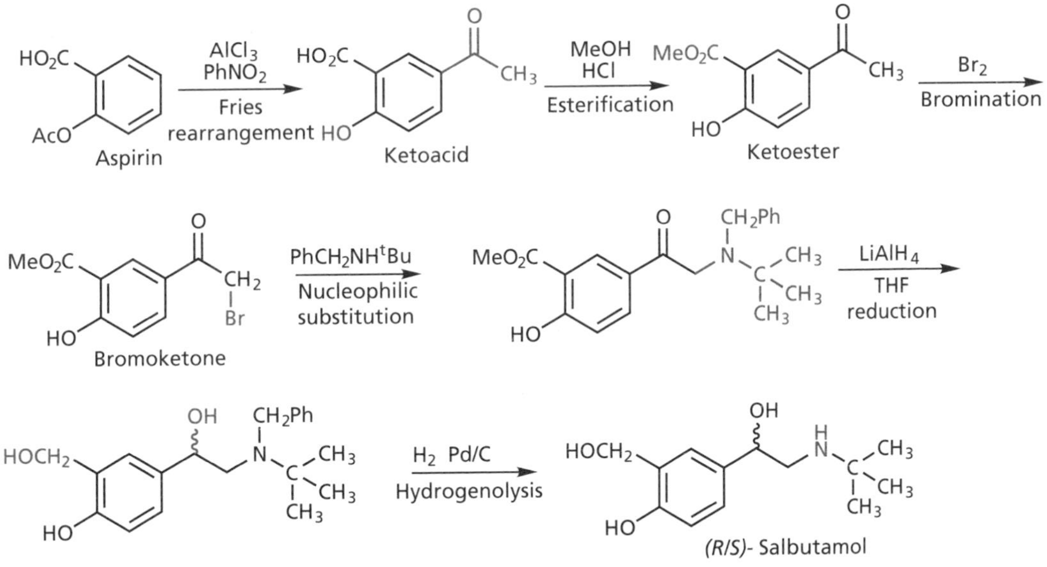

Synthesis of salbutamol.

has the same potency as isoprenaline, but is 2000 times less active on the heart. It has a duration of 4 hours, is not taken up by carrier proteins, and is not metabolized by COMT. Instead, it is more slowly metabolized to a phenolic sulfate. Thus, this modification proved successful in making the compound unrecognizable to COMT, while still being recognized by the adrenergic receptor (see also section 11.2.6) Salbutamol was marketed as a racemate and soon became a market leader in 26 countries for the treatment of asthma. The *R*-enantiomer is 68 times more active than the *S*-enantiomer. Furthermore, the *S*-enantiomer accumulates to a greater extent in the body and produces side effects. Consequently, the pure *R*-enantiomer (levalbuterol) was eventually marketed—an example of chiral switching (section 12.2.1).

Several analogues of salbutamol were synthesized. For example, analogues were prepared to test whether the *meta* CH$_2$OH group could be modified further. These and previous results demonstrated the following requirements for the *meta* substituent:

- It has to be capable of taking part in hydrogen bonding. Substituents such as MeSO$_2$NHCH$_2$, HCONHCH$_2$, and H$_2$NCONHCH$_2$ permitted this.

- Substituents with an electron-withdrawing effect on the ring have poor activity (e.g. CO$_2$H).

- Bulky *meta* substituents are bad for activity since they prevent the substituent adopting the necessary conformation for hydrogen bonding.

- The CH$_2$OH group can be extended to CH$_2$CH$_2$OH but no further.

Having identified the advantages of a hydroxymethyl group at the *meta* position, attention now turned to the *N*-alkyl substituents. Salbutamol itself has a bulky t-butyl group. *N*-Arylalkyl substituents were now added, which would be capable of reaching the polar region of the binding site described earlier (extension strategy; section 20.9.2). For example, salmefamol (Fig. 20.19) is 1.5 times more active than

salbutamol and has a longer duration of action (6 hours). The drug is given by inhalation, but in severe attacks it may be given intravenously.

Further developments were carried out to find a longer-lasting agent in order to cope with nocturnal asthma, a condition which usually occurs at about 4 a.m. (commonly called the morning dip). It was decided to increase the lipophilicity of the drug, since it was believed that a more lipophilic drug would bind more strongly to the tissue in the vicinity of the adrenoceptor and be available to act for a longer period. Increased lipophilicity was achieved by increasing the length of the *N*-substituent with a further hydrocarbon chain and aromatic ring. This led to **salmeterol** (Fig. 20.20) which has twice the potency of salbutamol and an extended action of 12 hours.

KEY POINTS

- The important binding groups in catecholamines are the two phenolic groups, the aromatic ring, the secondary alcohol and the ionized amine.

- Placing a bulky alkyl group on the amine leads to selectivity for β-receptors over α-receptors.

- Extending the *N*-alkyl substituent to include a hydrogen bonding group increases affinity for β-receptors.

- Agents which are selective for β$_2$-adrenoceptors are useful antiasthmatic agents.

- Early β$_2$-agonists were metabolized by catechol-*O*-methyltransferase. Replacing the susceptible phenol group

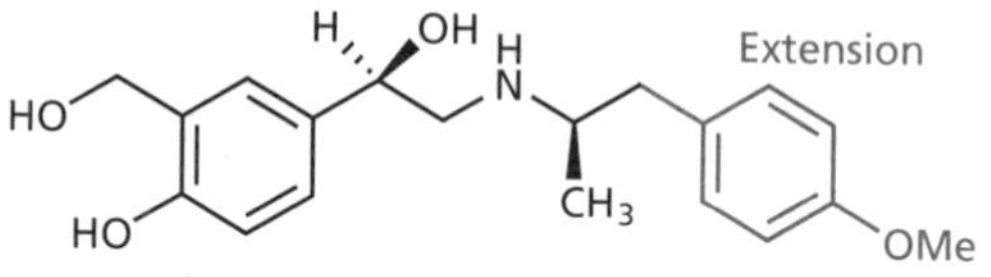

Figure 20.19 Salmefamol.

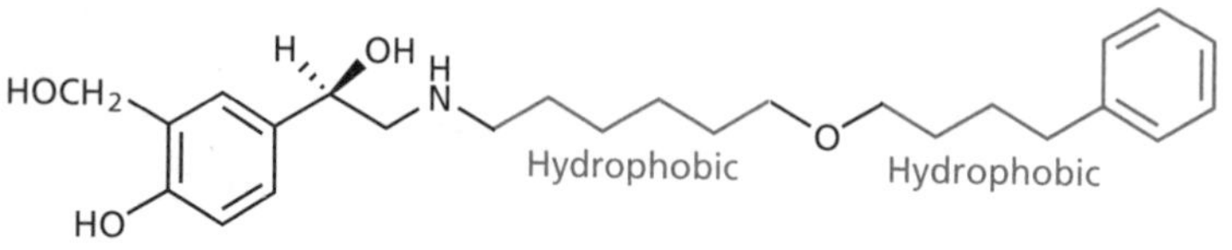

Figure 20.20 Salmeterol.

with a hydroxymethylene group prevented metabolism while retaining receptor interactions.

- Longer-lasting antiasthmatics have been obtained by increasing the lipophilic character of the compounds.

20.11 Adrenergic receptor antagonists

20.11.1 General α/β-blockers

Carvedilol and **labetalol** are agents which act as antagonists at both the α- and β-adrenoceptors (Fig 20.21). They have both been used as antihypertensives, and carvedilol has been used to treat cardiac failure.

20.11.2 α-Blockers

In general, α-antagonists have been limited to selective α_1-antagonists which have been used to treat hypertension or to control urinary output.

Prazosin (Fig. 20.22) was the first α_1-selective antagonist to be used for the treatment of hypertension, but it is short acting. Longer-lasting drugs such as **doxazosin** and **terazosin** are better, since they are given as once-daily doses. These agents relieve hypertension by blocking the actions of noradrenaline or adrenaline at the α_1 receptors of smooth muscle in blood vessels. This results in relaxation of the smooth muscle and dilatation of the blood vessels, leading to a lowering in blood pressure. These drugs have also been used for the treatment of patients with an enlarged prostate—a condition known as **benign prostatic hyperplasia**. The enlarged prostate puts pressure on the urinary tract and it becomes difficult to pass urine. The α_1-blockers prevent activation of the α_1-adrenoceptors that are responsible for smooth muscle contraction of the prostate gland, prostate urethra, and the neck of the bladder. This leads to smooth muscle relaxation at these areas, reducing the pressure on the urinary tract and helping the flow of urine. The agents are not a cure for the problem, but they relieve the symptoms.

α_2-Antagonists are being considered as antidepressants. Depression is associated with decreased release of noradrenaline and serotonin in the CNS, and antidepressants work by increasing the levels of one or both of these neurotransmitters. It may sound odd to consider an adrenergic antagonist as an antidepressant agent, but it makes sense when it is appreciated that the α_2-receptors are presynaptic adrenergic receptors, or autoreceptors (section 20.6.3). Activation of these results in a decrease of noradrenaline released from the nerve, so blocking the autoreceptor will actually increase noradrenaline levels. **Mirtazepine** (Fig. 20.23)

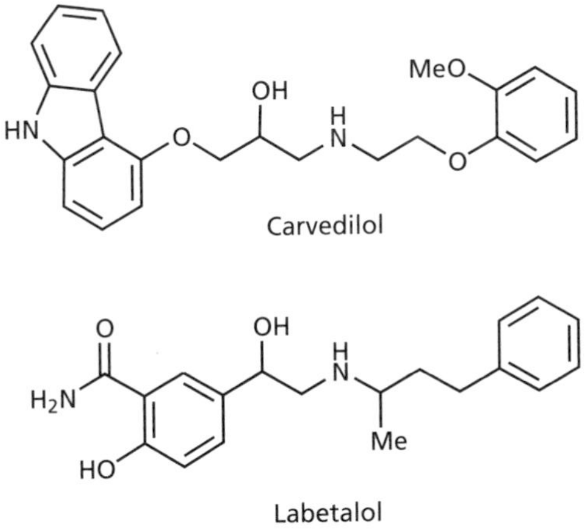

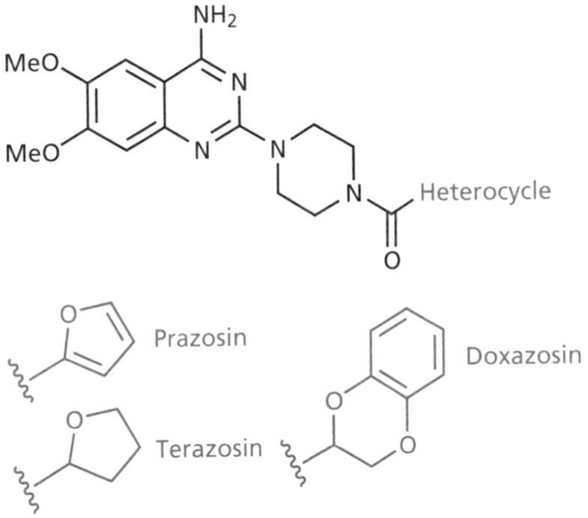

Figure 20.22 α_1-Selective antagonists.

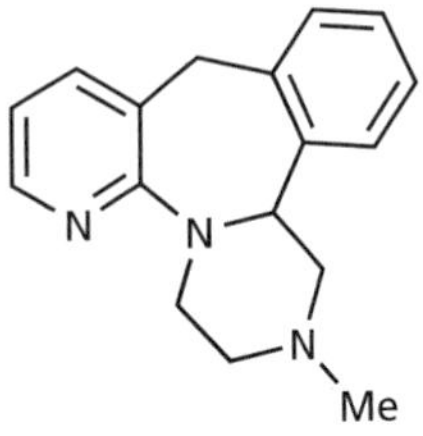

Figure 20.21 General α/β-blockers.

Figure 20.23 Mirtazepine.

is an antidepressant agent which blocks this receptor and increases the level of noradrenaline released. The α_2-receptor also controls the release of serotonin from serotonin nerve terminals, and blocking it increases the level of serotonin as well. Older antidepressants are known to take several weeks before they have an effect and it has been shown that this is due to an eventual desensitization of the presynaptic receptors.

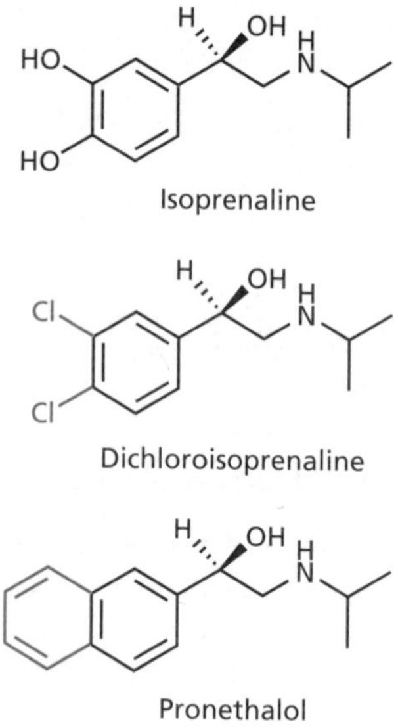

20.11.3 β-Blockers as cardiovascular drugs

20.11.3.1 First-generation β-blockers

The most useful adrenergic antagonists used in medicine today are the β-blockers, which were originally designed to act as antagonists at the β_1-receptors of the heart.

The first goal in the development of these agents was to achieve selectivity for β-receptors with respect to α-receptors. **Isoprenaline** (Fig. 20.24) was chosen as the lead compound. Although this is an agonist and not an antagonist, it was active at β-receptors and not α-receptors. Therefore, the goal was to take advantage of this inherent specificity and modify the molecule to convert it from an agonist to an antagonist.

The phenolic groups are important for agonist activity, but this does not necessarily mean that they are essential for antagonist activity, since antagonists can often block receptors by binding in different ways. Therefore, one of the early experiments was to replace the phenol groups with other substituents. Replacing the phenolic groups of isoprenaline with chloro substituents produced **dichloroisoprenaline** (Fig. 20.24). This compound was a partial agonist. In other words, it has some agonist activity, but it was weaker than a pure agonist. Nevertheless, dichloroisoprenaline blocks natural chemical messengers from binding and can therefore be viewed as an antagonist because it lowers adrenergic activity.

The next stage was to try to remove the partial agonist activity. A common method of converting an agonist into an antagonist is to add an extra aromatic ring. This can sometimes result in an extra hydrophobic interaction with the receptor, which is not involved when the agonist binds. This in turn means a different induced fit between the ligand and the binding site, such that the ligand binds without activating the receptor. Therefore, the chloro groups

Figure 20.24 Partial β-agonists.

of dichloroisoprenaline were replaced by an extra benzene ring to give a naphthalene ring system. The product obtained (**pronethalol**; Fig. 20.24) was still a partial agonist, but was the first β-blocker to be used clinically for angina, arrhythmia, and high blood pressure.

Research was now carried out to see what effect extending the length of the chain between the aromatic ring and the amine would have. One of these projects involved the introduction of various linking groups between the naphthalene ring and the ethanolamine portion of the molecule (Fig. 20.25). At this stage an element of good fortune came into play (Box 10.4) which led to propranolol rather than the original target structure (Fig. 20.26). Propranolol was found to be a pure antagonist having 10–20 times greater activity

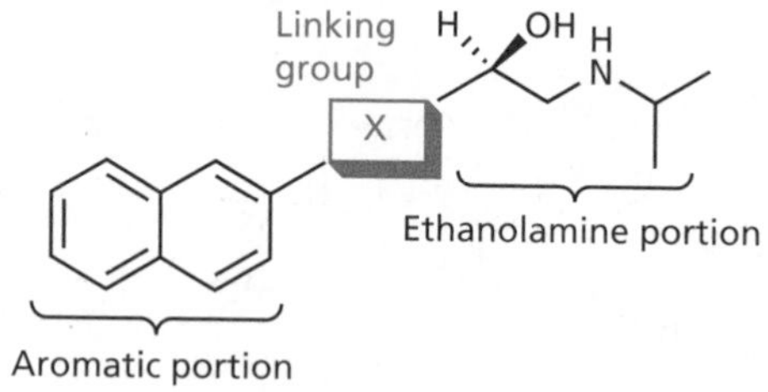

Figure 20.25 Chain extension tactics.

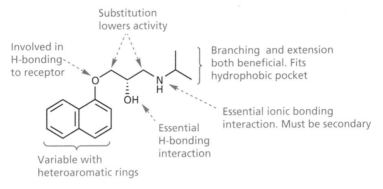

Figure 20.26 Propranolol and original target structure.

Substitution
lowers activity

Involved in
H-bonding
to receptor

Branching and extension
both beneficial. Fits
hydrophobic pocket

Essential ionic bonding
interaction. Must be secondary

Essential
H-bonding
interaction

Variable with
heteroaromatic rings

Figure 20.27 Structure–activity relationships of aryloxypropanolamines.

than pronethalol. It was introduced into the clinic for the treatment of angina and has since become the benchmark against which all β-blockers are rated. The S-enantiomer is the active enantiomer,[1] although propranolol is used clinically as a racemate. Later, when the original target structure was synthesized, it was found to be less active.

20.11.3.2 Structure–activity relationships of aryloxypropanolamines

Propranolol is an example of an aryloxypropanolamine structure (see Box 20.2). A large number of aryloxy-propanolamines have been synthesized and tested, demonstrating the following SAR (Fig. 20.27):

- Branched bulky N-alkyl substituents such as iso-propyl and t-butyl groups are good for β-antagonist activity, suggesting an interaction with a hydro-phobic pocket in the binding site (compare β-agonists).

- Variation of the aromatic ring system is possible and heteroaromatic rings can be introduced such as those in pindolol and timolol (Fig. 20.28).

- Substitution on the side chain methylene group increases metabolic stability but lowers activity.

- The alcohol group on the side chain is essential for activity.

- Replacing the ether oxygen on the side chain with S, CH_2 or NMe is detrimental, although a tissue-selective β-blocker has been obtained replacing O with NH.

- N-Alkyl substituents longer than isopropyl or t-butyl are less effective (but see next point).

- Adding an N-arylethyl group such as –$CHMe_2$–CH_2Ph or $CHMe$–CH_2Ph is beneficial (extension).

- The amine must be secondary.

[1] Although the S enantiomer of aryloxypropanolamines and the R enantiomer of arylethanolamines are the active forms, the absolute configuration at the asymmetric centre is the same for both classes of compound.

BOX 20.2 SYNTHESIS OF ARYLOXYPROPANOLAMINES

Propranolol is a first-generation β-blocker and acts as an antagonist at β-adrenoceptors. The synthesis of propranolol is relatively simple and can be easily adapted to produce a large number of analogues. A phenol is reacted with 2-chloromethyloxirane such that nucleophilic substitution of the alkyl chloride takes place, rather than opening of the epoxide ring—a regioselective reaction. The resulting product is then treated with an amine to ring-open the epoxide. This introduces the amine and generates the secondary alcohol at the same time. An asymmetric centre is created in the process and the final product is a racemate. Because of the nature of the nature of the synthetic route, a huge variety of phenols and amines can be used to produce different analogues.

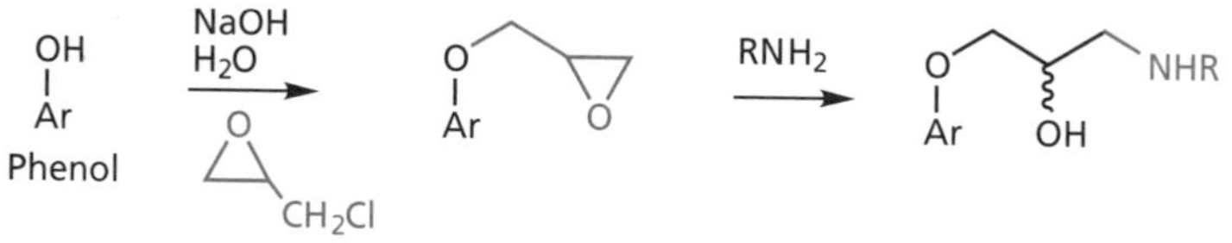

Synthesis of aryloxypropanolamines.

Pindolol

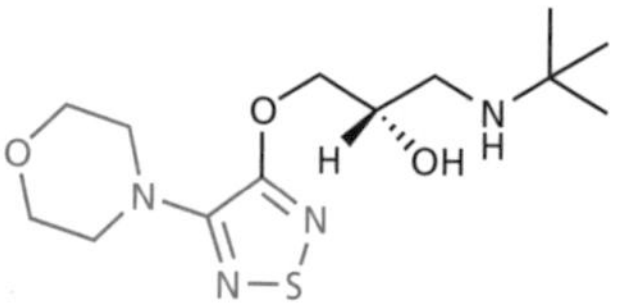

Timolol

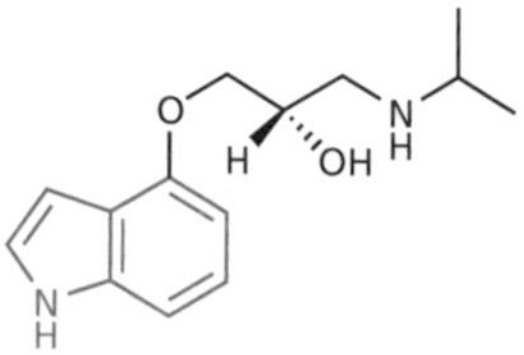

Figure 20.28 β1-Antagonists containing heteroaromatic ring systems.

20.11.3.3 Clinical effects of first-generation β-blockers

The effects of propranolol and other first-generation β-blockers depends on how active the patient is. At rest, propranolol causes little change in heart rate, output, or blood pressure. On the other hand, if the patient exercises or becomes excited, propranolol reduces the resulting effects of circulating adrenaline. The β-blockers were originally intended for use in angina since they were targeted on the heart, but an unexpected bonus was the fact that they also had antihypertensive activity (lowering of blood pressure). Indeed, the β-blockers are now more commonly used as antihypertensives rather than for the treatment of angina. The antihypertensive activity arises from a number of factors resulting from the following effects at various parts of the body:

- Action at the heart to reduce cardiac output.

- Action at the kidneys to reduce renin release. Renin catalyses formation of **angiotensin I** which is quickly converted to **angiotensin II**, which is a potent vasoconstrictor.

- Action in the CNS to lower the overall activity of the sympathetic nervous system.

These effects override the fact that β-blockers block the β-receptors on blood vessels. This would normally cause a constriction of the blood vessels and lead to a rise in blood pressure.

The side effects of first-generation β-blockers include the following:

- bronchoconstriction in asthmatics; such patients need β2-receptor stimulation during treatment with β-blockers

- tiredness of limbs due to reduced cardiac output

- CNS effects (dizziness, dreams, and sedation) especially with lipophilic β-blockers such as propranolol, pindolol, and oxprenol, all of which can cross the blood–brain barrier

- heart failure for patients on the verge of a heart attack (The β-blockers produce a fall in the resting heart rate and this may push some patients over the threshold.)

- inhibition of noradrenaline release at synapses.

20.11.3.4 Selective β_1-blockers (second-generation β-blockers)

Propranolol is a non-selective β-antagonist which acts as an antagonist at β_2-receptors as well as β_1-receptors. Normally this is not a problem, but it is serious if the patient is asthmatic as the propranolol could initiate an asthmatic attack by antagonizing the β_2-receptors in bronchial smooth muscle. This leads to contraction of bronchial smooth muscle and closure of the airways.

Practolol (Fig. 20.29) is not as potent as propranolol, but it is a selective cardiac β_1-antagonist which does not block vascular or bronchial β_2-receptors. It is much safer for asthmatic patients and, since it is more polar than propranolol, it has much fewer CNS effects.

Practolol was marketed as the first cardioselective β_1-blocker for the treatment of angina and hypertension, but after a few years it had to be withdrawn because of unexpected but serious side effects in a very small number of patients. These side effects included skin rashes, eye problems, and peritonitis. Further investigations were carried out and it was demonstrated that the amido group had to be in the *para* position of the aromatic ring rather than the *ortho* or *meta* positions if the structure was to retain selectivity for the cardiac β_1-receptors. This implied that there was an extra hydrogen bonding interaction taking place with the β_1-receptors (Fig. 20.30) which was not taking place with the β_2-receptors.

Replacement of the acetamido group with other groups capable of hydrogen bonding led to a series of cardioselective β_1-blockers which reached the market. These included **acebutolol**, **atenolol**, **metoprolol**, and **betaxolol** (Fig. 20.31).

20.11.3.5 Third-generation β-blockers

The *N*-alkyl groups in first- and second-generation β-blockers are normally isopropyl or *t*-butyl groups. Extension tactics involving the addition of arylalkyl groups to the nitrogen atom resulted in a series of third-generation β-blockers which bind to the β_1-receptor using an additional hydrogen bonding interaction. These include **epanolol** and **primidolol** (Fig. 20.32). Related to these structures is **xamoterol** which is a very selective β_1-partial agonist used in the treatment of mild heart failure. As an agonist it provides cardiac stimulation when the patient is at rest, but it acts as a β-blocker during strenuous exercise when greater amounts of adrenaline and noradrenaline are being generated.

Figure 20.29 Practolol.

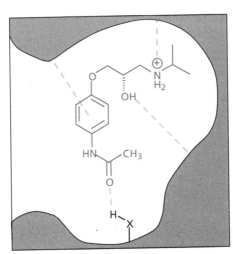 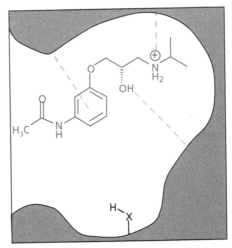

para substitution
Extra H-bonding interaction

meta substitution

Figure 20.30 Binding interactions with β_1-receptors.

20.11.3.6 Other clinical uses for β-blockers

β-Blockers have a range of other clinical uses apart from cardiovascular medicine. They are used to counteract overproduction of catecholamines resulting from an enlarged thyroid gland or tumours of the adrenal gland. They can also be used to alleviate the trauma of alcohol and drug withdrawal, as well as relieving the stress associated with situations such as exams, public speaking, and public performances. Timolol and betaxolol are used in the treatment of glaucoma, although their mechanism of action is not clear, while propranolol is used to treat anxiety and migraine.

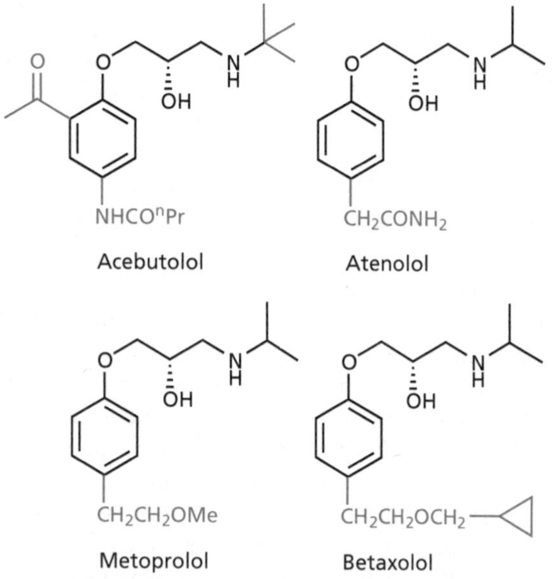

KEY POINTS

- Antagonists of β-adrenoceptors are known as β-blockers.
- Replacing the catechol ring with a naphthalene ring changes an agonist into a partial agonist.
- Variation of the linking group between naphthalene and the ethanolamine moiety resulted in the first β-antagonists.
- SAR of aryloxypropanolamines reveal the importance of the ionized amine, the side chain alcohol, and the ether linkage. Substituents on the nitrogen can be varied. The naphthalene ring can be replaced by various heterocyclic rings.
- First-generation β-blockers inhibits all β-receptors and can induce asthma in susceptible patients.
- Second-generation β-blockers show selectivity for β_1-receptors over β_2-receptors. Aryloxypropanolamines bearing a hydrogen bonding group at the *para* position of an aromatic ring show β_1-selectivity.
- Third-generation β-blockers bear an extended *N*-substituent which includes a hydrogen bonding group capable of an extra hydrogen bonding interaction with the β_1-adrenoceptor.

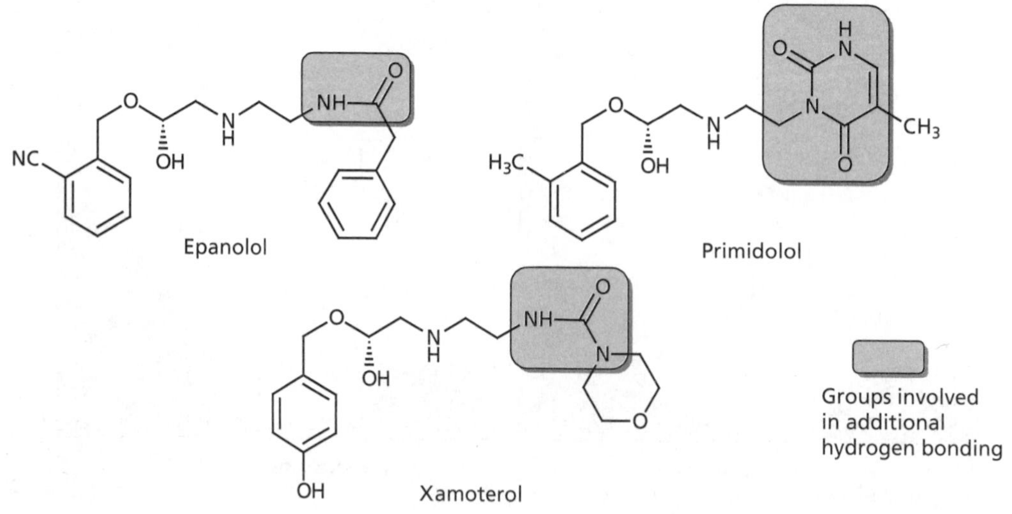

Figure 20.31 Second-generation β-blockers.

Figure 20.32 Third-generation β-blockers.

20.12 Other drugs affecting adrenergic transmission

In the previous sections, we discussed drugs which act as agonists or antagonists at adrenergic receptors. However, there are various other drug targets involved in the adrenergic transmission process which are important in controlling adrenergic activity. In this section, we briefly cover some of the most important aspects of these.

20.12.1 Drugs that affect the biosynthesis of adrenergics

In section 20.4, we identified **tyrosine hydroxylase** as the regulatory enzyme for catecholamine biosynthesis. As such, it is a potential drug target. For example, **α-methyltyrosine** (Fig. 20.33) inhibits tyrosine hydroxylase and is sometimes used clinically to treat tumour cells which overproduce catecholamines.

It is sometimes possible to 'fool' the enzymes of the biosynthetic process into accepting an unnatural substrate such that a false transmitter is produced and stored in the storage vesicles. For example, **α-methyldopa** is converted and stored in vesicles as

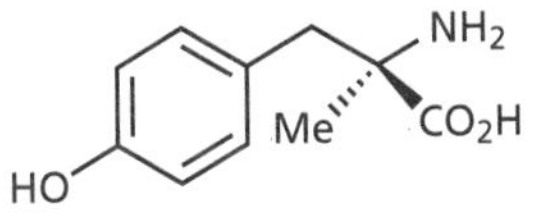

Figure 20.33 α-Methyltyrosine.

α-methylnoradrenaline (Fig 20.34) and displaces noradrenaline. Such false transmitters are less effective than noradrenaline, so this is another way of down-regulating the adrenergic system. The drug has serious side effects, however, and is limited to the treatment of hypertension in late pregnancy.

A similar example is the use of **α-methyl-m-tyrosine** in the treatment of shock. This unnatural amino acid is accepted by the enzymes of the biosynthetic pathway and converted to **metaraminol** (Fig. 20.35).

20.12.2 Uptake of noradrenaline into storage vesicles

The uptake of noradrenaline into storage vesicles can be inhibited by drugs. The natural product **reserpine** binds to the carrier protein responsible for transporting noradrenaline into the vesicles and so noradrenaline accumulates in the cytoplasm where it is metabolized by MAO. As noradrenaline levels drop, adrenergic activity drops. Reserpine was once used as an antihypertensive agent but has serious side effects, such as depression, and is no longer used.

20.12.3 Release of noradrenaline from storage vesicles

The storage vesicles are also the targets for the drugs **guanethidine** and **bretylium** (Fig. 20.36). Guanethidine is taken up into nerves and storage vesicles by the same carrier proteins as noradrenaline, and it displaces noradrenaline in the same way as reserpine. The drug also prevents exocytosis of the vesicle and so prevents

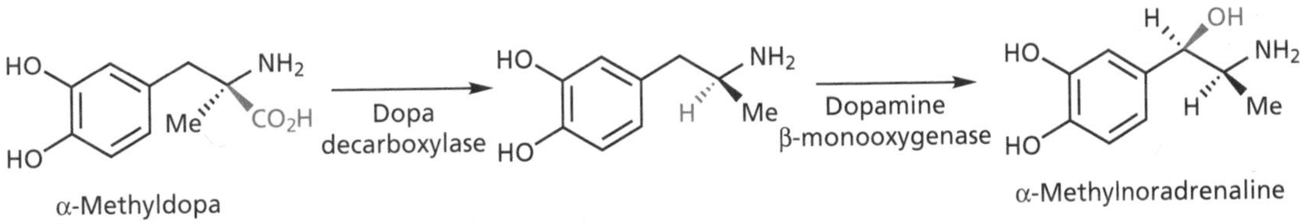

Figure 20.34 A false transmitter: α-methylnoradrenaline.

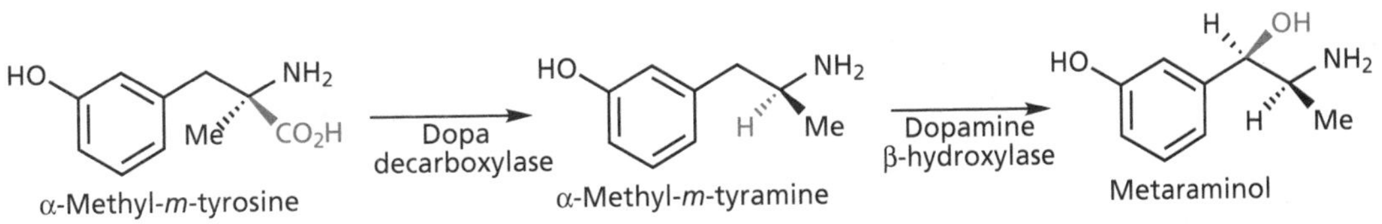

Figure 20.35 A false transmitter: metaraminol.

release of the vesicle's contents into the synaptic gap. Guanethidine is an effective antihypertensive agent, but is no longer used clinically because of side effects resulting from such a non-specific inhibition of adrenergic nerve transmission. Bretylium works in the same way as guanethidine and is sometimes used to treat irregular heart rhythms.

20.12.4 Uptake of noradrenaline into nerve cells by carrier proteins

Once noradrenaline has interacted with its receptor, it is normally taken back into the presynaptic nerve by its carrier protein. This carrier protein is an important target for various drugs which inhibit noradrenaline uptake and thus prolong adrenergic activity.

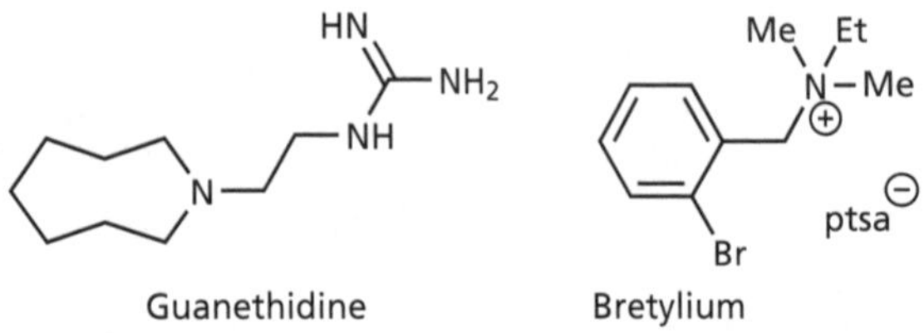

The tricyclic antidepressants **desipramine, imipramine**, and **amitriptyline** (Fig. 20.37) work by inhibiting noradrenaline reuptake in the CNS. Increased noradrenaline levels are thought to have an antidepressant action by eventually desensitizing presynaptic α_2-adrenoceptors, resulting in higher levels of noradrenaline and serotonin being released by presynaptic nerves.

It has been proposed that the tricyclic antidepressants are able to act as inhibitors because they are in part superimposable on noradrenaline. This can be seen in Fig. 20.38, where the aromatic ring and the nitrogen atoms of noradrenaline are overlaid with the nitrogen atom and one of the aromatic rings of desipramine. Note that the tricyclic system of desipramine is V-shaped, so that when the molecules are overlaid the second aromatic ring is held above the plane of the noradrenaline structure. Planar tricyclic structures would be expected to be less active as inhibitors, since the second aromatic ring would then occupy the space required for the amine nitrogen.

Reboxetine (Fig. 20.39) is a new antidepressant which selectively inhibits noradrenaline uptake, but unlike the tricyclics has no appreciable action on cholinergic or α_1-adrenergic receptors. It also rapidly desensitizes presynaptic α_2-adrenergic receptors, which further enhances its activity. **Bupropion** (Zyban; Fig. 20.39) inhibits the reuptake of both noradrenaline

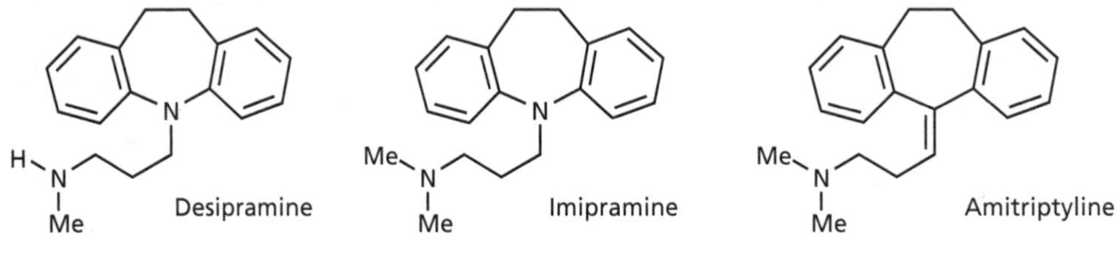

Figure 20.36 Agents which affect adrenergic activity (ptsa = *para*-toluenesulfonate).

Figure 20.37 Tricyclic antidepressants.

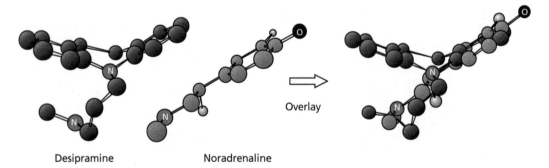

Figure 20.38 Overlay of desipramine and noradrenaline.

and dopamine and has been used for the treatment of depression and as an aid to giving up smoking.

Stimulants acting as noradrenaline reuptake inhibitors have been used for the treatment of **attention deficit hyperactivity disorder**. This is the most commonly diagnosed childhood behavioural disorder and is associated with inattention, hyperactivity, and impulsivity. **Methylphenidate** (Ritalin; Fig. 20.39) is the most commonly prescribed medication for this disorder, and **atomoxetine** (Fig. 20.39) was approved in 2002. Both agents lead to increased levels of noradrenaline and dopamine in the brain.

Cocaine also inhibits noradrenaline uptake when it is chewed from coca leaves, but this time the inhibition is in the peripheral nervous system rather than the CNS. Chewing coca leaves was well known to the Incas as a means of increasing endurance and suppressing hunger, and they would chew the leaves whenever they were faced with situations requiring long periods of physical effort or stamina. When the coca leaves are chewed, cocaine is absorbed into the systemic blood supply and acts predominantly on peripheral adrenergic receptors to increase adrenergic activity. Nowadays, cocaine abusers prefer to smoke or snort the drug, which therefore enters the CNS more efficiently. There, it inhibits the uptake of dopamine rather than noradrenaline, resulting in its CNS effects.

Some amines such as **tyramine, amphetamine**, and **ephedrine** (Figs. 20.10 and 20.14) closely resemble noradrenaline in structure and are transported into the nerve cell by noradrenaline's carrier proteins. Once in the cell, they are taken up into the vesicles. Since these amines are competing with noradrenaline for carrier proteins, noradrenaline is more slowly reabsorbed into the nerve cells. Moreover, as the foreign amines are transported into the nerve cell, noradrenaline is transported out by those same carrier proteins. Both of these facts means that more noradrenaline is available to interact with its receptors. This means that amphetamines and similar amines have an indirect agonist effect on the adrenergic system.

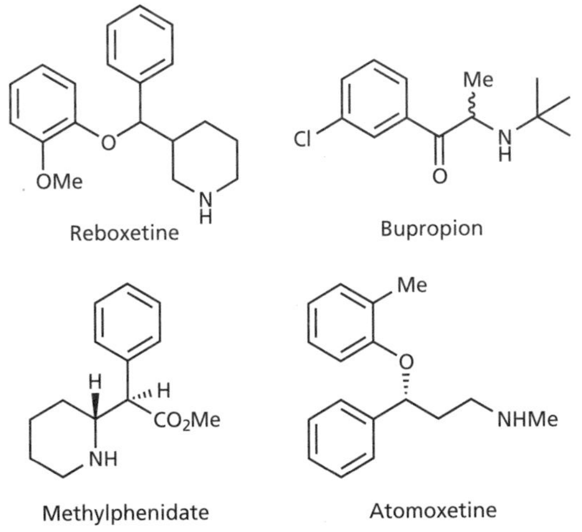

20.12.5 Metabolism

Inhibition of the enzymes responsible for the metabolism of noradrenaline should prolong noradrenaline activity. We have seen how amines such as tyramine, amphetamine, and ephedrine inhibit the re-uptake of noradrenaline into the presynaptic nerve. These amines also inhibit MAO, one of the important enzymes involved in the metabolism of noradrenaline. This in turn leads to a build-up in noradrenaline levels and an increase in adrenergic activity.

Monoamine oxidase inhibitors (MAOIs) such as **phenelzine, iproniazid**, and **tranylcypromine** (Fig. 20.40) have been used clinically as antidepressants, but other classes of compound such as the tricyclic antidepressants are now more favoured as they have fewer side effects. It is important to realize that the MAOIs affect the levels of all neurotransmitters that are normally metabolized by these enzymes. In particular, dopamine and serotonin levels are both increased by MAOIs. As a result of these widespread effects, it is difficult to be sure what mechanism is involved in the antidepressant activity of these agents.

Another serious problem associated with MAOIs is their interaction with other drugs and food.

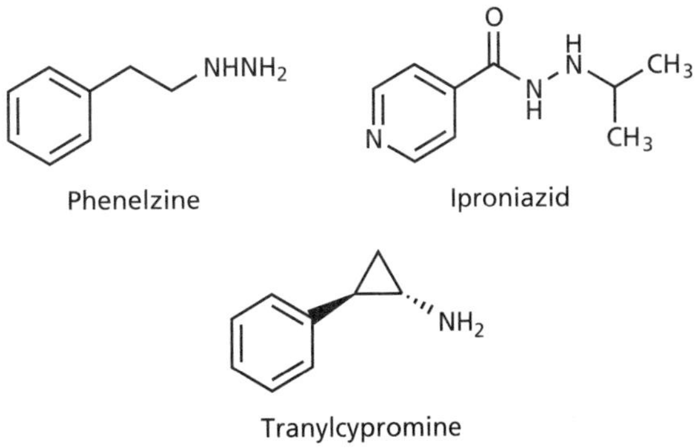

Figure 20.39 Adrenergic agents acting in the central nervous system.

Figure 20.40 Monoamine oxidase inhibitors.

A well-known example of this is the 'cheese reaction'. Ripe cheese contains **tyramine**, which is normally metabolized by MAOs in the gut wall and the liver, and so never enters the systemic circulation. If the MAOs are inhibited by MAOIs, tyramine is free to circulate round the body, enhancing the adrenergic system and leading to acute hypertension and severe headaches.

- The uptake and release of noradrenaline from storage vesicles can be inhibited by certain drugs.

- The tricyclic antidepressants inhibit the reuptake of noradrenaline into presynaptic nerves by blocking carrier proteins. Adrenergic activity is increased in the CNS.

- Cocaine increases peripheral adrenergic activity by blocking noradrenaline reuptake. In the CNS it inhibits the reuptake of dopamine.

- Amphetamines compete with noradrenaline for the carrier proteins responsible for transporting noradrenaline back into the presynaptic nerve. Adrenergic activity is increased in the CNS.

- Monoamine oxidase inhibitors inhibit the metabolic enzyme monoamine oxidase and result in increased levels of noradrenaline and other catecholamines.

KEY POINTS

- Inhibitors of catecholamine biosynthesis affect adrenergic activity.

- Drugs that are similar to tyrosine may be converted by the catecholamine biosynthetic pathway to structures that act as false transmitters and lower adrenergic activity.

QUESTIONS

1. How would you synthesize the following structures to test their adrenergic agonist activity?

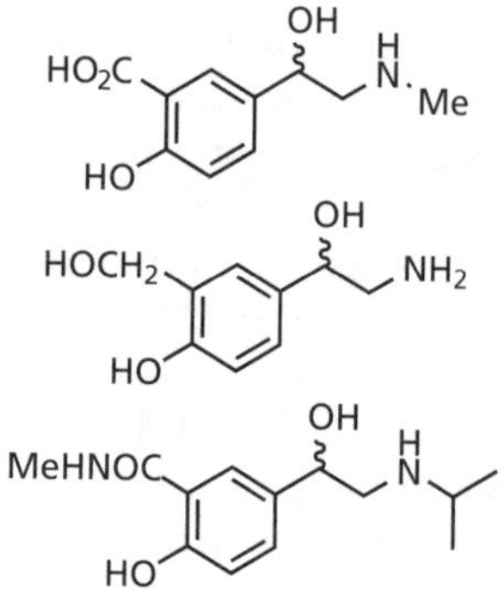

2. Suggest how you might synthesize the adrenergic antagonist pindolol (Fig. 20.28).

3. Suggest whether the following structures are likely to have good or bad activity as β-blockers.

4. The catechol system is important for the binding of adrenergic agonists, yet is not required for adrenergic antagonists. Why might this be the case?

5. How would α-substitution affect the metabolism of adrenergic agents and why?

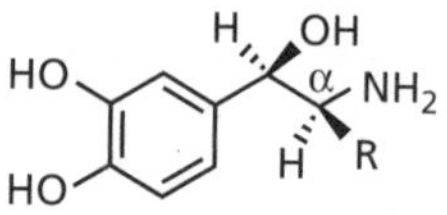

6. What synthetic complication arises from introducing an α-substituent as described in Question 5?

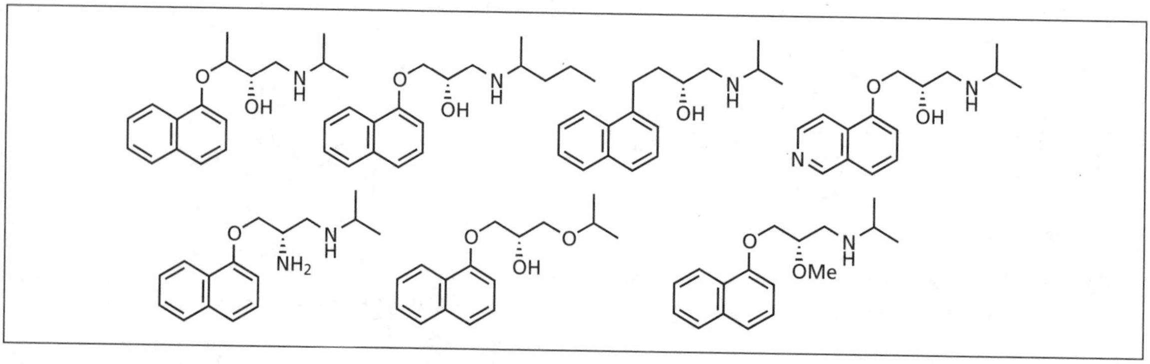

FURTHER READING

Abraham, D. J. (ed.) (2003) Adrenergics and adrenergic-blocking agents. Chapter 1 in: *Burgers' medicinal chemistry and drug discovery*, 6th edn, Vol. 6, John Wiley and Sons, New York.

Ganellin, C. R. and Roberts, S. M. (eds.) (1994) Salbutamol: a selective β_2-stimulant bronchodilator. Chapter 11 in: *Medicinal chemistry—the role of organic research in drug research*, 2nd edn. Academic Press, New York.

Ganellin, C. R. and Roberts, S. M. (eds.) (1994) Beta blockers. Chapter 10 in: *Medicinal chemistry—the role of organic research in drug research*, 2nd edn. Academic Press, New York.

O'Driscoll, C. (2001) Attack on asthma. *Chemistry in Britain*, September, 40–42.

Williams, D. A. and Lemke, T. L. (eds.) (2002) Drugs affecting adrenergic neurotransmission. Chapter 11 in: *Foye's principles of medicinal chemistry*, 5th edn. Lippincott Williams and Wilkins, Baltimore.

Titles for general further reading are listed on p. 711.

21 The opium analgesics

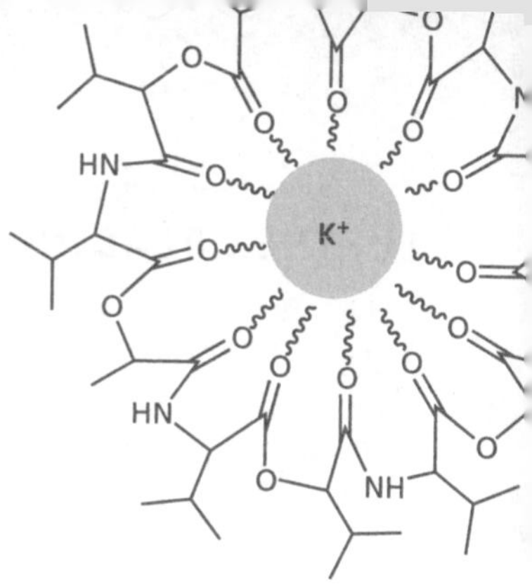

21.1 History of opium

The search for a safe, orally active, and non-addictive analgesic based on the opiate structure is one of the oldest fields in medicinal chemistry, yet one where true success has proved elusive.

It is important to appreciate that the opiates are not the only compounds which are of use in the relief of pain: there are several other classes of analgesic, including **aspirin.** These compounds, however, operate by different mechanisms from those employed by the opiates and are effective against different types of pain.

The term **opium alkaloids** has been used rather loosely to cover all narcotic analgesics, whether they be synthetic, semi-synthetic, or extracted from plant material. To be precise, we should only use the term for those natural compounds which have been extracted from opium—the sticky exudate obtained from the opium poppy (*Papaver somniferum*). The term alkaloid refers to a natural product which contains a nitrogen atom and is basic in character. Several thousand alkaloids have been extracted and identified from various plant sources. These compounds provide a vast 'library' of biologically active compounds which can be used as lead compounds in many fields of medicinal chemistry.

The opiates are perhaps the oldest drugs known to humanity. The use of opium was recorded in China over 2000 years ago, and was known in Mesopotamia before that. Because of its properties, the Greeks dedicated the opium poppy to Thanatos (the god of death), Hypnos (the god of sleep), and Morpheus (the god of dreams). Later physicians prescribed opium for a whole range of afflictions including chronic headache, vertigo, epilepsy, asthma, colic, fevers, dropsies, leprosies, melancholy, and 'troubles to which women are subject'.

As a result of this, its fame spread and by AD 632, knowledge of opium had reached Spain in the west, and Persia and India in the east. The poppy was cultivated in Persia, India, Malaysia, and China, and opium was used primarily as a sleeping draught (sedative) and as a treatment for diarrhoea. Its use in medicine is quoted in a twelfth-century prescription:

Take opium, mandragora, and henbane in equal parts and mix with water. When you want to saw or cut a man, dip a rag in this and put it to his nostrils. He will sleep so deep that you may do what you wish.

Its use as an analgesic came much later, in the sixteenth century, when Paracelsus introduced preparations of opium known as **laudanum.** However, problems were also reported with its use. Doctors stated that stopping the drug after long-term use led to 'great and intolerable distresses, anxieties and depression of the spirit'. These were the first reports of addiction and withdrawal symptoms.

Opium was first marketed in Britain by Thomas Dover—a one-time pirate who had taken up medicine.[1] Dover prepared a powder containing opium, liquorice, saltpetre, and **ipecacuanha.** The last-named compound is an emetic and had the advantage of making the consumers sick should they take too much of the concoction.

---

[1] Dover had another claim to fame. During his seafaring days, he rescued the marooned Alexander Selkirk from an uninhabited island. This was the inspiration for Defoe's *Robinson Crusoe*.

Another popular remedy dating from the eighteenth century was 'Godfrey's cordial', which contained opium, black treacle, and sassafras. This was used as a teething aid, for rheumatic pains, and for diarrhoea. These preparations were freely available in grocery shops without prescription or restriction, despite the fact that many people became addicted to them.

It has to be appreciated that in those days opium was considered as legitimate as tobacco or tea, and that this view continued right up to the twentieth century. Indeed, during the nineteenth century, the opium trade led directly to a war between the United Kingdom and China.

During the nineteenth century, China was ruled by an elite class who considered all foreigners as nothing better than barbarians and wanted nothing to do with them. As a result, trade barriers were set up against all foreign imports and China strove to be totally self-sufficient. Many nations felt aggrieved over this, including the British who were buying tea from China, and were not allowed to trade in return. Eventually, Britain thought it had the answer in opium.

Until the early seventeenth century, China had grown its own opium for use as an ingredient in cakes and as a medicine, but strangely enough it was the introduction of tobacco which changed all that. Tobacco was introduced to Europe from the Americas in the fifteenth century, and European sailors soon introduced the habit into the Far East. In China, smoking became so widespread that the Emperor Tsung Chen banned it in 1644. Deprived of tobacco, the population started smoking opium instead! By the end of the century about a quarter of the population was using the drug and the local crops were insufficient to keep up with demand. As India was a major producer of opium, the British East India company saw this as an opportunity to import the drug into China, and by the 1830s the company was supplying £1 million worth of opium per year. Due to China's embargo, however, most of this had to be smuggled in via the port of Canton by British and American merchants.

Eventually, the Chinese authorities decided to act. They seized and burnt a shipload of opium, then closed the port of Canton to the British. The British traders were outraged and appealed to Lord Palmerston, the British Foreign Secretary at the time. Relations between the two countries steadily deteriorated and led to the Opium War of 1839–42. China was quickly defeated and was forced to lease Hong Kong to Britain as a trading port. They were also forced to accept the principles of free trade and to pay reparations of £21 million. It may seem odd now, but at the time the British saw the Opium War as a just war aimed at defending the principles of free trade. Furthermore, China was seen as a tyrannical regime where justice was harsh and penalties were even harsher (e.g. death by a thousand cuts). The Opium War was first and foremost a trade war, and it wasn't long before other European nations and the USA picked their own fight with China and imposed their own trade settlements and trading ports.

In the mid nineteenth century, opium was smoked in much the same way as cigarettes are today, and opium dens were as much a part of London society as coffee shops. These dens were used by many of the romantic authors of the day including Thomas de Quincy, Edgar Allan Poe, and Samuel Taylor Coleridge. De Quincy even wrote a book recording his opium experiences (*Confessions of an English Opium Eater*) and Coleridge was consuming around 4 pints of laudanum a week when he wrote the 'The Rime of the Ancient Mariner'. A later poem called 'Dejection' may have been inspired by his experience of withdrawal symptoms.

Towards the end of the nineteenth century, doubts were beginning to grow about the long-term effects of opium and its addictive properties. There were many who leapt to its defence, and economic arguments were as powerful a weapon then as they are today. Indeed, in an 1882 parliamentary report, it was stated, 'If Indian opium was stopped at once it would be a very frightful calamity indeed. I should say that one third of the adult population of China would die for want of opium'. Nevertheless, doubts persisted and a motion was put forward in parliament in 1893 stating that the 'opium trade was morally indefensible'. The motion was heavily defeated.

It was not until Chinese immigrants introduced opium on a large scale to the USA, Australia, and South America that governments really cracked down on the trade. In 1909, the International Opium Commission was set up and by 1914, 34 nations had agreed to curb opium production and trade. By 1924, 62 countries had signed up and the League of Nations took over the role of control, requiring countries to limit the use of narcotic drugs to medicine alone. Unfortunately many farmers in India, Pakistan,

Afghanistan, Turkey, Iran, and the Golden Triangle (Burma, Thailand, and Laos) depended on the opium trade for survival. As a result, the trade went underground and has continued to this day.

21.2 Morphine

21.2.1 Isolation of morphine

Opium contains a complex mixture of over 20 alkaloids. The principle alkaloid in the mixture, and the one responsible for analgesic activity, is **morphine**, named after Morpheus. Although pure morphine was isolated in 1803, it was not until 1833 that chemists at Macfarlane and Co. (now Macfarlane-Smith) in Edinburgh were able to isolate and purify it on a commercial scale. Because morphine was poorly absorbed orally, it was little used in medicine until the hypodermic syringe was invented in 1853, allowing doctors to inject the drug directly into the blood supply.

Morphine was then found to be a particularly good analgesic and sedative, and far more effective than crude opium. But there was also a price to be paid. Morphine was used during the American Civil War (1861–65) and the Franco-Prussian war. However, there was poor understanding about safe dose levels, the effects of long-term use, and the increased risks of addiction, tolerance, and respiratory depression. As a result, many casualties were either killed by overdoses or became addicted to the drug.

21.2.2 Structure and properties

Morphine was an extremely complex molecule by nineteenth-century standards, and identifying its structure posed a huge challenge to chemists. By 1881 the functional groups on morphine had been identified,

but it took many more years to establish the full structure. In those days the only way to find out the structure of a complicated molecule was to degrade the compound into simpler molecules that were already known and could be identified. For example, the degradation of morphine with strong base produced methylamine, which established that there was an N–CH$_3$ fragment in the molecule. From such evidence, chemists would propose a structure. This is like trying to work out the structure of a bombed cathedral from the rubble.

Once a structure had been proposed, chemists would then attempt to synthesize it. If the properties of the synthesized compound were the same as those of the natural compound, then the structure was proven. This was a long-drawn-out process, made all the more difficult because chemists then had few of the synthetic reagents or procedures available today. As a result, it was not until 1925 that Sir Robert Robinson proposed the correct structure of morphine. A full synthesis was achieved in 1952, and the structure proposed by Robinson was finally proved by X-ray crystallography in 1968 (164 years after the original isolation). The molecule contains five rings numbered A–E and has a pronounced T shape. It is basic because of the tertiary amino group, but it also contains a phenol, alcohol, aromatic ring, ether bridge, and double bond.

Morphine (Fig. 21.1) is the active principle of opium and is still one of the most effective painkillers available to medicine. It is especially good for treating dull, constant pain rather than sharp, periodic pain. It acts in the brain and appears to work by elevating the pain threshold, thus decreasing the brain's awareness of pain. Unfortunately, it has a large number of side effects, which include:

- depression of the respiratory centre
- constipation
- excitation
- euphoria

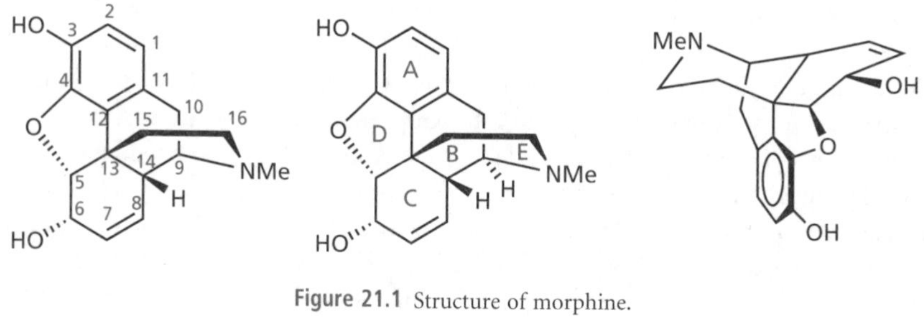

Figure 21.1 Structure of morphine.

- nausea
- pupil constriction
- tolerance and dependence.

Some side effects are not particularly serious, and some can even be advantageous. Euphoria, for example, is a useful side effect when treating pain in terminally ill patients. Others, such as constipation, are uncomfortable but can give clues to other possible uses for opiate-like structures. For example, opiate structures are widely used in cough medicines and the treatment of diarrhoea.

The dangerous side effects of morphine are those of tolerance and dependence, allied with the effects morphine can have on breathing. In fact, the most common cause of death from a morphine overdose is suffocation. Tolerance and dependence in the one drug are particularly dangerous, and lead to severe withdrawal symptoms when the drug is no longer taken. Withdrawal symptoms associated with morphine include anorexia, weight loss, pupil dilatation, chills, excessive sweating, abdominal cramps, muscle spasms, hyperirritability, lacrimation, tremor, increased heart rate, and increased blood pressure. No wonder addicts find it hard to kick the habit!

21.2.3 Structure–activity relationships

Morphine has several functional groups, and a study of structure–activity relationships (SAR) is important in identifying those groups that are important for activity and those that are not. The SAR story is presented here in a logical fashion for reasons of clarity. However, this was not how these studies were tackled at the time. Different compounds were made in a random fashion depending on the ease of synthesis, and the logic followed on from that. Another important aspect of SAR involves the methods used to test activity. We shall see later that morphine interacts with an analgesic

receptor. SAR studies can be carried out on receptors *in vitro* to see how well an analogue binds. Alternatively, they can be carried out on animals *in vivo* to test analgesic activity. Different results may be obtained depending on the testing methods used.

The first and easiest morphine analogues that were made were those involving peripheral modifications of the molecule; that is, changes to different functional groups.

21.2.3.1 Functional groups of morphine analogues

The phenolic OH

Codeine (Fig. 21.2) is the methyl ether of morphine and is also present in opium. It is used for treating moderate pain, coughs, and diarrhoea. Methylating the phenolic OH causes a drastic drop in receptor binding, and the binding affinity for codeine is only 0.1% that of morphine. This drop in affinity is observed in other analogues containing a masked phenolic group, so we can see that a free phenolic group is crucial for analgesic activity. Nevertheless, if codeine is administered to patients, its analgesic effect is 20% that of morphine—much better than expected. Why is this?

The answer lies in the fact that codeine is metabolized by *O*-demethylation in the liver to give morphine. Thus, codeine can be viewed as a prodrug for morphine. Further evidence comes from the fact that codeine has no analgesic effect when it is injected directly into the brain. Injecting codeine in this manner means that it does not pass through the liver and cannot be demethylated.

The 6-alcohol

The results in Fig. 21.3 show that masking or losing the alcohol group does not decrease analgesic activity. In fact, it often improves activity as a result of improved pharmacokinetic properties. In other words, the drugs concerned reach the target receptor more

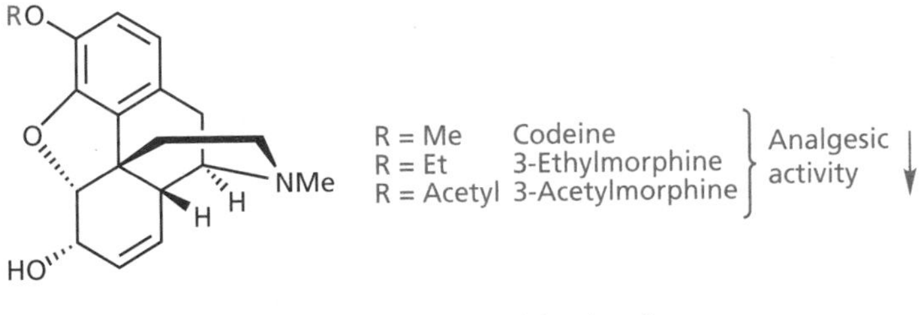

R = Me	Codeine	} Analgesic activity
R = Et	3-Ethylmorphine	
R = Acetyl	3-Acetylmorphine	

Figure 21.2 The importance of the phenolic group.

easily. In this case, the morphine analogues shown are less polar than morphine because a polar alcohol group has been masked or lost. As a result they cross the blood–brain barrier (section 8.3.5) into the central nervous system (CNS) more efficiently and accumulate at the target receptors in greater concentrations; hence the better analgesic activity.

It is interesting to compare the activities of morphine, **6-acetylmorphine** (Fig. 21.3) and **diamorphine**

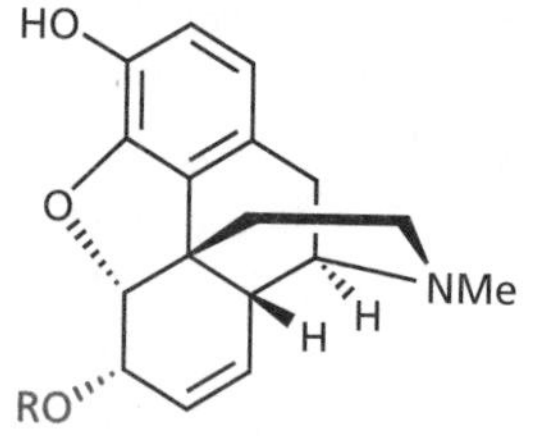

R		Analgesia with respect to morphine
Me	Heterocodeine	5×
Et	6-Ethylmorphine	greater
Acetyl	6-Acetylmorphine	4×

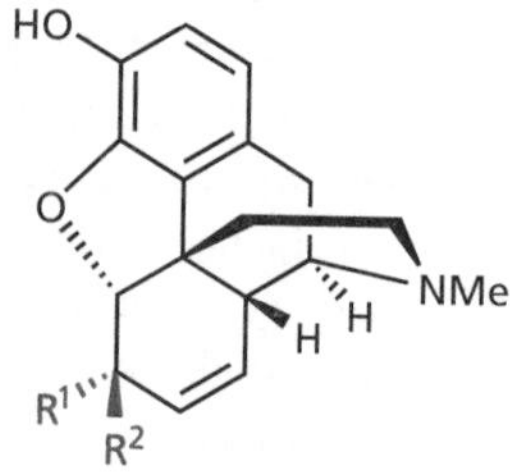

R¹	R²	Analgesia with respect to morphine
H	OH	Increased
H	H	or
	Ketone	similar

Figure 21.3 Effect of loss of alcohol group on analgesic activity.

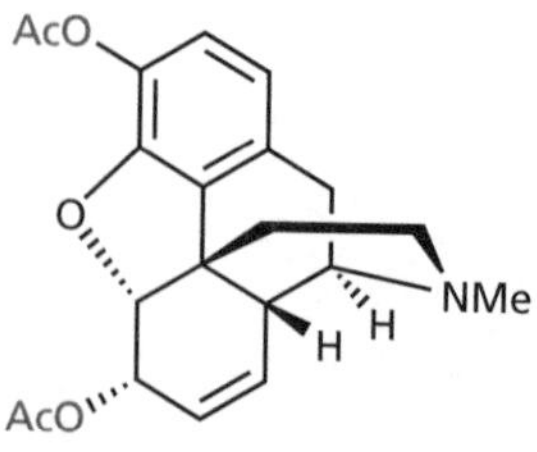

Diamorphine (heroin)   Dihydromorphine

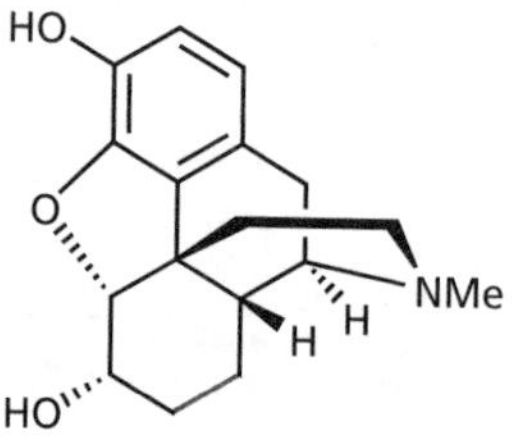

Figure 21.4 Diamorphine (heroin) and dihydromorphine.

(heroin) (Fig. 21.4). The most active (and the most dangerous) compound of the three is 6-acetylmorphine, which is four times more active than morphine. Heroin is also more active than morphine by a factor of two, but less active than 6-acetylmorphine. How do we explain this?

6-Acetylmorphine is less polar than morphine and will enter the brain more quickly and in greater concentrations. The phenolic group is free and therefore it will interact immediately with the analgesic receptors.

Heroin has two polar groups which are masked and is therefore the most efficient compound of the three in crossing the blood–brain barrier. Before it can act at the receptor, however, the 3-acetyl group has to be removed by esterases in the brain. This means that it is more powerful than morphine because of the ease with which it crosses the blood–brain barrier, but less powerful than 6-acetylmorphine since the 3-acetyl group has to be hydrolysed.

Heroin and 6-acetylmorphine are both more potent analgesics than morphine. Unfortunately, they also have greater side effects as well as severe tolerance and dependence characteristics. Heroin is still used to treat terminally ill patients suffering chronic pain, but 6-acetylmorphine is so dangerous that its synthesis is banned in many countries.

To conclude, the 6-hydroxyl group is not required for analgesic activity and its removal can be beneficial to analgesic activity.

The double bond at 7–8

Several analogues including **dihydromorphine** (Fig. 21.4) have shown that the double bond is not necessary for analgesic activity.

The N-methyl group

The N-oxide and the N-methyl quaternary salt of morphine are both inactive when tested *in vivo*, which might suggest that the introduction of charge destroys analgesic activity (Fig. 21.5). We have to remember, though, that these experiments were done on animals and it is hardly surprising that no analgesia is observed, as a charged molecule has very little chance of crossing the blood–brain barrier and reaching it target receptor. If these same compounds are injected directly into the brain, both compounds actually have a similar analgesic activity to morphine. The fact that neither compound can lose its charge shows that the nitrogen atom of morphine must be ionized when it binds to the receptor.

N-Demethylation to give normorphine reduces activity but does not eliminate it. The secondary NH group is more polar than the original tertiary group and so normorphine is less efficient at crossing the blood–brain barrier, leading to a drop in activity. The fact that significant activity is retained shows that the methyl substituent is not essential to activity. The nitrogen itself is crucial, however. If it is replaced by carbon, all analgesic activity is lost. To conclude, the nitrogen atom is essential to analgesic activity and interacts with the analgesic receptor in the ionized form.

X		Analgesic activity with respect to morphine
NH	Normorphine	25%
N⁺–Me, O⁻	*N*-Oxide	0%
N⁺–Me, Me	Quaternary salt	0%

Figure 21.5 Effect of the introduction of charge on analgesic activity.

The aromatic ring
The aromatic ring is essential. Compounds lacking it show no analgesic activity.

The ether bridge
As we shall see later, the ether bridge is not required for analgesic activity.

21.2.3.2 Stereochemistry
At this stage, it is worth making some observations on stereochemistry. Morphine is an asymmetric molecule containing several asymmetric centres, and exists naturally as a single stereoisomer. When morphine was first synthesized, it was made as a racemic mixture of the naturally occurring stereoisomer plus its mirror image. These were separated and the unnatural mirror image was tested for analgesic activity. It turned out to have no activity.

This is not particularly surprising if we consider the interactions that must take place between morphine and its receptor. We have identified that there are at least three important interactions involving the phenol, the aromatic ring, and the amine. Consider a diagrammatic representation of morphine as a T-shaped block with the three groups marked as shown in Fig. 21.6. The receptor has complementary binding groups placed in such a way that they can interact with all three groups. The mirror image of

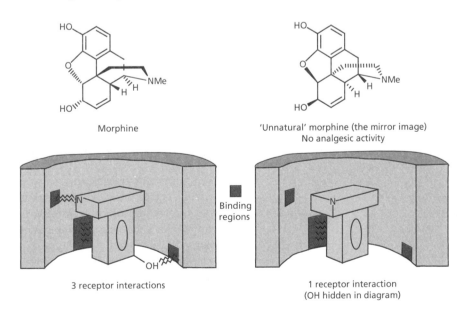

Morphine

'Unnatural' morphine (the mirror image)
No analgesic activity

3 receptor interactions

1 receptor interaction
(OH hidden in diagram)

Binding regions

Figure 21.6 Morphine and 'unnatural' morphine.

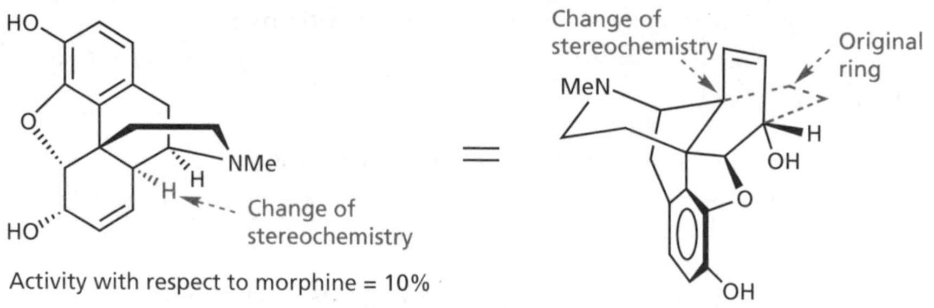

Figure 21.7 Epimerization of a single asymmetric centre.

Activity with respect to morphine = 10%

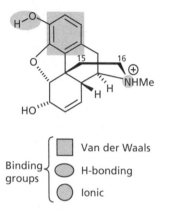

Figure 21.8 Important functional groups for analgesic activity in morphine.

morphine, on the other hand, can interact with only one binding region at any one time.

Epimerization of a single asymmetric centre such as the 14-position (Fig. 21.7) is not beneficial either, since changing the stereochemistry at even one asymmetric centre can result in a drastic change of shape, making it difficult for the molecule to bind.

To sum up, the important functional groups for analgesic activity in morphine are the phenol, aromatic ring and amine (Fig. 21.8). The stereochemistry of the molecule is crucial in ensuring that these groups are in the correct position for binding.

KEY POINTS

- Morphine is extracted from opium and is one of the oldest drugs used in medicine.

- Morphine is a powerful analgesic but has various side effects, the most serious being respiratory depression, tolerance, and dependence.

- The structure of morphine consists of five rings forming a T-shaped molecule.

- The important binding groups on morphine are the phenol, the aromatic ring, and the ionized amine.

- Analgesic activity of morphine analogues requires the presence of the important binding groups as well as the ability to cross the blood–brain barrier.

- Various opiates such as codeine and diamorphine act as prodrugs for morphine.

21.3 Morphine analogues

Considering the problems associated with morphine, there is a need for novel analgesic agents which retain the analgesic activity of morphine, but which have fewer side effects and can be administered orally. The following sections illustrate how many of the classical drug design strategies described in Chapter 10 were effective in obtaining novel analgesic structures.

21.3.1 Variation of substituents

A series of alkyl substituents were placed on the phenolic group, but the resulting compounds were inactive or poorly active. We have already identified that the phenol group must be free for analgesic activity.

The removal of the N-methyl group to give normorphine allowed a series of alkyl chains to be added to the basic centre. These results are discussed in the next section (section 21.3.2).

21.3.2 Drug extension

The strategy of drug extension described in section 10.3.2 involves the addition of extra functional groups to a lead compound in order to probe for extra binding regions in a binding site. Are extra binding regions likely in the analgesic receptor? Perhaps morphine is an ideal ligand for the binding site and already uses all the binding regions which are present? Perhaps morphine is produced naturally in the body and there is a receptor specifically for it?

All of these scenarios are certainly possible, but rather unlikely. Alkaloids such as morphine are secondary metabolites which are not essential to the growth of the plant and are only produced once the plant is mature. They could be looked upon as 'luxury items' which are not widespread in the chemistry of life and have evolved in individual species. It seems highly unlikely that the ability to synthesize morphine has evolved separately in poppies and humans. It also seems highly unlikely that morphine evolved in the poppy to act as an analgesic, so we have to conclude that it is a happy coincidence that morphine is able to interact with an analgesic receptor in the body. For

those reasons, it is unlikely that morphine is an ideal fit for the receptor binding site, and a search for further binding regions could be productive.

Many analogues of morphine containing extra functional groups have been prepared (Box 21.1). These have rarely shown any improvement. There are two exceptions, however. The introduction of a hydroxyl group at position 14 (Fig. 21.9) increases activity and suggests that there might be an extra hydrogen bond interaction taking place with the binding site.

The other exception involves the variation of alkyl substituents on the nitrogen atom. As the alkyl group is increased in size from a methyl to a butyl group, the

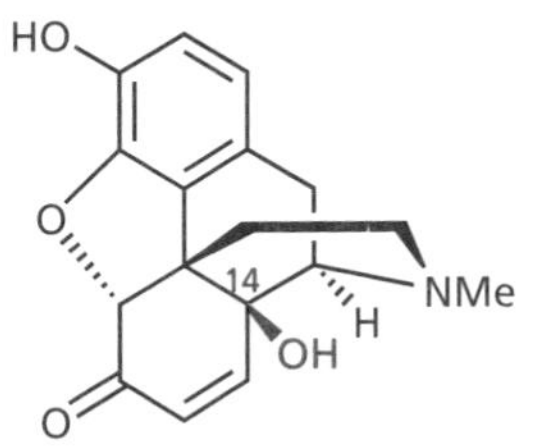

Oxymorphine
(2.5× activity of morphine)

Figure 21.9 Oxymorphine (2.5× activity of morphine).

BOX 21.1 SYNTHESIS OF N-ALKYLATED MORPHINE ANALOGUES

The synthesis of *N*-alkylated morphine analogues is easily achieved by removing the *N*-methyl group from morphine to give **normorphine,** then alkylating the amino group with an alkyl halide. Removal of the *N*-methyl group was originally achieved by a von Braun degradation with cyanogen bromide, but is now more conveniently carried out using a chloroformate reagent such as vinyloxycarbonyl chloride. The final alkylation step can sometimes be profitably replaced by a two-step process involving an acylation to give an amide, followed by reduction.

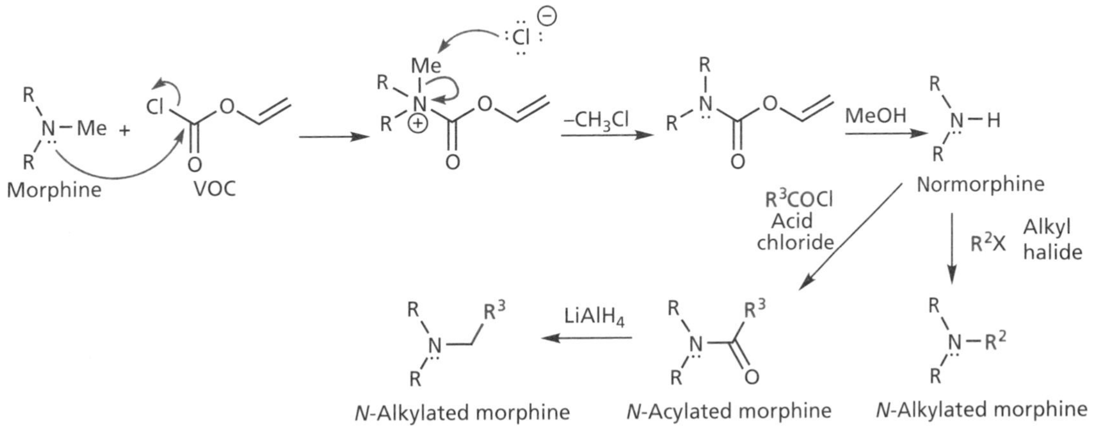

Demethylation and alkylation of the basic centre.

activity drops to zero (Fig. 21.10). With a larger group such as a pentyl or a hexyl group, activity recovers slightly. None of this is particularly exciting, but when a phenethyl group is attached, the activity increases 14-fold relative to morphine—a strong indication that a hydrophobic binding region has been located which interacts favourably with the new aromatic ring (Fig. 21.11).

To conclude, the size and nature of the group on nitrogen is important to the activity spectrum. Drug extension can lead to better binding by making use of additional binding interactions.

Before leaving this subject, it is worth describing important results which occurred when an allyl or a cyclopropylmethylene group is attached to nitrogen (Fig. 21.12). **Naloxone** and **naltrexone** have no analgesic activity at all, and **nalorphine** retains only weak analgesic activity. Not very exciting, you might think. What *is* important is that they act as antagonists to morphine, that is they bind to the analgesic receptors without switching them on. Once they have bound to

the receptors, they block morphine from binding. As a result, morphine can no longer act as an analgesic. One might be hard pushed to see an advantage in this, and with good reason. If we are just considering analgesia, there is none. However, the fact that morphine is blocked from all its receptors means that none of its side effects are produced either, and it is the blocking of these effects that makes antagonists extremely useful. For example, accident victims have sometimes been given an overdose of morphine. If this is not treated quickly, then the casualty may die of suffocation. Administering nalorphine means that the antagonist can displace morphine from its receptor and bind more strongly, thus preventing morphine from continuing its action.

Naltrexone is eight times more active than naloxone as an antagonist and is given to drug addicts who have been weaned off morphine or heroin. Naltrexone blocks the opiate receptors, blocking the effects that addicts seek if they restart their habit and making it less likely that they will do so.

Figure 21.10 Change in activity with respect to alkyl group size.

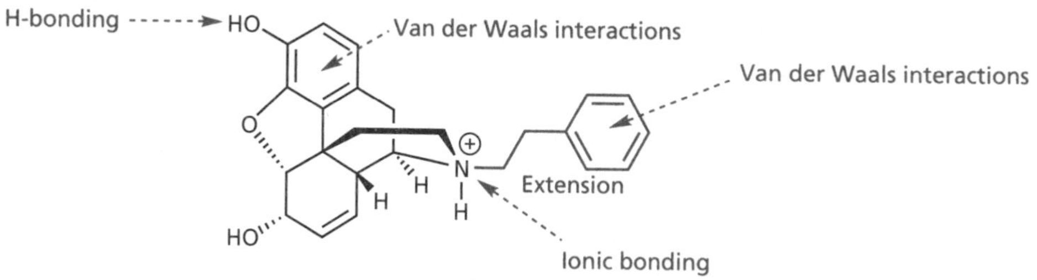

Figure 21.11 N-Phenethylmorphine.

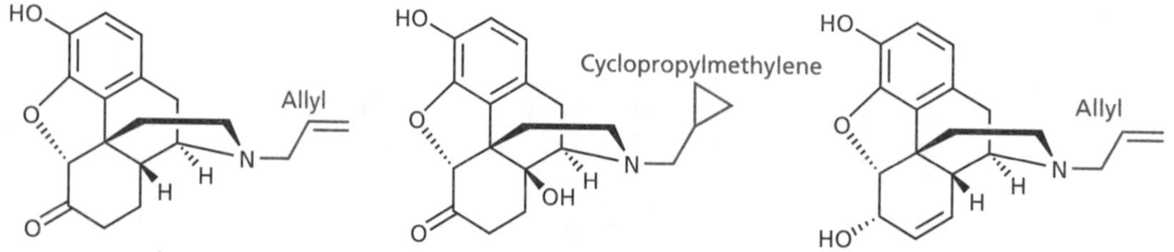

Figure 21.12 Antagonists to morphine.

There is another interesting observation related to these antagonists. For many years, chemists had been trying to find a morphine analogue without serious side effects. There had been so little success in this search that many believed that the analgesic and side effects were directly related, perhaps through the same receptor. The fact that the antagonist naloxone blocks both the analgesic and side effects of morphine did nothing to change that view. However, the properties of nalorphine offered a glimmer of hope. Nalorphine is a strong antagonist and blocks morphine from its receptors. Therefore, no analgesic activity should be observed. But a very weak analgesic activity *is* observed and what is more, this activity appears to be free of the undesired side effects. This was the first sign that a non-addictive, safe analgesic might be possible.

How can a compound act as an antagonist, but also act as an agonist and produce analgesia? If it is acting as an agonist, why is the activity so weak and why is it free of the side effects?

As we shall see later, there are several types of analgesic receptor rather than just one. Multiple receptors are common (Chapter 6) and we have already seen this in Chapters 19 and 20 with cholinergic and adrenergic receptors. In the same way, there are at least three types of analgesic receptor. The differences between them are so slight that morphine activates them all, but in theory it should be possible to find compounds which are selective for one type of analgesic receptor over another and choose a receptor that is safer than another. This is not the way that nalorphine works, however: it binds to all three types of analgesic receptor and blocks morphine from all three. It acts as a true antagonist at two of the receptors, but at the third it acts as a weak or partial agonist (section 5.8). In other words, it activates the receptor weakly. We could imagine how this might occur if the third receptor is controlling something like an ion channel (Fig. 21.13).

Morphine is a strong agonist and interacts strongly with this receptor, leading to a change in receptor conformation which fully opens the ion channel. Ions flow in or out of the cell, resulting in the activation or deactivation of enzymes. Naloxone is a pure antagonist. It binds strongly, but does not produce the correct change in the receptor conformation. Therefore, the ion channel remains closed. Nalorphine binds to the third receptor and changes the tertiary structure of the receptor very slightly, leading to a slight opening of the

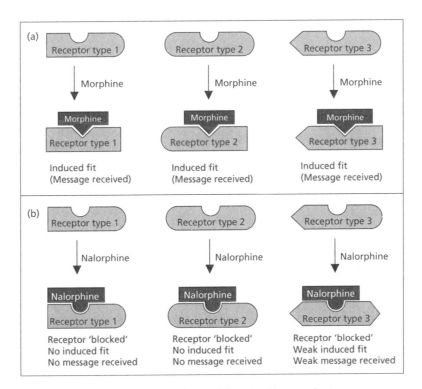

Figure 21.13 Action of (a) morphine and (b) nalorphine at analgesic receptors.

ion channel. It is therefore a weak agonist at this receptor, but it is also an antagonist since it blocks morphine from switching the receptor on fully.

The results observed with nalorphine show that activation of this third type of analgesic receptor leads to analgesia without the undesirable side effects associated with the other two analgesic receptors. Unfortunately, nalorphine has hallucinogenic side effects resulting from the activation of a non-analgesic receptor, and it is therefore unsuitable as an analgesic, but for the first time a certain amount of analgesia had been obtained without the side effects of respiratory depression and tolerance.

21.3.3 Simplification or drug dissection

We turn now to more drastic alterations of the morphine structure and ask whether the complete carbon skeleton is really necessary. If the molecule could be simplified, it would be easier to synthesize analogues. The structure of morphine has five rings (Fig. 21.14) and analogues were made to see which rings could be removed.

21.3.3.1 Removing ring E
Removing ring E leads to complete loss of activity. This emphasizes the importance of the basic nitrogen to analgesic activity.

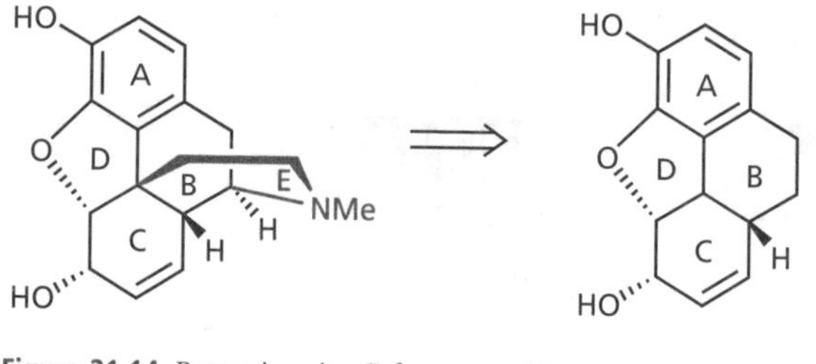

Figure 21.14 Removing ring E from morphine.

21.3.3.2 Removing ring D
Removing the oxygen bridge gives a series of compounds called the **morphinans** (Fig. 21.15), which have useful analgesic activity. This demonstrates that the oxygen bridge is not essential.

N-Methylmorphinan was the first such compound tested, and is only 20% as active as morphine, but since the phenolic group is missing, this is not surprising. The more relevant **levorphanol** structure is five times more active than morphine and, although side effects are also increased, levorphanol has a massive advantage over morphine in that it can be taken orally and lasts much longer in the body. This is because levorphanol is not metabolized in the liver to the same extent as morphine. As might be expected, the mirror image of levorphanol (**dextrorphan**) has insignificant analgesic activity.

The same strategy of drug extension already described for the morphine structures was tried on the morphinans with similar results. For example, adding an allyl substituent on the nitrogen gives antagonists. Adding a phenethyl group to the nitrogen greatly increases potency. Adding a 14-hydroxyl group also increases activity.

To conclude:

- Morphinans are more potent and longer acting than their morphine counterparts, but they also have higher toxicity and comparable dependence characteristics.

- Modifications carried out on the morphinans have the same biological results as they do with morphine. This implies that morphine and morphinans are binding to the same receptors in the same way.

- The morphinans are easier to synthesize since they are simpler molecules.

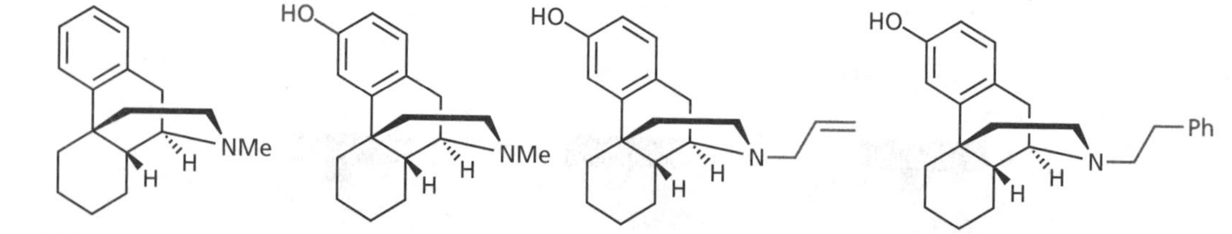

| *N*-Methylmorphinan (20% activity of morphine) | Levorphanol (5× more potent than morphine) | Levallorphan (Antagonist 5× more potent than nalorphine) | *N*-Phenethyllevorphanol (15× more potent than morphine) |

Figure 21.15 Examples of morphinans.

21.3.3.3 Removing rings C and D

Opening both rings C and D gives an interesting group of compounds called the **benzomorphans** (Fig. 21.16) which retain analgesic activity. One of the simplest of these structures is **metazocine** which has the same analgesic activity as morphine. Notice that the two methyl groups in metazocine are *cis* with respect to each other and represent the remnants of the C ring.

The same chemical modifications carried out on the benzomorphans as described for the morphinans and morphine, produce the same biological effects, implying a similar interaction with the analgesic receptors. For example, replacing the *N*-methyl group of metazocine with a phenethyl group gives **phenazocine** which is four times more active than morphine and is the first compound to have a useful level of analgesia without dependence properties.

Further developments led to **pentazocine** (Fig. 21.16) which has proved to be a useful long-term analgesic with a very low risk of addiction. A newer compound (**bremazocine**) has a longer duration, has 200 times the activity of morphine, appears to have no addictive properties, and does not depress breathing. These compounds appear to be similar in their action to nalorphine in that they act as antagonists at two of the three types of analgesic receptors, but act as agonists at the third. The big difference between nalorphine and compounds like pentazocine is that the latter are far stronger agonists, resulting in a more useful level of analgesia. Unfortunately, many of these compounds have hallucinogenic side effects due to interactions with a non-analgesic receptor.

We shall come back to the interaction of benzomorphans with analgesic receptors later. For the moment, we can state the following conclusions about benzomorphans:

- Rings C and D are not essential to analgesic activity.
- Analgesia and addiction are not necessarily co-existent.
- 6,7-Benzomorphans are clinically useful compounds with reasonable analgesic activity, less addictive liability, and less tolerance.
- Benzomorphans are simpler to synthesize than morphine and morphinans.

21.3.3.4 Removing rings B, C, and D

Removing rings B, C, and D gives a series of compounds known as 4-phenylpiperidines. The analgesic activity of these compounds was discovered by chance in the 1940s when chemists were studying analogues of cocaine for antispasmodic properties. Their structural relationship to morphine was only identified when they were found to be analgesics, and this is evident if the structure is drawn as shown in Fig. 21.17. Activity can be increased six-fold by introducing the phenolic

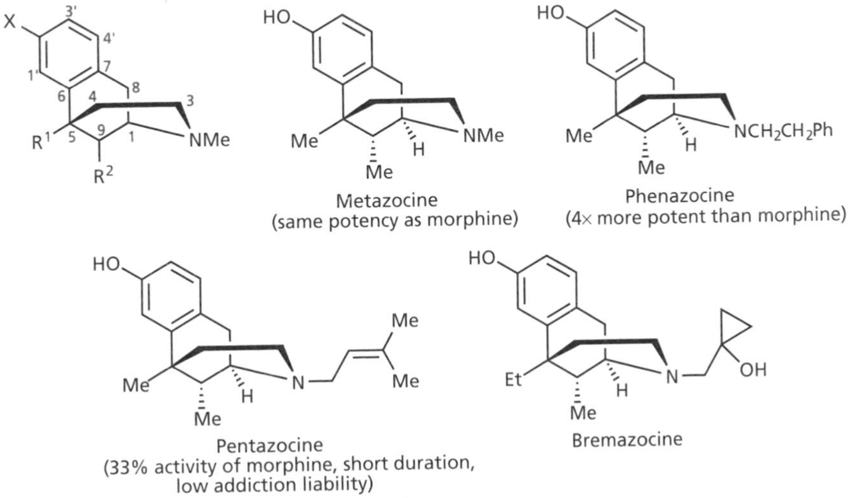

Figure 21.16 Benzomorphans.

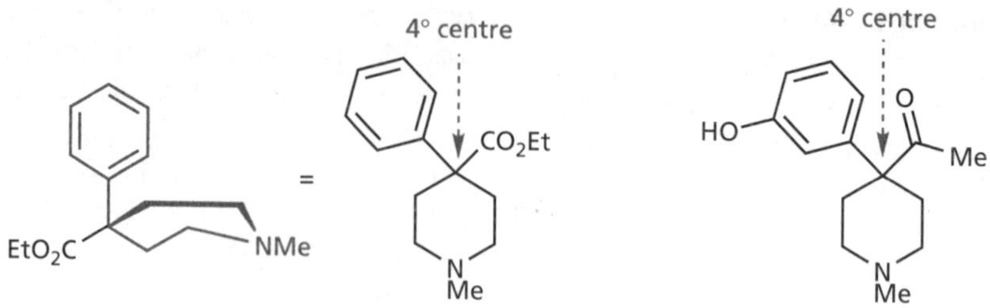

Figure 21.17 4-Phenylpiperidines.

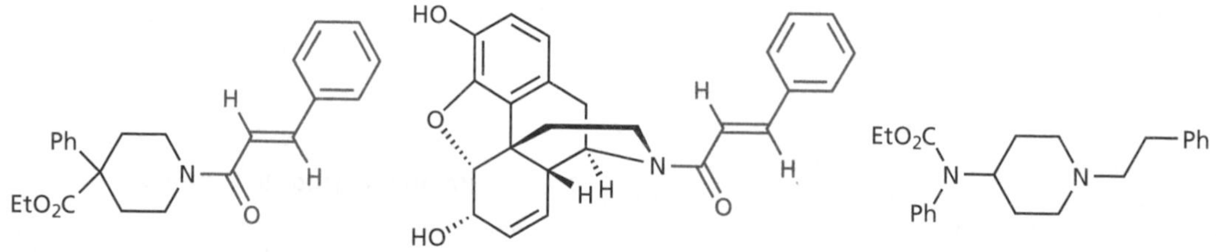

Figure 21.18 Effect of addition of a cinnamic acid residue on meperidine and morphine.

group and altering the ester to a ketone to give **ketobemidone**.

Pethidine (meperidine) is a weaker analgesic than morphine, but shares the same undesirable side effects. On the plus side, it has a rapid onset and a shorter duration of action. As a result, it has been used as an analgesic in childbirth. The rapid onset and short duration of action mean that there is less chance of the drug depressing the baby's breathing once it is born.

The piperidines are more easily synthesized than any of the previous groups and a large number of analogues have been studied. There is some doubt as to whether they act in the same way as morphine at analgesic receptors, since some of the chemical adaptations we have already described do not lead to comparable biological results. For example, adding allyl or cyclopropyl groups does not give antagonists. The replacement of the methyl group of pethidine with a cinnamic acid residue increases the activity by 30 times, whereas putting the same group on morphine eliminates activity (Fig. 21.18). These results might have something to

do with the fact that the piperidines are far more flexible molecules than the previous structures and are more likely to bind with receptors in different ways.

One of the most successful piperidine derivatives is **fentanyl** (Fig. 21.18), which is up to 100 times more active than morphine. The drug lacks a phenolic group, but is very lipophilic. As a result, it can cross the blood–brain barrier efficiently.

To conclude:

- Rings C, D, and E are not essential for analgesic activity.

- Piperidines retain side effects such as addiction and depression of the respiratory centre.

- Piperidine analgesics are faster acting and have shorter duration.

- The quaternary centre present in piperidines is usually necessary (fentanyl is an exception).

- The aromatic ring and basic nitrogen are essential to activity, but the phenol group is not.

- Piperidine analgesics appear to bind with analgesic receptors in a different manner to previous groups.

21.3.3.5 Removing rings B, C, D, and E

The analgesic **methadone** (Fig. 21.19) was discovered in Germany during the Second World War and has proved to be a useful agent, comparable in activity to morphine. It is orally active and has less severe emetic and constipation effects. Side effects

---

BOX 21.2 OPIATES AS ANTIDIARRHOEAL AGENTS

One of the main aims in drug design is to find agents that have minimal side effects, but occasionally it is possible to take advantage of a side effect. For example, one of the side effects of opiate analgesics is constipation. This is not very comfortable, but it is a useful property if you wish to counteract diarrhoea. The aim then is to design a drug such that the original side effect becomes the predominant feature. **Loperamide** is a successful antidiarrhoeal agent which is marketed as **Imodium**. It can be viewed as a hybrid molecule involving a 4-phenylpiperidine and a methadone-like structure. The compound is lipophilic, slowly absorbed, and prone to metabolism. It is also free from any euphoric effect, since it cannot cross the blood–brain barrier. All these features make it a safe medicine, free from the addictive properties of the opiate analgesics.

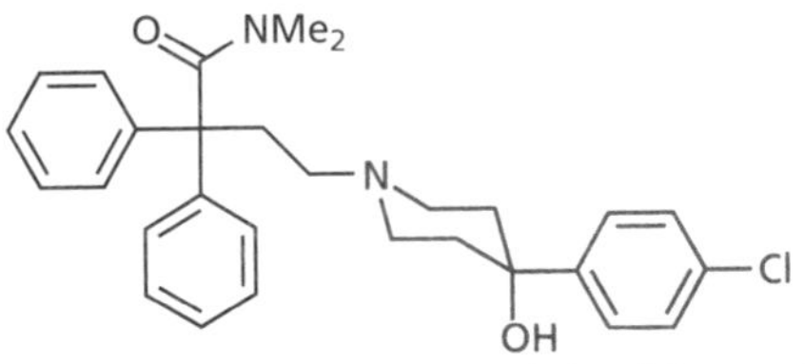

Methadone-like skeleton   4-Phenylpiperidine moiety

---

such as sedation, euphoria, and withdrawal symptoms are also less severe and therefore the compound has been given to drug addicts as a substitute for morphine or heroin in order to wean them off these drugs. This is not a complete cure, as it merely swaps an addiction to heroin or morphine for an addiction to methadone. This is considered less dangerous, however.

The molecule has a single asymmetric centre and when the molecule is drawn in the same manner as morphine, we would expect the *R*-enantiomer to be the more active enantiomer. This proves to be the case, with the *R*-enantiomer being twice as powerful as morphine, whereas the *S*-enantiomer is inactive. This is quite a dramatic difference. Since the *R*- and *S*-enantiomers have identical physical properties and lipid solubility, they should both reach analgesic receptors to the same extent, and so the difference in activity is most probably due to receptor–ligand interactions.

Many analogues of methadone have been synthesized, but with little improvement over the parent drug.

21.3.4 Rigidification

The strategy of rigidification is used to limit the number of conformations that a molecule can adopt. The aim is to retain the active conformation for the desired target and eliminate alternative conformations that might fit different targets (section 10.3.9). This should increase activity, improve selectivity and decrease side effects. The best examples of this tactic in the analgesic field are the **oripavines**, which often show remarkably high activity. A comparison of these structures with morphine shows that an extra ring sticks out from what used to be the crossbar of the T-shaped morphine (Fig. 21.20).

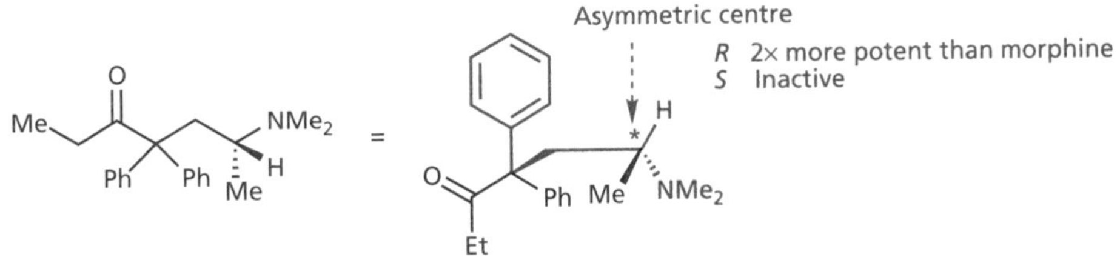

Figure 21.19 Methadone.

BOX 21.3 SYNTHESIS OF THE ORIPAVINES

The oripavines are synthesized from an alkaloid called **thebaine** which is extracted from opium along with codeine and morphine. Although similar in structure to both these compounds, thebaine has no analgesic activity. There is a diene group present in ring C and when thebaine is treated with methyl vinyl ketone, a Diels–Alder reaction takes place to give an extra ring and increased rigidity to the structure (Fig. 1).

As a ketone group has been introduced, it is now possible to try the strategy of drug extension by adding various groups to the ketone via a Grignard reaction. It is noteworthy that this reaction is stereospecific. The Grignard reagent complexes to both the 6-methoxy group and the ketone, and is then delivered to the less-hindered face of the ketone to give an asymmetric centre (Fig. 2).

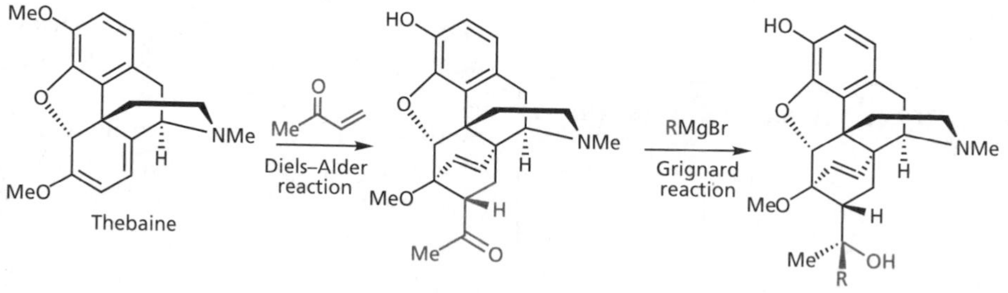

Figure 1 Formation of oripavines.

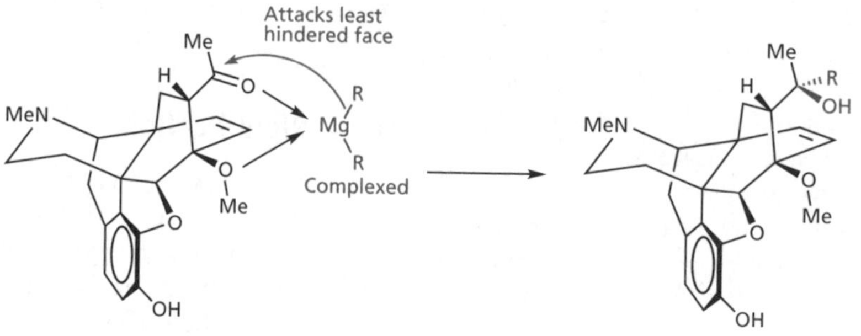

Figure 2 A Grignard reaction leads to an asymmetric centre.

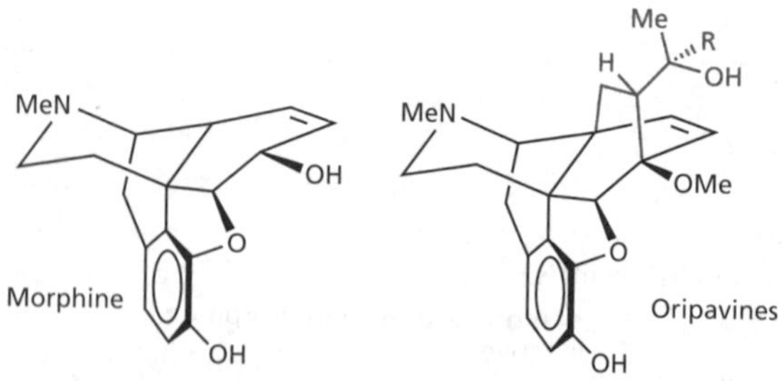

Figure 21.20 Comparison of morphine and oripavines.

Some remarkably powerful oripavines have been obtained (Box 21.3). **Etorphine** (Fig. 21.21), for example, is 10 000 times more potent than morphine.

This is a combination of the fact that it is a very hydrophobic molecule and can cross the blood–brain barrier 300 times more easily than morphine, as well as having 20 times more affinity for the analgesic receptor site due to better binding interactions. At slightly higher doses than those required for analgesia, it can act as a knock-out drug or sedative. It has a considerable margin of safety and is used to immobilize large animals such as elephants. Since the compound is so active, only very small doses are required and these can be dissolved in such small volumes (1 ml) that they can be placed in crossbow darts and fired into the hide of the animal.

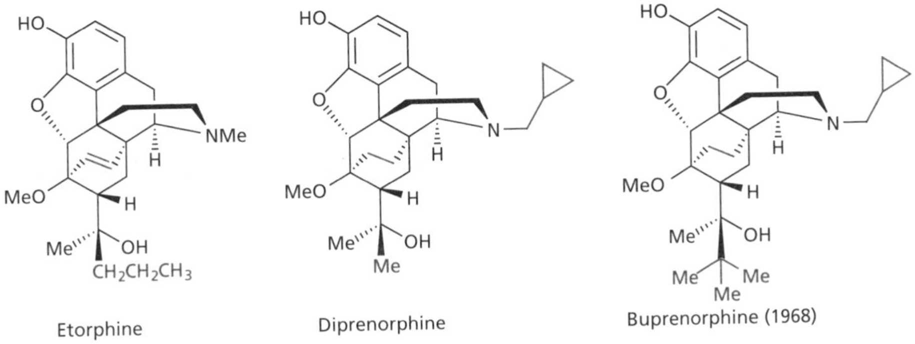

Figure 21.21 Etorphine and related structures.

The addition of lipophilic groups (R in Fig. 21.20) is found to improve activity dramatically, indicating the presence of an extra hydrophobic binding region in the receptor binding site.[2] The group best able to interact with this region is a phenethyl substituent, and the product containing this group is even more active than etorphine. As one might imagine, these highly active compounds have to be handled very carefully in the laboratory.

Because of their rigid structures, these compounds are highly selective agents for the analgesic receptors. Unfortunately, the increased analgesic activity is also accompanied by unacceptable side effects. It was therefore decided to see whether nitrogen substituents, such as an allyl or cyclopropyl group, would give antagonists as described in section 21.3.2. If so, it might be possible to obtain an oripavine equivalent of a pentazocine or a nalorphine—an antagonist with some agonist activity and with reduced side effects.

Adding a cyclopropyl group gives a very powerful antagonist called **diprenorphine** (Fig. 21.21), which is 100 times more potent than nalorphine and can be used to reverse the immobilizing effects of etorphine. Diprenorphine has no analgesic activity.

A similar compound is **buprenorphine** (Fig. 21.21) which has similar properties to drugs like nalorphine and pentazocine, in that it has analgesic activity with a very low risk of addiction. It is 100 times more active than morphine as an agonist and 4 times more active than nalorphine as an antagonist. It is a particularly safe drug since it has very little effect on

respiration and what little effect it does have actually decreases at high doses. Therefore, the risks of suffocation from a drug overdose are much smaller than with morphine. Buprenorphine has been used in hospitals to treat patients suffering from cancer and also after surgery. Its drawbacks include side effects such as nausea and vomiting, as well as the fact that it cannot be taken orally. A further use for buprenorphine is as an alternative means to methadone for weaning addicts off heroin.

The properties of buprenorphine appear to be related to its slow onset of action and slow departure from receptors. Since buprenorphine is the most lipophilic compound in the oripavine series of compounds and enters the brain very easily, the slow onset has nothing to do with how easily it reaches the receptor. Instead, it is a result of slow binding then slow departure from the binding site. Effects are gradual, so there are no sudden changes in transmitter levels. Although buprenorphine binds slowly, it binds very strongly once it is bound. As a result, less buprenorphine than morphine is required to interact with a certain percentage of analgesic receptors. On the other hand, buprenorphine is only a partial agonist and is less efficient at switching the analgesic receptors on. This means that it is unable to reach the maximum level of analgesia which can be produced by morphine. Overall, buprenorphine's stronger affinity for analgesic receptors outweighs its relatively weak analgesic action, such that buprenorphine can produce analgesia at lower doses than morphine. However, if the pain levels are high, buprenorphine cannot reach the analgesic levels required and morphine has to be used. Nevertheless, buprenorphine provides another example of an opiate analogue

---

[2] It is believed that the phenylalanine aromatic ring on enkephalins (see later) interacts with this same binding region.

where analgesia has been separated from dangerous side effects.

It is time to look more closely at the receptor theories that are relevant to the analgesics.

KEY POINTS

- The addition of a 14-hydroxyl group or an *N*-phenethyl group usually increases activity as a a result of interactions with extra binding regions.

- *N*-Alkylated analogues of morphine are easily synthesized by demethylating morphine to normorphine, then alkylating with alkyl halides.

- The addition of suitable *N*-substituents results in compounds which act as antagonists or partial agonists. Such compounds can be used as antidotes to morphine overdose, as treatment for addiction or as safer analgesics.

- The morphinans and benzomorphans are analgesics which have a simpler structure than morphine and interact with analgesic receptors in a similar fashion.

- The 4-phenylpiperidines are a group of analgesic compounds which contain the analgesic pharmacophore present in morphine. They may bind to analgesic receptors slightly differently from analgesics of more complex structure.

- Methadone is a synthetic agent which contains part of the analgesic pharmacophore present in morphine. It is administered to drug addicts to wean them off heroin.

- Thebaine is an alkaloid derived from opium which lacks analgesic activity. It is the starting material for a two-stage synthesis of oripavines.

- Oripavines are extremely potent compounds due to enhanced receptor interactions and an increased ability to cross the blood–brain barrier.

- The addition of *N*-cycloalkyl groups to the oripavines results in powerful antagonists or partial agonists which can be used as antidotes, for the treatment of addiction or as safer analgesics.

21.4 Receptor theory of analgesics

The identification of analgesic receptors came relatively late on in the opiate story (1973). There are at least four different receptors with which morphine can interact. Three of these promote analgesia. The initial theory on

receptor binding (the Beckett–Casy hypothesis) assumed a single analgesic receptor, but this does not invalidate many of the proposals which were made.

21.4.1 Beckett–Casy hypothesis

In the Beckett–Casy hypothesis, it was assumed that there was a rigid binding site into which morphine and its analogues could fit in a classic lock-and-key analogy. Based on the results already described, the following binding interactions were proposed (Fig. 21.22):

- There must be a basic nitrogen which is ionized at physiological pH to form a positively charged group. This group then forms an ionic bond with a comparable anionic group in the receptor binding site. Consequently, analgesics should have a pK_a of 7.8–8.9 such that there is an approximately equal chance of the amine being ionized or un-ionized at physiological pH. This is necessary as the analgesic has to cross the blood–brain barrier as the free base, then ionize in order to interact with the receptor. The pK_a values of useful analgesics all match this prediction.

- The aromatic ring in morphine has to be properly orientated with respect to the nitrogen atom to allow a van der Waals interaction with a suitable hydrophobic location in the binding site. The nature of this interaction suggests that there has to be a close spatial relationship between the aromatic ring and the surface of the binding site.

- The phenol group is probably hydrogen bonded to a suitable residue in the receptor binding site.

- There might be a 'hollow' just large enough for the ethylene bridge of C-15 and C-16 to fit. Such a fit would help to align the molecule and enhance the overall fit.

This first theory fitted in well with most results. There can be no doubt that the aromatic ring, phenol, and amine are all important, but there is some doubt about the ethylene bridge as there are several analgesics which lack it, for example fentanyl. There is also the extra binding region discovered by drug extension (section 21.3.2). Although the theory can be easily adapted to take account of extra binding regions, other anomalies exist and a simple one receptor theory does not adequately address these.

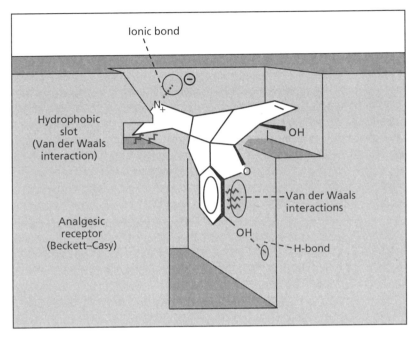

Figure 21.22 Beckett–Casy hypothesis.

21.4.2 Multiple analgesic receptors

The Beckett–Casy theory tried to explain analgesic results based on a single analgesic receptor. It is now known that there are three different analgesic receptors which are associated with different types of side effects. It is also known that several analgesics show preference for some of these receptors over others. This helps to explain many of the aforementioned anomalies. Nevertheless, it is important to appreciate that the main points of the original theory still apply for each of the analgesic receptors now to be described. The important binding groups for each receptor are the phenol, the aromatic ring, and the ionized nitrogen centre. Beyond that, there are subtle differences between the receptors. As a result, some analgesics show preference for one analgesic receptor over another or interact in different ways.

There are three analgesic receptors which are activated by morphine.

21.4.2.1 The μ-receptor

Morphine binds strongly to this receptor and produces analgesia. Receptor binding also leads to the undesired side effects of respiratory depression, euphoria, and addiction. We can now see why it is so difficult to remove the side effects of morphine and its analogues, since the receptor to which they bind most strongly inherently produces these side effects.

21.4.2.2 The κ-receptor

Morphine binds less strongly to this receptor. The biological response is analgesia with sedation and none of the hazardous side effects. It is the κ-receptor that provides the best hope for the ultimate safe analgesic. The earlier results obtained from nalorphine, pentazocine, and buprenorphine can now be explained.

Nalorphine acts as an antagonist at the μ-receptor, thus blocking morphine from acting there. However, it acts as a weak agonist at the κ-receptor and so the slight analgesia observed with nalorphine is due to the partial activation of the κ-receptor. Unfortunately, nalorphine has hallucinogenic side effects. This is caused by nalorphine binding to a completely different, non-analgesic receptor in the brain called the sigma receptor (σ) (section 21.7.4), where it acts as an agonist.

Pentazocine interacts with the μ- and κ-receptors in the same way, but is able to switch on the κ-receptor more strongly. It too suffers from the drawback that it switches on the σ-receptor.

Table 21.1 Relative activities of analgesics

Receptor	Effect	Morphine	Nalorphine	Pentazocine	Enkephalins	Pethidine	Naloxone
μ	Analgesia Respiratory depression Euphoria Addiction	+ + +	−	−	+	+ + +	−
κ	Analgesia Sedation	+	+	+	+	+	−
σ	Hallucinogenic		+	+			
δ	Analgesia	+ +	−	−	+ + +	+	−

Figure 21.23 New analgesic structures. Nalbuphine has the same activity as morphine, low addiction liability, no psychomimetic activity, but is orally inactive. Butorphanol is also orally inactive.

Buprenorphine is slightly different. It binds strongly to all three analgesic receptors, acting as an antagonist at the δ (see below) and κ-receptors, but acting as a partial agonist at the μ-receptor. This might suggest that buprenorphine should suffer the same side effects as morphine. The fact that it does not is related in some way to the rate at which buprenorphine interacts with the receptor. It is slow to bind, but once it has bound, it is slow to leave.

21.4.2.3 The δ-receptor

The δ-receptor is where the brain's natural painkillers interact—the **enkephalins** (section 21.6). Morphine can also bind quite strongly to this receptor.

Table 21.1 shows the relative activities of morphine, nalorphine, pentazocine, enkephalins, pethidine, and naloxone. A plus sign indicates that the compound is acting as an agonist. A minus sign means that it acts as an antagonist. No sign means that there is no activity or minor activity.

There is now a search going on for orally active opiate structures which can act as antagonists at the μ-receptor and agonists at the κ-receptor, and have no

activity at the σ-receptor. Some success has been achieved, especially with the compounds shown in Fig. 21.23, but even these compounds still suffer from side effects, or lack the desired oral activity.

21.5 Agonists and antagonists

We return now to look at a particularly interesting problem regarding the agonist/antagonist properties of morphine analogues. Why should such a small change as replacing an N-methyl group with an allyl group result in such a dramatic change in biological activity, such that an agonist becomes an antagonist? Why should a molecule such as nalorphine act as an agonist at one analgesic receptor and an antagonist at another? How can different receptors distinguish between such subtle changes in a molecule?

We shall consider one theory proposed by Snyder and co-workers (Feinberg et al. 1976) which attempts to explain how these distinctions might take place, but it is important to realize that there are alternative theories. In this particular theory, it is suggested that there are two accessory hydrophobic binding regions present in an analgesic receptor. It is then proposed that a structure will act as an agonist or as an antagonist depending on which of these extra binding regions is used. In other words, one of the hydrophobic binding regions is an agonist binding region, while the other is an antagonist binding region (Figs. 21.24–21.26). In the model, the agonist binding region is further away from the nitrogen and positioned axially with respect to it. The antagonist region is closer and positioned equatorially.

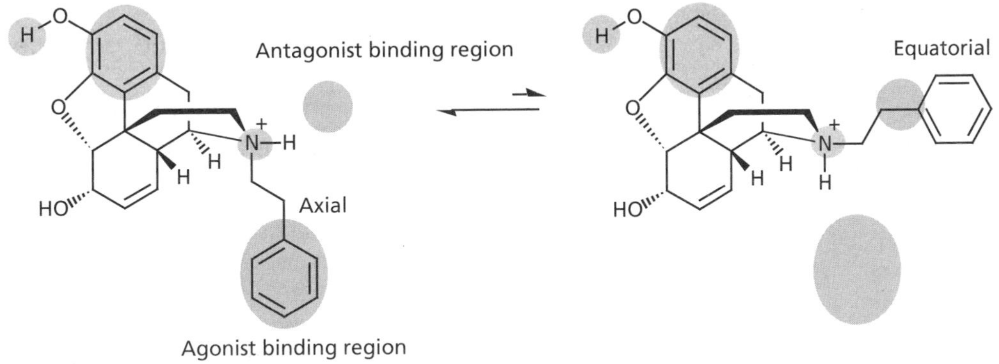

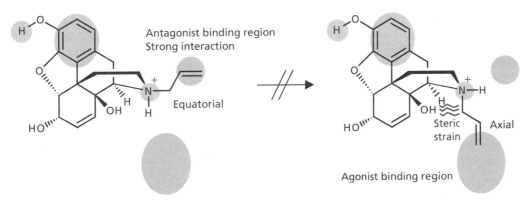

Figure 21.24 *N*-Phenethylmorphine binding interactions.

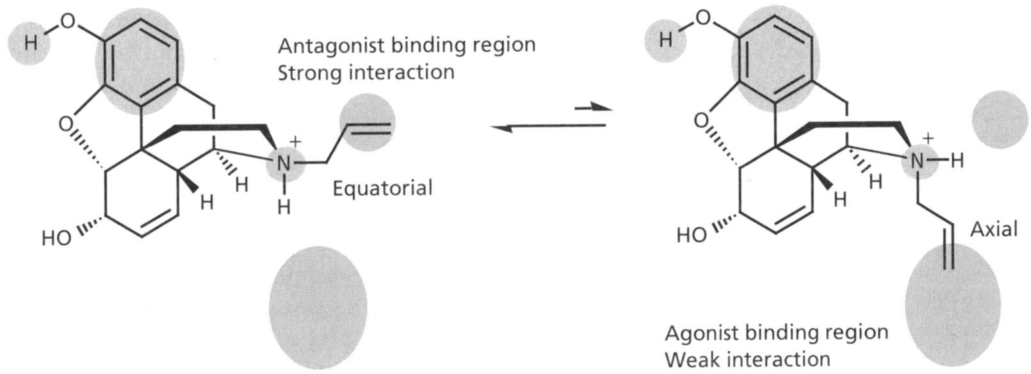

Figure 21.25 *N*-Allylmorphine binding interactions.

Figure 21.26 Influence of 14-OH on binding interactions.

Let us now consider *N*-phenethylmorphine (Fig. 21.24). Like morphine, it binds using its phenol, aromatic, and amine functional groups. If the phenethyl group is in the axial position, the aromatic ring is in the correct position to interact with the accessory agonist binding region. If the phenethyl group is in the equatorial position, the aromatic ring is placed beyond the antagonist binding region and cannot bind to either of the accessory regions. The overall result is that the molecule binds with the phenethyl group axial, resulting in increased agonist activity.

Now consider what happens if the phenethyl group is replaced by an allyl group (Fig. 21.25). In the equatorial position, the allyl group binds strongly to the antagonist binding region, whereas in the axial position it barely reaches the agonist binding region, resulting in a weak interaction.

It is proposed that a molecule such as phenazocine with a phenethyl group acts as an agonist because it can only bind to the agonist binding region. A molecule such as nalorphine with an allyl group can bind to both agonist and antagonist regions and therefore acts as an agonist at one receptor and an antagonist at another. The ratio of these effects would depend on the relative equilibrium ratio of the axial and equatorial substituted isomers.

A compound which is a pure antagonist would be forced to have a suitable substituent in the equatorial position. It is believed that the presence of a 14-hydroxyl group sterically hinders the isomer with the axial substituent, and forces the substituent to remain equatorial (Fig. 21.26).

21.6 Endogenous opioid peptides

21.6.1 Enkephalins, endorphins, dynorphins, and endomorphins

Morphine is an alkaloid which relieves pain and acts in the CNS. This implies that there are analgesic receptors in the CNS, and that there are endogenous chemicals which interact with these receptors. Morphine itself is not present in humans and therefore the body must be using a different chemical as its endogenous painkiller.

The search for this natural analgesic took many years, but ultimately led to the discovery of the enkephalins and the endorphins. The term enkephalin is derived from the Greek, meaning 'in the head', and that is exactly where the enkephalins are produced. The first enkephalins to be discovered were the pentapeptides **Met-enkephalin** and **Leu-enkephalin** (Fig. 21.27). These have a preference for the δ-receptor, and it has been proposed that enkephalins are responsible for the analgesic effects of acupuncture.

At least 15 endogenous peptides have now been discovered (the enkephalins, dynorphins, and the endorphins), varying in length from 5 to 33 amino acids. These compounds are thought to be neurotransmitters or neurohormones in the brain and operate as the body's natural painkillers, as well as having a number of other roles. They are derived from three inactive precursor proteins—**proenkephalin**, **prodynorphin**, and **pro-opiomelanocortin** (Fig. 21.28).

All the above compounds are found to have either the Met- or the Leu-enkephalin skeleton at their N-terminus, which emphasizes the importance of this pentapeptide structure towards analgesic activity. It has

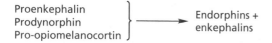

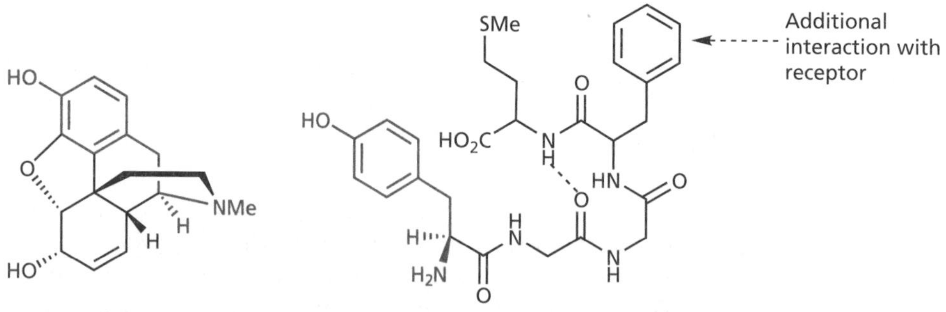

H-Tyr-Gly-Gly-Phe-Met-OH   H-Tyr-Gly-Gly-Phe-Leu-OH

Met-enkephalin   Leu-enkephalin

Figure 21.27 Enkephalins.

Proenkephalin
Prodynorphin   } ⟶ Endorphins +
Pro-opiomelanocortin   enkephalins

Figure 21.28 Production of the body's natural painkillers.

Morphine

Met-enkephalin

Figure 21.29 Comparison of morphine and Met-enkephalin (dashed line is a H-bond).

also been shown that tyrosine is essential to activity, and much has been made of the fact that there is a tyrosine skeleton in the morphine skeleton (Fig. 21.29).

If the crucial part of these molecules is the N-terminal pentapeptide, why should there be so many different peptides carrying out the same task? One suggestion is that the remaining peptide chain of each molecule is responsible for targeting each peptide to particular types of analgesic receptor. It is known that enkephalins show preference for the δ-receptor, whereas dynorphins show selectivity for the κ-receptor, and β-endorphins show selectivity to both the δ- and μ-receptors.

The most recently discovered opioid peptides are the endomorphins (Fig. 21.30), which are unlike previous opioid peptides. For a start, they are tetrapeptides, whereas all other opioid peptides are pentapeptides or larger. Secondly, the second and third amino acids in their skeleton differ from glycine, another break from convention. Finally, they have a primary amide functional group at the C-terminus. They do, however, have the mandatory tyrosine and phenylalanine residues that are present in all other opioid peptides. These peptides have a strong preference for the μ-receptor.

21.6.2 Analogues of enkephalins

SAR studies on the enkephalins have shown the importance of the tyrosine phenol ring and the tyrosine amino group. Without either, activity is lost. If tyrosine is replaced by another amino acid, activity is also lost (the only exception being D-serine).

It has been found that the enkephalins are easily inactivated by peptidase enzymes *in vivo*, with the most labile bond being the peptide link between tyrosine and glycine. Efforts have been made to synthesize analogues

which are resistant towards this hydrolysis. It is possible to replace either or both of the glycine units with unnatural D-amino acids such as D-alanine. Since D-amino acids do not occur naturally in the human body, peptidases do not recognize the structure and the peptide bond is not attacked. The alternative tactic of replacing L-tyrosine with D-tyrosine is not possible, as it completely alters the relative orientation of the tyrosine aromatic ring with respect to the rest of the molecule. As a result, the analogue is unable to bind to the analgesic receptor and is inactive.

N-Methylating the peptide link also blocks peptidase hydrolysis. Another tactic is to use unusual amino acids which are not recognized by peptidases or prevent the molecule from fitting the peptidase active site (Fig. 21.31). Unfortunately, the enkephalins also have some activity at the μ receptor and so the search for selective agents continues.

21.6.3 Inhibitors of peptidases

An alternative approach to pain relief is to enhance the activity of natural enkephalins by inhibiting the peptidase enzymes which metabolize them. Studies have shown that the enzyme responsible for metabolism has a zinc ion present in the active site, as well as a hydrophobic pocket which normally accepts the phenylalanine residue present in enkephalins. A dipeptide (Phe-Gly) was chosen as the lead compound and a thiol group was incorporated to act as a binding group for the zinc ion. The result was a structure called **thiorphan** (Fig. 21.32) which was shown to have analgesic activity. It remains to be seen whether agents such as these will prove useful in the clinic.

21.7 Receptor mechanisms

In general, receptors are situated in cell membranes and act as communication centres for messages sent by neurotransmitters or hormones—the ligands.

H-Tyr-Pro-Trp-Phe-NH₂ H-Tyr-Pro-Phe-Phe-NH₂

Endomorphin-1 Endomorphin-2

Figure 21.30 Endomorphins.

H-L-Tyr-Gly-Gly-L-Phe-L-Met-OH	Met-Enkephalin—δ agonist and some μ activity.
H-L-Tyr-D-AA-Gly-NMe-L-Phe-L-Met-OH	Resistant to peptidase. Orally active.
N,N-Diallyl-L-Tyr-aib-aib-L-Phe-L-Leu-OH	Antagonist to δ receptor (aib = α-aminobutyric acid).
Longer enkaphalins/endorphins	Increase in κ activity. Slight increase in μ activity.

Figure 21.31 Tactics to stabilize the bond between the tyrosine and glycine residues.

Receptor–ligand binding causes the receptor to change shape, resulting in further changes such as the opening of an ion channel or the activation/deactivation of enzymes (see Chapters 5 and 6). Ultimately, there is a biological effect which in the case of the analgesics is the inhibition of pain signals between nerves. We shall now look at the analgesic receptors in a little more detail.

21.7.1 The μ-receptor

As Fig. 21.33 demonstrates, morphine binds to the μ-receptor and induces a change in shape. This change in conformation opens up an ion channel in the cell membrane and, as a result, potassium ions can flow out of the cell. This flow hyperpolarizes the membrane potential and makes it more difficult for an action potential to be reached (see Appendix 4). Therefore, the frequency of action potential firing is decreased, which results in a decrease in neuron excitability. This increase in potassium permeability has a further indirect effect, as it decreases the influx of calcium ions into the nerve terminal. This in turn reduces neurotransmitter release.

Both effects 'shut down' the nerve and block the pain messages. Unfortunately, this receptor is also associated with the hazardous side effects of narcotic analgesics. There is still a search to see if there are

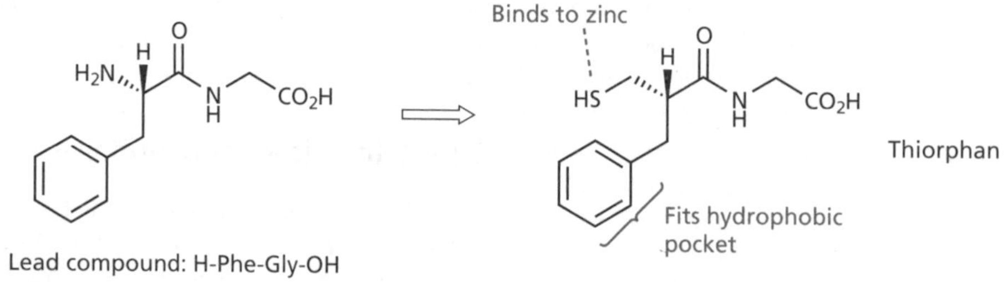

Figure 21.32 Development of thiorphan.

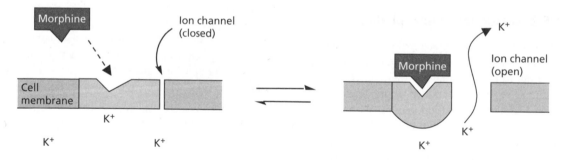

Figure 21.33 Morphine binding to μ receptor.

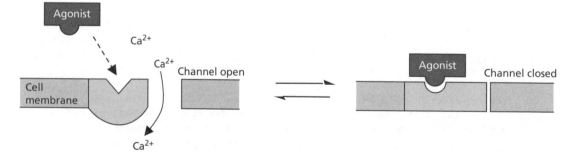

Figure 21.34 The association of the κ receptor and calcium channels.

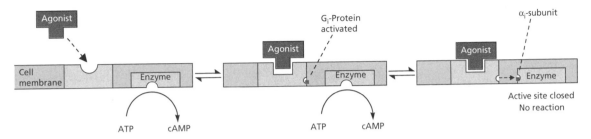

Figure 21.35 The delta receptor (δ).

different subtypes of the μ-receptor, one of which might be free of side effects.

21.7.2 The κ-receptor

The κ-receptor is directly associated with a calcium channel (Fig. 21.34). When an agonist binds to the κ-receptor, the receptor changes conformation and the calcium channel (normally open when the nerve is active) is closed. Calcium is required for the release of neurotransmitters and so the nerve cannot pass on pain messages.

The nerves affected by the κ-receptors are those related to pain induced by non-thermal stimuli. This is not the case with the μ-receptor, where all pain messages are inhibited. This suggests a difference in distribution between κ-receptors and μ-receptors.

21.7.3 The δ-receptor

Like the μ-receptor, the nerves containing the δ-receptor do not discriminate between pain from different sources. The δ-receptor is a G-protein-linked receptor (Fig. 21.35) (see Chapter 6). When an agonist binds, the receptor changes shape and triggers the fragmentation of a messenger G_i-protein. The α_i-subunit of the G-protein inhibits the membrane-bound enzyme adenylate cyclase and prevents the synthesis of cAMP (section 6.5.3). The transmission of the pain signal requires cAMP to act as a secondary messenger, and so inhibition of the enzyme blocks the signal.

21.7.4 The σ-receptor

This is not an analgesic receptor, but it can be activated by opiate molecules such as nalorphine to produce hallucinogenic effects. The σ-receptor may

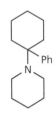

Figure 21.36 Phencyclidine (PCP) or 'angel dust'.

be the one associated with the hallucinogenic and psychotomimetic effects of **phencyclidine** (PCP), otherwise known as 'angel dust' (Fig. 21.36).

21.8 The future

There is still a need for analgesic drugs with reduced side effects. The following approaches are feasible in the field of opiates.

- **κ-Agonists:** Such compounds should have reduced side effects. However, a completely specific κ-agonist has not yet been found. Examples of selective agonists are shown in Fig. 21.37.

- **Selectivity between μ-receptor subtypes:** There may be subtypes of the μ receptor, one of which may lack the serious side effects. An agent showing selectivity would prove such a theory and be a useful analgesic.

- **Peripheral opiate receptors:** Peripheral opiate receptors have been identified in the ileum and are responsible for the antidiarrhoeal activity of opiates. If peripheral sensory nerves also possess opiate receptors, drugs might be designed to target these

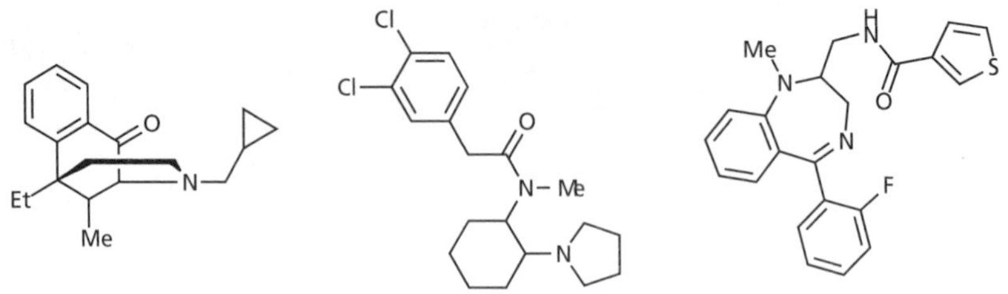

Figure 21.37 Examples of selective agonists.

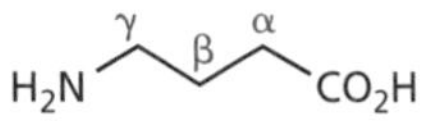

Figure 21.38 γ-Aminobutanoic acid (GABA).

sites and would not need to cross the blood–brain barrier.

- **Blocking postsynaptic receptors:** Opiates act on analgesic receptors on sensory nerves to modify the transmission of pain within a nerve or the release of a transmitter from a nerve. Another approach would be to block the postsynaptic receptors involved in the transmission of a pain signal from one nerve to another. This would involve finding selective antagonists for the various neurotransmitters involved. One promising lead is the neurotransmitter **γ-aminobutyric acid** (**GABA**) (Fig. 21.38), which appears to have a role in the regulation of enkephalinergic neurons and as such affects pain pathways. Another chemical messenger involved in the role of pain mediation is an undecapeptide structure called **substance P**, which is an excitatory neurotransmitter.

- **Agonists for the cannabinoid receptor:** Cannabinoid agonists may have a role to play in enhancing the effects of opiate analgesics, and may allow less opiate to be administered.

KEY POINTS

- The Beckett–Casy receptor theory identifies the importance of a hydrogen bonding interaction via a phenol group, an ionic interaction via a charged amine, and van der Waals interactions involving the aromatic ring. Modifications to the theory can account for extra binding interactions as a result of drug extension.

- There are three different analgesic receptors (μ, κ, and δ). All require the presence of a pharmacophore involving the phenol, aromatic ring, and ionized amine.

- The μ-receptor is responsible for the opening of a potassium ion channel. Morphine binds most strongly to the μ-receptor. This receptor is responsible for the serious side effects associated with morphine.

- The κ-receptor is responsible for analgesia and sedation, and lacks serious side effects. Receptor–ligand binding results in the closure of a calcium ion channel.

- The δ-receptor is favoured by the enkephalins. Receptor ligand binding results in the activation and fragmentation of a G_i-protein. The resulting α_i-subunit inhibits the enzyme adenylate cyclase and prevents the synthesis of cAMP.

- It is proposed that there are two accessory hydrophobic binding regions in the receptor binding site. An agent will act as an agonist or antagonist depending on which of these regions it can access.

- Enkephalins, dynorphins, endomorphins and endorphins are peptides which act as the body's natural painkillers. The presence of an N-terminal tyrosine is crucial to activity.

- Analogues of enkephalins have been designed to be more stable to peptidases, by the inclusion of unnatural amino acids, D-amino acids or N-methylated peptide links.

- The σ-receptor is responsible for the hallucinogenic properties of some analgesics.

QUESTIONS

1. Morphine is an example of a plant alkaloid. Alkaloids tend to be secondary metabolites that are not crucial to a plant's growth and are produced when the plant is mature. If that is the case, what role do you think these compounds have in plants, if any?

2. The synthesis in Box 21.1 shows that *N*-alkylated analogues can be synthesized by *N*-alkylation directly or by a two-stage process involving *N*-acylation. Why might a two-stage process be preferred to direct *N*-alkylation? What sort of products could not be synthesized by the two stage process? Is this likely to be a problem?

3. Show how you would synthesize nalorphine (Fig. 21.13).

4. Pethidine has been used in childbirth as it is short-acting and less hazardous than morphine in the newborn baby. Several drugs taken by the mother before giving birth can prove hazardous to the child after birth, but less so before birth. Why is this?

5. Show how you would synthesize diprenorphine and buprenorphine.

6. Why is buprenorphine considered the most lipophilic of the oripavine series of compounds.

7. Identify the potential hydrogen bond donors and acceptors in morphine. Structure–activity relationships reveal that one functional group in morphine is important as a hydrogen bond donor or as a hydrogen bond acceptor. Which group is that?

8. Propose the likely analgesic activity of 3-acetyl morphine relative to morphine, heroin, and 6-acetylmorphine.

9. Describe how you would synthesize the *N*-phenethyl analogue of morphine.

10. The *N*-phenethyl analogue of morphine is known as a semi-synthetic product. What does this mean?

11. Explain whether you think the following structures would act as agonists or antagonists.

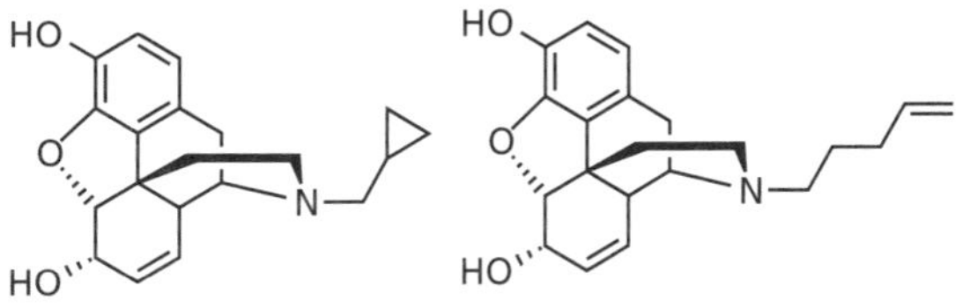

A   B

12. Thebaine has no analgesic activity. Suggest why this might be so.

13. Morphine is the active principle of opium. What is meant by an active principle?

14. Identify the asymmetric centres in morphine.

15. How could heroin be synthesized from morphine, and what problems does this pose for drug regulation authorities?

FURTHER READING

Abraham, D. J. (ed.) (2003) Narcotic analgesics. Chapter 7 in: *Burgers' medicinal chemistry and drug discovery*, 6th edn, Vol. 6. John Wiley and Sons, New York.

Feinberg, A. P., Creese, I., and Snyder, S. H. (1976) The opiate receptor: a model explaining structure–activity relationships of opiate agonists and antagonists. *Proceedings of the National Academy of Sciences of the USA*, **73**, 4215–4219.

Hruby, V. J. (2002) Designing peptide receptor agonists and antagonists. *Nature Reviews Drug Discovery*, **1**, 847–858.

Kreek, M. J., LaForge, K. S., and Butelman, E. (2002) Pharmacotherapy of addictions. *Nature Reviews Drug Discovery*, **1**, 710–725.

Pouletty, P. (2002) Drug addictions: towards socially accepted and medically treatable diseases. *Nature Reviews Drug Discovery*, **1**, 731–736.

Roberts, S. M. and Price, B. J. (eds.) (1985) Discovery of buprenorphine, a potent antagonist analgesic. Chapter 7 in: *Medicinal chemistry—the role of organic research in drug research*. Academic Press, New York.

Williams, D. A. and Lemke, T. L. (eds.) (2002) Opioid analgesics. Chapter 19 in: *Foye's principles of medicinal chemistry*, 5th edn. Lippincott Williams and Wilkins, Philadelphia.

Titles for general further reading are listed on p. 711.

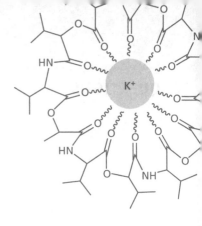

22 Antiulcer agents

22.1 Peptic ulcers

22.1.1 Definition

Peptic ulcers are localized erosions of the mucous membranes of the stomach or duodenum. The pain associated with ulcers is caused by irritation of exposed surfaces by the stomach acids. Before the appearance of effective antiulcer drugs in the 1960s, ulcer sufferers often suffered intense pain for many years, and if left untreated, the ulcer could result in severe bleeding and even death. For example, the film star Rudolph Valentino died from a perforated ulcer in 1926 at the age of 31.

22.1.2 Causes

The causes of ulcers have been disputed. Stress, alcohol, and diet have been considered important factors, but there is no clear evidence for this. Scientific evidence indicates that the two main culprits are the use of non-steroidal anti-inflammatories (NSAIDS) or the presence of a bacterium called *Helicobacter pylori*. As far as the NSAIDS are concerned, agents such as **aspirin** inhibit the enzyme **cyclooxygenase 1 (COX-1)**. This enzyme is responsible for the synthesis of prostaglandins that inhibit acid secretion and protect the gastric mucosa. Once an ulcer has erupted, the presence of gastric acid aggravates the problem and delays recovery.

22.1.3 Treatment

Antiulcer therapy has been a huge money spinner for the pharmaceutical industry, with drugs such as **cimetidine, ranitidine**, and the proton pump inhibitors (PPIs). None of these drugs were available until the 1960s, however, and it is perhaps hard for us now to appreciate how dangerous ulcers could be before that. In the early 1960s, the conventional treatment was to try to neutralize gastric acid in the stomach by administering antacids. These were bases such as sodium bicarbonate or calcium carbonate. The dose levels required for neutralization were large and caused unpleasant side effects. Relief was only temporary, and patients were often advised to stick to rigid diets such as strained porridge and steamed fish. Ultimately, the only answer to severe ulcers was a surgical operation to remove part of the stomach.

The first effective antiulcer agents were the H_2 histamine antagonists which appeared in the 1960s. These were followed in the 1980s by the PPIs. The discovery of *H. pylori* then led to the use of antibacterial agents in antiulcer therapy. The current approach for treating ulcers caused by *H. pylori* is to use a combination of drugs which includes a PPI such as **omeprazole**, and two antibiotics such as **amoxicillin** and **metronidazole**.

22.1.4 Gastric acid release

Gastric juices consist of digestive enzymes and hydrochloric acid designed to break down food. Hydrochloric acid is secreted from parietal cells, and the stomach secretes a layer of mucus to protect itself from its own gastric juices. Bicarbonate ions are also released and are trapped in the mucus to create a pH gradient within the mucus layer.

The H_2 antagonists and PPIs both work by reducing the amount of gastric acid released into the stomach by the parietal cells lining the stomach wall (Fig. 22.1). These parietal cells are innervated with nerves (not shown on the diagram) from the autonomic nervous system (sections 19.1–19.2). When the autonomic nervous system is stimulated, a signal is sent to the parietal cells culminating in the release of the neurotransmitter **acetylcholine** at the nerve termini. Acetylcholine activates the cholinergic receptors of the parietal cells leading to the release of gastric acid into the stomach. The trigger for this process is provided by the sight, smell, or even the thought, of food. Thus, gastric acid is released before food has even entered the stomach.

Nerve signals also stimulate a region of the stomach called the antrum which contains hormone-producing cells known as G cells. The hormone released is a peptide called **gastrin** (Fig. 22.2) which is also released

when food is present in the stomach. The gastrin moves into the blood supply and travels to the parietal cells further stimulating the release of gastric acid. Release of gastric acid should therefore be inhibited by antagonists blocking either the cholinergic receptor or the receptor for gastrin.

Agents which block the cholinergic receptor are known as anticholinergic drugs (section 19.11). These agents certainly block the cholinergic receptor in parietal cells and inhibit release of gastric acid. Unfortunately, they also inhibit acetylcholine receptors at other parts of the body and cause unwanted side effects.

The local hormone **histamine** also stimulates the release of gastric acid by interacting with a specific type of histamine receptor called the H_2 receptor. Thus, histamine antagonists have proved to be important antiulcer drugs although they have now largely been

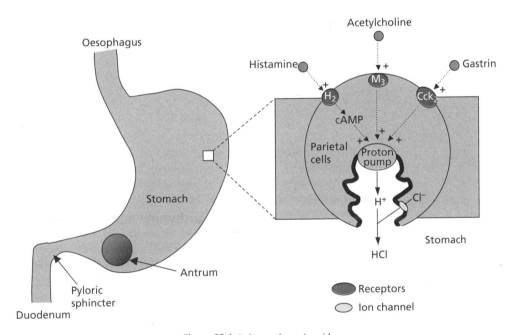

Figure 22.1 Release of gastric acid.

Figure 22.2 Gastrin.

superseded by the PPIs, which block the mechanism by which hydrochloric acid is released from parietal cells.

22.2 H₂ antagonists

The first breakthrough in antiulcer therapy came with the design of the H₂ antagonist **cimetidine** (Tagamet) (Fig. 22.32), produced by the company SmithKline and French (SKF). The cimetidine programme started in 1964, and was one of the early examples of rational drug design. Up until that time, many of the successes in medicinal chemistry involved the fortuitous discovery of useful pharmaceutical agents from natural sources, and the study of analogues often synthesized on a trial-and-error basis. Although this approach yielded a large range of medicinal compounds, it was wasteful in terms of the time and effort involved. Nowadays, the emphasis is on rational drug design using the tools of X-ray crystallography, molecular modelling, and genetic engineering (Chapters 10 and 15). Unfortunately, such tools were not available in the 1960s and the story of cimetidine is a good example of how to carry out rational drug design when the target has not been identified or isolated.

The remarkable aspect of the cimetidine story is that at the onset of the project there were no lead compounds and it was not even known if the necessary histamine receptor existed! In 1964, the best hope of achieving an antiulcer agent appeared to be in finding a drug which would block the hormone gastrin. Several research teams were active in this field, but the research team at SKF decided to follow a different tack altogether.

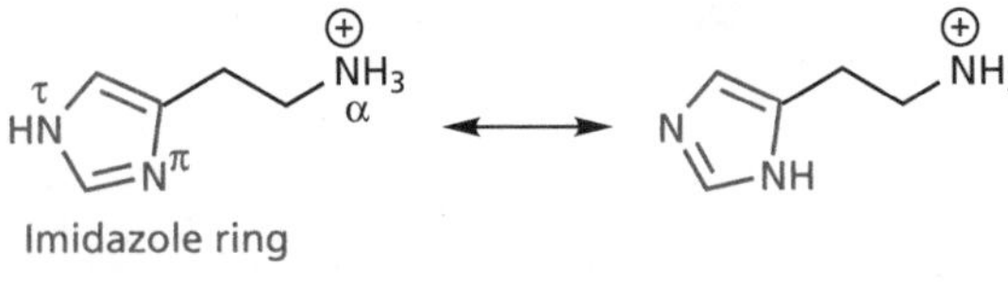

It was known experimentally that histamine (Fig. 22.3) stimulated gastric acid release *in vitro*, so the SKF team proposed that an antihistamine agent might be effective in treating ulcers. At the time, this was a highly speculative proposal as it was by no means certain that histamine played any significant role *in vivo*. Many workers at the time discounted the importance of histamine, especially when it was found that conventional antihistamines failed to inhibit gastric acid release. This suggested the absence of histamine receptors in the parietal cells. The fact that histamine had a stimulatory effect was explained away by suggesting that histamine coincidentally switched on the gastrin or cholinergic receptors. Even if a histamine receptor was present, opponents argued that blocking it would have little effect since the receptors for acetylcholine and gastrin would remain unaffected and could still be activated by their respective messengers. Initiating a project which had no known target and no known lead compound was unprecedented, and represented a massive risk. Indeed, for a long time little progress was made and it is said that company accountants demanded that the project be terminated. It says much for the scientists involved that they stuck to their guns and eventually confounded their critics. Why did the SKF team persevere in their search for an effective antihistamine? What was their reasoning? Before answering that, let us look at histamine itself and the antihistamines available at that time.

22.2.1 Histamine and histamine receptors

Histamine is made up of an imidazole ring which can exist in two tautomeric forms as shown in Fig. 22.3. Attached to the imidazole ring is a two-carbon chain with a terminal α-amino group. The pK_a of this amino group is 9.80, which means that at a plasma pH of 7.4, the side chain of histamine is 99.6% ionized. The pK_a of the imidazole ring is 5.74 and so the ring is mostly un-ionized at pH 7.4 (Fig. 22.4). Note that the lower the pK_a value, the more acidic the proton. It is also

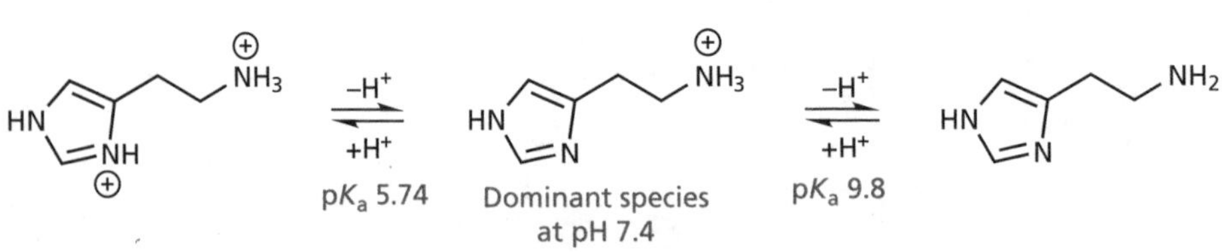

Figure 22.3 Histamine.

Figure 22.4 Ionization of histamine.

useful to remember that 50% ionization takes place when the pH is the same value as the pK_a.

Whenever cell damage occurs, histamine is released and stimulates the dilatation and increased permecability of small blood vessels. This allows defensive cells such as white blood cells to be released from the blood supply into an area of tissue damage and to combat any potential infection. Unfortunately, the release of histamine can also be a problem. Allergic reactions and irritations are caused by release of histamine when it is not really needed.

The early antihistamine drugs were therefore designed to treat conditions such as hay fever, rashes, insect bites, or asthma. Two examples of these early antihistamines are **mepyramine** and **diphenhydramine** ('Benadryl') (Fig. 22.5), neither of which has any effect on gastric acid release.

Bearing this in mind, why did the SKF team persevere with the antihistamine approach? The main reason was the fact that conventional antihistamines failed to inhibit all the then known actions of histamine. For example, they failed to fully inhibit the dilatation of blood vessels induced by histamine. The SKF scientists therefore proposed that there might be two different types of histamine receptor, analogous to the two types of cholinergic receptor mentioned in Chapter 19. Histamine—the natural messenger—would switch both on equally effectively and would not distinguish between them, whereas suitably designed antagonists might be capable of making that distinction. By implication, this meant that the conventional antihistamines known in the early 1960s were already selective in inhibiting the histamine receptors involved

in the inflammation process (classified as H$_1$ receptors), rather than the proposed histamine receptors responsible for gastric acid secretion (classified as H$_2$ receptors).

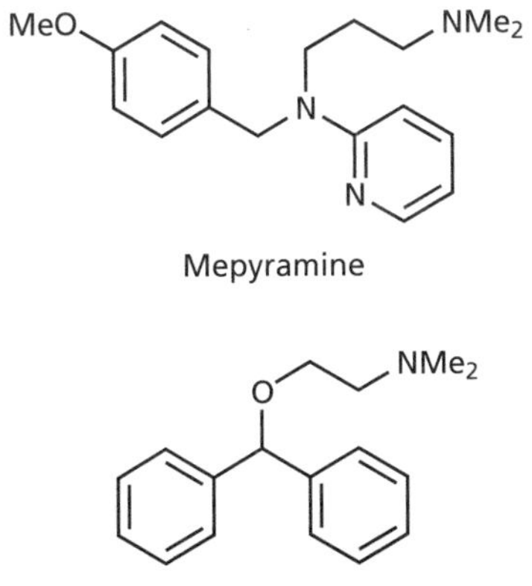

It was an interesting theory, but the fact remained that there was no known antagonist for the proposed H$_2$ receptors. Until such a compound was found, it could not be certain that the H$_2$ receptors even existed.

22.2.2 Searching for a lead

22.2.2.1 Histamine

The SKF team obviously had a problem. They had a theory, but no lead compound. How could they make a start? Their answer was to start from histamine itself. If the hypothetical H$_2$ receptor existed, then histamine must bind to it. The task then was to vary the structure of histamine in such a way that it would bind as an antagonist rather than an agonist. This meant exploring how histamine itself bound to its receptors. Structure–activity studies on histamine and histamine analogues revealed that the binding requirements for histamine to the H$_1$ receptors were as follows:

- The side chain had to have a positively charged nitrogen atom with at least one attached proton. Quaternary ammonium salts which lacked such a proton were extremely weak in activity.

- There had to be a flexible chain between the above cation and a heteroaromatic ring.

- The heteroaromatic ring did not have to be imidazole, but it did have to contain a nitrogen atom with a lone pair of electrons, *ortho* to the side chain.

For the proposed H$_2$ receptor, structure–activity (SAR) studies were carried out experimentally by determining whether histamine analogues could stimulate gastric acid release. The essential structure–activity requirements were the same as for the H$_1$ receptor except that the heteroaromatic ring had to contain an amidine unit (HN–CH–N:).

The results are summarized in Fig. 22.6 and appear to show that the terminal α-amino group is involved in

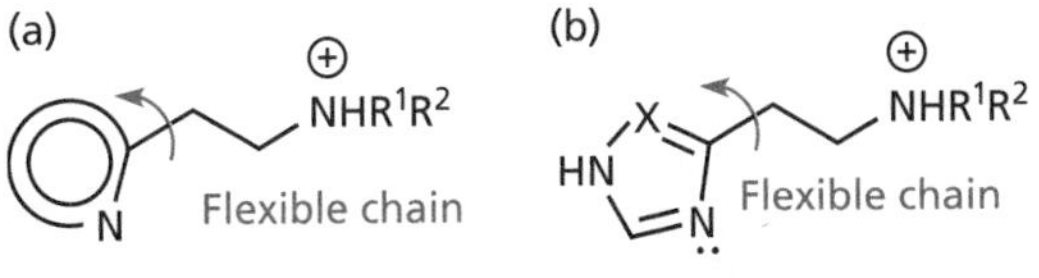

Mepyramine

Diphenhydramine

Figure 22.5 Early antihistamines.

Figure 22.6 Summary of SAR results: (a) SAR for agonist at H$_1$ receptor; (b) SAR for agonist at proposed H$_2$ receptor.

a binding interaction with both types of receptor via ionic or hydrogen bonding, while the nitrogen atom(s) in the heteroaromatic ring interacts via hydrogen bonding, as shown in Fig. 22.7.

22.2.2.2 N^α-guanylhistamine

Having gained a knowledge of the SAR for histamine, the task was now to design a molecule that would be recognized by the proposed H_2 receptor, but would not activate it. In other words, an agonist had to be converted to an antagonist. This meant altering the way in which the molecule bound to the receptor.

Pictorially, one can imagine histamine fitting into its binding site and inducing a change in shape which 'switches on' the receptor (Fig. 22.8). An antagonist

can often be found by adding a functional group which binds to an extra binding region in the binding site and prevents the change in shape required for activation.

This was one of several strategies tried out by the SKF workers. To begin with, the structural differences between agonists and antagonists in other areas of medicinal chemistry were identified and similar alterations were tried on histamine. Once again, analogues were tested to see whether they stimulated gastric acid release, the assumption being that an H_2 receptor would be responsible for such an effect.

Fusing an aromatic ring on to noradrenaline had been a successful tactic used in the design of adrenergic antagonists (see section 20.11.3). This same tactic was tried with histamine to give analogues such as the one

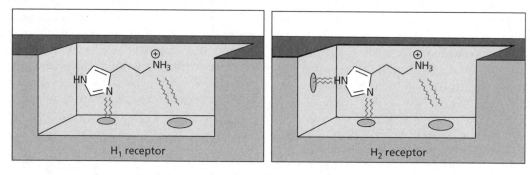

Figure 22.7 Binding interactions for the H_1 receptor and the proposed H_2 receptor.

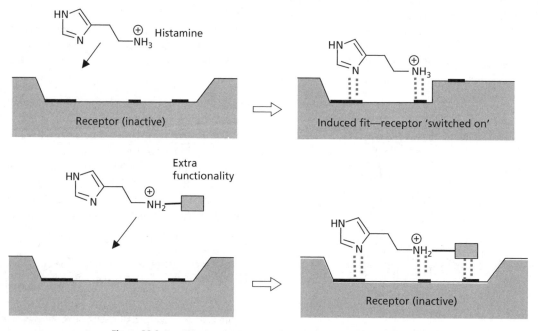

Figure 22.8 Possible receptor interactions of histamine and an antagonist.

shown in Fig. 22.9, but none of these compounds proved to be an antagonist.

Another approach which had been used successfully in the development of anticholinergic agents (section 19.11.2) had been the addition of non-polar, hydrophobic substituents. Similar substituents were attached to various locations of the histamine skeleton but none proved to be antagonists.

Nevertheless, there was one interesting result which proved relevant to later studies. It was discovered that **4-methylhistamine** (Fig. 22.10) was a highly selective H₂ agonist. In other words, it stimulated gastric acid release in the test assay, but had weak activity for all the other actions of histamine. How could this be?

4-Methylhistamine (like histamine) is a highly flexible molecule, because of its side chain, but structural studies show that some of its conformations are less stable than others. In particular, conformation I in Fig. 22.10 is disallowed as a a result of a large steric interaction between the 4-methyl group and the side chain. The selectivity observed suggests that

4-methylhistamine (and by inference histamine) has to adopt two different conformations in order to fit the H₁ and putative H₂ receptor. Since 4-methylhistamine is more active at the hypothetical H₂ receptor, it implies that conformation II is required for the H₂ receptor and conformation I is required for the H₁ receptor.

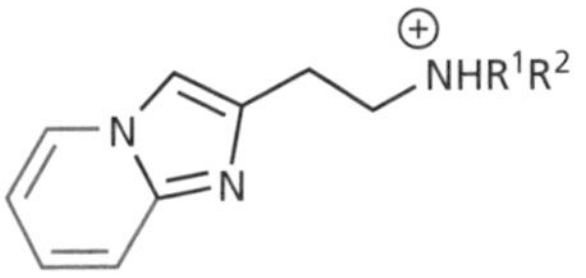

Despite this interesting result, the SKF workers were no closer to an H₂ antagonist: 200 compounds had been synthesized, and not one had shown a hint of being an antagonist. Research up until this stage had concentrated on adding hydrophobic groups to search for an additional hydrophobic binding region in the proposed receptor binding site. Now the focus switched to study the effect of varying polar groups on the molecule. In particular, the terminal α-NH₃⁺ group was replaced by different polar functional groups, the reasoning being that such groups could bond to the same binding region as the NH₃⁺ group, but that the geometry of bonding might be altered sufficiently to produce an antagonist. This led to the first crucial breakthrough, with the discovery that N^α-guanylhistamine (Fig. 22.11) was a weak antagonist of gastric acid release.

This structure had in fact been synthesized earlier in the project, but had not been recognized as an antagonist. This is not too surprising, since it acts as an agonist! It was not until later pharmacological studies were carried out that it was realized that N^α-guanylhistamine was also acting as an antagonist. In other words, it was a partial agonist (section 5.8). N^α-Guanylhistamine activates the H₂ receptor, but not to the same extent as histamine. As a result, the amount of gastric acid released is lower. More importantly, as long as N^α-guanylhistamine is bound to the receptor, it prevents histamine from binding and thus prevents complete receptor activation. This was the first indication of any sort of antagonism to histamine, but did not prove the existence of the H₂ receptor.

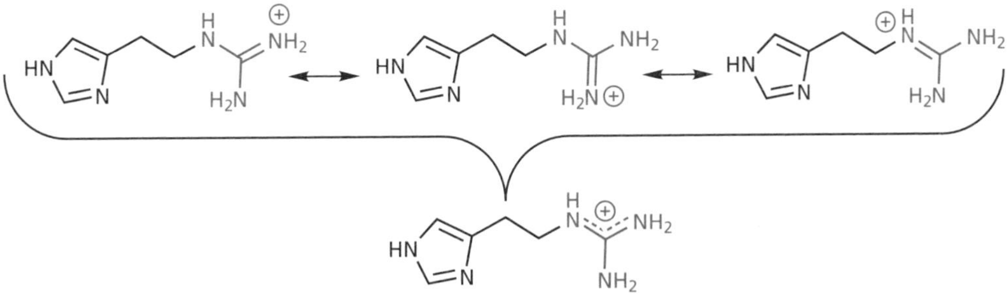

Figure 22.9 Histamine analogue—not an antagonist.

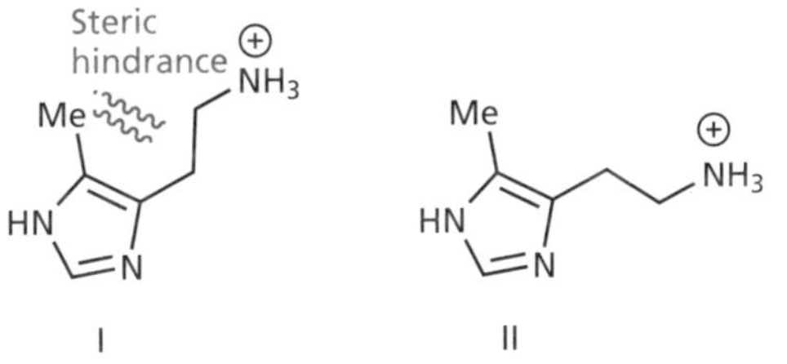

Figure 22.10 4-Methylhistamine.

Figure 22.11 N^α-Guanylhistamine.

The question now arose as to which parts of the N^α-guanylhistamine skeleton were really necessary for this effect. Various guanidine structures were synthesized that lacked the imidazole ring, but none had the desired antagonist activity, demonstrating that both the imidazole ring and the guanidine group were required.

The structures of N^α-guanylhistamine and histamine were now compared. Both structures contain an imidazole ring and a positively charged group linked by a two-carbon bridge. The guanidine group is basic and protonated at pH 7.4, so the analogue has a positive charge similar to histamine. However, the charge on the guanidine group can be spread around a planar arrangement of three nitrogens and can be further away from the imidazole ring (Fig. 22.11). This leads to the possibility that the analogue could be interacting with an extra polar binding region on the receptor which is 'out of reach' of histamine. This is demonstrated in Figs. 22.12 and 22.13. Two alternative binding regions might be available for the cationic group—an agonist region where binding leads to activation of the receptor and an antagonist region where binding does not activate the receptor. In Fig. 22.13, histamine is only able to reach the agonist region, whereas the analogue with its extended functionality is capable of reaching either region (Fig. 22.12).

If most of the analogue molecules bind to the agonist region and the remainder bind to the antagonist region, then this could explain the partial agonist activity. Regardless of the mode of binding, histamine would be prevented from binding and an antagonism would be observed due to the fraction of N^α-guanylhistamine bound to the antagonist region.

22.2.3 Developing the lead: a chelation bonding theory

The task was now to find an analogue which would bind to the antagonist region only. The isothiourea (a in Fig. 22.14) was synthesized, since the positive charge would be restricted to the terminal portion of the chain and should interact more strongly with the more distant antagonist binding region. Antagonist

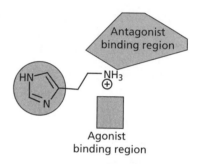

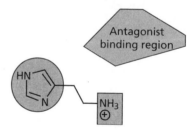

Figure 22.13 Binding of histamine: agonist mode only.

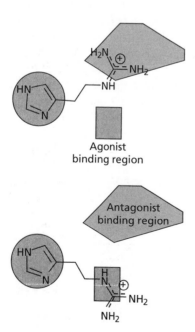

Figure 22.12 Possible binding modes for N^α-guanylhistamine.

(a)

(b) X = SMe, Me

Figure 22.14 Analogues.

activity did increase, but the compound was still a partial agonist, showing that binding was still possible to the agonist region.

Two other analogues were synthesized, where one of the terminal amino groups in the guanidine group was replaced by a methylthio group or a methyl group (b in Fig. 22.14). Both these structures were partial agonists, but with poorer antagonist activity.

From these results, it was concluded that both terminal amino groups were required for binding to the antagonist binding site. It was proposed that the charged guanidine group was interacting with a charged carboxylate residue on the receptor via two hydrogen bonds (Fig. 22.15). If either of these terminal amino groups were absent, then binding would be weaker, resulting in a lower level of antagonism.

The chain was now extended from a two-carbon unit to a three-carbon unit to see what would happen if the guanidine group was moved further away from the imidazole ring. The antagonist activity increased for the guanidine structure (Fig. 22.16), but strangely enough, decreased for the isothiourea structure (Fig. 22.16). It was therefore proposed that with a chain length of two carbon units, hydrogen bonding to the receptor involved the terminal NH₂ groups, but with a chain length of three carbon units, hydrogen bonding to the same carboxylate group involved one terminal NH₂ group along with the NH group within the chain (Fig. 22.17). Support for this theory was provided by the fact that replacing one of the terminal NH₂ groups in the guanidine analogue (Fig. 22.16) with SMe or Me (Fig. 22.18) did not adversely affect antagonist activity. This was completely different from the results obtained when similar changes were carried out on the C₂ bridged compound. These bonding

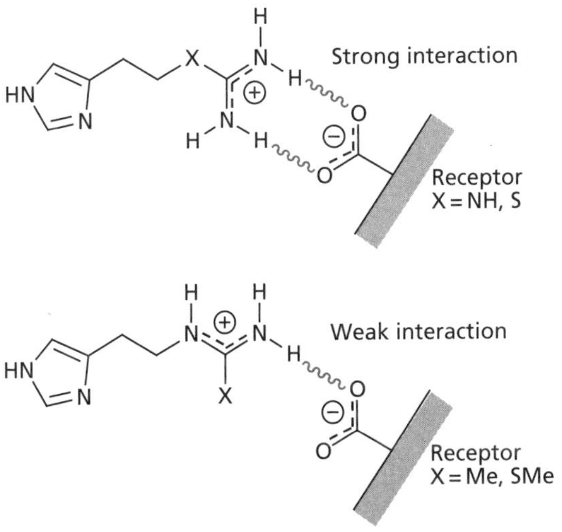

Figure 22.15 Proposed hydrogen bonding interactions.

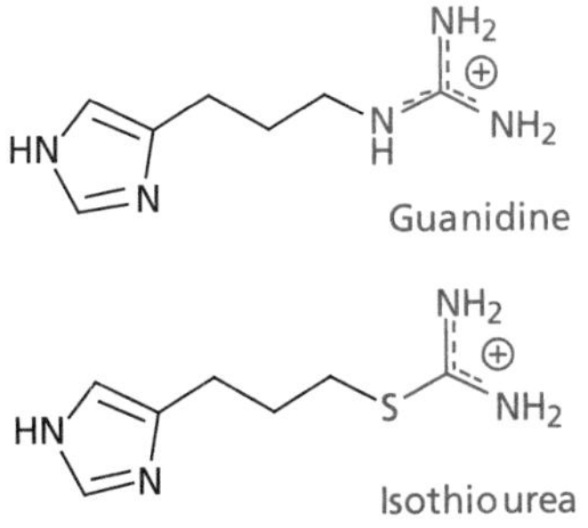

Figure 22.16 Guanidine and isothiourea structures.

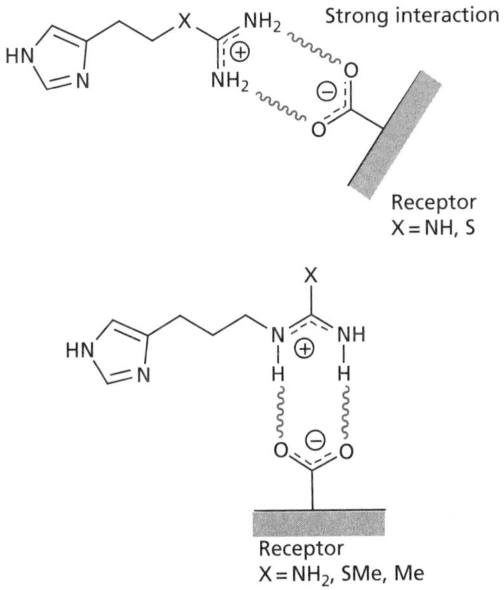

Figure 22.17 Proposed binding interactions for analogues of different chain length.

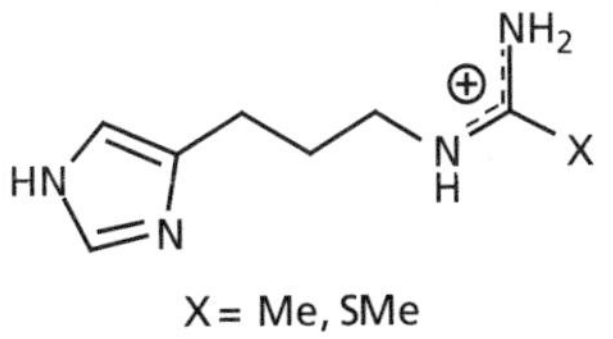

Figure 22.18 Guanidine analogue with X = SMe or Me.

interactions are represented pictorially in Figs. 22.19 and 22.20.

22.2.4 From partial agonist to antagonist: the development of burimamide

The problem was now to completely remove the agonist activity to get a pure antagonist. This meant designing a structure which would differentiate between the agonist and antagonist binding regions. At first sight this looks impossible, as both regions appear to involve the same type of bonding. Histamine's activity as an agonist depends on the imidazole ring and the charged amino function, with the two groups taking part in hydrogen and ionic bonding, respectively. The antagonist activity of the partial agonists described so far also appears to depend on a hydrogen bonding imidazole ring and an ionic bonding guanidine group.

Fortunately, a distinction can be made between the charged groups. The structures which show antagonist activity are all capable of forming a chelated bonding structure, as shown in Fig. 22.17. This interaction involves two hydrogen bonds between two charged

species, but is it really necessary for the chelating group to be charged? Could a neutral group also chelate to the antagonist region by hydrogen bonding alone? If so, it might be possible to distinguish between the agonist and antagonist regions, especially since ionic bonding appears mandatory for the agonist region.

It was therefore decided to see what would happen if the strongly basic guanidine group was replaced by a neutral group, capable of interacting with the receptor by two hydrogen bonds. There are many such groups, but the SKF workers limited the options by adhering to a principle which they followed throughout their research programme. Whenever they wished to alter a specific physical or chemical property, they strove to ensure that other properties were changed as little as possible. Only in this way could they rationalize any observed improvement in activity. Thus, it was necessary to ensure that the new group was similar to guanidine in terms of size, shape, and hydrophobicity.

Several functional groups were tried, but success was ultimately achieved by using a thiourea group to give **SKF 91581** (Fig. 22.21). The thiourea group is neutral at physiological pH because the C=S group has an electron-withdrawing effect on the

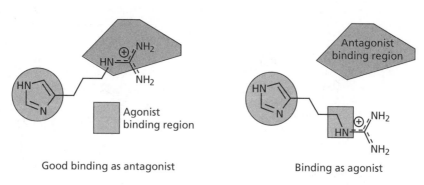

Figure 22.19 Binding interactions for 3C bridged analogue.

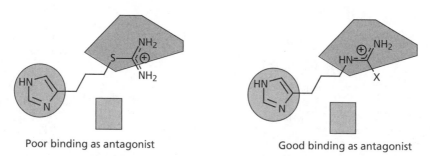

Figure 22.20 Effect of varying the guanidine group on binding to the antagonist region.

neighbouring nitrogens, making them non-basic and more like amide nitrogens. Apart from basicity, the properties of the thiourea group are very similar to the guanidine group. Both groups are planar, similar in size, and can take part in hydrogen bonding. This means that the alteration in biological activity can reasonably be attributed to the differences in basicity between the two groups.

SKF 91581 proved to be a weak antagonist with no agonist activity, establishing that the agonist binding region involves ionic bonding, whereas the antagonist region involves hydrogen bonding.

Further chain extension and the addition of an *N*-methyl group led to **burimamide** (Fig. 22.21) which was found to have enhanced activity, suggesting that the thiourea group has been moved closer to the antagonist binding site. The beneficial addition of the *N*-methyl group is due to an increase in hydrophobicity, and a possible explanation for this will be described in section 22.2.8.2 (desolvation).

Burimamide is a highly specific competitive histamine antagonist at H₂ receptors, and is 100 times more potent than N^α-guanylhistamine in inhibiting

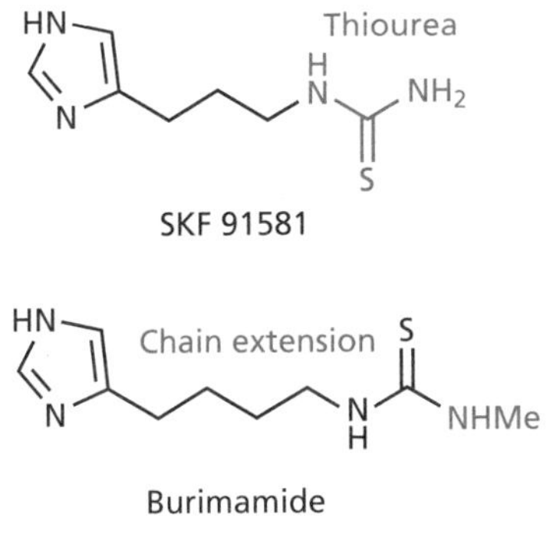

Figure 22.21 SKF 91581 and burimamide.

gastric acid release induced by histamine. Its discovery finally proved the existence of H₂ receptors.

22.2.5 Development of metiamide

Despite this success, burimamide was not suitable for clinical trials since its activity was still too low for oral administration. Attention was now directed to the imidazole ring of burimamide and, in particular, to its possible tautomeric and protonated forms. It was argued that if one of these forms was preferred for binding with the H₂ receptor, then activity might be enhanced by modifying the burimamide structure to favour that form.

At pH 7.4, it is possible for the imidazole ring to equilibrate between the two tautomeric forms (I and II) via the protonated intermediate (III) (Fig. 22.22). The necessary proton for this process is supplied by water or by an exchangeable proton on a suitable amino acid residue in the binding site. If the exchange is slow, then it is possible that the drug will enter and leave the receptor at a faster rate than the equilibration between the two tautomeric forms. If bonding involves only one of the tautomeric forms or the protonated form, then clearly antagonism would be increased if the structure was varied to prefer that form over the other. Our model hypothesis for receptor binding shows that the imidazole ring is important for the binding of both agonists and antagonists. Therefore, it is reasonable to assume that the preferred imidazole form is the same for both agonists and antagonists. If so, then the preferred form for a strong agonist such as histamine should also be the preferred form for a strong antagonist.

Figure 22.22 shows that the imidazole ring can exist as two un-ionized tautomers and one protonated form. Is the protonated form likely?

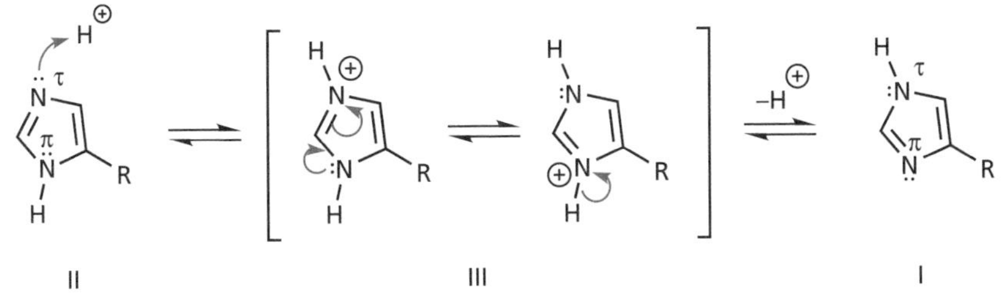

Figure 22.22 Imidazole ring can equilibrate between tautomeric forms (I and II) via the protonated intermediate (III).

We have already seen that the pK_a for the imidazole ring in histamine is 5.74, meaning that the ring is a weak base and mostly un-ionized at physiological pH. The pK_a value for imidazole itself is 6.80 and for the imidazole ring in burimamide it is 7.25, showing that these rings are more basic and more likely to be ionized. Why should this be so?

The explanation is that the side chains have an electronic effect on the imidazole ring that affects the basicity of the ring. A measure of the electronic effect of the side chain can be worked out by the Hammett equation (section 13.2.2):

$$pK_{a(R)} = pK_{a(H)} + \rho\sigma_R$$

where $pK_{a(R)}$ is the pK_a of the imidazole ring bearing a side chain R, $pK_{a(H)}$ is the pK_a of the unsubstituted imidazole ring, ρ is a constant, and σ_R is the Hammett substituent constant for the side chain R.

From the pK_a values, the value of the Hammett substituent constant can be calculated to show whether the side chain R is electron withdrawing or electron donating. In burimamide, the side chain is slightly electron donating (of the same order as a methyl group). Therefore, the imidazole ring in burimamide is more likely to be ionized than in histamine, where the side chain is electron withdrawing. At pH 7.4, 40% of the imidazole ring in burimamide is ionized, compared to approximately 3% in histamine. This represents quite a difference between the two structures and since the binding of the imidazole ring is important for both antagonist and agonist activity, it suggests that a pK_a value closer to that of histamine might lead to better binding and to better antagonist activity.

It was necessary, therefore, to make the side chain electron withdrawing rather than electron donating. This can be done by inserting an electronegative atom into the side chain—preferably one which causes minimum disturbance to the rest of the molecule. In other words, an isostere for a methylene group is required; one which has an electronic effect, but which has approximately the same size and properties as the methylene group.

The first isostere to be tried was a sulfur atom. Sulfur is quite a good isostere for the methylene unit, since both groups have similar van der Waals radii and similar bond angles. On the other hand, the C–S bond is slightly longer than a C–C bond, leading to a slight extension (15%) of the structure.

The methylene group replaced was next but one to the imidazole ring. This site was chosen, not for any strategic reasons, but because a synthetic route was readily available to carry out that particular transformation. As hoped, the resulting compound, **thiaburimamide** (Fig. 22.23), had a significantly lower pK_a of 6.25 and was found to have enhanced antagonistic activity, supporting the theory that the un-ionized form is preferred over the protonated, ionized form.

Thiaburimamide favours the un-ionized imidazole ring over the ionized ring, but there are two possible un-ionized tautomers. The next question is whether either of these are preferred for receptor binding.

Let us return to histamine. If one of the un-ionized tautomers is preferred over the other, it would be reasonable to assume that the preferred tautomer is the favoured tautomer for receptor binding, since it is more likely to be present. The preferred tautomer for histamine is tautomer I (Fig. 22.22) where Nτ is protonated and Nπ is not. This implies that Nτ in tautomer II is more basic than Nπ in tautomer I. This might not appear to be obvious, but we can rationalize it as follows. If Nτ in tautomer II is more basic than Nπ in tautomer I, it is more likely to become protonated to form the ionized intermediate (III). Moreover, deprotonation of III is more likely to give the weaker base which would be Nπ in tautomer I. Therefore the equilibrium should shift to favour tautomer I.

This is all very well, but why should Nτ (tautomer II) be more basic than Nπ (tautomer I)? The answer lies in the side chain R. The side chain on histamine has a positively charged terminal amino group, which means that the side chain has an electron-withdrawing effect on the imidazole ring. Since this effect is inductive, the strength of the effect will decrease with distance round the ring, which means that the nitrogen atom closest to the side chain (Nπ) experiences a greater electron-withdrawing effect than the one further away (Nτ). As a result, the closer nitrogen (Nπ)

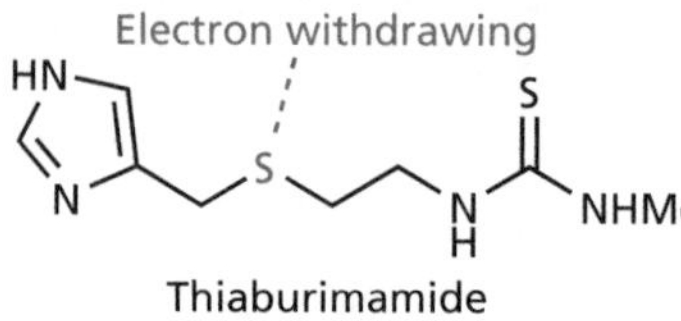

Thiaburimamide

Figure 22.23 Thiaburimamide.

is less basic, and is less likely to bond to hydrogen (Fig. 22.24). Since the side chain in thiaburimamide is also electron-withdrawing, then tautomer I will be favoured here as well.

It was now argued that tautomer I could be further enhanced if an electron-donating group was placed at position 4 of the imidazole ring. At this position, the inductive effect would be felt most strongly at the neighbouring nitrogen (Nτ), further enhancing its basic character over Nπ. At the same time, it was important to choose a group that would not interfere with the normal receptor binding interactions. For example, a large substituent might be too bulky and prevent the analogue fitting the binding site. A methyl group was chosen because it was known that 4-methylhistamine was an agonist that was highly selective for the H₂ receptor (section 22.2.2.2). This resulted in **metiamide** (Fig. 22.25) which was found to have enhanced antagonist activity, supporting the proposed theory.

It is interesting to note that the percentage increase in tautomer I outweighs an undesirable rise in pK_a. By adding an electron-donating methyl group, the pK_a of the imidazole ring rises to 6.80 compared to 6.25 for thiaburimamide. Coincidentally, this is the same pK_a as for imidazole itself, which shows that the electronic effects of the methyl group and the side chain cancel each other out as far as pK_a is concerned. A pK_a of 6.80 means that 20% of metiamide exists as the protonated form (III), but this is still lower than the corresponding 40% for burimamide. More importantly, the effect on activity due to the increase in tautomer I outweighs the detrimental effect caused by the increase in the protonated form III.

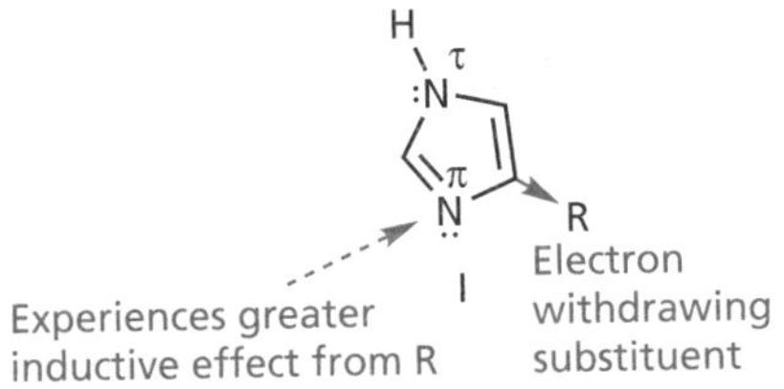

4-Methylburimamide (Fig. 22.26) was also synthesized for comparison. Here, the introduction of the 4-methyl group does not lead to an increase in activity. The pK_a is increased to 7.80, resulting in the population of ionized imidazole ring rising to 72%. This demonstrates the importance of rationalizing structural changes. Adding the 4-methyl group to thiaburimamide is advantageous, but adding it to burimamide is not.

The design and synthesis of metiamide followed a rational approach aimed at favouring one specific tautomer. Such a study is known as a **dynamic structure–activity analysis**.

Strangely enough, it has since transpired that the improvement in antagonism may have resulted from conformational effects. X-ray crystallography studies have indicated that the longer thioether linkage in the chain increases the flexibility of the side chain and that the 4-methyl substituent in the imidazole ring may help to orientate the imidazole ring correctly for receptor binding. It is significant that the oxygen analogue **oxaburimamide** (Fig. 22.26) is less potent than burimamide, despite the fact that the electron-withdrawing effect of the oxygen-containing chain on the ring is similar to the sulfur-containing chain. The bond lengths and angles of the ether link are similar to the methylene unit, and in this respect it is a better isostere than sulfur. The oxygen atom is substantially smaller than sulfur. It is also significantly more basic and more hydrophilic than either sulfur or methylene. Oxaburimamide's lower activity might be due to a variety of reasons. For example, the oxygen may not allow the same flexibility permitted by the sulfur atom.

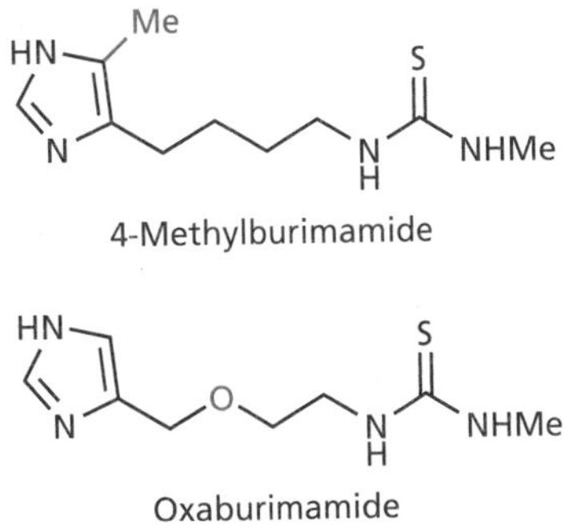

Figure 22.24 Inductive effect of the side chain on the imidazole nitrogens.

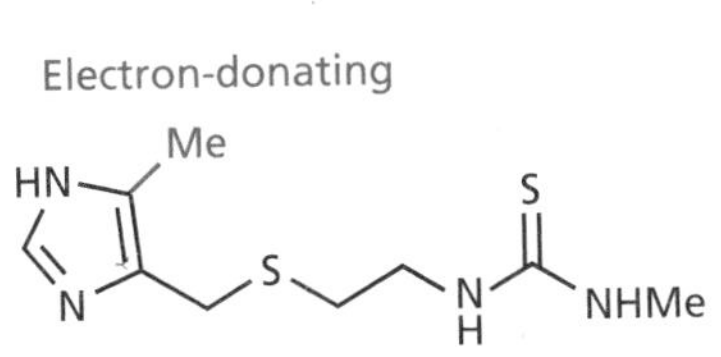

Figure 22.25 Metiamide.

Figure 22.26 4-Methylburimamide and oxaburimamide.

Alternatively, the oxygen may be involved in a hydrogen bonding interaction, either with the receptor or with the imidazole ring, resulting in a change in receptor binding interaction. Another possibility is the fact that oxygen is more likely to be solvated than sulfur, and there is an energy penalty involved in desolvating the group before binding.

Metiamide is 10 times more active than burimamide and showed promise as an antiulcer agent. Unfortunately, a number of patients suffered from kidney damage and granulocytopenia—a condition which results in the reduction of circulating white blood cells,

and makes patients susceptible to infection. Further developments were now required to find an improved drug without these side effects.

22.2.6 Development of cimetidine

It was proposed that metiamide's side effects were associated with the thiourea group—a group which is not particularly common in the body's biochemistry. Therefore, consideration was given to replacing the thiourea with a group which was similar in property but which would be more acceptable in a biochemical context. The urea analogue (Fig. 22.27) was tried, but found to be less active. The guanidine analogue (Fig. 22.27) was also less active, but it was interesting to note that this compound had no agonist activity. This contrasts with the C_3-bridged guanidine (Fig. 22.16), which is a partial agonist. Therefore, the guanidine analogue (Fig 22.27) was the first example of a guanidine having pure antagonist activity.

One possible explanation for this is that the longer C_4 chain extends the guanidine binding group beyond the reach of the agonist binding region (Fig. 22.28), whereas the shorter C_3 chain still allows binding to both agonist and antagonist regions (Fig. 22.29).

Figure 22.27 Urea and guanidine analogues.

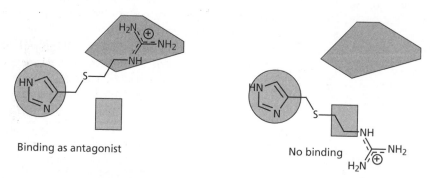

Figure 22.28 Binding of the guanidine analogue with four-atom linker.

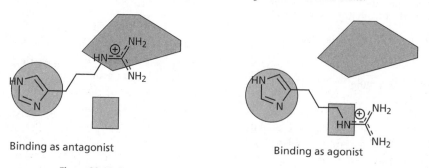

Figure 22.29 Binding of the guanidine analogue with three-atom linker.

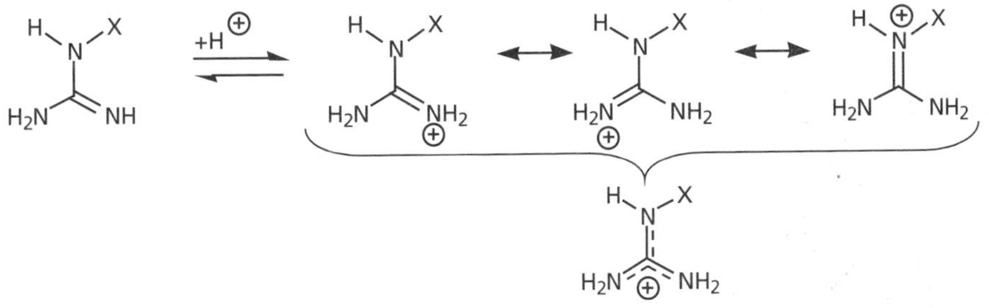

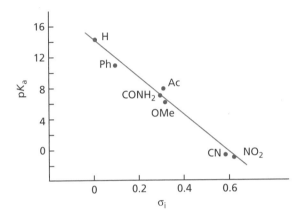

Figure 22.30 Ionization of monosubstituted guanidines.

Figure 22.31 pK_a vs. inductive substituent constants (σ_i).

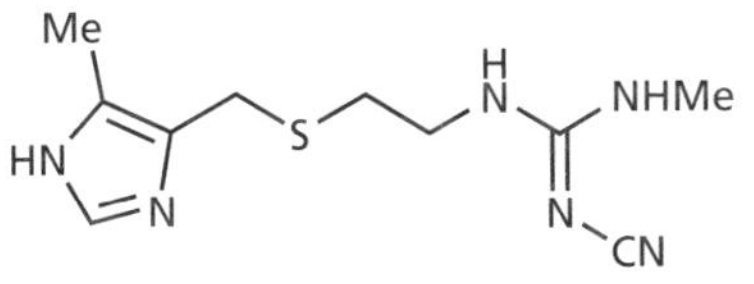

Figure 22.32 Cimetidine.

(*t*-Boc)N-β-Ala-Trp-Met-Asp-Phe-NH$_2$

Figure 22.33 Pentagastrin.

The antagonist activity for the guanidine analogue (Fig. 22.27) is weak, but it was decided to look more closely at this compound, as it was thought that the guanidine unit would lack the toxic side effects of the thiourea unit. This is a reasonable assumption since the guanidine unit is present naturally in the amino acid **arginine** (Appendix 1). The problem now was to retain the guanidine unit, but to increase activity. It seemed likely that the low activity was because the basic guanidine group would be essentially fully protonated and ionized at pH 7.4. The problem was how to make this group non-basic—no easy task,

considering the fact that guanidine is one of the strongest neutral organic bases in organic chemistry.

Nevertheless, a search of the literature revealed a useful study on the ionization of monosubstituted guanidines (Fig. 22.30). A comparison of the pK_a values of these compounds with the inductive substituent constants σ_i (section 13.2.2) for the substituents X gave a straight line as shown in Fig. 22.31, showing that pK_a is inversely proportional to the electron-withdrawing power of the substituent. Thus, strongly electron-withdrawing substituents make the guanidine group less basic and less ionized. The nitro and cyano groups are particularly strong electron-withdrawing groups. The pK_a^s for cyanoguanidine and nitroguanidine are 0.4 and 0.9 respectively (Fig. 22.31)—similar values to the pK_a for thiourea itself (-1.2).

Both the nitroguanidine and cyanoguanidine analogues of metiamide were synthesized and found to have comparable antagonist activities to metiamide. The cyanoguanidine analogue (**cimetidine**; Fig. 22.32) was the more potent analogue and was chosen for clinical studies. Its synthesis is described in Box 22.1.

22.2.7 Cimetidine

22.2.7.1 Biological activity

Cimetidine inhibits H$_2$ receptors and thus inhibits gastric acid release. The drug does not show the toxic side effects observed for metiamide and has been shown to be slightly more active. It has also been found to inhibit **pentagastrin** (Fig. 22.33) from stimulating release of gastric acid. Pentagastrin is an analogue of gastrin and the fact that cimetidine inhibits it suggests some relationship between histamine and gastrin in the release of gastric acid.

Cimetidine was first marketed in the UK in 1976 under the trade name of **Tagamet** (derived from an<u>t</u>agonist and <u>ci</u>me<u>t</u>idine). It was the first really

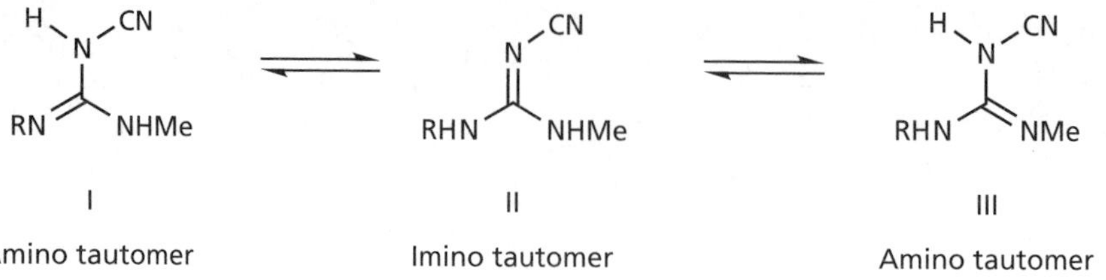

Figure 22.34 Three tautomeric forms of guanidine unit.

effective antiulcer drug, doing away with the need for surgery. For several years, it was the world's biggest selling prescription product, until it was pushed into second place in 1988 by **ranitidine** (section 22.2.9).

22.2.7.2 Structure and activity

The finding that metiamide and cimetidine are both good H_2 antagonists of similar activity shows that the cyanoguanidine group is a good bioisostere for the thiourea group. Three tautomeric forms (Fig. 22.34) are possible for the guanidine group, with the imino tautomer (II) being the preferred tautomer. This is because the cyano group has a stronger electron-withdrawing effect on the neighbouring nitrogen compared to the two nitrogens further away. As a result, the neighbouring nitrogen is less basic and less likely to be protonated. Moreover, tautomer II has an extra stabilization due to the conjugation of the double bond and the cyano group.

Since tautomer II is favoured, the guanidine group bears a close structural similarity to the thiourea group. Both groups have a planar π electron system with similar geometries (equal C–N distances and angles). They are polar and hydrophilic with high dipole moments and low partition coefficients. They are weakly basic and also weakly acidic, such that they are un-ionized at pH 7.4.

22.2.7.3 Metabolism

It is important to study the metabolism of a new drug in case the metabolites have biological activity in their own right. Any such activity might lead to undesirable side effects. Alternatively, a metabolite might have enhanced activity of the type desired and give clues to further development. Cimetidine itself is metabolically stable and is excreted largely unchanged. The only metabolites that have been identified are due to oxidation of the sulfur link or oxidation of the ring methyl group (Fig. 22.35).

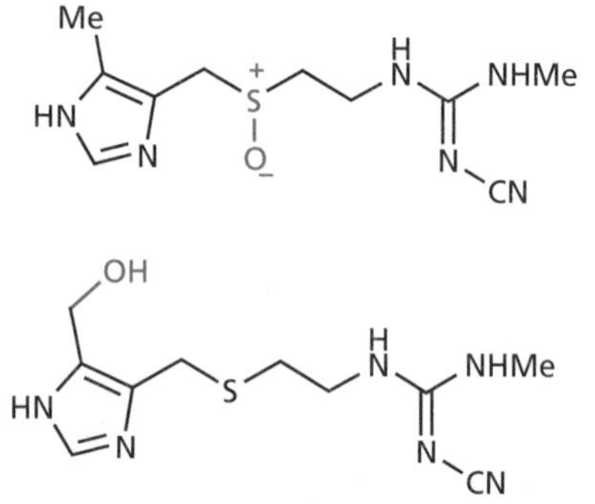

Figure 22.35 Metabolites of cimetidine.

It has been found that cimetidine inhibits the cytochrome P450 enzymes in the liver (section 8.4.2). These enzymes are involved in the metabolism of several clinically important drugs, and inhibition by cimetidine may result in toxic side effects as a result of increased blood levels of these drugs. In particular, caution is required when cimetidine is taken with drugs such as **diazepam, lidocaine, warfarin,** or **theophylline.**

22.2.8 Further studies of cimetidine analogues

22.2.8.1 Conformational isomers

A study of the various stable conformations of the guanidine group in cimetidine led to a rethink of the type of bonding that might be taking place at the antagonist binding region. Up until this point, the favoured theory had been a bidentate hydrogen interaction as shown in the top diagram of Fig. 22.15, where the two hydrogens involved in hydrogen bonding are pointing in the same direction. In order to achieve this kind of bonding, the guanidine group in cimetidine would have to adopt the Z,Z conformation shown in Fig. 22.36. (The Z and E nomenclature is

BOX 22.1 SYNTHESIS OF CIMETIDINE

The synthesis of cimetidine was originally carried out as a four-step process, where lithium aluminium hydride was used as the reagent for the initial reduction step. Subsequent research revealed that this reduction could be carried out more cheaply and safely using sodium in liquid ammonia, and so this became the method used in the manufacture of cimetidine.

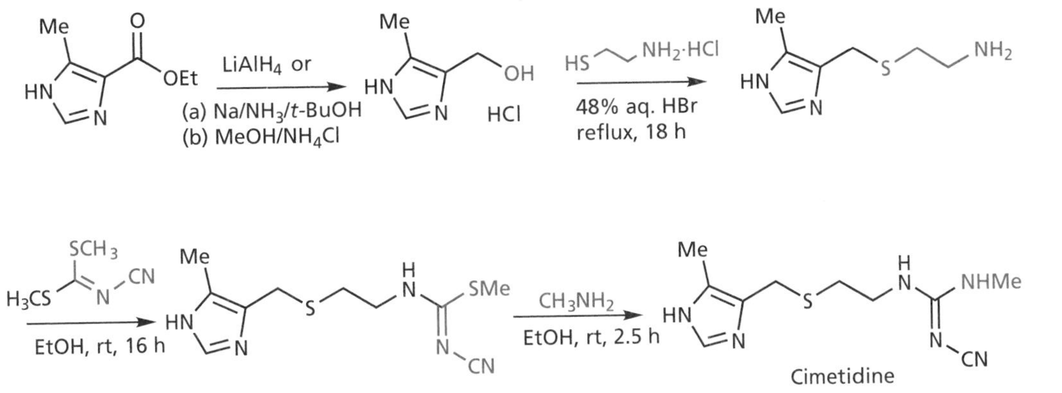

Synthesis of cimetidine.

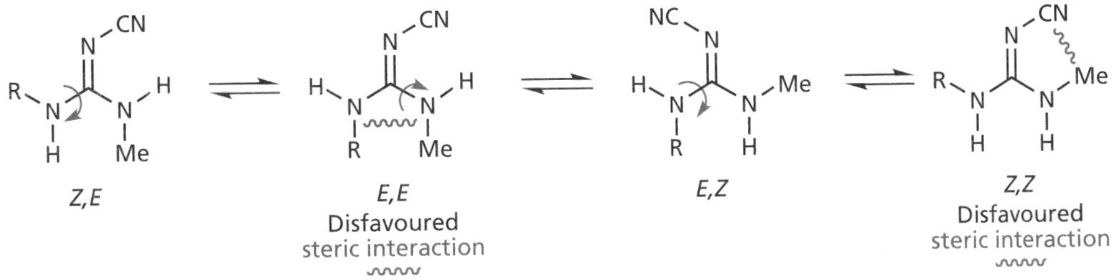

Figure 22.36 Conformations of the guanidine group in cimetidine.

relevant here, as there is double bond character in all the N–C bonds of the guanidine unit.)

However, X-ray and NMR studies have shown that cimetidine exists as an equilibrium mixture of the E,Z and Z,E conformations. Neither the Z,Z nor the E,E form is favoured, because of steric interactions. If either the E,Z or Z,E form is the active conformation, then it implies that the chelation type of hydrogen bonding described previously is not taking place. An alternative possibility is that the guanidine unit is hydrogen bonding to two distinct hydrogen bonding regions rather than to a single carboxylate group (Fig. 22.37). Further support for this theory is provided by the weak activity observed for the urea analogue (Fig. 22.27).

This compound is known to prefer the Z,Z conformation over the Z,E or E,Z conformations, and would therefore be unable to bind to both hydrogen bonding regions.

If this bonding theory is correct and the active conformation is the E,Z or Z,E form, restricting the group to adopt one or other of these forms may lead to more active compounds and the identification of the active conformation. This can be achieved by incorporating part of the guanidine unit within a ring—a strategy of rigidification (section 10.3.9). For example, the nitropyrrole derivative (Fig. 22.38) has been shown to be the strongest antagonist in the cimetidine series, implying that the E,Z conformation is the active conformation.

The isocytosine ring (Fig. 22.39) has also been used to 'lock' the guanidine group, limiting the number of conformations available. The ring allows for further substitution and development, as described below.

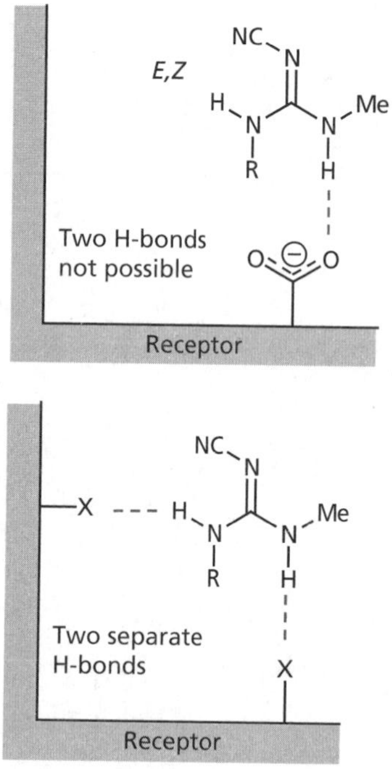

Figure 22.37 Alternative theory for cimetidine bonding at the agonist region.

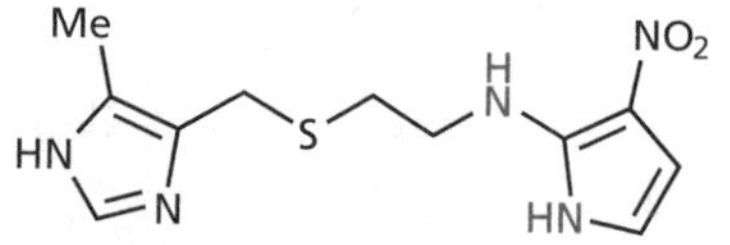

Figure 22.38 Nitropyrrole derivative of cimetidine.

22.2.8.2 Desolvation

The guanidine and thiourea groups, used so successfully in the development of H_2 antagonists, are polar and hydrophilic. This implies that they are likely to be highly solvated and surrounded by a water coat'. Before hydrogen bonding can take place to the receptor, this water coat has to be removed, and the more solvated the group is, the more difficult that will be.

One possible reason for the low activity of the urea derivative (Fig. 22.27) has already been described above. Another possible reason could be the fact that the urea group is more hydrophilic than thiourea or cyanoguanidine groups and is therefore more highly solvated. The energy penalty involved in desolvating the urea group might explain why this analogue has a lower activity than cimetidine, despite having a lower partition coefficient and greater water solubility.

Leading on from this, if the ease of desolvation is a factor in antagonist activity, then reducing the solvation of the polar group should increase activity. One way of achieving this would be to increase the hydrophobic character of the polar binding group.

A study was carried out on a range of cimetidine analogues containing different planar aminal systems (Z) (Fig. 22.40) to see whether there was any relationship between antagonist activity and the hydrophobic character of the aminal system (HZ).

This study showed that antagonist activity was proportional to the hydrophobicity (log P) of the aminal unit HZ (Fig. 22.41) and supported the desolvation theory. The relationship could be quantified as follows:

$$\log(\text{activity}) = 2.0 \log P + 7.4$$

Further studies on hydrophobicity were carried out by adding hydrophobic substituents to the isocytosine analogue (Fig. 22.39). These studies showed that there was an optimum hydrophobicity for activity

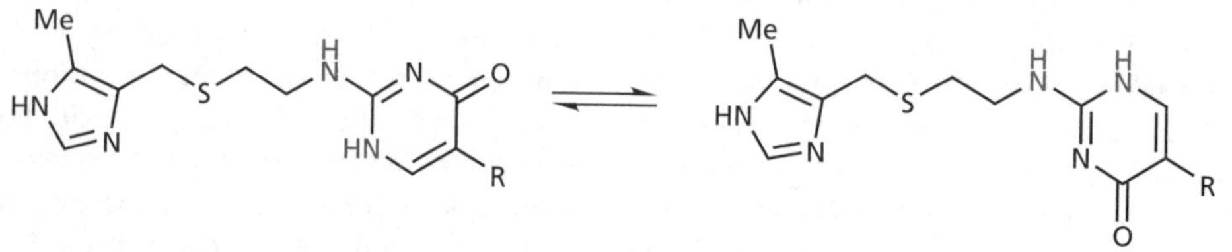

Figure 22.39 Isocytosine ring.

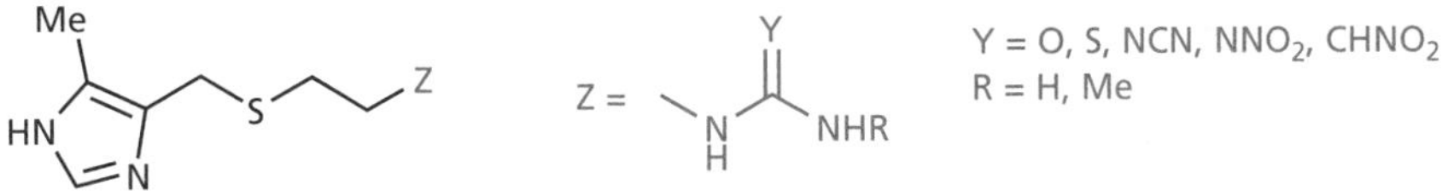

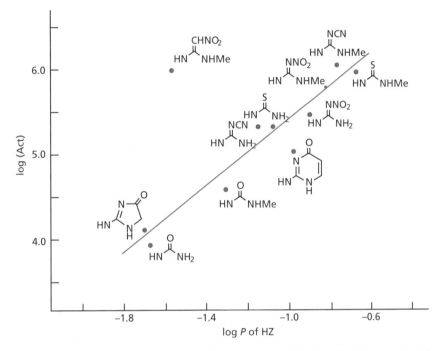

Figure 22.40 Cimetidine analogue with planar aminal system (Z).

Figure 22.41 Antagonist activity is proportional to the hydrophobicity (log P) of the aminal unit Z.

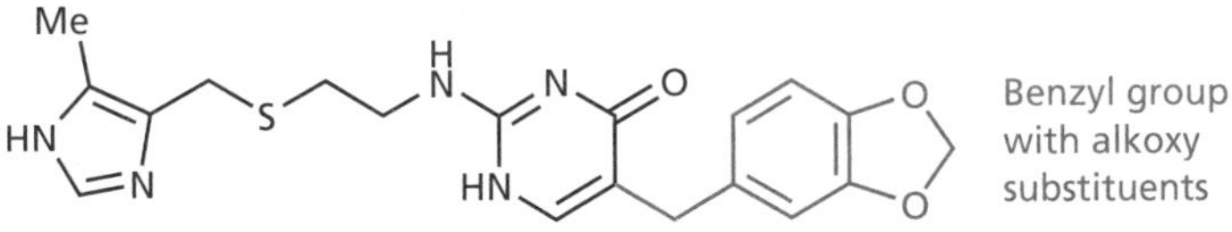

Figure 22.42 Oxmetidine.

corresponding to the equivalent of a butyl or pentyl substituent. A benzyl substituent was particularly good for activity, but proved to have toxic side effects. These could be reduced by adding alkoxy substituents to the aromatic ring and this led to the synthesis of **oxmetidine** (Fig. 22.42) which had enhanced activity over cimetidine. Oxmetidine was considered for clinical use, but was eventually withdrawn as it still retained undesirable side effects.

22.2.8.3 Development of the nitroketeneaminal binding group

As we have seen, antagonist activity increases if the hydrophobicity of the polar binding group is increased. It was therefore decided to see what would happen if the polar imino nitrogen of cimetidine was replaced by a non-polar carbon atom. This would result in a keteneaminal group, as shown in Fig. 22.43. Unfortunately, keteneaminals are more likely to exist as their

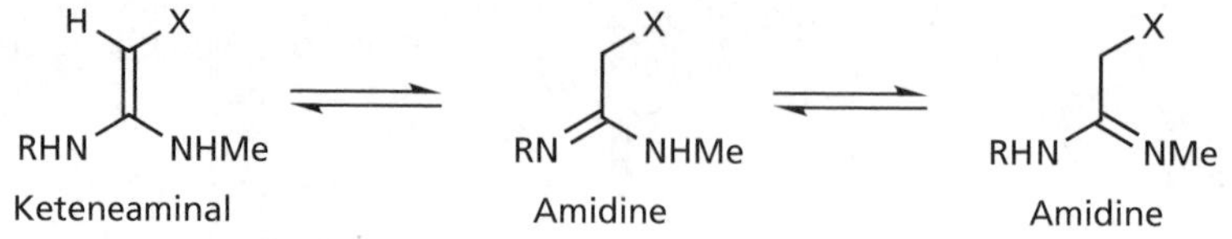

Figure 22.43 The keteneaminal group and amidine tautomers.

amidine tautomers unless a strongly electronegative group (e.g. NO_2) is attached to the carbon atom.

A nitroketeneaminal group was therefore used to give the structure shown in Fig. 22.44. Surprisingly, there was no great improvement in activity, but when the structure was studied in detail, it was discovered that it was far more hydrophilic then expected. This explained why the activity had not increased, but it highlighted a different puzzle. The compound was *too* active. Based on its hydrophilicity, it should have been a much weaker antagonist (Fig. 22.41).

It was clear that this compound did not fit the pattern followed by previous compounds, as its antagonist activity was 30 times higher than predicted. Nor was the nitroketeneaminal the only analogue to deviate from the expected pattern. The imidazolinone analogue (Fig. 22.44), which is relatively hydrophobic, had a much lower activity than would have been predicted from the equation. Findings like these are particularly exciting, as any deviation from the normal pattern suggests that some other factor is at work which may give a clue to future development.

In this case, it was concluded that the polarity of the group might be important in some way. In particular, the orientation of the dipole moment appeared to be crucial. In Fig. 22.45, the orientation of the dipole moment is defined by φ—the angle between the dipole moment and the N–R bond. The cyanoguanidine, nitroketeneaminal, and nitropyrrole groups all have high antagonist activity and have dipole moment orientations of 18°, 33°, and 27° respectively (Fig. 22.46). The isocytosine and imidazolinone groups have lower activity and have dipole orientations of −2° and −6°, respectively. The strength of the dipole moment (μ) does not appear to be crucial.

Why should the orientation of a dipole moment be important? One possible explanation is as follows. As the drug approaches the receptor, its dipole interacts with a dipole on the receptor surface such that the dipole moments are aligned. This orientates the drug in a specific way before hydrogen bonding takes place and determines how strong the subsequent hydrogen

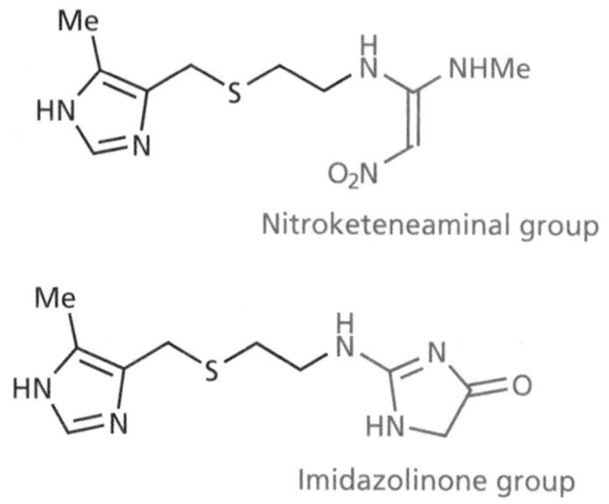

Figure 22.44 Cimetidine analogue with unexpected levels of activity.

Figure 22.45 Orientation of dipole moment.

bonding will be (Fig. 22.47). If the dipole moment is correctly orientated as in the keteneaminal analogue, the group is correctly positioned for strong hydrogen bonding and high activity will result. If the orientation is wrong as in the imidazolinone analogue, then the bonding is less efficient and activity is weaker.

QSAR studies (Chapter 13) were carried out to determine what the optimum angle φ should be for activity. This resulted in an ideal value for φ of 30°. A correlation was worked out between the dipole moment orientation, partition coefficient, and activity as follows:

$$\log A = 9.12 \cos \theta + 0.6 \log P - 2.71$$
$$(n = 13, r = 0.91, s = 0.41)$$

where A is the antagonist activity, P is the partition coefficient, and θ is the deviation in angle of the dipole moment from the ideal orientation of 30° (Fig. 22.48).

The equation shows that activity increases with increasing hydrophobicity (P). The $\cos \theta$ term shows

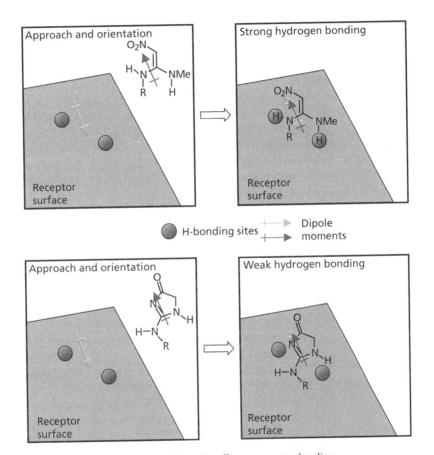

Figure 22.46 Dipole moments of various antagonistic groups.

Figure 22.47 Orientation effects on receptor bonding.

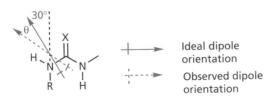

Figure 22.48 Definition of the angle θ.

that activity drops if the orientation of the dipole moment varies from the ideal angle of 30°. At the ideal angle, θ is 0° and cos θ is 1. If the orientation of the dipole moment deviates from 30°, then cos θ will be less than 1 and will lower the calculated activity. The nitroketeneaminal group did not result in a more powerful cimetidine analogue, but we shall see it again in ranitidine (section 22.2.9.1).

22.2.9 Further H$_2$ antagonists

22.2.9.1 Ranitidine

Further studies on cimetidine analogues showed that the imidazole ring could be replaced by other nitrogen-containing heterocyclic rings. Glaxo moved one step further, however, and replaced the imidazole ring with a furan ring bearing a nitrogen-containing substituent. This led to the introduction of ranitidine (Zantac) (Fig. 22.49). Ranitidine has fewer side effects than cimetidine, lasts longer, and is 10 times more active. SAR results for ranitidine include the following:

- The nitroketeneaminal group is optimum for activity, but can be replaced by other planar π systems capable of hydrogen bonding.
- Replacing the sulfur atom with a methylene atom leads to a drop in activity.
- Placing the sulfur next to the ring lowers activity.
- Replacing the furan ring with more hydrophobic rings such as phenyl or thiophene reduces activity.
- 2,5-Disubstitution is the best substitution pattern for the furan ring.
- Substitution on the dimethylamino group can be varied, showing that the basicity and hydrophobicity of this group are not crucial to activity.
- Methyl substitution at carbon-3 of the furan ring eliminates activity, whereas the equivalent substitution in the imidazole series increases activity.
- Methyl substitution at carbon-4 of the furan ring increases activity.

The last two results imply that the heterocyclic rings for cimetidine and ranitidine are not interacting in the same way with the H$_2$ receptor. This is supported by the fact that a corresponding dimethylaminomethylene group attached to cimetidine leads to a drop in activity. Ranitidine was introduced to the market in 1981 and by 1988 had taken over from cimetidine as the world's biggest-selling prescription drug. Over

a 10-year period, it earned Glaxo profits of around £4 billion ($7 billion), and at one time was earning profits of £4 million ($7 million) per day.

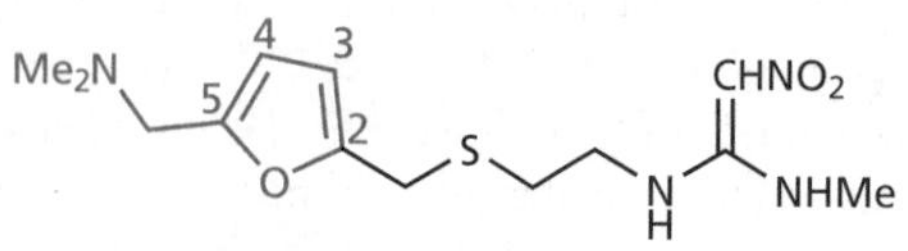

22.2.9.2 Famotidine and nizatidine

During 1985–87, two new antiulcer drugs were introduced to the market—famotidine and nizatidine (Fig. 22.50).

Famotidine (Pepcid) is 30 times more active than cimetidine *in vitro*. The side chain contains a sulfonylamidine group and the heterocyclic imidazole ring of cimetidine has been replaced by a 2-guanidinothiazole ring. SAR studies gave the following results:

- The sulfonylamidine binding group is not essential and can be replaced by a variety of structures as long as they are planar, have a dipole moment, and are capable of interacting with the receptor by hydrogen bonding. A low pK_a is not essential, which allows a larger variety of planar groups to be used than is possible for cimetidine.
- Activity is optimum for a chain length of four or five units.
- Replacement of sulfur by a CH$_2$ group *increases* activity.
- Modification of the chain is possible with, for example, inclusion of an aromatic ring.

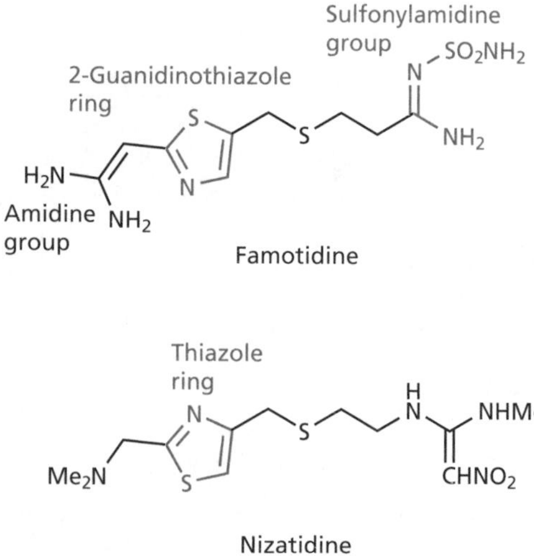

Figure 22.50 Famotidine and nizatidine.

Figure 22.49 Ranitidine.

- A methyl substituent on the heterocyclic ring, *ortho* to the chain leads to a drop in activity (unlike the cimetidine series).

- Three of the four hydrogens in the two amidine NH_2 groups are required for activity.

There are several results here which are markedly different from cimetidine, implying that famotidine and cimetidine are not interacting in the same way with the H_2 receptor. Further evidence for this is the fact that replacing the guanidine of cimetidine analogues with a sulfonylamidine group leads to very low activity.

Nizatidine (Fig. 22.50) was introduced into the UK in 1987 by the Lilly Corporation and is equipotent with ranitidine. The furan ring in ranitidine is replaced by a thiazole ring.

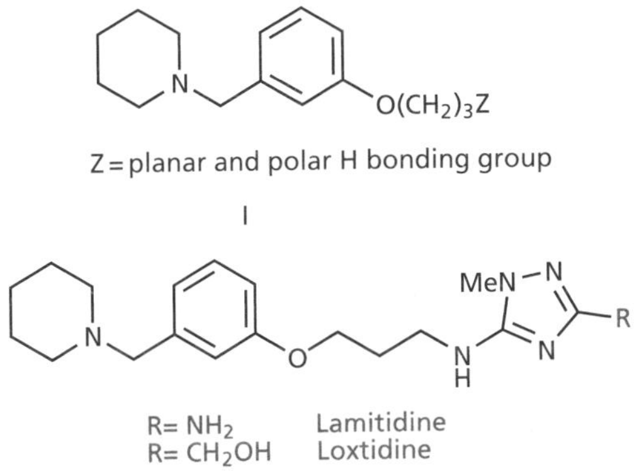

22.2.9.3 H₂ antagonists with prolonged activity

Glaxo carried out further development on ranitidine by placing the oxygen of the furan ring exocyclic to a phenyl ring and replacing the dimethylamino group with a piperidine ring to give a series of novel structures (I in Fig. 22.51). The most promising of these compounds were **lamitidine** and **loxtidine** (Fig. 22.51) which were 5–10 times more potent than ranitidine and three times longer lasting. Unfortunately, these compounds showed toxicity in long-term animal studies, with the possibility that they caused gastric cancer, so they were subsequently withdrawn from clinical study. The relevance of these studies has been disputed.

22.2.10 Comparison of H₁ and H₂ antagonists

The structures of the H_2 antagonists are markedly different from the classical H_1 antagonists, so it is not surprising that H_1 antagonists failed to antagonize the H_2 receptor. H_1 antagonists, like H_1 agonists, possess an ionic amino group at the end of a flexible chain. Unlike the agonists, they possess two aryl or heteroaryl rings in place of the imidazole ring (Fig. 22.52). Because of the aryl rings, H_1 antagonists are hydrophobic molecules having high partition coefficients.

In contrast, H_2 antagonists are polar, hydrophilic molecules having high dipole moments and low partition coefficients. At the end of the flexible chain they have a polar π electron system which is amphoteric and un-ionized at pH 7.4. This binding group appears to be the key feature leading to antagonism of H_2 receptors

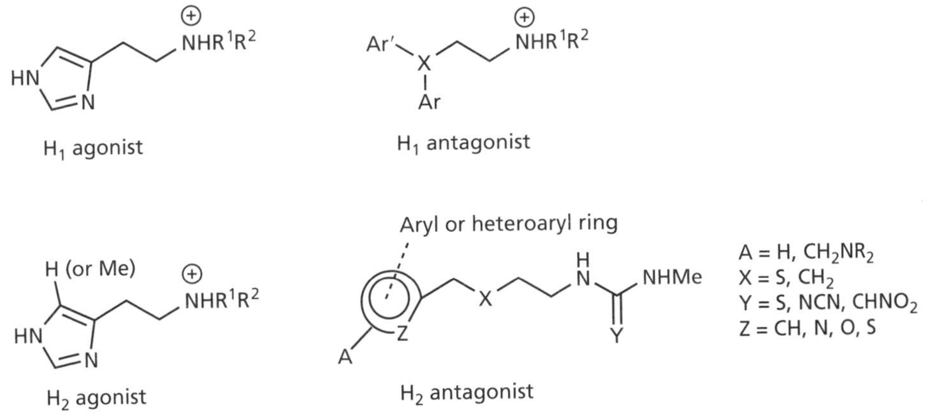

Z = planar and polar H bonding group

I

R = NH₂   Lamitidine
R = CH₂OH   Loxtidine

Figure 22.51 Long-lasting antiulcer agents.

H₁ agonist

H₁ antagonist

Aryl or heteroaryl ring

A = H, CH₂NR₂
X = S, CH₂
Y = S, NCN, CHNO₂
Z = CH, N, O, S

H₂ agonist

H₂ antagonist

Figure 22.52 Comparison of H₁ agonists, H₁ antagonists, H₂ agonists, and H₂ antagonists.

(Fig. 22.52). The heterocycle generally contains a nitrogen atom or, in the case of furan or phenyl, a nitrogen-containing side chain. The hydrophilic character of H_2 antagonists helps to explain why H_2 antagonists are less likely to have the central nervous system side effects often associated with H_1 antagonists.

22.2.11 H_2 receptors and H_2 antagonists

H_2 receptors are present in a variety of organs and tissues, but their main role is in acid secretion. As a result, H_2 antagonists are remarkably safe and mostly free of side effects. The four most-used agents on the market are cimetidine, ranitidine, famotidine, and nizatidine. They inhibit all aspects of gastric secretion and are rapidly absorbed from the gastrointestinal tract with half-lives of 1–2 hours. About 80% of ulcers are healed after 4–6 weeks. Attention must be given to possible drug interactions when using cimetidine, because of inhibition of drug metabolism (section 22.2.7). The other three H_2 antagonists mentioned do not inhibit the P450 cytochrome oxidase system and are less prone to such interactions.

KEY POINTS

- Peptic ulcers are localized erosions of the mucous membranes which occur in the stomach and duodenum. The hydrochloric acid present in gastric juices results in increased irritation and so drugs which lower the amount of hydrochloric acid released act as antiulcer agents. Such agents relieve the symptoms rather than the cause.

- The chemical messengers histamine, acetylcholine, and gastrin stimulate the release of hydrochloric acid from stomach parietal cells by acting on their respective receptors.

- H_2 antagonists are antiulcer drugs that act on H_2 receptors present on parietal cells and reduce the amount of acid released.

- The design of H_2 antagonists was based on the natural agonist histamine as a lead compound. Chain extension accessed an antagonist binding region, and the replacement of an ionized terminal group with a polar, un-ionized group capable of hydrogen bonding led to pure antagonists.

- The design of improved H_2 antagonists was aided by dynamic structure–activity analysis where changes were made to favour one tautomer over another.

- The orientation of dipole moments between a drug and its binding site plays a role in the binding and activity of H_2 antagonists. Desolvation of polar groups also has an important effect on binding affinity.

22.3 Proton pump inhibitors

Although the H_2 antagonists have been remarkably successful in the treatment of ulcers, they have been largely superseded by the PPIs. These work by irreversibly inhibiting an enzyme complex called the proton pump, and have been found to be superior to the H_2 antagonists. They are used on their own to treat ulcers that are caused by NSAIDs, and in combination with antibacterial agents to treat ulcers caused by the bacterium *H. pylori* (section 22.4).

22.3.1 Parietal cells and the proton pump

When the parietal cells are actively secreting hydrochloric acid into the stomach, they form invaginations called **canaliculi** (Fig. 22.53). Each canaliculus can be viewed as a sheltered channel or inlet that flows into the overall 'ocean' of the stomach lumen. Being a channel, it is not part of the cell but it penetrates 'inland' and increases the amount of 'coastline' (surface area) available to the cell, across which it can release its hydrochloric acid. The protons required for the hydrochloric acid are generated from water and carbon dioxide, catalysed by an enzyme called

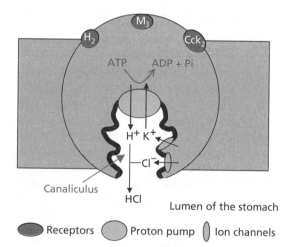

Figure 22.53 Role of the proton pump in secreting hydrochloric acid.

carbonic anhydrase (Fig. 22.54). Once the protons have been generated, they have to be exported out of the cell rather than stored. There are two reasons for this. First, a build-up of acid within the cell would prove harmful to the cell. Second, the enzyme-catalysed reaction which generates the protons is reversible, and so a build-up of protons within the cell would encourage the reverse reaction and slow the production down. The export of protons from the parietal cell is achieved by an enzyme complex called the **proton pump** or H^+/K^+-ATPase.

The proton pump is only present in the canalicular membranes of parietal cells and is crucial to the mechanism by which hydrochloric acid is released into the stomach. It is called an H^+/K^+-ATPase because it pumps protons out of the cell into the canaliculus at the same time as it pumps potassium ions back in.

$$H_2O + CO_2 \; \underset{\text{Carbonic anhydrase}}{\rightleftharpoons} \; H_2CO_3 \; \rightleftharpoons \; H^+ + HCO_3^-$$

Figure 22.54 Enzyme-catalysed generation of protons in the parietal cell.

$$ATP \; \underset{\text{ATPase}}{\rightleftharpoons} \; ADP + phosphate$$

Figure 22.55 Enzyme-catalysed hydrolysis of ATP.

Energy is required for this process, as both the protons and the potassium ions are being moved against their concentration gradients. In fact, the ratio of protons inside the cell to protons in the canaliculus is 1 to 10^6! The energy required to carry out this exchange is obtained by the hydrolysis of ATP (Fig. 22.55)—hence the term ATPase.

The pump is not responsible for the efflux of chloride ions: these depart the cell through separate chloride ion channels. This outflow closely matches the efflux of protons, such that a chloride ion is released for every proton that is pumped out. As a result, hydrochloric acid is formed in the canaliculus rather than inside the parietal cell.

As each chloride ion departs the cell, it is accompanied by a potassium ion which flows through its own ion channel. No energy is required for this outflow, since the potassium ion is flowing down a concentration gradient. The potassium ion acts as a counterion for the chloride, and once it is in the canaliculus it is pumped back into the cell by the proton pump as described previously. Consequently, potassium ions undergo a cyclic movement in and out of the cell.

22.3.2 Proton pump inhibitors

There are four PPIs in clinical use: **omeprazole, lansoprazole, pantoprazole,** and **rabeprazole** (Fig. 22.56).

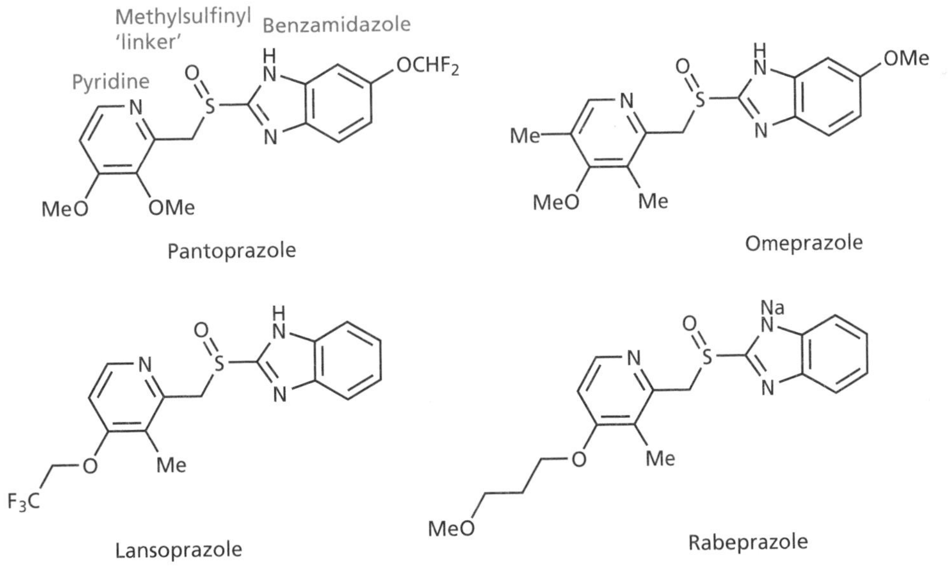

Figure 22.56 Proton pump inhibitors (PPIs).

The *S*-enantiomer of omeprazole has also been recently approved. All the PPIs have a pyridyl methylsulfinyl benzamidazole skeleton and act as prodrugs, as they are activated when they reach the acidic canaliculi of parietal cells. Once activated, they bind irreversibly to exposed cysteine residues of the proton pump and block the pump, preventing further release of hydrochloric acid.

There is a big strategic advantage in inhibiting the proton pump rather than blocking histamine or cholinergic receptors. For example, H_2 antagonists block the histamine receptors and in doing so block the stimulatory effect of histamine, but this does not block the receptors for acetylcholine or gastrin, and so it is still possible for the parietal cell to be activated towards secretion by these routes. The proton pump is downstream of all these targets and operates the final stage of hydrochloric acid release. Blocking it prevents the release of hydrochloric acid, regardless of what mechanisms are involved in stimulating hydrochloric acid secretion.

22.3.3 Mechanism of inhibition

The PPIs are weak bases, having a pK_a of about 4.0. As a result, they are free bases at blood pH (7.4) and are only ionized in strongly acidic environments where the pH is less than 4. These are conditions found only in the secretory canaliculus of the parietal cell, where the pH is 2 or less. The drugs are taken orally and are absorbed into the blood supply where they are carried round the body as fairly innocuous passengers until they reach the parietal cells. Since they are un-ionized weak bases at this stage, and are also lipophilic in nature, they are able to cross the cell membrane of the parietal cell into the strongly acidic conditions of the canaliculi. Here, the drugs undergo a personality change and become particularly vicious! The canaliculus is highly acidic, so the drug becomes protonated. The consequences of this are twofold:

- The ionized drug is too polar to cross back into the cell through the cell membrane. This leads to a 1000-fold accumulation of the drug in the canaliculi where it is intended to act.

- Protonation triggers an acid-catalysed conversion, as shown in Fig. 22.57, which activates the drug.

Protonation takes place on the benzimidazole ring of the drug. The nitrogen of the pyridine ring then acts as a nucleophile and uses its lone pair of electrons to form a bond to the electron deficient 2-carbon of the

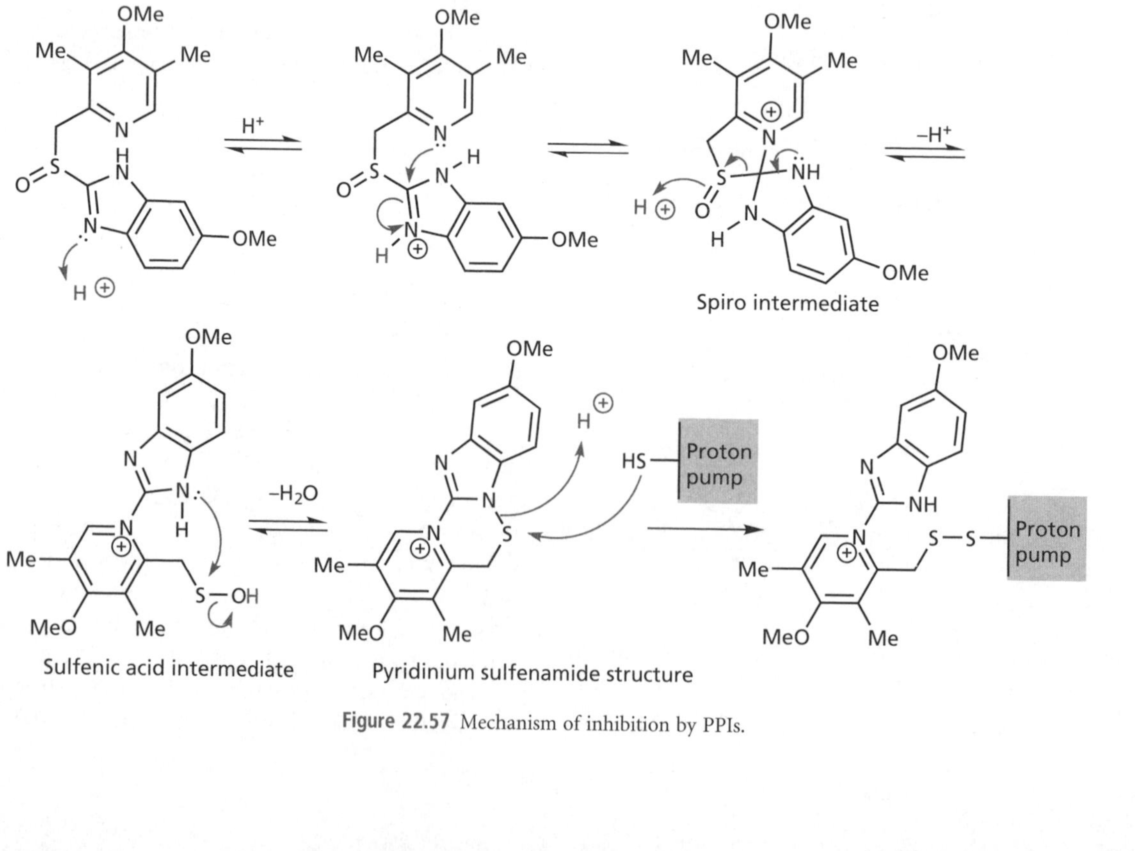

Figure 22.57 Mechanism of inhibition by PPIs.

benzimidazole ring to form a spiro structure. By doing so, the aromatic character of the imidazole portion of the ring is lost and so there is a high tendency for this ring to re-aromatize. This can be achieved by a lone pair of electrons from nitrogen reforming the double bond and cleaving the S–C bond to form a sulfenic acid. Sulfenic acids are highly reactive to nucleophiles and so a rapid reaction takes place involving an intramolecular attack by the NH group of the benzimidazole on the sulfenic acid to displace the hydroxyl group. A cationic tetracyclic pyridinium sulfenamide is formed which acts as an irreversible enzyme inhibitor (Fig. 22.57). It does so by forming a covalent bond to an accessible cysteine residue on the proton pump. There are three such accessible cysteine residues (Cys-813, Cys-892, and Cys-821) and it has been found that the specific cysteine residues attacked depend on which PPI is involved. For example, omeprazole binds to two of the accessible cysteine residues (Cys-813 and Cys-892), lansoprazole binds to all three, and pantoprazole only binds to one (either Cys-813 or Cys-822). Cys-813 is the only cysteine residue which appears to be bound by all the PPIs.

As acid conditions are required to activate the PPIs, they are most active when parietal cells are actively secreting hydrochloric acid, and show little activity when the parietal cells are in a resting state. Since a covalent disulfide bond has been formed between the inhibitor and the proton pump, inhibition is irreversible and so PPIs have a long duration of action. The duration depends on how quickly new pumps are generated by the cell.

PPIs also have very few side effects, because of their selectivity of action. This can be put down to several factors:

- The target enzyme (H^+/K^+-ATPase) is only present in parietal cells.

- The canaliculi of the parietal cells are the only compartments in the body which have such a low pH (1–2).

- The drug is concentrated at the target site due to protonation and is unable to return to the parietal cell or to the general circulation.

- The drug is rapidly activated close to the target.

- Once activated, the drug reacts rapidly with the target.

- The drug is inactive at neutral pH.

22.3.4 Metabolism of proton pump inhibitors

PPIs are metabolized by cytochrome P450 enzymes, particularly **S-mephenytoin hydroxylase** (CYP2C19) and **nifedipine hydroxylase** (CYP3A4). As a result of genetic variations, about 3% of white people of European descent are slow metabolizers of PPIs. Pantoprazole, in contrast to omeprazole and lansoprazole, is also metabolized by the conjugating enzyme **sulfotransferase**.

22.3.5 Design of omeprazole and esomeprazole

Omeprazole was the first PPI to reach the market in 1988 and was marketed as **Losec**. In 1996 it became the biggest-selling pharmaceutical ever. The story of how omeprazole was developed can be traced back to the 1970s. The lead compound for the project was a thiourea structure (CMN 131 in Fig. 22.58). This had originally been investigated as an antiviral drug, but general pharmacological tests showed that it could inhibit acid secretion. Unfortunately, toxicology tests showed that the compound was toxic to the liver and this was put down to the presence of the thioamide group. Various analogues were made to try to modify or disguise the thiourea group, which included incorporating the thiourea skeleton within a ring or modifying the structure of the functional group. This led eventually to the discovery of H77/67, which was also found to inhibit acid secretion. A variety of analogues having the general structure (heterocycle–X–Y–heterocycle) were synthesized, which demonstrated that the pyridine ring and the bridging CH_2–S group already present in H77/67 were optimal for activity. However, activity was increased by replacing the imidazole ring of H77/67 with a benzimidazole group to give H124/26. At this stage, drug metabolism studies revealed that a sulfoxide metabolite of H124/26 was formed *in vivo* and this structure was more active than the original structure. The metabolite was called **timoprazole** and was the first example of a pyridinylmethylsulfinyl benzimidazole structure. It went forward for preclinical trials, but toxicological studies revealed that it inhibited iodine uptake by the thyroid gland and so it could not go on to clinical trials.

Analogues were now synthesized to find a structure which retained the antisecretory properties but did not

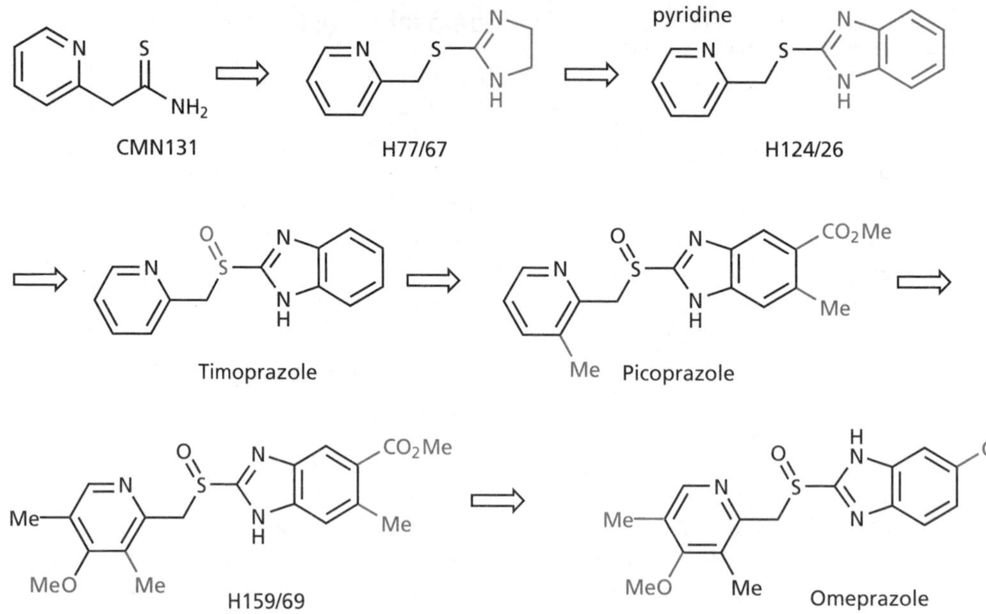

Figure 22.58 Development of omeprazole.

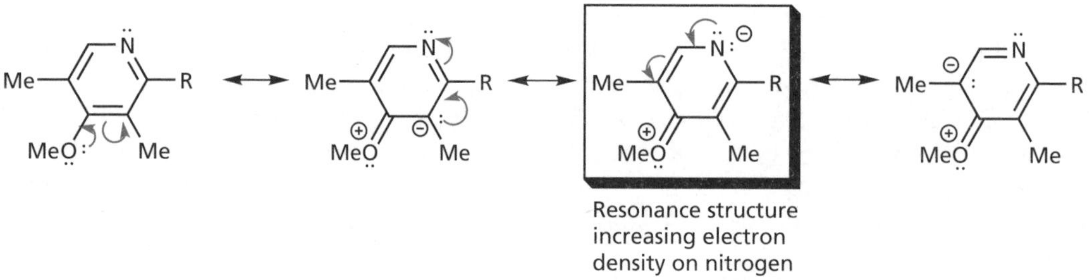

Resonance structure
increasing electron
density on nitrogen

Figure 22.59 Influence of the methoxy substituent.

inhibit iodine uptake. Eventually, it was found that the two effects could be separated by placing suitable substituents on the two heterocyclic rings. This led to **picoprazole** which showed potent antisecretory properties over a long period without the toxic side effect on the thyroid. Animal toxicology studies showed no other toxic effects and the drug went forward for clinical trials, where it was found to be the most effective antisecretory compound so far tested in humans. At this point (1977), the proton pump was discovered and this was identified as the target for picoprazole. Further development was carried out with the aim of getting a more potent drug by varying the substituents on the pyridine ring.

It was discovered that substituents which increased the basicity of the pyridine ring were good for activity. This fits in with the mechanism of activation (Fig. 22.57) where the nitrogen of the pyridine ring acts as a nucleophile. In order to increase the nucleophilicity of the pyridine ring, a methoxy group was placed at the *para* position relative to the nitrogen, and two methyl groups were placed at the *meta* positions. The latter have an inductive effect which is electron donating and increases the electron density of the ring. The methoxy substituent was added at the *para* position so that electron density could be increased on the pyridine nitrogen by the resonance mechanism shown in Fig. 22.59.

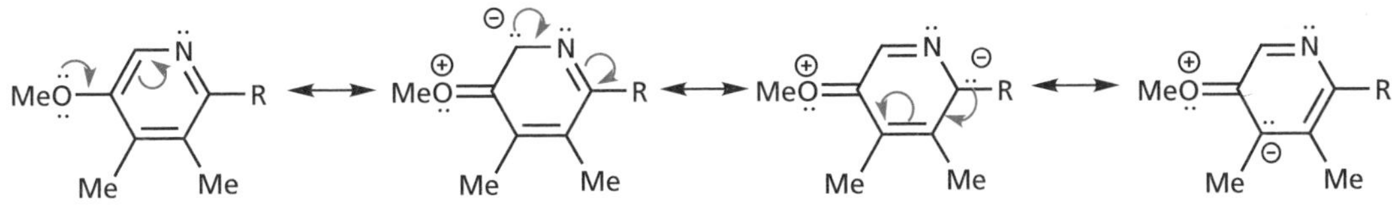

Figure 22.60 Possible resonance structures for methoxy substitution at the *meta* position.

It is noticeable that all the PPIs shown in Fig. 22.56 have an alkoxy substituent at the *para* position of the pyridine. The position of the substituent is important. If the substituent were at the *meta* position, none of the possible resonance structures would place the negative charge on the nitrogen atom (Fig. 22.60). Finally, if the methoxy substituent was at the *ortho* position it would be likely to have a bad steric effect and hinder the mechanism.

The introduction of two methyl groups and a methoxy group led to H159/69 (Fig. 22.58), which was extremely potent but chemically too labile. Further analogues were synthesized where substituents round the benzimidazole ring were varied in order to get the right balance of potency, chemical stability, and synthetic accessibility. Finally, omeprazole was identified as the structure having the best balance of these properties.

Omeprazole was launched in 1988 and became the world's highest-selling drug, earning its makers vast profits. For example, worldwide sales in 2000 were $6.2 billion (£3.6 billion). The patents on omeprazole ran out in Europe in 1999 and in the USA in 2001, but its makers (Astra) had already started a programme to find an even better compound. In particular, they were looking for a compound with better bioavailability.

Substitution was varied on both the pyridine and benzimidazole rings, but the best compound was eventually found to be the *S*-enantiomer of omeprazole—**esomeprazole** (**Nexium**; Fig. 22.61). At first sight, it may not be evident that omeprazole has an asymmetric centre. In fact, the sulfur atom is an asymmetric centre as it has a lone pair of electrons and is tetrahedral. Unlike the nitrogen atoms of amines, sulfur atoms do not undergo pyramidal inversion and so it is possible to isolate both enantiomers. The *S*-enantiomer of omeprazole was found to be superior to the *R*-enantiomer in terms of its potency and pharmacokinetic profile, and was launched as esomeprazole in Europe in 2000 and in the USA in 2001.

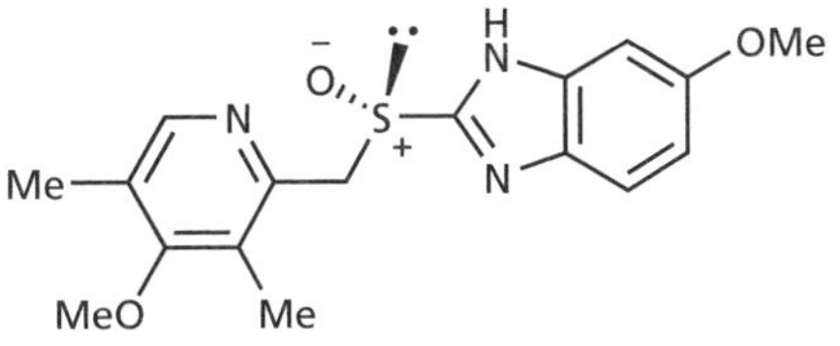

Figure 22.61 Esomeprazole.

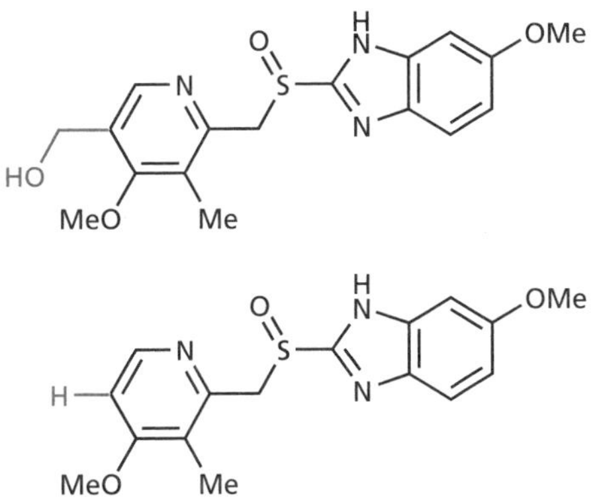

Figure 22.62 Metabolites of esomeprazole.

The story of esomeprazole is an example of **chiral switching** (section 12.2.1) where a racemic drug is replaced on the market with its more active enantiomer. Because of its better pharmacokinetic profile, it is possible to use double the dose levels of esomeprazole compared to omeprazole, resulting in greater activity. Esomeprazole is metabolized mainly by CYP2C19 in the liver, to form the hydroxy and desmethyl metabolites shown in Fig. 22.62. However, it undergoes less hydroxylation than the *R* isomer and has a lower clearance rate. A sulfone metabolite is formed by CYP3A4. Due to differences in metabolism, higher plasma levels of the *S*-enantiomer are achieved compared to the *R*-enantiomer. Synthesis of omeprazole and esomeprazole is described in Box 22.2.

22.3.6 Other proton pump inhibitors

The other PPIs shown in Fig. 22.56 retain the pyridinylmethylsulfinyl benzimidazole structure of omeprazole. They also share the alkoxy substituent at the *para* position of the pyridine ring. Variation has been limited to the other substituents present on the heterocyclic rings. These play a role in determining the lipophilic character of the drug as well as its stability. As far as the latter is concerned, there has to be a

---

BOX 22.2 SYNTHESIS OF OMEPRAZOLE AND ESOMEPRAZOLE

The synthesis of omeprazole appears to be relatively simple, involving the linkage of the two halves of the molecule through a nucleophilic substitution reaction. The benzimidazole half of the molecule has a thiol substituent which is treated with sodium hydroxide to give a thiolate. On reaction with the chloromethylpyridine, the thiolate group displaces the chloride ion to link up the two halves of the molecule. Subsequent oxidation of the sulfur atom with *meta*-chloroperbenzoic acid gives omeprazole. What is not obvious from the scheme is the effort required to synthesize the required chloromethylpyridine starting material. This is not the sort of molecule that is easily bought off the shelf, and its synthesis involves six steps.

The same route can be used for the synthesis of esomeprazole (the *S* enantiomer of omeprazole) by employing asymmetric conditions for the final sulfoxidation step. Early attempts to carry out this reaction involved the Sharpless reagent formed from Ti(O-*i*Pr)$_4$, the oxidizing agent cumene hydroperoxide (Ph(CH$_3$)$_2$OOH), and the chiral auxiliary (*S*,*S*)-diethyl tartrate. Although sulfoxidation took place, it required almost stoichiometric quantities of the titanium reagent, and there was little enantioselectivity. The reaction conditions were modified in three ways to improve enantioselectivity to over 94% enantiomeric excess, and which required less of the titanium reagent (4–30 mol%).

- Formation of the titanium complex was carried out in the presence of the sulfide starting material.

- The solution of the titanium complex was equilibrated at an elevated temperature for a prolonged period of time.

- The oxidation was carried out in the presence of an amine such as *N*,*N*-diisopropylethylamine. The role of the amine is not fully understood, but it may participate in the titanium complex.

The enantiomeric excess can be enhanced further by preparing a metal salt of the crude product and carrying out a crystallization which boosts the enantiomeric excess to more than 99.5%.

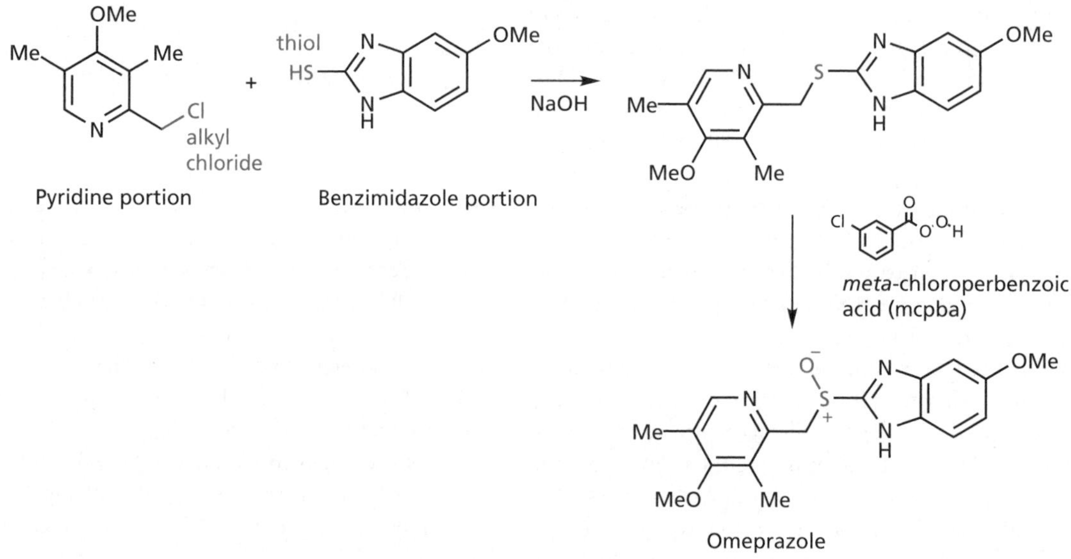

Synthesis of omeprazole.

balance between the drug being sufficiently stable and un-ionized at neutral pH to reach its target unchanged, and its ability to undergo rapid acid-induced conversion into the active sulfenamide when it reaches the target. Stability to mild acid is important to avoid activation in other cellular compartments such as lysosomes and chromaffin granules. Drugs which undergo the acid-induced conversion extremely easily are more active, but they are less stable and are more likely to undergo transformation in the blood supply before they reach their target. Drugs which are too stable are less reactive under acid conditions and react slower with the target.

The various PPIs all work by the same mechanism, but have slightly different properties. For example, pantoprazole is chemically more stable than omeprazole or lansoprazole under neutral to mildly acidic conditions (3.5–7.4), but it is a weaker, irreversible inhibitor under strong acid conditions. Rabeprazole is the least stable at neutral pH and is the most active inhibitor.

KEY POINTS

- The proton pump is responsible for pumping protons out of the parietal cell in exchange for potassium ions which are pumped in. The process involves the movement of protons against a concentration gradient, so it requires energy which is provided by the hydrolysis of ATP.

- PPIs prevent the proton pump from functioning. They offer a strategic advantage over H_2 antagonists since they act on the final stage of hydrochloric acid release.

- PPIs are prodrugs which are activated by the acidic conditions found in the canaliculi of parietal cells. They undergo an acid-catalysed rearrangement to form a reactive tetracyclic pyridinium sulfenamide which acts as an irreversible inhibitor. Reaction takes place with accessible cysteine residues on the proton pump to form a covalent disulfide bond between cysteine and the drug.

- PPIs need to be reactive enough to undergo acid-catalysed interconversion in the canaliculi of parietal cells, but stable enough to survive their journey through the bloodstream.

22.4 *H. pylori* and the use of antibacterial agents

One of the problems relating to antiulcer therapy both with H_2 antagonists and with PPIs is the high rate of ulcer recurrence once the therapy is finished. The reappearance of ulcers has been attributed to the presence of a microorganism called *H. pylori* which is naturally present in the stomachs of many people, and can cause inflammation of the stomach wall. As a result, patients who are found to have *H. pylori* are currently given a combination of three drugs—a PPI to reduce gastric acid secretion, and two antibacterial agents (such as nitroimidazole, clarithromycin, amoxicillin or tetracycline) to eradicate the organism.

It was once considered unthinkable that a bacterium could survive the acid conditions of the stomach. In 1979, however, it was shown that *H. pylori* can do just that. The organism is able to attach to a sugar molecule on the surface of the cells that line the stomach wall, and use the mucus layer which protects the stomach wall from gastric juices as its own protection. Since there is a pH gradient across the mucus layer, the organisms can survive at the surface of the mucus cells where the pH is closer to neutral (Fig. 22.63). *H. pylori* is a spiral curved bacterium which is highly motile and grows best in oxygen concentrations of 5%, matching those of the mucus layer. The bacterium also produces large amounts of the enzyme **urease** which catalyses

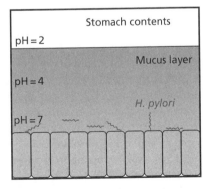

Figure 22.63 *H. pylori* attached to stomach cells.

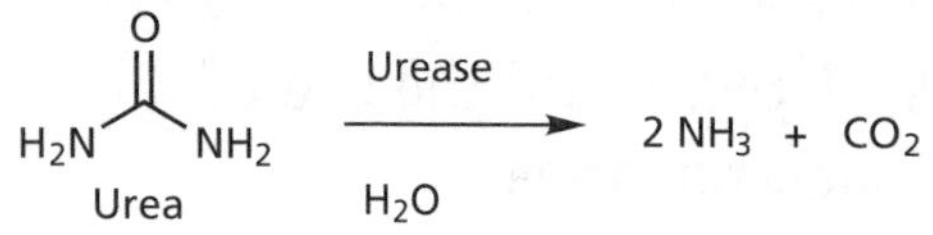

Figure 22.64 Action of urease.

the hydrolysis of urea to ammonia and carbon dioxide, thus neutralizing any acid in the local environment (Fig. 22.64). The organisms can contribute to the formation of stomach ulcers, as they secrete proteins and toxins that interact with the stomach's epithelial cells, leading to inflammation and cell damage.

22.4.1 Treatment

As mentioned earlier, *H. pylori* is treated with a triple therapy of a PPI and at least two antibacterial agents. A PPI is administered because the antibiotics used work best at higher pH levels than those normally present in the stomach. The combination of omeprazole, amoxicillin, and metronidazole is frequently used, but combinations involving other antibacterial agents such as clarithromycin and tetracycline are also possible. Bismuth chelate (bismuth subcitrate and tripotassium dicitratobismuthate) is present in some combination therapies. This preparation has a toxic effect on *H. pylori* and may help to prevent adherence to the mucosa. Other protective properties include an enhancement of local prostaglandin synthesis, a coating of the ulcer base, and an adsorption of pepsin.

Combination therapy has been shown to eradicate *H. pylori* in over 90% of duodenal ulcers and significantly reduce ulcer recurrence. Similar treatment is recommended for *H. pylori* related stomach ulcers.

Eradication of *H. pylori* can be difficult because of the emergence of resistant strains and the difficulty in delivering the antibacterial agents at the required therapeutic concentration. *H. pylori* can also assume a resting coccoid form that is more resistant to therapy.

It has been found that PPIs have an inherent anti-*H. pylori* action and it has been suggested that they inhibit urease, possibly by linking to exposed cysteine residues. However, the PPIs also inhibit strains of *H. pylori* which do not have urease, so this is not the full story. This antibacterial activity is sufficient to suppress the organism but not eradicate it, so traditional antibacterial agents are still required.

Research has been carried out into the design of drugs which act as sugar decoys to prevent *H. pylori* binding with stomach cells in the first place.

22.5 Traditional and herbal medicines

Several herbal remedies have been used for the treatment of ulcers.

- Liquorice has been reported to have a variety of medicinal properties and has been used as a medicine for several thousand years. It is reported to have antiulcer properties and this has been attributed to a component called **glycrrhetinic acid**—the aglycone of **glycyrrhizin**. **Carbenoxolone** is a derivative of glycrrhetinic acid and has been used in ulcer therapy. It is thought to have a mucosal protective role by increasing mucus production, and has some antibacterial action against *H. pylori*.

- **Silymarin** is a mixture of compounds (mainly silibinin, silichristin, and siliianin) obtained from the fruit of the milk thistle (*Silybum marianum*) and has antiulcer activity. It has been shown to reduce histamine concentrations in rats.

- Extracts from the neem tree (*Azadirachta indica*) have been used extensively in India as a medicine for a variety of ailments. The aqueous extract of neem bark has been reported to have antiulcer effects. Possible mechanisms include proton pump inhibition or antioxidant effects in the scavenging of OH radicals.

- Other herbal medicines include comfrey and marshmallow.

KEY POINTS

- *H. pylori* is a bacterium which is responsible for many ulcers. It can survive at the surface of mucus cells and produce proteins and toxins which damage epithelial cells.

- Ulcers which are caused by *H. pylori* are treated with a combination of drugs which include a PPI and at least two antibiotics.

- Several traditional and herbal remedies are used in the treatment of ulcers.

QUESTIONS

1. Omeprazole is administered orally as a galenic formulation to protect it from being activated during its journey through the acidic contents of the stomach. Once it is released in the intestines, it is absorbed into the blood supply and carried to the parietal cells where it crosses the cell membrane into the canaliculi and is activated. Since the canaliculi lead directly into the lumen of the stomach, why is omeprazole not orally administered directly to the stomach?

2. In the development of omeprazole, the methoxy and methyl groups were added to the pyridine ring to increase the pK_a. Subsequently, it was found that analogue (I) with only one of the methyl groups had a higher pK_a than omeprazole. Suggest why this might be the case.

3. Suggest whether you think structure (I) would be a better PPI than omeprazole.

4. The acid-catalysed activation of PPIs requires pyridine to be nucleophilic, which is why two methyl groups and a methoxy group are present in omeprazole. Suggest whether the addition of an extra methyl group (structure II) would lead to a more potent PPI.

5. The phenol (III) is a very difficult compound to synthesize and is unstable at neutral pH. Suggest why this might be the case.

6. Suggest what types of metabolite might be possible from omeprazole.

FURTHER READING

Agranat, I., Caner, H. and Caldwell, J. (2002) Putting chirality to work: the strategy of chiral switches. *Nature Reviews Drug Discovery*, 1, 753–768.

Baxter, G. F. (1992) Settling the stomach. *Chemistry in Britain*, May, 445–448.

Carlsson, E., Lindberg, P., and von Unge, S. (2002) Two of a kind. *Chemistry in Britain*, May, 42–45 (PPIs).

Ganellin, R. (1981) Medicinal chemistry and dynamic structure-activity analysis in the discovery of drugs acting at histamine H_2 receptors. *Journal of Medicinal Chemistry*, 24, 913–920.

Ganellin, C. R. and Roberts, S. M. (eds.) (1994) Discovery of cimetidine, ranitidine and other H_2-receptor histamine antagonists. Chapter 12 in: *Medicinal chemistry—the role of organic research in drug research*, 2nd edn. Academic Press, New York.

Hall, N. (1997) A landmark in drug design. *Chemistry in Britain*, December, 25–27 (cimetidine).

Lewis, D. A. (1992) Antiulcer drugs from plants. *Chemistry in Britain*, February, 141–144.

Lindberg, P. *et al.* (1990) Omeprazole: the first proton pump inhibitor. *Medicinal Research Reviews*, 10, 1–54.

Olbe, L., Carlsson, E. and Lindberg, P. (2003) A proton-pump inhibitor expedition: the case histories of omeprazole and esomeprazole. *Nature Reviews Drug Discovery*, **2**, 132–139.

Saunders, J. (2000) Antagonists of histamine receptors (H$_2$) as anti-ulcer remedies. Chapter 4 in: *Top drugs: top synthetic routes*. Oxford University Press, Oxford.

Saunders, J. (2000) Proton pump inhibitors as gastric acid secretion inhibitors. Chapter 5 in: *Top drugs: top synthetic routes*. Oxford University Press, Oxford.

Young, R. C. *et al.* (1986) Dipole moment in relation to H2 receptor histamine antagonist activity for cimetidine analogues. *Journal of Medicinal Chemistry*, **29**, 44–49.

Titles for general further reading are listed on p. 711.

APPENDIX 1

Essential amino acids

Non-polar (hydrophobic)

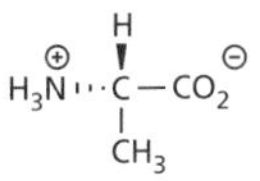

Alanine
(Ala or A)

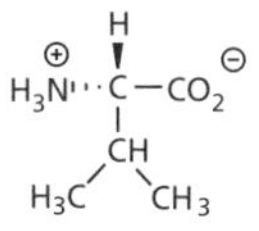

Valine
(Val or V)

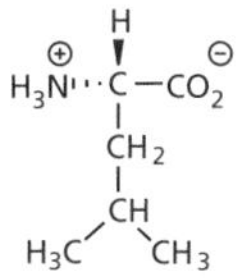

Leucine
(Leu or L)

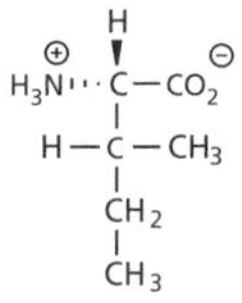

Isoleucine
(Ile or I)

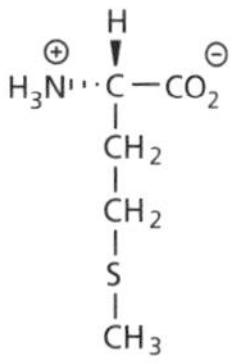

Methionine
(Met or M)

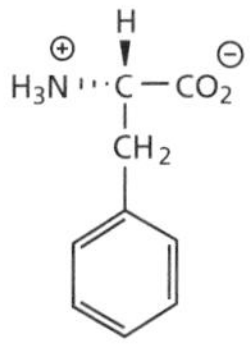

Phenylalanine
(Phe or F)

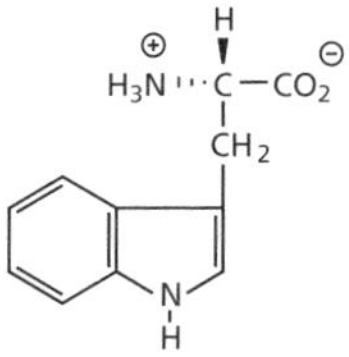

Tryptophan
(Try or W)

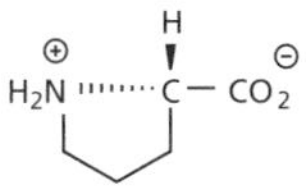

Proline
(Pro or P)

Polar

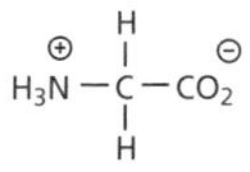

Glycine
(Gly or G)

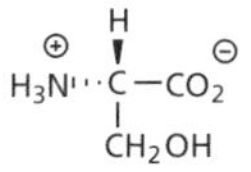

Serine
(Ser or S)

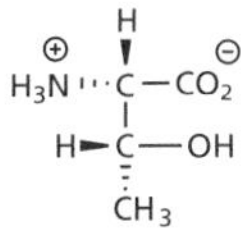

Threonine
(Thr or T)

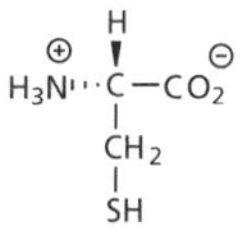

Cysteine
(Cys or C)

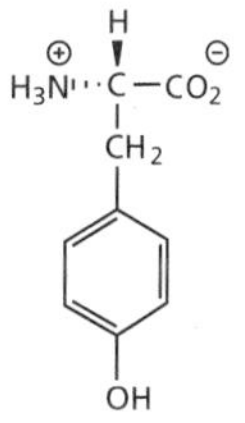

Tyrosine
(Tyr or Y)

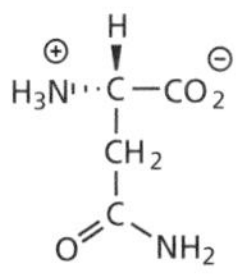

Asparagine
(Asn or N)

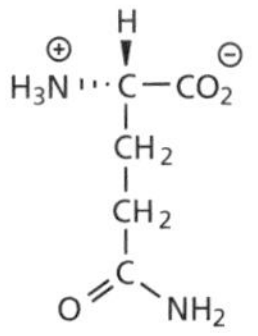

Glutamine
(Gln or Q)

Ionized

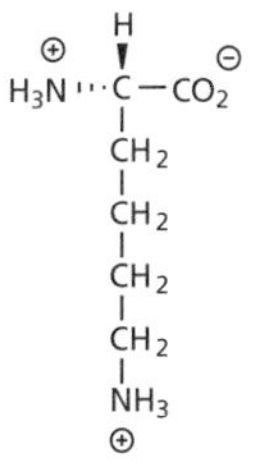

Lysine
(Lys or K)

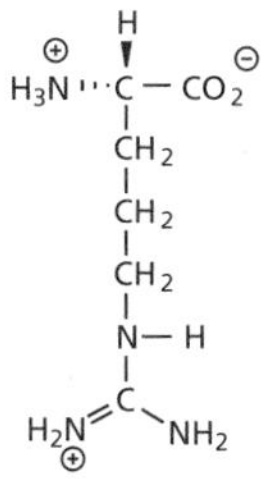

Arginine
(Arg or R)

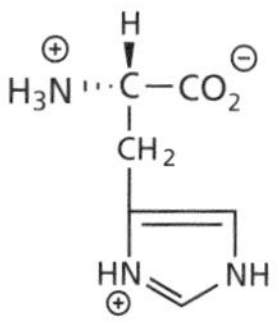

Histidine
(His or H)

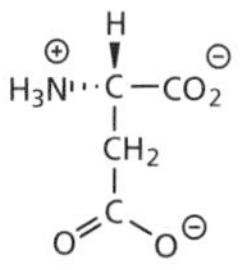

Aspartate
(Asp or D)

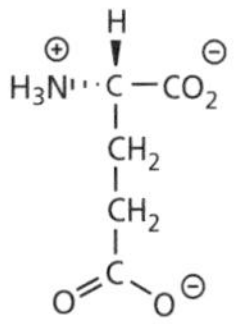

Glutamate
(Glu or E)

The standard genetic code

UUU	Phe	UCU	Ser	UAU	Tyr	UGU	Cys
UUC	Phe	UCC	Ser	UAC	Tyr	UGC	Cys
UUA	Leu	UCA	Ser	UAA	Stop	UGA	Stop
UUG	Leu	UCG	Ser	UAG	Stop	UGG	Trp
CUU	Leu	CCU	Pro	CAU	His	CGU	Arg
CUC	Leu	CCC	Pro	CAC	His	CGC	Arg
CUA	Leu	CCA	Pro	CAA	Gln	CGA	Arg
CUG	Leu	CCG	Pro	CAG	Gln	CGG	Arg
AUU	Ile	ACU	Thr	AAU	Asn	AGU	Ser
AUC	Ile	ACC	Thr	AAC	Asn	AGC	Ser
AUA	Ile	ACA	Thr	AAA	Lys	AGA	Arg
AUG	Met	ACG	Thr	AAG	Lys	AGG	Arg
GUU	Val	GCU	Ala	GAU	Asp	GGU	Gly
GUC	Val	GCG	Ala	GAC	Asp	GGC	Gly
GUA	Val	GCA	Ala	GAA	Glu	GGA	Gly
GUG	Val	GCG	Ala	GAG	Glu	GGG	Gly

Statistical data for QSAR

To illustrate how statistical terms such as r, s, and F are derived and interpreted, the following numerical data will be used. There are 6 compounds in the study ($n = 6$). Y_{exp} is the logarithm of the observed activity for each of the compounds and X is a physicochemical parameter. The QSAR equation derived from the data is

$$\log(\text{activity}) = Y_{calc} = k_1 X + k_2 = -0.47X - 0.022$$

The slope of the line is -0.47 and the intercept with the y-axis is -0.022.

The correlation coefficient r for the above QSAR equation is calculated using equation (1):

$$r^2 = 1 - \frac{SS_{calc}}{SS_{mean}} \qquad (1)$$

SS_{calc} is a measure of how much the experimental activity of the compounds varies from the calculated value. For each compound, the difference between the experimental activity and the calculated activity is $Y_{exp} - Y_{calc}$ (Fig. A3.1). This is then squared and the values are added together to give the sum of the squares (SS_{calc}).

SS_{mean} is a measure of how much the experimental activity varies from the mean of all the experimental activities and represents the situation where no correlation with X has been attempted (Fig. A3.1).

If there is a correlation between the activity (Y) and the parameter (X), the line of the equation should pass closer to the data points than the line representing the mean. This means that SS_{calc} should be less than SS_{mean}. For a perfect correlation, the calculated values for the activity would be

Table A3.1

Compound ($n = 6$)	Physicochemical parameter (X)	$\log(\text{act.})_{exp}$ Y_{exp}	$\log(\text{act.})_{calc}$ Y_{calc}	$Y_{exp} - Y_{calc}$	Square of $Y_{exp} - Y_{calc}$	$Y_{exp} - Y_{mean}$	Square of $Y_{exp} - Y_{mean}$
1	0.23	0.049	−0.129	0.178	0.0317	0.263	0.0692
2	0.23	0.037	−0.129	0.166	0.0276	0.251	0.0630
3	−0.17	0.000	0.057	−0.057	0.0032	0.214	0.0458
4	0.00	−0.155	−0.022	−0.133	0.0177	0.059	0.0035
5	1.27	−0.468	−0.613	0.145	0.0210	−0.254	0.0645
6	0.91	−0.745	−0.445	−0.300	0.0900	−0.531	0.2820
		Mean value Y_{mean} −0.214			Sum of squares SS_{calc} 0.1912		Sum of squares SS_{mean} 0.5279

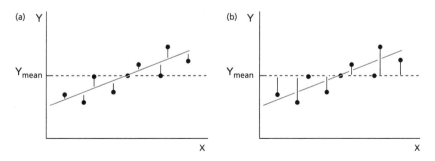

Figure A3.1 (a) Black lines show $Y_{exp} - Y_{calc}$. (b) Black lines show $Y_{exp} - Y_{mean}$.

the same as the experimental ones and so SS_{calc} would be zero. This would make $r^2 = 1$.

For the figures shown in the above table, the value of r works out as follows:

$$r^2 = 1 - \frac{SS_{calc}}{SS_{mean}} = 1 - \frac{0.1912}{0.5279} = 1 - 0.3622 = 0.638$$

This indicates that only 64% of the variability in activity is due to the parameter X. This is much lower than the minimum acceptable figure of 80% and so the equation is not a particularly good one. Nevertheless, it is possible that X may have some influence on the activity. To check whether there is any significance to the equation a statistical test called an **F-test** can be carried out. The equation used for this specific example is as follows:

$$F_{p_2 - p_1, \, n - p_2} = \frac{SS_{mean} - SS_{calc}}{SS_{calc}} \times \frac{n - p_2}{p_2 - p_1}$$

where p_2 is the number of parameters involved in the derived QSAR equation (Y and X) and p_1 is the number of parameters involved in the reference equation (Y only in this example). n, SS_{mean}, and SS_{calc} are as described above. This gives the following:

$$F_{2-1, \, 6-2} = \frac{0.528 - 0.1912}{0.1912} \times \frac{6 - 2}{2 - 1}$$

or

$$F_{1, \, 4} = \frac{0.528 - 0.1912}{0.1912} \times \frac{6 - 2}{2 - 1} = 1.7615 \times 4 = 7.05$$

$F_{1,4}$ is now compared against tables of F values which indicate the probability level of a significant correlation.

For $F_{1,4}$ the tables show that a value of 4.54 would indicate a probability level of 0.9, whereas 7.71 represents a probability level of 0.95. A value of 21.2 represents a probability level of 0.99. The higher the value of $F_{1,4}$, the closer the probability level approaches 1. The calculated value of 7.05 shows that the probability level is between 0.9 and 0.95.

The standard deviation (s) for the equation is calculated by using equation 2 and is dependent on the number of compounds (n) tested.

$$s^2 = \frac{SS_{calc}}{n - 2} \tag{2}$$

This gives a value of 0.218 for the data provided in Table A3.1. The value of s should be as small as possible, but not smaller than the standard deviation of the experimental data.

A QSAR equation could now be derived to see whether the biological activity matches a different physicochemical parameter. The Table A3.2 shows values for a different parameter (Z). In this case, the derived equation is:

$$Y_{calc} = 0.33Z - 0.62$$

The statistical analysis of this gives the following:

$$n = 6, \quad r = 0.840, s = 0.199, \quad F_{1, \, 4} = 9.6$$

All these results are better than the previous ones, showing that the parameter Z is more important than X in explaining the variation in activity. r is still less than 0.9, however, and further improvements are necessary.

If both of the above parameters are included in the analysis, the equation becomes:

$$Y_{calc} = -0.34X + 0.25Z - 0.38$$

Table A3.2

Compound ($n=6$)	Physicochemical parameter (Z)	log(act.)$_{exp}$ Y_{exp}	log(act.)$_{calc}$ Y_{calc}	$Y_{exp} - Y_{calc}$	Square of $Y_{exp} - Y_{calc}$	$Y_{exp} - Y_{mean}$	Square of $Y_{exp} - Y_{mean}$
1	2.03	0.049	0.0499	−0.0009	0.0000	0.263	0.0692
2	1.83	0.037	−0.0161	0.0531	0.0028	0.251	0.0630
3	1.38	0.000	−0.1646	0.1646	0.0271	0.214	0.0458
4	0.90	−0.155	−0.3230	0.1680	0.0282	0.059	0.0035
5	1.40	−0.468	−0.1580	−0.3100	0.0961	−0.254	0.0645
6	−0.26	−0.745	−0.7058	−0.0392	0.0015	−0.531	0.2820
		Mean value			Sum of squares		Sum of squares
		Y_{mean}			SS_{calc}		SS_{mean}
		−0.214			0.1558		0.5279

Table A3.3

Compound (n = 6)	Physicochemical parameter (X)	Physicochemical parameter (Z)	Log(act.)$_{exp}$ Y_{exp}	Log(act.)$_{calc}$ Y_{calc}	$Y_{exp} - Y_{calc}$	Square of $Y_{exp} - Y_{calc}$	$Y_{exp} - Y_{mean}$	Square of $Y_{exp} - Y_{mean}$
1	0.23	2.03	0.049	0.0493	−0.0003	0.0000	0.263	0.0692
2	0.23	1.83	0.037	−0.0007	0.0377	0.0014	0.251	0.0630
3	−0.17	1.38	0.000	0.0228	−0.0228	0.0005	0.214	0.0458
4	0.00	0.90	−0.155	−0.155	0.0000	0.0000	0.059	0.0035
5	1.27	1.40	−0.468	−0.4618	−0.0062	0.0000	−0.254	0.0645
6	0.91	−0.26	−0.745	−0.7544	0.0094	0.0001	−0.531	0.2820
			Mean value Y_{mean} −0.214			Sum of squares SS_{calc} 0.0021		Sum of squares SS_{mean} 0.5279

The corresponding results are shown in Table A3.3. The statistical results are $n = 6$, $r = 0.998$, $s = 0.028$ and $F_{1,3} = 230.3$. Note that there are three parameters in the QSAR equation and so the F term is $F_{1,3}$ rather than $F_{1,4}$. Comparison with tabulated $F_{1,3}$ values shows that the probability level for this equation is 0.999.

A final check has to be made to ensure that the values for the two parameters (X and Z) are not related in any way. An equation attempting to relate X and Z is derived and assessed statistically. For the values shown, $r^2 = 0.122$, which shows that there is little correlation between X and Z. The final equation is therefore validated.

QSAR equations may also include terms in parenthesis. For example, taking the previous equation,

$$Y_{calc} = -0.34(\pm0.08)X + 0.25(\pm0.05)Z - 0.38(\pm0.09)$$

The numbers in parenthesis represent the 95% confidence limits for the various parameters. For example, there is 95% confidence that the coefficient for Z lies between the values 0.20 and 0.30. If the number in parenthesis is smaller than the coefficient, it means the parameter is statistically significant in the F-test.

APPENDIX 4

The action of nerves

The structure of a typical nerve is shown in Fig. A4.1. The nucleus of the cell is found in the large cell body situated at one end of the nerve cell. Small arms (dendrites) radiate from the cell body and receive messages from other nerves. These messages either stimulate or destimulate the nerve. The cell body 'collects' the sum total of these messages.

It is worth emphasizing the point that the cell body of a nerve receives messages, not just from one other nerve, but from a range of different nerves. These pass on different messages (neurotransmitters). Therefore, a message received from a single nerve is unlikely to stimulate a nerve signal by itself unless other nerves are acting in sympathy.

Assuming that the overall stimulation is great enough, an electrical signal is fired down the length of the nerve (the axon). The axon is covered with sheaths of lipid (myelin sheaths) which act to insulate the signal as it passes down the axon.

The axon leads to a knob-shaped swelling (synaptic button) if the nerve is communicating with another nerve. Alternatively, if the nerve is communicating with a muscle cell, the axon leads to what is known as a neuromuscular endplate, where the nerve cell has spread itself like an amoeba over an area of the muscle cell.

Within the synaptic button or neuromuscular endplate there are small globules (vesicles) containing the neurotransmitter chemical. When a signal is received from the axon, the vesicles merge with the cell membrane and release their neurotransmitter into the gap between the nerve and the target cell (synaptic gap). The neurotransmitter

binds to the receptor as described in Chapter 5, and passes on its message. Once the message has been received, the neurotransmitter leaves the receptor and is either broken down enzymatically (e.g. acetylcholine) or taken up intact by the nerve cell (e.g. noradrenaline). Either way, the neurotransmitter is removed from the synaptic gap and is unable to bind with its receptor a second time.

To date, we have talked about nerves 'firing' and the generation of 'electrical signals' without really considering the mechanism of these processes. The secret behind nerve transmission lies in the movement of ions across cell membranes, but there is an important difference in what happens in the cell body of a nerve compared to the axon. We shall consider what happens in the cell body first.

All cells contain sodium, potassium, calcium, and chloride ions and it is found that the concentration of these ions is different inside the cell compared to the outside. The concentration of potassium inside the cell is larger than the surrounding medium, whereas the concentration of sodium and chloride ions is smaller. Thus, a concentration gradient exists across the membrane.

Potassium is able to move down its concentration gradient (i.e. out of the cell), since it can pass through the potassium ion channels (Fig. A4.2). But if potassium ions can move out of the cell, why does the potassium concentration inside the cell not fall to equal that of the outside? The answer lies in the fact that potassium is a positively charged ion and as it leaves the cell an electric potential is set up across the cell membrane. This would not

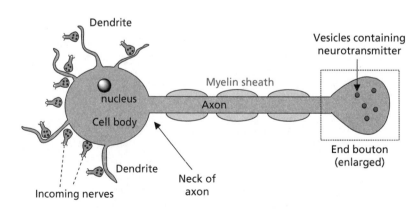

Figure A4.1 Structure of a typical nerve cell.

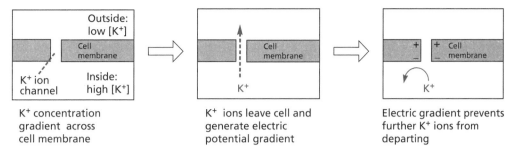

Figure A4.2 Generation of electric potential across a cell membrane.

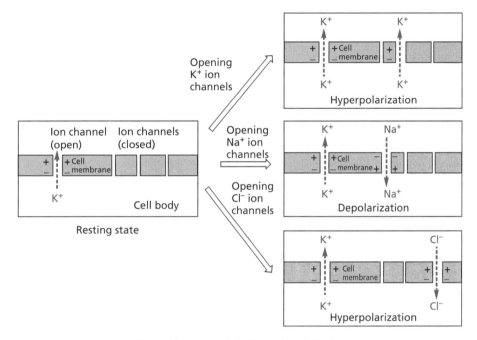

Figure A4.3 Hyperpolarization and depolarization.

happen if a negatively charged counterion could leave with the potassium ion. However, the counterions in question are large proteins which cannot pass through the cell membrane. As a result, a few potassium ions are able to escape down the ion channels out of the cell and an electric potential builds up across the cell membrane such that the inside of the cell membrane is more negative than the outside. This electric potential (50–80 mV) opposes and eventually prevents the flow of potassium ions.

But what about the sodium ions? Could they flow into the cell along their concentration gradient to balance the charged potassium ions that are departing? The answer is that they cannot, because they are too big for the potassium ion channels. This appears to be a strange argument,

as sodium ions are smaller than potassium ions, but it has to be remembered that we are dealing with an aqueous environment where the ions are solvated (i.e. they have a 'coat' of water molecules). Sodium, being a smaller ion than potassium, has a greater localization of charge and is able to bind its solvating water molecules more strongly. As a result, sodium along with its water coat is bigger than a potassium ion with or without its water coat.

Ion channels for sodium do exist and these channels are capable of removing the water coat around sodium and letting it through. However, the sodium ion channels are mostly closed when the nerve is in the resting state. As a result, the flow of sodium ions across the membrane is very small compared to potassium. Nevertheless, the presence of

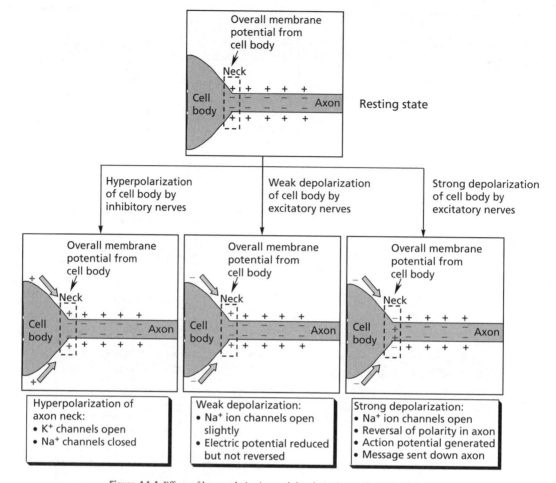

Figure A4.4 Effects of hyperpolarization and depolarization at the neck of the axon.

sodium ion channels is crucial to the transmission of a nerve signal.

To conclude, the movement of potassium across the cell membrane sets up an electric potential across the cell membrane which opposes this flow. Charged protein structures are unable to move across the membrane, while sodium ions cross very slowly, and so an equilibrium is established. The cell is polarized and the electric potential at equilibrium is known as the resting potential.

The number of potassium ions required to establish that potential is of the order of a few million compared to the several hundred billion present in the cell. Therefore the effect on concentration is negligible.

As mentioned above, potassium ions are able to flow out of potassium ion channels, but not all of these channels are open in the resting state. What would happen if more were to open? The answer is that more potassium ions would flow out of the cell and the electric potential across the cell membrane would become more negative to counter this increased flow. This is known as hyperpolarization and the effect is to destimulate the nerve (Fig. A4.3).

Suppose instead that a few sodium ion channels were to open up. In this case, sodium ions would flow into the cell and as a result the electric potential would become less negative. This is known as depolarization and results in a stimulation of the nerve.

If chloride ions channels are opened, chloride ions flow into the cell and the cell membrane becomes hyperpolarized, destimulating the nerve.

Ion channels do not open or close by chance. They are controlled by the neurotransmitters released by communicating nerves. The neurotransmitters bind with their receptors and lead to the opening or closing of ion channels. For example, acetylcholine controls the sodium

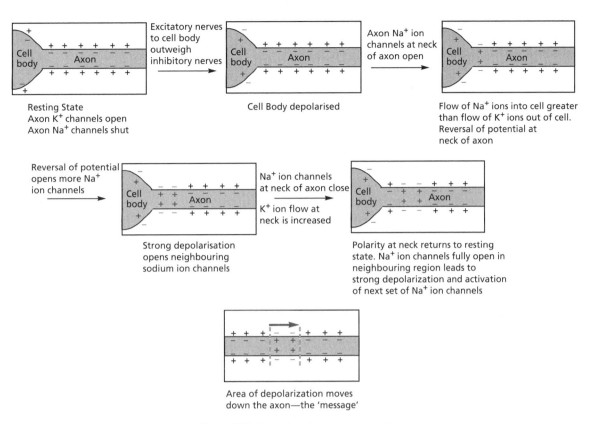

Figure A4.5 Generation of an action potential.

ion channel, whereas GABA and glycine control chloride ion channels. The resulting flow of ions leads to a localized hyperpolarization or depolarization in the area of the receptor. The cell body collects and sums all this information such that the neck of the axon experiences an overall depolarization or hyperpolarization depending on the sum total of the various excitatory or inhibitory signals received.

We shall now consider what happens at the axon of the nerve (Fig. A4.4). The cell membrane of the axon also has sodium and potassium ion channels but they are different in character from those in the cell body. The axon ion channels are not controlled by neurotransmitters, but by the electric potential of the cell membrane.

The sodium ion channels located at the junction of the nerve axon with the cell body are the crucial channels, since they are the first channels to experience whether the cell body has been depolarized or hyperpolarized.

If the cell body is strongly depolarized then a signal is fired along the nerve. A specific threshold value has to be reached before this happens, however. If the depolarization from the cell body is weak, only a few sodium channels open up and the depolarization at the neck of the axon does not reach that threshold value. The sodium channels then reclose and no signal is sent.

With stronger depolarization, more sodium channels open up until the flow of sodium ions entering the axon becomes greater than the flow of potassium ions leaving it. This results in a rapid increase in depolarization, which in turn opens up more sodium channels, resulting in very strong depolarization at the neck of the axon. The flow of sodium ions into the cell increases dramatically, such that it is far greater than the flow of potassium ions out of the axon, and the electric potential across the membrane is reversed, such that it is positive inside the cell and negative outside the cell. This process lasts less than a millisecond before the sodium channels reclose and sodium permeability returns to its normal state. More potassium channels then open and permeability to potassium ions increases for a while to speed up the return to the resting state.

The process is known as an action potential and can only take place in the axon of the nerve. The cell membrane of the axon is said to be excitable, unlike the membrane of the cell body. The important point to note is that once an action potential has fired at the neck of the axon, it has reversed the polarity of the membrane at that point. This in turn has an effect on the neighbouring area of the axon and depolarizes it beyond the critical threshold level. It too fires an action potential and so the process continues along the whole length of the axon (Fig. A4.5). The number of ions involved in this process is minute, such that the concentrations are unaffected. Once the action potential reaches the synaptic button or the neuromuscular endplate, it causes an influx of calcium ions into the cell and an associated release of neurotransmitter into the synaptic gap. The mechanism of this is not well understood.

Microorganisms

Bacterial nomenclature

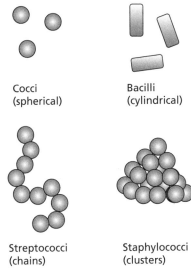

| Cocci (spherical) | Bacilli (cylindrical) |

| Streptococci (chains) | Staphylococci (clusters) |

Figure A5.1 Bacterial nomenclature.

Some clinically important bacteria

Table A5.1 Some typical infections

Organism	Gram	Infections
Staphylococcus aureus	Positive	Skin and tissue infections, septicaemia, endocarditis, accounts for about 25% of all hospital infections
Streptococcus	Positive	Several types—commonly cause sore throats, upper respiratory tract infections, and pneumonia
Escherichia coli	Negative	Urinary tract and wound infections, common in the gastrointestinal tract and often causes problems after surgery, accounts for about 25% of hospital infections
Proteus species	Negative	Urinary tract infections
Salmonella species	Negative	Food poisoning and typhoid
Shigella species	Negative	Dysentery
Enterobacter species	Negative	Urinary tract and respiratory tract infections, septicaemia
Pseudomonas aeruginosa	Negative	An opportunist pathogen, can cause very severe infections in burn victims and other compromised patients, e.g. cancer patients; commonly causes chest infections in patients with cystic fibrosis
Haemophilus influenzae	Negative	Chest and ear infections, occasionally meningitis in young children
Bacteroides fragilis	Negative	Septicaemia following gastrointestinal surgery

The Gram stain

A staining procedure of great value in the identification of bacteria. The procedure is as follows:

1. Stain the cells purple.

2. Decolourize with organic solvent.

3. Stain the cells red.

This test will discriminate between two types of bacterial cell:

- Gram-negative bacteria—these bacteria cells are easily decolorized at stage two and will therefore be coloured red

- Gram-positive bacteria—these resist the decolourization at stage 2 and will therefore remain purple.

Classifications

Bacteria can be classified as being Gram positive or Gram negative depending on what colour they retain on treatment with the Gram stain procedure. They can also be classed as aerobic or anaerobic depending on their dependency on oxygen. Aerobic organisms grow in the presence of oxygen whereas anaerobic organisms do not.

Definitions of different microorganisms

- **Bacteria** are unicellular organisms that have a prokaryotic cell structure. They are diverse in nature and some can carry out photosynthesis. Examples of typical infections are given in Table A5.1.

- **Blue green algae** are made up of prokaryotic cells that can form multicellular filaments and carry out photosynthesis in the same manner as the eukaryotic algae.

- **Algae**, with the exception of the blue green algae, are made up of eukaryotic cells and can perform oxygen evolving photosynthesis. Some are unicellular and some are multicellular. The latter have little or no cell differentiation, which sets them apart from higher multicellular organisms such as plants and animals.

- **Protozoa** are unicellular eukaryotic organisms that are unable to carry out photosynthesis. They are responsible for diseases such as malaria, African sleeping sickness, Chagas' disease, leishmaniasis, and amoebic dysentery.

- **Fungi** are multicellular eukaryotic organisms with little or no cell differentiation, which can form long filaments of interconnected cells called mycelia. They too are unable to carry out photosynthesis. Fungi are responsible for infections such as athlete's foot, ringworm, aspergillosis, candidiasis, and histoplasmosis.

Drugs and their trade names

Drug name (Trade name)

abacavir (Ziagen)

acebutolol (Sectral)

aciclovir (Virovir, Zovirax)

aclacinomycin A *see* aclarubicin

aclarubicin (Aclacin, Aclaplastin)

actinomycin D *see* dactinomycin

acyclovir *see* aciclovir

adalimumab (Humira)

adefovir dipivoxil (Hepsera)

adriamycin *see* doxorubicin

albuterol *see* salbutamol

aldesleukin (Proleukin)

alemtuzumab (MabCampath)

allopurinol (Zyloric)

amantadine (Lysovir, Symmetrel)

aminoglutethimide (Orimeten, Cytadren)

amoxicillin (Amoxil)

amoxicillin with clavulanic acid (Augmentin)

amoxycillin *see* amoxicillin

amphotericin (Fungilin, Fungizone)

ampicillin (Penbritin)

amprenavir (Agenerase)

amsacrine (Amsidine)

anastrazole (Arimidex)

arsenic trioxide (Trisenox)

atazanavir (Reyataz)

atenolol (Tenormin)

atomoxetine (Strattera)

atracurium (Tracrium)

azathioprine (Imuran)

azidothymidine *see* zidovudine

azithromycin (Zithromax)

AZT *see* zidovudine

aztreonam (Azactam)

benzatropine (Cogentin)

benzhexol *see* trihexyphenidyl

benztropine *see* benzatropine

benzylpenicillin (Crystapen)

betaxolol (Betoptic)

bethanechol (Myotonine)

bevacixumab (Avastin)

bortezomib (Velcade)

bupivacaine (Marcaine)

buprenorphine (Subutex, Temgesic, Transtec)

bupropion (Zyban)

busulfan (Busilvex, Myleran)

busulphan *see* busulfan

captopril (Capoten)

carbenoxolone (Pyrogastrone)

carbidopa with levodopa *see* co-careldopa

carboplatin (Paraplatin)

carmustine (BCNU, BiCNU, Giladel)

cefalexin (Ceporex, Keflex, Keftab, Biocef)

cefalothin (Keflin)

cefotaxime (Claforan)

cefoxitin (Mefoxin)

cefpirome (Cefrom)

ceftazidime (Fortum, Kefadim)

ceftriaxone (Rocephin)

cefuroxime (Zinacef, Zinnat, Kefurox)

celecoxib (Celebrix)

cephalexin *see* cefalexin

cephalothin *see* cefalothin

cetuximab (Erbitux)

chlorambucil (Leukeran)

chloramphenicol (Kemicetine, Chloromycetin)

chlordiazepoxide (Librium)

chloroquine (Avloclor, Nivaquine)

chlorpromazine (Largactil)

cholestyramine *see* colestyramine

ciclosporin (Neoral, Sandimmun)

cidofovir (Vistide)

cilastatin with imipenem (Primaxin)

cilazapril (Vascace)

cimetidine (Dyspamet, Tagamet)

ciprofloxacin (Ciproxin, Ciprobay, Cipro, Ciproxan)

clarithromycin (Klaricid)

clavulanic acid with amoxicillin (Augmentin)

clindamycin (Dalacin C)

clonidine (Catapres, Dixarit)

clozapine (Clozaril)

co-amoxiclav (Augmentin)

co-careldopa (Sinemet, Stalevo)

colestyramine (Questran)

co-trimoxazole (Septrin)

crisantaspase (Erwinase)

cyclopentolate (Mydrilate)

cyclophosphamide (Endoxana)

cyclosporin *see* ciclosporin

cyproterone acetate (Cyprostat)

dacarbazine (DTIC-Dome)

dactinomycin (Cosmegen Lyovac)

dalfopristin with quinupristin (Synercid)

daunorubicin (DaunoXome)

delavirdine (Rescriptor)

diazepam (Valium)

didanosine (Videx)

digoxin (Lanoxin)

diphenhydramine (Benadryl)

disulfiram (Antabuse)

dobutamine (Dobutrex, Posiject)

docetaxel (Taxotere)

donepezil (Aricept)

doxazosin (Cardura)

doxorubicin (Rubex, Doxil)

doxycycline (Vibramycin)

efavirenz (Sustiva)

enalapril (Innovace)

enfuvirtide (Fuzeon)

epirubicin (Pharmorubicin)

ertapenem (Invanz)

erythromycin (Erymax, Erythroped)

esomeprazole (Nexium)

estramustine (Estracyt)

etanercept (Enbrel)

etoposide (Etopophos, Vepesid)

famciclovir (Famvir)

famotidine (Pepcid)

fentanyl (Sublimaze, Actiq, Durogesic)

fexofenadine (Allegra, Telfast)

filgrastim (Neupogen)

flucloxacillin (Floxapen)

fluconazole (Diflucan)

fludarabine (Fludara)

fluorouracil (Efudix)

fluoxetine (Prozac)

fluphenazine decanoate (Modecate)

flutamide (Drogenil)

fomivirsen (Vitravene)

foscarnet (Foscavir)

fulvestrant (Faslodex)

fusidic acid (Fucidin)

galantamine (Reminyl)

ganciclovir (Cymevene)

gefitinib (Iressa)

gemcitabine (Gemzar)

gemtuzumab (Mylotarg)

gentamicin (Cidomycin, Genticin)

goserelin (Zoladex)

granisetron (Kytril)

guanethidine (Ismelin)

hexamine *see* methenamine

hyoscine (Scopoderm TTS)

ibritumomab (Zevalin)

idarubicin (Zavedos)

idoxuridine (Herpid)

ifosfamide (Mitoxana)

imatinib (Glivec)

imipenem and cilastatin (Primaxin)

imipramine (Tofranil)

imiquimod (Aldara)

indinavir (Crixivan)

indometacin (Rimacid)

indomethacin *see* indometacin

infliximab (Remicade)

α-interferon (IntronA, Roferon-A, Viraferon)

γ-interferon (Immukin)

ipratropium (Atrovent, Ipratropium Steri-Neb, Respontin)

irinotecan (Campto, Camptosar)

lamivudine (Epivir, Zeffix)

lansoprazole (Zoton)

L-dopa *see* levodopa

letrozole (Femera)

levalbuterol (Xopenex)

levobupivacaine (Chirocaine)

levodopa with carbidopa *see* co-careldopa

lidocaine (Xylocaine)

lignocaine *see* lidocaine

linezolid (Zyvox)

lisinopril (Carace, Zestril, Prinivil)

lithium carbonate (Camcolit, Liskonum, Priadel)

lomustine (CCNU)

loperamide (Imodium)

lopinavir with ritinavir (Kaletra)

losartan (Cozaar)

malathion (Derbac-M, Prioderm, Quellada M, Suleo-M)

medroxyprogesterone acetate (Farlutal, Provera)

megestrol acetate (Megace)

melphalan (Alkeran)

meperidine *see* pethidine

mercaptopurine (Puri-Nethol)

meropenem (Meronem)

mesna (Uromitexan)

methadone (Methadose)

methenamine (Hiprex)

methyldopa (Aldomet)

methylphenidate (Ritalin)

metoclopramide (Maxolon)

metoprolol (Betaloc, Lopresor)

metronidazole (Flagyl, Metrolyl)

mirtazepine (Zispin)

mitoxantrone (Novantrone, Onkotrone)

mivacurium (Mivacron)

morphine (Oramorph, Sevredol, Morcap, Morphgesic, MST Cintinus, MXL, Zomorph)

moxifloxacin (Avelox)

mupirocin (Bactroban)

nalidixic acid (Mictral, Negram, Uriben)

naloxone (Narcan)

naltrexone (Nalorex)

nelfinavir (Viracept)

nevirapine (Viramune)

nicotine (Nicorette, Nicotinell, NiQuitin CQ)

nitrofurantoin (Furadantin, Macrobid, Macrodantin)

nizatidine (Axid)

oblimersen (Genasense)

olanzapine (Zyprexa)

omalizumab (Xolair)

omeprazole (Losec, Prilosec)

ondansetron (Zofran)

oseltamivir (Tamiflu)

oxaliplatin (Eloxatin)

oxytocin (Syntocinon)

paclitaxel (Taxol)

palivizumab (Synagis)

pantoprazole (Protium)

pegademase PEG-adenosine deaminase (Adagen)

pegaspargase PEG-asparaginase (Oncaspar)

pegfilgrastim (Neulasta)

peginterferon α2a (PEG-IFN-α-2a) (Pegasys)

peginterferon α2b (PEG-IFN-α-2b) (Peg-Intron)

pegvisomant (Somavert)

penciclovir (Vectavir)

penicillin G *see* benzylpenicillin

penicillin V *see* phenoxymethylpenicillin

pentazocine (Fortral)

pentostatin (Nipent)

pethidine (Pamergan P100)

phenelzine (Nardil)

phenoxymethylpenicillin (Apsin)

pilocarpine (Pilogel)

pleconaril (Picovir)

podophyllotoxin (Condyline, Warticon)

prazosin (Hypovase)

probenecid (Probecid, Benuryl)

promethazine (Phenergan)

propranolol (Inderal)

propantheline bromide (Pro-Banthine)

pyridostigmine (Mestinon)

quinupristin with dalfopristin (Synercid)

rabeprazole (Pariet)

raltitrexed (Tomudex)

raloxifene (Evista)

ranitidine (Zantac)

reboxetine (Edronax)

ribavirin (Copegus, Rebetol, Virazole)

rifampicin (Rifadin, Rimactane, Rifater, Rifinah, Rimactazid)

risperidone (Risperdal)

ritonavir (Norvir)

rituximab (Rituxan, MabThera)

rivastigmine (Exelon)

rofecoxib (Vioxx)

salbutamol (Ventmax, Ventolin, Volmax, Airomir, Asmasal Clickhaler, Salamol Easi-Breathe, Ventodisks)

salmeterol (Serevent)

saquinavir (Fortovase, Invirase)

selegiline (Deprenyl, Eldepryl, Zelapar)

sildenafil (Viagra)

simvastatin (Zocor)

stavudine (Zerit)

sumatriptan (Imigran)

suxamethonium (Anectine)

tacrine (Cognex)

tamoxifen (Nolvadex)

tazobactam with piperacillin (Tazocin or Zosyn)

teicoplanin (Targocid)

temoporfin (Foscan)

teniposide (Vumon)

tenofovir (Viread)

terazosin (Hytrin)

testosterone propionate (Sustanon 250, Virormone)

theophylline (Nuelin, Slo-Phyllin, Uniphyllin Continus)

thioguanine *see* tioguanine

ticarcillin with clavulanic acid (Timentin)

timolol (Betim)

tioconazole (Trosyl)

tioguanine (Lanvis)

tipifarnib (Zarnestra)

tobramycin (Nebcin, Tobi)

topotecan (Hycamtin, Hycamptin)

toremifene (Fareston)

tositumomab (Bexxar)

trastuzumab (Herceptin)

tretinoin (Vesanoid)

trihexyphenidyl (Broflex)

trimethoprim (Monotrim, Trimopan)

tropicamide (Mydriacyl)

valaciclovir (Valtrex)

valdecoxib (Bextra)

valganciclovir (Valcyte)

vancomycin (Vancocin)

vasopressin (Pitressin)

vecuronium (Norcuron)

verapamil (Cordilox, Securon)

vinblastine (Velbe)

vincristine (Oncovin)

vindesine (Eldisine)

vinorelbine (Navelbine)

zalcitabine (Hivid)

zanamivir (Relenza)

zidovudine (Retrovir)

Trade name (Drug name)

Aclacin (aclarubicin)

Aclaplastin (aclarubicin)

Actiq (fentanyl)

Adagen (pegademase)

Agenerase (amprenavir)

Airomir (salbutamol)

Aldara (imiquimod)

Aldomet (methyldopa)

Alkeran (melphalan)

Allegra (fexofenadine)

Amoxil (amoxicillin)

Amsidine (amsacrine)

Anectine (suxamethonium)

Antabuse (disulfiram)

Apsin (phenoxymethylpenicillin)

Aricept (donepezil)

Arimidex (anastrazole)

Asmasal Clickhaler (salbutamol)

Atrovent (ipratropium)

Augmentin (amoxicillin with clavulanic acid)

Augmentin (clavulanic acid with amoxicillin)

Augmentin (co-amoxiclav)

Avastin (bevacixumab)

Avelox (moxifloxacin)

Avloclor (chloroquine)

Axid (nizatidine)

Azactam (aztreonam)

Bactroban (mupirocin)

BCNU (carmustine)

Benadryl (diphenhydramine)

Benuryl (probenicid)

Betaloc (metoprolol)

Betim (timolol)

Betoptic (betaxolol)

Bextra (valdecoxib)

Bexxar (tositumomab)

BiCNU (carmustine)

Biocef (cefalexin)

Broflex (trihexyphenidyl)

Busilvex (busulfan)

Camcolit (lithium carbonate)

Campto (irinotecan)

Camptosar (irinotecan)

Capoten (captopril)

Carace (lisinopril)

Cardura (doxazosin)

Catapres (clonidine)

CCNU (lomustine)

Cefrom (cefpirome)

Celebrix (celecoxib)

Ceporex (cefalexin)

Chirocaine (levobupivicaine)

Chloromycetin (chloramphenicol)

Cidomycin (gentamicin)

Cipro (ciprofloxacin)

Ciprobay (ciprofloxacin)

Ciproxan (ciprofloxacin)

Ciproxin (ciprofloxacin)

Claforan (cefotaxime)

Clozaril (clozapine)

Cogentin (benzatropine)

Cognex (tacrine)

Condyline (podophyllotoxin)

Copegus (ribavirin)

Cordilox (verapamil)

Cosmegen Lyovac (dactinomycin)

Cozaar (losartan)

Crixivan (indinavir)

Crystapen (benzylpenicillin)

Cymevene (gangciclovir)

Cyprostat (cyproterone acetate)

Cytadren (aminoglutethimide)

Dalacin C (clindamycin)

DaunoXome (daunorubicin)

Deprenyl (selegiline)

Derbac-M (malathion)

Diflucan (fluconazole)

Dixarit (clonidine)

Dobutrex (dobutamine)

Doxil (doxorubicin)

Drogenil (flutamide)

DTIC-Dome (dacarbazine)

Durogesic (fentanyl)

Dyspamet (cimetidine)

Edronax (reboxetine)

Efudix (fluorouracil)

Eldepryl (selegiline)

Eldisine (vindesine)

Eloxatin (oxaliplatin)

Enbrel (etanercept)

Endoxana (cyclophosphamide)

Epivir (lamivudine)

Erbitux (cetuximab)

Erwinase (crisantaspase)

Erymax (erythromycin)

Erythroped (erythromycin)

Estracyt (estramustine)

Etopophos (etoposide)

Evista (raloxifene)

Exelon (rivastigmine)

Famvir (famciclovir)

Fareston (toremifene)

Farlutal (medroxyprogesterone acetate)

Faslodex (fulvestrant)

Femera (letrozole)

Flagyl (metronidazole)

Floxapen (flucloxacillin)

Fludara (fludarabine)

Fortovase (saquinavir)

Fortral (pentazocine)

Fortum (ceftazidime)

Foscan (temoporfin)

Foscavir (foscarnet)

Fucidin (fusidic acid)

Fungilin (amphotericin)

Fungizone (amphotericin)

Furadantin (nitrofurantoin)

Fuzeon (enfuvirtide)

Gemzar (gemcitabine)

Genasense (oblimersen)

Genticin (gentamicin)

Glivec (imatinib)

Hepsera (adefovir dipivoxil)

Herceptin (trastuzumab)

Herpid (idoxuridine)

Hiprex (methenamine)

Hivid (zalcitabine)

Humira (adalimumab)

Hycamptin (topotecan)

Hycamtin (topotecan)

Hypovase (prazosin)

Hytrin (terazosin)

Imigran (sumatriptan)

Immukin (γ-interferon)

Imodium (loperamide)

Imuran (azathioprine)

Inderal (propranolol)

Innovace (enalapril)

Intron A (α-interferon)

Invanz (ertapenem)

Invirase (saquinavir)

Ipratropium Steri-Neb (ipratropium)

Iressa (gefitinib)

Ismelin (guanethidine)

Kaletra (lopinavir with ritinavir)

Kefadim (ceftazidime)

Keflex (cefalexin)

Keflin (cefalothin)

Keftab (cefalexin)

Kefurox (cefuroxime)

Kemicetine (chloramphenicol)

Klaricid (clarithromycin)

Kytril (granisetron)

Lanoxin (digoxin)

Lanvis (tioguanine)

Largactil (chlorpromazine)

Leukeran (chlorambucil)

Librium (chlordiazepoxide)

Liskonum (lithium carbonate)

Lopresor (metoprolol)

Losec (omeprazole)

Lysovir (amantadine)

MabCampath (alemtuzumab)

MabThera (rituximab)

Macrobid (nitrofurantoin)

Macrodantin (nitrofurantoin)

Marcaine (bupivicaine)

Maxolon (metoclopramide)

Mefoxin (cefoxitin)

Megace (megestrol acetate)

Meronem (meropenem)

Mestinon (pyridostigmine)

Methadose (methadone)

Metrolyl (metronidazole)

Mictral (nalidixic acid)

Mitoxana (ifosfamide)

Mivacron (mivacurium)

Modecate (fluphenazine decanoate)

Monotrim (trimethoprim)

Morcap (morphine)

Morphgesic (morphine)

MST Cintinus (morphine)

MXL (morphine)

Mydriacyl (tropicamide)

Mydrilate (cyclopentolate)

Myleran (busulfan)

Mylotarg (gemtuzumab)

Myotonine (bethanechol)

Nalorex (naltrexone)

Narcan (naloxone)

Nardil (phenelzine)

Navelbine (vinorelbine)

Nebcin (tobramycin)

Negram (nalidixic acid)

Neoral (ciclosporin)

Neulasta (pegfilgrastim)

Neupogen (filgrastim)

Nexium (esomeprazole)

Nicorette (nicotine)

Nicotinell (nicotine)

Nipent (pentostatin)

NiQuitin CQ (nicotine)

Nivaquine (chloroquine)

Nolvadex (tamoxifen)

Norcuron (vecuronium)

Norvir (ritonavir)

Novantrone (mitoxantrone)

Nuelin (theophylline)

Oncaspar (pegaspargas PEG-asparaginase)

Oncovin (vincristine)

Onkotrone (mitoxantrone)

Oramorph (morphine)

Orimeten (aminoglutethimide)

Pamergan P100 (pethidine)

Paraplatin (carboplatin)

Pariet (rabeprazole)

Peg-Intron (peginterferon α2b)

Pegasys (peginterferon α2a)

Penbritin (ampicillin)

Pepcid (famotidine)

Pharmorubicin (epirubicin)

Phenergan (promethazine)

Picovir (pleconaril)

Pilogel (pilocarpine)

Pitressin (vasopressin)

Posiject (dobutamine)

Priadel (lithium carbonate)

Prilosec (omeprazole)

Primaxin (cilastatin with imipenem)

Prinivil (lisinopril)

Prioderm (malathion)

Pro-Banthine (propantheline bromide)

Probecid (probenecid)

Proleukin (aldesleukin)

Protium (pantoprazole)

Provera (medroxyprogesterone acetate)

Prozac (fluoxetine)

Puri-Nethol (mercaptopurine)

Pyrogastrone (carbenoxolone)

Quellada M (malathion)

Questran (colestyramine)

Rebetol (ribavirin)

Relenza (zanamivir)

Remicade (infliximab)

Reminyl (galantamine)

Rescriptor (delavirdine)

Respontin (ipratropium)

Retrovir (zidovudine)

Reyataz (atazanavir)

Rifadin (rifampicin)

Rifater (rifampicin)

Rifinah (rifampicin)

Rimacid (indometacin)

Rimactane (rifampicin)

Rimactazid (rifampicin)

Risperdal (risperidone)

Ritalin (methylphenidate)

Rituxan (rituximab)

Rocephin (ceftriaxone)

Roferon-A (α-interferon)

Rubex (doxorubicin)

Salamol Easi-Breathe (salbutamol)

Sandimmun (ciclosporin)

Scopoderm TTS (hyoscine)

Sectral (acebutolol)

Securon (verapamil)

Septrin (co-trimoxazole)

Serevent (salmetrol)

Sevredol (morphine)

Sinemet (co-careldopa)

Slo-Phyllin (theophylline)

Somavert (pegvisomant)

Stalevo (co-careldopa)

Stratter (atomoxetine)

Sublimaze (fentanyl)

Subutex (buprenorphine)

Suleo-M (malathion)

Sustanon 250 (testosterone propionate)

Sustiva (efavirenz)

Symmetrel (amantadine)

Synagis (palivizumab)

Synercid (dalfopristin with quinupristine)

Syntocinon (oxytocin)

Tagamet (cimetidine)

Tamiflu (oseltamivir)

Targocid (teicoplanin)

Taxol (paclitaxel)

Taxotere (docetaxel)

Tazocin (tazobactam with piperacillin)

Telfast (fexofenadine)

Temgesic (buprenorphine)

Tenormin (atenolol)

Timentin (ticarcillin with clavulanic acid)

Tobi (tobramycin)

Tofranil (imipramine)

Tomudex (ralitirexed)

Tracrium (atracurium)

Transtec (buprenorphine)

Trimopan (trimethoprim)

Trisenox (arsenic trioxide)

Trosyl (tioconazole)

Uniphyllin Continus (theophylline)

Uriben (nalidixic acid)

Uromitexan (mesna)

Valcyte (valganciclovir)

Valium (diazepam)

Valtrex (valaciclovir)

Vancocin (vancomycin)

Vascace (cilazapril)

Veasnoid (tretinoin)

Vectavir (penciclovir)

Velbe (vinblastine)

Velcade (bortezomib)

Ventmax (salbutamol)

Ventodisks (salbutamol)

Ventolin (salbutamol)

Vepesid (etoposide)

Viagra (sildenafil)

Vibramycin (doxycycline)

Videx (didanosine)

Vioxx (rofecoxib)

Viracept (nelfinavir)

Viraferon (α-interferon)

Viramune (nevirapine)

Virazole (ribavirin)

Viread (tenofovir)

Virormone (testosterone propionate)

Virovir (aciclovir)

Vistide (cidofovir)

Vitravene (fomivirsen)

Volmax (salbutamol)

Vumon (teniposide)

Warticon (podophyllotoxin)

Xolair (omalizumab)

Xopenex (levalbuterol)

Xylocaine (lidocaine)

Zantac (ranitidine)

Zarnestra (tipifarnib)

Zavedos (idarubicin)

Zeffix (lamivudine)

Zelapar (selegiline)

Zerit (stavudine)

Zestril (lisinopril)

Zevalin (ibritumomab)

Ziagen (abacavir)

Zinacef (cefuroxime)

Zinnat (cefuroxime)

Zispin (mirtazepine)

Zithromax (azithromycin)

Zocor (simvastatin)

Zofran (ondansetron)

Zoladex (goserelin)

Zomorph (morphine)

Zosyn (tazobactam with piperacillin)

Zoton (lansoprazole)

Zovirax (aciclovir)

Zyban (bupropion)

Zyloric (allopurinol)

Zyprexa (olanzapine)

Zyvox (linezolid)

GLOSSARY

3D QSAR QSAR studies which relate the biological activities of a series of compounds to their steric and electrostatic fields determined by molecular modelling software.

Abzyme An antibody with catalytic properties.

ACE inhibitors Drugs which inhibit the enzyme angiotensin converting enzyme. Inhibition prevents the synthesis of a powerful vasoconstrictor and so ACE inhibitors are used as antihypertensive agents.

Acetylcholinesterase An enzyme that hydrolyses the neurotransmitter acetylcholine.

Action potential Refers to the reversal in membrane potential as a signal travels along the axon of a nerve.

Activation energy The energy required for a reaction to reach its transition state.

Active conformation The conformation adopted by a compound when it binds to its target binding site.

Active principle The single chemical in a mixture of compounds which is chiefly responsible for that mixture's biological activity.

Active site The binding site of an enzyme where a reaction is catalysed by the enzyme.

ADAPT Antibody-directed abzyme prodrug therapy.

Addiction Addiction can be defined as a habitual form of behaviour. It need not be harmful. For example, one can be addicted to eating chocolate or watching television without suffering more than a bad case of toothache or a surplus of soap operas.

ADEPT Antibody-directed enzyme prodrug therapy.

ADME Refers to drug absorption, drug distribution, drug metabolism, and drug excretion.

Adrenal medulla A gland that produces adrenaline.

Adrenergics Refers to compounds that interact with the receptors targeted by adrenaline and noradrenaline.

Adrenoceptors Receptors that are activated by adrenaline and noradrenaline.

Adsorption Refers to the situation where a molecule or a structure adheres to a surface. In virology it refers to a virus binding to the surface of a host cell.

Aerobic bacteria Bacteria that grow in the presence of oxygen.

Affinity Affinity is a measure of how strongly a ligand binds to its target binding site.

Affinity screening A method of screening compounds based on their binding affinity to a target.

Agonist A drug that produces the same response at a receptor as the natural messenger.

AIDS Acquired immune deficiency syndrome.

Alchemy A molecular modelling software package.

Alkaloids Natural products extracted from plants that contain an amine functional group.

Alkylating agents Agents which act as electrophiles and form irreversible covalent bonds with macromolecular targets. These agents are classed as cytotoxic and are used as anticancer agents.

Allosteric Refers to a protein binding site other than the one used by the normal ligand, and which affects the activity of the protein. An allosteric inhibitor binding to an allosteric binding site induces a change of shape in the protein which disguises the normal binding site from its ligand.

Aminoacridines A group of synthetic antibacterial agents that target bacterial DNA.

Aminoacyl tRNA synthetases Enzymes that catalyse the attachment of an amino acid to tRNA. Potentially useful targets in antibacterial therapy.

Aminoglycosides A group of antibacterial agents that contain sugar components and a basic amino function.

Anaerobic bacteria Bacteria that grow in the absence of oxygen.

Anchor see **linker**

Androgens Hormones that are used in anticancer therapy.

Angiogenesis The process by which new blood vessels are formed.

Angiostatin An endogenous compound that inhibits angiogenesis.

Anomers Cyclic stereoisomers of sugars that differ only in their configurations at the hemiacetal (anomeric) carbon.

Antagonist A drug which binds to a receptor without activating it, and which prevents an agonist or a natural messenger from binding.

Antedrugs see **soft drugs**

Anthracyclines A group of antibiotics that are important in anticancer therapy.

Anti-androgens Anticancer agents that block the action of androgens at their receptors.

Antibacterial agent A synthetic or naturally occurring agent which can kill or inhibit the growth of bacterial cells.

Antibiotic An antibacterial agent derived from a natural source.

Antibody A Y-shaped glycoprotein generated by the body's immune system to interact with an antigen present on a foreign molecule. Marks the foreign molecule for destruction.

Antibody–drug conjugates Refers to antibodies with drugs covalently linked to their structure.

Anticholinesterases Agents which inhibit the enzyme acetylcholinesterase.

Anticodon A set of three nucleic acid bases on tRNA that base pair with a triplet of nucleic acid bases on mRNA during translation. The amino acid linked to tRNA is determined by the anticodon that is present.

Antiemetic A drug used to prevent nausea and vomiting.

Antigen A region of a molecule that is recognized by the body's immune system and which will interact with antibodies targeted against it.

Antigenic drift The process by which antigens gradually vary in nature.

Antigenic shift Refers to a large alteration in the nature of antigens.

Antigenic variation A property of some viruses which are able to vary the chemical structure of antigens on their surface through rapid mutations.

Antimetabolites Agents which inhibit enzymes that are crucial to the normal metabolism of the cell. Used in antibacterial and anticancer therapy.

Anti-oestrogens Agents which bind to oestrogen receptors and block oestrogen. Used in anticancer therapy.

Anti-oncogenes Genes that code for proteins which check the 'normality' of a cell and which induce cell death if abnormalities are present. Important in preventing the birth of a cancer cell.

Antisense therapy The design of molecules which will bind to specific regions of mRNA and prevent mRNA acting as a code for protein synthesis.

Aorta The principle artery carrying blood away from the heart.

Apaf-1 A scaffolding protein for apoptosomes.

Apoptosis The process by which a cell commits suicide.

Apoptosome A protein complex that destroys the cell's proteins and leads to apoptosis.

Aquaporins Membrane bound proteins containing a pore that allows water to pass through the membrane.

Aromatase An enzyme that catalyses an aromatization reaction in oestrogen synthesis. Aromatase inhibitors are used as anticancer agents.

Arteries Blood vessels taking blood away from the heart.

Aspartyl proteases Enzymes that catalyse the hydrolysis of peptide bonds in protein substrates and which contain aspartate residues in the active site that take part in the hydrolysis mechanism.

Asymmetric centre An atom with four different substituents that frequently results in asymmetry for the whole molecule.

Asymmetric synthesis A synthesis which shows selectivity for a particular enantiomer or diastereomer of an asymmetric compound.

Attention deficit hyperactivity disorder A disorder associated with children—the name speaks for itself.

AUC The area under the plasma drug concentration curve. It represents the total amount of drug that is available in the blood supply during a dosing regime.

Autonomic motor nervous system Nerves carrying messages from the central nervous system to smooth muscle, cardiac muscle, and the adrenal medulla.

Bacilli Bacterial cells that are rod shaped.

Bacteriostatic Bacteriostatic drugs inhibit the growth and multiplication of bacteria, but do not directly kill them.

Bactericidal Bactericidal drugs actively kill bacterial cells.

Bacteriophage A virus that invades bacterial cells.

Bad A protein that promotes apoptosis.

Barbiturates A series of synthetic compounds with sedative properties.

Bax A protein that promotes apoptosis.

Bcl-2 and Bcl-X Proteins that suppress apoptosis.

Benign cancer or tumour A localized tumour that is not life threatening.

Benign prostatic hyperplasia The term for an enlarged prostate, resulting in difficulty in passing urine.

Beta-blockers Compounds that block or antagonize β-adrenoceptors. Particularly useful in cardiovascular medicine.

Beta-lactamases *see* lactamases

Bile duct A duct leading from the liver to the intestines. Some drugs and drug metabolites are excreted through the bile duct, but can be reabsorbed from the intestines.

Binding region A region within a binding site that is capable of interacting with a drug or an endogenous molecule by intermolecular bonding.

Binding site The location where an endogenous molecule or drug binds on a macromolecule. Normally a hollow or cleft in the surface of the macromolecule.

Bioavailability Refers to the fraction of drug that is available in the blood supply following administration.

Bioequivalence studies Studies carried out to ensure that the bioavailability of a drug remains the same should there be any alteration to the manufacture or formulation of the drug.

Bioisostere A chemical group which can replace another chemical group without affecting the biological activity of the drug.

Bioterrorism The use of toxic infectious agents by terrorist groups.

Bleomycins A group of naturally occurring glycoproteins used as anticancer agents.

Blood–brain barrier Blood vessels in the brain are less porous than blood vessels in the periphery. They also have a fatty coating. Drugs entering the brain have to be lipophilic in order to cross this barrier.

Boc A shorthand term for the protecting group t-butyloxycarbonyl.

Bronchodilator An agent which dilates the airways and can combat asthma.

Bump filter Used in conformational analysis *in silico* to reject conformations that have bad steric interactions.

Cache A molecular modelling software package.

Calmodulin A calcium-binding protein that activates several protein kinases.

Camptothecins A group of naturally occurring alkaloids and semi-synthetic derivatives used as anticancer agents.

Canaliculae Invaginations or channels formed by parietal cells which connect with the lumen of the stomach.

Capillaries Small blood vessels.

Carboxypenicillins A family of penicillins having a carboxylic acid substituent at the α-position (largely superseded).

Capsid A protein coat that encapsulates the nucleic acid of a virus.

Carbapenems A group of β-lactam antibacterial agents, so called because they lack a sulfur atom.

Carboxypeptidases Enzymes that hydrolyse the final peptide link at the C-terminal end of a peptide chain.

Carcinogenesis The birth of a cancer.

Carrier protein A protein in the membrane of a cell which is capable of transporting specific polar molecules across the membrane. The molecules transported are too polar to cross the membrane themselves and are crucial to the survival and functions of the cell.

Caspases Enzymes which have important roles to play in the ageing process of cells.

Catecholamines Compounds that contain a basic amino group and a catechol ring. The catechol ring consists of an aromatic ring with two phenolic groups, *ortho* to each other.

Cell cycle Refers to recognizable phases of cell growth, DNA synthesis, and cell division.

Cell membrane A phospholipid bilayer surrounding all cells that acts as a hydrophobic barrier.

Central nervous system The nervous tissue of the brain and the spinal column.

Centroid A dummy atom used in molecular modelling to define the centre of an aromatic or heteroaromatic ring. Has also been used to mean the scaffold of a molecule.

Cephalosporinases *see* lactamases

Cephalosporins A group of β-lactam semi-synthetic antibacterial agents that target the bacterial transpeptidase enzymes.

Cephamycins A family of cephalosporins having a methoxy substituent at the 7-position.

Chain cutters Agents that interact with DNA, leading to the splitting of the DNA backbone. Generally operate by producing radicals. Used as anticancer agents.

Chain contraction/extension strategy The variation of chain length in a drug to optimize the separation between different binding groups.

Chem-3D A molecular modelling software package.

ChemDraw A chemical drawing software package.

ChemWindow A chemical drawing software package.

Chem-X A molecular modelling software package.

Chemotherapeutic index A comparison of the minimum effective dose of a drug with the maximum dose which can be tolerated by the host.

Chimeric antibodies Antibodies that are part human and part mouse (or other species) in nature.

Chirality The property of asymmetry where the mirror images of a molecule are non-superimposable.

Chiral switching The replacement of a racemic drug on the market with its more active enantiomer or stereoisomer.

CHO cells Chinese Hamster ovarian cells. Commonly used to express a cloned receptor on their surface for *in vitro* tests.

Choline acetyltransferase An enzyme that catalyses the synthesis of acetylcholine.

Cholinergics Refers to compounds that interact with cholinergic receptors.

Cholinergic receptors Receptors that are activated by acetylcholine.

Chromatin A structure consisting of DNA wrapped round proteins such as histone.

Cloning The process by which identical copies of a DNA molecule or a gene are obtained.

CMV *see* cytomegalovirus

Cocci Bacterial cells that are spherical in shape.

Coenzyme A small organic molecules acting as a cofactor.

Cofactor An ion or small organic molecule (other than the substrate) which is bound to the active site of an enzyme and takes part in the enzyme-catalysed reaction.

Combinatorial libraries A store of compounds that have been synthesized by combinatorial synthesis.

Combinatorial synthesis A method of synthesizing large quantities of compounds in small scale using automated or semi-automated processes. Normally carried out as solid phase syntheses.

Combretastatins Naturally occurring anticancer agents that inhibit tubulin polymerization.

Comparative molecular field analysis (CoMFA) A method of carrying out 3D-QSAR that was developed by the company Tripos.

Competitive inhibitors Reversible inhibitors that compete with the normal substrate for an enzyme's active site.

Compound banks or libraries A store of synthetic compounds that have been produced by traditional methods or by combinatorial syntheses.

Conformational analysis A study of the various conformations permitted for a molecule. Conformations are different three dimensional shapes arising from single bond rotations.

Conformational blockers Groups that are added to molecules to prevent them adopting certain conformations.

Conformational space The 3D space surrounding the scaffold of a molecule.

Conjugation In the chemical sense it refers to interacting systems of π bonds. In the microbiological sense, it refers to the process by which bacterial cells pass genetic information directly between each other.

Conjugation reactions *see* phase II reactions

Convergent evolution In a biochemical sense, refers to receptors which are in different branches of the receptor evolutionary tree, but which have converged to recognize the same endogenous ligand.

Correlation coefficient *see* regression coefficient

Cotransmitters Chemical messengers which are released from nerves along with the major neurotransmitter and which have a fine-tuning effect on the signal received.

Craig plot A plot which compares the values of two physicochemical parameters for different substituents.

Cross-validated correlation coefficient (q^2) A measure of the predictability of a 3D QSAR equation.

Cross-validation Used in 3D QSAR in order to obtain a QSAR equation. Tests how an equation predicts the activity of a test compound that has not been included in the derivation of the equation.

Cryptophycins Naturally occurring anticancer agents that inhibit tubulin polymerization.

Cyclases Enzymes that catalyse cyclization reactions such as the formation of cyclic AMP from ATP.

Cyclin dependent kinases Enzymes that are activated by cyclins and which catalyse phosphorylation reactions that control the cell cycle.

Cyclins A group of proteins that are important in the control and regulation of the cell cycle.

Cyclo-oxygenases Enzymes that are important in the production of prostaglandins.

CYP Shorthand terminology for cytochrome P450 enzymes; for example CYP3A4.

Cytochrome C Released by mitochondria to promote apoptosis.

Cytochrome P450 enzymes Enzymes that are extremely important in the metabolism of drugs. They catalyse oxidation reactions.

Cytomegalovirus A virus that causes eye infections and blindness.

Cytoplasm The contents of a cell.

Cytotoxic agents Anticancer agents that are generally toxic to cells by a number of mechanisms.

Database mining The use of computers to automatically search databases of compounds for structures containing specified pharmacophores.

De novo drug design The design of a drug or lead compound based purely on molecular modelling studies of a binding site.

Death activator proteins Chemical messengers that trigger a cell to commit suicide.

Deconvolution The isolation and identification of an active compound in a mixture of compounds obtained from a combinatorial synthesis.

Dependence A compulsive urge to take a drug for psychological or physical needs. The psychological need is usually why the drug was taken in the first place (to change one's mood) but physical needs are often associated with this. This shows up when the drug is no longer taken leading to psychological withdrawal symptoms (feeling miserable) and physical withdrawal symptoms (headaches, shivering, etc.) Dependence need not be a serious matter if it is mild and the drug is non-toxic (e.g. dependence on coffee). However, it is serious if the drug is toxic and/or shows tolerance. Examples: opiates, alcohol, barbiturates, and diazepams.

Desensitization The process by which a receptor becomes less sensitive to the continued presence of an agonist.

Desolvation A process that involves the removal of surrounding water from molecules before they can interact with another; for example a drug with its binding site. Energy is required to break the intermolecular interactions involved.

Differentiation The ability of cells to become specialized in a multicellular organism.

Dihydrofolate reductase An enzyme involved in generating tetrahydrofolate—an important enzyme cofactor. Dihydrofolate reductase inhibitors prevent the synthesis of nucleic acids and are used as antibacterial and anticancer agents.

Dipole–dipole interactions Interactions between two separate dipoles. A dipole is a directional property and can be represented by an arrow between an electron rich part of a molecule and an electron deficient part of a molecule. Different dipoles align such that an electron rich area interacts with an electron deficient area.

Discovery Studio Pro A molecular modelling software package.

Displacer A test compound that competes with a radioligand for the binding site of a receptor.

Divergent evolution Receptors that diverged early in evolution have greater differences in their binding sites and ligand preferences.

DNA Deoxyribonucleic acid.

DNA ligase An enzyme that repairs breaks in the DNA chain.

DNA polymerases Enzymes that catalyse the synthesis of DNA from a DNA template.

DNA viruses Viruses that contain DNA as their nucleic acid.

DOCK A software program used for docking molecules into target binding sites.

Docking The process by which a molecular modelling program fits a molecule into a target binding site.

Dose ratio The agonist concentration required to produce a specified level of effect when no antagonist is present, compared to the agonist concentration required to produce the same level in the presence of an antagonist.

Drug–drug interactions Related to the effect one drug has on the activity of another if both drugs are taken together.

Drug load The ratio of active drug in the total contents of a dose.

Drug metabolism The reactions undergone by a drug when it is in the body. Most metabolic reactions are catalysed by enzymes, especially in the liver.

dsDNA Double-stranded DNA. A term used in virology.

dsRNA Double-stranded RNA. A term used in virology.

Dummy atom *see* centroid

Dynamic combinatorial chemistry The generation of a mixture of products from a mixture of starting materials in the presence of a target. Products are in equilibrium with starting materials and the equilibrium shifts to products binding to the target.

Dynamic structure–activity analysis The design of drugs based on which tautomer is preferred for activity.

Dynorphins Endogenous polypeptides that act as analgesics.

E_s *see* Taft's steric factor

EC$_{50}$ The concentration of drug required to produce 50% of the maximum possible effect.

ED$_{50}$ The mean effective dose of a drug necessary to produce a therapeutic effect in 50% of the test sample.

Efficacy A measure of how effectively an agonist activates a receptor. It is possible for a drug to have high affinity for a receptor (i.e. strong binding interactions) but have low efficacy.

Electronic screening *see* database mining

Electrostatic interactions *see* ionic interactions

EMEA *see* European Agency for the Evaluation of Medicinal Projects

Enantiomers The mirror image forms of an asymmetric molecule having one asymmetric centre.

Endocytosis The process by which a segment of cell membrane folds inwards and is 'nipped off' to form a vesicle within the cell.

Endogenous compounds Chemicals which are naturally present in the body.

Endomorphins Endogenous tetrapeptides that act as analgesics.

Endorphins Endogenous polypeptides that act as analgesics.

Endoplasmic reticulum Folds of membrane within eukaryotic cells. Endoplasmic reticulum can be defined as smooth or rough according to its appearance under the electron microscope. Rough endoplasmic reticulum has ribosomes attached to it and is where protein synthesis takes place.

Endpoint Some form of measurable effect. Used in clinical trials to determine whether a drug is successful or not.

Energy minimization An operation carried out by molecular modelling software to find a stable conformation of a molecule.

Enkephalinases Enzymes which hydrolyse enkephalins.

Enkephalins Endogenous peptides which act as analgesics.

Enteric nervous system Located in the walls of the intestine. Responds to the autonomic nervous system and local hormones.

Enzyme A protein that acts as a catalyst for a reaction.

Epimerization The inversion of an asymmetric centre.

Epitopes Small molecules that bind to part of a binding site and do not produce a biological effect as a result of binding.

Epothilones Naturally occurring anticancer agents that inhibit tubulin depolymerization.

Eukaryotic cell The cells that are present in plants, animals, and multicellular organisms. They contain a membrane bound nucleus and organelles.

European Patent Convention (EPC) A group of European countries for which patents can be drawn up based on a European patent.

European Agency for the Evaluation of Medicinal Projects (EMEA) The European regulatory authority for the testing and approval of drugs.

European Patent Office (EPO) Issues European patents.

Exocytosis The process by which vesicles within a cell fuse with a cell membrane and release their contents out of the cell.

Extension strategies The addition of functional groups to a drug with the aim of achieving a further binding interaction with another binding region in the binding site.

F A symbol used in pharmacokinetic equations to represent oral bioavailability. Alternatively, a symbol used in QSAR equations to represent the inductive effect of a substituent.

Farnesyl transferase An enzyme that attaches a farnesyl group to the Ras protein to allow membrane attachment.

Fast tracking A method of pushing a drug through clinical trials and the regulatory process as quickly as possible. Applied to drugs that show distinct advantages over current drugs in the treatment of life-threatening diseases, or for drugs that can be used to treat diseases that have no current treatment.

FDA see Food and Drug Administration

Feedback control The process by which the product of an enzymatic reaction or a series of enzymatic reactions controls the level of its own production.

FGF see fibroblast growth factor

Fibroblast growth factor A growth factor that stimulates angiogenesis.

Fight or flight response Refers to the reaction of the body to situations of stress or danger, and which involves the release of adrenaline and other chemical messengers that prepare the body for physical effort.

First pass effect The extent to which an orally administered drug is metabolized during its first passage through the gut wall and the liver.

Fischer's lock and key hypothesis see lock and key hypothesis

Fisher's F-test A statistical test used to assess the significance of coefficients in a QSAR equation.

Flagellum A tail-like structure used by some microorganisms as a method of propulsion.

Fluoroquinolones A group of synthetic antibacterial agents.

Fmoc A shorthand term for the protecting group 9-fluorenylmethoxycarbonyl.

Food and Drug Administration (FDA) The drugs regulatory authority in the US.

Force field Relevant to molecular modelling. Refers to the calculation of the interactions and energies between different atoms resulting from bond stretching, angle bending, torsional angles, and non-bonded interactions.

Free–Wilson approach A QSAR equation which uses indicator variables rather than physicochemical parameters.

G-proteins Membrane-bound proteins consisting of three subunits which are important in the signal transduction process from activated G-protein-coupled receptors.

G-protein-coupled receptors Membrane-bound receptors that interact with G-proteins when they are activated by a ligand.

Gastrointestinal tract Consists of the mouth, throat, stomach, and upper and lower intestines.

Gating The mechanism by which ion channels are opened or closed.

GCP see good clinical practice

GDEPT Gene-directed enzyme prodrug therapy.

Genomics The study of the genetic code for an organism.

Glomerulus A knotted arrangement of blood vessels which fits into the opening of a nephron and from which water and small molecules are filtered into the nephron.

Global energy minimum The most stable conformation of a molecule.

GLP see good laboratory practice

Glucocorticoids Hormones that are used in anticancer therapy.

Glycoconjugate The general term for macromolecules that are linked to carbohydrates.

Glycolipid A lipid molecule linked to one or more carbohydrates.

Glycomics The study of carbohydrates.

Glycopeptides and glycoproteins Peptides and proteins that are linked to one or more carbohydrates.

Glycopeptide antibacterial agents Glycopeptides with antibacterial properties, the most important being vancomycin.

Glycosidases Enzymes that catalyse the hydrolysis the glycosidic bond between carbohydrate groups.

Glycosphingolipids Glycoconjugates which are thought to be important in the regulation of cell growth. Includes the molecules responsible for labelling blood cells.

GMP *see* good manufacturing practice

Gonadotrophin-releasing hormone *see* luteinizing hormone-releasing hormone

Good clinical practice (GCP) Scientific codes of practice that apply to clinical trials and which are monitored by regulatory authorities.

Good laboratory practice (GLP) Scientific codes of practice that apply to a pharmaceutical company's research laboratories and which are monitored by regulatory authorities.

Good manufacturing practice (GMP) Scientific codes of practice that apply to a pharmaceutical company's production plants and which are monitored by regulatory authorities.

Granzyme An enzyme introduced into defective cells by T-lymphocytes and which induces apoptosis.

GRID A molecular modelling software program that maps the nature of binding regions within a binding site.

Group shifts The transposition of a group within a molecule to make it unidentifiable to metabolic enzymes but not to target binding sites.

Growth factors Hormones that activate membrane-bound receptors and trigger a signal transduction pathway leading to cell growth and division.

HAART Highly active antiretroviral therapy. A combination therapy used to treat HIV infections.

Half-life The time taken for the plasma concentration of a drug to fall by half.

HAMA response Human anti-mouse antibodies are antibodies that are produced against monoclonal antibodies which have been derived from a mouse source, and are recognized as foreign by the body's immune system.

Helicases Enzymes that catalyse the coiling and uncoiling of DNA.

Hammett substituent constant (σ) A measure of whether a substituent is electron withdrawing or electron donating and to what extent.

Hansch equation A QSAR equation involving various parameters.

Hard drugs Drugs that are resistant to metabolism.

HBA *see* hydrogen bond acceptor

HBD *see* hydrogen bond donor

Helicobacter pylori An organism that can survive in the stomach and cause damage to the stomach lining, leading to ulcers.

Hemagglutinin A glycoprotein on the surface of the flu virus that is crucial to the infection process.

Henderson–Hasselbalch equation An equation that is used to determine the extent of ionization of an ionizable drug at a particular pH.

Herpes Viruses responsible for cold sores and other herpes infections.

High-throughput screening An automated method of carrying out a large number of *in vitro* assays on small scale.

Histone acetylase and histone deacetylase Enzymes that acetylate and deacetylate the lysine residues of the structural protein, histone. Important in the control of gene expression.

HIV Human immunodeficiency virus.

HOMO Highest occupied molecular orbital.

Homology models A term used in molecular modelling for the construction of a model protein or binding site based on the structure of known proteins or binding sites.

Hormones Endogenous chemicals that act as chemical messengers. They are typically released from glands and travel in the blood supply to reach their targets. Some hormones are local hormones and are released from cells to act in the immediate area around the cell.

Houghton's teabag procedure A manual method of carrying out combinatorial synthesis.

HRV *see* human rhinoviruses

Human Genome Project The sequencing of human DNA.

Human rhinoviruses RNA viruses responsible for the common cold.

Hybridization The mixing of atomic orbitals to form hybridized atomic orbitals. With atoms such as carbon, nitrogen and oxygen, it involves the mixing of 2s and 2p orbitals to produce sp, sp^2, or sp^3 hybridized orbitals. This is important in determining whether the atoms concerned can form π bonds.

Hybridomas Cells that are formed from the fusion of B-lymphocytes with immortal B-lymphocytes in the production of monoclonal antibodies.

Hydrogen bond A non-covalent bond that takes place between an electron-deficient hydrogen and an electron-rich atom, particularly oxygen and nitrogen.

Hydrogen bond acceptor A functional group that provides the electron rich atom required to interact with a hydrogen in a hydrogen bond.

Hydrogen bond donor A functional group that provides the hydrogen required for a hydrogen bond.

Hydrolases Enzymes that catalyse hydrolysis reactions.

Hydrophilic Refers to compounds that are polar and water soluble. Literally means water loving.

Hydrophobic Refers to compounds that are non-polar and water insoluble. Literally means water hating.

Hydrophobic interactions Refers to the stabilization that is gained when two hydrophobic regions of a molecule or molecules interact and shed the ordered water 'coat' surrounding them. The water molecules concerned become less ordered, resulting in an increase in entropy.

Hyperchem A molecular modelling software package.

Hypoglycaemia Lowered glucose levels in the blood.

IC$_{50}$ The concentration of an inhibitor required to inhibit an enzyme by 50%.

IND *see* Investigational Exemption to a New Drug Application

Immunomodulators Agents that either suppress or enhance the immune system.

Immunosuppressants Drugs that inhibit the immune response. Useful in the treatment of autoimmune disease and in reducing the chances of rejection following organ transplants.

Impurity profiling The study of drug batches to identify and quantify any impurities that might be present.

Indicator variables A variable used in QSAR equations which is given the value of 1 or 0 depending on whether a substituent is present or not.

Induced dipole interactions The situation where a charge or a dipole on one molecule induces a dipole in another molecule to allow an ion–dipole interaction or a dipole–dipole interaction respectively. An induced dipole normally requires the presence of π electrons.

Induced fit The alteration in shape that arises in a macromolecule such as a receptor or an enzyme when a ligand binds to its binding site.

Inhibitor An agent that binds to an enzyme and inhibits its activity.

In silico Refers to procedures that are carried out on a computer.

Institutional Review Board (IRB) A regulatory body in the USA that grants approval to clinical trials at a particular site.

Integrins Molecules displayed on the cells of new blood vessels to protect them from apoptosis.

Investigational Exemption to a New Drug Application A document required by the FDA before clinical trials on a drug can begin.

Iontophoresis A means of encouraging topical absorption of a drug by applying a painless pulse of electricity to increase skin permeability.

Integrase A viral enzyme present in HIV that catalyses the insertion of viral DNA into host DNA.

Integrins Molecules that are involved in anchoring cells to the extracellular matrix.

Intercalating agents Agents containing a planar moiety that is capable of slipping between the base pairs of DNA. Important anticancer and antibacterial agents.

Interferons Endogenous proteins that are part of the body's defence system against viral infections. They work by inhibiting the metabolism of infected cells.

Interleukin-6 A protein that stimulates metastasis.

Intermolecular bonds Bonding interactions that take place between two separate molecules.

International Preliminary Examination Report (IPER) A report on a patent application that can be used when applying for patents to individual countries.

International Search Report (ISR) A report on a patent application that can be used when applying for patents to individual countries.

Intramolecular bonds Bonding interactions other than covalent bonds that take place within the same molecule.

Intramuscular injection The administration of a drug by injection into muscle.

Intraperitoneal injection The administration of a drug by injection into the abdominal cavity.

Intrathecal injection The administration of a drug by injection into the spinal column.

Intravenous injection The administration of a drug by injection into a vein.

Inverse agonist A compound which acts as an antagonist, but which also decreases the 'resting' activity of target receptors (i.e. those receptors which are active in the absence of agonist).

In vitro procedures Testing procedures carried out on isolated macromolecules, whole cells, or tissue samples.

In vivo procedures Studies carried out on animals or humans.

Ion channels Protein complexes in the cell membrane which allow the passage of specific ions across the cell membrane.

Ion channel disrupters A term used to describe a group of antiviral agents that act against the flu virus by disrupting ion channels.

Ion–dipole interactions A non-covalent bonding interaction that takes place between a charged atom and a dipole moment, such as the interaction of a positive charge with the negative end of the dipole.

Ionic interaction A non-covalent bonding interaction between two molecular regions having opposite charges.

Ionophores Agents which act on a cell membrane to produce an uncontrollable passage of ions across the membrane.

Irreversible inhibitor An enzyme inhibitor that binds so strongly to the enzyme that it cannot be displaced.

IsisDraw A chemical drawing software package.

Isomerases Enzymes that catalyse isomerizations and intramolecular group transfers.

Isostere A chemical group which can be considered to be equivalent in physical and chemical properties to another chemical group.

Isozymes A series of enzymes that catalyse the same chemical reaction but which differ in their amino acid composition.

K_d The dissociation binding constant.

K_i The inhibitory or affinity constant.

Kinases Enzymes which catalyse the phosphorylation of alcoholic or phenolic groups present in a substrate. The substrate is normally a protein.

Koshland's theory of induced fit *see* induced fit

β-Lactamase inhibitors Agents which inhibit the β-lactamase enzymes.

Lactamases Bacterial enzymes that hydrolyse the β-lactam ring of penicillins and cephalosporins.

LD$_{50}$ The mean lethal dose of a drug required to kill 50% of the test sample.

Lead compound A compound showing a desired pharmacological property which can be used to initiate a medicinal chemistry project.

LHRH *see* luteinizing hormone-releasing hormone

Ligand Any molecule capable of binding to a binding site.

Ligand-gated ion channels Ion channels that are under the control of a chemical messenger or ligand.

Ligases Enzymes that join two substrates together at the expense of ATP hydrolysis.

Lincosamides A group of antibiotics acting against protein synthesis.

Lineweaver–Burk plots Plots which can be used to determine whether an enzyme inhibitor is competitive or non-competitive.

Linker A term used in combinatorial chemistry for a molecule that is covalently linked to a solid phase support and contains a functional group to which another molecule can be attached for the start of a synthesis.

Lipinski's rule of five A set of rules obeyed by the majority of orally active drugs. The rules take into account the molecular weight, the number of hydrogen bonding groups, and the hydrophobic character of the drug.

Lipophilic Refers to compounds that are fatty and non polar in character. Literally means fat loving.

Liposomes Small vesicles consisting of a phospholipid bilayer membrane. Used to encapsulate drugs for drug delivery.

Local energy minimum Refers to the nearest stable conformation reached where energy minimization is carried out on a molecule by molecular modelling software.

Lock and key hypothesis The now redundant theory that a ligand fits its binding site like a key fitting a lock.

log *P* *see* partition coefficient

LUDI A software program used for *de novo* drug design.

LUMO Lowest unoccupied molecular orbital.

Luteinizing hormone-releasing hormone Hormones that are used in anticancer therapy.

Lyases Enzymes that catalyse the addition or removal of groups to form double bonds.

Lysis The process where a cell loses its contents due to weakening of a cell wall or cell membrane.

Lysosomes Membrane-bound structures within eukaryotic cells that contain destructive enzymes.

MAA *see* Marketing Authorization Application

Macrolides Macrocyclic structures that act as antibacterial agents. Erythromycin is the most used example of this class of agents.

Macromolecule A molecule of high molecular weight such as a protein, carbohydrate, lipid, or nucleic acid.

Magic bullet *see* principle of chemotherapy

Marketing Authorization Application (MAA) A document provided to the EMEA in order to receive marketing approval for a new drug.

Malignant cancers or tumours Life-threatening tumours that are undergoing metastasis and setting up secondary tumours elsewhere in the body.

Matrix metalloproteinases Enzymes that catalyse the hydrolysis of the proteins making up basement membranes. A target for new anticancer drugs called matrix metalloproteinase inhibitors.

MDRTB Multi-drug resistant tuberculosis.

Merrifield resin A resin used in solid phase peptide synthesis.

Messenger RNA (mRNA) Carries the genetic code required for the synthesis of a specific protein.

'Me too' drugs Drugs which have been modelled as variations of an existing drug.

Metabolic blockers Groups added to a drug to block metabolism at a particular part of the skeleton.

Metalloproteinases Enzymes that catalyse the hydrolysis of peptide bonds in protein substrates and which contain a metal ion as a cofactor in the active site.

Metastasis Refers to the breaking away of individual cancer cells from an established tumour such that they enter the blood supply and start up new tumours elsewhere in the body.

Methylene shuffle A strategy used to alter the hydrophobicity of a molecule. One alkyl chain is shortened by one carbon unit, while another is lengthened by a one carbon unit.

Michaelis constant The substrate concentration when the reaction rate of an enzyme catalysed reaction is half of its maximum value.

Microfluidics The manipulation of tiny volumes of liquids in a confined space.

Microspheres Small spheres made up of a biologically degradable polymer. Used in drug delivery.

Microtubules Small tubules that are formed in cells by the polymerization of a structural protein called tubulin. Important for cell division and as targets for anticancer drugs.

Mitochondria Organelles within eukaryotic cells that can be viewed as the cell's energy generators. They also play a role in cell apoptosis.

Mitosis The process of cell division.

Mix and split The procedure involved when synthesizing mixtures of compounds by combinatorial synthesis.

Modulator An agent that binds to the allosteric binding site of a target and modulates the activity of that target.

Molar refractivity (*MR***)** A measure of a substituent's steric influence in a QSAR equation.

Molecular dynamics A molecular mechanics program that mimics the movement of atoms within a molecule.

Monoamine oxidase inhibitors Compounds which inhibit the metabolic enzyme monoamine oxidase. Have been used as antidepressants but are less favoured now since they have side effects.

Monoclonal antibodies Refers to antibodies that are cloned and are identical in nature.

Monosaccharides The carbohydrate or sugar monomers that make up a polysaccharide.

Motor nerves Nerves carrying messages from the central nervous system to the periphery.

MR see **molar refractivity**

MRSA Methicillin-resistant *Staphylococcus aureus*: strains of *S. aureus* that have acquired resistance to methicillin (a penicillin).

Multidrug resistance Refers to the situation where a cancer cell acquires resistance to a range of drugs other than the one it was exposed to. Related to the overexpression of P-glycoprotein which expels drugs from the cell.

Murine antibodies Refers to monoclonal antibodies that were originally isolated from mice.

Mutation An alteration in the nucleic acid base sequence making up a gene. Results in a different amino acid in the resultant protein.

Muscarinic receptors One of the two main types of cholinergic receptor.

Mutagen A chemical or substance that induces a mutation in DNA.

NCE see **New Chemical Entity**

NDA see **New Drug Application**

Neighbouring group participation A mechanism by which a functional group in a molecule assists a reaction without being altered itself.

Nanotubes Tubular structures on the molecular scale which are being considered as possible antibacterial agents.

Neoplasm The proper term for a cancer or tumour. Means new growth.

Nephrons Tubes that collect water and small molecules from the glomeruli and carry these towards the bladder. Much of the water, along with hydrophobic molecules, is reabsorbed into the blood supply from the nephrons and does not reach the bladder.

Neuraminidase An enzyme present in the flu virus that catalyses the hydrolysis of a sialic acid molecule from host glycoconjugates and which is crucial to the infection process.

Neuropeptides Peptides that act as neurotransmitters.

Neurotransmission The process by which nerves communicate with other cells.

Neurotransmitter A chemical released by a nerve ending that acts as a chemical messenger by interacting with a receptor on a target cell.

New Chemical Entity (NCE) A novel drug structure.

New Drug Application A document provided to the FDA in order to receive marketing approval for a new drug.

New Molecular Entity see **New Chemical Entity**

Nicotinic receptors One of the two main types of cholinergic receptor.

Nitrogen mustards Alkylating agents used in anticancer therapy.

NME see **New Molecular Entity**

Non-nucleoside reverse transcriptase inhibitors (NNRTI) A group of antiviral agents that target an allosteric binding site on the viral enzyme reverse transcriptase.

NRTI see **nucleoside reverse transcriptase inhibitors**

Nuclear hormone or transcription receptors see **transcription factors**

Nucleases Enzymes that hydrolyse oligonucleotides and nucleic acids.

Nucleic acids RNA or DNA macromolecules made up of nucleotide units. Each nucleotide is made up of a nucleic acid base, sugar, and phosphate group.

Nucleocapsid Consists of a viral capsid and its nucleic acid contents. Viral enzymes may be present.

Nucleoside A building block for RNA or DNA that consists of a nucleic acid base linked to a sugar molecule.

Nucleoside reverse transcriptase inhibitors A group of antiviral agents that mimic nucleosides and target the viral enzyme reverse transcriptase.

Nucleosomes Repeating units of histone proteins within a chromatin structure.

Nucleotide A molecule consisting of a nucleoside linked to one, two, or three phosphate groups.

NVOC The nitroveratryloxycarbonyl protecting group.

Oestrogens Female hormones that are used in anticancer therapy.

Oligonucleotides A series of nucleotides linked together by phosphate bonds. Smaller versions of nucleic acid.

Olivanic acids A group of agents which inhibit β-lactamases.

Oncogenes Genes which normally code for proteins involved in the control of cell growth and division, but which have undergone a mutation such that they code for rogue proteins, resulting in the uncontrolled growth and division of cells.

Opportunistic pathogens Pathogens which are normally harmless but which cause serious infection when the immune system is weakened.

Organelles Identifiable structures within the cytoplasm of a eukaryotic cell.

Oripavines Complex multicyclic analogues of morphine which have powerful analgesic and sedative properties.

Organophosphates Agents that inhibit the acetylcholinesterase enzyme and which are used as nerve gases, medicines, and insecticides.

Orphan drugs Drugs that are effective against rare diseases. Special financial incentives are given to pharmaceutical industries to develop such drugs.

Orphan receptors Novel receptors for which the endogenous ligand is unknown.

Oxazolidinones A group of synthetic antibacterial agents that act against protein synthesis.

Oxidases Enzymes that catalyse oxidation reactions.

Oximinocephalosporins A group of second- and third-generation cephalosporins.

P1 or P1' Nomenclature use to label the substituents of a substrate that can fit into the binding subsites of an enzyme. P1, P2, P3, etc. refer to substituents on one side of the reaction centre, and P1', P2', P3' to substituents on the other side.

p53 protein An important protein that monitors the health of the cell and the integrity of its DNA. Important to the apoptosis process.

Papillomavirus A DNA virus responsible for genital warts.

Parasympathetic nerves Nerves of the autonomic motor nervous system that use acetylcholine as neurotransmitter.

Parietal cells Cells lining the stomach which release hydrochloric acid into the stomach.

Partial agonist A drug which acts like an antagonist by blocking an agonist, but which retains some agonist activity of itself.

Partition coefficient (P) A measure of a drug's hydrophobic character. Usually quoted as a value of log P.

Patent Cooperation Treaty (PCT) A treaty to which about 122 countries have signed up.

PEGylation Covalently linking molecules of polyethylene glycol to macromolecules.

Penicillins Natural and semi-synthetic antibacterial agents that are bactericidal in nature.

Peptidases Enzymes which hydrolyse peptide bonds.

Peptidomimetics Agents that have been developed from peptide lead compounds such that their peptide nature is removed or disguised in order to improve their pharmacokinetic properties.

Peptoids Peptides which are partly or wholly made up of non- naturally occurring amino acids. As such, they may no longer be recognized as peptides by the body's protease enzymes.

P-glycoprotein A protein that expels toxins and drugs from cells. Plays an important role in drug resistance in the anticancer field when cancer cells mutate and produce increased levels of the protein.

Phage *see* bacteriophage

Pharmacodynamics The study of how ligands interact with their target binding site.

Pharmacokinetics The study of drug absorption, drug distribution, drug metabolism, and drug excretion.

Pharmacophore The atoms and functional groups required for a specific pharmacological activity, and their relative positions in space.

Pharmacophore triangle A triangle connecting three of the important binding centres making up the overall pharmacophore of a molecule.

Partial charges A measure of the partial charge on each atom of a molecule calculated by molecular modelling software.

Partial least squares A statistical method of reaching a QSAR equation in 3D QSAR.

PDGF Platelet-derived growth factor.

Penicillanic acid sulfone derivatives A group of agents which inhibit β-lactamases.

Penicillinases *see* lactamases

Phase I metabolism Reactions undergone by a drug which normally result in the introduction or unmasking of a polar functional group. Most phase I reactions are oxidations.

Phase II metabolism Conjugation reactions where a polar molecule is attached to a functional group that has often been introduced by a phase I reaction.

Phosphatase An enzyme that catalyses the hydrolysis of phosphate bonds.

Phosphodiesterases Enzymes which are responsible for hydrolysing the secondary messengers, cyclic AMP and cyclic GMP.

Phosphorylase An enzyme that catalyses the hydrolysis of phosphate bonds.

Photodynamic therapy The use of light to activate a prodrug in the body. Used in cancer therapy.

Photolithography A method of combinatorial synthesis involving the synthesis of products on a solid surface. Reactions only occur on those areas of the surface where photolabile protecting groups have been removed by exposure to light.

π (pi) bond A weak covalent bond resulting from the 'side on' overlap of p-orbitals. Only occurs when the atoms concerned are sp or sp^2 hybridized, and when the bond between the atoms is a double bond or a triple bond.

π (pi) bond cooperativity A situation which can arise in conjugated systems where a hydrogen bond donor and a hydrogen bond acceptor enhance their respective hydrogen bonding strengths by a resonance mechanism involving π bonds.

Picornaviruses A family of viruses that include polio, hepatitis A, cold viruses, and foot and mouth viruses.

Pinocytosis A method by which molecules can enter cells without passing through cell membranes. The molecule is 'engulfed' by the cell membrane and taken into the cell in a membrane bound vesicle.

p*K*a A measure of the acid–base strength for a drug or a functional group.

Placebo A preparation that contains no active drug, but should look and taste as similar as possible to the preparation of the actual drug. Used to test for the placebo effect where patients improve because they believe they have been given a useful drug, regardless of whether they received it or not.

Placental barrier Membranes that separate a mother's blood from the blood of her fetus. Some drugs can pass through this barrier.

Plasma proteins Proteins in the plasma of the blood. Drugs which bind to plasma proteins are unavailable to reach their target.

Plasmid Segments of circular DNA that are transferred naturally between bacterial cells. Useful in cloning and genetic engineering.

Podophyllotoxins A group of natural and semi-synthetic agents used as anticancer agents.

Polymerases Enzymes that catalyse the polymerization of molecular units to form macromolecules.

Polysaccharides Macromolecules consisting of carbohydrate (sugar) monomers.

Porins Protein structures that create pores in the outer membrane of Gram-negative bacteria through which essential nutrients can pass. Some drugs can pass through these pores if they have the correct physical properties.

Potency The amount of drug required to achieve a defined biological effect.

pRB A powerful growth-inhibitory molecule that binds to a transcription factor to inactivate it.

Presynaptic control systems Receptors on the ends of presynaptic nerves that affect the release of neurotransmitter from the nerve.

Principle of chemotherapy The principle where a drug shows selective toxicity towards a target cell but not a normal cell.

Privileged scaffolds Scaffolds that are commonly present in established drugs.

Procaspase 9 An enzyme that activates caspase enzymes to produce apoptosis.

Prodrug A molecule that is inactive in itself, but which is converted to the active drug in the body, normally by an enzymatic reaction. Used to avoid problems related to the pharmacokinetics of the active drug and for targeting.

Progestins Hormones that are used in anticancer therapy.

Prokaryotic cells Simple bacterial cells that contain no organelles or well-defined nucleus.

Prostaglandins Endogenous chemicals that play an important role as chemical messengers.

Prosthetic group A cofactor which is covalently linked to the active site of an enzyme.

Protease inhibitors A group of antiviral agents which inhibit the HIV protease enzyme.

Proteases Enzymes which hydrolyse peptide bonds.

Protein A macromolecule made up of amino acid monomers. Includes enzymes, receptors, carrier proteins, ion channels, hormones, and structural proteins.

Protein kinases *see* kinases

Proteoglycan A molecule consisting of a protein and a carbohydrate.

Protomers The protein subunits that make up a viral capsid.

Proteomics A study of the structure and function of novel proteins discovered from genomic studies.

Proto-oncogenes Genes which code for proteins involved in the control of cell growth and division, but which can cause cancer if they undergo mutation to form oncogenes.

Proton pump inhibitors A series of drugs which inhibit the proton pump responsible for releasing hydrochloric acid into the stomach.

q^2 *see* **cross-validated correlation coefficient**

quantitative structure–activity relationships (QSAR) Studies which relate the physicochemical properties of compounds with their pharmacological activity.

Quinolones A group of synthetic antibacterial agents, largely replaced by fluoroquinolones.

R A symbol used in QSAR equations to represent the electronic influence of a substituent due to resonance effects.

Racemate or racemic mixture A mixture of the various stereoisomers of a molecule. A molecule having one asymmetric centre would be present as both possible enantiomers.

Racemization A reaction which affects the absolute configuration of asymmetric centres to produce a racemic mixture.

Radioligand labelling The use of a radioactively labelled irreversible inhibitor to label a macromolecular target.

Ras protein A small G-protein that plays an important role in the signal transduction pathways leading to cell growth and division.

Receptor A protein with which a chemical messenger or drug can interact to produce a biological response.

Receptor-mediated endocytosis Refers to the process by which a virus binds to a host cell glycoprotein and induces endocytosis to enter the cell.

Recombinant DNA technology The process by which DNA is manipulated to produce new DNA. Involves the controlled splitting of DNA from different sources, followed by the formation or recombination of hybrid DNA.

Recursive deconvolution A method of identifying the constituents in a combinatorial synthetic mixture. The method requires the storage of intermediate mixtures.

Reductases Enzymes that catalyse reduction reactions.

Regression coefficient A measure of how well a QSAR equation explains the variance in biological activity of a series of drugs.

Relaxation time The time taken for excited nuclei to return to their resting state in NMR spectroscopy.

Renal Relating to the kidney.

Replication The process by which DNA produces a copy of itself.

Restriction enzymes Enzymes that are used in recombinant DNA technology to split DNA chains in a controlled fashion.

Reverse transcriptase A viral enzyme present in HIV that catalyses DNA from an RNA template.

Reverse transcriptase inhibitors A group of antiviral compounds that inhibit the viral enzyme reverse transcriptase.

Reversible inhibitors Enzyme inhibitors that compete with the substrate for the enzyme's active site and which can be displaced by increasing the concentration of substrate.

Ribosomes Structures consisting of rRNA and protein which bind mRNA and catalyse the synthesis of the protein coded by mRNA.

Ribosomal RNA (rRNA) Present in ribosomes as the major structural and catalytic component.

Ribozymes RNA molecules with an enzymatic property.

Rifamycins A group of antibiotics and semi-synthetic agents used as antibacterial agents.

Rigidification strategies Strategies used to limit the number of conformations that a drug can adopt with the aim of retaining the active conformation.

Ring contraction/expansion strategy The variation of ring size in a drug to optimize the relative positions of different binding groups.

Ring fusion or extension strategy The fusion of one ring onto another to enhance a drug's binding interactions.

Ring variation strategies The replacement of an aromatic, heteroaromatic, or saturated ring with a different ring system to obtain different structural classes of a drug.

Rink resin A resin used in combinatorial chemistry.

RNA Ribonucleic acid.

RNA-dependent RNA polymerase An enzyme that catalyses the synthesis of RNA from an RNA template.

RNA viruses Viruses that contain RNA as their nucleic acid.

S1 or S1' Nomenclature used to label binding subsites of an enzyme. The subsites accept the amino acid residues of a peptide substrate. S1, S2, S3, etc. refer to subsites on one side of the reaction centre, and S1', S2', S3' to subsites on the other side.

Safety catch linker An example of a linker in combinatorial chemistry on which two molecules can be constructed, one the target molecule and the other a tagging molecule.

SAR *see* **structure–activity relationships**

Sarcodictyins Naturally occurring anticancer agents that inhibit tubulin depolymerization.

SARS Severe acute respiratory syndrome. A viral infection.

Scaffolds The molecular core of a drug to which the important binding groups are attached as substituents.

Scatchard plot A plot used to measure the affinity of a drug for its binding site.

Schild analysis Used to determine the dissociation constant of competitive antagonists.

Screening A procedure by which compounds are tested for biological activity.

Scintillation proximate assay A visual method of detecting whether a ligand binds to a target by its ability to compete with a radiolabelled ligand that emits light in the presence of scintillant.

Secondary messenger A natural chemical which is produced by the cell as a result of receptor activation, and which carries the chemical message from the cell membrane to the cytoplasm.

Secondary metabolites Natural products that are not crucial to cell growth and division. Generally produced in mature cells.

Self-assembly The process by which molecular units assemble into a structure without the aid of enzymes or other structures; for example, the assembly of protomers to form a viral capsid.

Self-destruct drugs Drugs which are designed to be inactivated in the body through chemical or enzymatic mechanisms.

Semi-synthetic product A product that has been synthesized from a naturally occurring compound.

Sensitization The process by which a cell adapts to the continued presence of an antagonist, resulting in increased receptor sensitivity or the production of more receptors.

Sequential blocking Describes the situation where two agents inhibit two different enzymes in a biosynthetic pathway. Allows each agent to be administered in lower and safer doses.

Sequential release The release of the products of a combinatorial synthesis from the solid support in a specific sequence.

Serine-threonine kinases Enzymes which catalyse the phosphorylation of serine and threonine residues in protein substrates.

σ (sigma) bond A strong covalent bond taking place between two atoms. It involves strong overlap between two atomic orbitals whose lobes point towards each other.

Signal transduction The mechanism by which an activated receptor transmits a message into the cell, resulting in a cellular response.

Simplification strategies The simplification of a drug to remove functional groups, asymmetric centres, and skeletal frameworks that are not required for activity.

Small G-proteins Proteins that have an important role in signal transduction pathways. So called because they are similar to G-proteins but are a single protein.

Small nuclear RNA Small molecules of RNA that are in the nucleus and are a constituent of a spliceosome. They are important to the modification and splicing of mRNA following transcription.

Smart drugs Anticholinesterases that act in the central nervous system to increase levels of acetylcholine. They relieve the symptoms of Alzheimer's disease.

Soft drugs Drugs that are designed to undergo metabolism in a predictable manner to produce non-toxic, inactive metabolites that are excreted.

Somatic gene therapy The use of a carrier virus to smuggle a gene into a human cell which has a defective form of the gene.

Somatic motor nervous system Motor nerves carrying messages to skeletal muscle.

Specifications The tests that have to be carried out on a manufactured drug, and the standards of purity required.

Spider scaffolds Scaffolds which have binding group substituents placed round the whole scaffold.

Spindle The arrangement of microtubules that is formed in order to separate cells during cell division.

Spliceosome A structure made up of protein and small nuclear RNA. Serves to modify and splice mRNA following transcription.

ssDNA Single stranded DNA. A term used in virology.

ssRNA Single stranded RNA. A term used in virology.

Steady state concentration The concentration of a drug that is maintained in the blood supply following regular administrations.

Stereoelectronic modifications The addition or alteration of groups to protect labile functional groups by steric and electronic means.

Steric shields Groups that are added to molecules to protect vulnerable groups by nature of their size.

Streptogramins A group of macrocyclic antibiotics acting against protein synthesis.

Structure–activity relationships Studies carried out to determine those atoms or functional groups which are important to a drug's activity.

Structure-based drug design The design of drugs based on a study of their target binding interactions with the aid of X-ray crystallography and molecular modelling.

Subcutaneous injection The administration of a drug by injection under the surface of the skin.

Subsites Often refers to enzymes that accept peptides or proteins as substrates. The subsites are binding pockets that accept amino acid residues from the substrate.

Substituent hydrophobicity constant (π) A measure of a substituent's hydrophobic character.

Substrate A chemical which undergoes a reaction that is catalysed by an enzyme.

Suicide substrates Enzyme inhibitors which have been designed to be activated by an enzyme catalysed reaction, and which will bind irreversibly to the active site as a result.

Sulfonamides Synthetic antibacterial drugs that are bacteriostatic in nature.

Sulfotransferases Enzymes that catalyse conjugation reactions involving sulfate groups.

Supercoiling The process by which DNA coils into a compact shape.

Suppositories Drug preparations that are administered rectally.

Surface plasmon resonance An optical method of detecting the binding of a ligand with its target.

Sybyl A molecular modelling software package.

Sympathetic nerves Nerves of the autonomic motor nervous system that use noradrenaline as a neurotransmitter at target cells and which use acetylcholine as a neurotransmitter between nerves.

Synapse The small gap between a nerve and a target cell, across which a neurotransmitter has to travel in order to reach its receptor.

Synergy An effect where the presence of one drug enhances the activity of another.

Tadpole scaffold A scaffold where substituents acting as binding groups are located at one region of the scaffold.

Taft's steric factor (Es) A measure of a substituent's steric influence in QSAR equations.

Tagging A method of identifying what structures are being synthesized on a resin bead during a combinatorial synthesis. The tag is a peptide or nucleotide sequence which is constructed in parallel with the synthesis.

Tautomers The different structures that a conjugated system can adopt arising from the rearrangement of double bonds and hydrogen atoms.

Taxoids Naturally occurring and semi-synthetic anticancer agents that inhibit tubulin depolymerization.

Telomerase An enzyme that catalyses the construction of telomeres.

Telomeres Polynucleotide structures at the 3' ends of chromosomes that stabilize DNA.

Tetracyclines Tetracyclic antibiotics that are bacteriostatic in their action.

Teratogen A compound that produces abnormalities in a developing foetus.

TGF Transforming growth factor.

Therapeutic index or ratio The ratio of a drug's undesirable effects with respect to its desirable effects. The larger the therapeutic index, the safer the drug. The therapeutic index compares the drug dose levels which lead to toxic effects in 50% of cases studied to the dose levels leading to maximum therapeutic effects in 50% of cases studied.

Therapeutic window The range of a drug's plasma concentration between its therapeutic level and its toxic level.

Thymidylate synthase Catalyses the synthesis of an important building block for DNA. Inhibitors are used as anticancer agents.

Thrombospondin An endogenous compound that inhibits angiogenesis.

TNF and TNF-R Tumour necrosis factors and tumour necrosis factor receptors. Play a role in apoptosis or cell death.

Transdermal absorption Refers to the absorption of a drug through the skin.

Tolerance Repeat doses of a drug may result in smaller biological results. The drug may block or antagonize its own action and larger doses are needed for the same pharmacological effect. Alternatively, the body may 'learn' how to metabolize the drug more efficiently. Again, larger doses are needed for the same pharmacological effect, increasing the chances of toxic side-effects.

Topliss scheme A scheme used to determine which substituents should be introduced in order to get more active drugs. Useful when analogues are synthesized and tested one at a time.

Topoisomerases Enzymes that catalyse transient breaks in one or both strands of DNA to allow coiling and uncoiling of the molecule. These act as targets for several antibacterial and anticancer drugs.

Transcription The process by which a segment of DNA is copied to mRNA.

Transcription factors Complexes which bind to DNA and control the expression of specific genes.

Transduction The process by which plasmids are exchanged between bacterial cells.

Transfer RNA (tRNA) An RNA molecule that bears an amino acid which is specific for a particular triplet of nucleic acid bases.

Transferases Enzymes that catalyse transfer reactions.

Transgenic animals Animals that have been genetically modified such that they can be used for the *in vivo* testing of drugs.

Translation The process by which proteins are synthesized based on the genetic code present in mRNA.

Transition state A high energy intermediate that must be formed during an enzyme catalysed reaction. The energy required to reach the transition state determines the rate of reaction. It is proposed that an enzyme binds the transition state more strongly than the substrate or the product resulting in a stabilization of the transition state.

Transition state analogues or inhibitors Enzyme inhibitors which have been designed to mimic the transition state of an enzyme catalysed reaction.

Transition state isostere An arrangement of atoms that mimics the arrangement of atoms in a transition state, but which is more stable.

Transpeptidases Important bacterial enzymes that catalyse the final cross linking of the bacterial cell wall. Targeted by penicillins and cephalosporins.

Transport proteins *see* **carrier proteins**

Tricyclic antidepressants A series of tricyclic compounds that have antidepressant activity by blocking the uptake of noradrenaline from nerve synapses back into the presynaptic nerve.

Triplet code Refers to the fact that the genetic code is read in sets of three nucleic acid bases at a time. Each triplet codes for a specific amino acid.

Tumour suppression genes *see* **anti-oncogenes**

Tyrosine kinases Enzymes which catalyse the phosphorylation of tyrosine residues in protein substrates.

Ureidopenicillins A group of penicillins bearing a urea group at the α-position.

Vaccination The introduction of foreign antigens to prime the immune system such that it will work more effectively against later infections.

van der Waals interactions Weak interactions that occur between two hydrophobic regions and which involves interactions between transient dipoles. The dipoles arise from uneven electron distributions with time.

Varicella zoster viruses Viruses responsible for chickenpox and shingles.

Vascular endothelial growth factor A growth factor that stimulates angiogenesis.

Vectors A process by which a molecule can be taken into a cell. Particularly important to gene therapy.

VEGF *see* **vascular endothelial growth factor**

Veins Blood vessels carrying blood back to the heart.

Verloop steric parameter A measure of a substituent's steric properties. Used in QSAR equations.

Vesicle A membrane-bound 'bubble' within the cell. Neurotransmitters are stored within vesicles prior to release.

Vinca alkaloids Naturally occurring compounds that inhibit tubulin polymerization and are used as anticancer agents.

Virion The form that a virus takes when it is not within a host cell.

Viruses Non cellular infectious agents consisting of DNA or RNA wrapped in a protein coat. Require a host cell to multiply.

Voltage-gated ion channels Ion channels that are controlled by the potential difference across the cell membrane. Important to the mechanism of transmission in nerves.

VRE Vancomycin-resistant enterococci.

VRSA Vancomycin-resistant *Staphylococcus aureus*.

VZV *see* **varicella-zoster viruses**

Wang resin A resin used in combinatorial chemistry.

Withdrawal symptoms The symptoms that arise when a drug associated with physical dependence is no longer taken.

Z A shorthand term for the protecting group benzyloxycarbonyl.

Zinc finger domains Refers to a region of a steroid receptor that is rich in cysteine residues and zinc cofactors. Involved in binding to DNA when the receptor is part of a transcription factor.

GENERAL FURTHER READING

It is recommended that you read round various topics as much as you can. There are strengths and weaknesses in every publication and if you find an explanation difficult to follow in one textbook or article, you may find a clearer one in a different publication. Textbooks also differ in the breadth and detail of coverage given to the various topics of medicinal chemistry. For example, this textbook has chapters concentrating in some detail on topics such as antibacterial, antiviral, and anticancer agents, but does not cover cardiovascular or anti-inflammatory agents to the same extent. The following are useful reference and general texts. You will also find references to articles and books at the end of each chapter which are more specific to the topics covered in that chapter.

REFERENCE WORKS

Abraham, D. J. (ed.) (2003) *Burger's medicinal chemistry*, 6th edn, Vols 1–6. John Wiley and Sons, New York.

British national formulary (BNF), British Medical Association and Royal Pharmaceutical Society of Great Britain, Pharmaceutical Press, London (twice-yearly publication).

Hansch, C., Emmett, J. C., Kennewell, P. D., Ramsden, C. A., Sammes, P. G., and Taylor, J. B. (eds.) (1990) *Comprehensive medicinal chemistry*, Vols 1–6. Pergamon Press, Oxford.

Medicines compendium (2003) Pharmaceutical Press, London.

O'Neil, M. J. (ed.) (2001) *The Merck index: an encyclopaedia of chemistry, drugs and biologicals*, 13th edn. Merck and Co., Whitehouse Station, NJ.

GENERAL TEXTBOOKS ON MEDICINAL CHEMISTRY

Drews, J. (1999) *In quest of tomorrow's medicines*. Springer-Verlag, New York.

Ganellin, C. R. and Roberts, S. M. (eds.) (1994) *Medicinal chemistry—the role of organic research in drug research*, 2nd edn. Academic Press, London.

Hardman, J. G., Limbird, L. E., and Goodman Gilman, A. (eds.) (2001) *Goodman and Gilmans' the pharmacological basis of therapeutics*, 10th edn. McGraw-Hill, New York.

King, F. D. (ed.) (2002) *Medicinal chemistry, principles and practice*, 2nd edn. Royal Society of Chemistry, Cambridge.

Krogsgaard-Larsen, P., Liljefors, T., and Madsen, U. (eds.) (1996) *A textbook of drug design and development*. Harwood, Amsterdam.

Le Fanu, J. (1999) *The rise and fall of modern medicine*. Little Brown and Co., London.

Patrick, G. (2001) *Instant Notes medicinal chemistry*. Bios Scientific, Oxford.

Silverman, R. B. (2004) *The organic chemistry of drug design and action*, 2nd edn. Academic Press, San Diego.

Sneader, W. (2004) *Drug discovery: past, present and future*. John Wiley and Sons, Chichester.

Thomas, G. (2000) *Medicinal chemistry: an introduction*. John Wiley and Sons, Chichester.

Wermuth, C. G. (ed.) (2003) *The practice of medicinal chemistry*, 2nd edn. Academic Press, London.

Williams, D. A. and Lemke, T. L. (eds.) (2002) *Foye's principles of medicinal chemistry*, 5th edn. Lippincott, Williams and Wilkins, Philadelphia.

GENERAL TEXTBOOKS ON RELATED AREAS

Berg, J. M., Tymoczko, J. L., and Stryer, L. (2002) *Biochemistry*, 5th edn. W. H. Freeman and Co., New York.

Cairns, D. (2003) *Essentials of pharmaceutical chemistry*, 2nd edn. Pharmaceutical Press, London.

Page, C., Curtis, M., Sutter, M., Walker, M., and Hoffman, B. (2002) *Integrated pharmacology*, 2nd edn. Mosby, St. Louis.

Rang, H. P., Dale, M. M., Ritter, J. M., and Moore, P. K. (2003) *Pharmacology*, 5th edn. Churchill Livingstone, Edinburgh.

JOURNALS

Advances in Drug Research

Advances in Medicinal Chemistry

Annual Reports in Medicinal Chemistry

Antimicrobial Agents and Chemotherapy

Bioorganic and Medicinal Chemistry

Bioorganic and Medicinal Chemistry Letters

Chemical and Pharmaceutical Bulletins

Chemistry in Britain

Current Medicinal Chemistry

Current Opinion in Drug Discovery and Development

Drug Design and Delivery

Drug Discovery Today

Drug News and Perspectives

Drugs

Drugs of the Future

Drugs Today

European Journal of Medicinal Chemistry

Journal of Combinatorial Chemistry

Journal of Computational Chemistry

Journal of Computer-aided Molecular Design

Journal of Medicinal Chemistry

Medicinal Chemistry Research

Medicinal Research Reviews

Nature

Nature Reviews Drug Discovery

Pharmacochemistry Library

Progress in Drug Research

Progress in Medicinal Chemistry

QSAR

Science

Scientific American

Trends in Pharmacological Sciences

INDEX

H

radioactive isotopes 550
  as anticancer drugs 549
radioimmunotherapy 550
radioligand labelling 86, 707
radiotherapy 494, 497, 511
radishes 7
Raf protein 110–11, 513
raloxifene 113–14, 521, 689
raltitrexed 514, 516–17, 689
Ran protein 110
ranitidine 642, 656, 661–4, 689
Ras protein 110–11, 490–2, 528, 531, 707
ras gene 490, 528
rashes 645
rate constants 171
Rebetol 692
reboxetine 612–13, 689
receptor 66, 68–9, 72, 75, 707
receptor mediated endocytosis 471, 707
recombinant DNA technology 129–31, 707
  HIV-protease production 462, interferons 486
  protein synthesis 36–7, 129, 131, 158, 245
  synthesis of monoclonal antibodies 38–9
  thymidylate synthase 353
recursive deconvolution 307, 309–10, 707
reduced folate carrier 515
reductases 48, 139, 143, 522, 707; *see also* dihydrofolate reductase, HMG-CoA reductase, nitroreductase, ribonucleotide reductase
regression coefficient; *see* correlation coefficient
regulatory affairs 183, 257–60, 263, 544
relaxation time (nmr) 169, 707
Relenza 19, 475, 692
Remicade 692
Reminyl 692
renal artery 149
renal cancer 545
renin 55, 456–7, 608
renin inhibitors 55–6, 457–8
replication 119–21, 445–6, 494, 707
  amplification of oligonucleotides 312
  inhibition of; *see* adamantanes, alkylating agents, antimetabolites, DNA polymerase inhibitors, intercalators, interferons, reverse transcriptase inhibitors, topoisomerase poisons, viral DNA polymerase inhibitors
  inhibition and the cell cycle 490–1
  viral 441–2, 451
Rescriptor 692
reserpine 174, 611

resonance effect (*R*) 278, 707
resonance-assisted hydrogen bonding 197
respiratory syncytial infection 486
respiratory syncytial virus 485
respiratory tract infections 685
Respontin 692
restriction enzymes 129–30, 707
restriction point 491
reticuloendothelial system 157, 554
retinal 349
retinal inflammation 448–9
retinoids 112
Retrovir 692
retroviruses 450, 552
reverse transcriptase 61, 165, 450–3, 455, 707
  inhibitors 456, 461, 707
reversible inhibitors 707
revimid 545
Reyataz 692
rheumatic pains 617
rheumatoid arthritis 37–8, 59, 183
rhinovirus 483
Rho protein 110
rhodopsin 95–6, 350
rhodopsin like receptors 95
rhodopsin receptor 96
rhubarb 174
ribavarin 485–6, 689
ribonucleic acid (RNA) 122, 442, 518, 707; *see also* messenger RNA, ribosomal RNA, small nuclear RNA, transfer RNA, RNA viruses, ribozymes, transcription, translation)
  disruption of synthesis 419, 431; *see also* antiviral agents
  methylation 510
ribonucleotide reductase 517–18
  inhibitors 516–17
ribose 17–18, 122
ribosomal RNA (rRNA) 122, 124–5, 424, 707
ribosomes 28, 122, 124–5, 707
  bacterial 122, 381, 424
  targets for antibacterial agents 127, 381, 424–8
ribothymidine 123
ribozymes 486, 707
ricin 234, 549
rifabutin 456
Rifadin 692
rifampicin 425, 431–2, 689
rifampin 456
rifamycins 172, 381–2, 431–3, 707
Rifater 692
Rifinah 692

rigidification strategy 210–13, 707
  design of analgesics and sedatives 629–31
  design of antihypertensives
  design of antiulcer agents 657–8
  design of matrix metalloproteinase inhibitors 542
  design of oxamniquine 219, 221
  design of serotonin antagonists 364–5, 372
Rimacid 692
Rimactane 692
Rimactazid 692
rimantadine 473, 485
ring contraction strategy 204, 707
ring expansion strategy 204–5, 707
ring fusion or extension strategy 206, 707
ring variation strategy 204, 207, 232, 364–5, 372–3, 707
ringworm 686
Rink resin 303–4, 707
Risperdal 692
risperidone 85, 689
Ritalin 613, 692
ritonavir 688–9
  combination therapies 242, 452, 464, 466
  design 15, 230, 462–5,
  resistance 466
Rituxan 692
rituximab 548, 689
rivastigmine 589–90, 689
RNA; *see* ribonucleic acid
RNA polymerase 471–3
  DNA-dependant RNA polymerase 431, 500
  RNA-dependent RNA polymerase 441–2, 485, 707
RNA viruses 441, 444, 485, 707; *see also* cold virus, HIV, influenza virus
  antiviral agents 450–84
Ro41-0960 61
Robinson, Sir Robert 618
Robinson Crusoe 617
Rocephin 692
Roche 459, 481
rofecoxib 59, 689
Roferon-A 692
rohitukine 538
root mean square distance 339
roscovitine 539
Rous sarcoma virus 489
rRNA; *see* ribosomal RNA
rubella 440
Rubex 692
rubidomycin 502
Russian flu 471

somatic peripheral system 561
somatostatin 519
Somavert 692
soterenol 603–4
soybean curd 379
Spanish influenza 471
specifications 262–3, 708
sphingosine 20
spider scaffolds 708
spinach 7
spindle 36, 523, 708
spliceosome 126, 708
sponges 172
spongistatin 1, 524
ssDNA 708
ssRNA 708
St. John's wort 148, 183
stability tests 253
Stalevo 692
standard deviation (s) 272, 677, 678
standard operating procedures
     (SOP's) 260
staphylococci 685; see also Staphylococcus
     aureus
   drug resistance 434
   treatment with cephalosporins 409
   treatment with erythromycin 427
   treatment with penicillins 390, 400
   treatment with rifamycins 431
   treatment with sulfonamides 385
Staphylococcus aureus (S. aureus) 685
   drug resistance 379, 393, 395, 399, 410,
        433–5
   treatment with cephalosporins 407,
        410
   treatment with fluoroquinolones 429
   treatment with mupirocin 435
   treatment with penicillinase-resistant
        penicillins 399–400
   treatment with vancomycin 415
   vancomycin resistant (VIRSA) 418,
        433
starch 17
statines 459
statones 459
staurosporine 539
stavudine 452–3, 689
steady state concentration 155, 709
stearic acid 20
stem cell factor 537
stepwise bond rotation 337
stereoelectronic modifications 230, 709
steric energy 330
steric factors 279
steric fields 291–2, 296–7, 371–2
steric shields 229–30, 709
   in cephalosporins 407
   in cholinergics 568–9

in farnesyl transferase inhibitors 531
in HIV protease inhibitors 464
in local anaesthetics 230
in matrix metalloproteinase
     inhibitors 542
in penicillins 399–400, 402
in peptidomimetics 246
to block cytochrome P450
     inhibition 371
vancomycin 417
sterilising agent 379
Sterimol 280
stimulants 4, 601, 613
stomach cancer 489, 511, 537
Stratter 692
streptococcal infections 380, 407
streptococci 393, 399–400, 403, 409–10,
     685
Streptococcus pneumoniae 390, 427, 429,
     431, 433
Streptococcus pyogenes 390
streptogramins 428, 708
Streptomyces 420, 500
Streptomyces achromogenes, 509
Streptomyces antibioticus 518
Streptomyces aureofaciens 425
Streptomyces cattleya 411
Streptomyces clavuligerus 408, 412
Streptomyces erythreus 427
Streptomyces garyphalus 414
Streptomyces griseus 424
Streptomyces lincolnensis 428
Streptomyces mediterranei 431
Streptomyces nodosus 17
Streptomyces olivaceus 414
Streptomyces orientalis 415
Streptomyces peucetius 501
Streptomyces pristinaespiralis 428
Streptomyces roseosporus 423
Streptomyces venezuela 426
Streptomyces verticillus 503
streptomycin 19, 380–1, 424–5, 433–4
streptozotocin 508–9
stress 593, 610, 642
stroke 59–60, 86, 164
stromelysins 540
structural databases 180, 328, 343, 349–
     50, 360
structural proteins 25, 35–6; see also
     tubulin
   viral 451
structure-activity relationships
     (SAR) 185, 709
   Abelson tyrosine kinase inhibitors
        536–7
   antiviral agents for flu 478
   anthracenediones anticancer
        agents 502–3

aryloxypropanolamines 606–7
atropine 572
carboxamides 478
catecholamines 599–601
cephalosporin C 405
cholinergics 565
EGF-R kinase inhibitors 534
enkephalins 637
ligands for the serotonin receptor 352
local anaesthetics 338
famotidine 662–3
farnesyl transferase inhibitors 532
histamine 645–6
morphine analogues 619–22
penicillins 397
physostigmine 584
role of combinatorial synthesis 300
sarcodictyins 527
serotonin antagonists 366
sulfonamides 383
testing procedures 197–8
structure-based drug design 79, 213–17,
     709; see also non nucleoside
     reverse transcriptase inhibitors
   of capsid binding agents 484
   of farnseyl transferase inhibitors 531
   of HIV-protease inhibitors 246–7,
        456–69
   of peptidomimetics 246–7
strychnine 5
SU 5416 539
SU 6668 539
subcutaneous injection 152–3, 708
suberoylanilide 544
suberoylanilide hydroxamic acid 544
Sublimaze 692
sublingual administration 151
subsites 708
substance P 563, 640
substituent hydrophobicity constant
     (π) 275–6, 708
substituent variation 364
substrate 42–7, 709
Subutex 692
succinyl choline 572
succinyl proline 216
succinyl sulfathiazole 385
sugar 5, 7
suicide substrates 56–8, 60, 413–14,
     515–16, 522, 709
sulbactam 413–14
Suleo-M 692
sulfa drugs 380, 382–3, 386; see also
     sulfonamides
sulfadiazine 384
sulfadoxine 384
sulfamethoxazole 387
sulfanilamide 176, 382–3